The Purple Island and Anatomy in Early Seventeenth-Century Literature, Philosophy, and Theology

The Purple Island and Anatomy in Early Seventeenth-Century Literature, Philosophy, and Theology

Peter Mitchell

Madison • Teaneck
Fairleigh Dickinson University Press

Associated University Presses
2010 Eastpark Boulevard
Cranbury, NJ 08512

The paper used in this publication meets the requirements of the American National Standard for Permanence of Paper for Printed Library Materials Z39.48-1984.

Library of Congress Cataloging-in-Publication Data

Mitchell, Peter, 1968–
The Purple island and anatomy in early seventeenth-century literature, philosophy, and theology / Peter Mitchell.
p. ; cm.
Includes bibliographical references and index.
ISBN-13: 978-0-8386-4018-0 (alk. paper)
1. Fletcher, Phineas, 1582–1650. Purple island. 2. Human anatomy in literature. 3. Body, Human, in literature. 4. Allegory. 5. Metaphor in literature. 6. Human anatomy—History—17th century. I. Title.
[DNLM: 1. Fletcher, Phineas, 1582–1650. Purple island. 2. Anatomy—history. 3. Medicine in Literature. 4. History, 17th Century. 5. Poetry—history. WZ 330 M682p 2006]
PR2274.P73M58 2007
821′.3—dc22 2006007972

PRINTED IN THE UNITED STATES OF AMERICA

To

PARK MIYOUNG

Without whom and whose understanding, the extremity would be less bearable; without whose faith and encouragement, the ascent might inspire only fear; and without a commitment to whom, I would have less motivation for the attempt.

Contents

Illustrations

Acknowledgments

THIS BOOK BEGAN SOME YEARS AGO AS A DOCTORAL DISSERTATION, AND IN RECOGNITION of how it was made possible, and how it has been shaped and informed by divers persons, bodies, and foundations, I herein wish formally and sincerely to express my gratitude to those following. Foremost among beneficial influences has been that exercised by the supervisor of my doctoral research, Professor John Manning, the depth of whose knowledge of this and adjacent fields of scholarship I have rarely succeeded in sounding. Careful to illuminate obscurity and confusion, he has been liberally generous with his knowledge, ever attentive, ever respectful; and under pressure he remained jovial and helpful in every way.

Initially, Dr. Lawrence Normand was my principal and later my joint supervisor, and remained so for the best part of the period of my doctoral research, continuing to comment on drafts until very nearly the end, and to offer hospitality to an impoverished scholar visiting London to conduct research on occasions since then. He contributed significantly to the incipient formation of my idea of the problem with which this book is concerned, and by gently insisting on the definition of the scope of the project, he wisely (if not always successfully) sought to bring focus and purpose to my digressive pursuits over interdisciplinary borders, following forever receding horizons. For the professionalism in which my *Viva voce* examination was conducted and for their acuity and unreserved encouragement, I should like to thank the examiners of my PhD dissertation, Professor Jonathan Sawday and Dr. Gordon Williams; the former especially also for the benefit of his erudition in our shared research field. All of the above and many I have yet to mention have my gratitude for their continuing commitment to the realization of my career prospects as an academic scholar despite a steady stream of setbacks.

The outcome of a demanding, not infrequently high-pressure, long-term research project, it is not surprising that this book should have accumulated numerous debts of gratitude. I owe the most heartfelt thanks to my wife, to whom this book is dedicated; to my parents, Dee and Frank Mitchell, for their constant encouragement and for the satisfaction they derive from my fulfillment as a scholar; and to them and my sister, Lucy, brothers, Stephen and Jonathon, late grandmother, Nancy Danby, and other members of my family, for their perseverance through long periods when the consuming entanglements of scholarly research have rendered me more remote than either they or I would have wished. For their enduring commitment to their son-in-law over approximately the same period in which this book was researched, written, and revised, I am grateful to Park Joon Taek and Choe Ki Sook.

My postdoctoral research has been supported by my affiliation as honorary research fellow and lecturer, with the Department of English at the University of Wales, Lampeter, among whose members, past and present, I extend the hand of gratitude to Professor Alcuin Blamires, Dr. Helen Vella Bonavita, Barbara Dennis, Dr. Rebecca Ferguson, John Hughes, Dr. William Marx, Peter Miles, Dr. Allen Samuels, and Harrison and Herb Solow, besides those already mentioned. From time to time, my research has enjoyed the benefit of their knowledge, intellectual engagement, keen analytic and scholarly scrutiny, or wondering curiosity. I have been the fortunate recipient of their continuous professional advice, strong encouragement and support, comradeship, and in some cases affectionate friendship. Beryl Doyle I thank for her administrative support, and for her uncanny combination of cordiality, tact, good-natured forbearance, and professionalism. I am also beholden to Peter Miles for the wholly unexpected gift of a copy of Frederick S. Boas's 1908–9, two-volume edition of the works of Giles and Phineas Fletcher.

The interest and encouragement of many others has been much appreciated, and I thank particularly Dr. Sarah Jane Boss, Professor Jonathan Dollimore, Dr. David Hillman, Dr. A. Doug Lee, Professor Michael Neill, Dr. Martin O'Kane, Professor Michael Schoenfeldt, and Professor Luke Wilson. It is to be regretted that the liberality with which the late Dr. Gareth Roberts gave of his experience, knowledge, and time produced results only belatedly; though I remain enduringly grateful to his memory. I thank Dr. Cindy Klestinec for conversations in which I, a specialist, feel like a generalist as our knowledge meets across several disciplines in our shared field. For his palaeographical skills and expertise in early modern literary manuscripts, I should like to thank Dr. Andrew Zurcher; for the generous application of his expertise in digital imaging, I am grateful to Ian Clewes.

I would like to thank Dr. Andy Gordon, Professor Ralph Houlbrooke, Dr. Claire Jowitt, Professor Bill Overton, Mathias Pozsgai, Emma Robertson, Dr. Ceri Sullivan, and Professor Diane Watt, for their provision of such opportunities as allowed this work at various stages of progress a platform or public output, and I am grateful for the incisive questions and comments of audiences and readers from which this book certainly ought to have benefited. For all that, errors and infelicities doubtless remain; for them only I am responsible.

My doctoral research was supported financially by the award of successive University of Wales Postgraduate Studentships. My postdoctoral research and the work of revising the manuscript for publication have been aided by the generous grant of a Leverhulme Research Fellowship (2004–2005). The cost of securing the reproduction rights of illustrations was met by the award of a British Academy Research Grant.

Research has been facilitated, often in the most conducive and comfortable of circumstances, by the advice, learning, and communication and organizational skills of librarians, archivists, and staff of the following libraries and archives, with whom I have liaised, in some cases over extended periods of time: the main library and Founders' Library of the University of Wales, Lampeter; The Hugh Owen and Thomas Parry Libraries of the University of Wales, Aberystwyth; The National Library of Wales; Cambridgeshire Record Office; Cambridge University Library; Gonville & Caius College Library and Archive, Cambridge; King's College Library and Archive Centre, Cambridge; Norfolk Record Office; The Department of Manuscripts and Special Collections, Hallward Library, Nottingham; The British Library; The Guildhall Library; and The Wellcome Library for the History and Understanding of Medicine. Further debts of gratitude for specific help are acknowledged in the notes to the book at the appropriate places.

For permission to reproduce illustrations, I thank Llyfrgell Genedlaethol Cymru / The National Library of Wales, The Wellcome Library for the History and Understanding of Medicine, and The Wellcome Trust Medical Photographic Library. Permission to quote copyright material has been sought from and granted by many. Excerpts from *The Anatomical Renaissance: The Resurrection of the Anatomical Projects of the Ancients* (Aldershot: Scolar Press, 1997) are used by kind permission of Dr. Andrew Cunningham; excerpts from *Andreas Vesalius' First Public Anatomy at Bologna 1540: An Eyewitness Report by Baldasar Heseler Medicinae Scolaris together with his notes on Matthaeus Curtius' Lectures on Anatomia Mundini,* edited by Ruben Eriksson (Uppsala and Stockholm: Almquist and Wiksell, 1959), are used by kind permission of Professor Karin Johannisson; excerpts from "Bodies by Art Fashioned: Anatomy, Anatomists, and English Poetry 1570–1680" (PhD thesis, University of London, 1988), are used by kind permission of Professor Jonathan Sawday; excerpts from *The Body Emblazoned: Dissection and the Human Body in Renaissance Culture,* copyright © 1995 by Jonathan Sawday, are used by kind permission of Professor Jonathan Sawday and Taylor & Francis Books Ltd.; excerpts from *Dissection and Vivisection in the European Renaissance,* copyright © by Roger French, 1999, are used by kind permission of Ashgate Publishing Limited; excerpts from *William Harvey's Natural Philosophy,* copyright © 1994 by Cambridge University Press, are reprinted with the permission of Cambridge University Press; excerpts from "William Harvey's *Prelectiones:* The Performance of the Body in the Renaissance Theater of Anatomy," *Representations* 17 (1987), 62–95, are used by kind permission of Professor Luke Wilson and The University of California Press; excerpts from *The Epitome of Andreas Vesalius,* translated from the Latin and edited by L. R. Lind (Cambridge, Mass.: M.I.T. Press, 1969), are used by kind permission of Professor L. R. Lind; excerpts from *Metaphor and Art: Interactionism and Reference in the Verbal and Nonverbal Arts,* copyright © 1989 by Cambridge University Press, but since reverted to the author, are used by kind permission of Professor Carl Hausman; excerpts from *Art and Anatomy in Renaissance Italy,* copyright © 1985, 1982 by Bernard Schultz, are used by kind permission of Professor Bernard Schultz; excerpts from "Such Nothing is Terrestriall: Philosophy of Mind on Phineas Fletcher's *Purple Island,*" by Lana Cable, *Journal of the History of the Behavioural Sciences* 19 (1983), 136–52, are used by kind permission of John Wiley & Sons, Inc.; excerpts from *The Double Face of Janus and Other Essays in the History of Medicine,* by Owsei Temkin, pp. 271–81, © 1977 The Johns Hopkins University Press, are reprinted with permission of The Johns Hopkins University Press; excerpts from Andrea Carlino, *Books of the Body: Anatomical Ritual and Renaissance Learning,* translated by John Tedeschi and Anne C. Tedeschi, copyright © 1999 by The University of Chicago, are used by permission of the publisher, The University of Chicago Press; excerpts from "William Harvey and the Purpose of Circulation," by Walter Pagel, *Isis* 42 (1951), 22–38, copyright © 1951 by The History of Science Society, are used by permission of the publisher, The University of Chicago Press; excerpts from *The Language of Allegory: Defining the Genre,* by Maureen Quilligan, copyright © 1979 by Cornell University Press, are used by permission of the publisher, Cornell University Press; excerpts from *Phineas Fletcher: Man of Letters, Science and Divinity,* by Abram Barnett Langdale, copyright © 1937 by Columbia University Press, are reprinted with permission of the publisher; and excerpts from "Metaphor and the Main Problem of Hermeneutics," by Paul Ricoeur, *New Literary History* 6 (1974), 95–110, are used by kind permission of the Assistant to the Editor of *New Literary History.*

In respect of *William Harvey and the Circulation of the Blood,* by Gweneth Whitteridge (London: Macdonald; New York: American Elsevier, 1971), copyright © 1971 by Gweneth Whitteridge, and *The Anatomical Lectures of William Harvey: Prelectiones Anatomie Universalis, De Musculis,* edited by Gweneth Whitteridge (Edinburgh and London: Published for the Royal College of Physicians, London, by E. & S. Livingstone, 1964), I have read and abide by *The Chicago Manual of Style,* 15th ed. (Chicago: University of Chicago Press, 2003), and all attempts to locate the author's/editor's estate have been made unsuccessfully. If any proper acknowledgement has not been made, we would invite copyright holders to inform us of the oversight.

Personal names are given in their native, vernacular form unless English or Latinized forms are generally accepted and widely adopted in modern English, as in the case of "Vesalius," for example. In conformity with modern scientific publications Latin scientific names are not italicized (whereas other Latin terms are italicized). This is because the nomenclature has become standardized. However, where a Latin or Greek term was current in the early modern period, when terms were only in the early stages of standardization, and many were interchangeable with common, popular English terms, I have elected to italicize the Latin or Greek term. With regard to quotations of sixteenth- and seventeenth-century printed books, all contractions have been silently expanded (as is conventionally the case); u's, v's i's and j's appear unaltered consistently (so far as I am aware); italicized words are unaltered except where stated, though completely capitalized words (with a few significant exceptions) are changed into lower case (though the initial letter remains in upper case where appropriate). The first reference to a sixteenth- or seventeenth-century work that I have consulted includes full details of publication, including either the publisher/printer or person for whom the work was printed, preferring always when available the former to the latter. The first reference to a work of this period that I have not personally consulted includes details of publication minus the publisher/printer or person for whom the work was printed.

The Purple Island and Anatomy in Early Seventeenth-Century Literature, Philosophy, and Theology

Introduction: Figuring Anatomy in *The Purple Island*

ON AUGUST 25, 1645, A PRISONER IN FLEET PRISON NAMED JAMES HOWELL WROTE TO Edward Benlowes (1603?–76) thanking him for his visits and for "other fair respects you show me." One such "fair respect" had been the gift or loan of a copy of an allegorical poem entitled *The Purple Island.* The poem's author, Phineas Fletcher (1582–1650), had Benlowes as his Maecenas and Benlowes was the poem's dedicatee. In his letter Howell goes on to say, "you have enlarg'd my quarters 'mong these melancholy walls by sending me a whole Isle to walk in, I mean that delicate *purple Island* I received from you, wher I meet with *Apollo* himself and all his daughters, with other excellent society; I stumble also ther often upon my self, and grow better acquainted with what I have within me and without me: Insomuch that you could not make choice of a fitter ground for a Prisoner, as I am, to passe over, than of that *purple Isle,* that *Isle of man* you sent me, which as the Ingenious Authour hath made it, is a far more dainty soil than that *Scarlet* Island which lies near the *Baltic* Sea."[1] Visitors to the shores of *The Purple Island* since (and in some cases before) its publication in 1633, have included readers like Howell curious to explore the reciprocity, or hermeneutic circle, between interpretation of the subject matter of the poem's story, and interpretation of themselves and their world as represented therein. Visiting literary critics have typically been concerned to discern and evaluate the effect of Apollo as god of poetry on *The Purple Island*'s language, style, generic character, formal arrangement, and conception or design. Historians of anatomy, physiology, and psychology have embarked on textual expeditions to discover the provenance or (more frequently) the character of the structure of the "*Isle of Man*" and its internal processes; and, recently, historians and critics of culture and ideas have chosen to deduce the derivation, literary context, and intellectual and political character of *The Purple Island*'s negotiations and settlements of subjectivity, nationhood, religion, natural philosophy, and poetry.

The risk to my readers' understanding of the nature and character of the island of which I write is not merely facetious; it stems from an allegorical and rhetorical complexity of the work itself. Rhetorically, the work gestures towards an identification with its subject matter, thereby problematizing the traditional critical distinctions between form and content, and between text and context. Cantos 2 to 6 are about the physical and (to a lesser extent) the mental anatomy of Man as a general, abstract, and purportedly universal concept. The rhetoric of anatomical observations sometimes creates the effect of an imaginary presence of the anatomized body, as though the poetic description itself were methodically consistent with or even performing an anatomical dissection. Moreover, the subject matter is represented allegorically, and so, in addition and *at the same time,* Cantos 2 to 6 are about a dynastic realm or nation, specifically Britain. Although interpretation may at any point of the text favor the anthropological signified over the societal, or vice versa, it is in practice impossible to separate the anatomical and psychological from the geographical and political meanings; and therefore, while the poem engages its implied British reader hermeneutically in reflection on her or his subjectivity and nationality, much of the criticism the poem has attracted from one critical or historical discipline or another has failed fully to appreciate the complex structure of its plurisignification.

It is frequently remarked by literary critics, though apparently not begrudged by Howell in his incarceration, that the persistence with which Cantos 2 to 5 of *The Purple Island* describe the structure and function of nearly the whole of the anatomy of the "*Isle of Man,*" and its natural, human-built, and sociopolitical environments, over 184 seven-line stanzas, requires considerable concentration on the part of the reader, as well as deductive and inductive interpretation. Grierson has complained that "the soul of the reader faints under the strain of such sustained and relentless allegory."[2] And interpretation to date has too frequently as a result shown signs of only cursory reading or incomplete attention, to redress which this study sets out initially to figure out the coherence and consistency in the mass of detail.

Fletcher was not the only Renaissance or early modern poet to look to anatomy for poetic invention. There was generally renewed intellectual interest in the human body in the late sixteenth and seventeenth centuries, and this book is concerned with its impact on English literary culture. What were imagined to be the anatomical projects of Aristotle and Galen were revived in the sixteenth century, and the practice of dissection developed. As the inherited authoritative knowledge of Galen, Aristotle, and their medieval and Arabic commentators was reassessed, the intellectual tools of anatomy, its increasingly specialized nomenclature, method, descriptive criteria, and organization of knowledge, became available to literary writers, whose response was to explore the emerging science imaginatively as well as intellectually. In *The Faerie Queene,* Edmund Spenser conceived of the temperate body as a feudal castle, and Fletcher imagined fallen yet redeemed anatomy as a dynastic realm or emergent nation. Although *The Purple Island* is the most persistent poetic anatomy of the period, and long dismissed by critics as defective of taste on account of this, it is also probably the most consistent and coherent, and this achievement alone, which is an allegorical achievement, deserves attention. The imaginative, literary, and anatomical excursions of Spenser, Fletcher, and others, not only embody anatomical knowledge differently, they invite the interdisciplinary critic to reveal the light they shed upon or that is shed upon them by the anatomy of the schools, and measure the extent to which they share its figurality, rationality, coherence, and systematicity. Studying literary works in conjunction with anatomical texts and illustrations from the Renaissance provides a distinct vantage point from which, firstly, to survey the culturally changed relationships between the body and the soul, mind, and person in Western culture, and secondly, to refocus the literary output of the period. The work of a relatively minor Jacobean poet, because outstanding as an imbrication of poetry and anatomy, achieves a not undue prominence. Accordingly, this is a study of Fletcher, and those of his immediate predecessors and contemporaries, who responded to the intellectual and artistic challenges posed by the rapidly developing field of anatomy.

To the reader who is unfamiliar with early modern anatomical terminology, its deployment well beyond the usual aesthetic, affective, and intellectual sites of literary interest (such as external anatomy, the heart, and the brain), with a systematicity probably unparalleled by any other literary work of its period, may simply suggest that Fletcher had an otherwise educated reader in mind, which in a specifically historical and biographical sense he did. It is in respect of the poem's descriptive fidelity to the order of anatomical knowledge that it may in the first instance be considered as describing a "*Purple Island*": in the veins "luke-warm waters di'd in porphyr hue, / Sprinkle this crimson Isle with purple-colour'd dew" (*P.I.* 2.10.6–7).[3]

Curiously, from a modern perspective (some have assumed, from *any* perspective), the anatomy in *The Purple Island* is presented, embellished, and developed along literary lines by Fletcher's "all-purpling Imagination."[4] Despite being on occasion "violently protestant in sympathy and [a] self-proclaimed follower . . . of [Guillaume de Salluste, Sieur] Du Bartas[,] as well as of Spenser," both of whom were Protestant, Fletcher's poetry has been described as "strongly suggestive of the baroque art of the counter-Reformation";[5] yet the "curious and interesting . . . workmanship" of the poem is not only stylistic.[6] It also creates of the Isle of Man a fictional geographical location, with characters that are simultaneously the virtues or faculties of bodily organs, and which dwell in demographic centers, and engage in love, governance, and epic warfare, in a story narrated by a shepherd named Thirsil. Correspondingly linguistically complex, rather than only anatomical, the poem has caused some literary critics much consternation.[7] Yet, the poem's application of anatomical descriptive criteria to the geopolitical realm is systematic and coherent, as this study will demonstrate.

Critics have been divided over the poetry's stylistic quality, its apparent preoccupation with style at the expense of "any real engagement" with society, and its dependence upon the poetry of Spenser for style, vocabulary, imagery, poetic theory, and philosophy.[8] But for critics since the seventeenth century until very recently, an overwhelmingly representative judgment was that of George MacDonald, who considered *The Purple Island* an "incongruous dragon of allegory."[9] The poem's purported incongruity has been thought to extend to questions of form, generic character, linguistic complexity, and overall conception or design.[10] Some critics have been unable or unwilling to conceal their puzzlement at or even distaste for the choice of subject matter for Cantos 2 to 5, and the "extreme persistence" and even "sensuousness" with which the anatomical description of the human body is pursued.[11] MacDonald described the poem in 1868 as "like a well-shaped house, built of mud, and stuck full of precious stones."[12] Jocularity aside, it is not difficult in cases like this to discern a motive to denigrate the "curious work" of the island-body (*P.I.* 2.7.1), as poem, but more significantly, as physical subject matter; a desire which seems to correspond to the assumption that anatomy is an "unseemly theme" for "the embroideries of imaginative writing."[13] These and other responses have evinced and continue to evince convictions not only about what is and what is not appropriate to poetry, but also about the phenomenon of the body.

In arriving at an independent judgment, a self-critical historicist literary criticism should seek both to account for and to evade the accretion of layers of meaning and evaluation over three centuries of intermittent critical attention.[14] *The Purple Island* was written in a very different context from that in which MacDonald described Fletcher's poem in 1868, and the intellectual climate has changed again since. Several intellectual strands contributed to the erosion of the binary opposition between mind and body during the last century. Some have exercised an influence on literary studies, not least Nietzsche's assault on philosophical rationalisms and idealisms, and Foucault's analysis of techniques of bodily management, presentation, and control as integral to the constitution of social power and cultural meaning. Since the late 1970s, literary critics and social and cultural historians have turned their attention to the loving, joyous, yearning, dying, or dismembered body, beginning perhaps with the influence of Bakhtin, or with Leonard Barkan's *Nature's Work of Art* and Philippe

Ariès' *L'Homme devant la mort,* although any selection of originary texts is somewhat arbitrary.[15] More recently cognitive and neurological sciences have also begun to influence literary theory and critical practice.[16] The subsequent lively interest in all things bodily in current historical and literary studies is the context in which this book is written and positioned, not only to take advantage of the interest, but to offer its readers an interdisciplinary method of historical and critical analysis of textual and visual representations of human anatomy, like the poetic allegory of *The Purple Island.*

Convictions about the body of an epistemological and evaluative kind come from outside the field of biology, though now but even more so in the early modern period, they inform its language, often through figurality. With this in mind, this study will draw on Renaissance and early modern poetic and rhetorical theory as well as modern aesthetic, literary, linguistic, and cognitive theory, to examine signification and figurality in *The Purple Island.* The degree to which signification involves either a correspondence between anatomy and other ideas, or understanding one thing—human anatomy—in terms of another that comes from outside, are variable, yet both are found in the work. The latter, which is metaphoricity, is often but not always conceptual. When conceptual it produces of and by itself the entities it signifies which may have no equivalence in nature or in ideas existing independently, because the truth or falsity of the cognitive content is dependent on conceptual systems which are themselves largely metaphorical. Comparative methods will also be employed herein, in order to reach a nuanced conclusion as to the role of figurality in *The Purple Island,* and what it may share with related literary and anatomical works. It is also important to ascertain the extent to which figures may be supposed to have required cognitive processing in the mind of an early seventeenth-century reader. Without this, we have a less reliable conception of a figure's conventionality or innovativeness. With a view to determining the innovation and conservatism of *The Purple Island,* the book explores the literary intertextuality, context, and form of the poem, the epistemology and hermeneutics of its language and figurality, and the relationship between ingenuity and the "conceit."

A question raised in an essay by Paul de Man is answered insofar as possible within the limits of this study: whether the entire semantic, semiological, and performative field of (anatomical) language can be said to be covered by tropological models. Analysis of this kind involves what de Man calls reading "semiotically" as opposed to semantically.[17] However, it is not my intention in this study to treat as suspect the difference between signifier and signified, or to place in doubt either the historically contingent meanings of anatomical terms, or the ability of the current of anatomical discourse to overcome the disruptive introduction of figures.[18] Rather, my intention is to explore the anatomical rationality of figurality. This is a topic that on the macro-level of paradigmatic models, metaphors, and analogies has received close attention in recent history and philosophy of science, but has barely been broached on the micro-level of textual analysis of anatomy textbooks. Significant work has been done specifically on the relationship between anatomy and architecture in the Renaissance, and this will be discussed; yet to my knowledge, the conceptual character and function of the comparison has rarely been analyzed in relation to theory of analogy or metaphor, or to cognition. Extensive attention is given in this study to the occurrence in anatomical and literary works of language informed by analogical or metaphorical concepts of the body as a microcosm, as constructed like a building or a city, by Nature or by God, as sharing society's stratification, or as participating in the vegetative growth of plants and sharing the features of the terrestrial globe. Moreover, this study attempts to reconstruct the rationality of the appearance of such figures in anatomy textbooks as the teleological understanding of morphology and function in terms of the faculties of the soul. More than any other literary work of the period, *The Purple Island* explores hidden aspects of this figurality and creatively develops it.

It is not without some disappointment that in turning to the responses to *The Purple Island* of historians of science and natural philosophy, one finds incidental observations of anatomical character and speculations as to provenance, but almost invariably a lack of interest in the occurrence of either anatomical figurality or literary modification. These visitors to the island's shores do not generally converse with their literary counterparts, which unfortunately is not conducive to understanding the imbrication of figurality and anatomy in Fletcher's poem. The exception to the rule is the only full-length biography of Fletcher, by Abram Barnett Langdale, although this too treats the allegory as superficial.[19] Attempts to extract the anatomy from figurality have given rise to contradictory conclusions as to character and provenance. That it was in part derived directly or indirectly from Galen, Andreas Vesalius, Realdo Colombo, and William Harvey, have all been suggested on either reasonable or dubious evidence.[20] From this it can be inferred, and will be argued in this study, that the allegory is far from superficial, adding to the considerable and sustained effort required of the reader to interpret the anatomy. Unsurprisingly, some critics have come to premature judgments on the assumption that the very density of figurality can only mean that the anatomy is conservative.

Comparatively recently, some critics have come by way of the didactic and religious design or perspective of the poem, by which the anatomy is justified, to understand better the combination of anatomy and poetic figurality. A significant contribution to Fletcher criticism was made when R. G. Baldwin advanced the theory that: "Far from being eccentric and merely fantastic as the early critics would have it, or an unimportant source of occasional ideas as most modern students take him to be, or the radically new apostle of science that Langdale creates, Phineas Fletcher is in fact as conservative, practical, steady a voice as can be found in the seventeenth century."[21] Baldwin offers insight into the central poetic figure—the "*Isle of Man*" (*P.I.* 1.48.7)—as combining "firmly established" analogical strands in philosophy, theology, and literature, between the lesser world of the human body as a microcosm or as microchristus, on the one hand, and the globe as a geocosm, the greater world or universe as a macrocosm, and the cosmic Christ, on the other hand.[22] His reading accounts for the poem's "epic aspirations" and, when extended, is capable of accounting for its analytical and aggregative, almost encyclopedic, survey of its subject.[23] An adherent of the theory that the microcosm of Man is a "secondary Scripture" that reveals the Creator, Baldwin's Fletcher sees the didactic and declarative function of poetry as "a supreme source of information concerning divine mysteries," or, in Fletcher's words (cited below), as God's "sacred skill" in Man.[24]

The character of *The Purple Island*'s narrator, Thirsil, combines the classical, pagan, and sensuous tradition of pastoral (which had been Christianized by Spenser and other sixteenth-century poets) with the Christian, even puritan, identification of the pastor as the "under-shepherd to the Master-and-Owner Shepherd or 'Pastor.'"[25] Before commencing his narration, this pastor-poet invokes this "Great Prince of shepherds" with the rhythm of a prayer for the inspiration to figure forth anatomy in poetry:

> Thou first and last, inspire thy sacred skill;
> Guide thou my hand, grace thou my artlesse quill:
> So shall I first begin, so last shall end thy will.
>
> (*P.I.* 1.33.5–7)

The nature of this invocation and of others in Thirsil's narration is to gesture towards salvation as the "will" of God, which organizes events in divine history, and in which the poetic anatomy has a soteriological role. Although the attention of this study to *The Purple Island* is confined almost entirely to Cantos 1 to 5, in which the anatomy is conducted, the crucially significant events recounted in this story of divine providence throughout the poem, are the Creation, the Fall, the incarnation, death, and resurrection of Jesus Christ, the apocalypse and the general resurrection; and it is through the perspective of these events that significance is drawn from anatomy, the significance of the body's creation, mortality, and salvation. So in a third, religious sense the island can be said to be purple, the color of the robe with which Roman soldiers in the Passion narrative sought to humiliate Jesus Christ, which was subsequently, in ecclesiastical contexts (as in Fletcher's poem), associated with penitence,[26] a preparation for regeneration of will (as of venous blood) in the heart (*P.I.* 6.57–77). The significance of Christ's resurrection in anatomy has received insufficient attention even from historians who have emphasized either the relationship to mortality through the *vanitas* and *memento mori* traditions, or that to creation through classical concepts of design and teleology. This study consequently explores the extent to which anatomy, and poetic anatomy in particular, partakes of the revelation and soteriology of the incarnation.

The religious principle of *The Purple Island* through which even the least parts of the body are regarded as redeemed, enables us to see more clearly how the denigration of the anatomical theme by earlier critics exemplified a distortion of historical perspective, a process which has been subject to considerable analysis in the humanities since the 1980s.[27] Just as Fletcher's metaphorical grasp of the reality of the human body was determined in part by the conventional relationship between figurality and anatomy at his time, so his critics have had an equally historically determined critical point of view from which to interpret the literary treatment of anatomy in Fletcher's poem. The reader will encounter throughout this study a twofold endeavor, conceived in relation to some of the deeply problematic but intriguing questions raised within historicism. However vitiated my method by historical determinacy, motivated character, and the diachronic vicissitudes of signification, its aim is firstly, to reconstitute the intellectual and cognitive grasp of the reality of the body demonstrated by Fletcher's poem—its rational and religious anatomico-poetic epistemology—and secondly, to subject this to analysis such as is possible perhaps only with historical (and distorting) perspective.

It is the first aspect of this endeavor that causes one to suspect that despite the merits of his critical intervention, Baldwin's implicit suggestion that the religious, didactic narratorial perspective of the pastor-poet of *The Purple Island* skews the human anatomy it surveys, entails a historical distortion of the early seventeenth century. The period is distorted by the assumption that while anatomy may have interested scholars from other fields, nevertheless professional anatomy was conceptually and discursively distinct from religious poetry, as indeed it is for the most part now and was when Baldwin wrote. It is with this historical assumption that the most recent contribution to Fletcher criticism has taken issue.[28]

Much traditional history of science and medicine is almost entirely diachronic in a heavily retrospective way, idealistically judging the theories and practices of the past according to a positivist concept of progress, consequently isolating individual discoveries from the contexts in which they were justified and in which the experiments which led to them were conceived. One of the first and most influential historians of early modern medicine to rock the positivist boat was Walter Pagel, who argued for "historical understanding," by which he meant (in his words): "an attempt to develop a medical theory or discovery from the world of ideas cherished by its author and characteristic of his period, disregarding their scientific or non-scientific value; so that ultimately one tries to explain why each discovery was typical of the period which begot it."[29] Pagel's and more recent scholarship in the history and philosophy of Renaissance and early modern anatomy has recognized the immense informing influence of theology and the philosophy of the soul and its faculties on the contexts of project formulation and the justification of discoveries. Since classical times, anatomy had been a part of natural philosophy which studied the world as Creation, and Renaissance and early modern anatomists envisaged the projects they undertook in the context of Aristotelian, Galenic, or Platonic philosophy.[30] This goes some way in this study to explaining why the illustrations of anatomy textbooks and the exhibits in anatomy theaters insisted iconographically on the human body as the image or house of God, on the ephemerality of human life, and on salvation. More generally, this study opposes the binary opposition of science to religion as the effect of suppressing differences within the opposed terms, differences which at least in the early modern period gave rise to a less than clear distinction between natural (or rational) and revealed knowledge.

Moreover, at this time "anatomists began to build up large amounts of anatomical knowledge that had no direct use in the practice of medicine and surgery."[31] This knowledge has been recognized by some historians as such, "by virtue of people knowing it, irrespective of whether or not we now think it to be true."[32] In support of this claim, this study will employ the conceptualist or cognitive argument in contemporary metaphorology that by means of metaphor, understood as an imaginative rationality, we grasp both new and conventional concepts, and that our whole conceptual system is largely metaphorical in character.[33] It is in relation to this system, which varies historically, that metaphors can be considered true or false at any given time. The idealist historian's objection to this might be to postulate an ahistorical ontology of the object of investigation, the body-in-itself, against which all theories may in principle be tested, even if the lack of either an appropriate scientific method or an epistemologically transparent and neutral (nonideological) nomenclature prevented most Renaissance and early modern anatomists from discovering this ontology. But this study does not assume the existence of such an ontology. If this cultural-historical endeavor is not to be merely a history of ideas, records, and representations, however, we must insist with Ivan Illich, that there exists the body "parallel to but clearly distant from the body that was painted, sculpted, and described." But as Illich argues, this body is not a natural fact which stands outside the historian or critic's domain, accessible only to the scientist, for arguably its sensations are culturally mediated or even structured by abstract concepts like enjoyment and suffering,[34] and historically the cognitive content of ideas about it attend to how we experience it, and comprise answers, tentative or emphatic, to the *whys* and *wherefores*, as well as the *whats*. The "phenomenon of the living organism," as the medical historian Owsei Temkin sagely puts it, "is hard to define; our modern textbooks usually shy away from the task."[35] As I believe is implied by some cultural materialism, and by some poststructuralist feminist theory of sexual difference, there is at all times a putative ontology of the body the truth of which is always mediated by, or thought in relation to, discourse and representation, but not exhausted thereby, and may be thought particularly at the "most tenuous borders" of a discourse.[36]

The rejection of historical positivism need not entail the neglect of the material determinacy of history. The textuality of history—the textual nature of its evidence, the character of the practices of the past as always already interpreted—has unnecessarily given rise in new historicism to skepticism of causality in events and an overreliance upon "synchronic analysis of a historical moment."[37] Where early modern anatomists envisaged the projects they undertook as garnering and accumulating knowledge in the anatomical tradition of the classical philosophy they revived, the effects of their discoveries could not often be foreseen and were not always easily recuperated. History and anatomical discovery have a material determinacy, which should not be neglected. To accompany the synchronic digressions from *The Purple Island* in this study, there are also diachronic genealogies or trajectories which can be identified either as anticipated or as actualized in the historical materials examined. However, the pace of the collapse of correspondences like the body politic analogy has in the past been overestimated. It was probably not a sudden "breaking of the circle," but was slow and incremental. Evidence of this lies in John Milton's *Paradise Lost,* which incorporates much contemporary natural history, yet, as late as 1667, assumes a Galenic, humoral psychology from which to develop moral and physiological strategies to regulate appetite.[38]

It was only from the mid-seventeenth century that anatomy began to resemble the subject of modern standard works like *Gray's Anatomy,* as natural philosophers of the Royal Society sought to exclude theology, questions of the soul, metaphor, and analogy from the discourse of science and medicine. During the same period medicine itself changed in correspondence to a paradigmatic change in anatomy—the emergence into dominance of a different anatomical model (one which the natural philosophers of the Royal Society did not acknowledge to be a metaphor)—and physicians began to assume "that the body can be regarded as a machine whose protection from disease and its effects depends primarily on internal intervention."[39] Medicine became increasingly professionalized, and anatomy an increasingly specialized body of knowledge with its own standard nomenclature and particular use to medicine. Some historians have argued that not only pain, disease, disability, and dying, but the body itself (or "body-percept") has as a result become iatrogenic, that is, "induced by a physician or his treatment."[40] The body as a machine, capable of operating in the Cartesian sense without a soul, far less than the ensouled body that had preceded it, was the locus for redemption, and the religious, affective, and figurative associations of dissection were more narrowly focused on the soul. But that story is beyond the scope of this study.

Since *The Purple Island* is situated historically on the threshold of the emergence of the machine metaphor into dominance, my analysis of its figurality accordingly seeks to ascertain the philosophical or theological justification or rationale, the reason *why* literature and natural philosophy figured anatomy, and why it did so in the ways it did in the sixteenth and early seventeenth centuries. Natural philosophy and theology constitute a part of my theoretical apparatus for interpreting *The Purple Island,* particularly with regard to the poem's putative soteriological role and that rhetorical aspect which not only gestures towards an identification with its subject matter, but seems "endowed with a power to instil a supernatural degree of life" not only into the literary artifact (as was often claimed by artists, poets, and theorists)[41] but also the anatomised body-in-itself. Even where the grain of discourse is not theological or philosophical, the use of a figure or a hairline verbal fracture, like a change of register or tone, may function as an index to the basically theological and philosophical contexts in which all the works examined here were written and read.

The most facilitating of recent contributions for the debate continued in this study, have come from the interdisciplinary fields of cultural materialism and cultural history, from Thomas Healy and Jonathan Sawday, respectively.[42] This should be no surprise given the fundamental assumptions shared by many practitioners in these fields, as in new historicism, that expressive acts such as poetry are "embedded in a network of material practices" which are generated by and embody ideology or "master fictions," and that "literary and non-literary 'texts' circulate inseparably."[43] Sawday explores *The Purple Island*'s interaction with its cultural context, forging intertextual links, and concluding that the poem synthesizes the anatomy of textbook and anatomy theater with the body in poetry and religion. By contrast Healy argues that both anatomists and poets ignored the distinctions that at this time were already emerging between their respective disciplines, and that *The Purple Island* appropriates early modern sources of anatomy for an intellectually reactionary project. In some part this study is a negotiation of the impact of these authors on the interpreted text of *The Purple Island.* But in exploring the textual negotiations that occurred in the early modern period, it is intended that the usefulness of a variety of interdisciplinary and new-historicist methods and tools will be critically assessed, and that of concepts used in Renaissance poetic and rhetorical theory will be explored. It is hoped thereby that a contribution will have been made to the marking out of a distinct interdisciplinary space between early modern literary studies and history of medicine, natural philosophy, and theology.

To this assessment of recent criticism, my examination of the congruity of anatomy with figurality will be a necessary preliminary stage. Textual and linguistic analysis that requires close attention to the grain of a text does not necessarily give rise to a "closed" reading, as is often suggested by theoretically informed historicists. Throughout this study analyses of figurality and anatomy in *The Purple Island* will lead in digressive ways to analyses of related passages of anatomy textbooks, as well as to works which are literary, theological, religious, natural-philosophical, and medical. Moreover, analysis will not be confined to written texts but will extend to techniques, conventions, conceptual and pictorial content, iconography, and style in anatomical illustrations, design of anatomy theaters, and method of public anatomical demonstrations, as well as the form and function of anatomy therein displayed or demonstrated. The desire to erode the text/context, literature/historical background dichotomy in the interests of a more nuanced conception of works of literature as culturally emergent events, albeit physically embodied ones, but not objects, and without clearly defined boundaries between what is inside and what outside, has led the digressions in this study to be extensive, not invariably individually, but certainly *en masse.*[44] It is thought that this method is justified as more equipped to tackle the linguistic, conceptual, and formal complexities of *The Purple Island,* as well to analyse the rhetorical corporealization of the text and textualization of the body-as-referent, which erodes the distinction between text and context which formal literary criticism and literary history tend to preserve.

Between texts and practices connections of a traditionally historical and biographical kind will be sought, and perhaps the most significant departure in methodology from recent criticism of *The Purple Island* is into the provenance and conceptual character of its anatomy. The aim of this source-criticism is to ascertain how the intratextual relationships between words and connotations, in the special set of conditions that pertain to *The Purple Island*'s form and allegory, effects changes in the anatomy of its sources. Where biographical and source-criticism does not establish a connection, there will nevertheless be reason on occasion to postulate relationships that do not depend on such contiguous relations. While wishing to avoid arbitrary connectedness, consideration will be given in such cases to formal, conceptual, linguistic, figurative, and semiotic similarities between *The Purple Island* and other texts and practices, which raise problems too complex to be reduced to the formula of literature as a mirror of society or reality, and which in one highly significant case will be theorized in a way not dissimilar to the modes of exchange proposed by Stephen Greenblatt.[45] However, the exchanges brought into focus are more susceptible to analysis as instances of the "rhetorical *energia*" of mimesis (the concept that best describes Greenblatt's use of the image of the mirror invested "with a sense of pressure as well as form") than as negotiations that approximate to market relations.[46] Specific individual utterances (what poststructuralist linguists call *parole*), which include Fletcher's poem, may resist the modes of exchange that predominate in early capitalist market relations by adopting equally nuanced and adaptable theories of linguistic representation from theology, natural philosophy, rhetoric, and poetics, which are not always everywhere determined in the last instance by what happens in the economic base of society. Moreover, my examination will bear out the fidelity of *The Purple Island* to a mimetic strategy of representation rather than to one of appropriation.

Both Healy and Sawday place great emphasis on how the formal layout of the pages of *The Purple Island* can be understood to bear the burden of the historical change in the relationship between religious poetry and anatomy and represent the poem's strategy for negotiating this change. As an approach to the question of the congruity of anatomy and figurality, and to that of modes of cultural exchange, I find this helpful only to a point. At the last these questions require analyses of language and rhetoric at least as much as formal correspondence, which, notwithstanding, has profound consequences for the way in which *The Purple Island* is, was, and was intended to be read. Indeed, a question asked by recent critics, though as yet not satisfactorily answered, is what *The Purple Island* effectuates either by synthesizing anatomy and poetry, or by appropriating the former, or by performing whatever other strategy it may be interpreted as deploying. We can only speculate as to what Fletcher intended, not merely because there is no extant record other than the poem itself, but as pure concept intention remains inaccessible even where putative intention is declared; and even if it were forthcoming authorial intention ought not, and of course would not, foreclose acceptable interpretation.[47] But answers may be sought to how changes effected in the anatomy affected the perception of the work. Without a full consideration of both the work's intervention in the cultural negotiations of anatomical debate and consensus, and its effect, we would be writing only a narrowly implicative kind of literary criticism or a brief episode in the history of representations. Yet cultural critics have by and large omitted consideration of *The Purple Island*'s contemporary reception, and in order to redress this omission, this study will pay greater attention to contemporary responses such as that of Howell.

Finally, I should like to outline the limits of this study, which does not answer Illich's rallying call for a history of the body as "parallel to but clearly distant from the body that was painted, sculpted, and described in that historical moment."[48] The study will confine its analyses for the most part to the textual productions of high culture and the putative experience of one Fletcher who participated therein. Textual analysis will be combined with theoretically informed historical analysis of a predominantly intertextual kind. But the study does recognize "a distinctive awareness of the body as the primary *locus* [not only] of experience," but of systems of concepts both concrete and abstract,[49] which in turn create the context in which the body's metaphorical structure is rationalized according to natural-philosophical and theological precepts. In this way, the study is aligned with the work of cultural and social historians, sociologists, and cultural theorists, who have turned their attention to bodiliness as a condition of understanding the society or culture in question, either from within or from without.[50] This awareness informs my methodology principally through an understanding of the emergence of conceptual systems that are largely metaphorical from spatial-orientational and ontological metaphors, from the fact that we have bodies of the sort we have and that they function as they do in their physical environment and in relation to other physical objects in the world. The writings studied here return us to something vaguely Aristotelian in which the body "is both phylogenetically and ontogenetically one of the first and most basic entities the mind can grasp."[51] This book is not concerned only with what some philosophers and historians of ideas occasionally still regard as "disembodied knowledge." The Renaissance did have such a notion,[52] yet works like *The Faerie Queene* and *The Purple Island* abundantly testify to the significance of intellectuals' embodied experience for knowledge-making or -revealing processes.[53] Additionally, Renaissance knowledge not only has an embodied production, but is maintained and reproduced through texts which frequently, like *The Purple Island,* are modeled formally or imaginatively on human bodies. Some of these works can be interpreted as reconstituting the anatomy they describe, ensuring the embodiment of knowledge through a kind of mimesis of the body-as-referent.

In the digressive interdisciplinarity of this study, *The Purple Island* is the one fixed if semiotically unstable point, chosen because of its complex figurality and the fertile hermeneutic difficulties arising from the indeterminacy of its anatomical, literary, and religious signs. These signs are understood as negotiated in or by the work in relation to the constraints and possibilities of narrative, discourse, genre, doctrine, and the energy of historical change running through the whole. This is an examination of *The Purple Island* as an *index* to anatomy in Christian literature, natural philosophy, and theology. Like any index it is selective: "anatomy" in early modern England could mean many things, not all of them relevant to Fletcher's poem. This focus makes the study a work of literary criticism, and as such holds that a construction or interpretation of a literary work, "may be said more probable than another," or "most probable" when it "accounts for the greatest number of facts provided by the text, including potential connotations."[54] (This is not to construe the work as an object or a unity, but as embodied graphically in the extent of its signifiers, and materially in bookhood.) But as a literary criticism that treats meaning as culturally dispersed, it concentrates on the anatomical cantos (2–5) of Fletcher's poem, rather than those (7–12) that describe the moral and theological virtues and vices and the events of the apocalypse. Therein it owes its allegiance to interdisciplinary historical analysis of anatomy rather than to formal criticism. On account of its treatment of meaning as culturally dispersed and possibly conflicted, it may form part of a larger history of the "culture of dissection" (however qualified or nuanced that appellation needs to be), and of the body-poetics both of the early modern period and of the present.[55]

The "reciprocity between *text*-interpretation and *self*-interpretation" which characterizes Howell's account to Benlowes of reading *The Purple Island* in 1645, suggests to me, as a modern reader, "the capacity to let new [albeit historically past] worlds build our self-understanding."[56] This is especially so because from modern perspectives *The Purple Island* has appeared acutely vulnerable to imminent historical change, as epistemologically conflicted or as achieving a fragile synthesis at probably the last possible moment in history when anatomy and religious allegory could be synthesized. I do not altogether agree with any but the first of these assessments, but propose that by extending conventional, synchronous structures of analogy, Fletcher may have sought to shore up knowledge of the body that was already implicitly assumed to be true. Yet if *The Purple Island* is composed in any measure of innovative metaphoricity, it might be thought to interrogate and break the mold of these dominant structures, thereby helping modern concepts to emerge. Additionally, whether the poem's figurality is found to be congruous with or conflicted by its anatomy, depending on the character of the anatomy, this may affect our self-reflexive reading of *The Purple Island* either as confirming or interrogating modern assumptions about the relationship of scientific and religious knowledge pertaining to the body, and may indeed affect our view of the assumptions that might belong either to the early seventeenth or the early twenty-first century.[57]

I
The Figurative Design of *The Purple Island*

1
Conceptual Metaphors of the "*Isle of Man*"

MUCH OF THE CRITICISM THAT PHINEAS FLETCHER'S POEM *THE PURPLE ISLAND* HAS ATtracted has failed fully to appreciate the complex structure of its plurisignification. However, almost without exception, the most thorough studies have been concerned, perhaps unavoidably, with the work's combination of anatomical exposition and Christian allegory (with epic and pastoral elements).[1] In the only critical monograph on the writings of Phineas Fletcher, and his brother Giles (1585/86–1623), Frank Kastor deploys his criterion of aesthetic unity to evaluate disparagingly what he interprets as the incongruity and dissonance consequent upon the poem's discursive, disciplinary, and generic juxtapositions.[2] Kastor says that, in writing an "anatomy lesson in verse form," Fletcher encountered, but did not solve, a discursive incompatibility: "To translate anatomical language into common layman's terms is difficult enough, but to turn a technical vocabulary into poetry involves more: translating a highly precise but denotatively limited vocabulary into its opposite. For Fletcher, the problem was even greater: he had to translate anatomy (the Body) into a combination of geographical jargon (the Island) and esthetic language (poetry)".[3] Kastor's verdict on Fletcher's attempt to "translate" one "language" or discourse into two others, is that it resulted in "large-scale incomprehensibility," rather than greater accessibility.[4] The chapters in Part One are concerned primarily to answer the question of the congruity between the subjects and discourses of anatomy and poetry, in the course of which reasons which are poetic, historical, theological, and theoretical, are found for constructing a less negative evaluation of *The Purple Island*'s purportedly "strange" combination of anatomical theme and poetic mode and genres.

THE "*ISLE OF MAN*": ALLEGORY AND ANATOMY

Edwin Honig has observed an allegorical device in the general nature and function of which he perceives slight variation, and which he calls variously the "threshold symbol" or "emblem."[5] Honig says that "[t]he emblem outlines a concise picture in brief focus," while the "threshold symbol generally frames a brief preparatory action," usually an action of the hero; and according to Maureen Quilligan, this "image" or symbol itself signifies the structural concept which shapes the opening landscape or action of an allegory.[6] Principal among the components of *The Purple Island*'s main story, as the story's location and an actant or agent (rather than an action),[7] is the "*Isle of Man,* or *Purple Island*" (*P.I.* 1.48.7), which is the poem's "threshold image." In reference to the figurality of the linguistic and conceptual structure of this component of the text, it would be more analytic to call it a "figure" (although the second section of chapter 3 relates the poem's figurality to theory of images).[8]

As was said in the Introduction, this "threshold figure" signifies simultaneously two meanings. When the diegetic narrator, Thirsil,[9] introduces the isle in stanza 34 of Canto 1, both its meanings can be detected in germinal form:

Heark then, ah heark, you gentle shepheard-crue;
An Isle I fain would sing, an Island fair;
A place too seldome view'd, yet still in view;
Neare as our selves, yet farthest from our care;
  Which we by leaving finde, by seeking lost;
  A forrain home, a strange, though native coast;
Most obvious to all, yet most unknown to most:

(*P.I.* 1.34)[10]

The ensuing explanation of this riddle of apparent contradictions (i.e., paradoxes) is a description of the Isle of Man as a location, a narration of its own actions and the actions occurring within it, and a gloss or commentary on the text of the threshold figure (which, as we shall see, takes two forms: a poetic and a prose marginal form).[11] It is as if the allegory embodies within itself the total form of a work of commentary, including both gloss and text. The chapters in part 1 of this study are concerned with how the question of the congruity between the subjects and discourses of anatomy and poetry may be answered with reference to the descriptive or expository aspect of the poetic stanzas and marginal glosses, while the relationship between the stanzas and glosses is examined more thoroughly in chapter 6. Aspects of the narrative of the poem are analyzed throughout this study, but particularly in chapters 2 to 5, and in chapters 7 and 9.

The isle is the realm of Britain, uniting England (incorporating Wales), Scotland, and Ireland under one sovereign since 1603. These "kingdomes" were already envisaged by William Camden's *Britannia siue florentissimorum regnorum, Angliae, Scotiae, Hiberniae chorographica descriptio,* as united "*Naturally,* according to the course of rivers, and interpose of mountaines," "*Nationally,* according as the people inhabite them," and "*Diversly* and *Civilly,* according to the wils and jurisdiction of Princes."[12] Camden, in fact, discursively *divides* the "land" of these said kingdoms in these said ways; but, as this study will maintain at several places, in respect of several discourses, division is a means to, and simultaneous with, unification through the establishment of a relationship of part to whole.[13] In Camden's *Britain,* the ideal and ideology of national unity preceded its political establishment in 1603, although closer in method to *The Purple Island* is Michael Drayton's *Poly-Olbion,* which I will look at shortly. The state represented in *The Purple Island* is not, however, the nation-state but the dynastic realm, with its legitimation of sacral monarchy (*P.I.* 6.28) and its imagining of a sacred community of dependent members (*P.I.* 1.45); and yet it presages the nationalism that was to accompany the emergence of nationality and of the nation-state in the seventeenth to nineteenth centuries.[14] This realm is perceived by *The Purple Island*'s diegetic narrator, Thirsil, to have been overlooked by those intent on foreign exploration, and so, although "still in view," it is "farthest from our care." The same complaint is voiced in Michael Drayton's Preface "To the Generall Reader" of the 1613 edition of the First Part of his *Poly-Olbion,* where he says that the unusual tract before the reader may prove difficult to those "as had rather read the fantasies of forraine inventions, then to see the Rarities and Historie of their owne Country delivered by a true native Muse."[15]

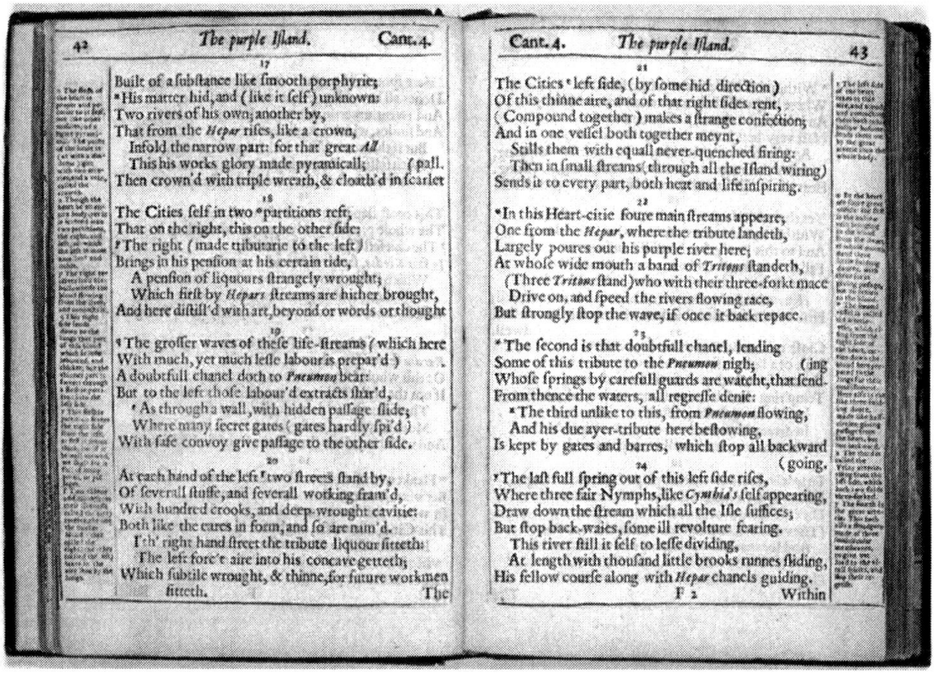

42 *The purple Iſland.* Cant. 4.

17

Built of a ſubſtance like ſmooth porphyrie;
*His matter hid, and (like it ſelf) unknown:
Two rivers of his own; another by,
That from the *Hepar* riſes, like a crown,
Infold the narrow part: for that great *All*
This his works glory made pyramicall;
Then crown'd with triple wreath, & cloath'd in ſcarlet (pall.

18

The Cities ſelf in two *partitions reft,
That on the right, this on the other ſide:
*The right (made tributarie to the left)
Brings in his penſion at his certain tide,
A penſion of liquours ſtrangely wrought;
Which firſt by *Hepars* ſtreams are hither brought,
And here diſtill'd with art, beyond or words or thought

19

*The groſſer waves of theſe life-ſtreams (which here
With much, yet much leſſe labour is prepar'd)
A doubtfull chanel doth to *Pneumon* bear:
But to the left thoſe labour'd extracts ſhar'd,
*As through a wall, with hidden paſſage ſlide;
Where many ſecret gates (gates hardly ſpi'd)
With ſafe convoy give paſſage to the other ſide.

20

At each hand of the left *two ſtreets ſtand by,
Of ſeverall ſtuffe, and ſeverall working fram'd,
With hundred crooks, and deep-wrought cavitie:
Both like the eares in form, and ſo are nam'd.
I'th' right hand ſtreet the tribute liquour ſitteth:
The left forc't aire into his concave getteth;
Which ſubtile wrought, & thinne, for future workmen fitteth.

The

Cant. 4. *The purple Iſland.* 43

21

The Cities *left ſide, (by ſome hid direction)
Of this thinne aire, and of that right ſides rent,
(Compound together) makes a ſtrange confection;
And in one veſſel both together meynt,
Stills them with equall never-quenched firing:
Then in ſmall ſtreams (through all the Iſland wiring)
Sends it to every part, both heat and life inſpiring.

22

*In this Heart-citie foure main ſtreams appeare;
One from the *Hepar*, where the tribute landeth,
Largely poures out his purple river here;
At whoſe wide mouth a band of *Tritons* ſtandeth,
(Three *Tritons* ſtand) who with their three-forkt mace
Drive on, and ſpeed the rivers flowing race,
But ſtrongly ſtop the wave, if once it back repace.

23

*The ſecond is that doubtfull chanel, lending
Some of this tribute to the *Pneumon* nigh;
Whoſe ſprings by carefull guards are watcht, that ſend-(ing
From thence the waters, all regreſſe denie:
*The third unlike to this, from *Pneumon* flowing,
And his due ayer-tribute here beſtowing,
Is kept by gates and barres, which ſtop all backward (going.

24

*The laſt full ſpring out of this left ſide riſes,
Where three fair Nymphs, like *Cynthia's* ſelf appearing,
Draw down the ſtream which all the Iſle ſuffices;
But ſtop back-waies, ſome ill revolture fearing.
This river ſtill it ſelf to leſſe dividing,
At length with thouſand little brooks runnes ſliding,
His fellow courſe along with *Hepar* chanels guiding.

F 2 Within

Phineas Fletcher, *The Purple Island; or, The Isle of Man* (Cambridge: University of Cambridge, 1633) [*STC* 11082], pp. 42–43, displaying Canto 4, stanzas 17 to 24, which describe the anatomy of the heart as the geography of Kerdia, the "Heart-citie." The formal arrangement of each of the pages of the first edition of *The Purple Island* consists of vertically arranged poetic stanzas and prose marginal notation. In Cantos 2 to 5, this notation is particularly full and densely anatomical. By permission of Lylfrgell Genedlaethol Cymru / The National Library of Wales.

Described as "neare as our selves," Fletcher's isle is also Man, though more specifically in Cantos 2 to 5, the human body. Questions of provenance and the conceptual character of the anatomy in *The Purple Island* are addressed directly in part 3 of this study, where the direct or indirect influence of Andreas Vesalius is proposed. When Charles Singer characterized Renaissance anatomy (and especially that of Vesalius) as "*living* anatomy," he explained that: "The parts are not the subject of morphological treatment, but of investigation as contributing to the existence of that complex vital unit we call a Man."[16] Whether what is called a Man, in respect of the definitions arising from *The Purple Island* and Renaissance anatomy, is, in fact, subject to a continuity between Renaissance and modern contexts, can only be answered in stages, and never adequately from anything other than a variety of discursive and disciplinary perspectives. Insofar as concepts of "Man" derive from metaphysics, ontology, epistemology, anatomy, theology, ethics, political philosophy, and ideology, they are of relevance to our subject, though a combination of my skepticism regarding totalizing significations of Man, and the immensity of argument required to move the debate in Renaissance studies in any direction whatsoever, persuades me that in the attempt erosion rather than onslaught is the better part of judgment, that is, to hold definition temporarily in abeyance under a skeptical gaze, while the effect of methodology gathers cumulatively.

Man as represented by Renaissance anatomy is presumed to represent male and female, and to do so in the form of the male except where sexual differences are expressly the object of inquiry. Some recent work in the field of gender in the early modern period tends to suggest that this is indicative of a hierarchical rather than oppositional view of female and male bodies. In *Making Sex,* Thomas Laqueur argues that classical and Renaissance anatomy and physiology produce an understanding of sexual difference as homologous, and that the homology rests upon an assumption that anatomical sexual differences are of degree (of the cardinal qualities "heat" and "perfection") rather than of kind.[17] These cardinal qualities differentiate between bodies that consist of essentially the same flesh, physiologies, and anatomies. There was in effect a "one-sex" view of difference, according to which it is as a result of a higher degree of heat, for instance, that the penis exists outside the body and the vagina within, not because those organs are essentially different from one another.[18] Heat was also thought to correspond to means of fluid elimination, so that female bodies, which were colder, retained more fluid. Thus sexual difference was also understood in terms of the kind, quality, and quantity of fluid.[19]

The fact that bodies could be and were generally divided into two categories is not because the difference between male and female bodies was regarded as essential, or because the flesh organized itself naturally into two distinct and opposite sexes, but because bodies approximate one or the other of two metaphysical and social categories. If anatomical sex was thought not to be ontological, it was believed to approximate to a similarly hierarchical, though much more stable, distinction between feminine and masculine. Although Laqueur encourages caution mapping the conceptual structure of the ancients and early moderns onto our modern concepts, he suggests that what we might call gender was regarded as real, whereas corporeal sex was an epiphenomenon. Anatomical sexual differences were not foundational and were never resorted to in an attempt to shore up evidence for essential, ontological differences.[20] So where we might consider an illustration or a description of a male body to be essentially and only of a male body, this in the Renaissance and early modern period would have been regarded as the more perfect form of the human body, not essentially distinct from the female body, but manifesting in matter the distinct form of masculinity. That it was routinely the male body that was used to express the principle to which Charles Singer alludes, that the parts contribute to the existence of Man, or the human being, is due to that fact that so far as male anatomy was regarded as the more perfect *human* anatomy, it was thought of as representative of the less perfect, female anatomy; though it ought to be stressed, it was *thought of* as representative, and as a result, anatomists neglected aspects of female anatomy, as the third section of this chapter shows.

Some qualification should be made to Laqueur's representation of the dominance of the "one-sex" conception of sexual difference in the early modern period. Whereas Laqueur argues that the special scrutiny to which female bodies were increasingly subject, focusing on their generative capacities, treated the female body as if it contained "necessary and sufficient reasons" for "medical problems and behavioural characteristics," because medical authors "wished to emphasise their specialist knowledge";[21] Susanne Scholz argues, that even if the "one-sex" paradigm was "still called upon to authorize 'medical' observation," and this tied in perfectly with a natural teleology in which the purpose for the sexual parts was reproduction, the special scrutiny "meant that female bodies were perceived, described, and finally constituted as different"; that the female body became "uterocentric."[22]

The subtleties and complexities of sexual difference in the period are exemplified by the anatomical and medical metaphor of the "wandering" womb, which may derive its principal qualities from classical authors. In his *Anatomica methodus,* de Laguna cites Plato's *Timaeus,* 91B, as authority for the notion of "a natural force . . . present in men, unruly and imperious, [which] like an animal . . . will not obey reason."[23] When Laqueur analyzes this passage in Plato, he argues that "Plato refers to *both* the male and female genital organs as animals prone to wander unless they are satisfied."[24] But the Renaissance anatomist de Laguna sexually corporealizes this animality, referring specifically to the "vulva and matrix in women," which "in the same manner as an animal eager for copulation, if it abstains unseasonably long from sexual relations, withstand the privation with difficulty and are greatly angered."[25] Although not regarded as exclusively feminine, the animality of desire is thereby construed as inhering disproportionately in the female body, making women more susceptible to the demands of concupiscence, and suggesting that the female body contains an organ for which there may be no homologue in the male, and which therefore may be ontologically female.

The animality of the womb gives rise to its purported capacity to wander throughout the abdomen. The uterus was ordinarily possessed of a "natural" faculty of voluntary movement, whereby the cervix "doth naturally open itselfe, attracting, drawing, and sucking into the Wombe the seede by a vehement and naturall desire."[26] Privation of the kind de Laguna refers to was often regarded by physicians as detrimental to health because it was thought to give rise to the "symptomatical" movement of the womb, which often included an interception of the ordinary processes of "the instrumental cause of respiration the midriffe [diaphragm]," "which consenting with the braine causeth the Animall facultie the efficient cause of respiration [in the brain] also to be intercepted," leading to a diagnosis of "the suffocation of the Mother," or hysteria, a kind of melancholy.[27]

In the etiology of hysteria, the movement of the womb was itself symptomatic. In the utero-centric model of the female body the emission of superfluity was regarded as heavily dependent upon menstruation. Robert Martensen explains that whereas "men could get rid of excess moisture by sweating, ejaculation, or extrude it to the skin's surface through their larger pores," where it would turn to hair on contact with the air, women's coolness "made such coctions, or transformations, unlikely."[28] For anatomical reasons of its shape, and its coldness relative to other organs, the uterus was regarded as an especially retentive organ. Retention was thought to cause the blood to corrupt and resolve into venemous vapor— "a distillation of variety of corrupt humours flowing from the whole body."[29] Françoise Mauriceau "lacked sufficient colors and odours to describe the 'filth' and 'refuse' in 'that cloaca' [sewer]."[30] Other menstrual etiologies of hysteria include the overdrying of the uterus through evacuation of the blood, but Edward Jorden was typical in thinking that, "as the want and scarcitie of bloud may procure this griefe, so the abundance & excesse thereof doth more commonly cause it."[31]

If the patient's menstruation was currently orderly, a diagnosis of hysteria could still have been reached, and with ethical imputation, because of an alternative etiology in the uterine retention of female "seed." This gives rise to a more venemous quality of humor than does blood because of the unnaturalness of seed's acquisition of a venemous quality. While it retains its nature, which is in fact refined blood, seed is more antithetical to venom.[32] An ethically neutral explanation of this retention was available in the uterus failing to release seed deposited within it by the female "testes." But a consideration of susceptible types of women makes evident the prevailing sexual-political and moral implications of the etiology. Widows were regarded as especially susceptible to the deposition and retention of seed, when formerly they had been sexually active with their husbands, and subsequently on occasion of being sexually tempted or aroused.[33] Any woman with "great store or abundance of seed, and [who] hath been used to the company of a man," the absence of whom leads the patient to "containe her selfe," or any woman who is unaccustomed to the "benefit of marriage" though is "apt thereunto," or any woman who refrains from marital sexual activity either through her own choice, for reasons of religious vocation, or through the choice of her husband, is regarded as liable to retain seed in her uterus.[34] The possible ethical inference of these patient histories is that insufficient marital sexual activity or inappropriate extramarital sexual excitement may result in a pathological condition.

In etiologies of hysteria, seed corrupts when not put to use, and in the administration of cures marital sexual relations arguably subserved the restoration of the effective, reproductive use of abject female bodies. It is intended that, when "the conjunct matter of the disease is scattered and wasted, the womb, and also the woman shall bee restored unto themselves againe."[35] As compared with Mauriceau's disgust at the inferentially diseased uterus, Thomas Raynalde admires that, at "the time of deliverance," the cervix "dilateth and openeth it selfe in such amplitude & largenesse, that it is wonderfull to speake of."[36] The uterus is delivered from the dangerous animality of its appetite, which is providentially reconfigured as having its cause in the divinely ordained purpose of procreation: "that man being mortall, and leaving his off-spring behind him, may become as it were immortall, and live in his posterity."[37]

Another explanation for the onset of hysteria lay in the "mola" or fibroid tumor.[38] The etiology of the mola involves the suppression or immoderate flux of the menses, together with the retention of female or male "seed" within the uterus. The mola was thought to be "a certain false conception" engendered either by the "mixture of the womans seed and menstrual blood," or by "the confluxion of the mans seed that is unfertile [possibly made so by an immoderate flux of menses], with the womans."[39] It is resultant upon masturbation, or sexual excitement which does not result in emission through sexual congress, in the first case, or "immoderate use of Venus" in the second.[40] There was some disagreement in the medical profession as to whether male seed was required, and Paré is insistent that his verdict be accepted, that male seed is required, otherwise "it would be no small cloake or cover to women to avoide the shame and reproach of their light behaviour."[41] According to this etiology, whether or not they lead to a diagnosis of hysteria, the formation of fibroid tumors in the uterus allows the imputation of immoderate sexual behavior in the patient, since masturbation is also regarded as excessive.[42]

Even if a patient's history did not admit of the menstrual or seed explanations, a plethora of uterine abnormalities could be held responsible. Paré throws out a wide diagnostic net, which classifies among the etiologies "the defluxion either of the seed or flowers, or of the womb or whites, or of some other humour, tumour, abscesse, rotten apostume, or some ill juice, putrefying, or getting or engendering an ill quality, and resolved into grosse vapours."[43] With the possible exception of some of Paré's more extravagant etiologies, and certainly with the exception of the suppression of the menses, the etiologies of hysteria infer in the patient either an insufficiency or an excess of sexual activity. Those etiologies which impute these disapproved levels of sexual activity are sufficiently diverse to catch a wide range of female patient histories.

In early seventeenth-century England, there was a virtual consensus of medical opinion that the etiology of hysteria was the "symptomatical" movement of the womb.[44] This medical diagnosis either in a living patient or via postmortem autopsy, as the cause of death, seems to have been exclusive to females.[45] Yet almost the entire constellation of hysteria's perceptible symptoms fluctuated and was observed in both men and women.[46] We should not therefore miss the religious symbolic significance of the principal symptoms when they occurred in women, whose bodies contained the "commodious and convenient place" assigned to the office of conception, wherein Nature "might at her leasure worke her divine feats about the seed once conceived," in order that the vitiating effects of the Fall might be vicariously, though providentially, overcome in procreation.[47] Besides "suffocation" and asphyxia, these symptoms are "hysterical sleep," or torpor, involving cutaneous anesthesia, deafness, cecity, and aphonia.[48] Recorded cases of revival after professional pronouncements of death as a result of hysterical suffocation, sometimes during autopsy, are not uncommon in the medical literature of the period, and became well known, inspiring "admiration and horror" in equal measure in the popular and medical imagination.[49]

The sexual specificity of the diagnosis of the suffocation of the Mother might have arisen from an imperative to support a providentially justified social division of gender roles, rather than from the view that the womb was a foundation of ontological anatomical sexual difference. With alarming facility, the condition's fluctuating symptomatology and highly flexible, widely applicable uterine etiology could be combined with an implicit imputation of sexual irregularity, to provide medicine with a powerfully coercive diagnostic tool that may have been commodious for the regulation of female sexual activity. But, however much the metaphor of the wandering womb, as an explanation of a social illness, became "profoundly bound up with extracorporeal meaning," Laqueur's suggestion that anatomical knowledge prevented its inherence in the body is incorrect.[50] It is true that by the seventeenth century, anatomy had dismissed the idea that the uterus passes via the diaphragm into the thorax, "worming its way up through the body and rising ultimately into the throat, producing the strangulating *globus hystericus* effect."[51] Instead, an explanatory basis was sought in the postures or connections of the womb: "It is joined to the heart by arteries [from the aorta], to the brain and spinal medulla by nerves and likewise to the diaphragm, breasts, kidneys, stomach, ilia, coxae, and pubis; all of these connections are perceptible in strangulations of the vulva or suffocations of the womb by which the breath is withdrawn, dizzy spells ensue, and women think the hair on the top of their heads is being pulled. They also feel pain in the loins, femora, groin, and stomach. . . . These connections are more apparent in the living body, but collapse in the dead person. . . . The uterus is certainly joined to the ilia with certain small nerves twisted back which are extended in pregnant women."[52] The fact that anatomy even demonstrated that the uterus was kept in place by ligaments, and separated from the throat by thick membranes and by the thorax, which was full of organs, was not consistently an impediment to the belief that the uterus actually moved voluntarily when "forced or provoked through the griefe of something contrary to nature that is contained therein," and its movement was thought of as effecting a concatenation of organic movement resulting in asphyxiation: "the wombe, swolne or puffed up by reason of the accesse of grosse vapours and humours that are contained therein and also snatched as it were by a convulsive motion, by reason that the vessels and ligaments distended with fulnesse, are so carried upwards against the midriffe and parts of the breast, that it maketh the breath to bee short, and often as if a thing lay upon the breast and pressed it."[53] Arguably, since the narrator of *The Purple Island,* Thirsil, omits to describe the sexual organs anatomically, except in terms of the desire which inheres therein, and their providential significance, there is insufficient evidence to align him with an anatomical or medical view of sexual difference. Notwithstanding this, the subtleties and complexities of early modern sexual difference provide an introduction to the not dissimilar complexities of knowledge of the body generally, and are relevant in matters which are various, and are to be discussed throughout this study: some of which are metaphysical, such as questions of the ontology of the body-in-itself and of what inheres therein and what is discursively or figuratively structured or defined; some of which are epistemological, such as the problem of rational inference, and the relative statuses of Classical authority and the evidence of the senses; matters which are theological and ethical such as the providentially configured cause of anatomical structure and function in divinely ordained purpose; matters which are political and ideological, such as the hierarchical organization of bodies and of members of the body; matters which are aesthetic, such as the denigration of the body as abject and mortal, and its idealization as Nature's work of art; and matters which are susceptible to narrative analysis, such as the transition from creation to death to revival or resurrection.

Laqueur argues that while the principle of heat remained a constant throughout the Renaissance, the particulars of the homology did change. As a seat of female sexual pleasure the clitoris emerged into prominence in Realdo Colombo's *De re anatomica.*[54] It was to supercede the vagina as the inferior homologue of the penis, initially due to its treatment in Gabriele Fallopia's *Observationes anatomicae.*[55] The conclusion that all women were in some sense hermaphrodites, as a result of their possession of another miniature penis in addition to a vagina, which formerly corresponded to the male penis, led to the corollary, drawn by some Italian writers, that hermaphrodites were really women with enlarged clitorises. "In this way, the discovery of the clitoris," according to Katharine Park, "far from being easily absorbed into the resilient earlier model of male and female as occupying different points on a vertically continuous hierarchy of complexional heat, as Laqueur has argued, seems rather to have contributed to its dissolution."[56] Besides, in Park's view, the early modern medical tradition was full of "complexities, confusions, and controversies" which are not preserved by the supposedly overarching structures of consensus upon which Laqueur's thesis depends. The difference between the Galenic theory of generation which conceptualized bodies as arranged along a continuum of male to female, and hence allowed for intermediates, and the Aristotelian model which was more highly dichotomized, however, is one of emphasis, not of kind. But dissenting voices do seem to have been heard. Andreas de Laguna could not "easily repress a smile" when he saw "Galen stoutly contending that [the female pudenda] are formed from the male genital parts, something which not even nature, the creatrix of all things, could ever accomplish."[57] Park differs from Laqueur in resisting treating medical discourse as though it were a master discourse in the sixteenth century, arguing that "gender was produced and maintained in many different sites."[58] This lends weight to Lyndal Roper's claim elsewhere that "what Laqueur is actually describing is the discourse of medical theory," and "[i]t is not apparent that it was by means of such theory that early modern people understood their bodies."[59] As far as learned discourse is concerned, however, this need not undermine Laqueur's account of the ontological priority of gender over sex, as a brief account of the provenance of the *Purple Island* sexual anatomy may indicate.

In exploring sexual difference, Sawday says that the natural philosophers of the Renaissance could shrug off neither the erotic dimension nor the scriptural tradition.[60] In the *Purple Island* representation of sexual difference in Canto 3, both of these influences are discernible. The sexual parts are introduced as the seat of desire—"here [Cupid] hides the fire with which each heart doth glow" (*P.I.* 3.22.7)—and the division of the sexes and the phenomenon of the "fire of generation" (*P.I.* 3.23.5) are treated in terms of the "great Providence" of God, who "their course foreseeing / Too eas'ly led into the sea of death," "from the first a fellow Isle he fram'd," and after the loss of their Edenic life at the Fall, "gave them a second being, / Which in their off-spring newly flourisheth" (*P.I.* 3.23.1, 1–2; 3.24; 3.23.3–4). None of this helps particularly to answer the question of the poem's representation of anatomical sexual difference, whether it is foundational and oppositional or hierarchical and indicative of essential metaphysical and social differences. Nor does the description of "*Arren,*" man, and "*Thelu,*" woman, as "Alike in all the rest," yet here "disagreeing," in the "sixt and last town in this region" of the abdomen, "Where *Venus* and her wanton have their being" (*P.I.* 3.24.3, 5; 3.22.1; 3.24.6). Disagreement does not imply essential difference. Although an understanding of the homology view might suggest an awareness of an underlying, essential agreement, it is unlikely that the choice of "disagreeing" would have been considered incompatible with it. The view expressed by Jesus in Matthew 19:4–5, that "he which made *them* at the beginning, made them male and female," and that "For this cause shall a man . . . cleave to his wife; and they twain shall be one flesh,"[61] is frequently reiterated in corporeal-geographical terms in Fletcher's poem:

> These two fair Isles distinct in their creation,
> Yet one extracted from the others side,
> Are oft made one by Loves firm combination,
> And from this unitie are multipli'd:
>
> (*P.I.* 3.26.1–4)

Elsewhere as well, the narrator Thirsil finds occasion to draw distinctions between the bodies of Arren and Thelu, where he asserts that the face "in the *Thelu* is the fairer sight," and that the hairs of the head "Much grace the town [the "skull" and "brain" within its "wall"], but most the *Thelu* gay" (*P.I.* 5.7.4; 5.9.6; 5.6 margin. c). But it is not clear that the poem represents these differences of corporeal sex as more than epiphenomena.

Fletcher also borrows a conception of sexual difference from Spenser's *Faerie Queene,* the ideal form or "perfected spiritual body" of which, in David Lee Miller's analysis, is the figure of the hermaphrodite.[62] One of Fletcher's sources for *The Purple Island* was the House of Alma in Cantos 9 to 12 of Book II of *The Faerie Queene* ("The Legend of Sir Guyon; or, Of Temperaunce").[63] According to Miller, Alma's castle refects this ideal form by being "metaphysically hermaphroditic," even while its description, like that of Fletcher's Isle of Man, omits the genital organs.[64] Fletcher, in fact, without referring directly to gender, as Spenser does, borrows from Spenser's architecture of the House of Alma precisely that part of the Pythagorean geometry which Miller argues represents it as hermaphroditic:

The frame thereof seemd partly circulare,
And part triangulare, O worke diuine;
Those two the first and last proportions are,
The one imperfect, mortall, fœminine;
Th'other immortall, perfect, masculine,

(*F.Q.* II.9.22.1–5)[65]

That Trine-one with himself in councell sits,
And purple dust takes from the new-born earth;
Part circular, and part triang'lar fits,

(*P.I.* 1.44.1–3)

The geometry of the Isle of Man hints at essential hierarchical differences of gender in *The Faerie Queene,* significantly, perhaps, without suggesting that these metaphysical differences may originate in differences in the body itself. The adherence of the *Purple Island* anatomy to an anatomical or medical view of sexual difference is inconclusive. Clearly though, the iteration of the "*Isle of Man*" signifies an anatomy that is assumed to be representative of both female and male. But whether the anatomy is male except where sexual differences are expressly the object of inquiry is unclear. When referring to the Isle as a nation, Fletcher adopts the conventional feminine pronoun (*P.I.* 1.35.6). But the signs are indeterminate on whether, if this is merely conventional, and if the anatomy of the Isle of Man is male it should be regarded accordingly as the more perfect human anatomy, and the female, when invoked to display inessential anatomical sexual differences, should be thought of as less perfect.

Crucially, not only Man but also the dynastic realm of Britain is implied when Thirsil describes the providence of "that great Wisdome" for the "Island": "being one, soon into two he fram'd it; / And now made two, to one again reclaim'd it" (*P.I.* 1.48.3–6). In geopolitical terms, the reclamation referred to is the unification of the realm of England and Scotland under James I/VI in 1603. As we shall see, *The Purple Island* exemplifies a creative exploration of the body politic analogy as analogy in which "concepts of the body . . . may be structured by *and* structure concepts of social organisation, even order those social structures themselves."[66] The hierarchical order of the body of Man is represented in *The Purple Island* with the brain, "the seat of the minde, and senses" (*P.I.* 5.6 margin. b), as the principal organ in the "head," which is "The third precinct," the "nobles[t] in frame, and office," the "highest set in place, as in degree" (*P.I.* 5.4.1, 7 & margin. a), in what is essentially a Galenic anatomical hierarchy. Although Galen (born *c.* AD 130) did not develop the political analogy of the body, two of his and the Renaissance's principal influences, Plato (427–347 BC) and Aristotle (384–322 BC), both used the analogy.[67] In his *Republic,* Plato compares the divisions of the body politic to the "three elements or 'parts' of the soul"—the rational, the irascible or agitative, and the appetitive or concupiscent—which in turn relate to a tripartite division of the body, although the body plays an ancillary role.[68] Aristotle's hylemorphic conception in his *Politics,* however, fuses body (matter) and soul (form) inseparably, in both the political and the human organism,[69] and his immense influence on later political literature is felt mostly after the middle of the thirteenth century until the end of the sixteenth.[70] In drawing analogies between parts of the body and parts of the polity, medieval and Renaissance political literature posits the sovereign as analogous to either the heart (after Aristotle) or the head, and sometimes to both, as in *De principis instructione* by

Giraldus Cambrensis (c. 1146–1220),[71] and in *The Purple Island,* wherein God frames the "Prince" "like himself," "A little living Sunne" (*P.I.* 1.45.4, 6, 7). This "Sunne" "moves" in the "dissected wreathes" of the thorax as the heart (*P.I.* 4.9.2), thereby drawing analogies between the political realm, human anatomy, and the cosmos or heaven.

Leonard Barkan suggests a three-stage development of the analogy of the body in political thought. The first stage is "purely anthropomorphic," and purportedly projects onto the commonwealth an understanding of unity in diversity derived from the body.[72] The second stage is a development and refinement of an organic conception of the commonwealth. Once an organic theory has evolved the original corporeal source dies imaginatively, but, in the third stage it is reborn or reactivated in order to defend some interpretation or version of the organic theory.[73] Such reactivations frequently occur when both analogous signifieds of what Fletcher calls this uncannily "strange, though native coast"—native soil and the human body—are in transition historically. This is the case in the late sixteenth and early seventeenth centuries, as is evident from Camden's etymological redefinition of the place-names of Britain's parts, and from the anatomical texts discussed in due course in this chapter. The analogues are in transition partly on account of each being an analogue for the other, and therefore implicated in the other's transformations, divisions, and reunifications.

As a mode and genre, the allegory of *The Purple Island* is, then, in some measure based upon analogy, which provides one of four possible grounds of metaphor in Aristotle's formulation, such that "there are four terms so related that the second (B) is to the first (A), as the fourth (D) to the third (C); for one may then metaphorically put D in lieu of B, and B in lieu of D.[74] This might be referred to as the "simple-analogy view" of metaphor, according to which, a proto-nationalistic aspect of the "truth" of the *Purple Island* metaphor or analogy of the Isle of Man might correspond to the perceived equivalence of two relationships, that between an island and that which surrounds it (the "sea" with the lands that lie beyond the sea), and that between Man and the knowledge "he" might obtain by trying "what e're is in the continent" (*P.I.* 1.51.1; 1.53.3). The superimposition of the geographical boundary on insularity with respect to sources of knowledge occurs in the statement and exclamation: "None knew the sea; (oh blessed ignorance!)" (*P.I.* 1.51.1). To be ignorant of the dynastic realm's boundaries, and those of Man, is "To be confin'd in bounds of happinesse" where and when "None sought new coasts, nor forrain lands descri'd" (*P.I.* 1.53.2; 1.51.6); whereas the entrenchment of the national boundary and psyche is only conceivable as a response to having permitted curiosity, in Thirsil's words, to "seek out ill, and search for wretchedness" (*P.I.* 1.53.4).

As the expository and narrative poetry unfolds to the reader of *The Purple Island,* the enwrapped analogues intermittently rise to become uppermost in the description.[75] At its creation out of "purple dust" by God (*P.I.* 1.44.2), the "Isle" seems to be human; for it is a microcosm, or "an Index" to the "summe of all" (*P.I.* 1.43.6, 7). However, God is said to have

> plac't it in the calm pacifick seas,
> . . . .
> Then peopled it with subjects apt to please
> So wise a Prince . . .
>
> (*P.I.* 1.45.1–4)

So the signification of the isle as a realm rises to the surface, only to submerge beneath the island's human signified when it is revealed that God did not make it "like other Isles; but gave it / Vigour, sence, reason, and a perfect motion" (*P.I.* 1.46.1–2). A moving island is not an unprecedented figure, and would be used again in later literature, but the innovation of its use in *The Purple Island* and its structure and significance at the heart of the allegory will become apparent in this and the following chapter.

The anatomical stanzas of Cantos 2 to 5 of the poem are accompanied by prose marginal glosses, without which, in Kastor's opinion, "most of the content [of those stanzas] is meaningless," and with which the reader becomes embroiled in a process of translation.[76] Kastor selects the following stanza and accompanying marginal notes to illustrate his point:

e The livers upper part rises
& swells gently; is very smooth,
and even; the lower in the
outside like to an hollow rock,
rugged & craggy.
f From it rise all the springs of bloud,
which runnes in the veins.

[e] Much like a mount it easily ascendeth;
The upper part's all smooth as slipperie glasse:
But on the lower many a cragge dependeth;
Like to the hangings of some rockie masse:
[f] Here first the purple fountain making vent,
By thousand rivers through the Isle dispent,
Gives every part fit growth and daily nourishment.
(*P.I.* 3.7)

The heavily terrestrial language of the stanza characterizes even the anatomical notes in the simile which compares the lower external surface of the liver to a hollow rock; but, contrary to Kastor's judgment, the stanza is neither incomprehensible nor meaningless. The liver has already been introduced in stanza 5 of the same canto as "Th'[a]Arch-citie *Hepar,*" from the Greek for "liver." The marginal note "a" reads: "Of all this lower region the Hepar, or liver, is the principall." Assembling this information, the reader deduces that Hepar is at once an organ *and* an important city situated upon a prominent geological feature. If the translations from Greek to English, and from margin to stanza, embroil the reader in linguistic complexity, it should nevertheless be acknowledged that there exists in the poetic stanzas some consistency and coherence, as a result of which anatomico-geographical information can be progressively assembled across cantos to form a coherent fictional location. For example: the "streams" of "purple-colour'd dew," which arise from the "purple fountain" described in the stanza cited above, are described formerly as among the "diff'ring streams" "Circling about, and wat'ring all the plain" of the isle (*P.I.* 2.10.1,7,1; 2.9.5). In order thoroughly to refute Kastor's claim of incomprehensibility, however, it is necessary that the extent and character of this metaphorical coherence and consistency in *The Purple Island* is explored, and this is done theoretically in several sections below and through close textual analysis of Fletcher's work in chapter 2.

Kastor's puzzlement in the face of the poem's semiotic complexity is shared by George Parfitt who, although primarily concerned with *The Purple Island*'s hybridity of genre, makes the observation that it is "a very odd poem indeed, for—all else apart—the style is constantly transforming the basic anatomy into self-conscious art."[77] Edmund Gosse was also concerned with poetic treatment but his response to this was more favorable, for although he considered the theme of Cantos 2 to 5 to be inappropriate for poetic writing, the poem is said to be "extremely ingenious, cleverly sustained, and adorned as tastefully as such an unseemly theme can be by the embroideries of imaginative writing."[78] By considering the poetic execution of the unusual plan of Cantos 2 to 5 to be ingeniously coherent and by implication comprehensible, Gosse is in a critical minority.[79] But he shares a critical paradigm with those who find the execution of the plan incomprehensible. Jonathan Sawday begins a chapter of his doctoral thesis by disclosing part of the assumption behind Gosse's criticism: "To say of [*The Purple Island*] that it is 'clever' is to see it as an exercise in the lumpy welding together of scientific ideas and imaginative writing," a view which assumes the inherent incongruity of the theme and language involved.[80] Cleverly sustained or incomprehensible: both views of the poetic modification of anatomy in *The Purple Island* regard Fletcher's choice of subject as inappropriate for poetry, at least for poetry of the highly decorous and imaginative kind to which *The Purple Island* belongs.[81] The cognitive content of such poetry is easily overlooked, and the other part of Gosse's assumption is that poetry and poetic metaphor in particular are ornamental whereas anatomy is a body of knowledge.

In his critical biography of Phineas Fletcher, Abram Barnett Langdale seems to modify Gosse's opinion that the admittedly extraordinary plan of *The Purple Island* is "cleverly sustained," by suggesting that the anatomical content and the poetic rendition of it are not inimical: "It seems to me that just one work [of Phineas Fletcher] is unspoiled by any structural weakness: Cantos I–V of *The Purple Island.* Here the material is original and here the author's technic is perfectly adapted to its presentation."[82] Langdale's response to the poem is otherwise strikingly different from Gosse's, because his critical interests and evaluative criteria are entirely different: "The reason why no one has suspected the real significance of *The Purple Island* is obvious enough. Nothing could be less scientific than the half-allegorical, half-philosophical, and entirely fanciful method chosen by Fletcher for presenting the cold facts of the laboratory."[83] Langdale is interested to extract the anatomy described by the poem, and to contextualize it in the history of science. As such although he does not regard the anatomical content and the poetic rendition of it as inimical, he does regard the poetic modification of anatomy rather as a hindrance than a saving grace. His approach to the poem's combination of anatomical theme and poetic language is of a type that is concerned with its literariness only insofar as it contributes to the interpretative difficulties of the poem's cleverly modified anatomical content. He is comparatively less concerned to judge the aesthetic success or failure of these modifications of material by technique. His evaluation corresponds to the relative conservatism or progressiveness of the anatomy, once it has been sufficiently extracted. Once it is clearly ascertained, he interprets it as predominantly Renaissance and anti-Galenist and derived in some measure from anatomical demonstrations, thus positioning it in a historical model of inquiry into nature which identifies the origins of modern science in the Renaissance.[84]

The possibility of interpretative or hermeneutic error is increased by such an extraction, however tempting, because the critic's isolation of the poem's anatomical content for analysis depletes it of its moorings in analogy, thereby risking a more precise (and therefore possibly more apparently scientific) interpretation than the allegorical context perhaps warrants. Although the appellative "scientist," which "only came into the language in the 1830s," need not have been coined at precisely that moment when the epistemological premises of inquiries into nature took a new turn, as an attempt to take stock of that change, David M. Knight's warning may need to be heeded here, that the whig interpretation of history is "particularly seductive to the historian of science."[85] And it may be found to be significant that the term "scientist" was coined by analogy with "artist," thereby arguably helping to create, or at least give expression to, a distinction that had not formerly, or perhaps formally, existed. Renaissance and much early modern anatomy was a science only in the classical Latin sense of *scientia,* "knowledge that could be perceived and understood through rational means." For Sir Kenelm Digby, in the later seventeenth century, Catholicism was a "science" in the same sense, for he found in the "perfection of faith expressed in Catholic doctrine" a "standard that all laws and principles in natural philosophy must meet."[86] It is Langdale's conviction, with regard to *The Purple Island,* however, that the "allegorical scheme of presentation is superficial, so that instead of penetrating the scientific facts, it hovers over and around them like a cloak."[87] Langdale does not expound this point further, but it will be done here, theoretically, and with a close textual analysis of the poem and a comparative study of anatomical texts.

Langdale's conviction that the allegory is superficial with regard to the anatomy in the poem suggests that he considers the anatomy to be, in conception, independent of the figurality with which it is presented by the poem. In order for this to be so, the language of the anatomy must be one of two types. It might be literal, that is, in the strictest sense recognized by George Lakoff, "directly meaningful language—not language that is understood, even partly, in terms of something else."[88] Langdale, however, does not rule out the possibility of the language of anatomy being figurative, but his claim that it is free of allegory implies that its figurality would be such that it is distinguishable from the figurality of allegory.

In order to decide on this question it is apposite to draw upon studies in theory of meaning and figurality, other than that of Aristotle who nevertheless remained the single most influential theorist in the sixteenth and early seventeenth centuries. Wayne Booth proposes that, "the many things that we call metaphor are not mutually compatible under a single determinate definition; at most they bear family resemblances to some other members of the family, and some of them are in fact essentially different from some others."[89] So it is important to be precise about what kind of figure is the threshold figure of *The Purple Island,* and this is the aim of the first section of chapter 2. A theory of allegorical, analogical, and metaphorical meaning is required to interpret *The Purple Island,* because, in the event that the allegory were not superficial in the reach of its language or its meaning, the risk of extraction of the kind Langdale describes is that critical methodology and interpretation could be caught in a self-fulfilling circularity, of which Langdale seems to be aware when he says: "Instead of a precise anatomical text, we are confronted with science in allegorical poetry, which offers a perilous footing, and where the highly colored phraseology will often suggest more than the reader has any right to infer."[90] The risk of this kind of misguided inference, of prematurely foreclosing meaning, may be especially acute in two circumstances. Firstly, when the figural aspect of representation is creative or novel; when the figure and its meaning are not consensually agreed, conventional, or pre-given. As this study in due course demonstrates, this is partly, but only partly, true of Fletcher's chosen figure, the Isle of Man. Secondly, this risk also increases with the extent of systematicity of the figuration; that is, the extent to which the entailments of the figuration are explored or detailed. This is the subject of part of chapter 2. Despite the indication that Langdale gives of his awareness of these risks, evidence can be produced from various quarters to suggest that Langdale's conviction may nevertheless determine his findings. This evidence tends to come from that part of the critical attention paid to *The Purple Island* that has been concerned with whether the figurality of the human body is characterized by the conventional analogy of the microcosm and macrocosm. Accordingly, the evidence will enable us in some part to establish the relative novelty of the poem's figurality, and so answer Langdale.

Microcosm and Microchristus

Whether or not *The Purple Island* represents the body as a microcosm is a point of contention between Langdale and R. G. Baldwin.[91] This is predictable because whereas Langdale's assessment is that the significance of *The Purple Island* lies in the Renaissance, anti-Galenist, and thus purportedly "modern" character of the anatomy, it is Baldwin for whom Fletcher's reliance on the conventional analogy of the microcosm-macrocosm is indicative of his intellectual conservatism. This difference of opinion features prominently in this study because our answer to the central question of the nature of Fletcher's intellectual interest in and poetic rendering of anatomy depends upon our understanding of the character of the anatomical, natural-philosophical, and theological ideas in *The Purple Island.* An analysis of the formation of the opinions of Langdale and Baldwin, however, may lead us to investigate the development of the theory of the microcosm, particularly in the Middle Ages and Renaissance, and so to understand differently the nature of their difference of opinion.

Although they were to be elaborated enormously, theories of analogy between the human body and the macrocosm extend historically from Empedocles' (fl. mid-fifth century BC) notion of the "*discordia concors*" of the four elements, ruled alternately by love or sympathy, which unites them, and strife or antipathy, which fragments them into separate concentric spheres.[92] In the diluted form in which the elements are encountered in the world, they are to their pure and essential form as the human body is to the macrocosm.[93] In successive periods there occurred an accretion of correspondences to the elements in Empedocles' theory. These include, most notably, Hippocrates' theory of the four forces or qualities of the human body, to which Galen adhered as one of a number of what he regarded as self-evident maxims;[94] and the development, possibly by Polybius, of the theory of humors, which were originally associated with the elements, but which subsequently lost their immediately cosmic significance and developed an "exclusively corporeal significance."[95]

The theory that emerged as a result of these accretions into Arabic philosophy and medieval scholasticism has become known as the chemical or elemental microcosm. However, the tradition of microcosmic speculation elaborated enormously, and Langdale's argument that the anatomy in *The Purple Island* is free from microcosmic theory rests upon a distinction between this and the (Renaissance) Platonic notion, derived in part from Plato's *Timaeus,* of the human—rather than only the human body—as an index to the whole cosmos, where the cosmos is hylozoic, a living being, which is in turn "a copy in the transitory world of 'becoming' of a divine original which exists in the world of 'being.'"[96] An infinite regress of similitudes, of *aemulatio,* exists between the human and God in this theory, which purportedly Fletcher upholds. Langdale contrasts this with the originally early Greek idea of the human body, as a microcosm, repeating "in miniature all the structural details of the greater world, the macrocosm," a multiplication of analogies which Plato purportedly does not suggest.[97] In each of these theories, knowledge may be construed as "a form of self-knowledge," and Man or the human body as an index or microcosm is "the mediator between himself and the universe; knowledge of one element in the microcosm-macrocosm analogy was analogical knowledge of the other."[98] Man or the human body is always half of a universal atlas.[99] While allowing that the two ideas are "connected" and even "remotely related," Langdale's distinction holds that it follows *only* from the idea of the human body as a microcosm "that the contemplation of God and his universe could serve as an adequate substitute for the study of the body."[100]

Plato does not say in the *Timaeus* that the cosmos is identical to all the organisms it contains, but that its nature is determined by them and of them it is a copy.[101] "Consequently, research into the constituent elements of man's body is a kind of first step toward understanding the more veiled mysteries of creator and cosmos."[102] But the direction of inference in this investigation is not necessarily reversible. Barkan claims that by linking two literally different elements, "[poets] could distort the truth about the concrete features of men and cosmos because they were concerned instead with abstract truths which could be drawn from concrete falsehoods."[103] This is the "figurative microcosm" in the Platonic tradition. But, says Barkan, "there is considerably less literal truth" when the microcosmic idea shifts from the human being to the human body alone, thereby assuming "an equivalence of man's body and the cosmos" and using this literal equivalence "as some sort of scientific key to the nature of the world and man."[104] Mazzeo agrees, saying that the analogical play often found in the works of Renaissance poets "is really indicative of their tendency to intuit nature directly rather than think about it."[105] Consequently, the effect of the "pernicious" metaphor of the microcosm on the history of observational and experimental anatomy "from the death of Galen to the birth of Mundinus" is understood by Langdale to have been inhibitive.[106] In the course of this and the following chapter, theoretical and historical aspects of the approach to analogy and metaphor represented by Barkan, Langdale, and Mazzeo will be challenged. These aspects include: the distinction between abstract and concrete truth and falsehood; the distinction between scientific (literal is implied) language and figurative or poetic language, particularly with regard to the cognitive content of utterances of either kind; and the role of the analogy of the microcosm in anatomy. It remains immediately, however, to resume discussion of the distinction between the theory of the microcosm and the connected Platonic theory.

Because in Langdale's view Fletcher "was far from subscribing to the microcosm-macrocosm speculation," he argues that *The Purple Island* does not systematically explore correspondences between Man and the universe.[107] Consequently, the "allegorical scheme of presentation," which relates Man to the world and universe, "is superficial," leaving the human body relatively free for anatomical inspection.[108] The question of the systematicity of the correspondences between Man and the universe is a topic for this and the following chapter, and cannot be answered simply. But, it is arguable that the distinction which Langdale makes between the (Renaissance) Platonic notion and that of the body as a microcosm is a rather specious one. Barkan holds that in Plato, the search for the proper understanding of one's place in the cosmos is the "sorting out and ordering chaos by means of analogies"; it "is the study of the ways in which the multiplicities of the world's body can be united with the multiplicities of man's."[109] If so, Plato left the door open for analogies between human and cosmic parts, and therefore to the microcosm theory (as understood by Langdale).[110]

Although it is not possible to collapse the difference between the two theories under discussion, it is possible to demonstrate that they were perceived as closer together than either Langdale or Mazzeo allows. One aspect of the figurative macrocosm as God (in a sense, the whole in the world of becoming as that in the world of being), is God's position in the great chain of being, wherein "there are the same number of beings in the whole of creation as may be found eminently contained in God himself."[111] God, who is related to the human by an infinite regress of similitudes, of *aemulatio,* in the Platonic theory, is thereby also linked by the resemblance of *convenientia* to all that He creates, that is, similitude as the visible effect of propinquity or proximity in the great chain of being. In a conventional theological formulation, God is all things in *complicatio,* and the world or universe is all things in *explicatio:* God is an image or text the world unfolds.[112] *The Purple Island*'s diegetic narrator Thirsil describes creation as "When that great Power" "Brought into act this undigested Ball, / Which in himself till then had onely room" (*P.I.* 1.39.1, 3–4). As a microcosm combining in equal weight the four elements and four qualities, moreover, the human is not only an emulation of the macrocosm but also of God. God's resemblance to "his" human creation, "his" "image," is therefore characterized by both *convenientia* and *aemulatio,* which combine to form analogy. Therefore, analogy is the form of resemblance which not only characterizes the multiple correspondences of the microcosm theory, but to a lesser extent informs the (Renaissance) Platonic notion of the whole human's correspondence both to the cosmos and to the divine original. It is only a matter of emphasis whether the whole human emulates the cosmos or only the human body forms the microcosmic analogue.

Baldwin disregards entirely Langdale's distinction between the theories, arguing that "the notion of the microcosm, in all its variety, inform[s] Fletcher's work at every point."[113] He thereby not only contests Langdale's interpretation of the figuration of the body in *The Purple Island,* but also the extent of systematicity of that figuration. In addition, he contests Langdale's argument that had Fletcher subscribed to the theory of the microcosm, he would have felt no need to study human anatomy. Langdale's view is that between Galen, in the second century A.D., and Mondino dei Luzzi, in the fourteenth, during which time there was widespread intellectual acceptance of the body's correspondence to the parts of the macrocosm, the textually transmitted microcosm theory may have had a stultifying effect on the development of the observation of physical objects and processes.[114] Baldwin disagrees, maintaining that for Fletcher: "it was in fact the very reason for his study of the body: a secondary Scripture, as it were, the Work reveals the Workman; and the key to an understanding of the Work is knowledge of the greatest Work of all, man, who is 'a short summarie / Of all that in this All is wide dispread'" (*P.I.* 5.8).[115] The logic with which Baldwin represents Fletcher's translation of an abstract and theoretical model (the microcosm) into "study of the body" requires some clarification and justification, not least because he fails to say which for Fletcher has epistemological priority—the "Work" or the "Workman" (who is the "All"). To understand how the model of the microcosm related epistemologically to knowledge of human anatomy it is necessary to look first at the natural-philosophical context of its origins in ancient Greece, and then to its development in relation to the natural philosophy of early medieval Platonism, scholasticism, and Renaissance Neoplatonism.

Greek natural philosophy advises conformity of the human mind to "the inherent pattern of Nature itself."[116] Thus knowledge or true belief is in conformity with the abstract concept of Nature or the mind of the cosmos.[117] This, however, did not preclude investigation into nature or the cosmos itself, because the idea of the presence of mind in Nature gave it and its organisms inherent order and purpose which made natural philosophical inquiry possible. Answers to "why" questions, rather than only "what" questions, were made possible by this conception of Nature. Therefore, inferences about either of the analogues, Man or nature, could be made from the other. By contrast, much medieval thought was opposed to the idea that either Man or the cosmos is in and of itself well ordered; rather "both take on a system and hierarchy when juxtaposed because they can indeed only be related by means of abstract conceptual order."[118] Correspondence is abstract insofar as it belongs only to the discourses of philosophical and theological rationality. From this point of view, inferences about Man or the cosmos are best made only from the abstract order, as those that are made about that order from investigation of Man or the cosmos are less persuasive. Abstract microcosmic speculation of this kind can be understood in relation to early medieval philosophy. "Before the 'new Aristotle' of the late twelfth century and its widespread adoption in the thirteenth, the Church's philosopher had been Plato." According to Roger French, "Plato, Platonists and Neoplatonists . . . all saw the soul as the important component of the man, the body being merely its workhouse in this world. Investigation into the body would have been an investigation into mere matter."[119] Consequently, after the ancients and prior to the twelfth century, dissection was not carried out in any routine way in the European West.

The adoption of Aristotle in the late twelfth and thirteenth centuries altered perceptions of the body and of dissection: "Aristotle's message about man was that the soul was the essence of human substance, in which inhered the accidents of matter. The form of the adult was the goal of development, in which the nature-of-man, the form or soul, found its full expression in arranging the matter of the body in a characteristic way. Mutilation of the body was material and accidental, not essential."[120] In his books on animals, which became available during the thirteenth century, Aristotle set out "the philosophical worth of knowledge" gained directly from dissections and vivisections. "At least as important for medical men" was Galen, however, and at the outset of dissection in the medieval West, at the medical "school" of Salerno, in southern Italy, Galen's experiments and anatomical descriptions are "held out as something worthy of emulation."[121]

These innovations should be seen in the context of the later medieval scholasticism, which flourished in the period between 1200 and 1500. Scholasticism's careful procedure of weighing the authorities of the works of Aristotle in natural philosophy and Galen in medicine with those of the Arab philosophers and physicians (all of them in Latin translations), with a view to the reconciliation of their apparently contradictory anatomical claims and theories, was applied to anatomy as early as the time of Pietro d'Abano (1257–1315?).[122] The scholastic Middle Ages inherited the ancient and Arabic theories of correspondence, and medieval anatomists subscribed to and even revived the microcosm theory.[123] In the fourteenth-century philosophy and medical theory of Gentile da Foligno, the Hippocratic notion of complexion retained its signification as the composition of any composite thing by four elements and four qualities; but it has been argued that much medieval speculation tended to center upon issues touching on divinity, such as the relative closeness to equality in weight (balance) of these elements in Man compared with other generated things, and Man's consequent closeness to the perfect balance of God.[124] This was because the procedure of reconciling authorities in natural philosophy and medicine was combined with the attempt to synthesize Classical natural philosophy, including anatomy, with Christian theology. This entailed, firstly, that when elucidating the human body, scholasticism "placed emphasis upon the rational justification of religious beliefs and the systematic presentation of those beliefs."[125] Furthermore, as Lind argues, "much early anatomical dissection, far down even into the sixteenth century, had as its immediate object to prove that what Galen and Avicenna or Mundinus had said about a given part of the body was correct."[126] As Andrea Carlino has said of fifteenth-century anatomy at Bologna and Padua, "dissections could only take place for didactic purposes," for "exclusively educational purposes": "Dissection provided an immediate means of learning anatomy by approaching the human body—the source of this knowledge—directly. It constituted, above all, an investigative tool, a technique for unveiling the secrets concealed by the exterior body covering without the mediation of texts. This was a tool that could resolve the contradictions and inconsistencies that these texts sometimes contained. The moment that dissection began to be permitted anatomy could have been rewritten and freed from the authority of Galen and the monopoly of the Galenists. This did not happen. . . ."[127] Teachers of medicine were concerned with demonstrating the truth of authoritative texts, or commenting on them, and even when they refuted Greek, Arabic, or scholastic authorities, they proceeded via scholastic argument and frequently sought authority in another such source.

There was a procedural reason for this: "the dissection itself was performed by unlearned 'anatomistae,'" or barbers or surgeons or even lay persons, while the line of incision in the corpse and the exposed anatomy was indicated or demonstrated by an *ostensor* (demonstrator) with a wooden pointer (the *anatomista,* who was also called the *sector* or *incisor* (dissector), and the *ostensor* were often both referred to as *censors*), and this typical pedagogical procedure required of the teacher (*praelector/lector*) or professor of anatomy only the delivery of "a lecture from Mundinus, one or other of Galen's works on anatomy, or from Avicenna," "literally *ex cathedra.*"[128] Mondino, however, did dissect cadavers himself, and the above procedure developed later, becoming traditional.[129] Made traditional, the procedure of demonstrating anatomy and its incorporation of an epistemological process of inference from authoritative text to cadaver, together with the abstract notion of the microcosm, would seem to suggest that inferences are most likely to have been made about anatomy from either the authoritative texts or from theological propositions about God, rather than about the truth or falsity of authoritative texts or about the nature of God from the human body.

The Renaissance augmented the Platonic and microcosmic theories. Neoplatonism continued the Platonic tradition of Man's correspondence to the All, while retaining something of the medieval emphasis upon God rather than Cosmos or Nature as Creator of (as well as containing) all created things in the figurative macrocosm.[130] In this way the Christian view of the relationship between God, microcosm, and macrocosm differs significantly from the Platonic view, in which the living being of the cosmos is interposed between the human and the divine in the infinite regress or hierarchy of similitudes. The Renaissance was a period of immense debate, but some historians have argued that in an attempt to supply the lack of proper method for studying the world in late medieval Nominalism, the Renaissance expanded and developed analogical reasoning in the Platonic tradition.[131] This continued tradition seems to have taken two different forms. Some, like the most influential exponent of Platonism in Italy during the fifteenth century, Marsilio Ficino, focused on the central position of Man, who shares with lower forms of life the powers of generation, nutrition, and sensation, and with the angels and God, contemplation, but whose distinctive and peculiar attribute is the rational or intellectual part of the soul, the operation and fulfillment of which in and through the world or universe, raising the world to itself and finding its unity in the "otherness" of the world, is Man's *appetitus naturalis* (natural movement or desire). Others, after Giovanni Pico della Mirandola, rejected as unsatisfactory the view that Man owes his distinction to his central position as a static microcosm, holding rather that "humans, unlike all other creatures, do not have a rigidly determined nature, a limited sphere of willing and acting . . . [but that] God formed them according to a general image," an "amalgam of those qualities which are assigned to the other orders," qualities in which humans have a power to share at will.[132] The two forms are closely related and may even be read as interdependent, for it is in the sense that "Man . . . contains all things" that he is by implication differentiable from any particular thing, distinguishable from the world he reflects as a kind of separate world.[133] Andrew Cunningham summarizes these views and develops them in a teleological direction: "Man is a little world, a *minor mundus;* he is every creature, in the sense that he contains all things; man is created in God's own image; man is the last thing created, and hence is the highest likeness of God and the summation of all other creation, and he is the end or goal of the creation process; man stands between visible creation and the things which can only be grasped by the intellect—he is the mean."[134]

Consciousness of himself as a microcosm distinguishes and distances Man from all of creation, granting him free will.[135] Don Parry Norford argues that in the Renaissance view, "the seed of alienation was sown"; Man can only become the whole universe *because* he is separated from it, "as a mirror can reflect . . . images only by being opposed to objects."[136] On the other hand, an important notion in theories of education was that reflection would facilitate identification with everything of which knowledge was sought. Identification combined with alienation from that which one shares with the universe—bodiliness—could lead, in theories like those of Pico and Ficino, to the view that the "incorporeall, immortall . . . or immutable" "soul of man receives and entertaines the generall and universall notions of things, free from all contagion or touch of Matter," and thereby "may be called the receptacle, promptuary, or store-house of all the species or kinds of things."[137] Despite this, it was not only the intellectual soul of Man that was thought to reflect the cosmos. Representation in the sixteenth century was, in Foucault's view, "posited as a form of repetition," and among the kinds of repetition were "*convenientia*" and "*aemulatio,*" of which analogy is composed.[138] And since it is analogy that characterizes the microcosm theory, Helkiah Crooke's *Microcosmographia* asserts, Man's "body is, as it were, a Magazine or Storehouse of all the vertues and efficacies of all bodies."[139] Moreover, "the whole body" is perfectly mixed, for there is "concord and agreement of the foure disagreeing qualities, and so just & equal a mixture of the elements, as that it is the very middle and meane amongst all living and animated things."[140] Thus, in addition to the Platonic theory, as Barkan argues, the literal microcosmic view of the human body as a "little world," which combines "total inclusiveness and miniature proportion," was essential and typical in the Renaissance and early modern period.[141] This literal microcosm could also suffer an alienation, an isolation from that which it reflects, and the reason is also its discontinuity with the classical view, at least "as *Empedocles* would have it," that Man "is in power (in a manner) *All things*" because "he" shares their "matter and substance." Rather, says Helkiah Crooke, "he" is "the Image and resemblance of the whole world" "[*a*]*nalogically* by participation or reception of the severall species or kinde of thinges."[142] Analogy rather than a continuity of composition or substance marks a shift of emphasis, and the beginnings of an alienation or isolation, the relation of which to *The Purple Island* this study gradually explores.

The abstract order to which Baldwin refers in his argument with Langdale is that whereby the microcosm and macrocosm assume the status of a written work of God. The immediate question is whether in the Renaissance inferences were made about God from the written work or about the microcosm from theological propositions about God. Part of the answer must lay in the nature of Renaissance anatomical knowledge, whether, like the knowledge of scripture, it was of the order of revealed as well as natural knowledge. This I shall return to in chapter 3. A further question about the formulation of the microcosm theory, to which Baldwin alludes, is the relationship between the actual human body and the theory of the microcosm as a written work of God. Could a belief in this idea of the microcosm also be an impetus to anatomize the body, as Baldwin suggests? And if so, would such dissections be merely demonstrative of opinions iterated in and pronounced from authoritative texts, or could they be investigative?

According to Lind, "[i]t is around the year 1500 when anatomy appears to spring once more to life as a form of research and publication, accompanied by a steadily increasing number of dissections, both public [often in the newly purpose-built anatomical theaters] and private."[143] But scholasticism and its methods continued into the early sixteenth century,[144] when it met with the Hellenist revival of Greek anatomical and medical terminology and theory—"the pure sources" of the intellectual paternity of Western medicine.[145] Unlike humanism, which "had some aims in common" with Hellenism, yet which, like the aggregative style of Gabriele de Zerbi's anatomy, "respected the Latin tradition of scholarship," the Hellenist revival occurred initially and predominantly outside the schools.[146] There was a notable exception in Alessandro Benedetti, who returned a convinced Hellenist from the Venetian territory of Candia and from Morea in Southern Greece, to teach at Padua in 1490. But Hellenism spread so that elsewhere too the "Arabs and their medieval commentators are brusquely rejected in favour of what Galen actually says."[147] Anatomical disputes were beginning to be resolved less by reconciling, and more by choosing authorities (especially Hippocrates, Galen, Plato, and Aristotle), and it was thought by Hellenists that "[s]uch medical errors as there were might be put right with new manuscripts to hand, or by emendation."[148] The humanists' and Hellenists' recovery and publication of Greek texts, followed by the Counter-Reformation's appropriation of the humanists' methods of textual scholarship, ensured that classical errors and assumptions about analogies and correspondences between the macrocosm and microcosm were enshrined in better edited texts.

In particular, some of the private dissections were experimental, however, involving both the formulation in "imagination" of rational hypotheses (however slight and not yet physiological) and careful observation.[149] The expressed motivations of Renaissance anatomists testify that a belief in the microcosm might indeed be an impetus to anatomize the body. In the Dedicatory Letter to Prince Philip (son to the Emperor Charles V, and later Philip II) prefacing Andreas Vesalius's *Epitome* of his seven books on the fabric of the human body, Vesalius warns that "man" should not be "completely unknown to himself," but recommends examination of "the finished product of a creation most perfected and nigh the most worthy of all."[150] And in the Preface to the *Fabrica* itself, addressed to the Emperor, Vesalius writes that this "most perfectly constructed of all creatures" "in many respects corresponds admirably to the universe and for that reason was called the little universe by the ancients."[151] Moreover, Crooke, among others, could invoke the inscription over the oracle of Apollo at Delphi, famously appropriated by, among others, the Neoplatonists, and claim that "whosoever dooth well know himselfe, knoweth all things, seeing in himselfe he hath the resemblances and representations of all things."[152]

Dissection that was purely demonstrative of analogies held to be true by those who, like Vesalius's dedicatees, were untrained in anatomy, however, posed little threat to the authority of textually transmitted anatomical knowledge of the microcosm, and may not have contributed to any discovery. Notwithstanding this, adherence to the microcosm theory in the first half of the sixteenth century, even by pioneering anatomists like Alessandro Benedetti and Andreas Vesalius, suggests that any incompatibility perceived by Langdale between this theory and the anatomical knowledge their investigations were producing was not apparent to them. Benedetti expresses a belief that "God, the creator of the universe, has in

His excellent reason joined with admirable union things which war among themselves."[153] These things are outlined in Benedetti's *Anatomice* (Bk. I, Ch. 4) as the four humors of the body (blood, phlegm, black bile or melancholy, and yellow bile or choler), with their qualities (possible combinations of hot, cold, moist, and dry), which correspond to the mutually striving four elements of the world and the macrocosm.[154] In comparing "skull to Earth, dura [mater] to Water, tenuis [pia mater] to Air, and brain substance to Fire," Vesalius remarks upon "the wisdom of the Great Craftsman [of the Universe, i.e., God] who, between Earth and Fire, as by nature most different, has placed Water and Air, so between skull and brain (which, by their substances, are as far apart) has interpolated the two membranes."[155] Adherence in early to mid-sixteenth-century anatomy to the chemical microcosm, that is to Empedocles' theory of elements in the world, and to their correspondence to the Hippocratic dogma of qualities (as self-evident in the wake of Galen's adherence) and the theory of humors in the body, is rather the rule than the exception, even among pioneering anatomists. There was no "drastic shift" from a humoral, microcosmic paradigm to a "Vesalian" model in the sixteenth century,[156] partly because Vesalius's anatomy was itself humoral and microcosmic. The dichotomy set up by Langdale is a false one, for the microcosm-macrocosm analogy between the parts of the human body and those of the cosmos should be acknowledged as having been a motivating concept in the study of Man.[157]

The examination of the plausibility of Baldwin's argument, that Fletcher's motivation for studying the body was a belief in the microcosm, requires further detailed analysis of the anatomical content of Fletcher's poem, and this part of the answer is deferred to part 3 of this study. However, the more immediate, and limited, question of whether or not the anatomical cantos of *The Purple Island* subscribe to the theory of the microcosm, and if so to what extent, can be answered here. There are three places in *The Purple Island* where the Isle of Man is predominantly figured as simultaneously a part of and a compendium or summary of all creation:

> He cast to frame an Isle, the heart and head
> Of all his works, compos'd with curious art;
> Which like an Index briefly should impart
> The summe of all; the whole, yet of the whole a part[.]
>
> (*P.I.* 1.43.4–7)

> Look as a scholar, who doth closely gather
> Many large volumes in a narrow place;
> So that great Wisdome all this All together
> Confin'd into this Islands little space
>
> (*P.I.* 1.48.1–4)

> For as this Isle is a short summarie
> Of all that in this All is wide dispread;
> So th' Islands face is th' Isles Epitomie,
> Where ev'n the Princes thoughts are often read:
> For when that *All* had finisht every kinde,
> And all his works would in lesse volume binde,
> Fair on the face he wrote the Index of the minde.
>
> (*P.I.* 5.8)

The first and especially the second of these quotations set up and extend a simile of the body as a book. In *Microcosmographia,* Crooke alludes to "ancient Writers" who "dignified the frame of mans body with the name & title of *The Booke of God*" for the reason that "in all men there appeareth certaine sparkes of a Naturall divinity, or divine nature."[158] The provenance of this simile for *The Purple Island* could be either the theory of the microcosm or the underpinning tenet of Renaissance Platonism.

The use of "*All*" in line 5 of the third quotation suggests a Neoplatonic conception, moderated by Christianity, because the "Great *All*" signifies the Christian God elsewhere in the poem (*P.I.* 4.17.5). Renaissance Neoplatonism is also responsible for the mind or intellect's capacity in the poem of "travelling all places, [though] changing none," of soaring "up to heav'n" or "down" to the "centre" and "*Dis* dark realm" (*P.I.* 6.31.1, 2, 3), thereby reflecting "within his compasse high / All heav'n and earth" (*P.I.* 6.30.3–4). The idea of the face as the "Index of the minde" probably also derives from this Neoplatonic tradition, such that the mind reflects the general and universal notions of things in the world, which is in Fletcher's poem sometimes identified with the "Isle" (*P.I.* 5.7.3), and is reflected in the face. But as "th' Isles Epitomie," neither the face nor the mind could be free from all touch of matter, as in the more idealistic Neoplatonic theories, for the Isle of Man in Cantos 2 to 5 is the anatomy of the body. Thus, it is unsurprising that the use of "whole" in the first of the quotations, and of "All" in both the second and third, signifies the macrocosm in the literal microcosm theory. Significantly the "All" which is the created macrocosm is distinguished from "that *All*" which is the Creator God, as the concepts which they name in Christianity should be. But a firm determination of the provenance of the conception of these textual instances remains elusive, because the poem nowhere expands upon or goes into further detail about these rather generalized microcosmic and Neoplatonic conceptions. The general impression is of a limited adherence to the Platonic and the literal microcosm theories.

A survey of other similar occurrences in the work produces the following results. One reference, astrological in conception, refrains from positively affirming analogical identity between organs and planets (in the scheme sometimes known as *melathesia*), and insists on the use of simile (*P.I.* 3.9).[159] Three further references draw analogies between the human heart and the sun. The first of these refrains from positing identity between the "ends, place, office, state" of the sun in the macrocosm and of the heart, in its guise as "Kerdia," "th'Isles Heart-city" (*P.I.* 4.3.4; 4.2.6). They are represented not as parallel but as "so nearly neare," which in the poem's context of analogies, and in that of widespread acceptance of the traditional analogy even as late as the early seventeenth century, is a significantly meticulous qualification (*P.I.* 4.3.4).[160]

There is no such equivocation, however, in the remaining two textual instances of analogy between the heart and the sun (*P.I.* 4.9, 26), because these are in fact metaphors in which the microcosmic term, heart, is supplanted by the metaphorical term, or subsidiary subject, "Sunne" or "Light" (*P.I.* 4.9.2; 4.26.1). They are instances of metaphor derived from analogy, but their correspondences are not primarily microcosmic. The first explicitly, and the second implicitly, pun on "Sunne, Sonne," and thereby evoke the incarnation, and the doctrine of cosmic redemption.[161] In the New Testament, the concept of Jesus as the Savior of the world "was at the very heart of the primitive Gospel."[162] Moreover, in and through the primitive church, as understood by Paul as Christ's body, all things return to Christ. "This view of the church as a body of cosmic magnitude enabled Paul to relate the 'powers' of the universe to Christ in the same way as human beings."[163] Just as Jesus Christ redeems

Man from within each individual conscience, so the Cosmic Christ redeems the universe, the cosmos; and the property of Christ and the efficacy of his redemptive action is seen on an analogy with the function of the life-giving human heart in the body and of the sun in the cosmos.[164] The body as microchristus, rather than microcosmus, carries the weight of meaning in these stanzas, which, combined with the former examples, possibly erode Baldwin's claim that the microcosm informs Fletcher's work at every point. That notion scarcely seems to be given full credence at those points in the poem at which it is explicitly alluded to, suggesting that the poem's figurality might be altogether more novel than conventional. Langdale's conviction, that the anatomy of *The Purple Island* is independent of metaphorical conceptualization, might be justified to some extent in respect of the conventional theory of the microcosm and macrocosm.

Baldwin, however, qualifies the meaning of the microcosm, in the light of the work of Marjorie Hope Nicolson, to include its correspondence with what she and others before her have called the "geocosm," the world or the terrestrial globe, as differentiated from the "macrocosm," or universe.[165] The two theories are closely related and consistent, for in the body as a microcosm of the cosmos, as in the body as a microcosm of the terrrestrial globe, the heart may be figured as "the Sunne of mans bodie."[166] "For even as the perpetuall and continuall motion of the Sun, and by the quickning and lively heat thereof, al things are cheered and made to flourish; the earth is decked and adorned, yea crowned with flowers, brings foorth great varietie of fruites"; even "so in like manner, by the perpetuall motion of the heart, and by the vitall heat thereof, this litle world is refreshed, preserved, and kept in vigor and good life."[167]

A string of natural philosophers, physicians, and writers from the classical to the Renaissance periods pursued chemical, numerical, and natural correspondences between microcosm and geocosm, among which and of special relevance to *The Purple Island* were those between the humors of the human body and the elements of the globe, and between sensually apprehensible objects in the world and features of the human body.[168] But exceeding all others in the Renaissance, by developing inwardly or "spiritually" perceived analogies between bodily elements and organs, on the one hand, and celestial bodies and all elements in nature, on the other, was Paracelsus;[169] though Cornelius Agrippa also provides examples of the multiplication of analogies, but his magical agenda are rather different from those of Paracelsus. In the classical theory, the four elements in the human body were regarded as diluted, and in the same relation to their pure and essential form as the human body is to the geocosm or macrocosm.[170] But in Paracelsus's "Spiritualist anatomy" there is complete correspondence, part by part, of macrocosm and microcosm, a "network of sympathetic relations" between the greater and lesser worlds, which elevates the position of Man and of the world, and gives them an underlying unity.[171] Barkan claims that in the work of Paracelsus, "Man is inscribed in a hyperorganized cosmos; but the very extent of analogous cosmic bodies stresses the multiplicity and fragmentation of man's body and condition over its unity."[172] Foucault's insight here is that knowledge produced by means of the construction of analogical models of this kind and on this scale is limitless; it proceeds by infinite accumulation of confirmation, because theoretical consistency demands that the whole world must be explored to justify one analogy.[173] However, as a general configuration of nature, Foucault argues, the microcosm poses real limits to the endless addition of analogy, because the whole of the macrocosm must be enclosed in the microcosm's dimensions.[174] Such knowledge as is gained is nevertheless, Foucault says, "a thing of sand," only allowing of addition, but always knowledge of the same thing and of the same kind;[175] without ever approximating firm knowledge of the kind wanted by a modern scientist, or (more to the point) perhaps an anatomist who privileges demonstration to the senses.

Some of this multiplicity of correspondences can be assumed to be encompassed in what Baldwin calls the microcosm's "variety," for the Isle of Man described in Cantos 1 to 5 of *The Purple Island* is quite obviously informed by the analogy between the human body and the island. It remains to unfold the entailments of this analogy and to discern either the degree of its systematicity and coherence, or, conversely, whether, as Kastor seems to imply, the amassed correspondences cause the poem to fragment like the Paracelsian human edifice. This task is undertaken in chapter 2. Before that, a digression through another aspect of the *Purple Island* figurality is necessary in order to explore some part of the multiplicity of correspondences between the geocosm and the microcosm, and in order to determine whether, as Langdale believes, the anatomy is essentially extractable from its analogical and allegorical context, or, as Baldwin argues, the anatomy is conceived wholly within a conceptual framework of analogy. This digression will take us into the figurality of Renaissance anatomy textbooks.

METAPHORICAL CONCEPTS IN ANATOMY

The geocosmic analogue for the body in *The Purple Island* is not simply a geological location, for it encompasses the built environment and human society, and is more accurately described as geopolitical. The body's analogy with this location is characterized by two more basic conceptualizations of the body: as constructed and as natural. The extent to which *The Purple Island*'s anatomical description is conceptualized in these two ways relates closely to the extent to which concepts of the anatomy of the body as a microcosm interpenetrate in the poem with those of the natural geocosm and the constructed environment of the island realm. Furthermore, determining the degree of this conceptual imbrication involves in part referring the poetically modified anatomy to figurations of the body as constructed, as natural, and as a microcosm, if they occur in anatomy textbooks. This should give an indication of the poem's indebtedness to the metaphorical language of scholastic or medical anatomy, against which we can assess the plausibility of extracting the *Purple Island* anatomy from the poem. The plausibility of such an extraction is therefore in inverse proportion to both the degree of systematicity of the analogy in the poem and the extent to which such figurality is shared with the language of the anatomy of the schools and medical profession.

If the anatomical language of *The Purple Island* cannot be found to be without analogy or metaphor, the tropes with which the anatomical concepts are articulated would need to conform to a distinct kind of figure, in order for the anatomy to be distinguishable from allegory. Questions of diachrony and historicity arise to problematize the generalizability of characterizations of typical scientific metaphors; yet, preliminary to an examination of the discursive and conceptual relationship between anatomy and poetry of the early seventeenth century, a characterization of typical scientific metaphors in the terms of the philosophy of science serves to introduce a further interpretative reference to our reading of anatomical metaphor in the sixteenth and seventeenth centuries, and thereby helps us to assess the level of interpretative difficulty faced by anyone who shares Langdale's project.[176] Scientific metaphors are usually in the form of models, which are "basic to the construction of scientific explanations."[177] They convey abstract essential information sometimes by means of concrete concepts. Additionally, they present the strange or mysterious in some familiar light. According to David Edge: "The process whereby we construe an uncertain, obscure, or puzzling area of experience in terms of one both familiar and apparently (in at least some respects) similar, the displaced pattern acting as a metaphorical redescription of the unfamiliar, has been shown to be central to many key scientific innovations."[178] In addition, scientific metaphors are typically unambiguous in meaning.[179] This is because they usually single out for purposes of knowledge an important characteristic or feature of their object. This leads directly to the next typical characteristic of scientific metaphors, which is that the feature that they highlight can be paraphrased in literal language, and as a consequence "a technical neologism is invented and assigned that designation."[180] Furthermore, scientific metaphors are inessential, that is, replaceable in principle by other tropes or a paraphrase equally useful in indicating the anatomical concept which the analogy, metaphor, or whatever trope it is, aims to grasp. Karsten Harries explains this final characteristic, saying that, in science and philosophy, "every particular occurrence of catachresis is in principle replaceable. To understand what is meant is also to know that the fact that it happens to have been expressed in just this way is inessential. This makes the metaphors of science and philosophy fundamentally different from those of poetry."[181] But Harries is speaking of modern science and poetry, and a comparative analysis of the figurality of anatomical texts and of *The Purple Island* will throw into relief the historical variability and perhaps the theoretical tenuousness of this and the aforesaid characteristics as typical of scientific metaphors during the "scientific revolution."

In this part of the study, the aim of these excursions into conceptualizations of the body as constructed, as natural, and as a microcosm, where they occur in anatomy textbooks, is not to follow intertextual links with *The Purple Island,* nor to suggest sources for its anatomical representations. These tasks are undertaken in parts 2 and 3. Prior to that, the aim is to demonstrate congruity where it exists between the languages, discourses, and ideas of *The Purple Island* and the anatomy to which the early seventeenth century was heir. In so doing, however, anatomical texts that could not have been at any time accessible to Fletcher, or to similarly educated English scholars and poets, are passed over.

There has been a tendency among some historians of anatomy to associate the return to the ancient Greeks' methods and assumed projects of dissection among early sixteenth-century anatomists and physicians with a desire to throw off the yoke of microcosmic analogy, which had developed inordinate influence in Arabic and medieval scholastic anatomy. Among pre-Vesalian anatomists, Alessandro Benedetti was especially concerned to abandon medieval scholastic and Arabic sources of anatomy, and, "in the spirit of the humanists," to return to and revive "the pure sources of medical knowledge which lay among the Greeks, beginning with Hippocrates."[182] He stayed for sixteen years both in Morea in Southern Greece and the Venetian domain of Candia in Crete, collecting information for the thirty volumes of his *De re medica.*[183] In Venice, he was the center of a "recognizable group of physicians . . . whose fondness for Greek and Greek learning . . . brought upon them the scorn of more traditionalist opponents."[184] The principal sources for the anatomy of Benedetti's self-consciously Hellenically entitled *Anatomice, siue historia corporis humani* were the anatomical doctrines of Aristotle (the doctrine of humors),[185] Galen (the doctrine of the similar composition of parts, or *homoiomereiai*), and Plato's *Timaeus;* though possibly because of the unavailability of Greek manuscripts and the scarcity of printed editions of Aristotle or Galen in Greek before the 1520s, he derived the Latin terminology from Julius Pollux's *Onomasticon,* as well as from Rufus, Aretaeus, and Paul of Aegina.[186] L. R. Lind says of Benedetti's *Anatomice:* "The return to the Greeks, [his] abandonment of detailed reference to the Arabs and of medieval scholastics as authorities, and a decisive reduction of the earlier attempts to harmonize conflicting doctrines marked the new anatomy which was to result in Vesalius's *Fabrica.*"[187]

The limits of philological discussions on physiology were reached early,[188] and the continuity between Benedetti's concerns with Greek culture and philology and the emphasis in Vesalius's work on evidence and argument is questionable.[189] But it can be said that the figurality of the anatomy of Benedetti's text is not obviously an elaboration upon medieval or Arabic microcosm theory. Nevertheless, the conceptualization of the body in *Anatomice* is composed to some extent metaphorically, along lines that are consistent with the concept and the figuration of the body as natural. This figuration is most suitably conveyed in the example of Benedetti's account of human conception and embryology, and so although *The Purple Island* does not treat of these subjects, the conclusions we draw from Benedetti's account can be deductively applied to Fletcher's Isle of Man. *Anatomice*'s description of the seminal vessels of women begins with a comparison between human conception in the uterus and fertilization in eggs, drawn largely from comparisons between vivaporous and oviparous generation in Aristotle's *Historia animalium.*[190] The overt comparison in these descriptions between different animal reproductive systems, embryos, and physiologies deliberately draws analogies, based on observable or interpreted similarities, yet potential sources of difference emerge too.[191] Analogies permit understanding of the obscure by reference to the familiar; they function like models, and thereby satisfy at least two of the characteristics of scientific metaphors. We will proceed to discover if their use in *Anatomice* is consistently thus, and if they resemble any further characteristics of scientific metaphors.

Subsequent to this comparison, in the same chapter in Benedetti's text, the analogy between fish and animal (including human) fertilization and embryology is extended to the process of generation in plants, thereby creating a three-way analogy: "[J]ust as there is something inside [the eggs of fish] called milky as in the seeds of plants, so in the material of animals there is an excrement of the constitution which is nourishment and the foetus grows through the umbilicus in the same way plants do by means of their roots and as happens also to animals themselves when they have been completely formed and released from that aliment in which they are contained."[192] In this passage, the overt analogy is apparent in phrases such as "just as there is" and "in the same way." In a later passage of the chapter, comparison is foregone in favor of a focused description of human embryological development. In this passage, overt analogy is superseded by the metaphorical use of the term "plant" for the omitted term "embryo": "On the tenth day the bubble of the heart swells with the boiling spirit within it. With the drying heat outside it the material of the secundinae is made firm since both plant and shell or covering are generated at the same time."[193] One term of the analogy, plant, has replaced the other, human embryo, thereby creating a metaphor, composed of "*two* distinct subjects—a 'principal' subject and a 'subsidiary' one [= the "focal word"]."[194] The metaphorical word, in this case plant, is the "focus," but "this focus requires the 'frame' of the [literal] sentence," which is animal and human embryology.[195] The similarities between the meanings of the two terms or subjects, which have been carefully described analogically earlier in the same chapter, are now implied, and the differences between the physiological processes and anatomical structures compared appear either to have been elided or overcast, such that one set of concepts (the nutrition, growth and structures of the embryo and fetus in the egg or uterus) is understood momentarily in terms of another set of concepts—the nutrition, growth, and structures of a plant. It is thus, "*understanding and experiencing one kind of thing in terms of another,*" that, according to David Edge's description (cited above), is descriptive of scientific metaphor, and that George Lakoff and Mark Johnson argue is the "*essence of metaphor.*"[196] Moreover, many cognitive scientists now consider this kind of metaphorical understanding to be guided by analogy, just as the occurrence of metaphor in Benedetti's *Anatomice* is.[197]

Lakoff and Johnson regard metaphor as conceptual, as "*metaphorical concept,*" in being not only a matter of the language used to speak or write about a concept, but rather in shaping, structuring, and defining that concept and our experience of its referent.[198] Lakoff and Johnson's theory aims to supplement what they refer to as "the myth of objectivism" in Western philosophy and linguistics, and opposes itself to objectivism's concerns. It describes these concerns as with the following: definitions for concepts as "senses" or fixed objective meanings of words corresponding to things that are inherent in the concepts, which in turn correspond to properties inherent in well-defined objects in the world;[199] secondly, what it means for something, a statement, actually to be true.[200] These concerns were very much those of the Royal Society, and characterize natural philosophy and science from the mid to late seventeenth century, though I should hesitate to identify them with the principle of "fallibilism" that is practiced by scientists nowadays.[201] According to Lakoff and Johnson, the myth of objectivism, which states that one can speak objectively, that is truthfully or correctly, about reality only if one uses "language that is clearly and precisely defined," "makes both myths and metaphors objects of belittlement and scorn: according to the objectivist myth, myths and metaphors cannot be taken seriously because they are not objectively true."[202] Conversely, the literal sense of metaphors is usually blatantly false (if true, it is usually banal), and therefore metaphors should be excluded from semantics and regarded as ornamental linguistic devices, the meaning of which may be translated into a literal paraphrase and thereby understood.[203] In contrast to each of these objectivist concerns, Lakoff and Johnson are concerned with "how human beings get a handle on" concepts, "how they understand . . . and function in terms of" their understanding of concepts;[204] and, secondly, with truth as relative to a conceptual system, defined in large part by metaphor, by virtue of which definition "metaphors can be true or false."[205] These concerns are justified because, as Johnson maintains, "*any adequate account of meaning and rationality must give a central place to embodied and imaginative structures of understanding by which we grasp the world.*"[206]

Whereas the remainder of this chapter will relate the conceptual character of the figurality of *The Purple Island* to that of anatomy textbooks, chapter 2 will attempt to establish the creative character of *The Purple Island*'s metaphoricity in terms of sixteenth-century understandings of the rhetorical and poetic functions of metaphor, and in terms of modern aesthetics, philosophy of literature, and metaphorology. The relevance of Lakoff and Johnson's theory to the occurrence of metaphor in sixteenth-century anatomy textbooks and *The Purple Island* is both historical and theoretical. Initially, only the theoretical relevance will be explored. In order to demonstrate this relevance, the rudiments of the theory require exposition.

Lakoff and Johnson propose that the defining or structuring concept of a conceptual or "structural metaphor" emerges directly or metaphorically from our bodily experience, and is grounded by virtue of systematic correlates within that experience.[207] Typically, say Lakoff and Johnson, we conceptualize the nonphysical in terms of the physical (although the physical is said to be more or less culturally mediated), such that, for example, by means of spatial-orientational metaphors, whole systems of concepts "arise from the fact that we have bodies of the sort we have and that they function as they do in our physical environment."[208] Additionally, our experiences with physical objects (especially our bodies) are said to provide the basis for ontological metaphors, which are similarly grounded within our bodily experience, and whereby more abstract phenomena, such as emotions or ideas, are understood as entities and substances.[209] In these ways, the grounding of our conceptual system is achieved, say Lakoff and Johnson (without "claiming that physical experience is in any way more basic than other kinds of experience"), by way of concepts which emerge directly, or partly or wholly metaphorically, from our physical, bodily experience.[210] This understanding is consistent with the "purely anthropomorphic" stage of the development of the political analogy of the body, and is basically in agreement with Barkan's view that the body "is both phylogenetically and ontogenetically one of the first and most basic entities the mind can grasp."[211] It is a vaguely Aristotelian view and Aristotle lay behind Renaissance teaching of rhetoric. The orientational or ontological metaphors that allow us to conceptualize abstract phenomena in more sharply defined terms are referred to by Lakoff and Johnson as "*emergent metaphors* and *emergent concepts*."[212] Once an anthropomorphic or organic theory or conceptualization of an abstract phenomenon, such as the body politic, has evolved, however, the original corporeal source dies imaginatively, and so a number of the metaphorical concepts of which Lakoff and Johnson write are actually conventional metaphors.

Accordingly Lakoff and Johnson's analysis is primarily of the "metaphorical structuring" by concrete, or "highly structured and clearly delineated concept[s]," of abstract or less clearly defined concepts, such that abstract concepts are thereby grounded in systematic correlations in our experience.[213] To the extent that Lakoff and Johnson's theory rests on the occurrence of metaphorical structuring in this direction, it appears unlikely at the outset to illuminate the metaphorical structure of the anatomy of the physical human body. Although the human body as referent is ontologically concrete rather than abstract, Lakoff and Johnson's theory does allow that less clearly delineated concepts of physical objects may be defined in terms of more clearly delineated concepts of other physical objects. This is illustrated when, for example, ontological metaphors project concepts of the object status and orientations proper to, and directly emergent from, our experience as physical beings onto other physical objects.[214] Analysis of the metaphorical structuring of the internal, and to some extent external, anatomy of the human body not only in sixteenth-century anatomy textbooks, but much more broadly in early modern culture, reveals it to be in several respects consistent with the structuring of abstract or less clearly delineated concepts in terms of concrete, more clearly delineated ones in Lakoff and Johnson's thesis. The reasons for this consistency will become apparent in the course of this chapter.

The acceptance of Lakoff and Johnson's thesis involves fending off objections from abstractionist views of metaphor, such as that of Barkan whose theory of the figurative microcosm of the Platonic tradition is probably abstractionist. Barkan argues that abstract concepts are drawn from the concrete falsehoods of metaphor:[215] embryos and fetuses are not literally identical to plants. This abstractionist view asserts an epistemological distinction between empirical or concrete truth, which might be had by using literal language, and abstract truth, referred to or signified by metaphor. If such a view allows that the use of the term plant for the embryo is a metaphor, a conventional one, it would claim that it is based on preexisting similarities in meanings which correspond to objectively inherent properties of each of the special cases of "plant": the embryo and the plant of the vegetable kingdom.[216] This view of meaning would therefore deny any direction of metaphorical structuring, that the human and animal embryo and fetus are understood in terms of plants. It would claim, moreover, that the preexisting similarities in the meanings of both concepts of "plant" may be translated into a literal paraphrase and thereby understood. The abstractionist view of literal language and of conventional metaphor as contributing to understanding only by making us see objective similarities, is compatible with, and understandably adopted by, objectivism.[217]

There are objections raised by Lakoff and Johnson to the abstractionist view. By asserting the existence of an abstract concept, such as "the field of nature," which is common to, and true of, both the uterus and the earth, abstractionist theories do not account for internal or external systematicity.[218] They do not account for the coherence between concepts which pertain to the physiology of reproduction, by referring them, as Lakoff and Johnson do, to the characterization of a system of metaphorical concepts by the most specific metaphorical concept and its entailments.[219] The metaphor of the embryo as a plant ought properly to be understood as emerging from the context of Benedetti's own overtly and deliberately comparative anatomy, but further instances of this analogy in other differently authored texts lend substance also to the view that such occurrences are systematic rather than incidental.[220] These occurrences give credence to Lakoff and Johnson's theory, which describes systematicity as the tendency to use language consistent with a conceptualization of a concept in speaking or writing about that concept. In this way, "metaphorical expressions in our language are tied to metaphorical concepts in a systematic way."[221]

Gerard Steen has observed that it follows from Lakoff and Johnson's theory that "the verbal aspect" or textual instance "of metaphor ought to be regarded as dependent on, or derived from, conceptual metaphor" because they argue "that conventional conceptual metaphors belong to the common knowledge of the language user, and that they are stored as conceptual units in the mind."[222] Although what Lakoff and Johnson have identified as conceptual metaphors have been widely accepted by theorists of metaphor as metaphors in terms of their linguistic structure, questions have been raised about Lakoff and Johnson's attribution of a cognitive function to all metaphorical elements and expressions in a language system from the perspective of language behavior and discourse processing.[223] Steen argues: "Many of the conceptual metaphors found by Lakoff and Johnson may be the conceptual repository of our cultural knowledge, but such knowledge is bound to have highly variable cognitive representations and effects at the level of individual minds."[224] It is also Steen's view that from "the synchronic, empirical perspective on *individual* processing, many of the conceptual metaphors identified by Lakoff and Johnson . . . may be regarded as instances of polysemy," and are therefore "dead and highly conventional metaphors" which are understood directly or nonmetaphorically.[225]

Although sixteenth- and early seventeenth-century anatomy textbooks (almost invariably in Latin and occasionally Greek or the vernacular) were owned by only the highly educated, their nomenclature and knowledge was accessible to all who were so educated, as the B.A. degree was the entry point for medical specialization, as for specialization in, for example, law or theology, and it was usual "to take the baccalaureate or magister artium or a degree in the faculty of theology before commencing the study of medicine."[226] The specialized body of knowledge accessed when reading early modern anatomy textbooks may have been in the full possession only of those philosophically and medically learned specialists, but some of the concepts and figurality of anatomical description has roots and branches spread more widely in early modern culture, in the general acceptance of analogy as inherent in the world. Nevertheless, the determination of any possible cognitive processing of anatomy textbooks requires some attempt to ascertain the intended and actual readership in addition to an assessment of the broad values in society relating to the role and meaning of anatomy.[227] We may deduce from the overt explanatory character of Benedetti's analogical comparisons in *Anatomice* that metaphorical meaning in this case of early sixteenth-century anatomy may not have been as directly accessible as literal meaning, polysemy, or highly conventional metaphor. We may also deduce that at the level of individual readers' minds the cognitive representations and effects of the metaphorical expressions that I have tied to the conceptual vegetative metaphor, may have required at least some analogical mapping from particular domains of plant to animal anatomy. There seems to be cognitive analogizing at work in the writer's inclusion of metaphorical expressions in the text, and in this case, then, the process of understanding "linguistic metaphors" "triggers a *process* of analogizing, which in turn *may* have recourse to *information structures* in long-term memory called conceptual metaphor."[228] But different processes may occur if a conceptual metaphor is highly conventional, or is neither available nor accessible (as we shall see when we look at novel metaphor).

According to Lakoff and Johnson, we conceive of whatever concept is structured metaphorically, in a way that is consistent with its systematic conceptualization in terms of a host of metaphorical concepts that are related in terms of entailment relationships.[229] That the nutrition, development, and structure of the embryo are understood in terms of those of a plant entails that in terms of physiology, spatially contiguous parts of the human reproductive system should be understood in a metaphorically consistent way. Detailed anatomical exposition makes clear that the human embryo bears no resemblance to a seedling in terms of anatomical structure, so "the most specific metaphorical concept of this system," that which characterizes the entire system, cannot be found in the metaphorical concept of the embryo.[230] Rather it concerns function, and it is that reproduction in humans is like reproduction in plants, as understood in pre-Linnaean botany. This concept characterizes an internally coherent system of metaphorical concepts of causally and temporally contiguous processes, which in turn characterize the spatially contiguous structures involved in those processes.[231] This systematicity, however, appears once again not to conform to the direction of metaphorical structuring in Lakoff and Johnson's theory of concrete structuring abstract concepts. Reproduction is a physical phenomenon, but conceptualization of the whole physiological process requires abstraction into stages, and is less readily observable than anatomical structure when described with only passing reference to function. The significance of this feature of metaphorical structuring in early modern anatomy will become clear in due course.

Physiologically central to this host of metaphorical concepts is the uterus, which, if the embryo is a plant, must be understood in terms of a role analogous to the "Earth," as indeed it is in Johann Veslingus's *The Anatomy of the Body of Man.*[232] In Jacopo Berengario da Carpi's *Isagogae breves,* the uterus is called "the field of nature."[233] It is similarly "the felde of mans generation," in Thomas Vicary's *The Anatomie of the Bodie of Man.*[234] And to demonstrate that these parts are understood on the model of reproduction in plants, and not vice versa, Veslingus argues in *The Anatomy of the Body of Man,* that the blood vessels adjoining the embryo from the placenta do not develop "to the Child from the Mothers Womb, and the Membranes adjacent"; rather the reverse is the case for Nature "doth the like in vegetables, whose roots come not from the Earth to the Plants, but the Plants send them to the Earth for nourishment, and this is abundantly proved in Bulbous roots that grow out of the Earth."[235] We may infer that "the field of nature" is not an abstract concept.

For these authors, the vegetative conception of the embryo and terrestrial conception of the uterus, in no sense detracts from their adherence to the anatomical homology of the "one-sex" view. Berengario, in particular, has been described as making "much use . . . of homology and analogy among the organs, as in the case of the uterus and penis."[236] But like the homology view, and the related idea that male anatomy could be representative of the less perfect, female anatomy, the consistent metaphorical structuring of the uterus and fetus led to the neglect of aspects of female anatomy by anatomists. In conceptual metaphor only an aspect of a concept is highlighted by, and understood in terms of, another, defining concept, so that some of the defined concept remains hidden.[237] Thus, in the metaphorical definition of the uterus as the earth, the internal location of the womb, its substance, shape, or form, even supposed movements, and construction from its parts are hidden,[238] while the concept of its providing the nourishment and environment necessary to the growth of the embryo and the latter's connection through the umbilicus to the uterus, like a plant to the earth, are highlighted. Whether that which remains hidden in this concept is understood in terms of inherent rather than metaphorical properties will be turned to later in this discussion.[239] The impression that the concepts of the embryo and uterus are insufficiently defined independently, finds support in the descriptions and illustrations of the female reproductive system and of embryological development prior to and even well into the sixteenth century.[240] It has been argued that "Berengario was the first to discover that the cavity of the uterus is not divided into two or seven compartments, as Galen and his followers believed, but 'has one single cavity or cell.'"[241] Yet, Book III of Charles Estienne's *De dissectione,* published in 1545, is typical for its espousal of the seven-cell theory of the uterus, which Estienne accepts ostensibly on the word of some French midwives.[242] Repeated illustrations attribute to the human uterus "two thick strong ligaments," often referred to by the appellation "horns of the womb."[243] The perpetuation of inaccuracies was encouraged by Vesalius's *Fabrica,* the female reproductive system of which has been described as "mediæval and full of errors" which are "set forth also in the figures of Vesalius, and through them they have been perpetuated in popular Anatomy."[244] Therein the uterus is represented as slightly bifid, and the account of the embryo has been described as "negligible."[245]

This evidence goes some way toward countering the abstractionist objection that there is an abstract concept of plant which applies to plants of the earth and to the embryo and fetus, such that the concepts of the latter are structured independently rather than in terms of the concept of plants of the earth.[246] And the example of the early sixteenth-century manual of midwifery, *The Birth of Mankinde,* effectively overcomes this objection at least as far as medical works popularizing anatomy are concerned.[247] At its most detailed, *The Birth of Mankinde* designates the vulva as the "passage port," the vagina as the "wombe passage" and the cervix as the "Wombe Port," thereby propagating the conceptualization of female anatomy as a land to be discovered and explored, both anatomically and by implication perhaps, sexually. The apertures of the sexual and reproductive organs of women become linguistically and functionally organized according to an economy of terms peculiar to navigation. This displacement of anatomy is a consequence of a prescientific understanding of the body, not as a distinct ontological entity with an independent structure, but as possessing an illustrative function within discourse.[248]

There is increasingly over the course of the sixteenth century, more detailed and precisely delineated reproductive anatomy and embryology, resulting from observation, the formation of anatomical projects to discover previously unmapped regions of human anatomy, and the standardization of nomenclature, especially in the work of Gabriele Falloppio, Giulio Cesare Aranzi, and Adriaan van der Spieghel.[249] Yet "in the production of a new map of the body, a new figure was also to be glimpsed—the scientist as heroic voyager and intrepid discoverer. The body was a remote and strange terrain into which the discoverer voyaged."[250] The assumption that the representation of the female reproductive system in particular is schematic, imprecise, figured on an analogy with plant development or foreign land, and characterized by repeated misconceptions, may be unwarranted, for there is reason to believe that "the earliest first hand drawing that has come down to us of any internal organ of the human body," "evidently drawn from the object, or at least not far removed from such a drawing," may well be that of the uterus in the figure of the pregnant woman in the *Fasciculo di medicina.*[251] The question arises of whether the entire semantic, semiological, and performative field of Renaissance and early modern anatomical language, and consequently the conceptual structure of all internal anatomy and physiology, is covered by tropological models, including the vegetative conceptual metaphor. It is a question which this study attempts in some measure to answer, and in so doing to counter the abstractionist objection to the conceptual theory of metaphor.

It is consistent with Lakoff and Johnson's theory that the metaphorical concepts of a system characterize in turn a corresponding coherent system of metaphorical expressions for the concepts,[252] among which are those of Berengario, Vicary, and the author or translator of *The Birth of Mankinde,* for the uterus. Berengario, like Zerbi, followed a method of commentary which "aims at completeness of quotation from the Arabic scientists," such as Avicenna, Averroes, Rhazes, and Haly Abbas, frequently using Arabic terms in preference over Latin and Greek; and he was more concerned than Benedetti to harmonize authorities, be they Greek, scholastic, or Arabic.[253] Berengario's most authoritative influence, however, was by his own admission Galen, and from him he derives his insistence upon careful, meticulous observation and touch in dissection.[254] While following "the high scholastic model . . . of the thirteenth and early fourteenth centuries closely," Berengario applied *demonstratio* to a proposition and "used it as 'proof' by exposing the structure in question to the sight and touch of his audience." The senses thereby became that which "provides the determination in a disputation."[255] So in his "per Carpum castigata" (1514) and his major anatomical work, the *Commentaria . . . super Anatomia Mundini* (1521), notwithstanding his tendencies to harmonize authorities, "Berengario insists upon the precedence of sense perception over [Galenic and Arabic] authority."[256] In the *Commentaria* he says emphatically, "let those who write books on anatomy also not trust in the authorities but in their sense perception as I do and shall do." Even Galen, whose opinion Berengario claims "I always maintain," gets short shrift "when sense-perception disagrees with him."[257] Moreover *Isagogae breves* "makes much use of . . . Aristotelian teleology, categories, and properties in describing the parts of the body."[258] Yet, apparently not in conflict with his anatomy of the senses, and consistent with this revived Aristotelian teleology is Berengario's use of the trope of the field of nature for the uterus.

This trope is developed and extended in Benedetti's description of the secluded garden of the uterine interior, wherein sprout seed leaves, "acetabula or cotyledons ["as the Greeks call them"]. By means of these through the umbilicus the foetus is caused to grow."[259] In human anatomy at this time the "cotyledons" are "the orifices of the veins [and arteries] that run from beneath and from above" the fetus in the uterus,[260] cleaving to the chorion.[261] "By means of these the fetus is attached to the uterus, and the menstrual blood is carried by them" to the fetus, which is the unborn offspring at a stage later than its embryonic period.[262] However, their metaphorical conceptualization is consistent with an apparently variable concept of the embryo or the fetus as the developing plant in different stages. During pregnancy, the cotyledons, or seed leaves, "suck out blood and spirit" from the placenta via the umbilicus "just as the fibres of roots do and carry it through the veins" of the developing fetus.[263] In this way, physiological function is rendered metaphorically coherent with anatomical structure in both animals and plants. The metaphorical conceptualization appears here slightly different from that in Benedetti's analogy of the embryo with the plant. If the cotyledons correspond to veins feeding the fetus with blood, then the fetus must be understood on an analogy with the flower or fruit of the plant, produced at a later stage of the plant's development. Although the cotyledonous uterus was refuted by Giulio Aranzi in 1564, this analogy is presented in the seventeenth century as verbal metaphor by Veslingus describing the "fruit . . . inclosed in its secundines," and as visual metaphor in one of the copper plates of Adriaan van der Spieghel's *De formato fœtv,* published in 1627.[264] A sleeping fully developed fetus is shown attached via the umbilical cord to the placenta, surrounding which like petals are the dissected parts of the chorion and the membranes of the uterus. The fetus lies in the center of this array as though emerging from a flower (as indeed fruit does developmentally).[265]

The apparent difference between the metaphorical models used by Benedetti does not amount to inconsistency. The first stage of the reproductive process, conception, is understood in Benedetti's *Anatomice* as the result of fertilization of the female genital humor or seminal fluid by male seed. Female genital humor, which is described as similar in color and consistency to egg white, is collected within the uterus, "just as happens in eggs in which the white and the yolk produce offspring from the female, with a constituent supplied in the seed from the male."[266] The analogy between generation in oviparous and viviparous mammals is rendered coherent with the metaphorical system by representing the male seed as sown in the field of nature.[267] In this way, the seed, which originates outside of the uterus, yet is incorporated into intra-uterine processes, is rendered coherent with the same metaphorically conceptualized system of reproduction, because the seed belongs to this system.[268] What emerges from this fertilization in the field of Nature is the young plant, the unborn offspring, the embryo in the early stages of its development. But at parturition, the cotyledons release the umbilicus "with its involucrum, that is, the secundinae, or part of the placenta," "like the stem of an apple already ripe falling of its own accord from the branch."[269] The implied and covert metaphorical concept in this description of birth, the final stage of reproduction, is that the neonate is the fruit, the formation of which is the goal or purpose of reproduction in plants and metaphorically in animals. In each case of the development of the fruit and of the animal offspring the end result is developed out of the material of the early stages.

Adriaan Van der Spieghel, *De formato fœtv liber singularis aeneis figvris exornatvs* (Frankfurt: Matthaeus Merian, 1631), Tab. IIII. A sleeping fully developed fetus is shown attached via the umbilical cord to the placenta, surrounding which like petals are the dissected parts of the chorion and the membranes of the uterus. The fetus lies in the center of this array as though emerging from a flower (as indeed fruit does developmentally). Wellcome Library, London.

It is the effect of the goal or purpose of reproduction in animals upon its metaphorical conceptualization that suggests a qualification to that aspect of Lakoff and Johnson's theory which deals with the direction of metaphorical structuring. The analogy between animal and plant reproduction operates not only on the levels of the description of anatomical structure and of physiological process, but is coherently extended to the purpose of reproduction, its reason for being, which is parturition and the "fruit of the womb." This final cause is what Aristotle, whose influence on Benedetti's *Anatomice* has been noted, terms *telos* (purpose, design). This abstract concept provides the rationale for the character of the metaphors which structure the comparatively concrete concept of the human or animal body. But before we proceed to elucidate Aristotelian teleology, we ought ideally to gauge the extent to which *The Purple Island* reproduces the vegetative conceptual metaphoricity of conception, embryology, and uterine anatomy. The simple answer to this, however, is not at all, for the reason already given, that the poem's diegetic narrator, Thirsil, chooses to "Flie then those parts, which best are undescri'd" (*P.I.* 3.25.5). But the answer is not this simple, for although the metaphorical conceptualization of anatomy in *Anatomice* along lines which are consistent with the concept and the figuration of the body as natural and vegetative is most suitably conveyed by the example of Benedetti's account of human conception and embryology, the *Purple Island* anatomy of the whole body is similarly figured in geographical and terrestrial terms.

Figured thus as an island, the body in Cantos 2 to 5 of *The Purple Island* ambivalently allegorizes the "heroic" voyage of the new figure of the anatomist into "the cavities and recesses, the interior secrets, of the body."[270] The allegory is ambivalent because the analogy is between the body and the dynastic realm of Britain, and the diegetic narrator, Thirsil, in Canto 1, stanza 34 (cited earlier in this chapter), complains that those intent on foreign exploration and colonization have "by seeking lost" knowledge of Man. "A forrain home, a strange, though native coast" (*P.I.* 1.34.6), the Isle of Man is not unambiguously "as dark as the interior of the continent of the newly 'found' americas."[271] The exploration that the implied reader of *The Purple Island* is invited to embark upon is partly announced by an epistle to the readers which prefaces the first edition.[272] The writer of this epistle, Daniel "Featly" (Daniel Fairclough), claims that the "way to God is by our selves: It is a blinde and dirty way; it hath many windings, and is easie to be lost: . . ." "This Poem will make thee understand that way," he says, in commendation of *The Purple Island.*[273] The writer is referring both to the exploration of bewilderingly intricate anatomical forms, connections, and functions, in pursuit of self-knowledge, and the constant vigilance required for moral self-amendment as well as personal and national salvation.[274] It takes the allegorical narrative of Cantos 6 to 12 fully to unfold this conception of knowledge of Man and its implications and purposes.

Besides the figurative theme of discovery, in some respects the vegetative metaphors employed in Cantos 2 to 5 of *The Purple Island* resemble closely those in anatomy textbooks; while in others they are at least superficially different. *The Purple Island* represents the flow of blood and humors from the liver to the body via the venous system, as a "Nile-like inundation" (*P.I.* 3.11.6).[275] Of the function of this fluvial distribution, it says the "Cava, the hollow vein," "So gives to all the Isle their food and vegetation" (*P.I.* 3.11.7 & margin. i). Despite the obvious differences in anatomical parts and figurality, this imagery is rationally consistent with the vegetative imagery of descriptions in anatomy textbooks of arterial and venous distribution through the seed leaves, acetabula, or cotyledons that run from beneath and from above the fetus in the uterus, feeding the fetus with blood and causing it to grow. There are two aspects to this rational consistency. Firstly, Fletcher actually provides in allegorical form the *telos,* the reason for the vegetative figurality of the textbooks. In the "geographical" context of the island, rivers from the "Arch-citie *Hepar*" (*P.I.* 3.5.3) nourish the land and so give rise to vegetation. In the anatomy textbooks, the very same nourishing function gives rise to the veins feeding the fetus. Secondly, in anatomy textbooks and *The Purple Island,* metaphorical conceptualization in terms of vegetative reproduction and growth is not confined to the anatomy of the reproductive organs and "fruit" of the womb. It extends to the vascular system, particularly the veins, and even beyond.

The character of the metaphors of vegetation and fluvial distribution in anatomy textbooks and *The Purple Island,* respectively, provide a means whereby the purposes of reproduction, nutrition, and growth are conveyed through description. In order to understand this and the first of the two aspects of rational consistency noted above, it is necessary to turn to Aristotelian teleology. According to Andrew Cunningham, anatomy for Aristotle was integral to the enterprise of philosophy: "Aristotle's anatomical project takes its beginnings from his concern with the soul," and in *De anima,* "the soul in a way is the principle of animals."[276] It is possible to formulate the rationale for Aristotle's project in four steps. Firstly, the particular form of structures of the body is understood to fulfill the goal or purpose of their function.[277] (This goal or purpose is referred to as the "final cause.")[278] In this regard, Aristotle states in *De partibus animalium* that "Nature never makes anything superfluous or in vain."[279] Secondly, although the goal, reason, purpose, or final cause of a function is distinguishable from its material or effective cause, that is the bodily material that enables the function to occur, nevertheless, the particular form of the structure which performs that function is understood to be fitted effectively by Nature to the function which achieves its purpose. It follows that "from the action or service of the organs you will learn the [meaning of the] arrangement and the nature of the parts themselves."[280] But, one must somehow learn the action or function of the parts, and the purpose of this action, and so to the third step. It is consistent with this teleological view of anatomy and the tradition that stems from Aristotle "takes it as axiomatic that we can gain reliable knowledge of the *function* the different parts of the body fulfil in health and disease from an investigation of their *structure*—by cutting bodies up, either dead (dissection) or alive (vivisection)."[281] It is thus by the appearance of a thing that one may rationally infer its function and its reason for being. Fourthly, and finally, the form of the whole body is understood to fulfill the function the purpose of which is to serve the soul. "Thus to investigate animated things is to investigate the soul, and reciprocally to investigate the soul is by definition to investigate the animated."[282] Aristotle was able to formulate such a theory because unlike Plato, for whom the soul was not the form of the body, he found form and intelligibility in orderly processes of nature.[283] It has been suggested that Aristotle holds that, as the form and final cause of the body, the incorporeal soul or essence "becomes something particular,"[284] and is bound to the body indissolubly.[285] But it has also been pointed out that this is the soul as form, "considered as motive or efficient force," that is, "the First Entelechy of a Natural Organic Body," and is not the same as the higher rational (portion of the) soul, which "has for its object pure form," but which is not itself form.[286] But Aristotle "thus embarked on the anatomizing of animals in order to discover the instruments (organa) that the soul [as form] of The Animal uses to express itself through the body."[287]

Connections between medicine and philosophy in the ancient world run deep, and as de

Laguna explains in the Preface to his *Anatomica methodus:* "There have been three sects of physicians: the empirical, methodical, and rational."[288] The skeptic's rejection of (Aristotelian) rational inference found its way into the empirical tradition in Greek medicine, founded by Serapion (fl. *c.* 225 BC).[289] Hellenistic skepticism "developed arguments to show or suggest that the evidence, reasons, or proofs employed as grounds for our various beliefs [or "offered to justify any proposition" "purporting to assert some knowledge about the world"] were not completely satisfactory."[290] The philosophical method adopted and developed by the Pyrrhonist Sextus Empiricus involves the "ability to place in antithesis, in any manner whatever, appearances and judgements."[291] Without an accepted, independent criterion "by which we shall be able to judge the dispute" between two opposing or contradictory propositions, argued Sextus, "there is always some doubt that any non-empirical or trans-empirical proposition is absolutely true, and hence constitutes real knowledge."[292] The Empiricist physicians proceeded to the administration of therapeutic measures via *autopsia* (personal experience) and careful assessment of *historia.* Drawing upon Galen's *De Sectis* (On Sects) in the Preface to his *Anatomica methodus,* de Laguna explains that "[t]he empirics believed that they became trained only through contact with actual things and by experience."[293] Such restriction in what counts as evidence is not acquiescence in the senses as indicating anything beyond appearances; it is in fact skeptical and the Empirics were undogmatic about therapeutic knowledge, preferring to say that observed sequential connections or disjunctions between certain things *may* ground a general theorem, and to consider agreement among authoritative sources a "Commemorative Sign," rather than a natural sign of the truth, acting in accordance with which *tends* to be beneficial. In *On Medical Experience,* Galen's Dogmatist argues that despite their alleged skepticism, the Empiricists must necessarily theorize or they would be unable to determine what are the salient similarities between cases which must always differ in some respects. Nevertheless, for Empiricists all explanation of why a therapy works will be epistemic rather than ontological: they refuse to speculate about etiology and underlying causes (such as the dispositions of patients), and in this they are essentially skeptical.[294]

For Aristotle sensory observation may give us the immediacy of something, a sense of its reality, but that thing is really defined by its function, use, utility, action, and its reason for being, to which sense perception was linked by rational inference.[295] "[T]he rationalists," asserts de Laguna, "obeyed their senses, it is true, but did not likewise acquiesce in them until they had consulted their reason more deeply and investigated it."[296] Reason is a "faculty in the soul by which one could go beyond the immediate data of experience" to the "underlying physical structures of things" to "form general conclusions" from "a theory of their interactions."[297] This is the rationale of dissection, and its justification as medically efficacious, even in the absence of medical application.[298] It is a rationale that the skeptical Empiricists did not share, a possible consequence of which, as Celsus points out in *De medicina,* is to contend that "the information acquired from anatomizing or vivisecting the human body simply cannot be used to infer how the body works while it is intact and alive."[299]

Aristotle divided the animal soul into three parts: the vegetative, the sensitive, and the motile; though the human animal possesses in addition a rational soul, the intellect.[300] "The distinction between the soul and body of an animal is, in Late Scholastic terminology, a distinction of reason, but not a real distinction, for the sensitive and vegetative faculties of the animal soul are immersed in corporeal organs, and are inseparable from them, functioning in effect as dispositional powers of organs of the body; sensitive and vegetative activities are corporeal processes. There is, though, a real distinction, not just a distinction of reason, between the intellect, a spiritual substance with activities performed by spirit alone, and the material body."[301] The "most basic aspect of the animal soul" in Aristotle's division is the nutritive or vegetative soul, which humans share with other animals and with plants, and "which [in humans and some animals] rules in the whole of the trunk of the body."[302] It is this aspect of the soul that Aristotle thought was possessed by the embryo; the sensitive, motile, and rational are acquired in later stages of development. I am suggesting therefore, that in Benedetti's *Anatomice,* the vegetative soul provides the reason for the metaphorical structure of the *concepts* of the reproductive parts of both animals and plants.[303] In itself the soul stands in a contiguous rather than metaphorical relation to the body: the soul resides in, uses as an instrument, and "express[es] itself through the body."[304] But the soul provides the rational principle, the reason why parts and functions of the bodies of animals and plants are described and understood in precisely the way they are. Within this theory, it remains possible that anatomical parts or physiological processes in animal and human reproduction gain their conceptual structure metaphorically from the more clearly delineated concept of reproduction in plants.[305] But the meaning of the embryo and the uterus follow not only from the natural history of plants, as one might be led in the application of Lakoff and Johnson's theory to expect, but from the natural philosophy of the vegetative soul.

Besides reproduction, the functions of the vegetative soul also include nutrition, and one can see from Matteo Corti's description of the mesaraic veins from the vena porta to the jejunum how the analogy with the roots of plants provides the concept of design by Nature with explanatory power. Asking why Nature made the jejunum straight, in the ninth of a series of lectures at the University of Bologna in 1540, Corti turns to Galen's *On the Use of the Parts* (Bk. V, Ch. 3), to explain that "as in plants when there are more branches and roots in one place, it also occurs in the body that more mesaraic veins go to one intestine than to another one."[306] When this kind of analogy develops into metaphor in anatomy textbooks, it is the vegetative soul that provides the reason for the metaphorical structure of the concepts of the nutritive parts of both animals and plants.

The texts of Benedetti on embryology and generation and Corti on nutrition both suggest that the Greek philosophical concern with teleology was adopted into Christian anatomy in the Renaissance—"the belief that everything in the universe has a purpose, which it strives to fulfil."[307] Likewise, the Spanish anatomist de Laguna, in his *Anatomica methodus,* is consistently at pains to explain how the body's internal organization is as it is for a reason.[308] Scientific metaphors aim to construe *how* something obscure occurs in terms of how something more familiar and known occurs. But the rational principle of the figurality in anatomy textbooks written at the dawn of the scientific revolution differs from this considerably: it aims to explain *why* an obscurely known process should be understood in terms of and as analogous to a more familiarly known process.

A determination that the figurality of anatomy is teleological on account of it being informed by Aristotle would be sufficient to analyze the *Purple Island* anatomy and figurality, were it not that the influence of Galen on early modern anatomy in matters not strictly philosophical was even greater than that of Aristotle. The French surgeon Ambroise Paré argues that Galen's *De anatomicis administrationibus* (On anatomical procedures / administrations) "laies the parts of mans body before our eyes, to the sense" "in the way of *knowing* them";

but he also concedes that Aristotle imparted teleological and rationalist anatomy to Galen, by whom the parts of "mans body" are subjected "to the minde in the faithfull *understanding* them."[309] According to Cunningham: "The thesis that Galen maintains in *On the use of parts* is that all the parts of the human body were made for a purpose; that they were made in the best way to fulfil that purpose."[310] Sixteenth-century Galenists like Corti are therefore concerned to understand why Nature has made the body in the ways it has.[311] Notwithstanding the similarities, there is a difference, for Galen derives his divisions of the body and soul not from Aristotle but from Plato, who distinguishes in his *Timaeus* between an immortal, "animal" soul, whose functions are rational or reasonable and sensory, and a mortal soul. The latter is divided into a higher, "vital" faculty, whose function is agitative, irascible, spiritual, or passionate; and a lower, "concupiscent" or "natural" faculty, whose functions are appetitive or desirous, the functions of generation, nutrition, and growth.[312] Galen departs from this scheme only when he attributes the functions of generation, nutrition, and growth to *physis,* or Nature.[313] Early modern Galenists after him tended nevertheless to attribute these functions to the natural faculty of the soul as well as to *physis,* or Nature, and some tabulated in great detail these and further divisions.[314] Galen also separates the "natural" functions from respiration, which in Aristotle had been a function of the vegetative soul, and locates them more precisely in the abdomen beneath the diaphragm, while simultaneously locating the vital faculty in the thorax and the rational in the head.[315] This difference in anatomical division and function gives rise to differing opinions on "the form and the purpose" of the organs; differences mainly concerning the efficient and final, rather than the material or effective, causes of function.[316] But the fact that Galen followed Aristotle's teleology suggests that Galenic and Galenist anatomy provide a reason (albeit *physis,* or Nature, rather than the vegetative soul) for the metaphorical structure of the concepts of the reproductive and nutritive parts of both animals and plants.

That the signs of nutrition and generative growth are to be found everywhere in the body is rationalized by Aristotle and Galen as the effect of the nutritive, generative, and auctive or developmental functions of the vegetative soul or *physis.*[317] In Galen, the liver produces the natural spirit (Greek, *pneuma physicon;* Latin, *spiritus naturalis*), a "vapor" which is distributed to the body with the natural blood and "humors" as nourishment via the venous system.[318] As regards Aristotle, by way of explanation, it is important to provide some indication of the difficulty scholars have experienced defining precisely what the soul's relation to organs of the body is. According to Cunningham, Aristotle argues that a bodily location, the heart, is the center of all aspects of the animal soul—the sensitive, the motile, and the rational, the latter of which belongs only to Man.[319] But B. F. C. Costelloe and J. H. Muirhead claim that for Aristotle, "*Nous,*" which means generally reason or thought, but specifically thought insofar as it deals with unmediated knowledge and supersensible reality, "has no bodily organ."[320] For our immediate purposes, however, in Aristotle's anatomy the heart is the center of the vegetative soul, so not only nutrition, generation, and growth fall within the functions of the vegetative soul, but also the respiratory function of the heart and lungs, and the heart's life and heat, which spread to the body via the arteries. The *Isagogae breves* of Berengario converts Aristotelian teleology into figurality in describing the heart as having blood vessels for roots.[321] Metaphorical conceptualization in terms of vegetative reproduction and growth is consequently not confined to the anatomy of the reproductive organs, but extends systematically outside them to the vascular system and other structures whose particular form is understood to fulfill the goal or purpose of their function.[322]

On occasion, the diegetic narrator of *The Purple Island* finds the need to adjust his "geographical," fluvial figuration to coincide with the vegetative figures of anatomy textbooks. The supposed movement of natural blood within "the hollow vein" (vena cava) is "from other streams much different; / For others, as they longer, broader grow" (*P.I.* 3.12.1–2). By contrast the veins in the direction of their flow divide and increasingly "runne in narrow banks impent" (*P.I.* 3.12.3). This failure of consistency gives rise to a change of metaphorical vehicle. According to Ambroise Paré, the blood and humors produced by the liver flow out via "the Hollow Vein [which goes] out of the gibbous part of the Liver, [and] which going forth like the body of a Tree, is divided into two great branches."[323] Just as Paré's *Works* present the "Hollow Vein" or vena cava going from the liver "like the body of a Tree," so *The Purple Island* describes this vein and the portal vein arising from the liver as:

> Much like a tree, which all his roots so guides,
> That all the trunk in his full body hides;
> Which straight his stemme to thousand branches subdivides.
>
> (*P.I.* 3.12.5–7)

The figuration of the body as an island could not sustain the consistent use of this vegetative metaphor, despite the explanation it provides in this instance, and so the description of veins reverts from branches to "streams" (*P.I.* 3.13.1). The full answer to the question of the internal and external systematicity of the figuration of the anatomy of Man in terms of the Isle—the coherence between metaphorical concepts which pertain to its functions, and the extent of the characterization of a host of these concepts by the most specific metaphorical concept and its entailments—is deferred to the next chapter. But we are beginning to gain a sense of how *The Purple Island* allegorically transforms the systematic vegetative figurality of anatomy into the geographical features of the Isle of Man. Among the copper plates of Giulio Casserio is one which presents the blood vessels arising from the liver as strikingly reminiscent of densely growing, highly divided branches and leaves of trees.[324] This conceit is extended to the entirety of the venous system of the whole body by the Spanish anatomist Juan Valverde da Hamusco. In each of his major anatomical works is a plate which contains the anterior and posterior aspects of two male écorches, over and through the musculature of each of which run veins like stems, roots, and tendrils, turning the bodies into herbaceous borders.[325] We need only imagine this profusion of plant growth as brooks, streams, and rivers and we have Fletcher's "*Purple Island*" (*P.I.* 1.48.7):

> [f] Nor is there any part in all this land,
> But is a little Isle: for thousand brooks
> In azure chanels glide on silver sand;
> Their serpent windings, and deceiving crooks
> Circling about, and wat'ring all the plain,
>
> f The whole body is as it were watered with great plenty of rivers, veins, arteries, and nerves.
>
> (*P.I.* 2.9.1–5)

Such figuration as we have found in anatomy textbooks and *The Purple Island* is capable of presenting carefully observed details of features, connections, and divisions within the venous system. In the *Epitome*'s description of the veins of the thorax, a branch of the axillary vein (known now as the volar interosseous vein) is said to "split into offshoots," which

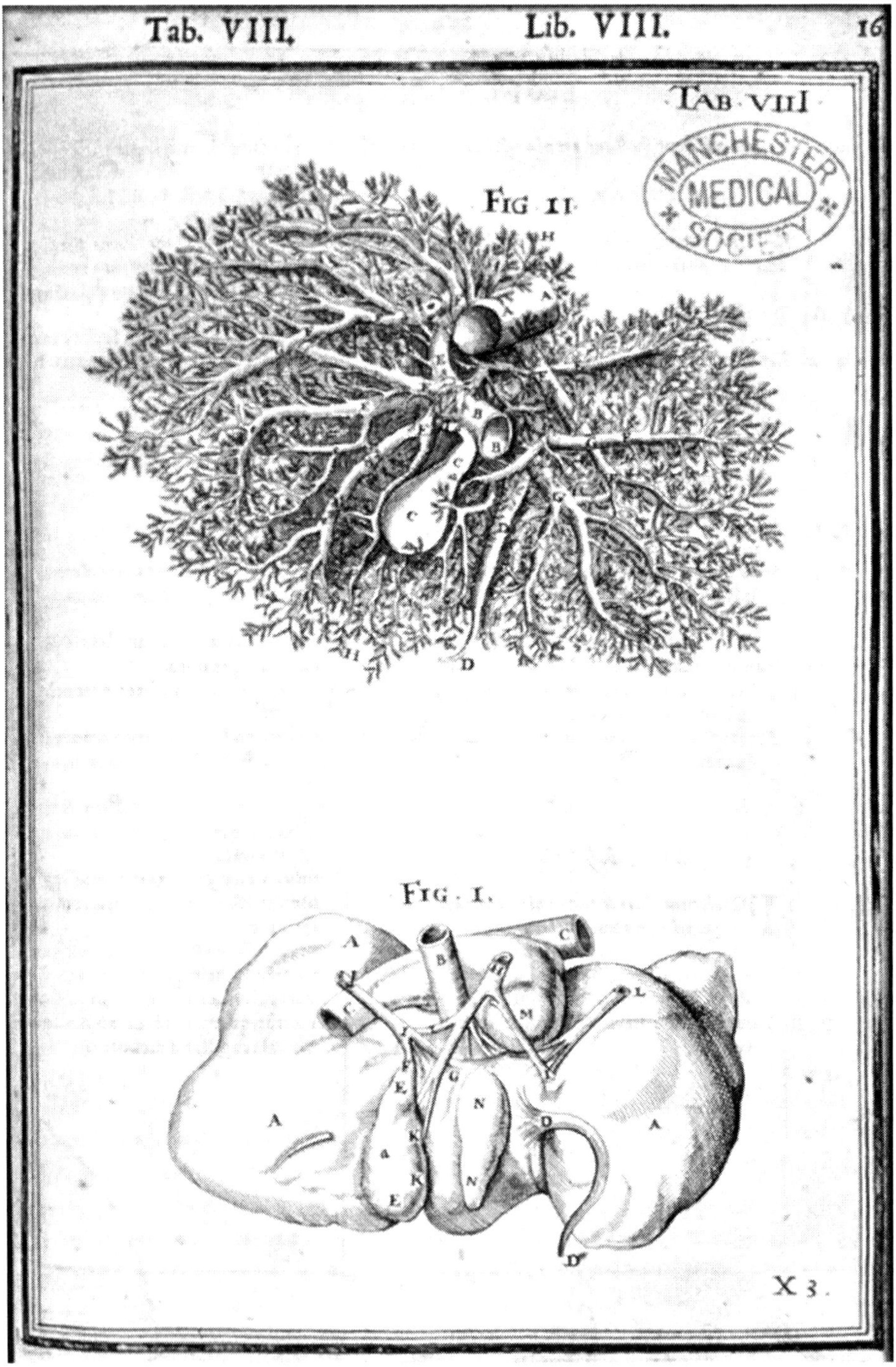

Giulio Casserio, *Tabulae anatomicae LXXIIX,* in Adriaan Van der Spieghel, *De humani corporis fabrica libri decem* (Frankfurt: Matthaeus Merian, 1632), Bk. VIII, Tab. VIII. The blood vessels arising from the liver are presented as strikingly reminiscent of densely growing, highly divided branches and leaves of trees. Wellcome Library, London.

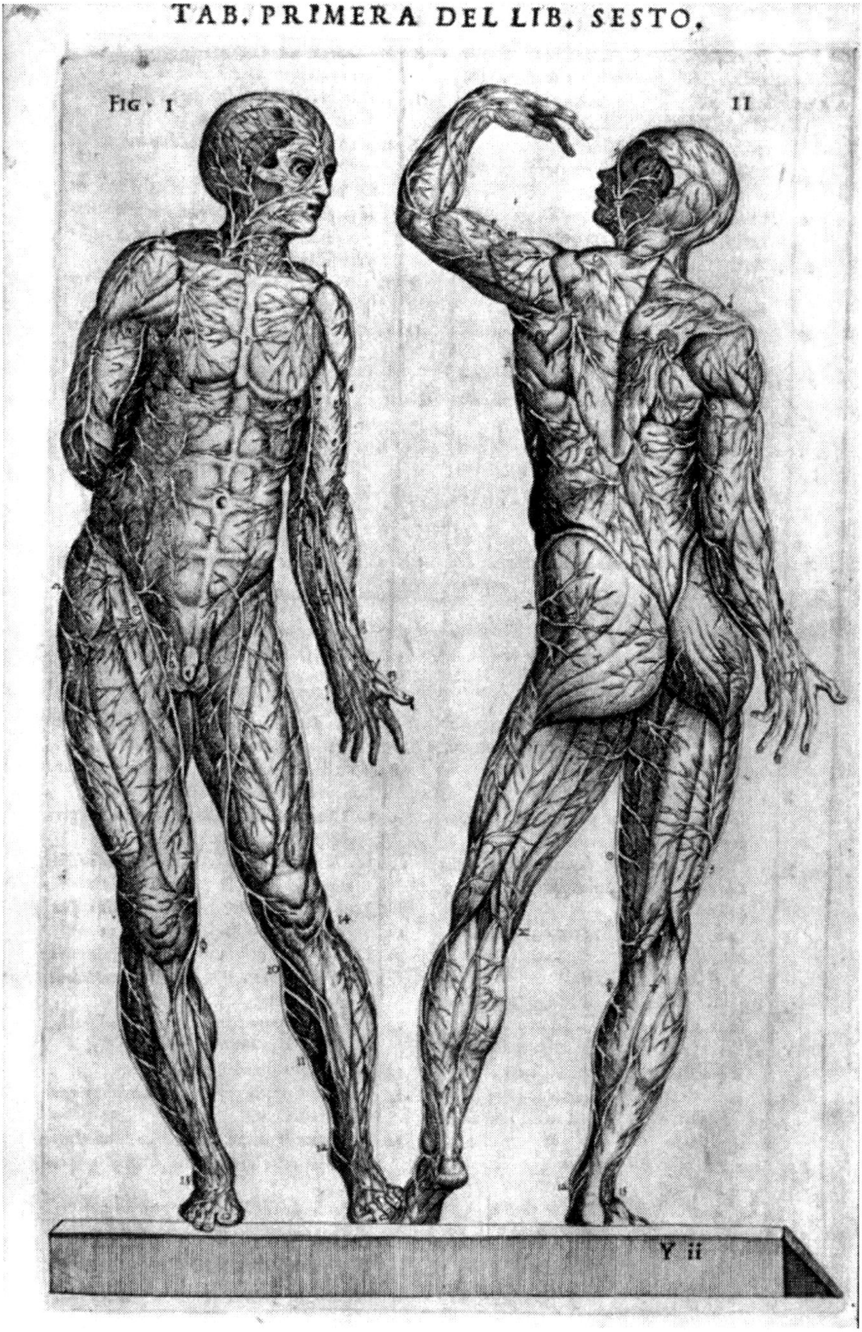

Juan Valverde da Hamusco, *Historia de la composicion del cuerpo humano* (Rome: Antonio Salamanca and Antonio Lafrery, 1556), Bk. VI, Tab. 1. The visual conceit of anatomy as vegetation extends to the entire venous system of the body in this plate which contains the anterior and posterior aspects of two male écorches, over and through the musculature of each of which run veins like stems, roots, and tendrils, turning the bodies into herbaceous borders. Wellcome Library, London.

are distributed "to the inner region of the fingers in such a way that it [the vein] offers two twigs to each of [the fingers]."[326] On describing, he thought for the first time, the delicate little membranes or *ostiolae* (little doors) in the cavity of the veins, which we now call the valves, Girolamo Fabrici of Acquapendente said that they have "a form like a node in a twig," and Plate II in his *De venarum ostiolis* visually demonstrates this on a comparison with figures of nodes such as those in the genus *Verbena.*[327] But to return to the first of the two aspects of the rational consistency of the figuration of Man in *The Purple Island,* the figuration, be it vegetative or fluvial, does not only present closely observed detail; it also conveys the meaning of venous distribution in the nourishment and growth of the organism, just as the analogy between animal and plant reproduction operates not only on the levels of the description of structure and function, but extends coherently to the purpose of reproduction, which is parturition and the "fruit of the womb." This final cause or *telos* provides the rationale for the character of the description of the uterus, the fetus, and that part of the anatomy of the whole adult human figured as vegetation or as fluvial distribution.

Whereas in *The Purple Island* the terrestrial and geographical figuration is allegorical throughout, in the descriptions in anatomy textbooks of parts and processes of the body external to the reproductive, this metaphorical conceptualization tends to carry less weight, and is frequently formulated in similes. There is creative exploration of vegetative figurality to be found in anatomy textbooks, however, not least those of Vesalius. In describing the arterial system, as with the venous system, Vesalius uses the vocabulary of vegetative growth.[328] The significance of Vesalius's creative exploration of vegetative figurality lies in his position at the forefront of innovation and in the history of anatomy. Vesalius is frequently regarded as the founder of modern anatomy. Therefore the figurative character of his anatomy indicates the depth of metaphorical structuring and the extent of its reach in the discipline. But as we shall also come to see, the significance of Vesalius's work for Fletcher is direct. Vesalius did not only criticize the Arabic anatomists and their scholastic commentators, as the Hellenists, including his immediate predecessor Benedetti, had done. While taking his inspiration from Galen's project of dissection, Vesalius also repeatedly draws attention to "what Galen had not known."[329] He was not, however, always without precedent in this; what he criticizes are "*particular* doctrines, claims, [and] statements" rather than broadly conceived projects or methods; indeed these he actually revives.[330] This is partly explained in terms of Vesalius's advocacy of observation as the only true guide to structure. This was freighted with a challenge to the established hierarchy of natural knowledge because, in contemporary (Aristotelian) terms, sensory observation revealed particulars. Natural philosophy sought rather to understand universals and first principles, but these are not observable, so there is in the *Fabrica* very little in the way of natural philosophy and the general principle of inference of cause and function from structure and morphology.[331] But this does not demonstrate a philosophical difference with Galen. Rather it was Vesalius's preoccupation with the amendment of Galen's morphology that led to his neglect of function.[332] The neglect was an effect of methodology, rather than a philosophical commitment. Like Aristotle, Galen, Benedetti, Berengario, and de Laguna, Vesalius fully appreciated "what valuable philosophical material is to be acquired from [anatomical] knowledge."[333] As Singer has argued, "Vesalius had always in his mind's eye" "the body as a whole"; his was a teleological conception of anatomy whereby the "parts he is examining must be fitted into the system as a whole," and placed in its relation to the soul, "the end for which man was made."[334] Although he often neglects (rather than avoids) to infer cause and function from

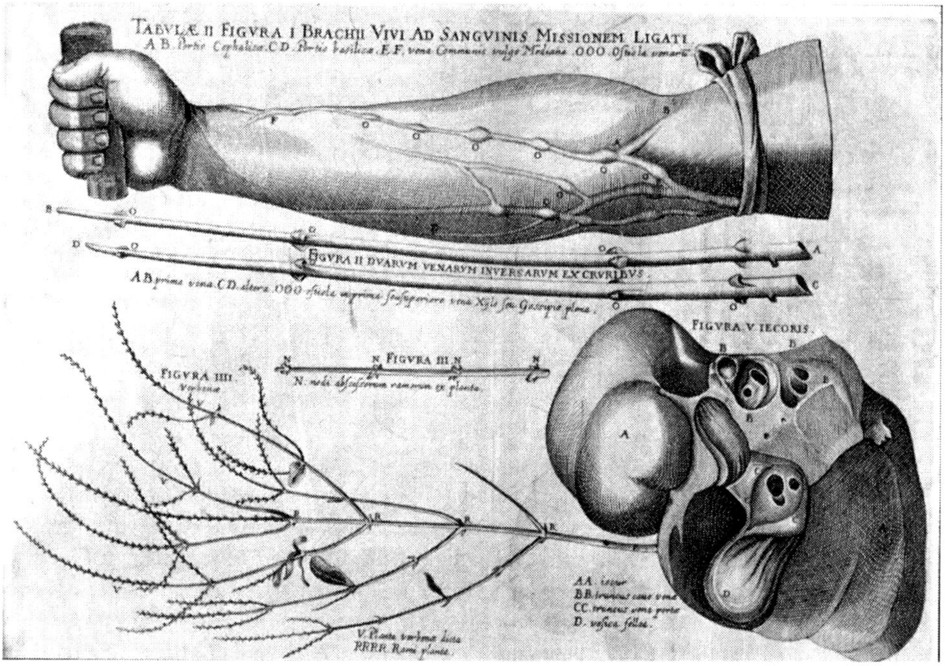

Girolamo Fabrici of Acquapendente, *De venarum ostiolis* (Padua: Lorenzo Pasquati, 1603); repr. in *Anatomices et chirurgiae . . . tractatus quatuor: Quorum, I, De formato fœtv, II, De lucutione et ejus instrumentis, III, De loquela brutorum, IV, De venarum ostiolis* . . . (Frankfurt: Jacques de Zetter, 1624), Tab. II. On describing, he thought for the first time, the delicate little membranes or *ostiolae* (little doors) in the cavity of the veins, which we now call the valves, Fabrici said that they have "a form like a node in a twig," and this plate visually demonstrates this on a comparison with figures of nodes such as those in the genus *Verbena.* Wellcome Library, London.

structure, he does not doubt in principle that function may be inferred thereby. He could not (in Singer's words) "shake off the yoke of Galenic teleology."[335] In his own return to the investigatory anatomy of Galen there are, moreover, a "large number of instances in which he accepts Galenic (or Aristotelian or [even] Arabic) doctrine without question."[336] Notwithstanding this, Vesalius's engagement with Galen differs somewhat from that of his immediate predecessors and most of his humanist and Hellenist contemporaries, whose concerns were philological and linguistic.[337] Partly through the number of dissections he carried out, and partly through "the infinite patience and detail with which he explored the human body and the marvellous powers of observation,"[338] upon which he was prepared during an anatomy to rely "without consideration to what earlier authorities claimed,"[339] Vesalius surpassed his predecessors in the number, significance, and collective impact of his refutations of authority. Yet this approach to anatomy did not appear to Vesalius as incompatible with an extensive and systematic conceptualization of the vascular system in terms of vegetative structure, growth, and function.

Philosophical consistency demands linguistic consistency, yet to understand the reason that the veins are "twigs," or, in *The Purple Island,* "brooks," is also to know that the fact that they happen to have been called "twigs" or "brooks" is inessential. Metaphorical expressions in the language of anatomical texts may in themselves be inessential, that is, replaceable in principle by other expressions equally useful in indicating the concept which the expression aims to grasp. But because these expressions are systematic and not merely incidental, they must be replaced by other metaphorical expressions with which they are consistent in order not to alter their meaning.[340] Any other descriptive rendering would have to be consistent with the idea that the parts in question are the result of several divisions of vessels in the venous system. Thus in the vegetative system, "branches" or "shoots" would be consistent replacements (and in fact Vesalius goes on to describe as "one shoot" one of the two "twigs" which extends to the external region of the hand, and "is distributed between the first internode of the thumb and the metacarpal bone supporting the index finger").[341] In this way, the most specific defining metaphor of a system—defining the animal in terms either of vegetative reproduction and growth or the fluvial distribution of nourishment—is indispensable to the language of anatomy textbooks and *The Purple Island.* It underlies the surface expressions and determines the knowledge the textbooks and poetic anatomy are capable of producing. Fletcher's allegory transforms the more conventional figuration in anatomy and by the same measure of consistency replaces branches and shoots with streams and brooks, retaining the metaphorical rationality of the former, which aims less like scientific metaphor to get a hold of the defined phenomenon, than to understand it in terms of the defining concept. In the cognitive process of understanding by means of a scientific metaphor, it has been argued that "the mapping itself may also create or affect the structure of the target [defined] domain."[342] But in the conceptual processing of sixteenth-century anatomical language, the most specific metaphor is replaced by another defining metaphor *always* with the effect that the defined concept is differently defined. It could not be otherwise. The language of anatomical texts varies in terms of metaphorical density, but by eschewing a premature distinction between "literal" and figurative language, we may avoid overlooking the embeddedness of much of the metaphor, its disguise as "literal" language, and the way in which covert embedded metaphors and abstract concepts from natural philosophy frequently seem to characterize overtly metaphorical expressions.[343]

Neither Vesalius's sensory approach to anatomy, nor his rationalist division of the functions of the body such that they correspond to the different parts of the soul, prevented him from extending the defining domain of the vegetative metaphor in this way, nor from extending its use from the reproductive and venous systems to the nervous system, which the *Epitome* describes as an intricate latticework of "roots," "tendrils," "branches," and "offshoots."[344] In the description of the division of the third pair of nerves ascribed to the brain (which is now referred to as the trigeminal nerve, the fifth cranial nerve), the vegetative metaphor is particularly vivid: "Also from this root [the inferior maxillary nerve] a branch twisted inward in the manner of a tendril is sent to the muscles just mentioned [the medial pterygoid which raises the lower jaw], and another branch to the upper teeth, another to the lower jaw and to the teeth fixed in it, and finally to the lower lip [the inferior alveolar nerve, the mylohyoid nerve and the mental nerve]."[345] From what is now known as the brachial plexus—"a varied plexus of nerves"—six nerves are said to "sprout forth into the arm, in addition to various offshoots dispersed to the hollow and to the gibbous part of the scapula."[346] Significantly, *The Purple Island* also figures the nervous system in a way that is consistent with its own figuration of the vascular system: "Three diff'ring streams from fountains different" (*P.I.* 2.10.1):

k The veins convey nourishment from the liver, the arteries life and heat from the heart, the nerves sense and motion from the brain.	k These with a thousand streams through th'Island roving, Bring tribute in; the first gives nourishment, Next life, last sense and arbitrarie moving: (*P.I.* 2.13.1–3)

The nerves arise from the brain and "Slide down to every part with their thick milky wave" (*P.I.* 2.12.7), which conveys the animal spirit [Greek, *pneuma psychicon;* Latin, *spiritus animalis*] necessary for the sensitive and motive functions of the animal faculty, to the organs of sense and the members responsible for voluntary motion.

This tendency in the *Epitome* and *The Purple Island* to use language consistent with the conceptualization of the vascular system in describing the nervous system is far from superficial and has two possible explanations. Firstly, both texts may be metonymically extending the defining concept from the domain in which it has been seen to function routinely as a metaphor into the conceptual and descriptive domain of the nervous system, a part of the body and its functions which lay beyond the domain ordinarily defined in this way. Vesalius does seem more inclined than some of his contemporaries stylistically to explore hidden aspects or possibilities of the conventional anatomical metaphor. In his descriptions of the vascular system, as we have seen or shall see, his metaphorical flourishes tend often to extend the ordinarily used parts of the defining concept or domain of vegetation to "tendrils," "twigs" and to parts of asexual reproduction. Such a possibility arises because, in Lakoff and Johnson's analysis, "there is always more in the defining concept than is carried over to the defined concept."[347] But, although Vesalius's accounts of the nervous system and the vascular system in the *Epitome* are very rich metaphorically, he is by no means alone in representing the nerves in this way. What does seem to emerge from the case of Vesalius is that his metaphorical flourishes are neither inconsistent with the language early modern anatomy routinely deploys, nor discontinuous with the conceptualization of the body it articulates.[348]

Consequently, a more likely explanation for the external systematicity between the metaphorical concepts of the reproductive, vascular, and nervous systems is that it possesses a rational consistency which rests upon the distribution of the vegetative faculty in the body. The systematic elaboration of this metaphorical concept in the *Epitome* and *The Purple Island* involves envisaging certain parts of the body interior as the ground upon which all this vegetative growth or fluvial distribution of nourishment takes place. In the *Epitome,* one branch of the carotid artery, having divided and subdivided several times, enters [as the cerebral arteries (anterior and middle)] the pia mater; "part creeps into the right ventricle of the cerebrum and forms a plexus" or network of small vessels "laid down in this ventricle" (the choroidal artery and choroid plexus).[349] The term "creeps" and the expression "laid down" are in the surrounding context of vegetative imagery highly suggestive of the activity of the runners of parent plants of species such as bramble, which reproduce in this asexual way. In this particular case, the right ventricle of the brain forms the field in which the runner lays down its roots. On this small scale in Vesalius's *Epitome* perhaps we have the suggestion that may have led to the conceptualization of the anatomy of the body in terms of the island in Fletcher's poem.

Generally internal organs like the uterus and the brain are metaphorically defined as parts of the ground or soil. But the *Epitome*'s description of the blood supply to the genital labia extends the metaphorical conceptualization of parts of the body as ground from internal to surface anatomy: "Extending downwards upon the thigh, the external branch [of the left "trunk" of the vena cava] sends a shoot to the skin of the pubis and to the hillocks of the female pudenda."[350] This extension of the vegetative metaphor may appear to suggest that even the concept of surface human anatomy is in some sense less clearly delineated than the concept of growth and reproduction in the earth, but a rationale may be that the pudenda are sometimes regarded as internal rather than external parts.[351] Elsewhere, however, offshoots of the axillary vein (the venous anastomosis around the elbow joint, and the cutaneous and muscular rami) "creep forward to the skin of the internal part of the hand"; and a "branch of the shoulder vein" (humerariae) "runs to the little finger and the ring finger."[352]

Consequent upon this systematic elaboration is a conceptualization of the whole of human anatomy, internal and surface, in anatomy textbooks, in terms of nutrition, growth, and generation in plants, and in *The Purple Island,* in terms of an island, the many geographical features of which are examined in the next chapter. A further consequence is a rationalization of this anatomy as formed in order to fulfill the purpose of these functions. This teleology relates the anatomy to the philosophical concepts of the vegetative soul or the faculty of Nature, which animals and plants share. The anatomy of the human and animal body is thereby conceptualized as natural. However, as with all metaphorical conceptualizations, this one is partial; it defines only one aspect of the concept of anatomy with a view to highlighting that aspect. In sixteenth- and seventeenth-century anatomy, there is another conceptualization which is equally systematic and similarly characterized by an abstract philosophical concept for a part of the soul.

In the eighteenth of the series of lectures given at Bologna in 1540, Corti cites *On Anatomical Procedures,* where Galen provides the authority for a metaphor of anatomy as "building" or architecture. According to Corti, Galen says, "that as the artisans in various ways connect and fasten their objects, such as stones, timber etc., so also Nature, whom they have imitated in their works, proceeds in joining the limbs to one another, some only by inserting them, e.g. the teeth, sometimes through sutures, e.g. the bones of the head, sometimes as with glue, which is the cartilage between the bones, sometimes with strings, e.g. the ligaments, the cords or the sinews, sometimes as with tissues, e.g. round the arteries and the veins etc."[353] This extended metaphor so defines the construction of the human body from its parts that the art of the architects is said to have imitated Nature's construction of anatomy, thus reversing the direction of metaphorical structuring we find in Corti's anatomical description.[354] A "consonance between architecture and the human figure . . . had been born in antiquity" and was "later nourished in fifteenth-century Italy."[355] Indeed the arts of anatomy and architecture entwine historically in various ways, and it is with the conceptual and theoretical aspects that I am principally concerned.[356]

In *De re aedificatoria,* Alberti notes that, "the most expert among the ancients . . . were of the opinion, that an edifice was like an animal, so that in the formation of it we ought to imitate nature."[357] Thus, the concept of an animal is prescribed for the purpose not only of conceptually structuring the concept of a building, but of dictating the building's architectural plan. Schultz remarks that the tradition of analogy between architecture and anatomy "developed along" both "a more literal and theoretical path," and in Leonardo da Vinci's anatomical drawings he finds "an interpretation that reaches beyond the traditionally held theoretical correspondence, for it touches a literalism";[358] by which I understand that the analogy becomes a metaphor whereby anatomy is, as it were, *literally* understood in terms of architecture, whereby the "correspondence" attains systematicity and considerable conceptual content. Leonardo's anatomical interests are known to have extended to dissection,[359] and his anatomical practices in Florence in the opening years of the sixteenth century have been described as pervasively defined by a "continuing dialogue between the textual authority of antiquity" and "an experimentally established knowledge of physical principles which in turn is governed by the ordained order of nature."[360] Whatever Leonardo's projects may have comprised in relation to the ancients of anatomy, their models and his own dissections, his contribution went unpublished and cannot be regarded as having had any measurable influence upon the development of either the texts or illustrations of anatomy textbooks.[361] It is to these textbooks that the present study must be confined because they represent the mainstream of anatomical concepts, language, debate, and consensus, from which Fletcher draws for the *Purple Island* anatomy.

In his *Isagogae breves,* Berengario provides two schemes of the knowledge acquired from anatomy and "building" or architecture.[362] Anatomy and "building" here both signify simultaneously the object, with a structure of parts, and the form of discourse or knowledge which pertains to the object, together with the method of procuring that knowledge.[363] Hence, of the discipline of anatomy, Crooke says, "the Groundworke of the building, and not onely so, but the whole frame was by the ancients reared up; and therefore now if any Ornaments be added, they must be fitted thereunto."[364] In his lectures on *Anatomia Mundini,* however, the Galenist Corti distinguishes between architecture, to which anatomy is analogous, and building, to which the art of dissection corresponds. The distinction reflects Corti's adherence to the authority of the written medicine and philosophy of Galen in contrast to Vesalius's appeal, in his anatomical demonstrations which accompanied Corti's lectures, to the epistemological priority of the skill of the hands and the demonstration to the senses. Corti explains that by sight and touch and good dissection we may learn the position, substance, connexion, and form of the parts, but the explanation of "why this organ has such a shape, position, connexions, number etc." is for the "philosophers and the doctors to give."[365] Corti's explanation articulates a position that the object in nature should not be effectively distinguished from knowledge of, or discourse on, the reason for the existence of that object. Such teleological language is imagined as coextensive or continuous with that to which it refers by means of its explanation. Since it purportedly offers the true reason in Nature for the object, such language is imagined as partaking of the object's identity and reality, arising directly from Nature. In his accompanying demonstrations, and in the *Fabrica,* Vesalius did not disagree that the proper way to understand anatomy is to discover the reason why a part is as it is; he disagreed with the division of labor. It is indicative of his methodological approach that, in the Preface to the *Fabrica,* Vesalius deplored the pedagogical procedure which required of the professor of anatomy only the delivery of a lecture from an authoritative text, and to stand over the *anatomistae "like architects,"*[366] and he argued strenuously that the art of medicine was "maimed by the neglect of that primary instrument, the hand, so that [the manual aspects of medicine] were relegated to ordinary persons wholly untrained in the disciplines subserving the art of medicine."[367]

In the first of Berengario's schemes, human bodies and buildings are both similarly divided into individual parts and defined according to the way in which they are put together. In this scheme, what makes anatomy different from "building," says Berengario, is the operation of the parts of the body.[368] In the second scheme, the life of the body distinguishes anatomy from architecture, and gives rise to the body's movements, innate bodily habits or complexions, influences and "passiones" (by which Berengario means illnesses or diseases).[369] Anatomy is therefore *partially* not wholly structured in terms of "building," as exemplified perhaps by a set of illustrations intended for Bartolommeo Eustachio's *De dissensionibus ac controverisiis anatomicis,* which went unpublished before its author's death, a fact that may explain his eclipse by Vesalius.[370] The illustrations were probably drawn by Eustachio himself and his relative and assistant, Pier Matteo Pini of Urbino, and engraved by Giulio de' Musi of Rome.[371] The engravings have been described as showing "all the elements of a developing observational science," and, with "their concern for measurement and for the study of variation," as "look[ing] forward to analytic biology."[372] "Eustachio's figures are maps of human anatomy, not representations from a single viewpoint"; and they have been described as "lifeless and stiff and not at all adequate representations of living parts."[373] They contrast remarkably with the figures of Vesalius, which will be discussed in chapter 3, and which "attempt to copy the natural appearance of anatomical structures."[374] One explanation is the engraver Giulio de' Musi, of whom "only architectural prints are known, apart from these engravings."[375] If so, it would seem that the analogy between architecture and anatomy, and the metaphor of anatomy as "building," may have facilitated the modeling of these anatomically accurate figures, in which case they consequently fall foul of the metaphor's inadequacy as identified by Berengario.

Some aspects of anatomy which exceed definition in terms of architecture are nevertheless rendered consistent with the body defined metaphorically in those terms. This systematicity is achieved via the philosophical purpose that is realized through "the ancient pneumatic physiological system, which explained muscular movement [and sensory perception] via the distension of the animal spirit, transported from the brain through the nerves":[376] "The brain, like the senses, and also like volitional movement, is made for the Reigning Soul."[377] The whole body is, according to Vesalius's Preface to his *Fabrica,* (combining Platonic and Aristotelian concepts of the body) "the temporary lodging and instrument of the immortal soul,"[378] which is the instrument's master.[379] Not only Vesalius, but Fletcher is clear that the Isle of Man is given the following attributes:

> Vigour, sence, reason, and a perfect motion,
> To move it self whither it self would have it,
> And know what falls within the verge of notion
>
> (*P.I.* 1.46.2–4)

While they cannot be derived from architecture proper, sensory perception and volitional movement are thereby rendered consistent with this metaphorical conceptualization of the body, in a contiguous relationship to the reason for its being.

The metaphorical concept of anatomy as "building" gives rise in *The Purple Island* as well as in anatomy textbooks to an exploration of marginal or downplayed aspects of the conventional metaphor, extending the ordinarily used part of the defining concept to provide detailed specifications of the body's architectural features.[380] In Canto 5, stanza 16, the third ventricle of the brain is described as "Vaulted by three fair arches," which the margin, barely modifying the figurality of the stanza, indicates belong to the "corpus Cameratum," "which with three arches bears up the whole weight of the brain." The "corpus Cameratum" is more usually in anatomy referred to as the fornix, which Vesalius similarly describes in his *Fabrica,* albeit with simile rather than metaphor, as "resembling a vault (fornix) or arched roof."[381] The fornix "is a longitudinally arched bundle of nerve fibres . . . connected with and adjacent to the basal ganglia. . . . It lies on the under surface of the corpus callosum, to which it is adherent."[382] The "inferior surface [of the corpus callosum] is concave and on both sides forms the roof of the lateral ventricles."[383] Fletcher's poem omits the corpus callosum, but very probably collapses the distinction between two anatomically distinct parts—the corpus callosum and the fornix. Vesalius, however, is precise and accurate, stating that "in its upper part, . . . the septum adheres along its whole length to the corpus callosum, and in its lower [part], . . . it adheres to the top of that portion of the brain which we shall describe as resembling a vault (fornix) or arched roof."[384] The same architectural function of support as alluded to by Fletcher's description is extended by Vesalius's *Fabrica* to the "septum [which] in its turn supports the tortoise [corpus callosum or fornix?] lest—by a catastrophe for the whole working of the brain—it might collapse and obliterate the common cavity [= third ventricle]."[385] Notwithstanding the difference in measure of precision between *The Purple Island* and Vesalius's *Fabrica,* the two architectural descriptions of this region of the brain and its functions of support share a continuity of metaphorical representation, which suggests that figurality and anatomy are not incongruous, either in anatomical texts or in *The Purple Island,* and further, that the figurality they share has conceptual content.

What yet remains beyond the scope of the metaphor in this description of the constructedness of the brain, and beyond the second of Berengario's schemes of the knowledge acquired from anatomy and "building" or architecture, and even beyond the philosophy that renders volitional movement and sensory perception consistent with the metaphor, is the life of the brain, and of the whole body, giving rise to innate bodily habits or complexions, influences, illnesses, or diseases, and the natural or vegetative functions of reproduction, nutrition, and growth. With these aspects of anatomy, the metaphor of the body as an abode and instrument of the soul is not consistent, yet the human body in the anatomy of the schools and medical profession, and the image or figure of the island for Man, encompass both constructed and natural, living features. The concept of anatomy is structured by two metaphors, and the remaining aspects of the life of the body which exceed definition in terms of architecture are metaphorically conceptualized as deriving from Nature or the vegetative soul. The two metaphors cannot replace each other and retain the same meaning, because they are not consistent. Part of what remains hidden by the concepts of the embryo as a plant and the uterus as a field of nature is understood in terms of inherent properties, but part is understood in terms of "building." Thus, whereas the metaphorical definition of the uterus as the field of nature highlights its reproductive function or operation and its connection through the umbilicus to, and influence upon, the fetus, the metaphor of "building" highlights that organ's interiority with respect to the body, its substance, size, shape, and its construction from its parts. Similarly, whereas the definition of the embryo as a plant highlights the concept of its growth and the operations associated with its connection through the umbilicus to the uterus, the metaphor of "building" highlights the internal location of the fetus, its substance, shape or form, construction, and volitional movement. Thus while each of these metaphors partly structures the concept of animal anatomy, they "jointly provide a coherent understanding of the concept as a whole," as both natural and constructed.[386]

In so doing the conceptual metaphors characterize a portion of the body of knowledge which sixteenth- and seventeenth-century anatomy was capable of producing and expressing to the reader of anatomy textbooks. They lead the reader to focus only on those aspects of anatomy that they jointly highlight, leading her or him to view the entailments as true.[387]

Perceptions and inferences may follow from the metaphors' acceptance, leading the reader to project metaphorically a structure and character, or metonymically a purpose, onto anatomy in order to understand it and, further, to judge the truth of a proposition in relation to it. True belief and knowledge are in this context relative to the conceptual systems of anatomy, which are defined in some fundamental respects by metaphor, and by its categories or divisions, which relate physiological systems and anatomical parts to Nature or to faculties, such as the nutritive, auctive, and generative.

In Benedetti's comparative anatomy, metaphorical identity between plant and animal parts, functions, and purposes begins in analogy, as does Berengario's metaphor of anatomy as "building."[388] Let us take firstly the former of these metaphors, in order to demonstrate their relationship to the analogies of the microcosm. In this way, a further vital piece of the jigsaw of *The Purple Island*'s figurative design may be put in place. The faculties of Nature and the vegetative soul are both internal to Man and external to "him" in the vegetative and animal world. They thereby provide a rationale for analogy between human anatomy and the world. The metaphorical conceptualization of human anatomy in terms of vegetation is therefore consistent with the analogy of the human body with the vegetative growth and "geological" terrain of the geocosm, which we find in *The Purple Island.* Corti is recorded by Baldasar Heseler as delivering a lecture drawing parallels between the arrangement of the poles of the earth and the arrangement of "man" as "Microcosmos."[389] The Hippocratic dogma of qualities and the theory of humors, both of which comprise the body's consistence as described by Benedetti's *Anatomice* (cited above), correspond in Empedocles' doctrine to the four elements in the macrocosm, and to those of the earth, the geocosm.[390] There is moreover a submerged microcosm-geocosm theory in the analogy between the fetus and the plant, because the fetus was thought to pass through various stages that mark the development of the soul from vegetative, sensitive, and motile to rational. Hence the soul reflects or participates in the vegetable and animal kingdoms, and can be seen to share some part of its life with them.

Related to and consistent with the analogy of the geocosm is the body as a microcosm of the heavenly macrocosm, or the cosmos. The "proportion," "composition," and "correspondency of the parts of mans body" suggested that as a whole "it may worthily be called a Litle world, and the patterne and Epitome of the whole universe."[391] With regard to the parts themselves, Berengario was far from alone in saying, "the dignity of the heart prevails over other parts since it is the chief of all the members and is called the sun of the microcosm because it illuminates the other members with its spirit."[392] Meanwhile the head corresponds to the "sphere of intelligentiae," to spirits and angels (the eyes can be said to "carry the same form the Stars do"),[393] and the "Inferior venter" or abdomen reflects the "sublunary" and "elementary" sphere.[394] Both of these analogies, of the geocosm and the macrocosm, share some of the connotations or implications of the metaphor of vegetation in anatomy, and both are therefore conceptual, playing a crucial role in the field of sixteenth-century knowledge, as a category of thought and a general configuration of Nature.[395] They imply that the body is natural, that it is continuous with the structure, growth, generation, and nutrition of plants and animals in the world.

But the geocosm also comprises the human-built and sociopolitical environment. So in the same way, the concept of anatomy as "building" is consistent with theories of the microcosm's correspondence with the dynastic realm, the body politic, and with constructed human environments and systems. These analogies share with the metaphor the idea that the body is constructed and is in this sense continuous with the material, political, and social fabrications made by humans. One frequently reads in anatomy textbooks that "[t]he Heart is the chief Mansion of the Soul," implying a body composed of many buildings rather than understood as one.[396] There is a rich vein of such figuration which stretches back at least to Plato's reference in *Timaeus* to the head as the "citadel," the thorax as the "council chambers," and the abdomen as the "manger."[397] And such analogical comparison between the hierarchies of the organism and the state could clearly serve the anatomist as well as the political theorist.[398]

The consistency found in anatomy textbooks between the metaphor of anatomy as vegetation and the analogies of the geocosm and macrocosm, and between the metaphor of the body as constructed and the analogy of the body politic, finds its way into the island-body analogy of *The Purple Island.* By means of its vegetation, fluvial distribution, "geological" formations, and built environment, the island unifies in a single image the coherent, but in anatomy inconsistent, metaphorical concepts which find their reasons for being either in the rational or the natural and vegetative faculties. In order to provide an example of how Fletcher's poem coalesces the metaphorical conceptualizations of the body and the analogies with which they are consistent in a systematic way unprecedented in the work of any sixteenth- or seventeenth-century anatomist, we can look to Galen's instructions on the order of dissection.

In *On Anatomical Procedures,* Galen indicates that not only should one conceptualize the body as a building, but that this conceptualization should determine the order of dissection, an order which Fletcher in *The Purple Island* follows. Although it is not the only order Galen employed,[399] he recommends that "one should first learn about the bones, because they play the same role in the human body as walls do in houses." The metaphor is reiterated in the first chapter, "De ossibus" (Of the Bones), of Realdo Colombo's *De re anatomica,* where the human osteological structure is comparatively related to architectural supports.[400] Galen says that for the next stage: "Then one should move on to the anatomy of the muscles for these two parts of the body underlie all the others, like the foundations of a building."[401] Foundations are stones in the earth, and the description of the island's "geological" formation, in *The Purple Island,* provides our example of how the metaphorical conceptualizations of the body and the analogies of the microcosm systematically coalesce.

It grounded lies upon a sure [a] foundation,
Compact, and hard; whose matter (cold and drie)
To marble turns in strongest congelation;

a The foundation of the body is the bones. Bones are a similar part of the body, most dry, or cold;

(*P.I.* 2.5.1–3)

Islands, bodies, and buildings are all imagined as having foundations or frames of similarly solid substance, which are located beneath the surface of each, and share in each case a complexion which is cold and dry. This literary exploration of the shared entailment is unprecedented by the figurality of anatomical texts where the metaphors are coherent but not consistent. Although, in anatomy textbooks, they jointly provide a coherent understanding of the concept of anatomy, each metaphor allows the reader to get a handle on different aspects of the concept being structured. The entailments they share, as has been said, are abstract and unclearly delineated; and unlike in *The Purple Island,* they provide no consistent image of anatomy.

The Purple Island thus coalesces the conceptual metaphors of anatomy and the analogies of the microcosm. In setting out to reconstruct and analyze the rationality of Fletcher's use of figurality, this study has so far demonstrated that *The Purple Island* shares figurative and rational coherence and consistency with descriptions of the human body in anatomy textbooks of the Renaissance and early modern period. Moreover the work's combination of anatomical exposition and allegory is congruous, and Kastor's verdict that Fletcher's design resulted in large-scale incomprehensibility should not deter the reader from the progressive assemblage of both geopolitical information about the dynastic realm of Britain and knowledge about Man, or human anatomy. This anatomy is nominally male but purports also to represent the female, and combines an erotic dimension, scriptural tradition, and metaphysical differences of gender, together with morphology and function, descriptively fashioned by conceptual anatomical and microcosmic figurality. As the allegory unfolds the threshold figure of the Isle of Man with consistency and coherence, the enwrapped analogues intermittently rise to become uppermost, though whether they are continuously superimposed is a question answerable only after a more exhaustive investigation in the following chapters. However, this chapter has reached the conclusion that the allegorical scheme is not a superficial means of presenting anatomy, as Langdale believed, and his critical methodology and interpretation can now be seen to have become caught in a self-fulfilling circularity.

This study has also begun to develop a theory of allegorical, analogical, and metaphorical meaning to interpret the figurality of *The Purple Island,* to reconstruct and analyze its rationality, to explore its congruity, and to assess the validity of the arguments of its critics, the most thorough of whom have been concerned with the combination of anatomical exposition and allegory, and hence with the status of figurality. The allegory is based upon analogy, broadly conceived within the traditions of the microcosm and microchristus. The crucial point of difference between those of Fletcher's critics, like Langdale, whose assessment is that the significance of *The Purple Island* lies in the purportedly "modern" character of the anatomy, and those, like Baldwin, for whom Fletcher's reliance on the conventional analogy is indicative of his intellectual conservatism, led us to investigate the development of the theory of the microcosm, and so to understand differently the nature of the question. By examining the development of the microcosm theory in natural philosophy, medieval Platonism, scholasticism, and Renaissance Neoplatonism, with reference to how the model related epistemologically to the development of observational anatomy, Langdale's argument was found to rest upon a less than wholly true distinction between the microcosm and the (Renaissance) Platonic notion derived in part from Plato's *Timaeus.* Not only were the Platonic theory and the literal microcosmic view of the human body essential and typical in the Renaissance, belief in the microcosm was even plausibly an impetus to anatomize. Hence, the distinction between abstract and concrete truth is also a misleading one with regard to the microcosm. In *The Purple Island,* the body as microchristus or as a microcosm of the terrestrial globe, as differentiated from the macrocosm, or universe, carries the weight of meaning. The geocosmic analogue for the body is a geological location which encompasses the built environment and human society, and the body's analogy with this geopolitical location is characterized by two more basic conceptualizations of the body: as constructed and as natural.

In order to determine whether the anatomy is essentially extractable from its allegorical context, or conceived wholly within a conceptual framework of the multiplicity of correspondences between the geocosm and the microcosm, this chapter has explored the Renaissance anatomy of the schools and medical profession for the extent to which it shares figurality with *The Purple Island.* We should expect from the theory of conceptual metaphor, that systems of metaphorical concepts relating to analogies like the body politic and the microcosm-geocosm arise from the nature of our bodily existence and experience, because our bodies are both phylogenetically and ontogenetically one of the first and most basic entities the mind can grasp. Notwithstanding this, we have discovered an unexpected feature of conceptual metaphor in early modern anatomy. Analysis of the metaphorical structuring of the anatomy of the human body in terms of vegetation and "building," in sixteenth-century anatomy textbooks and in *The Purple Island,* reveals it to be consistent with the structuring of abstract or less clearly delineated concepts in terms of concrete, more clearly delineated ones. Analogies between animal (including human) anatomy and that of plants were found to give rise to metaphorical expressions which were conceptually shaping, structuring, and defining of animal anatomy. Moreover, cognitive processing, whereby linguistic metaphors trigger a process of analogizing, which in turn has recourse to conceptual metaphor, seems to have been required of the early modern reader of anatomy. Each of the conceptual metaphors of vegetation and "building" partly structures the concept of human anatomy; they jointly provide a coherent understanding of the concept as a whole, as both natural and constructed. Human anatomy is systematically conceptualized in terms of a host of metaphorical concepts that are related in terms of entailment relationships to these conceptual metaphors, such that conceptual metaphor underlies surface metaphorical expressions and characterizes anatomical knowledge, eschewing the distinction between "literal," "scientific" language and purportedly ornamental figurative language, particularly with regard to the conceptual content of utterances of either kind. This corroborates Paul de Man's more general philosophical argument that because properties of substances will not totalize in any proper way, the empirical entity cannot generally be sheltered from what he calls tropological disfiguration.[402] And yet these conclusions seem not to be in conflict either with the sixteenth-century revival of the ancients' projects of dissection or the increasing emphasis upon sensory apprehension and observation of the body.

The *Purple Island* anatomy is similarly figured in vegetative, terrestrial, and architectural terms. Although it figures the vascular system as fluvial distribution rather than vegetative growth, the imagery is rationally consistent with descriptions in anatomy textbooks of arterial and venous distribution, osteology, and myology. Not undertaken in this chapter was a close examination of the internal and external systematicity of the figuration of the anatomy of Man in terms of the Isle. But we have begun to sense how *The Purple Island* allegorically transforms the systematic figurality of anatomy into the geographical and social features of the Isle of Man, and chapters 2 and 3 provide a more methodical and thorough examination of the entire semantic, semiological, and performative field of the *Purple Island* anatomical language, and its tropological models. We will see whether, as Kastor thought, the amassed correspondences cause the poem to fragment.

Since Renaissance anatomy was rationalist and teleological, the metaphorical conceptualization of morphology and function is coherently extended to the purpose of functions, their reason for being or final cause. This provides the rationale for the character of the metaphoricity. So, a further aspect to the rational consistency of *The Purple Island* is that it actually provides in allegorical form the *telos,* the reason for the vegetative figurality of the textbooks. The Aristotelian vegetative soul or Galenic *physis* provides the reason for the metaphorical structure of the concepts of the nutritive, auctive, and reproductive parts and functions of both animals and plants, and the animal, rational, or immortal soul provides the same for the sensitive and motive functions. Defining the human body in terms of the instrumentality of its construction to the rational soul, and the vegetative (or fluvial) distribution of nourishment for its growth and reproduction, is implicit to and embedded within the figurative language of anatomy textbooks and *The Purple Island.*

The digression into the anatomy of the schools and of medicine also reveals a coherence between the analogies of the body as a microcosm of the terrestrial globe or the macrocosm and the conceptual metaphors of anatomy, thereby further integrating the elements of Fletcher's figurative design. The consistency found between anatomy as vegetation and the analogies of the geocosm and macrocosm, and between anatomy as "building" and the analogy of the body politic, finds its way into the Isle of Man, which unifies in a single image coherent, but in anatomy inconsistent, metaphorical concepts. The threshold figure of the moving Isle of Man imposes an order on anatomy, and fulfills at least two of the main principles of metaphor in Elizabethan poetry, as described by Terence Hawkes. These are coherence and consistency.[403] The figure coalesces and unifies the conventional concepts of anatomy in a "normative," "reinforcing" way, and in this respect also conforms to the ordinary seventeenth-century poetic practice of constructing metaphor "from 'public' elements with 'established' ranges of relationships."[404] Hawkes says of metaphor in all societies: "It will be as much concerned with what we know as it is with what we don't know; it will retrench and corroborate as much as it will expand our vision."[405] It is now timely to see how the threshold figure of *The Purple Island* is exploratory and expansive of the vision of anatomy in the early seventeenth century. The next chapter will attempt to establish the creative and innovative character of the tropology of *The Purple Island* in terms of sixteenth-century rhetoric and poetics, and modern aesthetics, philosophy of literature, and metaphorology.

2

The Creative Metaphor of the "*Isle of Man*"

THE FIRST CHAPTER OF THIS STUDY BEGAN TO DEVELOP A THEORY OF ALLEGORICAL, ANAlogical, and metaphorical meaning in *The Purple Island,* to reconstruct and analyze the rationality of that figurality and to explore its congruity and systematicity. The allegory was found to be based upon analogy, broadly conceived within the traditions of the microcosm and microchristus. Additionally, however, the *Purple Island* anatomy is figured in vegetative, terrestrial, and architectural terms, and this figuration has been found to be rationally consistent with the anatomy of the schools and medical profession. The previous chapter's digression into this anatomy revealed consistency between the conceptual metaphor of anatomy as vegetation and the analogies of the geocosm and macrocosm, and between the conceptual metaphor of anatomy as "building" and the analogy of the body politic, as well as coherence between these consistent sets of conceptual metaphors and analogies. These investigations thereby provided several elements of Phineas Fletcher's figurative design but without fully elucidating how *The Purple Island* integrates the elements or explores the implications of that integration. In the second chapter, a close examination of the internal and external systematicity of the entire semantic, semiological, and performative field of the tropological models of the *Purple Island* anatomy will commence.

The Purple Island does not merely draw figurative ideas from anatomy and the tradition of microcosmic speculation, but allegorically transforms them into the geographical and social features of the Isle of Man. In anatomy the metaphorical concepts of the body as natural and as constructed cohere but are not consistent. The threshold figure of the Isle of Man, however, unifies the concepts in a single image, fulfilling the Elizabethan poetic principles of coherence and consistency. To what extent this image or figure retrenches and corroborates what Fletcher's seventeenth-century readers might already have known, and to what extent it might be thought to have expanded their vision, is discussed in this chapter, though the question of the creative and innovative character of the tropology of *The Purple Island* will exert a pressure on all the directions this study will take. In beginning our answer this chapter will turn to sixteenth-century rhetoric and poetics, and modern aesthetics, literary theory, and metaphorology.

The way in which the threshold figure is exploratory is intimately related both to the metaphorical concepts which more or less conventionally define anatomy and to the even more conventional analogues of the microcosm. These are the established ranges of relationships from which the Isle of Man figure is constructed. But the presence of coherence and consistency in the threshold figure enable *The Purple Island* systematically to explore unfamiliar parts of the geocosm and the body politic to a degree unparalleled by anatomy textbooks. There is, however, in de Laguna's *Anatomica methodus,* an example of a systematic exploration of an analogy which provides a preparation for the far greater imaginative systematicity to be found in *The Purple Island.* De Laguna's analogy is between the physiological anatomy of the organs of nutrition and the river systems of France with their merchant shipping routes. He says, "Hence from the intestines which are like large ships that carry a milky substance very many meseraic veins, like small boats or skiffs, bear away the purer juice and send it to the liver, similar to those tall ships which as soon as they have crossed the ocean come to Rouen with their cargoes on their way to Paris but transfer their cargoes at Rouen into small boats for the last stage of the journey up the Seine."[1] The River Seine in this description is the small membrane, the mesentery, which is "a continuation of the serous coat [the fourth and outermost coat of the alimentary canal beneath the diaphragm] and attaches the small and much of the large intestine to the posterior abdominal wall."[2] The anatomy of the Isle of Man superficially resembles this system of waterways, for, as Thirsil says, "Nor is there any part in all this land, / But is a little Isle: for thousand brooks in azure chanels glide on silver sand" (*P.I.* 2.9.1–3). But whereas the "brooks" dividing the Purple Island are the arteries, nerves, and veins, the analogy is less exacting in what is essentially de Laguna's teaching aid. Therein the "the ocean" corresponds to nothing anatomical, and the comparison of the intestines, meseraic veins, mesentery, and the humors with ships and rivers communicates only their carrying function and connectedness.[3] De Laguna's analogy, however, conveys not only the purpose of the anatomical structures, but also the active involvement of Nature in the maintenance of their functions, and consequently of health:

> And since no ship sails without a helmsman, exposed as it is to a thousand perils and shipwrecks, who, I ask, shall be that one who steers the ship, the kybernetes [steersman in Greek] who grasps the rudder? By Apollo, that one is Nature herself, who guides and disposes everything according to her will. But you ask, and properly too: 'What performs the function of a rudder, since a helmsman cannot operate without a rudder?' I answer: 'Those four first qualities of things, namely, heat, cold, moisture, and dryness.' Is there not an analogy here with the guidance of ships, that is, if the helm is turned farther than it should be toward one direction the course of the ship is changed? Does not the same thing happen in our human body? If some one of these qualities becomes greatly increased does not the entire body suffer wretchedly?[4]

Thus in the pedagogic context of a geocosmic and mercantile analogy, de Laguna calls on Hippocrates' theory of the four qualities to explain Nature's operation in nutrition, in an extended simile conceived within the philosophical context of the teleology of human anatomy. But, however much it draws on conventional analogies and conceptual metaphors, the extended simile or analogy is exceptional and isolated in de Laguna's text. By comparison, *The Purple Island* collapses the analogues which comprise these analogies into one single coherent and systematic image or figure, the Isle of Man, the anatomy of which is then capable of expansion or exposition over four cantos. That which I have until now frequently referred to as the island-body analogy, is in fact a metaphor, which the allegory extends. By thus creating a single metaphor out of two conceptual metaphors from anatomy—vegetation and "building"—and three conventional analogies of the microcosm—the geocosm or terrestrial globe, the macrocosm, and the body politic—*The Purple Island* opens out meanings, possibly creating new meanings, that can neither be connoted nor suggested in precisely the same way by another metaphor or analogy.

Creation of Meaning in Metaphor

The theoretical debate, early modern and modern, over whether metaphors extend, revitalize, discover, or actually create similarities is one that must be pursued in order to know what kind of metaphor the Isle of Man is. In order to demonstrate that the Isle of Man is creative, it is necessary to explore how ambiguity, or rather multiplicity, of meaning, or plurisignification, emerges from what is frequently referred to as the "area of interaction" in a metaphor.[5] This section is influenced throughout by the literary theory and philosophy of Max Black, Monroe Beardsley, Mark Johnson, George Lakoff, and Paul Ricoeur, among others. Despite the differences that emerge from the different disciplines in which they work, these theorists of metaphor share a view that, paradoxically, metaphor creates or extends similarities that connect two concepts which are disparate and incongruous when literally interpreted, and that metaphor thereby has "irreducible cognitive content" or "significance."[6]

In Max Black's "interaction view" of metaphor, the thoughts of a hearer or reader about the subjects of a metaphor are "active together" and "interact" to produce a multiplicity of meaning.[7] As with the conceptual or cognitive theory of metaphor, there is conceptual processing, and this is required precisely because of the second feature of creative metaphor, that it connects concepts that are literally incongruous. A metaphorical expression is not merely a substitute for some other literal expression as a "substitution view" supposes.[8] On a "substitution view" metaphors supply the want of a convenient abbreviation in the literal vocabulary and are therefore a species of catachresis, or they are merely stylistic, a decoration.[9] This view we have encountered as typical of objectivist views of metaphor, which exclude metaphor for its literal falsity from semantics. It is also typical of abstractionist views to maintain that metaphorical expressions may be substituted by a literal paraphrase. This study has argued, after Lakoff and Johnson, that metaphor is conceptual—an "*imaginative rationality*"—and from this it follows that poetic metaphors are not, or need not be, merely ornamental.[10]

It should be granted, though, that by arguing for a view of the Isle of Man metaphor as cognitively significant, one is not committed to deny that literary metaphor is ever merely ornamental. The sixteenth-century literary theorist George Puttenham had objectivist suspicions about the semantic falsity of metaphor and displays a naïve trust in the unfigurative transparency of "common vtterance," the "ordinary limits" of which figures are said to "passe": figures, he says, are "occupied of purpose to deceiue the eare and also the minde, drawing it from plainnesse and simplicitie to a certaine doublenesse, whereby our talke is the more guilefull & abusing, for what els is your *Metaphor* but an inuersion of sence by transport."[11] This view is also espoused in the influential (for the Renaissance) anonymous *Rhetorica ad Herennium* (c. 86 BC), which argues that metaphor has no real claim to meaning.[12] But Puttenham also writes plausibly of the possibility of dressing poetry: "This ornament we speake of is giuen to it by figures," he says, "disguising it no litle from the ordinary and accustomed."[13] Puttenham is in agreement with Quintillian, whose view that metaphor adorns, as "ornament," had considerable influence in the Renaissance.[14]

Because of misgivings like those of Puttenham and Quintillian about the cognitive content of metaphor, a question basic to all allegorical narrative is whether metaphor is "a thrust at truth or a lie."[15] But an effect of Peter Ramus's imposition upon the elaborate structure of traditional Aristotelian rhetoric, of a division between those parts of rhetoric belonging to Dialectic (i.e., logic) and the remaining parts of Rhetoric, was that metaphor could justifiably be regarded as cognitive and logical, "the arrangement of thought in an orderly manner." However emotive, metaphors could be regarded as akin to "conceptual statements which concern 'thinking.'"[16] One of the parts assigned to Dialectic was "invention," which could mean "a 'discovery,' a 'finding,' the 'faculty of discovery' but also the 'thing found.'"[17] Poetic invention therein derives its meaning from the meaning of *inventio* in (mainly Latin) rhetorical theory.[18] Although it is applied to the finding of subject matter, rather than words, invention is the "conception and the working up of material," and is consequently inextricably linked to choice of words.[19] This equivocation reflects what has been referred to as "one of the central problems of Renaissance aesthetics: 'What is the object, the material cause of poetry: reality or fiction?'"[20] Especially in the seventeenth century, under the influence of the Aristotelian poetics of Julius Caesar Scaliger, invention could also be used in a case where the subject matter is made up by the poet using her or his *fantasia* (fantasy) or *ingenium* or *ingegno* (in this context, wit), to resemble either forms in nature or the "intention" in the mind of either God or Nature.[21] In this respect invention is relevant to the fictional dimension or subject matter of the story in the narrative of *The Purple Island,* described by William Benlowes as one of Fletcher's "works of fancie."[22] Legitimate objects of invention were generally regarded to be things that are, that can be, or that the Ancients considered to be true, though tensions did exist between the concepts of mimesis, particularly with regard to the ontological status of that which is imitated.[23]

The threshold figure of *The Purple Island,* the Isle of Man is creative and cognitive, not merely ornamental. Because it is introduced as a riddle (see *P.I.* 1.34 cited above, in chapter 1), the figure requires active nonliteral analogical processing which gives rise to the creation or extension of similarities and a solution in the form of "new knowledge."[24] The knowledge of the Isle of Man figure may not be identical in kind with literal truth or falsity but it may be said to be true.[25] But because the allegory has been found to be based upon analogy, broadly conceived within the traditions of the microcosm and microchristus, it is not immediately obvious that the threshold figure is not in principle replaceable by "ordinary" language. On the "comparison view" of metaphor, the characteristic transforming function of metaphor is thought to be analogy or similarity.[26] This is similar to the "simple-analogy view" which was held to obtain in some cases of metaphor by Aristotle, as well as by Cicero, who says that metaphor is a "contracted" simile, giving decorous effect.[27] Hausman argues that "an analogy can be expressed as a complex simile": "The point of an analogy is to make a comparison."[28] The "simple-analogy view" of metaphor, which also corresponds to what Beardsley calls the "Literalist Theory," holds that metaphors are compressed, disguised, or elliptical similes:[29] "Since there is no fundamental difference, they can be understood the way similes are, by the ordinary rules of language, and present no special problems for explication."[30] As matters of language, rather than of thought, metaphors are regarded on the comparison view as capable of describing only preexisting isolated similarities, even if these similarities were formerly hidden, in which case we should perhaps say they were discovered.[31] Analogies and similes are usually, though not always, paraphraseable.[32]

But the view that the characteristic transforming function of the Isle of Man metaphor is analogy, as distinct from the view that the metaphor is based in some way on analogy, fails to account for the fact that the similarity or likeness between its component terms admits of degrees.[33] Beardsley notices how similes are "empty and uncontrolled without a context," or closed with a similarity in one particular respect.[34] By contrast, "complex," interesting metaphors are the linguistic site not simply of a single relation, but open out such that the subject *is* the predicate in more ways than one.[35] As a creative metaphor encompassing the similarities that emerge from conventional analogies and conceptual metaphors of anatomy, the Isle of Man figure gives rise in *The Purple Island* to the extension of the ordinarily used part of the defining concept or domain, a process we have also found in Vesalius's *Epitome.* In metaphorical extension of this kind, as Beardsley argues, "there is an important difference between a predication that is merely suggested and one that is overtly stated . . . and [the difference] is lost in the paraphrase."[36] We have already had cause, in chapter 1, to question Baldwin's reading of the microcosm-macrocosm analogy in *The Purple Island,* and to qualify it with the degrees to which the figure of the Isle of Man gives rise in the poem to each of the microcosm analogies. Unlike similes, metaphors may involve multiple potential points of similarity, they are "full and rich, apart from any context," although the context may "eliminate possible meanings,"[37] just as the context of the poetic work develops the geocosmic connotations of the Isle of Man more than the macrocosmic ones

In relation to Langdale's conviction that the allegory of *The Purple Island* is superficial with regard to the anatomy, chapter 1 argued that the risk of prematurely foreclosing meaning may arise in the interpretative situation as a result of a creative or novel figure, because the meaning of such a figure is not consensually agreed or wholly conventional. Fletcher does supply a classical precedent for the Isle of Man in "*Delos*" (*P.I.* 1.47 margin. *), a small island in the Cyclades, which in Greek mythology is represented as moving. But this does not provide the plurality of connotations emerging from that figure as it occurs in Fletcher's poem. The Isle of Man belongs to a kind of figure the object of which is to deepen and extend meaning, even indefinitely, yet in accordance with a notion of the world's intelligibility, rather than simply and precisely to denote how the world is understood (as plain language in objectivism), or to extend without coherent reference to that understanding (as metaphor in subjective or "free" association), or merely to adorn (as metaphor in objectivism).[38] It would be impossible to provide an exhaustive paraphrase in "literal" language of the metaphorical ramifications gained by the figure of Man as an island (although many are explored in the cantos of the poem). This assertion is not to deny clarity to the figure of the Isle of Man. There may be clarity in the profusion of metaphor, as well as in plainness, if the intelligibility of the profusion is in some sense systematic or logical and not freely associative.[39]

In order to appreciate the distinct conceptual process of interaction, which gives rise to and opens out multiple points of varying degrees of similarity, we may look more closely at the "interaction view" of Black. This is characterized by seven claims or principles, which Black says, pertain in their entirety only to interestingly complex metaphors.[40] In the "interaction view," the sense of a metaphor "emerges as the result of a certain contextual action."[41] In this action, the metaphorical word or subsidiary subject is the focus (in the threshold figure of *The Purple Island,* the Isle), but this focus requires the frame of the sentence, the principal subject (Man).[42] Black says: "This use of a 'subsidiary subject' to foster insight into a 'principal subject' is a distinctive *intellectual* operation (though one familiar enough through our experiences of learning anything whatever), demanding simultaneous awareness of both subjects but not reducible to any *comparison* between the two."[43] Paul Ricoeur argues that "this is why you cannot translate an interaction metaphor in plain language without 'a loss in cognitive content.'"[44] The "creative" metaphor is "not expendable"; it "is more than a mere substitution for another literal word [i.e., a kind of catachresis] which an exhausting paraphrase could restitute at the same place."[45] Black's claim is that: "It would be more illuminating in some of these cases to say that the metaphor *creates* the similarity than to say that it formulates some similarity antecedently existing."[46] Before they became conventional and prior to their deployment in *The Purple Island,* the similarities between anatomy and "building," and those between reproductive anatomy in plants and animals, were *created by* analogy and metaphor. Conceptual structuring in metaphors is in one direction only, and in this sense it is not merely a comparison, even where the metaphor is created out of a context of analogy, such as that between the animal body and architecture. As Lakoff says, though: "There is a big difference between having a metaphorical mapping exist as a unit in your conceptual system and putting together the same metaphor fresh for the first time."[47] It is apposite therefore to distinguish between those metaphors that create similarities which have become conventional, those that in some manner extend or rediscover those preexistent similarities, and those that are creative of novel similarities.

Black's second claim is that the principal and subsidiary subjects "are often best regarded as '*systems* of things', rather than "things.'"[48] He argues (making his fourth claim) that there exists a stock of commonplaces associated with the subsidiary subject, platitudes which are the common possession of the members of a speech community, which he terms the "*system of associated commonplaces.*"[49] Black's third claim is that drawn from the system of associated commonplaces is "a system of 'associated implications' characteristic of the subsidiary subject."[50] We are led by this system to construct a corresponding system of implications about the principal subject.[51] Concepts are brought together by the identity (*a* is *b*) of metaphor.[52] Associated with the subsidiary subject of the Isle of Man are aspects and organizations of the natural, human-built, and sociopolitical environments. The subsidiary subject, "island" or "isle," is for Kastor "a rather obvious nationalistic symbol."[53] Even though *The Purple Island* adheres to the conception of the state as a dynastic realm, the metaphorical term or focal word suggests a further system of political implications of a proto-nationalistic kind which identify the realm with reference to its borders, its shores.

In his fifth claim, Black explains how the corresponding set of implications about the principal subject arises. He states that the metaphor "selects, emphasizes, suppresses, and organizes features of the principal subject by *implying* statements about it that normally apply to the subsidiary subject."[54] These are, to use the term of Beardsley, the "significant connotations" of the metaphor.[55] Among the associated implications of the subsidiary subject of the Isle of Man figure is the concept of the isolation or insularity of the island or dynastic realm (specifically Britain). When the implied concept of geographical or national isolation is transferred from the associated implications of the subsidiary subject to the principal subject, it creates a corresponding associated implication, that a human being is alone, possibly alienated. As the metaphor suppresses some associated commonplaces of Man and Isle, it selects and emphasizes features of similarity as characteristic of the nation and the human being, like self-determination, thereby creating a corresponding sense of the human being as a self-determining, perhaps self-fashioning, individual. Other similarities suggest themselves, such as between a nation's xenophobia and a person's paranoia. A set of significant connotations are created which relate to the historically incipient and ideologically emergent concepts of individualism and nationalism, which are perhaps not reducible to extensions or discoveries of antecedently associated yet hidden connotations of the figure of Man as an island, and therefore must be distinguished from classical and medieval occurrences. The island realm of Britain is meanwhile redefined as anatomizable or analyzable, perhaps in the way a human body can be searched for causes of disease postmortem. These emergent concepts of Man and the nation conflict with those of the dynastic realm and the abstract or universal Man we find in Vesalian anatomy (to be discussed in chapter 9). But they also bring to fruition "the seed of alienation" that was sown in Renaissance microcosm theory. The Isle of Man metaphor is true insofar as it elucidates and deepens the meanings of Man in accordance with the world's intelligibility, but it also contains a truth that is the result of working out certain "modern" implications of representing Man as an island.[56]

It might be added to Black's analysis, that, between the meanings of the two "systems" of a metaphor is a similarity or equivalence (without which no metaphorical meaning could emerge) and a difference (without which there would be no metaphor, but simply identity); for a metaphor must mask or downplay aspects of the systems or domains of the defining and the defined terms in order to highlight other aspects of both domains and thereby discover similarity. James Deese argues: "The literal and figurative concept must have some features in common . . . but they had better be neither too numerous or salient nor too obscure and dependent upon a kind of erudition."[57] These criteria appear to be reliably apt for defining that part of the interaction in the threshold figure of *The Purple Island* which is based on the conceptual metaphors of anatomy and conventional analogy. Correspondences, such that the bones and muscles provide the body's architectural foundation, are not obscure; they are commonplace, at least in anatomy. Nor are they too salient or obvious: Man and islands do not share the same ontological category. But, like the "simple-analogy view" of metaphor, Deese's criteria seem insufficient to account for two aspects of complex or creative metaphors.

With regard to the first of these aspects, Beardsley argues that we do not have a metaphor "unless some fairly central characteristic associated with the word [modifier, subsidiary subject] is excluded by the context."[58] Ricoeur argues: "I agree with these authors [Beardsley and Black] on the fundamental issue: a word receives a metaphorical meaning in specific contexts within which they are opposed to other words taken literally; this shift in meaning results mainly from a clash between literal meanings, which excludes a literal use of the word in question and gives clues for the finding of a new meaning which is able to fit in the context of the sentence and to make sense in this context."[59] The "clash" of literal meanings is of course what the more or less conventional conceptual metaphors of anatomy tend to subdue, for there the principal subject does not generally resist definition in terms of the subsidiary subject, and that part of the domain or system of meanings of the defined term that is not so defined is ordinarily defined by another conventional metaphor or is quietly masked. If the metaphor is mundane or utilitarian, or highly conventional, tension diminishes and the similarity no longer surprises, because the difference is already assumed to be irrelevant to the metaphorical meaning, which is no longer emergent but emerged and paraphraseable.[60] The meaning may not even be recognized as metaphorical. By contrast, acknowledging the "clash" of literal meanings in creative metaphors, Hausman notes: "The divergence of meanings in metaphors and their characteristic way of intruding presents the interpreter with a special kind of tension" as well as a "fit."[61]

The word "conceit," often spelled "conceipt," deriving from the Italian *concetto,* carried "the general meaning of idea, concept, or thought."[62] But when it was used in Renaissance and seventeenth-century poetics to refer to a feature of a work, it signified a metaphorical manipulation of language: "Unlike the usual metaphor, where an obvious and close resemblance exists between vehicle and tenor, a conceit stresses a similarity between highly dissimilar elements that must be discovered through thought."[63] Hausman explains (in reference to I. A. Richards's theory of metaphor): "Disparities are as operative as similarities are. And the role of disparities opens the significance of metaphors so that they can act to bring out resultant meanings not given in the meanings of the terms they bring together."[64] "Having found an unforeseen relationship between two phenomena, the business of the concettist was to develop all possible similarities and to give them eloquent expression."[65] This gives rise to an "extended conceit," "one in which analogy is subject to detailed and ingenious development, often throughout an entire poem."[66] This suggests that the reason for the exposition of the Isle of Man in Cantos 1 to 5 of *The Purple Island* is to unfold similarities in order to "make the reader see what is not obvious; . . . that things that are dissimilar do resemble each other."[67] The same "self-reflexive tension between the literal and the metaphorical" is the "essence" of the interpretation of allegory.[68] Allegorical figurality simultaneously discovers meaning and hides it in shadows; the system of figures cloaks the discoverable meaning in the "sustained or continuous fictional activity or plot," which in literary criticism is often referred to as the "literal" meaning, but which in my view corresponds to the literal (or inertly metaphorical) meaning of the modifier or subsidiary subject of the metaphor; and where this meaning is taken for the meaning of an allegory, interpretation is not based in the interaction of metaphor, but is rather like understanding metaphor as ornament, which entails an avoidance of the tension of interaction.[69] The cloak of allegory hides or downplays the ordinary meanings of the principal subjects of the allegory, but the resultant tension acts as an impediment to comprehending the allegory in terms of the ordinary "literal" meanings of words used metaphorically, that is as subsidiary subjects in metaphor.

By assuming that the subjects of a metaphor must have some features in common, such as the relevant analogies already in existence as commonplaces, and which therefore may be revivified or extended, Deese's criteria seem insufficient to explain a second aspect of complex metaphors, the metaphorical creation or generation of meaning. "An analogy depends for its significance wholly on similarities common to antecedent significances. . . . If analogies attain their significance through their dependence on relations among common features, then they do not introduce new significance into the world."[70] In some intricate way, this feature is related to the tension of creative metaphors. Black and Beardsley suggest that the interaction may emphasise previously *unseen* aspects of the meanings of and similarities between the principal subject and of its modifier.[71] And this suggestion dovetails with the views of Lakoff and Johnson, who propose three subspecies of imaginative or creative metaphor.[72] One of these we have encountered in Vesalius's *Epitome*. His metaphorical flourishes tend often to extend the ordinarily used parts of the defining concept or domain. This may be said to extend or revivify similarities, which are nevertheless consistent with more or less conventional conceptual metaphors. The second subspecies is the use of the ordinarily unused part of the defining domain of a metaphor, which I think may be said to extend or even discover similarities which are nevertheless consistent with more or less conventional metaphors. And the third is the novel or new metaphor, which of all the above is the only metaphor that by putting aside, hiding, or downplaying some aspects of the defined concept and by virtue of giving "coherent structure" to a newly highlighted range of aspects, actually "*creates similarities of a new kind*" between the ranges of aspects of the defined and defining concepts.[73] To argue that the Isle of Man figure is creative, innovative, and reveals implications that are historically incipient and ideologically emergent, is not to claim that it is novel *per se.* Unlike in novel metaphor, the contextual effect perhaps goes no further than "actualization of the potential range of commonplaces or connotations" belonging to the system of associated commonplaces of the focal word.[74] The case is more like the modeling of anatomy on architecture in Leonardo's anatomical drawings, in which it has been suggested there is an interpretation that reaches beyond the traditionally held correspondence. The analogy becomes a metaphor in which the correspondence not only attains systematicity and considerable conceptual content, but is extended to encompass Leonardo's anatomical investigations, or, in Fletcher's case, newly emergent implications from within early modern British culture.

While much of the similarity in the threshold figure of *The Purple Island* is indebted to the systematicity within and coherence between anatomical concepts and conventional analogies, some is achieved as the threshold figure gives rise in Cantos 2 to 5 to a non-derivative exploration of the unused parts of the commonplace concepts of the geocosm and the body politic, which by analogy define the understanding of Man. By opening out the domain of the analogues of Man in the area of interaction, the Isle of Man exceeds or *extends* the microcosm's antecedent correspondences with the geocosm and the dynastic realm. It may even be that new connotations are created. This exploration or extension into unexplored aspects of the defining domain is Black's sixth claim for metaphor, that the interaction also "involves shifts in meaning of words belonging to the same family or system as the metaphorical expression."[75] *The Purple Island* was not the first text, nor the only one of its period, to systematize similarities with a view to anatomical detail (as part 2 will show), and so there is an "implied determinism by which [the] metaphorical insight is predictable in principle, or deducible in a system that contains the connections among similarities known and becoming known as the system is worked out."[76] These similarities are made explicit, we might say "discovered," and play a role in interpreting the Isle of Man metaphor.[77]

The exploration of the unused part of the domain of the defining term is more a creativity of discovery than of genuine innovation or novelty, however, and seems inadequate as an explanation of the combination in creative metaphor of tension and opening out of meanings. While he agrees "entirely with the 'interaction view'" of metaphor, Ricoeur argues that the working out of Black's theory (especially its third, fifth and sixth principles) relies heavily upon "trivial metaphors," which qualify the principal subject only with traits drawn from the system of associated commonplaces of the subsidiary subject, which Ricoeur defines, alike with Beardsley's similar concept of the potential range of connotations (in his "Controversion Theory"), as rules to which members of a speech community are committed, but not semantical and syntactical rules.[78] Ricoeur argues that Beardsley and Black are herein merely adding to the semantic polysemy of the subsidiary subject in the dictionary and to the semantic rules which govern the literal use of the lexical terms, the system of associated commonplaces or connotations and the cultural rules which govern their use. Ricoeur asks pointedly and rhetorically: "Is not this system something dead, or at least something already established?"[79]

Polysemy is, admittedly, important to allegory, as Quilligan remarks: "The 'other' named by the term *allos* in the word 'allegory' is not some other [meaning] hovering above the words of the text, but the possibility of an otherness, a polysemy, inherent in the very words on the page; allegory therefore names the fact that language can signify many things at once." Although Quilligan asserts that "[a] sensitivity to the polysemy in words is the basic component of the genre of allegory," she also argues that allegories "not only all pun, they also show many other shared concerns with the truthfulness of metaphor."[80] In the same emphatic vein as Ricoeur, Terence Hawkes states: "What happens in metaphor is that the "literal" or "dictionary" level at which words usually operate is systematically avoided, even violated."[81] Although it is fair to say that the Isle of Man figure is not a novel metaphor in this sense, metaphor rather than polysemy is the basic component of the allegory of *The Purple Island,* and analysis of metaphor in *The Purple Island* can suggest that exploration might be more than merely making known similarities that already exist if not in previous occurrences of the metaphor of an island for Man, then at least in early seventeenth-century culture, between the principal subject Man and the ordinarily unused part of the defining concepts of the geocosm and the body politic. For example, it can suggest that some of the associated implications of the subsidiary subject may be thought to suffer a metaphorical change of meaning in the process of their transfer to the principal subject.[82] From the interaction of the Isle of Man figure emerge similarities between the systems of the two subjects. Among these is the concept that the island realm, like the body, is composed of members among whom one is sovereign. When the implied concept of sovereignty is generated and transferred from the associated implications of the subsidiary subject to the principal subject, where it creates a corresponding associated implication, that the heart or head is sovereign in the body, it might be thought that it does not maintain its "literal" or conventionally metaphorical sense, but is metaphorically translated by features of the soul's dominion over and influence upon the human body from the heart or head. The transformation involves the implication that the sovereign exercises rational or divine control over the polity as its soul. The implication is not created new by *The Purple Island,* however, so hereby the Isle of Man has been merely "analyzed into a set of subordinate metaphors"—the polity as a body and human anatomy as "building" or as composed of (social) members—"so the account given is either circular or leads to an infinite regress," a situation which Black perceives as a challenge to his interaction view, because the primary or threshold metaphor then appears to be an analogy joining together subordinate metaphors, rather than a metaphor transforming the associated commonplaces through interaction.

Black counters this perceived challenge by "denying that *all* changes of meaning in the 'associated commonplaces' must be counted as metaphorical shifts. Many of them are best described as *extensions* of meaning, because they do not involve apprehended connexions between two systems of concepts."[83] It is important to maintain the transformative function of interaction as the cause of dilation of meaning in the definition of creative metaphor. But the infinite regress of subordinate metaphors which, as an analysis of creative metaphor, appears to threaten the interaction view, should not be surprising given what we know about the formation of conceptual metaphor. In Lakoff and Johnson's notion of systematicity, a concept is systematically conceptualized in terms of a host of metaphorical concepts that are related in terms of entailments. Amalgamating the two theories, in this limited respect, suggests that Black's system of associated commonplaces of the subsidiary subject may consist of metaphorical concepts even before they are transformed by interaction.[84]

There is another way in which analysis of metaphor in *The Purple Island* can suggest that exploration might be more than merely making known antecedently existing similarities between the principal and subsidiary subjects. It may identify something valid about the way in which readers have responded to *The Purple Island*'s purported strangeness, incongruity, even incomprehensibility, to affirm that the threshold figure at the heart of the allegorical design is not paraphrasable *even by other consistent metaphors,* and perhaps that its particular grammatical character is irreplaceable.[85] Perhaps linguistic and lexical structures, as well as rhetoric, have a role in cognitive processing in all occurrences of metaphor,[86] but many nonliterary metaphorical statements can in principle be said in other ways, and this is true of scientific metaphors, and also of the conceptual metaphors of anatomy as long as alternatives are metaphorically consistent, as was found to be the case in Vesalius's *Epitome.* The reason for the exceptional character of the Isle of Man, and perhaps of literary or at least poetic metaphor in general, is difficult to fathom, but it lies partly in the maximum of foregrounding of the utterance against a background of less actively metaphorical language.[87] In this foreground the contextual effect on metaphorical words is enhanced, giving rise to greater interaction or interanimation, and this effect spreads to make other words in the work lose the precision they formerly had and become dependent on one another (rather than their antecedently existing associations) for their meaning.[88]

Black's fourth claim about interaction metaphor provides not exactly a *differentia specifica* of literary metaphor, but the possibility of the literary work as a context or frame for a kind of metaphor which, while preserving the notion that the system of associated implications characteristic of the subsidiary subject is drawn from somewhere rather than created entirely anew, suggests that it may be drawn from a special system of associations as may be *created* by a poem or piece of sustained prose, in which, it is said, a writer can establish a *novel* pattern of implications for the literalness of key expressions, prior to using them as vehicles or subsidiary subjects for her or his metaphors, thereby creating "specially constructed systems of implications" which are not commonplaces.[89] This statement is distinct from his earlier claim that the interaction may emphasize previously *unseen* aspects of the associations of the principal subject. These points may not lead us to conclude that the interpretation of poetic metaphor is logically distinct from that of conventionally metaphorical and (so-called) literal language, but it does emphasize the importance of the relationship between individual occurrences of metaphor and the discourse context of the whole literary work.[90]

The Isle of Man figure in *The Purple Island* can be seen to establish a *novel* pattern of implications for the literalness of key expressions relating to Man and to the island realm, prior to Fletcher using them as vehicles for his metaphors, thereby creating specially constructed systems of implications which are not commonplaces. In this sense, *The Purple Island*'s proto-individualism and proto-nationalism are among the novel implications of the Isle of Man insofar as they are not connoted by earlier literary occurrences of the metaphor of an island for Man, and they interact in the foreground of the poetic work, giving rise to greater interaction or interanimation, which spreads to make other words which are associated with the Isle lose the precision they formerly had and become dependent on their newly acquired contextual meaning (rather than their antecedently existing associations). This can be seen in the entrenchment of the Isle's shores by the *mare mortuum,* or sea of death (*P.I.* 1.45, 54)—a "natural" division or border that serves to separate or set apart the saved Protestant nation of Britain and each of its people as the elect. But the metaphorical use of the novel pattern of proto-nationalistic implications becomes more pronounced in the unusual presentation of the anatomy and physiology of the closure of the eyelids against intruding objects. This takes on militant proportions as the objects become "rushing foes," the lids defensive "walls" of the twin "Towers" of the "castle," and the lashes the "pikes" of ranked "spearmen" (*P.I.* 5.27.3; 5.37.2; 5.27.7; 5.27.6). But the poem's realization of the nation-state and of a defensive nationalism and individualism is incomplete and ambiguous. The fortification of the island's "head" (*P.I.* 5.4, 11; 7.8), for the defense of its virtuous "Soveraigne," the "*Intellect*" and "understanding" (*P.I.* 6.28.2; 6.51.2 & margin. *), his "Queen," "*Voletta,*" "The will" (*P.I.* 6.57.3, 4 & margin. *), their daughter, "*Eclecta,*" the Church or elect (*P.I.* 11.6.3–11.8.7), and the king's counselors, of which he himself is one (the others being the "common sense" [*P.I.* 6.42 margin. *], "*Phantastes,*" or "The fancie" [*P.I.* 6.46.2 & margin. *], and "the *Pemptarchie,*" "The five senses" [*P.I.* 6.42.2 & margin. *]), against the invasion of vice, should not be regarded only as expressive of xenophobic nationalism and paranoia, although the former of these meanings is made explicit by the extended simile of conflict between "the *Britain* and *Iberian* fleet" (*P.I.* 6.23).[91] The fortification can also be understood as a reactionary entrenchment of the idea of the state as a dynastic realm, and the corresponding abstract concept of Man, both of which are ordered hierarchically and defined by internal centers, or heads, rather than by borders. Even as reactionary, however, the metaphor is not an analogy resting merely on antecedent isolated similarities, such as those between the head of the ideal or abstract human body and the sovereign of the state. Rather, the metaphor extends conventional similarities between Man and the state, and, ironically, in the very entrenchment of its reaction helps to forge the emergent concepts of individual and national insularity.

The implications of the threshold figure of the *Purple Island* allegory are in part antecedent and extended, emerging from between previously unobserved similarities and assumed differences between the subjects of the metaphor, and in part created or worked out. Many predications are merely suggested rather than stated. But in each case the "truth" of the Isle of Man figure is not reducible to the abstract. It connects, draws together, and associates; but its "ontological function" is not to be a self-sufficient whole, whether cognitive or aesthetic.[92] Poetic discovery or invention gives rise to the generation of a potentially infinite number of subordinate metaphors, metaphorical shifts of meaning, of inexhaustible cognitive significance. This condition of "open-endedness," "of the possibility of additional, accruing meaning," is of the character of creative and literary metaphor,[93] and it severely impacts upon the possibility desired by Langdale of finding where the figurative meaning of *The Purple Island* ends and where the "science" begins. The figure of the Isle of Man is intense: it is abundantly meaningful in a complex way; yet it is linguistically compact.

The Allegorical Extension of the "*Isle of Man*"

The purpose of this section is to explore how the allegory of Cantos 2 to 5 of *The Purple Island* may discover previously *unseen* aspects of the meanings of the principal subject of the creative metaphor, the Isle of Man, and of its modifier. Our discussion has revealed several ways in which this discovery, or even invention, can be made. One is to extend the ordinarily used part of the defining concept or domain, as Vesalius does. Another is to use the ordinarily unused part of that domain. Both of these ways may give coherent structure to a newly highlighted range of similarities between the defined and defining concepts. It will be for this section to begin to make a judgment about how non-derivative this exploration of Fletcher's is, and whether new connotations are created. The former section has pointed to a set of significant connotations which relate to the historically incipient and ideologically emergent concepts of individualism and nationalism, the implications of the interaction of which are perhaps not reducible to extensions or discoveries of antecedently associated yet *unseen* connotations of the figure of Man as an island, and the allegory's working out of these "modern" implications will be explored here too. It has also been argued that as a literary metaphor, the contextual effect is enhanced, giving rise to greater interaction, and this effect may spread throughout the work, thereby properly creating *new meaning*. The metaphorical use of key expressions in *The Purple Island* that have acquired *novel* implications creates fragments of what Paul Valéry calls "pure poetry embedded in the substance of a discourse," a background which is the poetic work.[94] The other aspect of the exploration undertaken in this section will be simultaneously to discern the systematicity in the profusion of metaphor, in order to assess how successfully and clearly Fletcher the concettist developed all possible similarities and gave them eloquent expression, how well he makes the reader see what is not obvious, that things that are ontologically dissimilar do resemble each other. A limitation of the previous section was that by working at a theoretical level on the threshold figure extracted from the poetic work, it was possible only to state potential implications or connotations. Now we turn to the metaphor's extension through the poetic allegory, and in order better to understand the relationship of the figure to the whole work, I want to look at Paul Ricoeur's investigation of the idea that we "may treat metaphor as a work in miniature."[95]

For Ricoeur, "discourse . . . implies the polarity of *sense* and *reference,* that is, the possibility to distinguish between *what* is said . . . and *about what* something is said."[96] He connects "the problems of explanation to the dimension of 'sense' . . . and the problems of interpretation to the dimension of 'reference' understood as the power of discourse to apply to an extralinguistic reality *about* which it says what it says."[97] He then theorizes the relationship between metaphor and the work to which it belongs: "From one standpoint the process of understanding a metaphor is the key for that of understanding larger texts, say, literary works; this point of view is that of explanation and develops only this aspect of the meaning which we called the *sense*—the immanent design of discourse. But, from one other standpoint, it is the understanding of a work as a whole which gives the key to metaphor; this other point of view is that of interpretation properly, which develops the second aspect of the meaning which we called the reference, that is, the intentional direction toward a world and the reflective direction toward a self."[98] Here Ricoeur is using a vertical model of textual analysis and interpretation, a pioneering example of which is that developed by Roland Barthes in his "Introduction to the Structural Analysis of Narratives," wherein the vertical, "paradigmatic" axis of narrative integrates the consequential and consecutive "grammatical units" of the narrative (called "functions") by means of "indices."[99] Indices proper refer to an implicit signified on the level of "actions" (including character-agents or "actants"),[100] and to the higher level of "narration," that is, the "narrative communication" and the "narrative situation." The latter refers the work to the world of social, economic, and ideological systems beyond and signified within the work. Whereas both "functions" and "indices" are distributed along the syntagmatic axis of the narrative, the former have only contiguous ('metonymic") relations to each other and to the indices, but indices operate as metaphors referring to the analysis or interpretation of the whole work.

Although on an abstract understanding, the paradigmatic axis may correspond to the integrative character of metaphor, Ricoeur's more precise distinction between interpretation and explanation accounts for how "some features of discourse—reference and self-reference—begin to play an explicit role" only when "discourse takes the form of a literary *work*."[101] Metaphor is "a paradigm for the explanation of a literary work" because the "decisive moment of explication [of a metaphor] is that of the construction of the network of interaction," which is "the point of intersection between several semantic lines; this construction is the means by which all the words taken together make sense."[102] Equally, only the integration of all the semantic lines of a work enables it to make sense as a work. Thus, the problems of explanation of the work are encountered in miniature in metaphor, which stands as an index or microcosm of the whole work. This is so with the threshold figure of *The Purple Island,* in which the human body also stands in this fashion in relation to the world of the isle, as both a part and a microcosm.

By way of contrast to the whole work, "the metaphor is too short a discourse to display th[e] dialectic between disclosing a world and understanding one's self in front of this world."[103] For this reason the interpretation of the whole poetic work is the key to unravel those connotations of the metaphor which fit the context. Ricoeur maintains, "there are no metaphors, in the sense of metaphorical *words,* without certain contexts"; equally it is not possible to interpret *actual* connotations which fit, without a wider context than that of the interaction of principal and subsidiary subjects.[104] The limitation which the context of the work imposes is necessary because, in principle, "any concept, as well as the lexical representation of that concept, can never be characterized by a limited set of features—the set is always indefinitely extended."[105] Beardsley argues, a good explication of a metaphor satisfies congruence *and* plenitude: "the metaphor means *all* the connotations that can fit [the subject]—except those that are further eliminated by, because they do not fit, the larger context."[106]

If the Isle of Man's implications are related systematically by means of entailment relationships, the poetic cantos may be described as unfolding or extending the entailments of the work's threshold figure, to reveal the conceptual and linguistic systematicity of the ways by which these entailments are related both to each other and to the threshold figure from which they emerge.[107] In this way, the poetic allegory can be seen as a kind of gloss on the text of its intrinsic threshold figure, which may be seen from one perspective as analogical, insofar as analogy "creates a new significance understood not only with reference to antecedent relations but also with reference to a special way in which terms foreign to one another are interlocked" in the text.[108] Langdale's assertion that the allegory is superficial would be wholly incorrect if the interpretation of Man were dependent on the context of the whole allegorical text, and the task of extraction would be further impeded by the extent of systematicity of the figuration; that is, the extent to which the entailments of the figuration are explored or detailed by the allegory. Systematicity knits the concepts of the principal and subsidiary subjects of the figure together in the "shadowes" of allegory, which by cloaking or masking must necessarily eliminate possible interpretations, but where extended or new connotations are discovered and worked out with logical consistency in the process of reading.[109]

By way of conclusion to his discussion of the theory of interpretation of metaphor in relation to that of the whole work, Ricoeur suggests: "We could even say that, if the interpretation of local metaphors is enlightened by the interpretation of the text as a whole and by the disentanglement of the kind of world it projects, the interpretation of the poem as a whole is controlled, reciprocally, by the explication of metaphor as a local phenomenon."[110] The linguistic creation of the Isle of Man serves the general project of *The Purple Island* to let a new world or creative imitation of the ontology and reality of Man emerge into the foreground from the interaction of conventionally agreed meanings of Man, the realm, and the world, with which the emergent meanings are continuous, if contradictory.[111] The relationship between the figure and the work understood in this way reflects the idea of the macrocosm as an unfolding of the meaning of the microcosm, and of the microcosm as a condensation of the whole of the macrocosm. The systematicity of this condensation and dispersal, or unfolding, lends neither closure nor unity to the threshold figure, but it may suggest its coherence. In addition, and as we shall see particularly in part 3, this relationship closely resembles part-whole relationships in anatomy. If each particular part of the anatomy of Man is characterized by the conceptual metaphors which undergird the whole of anatomy, and by the creative metaphor of Man as the island-realm of Britain, then it follows not only that the explication of the threshold figure is an index to the explication of the whole of Cantos 2 to 5, but also that the Isle of Man "contains" in miniature the whole of the anatomy of Man explored in Cantos 2 to 5.

As a way of testing the systematicity of the figuration of the body as a geopolitical realm throughout Cantos 2 to 5, it is necessary to introduce some criteria. It has been mentioned that *The Purple Island* deploys the concept of complexion in the description of anatomical parts. In so doing the poem's anatomy fulfills one of the descriptive criteria of the "anatomical *accessus*" employed in anatomical commentaries and textbooks from John of Alexandria's commentary of Galen's *De sectis,* to the *Fabrica* of Vesalius (although not overtly in his *Epitome*) and several textbooks published subsequent to this.[112] According to French, the anatomical *accessus* evolved out of several forms: a literary or philosophical *accessus,* considered mandatory by late classical and medieval philosophers in an introduction to any work of Aristotle, and applied in a literary scholastic form to medical literature from the postclassical Alexandrian period on; Aristotle's *Categories;* and the causality *accessus,* the Aristotelian apparatus of fourfold causality, with a material, efficient, formal, and final cause. The criteria or *occasiones* of the anatomical *accessus* came to dominate anatomical exegesis until the sixteenth century, by which time they vary only slightly in anatomical texts, and may be compiled as follows: i) substance, or composition;[113] ii) size, quantity, or magnitude; iii) number; iv) form, figure, or shape;[114] v) location, or position; vi) anatomical relations, posture, connexion, or a part's continuation or connectedness with other parts, which is called "junctures" by Berengario in his *Isagogae breves* (1522);[115] vii) complexion, which is called "innate bodily habits" and "influences" in *Isagogae breves,* "temperature" in Paré's *Works,* and "native temper," "innate propertie, and speciall vertue" in *The Purple Island* (5.42.5; 2.34 margin.), and can be defined as "the particular manner in which the four elements or four humours are mingled together and is to be distinguished from composition [i, above], which refers to the coarser structural nature";[116] viii) movements, uses, functions, or operations, which were taken from vivisection as an extension of earlier anatomical *occasiones;*[117] ix) passions, affections, or illnesses, which was also added to John of Alexandria's list by later commentators.

The concepts of (viii) functions and uses, and of (ix) illnesses, underwent radical reconceptualization in the late sixteenth century, in the former case, and in the seventeenth century, in the latter case. The former was eventually to develop into physiology and the latter came to be a vehicle for discussion of pathology, but before doing so both criteria were to figure far more significantly in the revival of the philosophical (teleological) anatomy of Aristotle at Padua in the late sixteenth century, in the work of Girolamo Fabrici (Hieronymus Fabricius ab Aquapendente) and others, and in the work of William Harvey in the early seventeenth century. In Cunningham's view, the goal of anatomy for Fabrici was "a *single, general, universal account* of the operation" in that single, general, universal creature "The Animal," and the organs which carry it out; an account which would "capture the essence of that particular faculty and instrument of the soul, and hence explain what it was and why it was as it was."[118] His description, or "*Historia,*" of a part, is consequently comparative, comparing it with the part used to the same purpose in other animals. No such comparison exists in the exclusively human anatomy of Fletcher's poem. After description, for Fabrici, is inquiry into "action," such as breathing, the function of which it is the part's purpose to perform. *The Purple Island* anatomy represents organs as the "instrument" of a function, such as "breathing" (*P.I.* 4.27 margin. a), but in no way attempts to provide an account of "use," such as Fabrici understands it, as a general account of why a part exists and why it is as it is in the "Animal." Although it is possible that Fletcher may have responded to the anatomy of the Aristotelian revivalists in particular ways the nature of which is discussed in part 3, his poem shows no such influence in its conception of the descriptive criteria or *occasiones* and the kind of body they aim to reveal.

By referring the poetically modified anatomy of *The Purple Island* to the standard descriptive criteria of the early sixteenth century, it is possible to test the systematicity of the figuration of the body as a geopolitical realm in the poem.[119] The extent to which the description of the realm (into which the body is modified) fulfills these criteria, and the extent to which the anatomical description is modified by its figuration, can be used to measure the systematicity of the analogy drawn between the entailment metaphors of the Isle of Man figure, that is, the metaphors consisting of subsidiary subjects drawn from the special set of associated implications of the Isle. In the lines of *The Purple Island* on the island's "geological" foundation (*P.I.* 2.5.1–3, cited in chapter 1), three of the descriptive criteria for anatomy are fulfilled for islands and buildings: substance, location, and complexion. The description of the complexion of the "isle" is humoral, and refrains from applying to the body the elemental composition of the terrestrial world—"Fire, Water, Earth, and Aire" (*P.I.* 1.41.1). The quoted description of the isle's foundation is taken from the first anatomical stanza of the poem, and lays, as it were, the bare bones or foundation of all the poem's further ramifications of this metaphorical complex. For Giovambattista Giraldi Cintio, the subject matter of the poem is "the bones which hold the flesh together," while the choice of subject matter derives from invention.[120] As such the said stanza also lays the skeletal frame of the anatomical exposition it introduces, conceived as resembling the bodiliness of its anatomical theme. In this way, *The Purple Island* appears as a body of words to replicate, or speak the truth of, the human body it depicts, thereby reversing the direction of metaphorical structuring in the concept of Man's body as the Book of God.

The representation of poetry as corporeal is widespread in the early modern period, and appears, for example, in Michael Drayton's attack on those poets "our Mimick Apes, out of their bragging pride, / That faine would seeme to be, what nature them denide; / Whose Verses hobling run, as with disjointed bones, / And make a viler noyse, then carts upon the stones" (*Poly-Olbion,* Song XXI, ll. 171–74). The metaphor is by no means confined to the early modern period and moreover entered critical vocabulary.[121] Its function is partly to convey the sense of presence created by the maximum of foregrounding of the utterance, yet the suggestion that a literary work on account of this notion is an aesthetic object runs into great difficulty, because of the elusiveness of an ideally convergent interpretation which discloses all of what is descriptively or perceptibly present but possibly hidden in a work.[122] Fletcher's use of the metaphor is notable, however, because his choice of anatomical theme entails that the poem's form and content reflect each other's materiality and textuality in a closer correspondence than is usual.

The two conceptualizations of the body as constructed and as natural inform *The Purple Island*'s choice of analogues and the subsidiary subject of the metaphor of the Isle of Man. This discussion of the range and systematicity of the defining domain of the metaphor, and of the metaphor's novel set of associated implications, will focus primarily on the connotations of constructedness, for reasons that will become clear as the discussion proceeds. The body's structured resemblance to a building ranges extensively so that it is compared to the island realm, which thereby takes the form of the body politic. As a whole, then, the body is compared to geological and political entities of which it is in fact a part, as well as an epitome. Cantos 2 to 5 systematically divide, or anatomize, the island realm, so that it is "parted in three regiments, / By three Metropolies is joyntly sway'd" (*P.I.* 2.14.1–2). These "metropolies," or geological "regions" (*P.I.* 3.5.4), are the venters of the body—"the lowest, or belly; the middle, or breast; the highest, or head"—and they are identically located in the body as in the island realm, thereby fulfilling that criterion of descriptive anatomy (*P.I.* 2.14. margin.).[123]

Anatomy is, according to Paré, "a perfect and absolute division, or artificial resolution of mans body into its parts,"[124] and *The Purple Island*'s tripartite division of the body conforms to that of Plato and Galen.[125] In *Timaeus* Plato provided a bodily residence for each of the three faculties of the soul. The concupiscent, appetitive, or desirous faculty in the liver; the agitative, irascible, spiritual, or passionate faculty in the heart; and the rational or reasonable faculty in the brain, with the motive faculty extending from the brain to the extremities.[126] Galen developed on Plato's division by specifying the natural, vital, and animal members, the principal one of each being the organs mentioned above.[127] The natural members are located in the lowest venter, or abdomen, separated by the diaphragm from the vital members in the middle venter, breast, or thorax, and the animal members are situated in the highest venter, or head.[128] This is closely related to that which is referred to by anatomists as the "common or vulgar" division of the body into the three "bellies."[129] Those in the lowest venter are concerned with the functions of digestion, growth, and generation; those in the thorax with respiration and the "vital action" of the heart and arteries; those in the head with voluntary motion, sensation, and the principal functions of imagination, reason, and memory.[130] From the principal members in each of these venters arose the spirits (Greek, *pneuma;* Latin, *spiritus*) of the body.[131] The liver produces the natural spirit (Greek, *pneuma physicon;* Latin, *spiritus naturalis*) which is distributed to the body with the natural blood via the venous system; the vital spirit (Greek, *pneuma zoticon;* Latin, *spiritus vitalis*) is produced by the heart and distributed with the vital blood by the arterial system; and the brain produces out of the vital spirit the animal spirit (Greek, *pneuma psychicon;* Latin, *spiritus animalis*).[132] The brain uses the animal spirit "partly for the divine operations of the Reigning Soul"; but partly "it distributes it continuously to the organs of sense and motion through the nerves, as through little tubes."[133]

In *The Purple Island,* among the associated implications of the Isle of Man are "streams" of "purple-colour'd dew," that arise from the "purple fountain" of Hepar (the liver, described in chapter 1), and which are the veins, whose form, or shape, is "long," and whose function is "to contein, concoct, and distribute" to the extremities as "nourishment" the natural spirit and blood produced by the liver (*P.I.* 2.10 margin. g; 2.13.2 & margin. k). They are among the "diff'ring streams," the "veins, arteries, and nerves," "Circling about, and wat'ring all the plain" of the isle (*P.I.* 2.10.1,7,1; 2.9.5 & margin. f). The arteries are "long, round, hollow, formed for conveyance of that more spritely bloud [and therefore "life and heat"], which is elaborate in the heart. This bloud is frothy [because hot], yellowish, full of [vital] spirits, therefore compast with a double tunicle, that it might not exhale or sweat out by reason of the thinnesse" (*P.I.* 2.11 margin. h; 2.13.3 & margin. k). The nerves are described as "spermaticall," that is composed of sperm, "Embracing as they runne, each with his brother," "none of them single," "carrying the animall spirits for sense and ["arbitrarie" or voluntary] motion and therefore doubly skinned as the brain" from which they arise (*P.I.* 2.12.3 & margin. i; 2.13.3 & margin. k). In each of the anatomical criteria of form or shape, location or position, posture or connectedness, functions or operations, substance or composition, complexion, and number, the anatomy of the vascular and nervous systems as described in Canto 2 of Fletcher's poem is consistent with that in anatomy textbooks, and in particular with Galenist anatomy.

Anatomical consensus in the sixteenth century concurred with Plato and Galen, although a few maintained Aristotle's location of reason and sensation in the heart, which Galen had disproved by demonstrating that the reversive nerves in the neck arise from the brain.[134] Thus Benedetti locates the origins of sensation in the brain, of life and reason in the heart and thorax, and of nutrition and reproduction in the natural members beneath the diaphragm and in the *sumen* or abdomen (which for Benedetti is beneath the umbilicus or navel).[135] De Laguna's reason for dismissing Aristotle's opinion is that which Galen provides in *De placitis Hippocratis et Platonis:* "For neither the natural nor animal faculty has its location in the heart since far more suitable members have been provided by nature for the accommodation of these powers."[136] Certain other conclusions based on less rigorous observation or inconsistent reasoning are frequently dismissed.[137]

While retaining the systematicity that characterizes the relationship of the poem's threshold figure to the whole of the anatomy in Cantos 2 to 5, which is demonstrated by the consistent use of the anatomical *accessus* and the simultaneously high degree of geopolitical figuration of that anatomy, *The Purple Island* goes beyond the conventional divisions of the body, to explore the ordinarily unexplored aspects of its defining domain. Each regiment, or region, provides the location for several cities and prominent geological features, the form, size, and number of which also correspond to those organs to which they are analogous, thereby fulfilling three descriptive criteria for anatomy. The description of the "Archcitie *Hepar,*" as "like a mount" which combines terrestrial and architectural features, may serve as an example (see *P.I.* 3.5–8). Beyond this, each city or region is systematically subdivided, so that the auricles of the heart or "Kerdia" become "streets" (*P.I.* 4.20.1) and the blood vessels "streams" (*P.I.* 3.11.1; 3.12.1). The structuredness of the former coalesces with the natural fluidity of the latter to produce "conduits" and "pipes" conveying "excrementall liquors" from the liver (3.16.1; 3.13.6. & margin.).[138]

Creation in *The Purple Island* is described in three stages. Firstly, "Light" and order replace "Chaos" and "darkness" (*P.I.* 1.40.1, 2). Thereafter Fletcher follows a Christian theocentric reworking of Empedocles' *discordia concors:*

Fire, Water, Earth, and Aire (that fiercely strove)
His soveraigne hand in strong alliance ti'd,
Binding their deadly hate in constant love

(*P.I.* 1.41.1–3)

In the third stage, the Isle of Man is created. Insofar as it is an island, a geological territory, it would be unsurprising if the humoral complexion of the Isle of Man were to derive directly from the elemental composition of the world. The "matter" which forms the island-body's "foundation" or "bones," is "Fram'd of fat earth, which fires together tie" (*P.I.* 2.5.2, 1, 4 & margin. a). But, as the marginal comment confirms, this "fire" is the "heat" of the "vertue generative" of the "seed," which in a sense bakes the bones into their particular forms, and not the elemental composition of the bones, which is not hot, but rather "cold and drie"(*P.I.* 2.5.2 & margin. a). Since it is heat rather than fire that knits the bones, the complexion is humoral rather than elemental, but although the anatomical criterion of complexion refrains from applying the elemental composition of the world to the body, the general tendency of the poem to detail systematically the correspondences between body and isle may be construed as compatible with the similar tendency in Paracelsian medicine.

The sociopolitical dimension of the conceptualization of the body as constructed, however, which Paracelsus does not describe, provides the poem with more scope for describing physiology and anatomical relations than that of the body as natural. This is demonstrated by the political conceptualization of the anatomical posture, or connectedness, of each of the venters, regiments, or regions, and of the organs, cities, or geological features within them, through the vascular and nervous systems. The "Arch-citie *Hepar*" is represented as:

c The liver is tied to the heart by arteries, to the head by nerves, and to both by veins dispersed to both.

[c] To th'Heart and to th'Head-citie surely ti'd
With firmest league, and mutuall reference:
His liegers there, theirs ever here abide,
To take up strife, and casuall difference:

(*P.I.* 3.6.1–4)

The political organization of bodily members and their relations entails that they are conceived as either ruling or subject to rule: the "Arch-citie *Hepar*" rules over the abdomen, "stretching her commands / To all within this lower region" (*P.I.* 3.5.3–4); "Kerdia," "th'Isles Heart-city spreads his large co[m]mands" throughout the "middle Province" (*P.I.* 4.2.6,5); but "[t]he third precinct," the head, "over all the rest [bears] rule and soveraigntie" (*P.I.* 5.4.1,7).[139] These jurisdictions and their rulers correspond to the hierarchy of noble organs in Galenic anatomy,[140] and the sovereignty of "Reason" in the "brain," "to which [de Laguna writes] all things appear to render homage as to the most noble emperor and highest monarch of the entire body,"[141] echoes the figurality of Renaissance and early modern anatomical texts, where the political figuration of the noble organs is either metaphorical, as in *The Purple Island,* or confined to analogy or simile.[142]

The attribution of political sovereignty to the hierarchy of noble organs extends from their figuration as cities to their allegorical personification as rulers. Situated in the fortified head of the island realm is its virtuous "Soveraigne," the "*Intellect*" and "understanding" (*P.I.* 6.28.2; 6.51.2 & margin. *). But the "Prince" of the geopolitical realm is also described as "A little living Sunne, Sonne of the living Light" (*P.I.* 1.45.4, 7) on an analogy with the property and function of the sun in the cosmos, the life-giving heart in the body, and Jesus Christ in both the cosmos and the heart. And just as both the heart and head wield dominion over the other members of the human body, the Isle of Man is itself "the heart and head / Of all [God's] works," "an Index" to "the whole, yet of the whole a part," integrated, not alienated, even as it reflects the "summe of all" (*P.I.* 1.43.4–5, 6). Of the two conceptualizations of the polity that emerge from *The Purple Island,* the dynastic realm with its legitimation of sacral monarchy is here in the ascendant, and consequently so is the idea of the reflection of the microcosm or microchristus as facilitating identification with everything of which knowledge is sought, rather than as alienating.

Relying on the assertion that other parts and powers of the body are created either for the brain or the heart, the argument in anatomy about supremacy necessarily diverted attention away from the interdependence implied by the "doctrine of the consent" of principal parts.[143] Political theorists justified different distributions of power and social formations as natural, or pre-given, by referring to various models within the spectrum of anatomical hierarchies and organizations.[144] The doctrine of consent was rarely attractive to anatomists, however, even though three of the most authoritative ancients differed in their views on superiority. That "Plato thought the liver was the noblest member" is often indicated in the sixteenth century and accompanied with such explanations as, "[i]f the liver is strong the members seem to grow with increased blood"; but ultimately it is an opinion that is dismissed.[145] Aristotle's belief "that the chief power of the soul was located nowhere else but in the heart" is no longer held as viable, even if, as de Laguna concedes, his reasoning is admissable: "Since the heart is the principal organ of the body and because it is the first to live and the last to die, we must pardon Aristotle to some extent. . . . If indeed from the heart alone rise anger or passion, fear, terror, and sadness; if from it alone spring shame, delight, and joy, why should I say more? In fact, almost all those perturbations which the Greeks call pathemata have their origin in the heart."[146] But Galen's observations of the reversive nerves descending from the brain, and his attribution of the natural and animal spirits to the liver and brain, respectively, meant that the heart would never retrieve its principality, nor would the faculties of reason or sensation ever be returned to its possession.[147] But, as a result of the Aristotelian revival at Padua in the late sixteenth century, and his own adherence to that philosopher, William Harvey saw in the necessity of the heart for animal life a reason to retrieve for it an ancient prominence, if not quite as equal in nobility to the brain, then nevertheless as more necessary.[148]

The Purple Island's personification of organs or their innate virtues as rulers or subjects further explores ordinarily unused parts of the defining domain of the threshold figure—the island realm—and in so doing, elaborates "one of the most trustworthy signals of allegory" to the purpose of emphasizing the very operation of metaphor and microcosmic analogy in the poem.[149] In the "fair town" of "Hepar,"

> the Isles great Steward dwells;
> His porphyre house glitters in purple die;
> In purple clad himself: from hence he deals
> His store to all the Isles necessitie:
>
> (*P.I.* 3.8.1–4)

The dispersal of nourishment from the liver to the rest of the body is socially conceptualized, by no means uniquely in the period, as a form of magnanimous dispensation from a ruler to his subjects, the effect of which, combined with "his sweet influence," is to "cheer" the isle, and to "slake" the "envious rage and endlesse difference" of dissentious and unruly members, such as the icy "*Splenion*" and "th'angry Gall" (*P.I.* 3.9.6,7,4).[150] The political modification of physiology thereby fulfills the anatomical desriptive criteria of use, or operation,[151] and complexion, or innate influence. Anatomy textbooks are free of such explorations of personification, although de Laguna's *Anatomica Methodus* is not without a concept of social justice: "But since (as I said earlier) the intestines not only contain the material which has been prepared by the stomach but also alter and change it they deserve to have the omentum, called epiploon by Galen, as a cushion and a support around them."[152]

The "*Splenion,*" or "the splene" (*P.I.* 3.17 margin. o), and "*Choledochus,*" or "the Gall" (*P.I.* 3.16.2 and margin. n), correspond to "*Saturn*" and "*Mars*" in the macrocosm (*P.I.* 3.9.2), and strive with each other, as do those planets, out of antipathy; but their discontent arises not merely from this relationship, but also from their natures and offices in the isle. The natures of these organs are intrinsically related to their functions or offices, to draw "excrement" or "ill humour" from "*fair* Hepar" (*P.I.* 3.15 margin. m; 3.17 margin. o; 3.14.1), and this is because their natures partake of the particular excrement it is their office to draw and contain, an office they hold not only in Fletcher's poem but in Hellenistic and academic anatomy.[153] The "ill stream" drawn "by divers drains unto the *Splenion*" is "dreery black and frightfull," just as the "splene" itself is "Dark, dolefull, deadly-dull," "the native seat of melancholie," and dwells "in earthy cell," "a little hell; / Where with him fright, despair, and thousand horrours dwell" (*P.I.* 3.17.1, 3, 2; 3.17 margin. o; 3.18.6, 5, 6–7 & margin. p).[154] It is made clear that the consequence of their failure to perform these offices affects "the whole body" (*P.I.* 3.18 margin.). The office of the "Gall" is described as follows:

> m The first excrement drawn from the liver to the gall is cholerick, bitter, like flame in colour; which were it not removed, and kept in due place, would fill all the body with bitternesse, and gnawing.
>
> Three pois'nous liquors from this purple well
> Rise with the native streams; [m] the first like fire,
> All flaming hot, red, furious, and fell,
> The spring of dire debate, and civile ire;
> Which wer't not surely held with strong retention,
> Would stirre domestick strife, and fierce contention,
> And waste the weary Isle with never ceas'd dissension.
> (*P.I.* 3.15)[155]

Conflict of interests is here tightly contained by the ordinary performance of physiology. It is restricted in text-time and whatever grievance might give rise to "dire debate" is apparently ameliorated.[156] But elsewhere in early seventeenth-century literature and politics, "civile ire" and "domestick strife" has as much imaginative force as order or unity has.[157] By considering such discourses where this is so, we open out *The Purple Island*'s containment, revealing it as ideological in its concealment of a contradiction in the meaning of the Steward's dispensation.

By providing a consistent representation of anatomy and physiology, figured in terms of a special set of associated implications of the island realm of Britain, the extension of the metaphor and analogy of the body politic leads readers to focus on those aspects that it highlights, leading readers to view the entailments as plausible or true.[158] Conversely, political literature employed the analogy of the body politic to describe the state, and theory could reactivate the anthropomorphic analogy of the state in order to "prove" some version of the organic theory and justify the action either of the sovereign or the state.[159] This mutual metaphorical definition was especially the case in the Renaissance and early modern period, when the questions that the analogy was designed to answer could be raised with specific reference either to the body or the state.[160]

During the Midlands Rising, James VI/I issued four proclamations, in May, June, July, and August of 1607, which deployed the political analogy of the body, to prove the paternal "care" and "clemency" of the sovereign and his "desire of the Publike good."[161] The exercise of paternal care, as an intrinsic aspect of the body politic, is nonetheless limited by its interpretation. Paul Archambault maintains that in political discourse, "the variation given the analogy provides a fairly accurate idea as to the author's political tendencies."[162] "The political writers of the late mediaeval and Renaissance periods did not consider it as an inert, neutral literary *topos,* but as capable of betraying certain political inclinations."[163] On a more confined discursive scale, the Royal Proclamation of June 28, 1607 articulates the need to distinguish which subjects qualify as recipients of such care: "We will prefer the saftie, quiet, and protection of our Subjects in generall, and of the body of our State, before the compassion of any such Offenders, bee they more or lesse, and howsoever misled."[164] This interpretation empowered James to dismiss the Rising as the insolence of "rebellious" members, guilty of "hainous Treasons" against the "politike body of our Realme."[165] This leads James to declare that the protesters have forfeited the "naturall clemency" of the sovereign, and hence to justify violent state oppression on the not unprecedented grounds of a medical analogy with amputation or venesection: "a sharpe remedy applyed to a small and infected part, to save the whole from dissolution and destruction."[166]

In William Shakespeare's *Coriolanus* (c. 1608–9) allusions to the plebeians as diseased bodies or infections within the body politic pepper the invective of Caius Martius (later Coriolanus): "What's the matter, you dissentious rogues / That, rubbing the poor itch of your opinion, / Make yourselves scabs?" (1.1.163–65); and turning exclamation into a curse in an incensed attempt to literalize the metaphor, "You herd of—boils* and plagues / Plaster you o"er, that you may be abhorr'd / Farther than seen, and one infect another / Against the wind a mile!" (1.4.31–34; * editor's modernization replaces "biles" with "boils").[167] Within the logic of the conception of the state as a body, with the patrician class as its head or physician, "domestick strife" (*P.I.* 3.15.6) may become disease (although *The Purple Island* refrains from this extreme invective and prefers the discourse of humoral imbalance), and violent state oppression may translate into "sharp remedy."

Typically versions of the body politic that draw an analogy between the ruler and the physician, in distancing the ruler from the sick or decaying body to which he administers, evince absolutist and authoritarian political sympathies.[168] It is so with Coriolanus who, in addressing the gentry and senators, says: "You,"

> that prefer
> A noble life before a long, and wish
> To jump a body with a dangerous physic
> That's sure of death without it—at once pluck out
> The multitudinous tongue
>
> (3.1.149–55)[169]

Coriolanus's proposed remedy is the removal of the tribunes, "The tongues o' th' common mouth" (3.1.22);[170] elsewhere (1.1.196–99) he expresses a desire for militant suppression.[171] Military and state suppression is precisely what the "offenders" in the Midlands Rising suffered on 8 June 1607 at Newton, Northamptonshire, when fifty of the protesters were killed outright, "a great many wounded, and many captured and later hanged and quartered, as exemplary punishment."[172]

True belief, knowledge, and judgment are in this context relative to the conceptual system of the political analogy of the body, although the analogy is open to different interpretations. In *Coriolanus,* Menenius insists that the Senate's "counsels and . . . cares" (1.1.149) for "the weal o"th'common" (1.1.150) are evident on analogy with the "good belly" (1.1.147):[173] he says to the citizens, "digest things rightly" and "you shall find / No public benefit which you receive / But it proceeds or comes from them to you" (1.1.149–52).[174] His pun on "digest" as also "understand" insinuates that not only the distribution of chyle to the members and their digestion of that nourishment, but also the correct interpretation of analogy, would slake the rage of unruly members. This dissemination of ideology makes the interpretation that is in the interests of the dominant class seem natural or inevitable.[175]

Allusions to the fable of the belly by anatomical texts do occur, and with reference to Livy's *History of Rome.* The language they employ is moreover political. In his chapter on "the Fruit in the Womb," Veslingus says: "The Bowels of the Abdomen which are allotted for *publique* digestion, do not *want private digestion,* but manifestly operate *for the common profit* of the Infant."[176] This description of the natural use of the belly is inseparable from its potential for political construction on the basis of an organic model of the state. And de Laguna's warning that when the stomach's "function is impaired, all other functions of the body are placed in the greatest danger,"[177] might easily furnish one with a plausible threat with which to coerce dissident or subversive elements.

On the perspective granted by *Coriolanus,* the distribution of nourishment in the dynastic realm of *The Purple Island* appears, not merely magnanimous, but as a form of condescension and control which implies that not only grievances arising from a dearth of food, but even constitutional or political grievances may be assuaged by filling the bellies of the discontented. In addition, by rendering the "*Splenion*" as innately (in fact, organically) "Where darknesse ever dwells, and melancholy fear," and the "Gall" as innately "angry," and consequently their "envious rage" as inevitable, the diegetic narrator of *The Purple Island* dismisses the possibility of any genuine or rational cause for discontent (*P.I.* 3.9.17.7; 3.9.4, 7). Whereas on July 24, 1607, James had denied the existence of food shortages, and had previously pointed out that "we have been careful to prevent such Enclosures, and Depopulations," later he tacitly admitted that the making of starch was irresponsible in a time "wherein the plentie of Corne shall not be very abundant," and had led to rising food prices.[178] This falsification on the part of James, combined with his selective application, further politicizes the Royal "care" that assumes his innocence as regards the causes of the Rising.

The diegetic narrator of *The Purple Island* unquestioningly assumes a lack of ambiguity and ideology in the "flowing substance" of "the Steward," which he is said to return "all the day" "to all the Isles necessitie" (*P.I.* 3.8.1, 6, 7, 4). But, in Shakespeare's play, in which "the popular voice, as [the plebeians] themselves speak it, has genuine grievances to express,"[179] the fable of the body finds the citizens unconvinced rather than ungrateful. Realizing the powerful eloquence of accepted analogy and suspecting its cognitive ideological deployment, the First Citizen protests to Menenius, "you must not think to fob off our disgrace with a tale" (1.1.92–93), and flatly refuses Menenius's interpretation of the innateness of patrician "care": "They ne'er cared for us yet" (1.1.78–79).[180] The First Citizen thus exposes ideology's concealment of contradictions in Menenius's interpretation of the conventional political analogy of the body.[181]

Yet, despite characterizing the "difference" between the "*Splenions* frost and th'angry Gall" as "endlesse" and unchangeable (*P.I.* 3.9.7, 4), these organs in *The Purple Island* are not only analogous to the common people of the realm. Anatomical function is intrinsically related to the natures of these, as all other, organs, and function or office is also position in a hierarchy: "If the splene should fail in his office, the whole body would be filled with melancholy fancies, and vain terrours" (*P.I.* 3.18 margin. p); similarly, "Where the splene flourishes, all the body decayes, and withers; where the splene is kept down, the body flourishes" (*P.I.* 3.19 margin. q). But the explicitly political analogy is not the social transgression of the multitude but the overweening power of a tyrant: "So when a tyrant raves, his subjects pressing, / His gaining is their losse, his treasure their distressing" (*P.I.* 3.19.6–7). *The Purple Island* therein draws on a literary tradition of use of the body politic analogy to admonish tyranny and limit monarchical authority.[182]

To an elaborate degree, anatomical or physiological operations in *The Purple Island* take various other socially integrated, personified forms, which are sometimes carefully and systematically delineated in tandem with the somatic processes upon which they are based. This is often an effect of the interaction of *The Purple Island*'s newly constructed set of implications, associated with the Isle as an emergent nation, with the somatic processes of the principal subject. Visual and audial perception is consistently described as the communication of intelligence to the head of a seemingly modern nation-state.[183] Light is represented as "sending / His beams and hundred forms" inward through the tunicles of the eye, to be "fully view'd" in the "Reticularis" or "netty tunicle," where "thousand nimble spies are round dispread," and where the forms are "caught" and brought to the "eye" of "Visus, or the sight"; whereupon the light "[f]rom thence with speedy haste is poasted to the minde," the "spies conveying" it along the pair of "Optick" nerves (*P.I.* 5.35.2–3; 5.35.6; 5.33 margin. k; 5.34.2, 3, 4; 5.25 margin. y; 5.35.7; 5.37.3 & margin. m). The spies whose duty is the detection and conveyance of information are a personification of the sensory (rather than the motive) faculty or virtue of the nerves.[184] This personification reoccurs in the "Cochlea, snail, or Periwincle," in the inner ear, "where the nerves of hearing plainly appeare," and "where ready waits / A nimble poast, who ne're his haste relenting, / Flings to the judgement-seat with speedy flight" (*P.I.* 5.46 margin. * & 5.46.4–6). The judge who is the recipient of this message is "*Auditus,*" or hearing, who, as "the Master of request, / Prefers petitions to the Princes eare," who is the "*Intellect*" or "understanding" (*P.I.* 5.48.1; 5.49.1–2; 6.28; 6.51 margin. *).The conveyance of secret information by the intelligence services is a feature of modern nation-states, including the proto-nation of Elizabeth I. By exploring emergent ideological formations and incipient state apparatuses, and drawing them into the metaphorical interaction of the Isle of Man, Fletcher then systematically figures Man as a nervous and defensive human individual.

The ear also provides a good example of how personification of anatomy occurs in the poem to a lesser or greater degree in proportion to observable anatomical postures and physiological operations. Of the organs, those are less personified that are less well supplied with arteries or veins, and therefore less endowed with either natural or vital spirit. A lessening of personification also occurs where the physiological activity is less pronounced or perceptible, such as in the bones, or in the "Portall," which is the "outward ear" (*P.I.* 5.39.1 & margin. p). Both these body parts are described as "hard and drie," which combines a composition and complexion generally unsuited to the anthropomorphic animation of anatomy; while the bones are in addition "cold," which depletes them further of the qualities of living tissue (*P.I.* 5.39.1; 2.5.2). Objectification also increases where the anatomical "members" are less well supplied with nerves than for example the "Periwincle," and therefore less endowed with animal spirit. An example from the inner ear is the drum "whose sounding head," while possibly endowing it with an anthropomorphic feature, is "strook" as an object by the "circling" waves of sound in the air (*P.I.* 5.40.4, 5). Where personification is diminished in the poem so is agency; but these less spiritual parts nevertheless perform their physiologico-social duties, albeit as "instruments" (*P.I.* 5.48) in the service of the body's physiology or the well-ordered running and defense of the state. As waves of sound "strike the warning drumme" of the inner ear, the "Hammer," "Stithe" and "Stirrup," mechanically respond with "fit motion," conveying the circles of sound toward the "judgement-seat" (*P.I.* 5.48.2; 5.43.1, 3, 5; 5.48.3; 5.46.5).

The descriptive attributes given by the poem to these organs of the inner ear—the "Hammer," "Stithe," and "Stirrup"—not only fulfill all (but illnesses) of the criteria of descriptive anatomy, but explain how each is especially conducive to the effective physiological operation of hearing and to the greater efficiency of secret intelligence services. In composition and complexion, these instruments, or informers, are "hard and drie, / Fitting the sound with their firm qualitie" (*P.I.* 5.42.5). Their "compact" size, number, and connection "[w]ith slender string knit to th'drummes inner side," enables them mechanically to convey sound (*P.I.* 5.42.3,4). Their form, location, and posture with respect to each other are configured ideally to produce the mechanical movement, described above, which conveys sound to Auditus:

> [t]The first an Hammer call'd, whose out-grown sides
> Lie on the drumme; but with his swelling end
> Fixt in the hollow Stithe, there fast abides:
> The Stithes short foot doth on the drumme depend,
> His longer in the Stirrup surely plac't;
> The Stirrups sharp side by the Stithe embrac't,
> But his broad base ti'd to a little window fast.
>
> (*P.I.* 5.43)

The "little window," being opened by the "Stirrup," admits "the sounds into the head," thereby efficiently fulfilling the function of these parts (*P.I.* 5.44. margin.).

The foregoing analysis demonstrates that Cantos 2 to 5 of *The Purple Island* explore and delineate systematically the entailments of a metaphor—the Isle of Man—which condenses an extensive analogy between the human body and the geopolitical realm as well as inventing new entailments or connotations and working out the implications of a "modern" proto-nationalist interpretation of the metaphor of an island for Man. Contrary to Langdale's opinion, the analogues are highly integrated throughout the unfolding of the entailments. Although Nicolson fails, in her work on the effect of science on seventeenth-century poetry, to acknowledge the political aspect of the geocosm, the persistence with which the correlations are sought between the analogues caused her to write: "Never did the ingenuity of allegorizing poets carry such similarities further than did Phineas Fletcher in *The Purple Island*, . . . it is the most extreme example in verse of the geocosm-microcosm analogy."[185] Furthermore, the analogy is delineated with a high degree of correlation between the two domains of connotations because the body and the geopolitical realm are metaphorically translated and modified by their involvement in the unfolding of a host of metaphorical entailments of the interaction of the principal subject with a novel set of associated implications of the Isle. Although the figuration of anatomy is far from superficial, its analogue or subsidiary subject is not allowed to take priority of meaning, for, although the figurality of anatomy is clearly extended beyond what usually occurs in anatomy textbooks, the poem painstakingly and consistently applies all but one of the descriptive criteria of anatomy textbooks to the description of the geopolitical realm, and regulates the degree to which personification of anatomy occurs in that realm, in relation to observable anatomical postures and physiological operations.[186] The consistency of the poem's technique with regard to this ought to ensure that, contrary to Kastor's critical judgment, the reader is seldom bemused or uncomprehending. Rather the reader is confronted with a meticulously imbricated allegory of the anatomy of Man.

The persistence with which *The Purple Island* details its anatomical content is the cause of an affront to the sensibility that characterizes much criticism prior to and including that of Edmund Gosse.[187] Having "a theme of unusual ugliness and aridity," the allegory is said to carry out its task of describing the corporeal and moral qualities of a human being "with extreme persistence, even where the imagery is most grotesque and inconvenient."[188] Gosse, however, ignores the fact that the execution of the poem's allegorical design requires an extraordinary fidelity to the details of each of its threshold figure's subjects or analogues. Some critics, such as Ezekial Sanford, have argued therefore that it is the poem's design that is ill-conceived. Sanford says that Fletcher "lavished on a bad subject the graces and ingenuity that would have made a fine poem on a good design. . . . [H]e tries to sweeten the language of anatomy by the flowers of poetry, and to support the wings of allegory by bodily instead of spiritual phenomena."[189] That some designs, or choices of theme, are inherently more appropriate for poetry seems to have been an assumption shared by "[s]uccessive generations of poetic readers [who] have been disappointed to find that *The Purple Island* is not some purpureal province of fairyland washed by 'perilous seas forlorn,' but the ruddy body of man, laced with veins of purple blood."[190] To assume this is to hold that a given mode of discourse properly possesses affinity with a given category or species of phenomena; it is to share Gosse's evident opinion that poetic writing is more suitably and tastefully employed in describing "the intellectual and moral qualities" of the mind, and to share his relief in arriving at Canto 6 of *The Purple Island.*[191]

The root of Sanford's displeasure is what appears to be his objectivism, for at the close of Canto 5 he reports that Fletcher "quits the dissecting table."[192] Sanford thereby obscures the possibility that what *The Purple Island* signifies is not a real dissection, though the rhetorical gestures of Fletcher's own text (to be discussed in part 2), together with the persistence with which Cantos 2 to 5 detail human anatomy, might be responsible for the confusion of categories. Certainly an objectivist could hold that a distinction between aesthetic and scientific language corresponds to a distinction between concepts as objective meanings of aesthetic and anatomical words, and to a distinction between properties inherent in well-defined aesthetic or anatomical objects in the world (referents) which is ontologically independent of the language used to speak about them. The objectivist response to *The Purple Island,* on perceiving a disorder of "natural" distinctions, might be distaste,[193] but even a structuralist may find it incongruous, a breaking of codes.

There are signifiers in Renaissance anatomy, as in any discourse or signifying system, that do not only or necessarily signify the phenomena that constitute the proper concepts or signifieds of that system, and the breaking of codes is widespread and inherent to discourse. The author of *The Purple Island* is particularly interested in these signifiers because they facilitate allegory and metaphorical extension of meaning. The anatomical term "*Aereolae*, or little altars" for the female nipples, which occurs in the anatomical marginalia to Canto 4, stanza 8, of *The Purple Island*, connotes sacred and erotic associations beyond the properly anatomical considerations of position or shape or use. In order to understand this theoretically, we can turn to structuralism, which takes greater account than objectivism of the relationship in the linguistic sign between the signifier and the concept that is signified, and admits that the conventionality of any signifying system is responsible for tying together its words and concepts, and that signification is determined both by the relations between terms in that system, and by that system's relation to other systems, rather than by the sign's correspondence to the phenomena to which it refers in the world.[194] On this view conventional agreement rather than objective correspondences may fix the reference of nomenclature. Conceptual or cognitive theory of metaphor suggests, moreover, that nomenclatures do not arbitrarily select signifieds; rather they are grasped metaphorically in ways commonly agreed by a historically and culturally specific speech community. Within these historical and cultural conventions, the foregrounding of language in poetry, and creative metaphor's condition of open-endedness, of the possibility of additional, accruing meaning, may be distinguishable from relatively inert metaphors, scientific metaphors, and more or less conventional conceptual metaphors and their creative extensions, but conceptual metaphor in early modern poetry is not categorically cognitively different from conceptual metaphor in early modern anatomy.

It is possible to demonstrate tropological disfiguration of referentiality, and the conceptual imperfection and historicity of linguistic and discursive distinctions, in anatomy, particularly well because in the early modern period anatomical nomenclature was in the process of being established, an activity in which many took part, including Benedetti (as we have seen) and Vesalius.[195] In most sixteenth-century anatomy textbooks, Greek, Latin, and Arabic terms were used; Arabic terms often only where they were found useful in the absence of the proper Greek or Latin ones found in Celsus's *De Medicina* or Pollux's *Onomasticon*.[196] "A standardization of names was [felt to be] so much the more urgent as the same term was often applied to several organs."[197] Such terms may be said to totalize only in an unreliable way, leaving the empirical referent unsheltered from tropological disfiguration. Additionally, an anatomical organ was selected by more than one signifier, and the areas of the body to which anatomical names refer sometimes overlapped as different linguistic systems divided and denoted in incompatible ways.[198] To add to the potential confusion, the vocabulary employed in popular medical works lagged behind that in academic works in the formation and standardization of nomenclature, a process which aimed to establish a more precise and transparent anatomical nomenclature, in an attempt to cut a swath through the nomenclatural confusion and impede the flow of phenomena, by selecting only those which distinguish the discipline and pertain to its object.

An example which illustrates all of these points is the referentiality of "womb," "mother," "matrix," and "uterus." Some authors, like Thomas Raynalde, insist that "three wordes, the Matrix, the Mother, and the Wombe, doe signifie but one thinge, that is to say, the place wherein the seede of man is conceived . . . conserved, nourished, and augmented, unto the time of deliveraunce, in Latine named Vterus et Matrix."[199] But, although Raynalde insists on the synonymity of all four terms, his philological exposition is insensitive to the ways in which the conventions of the discursive context effect signification and reference, and attract the "proper" use of one or other terms. Womb and mother are often employed in popular medical works, whereas matrix and uterus are conventionally used in academic textbooks. Moreover, in popular discourse, womb and mother, by transferring the property of containing, are capable of signifying the whole abdomen, whereas matrix and uterus more precisely isolate the female generative organ, "an official member, compounde and Nerveous."[200] This is in part because the term uterus was introduced in order to establish a more precise and transparent anatomical nomenclature. By contrast, the more broadly signifying, nonspecialist terms womb and mother, and the term matrix, retain more denotations and therefore greater poetic currency. Possibly because of the use by Arabic physicians and natural philosophers of "the name MOTHER" to express "the soft seat or covering of any thing," both mother and womb are more versatile grammatically.[201] John Davies of Hereford's *Mirum in modum* occasions the use of the term "enwombes" as a verb, to describe the dura mater's posture in relation to the contents of the skull, and the poem also attributes to the "*Pia mater,* or godly Mother" a "wombe," thereby conveying the surrounding posture and protective function of both dura mater and pia mater.[202] *The Purple Island* uses the yet more poetic appellation "*Venus* temple" (*P.I.* 3.27.4), the "sixt and last town in [the abdominal] region" of the Isle of Man being described with little of the anatomical precision of other regions, as chapter 1 indicated, owing to a moral delicacy which extends to some vernacular works of anatomy, including John Banister's *The Historie of Man.*[203]

The standardization of nomenclature was a process that proceeded at least until the late nineteenth century, when Basel hosted an international conference to further this aim,[204] and the etymology of nomenclature trailed in the wake of standardization. Etymology enhances the metaphoricity that has already been found in early modern anatomical writing and in *The Purple Island* by eroding the totalization of anatomical concepts which is the goal of the standardization of nomenclature.[205] But etymology also served a pedagogical function for Vesalius in his *Epitome,* where the derivation of anatomical names is often given by the device of descriptive comparison.[206] A survey of the historical and etymological accounts of various subjects and terms in medical dictionaries from Henri Estienne's *Dictionarium Medicum* of 1564 to Robley Dunglison's nineteenth-century medical dictionary is capable of impressing upon the reader that, in the course of arriving at an apparently neutral (nonideological) and transparent nomenclature, more recent dictionaries of anatomical, physiological, and medical science have elided vistas of metaphorical association in the nomenclature's formation and development.[207] *The Purple Island*'s exposition of the correlations between the anatomy of Man and the geography and built environment of the Isle demonstrates a clear interest in this etymology. The septum between the first and second, or lateral, ventricles of the brain is figured as "a wall built all of saphires shining, / As fair, more precious; hence it takes his name" (*P.I.* 5.16.1–2).[208] The marginal note reveals that name to be "Septum lucidum, or bright wall" (*P.I.* 5.16 margin. k).[209] But Fletcher's poetic, allegorical purpose should not be regarded as the only reason for his interest in etymology, which was more broadly intellectual. Grammarians like Ramus defined etymons not as words that gave rise to derivative or borrowed or later forms, whose original meanings are lost to those forms, but as the intrinsic properties of the current words to which they gave rise.[210]

The Purple Island's metaphorical extension of meaning through the exploration of etymology would have been regarded in his time as having recognizable cognitive content.

A theory of allegorical, analogical, and metaphorical meaning in *The Purple Island* has been significantly advanced in this chapter, and it now transpires that the anatomy is neither essentially extractable from its analogical and allegorical context in the poem, as Fletcher's biographer Langdale believed, or, as Baldwin argued, conceived wholly within a conceptual framework of analogy. It is true that the allegory has been found to be based upon analogy, broadly conceived within the traditions of the microcosm and microchristus, and that the analogies of the geocosm and macrocosm are consistent with the poem's figuration of anatomy in vegetative and terrestrial terms, and that the body politic is consistent with its figuration of anatomy in architectural terms. But the figure of the Isle of Man is a single image, an interactive, creative metaphor, not a compressed simile reducible to the antecedently existing analogical and conceptual-metaphorical strands which it unifies. It is abundantly meaningful, including in innovative ways, opening out degrees of similarity between the associated implications of its principal and subsidiary subjects. Equally, it has emerged from this chapter that the anatomy in the poem's "anatomical cantos" is indeed detailed and seemingly not in conflict with its tropological disfiguration.

Having arrived at an understanding of the creative character of the Isle of Man metaphor, this chapter has begun to undertake a methodical and thorough examination of the tropology of the semantic, semiological, and performative field of *The Purple Island.* This has revealed that significant connotations of the metaphor are unraveled in the allegory. This process is understood as an effect of the foregrounding of the metaphorical interaction of the Isle's specially constructed set of associated implications with the finely divided and detailed morphology and functions of the principal subject, Man, a process which gives rise to a corresponding set of implications of Man, some of which are novel. Since each of these metaphorical interactions gives rise at every part of the anatomy of the island-body to more connotations, the entire anatomical cantos may be understood as the unfolding of a host of entailment metaphors all systematically relating to the central, threshold figure of the "*Isle of Man*"—an index or microcosm of the whole in terms of the poem's explanation.

It is the understanding of the work as a whole, however, or at least the anatomical cantos, which gives the key to the interpretation of the threshold metaphor for the anatomy of Man, for this develops the work's reference, its intentional direction toward a world it creates and the reflective direction toward a self. This study has begun to build this understanding by exploring how the allegory may be said to have reproduced or developed conventional analogical and anatomical understanding and discovered previously *unseen* connotations of Man and realm, thereby accomplishing in some measure its intention to contrive a coherent world, and deepening and extending meaning in accordance with a notion of the intelligibility of the world of its readers. It has been found to do this by faithfully delineating and extending the conventional divisions of the body, and the ordinarily used parts of the metaphorical concept of the geopolitical island realm, and by using the ordinarily unused part of that domain to interact metaphorically with parts of the concept of Man, frequently giving rise to a clash of meaning as well as a fit. A full answer to the question of how derivative or novel this allegorical extension is must include an intertextual study of sources, influences, and comparable literature, and this task is allotted to part 2. But this chapter has begun to establish the creative character of the tropology, and therein this study has begun to address itself to the question of the extent to which the anatomical cantos of *The Purple Island* expanded the vision of its seventeenth-century readers.

In each of the poem's processes of discovery or invention, the anatomical cantos allegorically extend the threshold figure systematically and clearly, giving coherent structure not only to antecedently existing similarities but also to a newly highlighted range of similarities. Moreover, the systematicity extends to the working out of proto-nationalistic and individualistic implications of the Isle's characterization in terms of historically incipient and ideologically emergent concepts. But one of Fletcher's innovations is to take the geopolitical figuration of anatomy far beyond what occurs in anatomy textbooks while retaining consistent use of the anatomical *accessus* and regulating the degree of allegorical personification in relation to observable anatomical postures and physiological operations. For these reasons Fletcher's critics have lamented the "ill-conceived" design of *The Purple Island* or the persistence with which it executes that design, but their objections have been found to be the consequence of flawed reasoning. There are signifiers in Renaissance anatomy, as in any discourse or signifying system, that do not only signify the phenomena that constitute their proper concepts or signifieds within that system, and the breaking of codes is widespread and inherent to discourse. The reasons for this have been discussed in theoretical and historical terms, and this chapter has focused finally on one aspect of this signification—the etymology of nomenclature that trailed in the wake of the standardization of anatomical vocabulary.

The author of *The Purple Island* is particularly interested in the etymology of signifiers because it facilitates allegory and metaphorical extension of meaning by eroding the totalization of anatomical concepts which is the goal of the standardization of nomenclature. The early modern anatomical nomenclature, despite its developing adherence to precisely observed divisions of its object, only functioned as though it were referentially transparent insofar as it was shared by a group of individuals who utilized it in a specialized signifying context in a denotatively limited way.[211] Etymology enhances the metaphoricity that has already been found in early modern anatomical writing and in *The Purple Island.* But this does not yet go far enough in answering the question of whether the conceptual content of the anatomy in these cantos would have been in conflict with the poem's tropology had the anatomy not been transformed through Christian allegory, whether, for example, *The Purple Island*'s Christian theocentric reworking of Empedocles' *discordia concors* is consistent with Hellenistic, humanist, or scholastic anatomy, a question to which this study must now turn.

3

Creation, Mortality, and Salvation

THE TWO PRECEDING CHAPTERS HAVE TRIED TO ADDRESS THE NEED TO BE PRECISE ABOUT what kind of figure the "*Isle of Man*" is in Phineas Fletcher's *Purple Island.* Among the many things that we call metaphor or analogy, the figure was primarily acknowledged to be the threshold image or figure of the poetic allegory and a creative, interactive metaphor. In view of the role that the microcosmic analogy has been thought to have played in the history of observational and experimental anatomy, it was also necessary to establish whether Fletcher subscribed to the microcosm-macrocosm speculation, and if he did, whether *The Purple Island* allegory systematically explores correspondences between Man and the universe so that they act as some sort of key to the nature of world and Man that might have been thought to supplant the need for scientific inquiry. Chapter 1 related how the body as microchristus and as a microcosm of the terrestrial globe or geocosm contributes to a system of antecedently existing associated commonplaces of the threshold figure. The geocosmic analogue for the human body encompasses the imaginary built environment, human society, body politic, and geographical terrain of the island realm, Britain, and as such is characterized by two more basic conceptualizations of the body: as constructed and as natural. These two understandings also characterize the conceptual metaphors which *The Purple Island* draws from the anatomy of the schools and medical profession, which coalesce with the analogies of the microcosm among the significant connotations of the creative metaphor, the Isle of Man. The conceptual metaphors of anatomy as vegetation and as "building" were found jointly to provide a coherent understanding of the concept of anatomy, each metaphor allowing one to get a handle on different aspects of the concepts of morphology and function, notably as natural and as constructed of parts. Far from supplanting observation of human anatomy, the microcosm and conceptual metaphoricity provided Renaissance anatomists and the poet with a means to understand what they "discovered" through dissection and poetic invention, respectively.

Although metaphorical concepts and analogies like the body politic and the geocosm arise from our bodily experience and existence, the analysis of anatomy textbooks of the sixteenth and early seventeenth century in chapter 1 has revealed the metaphorical structuring of human anatomy to be consistent with the structuring of abstract or less clearly delineated concepts in terms of concrete, more clearly delineated ones, which are in this case vegetation and "building." This direction of metaphorical structuring and the metaphorical character of the structure and functions of the parts are rationalized by the coherent and systematic extension of metaphor from structure to function and thence to the purpose of function, its reason for being or final cause, such as growth, reproduction, or the instrumental use of the body by the rational or animal soul (from *anima,* soul), which produces the animal spirit. *The Purple Island* has been found to provide this *telos,* or purpose, in allegorical form, while in Renaissance anatomy, the faculties of the soul, Nature, or the Galenic *physis,* provide the rationale. This teleology provides the rationale also for the coherence which has been elucidated in this study between the metaphors in anatomy, and for the systematicity that obtains between them and the analogies of the microcosm, and the significant connotations of the Isle of Man which are unfolded throughout the anatomical cantos of Fletcher's poem.

In *The Purple Island,* however, the character of the allegory derives also from Christian theology. Cantos 7 to 10, and to a lesser extent, Cantos 11 and 12, provide a gallery of personifications of moral and religious virtues and vices, of which many, though not all, are portrayed in Cantos 11 and 12 engaged in apocalyptic combat with each other over the salvation or damnation of the inhabitants of the Isle, which are first described in Canto 7, and which include the "Soveraigne" prince or king, who personifies the "*Intellect*" and "understanding" (*P.I.* 6.28.2; 6.51.2 & margin. *), "Queen" "*Voletta,*" who personifies "The will" (*P.I.* 6.57.3, 4 & margin. *), their daughter, "*Eclecta,*" *ecclesia* or the elect (*P.I.* 11.6.3–11.8.7), and the king's counselors, as referred to at the end of an early section of chapter 2. The microcosm is also microchristus, and the poetry has been found to pun on "Sunne" and "Sonne," evoking the incarnation and the doctrine of cosmic redemption. Chapter 1 described how the property of Christ and the efficacy of his redemptive action are seen in an analogy with the function of the human heart in the body and of the sun in the cosmos. Although this study confines its attention for the most part to the anatomical cantos, in examining this final element in the figurative design of *The Purple Island,* the question which this chapter must address is twofold: Had the conceptual character of the anatomy in the poem not been transformed through metaphorical interaction, would it have been in conflict with the Christian allegory? And how does the allegory's Christian content relate to the systematically interwoven figurality of the microcosm, macrocosm, geocosm, and vegetative and architectural metaphors which characterize anatomical language and knowledge and which abound in Fletcher's poetic stanzas as a form of imaginative rationality?

Anatomical exposition is performed discursively by Thirsil in Cantos 2 to 5 as a part of his narration, but of the narrative dimension of *The Purple Island,* I have spoken only in broad terms, such as regard its main story line as divine or providential history. As such the theological narrative framework affects Thirsil's anatomy with eschatological delay or deferral of fulfillment, even perhaps of the full meaning of anatomy, until Canto 12. A suitable topic with which to start an exposition of the effect of this narrative is Fletcher's choice of threshold figure or image for *The Purple Island,* because that determines how the allegory unfolds and gives rise to the significant connotations of Man and the realm. The figure can be understood as facilitating the narrative design, as Baldwin explains: "[Fletcher] needed a figure that would permit him to communicate vividly man's relationship to God—that fluctuating relationship that involves man's moving away from God at the Fall, his subsequent wanderings in the seas of death, the sense of insecurity experienced by a troubled searcher, and the feeling of peace attendant upon completion of the mission. The mobile island that Fletcher fixes on is not completely successful: . . . But given the large number of requirements he had to meet, Fletcher has done well enough."[1] The threshold metaphor of *The Purple Island* purposefully, if obscurely, guides the reader to understand Man and the human body in the context of the providential plan which Thirsil's narration unfolds. The transition from human anatomy to the narration of providential history and personal salvation is facilitated by the fact that whereas the interpretation of the figurality of human anatomy required cognitive processing, whereby linguistic metaphors trigger a process of analogizing, which in turn has recourse to conceptual metaphor, "the importance given the poet in the Augustinian system imposed the obligation to write with the ultimate aim of promoting Christian doctrine through appeal to reason, mental stimulation," and, moreover, "[t]he Puritan dedication to spirituality, focusing upon the inner man imposed a similar obligation."[2]

The principal events of the main story line consist of the following, in the order in which they occur in the story, rather than the plot:[3]

1. The creation of the world (*P.I.* 1.39–42)
2. The creation of the isle (*P.I.* 1.43–44, 46), which is the dynastic realm (*P.I.* 1.45) and Man.
3. Sexual differentiation, represented as a binary division (*P.I.* 1.48), followed by a period of secluded innocence, which is both Edenic (*P.I.* 1.49–50) and British (*P.I.* 1.51, 53).
4. A temptation followed by commission of an act of sinful appetite, curiosity, or pride, which is the disobedience of Adam and Eve in Eden (*P.I.* 1.54), an exploration beyond Britain's shores (*P.I.* 1.53), and the usurpation of the "soveraigntie" of the Isle by "a false, foul, fiend-like companie" (*P.I.* 6.10.4, 2).
5. A fall into death from the timeless innocence of Eden and from national identity (*P.I.* 1.54–56).
6. The intervention or incarnation of a heavenly king, Jesus Christ (*P.I.* 1.57–58). This event consists of subevents, principally Christ the king's death and resurrection (*P.I.* 6.70–72).
7. A psychomachia, or contestation between vices and virtues for the body and island realm in its partially recovered state (*P.I.* 1.60; 6.20, 23; 11.5–12.53).
8. A second intervention by the king, followed by victory for the virtues, their regeneration or resurrection, and defeat and damnation of vice. This leads to the union of Eclecta (*ecclesia* or the elect) and her heavenly spouse (*P.I.* 12.54–86).

These events are historical, tropological, and eschatological, thereby consisting of the *sensus allegoricus* and *sensus anagogicus* of the allegory. These are two levels of allegory proposed by Dante's *Divine Comedy* as a method of biblical exegesis.[4] The *sensus allegoricus* is the historical meaning, though historical here should properly be termed typological, wherein, for example, Old Testament events prefigure the incarnation, life, death, or resurrection of Jesus Christ. The *sensus anagogicus* or anagogical meaning is the equivalent or corresponding eschatological, soteriological, or other religious event of individual or general import, the latter referring principally to the Last Judgment. Events 1 to 6 are repeated by Thirsil numerous times, at varying pace, always as an internal homodiegetic analepsis.[5] That is, these events lie within the scope (in story-time) of the story line; their narration invariably illuminates by providing information about a character, event, or part of the story, at that point in the text; and their narration always occurs at a point in the text after later events have been told. In addition, events 1 to 6 lead to that event of the story which is narrated in more detail at a decelerated pace. This is event 7, the subevents of which compose the contestation for the body and island, the fullest narration of which occurs between Canto 11, stanza 5 and Canto 12, stanza 6.[6] The meaning of events 1 to 6 in the earlier past of the story becomes intelligible when they are fulfilled by event 8. The retardatory device or "delay" of this eschatological narrative is thus past-oriented. But events 7 and 8 also open up a vista of eternal life in the future and an understanding for which Man strives in the earlier moral conflict, and so the delay is also future-oriented.[7] In respect of the past and the future, the retardatory devices are naturalized by reference, firstly, to a "reality model" which consists of belief in the Last Judgment and the Resurrection of the Dead, and secondly, to the literary model which assumes in its reader this belief.[8]

Because the narrative dimension, with its emphasis on soteriology and eschatology, has seemed to many critics of *The Purple Island,* including most recently Kastor, to constitute a rupture with the anatomy of Cantos 2 to 5, it is necessary to understand its effect on the conceptual character, coherence, systematicity, and rationality of anatomical figurality in the figurative design of Cantos 1 to 5. This is the main concern of this chapter. According to Lakoff and Johnson, when two or more metaphors jointly structure a defined concept, they usually share entailments so that there is an area of descriptive and conceptual overlap.[9] In Chapter 1, the analysis of the conceptual metaphoricity which *The Purple Island* shares with Renaissance anatomy suggests that something that is true of vegetation, which is also true of human anatomy due to its metaphorical structure in Renaissance anatomy textbooks, ought also to be true of "building." This is what defines the coherence that the two conceptual metaphors have been found to have. From a theological perspective, there are, in fact, three very significant shared entailments or sets of entailments, and these are the topic of this chapter: the concept of creation, with which events 1 to 3 of the main story line of *The Purple Island* are concerned; death and mortality or ruin, which concern events 4 and 5; and regeneration, rebuilding, and renewal, which extend semantically to redemption, with which events 6, 7, and 8 are concerned. Fundamental as these concepts are none is concrete or delineated enough to provide an image or analogy for the finely divided parts in anatomy.[10] That is why the system of related concepts and entailments characterizing the descriptions of anatomy derives metaphorically from vegetation and from "building," both of which are relatively clearly defined concepts that refer to concrete objects that are capable of division, and which connote processes either of construction or growth (although "building" of course refers also to the planning process). Moreover, creation, death, regeneration, and many of the others among the shared entailments are event concepts rather than object concepts (although "Creation," "ruin," and "mortality" are, or are also used as, nouns). This chapter argues that the metaphors of the body's constructedness and naturalness are coherent with the concepts of the body as created, mortal, and redeemed, and it further discusses the nature of this coherence and the knowledge it produces.

Creation and Mortality

A preoccupation with the related concepts of structure-function anatomy, teleology, and design distinguishes ancient and Renaissance anatomy from modern scientific anatomy, which "deals primarily with description and secondarily with origin," the latter only from an embryological, evolutionary, or genetic point of view.[11] In *On the use of the parts,* "[Galen] came to the conclusion that in the structure of any animal we have the mark of a wise workman or demiurge, and of a celestial mind; and that 'the investigation of the use of the parts of the body lays the foundation of a truly scientific theology which is much greater and more precious than all medicine.'"[12] That the architecture of the human brain is to be understood as created is implicit in the passage from Vesalius's *Fabrica* describing the wafer-thinness of the septum lucidum: "if the light be bright during dissection, or a candle be placed on one side, its brilliance shines through, as with those transparent stones cut thin and smoothed like crystal, which Italian architects fasten in the mullions of glimpses or windows."[13] There is something of the awe proper to a contemplation of creative ingenuity in this description, an ingenuity about which *The Purple Island* is explicit when describing the septum lucidum as "a wall *built* all of saphires shining" (*P.I.* 5.16.1, emphasis mine). The most immediate agent the existence of which the passage of the *Fabrica* assumes is Nature: elsewhere in the same chapter of the *Fabrica* the membrane which coats the skull externally is "a peculiarly cunning artifice,"[14] and Nature is to be commended for her "admirable diligence."[15] In 1540, Corti invites his audience of students at the University of Bologna to "make an experiment" with the duct of the gallbladder to demonstrate how "Nature has ingeniously made this duct" "so that not the least of air nor of bile could return or flood back."[16] The particular form of this duct, however small, is understood not to have been made in vain, but (in accordance with Aristotelian teleology) to be fitted effectively by Nature to fulfill the goal or purpose of its function.

In classical to early modern natural philosophy Nature does not only create the body with ingenuity, but participates through the vegetative soul or the natural faculty in the government of its functions. The conceptualization, discussed in chapter 1, of the embryo in the uterus as a plant in a fertile field is entirely consistent with the notion of Nature's actual coordination of the generative process. Moreover, the presence of mind in Nature gives it and its organisms inherent order and purpose: "Just as the human body consists of many parts, all in motion and performing different functions, yet held together in a delicately balanced harmony by the mind which controls and directs them all, so it is inferred that Nature itself, with its infinitely many parts performing their several operations, is constituted a harmonious whole by a cosmic mind whose inner principle is that of perfect regularity and orderliness."[17] According to de Laguna, the very end of dissection is to glean knowledge of the mind and creativity of Nature: "He who wishes to recognize accurately the care and industry of nature must present himself as dissector of even the repulsive bodily members and most diligently examine their location, shape, number, and consistency."[18] The ingenuity of even the muscular expulsion of the feces is admirably wondered at in the fifth of the series of lectures given by Corti at the University of Bologna in 1540.[19]

The philosophical perspective of de Laguna's anatomy reveals him to be a rationalist. As was said in the first chapter, the rationalist or dogmatist tradition springs from Aristotle, for whom as for de Laguna, "reason" relates to the "reason for being" of the anatomical form, location, shape, number, and so on; that is, their explanation in terms of function. And the proportional form of the whole body is regarded by the rationalist as fulfilling the harmonious functioning of its parts the purpose of which is to serve the soul, or the mind of Nature which made the body for the sake of the soul. The anatomy in *The Purple Island* is rational in this sense. Exquisitely formed and located, the hard, dry little mechanism of the three instruments of the inner ear—the "hammer," "stithe," and "stirrup"—is described as "compact by wondrous art" (*P.I.* 5.42.3) and perfectly formed for its function. From the "art," skill, or ingenuity of "artifice," or perfectly proportionate architecture, one is led a short distance to artistry.[20] Common are passages of anatomical works admiring the human form, its harmony, beauty, and proportion, as Nature's work of art, and such passages should be read not as exuberant flourishes but as possessed of conceptual content.[21] Benedetti's study of human embryology appeals by way of explanation directly to this concept of artistic creation: "The heart is created first, then the rest of the members are delineated ["sketched out"] and finally completed as if in a painting."[22] Not only painting, but sculpture functioned as the metaphor or simile through which to intimate the body's createdness.

Glenn Harcourt has argued that in the series of visceral figures of Book V of Vesalius's *Fabrica* the anatomy is presented as if through the dissection of antique sculpture.[23] Thus fragments of the object of the real anatomical demonstration are transformed through their representation as artifacts.[24] This I think is so, but it may be added that the "truth" of the body's having been created is one of the connotations produced by the visual metaphorical presentation.

The ingenious artist or artificer implied by the "wondrous art" of the *Purple Island* anatomy is not only, nor even primarily, Nature. Creative agency rests ultimately with God.[25] In the sixteenth century, this is also more or less true of Hellenist, humanist, and scholastic anatomy. Although works on anatomy would at times refer to "God" through "*Nature,*"[26] in Christian natural philosophy, God and Nature are neither identical nor equal. As Fletcher's diegetic narrator claims, God is "that All," yet "farre more then all" (*P.I.* 1.39.1). It is God who creates the cosmos or Nature, and via the agency of Nature fabricates the animal and human frame. Moreover, this theology was of greatest significance when the anatomical subject was human, because then the frame was the instrument of the rational soul. Understood teleologically the subject matter of anatomy was essentially the soul, and "the primary purpose for engaging in it was admiration of the Creator and his providence for man" in the purpose which the soul strives to fulfill through the body.[27] In the early seventeenth century, Crooke could boast a "farre more Divine and useful profit of Anatomy then [self-knowledge], proper and peculiar to us to whom the light of the Gospell hath shined, namely the knowledge of the immortall God," who "cannot be knowne but by his effects."[28] And, he adds, given "how wonderfull the Fabricke and structure of mans body is," "[w]ho is it therefore, that will not honor, reverence, and admire the author and workeman of so great a worke."[29]

Christian anatomy was, as Cunningham has maintained, "*religiously informed and constituted,*" and eulogies to the Creator are common in the prefaces and closing statements of anatomical works where conventional forms of submission to "the high Glory of God" are almost invariably followed.[30] In the dedication of Benedetti's *Anatomice* to Maximilian Caesar Augustus, the author speaks of examining "the marvelous and divine handiwork of God the creator, that is, the body which is (according to Plato) as it were the temporary vehicle of the soul."[31] But even in the midst of the densely anatomical text of a passage on the arteries, Paré offers an explicitly religious version of teleology: "God the Maker of the Universe, foreseeing this, made the coats of the vessels contrary to the consistence of the bodyes contained in them."[32] Paré's text performs this hairline verbal fracture, this shift of register, without an abrupt change of tone, suggesting an assumption, held by him and plausibly by his readers, that God's creativity is integral to anatomy. Crooke meanwhile devotes a whole chapter of *Microcosmographia* to "How profitable and helpefull Anatomy is to the knowledge of God."[33] While the awe found in Vesalius's description of the brain is proper to a contemplation of divine ingenuity, and implies that he had found therein the "obscure foot-steps" of God, Crooke, with less awe but more enthusiasm, finds that God has left "cleer arguments" leading from design to the architect of the built environment and landscape of the brain and spinal cord of the microcosm:[34]

> Enter thou whosoever thou art (though thou be an Atheist, and acknowledgest no God at all,) enter I beseech thee, into the Sacred Tower of *Pallas,* I meane the braine of Man, and behold and admire the pillars and arched Cloysters of that princely pallace, . . . the 4. arched Chambers [four

ventricles], the bright and cleare Mirrour [septum lucidum], the Labyrinthæan Mazes and web of the small arteries [choroid plexus], the admirable trainings of the Veines, the draining furrowes and watercourses, the living ebullitions and springings up of the sinnewes, and the wonderfull fœcundity of that white marrow of the back [animal spirit in the spinal cord], which the wiseman in the Book of the Preacher or *Ecclesiastes* calleth the Silver cord.[35]

Neither Crooke nor Paré was a pioneering anatomist; in fact Paré was a surgeon rather than a physician;[36] but Crooke was trained at Padua, and *Microcosmographia* draws extensively on the more obviously academic *Theatrum anatomicum* of Kaspar Bauhin.[37] Even for Vesalius, God is "the Great Craftsman,"[38] the "Artificer of all things,"[39] and "the Great Artist" or "Author of the human fabric."[40] Anatomical works are special among medical works in construing and expressing their project as the disclosure of the handiwork of God.[41]

Early modern natural philosophy, to which anatomy was subalternated,[42] was "about God. To study God through His works was the point of the whole enterprise. Natural Philosophy studied the world—that is, nature—as created by God: it was nature studied *as creation.*"[43] This religious conception of anatomy requires that to some extent it is read not according to a transparent nomenclature (the achievement of which was sought through the standardization of anatomical vocabulary), but metonymically as created and always in contiguous relation to its Creator. Thus the vegetative figurality gestures to the body as like the created world, and the conceptual metaphor of anatomy as "building" points to the body's artificial construction by the divine architect of the landscape and built environment of the body. The metaphors of "building" and vegetation, and the faculties of reason and Nature or the vegetative soul, are thereby reconciled and made coherent in and by the concept of the creative ingenuity of God.

Although the relationship between divine Creator and human creation is a relationship of contiguity rather than identity, God's resemblance to the image of God in Man has been said to be characterized by both *convenientia* and *aemulatio,* which combine together to form analogy, a far closer, reflective relationship. The upright form of Man is one of the ways, identified by C. A. Patrides, that Renaissance thought avers that "man's resemblance to God is in some way reflected in the human frame" in addition to the rational or intellectual soul that ideally controls it: "Man's upright form is foremost among the physical characteristics claimed as aspects of *imago Dei.*"[44] The reason of this "upright posture," the principal or final cause of it, distinguishable from either efficient or material causes, was derived out of "the function of the whole body" to enable "Man" to exercise his reason in "all arts."[45] For this, as Galen maintains in *On the use of the parts,* and as the Galenist Corti reiterates, "Man" must "amongst all Creatures [have] the Hand given him by God," and must be bipedal.[46] For Vesalius, the hand was indispensable to the role of anatomy and dissection in medicine; for Galen, it was persuasive of design in Nature; and for Aristotle it was the instrument of distinctive human activity.[47] In addition, there was a more emphatically religious explanation of upright posture, which in some measure, perhaps surprisingly, was drawn from Aristotle's *De partibus animalium.* Aristotle "maintains that man is participant of divinity" "in a fuller measure" than all other living beings, and "in him alone do the natural parts hold the natural position; his upper part being turned towards that which is upper in the universe."[48] The idea is developed to infer that the offices of the outward senses are not only for function but also contemplation, and are consequently "placed in the highest contabulation or Story of the body."[49] In *Anatomice,* Benedetti says: "The human body was created for the sake of the soul and stands erect among other animals, as established by divine nature and reason so that it might look upward more comfortably."[50] A heavenward propensity of anatomical structure was proposed, whereby the body of "Man" is itself teleologically predisposed to the contemplation of "the Heavens, his originall, to contemplate there Diuine occasions and permanent, leauing the Terrestriall as vaine."[51] However, Thomas Browne suggests that the "ground and occasion" for this mistaken cause "was a literal apprehension of a figurative expression in *Plato*"—"*Sursum aspicere*"—by which, Browne claims, Plato did not mean "to gape, or look upward with the Eye, but to have his thoughts sublime; and not only to behold, but speculate their Nature, with the Eye of the understanding."[52] Browne was not alone in his opposition to the literalization of Plato's expression, but he was in a minority.[53]

The upright posture of "Man" ensures that the hierarchical order of anatomical members has a spatial verticality to which it may correspond.[54] On observation, however, the correspondence was not always evident, for as Benedetti discovers when examining the colon, organs lower in nobility could be found located spatially higher than more noble members.[55] Benedetti and others concluded in such cases that the anomaly was only apparent and that the function of the organ necessitated its position. Such explanations of course defended the concept of design, which was itself implicitly hierarchical, the functions of some members being more noble or more "essential" than others to animal or human life.[56] The anatomy of the Isle of Man cannot be erect, for without serious threat to its credibility as an island realm and great inconvenience to its inhabitants, the geological distribution of the island must necessarily be more horizontal than vertical. For this reason Fletcher designed the island body as nonliterally upright, with the brain, "the seat of the minde, and senses" (*P.I.* 5.6 margin. b), as "the *Cephal* mount," elevated in the head (*P.I.* 5.33.6), which of the "three regions" of "this little world" is the "highest set in place, as in degree," thereby being "most like to heaven . . . in site" (*P.I.* 5.4.6 & margin. a). Thirsil implies in his description of the position and nobility of the head the contemplation that ought naturally to arise within, and he augments this suggestion by adding that in "figure" or shape also the head is "fram'd like heaven, sphericall" (*P.I.* 5.4.3 & margin. a). Thirsil's description is very probably drawn from the *Timaeus,* where Plato says the head being spherical is the "most perfect in shape" of the three regions of the body, and for this reason is "the seat of reason and intellect, the most perfect part of the soul."[57]

Apparently in these respects the description of the head and brain in *The Purple Island* conforms to notions informing the motif of the upright form of Man, "the excellent microcosm, the noblest work of God, whose body is a masterpiece of architecture and whose mind surveys the grandeur of the world," or contemplates the heavens.[58] But this is not entirely so with regard to the head's use of its elevated vantage point. Its eyes may be "Like heav'n in moving, like in heav'nly firing" (again after Plato), but they are "mounted high . . . on a lofty hill; / (For they the Princes best intelligence, / And quickly warn of future good, or ill)" (*P.I.* 5.24.5 & margin. x; 5.25.1–3). The eyes are "two watching towers" (*P.I.* 5.22.5). Through "the black window" (pupil) of each, "light and hundred shapes flock out & in apace" (*P.I.* 5.30.7 & margin. f), the "Crystalline" humor all the while subtly "Alt'ring it self to every shapes aspect" (*P.I.* 5.29.5 & margin. d). The "Reticularis" or "netty tunicle" is purpose-built "out of that milkie spring [animal spirit], / That from the *Cephal* mount his waves doth fling, / Like to a curious net his substance scattering" (*P.I.* 5.33.5–7); for therein "thousand nimble spies" spider-like entrap "The forms caught in this net" (*P.I.* 5.34.2, 3), in what can reliably be described as an efficient mechanism of intelligence-gathering for the government of the British nation. According to Sawday, *The Purple Island* interfuses "the providential salvation of England under James I, figured as the regeneration of the body-politic, with the regeneration of the soul and body in Christ." Moreover, "the body which is also England is an armed fortress," and the poem's "language of militant hostility towards the outside world—a hostility in which the body is created as a refuge—also represents the direct political programme of Fletcher's anatomy."[59] As a significant connotation of the Isle of Man figure, the "*Cephal* mount" may be drawn from the contemporaneous anatomical figuration of the head as the "Castle and tower of the soule,"[60] but it exceeds all parallels in the degree and systematicity of its elaboration as the intelligence-center of a government or human individual wary and suspicious of "what e're is in the continent" (*P.I.* 1.53.3), especially Roman Catholicism. In ways such as this the Christian allegory in *The Purple Island* maintains the systematicity that obtains between the metaphors in anatomy, the analogies of the microcosm, and the significant connotations of the Isle of Man, and extends that systematicity to those implications which arise from the interaction of newly emergent ideology and incipient nationhood, in order, as Sawday says, "to evoke a sense of continuous vigilance against the supposedly encroaching forces of Roman catholicism."[61]

Owing to the extensive analogies drawn between anatomy and the geocosm, macrocosm, and body politic, the anatomy of the human body as created is represented in *The Purple Island* and anatomy textbooks as mirroring all creation.[62] Vesalius claims that Nature used "equal diligence" in placing "Water and Air" "between Earth and Fire, [which] as by nature [are] most different," as when "between skull and brain (which, by their substances, are as far apart) [she] interpolated the two membranes [the pia mater and dura mater]."[63] In his second lecture given at Bologna in 1540, Corti draws attention to the place in *On the use of the parts* (Bk. III, Ch. 10) where Galen asserts that "man is called 'Microcosmos' . . . because in him is required as much intelligence as is required to build the great universe."[64] "If it had been possible . . . Nature would have man and all animals eternal," claims Corti in the fifteenth of these lectures. But this was not possible "due to the contrary elements of which we all consist."[65] It matters not whether the dominant concept of the body is Aristotelian—"The Animal, teleologically arranged"—or Platonic and Galenic—"the human body as divinely planned" by Nature—because in both cases Nature's "cunning artifice" is built of corruptible elements.[66] The constant counterbalancing of love and strife, of sympathy and antipathy, between the elements, gives rise not only to generation and growth, but to death.[67]

Renaissance anatomy, in which Nature was not the eternal creator it was for Aristotle, provided moral and theological argument not only from design, but from the demise to which "Man" as the chemical microcosm was inevitably subject. In Vesalius's *Epitome,* at the end of the chapter "Concerning the Brain and the Organs Formed for Furthering its Functions," the author contrives to encourage reflection in his reader on the noblest of organs: "In this manner, indeed, the great Creator of things has fashioned our body not only that it may live but that it may, subject to corruption, live properly."[68] The series of osteological plates in the *Fabrica* (Bk. I), unlike the myological plates (Bk. II), have no background scenes,[69] but rather "bits of rock and vegetation" drawn in the foregrounds, it has been claimed "as decoration for the skeletal figures."[70] By means of gesture and visual metaphor, however, these plates draw attention to the theme of mortality. The first of a sequence of three figures in Book I of the *Fabrica,* demonstrating the anterior, lateral, and

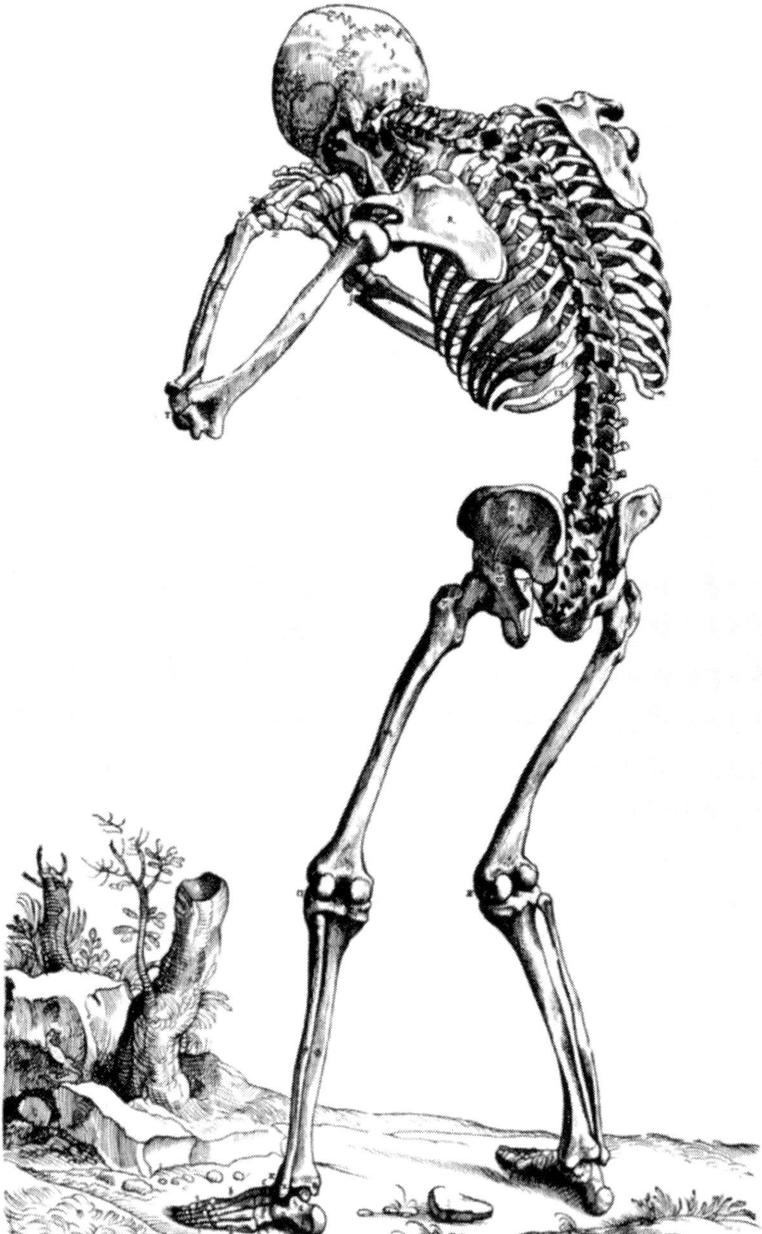

Andreas Vesalius, *Andreae Vesalii icones anatomicae* ([Munich]: [Bremen], [1934]), Bk. I, Fig. 99. The skeletal figure hunched beside a weathered tree stump has been interpreted as mourning its own demise, but might even be praying for redemption, or for the grace to live properly. "Just as it is natural for vegetation to wither, so it is natural for Man to die, whose complexion derives from the elements." Wellcome Library, London.

Andreas Vesalius, *De humani corporis fabrica libri septem* (Basle: Johann Oporinus, 1543), Bk. II, Fig. 2. Walking upright, eyes to heaven, gesturing in a way that has been interpreted as the *adlocutio,* a mode of imperial military address, modeled on the *Allocution of Alfonso d'Avalos, Marchese del Vasto,* completed by Titian in 1541, the figure engages in the arts of eloquence or rhetoric, representing human anatomy as a divine artistic creation, *imago dei.* With its right hand, the figure gestures to the "dust" to which the body will return, just as the ruined Roman arch in the background speaks of mutability in human affairs. Wellcome Library, London.

posterior aspects of the skeleton, is propped like "Death the Sexton" against a spade as though digging its own grave.[71] The second, which is also reproduced in the *Epitome,* parodies "the familiar *topos* of a Young Man with a Skull," and leans "like a Roman genius mortis" upon a stone pedestal contemplating its own mortality "with wittily redundant melancholy."[72] In other versions of this plate, the pedestal bears the inscription of a pair of funereal verses from Silius Italicus.[73] In the third of the sequence, the skeletal figure hunched in the foreground beside a weathered tree stump has been interpreted as mourning its own demise. The ruined Roman arch in the background of the one side-view figure in the *Fabrica*'s series of myological plates, speaks of mutability in human affairs.[74] Walking upright, eyes to heaven, gesturing in a way that has been interpreted as the *adlocutio,* a mode of imperial military address, modeled on the *Allocution of Alfonso d'Avalos, Marchese del Vasto,* completed by Titian in 1541,[75] the figure engages in the arts of eloquence or rhetoric, exemplifying human anatomy as a divine artistic creation, *imago dei,* and, with its right hand, gestures to the "dust" to which the body will return.[76] The aesthetically ambitious myological plates with their backgrounds and foregrounds such as those to which the anatomical figures were accustomed during life, with trees, shrubs, and small plants, and with a built environment, including a fortress or walled city and a ruin, function naturalistically to locate what Singer has called "*living* Anatomy."[77] But as visual metaphors, the ruin and the tree stump gesture towards the conceptual metaphors of anatomy which cohere in the concept of death or ruin.[78] Just as it is natural for vegetation to wither, for buildings to crumble with age, even for civilizations to pass away, so it is natural for Man to die, whose complexion derives from the elements. Not quite the motif of upright Man, "That glorious image" of God and summation of all other creation, which "was raz'd" after the Fall, "Then mortall made" (*P.I.* 2.3.1,5), Vesalius's strangely self-enclosed skeleton might even be praying for redemption,[79] or for the grace to "live properly."

Mortality is a theme not only of the figures of Vesalius, for other Renaissance anatomists elaborated the iconography of death in a variety of ways. Felix Plater's *De corporis humani structura* uses etched copper plates to present the anterior aspect with the skeletons of two children, the more fully developed of which carries a bow and arrow, symbols of the swift power of death.[80] Among the plates illustrating Guido Guidi's *De Anatome* is a skeleton with an enormous scythe, unmistakable tool of Death the Mower from the *memento mori* tradition,[81] a motif that is also found in the title page of the second edition of Vesalius's *Fabrica.* Although the woodcut for this page was recut without Vesalius's supervision from that of the first edition, the scythe blade being added by the cutter, Cunningham's contention that an original sketch of the first title-page reveals that "Vesalius did not initially intend there to be a skeleton at all in the picture," and that "Vesalius's gesture was one of simply pointing upwards—to God, to heaven," his message being "*personal experience of the human body, reveals God to us,*"[82] seems not to take into account either the iconography of the osteological plates discussed above, or the extremely close relationship between the artist(s) of the *Fabrica* and its author.[83] Whatever Vesalius's intentions regarding the title page, he was for moral and religious reasons not averse to presenting anatomy as revealing a God who has fashioned our body subject to corruption. The apparent pictorial extravagance of these anatomical plates has been explained from the point of view of artistic influence as the product of "an age inspired in much of its art by the conception of death."[84] Indeed the figures are based on models conceived upon the *vanitas* and *memento mori* traditions. But they also respond to "the vividly moralized context" of Vesalius's early osteological stud-

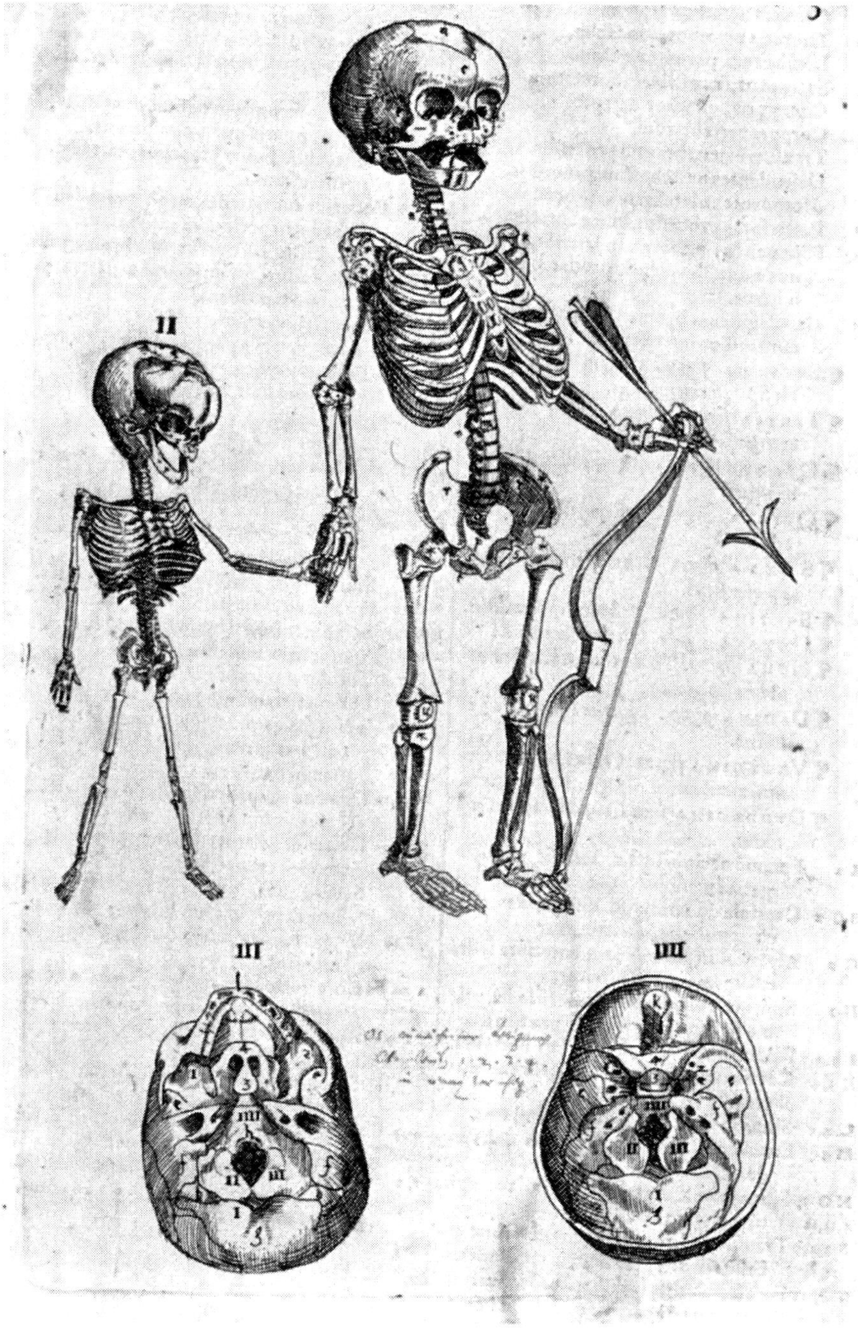

Felix Plater, ***De corporis humani structura et usu . . . libri III, tabulis methodice explicati, iconibus accurate illustrati*** **(Basle: Ambrosius Froben, 1583), Bk. III, Tab. III. The anterior aspect of the skeletons of two children, the more fully developed of which carries a bow and arrow, symbols of the swift power of death. Wellcome Library, London.**

ies in the great ossuary of Les Innocents, "the site of the original *Danse macabre*."[85] The osteological and myological plates of the *Fabrica* were reproduced in "at least twenty-five editions between the years 1543 and 1782," and copied, with or without Vesalius's permission, in numerous publications throughout the sixteenth and early seventeenth centuries;[86] and so they helped to fashion an expectation that the anatomy presented by illustration would serve a purpose in which the concept of death and the need for salvation are represented as integral to the understanding of anatomy within a Christian context.

Cunningham underestimates the extent to which death reveals God to the Christian anatomist, on account of his central thesis to demonstrate the resurrection of the projects of the ancients by the anatomists of the Renaissance.[87] Christian anatomists did not see in Nature only what was suggested to them by the concepts of the design and teleology of the body they inherited from Aristotle and Galen; they saw a Nature that was created but which with Adam and Eve had fallen, and was since the Fall particularly subject to corruption; a view that is shared by the diegetic narrator of *The Purple Island,* for whom the Isle of Man "still retain'd a naturall proclivitie / To ruine" (*P.I.* 1.35.3–4). The origin of this inclination is unmistakable on the title page of Valverde's *Vivae imagines partium corporis humani,* where a figure modeled on Adam holds aloft the apple of the tree of knowledge of good and evil. In that symbolic instrument of human transgression, the Creation of Man as *imago dei* and Man's mortality are the themes that characterize Christian anatomy, rendering the function of the whole body not only as instrumental to the rational soul but as of service in view of the prospect of Man's salvation.

Soteriology and the Anatomical Speaking Picture

Ancient natural philosophy studied nature as created, but philosophically and theologically, Christian anatomy differed not only in its emphasis on the coherence of creation and mortality in the fallen natural order: "while Galen might look upon nature or certain things in nature as a divine work, he would not admit any supernatural force in science or medicine, or anything bordering upon special providence."[88] This chapter has looked at how Vesalius and his contemporaries understood their anatomical project as the disclosure of the human body, understood as created by God to fulfill in action its purpose of serving the soul. We have also seen how anatomical figures could be based on models conceived upon the *vanitas* and *memento mori* traditions, and could communicate from their liminal state messages about human mortality and the need for salvation.[89] It seems to me surprising that no historian of anatomy has yet developed a view of the "living anatomy" of Man that takes seriously the effect of the Christian idea, articulated as early as the twelfth century by Hildegard of Bingen (1098–1179), the head of a nunnery at Bingen in Germany, and author of a number of visionary, religious, and medical works, that "God who created all things" so provides for human beings that in the human form "He" "introduces all those things allotted for the soul's salvation."[90] The case can be made more strongly, for in Christianity, the body is not merely the temporary vehicle of the soul, as it was for Plato, but is created, mortal, *and* redeemed; this notwithstanding the subtle knots of eschatology, in which the incorporeal risen body is both continuous with and transformed from the corporeal mortal body. The implication of the body's resurrection for Renaissance anatomy was that whatever is allotted for salvation must be related to the incarnation, and between Hildegard's twelfth-century intimation and Phineas Fletcher's seventeenth-century poetic anatomy, it had become self-evident to some that whatever these "things allotted for salvation" are, they cannot become clear to the human mind without the investigation of internal anatomy. Thus in Fletcher's paean of praise to the Creator, salvation is revealed by dissection:

> That mighty hand in these dissected wreathes,
> (Where moves our sunne) his thrones fair picture gives;
> The pattern breathless, but the picture breathes;
> His highest heav'n is dead, our low heav'n lives:
> Nor scorns that loftie one thus low to dwell;
> Here his best starres he sets, and glorious cell;
> And fills with saintly spirits, so turns to heav'n from hell.
>
> (*P.I.* 4.9)

The difficulties of this stanza diminish somewhat on close analysis of lines 3 and 4, each of which is comprised of carefully balanced indicative statements sharply divided by a *caesura.* In interpretation of line 3, Sawday says, "though the 'pattern' or copy of the creator (the human body) may be breathless, the picture, or image (which is the poem) 'breathes.'"[91] The term "picture" for poem is "an adaptation of the familiar Renaissance theoretical understanding of poetry, culled from the *Ars Poetica* of Horace, of creating a 'speaking picture.'"[92] The chiastic *poesia tacens, pictura loquens* (painting is mute poetry, poetry a speaking picture) was attributed by Pliny to Simonides of Chios, but, as Sawday's point implies, it came to be associated most closely with the *Ars Poetica,* as, in Christopher Braider's words, "Horace's defining tag."[93] Together with *ut pictura poesis* "('as is' or 'as in painting, so is' or 'so in poetry')," it "appears in virtually every treatise on art or poetry from the early Renaissance to the close of the Enlightenment," as "the central theme and presiding dogma of the theory and practice" of both arts alike.[94] Indeed, Braider claims there is even a "curious overdetermination at work in the treatment of the authorizing Horatian tag," a "pronounced misreading." "This yields, in place of Horace's careful comment that *it will sometimes happen,* in poetry as in painting, that one work is best viewed close to and another from a distance, the general stricture . . . it *will* be, always and essentially, in poetry as in painting."[95] Braider argues that, wrested from its proper context and misread, Horace's text and the accompanying adage ascribed to Simonides became what the Renaissance "*desired*" from them: "the authority for a universal theory of poetry and art based on values and goals quite foreign to anything the Augustan poet might have conceived for himself."[96]

The Renaissance may have desired a universal theory but there were very many cultural forms of the relationship between verbal sound and visual graph and image—emblems, hieroglyphs/hieroglyphics, *imprese,* types and symbols—each of which became highly specialized, generating theories sometimes deeply involved in their own technical vocabularies. As they juxtapose the meanings of words with the significance of images, some of these forms, like allegory, involve self-reflexive tension between the literal and the metaphorical, simultaneously discovering meaning and hiding it in shadows. Whereas Francis Quarles, in *Emblemes,* says "[a]n *Embleme* is but a silent Parable,"[97] referring therein to the image component, George Wither's use of "emblem" denotes either a single motif in the picture component of the visual-verbal amalgam, or (most often) a combination of motifs therein, or (occasionally) a combination of text and image, or (rarely) an allegorical or exegetical or emblematic mode of thought or interpretation.[98] The core of many of the relevant artistic and poetic theories is that art aims at "imitation of nature,"[99] a mimetic yet inevitably figurative project, like "invention" or "discovery" of subject-matter, drawing a similarity between nature and the artifact that stands at one remove from nature itself. Other, magical and esoteric theories and philosophies imagined a system of signs already ranged on the face of the world for the purpose of its legibility and recognition.[100] Bonaventura, for example, thought that analogies (of which the microcosm and macrocosm are constituted) reveal creatures, or parts of creatures, to be "not merely things (*res*) but signs (*vestigia*) as well, signs that reveal the marks of the Creator in His creation."[101] Mazzeo argues: "The analogical and symbolic potentiality of a thing is therefore not exterior or accidental to the thing but is derived from the very nature of creatures themselves and, indeed, is the law by which creatures were created."[102] Reconciling this semiotic ontology of creatures with his theory of emblems, Quarles writes, "indeed, what are the Heavens, the Earth, nay every Creature, but *Hieroglyphicks* and *Emblemes* of His Glory?"[103]

It is of particular importance to poetry that Man's words were thought ideally to resemble the signs within the things about which words are spoken or written, an idea which should not be confused, however, with an objectivist theory of language. Objectivism strives after language that transparently represents or reflects a world that nevertheless exists independently of language and representation. It is language shorn of figurality and disposed purportedly to serve, by revealing, "reality." But neither is the idea a conventionalist one, like structuralism. Renaissance theories of words and images tended by contrast to be "suprarealist."[104] They saw signs and pictures not only as coextensive with each other, but with what they depict, refer to, or signify. Signs are already inherent in the world and human language does not so much reflect the reality of those signs and images as interweave with the things of the world as another phenomenon of Nature.[105] At the same time, the sign or symbol has "the peculiar property of containing within itself in *alio esse,* in another mode of existence, that which is symbolized."[106] Words thereby, and particularly abstract nouns, were "perceived to be as real and as powerful as the things named," "to have a potency as solidly meaningful as physical fact," and were "sensed in terms of a nearly physical presence."[107] The analogues of an analogy constituted an adjacent similitude to that which is indicated between analogous things in Nature. The suprarealist view of language informed the concepts of discovery, invention, and creation, with which this study has associated its definition of metaphor. Without this view of language, and especially of metaphor, Fletcher would never have explored or described with such faithful deliberation and visual imagination the connotations of the Isle of Man figure, which for him exist in reality, some antecedently, some as etymons of anatomical nomenclature, and some in the working out of novel implications. The poem testifies to a belief that the entailments and associations unfolded allegorically over many cantos are inherent to a metaphor that has life and true meaning of its own, and not only in relation to empirical fact.

When in Canto 4, stanza 9 of *The Purple Island,* Thirsil says that when the pattern or copy, the human body, is breathless, the picture or image (which is the poem) breathes, Fletcher is drawing on the chiastic *poesia tacens, pictura loquens,* and *ut pictura poesis,* as well as emblematics and suprarealist theories of language. But his use of "picture" owes something also to the more classical Greek–oriented concept of descriptive anatomy that in words the body "may in the manner of an image be set before the eyes of those studious of the works of Nature."[108] While imaginatively exploring the metaphorical ramifications of the threshold figure of his poem, Fletcher's diegetic narrator also purports to be diligently describing the anatomy of the body as he sees it. In other words, the notion of words as forming a picture is a more broadly significant conception of language than any of the theoretical formulations of emblems, allegorical poetry, or descriptive anatomy may on their own suggest. The classical anatomical theory may rest uneasily with the emblematic theories, but in its simplest terms the understanding that both theoretical strands adhere to is that between the referent and the sign that is superimposed upon it there is the barest margin of displacement, an almost exact correspondence. That the philological labors of the anatomical and medical humanists, like Benedetti and Manardi, to achieve this correspondence were titanic, and that Quarles speculated that "[b]efore the knowledge of letters God was known by *Hieroglyphicks*,"[109] suggest not that correspondence was thought to be unnatural, merely conventional or impossible, but that how it could be found and whether human intellect was capable of finding it were by no means accepted as obvious to all.

The implication of line 3 of Canto 4, stanza 9, as Sawday rightly observes, is that "[e]ven when (or, rather, especially when) the body had been reduced to a scattering of decomposed parts, the divine footprints could be traced"; "[i]ndeed, the anatomical picture can only breathe, or reveal itself, when the body has ceased to breathe";[110] that is, when the body is dead and dissected (rather than "decomposed") it may be revealed as *imago dei,* the upright and whole, living and acting body. This accords entirely with the emphasis upon dissection over and above textual authority in anatomy after Berengario and Vesalius, and also conforms to what Cunningham calls "the odd assumption (*anatomia* = cutting up) that the functioning whole body which comes into existence as a whole and can only persist in existence as a whole, can nevertheless only properly be understood by cutting it up artificially into its supposed parts."[111] This assumption is of course that of the rationalist or dogmatist tradition in medicine, which springs from Aristotle.

There is, here, a parallel that I wish to draw, suggested by the pair of lines in stanza 9, each sharing a structure of balanced indicative statements. This parallel between lines 3 and 4 relates the foregoing discussion of Cantos 1 to 5 of *The Purple Island* as an anatomical speaking picture, to the poem's general theme of salvation. The significance of line 4 is that the eternal reconciliation of the human body and soul to God is only possible through the incarnation, death, and bodily resurrection of Jesus Christ. Paul says in his first epistle to the Corinthians (15:12–14), that by reason of faith in the resurrection of Jesus, Christians have faith in a general "resurrection of the dead"; moreover, "if there be no resurrection of the dead, then is Christ not risen. And if Christ be not risen, then is our preaching vain, and your faith is also vain." Paul repeats the statement in a more concise form in verses 16–17 to emphasize its fundamental centrality to Christian faith. On account of the parallel between lines 3 and 4, it seems that there may be a twin necessity of the cessation of the body's breath for the anatomical picture's demonstration of God's creative ingenuity, and, of the savior's death for the reconciliation of the fallen body to its Creator.

The reason this parallel may have been used by artists illustrating anatomical texts, like Berengario's *Commentaria* and the 1522 edition of his *Isagogae breves,* each of which contains a crucified anatomical figure, was perhaps in order to acknowledge, rather than vainly try to avoid, the anatomists' direct or symbolic involvement in judicial execution as the main source of cadavers for dissection, and yet to imply their victim's willingness to die for their own sins (rather than those of others), even in the cause of anatomical knowledge, and to grant her or, more usually, him a vicarious resurrection in the pages of anatomy textbooks.[112] But it is with the soteriological significance of the parallel for the theological rationality of *The Purple Island*'s speaking picture that we are here concerned. The title page of Johann Remmelin's *Catoptrum microcosmicum*, first published in 1619, confronts the reader with the words "MEMENTO MORI," in upper-case letters.[113] Inside is a plate which contains an anatomical Christ figure, or microchristus figure, and which employs the whole panoply of symbols from the *vanitas* tradition. Above the male figure an hourglass is accompanied by the text "*Nascentes morimur*" (Being born, we die), and beneath this in a crest is the Latin text of Psalm 144, verse 4: "*Homo vanitati*[.] *Similis factus est, dies eig* [*sic* (*suae?*)], *sicut umbra, praetereunt*" (Man is like to vanity: / His days are as a shadow that passeth away.)[114] The face of the Christ or microchristus figure, however, exudes benevolence and serenity, because defeated beneath his left foot is death, represented by a *memento mori* symbol, the skull, with sin as a serpent writhing within its cavities. The serpent's head is crushed by the foot of a crucifix which refers, via the keying mechanism of the illustration, to a Latin poem entitled "*Mediatoris Christi Solatium*" (The comfort [or consolation] of Christ the mediator). The plate does not cite 1 Corinthians, but just a few verses after Paul has twice declared the interdependence of Christ's resurrection and that of the dead, his epistle describes the time when Christ "shall have delivered up the kingdom to God, even the Father; when . . . he hath put all enemies under his feet" (1 Cor. 15. 24–25). "The last enemy that shall be destroyed is death" (1 Cor. 15:26). The iconography of the Christ or microchristus figure in Remmelin's work bears a close doctrinal relationship to Paul's exposition of the centrality to Christian faith of the doctrine of the resurrection.

The density of the religious iconography, symbolism, and text in this plate, and in the others in *Catoptrum microcosmicum*, persuaded Ludwig Choulant that "the book was to serve physico-theological purposes, and was intended for laymen."[115] But as he admits, "it contains altogether too much" anatomy for the untrained, even if its "anatomical value" to professionals has been described as "very slight."[116] Significantly, none of the iconography occludes the presentation of anatomy; on the contrary, iconography and anatomy visually combine. The skull symbolizing mortality outwardly presents the surface of the cranium with its sutures on what is of several layers or flaps on the page only the first. Once opened the flap reveals the dura mater with connecting blood vessels and the sinus meningis, which also are rendered on a flap. Opening this second flap sheds light onto the pia mater with blood vessels, which removed reveals on a fourth flap the cerebrum; and so on as by a series of flaps, some quite diminutive, the corpus callosum, medulla, cerebral cortex, fornix, anterior cerebral ventricles, and the pineal gland are disclosed to view. Finally, lifting one large remaining flap, in which smaller are inset, reveals the cerebellum.

We have already encountered the visual combination of *vanitas* iconography and anatomy in the osteological plates of the *Fabrica*, but without the overt reference we find in *Catoptrum microcosmicum* to the triumph over death and sin by the Savior. The introduction of salvation in the figure of Jesus Christ may seem conceptually incongruous were it not for the fact that the abdomen and thorax of the Christ or microchristus figure are themselves the objects of a dissection similarly composed of a series of unfolding flaps. Christ is evidently fully human as well as fully divine. Arguably this is the point of Paul's insistence in 1 Corinthians 15.16 (and in versus 13), that "if the dead rise not, then is not Christ raised." The general resurrection depends on the resurrection of Christ but the fullness of Christ's humanity means that he depends for his own resurrection upon the possibility of the general resurrection, or rather it effectively depends upon the divine person God the Son who

Johann Remmelin, *Catoptrum microcosmicum suis aere incisis visionibus splendens cum histo-ria et pinace de nova prodit* (Ulm: Johann Görlin, 1639), Fig. 2. An anatomical Christ figure, or microchristus figure, in a plate which employs the whole panoply of symbols from the *vanitas* tradition, yet which bears a close doctrinal relationship to Paul's exposition of the centrality to Christian faith of the doctrine of the resurrection. Wellcome Library, London.

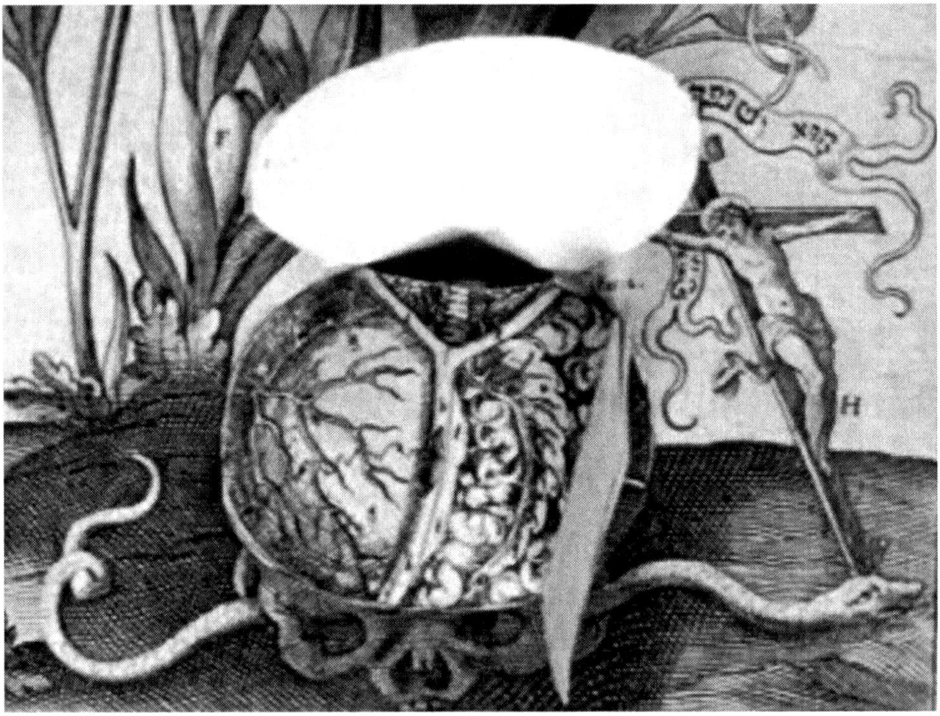

Johann Remmelin, *Catoptrum microcosmicum suis aere incisis visionibus splendens cum historia et pinace de nova prodit* (Ulm: Johann Görlin, 1639), Fig. 2 (detail). The skull symbolizing mortality outwardly presents the surface of the cranium with its sutures on what is of several flaps on the page only the first. The illustration shows that lifting the third flap reveals the corpus callosum, medulla, and cerebral cortex. In the anatomy textbook's approximation to the three-dimensional practice of dissection, Christian iconography and anatomy visually combine. Wellcome Library, London.

is capable of raising the dead and so of raising himself. *The Purple Island*'s diegetic narrator, in similar theological vein, addresses Jesus Christ, "Who reignedst in thy heav'n, yet felt'st our hell; / Who (God) brought'st man, whom man (though God) did sell" (*P.I.* 1.32.5–6).

The Purple Island and the plate in *Catoptrum microcosmicum* containing the anatomical Christ figure both demonstrate that the conceptual character of anatomy need not be in conflict with a soteriological perspective on Man. In Fletcher's poem, this perspective permits the coherence of vegetative and architectural conceptual metaphors through a combination of religious concepts about the body's creation and mortality, and the significance of the incarnation for its resurrection. As we have seen, arguments from design were of course used by anatomists to secure belief in the divine creation of human anatomy, but since Creation, humanity had fallen, and the Fall and the deluge had changed the face of the earth.[117] How could one know that the postlapsarian body was in any way commensurate with its prelapsarian counterpart, or that it is any longer contiguous with its Creator? Perhaps a Christian could genuinely justify theological arguments from design only because the *imago dei* had been restored in the human body by the incarnation of Jesus Christ, "Who th'deitie in-flesht, and mans flesh deifi'd" (*P.I.* 6.72.7). Despite mortality, Man's body becomes "God's dwelling place," and Man and God are thereby reconciled. The parallel that I sought to draw between the pair of lines in Canto 4, stanza 9, of *The Purple Island,* each sharing a structure of balanced indicative statements, now seems to suggest that the idea about human anatomy, that God "who first mad'st" (as the *Purple Island* diegetic narrator says) "never wilt forsake it" (*P.I.* 3.4.5),[118] is entirely dependent upon the incarnation, death, and resurrection of Jesus Christ; and that in dissection, which is of course entirely dependent upon human mortality, the cadaver is universalized and becomes microchristus, whereby this article of faith may be demonstrated. In microchristus, a personal copy of Christ, "we have a revelation of what "I" am in relation to God; of our destiny beyond death."[119] This, I suggest, is the congruity and the rationale of the plate containing the Christ or microchristus figure in *Catoptrum Microcosmicum,* and of the anatomical representation of microchristus in *The Purple Island.* Both are productions that can be seen in relation to the Protestant Christianizing of classical anatomy, which French describes as follows: "In place of the classicism of Vesalius' attack on Galen and the attempts to justify or negate that attack, popular anatomists now piled up praises of the human body as the image of God, or His temple on Earth: a more personal and more Christian view."[120]

The formulation of the doctrine of cosmic redemption in the epistles of Paul has been regarded as concluding that the human body "cannot be restored without the restoration of the whole natural order."[121] Furthermore, this created order was maintained by and subject to Jesus Christ, because in him dwelt all the fullness of the Godhead bodily.[122] Consequently, the reconciliation of Man to God brought about by the incarnation of Jesus Christ and "through the blood of his cross," could be understood to "reconcile all things" to God, "whether they be things in earth, or things in heaven" (Col. 1.20). In this way, the Fall of Man is reflected in a cosmic fall and the Resurrection of the Dead is a "reconciliation of the whole creation to God."[123] *The Purple Island*'s emblematic representation of the doctrine of the incarnation in Canto 4, stanza 9, gathers the allegorical strands of the microcosm, political geocosm, macrocosm, microchristus, and heaven, because all that applies to the human breast applies also to that region of the Isle where dwells the "Prince," "A little living Sunne" (*P.I.* 1.45.4,7). In Fletcher's poem the efficacy of Jesus Christ's redemptive work permeates creation in the way the sun is thought to affect the whole cosmos. In the first stanza the extradiegetic narrator says the "Sun" "began / To paint the world" (*P.I.* 1.1.1, 4), and, although the statement is preliminary to Thirsil's hypodiegesis of the Isle of Man, the context of analogy which that hypodiegesis is to provide suggests that the "Sun" metaphorically connotes also the "Son" of God, and the creative revivification of the earth in May with "ten thousand flowres" (*P.I.* 1.1.3) reflects human salvation or renewal.[124] This combines the idea of the sun as a life-giving heavenly or spiritual body with Pauline eschatology: "The idea of resurrection for Paul is completely inseparable from that of the quickening of our mortal bodies by the life-giving spirit."[125] Seasonal renewal and, moreover, the redemption it reflects, extend the vivifying Creation. The *discordia concors* of Empedocles is interpreted by Fletcher in this soteriological light when at the Creation, God is described as "Commanding strife and love should never cease" between the elements, "That by their peacefull fight, and fighting peace, / The world might die to live, and lessen to increase" (*P.I.* 1.41.5, 6–7).

Despite their perceived similarities, the sun and the spirit are not identical, however, and

Paul refers to man's unredeemed state in terms of "bondage under the elements of the world" (Gal. 4.3), into which sin and death have entered.[126] *The Purple Island*'s extradiegetic narrator adds in parentheses to the optimistic lines which begin the first stanza: "The world more aged by new youths accrewing" (*P.I.* 1.1.5). The unfallen status of all existence was symbolized in the myth of paradise, from which humanity in Adam and Eve has fallen.[127] "Man" is now "wretched" and pursues "this wretched world," "Which still grows worse by age, & older by renewing!" (*P.I.* 1.1.6, 7). There is notwithstanding, in the incarnation, a partial restitution and reconstitution of each of the analogues of the poem and the relationships between them. The meaning of the "picture" in line 2 of Canto 4, stanza 9 (cited above) is the image of God's throne, the sun of the macrocosm, and may be seen to correspond to the meaning of the same word in line 3, a "speaking picture" or emblem, because the poetic exposition in the canto to which this stanza belongs anatomizes the thorax, particularly the heart, which in the microcosm corresponds to the sun. The pun on "Sunne, Sonne" (*P.I.* 1.45.7) reoccurs by way of implication here, for the heart not only corresponds to the "sunne," but also to Christ, who is embodied and dwells therein, both in heaven and the human heart. As Fletcher's diegetic narrator says in Canto 1, it is Christ "who late didst deigne / To lodge thy self within this wretched breast" (*P.I.* 1.33.1–2). "In order for this 'low heav'n' to live immortally, the 'highest heav'n', which is the incarnate saviour, must die."[128] Christ "taught'st our death to live, thy life to die" (*P.I.* 1.32.3). By thus embracing human mortality as the wages of sin and the Fall, God sets "his best starres"—his son/ "sunne"—in the human microcosm, so glorifying it and redeeming it from eternal death or "hell." This providential care extends allegorically to the "Prince" of the geopolitical realm, whom God is said on the last day of Creation to frame "like himself, all shining bright; / A little living Sunne, Sonne of the living Light" (*P.I.* 1.45.4, 6–7). Before the Fall, the sovereign reflects the glory of the second person of the trinity and the providence of the first, whose representative he is on the isle.

Thus the allegory's Christian content relates coherently and systematically to the interwoven figurality of the microcosm, macrocosm, geocosm, body politic, and vegetative and architectural metaphors which characterize anatomical language and knowledge, and which abound in Fletcher's poetic stanzas as a form of imaginative rationality. As microchristus, or little copy of Christ, the dissected human body eloquently demonstrates not only the ingenuity of its Creator, but also the sacrificial and saving love of its Redeemer who is both Christ and "the King" (*P.I.* 1.57.4). From the coherence of the shared entailments of the metaphorical conceptualization of early modern anatomy—creation, mortality, and redemption—emerges an emphatically religious soteriological understanding of what anatomy may demonstrate. Buildings are capable of being rebuilt, and vegetation of renewal or regeneration. It is consistent therefore that by accepting mortality in humility and repentance, and by contemplating God's grace to "live properly," Man, and the world, may also find redemption. As Thirsil says, the Isle of Man, though retaining "a naturall proclivitie / To ruine," "Yet ever grows more prosp'rous by her crosses; / By with'ring springing fresh, and rich by often losses" (*P.I.* 1.35.3–4, 6–7).

Before we proceed to elucidate more fully how the soteriology of the *Purple Island* anatomy is purposively directed toward a soteriological effect through the poetic mode and anatomical method, there are two as yet unanswered questions relating to the character of the religiosity of anatomical knowledge which require that the topic is given further consideration.

The first is the question of how the natural knowledge of anatomy, which *reveals God,* yet remains a part of natural philosophy, related to the knowledge purportedly *revealed by God* either in Scripture or directly to individuals, groups of individuals, or the world. At what point did they meet or overlap and how is the process whereby one attains to the former kind of knowledge relate to that whereby one achieves the latter? The second question is related to this but has a very obvious moral dimension, and concerns ecclesiastical policy on anatomical knowledge and practice. Finding the answer to this second question seeks merely to place *The Purple Island*'s coherence of Christian allegory and rationalist anatomy in relation to the attitudes of its Christian cultural and social context towards dissection.

Notwithstanding anatomists" promotion of their own subject by referring to "How profitable and helpefull Anatomy is to the knowledge of God," it has been repeatedly suggested by medical historians that the Church was opposed to human dissection. Harcourt indicates that "the practice of anatomy can be castigated on no less an authority than that of St. Augustine [354–430]," who argued that whoever desires to estrange the body from the soul and the spirit ignores "the whole nature of man" and "acts without understanding."[129] The question is whether investigation of anatomy by the senses, combined with the rational inferences of teleology, does (or was thought to do) as Augustine's accusation suggests. Augustine himself had no doubt of this, and "he clearly condemns both the dissection and vivisection of human subjects"; and does so quite aside from any moral considerations.[130] Metaphysical anxiety about the potential of anatomy to "isolate certain judgements which record simple observations of fact from all the other types of judgement which normally form the background to those observations,"[131] is one reason why anatomical dissection might be neglected as a source of knowledge in the Middle Ages and Renaissance by Christian philosophers and even by anatomists themselves, as was suggested in chapter 1 by reference to Corti's contention that by dissection we may learn the position, substance, connexion, and form of the parts, but the explanation of why an organ has such a shape, position, connexions, number, and so on, is for the philosophers and the physicians to give.

If we leave what may be learned specifically from dissection to one side momentarily, and consider only knowledge pertaining to anatomy as subalternated to natural philosophy, we find that long after Thomas Aquinas (1226–74) had sought to maintain "a basic conformity between Aristotelian philosophy and theology," many natural philosophers assumed the compatibility of theology with investigation of Nature.[132] Nevertheless, a distinction was drawn clearly by Aquinas between the character of natural or rational knowledge and revealed knowledge, and between the processes whereby each is disclosed, although arguably Aquinas allowed that the means or process of revelation may be "standard uses of discursive reasoning." Sutherland specifies two different post-Thomistic religious uses of the term revelation: "one is to denominate *what* is revealed or disclosed, and the other is to denominate the *process* of disclosure or revelation."[133] Augustine believed that the process of dissection, for example, led to knowledge of a different order from that obtained either through discursive reasoning or through revelation (understood as a process). In terms of the processes, Aquinas distinguished between "an ascent, by the natural light of reason, through created things to the knowledge of God," and "a descent, by the mode of revelation, of divine truth which exceeds human intellect, yet not as demonstrated to our sight but as a communication delivered for our belief."[134] For Aquinas revelation "is not a matter of . . . intellectual ability or cognitive skill."[135] Discursive reasoning may provide a means of revelation, but because revelation is "premissed upon there being barriers to certain truths

or types of truth," and these barriers are "impenetrable to our normal or 'natural' way of seeking truth,"[136] the process or means of revelation could not be *identical with* standard uses of discursive reasoning. In Aquinas's view, "philosophy, in the sense of the independent use of human reason, was a proper handmaiden to theology," but could not be equal to it, for theology was "a science based on divine revelation."[137]

The truths of revelation are said to be distinctive in several ways:[138] what is revealed is true;[139] what is revealed is hidden and must be uncovered; what is revealed has been "hidden" because there are barriers which prevent its assimilation and appropriation; there is "news" in what is revealed. Arguably, in the matter of knowledge, sixteenth-century anatomy was capable of conceiving of itself as upholding each of the distinguishing criteria, except the third, which is also the principal Thomistic distinction between revealed and natural knowledge: "Revealed truths are not truths which are overlooked,"[140] and which are therefore immediately apparent to more attentive observers. There is nevertheless a visually striking comparatively early anatomical illustration which displays the anatomical use of the concept of revelation.[141] In Berengario's *Isagogae breves,* from the form of a self-dissecting male figure emanate rays of a revelatory light.[142] For sixteenth-century humanists and Hellenists, revelation in anatomy signifies the cognitive processes of removing the obstacles of Arabic or scholastic "error" and mistranslation of original Greek medical and anatomical texts,[143] to reveal the truth of Greek anatomical ideas, and to revive Greek anatomical projects which enlist the experience of the senses physically to uncover and newly bring to light that which was hidden.[144] Thus, anatomical "revelation" concerned both the kind of knowledge and the means of obtaining it.

Renaissance Neoplatonism also contributed to the definition and negotiation of the concept of revelation and to the erosion of its distinction from natural knowledge. Ficino inspired a new attitude toward Platonic material, including *Timaeus,* which chapter 1 discussed as central to Renaissance Platonic ideas of the microcosm. This treated Platonic material as "an authority comparable to that of the divine law," and the Platonic tradition as "divinely inspired," thereby raising the position of philosophy from handmaiden or rival to theology, to that of a rational manifestation of spiritual life and a means to revelation. A further contributing factor was that revelation had been, from the Middle Ages on, generally defined as "communicating a body of knowledge, some part at least of which could be independently obtained, or at least verified, by 'the light of reason and nature.'"[145] In other words, although Aquinas and most of his successors carried the distinction not only to the two ways of knowing, but to the respective areas of knowledge to which these were able to conduct us, "holding that there were many truths given by revelation to which reason could not attain at all," neither the distinction between the two processes of knowing nor that between the respective areas of knowledge was consistently upheld.[146] The anatomy of the human body could indeed be the object of revealed truths, especially in view perhaps of anatomy's concern with the soul and with the obviously religious subjects of creation, mortality, salvation, and ultimately God.[147] Furthermore, the interchange between natural philosophy and theology, between rational and revealed knowledge, and modes of knowing, was facilitated by the interdisciplinarity that was integral to the educational experience of would-be physicians and medical theorists.[148] During the sixteenth and early seventeenth centuries, scholars moved easily between the disciplines of theology, natural philosophy, including anatomy, and medicine, often teaching and sometimes publishing in several fields;[149] and, as was said in chapter 1, many took a degree in the faculty of theology before commencing

the study of medicine. Emerging from this educational climate, anatomists proceeded largely by means of observation and reason, as providing the process whereby divine truths about the anatomy of *imago dei* are revealed. Certainly for Crooke anatomy leads to divine worship, as we can see in his impassioned, enthusiastic invitation to the reader of *Microcosmographia:* "[W]ilt thou not cry out, though it bee against thy will, O admirable Architect! O unimitable workeman! And wilt thou not with the inspired Prophet sing unto the Creator this Hymne, *I praise thee (O Lord) because thou hast shewed the greatnesse of thy wisedome in fashioning of my body?*"[150] By no means out of place in this educational and intellectual climate is Fletcher's comparable response to the pudenda, which conventionally he feels required to be ashamed of and disparage:

> Which though we tearm the hidden parts of shame,
> Yet for the skill deserve no lesser praise
> Then they, to which we honour'd names impart.
> Oh powerfull Wisdome, with what wondrous art
> Mad'st thou the best, who thus has fram'd the vilest part!
>
> (*P.I.* 2.26.3–7)

The idea that knowledge of the human body is inspired or revealed knowledge had a tradition in natural philosophy as the revelation of God to human beings through a process of understanding this world more deeply than is permitted by the senses alone.[151] Magical and esoteric theories and philosophies in which analogies reveal creatures to be signs that reveal the marks of the Creator had some discernible influence on *The Purple Island,* as suggested above. The ancient philosophy that dignified the body with the title "The Book of God" finds its way to this island's shores, too, though in a Christian form in which the visible marks in God's works, having been understood as having the same nature as the words of ancient (Greek) texts (until reformers like Vesalius held them up to scrutiny and they became increasingly regarded as the image of truth at the second degree), came to constitute a second Bible, second only to the word of Scripture for bringing God before our eyes.[152] According to Baldwin, the poetic voice which Fletcher adopts is that of a pastor-poet, interpreting Man, "who, at once microcosm and microchristus, unlocks the universe and tends to the Godhead."[153] In so doing, Baldwin argues, the poet realizes in many respects the highest aspirations of both pastor and poet, "for, given the didactic function of poetry, second only to Scripture itself, as a supreme source of information concerning divine mysteries and the virtuous life, the office of the one is often very nearly identical with that of the other."[154]

Post-Reformation (especially Protestant) anatomy partly replaced strictly philosophical (often Aristotelian) justifications of anatomy, used by Vesalius and his contemporaries, with religious (more Platonic) notions of the nobility of the human body, as *imago dei* and God's dwelling place, and like Fletcher's poetry placed considerable emphasis upon the immediacy of the apprehension of God in his works.[155] Thus the order of anatomical knowledge became increasingly religious in tone as the suspicion developed that because the Christian Creator is omnipotent, not "bound by the laws of nature" which are merely God's "instruments,"[156] the knowledge sought of natural philosophy should be not even "the purely analogical resemblance between God and His creature," but rather the plan or "unity which underlies the multiplicity of the universe and which is God Himself."[157] As in the rational anatomy of Aristotle where purpose is constituted by the rational soul in the human body, the motile and sensitive souls of animals, or the vegetative soul in plants, for Paracelsus everything has a kind of soul, similarly "revealed by its structure, function and virtue,"[158] but for the latter the "mediator who bestows on things their aim, on material possibility, realization," is Christ, "the Logos."[159] The attainment of knowledge of the divine architectural plan for the materials of Nature, the builder, may be seen as a revelation on an analogy with the Word that was made flesh. In the work of Paracelsus in the sixteenth century, and in the "alchemistic mysticism" which became popular in England in the seventeenth century, "[k]nowledge of nature and religion represented an inseparable unit," and "natural science was rather revelation to the individual than science," which "is generally incontestable and objectively constraining."[160] There was a shift in epistemological method or process as well as in the kind of knowledge that "anatomy" was thought to reveal, and the former originated perhaps as a result of Paracelsus's "refutation of formal logic and reason as methods of natural research."[161] Paracelsus "put his trust in the Spirit, in the Word within, in holy inspiration," and believed that "only those who have direct, inner, experience of things (*res*) have real knowledge of things."[162]

The combination, originating in Paracelsus, of a disregard for formal logic and our "natural" way of seeking truth about natural things, with a view that "[b]y divine grace alone may the virtuous acquire knowledge,"[163] do not necessarily preclude a "bold incursion into the garden of nature," though this was "by means of unprejudiced experiment inspired by divine revelation,"[164] a "combination of religious irrationalism and vitalistic empiricism."[165] But for antirationalist reasons not too dissimilar to those offered by the empirical physicians of which Celsus and de Laguna wrote, and not uninfluenced by their skepticism, Paracelsus rejected the anatomical tradition of Galen, arguing against "the dead body as a ground of knowledge of what is useful for the living body" because "the essence, property, being and spiritual force [*Kraft*], which is the highest part of anatomy, have died and become corrupted."[166] The kind of knowledge that Paracelsus sought could not be supported by appealing to what can be observed or handled by everyone, in the way Vesalius invited support for his observations in the Bologna demonstrations.[167] For Paracelsus, " 'experience' is mystical experience, and the 'nature of things' that he seeks are mystical natures."[168] Paracelsus looked for a revealed system of "sympathetic relations" between microcosm and macrocosm, a "cosmic anatomy" which underlies and unites the multiplicity of the universe.[169] But it would be too neat a distinction to say, as Cunningham does, that "Vesalius saw with his bodily eyes, Paracelsus with his spiritual eyes."[170] As this study has shown, Vesalius also adhered to microcosmic theory, and when his *Epitome* metaphorically renders the vascular and nervous systems in terms of vegetation, the figurative language explores the ordinarily used parts of the domain of the defining subject, vegetation, while retaining precisely observed detail. Vesalius's text thereby deepens understanding of the body as a microcosm of the geocosm, with which the vegetative conceptualization is consistent.

The influence of Paracelsus upon the conceptual character of the *Purple Island* anatomy is considerable, though his skepticism of dissection suggests that his influence upon Fletcher's anatomical method is limited in this respect. The attitude to dissection that *The Purple Island* implies of its author may be discerned in the parallel that this section has been drawing between the necessity of the body's death for the anatomical demonstration, and the savior's death for the reconciliation of fallen Man to the Creator. Anatomical demonstration, which proceeds by means of observation and rational argument to conclusions about divine ingenuity and love, is not firmly distinguished from the revelation of the incarnation, that Jesus, a man, has embodied or manifested what is divine or eternal, and therefore, that God or divinity has conceptual content.[171] Aside from the general tendency to seek analogies between the microcosm and the world, and the Paracelsian character of some of the anatomy to be discussed in part 3, what may be found of Paracelsus in *The Purple Island* is the view, as expounded by Pagel, of knowledge as "a divine favour, science and research divine service, the connecting link with divinity."[172] "Grace from above" is indeed sought by the diegetic narrator of *The Purple Island,* to meet "human aspiration for knowledge from below."[173] For Thirsil the only way to secure the truth of rational knowledge is through revelation. Only God, to whom alone is ascribed full knowledge of the anatomy of the body, is qualified to be the narrator's guide:

> Thou Shepherd-God, who onely know'st it right,
> And hid'st that art from all the world beside;
> Shed in my mistie breast thy sparkling light,
> And in this fogge my erring footsteps guide;
> Thou who first mad'st, and never wilt forsake it:
> Else how shall my weak hand dare undertake it,
> When thou thy self ask'st counsel of thy self to make it?
>
> (*P.I.* 3.4)[174]

As the narrator here implores for light in his "mistie breast," he is praying both for divine inspiration in his own heart,[175] and for God to shed a light upon the anatomico-poetic exposition which the narrator's "hand," that primary instrument long neglected by the art of medicine until Vesalius, is about to undertake (*P.I.* 3.4.3,6).

It may be argued that the inconsistent and less than conceptually perfect distinction between rational and revealed knowledge, and between the processes whereby each is obtained, presented a way to justify dissection as a mode or process of revelation, and, though not through mere expediency, served well those assiduous investigators who were coaxing a recalcitrant and even fearful ecclesiastical establishment to accept what was yet in the sixteenth century a practice suspected of violating the sanctity of the human frame. Militating against such a view, however, is that a less than perfect distinction was also the resolution of many, Fletcher for one, with no professional dignity at risk. To what extent the *Purple Island* anatomy might be directly or indirectly derived from observation at anatomical demonstrations is a question for chapter 9. But the question of the Church's opposition, tolerance, or encouragement of anatomy has a direct bearing on our understanding of its Christian allegorical presentation in Cantos 2 to 5 of Fletcher's poem.

The persistence of beliefs even among some medical men and jurists in the residual life, sensitivity, even healing powers, of the cadaver,[176] together with "the dangerous and unseemly haste with which sixteenth-century anatomists appropriated fresh cadavers for dissection," their perceived appetite for bodies to dissect, and the dissemination of accounts of the revival of persons under the anatomist's scalpel, and "a haze of unsavoury stories" connecting famous anatomists with vivisection, blurred the distinction between dissection and vivisection in the sixteenth and seventeenth centuries, and gave rise to popular suspicion concerning dissection.[177] Opposition to human vivisection, which many believed the Alexandrian anatomists Herophilus of Chalcedon (c. 330–260 BC) and Erasistratus of Chios (c. 330–255 BC) had undertaken, was widespread, and partly because of the admiring references to the Alexandrians in the Preface to his *Fabrica,* the suspicion was formed that

Vesalius and even Berengario may also have practiced this.[178] Unlike Vesalius, who could at least see the very evident advantages to be had by basing a knowledge of human function on human vivisection, the physician and compiler of anatomy, Helkiah Crooke, argues that even though voluntary motions and senses in humans differ from those in other animals, "for the discerning of actions that belong to motion and sence, ther is no need of vivisection, for they are almost all of them apparent to the sences." Crooke favors vivisection only of non-human animals, arguing, with regard to the hidden and secret parts, that they "are all one in men and in beasts."[179] Unlike Augustine, however, in evidently not regarding anatomical investigations as missing the point, Crooke is typical of seventeenth-century physicians, for whom metaphysical anxiety is replaced with a moral recalcitrance generally reserved only for vivisection, and an attempt sometimes to find justification within anatomy itself for objecting to it: "it is not anatomy but butchery, to mangle the trembling members of mans body, and under I know not what slender and idle pretence of profit, or behoofe, to violate the sacred Law of nature, and of religion."[180] While condemning human vivisection, Crooke could still argue that by "the dissection of the body, and by Anatomy, wee shall easily attaine unto this knowledge [of our selves]."[181]

Historians have not succeeded in demonstrating what relationship the opposition of some religious men, physicians, or even schools of anatomy had to the attitude of the Church towards human dissection at any time. Historical references to dissections and postmortem autopsies begin in the thirteenth century.[182] Autopsy is performed for legal and medical reasons which are different from that of general anatomical knowledge which gives rise to private or public anatomical dissection; but, as Mary Niven Alston claims, "both processes involve considerable mutilation" and might be supposed to offend for the same moral or religious reason.[183] By the early fourteenth century "the technique of demonstrating . . . anatomy was already formalized" at Bologna;[184] until this time there is no evidence of any attempt by the Church deliberately to obstruct either postmortem autopsies or public or private anatomies.

On the issue of ecclesiastical policy towards dissection, canon law in this period furnishes only the bull *Detestande feritatis* (also known as "de Sepulturis"), issued on September 27, 1299 by Pope Boniface VIII.[185] In this bull Boniface passionately objects morally and theologically to a custom where, upon the death of "nobles and high-ranking dignitaries (and especially when such people died far from their native lands and the places they had elected for burial)," their bodies were disemboweled and the bones loosened from the flesh by boiling, thus allowing transportation of the bones to be laid to rest at the place reserved for interment.[186] Singer observes that although the bull "was not directed against the anatomists," nevertheless "it told against them" because boiling bones facilitated their prolonged and unobstructed study.[187] But if the bull did effectively prohibit osteological investigation, its effect was short-lived. By 1363, Guy de Chauliac writes "without any suggestion that such a thing had ever been prohibited."[188]

That the effect of the bull on dissection was minor seems to be suggested by various pieces of historical evidence, including a statute providing for an anatomy biennially from 1340, at the University of Montpellier,[189] and an increase in references to anatomical demonstrations and postmortems in late medieval Padua from the early fifteenth century.[190] *Detestande feritatis* was "aimed at a specific practice," not dissection; or at least it ceased to apply to dissection by 1491, as the burying of corpses used for dissection in consecrated ground proves.[191] Whatever problems the bull had posed anatomy were formally resolved by

Clement VII (pope 1523–24) in 1531.[192] Until this time, no evidence has been found either for "a general (or even common) prohibition concerning the opened corpse per se,"[193] or for the Church's encouragement of the practice of dissection.[194] But the evidence does suggest ecclesiastical tolerance,[195] which continued into the fifteenth century, when, in April 1482, Sixtus IV (pope 1471–84) issued a brief to the University of Tübingen which "permitted dissection following the approved authority of the Church."[196] This has been described by Bernard Schultz as "[p]robably in response to the uncertainty caused by Boniface VIII's bull of 1299." The recognition by Sixtus, who had himself, as Francesco della Rovere, studied philosophy and theology at Bologna and Padua, of the need to dissect conditional on permission of the ecclesiastical authorities, was confirmed later by Clement VII,[197] "a factor in the development of Anatomy at the end of the fifteenth century,"[198] for "coupled with the already accepted practice of university medical school dissections at centers like Bologna and Padua, [it] would have greatly facilitated access to the human body for anatomical research."[199]

While they imply ecclesiastical tolerance, the rules governing public anatomies of human corpses, which for the medical faculty of Bologna appear from at least as early as 1405,[200] do also suggest that an attempt was being made to address recalcitrance and perhaps even fear. A combination of factors which included "anthropological problems (the revulsion generated by contact with cadavers)"[201] (which partly explains the delegation of dissection to barbers and surgeons), the cultural association of public display and bodily disintegration with shame and dishonour,[202] and more specifically religious concerns about the delaying of burial and a belief in the nobility and dignity of the human body and "the sacredness of the human corpse," seem to have given rise to "a series of prudent strategies implemented to limit the circumstances under which anatomical demonstration could be legitimately allowed": "The *podestà* in Bologna, the *commissarius* in Pisa, the *praetors* in Padua, the cardinal vicar and senator or the governor at Rome acted as regulatory officers to prevent disturbances over the selection of unsuitable cadavers and to prevent physicians and students from recourse to such sacrilegious acts as exhuming the dead, or appropriating and dismembering bodies that did not deserve such ignominious treatment."[203] Those that were deemed to deserve such treatment were the cadavers of the condemned in public executions, the use of which had been tacitly accepted long before anatomical practice was governed by regulations written into university statutes.[204] It seems also to have been expedient to use the bodies of deceased poor foreigners, especially perhaps when they had been hospital patients, because there would likely have been no relatives nearby with an investment in a conventional and honorable funeral, for which the deceased themselves may not have been able to pay.[205] According to Carlino, though, "the moral quality of the body to be dissected had to be evaluated": "The statutes repeat with unusual emphasis such terms as 'criminals,' and 'delinquents,' to designate those who, having been found guilty in a terrestrial court of law and executed, might pass under the instruments of the *prosector.* Their bodies, punished and damned, would continue in their agony even beyond life, since their souls would pay in the hereafter for the sins they had committed."[206]

Notwithstanding the regulations surrounding the suitability (not from an anatomical but a social and religious point of view) of cadavers, even if the condemned was judged to be deserving of the additional ignominy and humiliation of public dissection as a reinforcement of her or his punishment, the anatomist might have been tainted by association with the infamy either of the criminal, accrued not only by the criminal deed for which he or she was condemned but also by public execution, or of "the executioner and his fellow workmen," who were "seen as uncomfortable members of the community; and considered for full participation in the life of that community." The latter may have been more likely where resistance to dissection as a further ignominy and shame arose because even "the most heinous sins, those that would induce an earthly court to promulgate a death sentence, could be forgiven [by God] to the extent that a profoundly repentant condemned person could even be admitted directly into God's presence in Paradise," and this may be reflected in the establishment by the statutes of special propitiatory provision for the dissected victims of execution, as, for example, in the number of Masses celebrated for the redemption of their souls.[207] Fifteenth- and sixteenth-century lay confraternities in Italy sought to persuade the condemned to contrition, and in so doing sometimes described the effect of penitence and acceptance of the sentence of death as the transformation of the gallows and execution block into "a holy martyrdom."[208] *Il manigoldo* (the executioner) was expected to play a "vicarious devil's role" in the symbolic drama of execution, in which the kneeling victim was imagined "as a pitiable soul struggling for salvation."[209] Using sources mainly from, and studies undertaken in the context of, England and northern Europe, Sawday's argument is that the anatomist's direct or symbolic implication in the exercise of judicial power, and unavoidable proximity to the infamy of the executioner, sat uncomfortably with his desire for the cultural inclusion and acceptability of his profession.[210]

In addition to the effect of these determining factors, although from at least the twelfth century the opening of the body was common funerary practice, it has been suggested that "anatomists might have considered that a disemboweled cut-up body, disarranged for a profane purpose (in contrast to the 'sacred' practice of plundering saints' bodies and the cult of relics) resulted in a higher dose of infamy."[211] According to Carlino: "In Rome especially, where the statutes bore the signature of the pontiff," but also elsewhere in Italy and at the other centers of anatomical practice in Western Europe, two contrasting demands required satisfaction in the organization of public anatomy lessons: "that of science, which required direct knowledge of the human body, and that of religion that sought respect toward the dead, their remembrance, and their bodies."[212] But as Park argues, the historical situation was not as polarized as this might suggest, and Carlino also notices that although the statutes established conventions "that safeguarded the afterlife of the dissected victims of execution while purifying the anatomist" they appear hardly ever to have been followed, especially after the mid-1570s. This, argues Carlino, is evidence of the anatomists" attitude that "knowledge of the human body as the foundation for all medical practice" "began to be seen as sufficient guarantee of the morality and lawfulness of their work." But, as he adds, at least in respect of Rome, which comprises the focus of his study, no authority involved in the organization of dissection reacted to any lapses, even though they must have been aware of them. "They never lamented any deficiency either of form or of content in those activities connected to medical research and teaching."[213] Had religious convictions about the human body, dead and alive, and the resistance of anthropological or cultural problems, necessarily and universally led in the Church to the view of "the use of the anatomist's scalpel" as "a sacrilege," as some historians like Foster have thought, the ecclesiastical authorities must surely have exercised power where they held it to prevent further lapses of adherence to the regulations.[214]

The mistake of these historians is to assume that the Church's response to the anatomical revival in the Renaissance and the increasing emphasis upon observation, combined with rationality, would have been to believe that the corpse was not used to demonstrate nobility or religious verities about creation, the Fall, mortality, and salvation, as anatomy may have done in the Middle Ages. This assumption is only part of the wider historiography in the Burckhardtian tradition, which understands the Renaissance as "an essentially *secularizing* movement," as "a movement which stepped aside and stood aloof from Christianity and matters of religious faith—and which succeeded because it did so";[215] succeeded, that is, in ceasing to see Nature "only through a glass darkly," and in distinguishing between science and religion as opposing each other.[216] The vantage point gained fails entirely to see that sixteenth-century anatomy was essentially concerned with the soul, not simply empirically as the sum of the parts of the whole body, but philosophically and theologically as its reason for being, and that, by the early seventeenth century, natural philosophy was also concerned with "things spiritual and things corporeal" as "emanations of the will of the Creator," "in all stages connected, and finally identical."[217] Anatomy's rational or natural knowledge, and indeed its purportedly revealed knowledge, might as easily have been produced and adopted as supportive of religious belief and piety, as rejected on either moral or religious grounds. So although many of the anatomist's acts, such as "opening and manipulation of the body," its public display and dismemberment, or the delaying of burial, were "religiously and anthropologically risky," dissection "was nevertheless able to evolve thanks to the establishment of a formal normative apparatus and to the mediation of judicial, political, and religious institutions."[218]

They may have exercised direction over anatomical studies or projects conducted under their auspices,[219] but "Roman ecclesiastical authorities did not disapprove of opening corpses."[220] While at Padua, Vesalius was invited to Bologna, "which was then a papal city," to give the anatomical demonstrations that were to accompany Corti's lectures.[221] And "Rome itself became something of an anatomical centre" with Bartolomeo Eustachio and Realdo Colombo as professors of anatomy at the Sapienza, the Papal University.[222] Meanwhile, in Protestant countries, ecclesiastical authority also aided dissection. Anatomy was taught at the University of Wittenberg, and later at other universities brought into existence or reformed through the Lutheran reformation.[223]

Augustine's attitude to human dissection had not been the dominant ecclesiastical one for over a hundred years when Fletcher came to write his poetic anatomy, so he would have had no reason to feel discouraged from doing so by contemporary ecclesiastical authority, and there is no sign in *The Purple Island* or its reception that he did. Rather, the emblematic representation of microchristus in Canto 4, stanza 9 has soteriological purpose because a part of the redemption is an epistemological "restoration of the 'knowledge of the Lord,'" which reveals as eternally true the analogy between human anatomy and its analogues.[224] By coherently and consistently combining the shared entailments of the metaphorical conceptualization of anatomy—that the body is created, mortal, and redeemed—with a religious soteriological understanding of what anatomy may demonstrate, Fletcher wrought the figurative design of *The Purple Island* with a politico-religious purpose, in the fulfillment of which he perhaps hoped that the Isle of Man, the realm or nation under James I, and the soul and body of his implied or intended readers, with whom he shared the nation, might be restored, regenerated, perfected, and preserved in Christ,[225] that it might grow "more prosp'rous by her crosses; / By with'ring springing fresh, and rich by often losses" (*P.I.* 1.35.6–7). In this design, however, neither the attribution of creative agency to God, nor the revelation or rational inference of salvation, explains fully how anatomy and figurality disclose the creative hand and saving grace of God to his readers. For this we must also look again, firstly, to the purpose of the poetic mode and then to the anatomical method or order of the poem.

In the suprarealist conception of language, with the authority of Augustine, the outward sounding of words or their sounds rehearsed in thought could be seen as a sign or revelation of the word that is inwardly luminous, the "locution of the heart," and may therefore be seen on an analogy with the incarnation of the Word of God as flesh.[226] Human language, thus understood, is sacred, created by God, and represents a mode or way of encountering God through revelation. *The Purple Island* appears to Baldwin as a didactic poem which teaches the knowledge of God and self, which was regarded as necessary to one intent upon salvation.[227] Didactic poetry includes major works such as Lucretius's *De rerum natvra libri sex,* and defenses of it traditionally accord to its poetic form an unrivaled capacity to "indicate through the provision of information the order and perfection of the universe as a whole," a subject which the tradition is especially concerned to represent.[228] A reason for the suitability of poetry for this role was seen in the following way (described here by Sawday): "The world was structured according to a gigantic act of repetition—a continual recapitulation—and that repetitive tendency was best mirrored in the compressed language and rhythms of poetry."[229] The poetic stanzas of Cantos 1 to 5 of *The Purple Island* may convey this "gigantic act of repetition" through their consistent seven-line *ababccc* rhyme scheme, and their regular iambic pentameter and hexameter. Nature and Man having fallen, however, Fletcher's poetic narrator, Thirsil, must try to revive perfection and help bring about salvation. Canto 6, stanzas 18 to 19, is just one of many examples of Thirsil imploring for inspiration, and it is significant that here he attributes to Phineas's brother, Giles Fletcher's *Christs Victorie, and Triumph,*[230] "speeches" taught by "Soft-hearted Mercy," second-born daughter of "th'Almighty King," which he describes not only as "So neare her heav'n they be, so farre from humane thought," but as having "calm'd thy father, and our desp'rate fears" (*P.I.* 6.19.6; 6.16.2; 6.13.2; 6.19.7,2). It would seem that it is also Fletcher's purpose, to induce divine mercy and thus salvation.

The world's repetition, and its deferred perfection, it might be argued, are also reflected or repeated in the poetry's assemblage of analogies between body and island, and between heaven and earth. Baldwin calls the poem's allegorical structure "a logical extension of Fletcher's didactic function."[231] Adding to the efficacy of language, by drawing the analogies of microchristus and microcosm, *The Purple Island*'s allegorical exposition of the compendium "volume" (*P.I.* 1.48.2) restores, or reveals as true, the relationships of *aemulatio* and *convenientia* between God and creation. According to Mazzeo, in the poetics of Tesauro the whole of Nature, or reality, is probed by metaphor to the end of a "revelation of little mysteries."[232] So, in the hands of Fletcher, allegorical structure purportedly partakes of the inspiration and revelation of the incarnation, the purpose of which is human and cosmic redemption. The poem purports to undertake a spiritual task of revelation and recovery. What it recovers is twofold: firstly, the character of Man as represented and dissected on the hypodiegetic level of the narrative, by the diegetic narrator Thirsil;[233] and secondly, and simultaneously, God's "art" of anatomy, knowledge of which is edifying but has like Man been "much defac'd" by the Fall (*P.I.* 2.3.3). In the limited intratextual context of its signification, *The Purple Island* becomes part of the saving work of redemption.

This work is augmented by the order or method of anatomical division. The anatomical cantos begin with a description of the foundation of the body and the terrestrial globe—

bones and stones. This order was advised by Galen and gathers significance as we reconsider the metaphor of God as architect. The analogy is drawn by John Banister in *The Historie of Man* between, on the one hand, the manner whereby an architect "hath a singular care, first in edifying his house, to be well aduised of the manner of his foundation, and heedy framyng of his Timberworke," and, on the other hand, the way that the anatomist should begin "before the partes supported . . . to consider the partes supporting."[234] Because God the Architect may also be seen as God the Artist, the order may be seen on an analogy with the artistic creation of the human figure in Renaissance art. According to David Summers: "The bones provided an order which was the beginning of the chief topic of art, the human body. Like proportion, or bilateral symmetry, the order of the bones was in most cases a beginning, and not the end toward which imagination laboured. The skeleton had to be clothed in flesh . . . and further, had to be enlivened. . . ."[235] In anatomy, this order is not only functionally necessary in the context of the concept of construction, but, according to Banister, aids understanding. Bones are a "similar part" of anatomy, composed of primordial elements and humors; and embryologically, "[a]t the first constitution of the parts the bones spring from the seminal excrement."[236] Usually it should not be inferred that the emulation of Nature's constructive, compositive, and embryological methods in the presentation of anatomy by anatomists implies that they were doing anything more than that—emulating Nature in order to understand her purpose for the function and structure of the body. But when the method of dissection in Fletcher's poem reveals and repeats the method employed by God and Nature in the creation of Man, structurally from the bones, compositionally from the elements, and embryologically, the method combines with the poem's adherence to an incarnational and emblematic conception of poetic language, and an interactive metaphorical mode of creation and discovery. The revelation of the method is, as it were, a continuation of the revelation of Creation and of incarnation in the poetic word.

Like the figure of the Isle of Man, which reflects in miniature the whole poetic work, the anatomy of microchristus in Canto 4, stanza 9 emerges from the analysis in this chapter as an index not only to the whole human body, which is the poem's theme or subject matter, but to the whole poem's soteriological design. Metaphor emerges as a way of understanding the incarnation, which combines in a way not too dissimilar to an interactive metaphor the revealed similarities and dissimilarities between its principal and subsidiary subjects—God and Man.[237] James Ussher asks at the outset in his *Immanuel:* "Where if it be demanded, how these things can stand together? that the *Son of Man* speaking upon *earth,* should yet at the same instant be *in heaven?*"[238] Like the intense, compact figure of the Isle of Man, the incarnation is for Ussher a "knot" capable of being "unfolded,"[239] the "untying" of which is to reveal not only an exposition of doctrine but a host of related analogues, often referred to in biblical studies as the "types" of the incarnation. These are the "*Tabernacle,*" "a figure of the humane nature of our Lord," and "the Temple of *Salomon,*" which, says Ussher, is "a Type likewise of the body of our *Prince of peace,*" filled with the Glory of God.[240] The meaning of types, moreover, is defined by the understanding of the incarnation itself: they are in effect the metaphorical entailments of the most specific metaphor of a coherent system which arises from biblical exegesis and doctrinal formulation, a system to which is appended the prince in the body politic, by both Ussher and Fletcher, and the sun in the macrocosm and geocosm, by Fletcher.

It would be inappropriate to extrapolate that Fletcher, even more so Ussher, understood the incarnation as nonliteral. Ussher clearly states: "The untying of this knot dependeth upon the right understanding of the wonderfull conjunction of the divine and humane Nature in the unity of the Person of our Redeemer."[241] "Right understanding" is here in line with all official theological formulations of the incarnation subsequent to and including the settlements reached at Nicaea (AD 325) and Chalcedon (AD 451), which took their prompt from the prologue to the Fourth Gospel—"And the Logos was made flesh" (John 1:14), "*sarx egeneto,* Latinized as *incarnatus*"—and did not intend incarnation as a metaphor, "but as shorthand for the doctrine that Jesus was God the Son living a human life, being both 'truly God' and 'truly man,' *vere Deus, vere homo.*"[242] Notwithstanding this, I suspect that the notion that the incarnation is both so defined *and* a metaphor is a contradiction only from a perspective informed by objectivist philosophy and linguistics. The conceptual metaphoricity of the Isle of Man neither so separates literality from metaphor, nor reserves truth only for nonmetaphorical language; indeed its whole approach to meaning, to truth, and to revelation is via metaphor, and the incarnation shares something of that figurality. Fletcher's concerns with metaphor and with the incarnation are intertwined in the design of *The Purple Island,* which uses the character of metaphor as combining subjects with dissimilarities, and as involving a cognitive revelation of something new, to shed light upon the paradox of the incarnation, to reveal the mystery more fully, anatomically, and persistently over the first five cantos.

This chapter began with two questions that had emerged from the analysis of the figurative design of *The Purple Island* in chapters 1 and 2. The first question was whether the conceptual character of the anatomy in the poem is in conflict with the Christian allegory. In the question it is necessary to recognize that the anatomy is transformed through metaphorical interaction, because the Isle of Man metaphor and all its entailment metaphors throughout Cantos 2 to 5 select, emphasize, suppress, and organize features of the principal subject by implying statements about it that normally apply to the subsidiary subject. But the question should also recognize that Fletcher's innovation was not confined to the creativity of the threshold figure, but extends to taking the geopolitical figuration of anatomy far beyond anything that occurs in anatomy textbooks while retaining consistent use of the anatomical *accessus* and regulating the degree of allegorical personification in relation to observable anatomical postures and physiological operations. So the question of what the *Purple Island* anatomy would have been like prior to its metaphorical, allegorical modification is answerable. Although a full answer is deferred to part 3, where the provenance and conceptual character of the anatomy is investigated, a partial answer has been given by this chapter.

The second question with which this chapter began was how does the allegory's Christian content relate to the systematically interwoven figurality of the microcosm, macrocosm, geocosm, and vegetative and architectural metaphors which characterize anatomical language and knowledge and which abound in Fletcher's poetic stanzas as a form of imaginative rationality? The chapter began to answer the first question with reference to the contemplation of creative ingenuity in anatomy, an ingenuity about which *The Purple Island* is explicit. In so doing it also established further ways in which *The Purple Island*'s Christian allegory maintains the systematicity that obtains between the metaphors in anatomy, the analogies of the microcosm, and the significant connotations of the Isle of Man, and extends that systematicity to those implications which arise from the interaction of newly emergent ideology and incipient nationhood. The vegetative figurality, for example, gestures to the body as like the created world, and the conceptual metaphor of anatomy as "building" points to the body's artificial construction by the divine architect of the body's landscape and built environment. The metaphors of "building" and vegetation, and the faculties of reason and Nature or the vegetative soul, are reconciled and made coherent in and by the concept of the creative ingenuity of God. But, in addition, while conforming in some respects to conventional notions informing the motif of the upright form of Man, whose body is a masterpiece of architecture and whose mind contemplates the heavens, *The Purple Island*'s allegory extends systematicity to the novel figuration of the structure and functions of the organs of sight as an efficient mechanism of intelligence-gathering for the government of the British nation.

The Christian allegory of *The Purple Island* is also rendered consistent with the anatomy in the poem through a further shared entailment of the two metaphorical concepts of vegetation and building. In addition to creation, the two defining concepts of anatomy share the entailment that they may die or be ruined. Renaissance anatomy provided moral and theological argument not only from design, but from the demise to which Man is inevitably subject, and developed an expectation that the anatomy presented by illustration would serve a purpose in which the concept of death and the need for salvation are represented as integral within a Christian context. The third shared entailment of the metaphors of anatomy explored in this chapter was that vegetation, buildings, and bodies may all be renewed in some sense, whether regenerated from seed or by asexual reproduction, or rebuilt from ruin, or redeemed from mortality by resurrection.

The shared entailment of redemption or salvation was explored in the context of Fletcher's use of the term "picture" for poem, an adaptation of the familiar Renaissance understanding of poetry as creating a speaking picture. The chapter suggested that Fletcher drew on the chiastic *poesia tacens, pictura loquens,* and *ut pictura poesis,* as well as emblematics and suprarealist theories of language, without which Fletcher would never have described, "discovered" or "invented" with such deliberation the significant connotations of the creative Isle of Man metaphor, which for him exist in reality, some antecedently, some as etymons of anatomical nomenclature, and others in the working out of novel implications. Examining closely the text of Canto 4 stanza 9 of *The Purple Island,* the chapter argued that a metrical parallel between lines 3 and 4 suggests a twin necessity of the cessation of the body's breath for the anatomical picture's demonstration of God's creative ingenuity, and, of the savior's death for the reconciliation of the fallen body to its Creator. *The Purple Island* demonstrates not merely that the conceptual character and imaginative rationality of anatomy need not be in conflict with a soteriological perspective on Man, and not merely that this perspective permits the coherence and systematicity of vegetative and architectural conceptual metaphors and the microcosm, political geocosm, macrocosm, microchristus, and heaven, through a combination of religious concepts about the body's creation and mortality and the significance of the incarnation for its resurrection; the poem also demonstrates that the idea about human anatomy, that God "who first mad'st" "never wilt forsake it" (*P.I.* 3.4.5), is entirely dependent upon the incarnation, death, and resurrection of Jesus Christ, and that in dissection, real or poetic, the body becomes microchristus and faith in the resurrection may be expressed. Moreover the emblematic representation of microchristus has soteriological purpose. As microchristus, the dissected human body eloquently demonstrates not only the ingenuity of its Creator, but also the sacrificial and saving love of its Redeemer who is both Christ and "the King" (*P.I.* 1.57.4).

Two questions were raised regarding the validity of the religious character of the anatomical knowledge presented by *The Purple Island.* The first was the question of how the natural knowledge of anatomy, which might reveal God, and the process whereby it was attained, related to the knowledge purportedly revealed by God and the process of revelation. The second question concerned ecclesiastical policy on anatomical knowledge and practice, and sought to situate *The Purple Island*'s coherence of Christian allegory and rationalist anatomy in relation to the attitudes of the poem's Christian cultural and social context towards dissection. Although metaphysical anxiety about the potential of anatomy to isolate judgments which record simple observations of fact from other types of judgment was found to be one reason why anatomical dissection might be neglected as a source of knowledge in the Middle Ages and Renaissance, and although Aquinas and most of his successors carried the distinction not only to the two ways of knowing, but to the respective areas of knowledge to which these were able to conduct us, neither distinction was consistently upheld; and anatomy was seen to be revelatory of structure, function, cause, creation, mortality, salvation, and ultimately God. This is why without fear of contradiction Fletcher could adopt the voice of a pastor-poet, interpreting Man as anatomical microcosm and microchristus.

The culture in which this was especially possible was Fletcher's own post-Reformation, Protestant culture of northern Europe, which placed great emphasis on notions of the nobility of the human body, as *imago dei* and God's dwelling place, and on the immediacy of the apprehension of God in his works. Here also Paracelsianism and related esotericism sought knowledge of Nature through revelation of sympathetic relations between microcosm and macrocosm, a cosmic anatomy. The influence of Paracelsus upon *The Purple Island* is considerable in respect of the notion of knowledge as a divine favor, science divine service or worship, the general tendency to seek microcosmic analogies, and the character of some of the anatomy to be discussed in part 3 of this study; but it was limited in respect of anatomical dissection. There is no reason to suppose, however, that by demonstrating Christian theology and faith through poetic dissection, Fletcher would have displeased a recalcitrant or fearful ecclesiastical establishment who suspected anatomy of violating the sanctity of the human frame. Anthropological problems, like the revulsion generated by contact with cadavers, the cultural association of public display and bodily disintegration with shame and dishonor, the taint of the infamy either of the condemned criminal or the executioner, and more specifically religious concerns about the delaying of burial and a belief in the nobility and dignity of the human body and the sacredness of the human corpse, all contributed to the formation of a series of prudent strategies which limited the circumstances under which anatomical demonstration could be legitimately allowed. But although it could have, ecclesiastical authority did not obstruct (at least intentionally) and latterly aided dissection, from as early as the fifteenth century in Roman Catholic contexts, and later in Reformed contexts.

In this generally conducive environment, Fletcher wrought the figurative design of *The Purple Island* with a politico-religious purpose, that the nation he shared with his implied readers might be renewed, regenerated, saved. By drawing on suprarealist theories of language, which may be Augustinian in origin, and the tradition of didactic poetry, and by restoring the relationship of Man to God through the poetry's assemblage of analogies to the microcosm and microchristus, and by adopting a method of dissection that reveals and repeats the method employed by God and Nature in the creation of Man, *The Purple Island* purportedly becomes part of the saving work of redemption. Although we can infer that this was Fletcher's purpose, regardless of what he consciously intended, our analysis in this chapter has revealed only that it might be thought of as having "taken effect" within the limited intratextual context of the poem's signification. Having thus far limited the scope of our analysis to the complex figurative design of the work, we are not yet in a position to measure the success of its purpose, or even assume that success in this case could be measured. What would it mean for the purpose to have been achieved? Of what significance could it be that from Fleet Prison, James Howell wrote, that reading *The Purple Island* "enlarg'd" his "quarters 'mong these melancholy walls," and that therein he encountered himself, growing "better acquainted with what I have within me and without me"? Of what significance could it be that he described Phineas Fletcher as "the Ingenious Authour" of the poem? As was remarked in the introduction to this study, critics have by and large omitted consideration of the contemporary reception of *The Purple Island,* and in order to redress this omission, the following chapter will explore the contemporary responses of its readers. And in order to gauge what the poem's likely effect was in early seventeenth-century literary culture, whether it would have retrenched and corroborated what Fletcher's readers might be expected already to have known, and to what extent it might be thought to have expanded their vision, part 2 of this study will place *The Purple Island* in the contexts of contemporary literature, including Fletcher's source-texts, in relation to which readers might be assumed to have gauged their own expectations when reading the allegory of the Isle of Man.

II
Ingenuity

4
Hermeneutics and Intertextuality

PART 1 OF THIS STUDY WAS CONCERNED PRIMARILY TO ANSWER THE QUESTION OF THE congruity between the subjects and discourses of anatomy and poetry in *The Purple Island,* and in so doing developed a theory of allegorical, analogical, and metaphorical meaning which took in questions of its congruity, systematicity, rationality, conservatism, and innovation. The conceptual metaphors of the human body in anatomy textbooks of the sixteenth and early seventeenth centuries, with which *The Purple Island* shares figurative and rational coherence and consistency, were found to coalesce with the analogies of the microcosm and microchristus in *The Purple Island.* Our conclusions on the figurative design of the poem have led this study to expose misconceptions in the judgment of some of *The Purple Island*'s more prominent and thorough critics. A crucial point of difference, for example, between those of Fletcher's critics, like Langdale, who thought that the significance of *The Purple Island* lies in the purportedly "modern" character of the anatomy, and those, like Baldwin, for whom Fletcher's reliance on the conventional analogy of the microcosm is indicative of his conservatism, led to an investigation of the microcosm analogy which concluded that not only were the Platonic and the literal microcosmic views essential and typical in the Renaissance, their distinction by Langdale was found to be false, and, moreover, a belief in the microcosm may even have been an impetus to anatomize and so was not necessarily conservative.

Nicolson and Baldwin both argue that the metaphoricity of the anatomical content of *The Purple Island* may be "more comprehensive and detailed" than anywhere else in literature, but that its conceptualization of the body in microcosmic, and (as an island) in geocosmic, terms is by no means novel or unique because already firmly established when Fletcher came to write.[1] Nicolson argues that the persistence with which the poem details the correlations of its central analogy can "show us how geology and physiology were retarded by the attempt of man to interpret the world in his own image."[2] In other words, Nicolson assumes that the character of the anatomy and physiology of the poem is dependant upon analogy for its conception. Her conviction of this is entirely consonant with her persuasion that the anatomical content of the poem represents no breaking of the circle of correspondence, and is intellectually conservative. Her evaluation of the poem is primarily of its anatomical content: "As poetry," she says, "*The Purple Island* is negligible; as a handbook of Elizabethan physiology, it has its historical interest."[3] Baldwin also emphasizes the critical significance of the poem for the history of ideas rather than literary accomplishment.[4] By taking issue with Langdale's conclusion that the anatomy is predominantly scientifically progressive, Baldwin reads *The Purple Island* as "a near-perfect handbook to the opinions of a period in the history of thought fast on the wane when he wrote; for Fletcher's particular genius was not to mirror the intellectual turmoil of the new age (as it is reflected for example in the work of Donne, who wrote at precisely the same time) but to fix—at very nearly the last moment they could be fixed—the firmest convictions and attitudes of the Renaissance."[5]

Although there is much to commend Baldwin's interpretation of *The Purple Island* as conservatively didactic, there are serious problems underlying both his and Nicolson's assessments of the intellectual and figurative character of the poem. Firstly, despite the typically didactic ways whereby the world's repetition is repeated in the poem's analogies between Man and the geopolitical realm, and between Man and heaven, the threshold figure of the allegory is a metaphor. It selects, emphasizes, suppresses, organizes, and extends features of the principal subject, in a way that analogies rarely do, by implying that statements true of the subsidiary subject are also true of the principal subject, and are so not merely on account of conventionally prescribed similarities. The figure is creative and allegorically transforms the systematic metaphors and analogies of anatomy into the geographical and social features of the Isle of Man. The significant connotations of this metaphorical interaction are in part antecedent, though the allegory extends them by discovering or inventing previously unobserved similarities, thereby deepening and extending meaning in accordance with a notion of the world's intelligibility. But the connotations include those which are related to the historically incipient and ideologically emergent concepts of nationalism and individualism. Moreover, as chapter 2 explained, the literary work enhances the contextual effect of metaphor, giving rise to greater interaction between the corresponding sets of associated implications of the principal and subsidiary subjects, and to a ripple effect throughout Cantos 2 to 5, wherein the associated implications of the subsidiary subjects of metaphors are already changed in "literal" meaning before they interact with their principal subjects, because of the specific, literary character of the "Isle" in Fletcher's poem, that distinguishes it from any other isle, realm, polity, or nation. Nicolson and Baldwin are mistaken to think that because they are bound up with analogy, the anatomy and physiology of the poem are therefore entirely dependent upon analogy for their conception. The emergent political, anthropological, and literary implications of the Isle of Man figure cannot by definition be "the firmest convictions" of the Renaissance.

The digression into the anatomy of the schools and of medicine in chapter 1 revealed coherence between the analogies of the body as a microcosm of the terrestrial globe or the macrocosm and the conceptual metaphors of anatomy, thereby further integrating the elements of *The Purple Island*'s figurative design. The digression also provided depth to the impression that the poem is indebted to the metaphorical and analogical language of Hellenistic, humanist, and scholastic anatomy. The poem consistently applies the descriptive criteria of anatomy textbooks—the anatomical *accessus*—to the description of the island realm, and regulates personification in relation to observable physiological operations. We may deduce from this not only that Langdale's project of extracting the anatomy from the poem unchanged by allegory is probably implausible, but also that Nicolson and Baldwin may be mistaken to think that because the anatomy and physiology of the poem are bound up with analogy, they are therefore conservative in conception. Admittedly, the figure of the Isle of Man coalesces and unifies the conventional metaphorical concepts and analogies of anatomy in a normative, reinforcing way, but the concepts are held to be true by compilers of anatomy, like Helkiah Crooke, who were Fletcher's contemporaries. The contribution which this study makes to this debate is not complete yet, however. The chapters in part 1 were concerned with the mainstream of anatomical concepts, language, debate, and consensus, insofar as Fletcher appears to have drawn from it in writing the anatomical cantos. These chapters sought to demonstrate congruity where it exists between the conceptual figurality of Fletcher's poem and the anatomy to which the early seventeenth century was heir, and to establish the creativity of the Isle of Man figure. In so doing the chapters refrained from examination of anatomical texts which could not have been at any time accessible to Fletcher or similarly educated scholars and poets in England; but they did not seek to follow specific intertextual links between *The Purple Island* and anatomical texts except for comparative purposes, never to demonstrate provenance. By contrast, part 3 will explore the anatomical intertextuality of *The Purple Island,* with a view to answering the question posed by Baldwin's and Nicolson's assessments of the conceptual character of anatomy.

As part 1 of this study developed from analysis of the threshold figure to analysis of its extension throughout Cantos 2 to 5 of *The Purple Island,* the questions of what the figurative design of the work was for, and what were its intentional direction toward a world it creates and its reflective direction toward a self, the self of its implied reader, have become ever more pressing. Baldwin's reading does provide an explanation for *The Purple Island*'s epic aspirations and, when extended, is capable of accounting for what chapter 5 of this study appreciates as the analytical and aggregative, almost encyclopedic, survey of its subject matter. An adherent to the theory that the microcosm of Man is a secondary Scripture that reveals the Creator, Baldwin's Fletcher sees the didactic and declarative function of poetry as a supreme source of information concerning divine mysteries, or, as God's sacred skill in Man. Although the allegory of *The Purple Island* derives from Christian theology, as chapter 3 has demonstrated, it is rendered consistent with the anatomy in the poem through the identity of the principal themes of its narrative with the shared entailments of the two metaphorical conceptualizations of anatomy as vegetation and "building." These themes are creation, mortality, and salvation, and Renaissance anatomy has been found to have provided moral and theological argument from the body's design, demise, and redemption. Therefore chapter 3 sought to understand how the Christian assurance of things hoped for does not merely shape the eschatological narrative of *The Purple Island* but informs its poetics and its desired effect. The chapter suggested that Fletcher drew on theory of poetry as a speaking picture, as well as emblematics and a suprarealist view of language, to produce a representation of microchristus that has soteriological purpose in eloquently demonstrating the creative ingenuity of God and the saving love of Christ (who is analogically related to the king of the Isle). But he also sought to renew the nation and the persons with whom he shared it. By drawing on the abovementioned theories, as well as possibly an Augustinian relationship between language and incarnation, and by restoring the knot of analogical correspondence in the Isle of Man and adopting a method of dissection that purportedly imitates Nature's method of creation, Fletcher designed for his work a restorative and saving effect on his readers.

The questions of the purpose of the work for its readers and the intentional and reflective directions toward a world and a self that emerge from an analysis of the literary work's design may be considered by situating the work in the contexts of readers" expectations and reception. Whereas the context of expectations, which the reader of *The Purple Island* might reasonably be assumed to have brought to the work in the seventeenth century, has received some recent critical attention, the context of the work's reception has suffered a lack of similar attention. The elucidation of questions by reference to these two contexts dictates the direction and scope of part 2 of this study. For example, there is the question of the work's conservatism and innovation. It was noted in the conclusion of chapter 2 that a full answer to the question of how derivative or novel the allegorical extension of the Isle of Man figure is must include an intertextual study of sources, influence, and comparable literature. This part of the study continues the concern of part 1 with conceptualizations of the anatomy of the human body as a microcosm, as microchristus, as constructed by God, as natural, as mortal, and as redeemed. Unlike the former part, however, the present is concerned to discover where such conceptualizations occur not only in *The Purple Island* and in anatomy textbooks, but in historically contemporaneous literary texts. There is also a shift, in the present chapter, toward an intertextual examination of literary works that may have exercised a specifiable influence on the writing of the anatomical cantos of *The Purple Island,* and an examination of the written responses of its seventeenth-century readers, with a view to answering the question of whether *The Purple Island* was or could have been received as a conservatively didactic poem which seeks to fix the firmest convictions and attitudes of the Renaissance. Although our analysis of the figurative design suggests that the answer to this is unlikely to be an unequivocable "yes," the meaning of the design is ever completed by the reader.[6]

The Contemporary Attribution of Ingenuity

To argue, as I did in chapter 2, that literary metaphor is a cognitive creation of language is to imply that it is a mental event rather than only a linguistic feature. But the event of literary metaphor may only be distinguishable from all other discourse in that it occasions the creation of a potentially unlimited number of new concepts, which may nonetheless be limited by the context of the whole work.[7] All discourse occurs as an event; this is what distinguishes it, as "*parole,*" from "*langue,*" the code or system of language: "[D]iscourse, as an act, may be considered from the point of view of the 'content' of the propositional act (it predicates such and such characters of such and such things) or from the point of view of what Austin called the 'force' of the complete act of discourse (the 'speech act' in his terms): what is *said* of the subject is one thing; what I 'do' *in* saying that is another thing: . . . Hence the polarity between the locutionary act (the act *of* saying) and the illocutionary act (that which I do *in* saying). . . ."[8]

Chapters 1 to 3 of this study were concerned largely to elucidate possible reasons why the anatomical content of *The Purple Island* has the figurative and theological character it has. This is a concern that may be expressed in terms of two critical criteria: the motivations that might be implied by the work about the author's design; and the locutionary act of the work as discourse—what is said by the work and how it is said. Chapter 3 was also concerned with the soteriological mode and method of the poem, and hence with what it does or aims to do in saying what it says in the way that it says it. This might in Austin's terms be called the poem's illocutionary act, though problematically, however, this act requires interpretation and as an effect of discourse is subject to individual and historical variation. Paul de Man criticizes Austin's model of discourse for giving insufficient allowance for this kind of variation. The fault as de Man sees it is that the model moves readily and without difficulty from grammar to speech acts.[9] The effect of discourse would in Austin's model commit the speaker, because it is the effect implied by the grammatical form; but there are also effects of speech acts that are not implied or intended and do not commit the speaker. This is because the sign must be interpreted; it is not a thing, a referent, but a meaning derived from the thing by a process—representation—that is not (despite objectivist claims to the contrary) simply transparent or generative, that is, dependent on a univocal origin.[10] Even in suprarealist conceptions of language, words form a secondary system continuous with but marginally displaced from the signs in the things or objects themselves. By contrast, a more consistent link between sign and meaning operates within grammatical patterns.[11]

The evidence of widespread metaphoricity in the language of Renaissance anatomy suggests that this hermeneutic process of apprehension through representation is likely in part to be a metaphorical process, but that such effects as are produced but not obviously implied are more pronounced in creative and literary metaphors because they are more open to divergent interpretations. Indeed, de Man is especially concerned to indicate that the effect of what he calls rhetoric, which he defines as "the study [and occurrence] of tropes and figures," is reduced in Austin's model to the effect that may be inferred from grammatical patterns.[12] De Man protests: "Rhetoric radically suspends logic and opens up vertiginous possibilities of referential aberration."[13] In order to rectify Austin's theory and to account for figures and tropes, de Man introduces a further term to identify this nuance in the concept of the speech act. He calls that which I effectuate in saying what I say, the perlocutionary act.[14] In response to these concerns and distinctions, and in addition to continuing to interpret the locutionary and illocutionary acts of *The Purple Island,* this section of chapter 4 is concerned with what is common or variable in contemporary critical responses to the effect of reading *The Purple Island,* and hence with the historically situated perlocutionary act of the work.

An early part of chapter 2 argued that we may treat metaphor as a work in miniature, an index to the whole literary work, not as in Barthes's model, but as a paradigm for the explanation of the work, because the decisive moment of explication of a metaphor is that of the construction of the area of interaction between its principal and subsidiary subjects. As in metaphor, where this construction is the means by which all the words taken together make sense, so in a literary work, the integration of all the semantic lines of the work, including any specially constructed systems of meaning, provides the means of the explication of the work. The explication of the threshold figure of *The Purple Island*—the Isle of Man—is therefore the key or index to the explication of the host of systematically related metaphorical entailments which are unfolded over Cantos 2 to 5. As when we were concerned to elucidate the locutionary and illocutionary acts of *The Purple Island,* we must also therefore consider the relationship of the whole work to the threshold figure when we consider its perlocutionary act. This does not imply that a metaphorical statement is simply a paraphrasable propositional speech act;[15] rather the perlocutionary act of a metaphor can be seen as an implied invitation to the reader to complete the process of semantic generation of similarities which the metaphor as a linguistic feature of the text prompts and the work begins to unravel.[16] The structure of metaphor, that is, its mode of functioning according to which its component subjects, their relations, and the metaphor itself, function in larger contexts, is examined from here on in this study neither as ornament nor as enclosed in the work, but as "something that contributes to the intelligibility of the world" in which the reader is situated.[17]

Although they cannot be regarded as representative of all contemporary responses to *The Purple Island,* the prose epistle to the readers and the commendatory verses, which originally prefaced probably all copies of the first edition, are significant seventeenth-century responses.[18] They are significant because there are few documented responses to Fletcher's poem, and because they tend to converge on certain points of interpretation. It was said in chapter 3 that Fletcher's poetic narrator, Thirsil, feels compelled to revive the perfect order of Creation and to help bring about salvation through the metrical rhythm and order of poetry; and that in Canto 6, stanzas 18 to 19, he attributes to Giles Fletcher's *Christs Victorie, and Triumph* a degree of success in having "calm'd" the wrath of "heav'ns dread King" (*P.I.* 6.19.2; 6.14.4). In that section it was argued that this was also an element of Fletcher's purpose. The enigmatic "A.C.," who may be the precocious Abraham Cowley, in a brief panegyric, judges the soteriological purpose or principle of the poem, the recreation and revelation of God's "much defac'd" art (*P.I.* 2.3.3), to have been successfully achieved:

> Henceforth let all the world thy verse admire
> Before that *Thracean Orpheus* charming lyre:
> He but enchanted *Beasts,* but thy divine
> And higher aires bring *Deities* to this *Isle* of thine.[19]

As with Orpheus, the way of singing, or in Fletcher's case writing, enchants. In Canto 5, Fletcher's narrator Thirsil refers directly to the myth of how "*Orpheus* wanne his lost *Eurydice*" (*P.I.* 5.61.1); the reference no doubt prompting A.C.'s verse. The occasion for Thirsil's digression into the myth is his description of "*Lingua,*" the "Tongue" (*P.I.* 5.56.1; 5.58 margin. c), from which it becomes clear A.C.'s boast is derived: "She heav'n to earth in musick often brings, / And earth to heaven" (*P.I.* 5.60.5–6). What Fletcher in the voice of Thirsil claims for the tongue, A.C. claims for Fletcher's poem.

If we are more specific about A.C.'s claim, relating it to the soteriological purposes or illocutionary act of the anatomical speaking picture of *The Purple Island* as discerned in chapter 3, the "divine aires," the song or aesthetic of Fletcher's poem, are in some measure identifiable as each of the following: the poetic mode of analogy, metaphor, and allegory, and the anatomical method, which in addition to a sequence of dissection involves a combination of examination by the senses, conceptual metaphor and analogy, rational inference, and revelation. The analysis of Canto 4, stanza 9, has suggested that in the scheme of the poem the incarnation "turns" the heart, the seat of the soul, and by extension the whole anatomical "picture," "to heav'n from hell," restoring analogy and *imago dei.* The "speaking picture" only "breathes" because the poem and the body it signifies partake of the effect of the incarnation, death and resurrection of Jesus Christ. Yet, it is by means of the arts of poetic figurality and anatomical demonstration, that God's "thrones fair picture"—the plan in the mind of God the Architect—is revealed amid the "dissected wreathes" of the microcosm and microchristus. By dint of what the poem aims to reveal, there is a suggestion of an aspiration for the poem to reflect or resemble the historical incarnation of the Word in human flesh. The poem may also aspire in its perlocutionary act to replicate the efficacy of the Cosmic Christ, for the designed effect of the poem's anatomy, sung "in rich graces key," is that Fletcher, like his Lingua, "tunes poor natures strings" (*P.I.* 5.60.7). A caveat to the poem's success in this case might be that A.C. describes it literally only as accomplishing its purpose within the intratextual context of its own signification: Fletcher is described as having brought "*Deities* to this *Isle* of thine," not this isle (body/nation) of ours.

Among the prefatory prose epistle and commendatory verses, Langdale argues that none finds any novelty or unusualness in Fletcher's poem;[20] which might seem a strange remark given that two in particular attribute ingenuity either to *The Purple Island* or its author. It might occur to one that this attribution militates against Baldwin's interpretation, for in the vein of modern critics like Gosse, the attribution of ingenuity or cleverness seems to imply the poetic overcoming of an inherent incongruity between aesthetic or allegorical language and anatomical theme. It does not appear, at least initially, to describe a reinforcement of settled convictions which (because of the role of ideology to suppress contradictions, for example in *Coriolanus*) might be assumed to appear to those who hold them as coherent and congruous. The attribution of ingenuity by seventeenth-century critics, who might be expected to know if the poem secured or challenged settled convictions, is on the face of it at odds with Baldwin's view of the poem's anatomy as conservatively conceptualized through a framework of the microcosm, rather than responding to recent anatomical projects and dissections.

But it is unclear whether the critical attribution of ingenuity by Fletcher's readers has the same conceptual content as implied when used by modern critics like Gosse. And so this chapter must raise again the question of the congruity of anatomy and poetic figurality, not as in the previous chapters with respect to the mainstream of anatomical concepts, language, debate, and consensus, but with respect to those contemporary theories of poetry and figurality that might have had an influence upon *The Purple Island.* Should ingenuity have been intended by the writers of prefatory material to signify innovative overcoming of innate difficulty, however, the creative metaphoricity of the Isle of Man figure may have been what caught their attention. Part 1 established the significant connotations of the Isle of Man figure as derived in some part from the conceptual metaphors of anatomy textbooks, and from conventional analogies of the microcosm, yet also as emerging from the interaction of the principal subject with a specially created set of associated implications of the Isle, an interaction involving the creation of similarities, but also the clash of dissimilarities. Cantos 2 to 5 not only give expression to the part of the domain of the subsidiary subject that is usually used in the metaphor of Man as an isle or a realm, but also explore the previously unused part of the domain of the subsidiary subject, particularly that relating to incipient nationhood, and thereby highlight a system of newly created or invented similarities with previously unused parts of the principal subject. In order to explore whether this metaphorical creation of similarities out of a clash of dissimilarities is the overcoming of inherent difficulty inferred by Fletcher's earliest critics, we must examine the context in which each attribution of ingenuity was made.

Because the first of the two verses of the anonymous Latin poem prefacing the first edition of *The Purple Island* contains what may be a self-referential appellation, "the bashful poet," I think this poem may have been written by a poet younger than Fletcher; one who found encouragement in Fletcher's poetry and friendship with its author. The first two lines of the verse refer to the fact that Fletcher withheld his identity as author of the first edition of *The Purple Island,* providing instead only the initials "P.F." on the title page.

Ingeniose tuo ne libro supprime nomen;
  Ingenio Authorem deteget ille suo.
Nempe verecundo memini te scribere vati,
  Quod pulchrè ingenio quadrat, amice, tuo.[21]

[Do not cleverly withhold your name from your book;
It reveals its author by its quality.
Certainly, I remember that you write to the bashful poet
That which accords beautifully with your talent, friend.][22]

The *Oxford Latin Dictionary* supplies for *ingenium* three sets of potentially relevant meanings: mental powers, natural abilities, talent, intellect; literary or poetic talent, inspiration; cleverness, skill, ingenuity.[23] These are of course meanings based on classical literature of the late Roman Republic and early Empire, and needless to say, there may have been semantic shifts by the seventeenth century. However, the second and third of these sets seem to fit perfectly well the terms in lines 2 and 1 of the Latin verse, respectively, that are grammatically related to *ingenium;* "Ingenio" perhaps translating as skill, or even ingenuity, in addition to literary quality.

The second of the two commendatory poems to commend Fletcher for his ingenuity consists of verses in English by Francis Quarles, which represent Fletcher as Quarles's "*ingenious Rivall.*"[24] Sixteenth-century English meanings of ingenious and of ingenuity include the first of the three sets of Latin meanings, notably "[h]igh or distinguished intellectual capacity; genius, talent, quickness of wit."[25] Seventeenth-century meanings of ingenuity also include the "[c]apacity for invention or construction; skill or cleverness in contriving or making something," and, as an attribute of the thing made, "[s]kilfulness of contrivance or design."[26] This set of connotations especially fits the use of "ingenio," in line 2 of the Latin commendatory verse, to infer the author of *The Purple Island* from the quality of the work; and probably also the use of "ingenious" by Quarles, given that Fletcher is extolled for his invention by the poem, also by Quarles, entitled "To my deare friend, the Spencer of this age," which was printed in probably all of the copies of the first edition.

In the Renaissance, as was said in chapter 2, the term invention could also mean a "discovery," a "finding," the "faculty of discovery" but also the "thing found." And as was said in that chapter, in the seventeenth century invention could also be used in a case where the subject matter is made up by the poet using her or his *fantasia* or *ingenium* or *ingegno.* Invention could mean "something close to 'imagination,' 'wit,' and positively or pejoratively a 'technique' or 'artifice,'"[27] which may refer to the figurative design of *The Purple Island.* Although "generally *inventio* concerns only subject-matter, not words, which are the province of *elocutio* [eloquence],"[28] a most significant development in the poetic theory of invention for the creative figurality of the Isle of Man is described by Ullrich Langer: "In the seventeenth century we observe in some quarters an increased emphasis on the faculty of the *ingenium,* as the capacity to manipulate language in such as way that novel relationships are created between things of the world. This emphasis displaces the focus of invention from the finding of things, or subject-matter, to the finding of words, especially metaphors."[29] Although applied to the finding of subject matter, invention is also inextricably related to choice of words, and was related in chapter 2 to creative metaphor as a cognitive process of discovery of similarities. Invention is in accord with the way in which creative metaphor uses previously unused parts of the defining domain, thereby bringing to light previously unconsidered or unexplored implications of the metaphors of anatomy and the analogies of the microcosm. Thus far, then, it may be said that ingenuity, as an attribution of *The Purple Island* or its author, used with emphasis upon its meaning as poetic invention, may connote the inspired discovery of *The Purple Island*'s anatomical subject matter as a vehicle for the execution of its author's purpose; the fanciful invention of an intellectually intriguing riddle of a location that is both the human body and the island realm of Britain; the imaginative contrivance of the poem's figurative design and the talent displayed by the witty, skillfully constructed and cleverly sustained allegorical artifice whereby the subject matter is presented.[30] But which meanings of ingenuity were pertinent for Fletcher's seventeenth-century readers may be deduced only uncertainly from the commendatory poems.

Fletcher is extolled for his ingenuity by other seventeenth-century readers.[31] As the introduction to this study indicated, James Howell, author of *Epistolæ Ho-Elianæ: Familiar Letters, Domestic and Forren,* wrote from Fleet Prison to Edward Benlowes, in 1645, extolling the qualities of "that delicate *purple Island,*" which provided him with "a whole Isle to walk in" imaginatively, thereby gesturing to the invention's skillful construction and cleverly sustained allegory. Howell draws attention to the poem's literary quality when he purports to "meet with *Apollo* himself and all his daughters," and to the way in which in the hermeneutic circle, the intriguing riddle of the figure is at the root of the work's intentional direction toward a world its reader apprehends, and its reflective direction toward the reader's self, when he acknowledges: "I stumble also ther often upon my self, and grow better acquainted with what I have within me and without me." For all these reasons he professes to consider Phineas Fletcher an "Ingenious Authour." Similarly William Winstanley's *The Lives of the Most Famous English Poets* describes Fletcher as: "This learned person, Son and Brother to two ingenious Poets, himself the third, not second to either . . . *Phineas Fletcher* . . . in Poetick fame exceeded his two Brothers, in that never enough to be celebrated Poem, entitled, *The Purple Island.*"[32] As with some of the other praise, this by Winstanley appears to be indeterminate in its use of "ingenuity." But Winstanley's attribution of ingenuity to both Giles Fletchers, elder and younger, may militate against certain specific readings of *The Purple Island*'s ingenuity, because neither wrote on an anatomical or natural philosophical subject. It is entirely possible that Winstanley's commendation of the ingenuity of these authors regards their literary or poetic talent and inventiveness rather than the clever overcoming of incongruity. But it is equally possible, and seems probable from Winstanley's *Lives,* that, for reasons that remain unclear in that text, *The Purple Island* presents for some readers an extraordinarily praiseworthy example of this inventiveness.

If the attribution of ingenuity to Phineas Fletcher and *The Purple Island* does imply the characteristics outlined above, it need not suggest that the author was unusual in being, or attempting to be, ingenious. To add another twist to the difficulty of establishing how unusual or innovative the poem was thought to be, we may turn to Grundy who says that, "ingenuity—*argutezza*—was, as we know, the aim of much seventeenth-century verse, especially of the kind designated 'baroque.'"[33] *Argutezza* or *acutezze* refers in literary criticism to the invention of new antitheses, paradoxes, witty metaphorical expressions, and conceits.[34] The attribution of ingenuity in this literary context posits a poem or metaphor as typical in mode; but typicality of poetic mode does not diminish the ingenuity of a particular figure. Those Renaissance and Baroque theorists like Emmanuele Tesauro who tried to theorize the conceit were endeavoring to distinguish *argutezza* or *ingegno,* and the poetic metaphor to which it gave rise, as "a non-discursive type of human understanding that discovers similitudes and creates metaphors," from ratiocinative intellect and from ornamentation, that is, from the *via philosophica* and the *via rhetorica.*[35] "Conceits, then, are a product of intellectual ingenuity, a capacity for finding likenesses in the apparently unlike."[36]

According to Langer, "[p]oetic invention allows, for [Philip] Sidney, a mimetic perfecting of the real, a homage to divine creation."[37] But the apparent ambiguity of the meaning of *inventio,* as applied to the finding of subject matter and as inextricably related to choice of words in creative metaphor where it signifies a cognitive process of discovery of similarities, has raised the question of the ontological status of that which is imitated. In respect of the representation of actions in Renaissance drama, Dollimore characterizes as mutually incompatible the explanations provided by Renaissance theorists for the rhetorical concept of mimesis. Dollimore notes the apparent inconsistencies between Sidney's justification of the didacticism of poetry, "to imitate to teach and delight," asking, "what, finally, is the ontological status of that which is imitated?"[38] On the one hand, Sidney describes the poet's "Idea or fore-conceit" of the work in ways that suggest God the artificer's plan of creation, and raises the poet as "maker" and possessor of a divine gift above Nature.[39] On the other hand, Sidney also says that the poet ranges "only within the zodiac of his own wit"; he "nothing affirms and therefore never lieth." Dollimore's analysis proceeds as follows: "In the context of Christian theology, morality depends ultimately on a metaphysical sanction for its prescriptive force; if it is accepted that what is being apprehended (and imitated) is a metaphysical ideal with real ontological status, then the prescriptive force of poetry is considerable; conversely, if the object of imitation is ideal in a fictive sense only, it cannot thus prescribe."[40] With regard to conceits that "invent" or apprehend some part of the ontology of the human body, or any other natural thing, as distinct from the political character and consequences of human action, the dichotomization is misleading, and would have appeared to Fletcher and many of his contemporaries as rather "puritanical" in respect of the imagination. Tesauro conceived *ingegno* as the equivalent faculty in Man to God's creative power, creating being in the poem where there was no being before,[41] just as the ingenuity of the threshold figure of *The Purple Island* necessitates cognitive processing and provides the whole work with its reason for being—to unfold a system of metaphorical entailments both in the allegorical poetry and in the minds of its readers. Creation and invention, here, are not logically distinct from discovery of what may already exist prior to its revelation, having been created purportedly by God. Fletcher's poetry is designed to unfold the mysteries of Creation, but the mysteries in part lay in the process of poetic creation of a work for which, it may be implied, a soteriological illocutionary act is intended; though, of course, a world that *has been created* definitively in the past only then to lapse and decay, such as that in which many Puritans believed, might be one in which fallen *human* creativity has no affinity with divine Creation.[42] But Fletcher does not appear to have believed in such a world devoid of correspondence between the creativity of Man and of God.

Since the creative metaphor requires interpretation, and since its structure and function can only be understood with reference to the perlocutionary act of the work in the larger hermeneutic context, "conceit" is not confined only to literary creation. The illustrator of *Poly-Olbion* explains his role as having illustrated that which the verse "in winding steps of Personating Fictions . . . so infolds, that suddaine conceipt cannot abstract a Forme of the clothed Truth."[43] Conceit and ingenuity could mean imagination,[44] and in this context denote the cognitive process required to apprehend and interpret that which the work disguises and reveals in allegory, fiction, or figure.[45] Howell's response to reading *The Purple Island* seems to evince a degree of gratified imaginative rationality in discovering the Isle himself.

Modern critics of seventeenth-century poetry sometimes seem to take their cue from Samuel Johnson (1709–84), who interpreted the wit of the poets whom he called metaphysical as unnatural and unjust, because, as he perceived it, it was "concerned less with the object itself than with the ingenious things [the poet] can find to say about it, with the relationships and analogies he can establish between it and things, ideas or experiences that would have been very unlikely to occur to anyone else."[46] Johnson contrasted "metaphysical wit" with a "conception of wit as that 'which is at once natural and new, that which, though not obvious, is, upon its first production, acknowledged to be just.'"[47] Between the early seventeenth and the late eighteenth centuries, the appreciation of the justice or naturalness of an analogy, just as the evaluation of its taste, had altered immensely. This was in part due to a considerable alteration in the concepts of poetry and nature, and hence of objects, their relations to each other, to language, and to the speaking, writing, and reading subject. In the early seventeenth century the concepts of natural objects, the human body included, were still substantially structured by a residue of (in some cases rejuvenated) interwoven figurative processes, including the drawing of extensive analogies between objects.[48] This residue of metaphorical processes was rendered coherent to early modern people by the Neoplatonic and Christian dignifying of mortal, concrete, earthly things, like human anatomical structure and function, by dint of comparatively abstract ideas about its nobility as created, as provided for by God, and finally perhaps as saved. Creative metaphor or conceit and ingenuity sought to involve individual minds more actively in the cognitive process of interpretation in order more deeply to understand the world thus constituted. When the interrelationship of these figurative processes is insufficiently appreciated by critics of seventeenth-century poetry, it is possible that figures, analogies, and correspondences, once firmly established but later eliminated by Baconianism, Cartesianism, and, later, neoclassical taste, are misunderstood as those that pushed at the forefront of conventional apprehensions, each appearing alike as lacking the natural justice that Johnson appreciated in poetic metaphors.[49] Nicolson has clearly made the point that modern critics have often mistaken seventeenth-century poetic images and metaphors, formed of commonplace Renaissance analogies or correspondences, for conceits, "produced, as Johnson believed, 'by a voluntary deviation from nature in pursuit of something new and strange.'"[50] At best such deviations appear as the curious "perception of occult [and hence, to the scientifically enlightened eighteenth century, unnatural] resemblances between things apparently unlike"; at worst, as the egotistic or forceful yoking together of heterogeneous material by the operation of the poet's mind.[51] Some modernist critics have reversed the Johnsonian evaluation, and exalted as the sign of true poetic individuality this perceived fusion of discordant or disparate elements into an aesthetic unity. Although their evaluation contrasts with Johnson's, in their interpretation of "metaphysical wit" there is no essential difference from his.

It is a historical error to consider it willfully individualistic or distasteful of a poet to "invent" the human body wittily and with considerable skill, in the imaginative and intellectual climate of the early seventeenth century, as an island realm.[52] Alexander Grosart is right to say that, in Sanford's judgment that the poem's plan is ill-conceived, and its theme distasteful, he, as well as Gosse, Kastor, and Henry Headley, "miss the deeper principle that lay in the Poet's mind when he made his selection of subject and the plan for his Poem, whereby no part of the wondrous human Body was to his eye and all-purpling Imagination without dignity or without insignia of creative wisdom, and power, and graciousness, and so none unworthy to be framed to song. . . ."[53] To the extent that the abovementioned critics perceive the ugliness or "deformity" of *The Purple Island*, and especially of Cantos 2 to 5, to derive from the anatomical knowledge and the persistence with which the theme is explored or "limme[d] forth," they miss the principle that lies in the fabric of the poem itself;[54] the principle shared by contemporary anatomy, that "nothing exists in nature that is not admirable."[55] The notion that "every minutest part of the Body had its wonder and interest, nothing ignoble, nothing 'unclean,'" because of its divine authorship and redemption,[56] survived in aesthetics in only a vestigial way after the seventeenth century, but motivates Thirsil's description of even "the vilest part," which, as was noted in chapter 3, was still recognized as having been made with "wondrous art" (*P.I.* 2.26.6–7). Admittedly, Thirsil is given a voice that is conscious of his "low verse" and he intimates that from the perspective of "Sacred *Thespio,*" the "holy Virgin" Mary, and the "Sunne" (Son) of God, the "vilest earth" is "low"; but these whom Thirsil implores for grace and inspiration all "deign'dst to dwell with mortals here beneath" (*P.I.* 6.18.3; 6.25.1,4,5,6; 4.9.2,5).

The distaste of Sanford, Gosse, Kastor, and Headley is the result of a historical incipience in the eighteenth century of taste as a literary and cultural critical guide to what was generally agreed to be good and what bad. It in no way reflects the "taste" of the 1630s. Seventeenth-century critics omit all judgment based upon any perceived distastefulness or inappropriateness of *The Purple Island*'s chosen theme. Disparaging judgments emerged not before the eighteenth century, when neoclassical canons of taste "found little of worth in the 'former age,'" and Fletcher was so neglected as to have been described as "hardly known."[57] Taste is wholly inappropriate as a category of critical evaluation for seventeenth-century poetry which displays excitement about the intellectual, soteriological, and poetic possibilities of an anatomy or allegory of the human body. The cultivation of taste attenuated poetic discourse in terms of its language, its cultural reference, and its applicability to the natural world, because it corresponded to a demarcation between scientific and literary discourses about the middle of the seventeenth century, each discourse affirming propriety over its respective facet of Renaissance Man. For even as evaluation of Fletcher's writings began to be once again more favorable from the late eighteenth century, *The Purple Island*'s perceived "gross deficiencies of judgment," "infelicity of subjects," and "false taste" condemned it.[58] Henry Headley, ordinarily an apologist for Fletcher, disparages the first five cantos, "in the course of which the reader forgets the poet, and is sickened with the anatomist. Such minute attention to this part of the subject was a material error in judgement. . . ."[59]

Although ingenuity as *argutezza* or *acutezze* signifies the invention of conceits, because of the interrelationship of figures constitutive of the world, *ingenium* also signifies for Fletcher's contemporaries the commendable achievement of "accuracy," or truthfulness of an accepted kind,[60] by means of intellect, skill, literary talent or wit, and invention. The achievement of truth is related closely to the first set of classical meanings of *ingenium*—mental powers, natural abilities, talent, and intellect. As such ingenuity is here considered an attribution that emphasizes the conservatism of *The Purple Island* as well as its innovation, even though it became closely associated in poetic theory with the interest in the conceit, the very object of Johnson's devaluation, though for reasons outlined immediately above, the aim of Johnsonians is not always true. On the face of it, the conceit seems innovative, rather than conservative, but wit or *ingegno* was not separated from "a traditional analogical habit of mind."[61] The conceit's function was to extend this habit, to discover hidden correspondences between things formerly believed unconnected, a purportedly higher and hitherto inaccessible unity, a reality that was the object as much of natural or occult philosophy as it was of poetry, for all were influenced through the Neoplatonic tradition, for example, by the Platonic doctrine of Ideal Forms.[62] This obscure unity was the world conceived as "a kind of divine and witty poem," the metaphors of which could be brought to light by a sufficiently insightful poet.[63] "The search for the 'far-fet' within this vast network of symbolic correspondences is probably what led [Johnson] to speak of the 'occult' when he describes the wit of the metaphysical poets as 'a kind of *discordia concors.*' "[64] And yet it was precisely as *discordia concors* that the universe provided a philosophical underpinning and mirror image not only for the Isle of Man figure, but for the general concept of Man in the Renaissance. And so perhaps in *ingegno* and the poetic of *concettismo,* or, as Mazzeo calls it, the "poetic of correspondences," we have the best contemporary literary-theoretical approximation to the Isle of Man figure and the whole work of *The Purple Island,* which aggregates and extends conventional knowledge, analogies, and metaphors, in a way that is entirely consistent with the way cultivated minds in early modern England may have run.[65]

The question of conservatism and innovation, or insightfulness, extends to erudition. Theorists often advised study in order to achieve "some depth of devise" in the invention and in the figures and narratives of poetic works.[66] Some seventeenth-century critics argued that "the more learned a poet the more likely his metaphors are to be conceits."[67] Yet the resemblances which *The Purple Island*'s allegory connotes are no less erudite, even exploratory, for being in some significant part conventional. The convergence of knowledge in the unfolding of the threshold figure involves, for example, some of the traditional strands of the theory of the microcosm. The Renaissance gave rise to what Barkan calls "the epistemological microcosm," which combines the chemical notion proper to the theory of the body as a microcosm—"man is made of the same elements as the cosmos"—with the idea proper to the Platonic and Neoplatonic figurative microcosm—knowledge of all is possible for the human mind because as the site of a *coincidentia oppositorum* it reflects or exemplifies God's infinity[68]—to produce a view that by knowing the cosmos "man . . . knows himself, makes of his brain the true microcosm."[69] Erudite correspondences in *The Purple Island* require the reader steadily to assemble in her or his mind a series of superimposed systems—anatomical microcosm, microchristus, macrocosm, and political geocosm—each of which becomes divided and subdivided as the poem proceeds to unfold the entailments, connotations, and ramifications of this purportedly universal knowledge. The ingenious aggregation of these corresponding systems of things and knowledges helps the reader to determine what her or his place in the world is to be, by establishing a world, where "world" means a space of meanings that assigns to things and to Man their proper places.[70] Several providers of prefatory material for *The Purple Island* responded to the poem's perceived erudition. Edward Benlowes commended *The Purple Island*'s "self-knowing art"; Francis Quarles did likewise the "well read lines [that] will teach . . . [future readers] how to be / The happie knowers of themselves."[71] Fletcher's "skill" is admired by "Lod. Roberts" (Lewis Roberts) for having "descri'd an *unknown World* at home," "which to search out, subdue, and till, / Is the best object of mans wit, strength, skill."[72] The brief panegyric, entitled "On the Most Accurate Poem, Inscribed *The Purple Island,*" is also an example of this evaluation. It claims that Fletcher's poem's erudition surpasses even that of the classical precedent for the metaphorical translation of the island realm's connotation of self-determination into volitional movement: "Henceforth let wandring *Delos* cease to boast / Herself the God of Learnings dearest coast."[73]

There was perhaps a consensus "that nothing absolutely new can be invented, that is, "conceived" or "imagined."[74] From this perspective, the ingenuity of Fletcher's conceit lies in its skill in combining in a single comparison—between island and body—those of the microcosm's conventional extensive analogues which focus on the body's constructedness, such as buildings, cities, and realms, with those that focus on its naturalness, such as the terrain of the geocosm. The accepted truthfulness of the analogy is not achieved by the novelty of Fletcher's particular choice of subsidiary subject for the body, but in its affirmation of the convention of the createdness and naturalness of both Man and the realm of Britain and through an eloquent and ingenious delineation in verse of correspondences already assumed implicitly to exist. Fletcher finds, but does not merely find, a new way to say old things, and the perlocutionary act of his work in this respect is to enjoin his readers as members of an erudite, literate, English-speaking, proto-nationalist, and staunchly Protestant (yet not puritanical) community, to complete the process of semantic generation in order to develop intimacy among members of the community.[75]

Evidence exists for this in the select persons among whom Fletcher may have circulated his manuscripts for a long period before Edward Benlowes encouraged the reluctant poet to take them to the press between 1627 and 1633.[76] Among his friends by 1633 were Francis Quarles, who may have met Fletcher at Cambridge before 1613, but certainly knew him in the 1620s; Izaak Walton, whose friendship followed Fletcher's ordination in 1615, and who, although not contributing to the prefatory material of *The Purple Island,* included seventeen lines from that poem in his *Compleat Angler;* and Quarles's friend Edward Benlowes, who in the 1620 decade ("about 1625" or "closer to 1630 than to 1620") came into Fletcher's life to become his dearest friend.[77]

However much the systematic division of the anatomical microcosm and geopolitical realm in *The Purple Island* demands a systematic correlation between analogues, Fletcher's conservative brand of ingenuity does not unfold the dangerous entailments of traditional metaphors to which John Donne's readers are exposed; *The Purple Island* holds "no snares for the timerous; no dangers for the bold," as the epistle of "Featly" to the readers maintains.[78] This is very much in line with Francis Quarles's aim in writing his *Divine Fancies,* where he says to the readers: "I heere present thee with a *Hive of Bees;* laden, some with *Wax,* and some with *Honey:* Feare not to approach; There are no *Waspes;* there are no *Hornets,* here."[79] The "strange" threshold figure or conceit of *The Purple Island* may arouse wonder, but not shock at the perverse or extravagant.[80] The intermittent rise and fall of anatomical and geopolitical signifieds occurs in *The Purple Island* in the context of its persistent pursuit of correlation; the logical end of which is not, as it is often in Donne, the moral shock ensuing upon the modification of a sacred term by its secular metaphorical vehicle. The end of Fletcher's ingenuity is the continuous signification of the Isle of Man as both body and island realm: an allegorical ideal of continuous superimposition of analogues, of opposites such as heaven and earth. *The Purple Island* seeks to recover God's art of anatomy and to help dispense the balm of salvation through allegory and metaphor. As with these, so with Quarles's epigrams, meditations, and observations: "There's none will sting thee, if thou strike not first: If any doe; she hath Honey in her *Bagge,* will cure thee too."[81]

Continuous superimposition, however, the poem does not achieve, despite the success with which the anatomy fully participates in the allegory's assemblage of superimposed analogous systems, and despite the encomiums it receives from Fletcher's friends and contemporaries. At moments in *The Purple Island*'s attempt to contrive an allegorical ideal of continuous superimposition, the assemblage of metaphorically compressed analogues teeters. Logically impossible narrative events are suggested by the ambiguous and multiple signification of the island. One of these is when Thirsil says that the "Serpent" (*P.I.* 1.54.1), who is Satan, "Vext with the glory of this happy Isle, / Allures it subt'ly from the peacefull shore" (*P.I.* 1.54.2–3). Sawday says that in these lines the allegory is inconsistent. He argues: "If the island, in a state of innocence, preserves its integrity through ignorance of any larger land-mass, what is the 'peaceful shore' from which it is lured by the serpent? It is as though Fletcher is confusing the island (transformed through the allegory into man) with the inhabitants of the island."[82] The logical meaning of the sentence in the poem refers to event 4 in the main story-line of the narrative, as analyzed in the introduction to the previous chapter. Event 4 simultaneously connotes two events, and requires the reader to interpret the meaning of "Isle" as metaphorical. Of the two events, the one is biblical—the temptation and Fall of Adam and Eve from grace and their subsequent exile from the Garden of Eden recounted in Genesis—and the other event is the sixteenth-century explorations of "forrain lands" by English sailors, merchants, and explorers lured from the "blessed ignorance" and "happinesse" of their existence in the isle (*P.I.* 1.51.6, 1; 1.52.1). The justification for this simultaneity of signification is that, as Adam and Eve are seduced into disobedience by Satan, so the realm's inhabitants also commit what the poem conceives as a sin of "curiosity," exploring "what e"re is in the continent" (*P.I.* 1.53.6, 3). The historicity of each episode is concealed by the poem's proto-nationalism and its past-oriented idealism, which equates Britain of the past with Eden.

Sawday's explanation of the inconsistency as he sees it is that "[t]he island is thus both a symbol of man in his unfallen state, and a garden of Eden which man exchanges for 'dead seas'" (see *P.I.* 1.54.5). But I venture to suggest that the figurality is more complex than symbolism. Using the most readily available model of metaphor for interpreting Isle in this case—the simple-analogy view of Aristotle—we may say that Adam and Eve are to Eden as the inhabitants of Britain are to the Isle of Britain, so the "Isle" of Eden and the Isle of Britain may supplant each other metaphorically. But in this case the transformation effected by the use of Isle in the syntax is metonymic and synecdochical; that is, the term used is that which names the geographical entity which is inhabited by the nation or realm, and it is used to stand collectively for the individuals who with that same geographical entity compose the nation or realm. This is not apparent on an analogical interpretation of the Isle of Man, which conspicuously fails to account for the transformations that occur.

By requiring synecdochical and metonymic transformations, this logically consistent reading avoids other possible readings, which while permitted by the allegory, when pursued in this context result in confusion or inconsistency. The shore is in fact not a greater land mass, but a synecdoche for the Isle. As such it signifies both a geopolitical location and Man, and in either case gives rise in the context of the stanza in question to a choice between either a contradictory statement or to a logically consistent synecdochical use of part for whole, such that, for example: Man (or part of Man) is lured from Man; or, the nation's inhabitants (or some of them), who are collectively the body of the geopolitical realm, are lured from their proper place in that realm which collectively they compose. As Man, the synecdoche "shore" is either both Adam and Eve or collectively and individually the inhabitants of the realm, whose anatomies the implied author, extradiegetic narrator, and diegetic narrator encourage the implied reader, narratee, and diegetic narratees, respectively, to return to and reexamine.[83] They are encouraged to examine that which is in fact their own self, or more specifically, in Cantos 2 to 5, the anatomies of their own bodies. Were Man and these bodies understood analogically, they would individually *reflect* the composition and division of the whole Isle. But as the principal subject of a creative metaphor they are *identified as* the whole Isle, which they are also (in a logical contradiction) represented as inhabiting. Since the Isle is used synecdochically and metonymically for the bodies of its inhabitants, it is logical to infer that Man may be a synecdoche and metonym not only for the dynastic realm or nation which comprises its inhabitants and the land, but also for that Isle or land of which Man is a part, and in which Man lives. The very metaphoricity of the Isle of Man figure gives rise to a tension which does not exist in Man's position *in* the Isle or in the Neoplatonic and microcosmic tradition of analogy, a tension which is avoided only by the transformation demanded by the context and the logical sense required of the text by the reader. If tested to extremes, then, the coherence of the allegory falters under the weight of its own ingenious and ideological artifice. Kastor could have mustered contradictions such as this in support of his argument that the poem's translation of anatomy into a combination of geographical jargon and aesthetic language results in incomprehensibility; although if he had chosen to do so, examples would have been few and far between.

I have thus far outlined both the innovativeness of the conceit and its deployment in Fletcher's conservative ingenuity. However, the conservative reading of the ingenuity and figurality of *The Purple Island* ought not to overlook a complicated feature of Fletcher's poetic practice that has the potential, if not the inevitable structural tendency, to run counter to his poetic design. In the sixteenth and early seventeenth century, "innovation," including "semantic innovation," "was not generally seen as desirable, even and especially by those who were most concerned to effect it." Innovation of any kind, including novel meaning in metaphorical creation, was often justified "by a discourse of purgation and return to an idealized prior state of things."[84] What *The Purple Island* implies about its author suggests that Fletcher was no different in his view. The semantic innovation of *The Purple Island* is inextricably related to just such a desire to arrest change, perceived as deterioration, not only in the language of poetry, but in the domains of Man, the realm, and the world "Which still grows worse by age" (*P.I.* 1.1.7).

Like the narrators created by other Spenserians, the diegetic narrator, Thirsil, complains about the lack of preferment of Edmund Spenser and the critical reception of his allegorical work, which as we shall see, profoundly influenced the figurative conception of *The Purple Island:* "Witnesse our **Colin;* whom though all the Graces, / And all the Muses nurst," "Yet all his hopes were crost, all suits deni'd; / Discourag'd, scorn'd, his writings vilifi'd" (*P.I.* 1.19.1–2, 5–6; margin. * indicates "*Spencer*"). Fletcher's innovation, which is a description of Man extended and transformed by creative metaphor, is by Thirsil justified as a rediscovery of what was "by seeking lost," "a strange, though native coast" (*P.I.* 1.34.5, 6). Fletcher's poem works to recover the knowledge of God's art of anatomy, against the flow of the narrative of degeneration since the Fall,[85] and does so by restoring the relationships of *aemulatio* and *convenientia* between God and creation, and by drawing the analogies of microchristus and microcosm. Furthermore, as chapter 3 showed, this restorative work is augmented by the order or method of dissection, which effects a kind of recreation of human anatomy, a return to Greek origins.

The very effort of shoring up these correspondences and parallels, and making them relevant to his contemporary readership, gives rise to the innovative aspect of the connotations of the figure of the Isle of Man, preserved from the deteriorating influence of otherness, and of the continent, by its individual, geographical, and national insularity. Fletcher imagined fallen yet redeemed anatomy as an emergent nation, yet this figure ironically helps to forge a language of nationalism and individuality which would in the history of discourse trace or help to bring about a collapse of such correspondences as *The Purple Island* endeavors to preserve. If we read the poem semiotically rather than semantically, we might conclude that the metaphorical sign undermines its pursued meaning.[86] But *The Purple Island* does not police its meaning as a philosophical text might. As chapter 2 has argued, the innovation is semantically generative as much because of a clash of meaning as because of a fit. It gives rise, for example, to the innovative physiological figuration of the person or nation-state's acquisition of knowledge or intelligence, the fact of which has a further irony, in that the individual or nation state's intelligence-gathering is a part of the apparatus of control, legitimation, or defense of meaning. By way of qualification, however, we should seriously consider the view that the clash of meaning was not experienced as great in the early seventeenth century. The emergent ideologies of English (or British) Protestant individualism and nationalism, which included the quasi-Paracelsian and alchemical notion of natural philosophy as individual revelation, were not inconsistent with the forging of correspondences.

One effect of understanding the contemporary attribution of ingenuity to Fletcher as the achievement of accepted truthfulness through creative figurality, is to reconcile two apparently contradictory responses to *The Purple Island:* as conservatively didactic and as strangely fanciful and inventive. While the definition helps to relieve the poem of some of the purported strangeness of its conception, which can now be seen as consistent with the poet's intellectual context, there is a further aspect to the purported strangeness. Most modern critical views agree with Kastor that, on account of Cantos 2 to 5, "*The Purple Island* qualifies easily as one of the most unusual poems in English."[87] Gosse regards it as a curiosity, ingenious yet grotesque, and like "none other in our language."[88] While it may have had an audience that was intellectually sympathetic to its plan, as the commendatory poems that preface *The Purple Island* and the responses of its seventeenth-century readers suggest, the view has been pervasive that *The Purple Island* did not have a historical context of comparable literature.

LITERARY SOURCES: THE *DEVINE WEEKES* AND *THE FAERIE QUEENE*

This chapter began by stating that the questions of the purpose and the intentional and reflective directions of *The Purple Island,* that emerge from an analysis of the work's design, may be inquired into by situating the work in the contexts of readers' expectations and reception. The first way in which this study considers the context of readers' expectations is to examine those literary works that exercised a specifiable influence on the writing of the anatomical cantos of *The Purple Island,* and may consequently have affected the expectations of *The Purple Island*'s readers and the intentions of its author. Despite the frequent critical characterizations of the poem as strange, it has long been acknowledged that the anatomical cantos in particular have two clearly identifiable literary sources. Historical intertextual studies of Cantos 1 to 5 of *The Purple Island* have tended to dwell on their reliance upon the second book of Spenser's *Faerie Queene,* and the *Devine Weekes* of the French Protestant poet Guillaume de Salluste, seigneur du Bartas (known as Du Bartas, 1544–90), as translated by Joshua Sylvester (1562/63–1618).[89] An intertextual entanglement exists here, which has been unraveled by A. H. Upham in the following way.[90] One of Spenser's sources for the House of Alma in Cantos 9 to 12 of Book II of *The Faerie Queene* ("The Legend of Sir Guyon; or, Of Temperaunce") was the anatomical description from the

Sixth Day of *La Semaine ou Création du Monde* (1578) by Du Bartas. Sylvester also translated *La Semaine,* and the subsequent *La Seconde Semaine* (1584), publishing them both in the collected edition entitled *Du Bartas His Devine Weekes and Workes;* but in doing so Sylvester imitated Spenser's "House of Alma." Having no French, Fletcher subsequently relied upon Sylvester's translation of Du Bartas, and Spenser's House of Alma.[91] The derivative passages in *The Purple Island,* according to Langdale, "regularly bear the marks of Sylvester, rather than of Du Bartas."[92] As to the respective weights of influence of Spenser and Sylvester upon the island-body of *The Purple Island,* it is generally agreed that Spenser was dominant.[93] Langdale details numerous textual echoes of Spenser's text in Cantos 1 to 5 of Fletcher's.[94]

This section will consider whether *The Purple Island*'s representation of the body in microcosmic, and (as an island) in geocosmic, terms was already firmly established by his own literary sources when Fletcher came to write, and whether the metaphoricity of the anatomical content of *The Purple Island* may be said to be more or less comprehensive and detailed than that of these sources. I will be looking at how the tropology of these works compares with *The Purple Island*'s allegorical transformation of the systematic metaphors and analogies of anatomy into the geographical and social features of the Isle of Man. As the previous chapters have shown, the implications of this metaphorical interaction are in part antecedent, though the allegory extends them by discovering or inventing previously unobserved similarities, and thereby deepening and extending truthfulness in accordance with a notion of the world's intelligibility. But the implications are also partly related to the historically incipient and ideologically emergent concepts of nationalism and individualism. This section thus performs the study's literary intertextual analysis, interpreting intertextuality as derivation and transformation, and accordingly seeks to know whether the threshold figure of the allegory of *The Purple Island* may be found elsewhere, in some shape or form, among the poem's source-texts, and to what extend their figurality and poetic are characterizable in terms of *ingegno* and the poetic of *concettismo.* The answer to this may affect our understanding of the poem's ingenuity, its innovation and conservatism, because in the context of an analysis of *The Purple Island*'s figurative design, the creative metaphor of the Isle of Man is said to select, emphasize, suppress, organize, and extend features of the principal subject, in a way that analogies and nonliterary metaphors rarely do, discovering unseen, emergent, or newly created similarities between the principal subject and the specially constructed set of associated implications of the subsidiary subject, which in many cases are not reducible or perhaps even traceable to conventionally prescribed similarities. This, as chapter 2 explained, is because the literary work enhances the contextual effect of metaphor, giving rise to greater interaction between the corresponding sets of associated implications of the principal and subsidiary subjects, and to a ripple effect throughout Cantos 2 to 5, wherein the associated implications of the subsidiary subjects of metaphors are already changed in "literal" meaning before they interact with their principal subjects. As was said above, this is because the specific, literary character of the "Isle" in Fletcher's poem distinguishes it from any other isle, realm, polity, or nation. But this interpretation may have to be qualified if the context of *The Purple Island*'s source-texts gives us reason to suppose that the threshold figure has a provenance elsewhere. Notwithstanding this, the comprehensiveness, systematicity, and coherence of the extension of meaning as it occurs in Cantos 2 to 5 of *The Purple Island* is to be compared with any similar metaphorical interaction and generation of meaning in the source texts.

As a further aspect of this unfolding of the Isle of Man figure, *The Purple Island* consistently applies the descriptive criteria of contemporary anatomy textbooks—the anatomical *accessus*—to the description of the island realm, and regulates personification in relation to observable physiological operations. We have deduced from this not only that Langdale's project of extracting the anatomy from the poem unchanged by allegory is probably implausible, but also that Nicolson and Baldwin are mistaken to think that because the anatomy and physiology of the poem are bound up with analogy, they are therefore conservative in conception. Both the allegorical figuration of anatomy and its fidelity to the order of knowledge in the schools and medical profession will be examined in each of Fletcher's literary sources, as will, as far as may be ascertained, what their figurative design was for, and what its intentional direction toward a world and its reflective direction toward a self. I will be gauging the extent to which the didactic, allegorical ways whereby the world's repetition and creation are repeated in *The Purple Island*'s anatomical sequence and analogies, may be said to have been derived from the works of Du Bartas and Spenser; and the extent to which a purpose shapes the narrative and informs the poetics in each case, and how this purpose relates to the soteriological purpose of Fletcher's poem. A dimension of this inquiry considered by the first section of this chapter in relation to *The Purple Island* will be considered here in relation to the *Devine Weekes* and *The Faerie Queene* insofar as Fletcher was obviously a reader of both works, and interpreted their locutionary, illocutionary, and perlocutionary acts as he drew on them for his own work.

Baldwin rightly indicates that what Fletcher borrows from Spenser and Du Bartas, regarding the representation of anatomy in *The Purple Island,* is the metaphor of "the body as edifice" or dwelling-place of the soul.[95] He borrows a lot more, but this device is conceptually foundational for the representation of anatomy in each of the three texts. It is also this figural aspect of the intertextuality which bears directly upon questions of the unusualness, ingenuity, and congruity of Cantos 2 to 5 of *The Purple Island.* Langdale observes that, in the commendatory verses which preface the first edition of *The Purple Island,* Quarles also proclaims to Fletcher that his own "Genius *jumpt with thine*"; and an untitled poem by Quarles, which also prefaces the first edition of *The Purple Island,* begins: "Mans Bodie's like a house."[96] Although it is not clear whether the writing was independent of the publication of *The Purple Island*—Quarles's commendatory poem was also printed a year earlier in the author's own *Divine Fancies*— the simile nevertheless mitigates any impression that in respect of the body as "building" metaphor, Fletcher's conception was wholly unique for poetry, even though its analogues for the constructedness of human anatomy extend beyond the simple home.

On "The Sixt Daie of the First Week" of Creation, in *Du Bartas His Devine Weekes,* newly created "Man" is referred to as "this Cittadel," whose "Fleshy Walls" are supported by "the Bones, which be the Pillars strong, / The beams and Rafters."[97] The poem therein appeals explicitly to the concept of Man and his body as created, and finds the metaphor of anatomy as "building" as efficacious as does *The Purple Island,* with a stronger emphasis upon carpentry and architecture. However, because of the lack of systematicity in the application of "building" to anatomy in Du Bartas's text, the tenor or principal subject of this metaphor is always and obviously the body. The effect is unlike the simultaneity of signification achieved by Fletcher's threshold figure. Although, in Fletcher's poem, the metaphorical term or focal word is almost invariably Island, Isle, or some metonym or synecdoche for these, the principal and subsidiary subjects act systematically as a metaphor or analogue for the other, having been compressed and transformed into a single image. Fletcher's poetic body is also more highly developed in extending the concept of its constructedness beyond the building, to the human-built, and the sociopolitical, environments. The *Devine Weekes,* however, does extend the "building" metaphor to include more than one object of construction, and makes one political point, evincing autocratic prejudice, and leaning for support, as does *The Purple Island,* on the hierarchy of organs in Galenic anatomy:

> [God] plantedst th'Intellectual Powr
> In th'highest stage of all this stately Bowr,
> That thence it might (as from a Cittadell)
> Command the members that too-oft rebell
> Against his Rule: . . .
>
> (*D.W.*, p. 158)

The Purple Island similarly rationalizes the form and location of the head (*P.I.* 5.4), but stands in the tradition of using the analogy of anatomy to warn against tyranny. Among the "members" of the head's body in the *Devine Weekes* are the "factious Commons, that still strive to gain / The golden Sceptre from their Soverain" (*D.W.*, p. 158). Personification occurs in the *Devine Weekes* only in the representation of digestion, and in this brief moral and political allegory in which "th'Intellectual Powr" of the head, or "Reason," and the seven deadly sins (who comprise a portion of the "factious Commons" or the body's rebellious members), bear some resemblance to allegorical figures in Cantos 6 to 12 of *The Purple Island* (*D.W.*, pp. 163, 158). The anatomical and physiological operations, which take various socially integrated, personified forms in *The Purple Island,* are entirely absent from the *Devine Weekes.*

This political lesson of the *Devine Weekes* is drawn from a comparison of Man and beasts, which distinguishes the body of Man literally as *imago dei:*

> Yet, not his face down to the Earth-ward bending
> (Like Beasts that but regard their belly, ending
> For ever all) but toward th'azure Skyes
> Bright golden Lamps lifting his louely Eyes;
> That through their nerves, his better part might look
> Still to that Place from whence her birth she took.
>
> (*D.W.*, p. 158)

"She" is specifically the animal or rational faculty of the soul, which is not only contemplative but in the brain is "The Treasurer of Arts, the Source of Sense" (*D.W.*, p. 162). The extent "of mans resemblance to his first Paterne, which is God" (*D.W.*, p. 169 margin.) extends from Man's upright posture to human ingenuity, exemplified by the conception and construction of mechanical contrivances made to resemble "flying Creatures (Artificial Fowls)" (*D.W.*, p. 167), the eagle and fly of John de Monte-Regio (*D.W.*, pp. 167–68), the "Engines of Archimedes," and "Admirable Dialls & Clocks" (*D.W.*, pp. 168–69). Invention characterizes the creative power of God "who the starry Sphears / Canst make to moue" (*D.W.*, p. 169); and so:

> This curious Lust [in Man] to imitate the best
> And fairest Works of the Almightiest,
> By rare effects bears record of thy Linage
> And high descent; and that his sacred Image
> Was in thy Soule ingrav'n, when first his Spirit
> (The spring of life) did in thy limms inspire-it.
>
> (*D.W.*, p. 170)

The word "conceit" was sometimes used in the sixteenth and early seventeenth centuries for an invention which might be a contrivance, device, or instrument of technology, such as wit or ingenuity gives rise to.[98] But invention is not confined by Du Bartas to the engineer or artificer. In fact the *Devine Weekes* draws closer to Fletcher's conception of invention, and to the Neoplatonic tradition, when it celebrates the swiftness of the "Soule: comprehending all things in Heav[en] and Earth" (*D.W.*, p. 166 margin.). This divinely imitative aspect of the tradition of *imago dei,* found in the character of ingenuity and invention, is not particularly explored by Thirsil in his anatomical description in *The Purple Island,* but it is central to understanding the metaphorical invention of the poem's author. In the remainder of this chapter and in the next, we will encounter other cases such as this wherein Fletcher's poetic method or mode carries out in practice what other poets describe discursively and theoretically.

Anatomy as natural terrain is evoked by both Du Bartas's and Fletcher's poems, and similarities exist between some of their descriptions. In the *Devine Weekes,* the liver is likened in a simile to:

> a fountain, whose divided Water
> It self dispersing into hundred Brooks,
> Bathes som fair Garden with her winding Crooks.
>
> (*D.W.*, p. 163)

The system of venous distribution in *The Purple Island* is conceived similarly as "thousand brooks," "Their serpent windings, and deceiving crooks / Circling about, and wat'ring all the plain" (*P.I.* 2.9.2, 4–5); and the liver in both works is figuratively a fountain (*P.I.* 3.7.5–7 cited above). Both metaphorical conceptualizations of the body—as constructed and as natural—coalesce in the *Devine Weekes* in much the same way as in *The Purple Island,* giving rise to human-built conduits: "the Blood (bred of good nourishment) / By divers Pipes to all the body sent" (*D.W.*, p. 163).[99] Unlike in *The Purple Island,* however, there is in the *Devine Weekes* no consistent boundaried "natural" analogue or subsidiary subject for the body like the Isle; consequently, as with the anatomy as "building" metaphor, there is no systematic unfolding of possible correlations and similarities between each term of comparison, either as the whole or the parts of which each is composed. Although the body in the *Devine Weekes* is highly figurative, it is loosely and inconsistently conceptualized in terms of its metaphorical vehicles or subsidiary subjects.

Differences and similarities in the microcosmic conception of the body in each poem extend to the descriptive criteria or *occasiones* of the anatomical *accessus*. In complexion, "the glorious goodly Frame of MAN" is said, in the *Devine Weekes*, to be composed of:

> Fire, Aire, Earth and Sea;
> Man's (in a word) the World's Epitome
> Or little Map: which heer my Muse doth try
> By the grand Pattern to exemplifie.
>
> (*D.W.*, p. 155)

Here, perhaps more than in *The Purple Island*, the complexion of the microcosm is felt to reproduce the elements of the geocosm, giving rise to the metaphor of Man as a "Map." But God, described as the "most peer-less learned Imager," when "Life to his lovely Picture" He conferred,

> Did not extract out of the Elements
> A certain secret Chymik Quint-essence:
> But, breathing, sent as from the lively Spring
> Of his Divinenss [*sic*] som small Riverling,
> Itself dispersing into every pipe
> Of the frail Engin of this earthen Type.
>
> (*D.W.*, p. 164)

The "Pattern" in the *Devine Weekes* is didactic poetry, wherein the anatomical content exemplifies, or copies, what the actual body is understood to be, an epitome. This epitome resembles that in *The Purple Island*, in being, not merely an elemental microcosm, but the "divinest Master-Piece of Art," and God its "Admired Artist, Architect divine" (*D.W.*, pp. 156, 157).[100] The express purpose of the "The Sixt Daie of the First Week" of the *Devine Weekes* of Du Bartas is "To paint" the "Image" of God's living "lovely Picture," so "That grave Spectators, in his face may spy / Apparent marks of [God's] Divinity" (*D.W.*, pp. 157, 164, 157). In this poetic design and in its religious didacticism, the *Devine Weekes* closely resembles *The Purple Island*, in which the face is an index not only to the mind but to the whole of Man, who in turn reflects the Isle and the world or "All" which is God or is contained by God. Moreover, the author of the *Devine Weekes* is therein imitating the inventiveness of God by trying to accomplish in poetry what is ascribed to the creative power of the Soul in the analogous activity of the painter when he or she fashions paintings that are taken for the real things they represent in paint (*D.W.*, p. 167).

In the application of complexion and other *occasiones* to the metaphorically translated body, the *Devine Weekes* is however far from being as systematic as *The Purple Island*. It fails not only to express the range of correlations suggested by the metaphors it employs, but also to detail anatomically those it does express. The form, location, posture, and operation of the body and its parts are only intermittently involved. Complexion is used, albeit only twice, in the instance quoted above and in the description of "The brains cold moisture" (*D.W.*, p. 162), and is preferred to composition which is almost entirely absent as a criterion for describing the material qualities of the body. Composition enters the description of the brain, which is the most anatomically detailed of any description in the poem, but only to compare the softness of the brain to the "hardnes" of "its boany Armor," the cranium (*D.W.*, p. 162). In this preference for complexion over composition Du Bartas retains the body's microcosmic qualities to the detriment of its empirical, structural ones. Unsurprisingly perhaps, the anatomical description of the poem is curtailed when the poetic voice interjects, to defer the work of pulling "in peeces all the Human Frame" to "Those skillful sons of Aesculapius" (*D.W.*, p. 163). But the *Devine Weekes* is also less developed than *The Purple Island* in specifically literary ways: in allegorical complexity and in personification. The anatomico-poetic unfolding of the body and its metaphorical vehicles in the *Devine Weekes* is far less systematic than that of the body and its geocosmic and political analogues in *The Purple Island.* Moreover, the *Devine Weekes* does not invent a creative metaphor such as the Isle of Man, and consequently espouses a more conservative political ideology than *The Purple Island.* To interpret the *Devine Weekes* as conservatively didactic is more appropriate than to do the same to Fletcher's poem.

While claims are made for the influence of Cantos 9 to 12 of Book II of Spenser's *Faerie Queene* on Cantos 1 to 5 of *The Purple Island,* the most tangible of these intertextual links exists between Canto 9 of the former and Cantos 1, 2, and 5 of the latter.[101] In copying Book II, Canto 9 of *The Faerie Queene,* Fletcher was indirectly reading the anatomical poetry of *La Première Semaine,* but two aspects of the figurality to some extent shared by these three texts place *The Purple Island* in proximity with each of these two literary sources in turn. Unlike Du Bartas, Fletcher chose the genre of allegory to represent human anatomy, and his choice is also that of Spenser who incorporated anatomy into the "continued Allegory, or darke conceit" of *The Faerie Queene.*[102] Secondly, Spenser's House of Alma is conceptualized almost entirely in terms of only one of the two structural metaphors employed by both Du Bartas and Fletcher: anatomy as "building."[103] *The Faerie Queene*'s omission of neural and vascular anatomy is perhaps symptomatic of this, for although a system of conduits might have sufficed, the metaphorical representation of neural and vascular anatomy in both the other poems, and in anatomy textbooks, is almost entirely in configurations either of terrestrial waterways or vegetative growth, or, occasionally (in textbooks), in terms of asexual reproduction in plants, as was shown in chapter 1.[104] Having said this, Book II is the most landscaped part of *The Faerie Queene,* and, as James Nohrnberg argues, its landscapes are "the most explicitly allegorical; they are very much the *objectum* of the temperament confronting them."[105] In this way what might be called the "internal environment of temperance" is reflected in or signified by the external and physical environment of the landscape.[106] Reading the allegory thus allegorically, the *sensus litteralis,* or surface narrative, can be seen partially to mask, but also to reveal and give way to the *sensus tropologicus,* the moral abstraction deriving from that narrative. The *sensus litteralis* and the *sensus tropologicus* are two further levels of allegory proposed by Danté Alighieri's *Divine Comedy* as a fourfold method of reading the Bible, the other two having been discussed in the previous chapter. Danté called the former the "letter," and in metaphorological terms this refers to the subsidiary subject of the extended metaphor or threshold figure.[107] The metaphor's principal subject in the relevant part of *The Faerie Queene* is the morally conceived anatomy of Man, the internal environment of temperance.

The body's analogue or subsidiary subject in Spenser's text is quite literally a building—"a goodly castle" (*F.Q.* II.9.10.3). Aspects of the architecture of anatomy in the *Devine Weekes* also feature fortification: the rational soul is seated in "a Cittadell" (*D.W.*, p. 158). By comparison with *The Purple Island,* however, the imagery of fortification in the *Devine Weekes* and in *The Faerie Queene* is conventional in a number of ways. The "Turret" or head of Alma's castle, "lifted high aboue this earthly masse, / Which it suruew'd, as hills doen lower ground" (*F.Q.* II.9.45.1, 3–4), bears considerable similarity to "The highest region in [the] little Isle" (*P.I.* 5.2.1). Both representations are in the tradition of *imago dei,* in which the head is "Like highest heauen compassed around" (*F.Q.* II.9.45.2). But the purpose of the turret of Alma's castle to "suruew" or survey the "earthly masse" is either contemplative or proprietorial.[108] Man is the natural philosopher of Creation or the descendent of Adam whose dominion Creation is. Neither meaning quite coincides with the function of the eyes or watchtowers in the Isle of Man, the instruments of surveillance.[109]

Because of the innovativeness of its emphasis upon national and individual defense, *The Purple Island* anatomy surpasses all comparison with either of its literary sources in the elaborate use of the imagery of defensive fortification, military operations, and installations of armaments. The poem's deployment of this vocabulary is also more consistent with the anatomical *accessus* than is its use in either Spenser's or Du Bartas's texts. Indeed the concept of the skeleton and myological structure as armature, which is systematic and pervasive in *The Purple Island,* is continuous with its use in anatomy textbooks. Paré remarks upon the situation of the heart "in the midst of the Chest": "It required that seat by the decree of Nature, because that region is the most safe and armed, and besides it is here on every side covered (as it were with the hands of the Lungs[ )]."[110] In a fertile figurative embellishment typical of the eloquence of mannerism, the second plate of Book III of Valverde's *Historia de la composicion del cuerpo humano,* published in 1556 at Counter-Reformation Rome, and reproduced in the Italian translation of 1560, the relatively unprotected abdomen of two figures sporting Roman military dress is dissected.[111] In each figure a series of incisions allows the peritoneum to be folded back to reveal as though through a rectangular window the stomach, duodenum, jejunum, ileum, cecum, and colon, parts of which are displayed in isolation at the figures" sides. The thorax of each figure meanwhile remains protected from the tools of the anatomist and artist by a breastplate, which, in this mannerist pictorial context of "the combination of real details in an imaginary framework," serves to give powerful expression to the concept of the chest as "most safe and armed."[112]

The Purple Island anatomy exceeds in variety of "building" and social metaphor any militarized description in anatomy textbooks, and extends the figuration to the comparatively vulnerable abdomen. Among the "common fences" which "round impile" the abdomen "and all the other Isle," is the skin (*P.I.* 2.20.5, 6), beneath which "The second rampier" is the "fat," which "outward enemies by yeelding most resisteth" (*P.I.* 2.19.1, 7 & margin. q). The third is "of more defence then all the rest"; this being the "fleshie pannicle" (*P.I.* 2.20.3 & margin. s). "Besides these three, 'two more appropriate guards / With constant watch compasse this government" of the abdomen (*P.I.* 2.21.1–2). The description in the margin of "the muscles of the belly-peece, which are eight, foure side-long, two right, and two crosse," becomes in the stanza that of "eight companies" distributed "in severall wards, / (To each his station in this regiment)" (*P.I.* 2.21.3–4). The other guard is the "Peritonæum," which the "lower region girts with strong defence" "with double-builded wall," "Admitting welcome friends, excluding enemies" (*P.I.* 2.22.2–3, 7 & margin. u). As with the espionage function of the sensory faculty of the optic nerves, to detect and convey information, there is an anatomical figuration of political processes in the function of the peritoneum to "Protect this coast from all conspiracies" (*P.I.* 2.22.6). Inland, the "Arch-citie *Hepar*" is "Fenc't with sure barres, and strongest situation; / So never fearing foreiners invasion" (*P.I.* 3.5.3, 5–6). Since the erection of these anatomical defenses in and around the abdominal organs depends

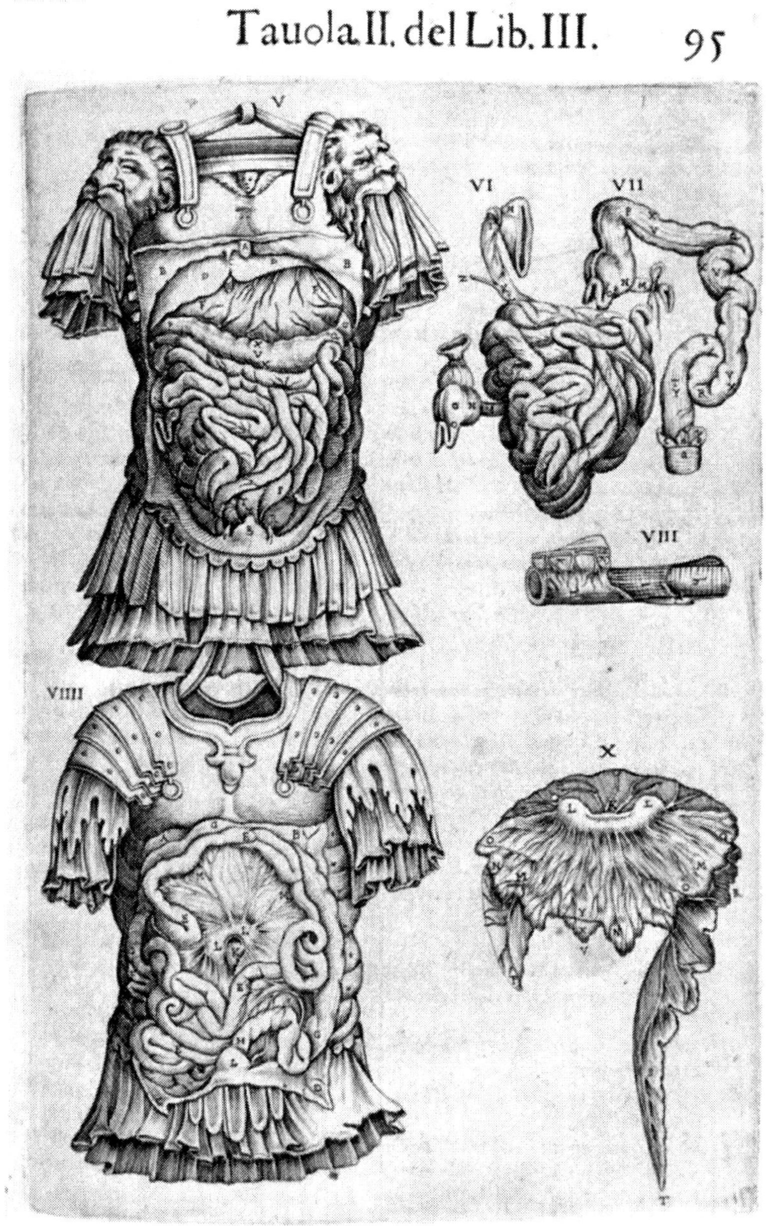

Juan Valverde da Hamusco, *Anatomia del corpo humano* (Rome: Antonio Salamanca and Antonio Lafrery, 1560), Bk. III, Tab. II. In each of the figures sporting Roman military dress a series of abdominal incisions allows the peritoneum to be folded back, while the thorax of each figure remains protected from the tools of the anatomist and artist by a breastplate. This fertile figurative embellishment typical of the eloquence of mannerism gives powerful expression to the concept of the chest as most safe and defended. Wellcome Library, London.

upon nutrition, food—"the customarie rent"—is made "Fit . . . for convoy, and this cities Arsenall," by the teeth, "twice sixteen Porters" (*P.I.* 2.30.2, 7, 1).

Thomas Healy interprets this figurative modification as a mechanism for the poem's appropriation of empirically founded anatomy for a politically reactionary agenda, which is anti-Catholic, parochial, and xenophobic.[113] But as surveillance and the conveyance of intelligence gesture to the modern nation state, so also *The Purple Island*'s emphasis upon coasts and shores, and their construction to resist invasion, characterizes the nation's insularity, specifically that of Britain. The pervasiveness of the defensive figuration of anatomy in Fletcher's poem, extending to all regiments and regions, to structures and to their operations, such as the closure of the eyelids (*P.I.* 5.27, quoted in chapter 2), unfolds a domain of previously *unseen* and even *novel* connotations of the threshold figure. Corresponding to an incipient nationhood, the national insularity to which it gives expression is among the *novel* connotations, however reactionary it may appear to a late twentieth-century or early twenty-first century reader.[114] By contrast, the imagery of fortification in the *Devine Weekes* and in *The Faerie Queene* is conventional, corresponding to a feudal and dynastic social order, in which subjects owe allegiance less to a head of state than to a local feudal lord, and both states and subjects are less self-defensively isolated or individualized.

In a chapter on "the Accurate Structure of Mans body," in *A New Method of Rosie Crucian Physick,* by John Heydon, published in 1658, the implications of Fletcher's proto-individualistic Isle of Man can be found not only to have continued into the second half of the seventeenth century, but to have become naturalized. Heydon exclaims of the eyes "Nature" has fabricated, how they are "necessarily liable to mischief from smallest matters"; "how safely they are *guarded*"; and how "fenced off by those two Wreaths of Hair which we call the Eye-brows."[115] But when he describes the eyelashes and eyelids, the resemblance to (but also the difference from) the *Purple Island*'s account becomes obvious: "the Eye-lids are fortified with little stiff bristles, as with Pallisadoes, against the assault of Flyes and Gnats, and such-like bold *Animalcula;* besides, the upper-lid presently claps down, and is as good a fence as a Port-Cullis against the importunity of the Enemy. . . ."[116] Fletcher's pikes of ranked spearmen have shed personification in favor of a more "scientific" description as "stiff bristles"; the walls of the castle of the head are now "as good as" a "Port-Cullis," but are no longer metaphorically identified as one; and the rushing foes of the Isle of Man—the moral, psychological, and national threat to the realm of Britain and its anxious, even paranoid, inhabitants—have turned into "Flyes and Gnats." Despite the loss of allegorical context, threat is still present, but it is a natural-historical threat, enlarged in hyperbolic phrases that would not be amiss in a mock-epic; suggesting perhaps that Heydon's text displaces moral, psychological, and perhaps even religious fears onto the world of microscopic (animalcular) creatures which natural philosophers in the later seventeenth century sought confidently and rationally to identify, categorize, and control.

The contestation between the dynastic realm and the nation-state in the late sixteenth and early seventeenth centuries characterizes both *The Purple Island* and *The Faerie Queene,* and despite the conventionality of the castle as representing the structure and function of anatomical members, *The Faerie Queene* addresses itself to the political task of undermining fealty to overlords and furthering allegiance to the English Tudor monarchy.[117] Like Camden's *Brittania,* Spenser's *Faerie Queene* appropriates histories of early feudal British monarchy, namely King Arthur, to legitimize the sovereignty of Elizabeth, who was on the throne until 1603 as the last of the royal Tudor line.[118] Indeed, the investiture of the natural body of Elizabeth with the "*corpus mysticum*" of the realm has been identified as the "founding trope" of *The Faerie Queene,* and is represented by the projected union of Arthur and Gloriana.[119] In David Lee Miller's analysis, the ideal form or "perfected spiritual body" of *The Faerie Queene* is the figure of the hermaphrodite that emerges from this marriage.[120] As was remarked in chapter 1, Alma's castle refects this ideal form by being "metaphysically hermaphroditic," even while its description, like that of Fletcher's Isle of Man, omits all mention of the genital organs;[121] and as was noted before, Fletcher borrows from Spenser's architecture of the House of Alma precisely that part of the Pythagorean geometry which Miller argues represents it as hermaphroditic. The geometry informing the Isle of Man is like that of the House of Alma, an index to the ideal form of the poem, the final cause or purpose, in order to reach which, the narrative of the poem is teleologically organized. But rather than a Eucharistic and matrimonial union of sovereign and realm, *The Purple Island*'s eschatological narrative culminates in Canto 12 with the union of the elect, personified as "*Ec*[*l*]*ecta*" (*P.I.* 11.8), and her heavenly spouse Jesus Christ. Whereas Spenser's concern is with the establishment of the Protestant English state, Fletcher's is with the salvation of the elect of the Protestant Church in England.

For Spenser, the innovation of Protestantism generally, and in relation to an emergent English nationality, was the reestablishment of the English Church as before papal dominance.[122] As in *The Purple Island,* innovation is justified as a return to pristine origins. Fletcher's poem was published during the reign of the Stuarts; yet it also finds in the divine and ordained creation of the Prince of the Isle, and the descent of the Isle's members and sovereign from its original inhabitants, the justification of the current Prince's sacral sovereignty. Although "this purple Islands nation" is composed of subjects "Not like those Heroes, who in better times / This happy Island first inhabited / In joy and peace," the new inhabitants "From their forefathers claim their right, & Islands due" (*P.I.* 6.6.3; 6.7.1–3; 6.21.7). The innovation of *The Purple Island* is to invent in its threshold figure a metaphor to concretize and give succinct yet infinitely expandable expression to the emergent ideologies of English (or British) Protestant individualism and nationalism.

Compared with Alma's castle, Fletcher's poetic body, as when compared with that of Du Bartas, is more highly developed in extending the concept of its constructedness figuratively beyond the feudal building, to the wider human-built and sociopolitical environments. However, perhaps because Spenser's concern is not so much the salvation of the individual person as the realization of the Christian state,[123] the anatomy of the House of Alma is no less social for being confined indoors. Spenser's text is closer to *The Purple Island* than to the *Devine Weekes* in this respect, and in one other: the ambiguity over whether Spenser's House of Alma is a building or a body is profound, whereas in the *Devine Weekes* the principal subject of the imagery of architecture and carpentry is never in doubt. When Guyon and Arthur approach Alma's castle, the sound of the Squire's horn causes the castle wall to shake "as it would fall"; a wall that is later described as "fensible withall, / Not built of bricke, ne yet of stone and lime, / But of thing like to that *Ægyptian slime*" (*F.Q.* II.9.11.5; II.9.21.3–5). Moreover, that this wall "was so high, as foe might not it clime" (*F.Q.* II.9.21.2) suggests, as A. C. Hamilton notes, "man's erect, unfallen stature" as *imago dei.*[124] But, furthermore, that this "goodly workemanship should not endure" is bemoaned as a "great pitty" (*F.Q.* II.9.21.8, 7); and although this reminder of eventual ruin is universally applied—"no earthly thing is sure" (*FQ* II.9.21.9)—the resonances of the traditions of *vanitas* and *memento mori* are unavoidable, as are the moral meditations on mortality so central to the

Christian understanding of anatomy. Nevertheless, the narrative of *The Faerie Queene,* unlike that of *The Purple Island,* is not providential history, and does not integrate the theme of death with the shared entailments of anatomy in the way Fletcher's poem does.

In Spenser's allegorical poetic, as in Fletcher's, neither analogue has priority over the other as far as the determination of meaning is concerned, and this is accounted for partly by the high degree of systematicity between the subjects of the metaphor in each work, and between the significant connotations each metaphor produces, including the social and moral. Figurality and anatomy are congruous, so that for both of these texts, for example, the "Rectum" is "the back-gate" "That cleped was *Port Esquiline,*" whereby "the excrement," the kitchen "filth" or "wast" is "throwne out priuily" (*P.I.* 2.43.4 & margin.; *F.Q.* II.9.32.7, 8, 1, 9), as was the city of Rome's common sewerage to the dump of the same name.[125] The multiple signification of the description and name, that is, the allegory, defamiliarizes the part that yet "those knights" Arthur and Guyon possess in their own bodies, causing them to gaze with "wonder," "For neuer had they seene so straunge a sight" (*F.Q.* II.9.33.2, 3, 4). This strangeness of looking anew at what is "too seldom view'd, yet still in view" is the central paradox of the Isle of Man, the "forrain home, a strange, though native coast" (*P.I.* 1.34.3, 6), that is simultaneously revealed and concealed in riddle and allegory. But there is on reflection less that is strange about the House of Alma than about the Isle of Man. Whereas a castle is architecturally and sociopolitically conventional as the subsidiary subject for the anatomy of Man,[126] and the body as an island is an ancient metaphor for what is in significant yet not dominant respects familiar; yet, nevertheless, as the Isle of Man the latter figure is developed in more innovative ways in *The Purple Island.*

It remains to be seen to what extent, if at all, the moral or sociopolitical significations of the anatomy of the House of Alma have priority over descriptively anatomical knowledge; and how the relative weight of emphases compares with that in the representation of the Isle of Man in *The Purple Island.* The physiological operations, which take various socially integrated, personified forms in *The Purple Island,* are present in less delineated form in Spenser's House of Alma, which, for example, has "a stately Hall," translated anatomically by Hamilton as "the passage to the gullet" or esophagus (*F.Q.* II.9.27.1).[127] Therein:

> sate, yclad in red
> Downe to the ground, a comely personage,
>
> . . . .
>
> He Steward was hight *Diet;* rype of age,
> And in demeanure sober, and in counsell sage.
>
> (*F.Q.* II.9.27.6–10)

According to Hamilton, Diet "personifies the temperate course of life . . . and also the parliamentary body."[128] The general location and the color of the back of the throat is the extent of the description's fidelity to the criteria of anatomy textbooks. With its greater adherence to the anatomical *accessus,* the *Purple Island* anatomy and physiology of the passage to the "Oesophagus, or meat-pipe" is comparatively more delineated (*P.I.* 2.31 margin. f). In the marginalia, "eight muscles" are described as serving the purpose of "swallowing"; in the stanza, although these are figured as "governours" whose duty is "To see the vict'als shipt at fittest tide" (*P.I.* 2.31.4, 5), the criteria of number, position, and function are maintained.

The imaginative conception governing the personification of anatomical parts in *The*

Faerie Queene is moral and political, and although the "internal environment" of the virtue with which Book II is preoccupied—temperance—is, like the external environment and the body itself, "particularly physical," description is not scrupulously anatomical in character.[129] The meaning of the language of fortification in both the House of Alma episode and *The Purple Island* is neither entirely that of proto-nationalism nor the shoring up of a dynastic realm. There are significant moral and psychological aspects to this imagery, as there are to the anatomical description of the Bower of Bliss in Spenser's poem. The reason for Spenser's moral emphasis is that the allegory of *The Faerie Queene* is a panegyric written in the context of a well-defined Renaissance theory of epideictic literature, and designed to illustrate a humanist ideal of moral conduct with the didactic purpose of portraying for emulation as well as glorification the ideal Renaissance sovereign.[130] Such a conception and purpose is borne out not only in Alma's castle: throughout the allegory the *sensus litteralis* and *sensus allegoricus* (the historical or typological meaning) are intertwined with the *sensus tropologicus* (the moral metaphorical meaning).[131]

Despite these differences of emphasis, the kind of personification is in Spenser's and Fletcher's texts similar, and the degree of personification is comparable. These points can be demonstrated by the physiology of digestion or nutrition, which in both texts is consistent with its metaphorical conceptualization in anatomy textbooks, though the textbook descriptions lack allegorical personification. Benedetti's metaphor of the stomach as a "cookshop" is a not unusual instance of overt figurative language: "Here as in a cookshop is the location of the dregs and excrement just as in buildings the architects avert from the eyes and nostrils of their masters whatever is necessarily regarded as foul and unclean."[132] The extended simile in this passage establishes a likeness between the functions of anatomical structures and the domestic activities taking place within buildings, in this case between concoction, or digestion, and cooking. The similarity forms the root of the metaphor for the anatomy of digestion in *The Faerie Queene,* where, "in the midst" of "the kitchin rowme" of the House of Alma, is "a caudron wide and tall, / Vpon a mighty furnace, burning whot," attended by "many Cookes accoyld, / With hookes and ladles" (*F.Q.* II.9.29.4; II.9.28.9; II.9.29.5–6; II.9.30.6–7). Spenser's literary text turns the similarity of simile to the identity of allegory and metaphor; an identity based on the definition of the concept of digestion in human anatomy in terms of the better defined concept of "building."

The anatomist Benedetti's comparison is also between the spatial distribution of anatomical parts and that of rooms in the plans of buildings. What de Laguna calls "sooty and smoky" food which could otherwise be too readily perceived, is confined in the stomach and intestines; yet vapors from the slow and weak (in humans) process of digestion (suitably figured therefore as "cooking") rise from the stomach and intestines "to the brain."[133] Vapors are superfluous by-products of digestion, but others rise as a result of the heat of the body, and among other means to "excrete" the "vaporous matter" is the growth of hair.[134] Vapors are one aspect of a general theoretical economy of fluids in the body, according to which any undue accumulation may result in disease: "A cataract results from vapor which descends from the head or ascends from the stomach. . . ."[135] The emission of vapors from the process of digestion is an entailment of the metaphor of human anatomy as "building," and hence, as an abode of the rational soul. And yet the nutritive process described is an operation of Nature or the vegetative soul, whose effects in the body are usually represented in the figurative terms of vegetation. But this figurative excursion that represents the involuntary movements of digestion as domestic activities thereby renders them consistent with the metaphorical concept of anatomy as "building," a concept which usually defines only voluntary movements, and consequently combines the two metaphorical conceptualizations of the body and their respective rationalizations.

This metaphorical consistency of nutrition, in which the two conceptual metaphors of vegetation and "building" are consistent rather than merely coherent, can be extended to two of their three shared metaphorical entailments: creation and mortality. Vesalius relates nutrition and growth to God's plan to raise Man in the face of mortality: "Since man has been unable to form the substance of an immortal being . . . the great Creator of things has carefully devised that man should live as long as possible. . . . In order that man might attain to the stature for which he was intended and that those elements upon which his innate heat is continually fed might be restored as quickly as possible, he possesses organs which serve to nourish him in many ways."[136] The spatial arrangement of the nutritive members, understood on an analogy with that of a building, is imagined as designed or planned by God the architect, so that the senses, which are made for the soul, can serve the purposes of their master without being interfered with by the kitchen staff. The indispensable part in this Galenic organization is the transverse septum or diaphragm, whose function as a panniculus is to restrict the "most unimportant part of the life-spirit to the lowest region of the central body cavity," to keep it and the vapors "of smells raised by the members of nutrition" "from interfering with the loftier operations of the life-spirit."[137] However, vapors from "sooty or smoky" food do rise to the brain, the seat of the rational faculty of the soul. Once at the brain they either pass out via the sutures of the cranium and are dissipated, or they are congealed by the brain's coldness, and pass out through the nostrils by means of their own weight.[138] Through the basilar bone, which is in the direction of the palate, at the place "commonly called the collatorium," "perforated like a sieve, the humid watery superfluity of the brain descends," growing thick.[139] Thus there is need of something in human anatomy akin to a chimney joining the nose to the stomach. In the House of Alma, arising from "the kitchin rowme," is "one great chimney, whose long tonnell thence / The smoke forth threw" (*F.Q.* II.9.28.9; II.9.29.3–4). This apparatus is the esophagus, but its figuration demonstrates the rational consistency of *The Faerie Queene*'s allegorical anatomy.

Hamilton describes the three processes of digestion in *The Faerie Queene* as follows: "The first, in which food is turned into chylus, [Spenser] names **Concoction,** the **maister Cooke,** from Lat. *Coquere,* to cook. The second, by which the chylus is turned into blood and distributed to the body, he names **Digestion,** from Lat. *Digerere,* to distribute. The third is elimination. . . ."[140] Principal in this physiology in both *The Faerie Queene* and *The Purple Island* is "Concoction": the "maister Cooke" of Alma's castle, and the "[Purple] Islands common Cook"; there is in addition a division of the labor of concoction between "many Cookes" in the former text and "many a groom" in the latter (*F.Q.* II.9.31.1; *P.I.* 2.33.2; *F.Q.* II.9.30.6; *P.I.* 2.36.1). The representations of concoction, however, differ in that, in *The Purple Island,* concoction is effected both in the stomach and the "thinne entrails" (*P.I.* 2.34 margin. k; 2.44 margin. t). Also in *The Purple Island,* a further concoction occurs in the veins, which is entirely omitted by the House of Alma episode, which of course does not represent the venous system (*P.I.* 2.10 margin. g). This elaboration of processes is one of the ways in which the physiology in Fletcher's poem is more systemic than that of the House of Alma. The distinction between processes in *The Purple Island* is less clear than in Alma's castle, where the process of "*Digestion*" is separately personified in the "kitchin Clerke," and this is because in *The Purple Island* the physiology of the whole system is less formulaic: "digestion" is not apart from but begun by "Concoction," who performs both his primary task and commences this distribution by sending chyle to the liver (*F.Q.* II.9.31.3; *P.I.* 2.34 margin. k; 2.35.7). Fletcher's account of the chyle is also more systemic: it is turned into blood only once it has been reduced of "excrementall liquors"—"gall" and black bile—and has reached the liver (*P.I.* 3.13 margin. k; 3.15 margin. m; 3.17 margin. o; 2.44 margin. u). Overall, the elaboration of digestion into processes is both less formulaic and more systemic in *The Purple Island;* while personification is either comparable or simplified.

The comparatively crude rendering of the physiology of digestion in *The Faerie Queene* is in part a consequence of the inherent limitation imposed upon it by the text's singular metaphorical conceptualization of the body (i.e., as a building). Without the enrichment of descriptive possibility that arises from extending the body's analogues and subsidiary subjects from buildings to streets, streams, cities, geological formations and regions, the metaphorical description of the body in Spenser's text is necessarily impaired and simpler than that in *The Purple Island.* But this comparative simplicity does not derive only from less metaphorical traffic; it also arises from (or gives rise to) Spenser's use of less precise, or more schematic, anatomy. The stomach in Spenser's text is "a caudron wide and tall, / Upon a mighty furnace [the liver], burning whot" in the "kitchin rowme [the abdomen]" (*F.Q.* II.9.29.5–6; II.9.28.9). This description admittedly encompasses all of the following anatomical descriptive criteria, however imprecisely: form, complexion, posture, and location. In Fletcher's text, by comparison, the stomach is the city "*Koilia,*" situated in the lower region or abdomen, "On this side border'd by the *Splenion,* / On that by soveraigne *Hepars* large commands"; its "form" "round-wise, and long, yet long-wise more," like a swollen "bag-pipe"; the "upper mouth" of which is a "port," a "haven," and a "creek," the "bottome" or pit a "market-place," in which "Concoction['s]" "furnace ever fumeth" with a combination of its own "innate propertie, and speciall vertue" of heat, and "the outward heat of parts [the liver and the gallbladder] adjoyning" (*P.I.* 2.27.3, 4–5; 2.28.1, 2 & margin. c; 2.31.7; 2.32.1, 4 & margin. g; 2.33.1, 2 & margin. i; 2.34.1 & margin. k). Much greater precision in the application of the anatomical descriptive criteria of location, posture, form, and complexion in *The Purple Island* is responsible to a large extent for the text's more highly developed physiology, or anatomical use.

Textual parallels repeatedly remind the reader of *The Purple Island* that Fletcher's anatomical cantos owe something of their conceptual and verbal character to Spenser's House of Alma[141]—a fact, about which Fletcher, who openly acknowledged his admiration for Spenser (and for Du Bartas, too),[142] apparently had no qualms.[143] An example is *The Purple Island,* Canto 6, stanza 51, lines 4 to 7, where Thirsil refers to *The Faerie Queene,* Book II, Canto 9, stanzas 47 to 58. The occasion for Thirsil's reference is his description of three counselors: "The common sense" (*P.I.* 6.42.4—6.45.7 [6.42 margin. *]); "*Phantastes*" or "The fancie" (*P.I.* 6.46.1—6.48.7 [6.48.1; 6.46 margin. *]); and "The understanding" (*P.I.* 6.51.1—6.56.7 [6.51 margin. *]). The passage corresponds to that in *The Faerie Queene* which describes the "three honorable sages" (*F.Q.* II.9.47.8):

> The first of them could things to come foresee:
> The next could of things present best aduise;
> The third things past could keepe in memoree
>
> (*F.Q.* II.9.49.1–3)

These are foresight, intelligence and memory. They correspond in *The Purple Island,* respectively, to "The fancie," "The understanding," and "*Eumnestes,*" "Father of memorie" (*P.I.* 6.48.7), who keeps the "Treasurie" of Phantastes (*P.I.* 6.48.6).[144] This trio is a personification of the tripartite division of the principal faculty of the animal faculty of the soul, which of course resides in the brain, and is described by Paré and other anatomists.[145] Fletcher acknowledges the debt to Spenser by putting in the mouth of his diegetic narrator these words:

> For these three late a gentle shepherd-swain
> Most sweetly sung, as he before had seen
> In *Alma's* house . . .
>
> (*P.I.* 6.51.4–6)

Despite Thirsil's gracious acknowledgment, his anatomy of this region of the human body typically expands upon that of the House of Alma by describing also the parts and functions of the eyes, ears, nose, tongue, teeth, brain, and cerebral ventricles. As a typical Spenserian poet, Fletcher thereby displays a derivativeness which "set [him] free to accomplish something new."[146]

To complete our understanding of the figurative design, the locutionary act, of *The Purple Island,* chapter 3 of this study sought to understand how the Christian assurance of things hoped for does not merely shape the eschatological narrative of the poem but informs its poetics and its desired effect, its illocutionary act. But in order to answer the question of whether the figurative design achieved its desired effect, and what this might mean, it was necessary in the present chapter to examine the work's intentional direction toward the world it creates, and its reflective direction toward a self, in two contexts. These are, firstly, the intertextual context of those works which as source texts may have influenced Fletcher's conception of the nature and purpose of the *Purple Island* world, and, secondly, the hermeneutic context of the historically situated perlocutionary act of the work, the responses of its earliest readers. In this chapter, intertextual study of literary works that exercised a specifiable influence on the writing of the anatomical cantos of *The Purple Island,* and an examination of the written responses of its seventeenth-century readers, have deepened our understanding of the figurative design of the work and our understanding of the work's conservatism and innovation.

On the evidence that exists, seventeenth-century readers of *The Purple Island* do not seem to have been divided like Langdale and Baldwin over whether Fletcher's use of the conventional analogies of the microcosm is indicative of conservatism. The most commonly occurring critical attribution made by Fletcher's contemporary readers is ingenuity, but its use does not have the same conceptual content as implied when used by modern critics like Gosse. As this chapter has stated, in the sixteenth and seventeenth centuries, ingenuity (*ingenium, ingegno, argutezza, acutezze*) could connote, among other concepts, intellect, wit, literary or poetic talent, inspiration, cleverness, skill in contriving or making something, and the capacity for invention, which was a part of dialectic, and which could mean *fantasia,* imagination, or discovery, the faculty or technique of discovery but also the thing found, the artifice. Applied to the finding of subject matter, invention was related also to the cognitive process of creating new antitheses, paradoxes, witty metaphorical expressions, creative metaphors, and conceits. Thus the attribution of ingenuity to either *The Purple Island* or its author recognizes the discovery and fanciful (re)invention of anatomical subject matter as an intellectually intriguing riddle; and the imaginative, witty contrivance of the poem's figurative design and skillfully, cleverly constructed and sustained allegorical artifice.

This chapter has argued that in addition the attribution of ingenuity recognizes the achievement of accuracy, or truthfulness of an accepted kind, since the obscure unity of the world was conceived as a kind of divine and witty poem, the metaphors of which could be brought to light by ingenuity and the poetic of *concettismo.* From the attribution it may therefore be inferred that Fletcher's seventeenth-century readers recognized in some sense what part 1 of this study recognized: that the threshold metaphor of the Isle of Man figure is creative and allegorically transforms the systematic metaphors and analogies of anatomy into the geographical and social features of the Isle of Man, selecting, emphasizing, suppressing, organizing, and extending features of the principal subject, in a way that analogies rarely do, giving rise to significant connotations that include those which are related to historically incipient and ideologically emergent concepts; but that it also aggregates, deepens, and extends conventional knowledge, analogies, and metaphors, in a way that is entirely consistent with the way cultivated minds in early modern England may have run. Moreover, the attribution recognizes that the contextual effect of the Isle of Man metaphor gives rise to a ripple effect of cleverly sustained allegory, wherein the associated implications of the subsidiary subjects of metaphors acquire specially constructed novel meanings before they interact with their principal subjects throughout Cantos 2 to 5. The recognition of these qualities of the work suggests that its earliest readers were engaged imaginatively in a cognitive process of apprehension and interpretation of that which the work disguises and reveals in allegory, fiction, and figure. That part of the perlocutionary act of the work that can be inferred from the figurative design is an implied invitation to the reader to complete the process of semantic generation prompted by the Isle of Man figure and begun by the allegorical extension in Cantos 2 to 5. Creative metaphor or conceit and ingenuity thereby sought to involve individual minds in this cognitive process of interpretation in order more deeply to understand the world as it was made intelligible by figures, analogies, and correspondences, later eliminated by Baconianism,, Cartesianism and neoclassical aesthetic taste. The commendatory poems that preface *The Purple Island* and the responses of its seventeenth-century readers suggest that the poem had a readership that was intellectually sympathetic to its plan.

The attribution of ingenuity and the question of conservatism and innovation have been extended in this chapter from the topic of the figurative design to that of erudition because several providers of prefatory material for *The Purple Island* responded to erudition displayed in the poem's eloquent and ingenious superimposition of correspondences already assumed implicitly to exist. The perlocutionary act of the work in this respect is to enjoin a small community of highly literate, erudite, proto-nationalist, Protestant readers, to complete the process of semantic generation in order to develop intimacy and self-knowledge, to recover God's art of anatomy and to help dispense the balm of salvation. In broad terms, this accords entirely with the purpose which chapter 3 attributed to *The Purple Island* in reference to the work's figurative design. But it differs considerably from *The Faerie Queene,* designed to illustrate a humanist ideal of moral conduct with the didactic purpose of portraying for emulation as well as glorification the ideal Renaissance sovereign. Within the context of constructing correspondences, Fletcher seems, like Tesauro, to have conceived of ingenuity as the equivalent faculty in Man to God's creative power, and the panegyric written by A.C. to preface *The Purple Island* suggests an aspiration to resemble the incarnation of the Word in human flesh. The allegorical ideal of continuous superimposition, however, is not achieved, despite the encomiums Fletcher's work received. This chapter has demonstrated that if tested to extremes, the coherence of the allegory falters under the weight of its own ingenious and ideological artifice. An additional effect of the figurative design on the fulfillment of the work's purpose is that Fletcher's shoring up of correspondences and analogies, and making them relevant to his readership, are what give rise to the Isle of Man figure preserved from the deteriorating influence of Roman Catholicism on the continent by its individualism and national insularity, the very connotations that are politically innovative. As is remarked above, *The Purple Island* ironically helps to forge an emergent language of nationalism and individuality that would help to bring about a collapse of such correspondences as it endeavors to preserve.

This chapter has considered whether *The Purple Island*'s representation of the body in microcosmic and geocosmic terms was already firmly established by Fletcher's own literary sources when he came to write, and whether the metaphoricity of the anatomical content of *The Purple Island* may be said to be more or less comprehensive and detailed than that of these sources. The section looked also at how the tropology of these works compares with *The Purple Island*'s allegorical transformation of the systematic metaphors and analogies of anatomy into the geographical and social features of the Isle of Man. This comparative work leads to the conclusion that *The Purple Island* develops the congruity of anatomy and figurality found in Du Bartas and Spenser, and is both less conventional and more innovative than either of the sections of their texts herein examined. *The Purple Island* differs from and (it is fair to say) exceeds the *Devine Weekes* and *The Faerie Queene* in several respects, all of which relate to the concept of ingenuity and the poetic of *concettismo.* In its intentional direction toward a world and its reflective direction toward a self, *The Purple Island* reflects the didacticism of the *Devine Weekes,* but differs from it in invention, especially the discovery of its anatomical subject matter and its witty, cleverly constructed allegorical representation, as well as the fanciful and mimetic invention of a strange, yet familiar location that is both the human body and the island realm. The House of Alma in *The Faerie Queene* shares *The Purple Island*'s invention, but the creative metaphor in *The Purple Island* uses to a greater extent and with greater systematicity previously unused parts of the defining domain, thereby bringing to light previously unconsidered or unexplored implications of the analogies and metaphors of anatomy.

The Christian providential allegory of *The Purple Island,* which contains panegyric, is not so very unlike the *Faerie Queene* panegyric as an example of Renaissance epideictic literature. But allegory in Fletcher's poem is rendered consistent with the anatomy in the poem through the identity of the principal themes of its narrative with the shared entailments of the two metaphorical conceptualizations of the body in Hellenistic, humanist, and scholastic anatomy. These conceptual metaphors of human anatomy are vegetation and "building," and the narrative themes are creation, mortality, and salvation. As an aspect of the unfolding of the Isle of Man figure, *The Purple Island* consistently applies the descriptive criteria of contemporary anatomy textbooks—the anatomical *accessus*—to the description of the island realm, and regulates personification in relation to observable physiological operations. Both the allegorical figuration of anatomy and its fidelity to the order of knowledge in the schools and medical profession have been examined in each of

Fletcher's literary sources in this chapter. The unfolding of the Isle of Man figure in *The Purple Island* exceeds the poetic exploration of the House of Alma in several ways in which it is proportionally more ingenious: in its extended range of analogues of anatomy (combined in the subsidiary subject, the geopolitical realm in *The Purple Island*) and metaphorical connotations; in the enhanced systematicity between human anatomy and these analogues and this subsidiary subject; in a more precise application of anatomical descriptive criteria, the *occasiones* of the anatomical *accessus,* to the body, its analogues and subsidiary subject; and in its generally more precise, less schematic anatomy, and more systemic, less formulaic physiology.

If ingenuity is understood as the achievement of accuracy or accepted truthfulness through the invention of new combinations of objects in strange, innovative, and cleverly contrived figures or conceits, probably only Fletcher's of the three poetic sections compared in this chapter qualifies. Consequently less conventional, Fletcher's text is proportionally more ingenious than either of its source-texts' sections for the extent to which it exceeds them in the range of previously unseen similarities it draws between its principal and subsidiary subjects, and in the systematicity of its exploration of the significant connotations or satellite conceptual metaphors to which the metaphorical interaction gives rise. In so doing it has been regarded by Gosse, Kastor, and others as yoking together heterogeneous material from anatomy, geology, and politics, by means of almost occult resemblances unlikely to occur to, and certainly to be pursued by, anyone else. It is consequently no wonder that a comparison with either of these two of *The Purple Island*'s source-texts has been seen as incapable of diminishing the effects of incongruity and unconventionality, which arise from the persistence with which Cantos 2 to 5 of *The Purple Island* detail human anatomy.

5
The Literary Context

There are a number of ways in which the comparison is to be made in this chapter between *The Purple Island* and contemporaneous literary works. As we have seen, the process of understanding the threshold figure of the Isle of Man as a creative metaphor is the key for understanding Cantos 2 to 5 of *The Purple Island.* This point of view is that of explanation and develops the *sense,* the immanent design of the work. From this standpoint, comparative work must ask if this anatomical figure or the significant connotations unfolded by Cantos 2 to 5 are found elsewhere in contemporary literature, and whether such occurrences may have had any influence upon *The Purple Island.* The character of the figure of the Isle of Man has been categorized as a creative metaphor, the product of *ingegno,* or ingenuity, and the poetic of *concettismo;* the answer to our question may affect our understanding of the poem's ingenuity, its innovation and conservatism, because the creative metaphor of the Isle of Man has been said to select, emphasize, suppress, organize, and extend features of the principal subject discovering unseen, emergent, or newly created similarities between the principal and subsidiary subjects, some of which have been said not to be reducible nor perhaps even traceable to antecedently existing similarities. The poetic foregrounding of the metaphor in the literary work, which enhances the contextual effect, giving rise to greater interaction, was said to have depended on the literary work's specially constructed set of associated implications of the subsidiary subject that distinguish it from any other isle, realm, polity, or nation. The literary contextualization of *The Purple Island* is necessary to determine whether any among competing connotations and ideologies of Man and of the realm may be interpreted as newly emergent, as opposed to either residual or dominant.[1]

From another standpoint, it is the understanding of a work as a whole which gives the key to metaphor; according to Ricoeur, this other point of view is that of interpretation properly, which develops the work's intentional direction toward a world of its own and the reflective direction toward a self, perhaps that of the implied reader. The first of these of the work's directions has been analyzed in terms of how in the process of reading Cantos 1 to 5 of *The Purple Island,* the Isle of Man figure gives rise to a host of newly created similarities and associations from the interaction of the figure's principal and subsidiary subjects, to which Fletcher, as concettist, has given eloquent expression in an allegory consisting of a main story line of providential history comprising a *sensus allegoricus* and *sensus anagogicus.* As the hermeneutic circle is initiated, these similarities combine the subjects with consistency and coherence, as a result of which anatomical and geopolitical information can be progressively assembled across cantos to form a coherent fictional location, though the ingenious superimposition of correspondences occasionally falters under its own weight. The comprehensiveness, systematicity, and coherence of the extension of meaning are to be compared with any similar metaphorical interaction and generation of meaning in comparable literary works of the period. As with *The Purple Island,* this will involve measuring the extent to which anatomical description relates closely to concepts of anatomy as vegetation and as "building" and how deeply it interpenetrates with concepts of the body as a microcosm of the natural geocosm or the constructed environment. Determining the degree of this conceptual imbrication will involve referring the poetic or prose figurations of the body once again to the anatomical *accessus* (which *The Purple Island* painstakingly applies to the Isle of Man) and, though to a lesser extent, to anatomy textbooks.

In *The Purple Island,* the conceptual metaphors of anatomy and the analogies of the microcosm and microchristus coalesce, as the anatomy is figured in vegetative, terrestrial, and architectural imagery that is rationally consistent with descriptions in anatomy textbooks, and which provides in allegorical form the *telos,* the final cause, of structure and function in some purpose of the soul, which is embedded within and implicit to the figurative language. To what extent this language is shared by or is as consistent in comparable literary works is for this chapter to discover; as, consequently, is the question of the derivativeness or innovativeness of the *Purple Island*'s allegorical extension of its threshold figure. Comparison will extend also to the relationship, whether congruous or incongruous, between literary genre and anatomy in the comparable works, involving an assessment of the effect of genre on the conceptual character, coherence, systematicity, and rationality of anatomical figurality and content. In *The Purple Island* the conceptual metaphors of anatomy and the faculties of the soul are reconciled and made coherent in and by their shared entailments of construction or creation, demise or mortality, and regeneration or salvation, which are also the thematic significance of the events of the main story line of the Christian allegory.

This chapter will also examine how *The Purple Island* and other literary works are modeled formally or imaginatively on human bodies, and gesture toward identification with their subject matter, rhetorically substantiating the accomplishment of their intentional direction toward a world that comprises the anatomy of the human body, a microcosmic world. These works hereby additionally problematize the critical distinctions between form and content, and between text and context. The chapter will seek to find out whether these comparable works relate their rhetorical reconstitution of the body, their illocutionary (and perhaps perlocutionary) act, to Christian soteriology, as does *The Purple Island.* As chapter 3 concluded, Fletcher wrought the figurative design of the anatomical speaking picture of *The Purple Island* with a politico-religious purpose, to demonstrate the creative ingenuity of God and the saving love of the "Sunne," or "Sonne." Additionally, by drawing a parallel between the dependence of resurrection on that of Jesus Christ and the reconstitution of the body on the incarnational, anatomical poetics of *The Purple Island,* he also sought as the perlocutionary act of his work the regeneration of the selves of his readers and the body of his nation, and he did so by drawing on theories of poetry, emblematics, suprarealist theories of language, and the tradition of didactic poetry, as well as by restoring the relationship of Man to God through the poetry's assemblage of analogies with the microcosm and microchristus, and by adopting a method of dissection that purports to imitate Nature's act of creation.

In the course of this chapter, it should become clear that it is ingenuity that is purportedly capable of this regeneration. *The Purple Island*'s reflective direction toward a self has been explored particularly in relation to two aspects of the work: the concept of ingenuity, attributed to the work and its author in the seventeenth century; and the soteriological purpose, the illocutionary act, of the poetic mode and anatomical method of the work. Hereby this study has begun to investigate aspects of the rhetorical and intentional complexity of *The Purple Island,* as theoretically distinct from though deeply involved in its figurative design. The Neoplatonic and Christian poetic aspects of the understanding of ingenuity discussed in the previous chapter will be developed in the present in order to elucidate the epistemological and hermeneutic contexts of the reconstitution of the body in or as the work. This development, a kind of literalization of concepts such as a body of knowledge, returns us to the notion of the body discussed earlier in this study as both phylogenetically and ontogenetically one of the first and most basic entities the mind can grasp. It will be of further significance that, in the previous chapter, ingenuity was defined in relation to *The Purple Island* as the commendable, restorative achievement of accuracy, or truthfulness of an accepted kind, by means of intellect, skill, literary talent or wit, and invention. Neither of Fletcher's source-texts was found to be equal to *The Purple Island* in the range or innovativeness of the similarities it discovers and creates between Man and "his" subsidiary subject, or in the systematicity of its exploration of these significant connotations, both of which achievements relate to ingenuity. It remains to be seen whether *The Purple Island* is exceeded in these respects by comparable literary works of the period. It is hoped that this comparative and interdisciplinary study will be found to have built understanding of the expectations that Fletcher's readers might reasonably be assumed to have brought to *The Purple Island.* It is expected that the study will be found to have given prominence to works by relatively minor authors, the reason being that they engaged aesthetically and intellectually with the challenging and by this time rapidly developing field of anatomy. The study will perhaps have contributed in some measure to a reconstruction of the intellectual and cognitive grasp of the body among educated British authors of the early seventeenth century, and will at least have subjected this to critical, historical, and regrettably but unavoidably distorting, analysis.

Encyclopedic Poetry and Expository and Fictional Anatomy

In the critical history of *The Purple Island,* because comparison with either of its source texts, discussed in chapter 4, has not generally been seen as capable of diminishing the effects of incongruity and unconventionality, which arise from the persistence with which Cantos 2 to 5 of *The Purple Island* detail human anatomy, Langdale was led in his critical biography of Fletcher to other poems composed within a century of Fletcher's day, the subjects of which are natural-philosophical, anatomical, and medical; and a handful of critics since—Baldwin and Nicolson at times, but most assuredly Sawday—have helped to reduce *The Purple Island*'s appearance of uniqueness by disclosing further aspects of its literary anatomical context.[2] Sawday explores the poem's cultural context and effectively reconstructs a congruity between the discourses of anatomy and religious poetry at the beginning of the seventeenth century, which he traces in *The Purple Island* and in other texts. Secondly, he attends to *The Purple Island*'s formal division into poetic stanzas and prose anatomical marginalia, examining how this design can be related both to other works and to the historical change in the relationship between poetry, on the one hand, and (what he calls) scientific discourses on the other. The latter of these approaches to *The Purple Island* is discussed in the ensuing chapter on the relationship between the stanzas and the marginalia, while the former is immediately relevant.

Several of the contemporaneous works with which Sawday relates *The Purple Island* are encyclopedic poems which include anatomical passages, some of which, like David Kynloch's *De hominis procreatione,* offered "popular accounts of scientific discovery."[3] "The beginnings of the literature of 'popular science' . . . is evident in many Elizabethan works," as Louis B. Wright argues, and "[f]rom the Middle Ages, encyclopedic works had preserved a large amount of miscellaneous information which was absorbed into the common stock of learning and reappeared in Elizabethan outlines of knowledge."[4] Some sixteenth- and early seventeenth-century encyclopedic poems in English, such as *Mirum in modum, Microcosmos,* and *Summa Totalis,* by the Welsh poet John Davies of Hereford, consist of an erudite "mass of scientific, quasi-scientific, and theological or philosophical speculation and observation," thereby reflecting the erudition for which *The Purple Island* was admired.[5] In *The Purple Island,* the knowledge to be gained is that of Man, "his" physical and mental anatomy, and the virtues and vices to which his moral composition is subject. Although not all "anatomies" are encyclopedic, all do tend to be learned, though they may deploy their erudition earnestly or satirically.

Encyclopedic and expository works could be either poetic or prose, and the prose form has a fictional counterpart that is, according to Northrop Frye, a species, or rather subspecies, of the Menippean satire or "anatomy," one of the four forms or chief strands to which Frye allocates all prose fiction. According to Frye: "The Menippean satirist, dealing with intellectual themes and attitudes, shows his exuberance in intellectual ways, by piling up an enormous mass of erudition about his theme or in overwhelming his pedantic targets with an avalanche of their own jargon."[6] Frye includes in this formal category *The Compleat Angler* by Fletcher's friend Izaak Walton, who included lines from *The Purple Island* therein. Among other reasons, Walton's work is an anatomy "because of its mixture of prose and verse," a feature shared by *The Purple Island;* because of "its rural *cena* setting," resembling in some measure the pastoral diegesis in which Thirsil addresses the "lovely crue / Of Nymphs & shepherd-boyes" (*P.I.* 2.1.6–7); and because of its dialogue form, which Fletcher's poem, having an extradiegetic and a diegetic narrator and only local dialogue, does not possess.[7] Whereas *The Compleat Angler* rails gently in Menippean fashion, *The Purple Island*'s satire, insofar as the lively personification of vices in Cantos 7 and 8 is satire, is both moral and fantastic.[8] For these reasons, though by no means securely classifiable in Frye's terms, *The Purple Island* is a fictional anatomy, but it is nonetheless an allegory for being so.

In "anatomies," learning may be communicated in fiction or nonfiction, exposition or narrative, poetry or prose. Like Wright, Frye tends to the view that the results of such expressions of "a magpie instinct to collect facts" are farraginous, a medley or hotchpotch. If they are fiction, the narrative "relies on the free play of intellectual fancy," and the work tends "to expand into an encyclopaedic farrago";[9] if they are expository, what they garner constitutes a miscellany.[10] But Frye also rightly acknowledges that, "[a]t its most concentrated the Menippean satire presents us with a vision of the world in terms of a single intellectual pattern"; as an example: "This creative treatment of exhaustive erudition is the organizing principle of . . . Burton's *Anatomy of Melancholy*": "Here human society is studied in terms of the intellectual pattern provided by the conception of melancholy."[11] While works like Robert Burton's (1577–1640) *Anatomy* study vast swathes of lived experience in terms of an intellectual pattern, others, like Jeremy Shakerley's *Anatomy of Urania Practica,* focus upon a subject of comparatively slender dimensions, in his case the mathematical "*Errors and*

Impertinencies Delivered in a Treatise Lately Published."[12] Another early seventeenth-century text which aspires to be encyclopedic and to focus all that it touches upon through the lens of "the change and alteration of things" is John Norden's *A Store-Hovse of Varieties.*[13] In these and other works, the piling up a mass of facts and a mass of authorities is neither unsystematic nor conceived as piecemeal, but rather like the description of anatomical parts and members according to the *accessus* in the teleological anatomy of the sixteenth century, it organizes parts in relation to one another and recuperates individual parts to the defining analytical perspective, of which more in due course.[14]

For a further reason, that has a bearing on our understanding of *The Purple Island*'s ingenious kind of analysis, which assembles correspondences and explores the connotations of metaphor and the etymons of conventional metaphorical words, these compilations of knowledge are the result of a more coherent design than is suggested by their characterization as miscellanies. This study has maintained the view that in the suprarealist conception of language, knowledge in the sixteenth century was thought to be continuous with, or only slightly displaced from, the objects whose teleological meaning it expresses and discloses to the knower. Because knowledge for many scholars, as well as less educated readers, consisted in, as it were, making words and things speak, rather than in objective demonstration or observation, the compilation of a commentary or "history" of a thing in nature consisted of amassing all forms of writing or discourse on that thing, all that has been written about it, thereby fabricating "an unbroken tissue of words and signs, of accounts and characters, of discourse and forms," some of which are moral, some philosophical, some anatomical, some theological, and so on.[15] Such compilations are not comprised of what Wright maintains is "miscellaneous information." Rather, as Foucault argues with regard to the sixteenth century: "When one is faced with the task of writing an animal's *history* [for instance, that of Man], it is useless and impossible to choose between the profession of naturalist and that of compiler: one has to collect together in one and the same form of knowledge all that has been *seen* and *heard,* all that has been *recounted,* either by nature or by men, by the language of the world, by tradition, or by the poets."[16] The knowledges of natural philosophy "were to be integrated with what [was known] from other sources, not to stand apart as a separate, independent, dissenting discipline."[17]

Besides erudition, the use of the concept of anatomy in the titles, texts, or designs of works connotes a concept of disclosure or discovery. What the works by Burton and Shakerley have in common, despite their obvious difference in scope of subject matter, is the "*Laying Open*" of their subject matter, in the words of Shakerley's subtitle. That of Burton's *Anatomy* indicates how melancholy is to be "*Philosophically, Medicinally, Historically Opened & Cut Up.*" Anatomies do not merely amass knowledge, but claim to throw the light of truth on it, doing so by appealing to the anatomical concept of *demonstratio* (demonstration).[18] The idea is variously expressed in the subtitles of many relevant works: Richard Ward's *The Anatomy of Warre* claims in its subtitle to expound upon its subject "*Laid out to the Life*";[19] Pierre Du Moulin's *The Anatomy of Arminianisme* is sub-titled "*The Opening of the Controversies of These Times*";[20] and in similar semantic vein, Denne's "anatomy" claims to have "*Discovered*" the Levellers' design.[21] Discovery is also invention, in the Renaissance theoretical sense of the choice of subject matter and the linguistic, metaphorical means of its expression. Fletcher's *The Way to Blessednes,* which describes Psalms as the divine speaking picture of the inward man, "lively drawne," follows a method or design that is proposed and introduced as a kind of anatomy: "In all Sciences whether liberall, or manuall, that Scholler makes the most happy progresse, whose Master doth first by instruction lay downe before him the grounds of his Art, and then further opens them by practise, and example."[22] Occasionally in literary anatomy, laying open a subject was understood in an additional sensory way as "handling" it. Ephraim Huit's *The Anatomy of Conscience* bears the subtitle, "*Wherein Are Handled the Places of Conscience, Worship, and Scandall*";[23] and Du Moulin's *The Anatomy of Arminianisme* claims in its subtitle that the controversies it "opens" were "*Formerly Handled in the Low-Countries.*"

Anatomy in the context of natural philosophy can be interpreted as a dissection or cutting up,[24] a "reiterated incision," and "a perfect and absolute division . . . of mans body into its parts."[25] According to Sawday's seminal historical investigation of what he calls the Renaissance "culture of dissection," that part of the epistemological method of anatomy which might be called division, particularization, or partition, was extended pervasively from the domain of anatomy proper: "[P]artition stretched into all forms of social and intellectual life: logic, rhetoric, painting, architecture, philosophy, medicine, as well as poetry, politics, the family, and the state were all potential subjects for division. The pattern of all these different forms of division was derived from the human body. It is for this reason that the body must lie at the very centre of our enquiry. And it is in this urge to particularize that 'Renaissance culture' can be termed the 'culture of dissection.'"[26] Within the field of encyclopedic and expository anatomy, the subtitle of Burton's *Anatomy* indicates that melancholy is "*Opened & Cut Up*" "*In Three Partitions*"; and within the field of fictional anatomy, the subtitle of John Lyly's *Euphues* divides its discourse into two aspects, wit in youth and in age.[27] Ward's *The Anatomy of Warre* not only claims in its title to lay out to the life the anatomy of its subject, but enumerates the parts: "*1. What Warre Is; 2. The Grounds, and Causes of Warre; 3. The Things Requisite in War,*" and so on. This enumeration suggests less the empirical division of the object or activity under discussion, however, than a division of the questions asked in the course of investigation. In this it resembles the method of anatomical investigation according to the *occasiones* of the anatomical *accessus,* or the use of other forms of *accessus* in scholarship and research as they derive from the Aristotelian categories of substance, quantity, quality, and so on.[28] Victorinus Strigelius (1524–69) categorized such methods generally as the "anatomical method," by which precepts are disposed.[29]

Mark S. R. Jenner has argued that Sawday's account may only demonstrate that dissection became a paradigm or pattern in "high" culture,[30] but its representation requires a further qualification even within the context of that culture. Anatomy was an "artificial resolution of mans body into its parts";[31] artificial because in Nature the parts cannot perform their function in isolation, and because anatomy's purpose is to describe the body as Nature created it. One function of rational analysis was therefore to understand the form and function of the parts in relation to each other and to the whole, as discussed in chapter 1, and another was to devise a system of organization, like the anatomical *accessus* or "method," an architectural plan of the parts it isolates for the purposes not merely of identification or enumeration, but to answer questions raised by the disposition of the precepts.[32] Not only a work's division of its subject matter is relevant, but also its analysis.[33] In a theoretically astute and historically and textually sensitive essay on the active and passive roles of the hand in dissection, Katherine Rowe observes that in Crooke's *Microcosmographia,* the hand (defined anatomically as pertaining to the arm from shoulder to finger tips) is revealed by anatomy as the "Judge and discerner of touch"; it is therein endowed with efficacy as the instrument of the rational soul, enacting a "continuity of action" which combines with Crooke's apparently opposing interest in the particularity and contiguity of bones and tissues.[34] Moreover, to "discern of" in the sixteenth and seventeenth centuries had a transitive sense—"to separate as distinct, to distinguish and divide"—with legal connotations of judicious interpretation. Hence, the anatomist's hand performs a morally and spiritually discriminating act, rather than merely a division. Other anatomists and natural philosophers concurred with Crooke in likening the action of the hand in dissection to the "most perfect action of the intellect," and to the composition of a whole body of knowledge from the scattered parts of the cadaver. Rowe remarks that anatomy possesses "marvellous recursive logic": "The hand that actually cuts and orders embodies the rational design it produces: unity in division."[35]

The part-whole relationship integral to rationalist anatomical inquiry was maintained in many other works entitled "anatomy" by a similar organization of the parts isolated by dissection, particularization, or cutting up. Often this is implied in the enumeration of the parts into which the subject is divided, as, for example, youth and old age form parts both of a life and of the manifestation of wit in the mind and behavior of the hero of *Euphues*. Sometimes, as in Burton's immense work, where melancholy is opened and cut up "*Philosophically,*" recuperation of part to whole is implied by the governing perspective of division. But discernment or analysis need not be of the Aristotelian, teleological kind, and may be subsumed under a religious or moral perspective. This too, however, relates at least some of the philosophical and medical anatomy textbooks of the schools more closely to their literary counterparts, and to the illocutionary act of *The Purple Island,* to administer the balm of salvation through an imitation of God's ingenuity, the incarnation, and the anatomical method of Nature; for anatomy as a discerning act of analysis has this further dimension, which is theological: "To take in hand the rational work of dissection . . . is to do . . . 'God's handy worke': both to reveal and to perform God's will in the flesh."[36] Contemporary representations of the physician's hand, according to Rowe, typically "integrate the actions of God's hand and the human hand in the same inscription." So analogies were drawn between the internal structures that facilitate grasp, the hand of the anatomist which facilitates the demonstration, and the spiritual authority of God's hand which inhabits both.[37] The epistle of Daniel "Featly" to the reader of *The Purple Island* suggests that self-knowledge and analysis are understood in the context which leads to knowledge of God: "He that would learn *Theologie,* must first studie *Autologie.*"[38]

Equally, however, the aesthetic perspective of proportion may dictate the division or aggregation of parts. This is the case in the declared aim of Lyly's dedication of *Euphues* to Sir William West, Lord de la Warre, to fulfill the "necessity of the Historie" and "giueth every peice a iust proportion." Lyly explicitly relates the method of his "invention" to that of the painter and anatomist: "For as euery *Painter* that shadoweth a man in all parts, giueth euery peice a iust proportion: so he that decyphereth the qualities of the minde, ought as well to shew euery humor in his kind, as the other doth euery part in his colour. The Surgion, that maketh the *Anatomy,* sheweth as well the muscles in the heele, as the veines of the heart. If then the first sight of *Euphues* shal seem too light to be read of the wise, or too foolish to be regarded of the learned, they ought not to impute it to the iniquity of the Author, but to the necessity of the Historie."[39] Lyly's justification of aggregation of detail in terms of just proportion and "the necessity of the Historie" echoes the explanation this study has provided for the persistence with which *The Purple Island* describes human anatomy, though the latter is overtly theological. Though they do not constitute a firmly knit family of works, each of the encyclopedic, fictional, and expository anatomies adheres to at least some of four interrelated concepts. These are erudition and self-knowledge, opening or discovery, division or dissection, and analysis or justification. It may be that these are the conventional, literary associations and commonplaces of anatomy.

Popular encyclopedic sources of natural philosophy in the sixteenth and early seventeenth centuries included Pierre de La Primaudaye's *The French Academie,* the first section of which was translated and published by "T.B." in 1586. Wright says of this work: "Beginning with the new portions of the third edition in 1594, natural philosophy takes a prominent place in the work. The second part of the *Academie* is concerned with 'a naturall historie of the bodie and soule of man.'"[40] The French academicians' anatomy was taken from various sources, and had primarily contemplative ends. Some Academic texts were the products of Protestant humanism, and sought to synthesize Protestant natural theology and Neoplatonic philosophy.[41] But *The French Academie* is strongly influenced by theology, and "[w]hen the sixth edition was printed in 1618, it was concluded with a theological and devotional section, which would have saved the work from any taint of worldliness even had the author not persistently moralized elsewhere."[42] The intellectual pattern providing the lens through which the anatomy is perceived is that of moral anatomy. As Wright maintains, many encyclopedic poems were additionally concerned to expound the things of the natural world as created: "So pertinent a subject as the creation was the place for a discussion of man, plants, and animals, and the world in which they live."[43] Several of these poems brought scientific "discoveries into harmony with the revealed truth of scripture" (as does *The Purple Island*) in order "to praise God through a poetic summary of his works."[44] In this respect, encyclopedic poems approach the Bible, albeit in a different mode, and become an "analogy of revelation."[45] Philippe de Mornay's *Vérité de la Religion Chrestienne,* translated by Sir Philip Sidney and Arthur Golding in 1587, "follows in its main outline an order found in theological *summae* of the middle ages and in most Protestant theological works": "following the 'Chain of Being,' it treats in turn the existence of God, the properties of God, the creation, the nature of men before and after the Fall, and finally the Christian means by which man may return to God."[46] This work forms the basis of "[s]ubstantial portions of the conventional scholastic divinity" of Davies of Hereford's *Mirum in modum.*[47]

A further example evincing this design is Henoch Clapham's *Ælohim-Triune,* the poetic stanzas of which expose God's "*worke*" on each of the days of Creation, "*according to* collection from sacred *Scripture.*"[48] Clapham is no Menippean satirist, but his poem presents us with a vision of the world in terms of a single intellectual pattern, that of the Creation. Nor is he an allegorist, but the organizing principle is assisted by a kind of threshold text, what Clapham calls "The Poemes text," which is unfolded by the stanzas.[49] As with the vast majority of medieval and Renaissance allegories, the threshold text is scriptural,[50] and unsurprisingly given the theme, it is Genesis 1:1–5. Du Bartas's encyclopedic and religious poem, translated by Sylvester, published in *Du Bartas His Devine Weekes and Workes,* and read by Fletcher, is also typical of the expository version of the encyclopedic poetic form.[51] Again, Du Bartas, who was a member of the Navarre Academy, was no Menippean satirist, though Sylvester replaced the mild and liberal Protestantism of the original work by "a dogmatic and uncompromising Puritanic spirit, that hardens the moralizing and places parts of the work very near the boundaries of satire."[52] *The Purple Island* is in some measure a narrative version of the same form, situating its anatomical exposition of prelapsarian and postlapsarian Man and its emblematic, incarnational revelation of the means for salvation in the anatomy of Man, amid the events of the Creation, the Fall, the incarnation, the apocalypse, and the general resurrection.

In his study of the literary anatomical context of *The Purple Island,* Sawday concludes, saying that: "placed along side the work of Kynaloch [*sic*], Norden, Hagthorpe, Clapham, Davies of Hereford, and "Ro. Vn.", themselves representatives of what might be termed a populist approach to contemporary religious and scientific ideas, *The Purple Island* appears not so much as an 'incongruous dragon' as a sophisticated representative of those poetic texts which took the human frame as their subject."[53] The work of each of these authors is capable of providing illuminating comparisons with *The Purple Island,* of the kind outlined in the introduction to this chapter, and of disclosing aspects of a literary-historical context in which *The Purple Island* may be read. Many such works are discussed in following sections of the present chapter, but for reasons that will become apparent, two of these works are selected for especially close analysis: *Mirum in modum* and *Microcosmos* by Davies of Hereford.[54] Before proceeding, it is noteworthy at the outset that Fletcher's poem is not in any straightforward manner a "popular account" of anatomy. Fletcher "wrote to provide courteous little offerings to his friends, to gain the respect of his university associates, and to better his academic position"; and his reluctance to publish does not mark him out as an obvious popularizer.[55] Even the soteriological purpose of the figurative design of *The Purple Island* is primarily for the benefit of a small circle of readers who were linked through friendship as members of an erudite, literate, English Protestant community. Moreover, most popularizers were "content with the facts as set forth in the established authorities," such as Plato, Aristotle, and Galen.[56] Part 1 of this study has demonstrated Fletcher's indebtedness to the mainstream of anatomical concepts, language, debate, and consensus in the sixteenth century, and Part Three will argue that the conceptual character and provenance of the anatomy in *The Purple Island* are not entirely traditional and uncontroversial, even though they undoubtedly remain inconclusive. These initial qualifications aside, the objective remains in the following pages to compare Fletcher's work with encyclopedic poetry and popular, as well as erudite, expository, and fictional anatomy, partly in order to decide to what extent *The Purple Island* may be defined by this context.

Allegory, Analogy, Metaphor, and the Anatomical *Accessus*

As with the *Devine Weekes* and *The Purple Island,* the conceptually mobilizing metaphorical words or subsidiary subjects in the anatomy in Davies of Hereford's *Mirum in modum* and *Microcosmos* are natural terrain and "building."[57] The body's correspondence to the geocosm is underpinned by the poems" adherence to a version of the theory of the microcosm, in which "[t]he *Body* in the Elements is cloz'd."[58] The obscurity of "cloz'd," which may signify spatial enclosure or "united together,"[59] is diminished in *Microcosmos,* where "*Humors*" are said to "proceede" from the "Elements," sharing the latter's innate "qualities."[60] This apparently literal derivation of humors from elements in *Microcosmos* is rationalized physiologically as the digestion of "*foode*" into "*bloode,*" wherein the "*humors*" "hath dominion chiefly."[61] In Norden's *A Store-Hovse of Varieties,* although created "things" are said to be "compounded of the *Elements,*" the "like compounded faculties" in living bodies are "onely like, and not indeede the same" as elements.[62] Distinguishing these faculties from elements is the process of creation, whereby the "*Power* aboue" creates "formall bodie" from the "elements," giving them shape, while "*Nature*" gives to them "life."[63] Notwithstanding their creation, these faculties of bodies "partake" of the "*vertue*" and "*contents*" of the elements.[64] The "*discord*" in "humane *bodies*" between the humors—"*Choler* and *bloud, fleame* and melancholie"—and the "foure *complections*" or Hippocratic qualities these "beget"—"*Hot, cold*[,] *drie,* [and] *moyst*"—reflects exactly the repugnance felt by the "*Elements*" for each other.[65] For Norden, as for anatomists, this conflict, and the fact that the qualities or faculties "Hath not selfe-*nature*" (because "each partaketh of the others *lot*"), are at the root of human mortality and corruptibility, the general law of change to which created bodies—"al the earthly host"—are subject.[66] John Hagthorpe's *Visiones Rervm* claims that "This lesser World," which is "the great ones map," is "Of fire and Water, Earth and Heauen composde";[67] but his *Divine Meditations, and Elegies* avoids the elemental correspondences between the microcosm and the geocosm and macrocosm, and provides instead Paracelsian analogies, some of which may also be found in *The Purple Island* and have been detailed in chapter 2.[68] Among the ones in Hagthorpe's catalog which occur in similar form in *The Purple Island* are those between "Our flesh" and the "Earths superficies"; between "haire" and the "Woods and Groues" ("Earths like ornament"); between "fruitfull Veines, and panting Arteries" in our bodies, and the earth's "christall streames"; and finally, between "Natures Magazines of humiditie"—"The two great Seas"—and "the heart, and Liuer" in our frames.[69]

The complexion of the body in *Microcosmos* also relates chronologically with the planets—"So our *Adolescence* / Is swaid by *Wit*-infusing *Mercury* / Being *hot* and *moist*"—thereby extending the microcosm's analogues from the terrestrial world to the astrological representation of the cosmos.[70] But, rather than Davies of Hereford's poems, Hagthorpe's *Divine Meditations, and Elegies* provides an example of how in poetry the correspondences between the microcosm and the macrocosm may be summarized and explored. Chapter 37, entitled "*The Comparison of the great and little Worlds,*" presents the microcosm-macrocosm analogy as heretofore discussed in chapter 1, combining *aemulatio*—resemblances across space between the firmament and Man—and *convenientia*—similitudes arising from propinquity, adjacencies, bonds, or joints. So, whereas in emulation of the macrocosm "this lesse Worlds frame" is "composde" "Of Heauen and Earth"; within the context of adjacencies in the macrocosm, "The Sunne illuminates the Heauens all, / And giues earth life," and by analogy "in these bodies small, / The soule performes as much."[71] As the "Sunne transmits / His influence, his light and benefits, / Through the tralucent bodies interposed," the adjacent "triple Ayre";

> Even so the sentient Soule likewise susteines
> Both moues and gouerns, as with certaine reines,
> On th'ayrie wings of (b) threefold Spirits sent,
> Each facultie of this her instrument.[72]

The relationship of *convenientia* and of instrumentality between the rational soul and "her instrument" the body is here far less systematically expounded than in textbook versions, but it is nevertheless summarized in broadly teleological terms as that between a horse and its rider.[73]

Within the context of analogy the microcosm and macrocosm (or the geocosm) may in principle either function as principal or subsidiary subjects in a metaphor. For this it is necessary that there is difference as well as similitude between the subjects. Thus Hagthorpe introduces his "comparison" asserting (notwithstanding the grammatical form of the question in which the assertion is rhetorically couched): "How much vnlike this great World seemes to be / Vnto this little World in quantitie?"[74] Yet it is "qualitie" or complexion in which the soul or essential form is often thought to dwell, and "in qualitie how neere" the great and little worlds do "come."[75] And so after his list of Paracelsian analogues, Hagthorpe writes, "How like are these? yet how vnlike againe?"[76] In expository and fictional works on human anatomy, such as this study has focused upon, the usual direction of metaphorical structuring is that whereby the defined term is human anatomy. On occasion, however, in Hagthorpe's *Visiones Rervm* and Davies of Hereford's poetry this direction is reversed, and the body's anatomy becomes the subsidiary subject for the globe: changes are said to occur "*in the intrals of the Earth, her Mines and Treasures*" (*V.R.* I.40 note 1); and "The *Sea* through vaines and Arteries of the *Earth,* / Creeps through her *Corpes,* to fix her droughty dust."[77] Norden also reverses the direction of conceptual structuring, saying that seas are "Linckt in the *earth* with *veynes* and concave *springs.*"[78]

In Davies of Hereford, the earth's corporeal figuration is supported by the poetry's representation of its analogical scheme for natural and divine philosophy (or theology):

> God is a sp'rite, the World a Body is,
> Both which in Man are plaine Epitomiz'd,
> Of God hee's Abstract in that soule of his;
> And in his Corps the World is close compriz'd:[79]

The body and the world are analogues of, and metaphors for, each other, as are the soul and God—understood here as the Platonic "World-soul."[80] The same conception is articulated by Hagthorpe. In his meditation on the microcosm and macrocosm "the supreame irradiance of the mind" is described as "Farre liker to the Worlds high Soule" than is the heart:

> Both incorporeall essences, and high,
> Vnbounded by (c) Time, Place, or Quantitie
> And as the Worlds high Soule, containing all,
> Is not contained.[81]

Since this "Worlds high Soule" is God, who "hath supreame powre / In all," "so by Creation's right" the rational soul has power "Ore this her petty Kingdome."[82] Hagthorpe also adheres to the idealism of Neoplatonism, holding that, "as God is by his infinitie / Of essence euery thing," so are human souls "by apprehending all / Materiall things, and immateriall."[83]

As with the *Devine Weekes* and *The Purple Island,* the poetic form and analogical and metaphorical representational mode of Hagthorpe's *Divine Meditations, and Elegies,* Norden's *A Store-Hovse of Varieties,* and Davies of Hereford's *Mirum in modum* and *Microcosmos* "indicate . . . the order and perfection of the universe as a whole."[84] However, as with the *Devine Weekes,* the correspondence between body and world does not persistently or systematically inform their description in either of these of Davies of Hereford's poems, or in the works of Norden and Hagthorpe, which consequently differ from *The Purple Island.* A reason for the lack of systematicity might be found by comparing in general terms the figurality of Davies of Hereford's poetry with that of *The Purple Island.* The threshold figure of *The Purple Island* is the most specific metaphorical concept of a system of entailment metaphors, the connotations of which emerge from the interaction of their principal subjects (a part of the structure or function of human anatomy) with a specially constructed set of associated implications of the parts and social life of the island realm; and the design of Cantos 2 to 5, unlike that of Davies of Hereford's poems, is to unfold the entailment metaphors. Although Davies of Hereford shares a predilection for complicated figures,[85] neither the world nor the cosmos may constitute for Davies of Hereford, with facility equal to that of the Isle, an analogue that consists of spatial boundaries; and so they discourage a systematic unfolding of the entailments of the figurality that the poems philosophize. Unlike the metaphors and analogues used in *Mirum in modum* and *Microcosmos,* the Isle of Man is unusually both intense and plurisignificant.

The concept of "building" is more systematically applied to the description of the body in Davies of Hereford's poems, although like the imagery of natural terrain it too is restricted by the scope of the anatomy described in each. *Mirum in modum* describes the anatomy of the cranium and the brain, including the ventricles, or "sev'rall caverns of the *Braine,*" while *Microcosmos* discloses the anatomy of the thorax, especially the heart and lungs. *Mirum in modum*'s description of the anatomy of the cranium, brain, and cerebral meninges occurs within a conceptual metaphorical context of the head's figuration as a house, initially by means of simile, and then metaphor:

> The Head is like a House, that is within
> Too many rowmes, or chambers subdivided,
> Vaulted with *Bone,* and with *Bone* likewise sided;
> The skinne that rafters, or else lines the roofe,
> Is hard, for durance, and thicke, to enwall,
>   Which is the skinne of Skinnes; a skinne of proofe
>   That Dura mater loe, the Latines call,
> For it enwombes the rest from dangers all.
>
> The use whereof, is to preserve the *Braine,*
>   (When it doth move) from hardnesse of the Scull;[86]

A narrative fictional poem entitled *The Little World,* the author of which is known only from the title page as "Ro. Vn," is more emphatically and systematically metaphorical than this. The intradiegetic narrator of the story describes "the partes and properties of Man" entirely as "*Houses*":

> These *Houses* they were mooueable,
> and often did remooue
> To places where the *Owners* thought
> it fit for their behove.[87]

In metaphorically translating Man thus, both Davies of Hereford and "Ro. Vn." draw on the same conventional metaphor which informed Quarles's untitled prefatory poem to *The Purple Island,* and the conception of the House of Alma. In *The Little World,* however, the metaphor so informs the partially allegorical work that it becomes a threshold figure, like the Isle of Man. As with anatomy textbooks, following the order or method advised by

Galen, *The Little World* begins by anatomizing "The Bones," "The *Pillars* and the *Timber-worke,*" "framed all of Boane." After a brief look at "The Sinewes," each house is found to have a "*Chimney,*" which like that of the House of Alma rises to "*The noctrils.*"[88]

As we know, *The Purple Island* extends figuration to include the human-built environment of the dynastic realm. Similarly, *The Little World* describes "A thousand thousand *Houses*" in "A *Cittie* large, of bignes such, / as it the World had beene" (the marginal comment confirms that this city should be interpreted as "*the world*").[89] But that which is a part of this world, city, or "Common-wealth" may well be compared to it, because, according to *The Little World*'s narrator, the "ordering" and "government" of anatomy "Even from the state of Common-wealths, do differ not a whit."[90] As a "Symilie" of the body, the features of the "Common-wealth," "Cittie," or "Corporation," consist of "a Mayer, / And many other Officers," "Commons too" and "gouernment."[91] In the person or subject, the "*Hart* of man" is "*The Court of Conscience,*" the "Ecclesiasticall" Court, where this "Iudge," "chiefe Iustice" or "*Magistrate*" is "*A good Conscience,*" or "the *Sperit of Grace,*" which may be supplanted by "*A badde Conscience.*"[92] There is also in the skull the Court "of the Common Lawes," which "houlden be, / for matters that depende / Twixt man and man."[93] The rules of these two courts derive, respectively, from "The Gospell, and the Morall law," or "Commandements."[94] Besides these, the commonwealth is composed of "*The Commons, or communaltie,*" the "men of Trade, / Artificers, and Laborers," which "Societie maintaine"; while "in euery *Body,* or / in euery *House*" are said to be "Such partes and members" "as may compared bee / To *Commons.*"[95]

The narrator of *The Little World* provides an organicist explanation of the comparison between body and polity in the origin of the "Realme," the "Common-wealth," and the "Cittie" in anatomy: "all Common-wealths / do of our *Bodyes* come."[96] Though he does not subject his assertion to analysis, it is consistent with the view that both phylogenitically and ontogenetically the body is one of the first and most basic entities the mind can grasp. As it has been said, however, the realm could also reflect the cosmos.[97] Those comparisons as are described in *The Little World* are merely functional rather than morphological, and unlike the correspondences in *The Purple Island* they deploy few if any anatomical *occasiones.* The systematicity of the entailment metaphors of the threshold figure in Fletcher's poem has been exemplified by the poem's descriptions of the anatomical structures and functions of the visual and auditory apparatuses. These descriptions also demonstrate how anatomical or physiological operations take various socially integrated personified forms, and how these personifications are integrated with the somatic processes upon which they are based. The comparative lack of such systematicity and integration in *The Little World* is abundantly evident from the following description of the same sensory apparatuses of the body on a similar analogy with intelligence-gathering in the commonwealth:

> It hath Intelligencers, which
>   seeke out to see and heare
> What newes abroad, both good and bad,
>   and then the same they beate
> Vnto the owner of the House,
>   and him thereof doth warne,
> That he the better may prevent
>   the things that might him harme.[98]

Notwithstanding its lack of systematicity and of anatomical specificity, this description contains the essential idea of the metaphor of defense and surveillance. There are other textual parallels between the two poems, but if *The Little World* provided a source text for Fletcher, its relatively rudimentary figurative ideas were elaborated and extended in *The Purple Island* in an anatomically consistent way to produce new similarities.

The Isle of Man by Richard Bernard also represents the body as an epitome of that of which it is also a part. Its promising title is a misleading guide to the work's content, however, and it is unlikely to be a source text for Fletcher's poem. The metaphor of the body as an island is not originally an early modern one, after all; it occurs in ancient and medieval didactic literature. But the anatomy of *The Isle of Man* is concerned throughout with *abstract* Man, rather than with the self-determining or self-defensive, paranoid *individual.* Moreover the proto-nationalism and proto-individualism which is represented in *The Purple Island* by the entrenchment of the Isle's borders by the *mare mortuum,* finds no such geographic or semiotic foothold in *The Isle of Man,* which is better represented by the subtitle: *The Legall Proceeding in Man-Shire Against Sinne.* Indeed the work figures Man as "the County Towne" and proceeds to describe the "chiefest Seate there, called *Soule,*" or "*Soules-towne,*" the "Common Inne" of which is "the *Heart.*"[99] Despite its title, Bernard's work in no way anticipates the set of associations relating to the historically incipient and ideologically emergent concepts of individualism and nationalism, which are connoted by the Isle of Man in *The Purple Island.*

Davies of Hereford's *Microcosmos* describes the thorax or "*Brest*" as resembling a "*Shoppe.*"[100] Yet it also extends its comparison to incorporate not only the material defenses, but also the social organization, of a feudal castle, wherein:

> The Hart, the Lunges, with al Life's incidents
> In region of the Brest, doe hold their States,
> Whose Bulke them Bulwarkes from what ruynates:[101]

The pre-Harveian model of the "double motion" of the heart, its dilation and "restraint," it is said in *Microcosmos* "hath two double uses," by means of the first of which, air is drawn from, and "smoky *Excrements*" sent to, the lungs, and by means of the second, blood is drawn from the liver, and sent to the lungs.[102] This physiology is used to demonstrate how the heart and lungs" "mutuall love" and "interchange" of "*kindnesse*" should govern social relations; in addition to which the heart's anatomical substance and location, and its physiological reliance upon the "helps" of "*members,*" should induce monarchs to be "most firme" and just, though "not to tyranize, / But, like good *Hartes, life's*-pow'r to exercize."[103] A footnote summarizes the strategy of this verse passage of the poetry: "A motiue to brotherly loue taken from the disposition of the Members."[104] The *occasiones* of the anatomical *accessus* are thereby deployed in the service of a socially prescriptive model, which is akin to that in *The Purple Island,* and also warns against tyranny. In Norden, "humane *bodies*" and the "body *politike,* or publike *state*" are also analogous, the latter having "like dissents, which yet assenting stands," because although, like the human body, it is "Dislike in parts," these parts "in *consort* agree."[105] Similarly in *The Little World:*

> And as a Realme, a Common-wealth,
> a Cittie, or a Towne,

Cannot continue long, but is
  in state of falling downe,
If that the Common-people, which
  within them do remaine,
Do not together ioyne as one,
  their state for to maintaine.

. . . .

Even so it with our *Bodyes* stands:
  for if it so befall
The Eye, the Eare, the Hand, the Foote,
  the parts and members all,
Ioyne not together all as one,
  the *Body* to sustaine,
Wee see by dayly proofe, the *Body*
  cannot long remaine.[106]

The anatomy of Man in *Microcosmos* and *Mirum in modum* is organized around the principle of *nosce teipsum* (know yourself), rather than *discordia concors,* but both organizing principles function ideologically in their respective works to legitimize as consensual the maintenance of social hierarchy and inequality.[107]

The location of the "interchange" in the thoracic castle of *Microcosmos* is reminiscent of Alma's castle in *The Faerie Queene,* though the latter comprises the whole body. The "*Turrets*" of the houses in *The Little World* by "Ro. Vn." are in some respects more reminiscent of Alma's castle and *The Purple Island,* for the turrets are "*Their heads,*" in the façades of which are "two *Windowes*" for "*The Eyes,*" and the narrator, like Guyon and Arthur, enters the body-house via "a *Dore*" that is "*The mouth.*"[108] But in division and ramification of structure, the anatomy in *The Little World* falls short of that in *The Faerie Queene* (II.9.23–26), which describes the "gate" of the mouth, through which Guyon and Arthur pass in entering the "porch" of "hewen stone," or jaw; further within which the knights encounter the "Porter" sat in "the Barbican" or oral cavity, from where they see the "Twise sixteen warders" or teeth (*F.Q.* II.9.23.1; II.9.24.1; II.9.25.1; II.9.26.2). The narrator of *The Little World,* by contrast, takes in at a glance "a *Mill,*" which the marginal note identifies as "*The Teeth,*" before mounting the "chamber" of the cranium.[109] Those parts of human anatomy described as in Alma's edifice, although not by any means comprising a whole anatomy, are nevertheless simultaneously more metaphorically transformed and more anatomically detailed.

Whereas in Alma's castle, the physiological operation of concoction takes a highly personified, socially integrated form, in *Microcosmos* the personification of the heart and lungs is comparatively slight. One reason for this is that, in *Mircrocosmos,* the anatomy and physiology of the thorax constitute a prescriptive corporeal model of society, for which a sufficient, yet not distracting degree of personification is required. To make a local comparison between figurations of the heart in *The Faerie Queene* and *Microcosmos* is to discover further why the feudal thorax of the later poem is comparatively bereft of figurality. In her castle, Alma, a personification of "the rational soul," in a situation consistent with Aristotelian anatomy, holds state within the heart, or "goodly parlour . . . / That was with royall arras richly dight," where dwells "[a] lovely bevy of faire Ladies," such as "Prays-desire" and "Shamefastnesse," who personify the heart's affections (*F.Q.* 2.9.33.6–7; 2.9.34.2;

2.9.39.8; 2.9.43.9).[110] In the text of *Microcosmos,* the exposition of the soul's three kinds of appetite precedes, and is resumed after, that of the heart and thorax, which in turn precedes the poem's disquisition on pleasures, affections, policy, and the soul's composition, location, origin, and end.[111] Consequently there is in *Microcosmos* comparatively less textual integration of the spiritual or psychical with the somatic.

This feature of the whole work reflects on the level of individual linguistic figurality the lack of an intense metaphor in *Microcosmos* that may be compared with the Isle of Man, and which may be unfolded poetically to reveal a systematic integration of its subjects. The lack of integration entails less personification of members, and less fictionalization of anatomy and physiology in general, for in Aristotelian teleology the soul not only enforms but invigorates or actualizes the otherwise lifeless body. This is why in *The Purple Island,* as has been said in chapter 2, personification occurs in proportion to an organ's endowment with either natural, vital, or animal spirit (which of course arise at the organic sites of the soul in the body), and to the pronouncement or perceptibility of physiological activity. The textual arrangement of *Microcosmos* is repeated with the same effects in *Mirum in modum,* where a formal division of the "*Animall, Vitall,* and the *Naturall*" "*Faculties*" of the soul precedes the poetic dissection of the skull and brain, after which the poetic voice moves on "to prosecute the rest," which consists of a further exploration of the "*Animall*" faculties's subdivisions, introduced formerly, much of which is explanatory of involved theological suppositions and philosophical principles.[112] "In scope," both of Davies of Hereford's poems "mirror the programme undertaken by the French Encyclopediasts," especially de Mornay and La Primaudaye,[113] but they lack the concentration possessed of some Menippean satire of a single intellectual pattern or organizing principle. This principle belongs to *The Purple Island* on account of the existence on the level of individual linguistic figurality of an intense metaphor, which both of Davies of Hereford's poems lack.

A substantial portion of *Microcosmos* is devoted to a theoretical and historical digression on "*Policie*" and the body politic.[114] During this, the poetic voice declares that in the lurid exposition of "the last *Richard* [III] may be lively seene / *Ambition* really *anatomiz'd,*" thereby identifying the foregoing history of the kings of England as a moral dissection or analysis.[115] Furthermore, in the digression, the polity is intermittently described in conditions of dismemberment and disease, while policy is conceptualized as medicine.[116] Unlike the geopolitical realm of *The Purple Island,* however, the model of the body politic in *Microcosmos* is only inconsistently informed by its corporeal analogue or subsidiary subject, and is expounded for the most part without even oblique reference to anatomy. The poem's textual distinction between anatomical and political expositions is not designed to unfold correlations which, in *The Purple Island,* are delineated between human anatomy and its geopolitical subsidiary subject.

The anatomy in *Mirum in modum* and *Microcosmos* is less personified, and perhaps less figurative in general, than that of either the House of Alma or *The Purple Island,* yet is more detailed and less schematic than that of either the *Devine Weekes, The Faerie Queene,* or the works of Hagthorpe and Norden. Sawday remarks: "Given the importance attached to anatomy in the Encyclopedic programme, the presence of the subject in Davies of Hereford's poetry should be no surprise," but what is surprising is "the detail with which Davies of Hereford is willing to pursue his anatomic investigations."[117] In Sawday's view, the "discussion, in *Microcosmos,* of the structure and function of the heart is exhaustive—as though the author envisaged his work being read as a medical text rather than a poem."[118] The discussion certainly exhausts all, but one, of the ordinary *occasiones* of the anatomical *accessus:* the composition of the heart's "flesh" is said to be "firmer, then the flesh / Of all the *parts* the *Body* hath besides";[119] the locations in the thorax of the "*Midriff,*" or diaphragm, the thoracic cavity's "*Tunicle,*" and the lungs bear relation to, and help indirectly to reveal, the specific location of the heart, enclosed within its "*Pericardion,*" the fibrous bag or "*Caule.*"[120] Yet more precise still is the description of the heart's situation, for which anatomical reasons are provided:

> And in the *Bulke* it is so situate
> As that its *Base* is *Center* of the *Brest;*
> The end whereof (where *greatnesse* doth abate)
> Leanes to the *left*-side more then all the rest;
> . . . .
> Yet leanes the *Hart* so, for two causes great;
> One, that the *Brest*-bone should it not infest,
> The other, that it should the left-side heate,
> Sith on the right, the *Liver* doth that feate.[121]

In this description of position is implied the heart's quality or complexion (in the Hippocratic tradition), that is, its "Naturall heate," of which the heart's form or shape—"Ovall round, / Or like a *Flame*"—is held to bear the signature, and thereby "[t]o shew that vitall fire is there repos'd."[122] Although it makes no obvious distinction between the auricles and ventricles of the heart,[123] the description of the "left concave['s]" posture in relation to the curve of the aorta, where it divides into the abdominal aorta and the vessels going to the head and arms, is clearly represented, as is, to a lesser degree, the heart's connections with the pulmonary artery, or "*Arteriall Veine,*" and pulmonary vein, or "veyny-*artire.*"[124] Treated in anatomical detail over two stanzas, prior to its employment as a prescriptive model for "how we should our selves behave," is the pre-circulatory function of these vessels and the "double motion" of the heart.[125] Although the poetic voice concedes that, "[t]he hart (besides) hath many *members* more, / Which are distinguisht by *Anatomists,*" those it does describe are rendered with a precision unparalleled by either the *Devine Weekes* or *The Faerie Queene,* and only surpassed by *The Purple Island.*[126]

Describing individual humans as houses, the narrator in the narrative of *The Little World* remarks upon the "fashion, stuffe, and workemanship," and the "order of these *Houses.*" Their substance is "of mixed stuffe," and "deriued from the *Fire, / Water, Earth,* and *Ayre.*"[127] There is no suggestion here of the Hippocratic qualities or the humors, or indeed of the rational anatomy of Aristotle, de Laguna, and Vesalius. The soul does not reside mysteriously in the anatomical criterion of complexion, as it does for overtly religious anatomists like Crooke; nor does it enform but rather owns its house, in the construction of which elements themselves are "Compacted by . . . cunning skill" or ingenuity.[128] Intermittently amid rather cursory and highly figurative glances at the structure and function of the parts, *The Little World*'s narrator decelerates the pace of his narration as against events in the story to admit into his description a more thorough application of the anatomical *accessus* than he generally allows.[129] These decelerations of pace correspond to significance in the story as in the anatomy.[130] One such account is of the heart and includes a description of its size as "a little thing," its shape or "fashion like a *Pine-aple*" with a "hollownes in it," and its "cullour somewhat redd," though with "the greater end" overspread with "a Cannôpie" of its own "fatnes,"

"That was of cullour very white."[131] Color is frequently in anatomy related to complexion, though from the combination of white and red "Ro. Vn." draws a personal and social moral:

That he that sits therein as Iudge,
must execute with speede
True *Iustice:* yea and *Mercie* both:[132]

The moral and social application of anatomical description far outweighs in *The Little World* anything approaching the preoccupation with detail we find in Davies of Hereford, or Thirsil's enthralled observation of the exquisite craftsmanship displayed by each anatomical part's fitness for its function. Consequently it is a relatively impoverished anatomical *accessus* that guides the observations of *The Little World*'s narrator.

Although the precision of the anatomy of the heart is unequaled elsewhere in Davies of Hereford's poetry, the description of the house of the head (cited above, from *Mirum in modum*), in the midst of its deployment of simile and metaphor, applies four anatomical *occasiones:* the location of the bone of the cranium; the posture, or connectedness, of the skull and dura mater; the latter's "hard" substance; and its "use." It might also be argued that a notion of the skull's shape, albeit imprecise, is conveyed to the reader by the terms "vaulted" and "sided," which extend from its metaphorical vehicle. This anatomy of the brain and cranium, however, is rather more schematic than that found in *The Purple Island* (Canto 5, stanzas 11 to 21), which is discussed in chapter 4. Elsewhere, in *Mirum in modum,* the criteria of size and number are included in the anatomical descriptions.[133] And, in *Microcosmos,* the description of concoction, although not detailed according to the anatomical *accessus,* is less formulaic than that in *The Faerie Queene,* and in its elaboration of processes resembles *The Purple Island,* including the third concoction, that of the "*Masse Sanguiner,*" or blood, which occurs "through al the *bodie,* where / *Humors* are made, that *Meate* and *Chyle* first were."[134] Although not consistently more detailed or precise than *The Purple Island,* Davies of Hereford's anatomy and physiology is clearly less schematic and formulaic than either Du Bartas's or Spenser's.

Sawday suggests two ways of accounting for this anatomical detail: "In part, Davies is responding to the example of the Encyclopedic authors. In part, too, it is a literal response to the patterns of enquiry suggested by Davies of Hereford's namesake, Sir John Davies, in his *Nosce Teipsum* (1599)."[135] While it is widely acknowledged that doubt and uncertainty pervade cognition of the body in poetry by Sir John Davies (1569–1626) and John Donne (1572–1631),[136] the character of each of the Davies's responses to the encounter with human anatomy is significantly different, as Sawday himself acknowledges: while "it is the potential strangeness of the human frame which has caught Sir John Davies's imagination," and caused him to "evoke this foreign element in the understanding of the body as itself a metaphor for human ignorance, Davies of Hereford is anxious to dispel doubt and uncertainty. Hence his attention to detail," and (it might be added) to the deployment of anatomical nomenclature and the *occasiones* of the anatomical *accessus.*[137] Although anxiety may signify an encroaching doubt, the poetic anatomies of Davies of Hereford apparently demonstrate a decidedly less skeptical attitude towards natural knowledge than does the work of his namesake. The significance of *Nosce Teipsum* for *Mirum in modum* may further be reduced if the intertextual influence is less than was formerly supposed. It had been thought by many that Sir John Davies's *Nosce Teipsum* formed the basis of *Mirum in modum,* and that the *Devine Weekes* of Du Bartas, as translated by Sylvester, "provided a model" for *Microcosmos.*[138] It was argued by Ruth L. Anderson, however, that "John Davies of Hereford drew his theory of the soul directly from La Primaudaye and that *Mirum in modum* and the *Microcosmos* are much closer to *The French Academie* in both thought and language than is *Nosce Teipsum.*"[139]

Sawday also suggests that the attention to detail in Davies of Hereford's work meets the challenge to anatomists made, in 1605, by Francis Bacon (1561–1626): "they inquire of the *parts,* and their *substances, figures,* and *collocations;* but they inquire not of the *diversities of the parts,* the *secrecies of the passages,* and the *seats or nestlings of the humours,* nor much of the *footsteps and impressions of diseases.* . . ."[140] The attributes of parts, to which, Bacon is saying, anatomists do attend, are the anatomical *occasiones* of substance, form, and posture, the latter possibly combined with location. At its most detailed, Davies of Hereford's anatomy abides by all the usual descriptive criteria of sixteenth-century anatomy; but, he does not attend to the "diversities" or "secrecies" of the parts, nor to the anatomical "seats" of the physiologically significant humors, nor to the etiologies and symptomatologies of diseases. Although they were to become immensely important as subjects in their own right in the seventeenth century, physiology and pathology until the late sixteenth century are confined to just two of the anatomical *occasiones:* a part's function, movements, use or operation; and its illnesses, affections, or passions. Although the description of the function of the heart and lungs is, as Sawday claims, exhaustive, and that of concoction as systemically developed as that in *The Purple Island,* the physiology of *Microcosmos,* and of *Mirum in modum,* is otherwise confined to comparatively slight descriptions, and none of these poems ventures any pathology.

Furthermore, Davies of Hereford, as Sawday observes, does not describe "the anatomy [or physiology] of the complete body."[141] His work in this respect differs from *The Little World* of "Ro. Vn.," which after describing the bones, albeit summarily, proceeds to "the *kitchen,*" the "nethermost of all" the venters, fitted with "*Gutters, Holes,* and *Channels*" to "convay" "bad" and "noysome" things.[142] This figurative conception of concoction as cooking is close in some cases to Alma's house, and in others to *The Purple Island,* as is the description of functions and parts. In the midst of the kitchen in *The Little World,* "A *Pott* hunge boyling," and "under it a *Fountaine* was" ("*The Lyver*") "out of which did flow, / A lickuor very like to *redd* / or *Claret Wine* in show"—clearly a Galenic account of the function of the liver to produce the blood and natural spirit distributed in the "*Vaines.*"[143] The fountain in *The Little World* and *The Purple Island,* as we have seen, is a furnace in *The Faerie Queene.* In the order or method of discipline prescribed by de Laguna as practically necessary for demonstration and consistent with the tripartite division of the body according to the functions of the natural, vital, and animal faculties, the *Little World* narrator's attention moves to "a large and a faire *Hall,*" which corresponds to the "goodly Parlour" in Alma's House (*F.Q.* II.9.33.6), and as the thorax or middle venter it contains "a Throne of Maiestie," "a Chaier of estate," which the marginalia tells the reader is "*The Heart.*"[144] Finally the narrator enters by the "*Dore*" of the mouth the "*Turret*" of the head, the highest venter, containing in the cranial "chamber," a "*Bed*" which is "*The Braine.*"[145]

At the midway point of the text of *The Little World,* "*Philotimus,*" or Truth, says that he has heard of one "that tooke in hand / For to describe in every part, the Body of a Man," thereby summarizing (somewhat disingenuously) the achievement of the poem in having described "Both inwardly and outwardly, / from topp to toe," human anatomy, which it has compared "vnto a House," "well gouerned / by one of worth," applying "euery part which in / the Body wee do see, / Vnto some necessarie thing, / which in an House should bee."[146] In fact vast swaths of members and parts are omitted from this description, but unlike the anatomy surveyed by Davies of Hereford, that in *The Little World* belongs to the lowest as well as the middle and upper venters.

The cursory treatment of some of the physiology, and the anatomical and physiological omissions in Davies of Hereford's poems are related to, and perhaps an effect of, the anatomy's dominant metaphorical conceptualization. This may be observed, when, towards the end of *Microcosmos,* the expository poetic voice refrains from describing the abdomen, "the third *Wombe* situate below / The *Midrife,*" deferring the task:[147]

> unto *Anatomists,*
> Who marke each *Mortesse* of the *Bodie's* frame,
> The *Pynns,* the *Tenons, Beams, Bolts, Windings, Lists,*
> All which they marke when they doe it unframe:
> To these *Craft's-masters,* I referre the same.[148]

In what may appear to be a gracious gesture to the professional anatomists, the poetic voice concedes that the complexity of the structure of the body is beyond the scope of the poem's encyclopedic project. This complexity is evoked by an accumulation of terms emerging from anatomy's conceptualization as "building," because, as Sawday explains:

> Anatomy, like carpentry, is a specialized craft. The timber construction, analogous to the human frame, may be taken apart methodically by specialists versed in the complexities and vocabulary of their trade. It is as though Davies of Hereford wants to emphasize the specialist nature of the undertaking by invoking a language ('mortresse' [*sic*], 'tenon', 'lists') appropriate to a trade less foreign to his lay readership.[149]

But this is no gracious gesture. By deploying the vocabulary of carpentry, as does the *Devine Weekes* to similar effect, this stanza of *Microcosmos* creates a palpably false impression, that the complexities and details of anatomy and carpentry are comparable, both in scope and kind, something which Berengario's comparison between anatomy and "building" does not do. In anatomy textbooks the involuntary movements, however formulaic, rarely fail to demonstrate this metaphor's inadequacy if it is not combined coherently with the metaphor of anatomy as vegetation. This is unless the system of associated commonplaces or connotations of "building" are not developed beyond material structures to social practices, as is the case with the heart and lungs in *Microcosmos,* and to a greater degree with concoction in *The Faerie Queene* and with most of the physiology in *The Purple Island.* Moreover, the vocabulary of the domain of the subsidiary subject or defining term can only vaguely represent a principal subject that is omitted entirely from description (as is the abdomen in this case).

Mirum in modum and *Microcosmos* are not designed to comprise "an exhaustive treatment of the anatomy of the complete body," and not only because "the abdomen is not dealt with."[150] The skeleton is represented entirely in terms of its metaphorical translation into the walls, vaulted roof, and fortification of a building, and the limbs, ears, eyes, and all external anatomy goes without dissection. The reason for the remarkable (by contrast) adherence to the anatomical *accessus* and nomenclature with respect to the brain, the heart, and the members in their vicinity, is that their exposition bears directly upon the soul's embodiment in what Neoplatonic and Protestant anatomy in particular regarded as "the image of God, or His temple on Earth."[151] In complete accord with the Galenic hierarchical organization of the members, the brain is for Davies of Hereford the seat of the "*Animall*" faculty of the soul, the "*Principall*" parts or faculties of which are "*Imagination, Reason, Memory,*" just as they are in Paré's detailed division of the faculties and functions of the soul, and they correspond to Spenser's three honorable sages; and the heart is the seat of the "*Vitall*" Faculty (as it is in Paré) from which proceeds "Life," and in which exists "*Ire.*"[152] The "*Naturall*" faculty of the soul, or "*Concupiscence,*" which consists of the functions of "*Feeding, Growing,* and *Ingendering*" (the nutritive, auctive, and generative faculties and functions in Paré),[153] resides in the "*Belly*" which Davies of Hereford also calls the "wombe," thereby flexibly extending that term's anatomical applicability in the manner of popular anatomical and medical works to the whole abdomen of either sex.[154] But whereas the "*Brayn-wrights* skill, / And wisdome infinite do most appeare" in the "middle *Ventricle*" of the "*Brayn,*"[155] the lowest region or venter of the body is felt in *Microcosmos* to be:

so farre remov'd from thence
From whence the *Soule* doth her *arch-wonders* show,
(Namely the *Seate* of the *Intelligence*)
Wee'l balke the same for its *impertinence.*[156]

To "balke" (otherwise "baulk" or "balk") has from the fifteenth century meant, among other things, to ignore.[157] Thereby contemptuously dismissing the abdomen, and foregoing osteology and the anatomy of the extremities, Davies of Hereford implies not only that the body should be understood above all Neoplatonically and teleologically as a building for, or denigrated as the "*muddy* walls" of, the mental and affective soul; but that it is safe presumptuously to assume that anything anatomists may find is sure to support this understanding.[158] The accumulation of terms of carpentry, as Sawday rightly suggests, "signals more than a sense of the poem's growing unease with anatomical language."[159] In addition, it is symptomatic of an epistemological anxiety to dispel doubt and uncertainty about the body's reason for being, its role in Christian Neoplatonic and Aristotelian thought as the temporary lodging and instrument of the immortal soul. Among other rhetorical strategies, that may have the consequence of allaying epistemological anxiety of this kind, this will be examined more closely in the following section of the present chapter, where it will be argued that the epistemology and hermeneutics of the poetry of Du Bartas, Davies of Hereford, and Fletcher are best understood in relation to the concept of ingenuity.

From the standpoint of the explanation of the works that have been compared with the figurative design of *The Purple Island* in this chapter, Clapham's *Ælohim-Triune* possesses a single intellectual pattern, not of Man but of the Creation, and although it is not allegory it has a threshold text which is unfolded by the stanzas. *The Little World,* moreover, is allegorical and its threshold figure is closer to the Isle of Man in conception. As in *The Purple Island* the unfolding of its significant connotations follows an order or method of dissection advised by Galen, beginning with the bones. None of the other works contains a threshold figure but many use metaphorical words or subsidiary subjects derived from the discourses of natural terrain and "building," and from the geocosm, macrocosm, and body politic. Thus the body is compounded of faculties like the elements, or consists of Paracelsian analogies, or gives rise to the conceptualization of the realm, the commonwealth, and the city, some of which connotations are also found in *The Purple Island.* The root of this figurality in conventional analogies of the microcosm is evident when the direction of metaphorical structuring is reversed, and the body's anatomy becomes the subsidiary subject for the terrestrial globe. Sometimes figurality is rationalized in terms of natural and divine philosophy, or Neoplatonism, or organized around the principle of *nosce teipsum* or *discordia concors.* Despite many similarities with the figurative design of *The Purple Island,* none of these works has been found to represent a comparable selection and organization of features of the principal subject, Man, or explore a comparable range of previously unseen or newly emergent similarities and significant connotations. In most cases, the significant connotations of the figurality of these works are traceable to antecedently existing similarities in analogy.

Though they do not constitute a firmly knit family of works, each of the encyclopedic, fictional, and expository anatomies we have compared with *The Purple Island* adheres to at least some of four interrelated concepts: erudition and self-knowledge, opening or discovery, division or dissection, and analysis or justification. These are the conventional, literary associations and commonplaces of anatomy. Several works consist of an erudite mass of scientific, quasi-scientific, and theological or philosophical speculation and observation, thereby reflecting the erudition for which *The Purple Island* was admired. But only some present us with a vision in which this erudition is integrated and organized in terms of a single intellectual pattern, as in Fletcher's poem, where the metaphorical concept of the Isle of Man is the organizing principle for an assemblage of correspondences and for a systematic exploration of metaphorical connotations and etymons. Besides erudition, anatomy in the titles, texts, or designs of works connotes a concept of disclosure, opening, or "handling" a subject. But it also takes from anatomy the notion of a reiterated incision, a division into several parts. Farraginous, full of detail, and subdivided, many of these works nevertheless share with *The Purple Island* some mode or other of rational analysis or discrimination. All relate in some measure to anatomy of the philosophical and medical textbooks of the schools in this discerning act of analysis, which has a theological dimension, in which to take in hand the rational work of dissection is to do "God's handy worke": both to reveal and to perform God's will in the flesh. Though they concur in the profit to be had by anatomy, and rarely isolate merely for the purposes of identification or enumeration, they are inconsistent in their application of the anatomical *accessus* or "method" to the human body (the observations of *The Little World*'s narrator being especially impoverished in this respect though more comparable with *The Purple Island* in others).

Although not generally more detailed or precise than *The Purple Island,* Davies of Hereford's anatomy and physiology is clearly more detailed and less schematic and formulaic than all the other works compared, including those of Clapham, "Ro. Vn.," Hagthorpe, and Norden, and those of Du Bartas and Spenser, which were compared in the previous chapter. But an effect of the anatomy's dominant metaphorical conceptualization is that the anatomy of function (physiology) in *Microcosmos* and *Mirum in modum* is comparatively slight, and the anatomy is not that of the complete body. Davies of Hereford's work in this respect differs from *The Little World* of "Ro. Vn.," though that work omits vast swaths of detail. Unlike the geopolitical realm of *The Purple Island,* the model of the body politic in these other works is in almost all cases, including *Microcosmos,* only inconsistently informed by its corporeal analogue or subsidiary subject. In the one case of the figuratively complex conceptualization of concoction as cooking, significant similarities exist between some of these works, but none compares favorably with *The Purple Island* in the extent to which its intention to ingeniously contrive a world consisting of figurative anatomy is achieved. No work's extension of meaning compares with that of *The Purple Island* for comprehensiveness, systematicity, and coherence. The correspondence between body and world does not persistently or systematically inform the description in either of Davies of Hereford's poems, or in the work of Norden, or that of Hagthorpe. Unlike the metaphors and analogues used in *Mirum in modum* and *Microcosmos,* and most of the works discussed, the Isle of Man is unusually both intense and plurisignificant.

In *The Purple Island,* anatomical or physiological operations take various socially integrated personified forms, and these personifications are integrated with the somatic processes upon which they are based. The comparative lack of such systematicity and integration in any of the compared works is notable, and entails less personification of members and less developed fictionalization of anatomy and physiology in general. More comparable in other respects, the anatomy in the didactic poetry of Davies of Hereford is less personified and perhaps less figurative in general, than that of either the House of Alma or *The Purple Island.* On the question of innovation, however, in *The Little World* a description of the visual and auditory apparatuses on a similar analogy with intelligence-gathering in the commonwealth might suggest that some of those of the significant connotations of the Isle of Man figure formerly thought not to be reducible to antecedently existing similarities, notably those of defense and surveillance, may not be novel. Though *The Little World* shares this set of connotations, it does so in a relatively rudimentary and unsystematic way, in which the anatomy is unspecific. By contrast the ideas were elaborated and extended in *The Purple Island* in an anatomically consistent way to produce new similarities. Additionally, *The Little World* does not in proto-nationalistic fashion identify its principal or subsidiary subject with its borders, as the Isle of Man is identified in terms of its shores.

Several works are comparable with *The Purple Island* in bringing developments of knowledge in natural philosophy into harmony with the revealed truth of scripture in order to praise God through a poetic summary of his works. Some are organized as expositions of Creation and designed to instill faith in the resurrection, as in similar fashion the main story line of *The Purple Island*'s hypodiegetic level is providential history: it represents anatomy as created and redeemed, and its illocutionary act is to dispense the balm of salvation through the eloquent exposition of divine ingenuity and saving grace. In this respect, and on account of the conceptually imperfect differentiation between natural and revealed knowledge and ways of knowing, discussed in chapter 3, literary anatomies become an "analogy of revelation." As with the *Devine Weekes* and *The Purple Island,* the poetic form and analogical and metaphorical representational mode of Hagthorpe's *Divine Meditations, and Elegies,* Norden's *A Store-Hovse of Varieties,* and Davies of Hereford's *Mirum in modum* and *Microcosmos* didactically indicate the order and perfection of the universe as a whole. No comparable works, however, share *The Purple Island*'s complex figurative integration of providential narrative and Christian anatomy with scholastic and humanist anatomical exposition and method, and emblematic, "Horatian" poetic theory, suprarealist Augustinian incarnational language, and quasi-Spenserian allegory. This figural and intentional complexity and sophistication of *The Purple Island* do mean that, as literature, it stands apart from much encyclopedic poetry and popular, as well as erudite, expository, and fictional anatomy, but our understanding is enhanced by if not determined by this context.

THE SOTERIOLOGICAL EPISTEMOLOGY AND HERMENEUTICS OF INGENUITY AND ELOQUENCE

In setting out the aims of this chapter, the introduction stated that in all respects the figurative design and soteriological purpose, or illocutionary act, of *The Purple Island* would be compared with other literary works of the period. Though discussion is incomplete with regard to those aspects of the design and purpose that have been compared, there remains an aspect of the design that has not yet even been broached in the context of comparison. When the conceptual metaphors of anatomy and the analogies of the microcosm and microchristus cohere and coalesce in the vegetative, terrestrial, and architectural figurality in *The Purple Island,* they are consistent with, and provide in allegorical form, the teleology of rationalist Renaissance anatomy, embedding the final cause of structure and function within the figurative language as a significant connotation. For Fletcher, as for Renaissance anatomists, this teleology was Christian in that the rational inference of function and cause from structure purportedly demonstrates the ingenuity not principally of Nature but of God. It is in part the aim of the present section to discern the teleology of rationalist Christian anatomy where it exists amid the texts compared with *The Purple Island* above, though particular attention is given to the works of Davies of Hereford because of their philosophical underpinning of anatomy and figurality. Initially this will take the form of an examination of epistemology, and the role of sense perception therein, for which a reminder of the relevant points of our discussion of Aristotelian teleology in Chapter 1 may be helpful.

In chapter 1, where the rationality of Aristotelian, Galenic, and Renaissance anatomy was under discussion, it was argued that according to this tradition, it is by the appearance of a thing that one may rationally infer its function and its reason for being, and that the form of the whole body is understood to fulfill the function the purpose of which is to serve the soul. This rational inference, it was said, was rejected by skepticism, which found its way into the empirical tradition in Greek medicine, the adherents to which tended to proceed to the administration of therapeutic measures via *autopsia* (personal experience) and careful assessment of *historia.* According to Sextus Empiricus, the skeptic does not assent to the rational inference of nonevident objects of inquiry from appearances, and regards such assent as a positive epistemological dogmatism (of which positive ontological dogmatism is an implication, because to make any nonempirical or transempirical proposition on the basis of a rational inference from appearance implies that the object of the proposition is real and does exist).[160] For Aristotle, by contrast, sensory observation must be linked via rational inference to that which for Nature is real, and consequently really defines something, that is a thing's use, utility, or action, which discloses its form, its reason for being.[161] While retaining its rationalist basis and the "broadly teleological" conception of structure, however, Renaissance anatomy was increasingly seen by its practitioners as "a well-defined territory with a special sensory epistemology."[162]

At the beginning of the sixteenth century, such questions as had been resolved by choosing or reconciling authorities, such as those which Andrés de Laguna cites to "prove" his anatomy—Hippocrates, Galen, Celsus, Plato, Aristotle, Pliny, and Alexander of Aphrodisias[163]—were beginning to give rise, in Alessandro Achillini's *Annotationes anatomicae,* for example, to yet still hesitant recourse to demonstration as a legitimate resolution of debate.[164] Even the exaggerated Hellenism of Benedetti did not prevent him from acknowledging that "Aristotle has had so much authority for so many centuries that even those things which (physicians) have not seen they will affirm to exist, even without experiment," by which he meant human anatomical dissection.[165] Berengario da Carpi "deliberately excluded from anatomy what could not be seen . . . and just as deliberately drew a line between 'philosophy' and anatomy.[166] Moreover, he derived *demonstratio* (demonstration) from the simple *ostensio* of the public anatomies, but used it in a way that we might call "physical logic": he applied it to a proposition and "used it as "proof" by exposing the structure in question to the sight and touch of his audience."[167] Questions and difficulties not solved by Averroes and other commentators on the anatomical work of Aristotle and Galen made Niccolò Massa, in his own words, "a sedulous investigator of the parts of the human body."[168] French claims that: "While Berengario argues with all the learning and most of the apparatus of the schools for a separate territory for an *anatomia sensibilis,*" Massa "went a stage further and wrote an anatomy of structure that had actually been seen, *anatomia sensata,*" which "confidently inhabits" the territory for which Berengario argued.[169] As early as his 1540 demonstrations at Bologna, Vesalius was discounting the truth of authoritative claims and appealing to his audience's powers of observation to do the same;[170] and by 1543, he had become convinced that Galen's inaccuracies were "learned rather from copious discussion than from dissection, for these things are so obvious that it is superfluous to warn against them those who dissect the brain themselves or watch others dissecting, and do not trust merely in books."[171]

This special sensory epistemology, at least in Vesalius, has been construed as a challenge to the established hierarchy of natural knowledge;[172] but as was argued in chapter 1, though Vesalius's preoccupation with the amendment of Galen's morphology led to his relative neglect of function, this does not demonstrate a philosophical difference with Galen. Vesalius was a rationalist, not a skeptic: he did not doubt that things really have the properties they appear to have. His anatomical demonstrations and his visual aids, as well as the technically astute and naturalistic illustrations of the *Fabrica,* were not intended merely to represent the body as it appears, but rather to demonstrate that it really is as it appears in dissection and not (at least consistently) as in the words of Galen. While deploying all the descriptive criteria of anatomy in their graphic display, the osteological and the myological figures, in Books I and II of the *Fabrica,* are designed to manifest the use and function of the parts.[173] From the action of each of the bones and muscles the figures are intended to teach the arrangement and the nature or reason of these parts. But the myological figures do not only display the function which the particular form of each demonstrated muscle is effectively fashioned by Nature to perform. Rather, as Harcourt argues, the correlation between structure and function is "invoked with respect to the muscles of the body as a whole."[174] Moreover, function in the figures "is not simply a matter of bearing, articulation, flexion, and so on: it is a structural potential that expresses itself essentially as action." The action of the whole body is a "perfect commensurability of all the body's parts,"[175] the purpose of which is to serve the soul.[176] Furthermore, the tendency of the illustrations to understand the function of each particular muscle in terms of the way it operates in the functional context of the other parts to which it is connected is far from being absent from the texts of the *Fabrica* and the *Epitome.*[177] The figurality that in the *Epitome* explores unused parts of the defining domain of the conceptual metaphor of vegetation for anatomy, as discussed in chapter 1, not only indicates that the conceptual metaphors of anatomy were not systematically undermined by the increasing emphasis upon observation (as do, incidentally, the plates intended for Eustachio's *De dissensionibus ac controverisiis anatomicis,* which were discussed in the same section), but that Vesalius's descriptive anatomical vocabulary is shaped by the teleological notion that the distribution of blood, humors, and natural spirit in the veins serves the function of growth and nourishment, which in turn serve the faculties of the vegetative soul or Nature or *physis.*

Some, including Vesalius's former teacher at Paris, Jacobus Sylvius, were outraged by Vesalius' preoccupation with the amendment of Galen's morphology, and deplored his "lack of 'piety' and 'faith'" in Galen.[178] As French has argued, "in turning so readily to a language of *odium theologicum* Vesalius's opponents saw a parallel between philosophical and religious dissent." Like a religious tradition, Renaissance medicine had "its divine and revealed sources of knowledge and procedures," "its sacred texts," "a history of commentary, authorities, and interpretors"; its practitioners had "observances," "mysteries to be understood and kept," and "a piety and faith binding the whole together."[179] Contributing to the chorus of condemnation was Eustachio, who defended Galenic anatomy "most vigorously against newer investigations, particularly those of Vesalian anatomy," though this did not prevent him from enriching anatomy "by exact investigations" particularly concentrated on the auditory apparatus.[180] Eustachio worked in Catholic Rome and his defence of Galen against Vesalius took a Counter-Reformation form. He realized Galen could be in error, but argued that he was not deceitful and pleaded for "'indulgence', *venia,*" in view of the faith that was owed him. Furthermore, Eustachio asserted that "to sin, *peccare,* is Vesalius's vice, not Galen's, and is constituted in having bad faith rather than good faith, *mala* rather than *bona fides,* or no faith at all, *nulla fides.*"[181]

Vesalius, for his part, revealed just how indispensable religious justification was for anatomy by elevating "the investigation of nature" with the hands as an integral part of "this very holy art" of medicine, and launching a broadside assault on those trained in the "liberal arts" who "shamefully permit" such to be torn away from them.[182] For Vesalius, accuracy of investigation was quasi-religious, understood principally in the context of creation, and false information leads one into a state of "impiety." In the *Fabrica* he reveals his astonishment at "the host of contemporary philosophers and even theologians who detract ridiculously from the divine and most wonderful contrivance (*admirabile machinae*) of man's brain."[183] In similar vein, Park remarks on Berengario's repeated dissections of the human head: "Only by dint of such varied and repeated observations could the anatomist truly come to understand the divine craftsmanship with which the human body had been created."[184] Wiping the anatomical slate clean of the *accessus,* and abolishing macrocosm-microcosm speculations and teleology as the projection by humans of their own image onto the creatures and works of God, was for Francis Bacon an expression of a conviction that was both moral and epistemological, and justified by religion. According to Charles Taylor, "Bacon has no doubt that the root of this momentous error is pride. The old science amounts to 'prescribing laws to nature.'"[185] Bacon advises that the inductive method would "lead to piety by revealing God in Nature,"[186] and, in *Novum Organum,* he opposes the lack of utility of contemplative methods saying: "Just as in religion we are cautioned that faith is shown by works, the same well applies to philosophy."[187]

The implications of religious justification persuade Cunningham that the Renaissance, essentially a secular category, is empty of historical explanatory power for sixteenth-century anatomy. He proposes instead the Reformation "as a potentially better explanatory category for the dynamic changes in anatomical practice,"[188] and finds in Vesalius's encouragement of students not only to see the anatomy being demonstrated but also to dissect at home, urging them to take "the objects (*subjecta*) into your own hands," a parallel with Luther's insistence on "*the duty of every person to read God's Word (the Bible) for himself.*"[189] By long tradition held to be analogous, the body became the "true bible" for Vesalius.[190]

Before commencing to detail the anatomy of the three venters, the intradiegetic narrator of *The Little World* remarks upon the "wondrous skill, / and Wisdome" of the Creator who "By mixing of the Ellements" made the "*Houses*" of the story. Such "admyrable workemanshipp" displays God's "wisedome, power, and glorie." Knowledge of this skill entices the narrator to desire to look beyond or within the surface or outer walls of the body: "Those things, the which we cannot see" "Must nedes exceede these outward things / and go beyond them farre."[191] The narrator had commenced his narration by saying that his story is a "Revelation," "Vision," or "Dreame," and as is often the way with dreams, desire is met with its immediate fulfillment in action:

> I looked then,
> and loe, I did behold
> The *Houses* all were open, and
> each did it selfe vnfold,
> That I might see all things in them,
> which was a thing most rare,
> To marke how thinges within the same,
> in order placed were.[192]

The epistemology of *Mirum in modum* and *Microcosmos* is not so naïve. Aside from the omission of the "least parts," which is as a result of presumption rather than caution, the perception of the order, design, and teleology of anatomy by means of the soul's material instruments is knowingly problematized by Davies of Hereford's poetic voice because of his adherence to the theology of the Fall. This is a skepticism that was given an extra lease of life by the Reformers' theory of the substantial corruption of human nature and the vitiation of "man's powers for good" by original sin.[193] It shares perhaps a little of the "Puritanic spirit" with which Sylvester replaced the liberal Protestantism of *La Semaine* and *La seconde Semaine* of Du Bartas. Hagthorpe's *Divine Meditations, and Elegies* shares this pious skepticism.

Enjoining Neoplatonic analogy to the functions of the animal soul and the vital faculty of the mortal (if Platonic) or vegetative (if Aristotelian) soul, Hagthorpe discloses the similarity between how the "Worlds high Soule" "Doth this great World two waies illuminate, / by (e) corporall and incorporeall meanes," and how the "soule" powers "forth two-fold beames."[194] The first of these beams "this dead earth viuificate"; the second "this darke sence illuminate."[195] Such ideally is the relationship between the senses and the rational soul, for which the senses, the brain, and volitional movement are made, according to Vesalius; but the rational or animal faculty of the soul may not reign consistently. The senses may bring to mind marred impressions, as Nicholas Culpeper explains in *The Anatomy of the Body of Man:* "Now if these Messengers of the Mind, or the places of the Body where they lodg, be foul, or gross, or thick, they give either dull, or false information to the Mind; a looking-glass if it be so corrupted will do the like to the face of the Body: . . ."[196] For Davies of Hereford, the external senses belong (as they do in the anatomy of the schools) to the "*Sensitive*" power of the animal faculty of the soul, but they are represented similarly as subject to delusion.[197] Although they are themselves omitted from anatomical inspection, the senses are crucial philosophically and theologically to understanding Davies of Hereford's natural philosophy, because they impinge in an Aristotelian way upon the method of the soul's inquiry into its own embodiment and the world in which it lives. Thus the remark that "nought that *Understanding* doth digest, / But first on it the outward *Senses* feedes" is indebted to Aristotle; but the senses have been "[u]nhallowed" by original sin, "drown'd in that damned juice, / (Synne's Syder) from Eave's fatall Apple bruiz'd."[198] Consequently gross and dull, they "are often deceiv'd. Therefore all sciences which are deriv'd & fast rooted in the senses are uncertaine, & deceiptfull."[199] In *Visiones Rervm,* Hagthorpe enlarges upon this with the consequences for the concept of *imago dei* and the generally Neoplatonic ideas about the microcosm of Man's mind. Nature imparted "wit, and memorie to know / Both things above, and in the earth below" (*V.R.* I.54.6–7); but by the tumultuous Fall Man was "disrobde of all this excellence" (*V.R.* I.56.2):

> So chang'd, I say (by misterie diuine)
> For our first Parents haplesse high offence,
> To wretchednes, and depreuated sence.
>
> (*V.R.* I.56.3–5)

Doubt is not unusual in the early seventeenth century, when many questioned the possibility of human participation in knowledge, when knowledge is in truth infinite,[200] but the theological skepticism of Hagthorpe and Davies of Hereford is distinguishable from the Pyrrhonian skepticism of those like Walter Ralegh who suspected that the human senses are finite and limited, rather than systematically deceptive. In his essay, *Sceptick,* Ralegh's skepticism is limited to providing Pyrrhonian reasons for doubting (that one can infer the reality of something from) the evidence of the senses, as it is an incomplete translation of the first three Pyrrhonic "Modes" or arguments and part of the seventh of Sextus Empiricus's *Pyrrhonean Hypotyposes* (Outlines of Pyrrhonism), those modes being concerned with the unreliability of sense perception.[201]

Although consistent with Thirsil's invocation for inspiration for his poetic anatomy, the religiosity of pious skepticism contrasts with *The Purple Island*'s scholastic and rationalist account of knowledge and perception, in which images of light are "poasted to the minde" and Auditus, "the Master of request, / Prefers petitions to the Princes ear," thereby submitting perception to the judgement of understanding or intellect (*P.I.* 5.35.7; 5.49.1).[202] Also involved, however, is imagination, which is personified as Phantastes, one of three counselors, who fits all shapes of perception "unto the noble Princes sight" (*P.I.* 6.48.4). Phantastes is "the fount of speedy apprehension, / Father of wit, the well of arts, and quick invention" (*P.I.* 6.46.2, 6–7)—a definition that identifies or at least allies imagination or "The fancie" (*P.I.* 6.46 margin. *) with ingenuity, the faculty that discovers, contrives, or makes creative conceptual metaphors and conceits and eloquently extends their significant connotations. Some other accounts also include imagination or fantasy in the functioning of perception. That in Bernard's *The Isle of Man* describes sensations or sins entering the "Fiue doores" of the senses, whereupon they "come into a *Hall,* where attendeth *Common Sense,*" and from here they "goe into a *Parlour,* a more inner roome, and there stayeth *Fantasie* to entertaine them."[203] Beyond and above is "an upper *Chamber,*" where sensations or sins are received by "*Intelligence,* who presently acquainteth Mistrisse *Heart,* the Mistresse of the house with it, which is in her *Dyning-roome,* what are the company and number of her guests come in."[204]

Bernard's work is a moral, satirical anatomy, and Mistrisse Heart as a "Harlot" who owns a "*Common Inne,*" is a personification of vice.[205] The model of sense perception the allegory presents implies that, even if the communication of sense impressions to the intelligence is indirect, it is a dereliction of duty for intelligence to defer judgment. Ideally intelligence is the ultimate arbiter, but only in the morally virtuous. *The Purple Island* also registers doubt in the reliability of imagination, but this is perhaps better understood with reference to early modern Pyrrhonian skepticism. As a result of differences in perception, Ralegh argues that different kinds of animal and different human beings receive different impressions in their "Imagination," which Ralegh doubts correspond to the object-in-itself, that is, the object as it really is or should be understood. There seems to be in Ralegh's use of the term "imagination," a very subtle subjectification of the concept of appearance as it is in Sextus. Hellenistic Skepticism does not doubt the existence of intentional objects ("*phainomena*") or deny "those things which, in accordance with the passivity of our sense impressions ["*phantasia*"], lead us involuntarily to give our assent to them." In fact, according to Sextus, "the *criterion* of the Skeptic discipline is the appearance, and it is virtually the sense-perception to which we give this name." "[T]he question is whether [a thing] is in reality as it appears to be," for what is a matter of doubt is "the proper *explanation* for things wearing the phenomenal dress they do." The skeptical inquiry is "a question of what is predicated of [the appearance], and this is a different thing from investigating the fact of the appearance itself."[206] Whereas Greek skepticism thus construed the question of appearance and impression as metaskeptical, early modern skepticism (especially after Descartes) was more inclined to consider the contents of our own consciousness as things properly so called, and consequently to regard appearances as private, internal phenomena. Subjective states are thereby rendered like objects as candidates for being real or otherwise.[207]

This modern situation is suggested by Fletcher's allegory of Phantastes as engrossed "in his private thoughts and busy brain" with "Thousand thinne forms, and idle fancies," "Which in the world had never being yet" (*P.I.* 6.47.1, 2, 4). Fletcher's description of Phantastes (*P.I.* 6.46.1–6.48.7) draws upon that of the first honorable sage in Spenser's *Faerie Queene* (II.9.49.1), but Fletcher's "fancies" that "had never being yet" is perhaps a more skeptical rendering of Spenser's allegory of foresight who "could things to come foresee." Ralegh avoids the self-delusion suggested in the portrait of Phantastes by refusing to prefer his own "conceit and imagination, in affirming that a thing is thus, or thus, in its own nature, because it seemeth to *me* to be so."[208] But unlike Pyrrhonian skepticism, Fletcher's physiology and philosophy of sensory perception does not abandon the immediacy with which reason may affirm or deny the evidence of the senses, and so apprehend the world without the unreliable intermediacy of imagination.[209] This difference alone is sufficient to account for why the ingenuity of *The Purple Island* affirms accepted truths about Man and the world, while *Sceptick* grants to the desire for certainty no recourse either to reason or inspiration, and remains defiantly uncommitted.

The representation of sense perception and reason in *Mirum in modum* and *Microcosmos* is closer to that of Fletcher than to Pyrrhonian skepticism, for Davies of Hereford's pious skepticism with regard to the senses does not appear to extend either to the rationality of his anatomy or the role of reason in his epistemology.[210] The meaning which Davies of Hereford's poetry attempts to retain for anatomy is teleological, in the Neoplatonic and the Aristotelian senses. In an echo of Aristotle's *De partibus animalium* (Bk. III, Ch. 1, 661^{b}), Nature's "tender providence" is said to ensure that nothing in the body is made "in vaine," "[f]or ev'ry thing that *Nature* doth produce, / (As by experience is most evident) / She doth direct unto some end and use."[211] The whole body is thus the instrument as well as the abode of the immortal soul. In addition, the teleology is Christian, but Davies of Hereford does not concur with the earlier "Christianized Aristotelian doctrine" of "soul as form, that is, as the first actuality of the living body."[212] Nevertheless, according to Neoplatonic anatomy, "[t]he Platonic soul, taken from its heavenly home and inserted into its earthly body, cannot exercise its faculties without material instruments, the parts of the body." So "by dissection we discover the instruments of the soul and from our knowledge of the instruments we come to know the soul," as well as our "relationship with God."[213] The design which subordinates the meaning and purpose of this "little *World*" to the soul that dwells within, for the sake of its redemption, is created by God, whose "rare existence" in the "least parts doth wholy stay" in order to be revealed, "felt (as t'were)" and seen; though this assertion goes uncorroborated because the "least parts" (hierarchically speaking) forego description.[214]

As to the role of reason in Davies of Hereford's epistemology, despite his rhetorical question—"how can humane *Wisdome* chuse but erre, / When all hir science comes from th'outward *Sences*"?[215]—rational epistemological method is defended as a reliable corrective to deluded sense, because, although "[e]very knowledge hath its beginning of the senses," "*Sence* deluded, *Reason* out faces," and is said to do so in two ways: by induction it "makes *rules* general / Of things particular," and it "by the *effect* the *cause* doth sho," thereby deductively disillusioning sense perception of false impressions derived from "*particulars.*"[216] Therefore the anatomist's inquiry must not be characterized by empiricism, but by reason (constructed as masculine), in the tradition of Aristotle among ancients, and de Laguna, Vesalius, and others in the Renaissance: "For *Natures* bosome nothing doth embowre, / That is not subject to his searching pow're."[217] But, whereas Vesalius entrusted the hand and the eyes with investigation of anatomy, Davies of Hereford is restless for deliverance of reason from the unreliability of the senses: "*Reason's* pleasures" "make the *Soule* forsake the *Bodie's* Trunck, / Leaving it *Joy*-tranc'd whilst shee takes her flight / Through *Nature's* workes to have her *Maker's* sight."[218] Such sight is a revelation of the architectural plan in the mind of God, conveyed purportedly without the intermediacy of the senses.

The *Devine Weekes, The Purple Island,* and *The Little World* attribute creative agency to God. Furthermore, the intradiegetic narrator of *The Little World* says piously of the body that "by his precious blood," Christ has "long since redeemed it," and as such it is "The Temple of the Holy Ghost";[219] but there is no evidence of a design comparable with that of Fletcher, to compose a speaking picture of microchristus, in which the subject of human anatomy, as well as its method, eloquently demonstrate not only the creative ingenuity, but also the sacrificial and saving love, of its Creator and Redeemer. Davies of Hereford's poetry, however, is more explicitly didactic than either *The Little World* or the *Devine Weekes* or *The Purple Island,* in developing an abstract, inductive, and deductive model for understanding anatomy, and thereby the Creator's design for the human body.

In the ideology of *The Little World,* and of *Mirum in modum* and *Microcosmos,* it is not enough merely to demonstrate knowledge of this design; the demonstration's end should be to "laude, / and prayse his holy name" whose creation so reveals his glory, wisdom and power.[220]

The *Divel's* knowledge to the cause extends,
Who enters *Nature's* Brest, and doth select
All secrets of the same, to secret *endes:*[221]

The reasoning "*Divel*" "th'*abysse* of *Causes* darke descendes, / And . . . / Those *Causes* to the *Causer* comprehendes," but his "secret *endes*" are, as ever, to thwart the purpose of that "*Causer,*" whom he nonetheless acknowledges.[222] The anatomical method in *Mirum in modum* and *Microcosmos,* even where the anatomy is most detailed and precise, is a demonstration and celebration of God's providence, rather than an empirical, inductive, or rational investigation. And precision is subordinated, much more consistently than in *The Purple Island,* to this theological didacticism.

Although indebted to the rationalist tradition of anatomy, both of reason's strategies in Davies of Hereford's poetry are deployed in such a way as to make them vulnerable to skepticism and to the development of a special sensory epistemology in anatomy: in the first case, because the criterion proposed, however implicitly, is not independent of the proposition which it must justify, and so the argument reduces itself to a form of circular reasoning; in the second case, because of the poetry's adherence to a narrow metaphorical scope in describing the body. As was discussed in the previous section, generalization from particulars, in *Microcosmos,* characterizes the inductive application of the concept of a building, from the cranial house and thoracic castle, to the abdomen. But, rather than being properly inductive, the abdomen, in this case, is represented as an effect, the principal cause of which resides in its divine architect. In practice, then, the abstract model of reasoning is deductive. It resembles how, in *The Purple Island,* anatomy is reconciled to God, not merely through the attribution of creative agency, but through the implication of anatomy's figurality of constructedness and naturalness, that the body is designed by God and built by Nature. In the absence of evidence from dissection or description (which is supplied by the more comprehensive anatomy in *The Purple Island*), however, the anatomy of the abdomen cannot but be overdetermined by a generalized and unnuanced conceptualization as "building." The poem seems to assume that if, as the above suggests, the soul provides the reason for the metaphorical structure of concepts of the morphology and function of the human body, then propositions regarding human anatomy could be judged as true or false according to whether they are consistent with the criterion that the human building is created by God as the instrument and abode of the soul.

Davies ignores the possible objection that all hypotheses may in principle be tested against the ontology of the body-in-itself by observation and experiment, for "there is an 'outside' to what is constructed by discourse." But, given the (possibly perennial) absence of an epistemologically transparent and neutral nomenclature to relate this putative ontology, which anyway "is not an absolute 'outside,' an ontological thereness that exceeds or counters the boundaries of discourse," but "a constitutive 'outside,'" that "can only be thought—when it can—in relation to that discourse, at and as its most tenuous borders," it is more significant that Davies also ignores the fact that a different metaphor for the body could replace that of "building."[223] Regardless of whether Davies of Hereford intended his abstract model of reasoning to be applied in precisely this way, the evidence of its poetic deployment allows the reader to infer that the poem's implied author holds that the body may be understood as though metonymically, through reasoned inference of contiguous cause from figurative effect, and, as though by synecdoche, through deductive generalization from metaphorically translated detail. In practice, then, Davies of Hereford's abstract model of reasoning is deductive from the efficient and final causes (God and service of the soul) and reduces itself to a self-confirmatory circularity.

The model of reason in Davies of Hereford's poetry may perhaps be distinguished further from the observation and rationality required of early modern anatomy, for it has "Witt's-wonder-working *Ingeny,*" ingenuity, or "inventiveness," with which Fletcher is attributed on account of *The Purple Island*'s purportedly commendable affirmation of conventional truthfulness through the invention of the creative cognitive metaphor of the Isle of Man and the intentional direction of its allegorical extension toward a fictional world. Although Davies of Hereford's figuration of the body is less inventive, it integrates, albeit to a more limited extent, those of anatomy's conceptual metaphors and conventional analogues that convey its constructedness, such as buildings and feudal communities, with those that focus on its naturalness, such as the terrestrial globe or earth. It thereby shares much of the conventionality but less of the innovation of the Isle of Man figure. Nevertheless, where Fletcher's writing actualizes ingenuity, Davies of Hereford theorizes it.

The Welsh poet finds the ingenuity of which he writes reflected in the first observers of the "workes of Nature," the "*Poets*":[224]

> They searcht the *causes* of things *generable,*
> With their *effects* and distinct *properties;*
> And made them (by their *skill*) demonstrable.[225]

This "*skill*" is "sweete *Eloquence,*" which, in the estimation of the poetic voice of *Microcosmos,* "[w]as the first *retorick* borne of *Sapience.*"[226] In rhetoric, "*elocutio*" or eloquence was thought to give expression to invention through correct and apt use of words, composition, and embellishment.[227] Although here ascribed to poets, Du Bartas attributes the same combination of incisive inquiry or discovery and persuasive use of language to:

> Those skilfull sons of *Æsculapius:*
> *Hippocrates;* or deep *Herophilus:*
> Or th'eloquent and artificiall Writ
> Of *Galen,* that renowned *Pergamite.*
>
> (*D.W.*, p. 163)

This same combination is nowhere better expressed in *Microcosmos* (albeit to illustrate its limitations) than in the short conditional narrative of "a *Child,* confin'de t'a *Dungeon* deepe / Vntil he had attainéd *Manhood's* yeares."[228]

> . . . while he's *Dungeon'd,* let the expert'st *Tonge*
> (That able were to create *Living wordes*)
> Paint out the *Earth* with quicke-*words,* great with *yonge,*
> And though that *Fry* againe like *Spawne* affords,
> And ev'ry one had pow'r to pierce like *Swords*
> Into the nature of these *Rarities,*
> To make him comprehend the highest *Lord's*
> Inferior'st *workes,* he could not well comprize
> The thousandth part of *grace* which in them lies.[229]

The metaphor of a text as a living and speaking picture has also been found in *The Purple Island* (especially *P.I.* 4.9.3), and unraveled in chapter 3 to reveal its composition from a variety of poetic, emblematical, art-theoretical, philosophical, theological, and esoteric strands regarding the relationship between words, images, and things in nature. The "painting" in *Microcosmos* not only lives and speaks, however, but gestates and "spawns."[230] In the hermeneutic circle it is pregnant and gives birth to meaning. As the reader of the speaking picture interprets the images and words, he or she "unfolds" them and thus dilates meaning in accordance with the intelligibility of the world toward which the work is intentionally directed.[231] The resultant meanings are interpretations, which are for reasons historical and theoretical comprised of words. Theoretically it is arguable that the interpretation of any sign is "another sign; it is a reading, not a decodage, and this reading has, in its turn, to be interpreted into another sign, and so on, ad infinitum."[232] In Renaissance and early modern theoretical discourse, the meaning of a thing in nature is a sign, of which the thing itself is possessed, and which discloses that thing's reason for being in Creation. For both of these reasons, the passage in *Microcosmos* has the "quicke-*words,* great with *yonge,*" give birth not to abstracted meaning, but to "*Fry*" that in their turn "againe like *Spawne*" produce.

That part of eloquence called "exoneration" or embellishment describes the use of figures of speech, the definition of which generally followed Quintilian: "A figure of speech is any artful deviation from a supposed flow of careless, everyday speech."[233] In cases of linguistic items of highly figurative language, which belong to tropes of words in Henry Peacham's categorization, and especially in cases of creative metaphor, the process by means of which "one sign gives birth to another" is greatly enhanced,[234] because metaphorical interaction unfolds a host of significant connotations, entailments, or metonymic implications, all signified by link upon link of signifying "fry." Although more clearly and evocatively articulated in *Microcosmos,* this figurative unfolding is pursued with more systematicity in the poetic stanzas of Cantos 2 to 5 of *The Purple Island,* where the hermeneutic circle is initiated by the implied invitation to the reader to complete the process of semantic generation. The reason for this systematicity is manifold, but we are now at a stage where we can begin more fully to appreciate it, in relation to Renaissance eloquence and rhetoric. Not only is the figurative design indebted to theories of suprarealism, emblematics, post-Horatian relationships between images and words, and the closeness of human language to the signs in natural things, it also embeds the anatomical account of teleology and is directed toward inspired or revelatory apprehension of the body's true nature. Being an allegory, which Peacham categorizes under tropes of sentences,[235] *The Purple Island* is literally (as it were) an extended or unfolded metaphor. The threshold figure that is thus unfolded, moreover, as we know, is creative and generates or gives birth to unseen and newly emergent similarities between its principal and subsidiary subjects *ad infinitum.*

As the ensuing signifiers accumulate, according to the theory in the narrative of the child confined to a dungeon, in *Microcosmos,* they individually and cumulatively "pierce" and expose the nature and truth of the subject matter of the speaking picture. They not only reproduce their subject but also get to the heart of the matter. This is mimesis, which in Peacham's scheme of figures is categorized under *descriptio* (description), which in turn comes under figures of amplification, which belong to schemes rather than tropes.[236] To the objectivist, and often Puritan, charge that literary or poetic figurality involves falsity, literature's apologists responded by stressing its mimetic function.[237] But as we shall see this was not yet in the early seventeenth century a "realist" mimesis. *Descriptio* is defined as when a thing is expressed so plainly and "lively," that it seems rather "painted in tables, then declared with words."[238] Both of Davies of Hereford's poems profess to "touch" the "tender parte[s]" of the body with their discourse on them, therein making an epistemological claim for poetic representation, that goes beyond the modern colloquial sense of touching upon a point or subject.[239] The heart of a matter is its "life" for Drayton, who in *Poly-Olbion* identifies the skill of poets with that of "Paynters," to "expresse things nearest to the life," to "touch the very poynt."[240]

Since the "poynt" in *Microcosmos* and *Mirum in modum* is the body, the poems touch "tender part[s]." Drayton's idea may be expanded with reference to Sidney, who uses the expression "figuring forth" to mean representation, "counterfeiting," "imitation," "*mimesis*," as derived from Aristotle.[241] But he also hooks Horace's defining tag onto the same phrase in such a way as to combine in poetry the meaning of incisive accuracy with figurality or imagery, thereby distinguishing poetry from philosophy as: "A perfect picture I say, for [the peerless poet] yieldeth to the powers of the mind an image of that whereof the philosopher bestoweth but a wordish description, which doth neither strike, pierce, nor possess the sight of the soul so much as that other doth."[242] Sidney evidently does not think that figurality falsifies; indeed he "concurs with Aristotle's judgement that poetry 'is more philosophical and more studiously serious than history,'" whereas, according to Dollimore, "Bacon asserts exactly the contrary, and the reversal results from the different ontological status accorded to the ideal world of poetry."[243] With this I agree, but it should also be said that, in reference to the ontology of natural objects rather than the political justice of actions or events, and as a development of imaginative rationality, Sidney's poetics are not dissimilar to the rationalism of the anatomists (among whom Bacon cannot be counted) insofar as both were concerned with the power of words figuratively to convey the final cause, or reason for being, which defines the "reality" of a thing. It is these qualities which Davies of Hereford calls ingenuity and eloquence, terms which may supplement those of discovery, invention, conceit, creation, and extension, with which this study has associated its definition of metaphor and allegory.

With Davies of Hereford, Du Bartas, and Fletcher, however, we must include in our definition of ingenuity a further sense in which the poem is said to "pierce" its subject matter. Like Davies of Hereford, Du Bartas at frequent turns in his discourse makes rhetorical claims to be dissecting the body of which his poem speaks. These claims are not the narration by an intradiegetic narrator of his own actions in a narrative, such as we find in *The Little World* by "Ro. Vn.," but rhetorical gestures about the power of the language to reproduce truly the meaning of things in nature. Whereas for Drayton and for Sidney this power is mimetic, in an Aristotelian sense, Du Bartas seems to be suggesting something further which has to do with the anatomical semantic content of the rhetorical gesture: "First, with my Launcet shall I make incision, / To see the Cells of the twin Brains division" (*D.W.*, p. 162). The verb, to anatomize, in its broadest sixteenth and early seventeenth-century senses, means to open, divide, or analyze. But when the verbs, to explain, to cleave, and to rip, are used in highly similar syntactical, verbal, and metrical contexts in a poetic discourse on the human body, the reader is inclined to suspect a context-driven semantic similarity between them. The reader may also suspect that an epistemological claim is being made with regard to the poetic discourse; and additionally, that the claim is being made for the relationship between the poetry and not merely the subject or discourse of anatomy, but the anatomy of the human body-in-itself (the body-as-referent):

Shall I the Hearts un-equall sides explain,
Which equall poiz doth equally sustain?

(*D.W.*, p. 162)

Or, shall I cleave the Lungs, whose motions light
Our inward heat doo temper day and night:

(*D.W.*, p. 162)

Or, shall I rip the Stomachs hollowness,
That ready Cook concocting every Mess

(*D.W.*, p. 163).

In his first lecture to the students at the University of Bologna, in 1540, Corti argues that "dissection can be performed in two different ways: in one way really or actually, in another way through description, e.g. in writing or lecturing. For also this is to dissect the body."[244] Later reassessments of the relative worth of practical and textual dissection or anatomy occurred as a result of Vesalian advocacy of "hands-on" anatomy,[245] but status was still accorded rational discourse in the pursuit of anatomical knowledge. Even so, the rhetorical gestures with which I am here concerned draw attention to the supposed reality of the speaker's rhetorically purported action in a way that distinguishes them from even the most textually committed of anatomists. Nor does this kind of gesture conform to the instructions offered in the imperative in anatomical manuals written in the form of a dissection, such as Achillini's *Annotationes Anatomicae.*[246] Du Bartas's rhetorical gestures describe or propose a course of action which the implied author or speaker is purportedly capable of undertaking in the course of writing the text. These gestures are understandable only within the context of theory of mimesis and suprarealist language (but even then the gestures seem almost to exceed the theory). The claim they make is that anatomical poetry reproduces or recreates the actuality, the life or the truth of the body-as-referent as the signs of its speaking picture, a claim that sheds new light on *The Purple Island.*

Fletcher's poem purports to be a body of words inspired to replicate and participate in the truth of the human anatomy it depicts. Accuracy is determined in part by the accuracy or accepted "truth" of the threshold figure which the poem unfolds or dilates. In the poetic theory of Tesauro, a conceit gives life, truth, and being to the poem as a whole work, and Fletcher's allegory which systematically imbricates anatomy and figurality, may be said, in Sidney's terms, to figure forth anatomy with greater accuracy, precision, and systematicity than any comparable text, in, or rather *as,* the body of its discourse. Davies of Hereford seems to draw on Plato for the concept of ingenuity, as Alexander Pope would later do in *An Essay on Criticism* for his related concept of literary judgment.[247] But he also draws on the Renaissance poetics of Tesauro and others, who conceived of *ingegno* as the poet's idea or fore-conceit, that is, the design or purpose of a work which in Sidney's poetics corresponds to God the artificer's plan of creation. We may recall how important in Vesalius were the muscles, sinews, and bones for the soul's articulation of its instrument, the body, when we consider how ingenuity was regarded as giving "sinews" to the unformed matter of a poem.[248] Where Sidney and Tesauro's conceptions were discussed earlier in this study, they were seen to raise the poet as "maker," and to tie the Renaissance poetic concept of the conceit to modern theory of creative metaphor. Now, however, the suggestion is that the divine poetic gift not only creates unusual similarities between the meanings of newly combined signifiers, creating being in the poem where there was no being before, but also reproduces or recreates the actuality, the life, or the truth of the body-as-referent as the signs of the text which the speaker opens and analyzes. With Du Bartas and Davies of Hereford, as with Fletcher, however, we must also include in our definition of ingenuity a sense in which the poem participates in and partakes of the nature of the thing it describes and actualises or reproduces as the poem's signifying "fry."

In Davies of Hereford's theorized mode of representation, the relation between referent and sign, between the object and knowledge of that object, is metaphorical. The signs of the poetic text assume a likeness to the body-as-referent, whereby those signs possess or acquire a materiality akin to the body they refer to and signify, or "paint." The signs figuratively stand for or become a body, a "brain-begotten Birth" (in the words of Nicholas Deeble) of what Davies of Hereford calls "*Wits* sweete words."[249] This compound expression combines two metaphors: of a text as a physical body and of literary creation as sexual generation. They are "potent and pervasive" in literature and philosophy and by no means only so in the early modern period.[250] But whereas usually their deployment metaphorically conceptualizes the creative relationship of the author to her or his work, for Du Bartas and Davies of Hereford the bodiliness of the book or text is primarily a means of understanding its relationship with its subject matter. On the face of it, the manner in which wit's words are said to "pierce like *Swords*" the earth, in the narrative of the confined child, doesn't imply that they acquire or recreate the earth's materiality. Words seem to anatomize or analyze; they are words of incisive reason. Though perhaps, in an almost Aristotelian sense, they generalize from particulars and disclose causes, giving form and actuality to matter. Indeed the body's final cause is inferred from the postulate that the body is an effect of the soul and of God's act of creation. This final cause provides the reason for the particular metaphorical character of the body's parts—the conceptual metaphor with or by which the body is *figured out.* Incising or analyzing the body is reconciled in an imaginary way with becoming or actualizing or giving form to the (reason for being of the) body. The body, which is figurative and semiotic, thereby acquires corporeality. Between the body-as-referent, which is ontologically material but understood as constitutionally textual, and the signs of the poetic text, which are ontologically semiotic and graphic but which acquire a rhetorical corporeality, exists a circle of correspondence tighter than that whereby *The Purple Island*'s form and theme reflect each other's materiality and textuality. The extent of empiricism or inductive method is consequently negligible.

Epistemologically this circle relates the body-as-referent to the poetic text, but the circle is also hermeneutic. Both the body-as-referent and the text can be imagined as pregnant with inherent meanings or signs that the anatomist, ingenious poet, or reader is invited to dilate, and the metaphor of birth for interpretation applies to both the body-as-referent of the text and the text-as-body. A double circularity is thus created between, firstly, the body-as-referent and the text, and secondly, between the text and the reader. What is most at issue for hermeneutics is the relationship between a work as embodied graphically in the extent of its signifiers and the intentional direction of that work toward a world the realization of which is initiated by the interpretation of the work's signification. Interpretation inquires into "the power of a work to project a world of its own and to initiate the hermeneutic circle between apprehension of those projected worlds and the expansion of self-understanding in front of these novel worlds."[251] Interpretation in the context of Renaissance poetic theory was described in chapter 4, as involving conceit, as imagination, in the hermeneutic context, to apprehend the world of the work, its significant connotations. But something of the reflexiveness of the hermeneutic circle is discernible in Sidney's *Defence of Poesy,* where the "speaking picture of Poesy" illuminates or figures forth wisdom when "the imaginative and judging power" of the reader, comprised of "inward conceits," is "witness to itself of a true lively knowledge."[252] Thus interpretation holds in tension a retrieval of what exists as connotation, implication, or merely suggestion, within a work (as its locutionary and illocutionary acts), and an imaginative production of meaning by the reader (the perlocutionary act of the work). Although it is correct, I think, to claim that nothing, in literature, is " 'given' or 'determinate', if by that is meant 'non-interpreted' "; nevertheless "texts belong to language as a whole, have intricate relations to other linguistic practices, however much they might also subvert and violate them."[253] A literary work, as Davies of Hereford's hermeneutic theory implies, both produces a potentially indefinite number of interpretations *and* constrains our interpretations of it on account of there being meaning that is "to some extent 'immanent' in it."[254]

A view of sexual reproduction held in the early modern period tackles the contradiction between the death of the agents of generation and the vicarious continuity of their identities through their progeny, and therefore presents itself as possibly having informed Davies of Hereford's view of words as "great with *yonge,*" and giving birth to "*Fry*" that in their turn "againe like *Spawne*" produce. The following extract from Du Bartas's "Epithalamie, or wedding song" for Adam and Eve, concisely articulates this view: "I admire how God did then / Make Two of One, and One of Two again" (*D.W.,* p. 172). It is repeated in allegorical form in *The Purple Island,* where Adam and Eve become "*Arren*" and "*Thelu*" (*P.I.* 3.24.3):

> These two fair Isles distinct in their creation,
> Yet one extracted from the others side,
> Are oft made one by Loves firm combination,
> And from this unitie are multipli'd:
>
> (*P.I.* 3.26.1–4)

Eve or Thelu finds her origin in Adam or Arren, and children find theirs in their biological parents. This provides for Davies of Hereford's hermeneutics of representation an imaginary and metaphorical resolution to the problem of interpretation. The meaning of the parent text (which is composed of signifying "fry") can be imagined to persist in the reader's interpretation (which is composed of signified "fry," which in their turn signify).

As theorized by Davies of Hereford, poems are products not only of eloquence, but of reason's "*Ingeny,*" which is said to be "[f]ixt to" the "*Eye*" of "understanding" and to the "soule's facultie[s]," and, as such, is sharply contrasted with the abdomen, which is "so farre remov'd" from "the *Seate* of the *Intelligence*" as to be unworthy of treatment, as was discussed in the previous section.[255] In the anatomical exposition, as was remarked above, the abdomen is brought to the surface of the poem's representation only to be repudiated, negated. The unintended irony is that, despite having provided the metaphor for the hermeneutic process, sexual reproduction, the purpose and function of the organs of generation, is considered to be beneath (situated in the abdomen) the value of the words that are interpreted in that process, and therefore inadequate to the very meaning that the metaphor produces.

One interpretation of the pervasiveness and potency in literature of the metaphors of writing as a physical body and of literary creation as sexual generation, is "as a poignant commentary on the imminent dissolution of texts, for once the text is conceived as a living body, it follows that it is equally subject to change and that its lifespan is limited."[256] As a consequence of the ephemerality of matter and of words, the poetic voice of *Mirum in modum* hopes that his words "of *Wisedomes* salt may taste, / Which can from crude *Conceit* corruption stay, / And make the same eternally to last."[257] On one level, this is a conventional expression of concern about the adequacy of the poem's expressed wisdom, and the vicissitudes of posterity. As salt is a preservative of flesh, so the sapience of eloquence preserves its relevance through time.[258] There is also, however, a soteriological development of the poetry's theory of representation, such that, although the body-as-signified is composed of poetic eloquence, which acquires corporeality from the mortal "*Corpes,*" it is hoped that "brain-begotten" ingenuity, which belongs to the only immortal part of Man—the rational soul or faculty—will stem the tide of corruptibility in this body of eloquence.[259] It is hoped, in other words, that ingenuity will not merely reproduce the body-as-referent as another natural body subject to imminent dissolution, but mimetically perfect it (as was said in chapter 4) and resurrect it in its ideal and immortal form. In an epistemological sense, the relationship of the word to the body-as-referent, as theorized by Davies of Hereford, is not only metaphorical, it is also hierarchical and soteriological.

Something of the nature of this immortal textual body may be gleaned from Hagthorpe's *Visiones Rervm,* which is composed of four poems, the second of which is entitled "*Cursus & Ordo rerum;* or, Art and Nature." The "Art" and "Nature" of the subtitle are personified as "Two Royall Queenes, whose persons did appeare / The types of beautie" in the intradiegetic narrator's "dream" (*V.R.* II.26, 5; II.3.3–4, 1). On the "kirtle" of Nature, the "lowly" one of the two, appear animals, birds, and plants (*V.R.* II.4.3); whereas Art personifies the products of "art," ingenuity, skill, eloquence, and poetry. On her "skirts" appear "Ten thousand Flowers," that is the flowers of poetry, and "Ten thousand neate conceits . . . / Strange workes and rare inuentions" (*V.R.* II.21.1, 2, 6–7).

If this contrast should be applied to the representational and hermeneutic theories of Davies of Hereford, or to the threshold conceit, strange work, and rare invention of *The Purple Island,* it would imply that whatever is reproduced as the signified of this poetry does not belong to Nature but to Art. It is after all God, not Nature, who creates the wondrous artifice of anatomy in *The Purple Island,* and whose creative act the poem is designed to emulate. For these Christian poets the rational soul is analogous to the mind of Nature, as in Greek natural philosophy, but only when Nature's mind is understood as God. The argument that the *Purple Island* anatomy preserves the teleology of the soul's instrument in an Aristotelian sense in which Nature is akin to the soul appears on this view simply not to hold water.

The essential similarities between the allegorical distinction of Art and Nature in *Visiones Rervm* and the epistemological or representational theory of Davies of Hereford and Fletcher extend to the corruptibility not only of the body-in-itself, but of Nature. In Hagthorpe's poem, "*Alethia,*" or "*Truth,*" declares:

> *Nature* is of her selfe but a rude masse,
> Which of it selfe each day to ruine tends:
> Whom *Art* still strives to beautifie and grace,
> And to preserve and forme it euer bends
> Her chiefe endeauour . . .
>
> (*V.R.* II.76.1–5)

But the contrast Hagthorpe draws between Nature and Art is for the sake of his allegorical purposes much starker than that in the work of Davies of Hereford and Fletcher: "Art she tills and formes the soules of men, / Giving them light, whom Nature hath made blind" (*V.R.* II.77.4–5).

Quarles provides an ironic and morally satirical application of this conception of enlivening poetic eloquence in his preface to *Hadassa; or, The History of Qveene Ester:* "But in these lewd times, the salt, and soule of a Verse, is obscene scurrility, without which it seemes dull, and liuelesse."[260] He then justifies his work to the reader within the same conceptual parameters by referring to its more moral aims, which are apparently also more modest than those of Davies of Hereford:

> And though the sacred History needs not (as humane doe) Poetry, to perpetuate the remembrance, (being by God's owne mouth blest with Eternity) yet Verse (working so neere vpon the soule, and spirit) will oft times draw those to haue a History in familiarity, who (purchance) before, scarce knew there was such a Booke.[261]

Without irony, Davies of Hereford's model of the soteriological function or effect of ingenuity is at once both a mimetic version of the suprarealist view of language, and a specific application of it. By polarizing immortal wisdom or ingenuity and mortal conceit or eloquence, this formulation firmly privileges the former as what might otherwise be regarded as grasping the spirit, while the latter is the letter, the expression in the flesh or matter of words. This differs from the suprarealist view of language which in Sidney's poetics retains the theoretical appellation, conceit, even for the inward process of conception in the rational imagination.[262]

Davies of Hereford's theory privileges ingenuity over eloquence, but rather than seeking transcendence by signifying abstract plans or forms of which all language and matter are a manifestation or unfolding, the theory suggests that ingenuity has the power to (re)create in a specific way. It may appear that this may be compared to the "visionary hermeneutics of Christian Neoplatonism" found in Edmund Spenser's *Fowre Hymnes,* wherein, argues David Lee Miller, disciplined meditation "offers itself as a methodical return from the secondary, derivative, merely apparent object of regard—the body, text, or world—to the source of its form, the logos, which the object is then assumed to signify by the metonymy of effect for cause."[263] This is because Spenser's Neoplatonic (partly Aristotelian) theory of causation claims that the soul "frames her house," "For of the soule the bodie forme doth take; / For soule is forme, and doth the bodie make."[264] Miller's analysis concludes that "this hermeneutic systematically misreads its own procedure, offering as an act of decoding what is in fact the work of production," a work which "derives the heavenly logos from an image of the human body."[265] The disingenuous misreading is comparable with that of Davies of Hereford who, in having provided the metaphor of sexual generation for the hermeneutic process, proceeds to devalue and reject the metaphor as inadequate to explain the origin of the body-as-signified which is purportedly produced by his ingenious text.

Davies of Hereford's theory is not, however, a Neoplatonic transcendent theory. It suggests instead that ingenuity has the power to (re)create by investing eloquence with immortality. Ingenuity thereby acquires Christian soteriological significance, as does *The Purple Island*'s anatomico-poetic exposition of microchristus, which eloquently demonstrates the saving love of the Redeemer in the breathless "pattern" or corpse, and recreates the "picture" of God in human anatomy—*imago dei*—as the speaking "picture" of the poem (*P.I.* 4.9.3, 2), with the aim of dispensing the balm of salvation. Davies of Hereford hereby draws upon the analogy between the operation of human and divine locutions, as discussed in chapter 3, suggesting that ingenious eloquence partakes in some manner of the efficacy of the incarnation of Jesus Christ.[266]

When the conclusion reached in the forementioned chapter, that *The Purple Island* aims to participate in the saving work of redemption, was carried forward to the discussion in chapter 4 of the work's reception, the soteriological efficacy was qualified as having been understood by "A.C." as pertaining possibly only to the limited intratextual context of its signification. The poem's participation in redemption may be compared to the words of "God's owne mouth blest with Eternity" in Quarles's preface to *Hadassa.* Therein God's locutions perpetuate the sacred history of Ester. But as Francis Bacon's translation of Psalm 104, dedicated to George Herbert, makes clear, God's "breath," his "word," does not only perpetuate the life of narratives or of any form of discourse, but "doth all things generate and feed":

> If thou withdraw'st it, then they cease to be,
> And straight return to dust and vanity;
> But when thy breath thou dost send forth again,
> Then all things do renew, and spring amain,
> So that the earth but lately desolate
> Doth now return unto the former state.[267]

Davies of Hereford's re-creative mode of representation and his regenerative application of the suprarealist conception of language to the hermeneutic context, suggest that he postulates the extension of this soteriological effect beyond the signifiers of the poetic text, and beyond their signifieds, to human anatomy-in-itself, which is also the anatomy of the reader. But how can the effect extend thus? Given that anatomy is an exposition of Man, that is of body and soul, the purpose of the function of the former being to serve the latter, the soteriological effect of poetic anatomy upon the reader may be the effect of learning on "our degenerate souls," for which Sidney regards poetry superior to any other form of writing.[268] But whereas Sidney, perceiving our souls as "made worse by their clay lodgings," sees the effect in Neoplatonic terms, the anatomy in *The Purple Island* and that in *Mirum and modum* and *Microcosmos* is Neoplatonic and Aristotelian in conception. In respect of the Aristotelian conception, regenerative effect cannot be envisaged as extended to the soul of the reader without also pertaining to the body.

Not only for this reason a different model of regeneration, preservation, and perfection to that which we find in Sidney, is required. The regenerative, soteriological efficacy of ingenuity in Davies of Hereford is represented as extending not only to the body of the reader, but beyond the hermeneutic context to the body (as referent) generally or universally. It was argued in chapter 3, that *The Purple Island*'s diegetic narrator draws attention in Canto 4, stanza 9, to the fact that in order for the poem to demonstrate God's creative ingenuity and the Redeemer's love, there is a necessity for the human body to die; there is a necessity of death and dissection to reveal the picture of anatomy as redeemed and reconciled to God as *imago dei.* It is necessary, furthermore, that "our *Bodyes* cease to bee," as *The Little World*'s narrator claims, and that by extension and analogy "a Realme, a Common-wealth, / a Cittie, or a Towne, / Cannot continue long," in order that our bodies can "arise / in glorie for to sitt / With Christ."[269] These considerations suggest that the eschatological doctrine of the Resurrection of the Dead may provide for the elaborate hermeneutics and epistemology in Davies of Hereford, a metaphorical resolution that develops upon the resolution already fielded by the concept of sexual reproduction.

This is a resolution to the problem of the difference between the ontology of the actual body and that of the text, and to the problem of the semantic difference between the body in the text and that in the interpretation. The doctrine of the Resurrection of the Dead implies that the corruption and decomposition of corpses is congruous with their exhumation at the Last Judgment,[270] and the congruity is said to lie in the continuity between the mortal and resurrected bodies. The narrator of *The Little World* insists that it is "that" same "Temple of the Holy Ghost" as we now inhabit, "that we shall . . . see / At latter day, for to arise / in glorie."[271] The Resurrection is thereby a defeat of death, as Paul argues in his discourse on resurrection in 1 Corinthians 15 (cited in chapter 3); and yet, precisely because of this, it is also a defeat of corporeality, which inherited mortality with original sin. That the body that is resurrected cannot therefore be conceived in terms of matter prompts the question—"How are the dead raised up? and with what body do they come?" (1 Cor. 15:35). As a defeat of death and, therefore, of corporeality, the resurrection must necessarily transform the mortal, earthly, "natural," "physical body" (*sōma psychikon*) into an immortal, "spiritual body" (*sōma pneumatikon*);[272] yet paradoxically, the immortal must remain identical with its mortal precursor, for the identity of the individual to be continuous. Because of the Christian inheritance of Jewish anthropology, when Christian eschatology struggles to overcome contradiction, as in Paul's epistle, it is attempting to reconcile the opposition between natural and spiritual bodies with the continuity in eternal life of "man's essential being," "characterised by existence in a body."[273] Paul's resolution is to propose that the "body in the sense of the 'I', the 'person', will survive death through the creative act of God."[274] He uses the analogy of the seed: "that which thou sowest is not quickened, except it die: and that which thou sowest, thou sowest not that body that shall be, but bare grain, it may chance of wheat, or of some other grain: but God giveth it a body as it hath pleased him, and to every seed his own body. . . . So also is the resurrection of the dead. It is sown in corruption; it is raised in incorruption . . . it is sown a natural body; it is raised a spiritual body" (1 Cor. 15:36–39, 42, 44).

Continuity of identity after death cannot be asserted without accounting in some recuperative way for the discontinuity between death, which the soul but not the body survives, and the resurrection of the body at the Last Judgment. In the view of the modern theologian, Siegfried Wibbing, the physical body represents Man's earthly existence and the spiritual body his post-resurrection life, yet the "continuity between the earthly body and the heavenly body does not rest upon a transformation" in Paul, for: "If that were the case, Paul would have stressed the temporal aspect of his concept of the 'spiritual body'. . . . The thought would then be of a body consisting of a substance *pneuma*. But Paul is not concerned with the description of a spiritual substance of which the heavenly *sōma* is constituted. His concern is with the fact that God determines this *sōma* through the Spirit as a power of God."[275]

Thomas Aquinas's hylomorphic analysis of the discontinuity between death and the resurrection of the body and of the problem of identity and transformation, in his comments on 1 Corinthians 15, produces the view that "the soul as a substantial form survives the death of the body, but the full person does not exist until body (matter) is restored to its form at the end of time."[276] Although Aquinas thus made body philosophically necessary, he also "telescoped body into form by holding that soul is enough to account for individual continuity and that soul is the *forma corporeitatis.* (In other words, it is soul that accounts for the 'whatness' of body. Thus, any matter which soul informs at the end of time will be its body.)"[277] Aquinas herein seems to have taken the Aristotelian efficient and final cause of the body, that is, its instrumental means and aim or end, which is the soul or form, which is dependent upon the body for its existence and activity, and must come into existence and perish with the body, even though it is theoretically separate from it,[278] and Aquinas combined this with incorporeal soul or spirit, which, according to Aristotle, is to be found only in the reason and in that part of it which is without taint of the lower activities of the soul—namely, the Active Nous, which, although wholly unaffected by the body, does not constitute a survival of personal identity or form.[279] In the sixteenth century, on finding that Aristotle's analysis of the soul is actually inconsistent with the soul's immortality, Pomponazzi concluded that "we know the immortality of the soul only by faith, not by reason."[280]

Neither the works of Davies of Hereford nor *The Purple Island* can be regarded as having resurrected the body-as-referent as the signified of the text, without a comparably heavy dose of Pauline eschatology and Aristotelian teleology. The intended effect of ingenuity is analogous to that of the soul in Aquinas's hylomorphic analysis. In order for the mortal body-as-referent to be resurrected as the poetic text, matter must be informed, or enformed, by ingenuity as the rational faculty of the soul, an imaginative rationality. Matter thereby partakes of the immortality of the rational soul from whence the poetic account comes. But in the poetic case, in the context of the locutionary act, the matter is that of the visible graph and poetic eloquence, which acquires corporeality mimetically from the mortal "*Corpes.*" The illocutionary act, the anticipated or intended effect of this provision of a rational, poetic account of human anatomy on the implied reader is that of contributing towards the salvation of her or his body and soul. This is done through the (implied) reader's imbibition of a fideistic understanding of human anatomy as microchristus and therefore as redeemed. But in these terms, it is conceivably more achievable by *The Purple Island* than by any of the comparable works, because the creative metaphor or conceit of the Isle of Man and the emblematic representation of Man as microchristus so deeply inform (or enform) that work and help to regenerate Man through the emergence of new significant connotations in the imagination (conceit) of the reader, who (ideally) reflects the redeemed Isle of Man and the redeemer about which he or she reads in the hermeneutic circle.

There is an iconographic, rhetorical aid to this regeneration in some of the literary works this section has examined. It was referred to earlier in the section as the kind of gesture that describes or proposes a course of action which the implied author or speaker is purportedly capable of undertaking in the course of writing the text. It was said there that the gestures claim that anatomical poetry reproduces or recreates the actuality, the life or the truth of the body-as-referent as the signs of its speaking picture, though it was not until now supposed that the recreation was conceived on the model of resurrection. The wounds or "red letters" of the risen Christ were interpreted, as Richard Crashaw reads them in his divine epigram "On the Still Surviving Marks,"[281] both as traces of how Christ's corporeal body was killed and as evidence that his risen, transformed body is nonetheless demonstrably his own, and, hence, as assurances of the resurrection. Similarly, the rhetorical gesture made by the *Devine Weekes,* by *Mirum in modum,* and by *Microcosmos,* where the poetic voice professes to pierce, cleave, and touch the tender parts of the body, does not only imply the poetry's purported mimetic accuracy with respect to the body it describes; it also testifies to the truth of that body's resurrection and preservation in and through the signs of the text. And, similarly, the immediacy of the anatomized body to the diegetic narrator of *The Purple Island* serves rhetorically to reproduce or resurrect the body-as-referent as the signified of the poem. How appropriate, then, that "*these* dissected wreathes" (*P.I.* 4.9.1, my italics) should occur in Canto 4, where the body is most overtly microchristus, "[w]here moves our sunne[/son]," which Christ's incarnation, death, and resurrection make demonstrably "glorious," and so turn "to heav'n from hell" (*P.I.* 4.9.2, 6, 7).

Despite testifying rhetorically to the truth of the body's resurrection in and through the signs of the text, there is a discernible equivocality in *Microcosmos* over whether its sapience has sufficiently salted its eloquence to resurrect the referent of the text into a living body of knowledge that preserves the body of the reader. Without this resurrection, however much the poet may "Paint out the *Earth* with quicke-*words,*" neither the comprehension of the fictional child confined to a dungeon, nor the regeneration of the reader by means of interpretation of the text is conceivable. The modesty of the poetic voice which narrates the story of the confined child does not allow even the "expert'st *Tonge,*" who tries "[t]o make him comprehend the highest Lord's / Inferior'st *workes,*" to "comprize / The thousandth part of grace which in them lies." "Grace" signifies, now and in the seventeenth century, both aesthetic and theological concepts, and here the two sets converge to produce a concept of the "divine regenerating and inspiring influence" of seeing the "elegant proportions" and beauty of creation "as t'were a Scroule, / Wherein in letters plaine, that none can misse, / God is enroulde."[282] Such influence is presented in *Microcosmographia* as having arrested Crooke in the midst of his anatomy of the human body (cited in chapter 3).

But the whole premise of resurrection in the rational poetics of ingenuity is that the rational soul is immortal. The modesty and wonder of Davies of Hereford's response is in order to avoid the intellectual pride that not only led to the fall of the senses, but also to the "Jointes" of "*man's* divinest part" becoming "out of frame"; it is a response that implies not only an acknowledgment of the failure of the poetry's vast project to immortalize the principal parts of the body, but also militates against its obsessive use of detail in deference to the ontology of the created body-in-itself.[283] Davies of Hereford is even close to Bacon in his avoidance of intellectual pride, though Bacon associated such pride with the macrocosm-microcosm speculations and teleological accounts of causes in natural philosophy to which Davies of Hereford adheres. Bacon's translation of Psalm 104 gives expression to an identical wonder and humility: "But who can blaze thy beauties, Lord, aright?"[284] In chapter 1 of Hagthorpe's *Divine Meditations, and Elegies,* it is not clear from the syntax of the relevant poetic stanza whether "we see, / Nothing but what is obvious to the eye" because many "for game circle rownde / The whole worlds frame," or if cause and effect are the reverse of this.[285] Whichever is the case, however, both phenomena may be read as part of an accumulation of symptoms or signs of the fact that, in words that seem to echo yet significantly differ from those of Davies of Hereford, "Our vnderstanding partes, in sence, are drownd."[286] This is not necessarily a comment on the epistemological veracity of sense perception, but a reaction against the senses *per se.* As a remedy to this and as a response to "The shortnesse of mans life," which is the subject of the chapter, Hagthorpe prescribes the measures advised by Thirsil in *The Purple Island:* "We live not to our selues. Those onely liue / That doe themselues to contemplation giue."[287] Yet, as with Davies of Hereford's poetry, *The Purple Island* does not suggest, in the way that Hagthorpe seems to, that contemplation is in itself redemptive.

As in Davies of Hereford's *Microcosmos,* so in Fletcher's *Purple Island,* the resurrection of the body-as-referent and the salvation of the reader through rational ingenuity are knowingly unaccomplished, again for narrative reasons of the Fall, which are expressed by the structure of Canto 4, stanza 9 (cited in chapter 3).[288] The possibility in Fletcher's poem of the resurrection of the body-as-referent is centered on human anatomy as redeemed by the incarnation. But this stanza consists of oppositional terms, which establish an antithesis between the analogous Christ and microchristus. Admittedly, analogy ensures that since Christ is in "heav'n," microchristus metonymically acquires a "heav'n"; but Christ's is "highest" and "our[s]" is "low" (*P.I.* 4.9.4). Notwithstanding the opposition, the "loftie one" scorns not "thus low to dwell" in his incarnation, and set "[h]ere" "his best starres," thereby making his earthly, microcosmic, "cell" "glorious" (*P.I.* 4.9.5, 6). What is accomplished thereby, though, is reconciliation, not identification, of opposites. The still mortal human heart "*turns to* heav'n from hell" (*P.I.* 4.9.7, my emphasis); it is not in itself "glorious" (*P.I.* 4.9.6). The narrative allegory of Fletcher's poem may be seen as a direct result of the subtle effects Christian theology has on notions of classical rhetoric, on the classical Greek *logos* (word or reason), and on notions of form in Nature.[289] Through its analogical poetic mode and anatomical method *The Purple Island* becomes part of the saving work of redemption, a continuation of the revelation of Creation and a perpetuation of the balm of salvation dispensed through the incarnation in the inspired poetic word, as was argued in chapter 3. But notwithstanding this soteriological role, the poem, like its implied reader, still awaits the general resurrection. Although the "Isle" is "pull'd from" the "*Mare mortuum*" and "setled" with the "bloud" of Christ, its restitution is to a "baser state then what was first assign'd" (*P.I.* 2.2.1; 1.54 margin; 2.2.3, 5).

Norden provides an expository rather than fictional summary of the same theological position, though without remarking upon its consequences for the regenerative efficacy of human locutions:

> Yet man is not free from his fatall foe,
> That first supplanted *Adam,* antique fire,
> And works to wrecke on *Adams* line, like woe:
> Beleeuers onely make this foe retire.
>   Sacred *Messiah* payde that deathfull hire,
>   Whereby wee liue, and haue *eternitie,*
>   Though here beclowded with the contrary.[290]

As a speaking picture of microchristus, the poetic dissection of *The Purple Island* provides a verbally constituted form of ocular evidence for believing in the divine promise of a full recovery of the prelapsarian "picture" of God, based on faith in the incarnation and the conviction that God "never wilt forsake" what "he" "mad'st"; but this recovery is deferred in anticipation of the Resurrection of the Dead, when "man in heav'n shall dwell" (*P.I.* 4.9.3; 3.4.5; 4.26.7).[291] The qualified rhetorical resurrection reveals itself to be a prolepsis that foreshadows the Resurrection of the Dead while it speaks of its own soteriological and eschatological significance.[292]

Among the works compared with *The Purple Island* in this chapter, *Mirum in modum* and *Microcosmos* are clearly more comparable, in some ways, with *The Purple Island* than are either of Fletcher's source texts, or the works by Norden, Hagthorpe, "Ro. Vn.," Bernard, and Clapham. The texts of Davies of Hereford and Fletcher share a complexity of anatomical description, and a precision in the application of anatomical descriptive criteria to at least some of the body, though besides *The Faerie Queene,* only *The Purple Island* extends this application to the anatomy's subsidiary subject, as a result of its metaphoricity and its pursuit of the allegorical ideal of continuous superimposition. *Microcosmos* and *The Purple Island* also share a detailed and systemic description of concoction, and of the operation of the heart and lungs. Anatomy in *Microcosmos,* however, is overdetermined by its teleology as a building for the soul, and its author consequently does not adequately respond to the challenge to anatomical method set by Bacon.

Fletcher's *Purple Island* is said by Sawday to differ from the work of Davies of Hereford, and that of the other poets with which he compares it, in that they, with the exception of "Ro. Vn.," refuse to organize their material formally, while "*The Purple Island* brings a poetic organization of its material to the fore. Firmly set in the pastoral mode, the rules of the pastoral govern each of its many closures."[293] In fact, they govern also the openings to each canto. The anatomical content of Cantos 2 to 5 is, therefore, framed by the narration of Thirsil, who is also a character in a pastoral diegesis; the pastoralism is judged by Kastor to be "essentially external and very superficial," and by Grundy as "merely perfunctory" and conventional rather than descriptive.[294] The hypodiegetic level of the narrative of *The Purple Island* does not only consist of the physical anatomy, but also covers a "moral-mental-spiritual anatomy of Man-Island" (Canto 6), a "gallery of allegorical portraits" of virtues and vices (Cantos 7 to 10), and a "war of virtues and vices in the Man-Island" (Cantos 11 and 12), each of which, according to Kastor, ill becomes the pastoralism.[295] The diegetic narrator sews his anatomical and moral "paintings" and the ensuing war into a narrative of the most significant events of divine history, which consequently "forms an important unifying aspect of this work"; however, Kastor and others maintain that the dissonances between "the anatomy theater," the allegorical figures, "the heroic, valorous combat," and the Christian mythology, prevail over unity.[296] Further to what has previously been said about this in this study, including the fact that the attribution in the seventeenth century of ingenuity does not imply any discordance or incongruity, it is defensible to argue that *The Purple Island* significantly differs from the *Devine Weekes, Mirum and modum, Microcosmos,* Drayton's *Poly-Olbion,* Norden's *A Store-Hovse of Varieties,* Hagthorpe's *Divine Meditations, and Elegies,* and Clapham's *Ælohim-Triune,* in that they are, despite their responses to the Fall, expository poems without a narrative arrangement. In this, and other specifically literary ways, such as allegorical complexity, personification, and fictional invention or fantasy, *The Purple Island* is more comparable with *The Faerie Queene. The Purple Island* may nevertheless be distinguished from the texts of Du Bartas, Spenser, Davies of Hereford, Clapham, Hagthorpe, "Ro. Vn.," Norden, and Bernard, because it is simultaneously persistently anatomical *and* highly poetic and literary.

It is nevertheless true that *The Little World* by "Ro. Vn." is also a narrative, with an intradiagetic narrator, and a "Revelation," "Vision," or "Dreame" as its story. And Sawday's remark is possibly unfair in respect of Hagthorpe's *Visiones Rervm.* Drawing upon the same medieval genre of dream-visionary poetry, the intradiegetic narrator begins his narration with the words: "I here intend to write / A Dreame or Vision . . . that me befell" (*V.S.* 2.1–2).

In addition, like *The Faerie Queene* and *The Purple Island,* Bernard's *The Isle of Man* is allegorical, though not perhaps an allegory, at least not as sophisticated an allegory as either that of Fletcher or Spenser. There is evidence of formal organization in Clapham's *Ælohim-Triune,* though the poetic form is "diuerse."[297] But no formal organization is as literary as that of *The Purple Island.* Aside from generic and formal organization, a thematic coherence is granted many of the works discussed in this chapter by their intellectual organization of facts or ideas, an organization that is obvious in Drayton's *Poly-Olbion,* Norden's *A Store-Hovse of Varieties,* Hagthorpe's *Visiones Rervm,* Bernard's *The Isle of Man,*[298] and even in Davies of Hereford's poetry which shares with Bernard's allegorical and expository prose the perspective on its material provided by the highly moral ancient precept *nosce teipsum;* while *Mirum in modum* focuses on the animal soul in the brain, and *Microcosmos* on appetite, pleasure, affection, and their location in the heart. There is nonetheless a tendency of some, like Hagthorpe's *Divine Meditations, and Elegies,* and like Norden's *A Store-Hovse of Varieties,* to expand into an encyclopedic farrago which some assume to be typical of the genres of encyclopedic poetry and expository and fictional anatomy to which many of these titles belong.

While it is a poetic anatomy, the literary sophistication of *The Purple Island* means that it stands apart from much encyclopedic poetry and expository and fictional anatomy, and while comparative study with reference to allegory, analogy, metaphor, and the anatomical *accessus* enhances our understanding of the literary context of *The Purple Island,* it is clear that Fletcher's poem was not determined by this context, and our understanding of the work's figural and intentional complexity has not been advanced greatly. Comparative study yielded much more when it was concerned to elucidate the epistemological and hermeneutic dimensions of the use of ingenuity and eloquence in a selection of the works, particularly as these works share a broadly teleological and soteriological epistemology and hermeneutics. For the authors of these works, as for Renaissance anatomists, this teleology was Christian in that the rational inference of function and cause from structure purportedly demonstrates the ingenuity of God. While retaining its rationalist basis and the broadly teleological conception of structure, however, Renaissance anatomy was increasingly seen by its practitioners as a well-defined territory with a special sensory epistemology. From a historiographical perspective, it is possible to say that this epistemology, particularly as advanced by Vesalius, has persuaded some historians that anatomy in the Renaissance was no longer teleological but empirical. Yet the tendency to understand the function of each particular muscle in terms of the way it operates in the functional context of the other parts to which it is connected characterizes Vesalian myological illustrations as well as the texts of the *Fabrica* and the *Epitome.* This structure-function anatomy, like the figurality in the *Epitome,* is shaped by teleological notions.

The epistemological field of early modern anatomy is uneven, nuanced, and combines in varying measures deductive and inductive reasoning and debate, the resolution of which could still be textual authority or experience and observation, though on the whole there was increasingly less recourse to the former, and entire anatomical projects could be based on anatomy that had been seen rather than inferred from the eye of reason. But as previous chapters have argued, especially in the late sixteenth and early seventeenth century, natural philosophy reveals an increasing reliance upon inspiration and the immediacy of spiritual apprehension, though this was less evident in the rationalist anatomy of the schools than in the medical profession broadly speaking, and the literary encyclopedic, fictional, and expository anatomies. Although this study is literary-critical in focus, its investigation of relevant works of anatomy and the standardization of nomenclature suggest that there may have been an emerging notion of an ahistorical ontology of the body, independent of the language used to speak about it. The objectivist assumption of this ontology in historians of science would give rise to the positivist historiographical concept of an anatomy discovered incrementally from the beginning of the "scientific revolution." But in the early modern period, the notion of such ontology was largely rhetorical, in the sense of gesturing artificially at something which was not, and anatomy was grasped by a combination of sense perception and an imaginative rationality, through conceptual metaphoricity, suprarealist language, and Christian theology. Religious justification was indispensable for Renaissance anatomy, elevating the investigation of nature with the hands as an integral part of the very holy art of medicine, the preservation of the health of the body. For Vesalius, as for Berengario, Bacon, and other anatomists and natural philosophers, accuracy of investigation was a kind of piety, understood principally in the context of creation. The "polarization of religion versus the investigation of Nature may well not have obsessed men of the *sixteenth* century . . . indeed, it may never have occurred to them."[299]

The literary anatomy varied greatly in its epistemology and theory of representation. The *Little World* narrator avoids all the deeply involving epistemological debates between skeptics and rationalists about inference from the appearance of structure, by presenting anatomical discovery as a revelation, a vision, or a dream. But for several authors, knowledge was made problematic by the doctrine of the Fall. This has been referred to as pious skepticism, distinguishable from the Pyrrhonian skepticism of those like Walter Ralegh who suspected that the human senses are finite and limited, rather than systematically deceptive. Although the *Purple Island* narrator Thirsil invokes God for inspiration, his account of knowledge is scholastic and rationalist, except where it yields to skepticism over subjective states, which like objects are candidates for being real or otherwise. The representation of sense perception and reason in *Mirum in modum* and *Microcosmos* is closer to that of Fletcher than to Pyrrhonian skepticism, for Davies of Hereford's pious skepticism of the reliability of the senses does not extend either to the rationality of his anatomy or the role of reason in his epistemology. Reason induces general rules from particulars, and deduces the character of particulars from their purported causes. Both of these strategies, however, were found to be vulnerable, firstly, to skepticism because the cause they propose is not independent of the proposition about particulars which it must justify, and so the argument reduces itself to a form of circular reasoning; and secondly, to the development of the sensory epistemology of anatomy, because the particulars are described according to a narrow metaphorical scope, quite unlike *The Purple Island.* Nevertheless, Davies of Hereford's poetic exposition of the epistemology and hermeneutics of reason in poetic representation has "Witt's-wonder-working *Ingeny,*" ingenuity, or "inventiveness," and is that from which ideas are drawn and applied to our interpretation of the ingenuity with which *The Purple Island* and its author were attributed on account of that poem's purportedly commendable affirmation of conventional truthfulness through the invention of the creative cognitive metaphor of the Isle of Man and the intentional direction of its allegorical extension toward a fictional world.

According to Davies of Hereford's epistemological and representational theory, as a poetic text generates connotations which the reader "conceits" or apprehends, these signifiers individually and cumulatively pierce, touch, and expose the nature, truth, or heart of the subject matter. In rhetorical theory, this is mimesis, categorized under *descriptio* by Peacham, which counterbalances the qualities of plainness and liveliness. The heart of the matter of the body is its life, and since the "poynt" in *Microcosmos* and *Mirum in modum* is the body, the poems touch the "tender parts," "figuring them forth," as Sidney would say, who hooks Horace's defining tag onto the same phrase in such a way as to combine in poetry the meaning of incisive accuracy, or ingenuity, with figurality or imagery, which is eloquence, thereby distinguishing poetry from philosophy as a "perfect picture" which strikes, pierces, and possesses the sight of the soul.

With Davies of Hereford, Du Bartas, and Fletcher, however, a further sense has been discerned in which a poem is said to "pierce" its subject matter. This involves a rhetorical claim made by an implied author or speaker to be dissecting the body of which the poem speaks, which is also an epistemological claim that not merely the subject or discourse of anatomy, but the anatomy of the human body-in-itself is dissected thus, and a representational claim that poetic anatomy reproduces or recreates the actuality, life, or truth of the body-in-itself as the signs of its speaking picture. The poetic text assumes or acquires a materiality akin to the body for which it figuratively stands. This perfectly exemplifies the circularity of Davies of Hereford's epistemological and representational theory. Words which anatomize do so by generalizing from particulars to disclose final causes, which are inferred from the postulate that the body is an effect of the soul and of God's act of creation, so the efficient cause (God) and the final cause (the soul) are not independent of the proposition about particulars which they must justify. Davies of Hereford's reconciliation of the body's ingenious incision or dissection, and its eloquent figuring forth in poetry can only be reconciled in an imaginary way, but the relatively higher degree of faithfully rendered anatomy in *The Purple Island* makes the possibility of reconciliation more plausible. A decisive conclusion must wait until the conceptual character and provenance of the *Purple Island* anatomy has been treated in part 3.

This chapter has described how Davies of Hereford's poetry claims even more for poetic ingenuity than is outlined under the epistemological and representational claims above. This further claim assumes that because the body, as signified by the poem, is composed of poetic eloquence, which acquires corporeality from the mortal body, it is subject to imminent dissolution. It is proposed that ingenuity, which belongs to the immortal rational soul, will stem the tide of corruptibility by mimetically perfecting and resurrecting it in its ideal and immortal form. It is implied by Davies of Hereford's poetry that the author intends this as the illocutionary act of his work, which is Christian and soteriological, like that of *The Purple Island*'s anatomico-poetic exposition of microchristus. In the work of both poets, ingenuity purportedly partakes of the efficacy of the incarnation of Jesus Christ, and the rhetorical gesture made by the *Devine Weekes*, by *Mirum in modum, Microcosmos,* and by *The Purple Island,* where the poetic voice professes to pierce, cleave, touch, and dissect the tender parts of the body, is designed to testify to the truth of the body's resurrection in and through the signs of the text.

The effects of ingenuity and eloquence extend to the hermeneutic context, to the relationship between a work as embodied graphically in the extent of its signifiers and the intentional direction of that work toward a world the realization of which is initiated by the interpretation of the work's signification, but the "inward conceit" or apprehension of which accompanies the expansion of self-understanding in front of the world. The picture of poetry as represented in *Microcosmos* is not only living and speaking, but gestating and "spawning" images and words, which the reader unfolds, dilating meaning in accordance with the intelligibility of the world toward which the work is intentionally directed. Davies of Hereford's theory has been found to be consonant with the rhetorical theory of Quintilian and Peacham, and this study has consistently found that despite objectivist elements in Renaissance dialectical, poetic, and rhetorical theory, it was widely acknowledged that in the case of creative metaphor, the dilation of meaning whereby one sign gives birth to another is greatly enhanced, and what is now sometimes referred to as the hermeneutic circle is initiated by the implied invitation to the reader to complete the process of semantic generation by using her or his own "conceit" or imagination.

Although theoretically articulated in *Microcosmos,* this invitation to the reader is that much more inviting in *The Purple Island* than in any of the comparable works. The reasons for this are manifold, and have incrementally gathered in the course of this study. Their cumulative effect has been argued for over the chapters, but to list them briefly, they include the work's figurative design, its conceptual and creative metaphoricity, its exploration of the unused parts of the defining term of its threshold figure, and of the unused part of the defined term, and consequently its discovery of previously unseen significant connotations of the Isle of Man metaphor, and also some novel ones. Reasons also include the work's indebtedness to theories of suprarealism, emblematics, post-Horatian relationships between images and words, and the closeness of human language to the signs in natural things. Furthermore, there is the work's integration of anatomical teleology with allegorical figurality, and of Christian providential history with the shared entailments of the conceptual metaphors of anatomy, and its cleverly contrived and sustained allegory that explores systematically and coherently many of the connotations implied by the metaphorical interaction of Man with the specially constructed implications of the Isle. Being an allegory, which Peacham categorizes under tropes of sentences, *The Purple Island* is literally (as it were) an extended or unfolded metaphor, though it continues with each new reader ad infinitum to generate unseen and newly emergent similarities between its principal and subsidiary subjects, the interpretation of which, however, is not only constrained by the context of the whole work, but holds in tension a retrieval of what is to some extent immanent in the work as connotation, implication, or merely suggestion (the work's locutionary and illocutionary acts) and the individual and historical variation implied by the perlocutionary act of the work.

Despite this variation Davies of Hereford postulates the extension of the soteriological effect of poetry beyond the signified of the poetic text, to human anatomy-in-itself, the anatomy of the reader, and beyond the hermeneutic context to the body (as referent) generally or universally. In so doing he seems to derive his idea of the relation of ingenuity to the materiality of the body from the Thomistic eschatological doctrine of the Resurrection of the Dead. As an imaginative rational faculty, ingenuity informs or enforms the mortal body-as-referent, resurrecting it as the poetic text, whereby the matter of the visible graph and of poetic eloquence partakes of the immortality of the rational soul. The anticipated or intended effect of this on the implied reader is to contribute towards the salvation of her or his body and soul, through her or his imbibition of a fideistic understanding of human anatomy as microchristus and therefore as redeemed.

My view that there is no evidence of a design comparable with that of Fletcher, to compose a speaking picture of microchristus, in which the subject of human anatomy, as well as its method, eloquently demonstrate not only the creative ingenuity, but also the sacrificial and saving love, of its Creator and Redeemer, suggests that this imbibition is conceivably better accomplished, if at all, by *The Purple Island,* the variation of reader response notwithstanding. This chapter has discerned equivocality and qualification in *Microcosmos* and *The Purple Island,* however, over whether their sapience has sufficiently salted their eloquence to resurrect the referent of the text into a living body of knowledge that preserves the body of the reader. In both cases, it has been suggested, that a wish to avoid the intellectual pride that led to the "Jointes" of "*man's* divinest part" becoming "out of frame," is part of the reason. In conclusion, the dissection of microchristus by the speaking picture of *The Purple Island* provides a verbally constituted form of ocular evidence for the assurance that God never will forsake what "he" made, but that this recovery of the prelapsarian *imago dei* is deferred in anticipation of the Resurrection of the Dead.

In this chapter we have seen how *The Purple Island* may be better understood in the literary context of comparable works of encyclopedic poetry and fictional and expository anatomy in the late sixteenth and early seventeenth centuries. The comparative study has developed our understanding of the figurative design of Fletcher's poem, and its illocutionary act, particularly regarding the sophisticated poetics it shares with the works of Davies of Hereford and Du Bartas. The implication of this poetics is that the epistemology and hermeneutics of ingenuity and eloquence are encountered under the aspect of soteriology. There is one feature of the design of *The Purple Island* that has escaped notice thus far in this part of the study, and yet it is a feature broadly speaking which the work shares with several of the other works discussed in this chapter. This is the formal juxtaposition, in Cantos 2 to 5, of the poetic stanzas and the prose anatomical marginalia. It is additionally important to attend particularly closely to this feature because, recently, answers to questions about *The Purple Island*'s unity have focused upon it.

6
The Form of Stanza and Margin

THE FORMAL ARRANGEMENT OF EACH OF THE PAGES OF THE FIRST EDITION OF *THE PURPLE Island* consists of vertically arranged poetic stanzas and prose marginal notation. In Cantos 2 to 5, this notation is particularly full and densely anatomical. It is the one feature of the design of the earlier cantos of the poem that has not been considered analytically thus far in this study, even virtually escaping notice. Yet there are a number of reasons why it is important to attend particularly closely to this feature of the work. Firstly, the formal juxtaposition bears directly on our assessment of the validity of the claim to be able to extract the anatomy from its shadow of allegory, because the existence of prose anatomical notation suggests anatomical content that is independent of the figurality with which it is presented in the poetic stanzas. More broadly, the formal disposition of the work has a bearing on our response to what has preoccupied the most thorough critics of *The Purple Island* almost throughout its critical history, that is, the combination of anatomical exposition and allegory, and hence with the status of figurality. Although the chapters in part 1 described the congruity between the subjects and discourses of anatomy and poetry with reference to the descriptive or expository aspect of the poetic stanzas and marginal glosses, the actual relationship between the stanzas and glosses has not been examined, though individual examples of how the marginal notes work in relation to the stanzas have occurred throughout this study. Consequently, there has been developed no consistent sense of how the language and figurality of the poetic stanzas relates to the marginalia, whether or not there is a linguistic or conceptual distinction.

The formal arrangement of vertically arranged poetic stanzas and prose marginal notation is another way in which *The Purple Island* brings the organization of its material to the fore; and yet equally interestingly, it is a feature which broadly speaking the work shares with several other works discussed in the previous two chapters, and which consequently may have comprised an aspect of the set of expectations readers may have brought to Fletcher's poem. In broad terms the form is shared by Du Bartas's *Devine Weekes,* Clapham's *Ælohim-Triune,* Drayton's *Poly-Olbion, The Little World* by "Ro. Vn.," and Kynloch's *De hominis procreatione.* The marginalia of the work by Kynloch is compared by Sawday to that of *The Purple Island.*[1] Whereas Davies of Hereford's poetic *Summa Totalis* and *Microcosmos* contain copious prose notes, *Mirum in modum* contains few. The forms of both Burton's *Anatomy of Melancholy* and the surgeon John Banister's work on human anatomy, *The Historie of Man,* are composed of a main text and marginalia, both prose, and Hagthorpe's *Visiones Rervm* provides a series of notes beneath the texts of the poetic stanzas to which they correspond. In addition, the notes of *Summa Totalis, Mirum in modum, Microcosmos, The Purple Island, Ælohim-Triune, Visiones Rervm* and *Poly-Olbion* are fixed to parts of the poetic text by means of a number-, symbol-, or letter-referencing system, or a combination of these.[2] A full answer to the question of how derivative or innovative the allegorical extension of the Isle of Man in *The Purple Island* is, I have already said, cannot be provided without an intertextual study of sources, influences, and comparable literature, but thus far this study has not regarded the relation of the work's formal design to other comparable works. And so this chapter continues comparative work of the kind undertaken in chapters 4 and 5. A closer look at the varieties of arrangement and function of notation, and particularly at the use of it, in *The Purple Island* is additionally important because recent answers to the question of that work's unity have tended to focus upon this aspect of its form. The present chapter therefore investigates ways by which the form of *The Purple Island* may affect our understanding of the congruity of anatomy and poetry in the work, the poem's figurative design and soteriological illocutionary act, and its intervention as both innovative and conservative in the context provided by the other comparable works that broadly share the same formal design of juxtaposing main text and marginalia.

Among the interpretations to which the formal arrangement of *The Purple Island*'s anatomical cantos has given rise is that of Langdale. In his endeavor to extract the anatomy from its poetic modification, he proposes that it is possible to avoid interpretative pitfalls, "because *The Purple Island* anatomy is clarified by concise prose marginals."[3] He infers or implements a categorical distinction between a decorous allegorical poetic in the stanzas and a concise unfigurative prose in the margin, which purportedly throws off the cloak of allegory. This distinction, and the consequent clarifying function of marginalia, holds more or less true for the marginalia of Du Bartas's *Devine Weekes:* "Two equall ranks of Orient Pearls" are rendered in the margin as "*The Teeth*" (*D.W.,* p. 160); the marginal note "The lippes" refers in the allegorical poetic text to "Two mooving Leaves of Corall soft and sweet" (*D.W.,* p. 160). Given its significant influence on the composition of *The Purple Island,* it would seem not unlikely that the nomenclatural function of the marginalia of the *Devine Weekes,* providing the anatomical term corresponding to the metaphorically translated description in the poetry, was carried into the formal design of Fletcher's poem. As has been remarked earlier, the "matter" which forms the island-body's "foundation" is "Fram'd of fat earth, which fires together tie" (*P.I.* 2.5.2, 1, 4), and the marginal comment (a) confirms that this "fire" is the "heat" of the "vertue generative" of the "seed," which may be said to bake the "bones" into their particular forms. Another example is where in the stanza the "eight companies" distributed "in severall wards" (*P.I.* 2.21.3), are translated in the margin (t) as "the muscles of the belly-peece, which are eight, foure side-long, two right, and two crosse." None of these examples, however, necessarily supports Langdale's view that the marginalia consist of unfigurative prose which throws off the cloak of allegory, a point to which this discussion will return.

The same confirmatory, clarifying function is held by many of the marginal notes in *The Little World* of "Ro. Vn." Where the narrator describes "A thousand thousand *Houses*" in "A *Cittie* large, of bignes such, / as it the World had beene," the marginal comment confirms that this city should be interpreted as "*the world.*"[4] In addition, the marginal notes frequently provide the anatomical name for the part which in the poetry is translated metaphorically into a part of the house, the subsidiary subject of the poem's conventional threshold metaphor. Thus "a *Mill*" is identified by the marginal note as "*The Teeth*"; and when the *Little World* narrator's attention moves to "a large and a faire *Hall,*" which corresponds to the "goodly Parlour" in Alma's House (*F.Q.* II.9.33.6), and which anatomically is the thorax or middle venter, we know that the "Throne of Maiestie" or "Chaier of estate" therein is "*The Heart,*" because the marginal note tells us so.[5]

How the marginalia of *The Purple Island* clarify the poetic allegory of the human body is a contentious and complicated issue. Lana Cable argues for two distinct discourses expressed and reflected by the formal juxtaposition of poetic stanzas and prose marginalia.[6] She claims "the features of the poet's anatomical landscape [in the poetic stanzas] are as vague and as familiar as anything in *The Faerie Queene,* coming to the foreground briefly as the poet requires them and then slipping away again, leaving no more question about where they have gone than whence they had come. If we attempt to read the verse as if from it we were to extract some literal sense of a portion of human anatomy, we immediately run into the difficulties Kastor and others complain of."[7] Here Cable astutely fathoms the reductionism behind Kastor's charge of incomprehensibility: "behind the feeling of 'losing track' lies the assumption that the truths of the poem are *located* in its facts—which can be followed only by way of the gloss."[8] She describes the process of reading the anatomical cantos thus: "[W]e are repeatedly, almost irresistably drawn to the marginal gloss in cantos 2–5, the prose. We are drawn not just for explanation of a term or allusion (this the gloss rarely does), nor even to acquire a scholiast's interpretation, but simply—sometimes desperately—to try to find out what is *really* going on. And the gloss *tells* us what we were after—again and again it *makes sense* of the poem for us."[9] Cable ventures to suggest further: "What we find in the gloss is reassuringly lucid and perfectly comprehensible. We find facts. Encouraged, we turn back to the verse, facts in hand, and try once again to follow the poetry—'translating,' as Kastor puts it—literally forcing the poetry to make sense." "[A]gain and again we turn back to the poetry to make *all* of it 'make sense.' And we fail." "[W]hat *can* be 'made sense' of, what can be rendered in purely empirical terms, has already been so rendered: that was the gloss. What cannot be so rendered must be lost if we stick to these terms. . . ."[10]

There is reason to believe, however, that there is less discursive or linguistic difference between the poetic stanzas and the prose marginalia than Cable's analysis suggests. One of the reasons that Kastor's charge of incomprehensibility is misjudged is because in the stanzaic descriptions in *The Purple Island* like that of "governours" whose duty is "To see the vict'als shipt at fittest tide" (*P.I.* 2.31.4, 5), the anatomical descriptive criteria, in this case number, position, and function, are maintained in both the stanza and the description in the corresponding marginal note, in this case (f), of the "eight muscles" as serving the purpose of "swallowing." As was established early in this study, Kastor misjudges this, but additionally, Cable's analysis is astute and accurate with regard to his assertion that while most of the content of the stanzas is meaningless without the marginal glosses, by referring between stanzas and marginalia the reader becomes embroiled in a process of translation that renders the stanzas substantially incomprehensible. The coherence and consistency of the poem's application of the anatomical *accessus,* however, suggests to me that Cable has overlooked how, assembling information, the reader may deduce that the objects described are at once members, organs or parts *and* geopolitical structures, like the city Hepar (the liver), situated upon a prominent geological feature.

The fact that *The Purple Island* marginalia serve to clarify poetry that is systematically anatomical and geopolitical, and provide anatomical nomenclature, militates somewhat against some views that have achieved a prominent place in the critical history of the poem. For reasons previously supplied by this study, the view of Nicolson that the metaphoricity of the anatomical content of *The Purple Island* is more comprehensive and detailed than anywhere else in English literature, at least in the Renaissance, is justified. And Baldwin's reading does provide an explanation for the epic aspirations of *The Purple Island* and, if supplemented with an understanding of the character of the literary anatomy, is capable of accounting for what chapter 5 of this study has referred to as the erudite, analytical, and aggregative, almost encyclopedic, survey of its subject matter. But this study has found reason to disagree with or qualify Baldwin's and Nicolson's views that *The Purple Island* is conservative and didactic, that its conceptualization of the body in microcosmic, and (as an island) in geocosmic, terms was already firmly established when Fletcher came to write, that the work is explicable entirely within the context of analogy, and that it fixes at very nearly the last moment they could be fixed, the firmest convictions and attitudes of the Renaissance. These views are not wholly supported by the evidence; they have needed and need further qualification.

The argument that Fletcher's reliance on the conventional analogy of the microcosm is indicative of conservatism led this study to investigate the microcosm analogy only to conclude that a belief in the frequently indistinguishable Platonic and literal microcosmic views, while typical in the Renaissance, was not necessarily conservative and may even have been an impetus to anatomize. In addition, Nicolson and Baldwin are mistaken to think that because the anatomy of the poem consists of analogy, it is entirely dependent upon analogy for its conception. Chapter 2 has described how, as a creative metaphor, the threshold figure of the allegory selects, emphasizes, suppresses, organizes, and extends features of the principal subject, in a way that analogies rarely do. The figure allegorically transforms the systematic metaphors and analogies of anatomy into the geographical and social features of the Isle of Man, and although the significant connotations of this metaphorical interaction are in part antecedent, by discovering or inventing previously unobserved similarities the allegory deepens and extends, rather than merely aggregates and fixes, meaning in accordance with a notion of the world's intelligibility. Moreover the connotations include those that are related to the political, anthropological, and literary implications of nationalism and individualism, which are not among the firmest convictions of the Renaissance because they are historically incipient and ideologically emergent. It is also the case that to argue that *The Purple Island* merely fixes convictions is to underestimate the contextual effect of metaphor in the literary work, which at least in this case gives rise to greater interaction between the corresponding sets of associated implications of the principal and subsidiary subjects, wherein the associated implications of the subsidiary subjects of the entailment metaphors throughout Cantos 2 to 5 are already changed in "literal" meaning before they interact with their principal subjects, precisely because the specific, literary character of the "Isle" in Fletcher's poem is not wholly the same as any other preexisting isle, realm, polity, or nation, and cannot therefore be wholly conservative.

On the evidence that exists, seventeenth-century readers of *The Purple Island* do not seem to have been divided like Langdale and Baldwin over whether Fletcher's use of the conventional analogies of the microcosm is indicative of conservatism. The contemporary attribution of ingenuity to Fletcher and *The Purple Island* does not undermine but militates against Baldwin's interpretation. The attribution recognizes the discovery and fanciful (re)invention of anatomical subject matter as an intellectually intriguing riddle; and the imaginative, witty contrivance of the poem's figurative design and skillfully, cleverly constructed and sustained allegorical artifice. But as an early section of chapter 4 has argued, it also recognizes the achievement of accuracy, or truthfulness of an accepted kind, since the work's metaphoricity, its poetic of *concettismo,* was perceived as having illuminated the obscure unity of the world conceived as a kind of divine and witty poem, thereby having brought "*Deities* to this *Isle*" (in the words of "A.C."). The poem was perceived as innovative in the way of achieving its purpose to deepen and extend conventional knowledge, analogies and metaphors, a way that is nonetheless consistent with the way highly educated, cultivated minds in early modern England may have run.

So although there is a strong conservative element to the religious, political, and philosophical strands of *The Purple Island,* the notion that the marginalia serve to clarify (though not exhaust the meaning of) poetry that is systematically anatomical as well as being geopolitical, and confirm the identity of anatomical content with anatomical nomenclature, does provide a further objection to an overall reading of the work as conservative and didactic in a traditional sense. Although the poetic form, allegorical structure, and analogical representational mode of the stanzas may "indicate . . . the order and perfection of the universe as a whole,"[11] the poetic voice which Fletcher adopts cannot be understandable wholly as that of a pastor-poet, interpreting Man, who, at once microcosm and microchristus, unlocks the universe and tends to the Godhead. Whereas the poetry requires the reader to approach the human body, and from there its creator, obliquely, through a reading of the metaphorical signs of the island realm, the marginalia are in prose, and tend towards particularization, as opposed to extensive analogy, as the way to proceed in order to understand Man.

The marginalia of the *Devine Weekes* do not only clarify and confirm; they also perform another function, which they share with Banister's *The Historie of Man.* They indicate the subject of the relevant passage of poetry. The note, "The Seat of the Soule," by accompanying a passage discussing the conflicting ideas about the soul's anatomical location, whether it be in "the Brain or Heart," or whether it "Be all in all, or all in every part," effectively performs the function of an index, a marginal index (*D.W.,* p. 165). Occasionally in this role the marginal renders an expression in the poetic text into a more familiar form. Thus the vernacular "Ther's under Sun (as *Delphos* God did show) / No better knowledge then *Our selfe to know,*" is rendered in the margin simply as the Latin, though more familiar, injunction "*Nosce te ipsum*" (*D.W.* [1641], p. 52). This function is also fulfilled by those of the marginal notes in Kynloch's *De hominis procreatione,* that "act as an index to [the medical subject that] the poem is discussing" and those that identify the types of literary mode employed by the poem.[12] Du Bartas also on occasion employs the marginalia of the *Devine Weekes* to indicate either an event described, such as the "Mariage" of Adam and Eve (*D.W.,* p. 172), or the generic mode which the poem adopts at the relevant point in the text. Thus after Adam and Eve's marriage, a few lines comprise what is called in the margin "Their Epithalamie, or wedding Song" (*D.W.,* p. 172).

The marginal notes of *The Purple Island,* however, are often more discursively expansive than these concise clarifying, nomenclatural, or indexical marginalia, and this has two consequences for my reading. Firstly, if having more scope, the marginalia's expanse contains figurality (which it will be my immediate task to determine) this suggests that Langdale's distinction warrants a guarded skepticism, because it exacerbates the methodological risk, already run by Langdale, of succumbing to a hermeneutic circularity. His expectation that the anatomy is sufficiently free of analogy and metaphor to be extracted, would appear to be confirmed by what appears to him to be the prior performance, or at least commencement, of this process of extraction by the parallelism of the vertical arrangement. Secondly, the expansivity of the prose marginalia of *The Purple Island,* as compared with many other works, suggests that they may perform an explicatory role like that of some of the originally marginal notes in *Microcosmos* and *Summa Totalis* or the marginal notes keyed by letters to the rhyming couplets of *Poly-Olbion.* A variation on the indexical function is to explicate the content or the rhetorical strategem of the poetic text. In Davies of Hereford's *Microcosmos* the notes sometimes summarize what the poetic passage performs. An example of this is cited in the previous chapter, where it is remarked that the "disposition of the Members" is deployed in the service of a socially prescriptive model. More often, however, they provide a pithy statement about the content of the related passage of poetry, sometimes expanding explanatorily or merely extracting from the poem in unambiguous and unpoetic prose what the poem is authorially intended to say.[13] This latter function to extract what might be called the prose pith of the poetic passage, might be thought to conform to a reductionist or substitution view of poetic metaphor (which is discussed early in chapter 2); but frequently the note reiterates in prose the linguistic and even occasionally the syntactical character of the verse, thereby repeating metaphor as it appears in the poem. The theological etiology of the heart's position as described in the following passage is shaved off in the note, yet the metaphorical use of the verb, to hang, remains unaltered:

> Lo, thus the *Highest* holy vpright hand
> By even counterpoise hath hang'd the *Heart*
> In the *Brest's* Center. . . .[14]

The corresponding note reads: "The Hart is hang'd in the Brest by even counterpoise."[15]

The indexical function is stretched beyond its proper bounds in the expansive notes and annotations of *Poly-Olbion.* In addition to letters, *Poly-Olbion* also employs asterisks as a part of its keying apparatus, and the "Illustrations" or annotations are also keyed to marginal notes with a symbol, letters, and asterisks. But the function of the notes in Drayton's work is to explain, sometimes expansively, a technical or obscure term in the poetry, a function shared by the notes of *Microcosmos* and *Summa Totalis,* and by the notes of Hagthorpe's *Visiones Rervm,* which are numbered or keyed by symbols.[16] In Fletcher's work, on the other hand, the technical nomenclature appears in the marginalia themselves because the poetic stanzas are thoroughly allegorical.

Some of the notes to the second poem in *Visiones Rervm*—"*Cursus & Ordo rerum;* or, Art and Nature"—share with *The Purple Island* both discursive expanse and a subject matter and vocabulary that fall within the fields of natural history and philosophy.[17] The subject of Hagthorpe's poem is not, however, anatomical, though the cataloging style with which, among other subjects, the natural history of animals, birds, and plants is accumulated qualifies the work in part as encyclopedic or expository anatomy. The notes appear beneath the stanza to which they correspond, and on several pages, they exceed in space that given to the stanzas. Indeed they become so extensive that at stanza 24, the "Reader" is told to "looke for the rest of the Notes at the latter end of the Poem, by reason of the largenesse of them." Such expansiveness raises the status of notation, and does so whether to flaunt erudition, or to clarify, index, or name allegorically translated detail, or to explain poetic or rhetorical strategy, or to extract pithy informing statements, or to explicate dark passages: whatever the function of the notes, and in *Visiones Rervm* it is principally explicatory and explanatory, they achieve in the work of Hagthorpe and Fletcher a status comparable with that of the poetry.

Combined with expansiveness the experience of reading the notes as almost a sequentially continuous commentary rather than a series of notes may give rise to the notion that the juxtaposed discourses of poetry and anatomy in Fletcher's poem, and possibly those of poetry and natural history among other subjects in that of Hagthorpe, are "two autonomous forms of discourse, rather than one (the 'poetic') supported by the other (the 'scientific')."[18] Among the most eloquent of exponents of this view of the formal disposition of *The Purple Island* are Cable and Sawday, whose response this is. Cable remarks correctly that "[i]t is not the poet's intention to administer an anatomy lesson in a poetic sugar coating." "Yet," she adds "he will explain the anatomical details of his poem by means of a gloss, because his purposes include complete acknowledgement of, and respect for, all the empirical facts of man's existence—an existence that, for Fletcher, ultimately transcends empirical facts."[19] Among the examples that Cable provides to support her argument for two distinct discourses, is Canto 4's description of the "windepipe" (*P.I.* 4.28 margin. c). Cable concentrates on stanza 29 and the first four lines of stanza 30, and limiting her interpretation purposely at the outset to the poetic stanzas, she comments: "The purpose of the pipe is clear enough ('for th'aiers safe purveiance' [*P.I.* 4.29.1]), as is the need for air in order to produce vocal sound. We also see that the windpipe is formed of several substances ('divers matter' [*P.I.* 4.29.3]). But soon the picture gets confused. We learn that the first of the substances is bent into a hundred circles, 'hard and drie,' [*P.I.* 4.29.5] which do not touch one another but come very near to touching, 'meeting very nigh.' [*P.I.* 4.29.7] But in the lines that follow, the proximity of the circles has changed: the second substance 'Fills up the *distant* circles *emptiness*'" (*P.I.* 4.30.3 [Cable's italicization]).[20] That the poetic stanzas are possessed of nuances that distort, depart from, and appear even to contradict the comparatively empirical, matter-of-fact priorities of the marginalia is well observed, and Cable proceeds not only to demonstrate why recourse to the marginalia to explain the poetic stanzas is frustrating for Kastor and other critics, but also to propose a distinction between the components of the formal design:

> As before, any attempt to make objective sense of such a passage is bound to fail. But we can now begin to see why such lines would be composed as they are. For the purposes of the poem, a distance is measured by its status in the imagerial construct. The windpipe has been established as a "safe" purveyor of air, a supplier toward "perfect" vocal sound [*P.I.* 4.29.2]. Thus the fitness of the "hundred circles" [*P.I.* 4.29.5] for their task is measured first by their hardness, their dryness, and their ability to function in concert—"meeting very nigh." But when the second substance—the connector of the circles—is introduced, *its* importance as a *connector* is measured by the distance it must span between the circles: it is "smooth and pliable made for extending, / Fills up the distant circles emptinesse." [*P.I.* 4.30.2–3] The prose gloss, as usual, conveys none of these subtleties: "The windpipe, which is framed partly of cartilage, or grisly matter . . . (these cartilages are compassed like a ring); partly of skin, which tie the grisles together." [*P.I.* 4.28 margin. c] Here, the description is simple and reasonably clear; there is no shifting of valuation, no "measuring" at all. The gloss is dealing with the world of objective, empirical fact. In the world of the poetry, however, to describe or even to allude to something is to grant it a place in the overall moral and ethical, as well as esthetic, structure of the poem.[21]

The subtleties of which Cable writes are indeed lacking or reduced in the marginalia, but the poetry's concern with the status of complexion, structure, position, posture (or connectedness), and function is not, strictly speaking, aesthetic or moral, as she maintains. These descriptive criteria are the *occasiones* of the anatomical *accessus,* and both marginalia and poetic stanzas integrate them into their descriptions, though the poetic stanzas place greatest emphasis on those which characterize the function of a part or member that most effectively displays its purpose, its reason for being, "*its* importance as a *connector,*" for example. A shifting perspective, from the fulfillment of the function of the rings of cartilage to that of the "muscles" (*P.I.* 4.30 margin. d), does not entail that gestures beyond the purportedly empirical facts cannot at the same time include them; that the facts must be confined to the margin while the poetry embraces *other* (rather than *more*) objectives. But the character of the criteria upon which the poetic stanzas place emphasis is teleological rather than strictly structural or compositional, to mention only two of the *occasiones.* Nor, however, is their character moral or ethical, though they do contribute to the fulfillment of the work's intentional direction toward a world that it creates, and thereby enhance its fictive literary or aesthetic quality.

The poetic stanzas tend to grant the reader a shifting perspective on the anatomized microcosmic world. In stanza 29, as Cable notes, the distance or space between the rings of cartilage is indeed specified, as "Not altogether close, yet meeting very nigh" (*P.I.* 4.29.7), yet after the shift of perspective to one from which the distance seems great, these "circles" are again described in the three lines of stanza 30 that Cable omits from her analysis, as "Not fully met," and the "muscles" as supplying "where the circles scanted" "what they have wanted" (*P.I.* 4.30.5–7). In other words, the stanzas disorientate only then to reorientate the reader, granting her or him a perspective on the whole of anatomy which is capable not only of taking in relative distances but of placing her or him in different positions or locations within the world of the microcosm, each with a vantage point on a particular organ or member or part that best displays its reason for being what it is and where it is in that world.

Cable's account exaggerates the discontinuity between moral-aesthetic and empirical perspectives of the parts of the formal disposition. But Sawday somewhat qualifies Langdale's distinction between stanza and margin in order to propose a similar kind of distinction: "The chief difference between the two texts is not so much that one is purely unfigural ('scientific') whilst the other is metaphorical ('poetic') since similes appear in the notes and scientific patterns of language in the poem. . . . Instead the difference lies in the terms of comparison which the two texts develop. In the margins, the system of comparison is, finally, a closed one. The body can be compared only to itself or to its own features. Thus the 'humour' which encloses the heart (note m) is itself of the texture of other bodily fluids—whey or urine [*P.I.* 4.16]. The poetic text, on the other hand, endeavours to expand the terms of the comparison outwards, to uncover the significance of the body in relation to the external world."[22] Such demarcation of distinct figurative characters does not affect "*Cursus & Ordo rerum.*" The creatures that appear on the kirtle of personified Nature are not thereby transformed metaphorically. They are in this unlike the human anatomical parts that do undergo such transformation as they become parts of the Isle in the poetic stanzas of *The Purple Island.* It is the purpose of this chapter, in part, to judge whether the notion that the *Purple Island* text consists of parallel and autonomous discourses of allegorical poetry and anatomical science is accurate as a description of the work. Without accepting Kastor's opinion that the juxtaposition of the stanzas and marginalia results in incomprehensibility, it remains the case that it nevertheless problematizes the status of *The Purple Island* as a conventional didactic poem.

Sawday's distinction between the figurality of the stanzas and the marginalia of *The*

Purple Island, however, requires qualification, for two reasons which qualify Langdale's distinction all the more, because it is the sharper of the two. Both distinctions underestimate the conceptual, figural, and verbal interdependence of the stanzas and marginalia; secondly, the differential in the spectrum of various stanza-margin relations, with regard to figurality, is broader than either distinction allows. Contrary to Sawday's claim, the system of comparison in the margin of *The Purple Island* is not always "closed." There are a few occasions where a metaphor, representing the body as the geocosm or the human-built environment, is reduced to a simile in the margin, such as that in Canto 2, stanza 9, which represents the veins, arteries, and nerves as like rivers.[23] The sense of how the marginalia function to turn poetic metaphor to simile or comparison may be deduced further from an example provided early in chapter 1 of this study: the heavily terrestrial language of Canto 3, stanza 7 characterizes even the anatomical notes in the simile that compares the lower external surface of the liver to a hollow rock.

It is not possible to make any reliable comparison or distinction between the discourses of the stanza and the margin without taking into account the occurrence of etymological metaphoricity, which compares the body to the external world in the proper nouns of anatomical nomenclature. As this study has demonstrated, the metaphorical concept of anatomy as "building" gives rise in *The Purple Island* to an exploration of marginal or downplayed aspects of the conventional metaphor, extending the ordinarily used part of the defining concept to provide detailed specifications of the body's architectural features. This often involves an opening out of the etymons of anatomical nomenclature, and in Canto 5, stanza 16 the third ventricle of the brain is described as "Vaulted by three fair arches," which the margin (m) indicates belong to the "corpus Cameratum," or fornix, "which with three arches bears up the whole weight of the brain." As was noted when this description was discussed earlier, whereas the marginalia thereby barely modify the figurality of the stanza, Vesalius similarly describes the fornix in his *Fabrica,* albeit with simile and comparison rather than metaphor, as "resembling a vault (fornix) or arched roof." There is one occasion when such stanzaic etymological metaphoricity is even more substantially reproduced in the margin. The stanzas use "caverns" for the lateral ventricles of the brain, and "cave" for the third ventricle; whereas the margin uses "cavitie," which is anatomically specific and nonfigurative, for the third ventricle, but reiterates "caverns" for the lateral ventricles (*P.I.* 5.14.2; 5.16.3 & margin l, k). These terms are punningly, etymologically, and semantically related, and, while expanding the signification in a way that is entailed by the Isle of Man metaphor, both stanza and margin are no more than merely exploring the metaphoricity inherent in the anatomical term "cavitie."[24] This limitation should not, however, detract from the genuine metaphoricity of the linguistic form and the exploration it commences and prompts in the mind of the reader.

Etymology served an important pedagogical function for Vesalius, as chapter 2 revealed, and so account should be taken of the etymological figurality in the stanzaic and marginal texts of *The Purple Island,* even if only to indicate that in producing their meanings, etymological metaphors may not consistently, like conceptual metaphors, define their principal subjects in terms of their subsidiary subjects, or rely as heavily as creative metaphors on the interaction between their principal and subsidiary subjects, though I am unaware of why the latter should be so in an allegorical poem. Indeed etymological figurality also deserves consideration because *The Purple Island* demonstrates a clear and serious suprarealist interest in its allegorical potentialities. It treats etymons as the intrinsic properties of the words in current use which descriptively fashion the anatomy of Man.[25] In suprarealist conceptions of language, words form a secondary system continuous with but marginally displaced from the signs in the things or objects themselves. So it is conceivable that rather than constructing two autonomous discourses, the one "poetic" and the other "scientific," Fletcher may have thought that he was unraveling in the poetic stanzas the etymological connotations that he thought were implicit in the anatomical terms or signs existing as though within the parts and members of the body, and which he provided in more concise form in the marginalia. There is no sharp division but an unfurling of potential connotations embedded within terms that are historically yet to become standardized, denotatively more limited and epistemologically transparent. When Canto 5, stanza 16 describes the septum lucidum metaphorically as "a wall built all of saphires shining," appending the comparative and qualifying phrase "As fair, more precious" (*P.I.* 5.16.1–2), the marginal note's translation, "bright wall" (k), does not need to indicate a metaphorical similarity because, in unavoidably reiterating the metaphoricity of the Latin name, similarity is implied anyway. The same metaphoricity is also found in the English names of the organs of the inner ear, the "Hammer," "Stithe," and "Stirrup," which are said, by the marginal note, to be taken "from their likenesse" (*P.I.* 5.43 margin. t).

The tendency of the marginalia to translate anatomical nomenclature, and to make a descriptive comparison, perhaps as a way of offering a (supra)realist explanation for the metaphoricity of nomenclature, is reflected in anatomy textbooks. It was found in chapter 2 that Vesalius's description, in his *Fabrica,* of the septum lucidum admitting candlelight through its membrane, "as with those transparent stones cut thin and smoothed like crystal . . . which Italian architects fasten in the mullions of glimpses or windows,"[26] employs imagery resembling that in Fletcher's stanza; but as with the fornix, Vesalius uses simile rather than metaphor, and offers a practical context in which the anatomical name might be understood as empirically descriptive of the part understood as created. It is, by contrast, typical of the poetic stanzas of *The Purple Island* subtly to develop the inherent figuration of the nomenclature of the body, as either constructed or natural, shading it into the built, natural, and social features of the geopolitical realm of the island and the cosmos above it, thereby incorporating those features into an extensive analogy.

Stanzaic figurality is reproduced in the margin, however, even where it is not inherent in the nomenclature. This is confined to the personification of anatomical parts, and the conceptualization of anatomical relations in terms of social ones. Both of these figurative representations can be seen in some instances to be omitted from the marginalia and in others reproduced. The familial relations between members in Canto 2, stanza 6 are omitted from the margin, whereas the concept of political "sovereignty" is reproduced in the marginal note (l) to stanza 14 of the same canto; and personification may be seen in the same canto to be omitted from the marginal (e) of stanza 30, and reproduced in the marginalia (d) and (l) of stanzas 29 and 37. The concept of sovereignty is closely allied to the anatomical concept of nobility or principality (discussed in chapter 2); and, by and large, personification in the margin is an extrapolation from anatomical nomenclature, though this is occasionally an imaginative or popular nomenclature, not to be found in anatomy textbooks: the "needfull groom, the Porter nam'd," who sits at the "second Port" of "*Koilia,*" is reproduced in the margin as the "lower orifice, or mouth of the stomack . . . [which] is called the (Janitor or) Porter, as sending out the food now concocted through the entrails" (*P.I.* 2.37.3, 1 & margin. l).

In conclusion, the differential in the various stanza-margin relations is broader than either Langdale's or Sawday's distinction allows, partly because neither examines etymological figurality. Although there are broad tendencies that distinguish the stanzas from the marginalia, these are only tendencies, which, together with the frequency with which the marginal notes reiterate the personification and socialization of anatomical members, mark the conceptually, linguistically, and discursively imperfect differentiation between stanzas and marginalia in Cantos 2 to 5, and, significantly for the direction this study takes in part 3, between anatomical nomenclature and its metaphorical and etymological extension and exploration in poetry. Consequently, *The Purple Island* is not, as Sawday has it, "a text which stands as an icon for the programme of linguistic reform which was to be undertaken in the later seventeenth century," as a result of "the post-Hobbesian proscription of poetic language in the languages of science."[27]

This proscription had a special effect on the standardization of anatomical nomenclature. Although etymology trailed in the wake of standardization, it was gradually elided in dictionaries of anatomical, physiological, and medical science. But this is only one aspect of the proscription, one which the evidence of Dunglison's *Dictionary of Medical Science* demonstrates lasted longer than other kinds of figurality, doing so because it was bound up with the formation and development of the terminology from which it was being expunged. The objects of more widespread and vociferous attacks, sometimes even from unlikely quarters, were "all deductions from Metaphors, Parables, Allegories,"[28] analogies, and teleology, as well as eloquence, all of which are integral to the figurative design or locutionary act, and the illocutionary act, of Fletcher's poem.

Among the earlier arguments in favor of such a proscription was Bacon's proposal that natural philosophy should disentangle words which in their innocence were equal to things, from the encrustation of their "exasperating, spiritually loaded, and therefore confusing histories."[29] It is evident from *The Purple Island* that a suprarealist interest in etymology, analogy, metaphor, and allegory was related to contemplation and speculation as a means of access to truth. Bacon argued that the advancement of learning would occur only when philosophers ceased to study words, and instead put words at the disposal of things, rather than spiritual truths, correspondences, or teleology of form and function.[30] So although motivated religiously by a desire to know the nature of things not as they have become in human minds and languages, but rather as they have been created by God, and (Bacon conjectured) as Adam had known them, Bacon sought to disentangle natural philosophy from the accretion of associated commonplaces and connotations derived from philosophy's context as handmaiden to divinity or theology.[31]

Bacon was an early advocate of empirical matter-of-factness shorn of such figure-laden knowledge, but among the most influential in this regard was Thomas Hobbes, who in *Leviathan* called for a new truth-seeking discourse of "Reason and Science" devoid of "Metaphors, Tropes, and other Rhetoricall figures."[32] *Leviathan* allows only the instrumental use of "some apt simile" when "the understanding have need to be opened"; metaphors should be "utterly excluded," they are "not to be admitted" "in reckoning [Hobbes's word for reasoning] and seeking of truth," for "they openly professe deceipt."[33] Hobbes's objectivist suspicions of metaphor were shared by the Restoration scientist Joseph Glanvill, whose *Vanity of Dogmatizing* reveals both the objectivist trust that ordinary language is unmetaphorical, and an objectivist and positivist postulation of the existence of ahistorical ontology about which a neutral and transparent nomenclature can speak the truth. He "warned against the habit of using words 'arbitrariously' or 'distorted from their common use,'" whereby "things plain and easy in their naked natures, are made full of disputable uncertainty."[34] Van Helmont also contributed to the sea-change with his *Oriatrike,* published in England in 1662 (having first appeared as *Ortus Medicinae* at Amsterdam in 1648), which demonstrates that since deductive argument about human anatomy from theories of the microcosm is a deduction from analogy, as well as being suspiciously contemplative and speculative, it too was a target of the mid-seventeenth-century linguistic reformers: "The name therefore of Microcosm or Little World is Poetical, heathenish, and Metaphorical, but not natural or true . . . the life of man is too serious, and also the medicine therof, that they should play their own part of a parable or similitude, and metaphor with us."[35]

The broad reform in which these Protestant reformers of their different philosophical fields shared a role was a search conceived in earnest "for a type of linguistic origin, a moment of Edenic purity of expression."[36] They are thus far removed from Fletcher, for whom microcosmic analogy and metaphor were of the prelapsarian order it was the illocutionary act of *The Purple Island* to try to help to restore. But despite his advocacy of this "purified" kind of language, Bacon's own writings in particular have been described as "sparkling" and to "avoid the matter-of-factness which one associates with the Baconian programme."[37] Nevertheless taking Bacon as well as Hobbes among its luminaries, England's Royal Society was founded in 1662, and sought ways "to degrade language to the same colorless symbolism which had proved so successful in mathematics."[38] Sprat's *History of the Royal Society* (1667) "articulated the same linguistic attitude, which demanded that 'eloquence be banished out of all civil societies,' and, more significantly, claimed that in works of natural philosophy rhetorical figure had no place."[39]

Quilligan argues that the "effect of such pressures on allegorical narrative can most immediately be sensed in Phineas Fletcher's self-consciously Spenserian allegory, *The Purple Island.*" She does not identify the effects of this linguistic and epistemological change in *The Purple Island*'s formal arrangement, but rather in the allegory: "Fletcher's numerous puns irradiate no underlying structure, being mere bits of verbal wit, bright moments of local color, but no more. Fletcher's poem is hardly allegory at all, not merely because it is not structured by considerations about language, but because Fletcher has to strain to make the physical world morally meaningful to his reader."[40] In a work the ideas of which have contributed significantly to this study's conception of allegory, and which is otherwise theoretically astute and sensitive to poetic language and allegorical design, this judgment is an isolated example of only cursory reading or incomplete attention, all too common in Fletcher scholarship.

Quilligan's judgment is twofold. Firstly, it is a judgment about the figurative design, or lack of it, and, secondly, it relates that lack symptomatically to the cultural and intellectual context. Contrary to the first part of the judgment, this study has attempted to reveal an underlying allegorical structure wherein the Isle of Man may be understood as a conceit in a poetic of correspondence, the threshold figure of the allegory, or the most specific metaphorical concept in a system. The creative metaphor of the Isle of Man extends the ordinarily used part of the defining concept or domain and uses the ordinarily unused part of that domain, giving coherent structure to a newly highlighted range of similarities. The significant connotations of the metaphor are unfolded or extended throughout the cantos in a process that I have described in several ways, each of which has implied coherence and systematicity. As the most specific metaphorical concept in a system, Man is systematically conceptualized in terms of a host of metaphorical concepts that are related in terms of entailments. This system has been identified also as the system of associated commonplaces of the subsidiary subject which may therefore consist of metaphorical concepts from anatomy, political theory, literature, and natural philosophy and theology even before they are transformed by interaction. This system of associated commonplaces is at least partly comprised of a literary, allegorical system of novel implications, because of which the interaction of the Isle of Man figure produces connotations which create further creative metaphors as they are unfolded and used throughout the cantos to describe parts of the body of Man. Although they are potentially indefinitely extended, these entailments or connotations are not arbitrary; they satisfy congruence as well as plenitude, and are sewn into the shadows of allegory with logical and conceptual consistency. The threshold figure functions as a microcosm, an index to the system of entailed metaphors, which therefore cannot be merely "local"; rather they are interrelated in the body of the poem in the way that parts are related to the whole of anatomy.

The Purple Island is structured by considerations about language not only because of this system or structure of metaphors, but because etymology is of central relevance to this system. Admittedly, these considerations do not extend as far into polysemy and homonymy as do many other allegories; but the poem's conceptual and creative metaphoricity, its incarnational, emblematic, and suprarealist language, and its soteriological poetic mode, epistemology, and hermeneutics, all indicate or imply in the author an earnest concern with the relationship between language and right or true belief and understanding with regard to Man. Principally, like most Renaissance allegorists, the implied author of *The Purple Island* is concerned with the inextricable relationship between human redemption and the redemption of language.

Quilligan's second point, that Fletcher has to strain to make the physical world morally meaningful, is a point about the poem's interaction with its cultural, intellectual, literary, and anatomical context. Conversely, both Thomas Healy and Sawday place great emphasis on how the formal layout, rather than the allegory as such, can be understood to bear the marks of engagement with historical change in the relationship between religious poetry and anatomy. Cable argues, correctly in my view, that Sprat's linguistic recommendations for a distinction between the languages of the rhetorician, the poet, the fabulist, and the religious enthusiast, on the one hand, and that of the scientist, on the other, do not apply to Fletcher's poetic, though her reason for arguing this, which differs from mine, is "not because he wrote too early to have heard of [the distinction]—his gloss is evidence enough that he knew the distinctions a Sprat could make—but because his purposes were not scientific, they were esthetic and moral."[41] That neither the categorical nor the linguistic distinctions proposed by Cable and Sawday, respectively, as ways of understanding the stanza-margin relationship, hold generally for Cantos 2 to 5, and that the allegory, analogy, and metaphoricity of both stanzas *and* marginalia require the reader to approach the human body obliquely through comparison and figuration, might have counted in favor of Baldwin's endeavor to understand the poem in the conservative tradition of didactic poetry, were it not for reasons against this interpretation which have already been given, including the reason that also tells against Cable's distinction between the stanzas and the marginalia, that the descriptive criteria of the anatomical *accessus* are formative of the descriptions of both stanza *and* margin. The anatomical *accessus* is the unifying feature of the formal design of Cantos 2 to 5.

Much of the evidence has been interpreted in this study as demonstrating that there was not in the early seventeenth century an absolute disjunction of the anatomical and the literary, and it would seem that the poetic voice of *The Purple Island* is not only that of a pastor-poet but also that of an anatomist. Quilligan specifically claims that "Fletcher versified Vesalius and Galen, not the modern physiology of William Harvey."[42] Preliminary to an examination of the work of Harvey (1578–1657) in chapters 8 and 9, I contend merely that the claim is freighted with assumptions perhaps arising from its author's beliefs regarding the formation of the theory of systemic circulation and the composition of *The Purple Island*. But as regards the more general thrust of Quilligan's second point, that Fletcher had to strain to make the physical world morally meaningful, it is clear from the intertextual and literary contextual studies of this and the previous two chapters that works representing human anatomy in allegorical, metaphorical, and analogical linguistic forms and discourses abounded in the early seventeenth century. The potential readership for such works I would suggest was wide, even if *The Purple Island*'s was comparatively limited until its publication in 1633. Conventional analogy and the more or less conventional conceptual metaphors of anatomy, moreover, persuade me that readers, or at least *The Purple Island*'s implied reader, would have understood the ingenious artifice Fletcher sought to create.[43] Admittedly, the assemblage of analogues in Fletcher's invention creeks and teeters, but it is generally upheld in the pursuit of a continuous superimposition, which results in a relatively coherent discourse, at least on an imaginary level. Cultural materialist studies of early modern literature and culture suggest that there is never such ideological dominance or hegemony in a work that the contradictions that it is designed to obscure do not break forth.[44] Such faultlines in Fletcher's work do not as Quilligan suggests erupt as an effect of later linguistic reforms, but as a result of the allegorical extension or development of a poetic of *concettismo* into an ideological artifice.

This is perhaps best explained with reference to the attribution of ingenuity by seventeenth-century critics, which does not imply, as it does for modern critics like Gosse, the poetic overcoming of an inherent incongruity between aesthetic or allegorical language and anatomical theme. But from the perspective of several recent critics, including Quilligan, Sawday, Nicolson, and Baldwin, *The Purple Island* has appeared acutely vulnerable to imminent historical change. The text stages its relation between anatomy and poetry, as Sawday says, at almost the moment "at which science and poetry finally struggle free of one another." And the design of the poem would appear to bear the traces of this "transition in the history of discourse" from correspondence to mechanism, "with the 'marginal' text appearing as a proto-scientific" description, "whilst the 'central' poetic text is the remnant of a 'pre-scientific' view of the world bounded by the theology of the microcosm."[45] Although this study has argued that the attribution of ingenuity by readers of the work demonstrates their perception that the poem secured settled convictions in a challenging, innovative way, the attribution does not suggest that the work was perceived as epistemologically, linguistically, or conceptually conflicted or incongruous. Rather by extending conventional, synchronous structures of analogy, Fletcher may have sought to shore up knowledge of the body that was already implicitly assumed to be true, but sought to do so in a way that simultaneously gave rise to significant metaphorical connotations that are ideologically consistent with formations of self and state that would supersede the universe of correspondences in which the allegory of the Isle of Man is conceived. As was said in the first chapter, reactivations of an analogy frequently occur when both analogous signifieds of what

Fletcher calls this uncannily "strange, though native coast"—native soil and the human body—are in transition historically. This is the case in the late sixteenth and early seventeenth centuries. The new ideologies had already emerged and, crucially for our understanding of the allegory and formal disposition of *The Purple Island,* they were nearer to being established than was the notion of a scientific language "purified" from the encrustation of figurality and religious association.

It seems to me, at this stage of our study, that in the political and anthropological ideological senses is the work conflicted, not in the epistemological, and there would seem to be no direct correspondence between this conflict and the division of the text of the work into poetic stanzas and anatomical marginalia. So the idea that the formal design represents the achievement of a fragile synthesis at probably the last possible moment in history when anatomy and religious allegory could be synthesized seems to me not only to place the burden of the work's engagement with historical change too firmly upon this one aspect of the figurative design; it also seems to attribute too strong a cultural position to the emergent notion of a scientific language devoid of figurality. Bacon's "sparkling" writing which belies his professed disinterest in words and rhetoric, Puttenham's objectivist suspicions of metaphor, and the imperfect or inconsistent distinctions made between images of life and images of wit by Fulke Greville, for example, are the germinal counterparts of the later, more thoroughgoing linguistic reforms of the middle and later part of the century.[46] Quilligan's view is that it is in the allegory, rather than the formal disposition of the text, that the effect of the pressures of discursive change can most immediately be sensed. But equally it may be said that rather than being a response to developments in philosophical discourse that occurred almost half a century after most of Fletcher's poem was probably written,[47] the "effect" of which Quilligan speaks is in fact the effect of a historiography that desires to see in the seventeenth century an epistemic change, composed of *coterminous* epistemological and political changes, and from which are deduced retrospectively the signs and strains of the epistemological change in literary works written *before* as well as during and after it.

As a direct approach to the question of the congruity of anatomy and figurality, I find the form of *The Purple Island* helpful only to a point. The lack of a consistent conceptual, discursive, and linguistic distinction between the poetic stanzas and the anatomical marginalia, however, need not preclude the possibility that the formal arrangement of the poem was designed to achieve an effect, and perhaps did achieve this, in its cultural context. The evidence of the responses of some of its modern readers would at least seem to suggest that the formal design has profound consequences for the way in which *The Purple Island* is read, though the lack of reference to the form among the responses of its contemporary readers gives no clues as to its effect on how the poem was or was intended to be read in the seventeenth century. Stephen Greenblatt argues that "specific, institutional interests [are] served . . . by the powerful if conceptually imperfect differentiation between the literary and the nonliterary."[48] A formal juxtaposition and differentiation between poetic stanzas and anatomical marginalia, however conceptually imperfect, such as is *The Purple Island*'s, may have been designed to negotiate and re-present as *distinction* or *division* the *relationship* between anatomy and religious poetry, such as it was; and to do so in order to serve a specific interest. It remains to be seen what *The Purple Island*'s negotiated marks of difference result in, though it is not a text that is merely conservative and theologically didactic, or one that augurs the later reformulations of linguistic practice in natural philosophy. But it would be unsurprising to discover that the answer lies in the attribute of ingenuity with which this study has come to identify the work, as characterizing its locutionary, illocutionary and perlocutionary acts.

As chapter 4 has described in more detail, in the sixteenth and seventeenth centuries, ingenuity connotes wit, literary or poetic talent, skill in contriving or making something, and the capacity for invention, which was a part of dialectic, and which could mean *fantasia*, imagination, or discovery, the faculty or technique of discovery but also the thing found, or the artifice. Although in the invention of *The Purple Island*'s allegorical island realm, Fletcher may have sought to shore up knowledge of the body that was already implicitly assumed to be true, the allegory simultaneously gave rise to significant metaphorical connotations of the Isle of Man that were ideologically emergent and historically incipient. As regards the formal disposition of *The Purple Island,* ingenuity creates a similar Janus-effect, but with a notably different explanation. The formal design in which the marginalia serve to clarify and confirm anatomical and geopolitical poetry with reference to anatomical nomenclature, a design which has given critics the impression of empirical verification, or has suggested the appearance of a proto-scientific description, is the very same curious design which is an expression of the Renaissance habit of forging correspondences between discourses, ideas, and objects, a habit which culminated in metaphysical poetry, or the poetic of *concettismo.* Applied to the finding of subject matter, invention and discovery were also related to the cognitive process of creating new paradoxes, like the riddle of the Isle of Man in Canto 1, stanza 34, and creative metaphors or conceits, like the Isle of Man figure itself. The formal juxtaposition of the work is understandable as indicative of the great imaginative lengths that Fletcher went to in order to secure the correspondences his poem asserts. The poem, and in this it resembles other baroque or metaphysical poetry, actually seems to accentuate differences as it tries to forge relationships: by exploring unseen aspects of the metaphor of the isle for Man, the poem revivifies it and at the same time estranges the conventional metaphor, foregrounding differences as well as highlighting similarities. The formal juxtaposition, which belies the imperfection of the linguistic and conceptual differentiation, may be said to serve the poem's aesthetic and stylistic interest.

The notable difference between this and the production of connotations of individualistic subjectivity and nationhood is that whereas these connotations were actually consistent with formations that were ideologically emergent and historically incipient at the beginning of the seventeenth century, the Janus-effect created by the aesthetic form is probably a result of retrospection. The incongruity to which the juxtaposition gives rise in the mind of the modern reader is causally belated, even as it may conceivably be causally linked to the text, and to modes of poetic practice affecting the text.[49] It is possible to see the poem's apparent incongruities as the curious effect of a discursive imbrication that functions in the poem as if the distinction between the discourses it employs was not only simultaneous with but even prior to and itself the cause of the incongruity.

The Purple Island is clearly a more culturally interactive and dynamic text than one would expect a "handbook to the opinions of a period in the history of thought fast on the wane" to be.[50] In developing ways of understanding the formal arrangement of Cantos 2 to 5 as designed to be culturally interactive, both Sawday and Healy concur in that it may initially appear (as it does to Langdale) that, "the empirically based descriptions in the surrounding commentary are employed to support [or "authorize"] the poetic organization of the body," perhaps by anchoring it in "scientific" observation; but, both modify this initial impression.[51] The initial impression is an understandable one, because it is a further principal function of the marginalia in Kynloch's *De hominis procreatione* and Banister's *The Historie of Man* to cite textual authority, as is often the function of marginal citation in the form of the printed sermon or homily.[52] Banister, for example, cites other anatomy textbooks. In addition, the indicative, pithy notes of *Microcosmos* are frequently proverbial, resting on conventional or commonsense wisdom, and both these and the notes of *Summa Totalis* cite scriptural or classical sources, thereby displaying their erudition.[53] Their function is to lend authority to the poem,[54] but it is also to explicate or clarify, a function they share with the marginalia of *The Purple Island,* and the authoritative and the explicatory functions of notes are not very dissimilar.

The texts which accompany the Christlike anatomical figure in Remmelin's *Catoptrum microcosmicum* (discussed in chapter 3) serve as well as an informing, interpretative function, the further purpose of citing biblical authority for the work's interpretation of anatomy. Thus the theme of mortality, emphasized by the iconography of the *vanitas* tradition, is given authority by the crested Latin text of Psalm 144, verse 4. The illustration also employs a keying mechanism whereby the crucifix in the illustration is referred to the foot of the illustration where the Latin poem entitled "*Mediatoris Christi Solatium* (The Consolation of Christ the Mediator)" appears. A late seventeenth-century English translation of Remmelin's work significantly carries out editorial cuts in conformity with the linguistic reforms of midcentury to late century.[55] From the plate of the Christlike figure, the *vanitas* text, *Nascentes morimur* (Being born, we die), disappears, as do the verse from Psalms, the serpent, the crucifix, and the Latin poem to which it refers and which was formerly at the foot of the illustration. Omissions from the work's other two illustrations that are taken from the original are consistently edited in the same way, to erase classical and theological interpretative schemes and authority.[56] These developments are typical after the mid-seventeenth century, although intermittently an explicit juxtaposition of the theological and the anatomical appeared, such as in Thomas Banks's *Anatomical Crucifixion,* "an *écorché* version of the *Crucified Christ* . . . made in 1800 from the body of a murderer."[57] Whether theological schemes of interpretation inform this anatomical representation in a residual way is not clear, but such conceptual processing does not appear in the declared scientific intention of the initiators of the project, which was to "'put . . . to a test' what really happened to a crucified body 'when . . . being warm' it naturally 'fell into the position that a dead body must fall into.'"[58]

Although *The Purple Island* deploys some marginal notes in the citation of authority,[59] both Healy and Sawday are aware that, by contrast, most of the elaborate marginalia in *The Purple Island* "correspond to the deployment of such glosses and notes in the production of scriptural texts."[60] An earlier, less developed version of this can be found in *Ælohim-Triune,* where the marginalia have an explicatory function. Rather than rendering an allegorical or metaphorical description in anatomical prose, because the poem is not allegorical or highly metaphorical, the marginalia in *Ælohim-Triune* tend to provide some extra information that the reader might reasonably be expected already to know, thereby simply acting as a reminder and prompt as to what to employ in her or his interpretation of the poem.[61] The marginal notes in *The Little World* are also interpretative, in this case of personifications and allegorical structures and their functions. A marginal informs the reader that "*God*" is the "*Maker of the house,*" which is metaphorically the body, and the function of the "Bellowes" of each house is to be understood as "*The offict* [*sic*] *of the Lounges* [*sic*]."[62] But the interpretative apparatuses brought formally to bear upon the poetic texts of both Clapham and "Ro. Vn." pale in comparison with that of *The Purple Island,* as do the interpretative feats they perform for the reader.

Healy's view of this is that, "by creating his own scholia in the margins, Fletcher is attempting to emphasize the poem's importance, and create the impression of a time-tested authoritative text."[63] In a sense this is entirely consistent with the view of human anatomy, represented and recreated as or by *The Purple Island,* as a second scripture, and with the soteriological purpose for which Fletcher designed the work. But as a consequence of this formal presentation of its anatomy, the *Purple Island* marginalia may plausibly undercut the reader's process of unraveling the elaborate allegory.[64] The riddle of the Isle of Man, introduced in Canto 1, stanza 34, requires an explanation and an interpretation, provided or initiated as a gloss or commentary on the text of this threshold figure, in the allegory of Cantos 2 to 5. This commentary takes the forms of the poetic stanzas and the prose marginalia. The notion that the allegory embodies within itself the total form of a work of commentary, including both gloss and text, is strengthened by the position of a significant part of this commentary or *scholia* in the margin. Healy is suggesting that *The Purple Island* thereby preempts the hermeneutic process, as opposed to merely prompting it. Rather than evolving a narrative which glosses its own threshold text, as does *The Faerie Queene,* thereby combining creative and critical processes, *The Purple Island* effects this combination through its formal juxtaposition.[65] The accretion of the anatomical aspect of allegorical meaning over the length of Cantos 2 to 5 accumulates not so much "in the self-consciousness of the reader, who gradually becomes aware, as he reads, of the way he creates the meaning of the text," but rather in the prose marginalia.[66] In a more pronounced way than with other allegories, *The Purple Island* does not need *allegoresis* because "the commentary . . . is already indicated by the [marginal] text."[67] Moreover, as the reader desists from interpretation, the poem impresses upon her or him its "primacy in revealing the truth of the body," and the supplementary character of the "ostensible empirical support" in the margin.[68]

Although this argument very clearly perceives the importance which *The Purple Island*'s soteriological illocutionary act places upon the allegorical artifice, it also seems to suggest wrongly that marginal commentary on authoritative texts can be exhaustive, and so not invite the reader to interpret further. Secondly, by firmly understanding the role of the marginalia as explicating the poetic text, like *scholia,* Healy does not appreciate how the expansiveness of the commentary can result in *a further text,* or how the juxtaposition may extend the poetic of *concettismo* from conceit and allegory to formal design. I will return to examine the next step, as it were, in Healy's argument in due course, but in relation to the first of these objections to Healy's argument I will examine a related argument by Sawday, not the one regarding the linguistic distinction between the stanzas and the marginalia, but one which considers the derivation of this formal design.

Sawday develops an interpretation, to which Healy is indebted, and in which the marginal anatomy's status is raised, as it tries to achieve a symbiotic relationship with the poetry.[69] The formal design's attempt at synthesis is modeled on the way by which anatomy textbooks, from the early sixteenth century, had combined text and illustration, particularly how they had included illustrations with accompanying expository text. Parallels are suggested by the formal layout of the pages of Remmelin's *Catoptrum microcosmicum,* and by the items of text therein, some of which are marginal, while some appear above or below the anatomical figure. These items of text serve either an informing, interpretative function or an authoritative function. These are two of the conventions available to artists and anatomists when combining illustration and text in anatomy textbooks.[70] The function of the marginalia in the formal design of *The Purple Island,* however, is for the most part interpretative and nomenclatural rather than authoritative, and therefore more in line with the widespread convention of division in anatomy textbooks between the descriptive anatomy of the marginal text and the pictorial anatomy of the illustrations. Sawday is therefore right to argue: "The poem becomes the equivalent of a drawing—the anatomical figure itself—and the [marginal] notes become the text or main written explanation of the figure. The poem is, in this sense, the dissected body, and the notes the anatomist's observations."[71]

For *The Purple Island* there were perhaps three significant aspects of anatomical illustrations themselves, besides their relationship with the text. Firstly there was a major contribution made by illustrations like those of the *Fabrica* and other works published up to about 1627, to the development of an observational science. In addition, Fletcher also responded to what Choulant has characterized as the development therein of an anatomical ideal or norm "as the expression of nature's highest endeavor" and "the ideal purpose which appeals to us as beauty in all the works of nature."[72] And thirdly, anatomical illustrations presented the human body as a microcosm, in which was represented or imitated the divine order of the universe. The elucidation of these aspects requires a brief digression into the historical development of the purposes and conventions of illustration in anatomy.

In all fields claimed in the Renaissance by the humanists book illustration was "extremely slow to assert itself as an independent form of scientific communication."[73] Although there already existed the graphic incunabula and fugitive sheets of anatomy, the first medical book to contain woodcut anatomical illustrations was the *Fasciculus medicinae* of 1491, and the first illustrated anatomy textbook was Berengario's *Commentaria . . . super Anatomia Mundini,* which did not appear until 1521.[74] Imprecision ailed much anatomical illustration. This was often the result either of a lack of knowledge or of deliberately schematic forms of representation which are conventional in certain diagrammatic demonstrations.[75] In chapter 3, we mentioned how Fletcher's description of his poetry as a "picture" owes something to the more Classical Greek–oriented concept of descriptive anatomy, advocated by Vesalius, that in words the body "may in the manner of an image be set before the eyes of those studious of the works of Nature."[76] Yet in precision illustration would appear potentially to have the advantage over discursive, verbal description.

According to Leonardo, the lack of succinctness of words, the paradoxical situation in which one has to use more words in order to be more precise, means that "the more minutely you describe, the more you will lead the reader away from the thing described."[77] The author of the popular medical work, *The Birth of Mankinde,* encounters this problem in trying to describe what he calls "the braded body" or "*Varicoso formim plexum,*" where branches of the vena cava (the spermatic veins) meet branches of the branches of the spermatic artery, from the aorta or "Great Artery" (they also meet the artery of the vas deferens and the cremasteric artery in the spermatic cord, although these are unmentioned by *The Birth of Mankinde*): "And at the Upper and smaller end where this vaine branche and arterie branche doe first meete, they begin to entermingle, enbrade, and enterlade each other in such infinit wise, writhing and deviding themselves in thousands of little branches, as it were hayres of the head, the one embracing, compassing, and overthwarting the other so confusedly, that no witte can expresse the right manner and order of their commixion."[78] This description belongs to a work that is popularizing medical anatomy for the benefit of an implied reader who is untrained in the observational skills and knowledge of that subject, but even Gray's *Anatomy* is descriptively challenged by the structures described as bewildering in *The Birth of Mankinde*.[79]

Leonardo claimed that representation in the imitative arts can transcend "the confusion of tongues and verbal categories."[80] More emphatically, and with more authority in the context of academic or medical anatomy, Vesalius says in the Preface to the *Fabrica:* "how greatly pictures assist the understanding of these matters and place them more exactly before the eyes than even the most precise language, no student of geometry and other mathematical disciplines can fail to understand."[81] Indeed the accumulating importance of illustrations in early sixteenth-century anatomy textbooks culminates in the priority given the illustrations of the *Fabrica* over the accompanying text, by Vesalius, whose scheme of publishing illustrations dates from early in his career.[82] In the Bologna demonstrations of 1540, the report that he drew with charcoal on the dissection table suggests to Cunningham that already he was thinking visibly.[83] The fugitive sheets of his *Tabulae sex* have been described as "on a more generous page than any of their predecessors," giving "a clearer visualization of the working and structure of the body than had yet been set forth."[84] When it came to the publication of his *Fabrica,* Vesalius told his printer, John Oporinus, Professor of Greek literature at Basel, "neglect nowhere the matter of the picture (even if you do occasionally omit the text on which the illustrations are based)."[85]

The significance of Vesalius's *Fabrica* for *The Purple Island* has already been suggested by this study in several ways, but the possibility of a more intertextual importance is argued for henceforth, though particularly in the following chapter. On the analogy in which the anatomical speaking picture of Cantos 2 to 5 of *The Purple Island* mimics the layout of text and illustrations in anatomy textbooks, the poetry of *The Purple Island* comprises the significantly larger and equally detailed part of the form of Cantos 2 to 5. So the function of the poetry may correspond to that of illustrations, like those of the *Fabrica,* that follow the convention of providing a comprehensive visual description, rather than those the convention of which is to provide schematic, illustrative points of reference in service of a text.[86] The priority of illustration over expository text corresponds to Healy's assessment of the supplementary as opposed to authoritative function of the marginal notes in *The Purple Island.* By means of mimesis, therefore, Fletcher employs not only the rhetorically acquired corporeality of the speaking picture, but also the representational status of anatomical illustration and the expository function of marginal notes. But Fletcher alludes to Horace's defining tag in such a way as to combine in poetry the meaning of pictorial imitative accuracy with figurality or poetic imagery, thereby in his own way trying to achieve what Vesalius tried to achieve in "pictures" that "assist the understanding" of anatomy by placing it "more exactly before the eyes than even the most precise language." Vesalius sought to persuade others, his printer included, of the advantages of illustration over the "wordish" description of natural philosophy, just as Sidney distinguished poetry as an "image" of that whereof the philosopher bestows but "a wordish description, which doth neither strike, pierce, nor possess the sight of the soul."[87]

Although not all physicians concurred with Vesalius's estimation of illustrations,[88] no writer on anatomy after the *Fabrica* could deny their usefulness,[89] and few anatomy textbooks appeared without them thereafter before the mid to late seventeenth century, when a new interest in physiology rendered precise morphological studies less immediately relevant.[90] Upon what then was Vesalius's estimation based? Firstly, it was the contribution of illustrations to the *demonstratio* of the anatomy he had so acutely and repeatedly observed and written about, placing it before the reader's eyes. Yet this observational interest was not separated by Vesalius from the development and beauty of an anatomical ideal: "Particular pains must be used in printing the engravings, since these are not made in the common and ordinary manner and as it were in outline only," Vesalius told Oporinus. "I should particularly desire this one thing," wrote the anatomist to the printer, "that in printing, you would follow as nearly as possible the printed copy sent by the engraver. . . . For this will ensure that no character, however much in the background, will escape the keen-sighted and attentive reader. . . ." But the strength of Vesalius's request arose simultaneously from a desire that "that feature which is most artistic about these pictures and to my eye exceedingly attractive, I mean the thickness of the lines, together with the nice shading, will be clearly apparent."[91] Observational detail is accompanied by *rilievo* (modeling or relief) and other artistic priorities.[92]

Not only linguistic but visual representation of human anatomy has limitations and needs necessarily to have a range of techniques and conventions. The various relationships between the illustration and the accompanying text that developed between the thirteenth and the sixteenth century conform to a set of conventions that respond to questions of priority, division of labor, and space in the printed book.[93] But some changes respond also to a tension that had arisen "between the ideal anatomization, which took place only in the pages of textbooks, and the practice of dissection which was situated in the anatomy theatre." Among these were the dissociation of text from image even as they are linked together by the keying mechanism, and the "growing division and specialization of knowledge, so that the [textual] understanding of each part of the body has itself to be anatomized—sectioned into its own discrete 'box.'"[94] These changes reflect the shifting emphasis in anatomy from merely demonstrating the truth of an authoritative text, to the investigation of human bodies as the basis for a new, purportedly correct morphology or an anatomy of what has been seen. In the first case illustration subserves or (figuratively, though occasionally literally) rests on the text, because the body is "close to language itself," occupying a space within discourse and meaning.[95] The second case initially appears to free the body-in-itself from its contexts of interpretation, though this freedom and raised status have been construed as a banishment of the body in the course of the seventeenth century "to the other side of the distance across which representation must now take place, expelled from the place of signification to the object status of what is signified."[96] In *The Purple Island* the text is linked to the speaking picture of the poetry via the keying mechanism, whereby the marginal notes are divided into discrete units. Composed of signs, however, the speaking picture of the body in *The Purple Island* is not expelled from the place of signification.

The tension between the ideal anatomization and the practice of dissection arose also because of the body's corruptibility, its volumetric proportions and its complexity of structures.[97] These added to the monumental task of transcription to the two-dimensional form of printed illustration. Corresponding to key conceptual shifts which took place in Europe in the early modern period, space, "the positioning of the body within a three-dimensional matrix," Sawday argues, was "the key to anatomical understanding."[98] But technical questions arose, such as whether the anatomy should be presented from a particular viewpoint or a combination of viewpoints, and whether at a particular stage of dissection or as a synthetic demonstration of, for example, the whole of the vascular system in isolation, as we find in the plates of Book III of the *Fabrica*.[99] Pictorial techniques related particularly to *designo* (design), *scorci* (foreshortening, a local application of *prospettiva* [perspective]), and *rilievo* were being developed among artists, and Leonardo's "entirely new" "*prospettiva* employed to record his observations," enabled him "to grant a *perspicuitas* [perspicuity] never before realized in medical illustration."[100] "This depicting of the human body," Leonardo argued, "will be as clear to you as if you had the natural man before you; and the reason is that if you wish thoroughly to know the parts of man, anatomically, you turn either him, or your eye, to see him from different aspects."[101] By Vesalius too, an enormous range of techniques, including systems in isolation, separate and unified forms, and varied viewpoints, are "exploited with an amazingly secure feeling for which technique is appropriate at each place in relation to the text."[102] It has been argued that Leonardo also developed "an exaggerated *contrapposto* that moved beyond the qualitative norm [of contrapositional movement] derived from antiquity," and so enhanced not only the *maniera,* but the *furia* of his figures, that is, their capacity to seem to move themselves.[103] Vasari described the figures of Leonardo as possessing "il moto ed il fiato" ("motion and breath").[104] As intrinsically related to the pneumatic physiological system, muscular movement was explained via the distension of animal spirit from the brain, and in artistic works it acted as an index to the life and affections of the soul.[105] But the engendering of life or being in an image of human anatomy also relates obliquely to the way in which the conceit was thought to give life or being to a poem.

Anatomy "became the testing-ground for that key experiment, which was the transmission of three-dimensional space onto the flat surface of canvas, wood, fresco, or copperplate."[106] For obvious reasons, Kemp says that the medium affects the information conveyed: the woodcut, for example, is less finely detailed than the copper plate.[107] So whereas Valverde "was able to take advantage of the finer detail possible in copper plates," the precision of the first illustrated anatomy textbook, Beregario's *Commentaria . . . super Anatomia Mundini,* was impeded by the limits of the woodcut, the dominant form of representation in anatomy textbooks until the set of anatomical plates was drawn and etched (possibly by Josias Maurer) for Giulio Casserio some time between 1600 and 1616.[108] "In many respects, Berengario's devotion to *anatomia sensibilis* is comparable to Leonardo's stance, and it is tempting to see his pioneering use of naturalistic illustration as a profound expression of his method of sensory scrutiny." But, as Kemp points out, "when we look at the nature and role of the illustrations in his text, their use seems to be more limited and pragmatic than his method might lead us to hope."[109] "There is," according to Kemp, "no suggestion in the way that he uses each illustration that it is to be regarded as a true *demonstratio* serving the end of *determinatio* [that is, determination of knowledge]."[110] "[T]his figure does not exhibit a true likeness of the vertebrae, except in their number," Berengario candidly advises his reader, whom he directs to the "dried vertebrae in the cemeteries."[111] Of Vesalius, whose plates became the models for anatomical illustration in woodcuts (as Casserio's plates were to become for anatomical illustrations in copper), Kemp writes: "He was fully aware that his 'pictures' were translations of the seen object via the particular conventions of the woodcut and was keen that the pictorial skills of his graphic artists should be realized as fully as possible within the limits of the medium."[112]

Once the various technical challenges are met, and the limitations of the medium acknowledged and worked with, "the kind of display used in naturalistic illustration relies heavily upon knowing how to read a series of conventions,"[113] like the page layout and the keying mechanism. One must learn, in addition, whether to read an illustration as presenting particular, individual or typical, exemplary anatomy. Some of the earliest anatomical figures in printed books in which an internal organ has been drawn from the object, such as that of the uterus in the figure of the pregnant woman in the *Fasciculo di medicina,* or the visceral figure in the fugitive sheets of Friesz's *Spiegel der Artzny,*[114] may have been derived from an individual body; but Vesalius's project to develop an universal, ideal anatomy involved the systematic dissection of several bodies of which the illustrations of the *Fabrica* are, as it were, composite representations. Sawday argues that "[t]he effectiveness of perspectival systems of representation depended upon the viewer's relationship to the picture plane; and this in turn depended upon "a philosophical dimension which bore directly onto the work of the anatomists."[115] One-point perspective granted the observer a central position analogous to the central position of Man in the Neoplatonic universe.

Understanding the form of Cantos 2 to 5 as mimetic of the layout of illustration and accompanying text in the anatomy textbook may throw light upon the poem's ingenuity, suggesting that it may be disingenuous. In his analysis of the conventions of visual presentation of anatomy in illustrations, Kemp argues that figures were even drawn in such a way as to suggest they were drawn "from life," that is, from the object, when in fact they were not.[116] We may now return to Healy's argument.

If, as Healy and Sawday agree, Fletcher does deploy the marginal notes "to act as a scientific exposition of his poetic depiction," in the way marginal scholia elucidate the scriptural text, or the explanatory labels of keying mechanisms elucidate anatomical illustration, then Healy might be right in saying that the implication is that, "the commentary's scientific accuracy depends on how well it describes what is taking place in the poem, not how it relates the poem to scientific discourse founded on observations of real corpses."[117] As we have already realized, the marginalia, in their process of elucidation, do not constitute an autonomous descriptive discourse, still less a "scientific" one. They reproduce the allegorical personification and social conceptualization of anatomy found in the stanzas, which, as a demonstration of what is said in the margin, extend and extrapolate from the metaphoricity often inherent in the marginal anatomical nomenclature. It would seem to follow, as Healy rigorously unfolds, that: "the poetic description [is presented] as though it were an empirically verifiable representation of the body which supposedly can be found replicated in each anatomy performed—whether in the anatomy theatre of the Royal College of Physicians, the pages of a Vesalian text-book or a reading of the poem."[118] The problem with this, as Healy sees it, is that the poetic allegory is a far cry from anatomy's "usual terms of reference [which] are to the actual human [body] dissected in the anatomy theatre." In Healy's estimation, the formal arrangement of Cantos 2 to 5 elides the epistemological differences between anatomical and poetic modes of inquiry, in order to promote (the) poetry's "primacy in revealing the truth of the body." This primacy, I would suggest, is of eloquence and ingenuity over descriptive anatomy, and the elision is for Healy a response to the increasing emphasis on observational methods in anatomy, rather than (as with Quilligan) the later linguistic reforms.

The problem with Healy's argument, however, is fourfold. Firstly, Healy is wrong to imply that marginal commentary on authoritative texts can be exhaustive; secondly, the commentary need not be merely supplementary "ostensible empirical support," for its expansiveness can result in *a further text* that relates the poem to anatomical (though not "scientific") discourse, which was partly, though not wholly, "founded on observations of real corpses"; thirdly, Healy fails to appreciate how the juxtaposition may extend the poetic of *concettismo* from conceit and allegory to formal design which thereby acquires an aesthetic explanation consistent with the work's literary context; and fourthly, anatomical illustration did not create a representation that is empirically verifiable in all respects. With this final observation, Sawday (as it were) preempts the charge of epistemological elision, as made by Healy, and robustly defends *The Purple Island* against it. Sawday argues that "prior to the "new science" of the late sixteenth and seventeenth centuries," categories bounded by theology and cosmology did not "admit the possibility of thinking about the body as a discrete entity."[119] Moreover, in the seventeenth century, anatomy continued to be understood in terms of an analogy with the naturalized organic theory of political organization. Francis Barker remarks: "Rather than an extra-historical residue . . . this body is . . . [t]he site of an operation of power, of an exercise of meaning." And meaning was generated in part in terms of the conceptual metaphors which issued from natural philosophy. Since this was how the body in early modern natural philosophy and literature was constituted, Barker is right to understand that, in anatomy, "[h]owever necessary it may be to isolate the body for analytic purposes, the body in question is . . . a relation in a system of liaisons which are material, discursive, psychic, sexual, but without stop or centre."[120]

Surviving lecture notes of 1619–20 from Marischal College, Aberdeen, where anatomy was taught in the fourth year of the arts course, provide a number of educational purposes of early seventeenth-century Protestant anatomy, and a useful summary of some of the general points made in this study about anatomy in the sixteenth century.[121] In so doing they give us a very clear idea of the context in which the illustrations of anatomy textbooks were placed. The first purpose—that the study of anatomy was a purpose on its own—would suggest the contrary of the argument immediately above, were it not that the notion of the independence of the subject is subordinated to the overall purpose to "know thyself," declared in the opening words of the first lecture. As this study has shown, knowing oneself implied an acknowledgment of one's creation, mortality, and redemption. A further educational purpose of anatomy is its use in grammar and rhetoric, "supplying a full vocabulary of its own terms and of metaphors and analogies." This use relates closely to the conceptual metaphoricity and analogy of anatomy, the standardization of nomenclature, and its etymology. Anatomy is also said to be useful to natural philosophy, "particularly that part that deals with the soul, for the parts of the body are the operative organs of the soul, whereby the powers of the soul are known."[122] In this way the Marischal College manuscript lends support to Sawday's point that theological and natural philosophical "belief in the presence of [an] essence, a belief, that is, in the existence of an *anima,* a soul or a thinking entity, necessarily informed any possible perspective of the body."[123] The lecture notes, which were written by John Moir, provide three further purposes of anatomy—its uses in medicine, ethics, and theology. In his discussion of the notes, French says that the anatomy is based on that of Kaspar Bauhin, and he is surprised that they make no mention of the overtly theological themes of "the body being the image of God, or a book of nature in which we can read the providence of the Creator."[124] The manuscript's declared purposes, and those from Bauhin which it omits, suggest that in the early seventeenth century, it was not usual anatomically to think of the human body independently of the philosophy, theology, and metaphorical concepts that provided its study with meaning and purpose.

It is unsurprising then that anatomical illustrations "sought to achieve more than just specific understanding of the structure of the human frame." Like the didactic allegory of *The Purple Island,* "they spoke of the divine order of the universe as it could be seen represented or imitated in the body."[125] This imitation occurred in various reflections, such as of the corruptibility of all natural things by the *vanitas* iconography in the osteological plates of Book I of the *Fabrica,* the teleology of the universe in that of anatomy in the myological plates of *Fabrica,* Book II, cosmic redemption in the redemption of human anatomy in the Christ or microchristus figure of Remmelin's *Catoptrum microcosmicum,* and the astrological conception of the cosmos in the "Zodiacal Man" of the *Fasciculo di medicina.*[126] These examples suggest that anatomical illustrations represent human anatomy not as a discrete ontological entity, but as possessing an illustrative function within analogical conceptual systems. There was undoubtedly an increase in observational and representational accuracy and precision, but both theoretically and historically, this bears no necessary correlation to a diminution of the capacity of illustrations to register on levels other than the strictly corporeal. The reflections were echoed in *The Purple Island*'s anatomico-poetic demonstration of the verities of the body's correspondences, conceptual metaphors, createdness, mortality, and redemption; these truths were already assumed implicitly to exist.

The purpose to present its poetry as "an empirically verifiable representation of the body" would have made little sense to the author of *The Purple Island* in this context. For several reasons provided in this study, theoretical and historical, it is at least strategically if not philosophically important to resist the temptation to make the commonsense objectivist assumptions that the early modern body possessed its own meanings objectively independent of the ways in which it is seen, visually represented, and written about. The conceptual metaphoricity, teleology, and sanctity of the human body were not objectively inherent in the body-as-referent, but arose in the relation between physicians and their anatomical object. However, although it was rationally inferred rather than observed, the body's purpose was thought to inhere *within* the body as discernible in figuratively and imaginatively understood structure and function as their reason for being. A belief in the possibility of discovering a body possessed of its own meanings, and an optimism about the formation of a nomenclature and set of visual techniques objectively to convey such meanings, are a part of the belief system and aim that in germinal form motivated early modern anatomists, in their terms, to uncover the body's truths hidden in its inner recesses, and to convey the knowledge gained thereby in text and illustration. The formally juxtaposed texts of *The Purple Island* did not need an empirical, inductive method to be actually in evidence in the illustrations of anatomy textbooks in order to gesture toward something like transparency of epistemological method. All that was required therein was a set of visual features and conventions that arose from techniques developed to overcome representational challenges, and which presented anatomical method, if not as epistemologically transparent in relation to its object, then at least as from the life, that is, from the object. Thus it is pertinent that Kemp should argue, as mentioned above, that figures were even drawn in such a way as to suggest they were drawn "from life" when in fact they were not.

In having taken a close look at some of these conventions and techniques, the paradoxical gesture toward an objectivism that did not actually exist (or even an absent object) may remind us of the "sparkling prose" wherein Bacon develops and argues for an inductive method. According to Devon Leigh Hodges, recognition of the way in which the anatomizing approach to nature "breeds despair about reassembling a fragmented world" turned Bacon "from the demolition of idols to the task of enunciating the optimistic, utopian side of his project." Bacon's justification for optimism was the inductive method, whereby objects are anatomized "in order to find the 'forms' of nature, such qualities of matter as 'dense, rare, hot, cold, heavy, light,' that are supposed by Bacon to constitute an 'alphabet' from which all natural compounds are made." But the alphabet does not constitute a neutral and referentially transparent nomenclature. Hodges finds that Bacon's practice shows that "the endeavor to discover 'forms' by anatomizing the world disrupts nature and generates a mass of contradictory information."[127] Both in terms of this negative induction and in terms of the aphorisms that he uses, for example, in *Novum Organum,* Bacon's method both "has the power to dispel illusion and to create the illusion, the dream, of a future in which mute objects will speak truths that will make all things possible."[128] According to Hodges, "[t]hese two poles of his project, scientific and alchemical, justify science and inspire faith in its telos."[129]

The abovementioned paradoxical gesture may also remind us of the modeling of Vesalius's finely observed anatomical illustrations on fine art and their having been conceived in relation to considerations of human creation, mortality, and salvation. There was, in addition, a supporting rhetoric of the realism of anatomical illustration. Vesalius's mastery of all the technical questions in response to which the conventions set out above arose,[130] may be inferred from the declaration in Raynalde's translation of *The Birth of Mankinde,* that in its Vesalian figures, the inward parts of women "may in maner be as exactly and clearly perceived, as though yee were present at the cutting open of Anathomy of a dead woman."[131] Clearly, the plates "acted powerfully on the senses" and "acted in practice like some sort of universal," demonstrating "the anatomy of 'the human body.'"[132] But we must acknowledge that the author of *The Birth of Mankinde*'s judgment is overwhelmed by the technological innovation's seeming capacity to represent the previously invisible truths of bodies in a direct and unmediated fashion. The methodology of this study aims to reconstitute the modes of anatomical knowledge and representation in the early seventeenth century. But not to resist the motivating assumptions of early modern belief or submit them to analysis would be to comply with them.

The combination of visual gesture toward the veracity of anatomical illustration and elaborate pictorial content is very much in evidence in the illustrations of Estienne's *De dissectione,* which was assembled between 1521 and 1545, yet remained unfinished when it was published in the latter year.[133] What is of interest here is that the visual gesture toward the veracity of anatomical illustration bears a functional similarity with a rhetorical gesture in *The Purple Island,* the *Devine Weekes* of Du Bartas, and the *Mirum in modum* and *Microcosmos* of Davies of Hereford. Where this rhetorical gesture was discussed in chapter 5, it was argued that the poetry makes rhetorical claims about the power of the ingenuity of poetic eloquence to reproduce truly the meaning of things in nature and the body of which it speaks. The anatomy in many of the illustrations of Estienne's *De dissectione* is quite unlike the anatomy in *The Purple Island,* which consistently involves the application of the descriptive criteria or method of anatomy and is persistently detailed, if allegorically translated: "The anatomy throughout [the illustrations] is pre-Vesalian and the figures of the abdominal viscera quite arbitrary and false. The figures of the thorax, the brain, and the eye are better. The plates generally represent the whole body with a great many nonessential elaborations, so that the rendering of the actual anatomic portion is small and indistinct."[134] The elaborations to which Choulant refers as "nonessential" are various adornments in the style of the School of Fontainbleau artists, the intellectual and social context in which Estienne, who was physician to Francis I, thought that his work could flourish.[135]

Arguably the elaborations of composition affect the anatomical interpretation of the plates, as do their history and their interpictorial and intertextual relationships. Although in many ways typical of the luxury of the court of Francis I, the elaborations can be considered nonessential only from beyond the seventeenth-century epistemological shift in anatomy, after which the body can provide its own justification for anatomical dissection (a purpose in and of itself), and, as Francis Barker argues, the body no longer itself signifies as an element or part of a whole context which it reflects, but can only be an object for discursive practice, present to discourse only across the distance of representation.[136]

Eight of the figures in a series of plates in Book III of *De dissectione* that present female sexual anatomy, some of which have a pictorial context of a bedroom scene, were not intended originally to illustrate an anatomy textbook, and are simply adaptations of some of the figures in *Gli amori degli dei,* a series of erotic prints depicting the loves or transformations of the gods, drawn by the Florentine Perino del Vaga and Rosso Fiorentino, and engraved by Giovanni Jacopo Caraglio. They were probably executed by François Jollat, whose work was well known in Paris between 1502 and 1550, though the plates have anatomical portions inset, and "the woodcut inserts which represent the internal anatomy are certainly the work of the surgeon Étienne de la Rivière."[137] According to Charles Estienne's introduction, *De dissectione* "is not specifically addressed to medical students and is not intended as an 'academic' book."[138] It may also be said that since the anatomical details are generally too "small and indistinct" to be of service to the anatomist or the physician, the original character of the drawings must mean something and function in the context of anatomy.

Published as anatomical plates, the erotic drawings stage a production of meaning from a formal combination of depictions of luxurious divine and human sexuality, on the one hand, and dissections designed to demonstrate the moral reason for the female sexual organs, on the other. But this is not an incongruous convergence. One reason the elaborations are essential to our interpretation of the plates is that they reflect Estienne's purpose for written and drawn anatomy as he outlines it in the introduction to *De dissectione,* to gratify and delight ("*pascere* and *oblectere* are the verbs he uses") "the soul as much as the eye." As Carlino summarizes, Estienne's "purpose of juxtaposing images with the text is thus not only to transmit knowledge untranslatable into words, nor does it have a merely mnemonic or synthesising function, as would a didactic or academic book"; the images "are the means by which the reader ["male" is implied] can bring together the intellectual pleasure of knowledge and aesthetic enjoyment."[139] On this view the appearance of architectural structures and ornamental elements, and the attributes and gestures of figures, "are not only decorative or appended elements . . . , but iconographical devices used to bolster an enjoyment of the anatomy, a mode of transmission of knowledge and meaning."[140]

With this being so, it is necessary to ask, what is the character or the content of the aesthetic enjoyment? Laqueur says that the plates "are about theater, about appearances, about erotic fetishes. . . . In short, these are anatomical pictures about gender and not about what we would call sex, or the structures in the body that mark male and female." On the homology of male and female sexual organs, "Estienne is thoroughly, indeed obsessively, Galenic."[141] The plates not only "perform" sexual difference as socially, culturally, and philosophically constructed, as Laqueur suggests, however; Valerie Traub has argued that they also represent strategies whereby anxieties about the terms of male and female embodiment are negotiated, managed, and displaced. These anxieties may originate in the development of the special sensory epistemology in anatomy, which attempts to reconcile the pursuit of knowledge with the ontology and materiality of the body-in-itself, thereby bringing into conflict gendered terms (intrepid masculine pursuit and discovery of knowledge / discovered and controlled, or domesticated, feminine matter, or Nature) that are otherwise produced as stable and distinct, as chapter 1 of this study argued.[142] Whatever the causes of anxiety—and they may have included the cultural associations of public display and bodily disintegration, or the anthropological problems and religious concerns to which the practice of dissection gave rise—the comparative psychic safety of pornography, or the *ars erotica* of the "emerging 'scientia sexualis'" of anatomy ("the representational field in which the uncovering of the female body was taken to its physical limit"), turns anxiety into arousal.[143] This happens as the objectified female figures, in theatrical demonstrations of availability combined with an anatomical revelation of sexual purpose, are "represented as taking exhibitionistic pleasure in the viewers' gaze."[144] For the implied heterosexual male viewer, the figures instrumentally embody or complete a circular, self-fulfilling, yet endlessly displacing, unfulfilling, structure of desire, fantasy, and aggressive sexual curiosity, in erotic solitude.[145]

Notwithstanding their "performance," amid the semiotically suffused pictorial content of the plates of female reproductive anatomy is a feature that ideally and optimistically gestures towards an epistemologically transparent form of representation. The device that gestures towards transparency of observation is the representation of anatomical incision, which visually reiterated becomes dissection. The engraved incision into the figure is one of the means whereby the illustration presents itself as derived from anatomical dissection. It suggests that the object depicted by the illustration is faithfully represented, even perhaps substantially present to its representation, as the cadaver is present to its dissection. The incision should be seen in the context of mimetic techniques, for mimesis is a kind of metaphor of representation for the thing-in-itself. The artists' tools substitute for those of the anatomist or *chirurgus,* and the graphic incision becomes a synecdoche for the general practice of dissection, which, it is assumed, produces a body of knowledge that faithfully represents the anatomy of the dissected body. The knowledge that one derives from the anatomy revealed in these plates by the incision, however, is not independent of the pictorial situation in which the figure is placed. In the first of the series of plates beginning on page 260 of *De dissectione,* the figure is pregnant and bears two incisions, one laterally crossing her rotund abdomen, the other outlining her rotundity and participating in the illustration's replication of circular architectural shapes. The spatial arrangement and pictorial context of the illustration draw upon the metaphorical conceptualization of anatomy as "building" and the conceptualization of God's agency, which is directly related to this. The shared circularity between abdomen, outlined by the incision, and the architecture seems to inform the peruser that pregnancy is the fulfillment for women of the architectural design in the mind of God.

The following plates in the series replicate this teleology, not only individually in their spatial arrangements and pictorial contexts, but sequentially as each plate represents a deeper level of dissection into the organs of generation. Despite being governed by tropes or gestures of pornography, the "progression of the plates" thereby "recapitulates the telos of heterosexual intercourse."[146] The sequence is, as it were, a progressive excavation of the final cause of female sexual anatomy, the purpose for which it is designed and constructed, and as such, it augments the sense of control that is desired and sought through pornography. The layers of skin and the peritoneum outlined by the incisions in the figure in the first plate await only the next plate in order to be folded back to reveal the anatomical interior.[147] By

Charles Estienne, *De dissectione partium corporis humani libri tres* (Paris: Simon de Colines, 1545), p. 260. The pregnant figure's incision outlines her rotundity and participates in the illustration's replication of circular architectural shapes, thereby drawing upon the metaphorical conceptualization of anatomy as "building" and the conceptualization of pregnancy as the fulfillment for women of the architectural design in the mind of God. Wellcome Library, London.

the third plate, the incision, neat and regular, no longer follows the contours of the pregnant body; instead it creates a square window, through which may be viewed the first proper tunic of the matrix or uterus, the veins shared with or woven into the chorion, from the external parts of the matrix, and the substance and connection of the chorion.[148] There is an absence of any architectural setting, but the window into the figure's abdominal interior nonetheless carries with it the metaphorical conceptualization and connotations of her anatomy as "building." In a subsequent plate in the series (p. 276), the female figure reclines slightly on a bed, her abdomen dissected so as to reveal as though through a window two fetuses. With her left hand, the figure gently lifts the placenta and umbilical cord of each into view. Although this plate does not represent the terminus of the sequence, it is possible to anticipate that the final plate, complete with birthing stool, represents the final cause of female sexual anatomy, the purpose for which it is designed and constructed.

The method of presentation of the anatomical portions of these plates, as though through a window, rather than the anatomy itself, which is often indistinct, carries the optimism that the knowledge they produce heralds an emergent epistemological mode for understanding the body, a mode that posits the true anatomy of the human body as existing independently of previously authoritative ideas about it, yet as accessible to anatomical investigation. Yet the successive graphic incisions, transforming into a window, are simultaneously freighted with the conceptual metaphoricity and teleology to which anatomical knowledge still adheres. The sequence of plates of female sexual anatomy seems almost to record an early modern historical process whereby the object of purportedly true anatomical knowledge, still replete with conceptual metaphoricity, was increasingly located within the confines of the body itself, yet the exploration of the anatomical interior is more rhetorical than actual, the incision perhaps an ingenious visual conceit, so innovatively does it combine figurality with the impression of incisive accuracy and the actual body's substantial presence.

Unlike the human anatomy that is presented visually in Estienne's plates, that which is presented linguistically in the concise nomenclatural explanations which accompany the plates is not replete with conceptual metaphoricity, but is nonetheless etymologically metaphorical. With this not only the mimetic formal juxtaposition of poetic stanzas and anatomical marginalia in *The Purple Island,* but also the linguistically imperfect differentiation, is comparable. Both works in their different ways juxtapose picture and words in order to accentuate difference while forging a relationship through a keying mechanism. Like the graphic incision in the plates of *De dissectione, The Purple Island*'s rhetorical gesture of exposing "these dissected wreathes" in the speaking picture implicitly claims to recreate a substantially present body with the immediacy of dissection. The gesture is from an empirical perspective disingenuous to the extent that it may be thought to justify science and inspire faith in its *telos,* in belief in the possibility if not inevitability of discovering a body possessed of its own meanings and of forming a nomenclature and set of visual techniques objectively to convey those meanings. But the possibility was as yet unrealized (it is perhaps unrealizable), and so the rhetorical gesture and the formal juxtaposition that provides the mimetic context for the transposition of this gesture from anatomical illustration to poetic text, are not mystifying about the poetry's representational status; they do not constitute an epistemological elision of what did not exist as a mode of knowledge acquisition even in anatomical illustration. What is elided by the form and rhetoric of *The Purple Island* is not the difference between anatomical and poetic modes of inquiry, as Healy has it, for they are not yet sufficiently distinguishable, but that between a rhetoric of unmediated access to

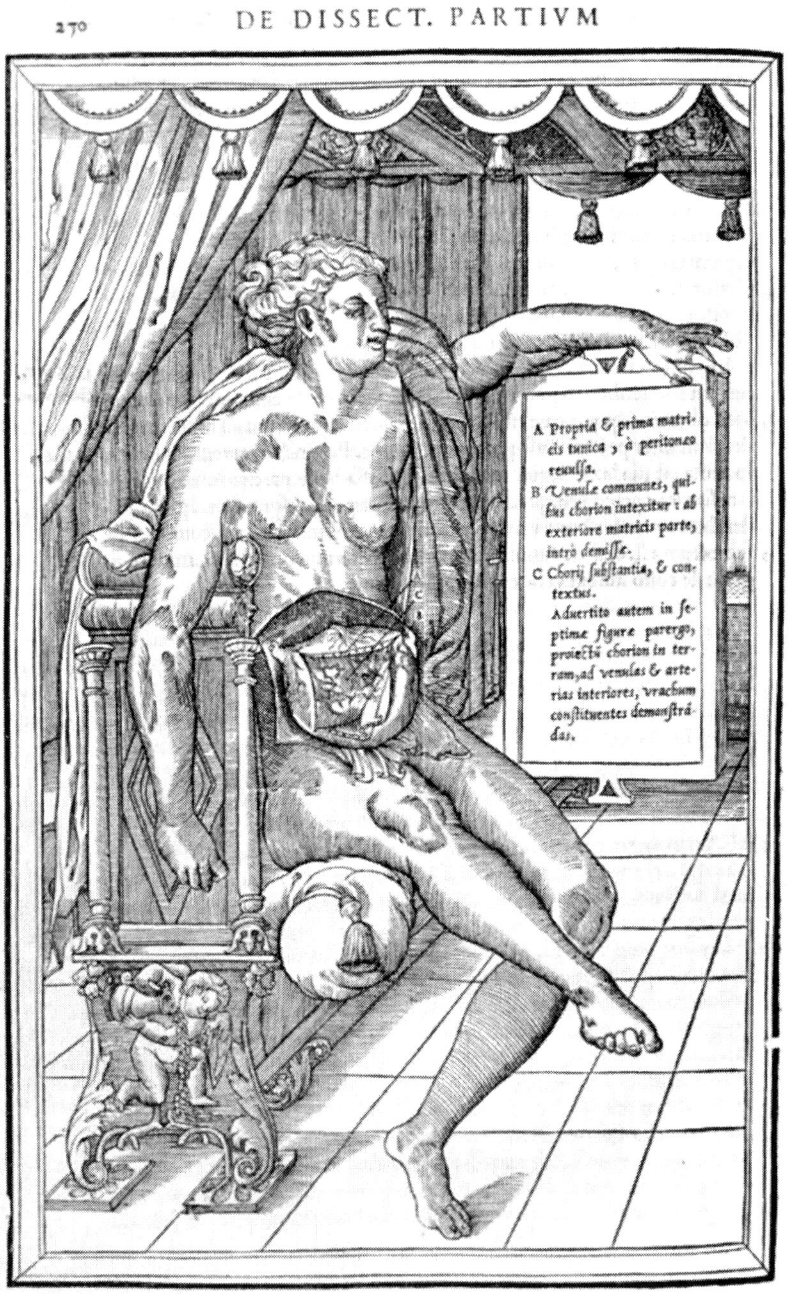

Charles Estienne, *De dissectione partium corporis humani libri tres* (Paris: Simon de Colines, 1545), p. 270. The incision in the figure no longer follows the contours of the pregnant body; instead it creates a square window, through which may be viewed the first proper tunic of the matrix or uterus, the veins shared with or woven into the chorion from the external parts of the matrix, and the substance and connection of the chorion. Wellcome Library, London.

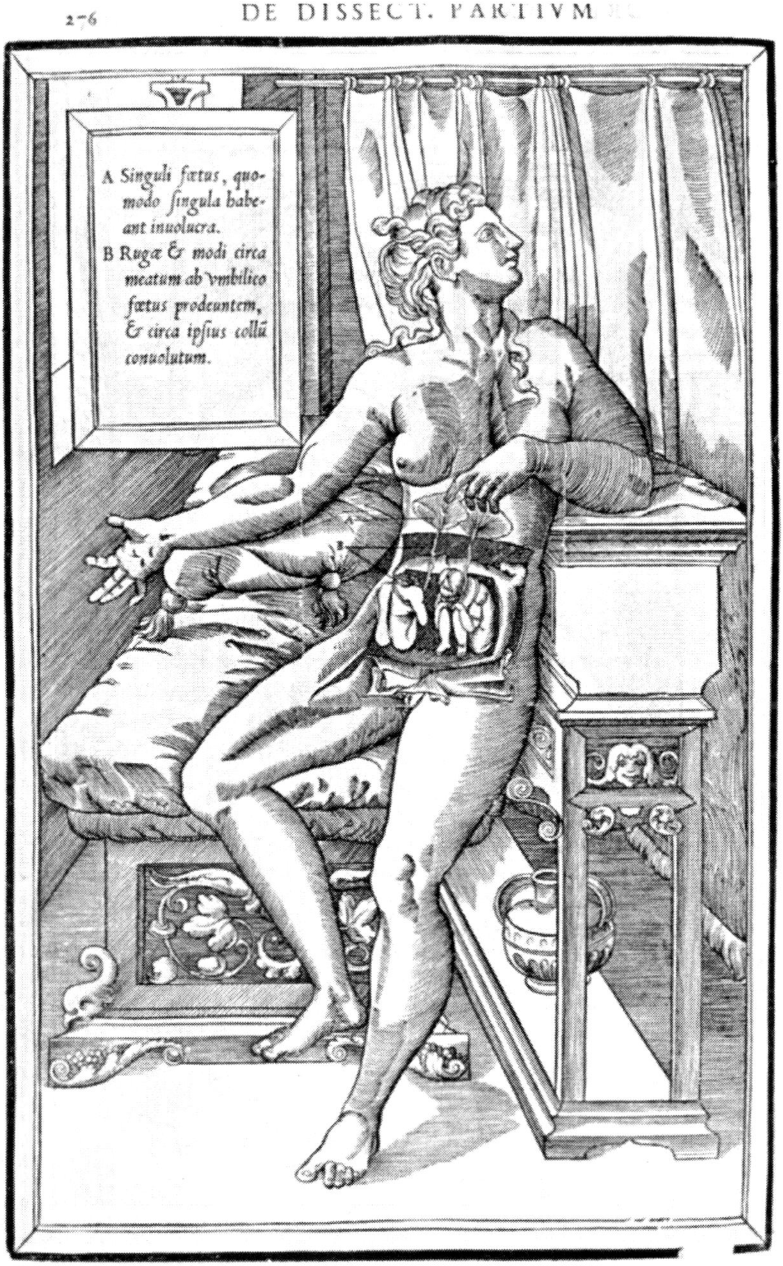

Charles Estienne, *De dissectione partium corporis humani libri tres* (Paris: Simon de Colines, 1545), p. 276. With her left hand, the figure gently lifts the placenta and umbilical cord of each fetus into view. Despite being governed by tropes or gestures of the *ars erotica* or pornography, the depictions of deeper levels of dissection into the organs of generation reveal the purported *telos* of heterosexual intercourse and the final cause of female sexual anatomy, the purpose for which it is designed and constructed. Wellcome Library, London.

a referent through words that are mimetic and suprarealist, on the one hand, and the theoretical incommensurability of the signifier with the signified and with (what might hesitantly be termed) the ontology or reality of the body-as-referent, on the other. Fletcher's invention reconstitutes the formal layout of anatomy textbooks and reproduces the dissected body-in-itself, not though as empirically verifiable because neither anatomy nor poetry was empirical.

It was suggested in chapter 3 that the figurative design and soteriological purpose of *The Purple Island* relate to emblematic theories of the relationship between visual image and verbal graph and sound. The emblematic character of the figurative design extends to the formal disposition of the text. Whereas emblems usually consist of a motto, a picture, and an epigram, *The Purple Island* consists of poetic stanzas that form a speaking picture and anatomical prose marginalia. In emblems and in Fletcher's poem the cloak of figurality or imagery in the form of an illustration or poetic "picture" illustrates and is illuminated by a linguistic form with less occulted meaning. Fletcher made the theory of emblems a more obviously relevant interpretative model for his implied reader by adding three emblematic plates to the second issue of the first edition of *The Purple Island.*[149] One emblematic plate has a picture of the sun, towards which bend a pansy and a marigold, and with an anagram on "Edward Benlowes" ("Sun-warde beloved"). With this is an epigrammatic poem of two stanzas which describes in echoes of the emblematic description of microchristus in Canto 4, stanza 9, how the poetic speaker turned to see the "Sun," whose "heavenly powers" are "Stampt on" the "Sun[n]es lesse Son[n]e," that is, the speaker's "new-born spirit," "his Image bright."[150] The soteriological efficacy of the incarnation, which Fletcher's poem purportedly perpetuates, is here extended to Fletcher by his "beloved" friend.

The Purple Island's reconstitution of the formal layout of text and illustration in anatomy textbooks may be seen as a function of ingenuity, in which the graphic symbolization of words, especially conceits, is seen on analogy with the incarnation of the Word of God. Poetry represents a mode of apprehending true anatomical knowledge through revelation, for which the diegetic narrator of *The Purple Island* prays. By purportedly truthfully reflecting human anatomy-in-itself, Fletcher's poem not only partakes of its ontological corporeality, but claims to resurrect the body-as-referent as the signs of its anatomical speaking picture, though it does so only proleptically. The speaking picture represents a verbally constituted form of the ocular evidence that the illustrations of anatomy textbooks provide independently of Fletcher's soteriological poem, for believing in the divine promise of a full recovery of the prelapsarian "picture" of God (*imago dei*) at the general resurrection.

The formal arrangement of Cantos 2 to 5 of *The Purple Island* into poetic allegorical stanzas and prose anatomical marginalia contributes to our understanding of the work's figurative design and its soteriological illocutionary act or purpose. Our understanding is enhanced partly because the formal juxtaposition is a feature shared by several other comparable works of the period, and consequently may have comprised an aspect of the set of expectations readers may have brought to Fletcher's poem, expectations of which the poet may have been aware when conceiving of the formal design of his work. The function of notes in their relation to the main poetic or prose text of a work is not always the same, however, as this chapter has made clear, though the notes in *The Purple Island* share the nomenclatural function of the marginalia of the *Devine Weekes,* providing the anatomical term corresponding to the metaphorically translated description in the poetry. The same confirmatory, clarifying function is held by the marginal notes of other works, like *The Lit-*

tle World of "Ro. Vn." It contributes to our understanding of the conservatism and innovation of *The Purple Island* that the marginalia serve to clarify poetry and confirm the identity of anatomical content with nomenclature, because this provides a further objection to an overall reading of the work as conservative and didactic in a traditional sense.

Although Fletcher could without fear of contradiction adopt the voice of pastor-poet to demonstrate the revelation of structure, function, cause, creation, mortality, salvation, and ultimately God, in anatomy, it is also true that his language becomes that of the anatomist in the marginalia of *The Purple Island,* and with more consistency than in the notes of any comparable work. *The Purple Island* is unremarkable for its period, however, in the deployment of marginalia, except for their expansiveness, their complexity, and the fact that they seem to achieve a status comparable with that of the poetry, thereby having given rise to the (albeit erroneous) notion that the juxtaposed discourses of poetry and anatomy are two autonomous forms of discourse, rather than one supported by the other. It is in respect of this expansiveness and complexity that the allegorical extension of the Isle of Man in *The Purple Island* might be thought not to have retrenched and corroborated readers' expectations but to have expanded their vision of the literary allegory, its formal design and its intentional direction toward a world. Whereas the poetic stanzas tend to grant the reader a shifting perspective on the anatomized microcosmic world, the marginalia tend to stabilize the world without reducing it of its meaning. This chapter has completed our comparative work within the literary context of *The Purple Island,* but it remains for Part Three to explore the context of anatomical works to which Fletcher may have been indebted in the composition of his poetic anatomy.

The character of the criteria upon which the poetic stanzas place emphasis is teleological. The aim that the reader should appreciate the function, as more important, causes apparent distortions of composition, magnitude, shape, position, and connexion, with the purpose that the reader's exploration of the world it is the work's intention to create should be characterized by a teleological understanding. The stanzas disorientate only to reorientate the reader, granting her or him a perspective on the anatomy which is capable not only of taking in relative distance but of placing her or him in different locations within the world of the microcosm, with vantage points on particular organs or members that best display their reason for being what they are and where they are in that world.

The function of the marginalia has been found sometimes to be to lend authority to the poem, and only rarely to offer an explanation of a term or allusion, but it is most often to explicate, clarify, or stabilize the geopolitical, anatomical world of the poetry, and most of the elaborate marginalia correspond to the deployment of such glosses in the production of scriptural texts. Healy suggests that *The Purple Island* thereby preempts the hermeneutic process, as opposed to merely prompting it (as this study has argued), and that the poem impresses upon the reader its primacy in revealing the truth of the body and the supplementary character of the ostensible empirical support in the margin, eliding the epistemological differences between anatomical and poetic modes of inquiry in order to do so. However, marginal commentary on authoritative, scriptural texts is never exhaustive, and despite the importance which *The Purple Island* places upon the soteriological purpose of its allegorical artifice, the threshold figure's generation of a potentially infinite number of subordinate metaphors, metaphorical shifts of meaning, of inexhaustible cognitive significance, remains limited only by the context of the whole literary work (stanzas and marginalia) and the "conceit" of the reader. Moreover the expansiveness of the commentary results in a *further,* though not an independent or autonomous, text, which the reader may in turn unravel, rather than merely supplementary ostensible empirical support the nature of which suggests no invitation is extended to the reader to complete the process of semantic generation.

One of the most important reasons for having attended closely to the formal disposition of Cantos 2 to 5 of the work is to discern the imperfect linguistic differentiation between stanza and margin, for it confirms our view that Langdale's claim to be able, with the aid of the prose marginalia, to extract anatomical content that is independent of the figurality with which it is presented in the poetic stanzas is a misleading one. It will be the task of part 3 to discern the provenance of the *Purple Island* anatomy, not as extracted from figurality, but partly to judge the extent and character of its own figuration. This chapter has developed a more consistent sense of how the figurality of the poetic stanzas relates to the marginalia, and a clearer view of the linguistic or conceptual distinction between the two parallel discourses of the work, such as it is. Distinctions previously proposed by Langdale and Sawday have required qualification, because they underestimate the conceptual, figural, and verbal interdependence of the stanzas and marginalia. The differential in the spectrum of various stanza-margin relations, with regard to figurality, is broader than either distinction allows. The figurality in the marginalia is often etymological metaphoricity, which compares the body to the external world in the proper nouns of anatomical nomenclature; and *The Purple Island*'s clear suprarealist interest in the allegorical potentialities of etymology suggests the importance of recognizing this. It is on the whole typical of the poetic stanzas to develop the inherent figuration of the nomenclature of the body, as either constructed or natural, shading it into the built, natural, and social features of the geopolitical realm of the island and the cosmos above it, including the personification of anatomical parts and the social conceptualization of anatomical relations. There are only broad tendencies that distinguish the stanzas from the marginalia, and these mark the conceptually, linguistically, and discursively imperfect differentiation. Underlying the descriptions in both of the component parts of the work's formal design is consistent use of the descriptive criteria of the anatomical *accessus.* The anatomical *accessus* is the unifying feature of the formal design of Cantos 2 to 5. Part 3 of this study will extend analysis of this kind to the more general differentiation between anatomical nomenclature and its metaphorical and etymological extension and exploration in poetry, by looking more closely at the anatomy of the late sixteenth and early seventeenth century in relation to *The Purple Island.*

This chapter has argued that the imperfection of the differentiation in *The Purple Island* implies that it is not anticipatory of the program of linguistic reform which was to be undertaken in the later seventeenth century, in which the objects of attack were deductions from metaphors, parables, allegories, analogies, and teleology. This chapter has provided only a brief introduction to a select few of the historical actors attributing rational validity to objectivist suspicions of metaphor, to trust in ordinary language, and to the objectivist and positivist postulation of the existence of ahistorical ontology about which a neutral and transparent nomenclature can speak the truth. The search conceived in earnest for a type of linguistic origin, a moment of Edenic purity of expression was far removed from the figurative design and soteriological purpose of *The Purple Island.* Significantly these developments in natural-philosophical discourse took place almost half a century after most of Fletcher's poem was probably written, and indeed after the other works representing human anatomy in allegorical, metaphorical, and analogical linguistic forms examined in this study. It has been argued here that, although *The Purple Island* has been thought, probably correctly, by Quilligan, Sawday, Nicolson, and Baldwin to have been acutely vulnerable to imminent historical change, such faultlines as appear in Fletcher's work do so not as an effect of later linguistic reforms, but as a result of the allegorical extension or development of a poetic of *concettismo* into an ideological artifice. The work of several critics seems to attribute too strong a position, both in *The Purple Island* and in early seventeenth-century culture more generally, to the emergent notion of a scientific language devoid of figurality. Rather than interpreting the formal juxtaposition in *The Purple Island* as a sign of the later epistemological change in natural philosophy, this chapter has provided an explanation that is consistent with the concept of ingenuity, with which the poem's early readers attributed the work, and with which this study has characterized the work's locutionary, illocutionary, and perlocutionary acts. The formal juxtaposition is here understood as an extension of ingenuity's invention or discovery of new paradoxes and conceits. While it seems to accentuate the differences between poetry and anatomy it actually explores new ways of presenting conventional correspondences, thereby foregrounding differences as well as highlighting similarities. Ingenuity can be understood to have shaped the formal disposition of the work as an expression of its formal design in terms of a poetic of *concettismo,* of creative metaphor.

All my analyses to this point suggest that there existed congruity between figurality and anatomy both generally and in both literary and anatomical works. This is insufficient, however, to counter Healy's estimation that there was "a *refusal* within the early seventeenth century by either scientific or poetic constituencies to acknowledge a division in the forms of discourse considered appropriate to their disciplines."[151] This refusal, were it to have occurred, would suggest that conceptual, discursive, and disciplinary congruity were residual rather than dominant, even if in the world intended by *The Purple Island* they remain dominant through a continuity of stanzaic figurality into the marginalia, and through formal mimesis and rhetoric. It is not possible, therefore, to examine the historical or diachronic continuity or discontinuity of this congruity by concentrating mainly upon the form of the poem, as does Sawday and, incidentally, Healy too. A determination of the precise contours and marks in *The Purple Island,* of the conceptual and discursive differentiation between anatomy and poetry, and consequently of how far they are congruous or synthesized, is possible only by resisting the collapse of this differentiation into that between the stanzas and marginalia. Rather we must examine the conceptual character of the anatomy and its relation to the poem's figurality. We must return to the problem the poem posed for Langdale. In order to judge how *The Purple Island* in this regard relates to anatomy in the early seventeenth century, whether in both there is a "refusal" to acknowledge the difference in the forms of discourse appropriate to their respective disciplines, we must examine early seventeenth-century anatomy, rather than (as has hitherto been the case) the anatomy to which the early seventeenth century was heir.

The analyses of this chapter have developed our understanding of *The Purple Island*'s formal relationship to anatomy textbooks, its marginal as well as stanzaic use of the anatomical *accessus,* and its rhetorical adoption of a visual pictorial device which in anatomical illustration bears the weight of an optimistic belief in the *telos* of scientific natural philosophy. The tendency of the marginalia to provide and translate anatomical nomenclature, and to make a descriptive comparison, as a way of offering an explanation for the metaphoricity of nomenclature, is one among several features of the formal organization of *The Purple Island* which lend weight to Sawday's view that while the poem is the equivalent of an anatomical figure, the marginal notes mimic the text or main written explanation of the figure. The consistent and coherent combination of similarities between the principal and subsidiary subjects characterizes the work's intention to assemble a microcosmic world which unfolds to the reader's "inward conceit" or apprehension as an anatomized body not unlike those found in illustrated anatomy textbooks, designed for the purpose of self-understanding. This is another measure of the fidelity of the work's design to the particulars of its anatomical subject.

In making his diegetic narrator Thirsil describe his poetry as a "picture," Fletcher was arguably alert to more than the concept of descriptive anatomy, advocated by Vesalius, that in words the body may in the manner of an image be set before the eyes of those studious of the works of Nature. The significance of Vesalius's *Fabrica* for *The Purple Island* has already been suggested by this study in several ways, and this chapter has further suggested that the priority of illustration over accompanying text in the *Fabrica* corresponds to the arguably supplementary as opposed to authoritative function of the marginal notes in *The Purple Island;* and comprising the significantly larger part of the form of Cantos 2 to 5, the poetry corresponds to illustrations like those in the *Fabrica,* that follow the convention of providing a comprehensive visual description. In the following chapter, the possibility of an intertextual connection between Vesalius's *Fabrica* and *The Purple Island* is argued for. This chapter has provided a brief overview of how the development of illustration responded to technical questions and limitations with a range of conventions and techniques, some of which drew extensively on fine art techniques such as *designo, scorci, prospettiva,* and *rilievo,* making anatomical illustration more aesthetic than it had ever been hitherto. Fletcher employs not only the rhetorically acquired corporeality of the speaking picture, discussed in chapter 5, but also combines in poetry the meaning of pictorial imitative accuracy with aesthetic figurality, thereby in his own "ingenious" way trying to achieve what Vesalius tried to achieve in actual pictures that assist the understanding of anatomy by placing it more exactly before the eyes than even the most precise language. Although the evidence is lacking, it is conceivable that seventeenth-century readers of *The Purple Island* recognized its formal imitation of the layout of illustrations with accompanying text in anatomy textbooks as a further instance of ingenuity, of imaginative, witty contrivance and skillful, cleverly constructed artifice. But it is equally probable that they could have seen how the emblematic character of the figurative design extends to the formal disposition of the text. Either way they would have encountered a speaking picture of microchristus, like that in Remmelin's *Catoptrum microcosmicum,* which demonstrates the creative ingenuity of God and the saving love of Christ.

This chapter has found conclusively that anatomical illustrations represent human anatomy not as a discrete ontological entity, but as possessing an illustrative function within analogical conceptual systems. Yet a belief in the possibility of discovering a body possessed of its own meanings, and optimism about the formation of a nomenclature and set of visual techniques objectively to convey such meanings, were a part of the belief system that in germinal form motivated early modern anatomists. The visual features and conventions that arose from the techniques developed to overcome representational challenges and answer technical questions, presented anatomical method as from life, and a supporting rhetoric perpetuated the realism of anatomical illustration. Additionally, amid the semiotically suffused pictorial content of anatomical plates is found a feature which many anatomical figures display, an incision, which in some plates optimistically gestures towards an epistemologically transparent form of representation. Visually reiterated, the incision becomes dissection, and in a sequence of plates, like those of female sexual anatomy in *De dissectione,* it seems to trace a process whereby purportedly true anatomical knowledge was increasingly located within the confines of the body itself, though because it was still replete with conceptual metaphoricity, the discovery was still at least as rhetorical as it was real. Like this visual conceit, *The Purple Island*'s rhetorical gesture of exposing "these dissected wreathes" in the speaking picture implicitly claims to recreate a substantially present body with the immediacy of dissection. The poem's rhetoric of unmediated access to a referent through words that are mimetic and suprarealist claims to overcome the incommensurability of the signified with the putative ontology of the body-as-referent, which however is not exhausted by representation. The poem implicitly claims to do so through ingenuity, and then demonstrates in the textual incision or dissection that this is done. In this way, simultaneously, the soteriological speaking picture represents a verbally constituted form of the ocular evidence that anatomical illustrations provide independently, in their own way, for believing in bodily resurrection.

This chapter has argued that the formal organization of *The Purple Island* neither elides the difference between anatomical and poetic modes of inquiry, nor represents the achievement of a fragile synthesis at probably the last possible moment in history when anatomy and religious allegory could be synthesized. Sawday's view that "the poetic body—the visual image—[which] speaks in the eloquent language of gesture via metaphors, similes, and the devices of figurative language" is in *The Purple Island* "harmonized with the revealed knowledge of the human form pursued in the anatomy theatre,"[152] places the burden of the work's engagement with epistemological historical change too firmly upon the formal design. Whether or not this harmony is achieved (and I defer judgment until after the final chapter's analysis of public anatomical demonstration), it must have been facilitated by the basic congruity of anatomy with figurality. What is more, it does not follow from the form of *The Purple Island,* as Sawday suggests. Even if we discount certain graphic features as embodying an epistemological ideal as yet unrealized (and possibly unrealizable), the formal juxtaposition or proximity in anatomy textbooks of illustrations and their marginal keys and texts does not imply that the visual content and the languages or discourses thereby engaged are not conflicted. My analysis in this chapter suggests that they are not conflicted, but we cannot infer this from their formal arrangement in books. Likewise, although Sawday contends that "[w]hat tends to appear to us [in Cantos 2 to 5 of *The Purple Island*] as a struggle between two discourses should more properly be understood as an attempt at synthesis," this synthesis is formal, and not necessarily, or by implication, conceptual or discursive.[153] We cannot conclude from the form alone that the allegorical aggregation or metaphysical forging of correspondences between discourses, ideas, and objects has been successful conceptually as well as linguistically and formally, with regard to anatomy.

III
Anatomy

7
Questions of Provenance and Conceptual Character

IN CANTO 3 OF *THE PURPLE ISLAND*, THE DIEGETIC NARRATOR, THIRSIL, IMPLORES HIS "Shepherd-God" to shed "sparkling light" upon the anatomico-poetic exposition which the narrator is about to undertake with his "hand," that primary instrument long neglected by the art of medicine until Berengario, Massa, Vesalius, and others revived the practice of dissection by physicians in the first half of the sixteenth century. In the same invocation, Thirsil prays for divine inspiration in his own "mistie breast," for a revelation that may "guide" his "erring footsteps" and "weak hand" through "this fogge" that conceals the "art," design or *scientia* of the microcosmic world that God, who "onely know'st it right," "first mad'st" and now "hid'st . . . from all the world beside" (*P.I.* 3.4.1,3,6,4,2,5). The invocation presents Thirsil as a curious combination of pious pastor-poet (or humble pastoral poet-shepherd) and rational anatomist, devoutly wishing to be divinely inspired, to be possessed of soteriological ingenuity and eloquence. The conceptual character of the anatomical content of *The Purple Island* should not be prejudged to conform to the character of the narrator who presents it, however, and in this and the following two chapters, which comprise the third part of this book, I will be concerned to discern more specifically than has been hitherto the case, the conceptual character and provenance of the anatomy that Thirsil presents.

Fletcher's biographer Langdale is convinced that the allegory of *The Purple Island* is superficial with regard to the anatomy. He considers the anatomy to be, in conception, independent of the figurality with which it is presented by Thirsil, a view with which this study has found reason to disagree. Conversely, Nicolson's conviction that the anatomical content is intellectually conservative rests on an assumption (though, I hasten to add, she would not articulate it in these terms), that notwithstanding the character of the anatomy upon which Fletcher drew for his poetic allegory, the principal and subsidiary subjects of the Isle of Man metaphor are substantially modified in conception by their involvement in analogy, rendering *The Purple Island* intellectually conservative in respect of its anatomy and physiology. There is circularity to this argument wherein the postulated cause of the conservatism—the presence of analogy in the anatomy—is not independent of the proposition that it is supposed to justify—that the *Purple Island* anatomy is conservative. Completing the circle is the assumption that analogy is necessarily a conservative mode of apprehension in early seventeenth-century natural philosophy. Assumptions like this have led Nicolson and other critics to the premature judgment that the very density of figurality can only mean that the anatomy in *The Purple Island* is conservative.

This set of assumptions applies specifically to the figurative design of *The Purple Island,* and simultaneously refers away from that literary work to the historical context of anatomical texts, illustrations, dissections, and public lectures and demonstrations. Though much that has already been said in previous chapters contributes to the conclusions that will be drawn from the discussions in this and the remaining chapters of this study, the assumptions will be tested in this part of the study, by asking if the poem unfolds the entailments of the microcosm-geocosm analogy, and the significant connotations of the allegorical threshold figure, to the detriment of anatomy, and how *The Purple Island* in this respect compares with Fletcher's anatomical sources and contemporaneous anatomical literature and practices.

Let us, by way of introduction, revisit some core conclusions regarding the figurative design of *The Purple Island,* and in a preliminary way work through their effect on the kind of answers that might be given to the questions raised by the disagreement most starkly represented by the opinions of Langdale and Nicolson. Although it is based on analogy, broadly conceived within the traditions of the microcosm and microchristus, the allegory of *The Purple Island* is an extension, an unfolding of a creative metaphor, rather than that of an analogy, though the analogies of the microcosm and microchristus are explored therein. Consequently, the anatomy corresponds to the creative threshold metaphor's reproduction or development of conventional analogical understanding, its extension of the ordinarily used part of the defining concept or domain, and its use of the ordinarily unused part of that domain, and consequent discovery, exploration, and coherent structuring of antecedently existing, yet previously unseen and newly highlighted similarities between its principal and subsidiary subjects. In this sense at least, the anatomy cannot be said to have been conceived wholly within a conceptual framework of analogy.

But the allegorical extension has its own consequences for the anatomy. Along with theories from philosophy and Renaissance rhetoric and poetics, this study has employed conceptual or cognitive theory of metaphor to analyze allegorical and anatomical metaphors, and this theory holds that the integration or interaction of the principal and subsidiary subjects, the defined and defining terms, in this case Man and the dynastic island realm or protonation, should result in a metaphorically figured, conceptually shaped anatomy of Man. Modification, or retardation as Nicolson calls it, of anatomy or of politics, depends upon the direction and degree of metaphorical conceptualization,[1] and our analysis of the metaphorical structuring of the anatomy of the human body in terms of vegetation and "building," in sixteenth-century anatomy textbooks, and in *The Purple Island,* has revealed it to be consistent with the structuring of abstract or less clearly defined or delineated concepts in terms of concrete, more clearly defined or delineated ones. However, according to Lakoff and Johnson's theory of conceptual metaphor, there is always more in the defining concept than is carried over to the defined concept, and other aspects of the latter concept are defined in terms of other defining concepts; thus in the anatomy of the schools and medical profession the conceptual metaphors of vegetation and "building" jointly provide a coherent understanding of the concept of anatomy, each metaphor allowing the reader to get a handle on different aspects of the concept being structured. The result of metaphorical structuring is not the same as the correspondence between analogous terms, such as the microcosm and the geocosm, or the former of these and the body politic.

Analogies correspond to the perceived equivalence of at least two, but not an unlimited number, of relationships, wherein each term is defined in terms of its contiguous relation to an adjacent term (*convenientia*) and an emulative relation to its analogue (*aemulatio*). Likewise, similes are empty and uncontrolled without a context, or closed with a similarity in one particular respect. By contrast systematicity in metaphorical structuring is often such that the defined concept is systematically conceptualized in terms of (its many similarities to) a host of metaphorical concepts that are related in terms of entailment relationships to the central conceptual metaphor of the system. Although the *Purple Island* allegory unfolds a creative metaphor, not an analogy, our definitions of metaphor, and of metaphorical structuring in particular, suggest, at least initially, that notwithstanding the theorized differences between analogy and metaphor, the anatomy of Man might be retarded by its involvement in allegory, as Nicolson argues.

In *The Purple Island,* however, since the Isle of Man is a creative literary rather than highly conventional metaphor, the concept of the Isle does not define the concept of Man directly, but the metaphorical interaction involves transformations in the associated commonplaces of the subsidiary subject, the Isle, resulting in the formation of a set of associated implications from which are drawn those that are pertinent to the principal subject, so that not only a newly highlighted range of similarities but also novel similarities between the terms of the metaphor emerge as a potentially unlimited number of significant connotations. So the anatomy also corresponds to the creative metaphor's invention or discovery of new similarities that are perhaps not reducible to extensions or discoveries of antecedently associated yet unseen connotations of the figure of Man as an island, though in some cases these similarities had been partially anticipated by earlier comparable works. In the hermeneutic process, unlike the interpretation of an analogy, unfolding the significant connotations of the creative metaphor is limited only by the context of the whole work and the "conceit" of the reader. In *The Purple Island* as a literary work, these connotations are subsequently used throughout the allegory (in Cantos 2 to 5) as the specially constructed set of associated implications of the subsidiary subjects of a host of other metaphors for the parts of Man, all of which relate like satellites to the Isle of Man metaphor in a set of entailment relationships. It is not clear from this that the anatomy is retarded by its involvement in allegory, or that the proper meanings of recent anatomy could not be among the significant connotations of the Isle of Man.

Contrary to the opinion of Langdale, though, the associated implications of the principal and subsidiary subjects of the metaphor of the Isle of Man are highly integrated. Perhaps it could be argued, then, that the creativity of the Isle of Man is directly proportional to its infidelity to the highly precise and denotatively limited character of the vocabulary of anatomy. As Kastor found to his dismay, and the analysis undertaken in chapter 2 confirmed, the descriptions of anatomy in the marginalia are modified by their involvement in the allegory; and if they are modified, it may consequently follow that they are retarded, as Nicolson maintains. One distinctive aspect of the hermeneutic process, as the result of the metaphorical figuration of Man, may be that having turned to the marginal notes to clarify or confirm an interpretative deduction relating to a passage in the poetic stanzas, the reader may turn again to the stanzas only to be disoriented by the distortion of perspective demanded by the allegorical integration of the human body with the geographical and sociopolitical features of the realm in the allegorical world created by the work.

But systematic integration in the poetic stanzas need not entail that they are discontinuous with the anatomy of the prose anatomical marginalia, because the differentiation between the poetic stanzas and the prose marginalia as regards figurality is less than conceptually, linguistically, or discursively perfect, as the previous chapter has shown. This is especially due to the figurality in *The Purple Island* being, in part, inherent to anatomical nomenclature, though unfolded or extended by the poem's allegory. The ongoing standardization of nomenclature amounted to a paring down of etymology, though individual anatomists might choose to employ inherent nomenclatural figurality for pedagogical and descriptive purposes.

Consequently, the extent to which etymological nomenclatural figurality was cognitively operative in anatomy textbooks is variable, and at its most pared down contrasts sharply with the plurisignification of the stanzaic figurality in *The Purple Island.* And consequently, the choice of nomenclature in *The Purple Island* may help to determine the provenance and conceptual character of the anatomy in the poem, but it may not be a sure guide.

The systematic integration of the human body with the geographical and sociopolitical features of the realm need not entail that the anatomy of the poem is dependent for conception upon the descriptive features of the political geocosm, for despite what is suggested by metaphorical structuring in the theory of conceptual metaphor, and the one-directional structuring of concepts in anatomy, the anatomy in *The Purple Island* may determine the character of those geographical and sociopolitical features. Even though, clearly, in the course of the allegory, the figurality of anatomy is extended far beyond what usually occurs in anatomy textbooks, the distortions of perspective nonetheless consistently and painstakingly maintain the *occasiones* of the anatomical *accessus* from the marginalia. As the intertextual and comparative analyses in this study have demonstrated, Fletcher's poem is exceptional in its application of the anatomical *occasiones* to the metaphorically translated body. Much greater precision and enhanced systematicity in their application in *The Purple Island* is responsible to a large extent for the less schematic anatomy and less formulaic, more highly developed and systemic physiology than that found in comparable literary works. In the poetic stanzas, though, the use of the *accessus* emphasizes functions as a way of ensuring that the reader's exploration of the microcosmic world should be characterized by a teleological understanding. As was argued in chapter 2, the degree to which allegorical personification occurs in the poetic world of the geopolitical realm is regulated in relation to observable anatomical postures and physiological operations. The specially constructed set of associated implications of the Isle consists of many implications that are *shaped by* the requirement that they accommodate the detailed structure and function of anatomy in terms of the descriptive criteria of the *accessus.* (As chapter 2 argued, the special literary construction of a set of associated implications of the subsidiary subject of a metaphor is not the same as analyzing the metaphor into a set of subordinate metaphors, so that the account given is either circular or leads to an infinite regress.)

In the hermeneutic process of negotiating the stanza-margin relationship, because the modifications in the poetic stanzas are always consistent with the *accessus,* disorientation usually precedes reorientation. Whereas the poetic stanzas grant the reader a shifting perspective on the structural and compositional features of the anatomized microcosmic world, their size, position, and connections, the marginalia tend to stabilize this perspective, making it more matter-of-fact without reducing the world of its meaning. As a consequence of this consistency of representation, it remains conceivable, despite what Grosart and Grundy have referred to as the "all-purpling Imagination" of *The Purple Island*'s author, that anatomy in the poetic stanzas is identifiable, and that it is identifiable not because it is conceptually independent of analogy or allegory for what might be regarded as its essential character, but rather because *the figuration of anatomy is consistent with its essential conceptual character,* and the figuration demonstrates this consistency through the application of the *occasiones* of the *accessus.* Since the anatomy thereby determines the allegory, rather than vice versa, at least some of the time, the possibility remains open that, contrary to Nicolson's belief, the anatomical content of *The Purple Island* consists of recent anatomy.

Although of course the element of necessity in Nicolson's assumption has been reduced to possibility or probability by what is said here regarding the consistency of the figuration of anatomy with its essential conceptual character in *The Purple Island,* this does not preclude the possibility that the *Purple Island* anatomy is nevertheless in fact (if not necessarily in principle) intellectually conservative. There are instances where the application of anatomical descriptive criteria does not imply familiarity with *any* anatomical text. As Baldwin remarks, knowledge of the redness of the liver "may not have been obtainable from books, but . . . from any butcher or housewife."[2] Additionally, it does not necessarily follow from the diligent application of anatomical descriptive criteria that the anatomy complies with contemporary anatomical ideas or texts. The detail and the accumulation of terms of carpentry in Davies of Hereford's *Microcosmos* were conjectured possibly to be symptomatic of a growing epistemological anxiety about the impact of innovative critiques of sixteenth-century anatomy on the body's correspondence to the macrocosm and its assumed reason for being, that is, the temporary lodging and instrument of the immortal soul. Yet despite the possible anxiety of their deployment by Davies of Hereford, the same anatomical descriptive criteria continued to play a dominant part in professional anatomical description until the developments in physiology that so changed anatomy's character in the later seventeenth century.

The anatomical exposition of the Isle of Man in Fletcher's poem was found in chapter 4 to exceed the anatomy in its literary source-texts in several ways: in its extended range of analogues and metaphors of anatomy; in the enhanced systematicity between human anatomy and these analogues and metaphorical terms; in a more precise application of anatomical descriptive criteria to the body and its analogues and subsidiary subject; and in its generally more precise, less schematic anatomy, and more systemic, less formulaic physiology. No comparable literary work discussed in chapter 5 extends its examination of anatomy as comprehensively, systematically, or coherently as the *Purple Island* allegory, though like some of these works, *The Purple Island* is modeled formally and imaginatively on the human body, through mimesis, and it adopts a method of dissection that purports to imitate Nature's act of creation; with ingenuity and eloquence it rhetorically gestures toward identification with and reconstitution or resurrection of its subject matter, as it is poetically dissected, thereby substantiating the accomplishment of its intentional direction toward an anatomized microcosmic world.

Though it shares some of these features with other comparable works, the anatomical content of *The Purple Island* is far in excess of what Fletcher may have gleaned from encyclopedic and fictional or expository literary sources. These observations, and the fact that the religious allegory of *The Purple Island* has been found to be congruous with the conceptual character and figurality of some sixteenth- and a very little seventeenth-century anatomy, may nevertheless seem initially not to tell against Nicolson's judgment that a combination of analogy with the *Purple Island* anatomy nevertheless produces an anatomy that is in fact (if not necessarily in principle) intellectually conservative in the early seventeenth century. By way of preamble to this particular question, let us recall the conclusions reached in this study regarding the character of the anatomy discussed so far.

During the period in which anatomical humanism and Hellenism thrived and the practice of dissection by physicians was revived—the first half of the sixteenth century—human anatomy was systematically conceptualized in terms of a host of metaphorical concepts that are related in terms of entailment relationships to the conceptual metaphors of vegetation and "building," such that (Aristotelian) analogy and the conceptual metaphors to which it gives rise underlie surface metaphorical expressions and characterize anatomical knowledge, eschewing the distinction between purportedly literal, "scientific" language and purportedly ornamental figurative language, with regard to the conceptual content of utterances of either kind. This study's digression into the anatomy of the schools and of medicine has also revealed coherence between the analogies of the body as a microcosm of the terrestrial globe or the macrocosm and the conceptual metaphors of anatomy, thereby further integrating the elements of Fletcher's figurative design. The *Purple Island* anatomy is also figured in vegetative, terrestrial, and architectural terms, and the imagery is rationally consistent with descriptions in anatomy textbooks, the figurality of which may frequently be distinguishable in kind from that of allegory, in that where *The Purple Island* uses etymological metaphor Vesalius might have used simile; but there is also a degree of continuity of figurative use or conception, for example, of the vascular system in terms of vegetation or fluvial distribution, and osteology and myology as the foundations or framework of a building and as analogous to the stones of the earth.

Furthermore, Renaissance anatomy was teleological and rational in the way in which the particular form and structure of parts or members of the body were analyzed and understood as having been designed for the particular function it is their purpose to fulfill. The metaphorical conceptualization of structure and function was coherently extended to the purpose or final cause of function, as this provided the rationale for the character of the figuration of generative, auctive, and nutritive processes as vegetative, and of volitional movement and sensory perception as the exercise of the rational or animal soul's functions over its constructed instrument—the body. The figurality of the microcosm, macrocosm, and geocosm, and the vegetative and architectural metaphors of anatomy, share entailments with the effect that human anatomy was understood as mortal, and as having been created and redeemed by God. Neither the special sensory epistemology of Renaissance anatomy nor its metaphorical conceptual character and imaginative rationality were in conflict with a soteriological perspective on Man; indeed to take in hand the rational work of dissection was piously to do "God's handy worke," and in dissection the body could become microchristus and demonstrate the creative ingenuity of God, the saving love of Jesus Christ, and hence faith in the resurrection. *The Purple Island* shares with anatomy textbooks the concept of design, rationality, teleology, Christian natural theology, and the two basic conceptualizations of the human body, as constructed and as natural. Indeed these conceptions more or less deeply inform the corporeal figurality in all the works discussed in this study, regardless of whether they are poetic, prose literary, encyclopedic, fictional, expository, dramatic, philosophical, anatomical, or theological.

Very often then, the figurality of anatomy textbooks of the sixteenth century is consistent conceptually and in kind with that of *The Purple Island,* though it is distinguishable from it. *The Purple Island*'s figurative design differs insofar as the interaction of the special associated implications of the subsidiary subject of the Isle of Man metaphor with those of the principal subject, gives rise to a literary exploration of the shared entailments and significant metaphorical connotations of islands, bodies, and buildings that is unprecedented in anatomy textbooks in its systematicity, coherence, and persistence. Whereas, for instance, in anatomy textbooks, the conceptual metaphors of anatomy as vegetation and as "building" jointly provide a coherent understanding of the concept of anatomy, *The Purple Island* by contrast coalesces the conceptual metaphors and the analogies of the microcosm in a consistent and coherent image, the threshold figure. *The Purple Island* also pursues peculiarly literary and rhetorical effects which include an intricate and innovative figurative design, an ingeniously contrived allegorical artifice, and a soteriological illocutionary act, for which it draws on emblematics, pictorial and poetic theories, the tradition of didactic poetry, theological and suprarealist views of language, and an imperfect and inconsistently upheld distinction between revelation and natural or rational knowledge.

Yet the deployment of Renaissance anatomical terminology in *The Purple Island,* well beyond the usual aesthetic, affective, and intellectual sites of literary interest, such as external or surface anatomy, the heart and the brain, with a systematicity unparalleled probably by any literary work of the period, may suggest that Fletcher implied a specifically qualified reader when he wrote his poetic allegory, someone perhaps who had graduated M.D. (*medicinae doctor*), or M.B. (bachelor of medicine), or at least M.A. (*magister artium*). Although some literary critics have responded with puzzlement or distaste to the anatomical invention or discovery of subject matter, and the persistence and not infrequent aesthetic sensuousness with which its description is pursued, these responses have been found to be rooted in historical and rational misjudgments regarding the distinctions between what is appropriate as a theme for imaginative writing and what is unseemly or inappropriately denotatively limited. The allegory of *The Purple Island* is facilitated by an etymology of anatomical nomenclature that was not standardized, and only functioned as though it were referentially transparent to the extent that it was becoming shared by humanist and Hellenist natural philosophers, philologists, and physicians who were increasingly utilizing it in a specialized signifying context in a denotatively limited way. But as was said earlier in this study, during the sixteenth and early seventeenth centuries, scholars moved easily between the disciplines of theology, natural philosophy, including anatomy, and medicine, often teaching and sometimes publishing in several fields; and at some institutions, particularly northern European Protestant institutions like Marischal College, Aberdeen, anatomy was taught in the arts course.

Moreover, the critical attribution of ingenuity by contemporary critics of *The Purple Island* does not refer to the poetical overcoming of an inherent incongruity between aesthetic taste and anatomy. Rather the attribution was to designate a combination of conservatism and innovation, an affirmation of a conventional if erudite body of analogical knowledge and truthfulness of an accepted kind through the invention of anatomical subject matter as an intellectually intriguing riddle, and through the imaginative, witty contrivance of the poem's figurative design and cleverly sustained allegory, particularly its threshold figure or conceit. The invention of conceits was thought to illuminate similarities that were already assumed to exist in the obscure unity of the world, by reproducing and developing them through allegorical extension, and thereby deepening conventional understanding. Even though dissimilarities are as operative as similarities in conceits or creative metaphors, giving rise to a clash of meanings as well as a fit, they highlight similarities that although new are thought for the most part to be extensions of conventional analogical understanding, and so consistent with the way cultivated minds may have run.

But there is a reason why the congruity between the religious allegory of *The Purple Island* and the conceptual character and figurality of the predominantly sixteenth-century anatomy that has been discussed so far may not seem initially to tell against the view that the combination of analogy with anatomy in Fletcher's poem *is* (even though it need not *necessarily be*) intellectually conservative in the early seventeenth century. The reason, I surmise, is that the analysis thus far leaves the question unanswered as to whether the allegory would remain congruous, coherent, and consistent if the anatomy were predominantly more recent than the first half of the sixteenth century. If Nicolson is correct, that not only the descriptive but also the conceptual character of the anatomy of *The Purple Island* is wholly derived from an earlier scholastic or humanist anatomy that was inextricably involved in the correspondences assumed to exist between the microcosm and the geocosm and macrocosm, then it might be argued that had the poem's method of inquiry been more anatomically specialized, the anatomy more faithful to observation, *demonstratio,* or a developed *anatomia sensata,* and the physiology or anatomy of function more mechanistic and modern, then the poem's allegory, which seeks resemblances between the body and the island and the cosmos, and which relies upon the social personification of anatomical features and physiological operations, could not have been as systematic as it is. For all its fidelity to the order of Renaissance anatomical knowledge and textbook presentation of illustrations, *The Purple Island* may be ideological in refusing to acknowledge a division in the forms of discourse appropriate to poetry and to anatomy.

Against this view, however, is the evidence that the analogy of the microcosm had provided Renaissance anatomists with a means to understand what they discovered through dissection, and the reasonable conjecture that their belief in the microcosm had been an impetus or motive to anatomize rather than an impediment. It remains possible, however, that by the seventeenth century, microcosmic theory had become residual rather than dominant in natural philosophy and anatomy. The evidence of Protestant anatomy at Marischal College, however, suggests that among the declared educational purposes of anatomy was not the inculcation of a view of the human body as represented by an epistemologically transparent and neutral nomenclature, but one that would contribute to grammar and rhetoric, supplying a full vocabulary of its own terms and of metaphors and analogies. Nor is the body understood as a machine, in the (Cartesian) sense of being capable of operation without a soul, but rather as the operative instrument of the soul, whereby the powers of the soul are known. The following chapters will ask how closely related are genuinely mechanistic anatomical ideas to linguistic developments in natural philosophy in the mid to late seventeenth century. In addition, as previous chapters have described, natural philosophy in the late sixteenth- and early seventeenth-century reveals an increasing reliance upon inspiration and the immediacy of spiritual apprehension. When ascertaining the character of early seventeenth-century anatomy, one ought also to consider the enduring legacy of the *Fabrica,* the significance and utility of Vesalian anatomy as disseminated by copious references, partial translations, and copies of its illustrations, well into the seventeenth century. Many anatomists of the second half of the sixteenth century aligned themselves with "the supporters of the anatomical revisionism championed by Vesalius against the Galenists."[3] It is with Vesalian anatomy, directly or indirectly received, that this chapter will presently be concerned.

Notwithstanding the avowed and enduring use of anatomy to natural philosophy on the soul, in the early seventeenth century, and the probable persistence of a lack of conceptually and linguistically perfect differentiations between poetry and anatomy and between revealed and rational knowledge, this study's analysis has been unable so far to ascertain the full extent to which the anatomy in *The Purple Island* is conceptualized in terms of its allegorical figuration, or additionally whether the anatomy is intellectually conservative or innovative, and whether or not this character is a result of or even relates to its figuration. This is because while the study has been concerned to demonstrate congruity where it exists between the conceptual figurality of Fletcher's poem and the mainstream of anatomical concepts, language, debate, and consensus to which the early seventeenth century was heir, those chapters did not seek to follow intertextual links between *The Purple Island* and anatomical texts of any period, links of the kind examined in chapter 4 with respect to Fletcher's literary source texts.

So this part of the study must raise again the question of the congruity of anatomy and poetic figurality, while continuing with the shift in chapter 4 toward an examination of works that may have exercised a specifiable influence on the composition of Cantos 2–5 of *The Purple Island.* The effect of *The Purple Island*'s form, which is, as Healy says, to render the marginalia's scientific accuracy dependent on how well it describes the poetic stanzas, not how it relates the poem to scientific discourse, should not be allowed to deflect from the consequences of our observations of how *The Purple Island* relates not merely formally, but discursively and conceptually, to particular anatomy textbooks, or to lectures and demonstrations. Surprisingly neither Baldwin nor Nicolson before him examines the *Purple Island* anatomy anything like exhaustively; yet neither could have arrived at the conclusion that "Phineas Fletcher is in fact as conservative, practical, steady a voice as can be found in the seventeenth century,"[4] without having dismissed Langdale's interpretation of the anatomical content of both stanzas and marginalia of *The Purple Island,* a significant portion of which he characterized as Renaissance and anti-Galenist, and some of which he claims to have identified as Harveian, and thus purportedly as "modern."

As the introduction to this study outlined, and as the present part will substantiate, the responses of historians of science and natural philosophy to *The Purple Island* have tended to involve incidental observations of anatomical character and speculations as to provenance, but almost invariably a lack of interest in the occurrence of either anatomical figurality or literary modification. Interested only in the isolation of anatomical content for the purposes of its identification as progressive or conservative, these historians have tended like Langdale to attempt to extract the anatomical content from its bed of allegory, with varying and contradictory conclusions, many of which are discussed in the course of the following chapters. The reason for this attempted extraction and rather limited, incidental interest may be that much traditional history of science eschews the "context of justification" for the "context of discovery,"[5] and organizes discoveries in a heavily diachronic and retrospective fashion, not synchronically as expressive or characteristic of, or as interacting with, their period, but idealistically as preliminary to further, later discoveries. This teleological arrangement of anatomical discovery is the consequence of a later, positivist (and perhaps idealist) concept of progress, a whig interpretation of history that is particularly seductive to the historian of science, but which fails fully to explain the contexts in which Renaissance and early modern anatomists conceived of their anatomical projects and justified or interpreted their discoveries in the construction of theses, commentaries, and disputations.

I think it is true, as Scholz argues, that "we cannot escape the teleologies involved in describing early modern approaches to the human body in the light of our own thought about our bodies," because "any gaze into the past can only render its otherness in terms and categories provided by our own location in history," but in contrast to those histories which see "only the "progressive" element in early modern discourses about the body, which then conduce to a view of Renaissance "science" as inevitably leading to our modern ways of describing the body,"[6] recent scholarship in the history and philosophy of Renaissance anatomy has recognized the immense informing influence of theology, microcosm-macrocosm speculation, and the philosophy of the soul and its faculties on the contexts of project formulation and the justification of discoveries, which anatomists related instrumentally to the soul. Although the view held by the earlier historians of science who demonstrated incidental interest in *The Purple Island* is no longer fashionable, there have not been any contributions to our understanding of *The Purple Island* from this quarter for nearly three quarters of a century.

In the introduction to this study I made the point that the rejection of historical positivism need not entail the neglect of the material determinacy of history in "scientific" *inventio* or discovery. Although Renaissance anatomists envisaged the projects they undertook as accumulating knowledge in the anatomical tradition of the classical rationalist anatomy they revived, figuring that which they discovered in terms that maintained the subalternation of anatomy to natural philosophy, or even theology, and further elucidated the relationship of body to soul or to the macrocosm, even so, the later positivist, idealist history of anatomy does echo the notion of a *telos* of science that was germinal in the sixteenth century, even if only as rhetoric which very often belied the more complex set of factors that determined questions of why projects were undertaken and how they were understood. Over the course of the sixteenth century, as was remarked in chapter 1, the formation of anatomical projects to discover and observe previously unmapped regions of human anatomy, resulted in the production of a new map of more detailed and precisely delineated anatomy and the coining and standardization of items of nomenclature. These developments corresponded to the emergence of a new figurative complex which *The Purple Island* ambivalently reiterates—the anatomist as heroic voyager and intrepid discoverer, and the body as a remote and strange terrain into which the discoverer voyaged. As this figuration suggests, the effects of discoveries could not often be foreseen and were not always easily recuperated, so the figurative conceptualization of anatomical parts had to change in accordance with their newly understood functions, shapes, positions.

In endeavoring to identify the anatomical ideational content of Fletcher's poem, and relate it to discoveries made in the sixteenth or early seventeenth centuries, consequently this part of the study is concerned to do so through an interpretation of figuration, in some cases providing for the anatomy a provenance, though this will remain tentative in most cases. Ascertaining the conceptual character of the anatomy in *The Purple Island* will involve identification of particular parts or members from descriptions of their substance, size, number, shape, location, posture, complexion, and function. The reason is that, as was said above, the stanzaic distortions of perspective consistently maintain these, the *occasiones* of the anatomical *accessus,* and the specially constructed set of associated implications of the Isle consists of many implications that are shaped by the requirement that they accommodate them. It is, therefore, the stanzaic deployment of the anatomical *accessus* that demonstrates that the figuration of anatomy is consistent with its essential character, and thus it is to the *accessus* that we must turn when ascertaining the conceptual character of the anatomy in *The Purple Island,* its relative conventionality or modernity, and its likely provenance.

We must raise the question of whether, notwithstanding precaution, the critical approach that tries to discern conceptual character and provenance unavoidably extracts the anatomy from its allegorical moorings and thereby distorts it. Because the geopolitical features of the Isle in Cantos 2 to 5 are specially constructed in order that the anatomical *accessus* may be applied to them, there is the strong possibility, already suggested here, that the anatomy retains its integrity despite its metaphorical interaction with the associated implications of the Isle. Our success in identifying the specific anatomical ideational content of *The Purple Island* would in this case be the result of the special conditions pertaining to its figurality. Not the least of these is the condition that the allegory extends a creative threshold metaphor rather than an analogy, and that the figuration of anatomy is consistent with its essential conceptual character. Notwithstanding our assurance of the latter, we must remain alert to the danger, mentioned in the first section of chapter 1, of misguided inference, of prematurely foreclosing meaning, of being caught in a self-fulfilling (hermeneutic) circularity. As was said in chapter 1, this danger is especially acute where the figural aspect of representation is creative or novel; when the figure and its meaning are not consensually agreed, conventional, or pre-given. This study has demonstrated that the Isle of Man figure is creative and innovative to some extent. Additionally the danger increases with the extent of the systematicity of the figuration, and as this study has demonstrated, particularly in the early part of chapter 2, the entailments or significant connotations of the figuration of Man are extensively and systematically explored and coherently and eloquently detailed in Cantos 2 to 5 of *The Purple Island.* Reason has been found to believe that despite his apparent awareness of the danger of inferring from the highly colored phraseology of the poetry more than the reader has any right to infer, Langdale's conviction that the allegorical scheme of presentation is superficial and hovers around the scientific facts like a cloak nevertheless determined his findings. This was probably the case with regard to his distinction between the conventional analogy of the microcosm-macrocosm and the (Renaissance) Platonic notion of the human—rather than only the human body—as an index to the cosmos.

The identification of the provenance and conceptual character of the *Purple Island* anatomy over this and the following two chapters will combine with biography, bibliography, and source criticism, but where evidence prohibits connections of this kind, or where evidence is insufficient or lacking, there may yet be reason to postulate relationships that depend instead on considerations of conceptual, linguistic, semiotic, and figurative similarities between *The Purple Island* and anatomy textbooks, lectures, and demonstrations. Concern with provenance may represent this study's most significant departure in methodology from recent criticism of *The Purple Island,* and yet this part of the study will assess more or less recently developed interdisciplinary methods and tools, such as the modes of exchange proposed by Stephen Greenblatt, for their usefulness to the theorization of intertextual, intracultural relations in the early modern poetics of anatomy. But it will also be asked whether the exchanges brought into focus may also, or alternatively, be susceptible to analysis as instances of the rhetorical *energia* of mimesis, since it is mimesis of the formal layout of illustrations with accompanying text in anatomy textbooks that has been found to characterize *The Purple Island*'s formal juxtaposition of allegorical poetic stanzas and prose anatomical marginalia.

The ingenious rhetorical reconstitution of the body-as-referent as the signified of the poetry's anatomical speaking picture is also a kind of suprarealist mimesis. It involves several claims all of which are partly mimetic in character. It involves a rhetorical claim by the diegetic narrator to be dissecting the body of which the poem speaks; an epistemological claim that the anatomy of the human body-in-itself is dissected therein; a representational claim that the poetic anatomy reproduces or recreates the actuality, life, or truth of the body-in-itself as the signs of its speaking picture; a soteriological claim, which is simultaneously qualified, that poetic ingenuity may mimetically perfect and resurrect poetic eloquence that has purportedly acquired corporeality and is consequently subject to imminent dissolution; and a hermeneutic claim to contribute towards the salvation of the body and soul of the implied reader through her or his imbibition of a fideistic understanding of human anatomy as microchristus and therefore as redeemed. We will accordingly investigate the possibility that *parole,* that utterances like that of Fletcher's *Purple Island,* resisted the modes of exchange that predominate in early capitalist market relations by adopting nuanced and adaptable theories of poetic and linguistic representation from sources like Thomistic and Augustinian theology, rationalist natural philosophy, emblematics, and Renaissance rhetoric and poetics. The suggestion is that discourse or *parole* is not always everywhere determined in the last instance by what happens in the economic base of society.

The Methods of Nature and Discipline

As Sanford V. Larkey has indicated, one of the most important factors in the dissemination of the anatomical knowledge of Vesalius, especially in England, was "the production of compendia of his work, and particularly those translated into the vernacular languages."[7] Before the second folio edition of Vesalius's *Fabrica* was published in 1555, the plates of the 1543 (first) edition were being copied in London, and were published in 1545, in Thomas Gemini's *Compendiosa totius anatomie delineatio.* According to Thornton and Reeves, this was "the first illustrated textbook of anatomy to be published in England, and the copper-plates with which it was illustrated were the first to be engraved there."[8] The *Compendiosa delineatio,* 1545, was "a definite plagiarism of Vesalius" (in the fullest sense of the word), "literally a compendium of the two great treatises of Vesalius"—the text of the *Epitome,* with some additions from the *Fabrica,* followed by two nude figures from the *Epitome* and then, thirty-nine plates with their descriptions, from the *Fabrica.*[9] More than any other single edition of Vesalius, this work brought about a dissemination of the new anatomy with a sensory epistemology and corrected morphology in England. But the compendium was in Latin, and most of the apprentice surgeons in England could not read Latin. When Geminus brought out the 1553 (second) edition of *Compendiosa delineatio,* he included the illustrations from the 1545 edition, and chose the dramatist and scholar Nicholas Udall (1504–56) to translate the accompanying descriptions of these into English, but instead of the text of the *Epitome,* Udall was to "re-order," for expediency, as Larkey surmises, Thomas Vicary's *A Profitable Treatise of the Anatomy of Mans Bodie,* probably, it has been conjectured, the missing edition of 1548.[10]

It has been suggested by Sawday that Fletcher used a text such as this, Geminus's 1553 "edition" of Vesalius.[11] By no means improbable, given the importance and popularity of the edition, the suggestion also satisfies the criterion of the relationship of illustration to accompanying expository text, because that in the *Fabrica* has been found to correspond best to the juxtaposition of allegorical poetic stanzas and anatomical prose marginalia in Fletcher's poem. But Sawday's suggestion is only partially satisfactory, because it avoids or assumes less likely other provenances that may have been more probable, and may be thought to have been so on account of the order in which anatomy is presented by Thirsil in Cantos 2 to 5 of *The Purple Island.*

It has been noticed by Langdale and Sawday that Thirsil's narration of the anatomy is arranged to correspond to methods of dissection properly belonging either to the anatomy theater or to anatomical theory.[12] Various methods or orders of dissection were available to sixteenth- and seventeenth-century anatomists, and so it is important to be clear which of these is influential in the narrative design of Fletcher's poem. The order of the dissection therein is to some extent "a theoretical rather than a practical dissection."[13] In part this is because it corresponds partly to that prescribed by de Laguna as the method of Nature or Creation, which was described in chapter 3 as augmenting *The Purple Island*'s soteriological work or purpose to reveal the Creation and the incarnation in the poetic word.[14] By beginning with a description of the foundation or supportive framework of the body and the terrestrial globe—bones and stones—the anatomical cantos emulate or replicate the creative ingenuity of Nature or God as Creator or architect. It is a structural order that was thought to be reflected in embryological development and in the composition of the parts, those which are "similar," like the bones, being composed of primordial elements and humors.

This method, which was described by Corti and taken from Galen's *On Anatomical Procedures,* is as follows: the bones; the muscles; the arteries, nerves, and veins; the inner organs, including the brain, liver, heart, and lungs; the intestines and glands.[15] This sequence was more or less adopted by Vesalius in his Bologna demonstrations of 1540, and later in his *Fabrica* in that the bones, the muscles, ligaments, and cartilage, the vascular system, and the nervous system are dealt with in Books I, II, III, and IV, respectively.[16] Larkey describes Vesalius's method as "virtually a building (*Fabrica*) of the human body. First he describes the scaffold—the bony skeleton—then the muscles, the vascular system, the nervous system. . . ."[17] But at the end of Book IV the method of the *Fabrica* diverges slightly from the method of Nature in precisely the way that *The Purple Island* order does. Book V of the *Fabrica* treats of the abdomen, Book VI of the thorax, and Book VII of the head. The order in Fletcher's poem is as follows: the bones, cartilage, ligaments, "flesh," veins, arteries, and nerves (Canto 2); the abdomen (Cantos 2–3); the thorax (Canto 4); finally, the head (Canto 5).[18] This order combines the method of Nature with the method of discipline or necessity, the latter applying to the sequence of anatomy from the abdomen to the head, and corresponding to the tripartite division according to the functions of the animal, vital, and natural faculties, used by Mondino. Crooke remarks upon the method of discipline, which he calls the order of "Preservation or continuance": "This order all Anatomists do observe in Schooles, and in all publique dissections, when as they desire to shew all, or the most part, in one and the same body." Paré provides the practical reason for this method: "It is convenient we begin our Anatomical Administration from this [the "lower belly" or abdomen] because it is more subject to putrefaction than the rest, both by reason of its cold and moist temperature, as also by reason of the feculent excrements therein contained."[19] But unlike in the *Fabrica* and *The Purple Island,* Mondino's sequence of abdominal viscera, thoracic viscera, brain, was followed by rather than preceded by the muscles, bones, and other parts of the body.

Therein lay the main reason against the idea of the *Purple Island* anatomy's provenance in Geminus's 1553 (second) edition of *Compendiosa delineatio,* the text of which has been conjectured to have been based on the missing 1548 edition of Thomas Vicary's *Profitable Treatise.* Throughout the sixteenth century, Vicary's *Profitable Treatise,* later published as *The Anatomie of the Bodie of Man,* remained a popular work on anatomy in England. As such it might have provided a provenance for the *Purple Island* anatomy, but despite similarities of phrasing and initial parallels in the methods of their anatomies, *The Purple Island* and Vicary's work diverge both in order and anatomical character, the latter following the anatomy of compound members, arranged from head to foot, which is the medieval scholastic way of proceeding.[20] When Udall came to "re-order" Vicary's anatomy, he did so for expediency, as Larkey surmises, which is to say, he arranged it according to the method of discipline, "the same arrangement as in the anatomy of Mondino, the most practical way to dissect a single specimen." In so doing, Udall created "a useful dissecting manual," albeit one in which "many of the descriptions of parts are incomplete or faulty."[21]

It might have been conceivable that when he wrote Cantos 2 to 5 of *The Purple Island,* rather than following the order of the text of Geminus's 1553 edition of *Compendiosa delineatio,* Fletcher followed that of the illustrations, which are taken predominantly from the *Fabrica,* but their arrangement varies in different copies, and in at least one copy they are in reverse order.[22] Larkey argues that Udall "realized the value of the illustrations of Vesalius, but also that the text of the *Epitome* was not suitable for teaching" practical anatomy or dissection, its whole aim being "to build up the body, beginning with the bones." "His rearrangement of the Vicary text along the lines of Mundinus was the form followed in a great many of the later anatomies. This type and the Vesalian *Compendia* following Geminus were the most popular in the sixteenth and seventeenth centuries."[23]

Accordingly, it might be thought to be more probable that the *Purple Island* anatomy has a provenance in Geminus's 1545 *Compendiosa delineatio,* which plagiarized the text of the *Epitome,* "word for word,"[24] with some additions from the *Fabrica,* and was illustrated by figures from both works. It is more plausible that through this Latin compendium, Fletcher would have imbibed the new sensory epistemology and corrected morphology of Vesalian anatomy. Moreover, the same combination of methods of Nature and of discipline as is used by Fletcher is utilized by Vesalius not only in his *Fabrica,* but also in his *Epitome,* though there he omits separate chapters for the vascular and nervous systems, and hence omits them from their place in the order which we find in both the *Fabrica* and *The Purple Island.* In the *Epitome* the division of the body into the anatomy of the natural members, or those associated with digestion (chapter 3), of the spiritual members, or contents of the thorax (chapter 4), and of the animal members and the nervous system (chapter 5), follows upon the osteology in chapter 1 ("Concerning the Bones and Cartilages") and the myology in chapter 2 ("Concerning the Ligaments of the Bones").[25] The *Epitome* deals with "the Organs which Minister to the Propagation of the Species" separately, as the subject of chapter 6. The discrepancy between this and the order or method of anatomy followed by Thirsil in Cantos 2 to 5 of *The Purple Island* makes Geminus's 1545 *Compendiosa delineatio,* in addition to his 1553 edition, an unlikely though not entirely improbable provenance for the *Purple Island* anatomy.

In the previous chapter it was remarked that despite its expansiveness the arguably supplementary as opposed to authoritative function of the marginal notes in *The Purple Island* might correspond to the priority of illustration over accompanying text in the *Fabrica.* Rather than constructing two autonomous discourses, the one poetic and the other "scientific," Fletcher in the poetic stanzas unraveled the etymological connotations that were implicit in anatomical nomenclature, which he provided in more concise form in the marginalia. Thereby in poetic terms providing an anatomical speaking picture, combining pictorial imitative accuracy with aesthetic figurality, the poetry in Cantos 2 to 5 ingeniously corresponds to illustrations like those in the *Fabrica,* that follow the convention of providing a comprehensive visual description which places anatomy more exactly before the eyes than ordinary language is capable of doing. Is it not conceivable that rather than drawing inspiration for his anatomical speaking picture from a Vesalian compendium, Fletcher took anatomical ideas and the idea of a "visual"-verbal juxtaposition from the *Fabrica* itself? The possibility presents itself because the same combination of methods of Nature and of discipline as is used by Fletcher is utilized also by Vesalius in his *Fabrica.*

From among other anatomical works a further possible provenance emerges in *The Historie of Man,* the author of which sought to take his method, a significant part of his work's anatomical content, and five of the plates from Vesalius's *Fabrica.* John Banister's *Historie of Man* is dedicated to "The Worshipfvll the Maister, Wardens, Assistantes, and fellowly Fraternitie of Chirurgians in London: and from them to all godly, true and zealous professours of Medicine, throughout this Realme of England."[26] In the dedication, Banister writes about the provenance of his own anatomy: "Omitting diuers old writers, whose workes had not all equall successe, I came at length to *Vesalius,* whose whole worke seemed as tedious as his Epitome ouerculled & short. But when I saw *Fuchsius* to haue extract a notable Epitome out of *Vesalius & Galen,* I had thought to looke no further till *Collumbus* appeared in my sight: whose labours then reuoluing, and seying him in some thynges use sufficient prolixitie, as in his Bookes of Bones and Muscles, & in other causes to be somewhat brief (because *Vesalius* had sufficiently handled them) as in the nutritiue partes, immediately I refused to bynde my selfe to any peculiar translation, chusing rather to picke a posie of the chiefest flowers from all their Gardens. . . ."[27] *The Historie of Man* shares the same combined method as the *Fabrica* and *The Purple Island,* though like Thirsil's anatomy, and like Vesalius's *Epitome,* it can be thought to have elevated anatomy of the "Cartilages" and the "Ligaments," unusually devoting a book to each as, for example, to the bones and the muscles, as one might expect. Fletcher surprisingly omits to treat of the muscles separately, but the cartilage and "ligaments" feature prominently in Canto 2, and the *Epitome* entitles a chapter "Concerning the Bones and Cartilages," and devotes a separate chapter to the "Ligaments of the Bones," as remarked above. The nine books of Banister's *Historie* divide human anatomy as follows: "Of the History of Man, the first Booke. Of Bones, the frame of the Body"; "The second Booke . . . , discoursing the Cartilages"; ". . . , the third Booke, of Ligamentes"; ". . . , the fourth Booke. Of Muscles to euery member"; ". . . , the fift booke describyng the instruments seruing to nourishment, . . . that is, *Of the nutritiue and naturall partes*"; ". . . the vj. Booke of the instruments seruyng to the propagation of mankind. That is to say, *The generatiue partes*"; ". . . the vij. Booke: of the making of the hart, and of the Instrumentes ministryng to his function: commonly called, *The Vitall or Spirituall partes*"; ". . . the viij. Booke, of the brayne, and Instrumentes seruyng to the Animall function"; ". . . the ix. Booke, discoursing the marey conteined in the Bones, Periostion, and the heares."[28]

The Purple Island follows the *Fabrica* order or method, in describing the veins and arteries after the bones, cartilage, ligaments, and muscles, and before the abdomen; in this both works differ from the *Epitome* and Banister's *Historie of Man.* Otherwise Banister places great emphasis on the Vesalian order, or more precisely, on the commencement of anatomy with the skeleton, which he likens to the way in which "the good and expert *Architect,* hath a singular care, first in edifying his house, to be well aduised of the manner of his foundation, and heedy framyng of his Timberworke."[29] Unfolding the now familiar simile, Banister explains: "so it behoueth us, in searchyng the true Secresie of this Misterious Science, and body of Man, first, before the partes supported, distinctly, and deliberatly, with diligent care, to consider the partes supporting." The reason that he supplies alludes to the necessity of following the method whereby the human body was constructed, in order to facilitate understanding, and to the body's dignity which is most apparent in its upright frame, without the diligent perusal of which, he advises, the reader may "finde the rest of this booke, not onely in readyng unsauery, but also to your sensible understanding uery obscure."[30] Banister's conception of the constructed human fabric, which is quintessentially Vesalian, extends from the anatomical exposition itself to his dedication, which is distinctive not for the customarily modest appraisal of the author's "inferiour" contribution to the accumulated knowledge of anatomy, which he compares to "the building of some riche tower, or strong castle," but for its invitation to the implied reader to "lay of the stones of my gatheryng to the buildyng of this famous tower of health." Banister expounds, saying, "I shall conceiue incredible felicitie, if I may hereafter, in uiewyng this renowned buildyng (for so I hope surely to see it) cast myne eye uppon those stones whiche my homely handes had so rudely congested, and see them cunnyngly wrought, and aptly applyed to the buildyng."[31] Though Thirsil's anatomy may not have been quite what Banister had in mind, it is not improbable that Fletcher may have cunningly, cleverly wrought those very stones that the surgeon and anatomist gathered.

There are other reasons for a possible provenance for the *Purple Island* anatomy in Banister's vernacular anatomy textbook and these include his disinclination to provide an anatomical description of female reproductive anatomy, and his circumspection when it comes to that of the male.[32] As was mentioned in chapter 2, the "sixt and last town in [the abdominal] region" of the Isle of Man is similarly described with little of the anatomical precision of other regions, owing to moral delicacy, though the similarity of the two works in this respect may not be intertextual in origin, but due merely to a moral anxiety and guardedness against prurience and the widening of access to sexual anatomical knowledge, that both works share with other vernacular works of anatomy, as has been suggested previously in this study. The text of Banister's *Historie* is divided into a main text and marginal notes, all of course prose. Although visually similar to the page layout of Cantos 2 to 5 of *The Purple Island,* with a comparable density of marginal notes, the function of these notes in the *Historie* is either to provide references to authoritative texts or anatomists, to comment on authorities like Aristotle, Galen, Vesalius, Realdo Colombo, and Jean Fernel (Fernelius, 1497–1558), or to summarize or act as an index to the content of passages of the main text. None of these functions predominates in Fletcher's work, as the previous chapter has indicated.

Although an expository "historie" of the anatomy of Man (with a title that reminds one of Fletcher's "*Isle of Man*"), there is one feature of *The Historie of Man* that is reminiscent of the way in which the *Purple Island* diegetic narrator Thirsil approaches the anatomy he is to present at each new stage of dissection with considered caution and appreciation of the elusiveness of explanation, interpretation, or even accurate description, with which to represent the formidably complex structures of the human body and their even less fathomable functions. Banister does not turn, as Thirsil does, to God for inspiration, but in his preamble to Book VII, "of the making of the hart, and of the Instrumentes ministryng to his function: commonly called, *The Vitall or Spirituall partes,*" he wrote that "hetherto my penne hath not wauered in the doubtfull balance of ambiguitie, but passed the playne pathe of truth," "till now at length, beyng come to a place somewhat ascendyng (as they that will attayne to the scope of their desire, must not refuse the hill with the dale). . . ."[33] This apprehension and communication of the difficulty of discovery is shared by Thirsil as he guides the reader of *The Purple Island* through the strange, though native features of the microcosmic world of the Isle of Man.

The evidence for a provenance for the *Purple Island* anatomy in either Vesalius's *Fabrica* or Banister's *Historie of Man* is not overwhelming or compelling, but, as we will see later in this chapter, the conceptual character of some anatomical features combined with those points already made here, do suggest a certain probability. Unlike the case of his literary sources, like Du Bartas and Spenser, Fletcher does not cite the authorities for his anatomy in the marginalia of Cantos 2 to 5. But the probability of one particular provenance rather than another may be enhanced by consideration of the evidence, such as it is, of what Fletcher may have been reading during the period of *The Purple Island*'s composition.

Fletcher graduated B.A. in 1604, and M.A. in 1607–8, and possibly afterwards B.D., all at King's College, Cambridge. He obtained a fellowship at the same college apparently in 1603. In 1596, John Banister had donated a copy of his *Historie of Man* to King's College Library. We may also learn from the book lists included in inventories presented for probate in the Vice-Chancellor's Court at Cambridge, that on January 27, 1588/89, an anonymous inventory of the stock of a bookseller was exhibited in court, and that item 245 was "banisters anatomy parchment," which may refer to a copy of Banister's *Historie of Man* bound in limp vellum (rather than printed on parchment), or it may refer to anatomical fugitive sheets; and that in 1591, Thomas Lorkyn (c. 1528–1591) bequeathed to the University Library a copy of the 1555 (second folio) edition of Vesalius's *Fabrica.* I have not discovered the destination of Banister's "anatomy parchment" in the university, but Lorkyn's will instructs its executors as follows:[34] "[A]ll my physicke bookes with thos that appertaine allsoe to phisicke to be kept in the inward Librarie in a gret Cubbard locked with two locks the librarie keeper to have on key and the phisicke Reader an other for him and his auditors Students of phisicke to looke one soe that the said vniuersity be at chardge to make convenient cubbords or presses or Cubbord or presse to keepe them in Locked within on yeare after my death."[35] What is not clear from this evidence is whether a Master of Arts and Fellow of King's with an interest in anatomy, like Fletcher, but who like him was neither the "phisicke Reader" (i.e., the Regius Professor), nor a Doctor or Bachelor or student of medicine, might have been able to negotiate access to the books within the cupboard, secreted as it was within Rotherham's library room ("the inward Librarie"), and to which access was specifically granted by Lorkyn's will only to the "librarie keeper" and the "phisicke Reader" and his students.[36] To this I have no definite answer, but because the first quarter of the seventeenth century is virtually a blank in the library's history, and during the early years of the century recorded accessions hardly exceed a dozen volumes, consequently, in a library in which so little interest was taken, the regulations probably were not very strictly observed, and Fletcher might well have obtained access to books bequeathed by Lorkyn.[37] This disinterest is perhaps reflected in the condition in Lorkyn's will that the cupboard was to be made at the university's expense within a year of his death, and that in the event of the university failing to provide the appropriate storage within the year, the books were to go to Peterhouse.[38] These precautions suggest a reasonable fear of their neglect even in the last decade of the sixteenth century.

The Purple Island was published in 1633, and even if Fletcher did have access, is there any reason to suppose that studies of nearly thirty years earlier could have borne fruit, or even been remembered? The exact date of the composition of *The Purple Island* is unknown. However, there is a strong case presented by William B. Hunter, among others, for arguing that "Phineas' otherwise undatable poem at least in part antedates that of his younger brother"—*Christs Victorie, and Triumph* (1610).[39] Kastor is bolder in claiming that Fletcher began writing *The Purple Island* "sometime about 1606–8," about the time of his graduation to M.A.; and Kastor adds that, by 1610, "the main design and some parts . . . had been written."[40] Phineas adopted the pastoral epithet of his diegetic narrator, Thirsil, around 1608. This, therefore, provides the earliest possible date for *The Purple Island*'s completion;[41] this suggests that he continued to work on the poem after his graduation as Master of Arts, and hence after he had gained access at least to the books of the University Library not secreted by Lorkyn. Kastor is confident that: "Internal evidence, including topical references and stylistic features, indicate, however, that the poem continued to be worked upon until publication in 1633."[42] According to Langdale, "Fletcher was following his usual method of composition—subjecting his work to a process of elaboration and adaptation that went on for years."[43] All of this at least allows the possibility that the anatomical cantos were composed in or shortly after the period in which Fletcher was a Master of Arts and Fellow of King's, or that the composition, if much later, relied upon notes (possibly from either Banister's *Historie of Man* or the 1555 (second folio) edition of Vesalius's *Fabrica*) taken during that period. That the composition of the anatomical cantos of *The Purple Island* may have relied upon the second folio edition of Vesalius's *Fabrica* is supported by the observation regarding the order or method in which the anatomy is presented in those cantos.

The anatomical method that Fletcher employs in Cantos 2 to 5 of *The Purple Island* follows neither the order of "Dignity" or nobility, which is the medieval scholastic way of proceeding and "requireth, that we shold begin with the brain, as with the most noble part," nor the order of "Scituation or dissection," which "requires that we should first demonstrate those parts which first appeare in view to the beholders."[44] Apart from the comparatively small mention that these two methods receive in Renaissance anatomy textbooks, what can be the significance of the methods that Fletcher chose? It has been argued that by reflecting the method whereby Nature creates the human body, the mimesis of *The Purple Island* is an attempt to partake of the creative and regenerative power of Nature or of God. The rhetorically, ingeniously acquired corporeality of the speaking picture and its clarification by the marginalia, a formal disposition that mimics the layout of anatomical illustration with accompanying text, is apparently enhanced by the narrative mimesis of theoretical anatomical method, that part of Thirsil's exposition that conforms to the sequence of the method of Nature. All of these facets of the work contribute to the "building" of anatomy as the subject matter of the poem, and to the rhetorical reproduction of the body-as-referent as its signified. Yet, as we have seen, the poem also recreates the presentation and order of practical anatomical demonstration. What precise contribution, if any, does the mimesis of practical method make to the corporeality of the speaking picture? This question is in part answerable in that the method of necessity, as French explains, "could be defended philosophically because the nutritive faculty, located in the abdomen, was taken to be a prerequisite for the existence of the vital faculty in the thorax and this faculty in turn was necessary for the existence of the rational faculty in the head. The anatomist was thus following the sequence adopted by nature in constructing the body."[45] Consequently, while apparently accommodating the necessity of responding to the exigencies of demonstrating the whole of anatomy in a single specimen, the entire Vesalian sequence is actually designed to replicate the way Nature or God fashions human bodies.

THE HEAD, BRAIN, AND *RETE MIRABILE*

The anatomical method of Thirsil's exposition of anatomy in Cantos 2 to 5 of *The Purple Island,* then, is Vesalian, though it may be derived from a source that is not the *Fabrica,* but a compendious work, the most probable being Banister's *Historie of Man.* But what of the anatomy itself, the conceptual character of the anatomical content? Langdale identifies in Fletcher's descriptions the vas breve or short branches of the splenic artery and the vermiform appendix, which were both first observed in the early modern period (*P.I.* 2.32 margin.; 2.42 margin.); and he claims further that "no equally accurate accounts of the gastrocolic omentum and the mesentery are to be found prior to those of Vesalius."[46] If they were to be firm, these identifications would add to the number of reasons for a Vesalian or post-Vesalian provenance for some of the *Purple Island* anatomy. But although *The Purple Island*'s application of anatomical descriptive criteria to the body is compatible with the allegory's assemblage of superimposed analogous systems—anatomical microcosm, microchristus, macrocosm, and political geocosm—none of the aforesaid anatomical parts participates deeply and integrally in the allegory of the poem. Consequently, while the descriptions of these parts might provide a reason for not concurring with the general tenor of Nicolson's view that the conceptual character of the anatomy in *The Purple Island* is wholly derived from an earlier scholastic or humanist anatomy (though of course Vesalius was a humanist), the descriptions do not contribute significantly to our answer to the question of whether the allegory of *The Purple Island* would remain congruous, coherent, and consistent if the anatomy were predominantly more recent than the first half of the sixteenth century; they may even suggest that where the anatomy is more faithful to observation, *demonstratio,* or a developed *anatomia sensata,* then the poem's allegory is not as inextricably involved in the correspondences assumed to exist between the microcosm and the geocosm and macrocosm. It may even be that *The Purple Island*'s fidelity to the order of Renaissance anatomical knowledge and textbook presentation of illustrations, is indeed inconsistent with its allegorical artifice, and the poem may be ideological in refusing to acknowledge a division in the forms of discourse appropriate to poetry and to anatomy.

These conclusions are premature, however, for the anatomy and physiology of the eye and inner ear (*P.I.* 5.23–49), which have been discussed in chapter 2 as fully participating in the figurality and personification required to substantiate the work's intentional direction toward a microcosmic island nation, provide an exemplary occasion to study the effect of allegory on recent anatomy.[47] According to Langdale, Fletcher "lists correctly the six ocular muscles [*P.I.* 5.26.1–7 & margin. z], his work comparing favourably with that of Vesalius, who erroneously described seven, the extra one being found only in animals."[48] Banister, following Colombo, describes only five.[49] The *Purple Island*'s stanzaic figuration of these muscles as "bands" "set to stirre the moving tower" of "*Visus*" (*P.I.* 5.26.1; 5.25.5), and the function of each, which simultaneously signifies physiologically and in terms of the mechanisms of the head city's defensive surveillance, confirm that even where, or rather because, the figuration of the body is highly systematic, the character of the anatomical content is capable of retaining conceptual integrity despite modification by its involvement in analogy. Similarly, *The Purple Island*'s figuratively rendered physiology of sight (*P.I.* 5.35–37, cited above, in chapter 2) does not deflect from the poem's location of "the mean of sight" in the "Reticularis" or "netty tunicle" (*P.I.* 5.33 margin.). In locating sight thus, Langdale proposes that *The Purple Island* possibly responds to Johann Kepler's *Ad Vitellionem, paralipomena,* and is in contradistinction to Galen and most Renaissance and early modern anatomists.[50] This further testifies to the compatibility of recent anatomy and its allegorical modification.

Another example of this compatibility is the description of the organs of the inner ear, including the stapes or "Stirrup," which was discovered by John Philipp Ingrassius in 1548.[51] The description in *The Purple Island,* which is discussed in chapter 2, explains how each of these organs is especially conducive to the greater efficiency of secret intelligence services (*P.I.* 5.42–43). All of these significations simultaneously of post-Vesalian anatomy and physiology, and the structures or processes of surveillance and intelligence, upset Nicolson's assumption that the involvement of anatomy and physiology in analogy determines their conservative character.

The brain and the heart, more than any other member or part, participate in all the allegory's assemblage of superimposed analogous systems, including the analogy between microchristus and Christ in heaven, whereby humanity participates in divine providence and redemption. Accordingly it is on these principal organs that I will focus my analysis in the remainder of this chapter and in the following chapter, respectively, determining the provenance and conceptual character of the anatomy in each case. Whether this character is conservative or "modern," the fact will help to determine whether what "particularly emerges for Fletcher's poem" is, as Healy argues, "the apparent difficulties for successfully negotiating a theory of correspondence in the early seventeenth century."[52]

The brain is indubitably understood in *The Purple Island* in terms of the created anatomical microcosm and its relationships with the macrocosm and political geocosm. Chapter 3 has argued that the elevated location of the brain in "the *Cephal* mount" (*P.I.* 5.33.6), the "Islands and Creatours glorie" (*P.I.* 5.2.2), corresponds to the principal causes of Man's upright form, as *imago dei,* in that on account of the nobility of the brain's position, contemplation ought naturally to arise within. Within the microcosm-macrocosm theory of Fletcher's poem, the figure or shape of the head is spherical like heaven, as it is in Plato's *Timaeus,* on account of being the seat of the rational soul. Within the political analogy of human anatomy, the head bears "rule and soveraigntie" "over all the rest" (*P.I.* 5.4.7), for in the "castle" or "Citie" of the brain (*P.I.* 7.12.6; 5.6.3) the Prince, the Intellect, resides with Voletta, the Queen, and in the face—the microcosm or "Index of the minde"—"the Princes thoughts are often read" (*P.I.* 5.8.7, 4). Davies of Hereford's description of the house of the head in *Mirum in modum* (cited in Chapter 5) applies several *occasiones* or descriptive criteria; but as was said in that section, this anatomy of the brain and cranium is rather more schematic than that found in *The Purple Island* (5.11–21). There are many aspects to the anatomy of the brain in Fletcher's poem, including the location of "common sense" "in midst of that high Tower," and of Phantastes or "fancie" "in the Castles front" (*P.I.* 6.42.4 and margin. *; 6.46.1 and margin. *),[53] and the morphology and nomenclature of the cerebral ventricles, septum lucidum and blood vessels. But that part of the anatomy of the cerebrum that demonstrates a progressive, Vesalian character is the location and physiology of the parts involved in the creation of animal spirit.

De Laguna provides a traditional account of the production of animal spirit from vital spirit in the *rete mirabile* ("marvelous net" or "wonderful network"). In this vascular plexus, said sometimes to consist of "veins," though on account of their purported function they ought logically to have been arteries, the vital spirits are "weakened in temperature and made cooler": "When they have attained such a degree of temperature they are scattered from that net-shaped plexus into the benign substance and ventricles of the brain so that they may be distributed thence as though through rivulets to the separate parts of the body by means of nerves."[54] Traditionally the *rete mirabile* was thought to be formed from "many very thin branches" of the internal carotid arteries, "marvelously united together one upon the other in the shape of a net," which was located above both the basilar bone and the mouth inside the bottom of the cranium, below the hard panniculus or dura mater.[55] Achillini claims to have seen it in 1503, in "a chamber in which it lies closed up with bones."[56]

Using "the experience of the senses" as his "chariot-driver," Berengario was unable to find the plexus in this position in human bodies, possibly because human bodies do not have a vascular plexus at this site (its existence having been inferred from its location in other animals), though the historian Lind identifies this "net" as the central branches of the Circle of Willis.[57] Believing Nature could more effectively transform vital into animal spirit elsewhere, Berengario placed the net further upward in the pia mater, where there are "very small branches of the arteries dispersed everywhere," "descending (ascending) upon the dura mater which adheres to the basilar bone and ascending through the pia mater up to the center of the brain."[58] In these branches, above the dura mater immediately below the medulla of the brain, is made "a certain membrane continuous and similar to the pia mater in whose branches," said Berengario, "the blood or vital spirit is rendered subtle" and (in one account) "made into animal spirit" or (in another account) "prepared so that in the substance of the brain and in the ventricles it may be made into animal spirit."[59] "Thus subtilized," the spirit "passes to the ventricles by means in part of the worms of the brain," the "vermis of the cerebellum."[60] Berengario's account of this process of making animal spirit seems to be anatomically precise about the parts involved, and it is reasonable to suppose, as Lind did, that Berengario suggests that the process occurs in the *plexus chorioideus ventriculi lateralis* (choroid plexus of the lateral ventricles).[61] The choroid plexus (sometimes referred to in the plural form) is "a vascular proliferation or fringe of the tela choroidea in the third, fourth and lateral ventricles"; it consists of "two membranous and vascular duplicatures of the pia mater, situated in the lateral ventricles" in the center of the brain, principally, though it also extends into the third and fourth ventricles (hence also the plural form),[62] and therefore fits Berengario's description of the structure he proposes as the site of the production of animal spirit.

Heseler claims that at the anatomical demonstrations at Bologna in 1540, "Vesalius accepted, as in [his] *Tabulae anatomicae,* the theory of rete mirabile in humans."[63] Two of Vesalius's demonstrations involved the dissection of regions of the anatomy of the brain contiguous to and including the contested parts. Baldasar Heseler makes the following note on Vesalius's twenty-fifth demonstration: "Eventually, he showed us the network of winding arteries around the rete mirabile in which the spiritus animales are produced, transmitted there from the heart as spiritus vitales. And I saw particularly those which run in coils around the base of the cranium, which are called rete mirabile."[64] The location of the object so evidently desired for inspection by Heseler at the base of the cranium would imply that Vesalius's view of the *rete mirabile* in 1540, or at least that which he was willing to espouse in public, was traditional, rather than Berengarian, were it not for the denial of the *rete mirabile* as a locus for the creation of animal spirit expressed in the following record by Heseler: "He [Vesalius] did not say that the spiritus animales are produced in the substance of the [basilar?] bone or in the rete mirabile, but in the network of arteries."[65] This clear denial of the *rete mirabile* as a locus for the creation of animal spirit is developed by the designation of another part as the *rete mirabile* and locus for production of animal spirit, in Vesalius's twenty-fifth demonstration: "At last he showed us the rete mirabile, *situated higher up in the middle of the cranium near where the arteries ascend,* and forming the plexus in which the spiritus animales are produced out of the spiritus vitales transferred there. And it was a reddish, fine, netlike web of arteries lying above the bones, which I afterwards touched with my hands, as I did with the whole head."[66] As early as 1540, Vesalius's view of the locus of the creation of *pneuma psychicon* was Berengarian, as is evident in the phrase I have emphasized with italics. It is in the middle of the cranium, and therefore above, not beneath, the dura mater, near where the arteries ascend into the cerebral ventricles, above the basilar bone. That both the fifteenth and twenty-fifth demonstrations involved the dissection of a sheep's brain and a human one, in order to compare their respective anatomies, should lead more decisively to the view that even if the traditional *rete mirabile* could be found in other animals, it was not thought by Vesalius to be the locus of the creation of animal spirits in human brains.[67] For Vesalius it would have been a crucial aspect of Berengario's discovery that in distinguishing between animal and human brain structure and function he was advancing upon the anatomy of Galen.[68]

Vesalius's denial of the traditional *rete mirabile* extends in the *Fabrica* from its function to its very existence in humans.[69] As in 1540, Vesalius shifts the function of the choroid plexus, but unlike then, or unlike Heseler's representation of Vesalius's twenty-fifth demonstration, Vesalius in the *Fabrica* does not shift the name *rete mirabile* from its traditional designation, but denies entirely the existence of the structure traditionally so designated in human beings. Singer says that in the *Fabrica* the lack of a *rete mirabile* in Man persuades Vesalius that vital spirit is transformed into animal spirit by the choroid plexus and ventricular cavity (the cerebral ventricles).[70] The account of the function of the choroid plexus is succinct and clear in the *Epitome,* and has been cited in part in chapter 1 on account of its vegetative figuration of the arterial system. One branch of the carotid artery (the anterior and middle cerebral arteries), having divided and subdivided several times, is said to disappear into the pia mater; "part [the choroidal artery] creeps into the right ventricle of the cerebrum and forms a plexus [the choroid plexus] laid down in this ventricle and compared to the outermost covering of the fetus [the chorion]." Vesalius says further, with a self-referential remark that emphasizes that he is indicating his opinion, that this "offshoot brings the vital spirit to the brain so that, as I shall now say, the animal spirit may be prepared by the function of the brain":[71] "From the vital spirit adapted in this plexus to the functions of the brain and from the air which we draw to the ventricles of the brain when we breathe in, the inborn force of the brain's substance creates the animal spirit."[72]

Langdale correctly states that *The Purple Island* omits the *rete mirabile,* and that "Fletcher's silence is significant because the *rete* was a picturesque detail admirably suited for allegorical treatment."[73] Not only does *The Purple Island*'s diegetic narrator omit the *rete mirabile,* he identifies "a knot of veins and arteries weaved together; by which the animall spirits are concocted, thinned and fitted for service" (*P.I.* 5.15 margin. i). This part significantly is not called a net, yet "knot" represents very aptly the choroid plexus, described in a standard modern medical dictionary as "infoldings of blood vessels of the pia mater covered by a thin coat of ependymal cells that form tufted projections into the third, fourth, and lateral ventricles of the brain."[74] The observable location of this "knot" in the brain of the Isle of Man "Hard by" "Two caverns" or "hollow places, like half moons, of much use for preparing the spirits," also suggests its identity as the choroid plexus, for these "caverns" "Within the centre" of the brain where "first are born the spirits animall" are the lateral cerebral ventricles (*P.I.* 5.15.1; 5.14.2, 1, 5 & margin. h).[75] It is the stanzaic maintenance of the following anatomical descriptive criteria that allows us to ensure the safety of our identification: location or position; form, figure, or shape; anatomical relations, posture, or connexion; and uses or functions.

Fletcher could have derived the *Purple Island* form, function, and location of the choroid plexus from either Berengario or Vesalius, though on considering the *Purple Island* mimesis of a form of anatomical illustration and accompanying text that approximates more closely the layout in the *Fabrica* than in either of Berengario's principle publications, and its mimesis of the combined Vesalian anatomical method of Nature and discipline, as well as the derivation of other aspects of the *Purple Island* anatomy such as the gastrocolic omentum and the mesentery from Vesalian anatomy, the evidence accumulates for Fletcher's possible use of the second folio edition of the *Fabrica* in Cambridge University Library. But Fletcher might have taken Vesalian anatomy from a Vesalian compendium, or from post-Vesalian works of anatomy, and he could have derived the choroid plexus as the site of the manufacture of animal spirit from Banister's *Historie of Man,* which describes the position and use of the appropriate part thus: "Through these upper ventricles of the brayne the reticular or net like foldes called *Coriformes* are caryed ["Plexus Coriformes"], whose uses doe belong to the generation of animall spirite." Notwithstanding Banister's description of the plexus as a "large and notable plicature, or weavyng," nouns from which Fletcher may easily have derived "knot," Banister's retention simultaneously of the term "marveilous nette" denotatively for the same anatomical feature implies that Fletcher's use of *The Historie of Man* would have been judicious and discerning, and probably not exclusive.[76]

Significantly for our assessment of the congruity of allegorical figurality with recent anatomy, notwithstanding its probable Vesalian or post-Vesalian provenance, the choroid plexus in Thirsil's account undergoes an allegorical transformation into "an hundred nimble workmen," "Lab'ring to make [the spirits] thinne, and fit to hand, / With never ended work, and sleeplesse caring" (*P.I.* 5.15.1, 3–4). The modification occurs entirely in accordance with the poem's general strategy of personification of anatomy to a greater degree when physiological activity is more pronounced or perceptible, and when the structures involved are imbued with significant quantities of either natural or vital or animal spirit. The operation of the vascular proliferation or knot takes a socially integrated, personified form without (it should be added) in any way detracting from the accuracy of the anatomical account.

Other aspects of cerebral anatomy do suggest a probable provenance in either Vesalius or Banister and the marginal and main texts at the places in both texts where they describe the third ventricle, the corpus callosum, and the fornix bear some comparison, such as was undertaken in chapter 1 between the *Purple Island* description at this place and that in Book VIII of Vesalius's *Fabrica.* The main text in Banister's *Historie of Man* proceeds as follows: "Above the third ventricle of the brayne, you may behold a portion supereminent, or appearyng ouer it sufficiently white, whiche is called *Callosum corpus,* a Callous body. Under this lyeth an other portion named *Fornix,* and *Psalloides,* in figure of a halfe circle, or rather an arche, or bendyng bridge. . . . Neither unprofitable is this body callous, or hard, or that after the fashion of a compassed arche made, but to the end that the wayght of the brayne might not ruine or fall upon the third cauitie."[77] Whereas in the main text of *The Historie of Man,* the ventricle is vaulted by the "*Callosum corpus*" and the "*Fornix,*" the marginalia of Thirsil's anatomical exposition conflates these into the "corpus Cameratum" (*P.I.* 5.16 margin. m). Vesalius similarly describes the fornix in his *Fabrica,* though of course in Latin, as "resembling a vault (fornix) or arched roof,"[78] and Banister's marginal use of the vernacular term "vaulted" to describe the third ventricle is also echoed by *The Purple Island*'s similar description of the ventricle as "Vaulted by three fair arches" (*P.I.* 5.16.6).[79] Like Banister, and with some similarity of phrasing, Fletcher's marginal note places emphasis upon the structural importance of the function of the "corpus Cameratum," "which with three arches bears up the whole weight of the brain" (*P.I.* 5.16 margin. m). In all three works compared here, the same architectural function of support applies to this function, and each shares a continuity of linguistic metaphorical representation. This metaphoricity is etymological, though *The Purple Island* omits to state the nomenclatural term (fornix), and in each case, figurality is cognitively operative, even though it is pared down in the anatomy textbooks as compared with the plurisignification of the stanzaic figurality in *The Purple Island.*

As regards the provenance and conceptual character of the anatomy in the poem, nomenclatural figurality of this kind is not a sure guide, precisely because it unfolds a standard vocabulary. But in Vesalius's works and in Banister's *Historie* may also be found figurality of a different kind, often informed by the functions of the parts and by teleology, as was argued in chapter 1. Book VIII of Banister's *Historie* might be thought to have provided Fletcher with a suggestion for his elaborate figuration of the optic and other nerves concerned with sensory perception, for therein the nerves are described briefly as "the diligent messengers and garders of the brayne."[80] Furnishing the reader with reasons for the preeminence or "primacie" of the brain, Banister refers to "the strong munimentes and straunge defences and couerynges thereto appertainyng," listing the hair, skin, fat, "*Membrana carnea, Pericranion, & Periosteon,*" and the skull as comparing favorably with the parts defensive of the heart.[81] Sufficiently akin in figurative conception to the key ideological determination of *The Purple Island*'s figural character, that the Isle of Man should be defended, Fletcher needed only to have politicized, elaborated, generalized, and extended that which he may have found in Banister's *Historie.*

A *Purple Island* marginal, which is keyed to stanza 18 of the same canto, describes a "passage" that "reaches" from the third ventricle or "cavitie" "to the fourth cavitie," and "yeelds a safe way for the spirits," as though (in the stanza) through a "vale" between or beneath "two round hills" (*P.I.* 5.18.1,3 & margin.).[82] Langdale regards this description to be "of considerable interest, because modern historians think that the cerebral passage was not known until many years after *The Purple Island* was written," in the mid-seventeenth century, when it was purportedly discovered by Franciscus Sylvius, and accordingly named the aqueduct of Sylvius.[83] However, Fletcher's narrator imagines the passage of animal spirit between the third and fourth ventricles as "through" a "vale" (*P.I.* 5.18.4, 1), rather than a conduit, pipe, or street—the figurations besides the stream that the poem usually applies to vessels of moving fluid, and might be thought to have applied to that which was to acquire in anatomical nomenclature the name "aqueduct." In the previous stanza, the passage of the animal spirit is in "milkie streets," whereas "rheugms" are discharged from the third ventricle in "Base noisome streams" (*P.I.* 5.17.2 & margin. n). It is more probable, therefore, that Fletcher had in mind or sight some such description of the passage of animal spirit from the third to the fourth ventricle as can be found in Vesalius's *Fabrica* or *Epitome:* "A not inconsiderable part of the animal spirit spreads from the third ventricle under the testes of the brain into the ventricle common to the cerebellum and the dorsal medulla."[84] Perhaps more similar to the *Purple Island* rendering is that of *The Historie of Man:* "Behynd this vaulted part in the extreme part of the brayne towardes *Cerebellum,* and in the upper part of the thyrd ventricle, Nature hath feyned certaine eminent partes, whiche in their upper partes, represent the likenes or Image of Testicles, and so called therfore of Anathomistes *Testes:* neare unto the which, two other particles yet somewhat greater are to be discerned, called according to their figure clunes, the haunches or buttockes. Betwene which lyeth ['a small hole goyng from the thyrd to this fourth cauitie now spoken of, being but litle, yet not so small but as may suffice the animall spirites to passe thereby from the thyrd, to the fourth Ventricle']."[85]

Significantly once again for our assessment of the congruity of allegorical figurality with recent anatomy, notwithstanding its probable provenance in Vesalius or Banister, the passage of animal spirit from the third to the fourth ventricle undergoes an allegorical transformation into a "vale" or "pleasant dale" between or beneath "two round hills," "Through which the spirits thither safe are past" (*P.I.* 5.18.1,3,4). The transformation distorts the anatomical descriptive criteria of size, with regard to the cerebral testes, but maintains number, form, location, and posture. Figuratively and pictorially the stanza conceptualizes the movement of the animal spirits from the third ventricle as a stream that culminates in "the fourth cave," where the spirits "refin'd their full perfection have; / And therefore close by this fourth wondrous cave / Rises that silver well, scatt'ring his milkie wave" (*P.I.* 5.18.1,5–7). The well is identified in the marginal note (p) as "the pith, or marrow, the fountain of these spirits"; and thus the stanzaic figurality ensures that the brain's function, to provide the body with voluntary movement and sense perception, and hence with life, is among the connotations the reader is invited to unfold. The same understanding is impressed upon the reader of Banister's *Historie of Man,* though there it is conveyed in what Sidney might have regarded as a more "wordish description," which does not so vividly impress, pierce, or possess the sight of the soul: "the brayne . . . is the author of sense and mouyng, without which the body were no otherwise but a stocke, or dead Image, altogether destitute of sense and motion."[86]

In this chapter's introductory paragraphs, which serve as preliminary to the whole of part 3, the problem of determining the conceptual character of the anatomy in *The Purple Island* was divided into theoretical and practical aspects. On the strength of conclusions reached in earlier chapters of this study, it has been possible to argue theoretically that because the unfolding of the significant connotations of the creative metaphor of the Isle of Man is limited only by the context of the whole work and the "conceit" of the reader, the anatomy need not be retarded by its involvement in allegory, and the proper meanings of recent anatomy could be among the significant connotations of the Isle of Man. The conditions permitting these possibilities include the special construction of a set of associated implications of the Isle consisting of many implications that are shaped by the requirement that in the poetic stanzas they accommodate the detailed structure and function of anatomy in terms of the descriptive criteria of the anatomical *accessus.* Hypothetically, it follows from the consistency of representation thereby achieved through Cantos 2 to 5, that anatomy in the poetic stanzas is identifiable, and that it is identifiable not because it is conceptually independent of analogy or allegory for what might be regarded as its essential character, but rather because the figuration of anatomy is consistent with its essential conceptual character.

In practice, this chapter has found that the figuration of anatomy in the poetic stanzas of *The Purple Island* does indeed demonstrate consistency of representation through the application of the *occasiones* of the *accessus;* though, as was anticipated, the stanzas emphasize functions as a way of ensuring that the reader's exploration of the microcosmic world should be characterized by a teleological understanding. The stanzaic maintenance of the descriptive criteria of number, form, location, and posture, despite the allegorical distortion of size, allows us to be sure of the safety of our identification of a "knot" in the brain as the *plexus chorioideus ventriculi lateralis* (choroid plexus of the lateral ventricles), and additional evidence has been found in the representation of this anatomical member for the view that allegorical personification in the poetic world of the geopolitical realm is regulated in relation to observable anatomical postures and physiological operations. Analyses such as this confirm that the stanzaic deployment of the anatomical *accessus* demonstrates that the figuration of anatomy is consistent with its essential character, and thus we must continue to turn to the *accessus* in the following chapter when ascertaining the conceptual character of the anatomy in *The Purple Island,* its relative conventionality or modernity, and its likely provenance. Some of the *Purple Island* anatomy studied in this chapter is identifiably post-Vesalian, and its integration in the structures and processes of the island realm or incipient nation upsets Nicolson's assumption that the involvement of anatomy and physiology in analogy determines their conservative character. The integration of this anatomy also suggests that what particularly emerges for Fletcher's poem is not so much the difficulties for successfully negotiating a theory of correspondence in the early seventeenth century, as Healy has thought, but Fletcher's willingness to allegorize very recent anatomical discoveries without being detrimental to the anatomy, a factor that will increase in importance in the following chapter.

This chapter has found a probable though tentative provenance for the method by which Thirsil organizes and undertakes his anatomical exposition, and for a significant part of the anatomy which he demonstrates, either in the second folio edition of Vesalius's *Fabrica,* a copy of which was bequeathed by Thomas Lorkyn to the University Library, in 1591, or in Banister's synthesis of Vesalius and Colombo, *The Historie of Man,* possibly the copy that was apparently included in an anonymous inventory presented on 27 January 1588/89 for probate in the Vice-Chancellor's Court at Cambridge, or more probably the copy at King's where Fletcher graduated B.A. in 1604, M.A. in 1607–8, possibly afterwards B.D, and obtained a fellowship apparently in 1603.[87]

Accordingly this study has begun to be able to ascertain the extent to which the anatomy in *The Purple Island* is conceptualized in terms of its allegorical figuration, whether the anatomy is intellectually conservative or innovative for the period when it was composed, from about 1608, and whether or not this character is a result of or even relates to the character or degree of its figuration. Langdale's interpretation of a portion of the anatomical content of *The Purple Island* as Renaissance and Vesalian, rather than either medieval scholastic or Galenist (like the anatomy of Matteo Corti), has found support here, though his claims to have identified some as Harveian, and thus purportedly as "modern," or seventeenth-century, remain to be tested in the following chapter. This study has sought to measure the conservatism or innovativeness of the anatomical content of Fletcher's poem using a method that recognizes that the effects of discoveries could not often be foreseen and were not always easily recuperated, so the descriptive vocabulary and the figurative conceptualization of anatomical parts had to change in accordance with their newly understood functions, shapes, positions, and postures. In this way our analysis of that part of the anatomy of the cerebrum that demonstrates a progressive, Vesalian character—the location and function of the parts involved in the creation of animal spirit—enables us to appreciate, firstly, Vesalius's innovation not only in placing the transformation of vital spirit into animal spirit in the *plexus chorioideus,* as Berengario did, but also in denying the existence of the traditional *rete mirabile* and putting its name into disuse, and secondly, the fact that in Fletcher's account, which is anatomically consistent with those of Berengario, Vesalius, and Banister, the stanzaic figurality ensures that the brain's function, to provide the body with voluntary movement and sense perception, and hence with life, is among the metaphorical connotations the reader is invited to unfold. Frequently we have found the seeds of the *Purple Island* figurality in the anatomical descriptions of Vesalius and Banister, seeds which are elaborated, extended, and politicized, in the allegorical context of the figurative design of Fletcher's work.

The discovery in this chapter of the precise delineations of the order or method adopted by Thirsil to demonstrate human anatomy in Cantos 2 to 5 of *The Purple Island,* provides further evidence of the role of the rhetorical *energia* of mimesis in the work's exchanges with textbook anatomy, evidence that combines with that of the work's mimesis of the formal layout of illustrations with accompanying text in anatomy textbooks, and with the work's ingenious rhetorical reconstitution of the body-as-referent as the signified of the poetry's anatomical speaking picture. The *Purple Island* method of anatomy corresponds partly to that known as the method of Nature or Creation, but diverges like that of the *Fabrica* from the method of Nature, in order to combine it with the method of discipline or necessity. This combined method could not have been drawn from either Thomas Vicary's work, or from the editions of Geminus's *Compendiosa delineatio* of 1545 or 1553, or indeed the reissue of 1559, making these works in my view unlikely though not entirely improbable provenances for the *Purple Island* anatomy. The rhetorically and ingeniously acquired corporeality of the speaking picture of *The Purple Island,* which is clarified and stabilized by the marginalia, a formal disposition that may also mimic formally if not functionally the division of the page of Banister's *Historie* into a main text and marginal notes, is a suprarealist mimesis that is apparently enhanced by the narrative mimesis of theoretical and practical anatomical method, the entire Vesalian sequence of which is designed to replicate the way Nature or God fashions human bodies. Thereby, perhaps, Fletcher accepted Banister's invitation to the implied reader of his *Historie* to lay the stones of his gathering to the building of a cleverly wrought tower of health, and in designing his work thus Fletcher sought to preserve the spiritual and physical health of his Protestant nation and its inhabitants, or at least that of a few friends, and sought by grace to restore and perfect their human nature, and to contribute to their salvation.

The figurative design and religious allegory of *The Purple Island* has been found in previous chapters to be congruous with the conceptual character and figurality of some sixteenth- and a little seventeenth-century anatomy, and although this may not in itself tell against Nicolson's judgment that combination with analogy in the *Purple Island* produces an anatomy that is in fact (if not in principle) intellectually conservative in the early seventeenth century, further evidence has accumulated in this chapter to suggest that neither the special sensory epistemology of Renaissance anatomy nor its imaginative rationality were in conflict with a soteriological perspective on Man. Vesalius himself, and the compilers of Vesalian compendia, as well as later sixteenth-century synthesizers like Banister, sought to take in hand the rational work of dissection in order piously to do "God's handy worke," and demonstrate the ingenuity of God who has created the artifice of human osteology and cranial architecture. Banister does not frequently refer to God, though he often explains function and structure by way of reference to Nature: this is explained at the very outset when he refers to "the mighty pleasure of God (whom we call *Nature*)."[88]

Although the analysis in previous chapters and in this leaves the question still partly unanswered as to whether the figurative design and religious allegory of *The Purple Island* would remain congruous, coherent, and consistent if the anatomy were predominantly more recent than the first half of the sixteenth century, this chapter has provided a clearer sense of the enduring legacy of Vesalius's *Fabrica* and *Epitome,* and the significance and utility of Vesalian anatomy as disseminated by copious references, partial translations and compendia, and copies of its illustrations. But because this chapter has been concerned almost exclusively with Vesalian anatomy, directly or indirectly received, the case is not yet closed on the validity of the argument that had the poem's method of inquiry been more anatomically specialized, the anatomy more faithful to *demonstratio,* and the physiology more modern, then the poem's allegory could not have been as systematic as it is. But we are beginning to develop a better understanding of how it is, in fact, the systematicity of the allegory in Fletcher's work that is responsible for the identifiableness of the anatomy, and this bears no direct relation either to *how* allegorical the poetry is, for provided it is anatomically systematic the anatomy will remain identifiable, or to the conservativeness or innovativeness of the anatomical content, because the allegory can be seen to adjust in relation to it, or to be determined by it, and hence leave the identity of the conceptual character of the anatomy fathomable.

8

The Heart, Veins, and Arteries

THE TASK THAT LAY BEFORE US AT THE OUTSET OF THE PREVIOUS CHAPTER WAS TO DIScern whether *The Purple Island* unfolds the entailments of the microcosm's analogies, and the significant connotations of the allegorical threshold figure, the Isle of Man, while ignoring contemporaneous anatomy (if Fletcher's choice of anatomical sources was conservative), or to the detriment of the anatomy of its sources (if they were more recent). This task has been only partly accomplished, and therefore must continue to be undertaken in this chapter, to which the previous chapter's introductory paragraphs will serve as preliminary. The evidence gathered in the previous chapter regarding the figuration of Vesalian anatomy in Fletcher's poem, that it is consistent with the essential character of the anatomy, may suggest that the task set out in this fashion risks unnecessarily conceptualizing the problem of determining the conceptual character of anatomy in the intellectual mold of Nicolson and others who assume that, in the late sixteenth or early seventeenth century, some anatomical inquiries and representations were devoid, or almost devoid, of figurality, and that had the anatomy in *The Purple Island* been more specialized, more reliant upon observation or *demonstratio,* and the physiology more mechanistic and modern, then the poem's allegory could not have been as systematic as it is. Ingrained in and structural to a yoke of assumptions that we are at this stage beginning to throw off only with difficulty is the view that analogy is necessarily a conservative mode of apprehension in late sixteenth- and early seventeenth-century natural philosophy.

In the previous chapter I began to demonstrate that these assumptions are misleading, but I chose to do so by allowing that they may be reliable, even perhaps by seeming to assume that they are so, the more forcefully thereby to demonstrate how erroneous they are. This is why the questions which this part of the study has set out to answer are framed in the way they are, as though allegory and observational anatomy may be inherently (and in the early seventeenth century were actually) incompatible, even while the framing of the questions allows also for the possibility that the anatomy of even *The Purple Island*'s most recent sources is in fact conceptually metaphorical in the way that early sixteenth-century and Vesalian anatomy has been found to be. Were this possibility in fact to be the case, our task in this chapter would be to develop an argument capable of discerning how *The Purple Island* figurative anatomy compares with that of Fletcher's most recent anatomical sources and with contemporaneous anatomical works, and how the *difference* in the *Purple Island* figurality modifies the figurative conceptual character of the anatomy of its sources.

Although still highly regarded by physicians, surgeons, and anatomists, Vesalius's anatomical work was largely done over half a century before Fletcher came to write the anatomy of the Isle of Man. Moreover, the analogy, conceptual metaphoricity, teleology, and philosophical and theological coherence of the representation of the human body in Vesalian anatomy has been established in previous chapters, and presents no great challenge to the religious purposes for which Fletcher divided the body in *The Purple Island.* The anatomy textbooks of the sixteenth century, and even those few from the seventeenth century that we have considered, are characterized by the metaphorical structuring of human anatomy by the more clearly delineated concepts of vegetation and "building." The direction of the structuring and the metaphorical character to which it gives rise are rationalized by the coherent and systematic extension of metaphor from structure to function and thence to the purpose of function, its reason for being or final cause, such as growth, reproduction, or the instrumental use of the body by the rational soul. Moreover, defining the coherence that the structural metaphors of anatomy have are three shared sets of entailments, which have been shown to be significant from a theological perspective. As chapter 3 has made clear, these are the human body's creation, its mortality, and its salvation. Fletcher's allegorical design involves identifying these closely with the themes of the main story line of *The Purple Island*'s narrative of providential history, thereby unifying the extreme ends of the chain of being, the cosmos, and the stones or "bones" of the earth, uniting all in Man the microcosm or microchristus. In significant part the figurative, allegorical design highlights correspondences already assumed to exist, many in Renaissance anatomy, and Langdale's claim that the anatomical character of the heart and vascular system in *The Purple Island* is influenced by Harvey's theory of systemic circulation, if it is justifiable, seems to imply that Fletcher met a much greater challenge to his soteriological purposes than the descriptive anatomy of Vesalius. But Langdale is of no help in establishing to what extent, if at all, Fletcher needed to break with analogical tradition and allegorical structure in order to accommodate Harveian physiology to the superimposed analogous systems of his poem, because Langdale misguidedly insists on the superficiality of the allegory throughout.

The previous chapter focused on the Vesalian anatomy of the head, brain, and *rete mirabile,* because the brain and the heart, more deeply and thoroughly than any other member or part, participate in all the allegory's assemblage of superimposed analogous systems, even the analogy between microchristus and Christ in heaven, whereby humanity participates in divine providence and redemption. Accordingly, and because the Harveian physiology that Langdale claims influenced Fletcher is cardiovascular, it is on the cardiovascular anatomy in *The Purple Island* that the analysis will focus in this chapter, determining its provenance and conceptual character, insofar as has been found possible.

The heart or "*Kerdia*['s]" dispensation of innately abiding "heat and life" to "all the Isle," its central location in the "middle coast" or thoracic cavity, and its command over this region, are all represented as corresponding "so nearly neare" to the "ends, place, office, state" of the sun in the macrocosm (*P.I.* 4.14.4, 4.25.2; 4.4.2, 1; 4.3.4), and to those of the "Prince" of the geopolitical realm (*P.I.* 1.45). Furthermore, the heart's form and posture are said to derive from the office of rulership it shares with Christ, the Prince and the sun: the vena cava and two coronary arteries are said to "Infold" the base or root of the heart "like a crown," "for that great *All* / This his works glory made pyramicall; / Then crown'd with triple wreath, & cloath'd in scarlet pall" (*P.I.* 4.17.5, 4, 5–7).[1] The heart's "motion fixt, not roving" is provided as demonstrative of the correctness of the opinion that the heart is "the seat of sure and active loving," in which it bears an idealized resemblance to the "Sunne," the "Sonne" of God, and the Prince, in each of their respective spheres of influence—the Cosmic, the divine, and the terrestrial and sociopolitical (*P.I.* 4.16.5, 7).

On account of figural complexities like these, involving a base of conventional analogy combined with the working out of the significant connotations of a metaphor and an allegorical personification, *The Purple Island* has rightfully been called "one of the last examples of 'sacred anatomy'—a system which was to collapse before the Cartesian construction of the mechanical body," a version of the model upon which some subsequent anatomy was to be based.[2] If Nicolson were correct that notwithstanding the original character of the anatomy upon which Fletcher drew for his poetic allegory, the principal and subsidiary subjects of the Isle of Man metaphor are substantially modified in conception by their involvement in analogy, this degree of integration of anatomical description in allegory ought to ensure that the meanings of both the brain and the heart are intellectually entirely enclosed within the superimposed circles of correspondence; and yet the anatomy of the brain is in significant part Vesalian, and although Vesalian anatomy may present no great obstacles to figuration for Fletcher the *concettist* and allegorist, it nevertheless remained influential in the universities and medical profession in the early seventeenth century. But in apparent contradiction of sacred anatomy, the heart's description incorporates physiology that has been identified by Langdale as Harveian, and as mechanistic.[3]

Our task in this chapter includes making a judgment as to whether Langdale's identification is correct. It has already been established that contrary to Nicolson's apparent assumption, the proper meanings of recent anatomy could *in principle* be among the significant connotations of the Isle of Man, and that the conditions permitting both this and the identification of this anatomy include the special literary construction of a set of associated implications of the Isle, many of which are shaped by the requirement that in the poetic stanzas they accommodate the structure and function of anatomical parts by means of the *occasiones* of the anatomical *accessus,* or criteria of methodical observables. The deployment of the *accessus* in the poetic stanzas demonstrates that the figuration of anatomy is consistent with its essential character, and so it is effectively through the figuration of anatomy in *The Purple Island* that the anatomy's character and provenance may begin to be determined. Were it in fact the case that cardiovascular anatomy in *The Purple Island* is Harveian, or has some other but similarly recent provenance, our task would be to develop an argument capable of discerning how the *Purple Island* figurative anatomy compares with that of Fletcher's Harveian or most recent anatomical sources, and with contemporaneous anatomical works, and how the *difference* in the *Purple Island* figurality modifies the conceptual character of the anatomy of its sources, detrimentally or otherwise, whether or not that character is originally figurative.

THE PROVENANCE OF CARDIOVASCULAR ANATOMY, THE MOVEMENT OF BLOOD, AND THE CONCOCTION OF VITAL SPIRIT

Langdale's case for the Harveian character of the heart rests on *The Purple Island*'s descriptions of the operation of the cardiac valves, the movement of the blood through the heart and the rest of the body, and the marginal statement: "The flesh of the heart is proper and peculiar to it self, not like other muscles" (*P.I.* 4.17 margin. n).[4] Taken alone, the marginal statement claiming that the heart is a muscle is insufficient to indicate a Harveian influence on the *Purple Island* cardiac anatomy and physiology. "Some anatomists, following Hippocrates, believed the heart to be a very strong muscle"; and the fact that Galen and Galenists like Matteo Corti, who ascribed to the heart a "pulsific quality," felt compelled to deny the Hippocratic position implies that this notion was discussed among Renaissance and early modern anatomists.[5] It may accrue significance, however, from the movement of the blood through the heart, veins, and arteries, and from the anatomical and physiological character of the valves, if these are found to be influenced by or consistent with Harveian or other late sixteenth- or early seventeenth-century anatomy.

The description of the veins and arteries, and of the movement of blood through them, occurs in Canto 2, and is thereby separated from the description of the structure and function of the heart, and the movement of blood through its ventricles, auricles, and adjoining vessels, in Canto 4. This separation adheres to the Vesalian method or order of dissection, and may, in itself, suggest a correspondingly pre-Harveian physiology, because in Harvey's published works, his way of approaching the body is like that of Fabrici, whose Aristotelianism Harvey imbibed at Padua. The approach centers on a physiological operation, such as the movement of the blood, treating the organs and their systems of vessels together according to their involvement in this process.[6] According to Cunningham, "[t]here is no room in this inquiry for the traditional 'three venter' division of the body, nor for the skeleton-muscles-arteries-veins-nerves sequence of Vesalius."[7] Besides the method, there are other Vesalian aspects of the anatomy and physiology of the veins and arteries in Canto 2 of *The Purple Island.* Although it has already been established that at least some of the anatomy in *The Purple Island* is Vesalian in character, these cardiovascular aspects should be attended to, because of the ambiguities, differences, and possible contradictions that may consequently arise between the character of the cardiovascular anatomy and physiology in this canto and that in Canto 4.

In the marginalia of Canto 2, the body is said to be "as it were watered with great plenty of rivers, veins, arteries . . ."; and in the stanza, these blood flows are a "thousand brooks" which, "Circling about, and wat'ring all the plain, / Emptie themselves into th'all-drinking main; / And creeping forward slide, but never turn again" (*P.I.* 2.9.2, 5–7 & margin. f).[8] Besides providing figuratively in allegorical form the *telos* or purpose of the vascular system, to provide nourishment and life to the parts of the body, as was remarked earlier in this study, these lines harbor the tension between the proposed models of physiology to which they refer, a tension that resides in the imagery and not only the technical aspect of the description.

To explain the movement of the blood, in Galenic theory, various analogies were used by anatomists. "It was likened to the movement of the tides which perpetually ebbed and flowed. . . . The tissues of the body were thought to use up the natural and vital spirits inherent in the venous and arterial blood as the fields used up supplies of rain."[9] This might account for Fletcher's imagery of the streams watering and emptying "themselves into th'all-drinking main" (from at least as early as the sixteenth century, "main" signifies elliptically "*main land,* Mainland," as well as "Main sea: The high sea, the open ocean");[10] but it fails to account for Fletcher's emphasis upon the one-directional flow of the blood. Most commonly in Galenic accounts, however, "the analogy applied to the veins and arteries was that of rivers": "The fact that in rivers the water does not ebb and flow did not prevent this analogy from becoming part of the common-place notions of the sixteenth and seventeenth centuries."[11] But the writers who popularized this analogy did not do as Fletcher does, in emphasizing the one-directional movement, because this would have literalized the analogy and undermined its application to the body as they understood it. Thus Canto 2, stanza 9 epitomizes the tensions of the poem's account of the heart and blood, which have given rise to divergent and contradictory readings, the phrases "Circling about" and "creeping forward slide, but never turn again" suggesting to some the influence of Harvey's theory of systemic circulation.

Entirely consistent with Galenic physiology is the differentiation in Canto 2 between the veins and arteries,[12] but then no Renaissance anatomist disagreed with Galen in the matter of their structural differences. The statement that they are "diff'ring streams" (*P.I.* 2.10.1), "Neither in nature nor in shape agreeing" (*P.I.* 2.10.2), is expanded to encompass their differences in color (*P.I.* 2.10.5, 6, 7 & margin. g; 2.11.7 & margin. h), in temperature (*P.I.* 2.10.6; 2.11.5), and in number of containing tunicles (*P.I.* 2.10.5 & margin. g; 2.11.6 & margin. h). Furthermore, the veins are said to be "appointed to contein, concoct and distribute the bloud" (*P.I.* 2.10 margin. g); whereas the arteries are "formed for conveyance of that more spritely bloud, which is elaborate in the heart" (*P.I.* 2.11 margin. h). Physiologically they also differ. The veins provide for the body "fruitfulnesse" or "nourishment," and the arteries give it "being" or "life" (*P.I.* 2.10.4; 2.13.2, 3). The physiological origin of blood in the liver is Galenic and pre-Harveian, despite the potentially misleading suggestion that the "streams" of the veins and arteries emerge "from fountains different" (*P.I.* 2.10.1): this derivation is explained because the blood in the arteries, "though from the same spring [the liver] first it rise, / Yet passing through another greater fountain [the heart], / Doth lose his former name and qualities" (*P.I.* 2.11.1–3).[13] Also in accordance with Galen and Galenists like Corti, as well as the Aristotelian Fabrici, but out of step with Aristotle, Vesalius, and Harvey, the liver, in *The Purple Island,* is embryologically the origin of the veins, as well as physiologically the origin of the blood (*P.I.* 3.7; 3.11 & margin.).[14] And the embryological priority of the liver over the heart (in Canto 3) is also from Galen "the physician," and defined in opposition to the opinion of the "Peripateticks," or followers of Aristotle, an opinion that was observationally supported by Harvey, as we shall see (*P.I.* 3.14 margin.).[15]

Fletcher's emphasis upon the direction of the flow of blood in the vascular system, and his suggestion that this flow is circular, are the only non-Galenic features of the physiology of the blood and containing vessels in Canto 2. But they should not on that account be ignored as anomalous; nor should they be explained prematurely as effects of the rivers of the island becoming uppermost in the poem's signification. The previous chapter has established that when the poem details early to mid-sixteenth-century anatomy and physiology, it is precise in number, location, posture, and other anatomical descriptive criteria, applied as they are to both the body and its geocosmic analogue or subsidiary subject. If the poem were in the case of vascular physiology to abandon such precision, this would threaten the carefully constructed superimposition of allegorical signification (which has thus far been found to falter on occasion, as has been said in chapter 4, because of the ideological character of the allegorical artifice, not because of its intellectual, anatomical character). But it is of course part of our task in this chapter to discern whether what applies to *The Purple Island*'s use of earlier anatomy applies also to its use of more recent, perhaps even seventeenth-century anatomy, and so the verdict on the effect of the conceptual character of recent anatomy on the success of its allegorical integration remains open for now.

The lack of critical attention paid to the analogy of rivers in early modern anatomy is partly owing to an ambiguity in the Galenic system about blood flow, especially in the veins. Galen had argued that "each part attracted and retained only enough blood for its immediate requirements and fresh supplies of this sufficiency were continually brought to it from the liver"; but "[o]ther writers seem to have imagined this as a continuous movement of ebb and flow in the veins, the blood returning to the liver for fresh supplies of natural spirit as this was used up by the parts."[16] Even if the venous blood could be construed as traveling "where and when it was required," and not back again, this movement was not thought to be applicable to the arteries, in which the blood "ebbed and flowed as the supplies of vital spirit which it conveyed to the parts were endlessly used up by these parts that they might continue to live."[17] Fabrici exemplifies the tendency among some early modern anatomists to postulate a difference between the movement of blood in the veins and that in the arteries. He believed, with Galen, "that the use of the veins was that of carrying crude blood, blood not vivified by the vital spirits, as nourishment from the [liver and the] heart to the tissues."[18] Fabrici's argument for the use of the valves in the veins, which he describes carefully in *De venarum ostiolis* (Concerning the little doors of the veins),[19] is that the valves "are formed by nature in order that they may to a certain extent delay the blood and so prevent the whole of it flowing at once like a flood" to the body's extremities, "and becoming collected there."[20] Such an explanation presumes at least a predominance of flow *from* the liver and heart *to* the rest of the body, and Fabrici's physiological explanation of the absence of valves in the arteries is to say that they "had no need of valves . . . to delay the blood" because "an ebb and flow goes on continually within them."[21] For Fletcher, however, there is no such difference between the movement of blood in the veins and arteries, for in both it is unequivocally one-directional. In all likelihood Fletcher does not derive this conceptual character of the blood from a Galenic account. Since the movement of blood in the arteries and veins is systemically conditional upon the understanding of the structure and function of the heart, whether in the Galenic or the later, Harveian model, we shall turn to the heart with the aim that its character in *The Purple Island* may shed light upon the character of the movement of blood therein.

It was mentioned in the previous chapter that when introducing the seventh book of his *Historie of Man,* John Banister wrote with caution and careful consideration of the elusiveness of explanation, interpretation, or accurate description with which to represent the formidably complex structures of the human heart and its even less fathomable operations. Banister's appreciation of the difficulty with which he was beset in that chapter arises from the context in which he compares the writings on the complex structure and elusive function of the heart by his forebears in medical and anatomical writing. The passage is as follows: "I freely protest, that hetherto my penne hath not wauered in the doubtfull balance of ambiguitie, but passed the playne path of truth, guided by the safe steppes of my forerunners: till now at length, beyng come to a place somewhat ascendyng (as they that will attayne to the scope of their desire, must not refuse the hill with the dale) I finde the steppes of some willingly yeldyng to so tedious a iourney: others, with senses unappalled, to have plowed a path directly ouer the mountaine: here I see one willingly walking a contrary way: and there an other, praysing their endeuours, but professyng him selfe onely to haue sight in the nearest way. So that here, no maruell, though I should stand amased, and as one plunged in the depth of contrary streames."[22]

Banister goes on to describe the opinion of "*Galen, Vesalius* & diuers others" on how blood and air combine to make vital spirit in the left ventricle of the heart. Vesalius, indeed, had provided a description, albeit an entirely conventional, Galenic one, of the processes that occur in the heart, with which that in *The Purple Island* may be compared: "The heart attracts this air and draws a large supply of blood from the right ventricle into the left ventricle. From the steamy vapor of that blood and from that air, by the inborn virtue of its own substance, the heart creates the spirit which the blood with a rushing flow distributes, thus accompanied and nourished, to the entire body through the great artery; the heart tempers the native heat of each part in the same way that the respiration restores the tinder of the innate heat to the heart."[23] In Galenic physiology, to which Vesalius here adheres, venous blood is brought by the (inferior) vena cava, from the liver via the hepatic veins to the right auricle, also known as sinus, atrium, or "ear" (*P.I.* 4.20 margin.), of the heart—a passage which *The Purple Island* describes as "by *Hepars* streams" (*P.I.* 4.18.6)—and thence to the right ventricle of the heart.[24] The function of this blood was not to nourish the heart, for "[b]efore it reached the heart, the vena cava gave off a branch, the coronary vein, which supplied the heart . . . with natural spirit and nutriment."[25] *The Purple Island* makes no mention of the coronary vein, however, and the description of the blood sent from the liver to the right auricle as Hepar's "pension" and "tribute" might suggest a nourishing function, though this is not explicit (*P.I.* 4.18.5; 4.20.5). In Galenic physiology, the blood entering the right ventricle is "concocted";[26] which action is described in *The Purple Island* variously as "concocts," "laboured," and as "distill'd with art, beyond or words or thought"; the latter of which, in terms of the plurisignification of the anatomical character, is perhaps the most interesting description and will bear closer examination in due course (*P.I.* 4.18 margin.; 4.19.2 & margin.; 4.18.7). By this action the "grosser waves" or "thicker" part (*P.I.* 4.19.1 & margin.) of the blood were "expelled through the pulmonary artery," "a doubtful chanel" (*P.I.* 4.19.3), "into the lungs ["*Pneumon*" (*P.I.* 4.19.3)] for their nourishment," while the thin remainder "sweats through" (*P.I.* 4.19 margin.) "the porosities in the interventricular septum ["a fleshie partition" (*P.I.* 4.19 margin.)] into the left ventricle."[27] Banister describes this septum with vegetative figurality as "the hedge betwene the ii. ventricles."[28]

The functions of the left side of the heart concern the production of vital spirit. For Galen, the "basic principle of life, the spirit or pneuma," existed in the air. Inspired into the lungs, it "passed through the ends of the bronchioles" and entered "the surcles and branches of the veniall Arterie,"[29] that is, the pulmonary vein, and, from this vein, entered the left auricle, or "ear," of the heart, and thereafter the left ventricle.[30] In the ventricle, or (in *The Purple Island*) the auricle, the air is "heated and refined" into "pneuma," "subtile wrought, & thinne, for future workmen fitteth" (*P.I.* 4.20.7).[31] Also, in the ventricle, the *pneuma* or "thinne air" (*P.I.* 4.21.2) "met the venous blood," the "right sides rent" (*P.I.* 4.21.2), which "had come through the septum, and by means of the innate heat of the heart they were concocted," or "Compound together" (*P.I.* 4.21.3) "into vital spirit [*pneuma zoticon*]; that is to say, the venous blood was concocted into arterial blood in which the vital spirit inhered."[32] In *The Purple Island,* the left ventricle, "concocting," "makes a strange confection": a term not used by either Berengario or Vesalius or Banister, used here in the stanza but not the marginal (t), and to which greater significance is attributed hitherto in the second section of this chapter. Then, almost as in Berengario's account, Thirsil describes how, "in one vessel [i.e. the aorta or "great arterie"] both [*pneuma* and thin laboured blood] together meynt, / [the ventricle] Stills [the pneuma and the blood] with never-quenched firing" (*P.I.* 4.21.4–5 & margin. t). Thence the arteries carry "in small streams (through all the Island wiring)" (*P.I.* 4.21.6) "vital life-giving blood," "both heat and life inspiring" (*P.I.* 4.21.7).[33] It will be seen in due course to be of significance that whereas Berengario, for example, speaks only of the blood being heated, a "distillation" occurs on both sides of the heart in *The Purple Island.*

Although items of anatomical vocabulary or nomenclature are patched into this summary account of the function of the right and left sides of the heart, as presented in Canto 4 of Fletcher's poem, it is possible to see how the poem's process of discovery of significant connotations of the Isle of Man extends the threshold figure systematically and clearly, giving coherent structure to (in this case) antecedently existing similarities. Although not all aspects of the cardiovascular anatomy described here have yet been assigned a provenance or even been found comparable to the accounts of Berengario, Vesalius, or Banister, much can be said to be conceptually consistent with these accounts. As was explained in chapter 2, the implications of the threshold figure are in part antecedent and extended, and even where the allegorical figurality is not derived from works of anatomy, it may in such cases be described as emerging from between previously unobserved similarities and assumed differences between the subjects of the metaphor. Many predications are merely suggested rather than stated, but it is not difficult from those that are stated to infer the proper anatomical terms for the parts involved in the movements and uses being described.

Until Canto 4, stanza 22, the vascular and cardiac physiology of *The Purple Island* is thoroughly Galenic, with the following exceptions: the one-directional, apparently circular, flow in the vascular system; the poem's use of the alchemical or chemical term "distillation" to describe the concoction of the venous blood in the right ventricle, and that of the resultant labored blood and *pneuma* in the left ventricle; and the use of the term "confection" to refer to this latter concoction in the left ventricle. These exceptions accumulate relevance as the text proceeds after stanza 21 of the fourth canto.

Stanza 22 describes the right atrioventricular orifice (between the right auricle and ventricle), which "lies anterior to the inferior vena caval opening and is guarded by the tricuspid valve."[34] In the hitherto Galenic physiology of *The Purple Island,* it is necessary, in order that the inferior vena cava "sheds" to the arms and head "his Nile-like inundation" (*P.I.* 3.11.6), that this blood passes through the right auricle or atrium of the heart into the superior vena cava. Traditionally, although this flow was "predominantly one-way," it was not exclusively so.[35] Blood divested of its nutriment passed back along its former path through the right auricle, along the inferior vena cava, where it would be met by backward flows from other veins such as the coronary vein which feeds the heart, and on to the liver where it would replenish its supply of nutriment and natural spirit. In addition, subscribers to this Galenic physiology traditionally described the inferior vena cava as carrying venous blood which contains nourishment and natural spirit *from the heart* to the more peripheral parts of the thorax, abdomen, and legs. So, for Fabrici, for example, the flow of blood to the parts was not directly from the liver, but to and "from the heart [along the inferior and superior venae cavae and other veins] to the periphery."[36] As one examines this traditional theory of blood "which flows back and forth many times in large quantity," an inherent problem looms into focus.[37] This is the omission of any consideration of the competence of the tricuspid valve which guards the right atrioventricular orifice of the heart. There is nothing in the writings of these Galenist accounts to suggest that the ebb and flow in and out of the right side of the heart is confined to the auricle; and, besides, the design of the valve is such that it may not prevent blood in the auricle from entering into the right ventricle. But the ambivalence of these writings is such that they do not profess precisely how far or even often how the vena cava "carries . . . blood into the heart,"[38] or how this is affected by the operation of the tricuspid valve.

The Purple Island is by contrast admirably clear about this and about the function of the tricuspid valve. The "tribute liquour" in the "hollow vein," or vena cava, travels in one direction only, from "*Hepar*" to the right "ear," or "right hand street," of the heart and thence to the right ventricle (*P.I.* 4.20.5; 4.22.2 & margin. u; 4.20.4, 5). As has been mentioned, conduits, pipes, streams, and streets are the figurations that the poem usually and consistently applies to vessels of moving fluid. The return passage of this "tribute liquour" is moreover prevented by the operation of the tricuspid valve, which is situated between the right auricle and ventricle, and which is figured as a "band of *Tritons*" who "strongly stop the wave" of blood from "the hollow vein" "if once it back repace" (*P.I.* 4.22.4, 7 and margin. u). The "three little folding doores, with three forks" are described emphatically as "giving passage, but no return to the bloud" which flows into the right ventricle from the inferior vena cava (*P.I.* 4.22 margin. u).

There are few examples that better illustrate how the anatomy in the poetic stanzas is identifiable not because it is conceptually independent of analogy or allegory for what might be regarded as its essential character, which the marginal notes reveal by throwing off the cloak of allegory, but rather because the figuration of anatomy is consistent with its essential conceptual character. Although fanciful in a way that is consistent with the concept of literary ingenuity, the trident-bearing tritons in the main "stream" which runs "from the *Hepar*" to "*Kerdia*," at the point at which it enters "th" Isles Heart-city," are shaped by the requirement that they accommodate the detailed structure and function of anatomy in terms of the descriptive criteria of the anatomical *accessus* (*P.I.* 4.22.1,2; 4.15.2; 4.2.6). The figuration of the right auricle is as a one-way street and simultaneously as a flow of water, the backward flow of which is prevented by the tritons, each of whom with their trident or "three-forkt mace" operates one of the three little doors like valves (*P.I.* 4.22.5). This figuration demonstrates consistency through the application of the anatomical descriptive criteria of size, number, function, position, posture or connexion, and plausibly even form.

Langdale remarks, albeit slightly misleadingly, that Fletcher "gives more prominence to the valves than they received in the Galenist texts, and he [i.e., Fletcher] specifically mentions the tricuspid, the mitral [a term which was introduced or stabilized by Vesalius],[39] and the semilunar," between the ventricles of the heart and the auricles and adjoining vessels.[40] The marginal descriptions of the tricuspid valve, the semilunar valve situated between the right ventricle and the pulmonary artery or "arterie-vein," the semilunar valve between the left ventricle and the aorta or "great arterie," as composed of "three little folding doors" (*P.I.* 4.22 margin. u; 4.23 margin. *), and the mitral valve between the left auricle and ventricle, "which hath two folds" (*P.I.* 4.23 margin. x), are all precise in terms of size or quantity, number, form or shape, position, anatomical relations or posture, and uses or functions; in short, six out of the nine anatomical *occasiones*. Such precision might suggest a provenance in Fabrici's observationally accurate descriptions of the valves in the veins in *De venarum ostiolis.* Although Fabrici "clearly recognized that the valves [in the veins] offered opposition to the flow of blood from the heart toward the periphery," and supported this with experiment, his Galenic interpretation of their use (described above), and of the blood vessels generally, might preclude his unmediated influence on Fletcher. More persuasive, however, is the fact that Fletcher's anatomical descriptions do not include the valves of the veins, but only the cardiac valves.[41] A more probable provenance is again Banister's *Historie of Man,* which is precise in attributing the number of folds of which each valve consists, though he consistently uses the proper anatomical term "membranes." Banister describes their structure as comprising also "certaine filmentes or threedes," which may correspond to the forks or tridents of Fletcher's tritons. *The Purple Island*'s marginal description of the shape of the folds or membranes of the "three-folding doore" of the semilunar valve of the pulmonary artery as "made like half-circles" (*P.I.* 4.23 margin. *) might easily have been derived from Banister's description of them as "like iii. of those letters which the Latins call C."

Galenism seems to recede even further from Fletcher's physiology when the actions of the other cardiac valves are considered together with that of the tricuspid. The passage from the lungs to the heart in the pulmonary vein or "Veiny arterie" is, in *The Purple Island,* also a one-way passage (*P.I.* 4.23 margin. *). This passage of the lungs" "ayer-tribute" to the left auricle of the heart is secured by "gates and barres"—the mitral valve with "two folds" "which stop all backward going" (*P.I.* 4.23.6, 7 & margin. x). This is in unambiguous contradiction to Galenic physiology, wherein: "Some part of the arterial blood [in the left ventricle] . . . did not go out through the aorta but back through the pulmonary vein into the lungs to supply them with vital spirit and arterial blood so that they could live. At the same time, the fuliginous vapours [of which *The Purple Island* makes no mention] which were the excrement of the concoction in the left ventricle . . . were driven back through the same pulmonary vein into the lungs and so breathed out of the body."[42] In *The Purple Island*'s physiology, two of the "four great vessels leaving the heart," which in Galenist physiology "were likened again and again to the four great rivers of Paradise," permit no flow *from* the heart.[43] Furthermore, the flow of blood from the heart to the lungs in the artery that was thought to be a vein is one-way: the "guards" of the "arterie-vein" or pulmonary artery "all regresse denie" to their sanguineous "tribute" (*P.I.* 4.23.3, 4, 2 & margin. *). And the "three fair Nymphs" of the semilunar valve of the aorta likewise "stop back-waies" flow of the blood vivified with "vitall spirits" (*P.I.* 4.24.4 & margin. y). The flow of blood (or air) through the valves at the venous and arterial orifices of the heart in Fletcher's poem is consistently one-directional, into the heart through the tricuspid and mitral valves, and out of the heart through the semilunar valves.

The operation of all of these valves is consistent with the one-directional flow in the veins and arteries, as represented in Canto 2. The operation of the tricuspid valve in ensuring that the liver's "tribute" enters the heart but does not leave by the vena cava is consistent with the flow in the veins from the liver to the extremities, in Canto 2. The artery-vein's derivation from the right ventricle of the heart, whence it carries blood, might appear to contradict the rule of direction of venous blood to the parts from the liver, but in the Galenic model, the right ventricle, and therefore the pulmonary artery, or artery-vein, was "one of the main branches of the venous system," transporting blood from the liver to the lungs, through the heart.[44] Banister states: "The arteriall veyne neither springeth from the hart, but from the liuer."[45] For this reason, and for carrying nourishment and impurities from venous blood in the heart to the lungs, the pulmonary artery was regarded as a vein, the artery-vein or "arteriall veyne,"[46] and the operation of its semilunar valve is entirely consistent, in *The Purple Island,* with the tricuspid valve of the right atrioventricular orifice (*P.I.* 4.23 margin. *).

Competence and consistency extend to the representation of the other cardiac valves. The operation of the aortic or semilunar valve is unambiguously consistent with the direction of *pneuma zoticon* from the heart to the periphery. And, finally, the operation of the mitral valve at the left atrioventricular orifice is perhaps unexpectedly consistent with the one-directional flow of blood in the arteries, despite allowing a flow only of air and that only toward the heart. The pulmonary vein, or "Veiny arterie" or "veniall arterie,"[47] was regarded as an artery for the same reason that the trachea was called the "rough artery"—for carrying air to be prepared into *pneuma*—and for its derivation from the left cavity of the heart from which the aorta (artery) arises. As Banister explains, "set to the left ventricle of the hart it is called an arterie, in that it serueth to spirites and arteriall bloud: but a veyne, because it hath the body or substaunce of a veyne."[48] As regards competence and consistency in the *Purple Island* account, the mitral valve's operation is consistent with a movement involving that of blood from the heart to the lungs in the "arterie-vein," and of air from the lungs to the heart in the "Veiny arterie."

The traditional theory of the ebb and flow of arterial blood required, on the part of its adherents, ignorance or disregard of Galen's observation that blood leaving the heart by the aorta and pulmonary artery cannot return along the same path because of the action of the cardiac valves.[49] By the time Harvey came to write, the action of these valves, in the orifices of the aorta and the pulmonary artery, "had been . . . recognized by Mondino, Leonardo, Berengar, Vesalius, and several later writers."[50] But none of those mentioned here had drawn from their observations hypotheses regarding the physiological consequences for the flow of blood in the aorta and its branches. Without attributing to Fletcher insight beyond that achieved by any of the abovementioned anatomists, it is implausible to conclude that, in ignoring the "ebb and flow" theory, Fletcher drew his physiology of the arteries entirely from the descriptions of the relevant valve by Galen or these other anatomists. It becomes necessary, therefore, to look elsewhere for the provenance of the combination of the competence of the cardiac valves and the movement of blood in the arteries in *The Purple Island.* But although Harvey was to draw physiological consequences from observations of the functions of the valves, it would be premature to conclude that the physiology of the heart, veins, and arteries, including the physiological structure of the valves, in *The Purple Island,* is Harveian, either in provenance or character. The effect of the operation of the semilunar valve of the pulmonary orifice on the movement of blood in the pulmonary artery had been described by the proponents of the pulmonary transit of the blood. Before Harvey, the lesser or pulmonary circulation of the blood from the right venticle of the heart to the lungs and back to the left auricle had been proposed by Michael Servetus, Realdo Colombo, and his pupil Juan Valverde da Hamusco, and described and opposed or advocated or questioned by Ambroise Paré, Guido Guidi, Carlo Ruini, Johann Guinther von Andernach, Felix Plater, Kaspar Bauhin, Giulio Cesare Aranzi, Andrea Cesalpino, Leonardo Botali, and John Banister (who follows Colombo).[51]

The observational spur to the formation of Colombo's theory and that of his predecessor Servetus was that, although, "nearly everyone thinks that there is a way open for the blood to pass [through the interventricular septum of the heart] from the right ventricle to the left," no pores can be observed to pass entirely through the septum.[52] *The Purple Island* does not conform to the rejection by Colombo, Paré, Guidi, Ruinus, Banister, and Harvey of the existence of these porosities in the septum.[53] This "porous" septum was the anatomical premise of one of the central tenets of Galenic cardiac and vascular physiology: that blood passes through minute foramina in the septum from the right ventricle to the left, or, according to Berengario, from the right sinus or auricle to the left.[54] "As far as is known, all anatomists writing before Vesalius, apart from the 'sedulous investigator' Nicholas Massa, concurred in the belief in the existence of the intraventricular pores, though many stressed the fact that they could not find them."[55] Keynes was not the first historian to suggest that "Galen saw the openings, particularly well developed in the heart of the ungulates, of the vessels supplying the substance of the septum, and interpreted these as pores through which the cavities of the two ventricles communicated."[56] Additionally, however, I think that the sponge-like appearance of the septum, caused by projecting ridges now known as trabeculae carneae, may have lent observational support to the belief that they offered a transventricular passage.[57] Vesalius, however, failed to thread bristles through these foramina. He was therefore skeptical about the septum's porosity and the transventricular movement of blood, because unlike the septum's sponge-like appearance, both are imperceptible to the senses. He did not, however, deny their existence; though his skepticism is bolder in the second folio edition of the *Fabrica;* but even so, therein he provided no alternative physiological explanation of the movement of blood from the right to the left ventricle, despite promising to do so.[58]

While admitting that a little blood may pass through the septum, Servetus did posit lesser or pulmonary circulation as the actual means whereby blood travels from the right ventricle to the left auricle and ventricle; but all but three copies of *Christianismi restitutio,* in which this theory appeared, were burned along with their author in 1553, though not for advocating pulmonary transit.[59] In *De re anatomica,* Colombo denied the transventricular passage of blood and the existence of the pores in the septum, and like Servetus, advanced a theory of the pulmonary transit of blood.[60] According to Whitteridge, Colombo's theory "does not seem to have had immediately great or widespread repercussions";[61] but a muted controversy ensued,[62] into which, according to Sawday, Davies of Hereford entered, by declaring that "bloud . . . / Like *sweate* . . . from the right [ventricle] doth proceede" to the left ventricle, through "a little *dore,*" which consists of "many a *pipe* so small . . . / That scarse *man's* eie can see how *each* exists."[63] Davies of Hereford enters into the debate by reiterating Galenic theory, which, by the end of the sixteenth century, had been rejected in favor of the pulmonary transit by many prominent anatomists,[64] and even by some surgeons like Banister, who also conducted anatomies; though pulmonary transit had not yet "won any kind of general acceptance," for "Galenic theory was too firmly ingrained."[65] Davies of Hereford may even conflate Galen's theory with that of Massa, who, in his *Introductorius anatomicae,* denied the existence of the pores in favor of "a passage which he had found at the base of the heart," which, Whitteridge speculates, was "presumably a patent foramen ovale," and through which Massa thought the blood could pass.[66] In his *Historie of Man,* Banister presents an attempted synthesis of Vesalius and Colombo, though on the heart and blood vessels he is usually in favor of pulmonary transit, with a "mixture" of blood and air entering the left ventricle via the pulmonary vein or "veniall arterie," whose utility "for the coolyng, and refrigeration of the hart" is nevertheless maintained, "this effect beyng wrought, by the bringyng of cold ayre unto it." After Colombo, he denies the porosity of the interventricular septum.[67]

Fletcher on this matter apparently follows Galen more faithfully than does either Banister or Davies of Hereford, figuring the foramina as "many secret gates (gates hardly spi'd)," through which "the thinner part" of the blood "sweats"(*P.I.* 4.19.6 and margin. q). Fletcher even informs his reader that "if [the septum] be well viewed, we shall see it full of many pores, or passages" (*P.I.* 4.19 margin. r). In this Fletcher may follow Kaspar Bauhin (1560–1624), who in his *De corporis humani fabrica: Libri IIII* of 1590 had "pointed out that there was no need to suppose the existence of pores in the interventricular septum of the heart," and proposed a theory of pulmonary transit for which he gives no authorities. As Whitteridge says, "it is open to conjecture whether he formulated it independently." But in the revised and enlarged version of all his previous works of anatomy, published as *Theatrum anatomicum,* in 1605, Bauhin claims: "These pores (*spiracula*) are very conspicuous in the heart of an ox after it has been boiled for a long time."[68]

It is clear from this that while the attention to "barely visible" detail in both *Microcosmos* and *The Purple Island* shades into belief in nonvisible things and the inference of the existence of things in human anatomy through the eye of reason, it might but need not also imply their sensitivity to contemporary anatomical controversy and their receptiveness to recent or contemporary ideas. Although the *Theatrum anatomicum* "soon acquired the reputation of being the best anatomical textbook available," it does not seem likely that Fletcher could have read a copy in any Cambridge library during the periods he was resident at King's, nor could he have read there a copy of Crooke's *Microcosomographia* which is a conflation of an English translation of *Theatrum anatomicum* with the textbook of Laurentius.[69] The insistence in *The Purple Island* upon ocular evidence for the intraventricular pores arises from the necessity of their existence for physiological coherence and rational consistency, given that blood in *The Purple Island* system does not reach the left side of the heart by the pulmonary vein which carries only air or *pneuma,* and *The Purple Island* proposes no substitute means, like Botali's intra-auricular *vena nutrix* or Massa's intraventricular passage, whereby blood (which is thought to be produced by the liver) could reach the left ventricle of the heart.[70]

Notwithstanding Fletcher's adherence to Galen, or perhaps more specifically Bauhin, on the matter of the interventricular septum, the physiology of the cardiac valves in *The Purple Island* is not traditionally Galenic. The first to write of the physiological function of all the cardiac valves as precluding a returning motion was the Alexandrian anatomist Erasistratus, to whom the tricuspid valve owes its name.[71] Histories of anatomy, including those of prominent and influential historians like Charles Singer, in which Vesalius and Harvey loom large, as they do in the present work, frequently place undue emphasis upon the continuity between Renaissance and modern conceptions of anatomy, so that, for example, the Alexandrian Erasistratus is said to have come "very near discovering the circulation of the blood."[72] Furthermore, Singer says: "He regarded the heart as the source of both arteries and veins, a matter in which he was not only ahead of his time, but was ahead of all opinion until Harvey."[73] It has been remarked hitherto in this study that such retrospective organizations of anatomical discoveries are frequently positivist in presuming or arguing for continuity composed of the scientific search for and accumulation of objectively verifiable true knowledge by "scientists" held back from discovering the truth merely by the ignorance of their age and the clutter of their nonscientific ideas; they are often philosophically objectivist in postulating the existence of ahistorical ontology about which a neutral and transparent nomenclature, shorn of figurality, can speak the truth; and they are not uncommonly idealist in seeing "science and its individual doctrines as having an in-built *internal development* which unfolds over the centuries, in accordance with the actions of the [Hegelian] 'Spirit of the Age.'"

According to Cunningham, "both the Idealists and the positivists thought that *progress* was of the essence of science, and that the historian of science should therefore seek out and show the progress of knowledge over the centuries."[74] But although Renaissance anatomists may have developed or even generated a putative ontology of the body, such that by dissection they might discover the truths of structure and function, the projects upon which Renaissance anatomists were engaged were not yet concerned with objectivist accounts of truth but with (equally dogmatic) rationalist conceptions of Nature and the soul inherited either from Aristotle or Plato and Galen, and with microcosm-macrocosm speculations about the fabric of the world, and other aspects of natural philosophy and theology, and it is these conceptions that still determined what anatomists sought to discover, and often how they interpreted their discoveries, constructing for them a context of justification comparable to the way in which Thirsil provides the context of the allegorical figurative design and narrative of divine history as a justification of the anatomical "discoveries" of Cantos 2 to 5 of *The Purple Island.* Even where the inspiration for Renaissance anatomy came from the comparatively empiricist human anatomy of the Alexandrians, it cannot be maintained that the empiricism was objectivist; rather it was mildly skeptical, for at least Herophilus had skeptical tendencies.[75] This is not to claim that where anatomists placed their allegiance, what they claimed their method to be and what they actually did need necessarily coincide.[76] But when the natural philosophy of Aristotle is used to justify a mode of procedure, the knowledge to which it gives rise, even if new, may yet often be validated according to the precepts and projects of Aristotelian philosophy, teleology, and even microcosmic theory. In other words, the "context of discovery" may be distinct from the "context of justification," and if the knowledge is often already interpreted ("justified") differently, there are consequently no "timeless standards by which to judge the validity of claims to scientific knowledge."[77]

Having said this, I ought to reiterate the point made earlier in this study that while its methodology rejects historical positivism, this study does not thereby deliberately neglect the material determinacy of history by "scientific" discoveries or *inventio.* For Renaissance and early modern anatomists, the context of justification may have been such that they envisaged the projects they undertook as garnering and accumulating knowledge in the tradition of Aristotle, Galen, Erasistratus, or Mondino, thereby sustaining the body's relationship to the macrocosm, the geocosm, or the soul, but the effects of their discoveries, or even their hypotheses, could not always be foreseen and were not always easily recuperated. This is not the same as saying that Renaissance and early modern anatomists adhered to the principle of "fallibilism," discussed in chapter 1, for they had not developed a method of testing hypotheses by trial and error, and were still largely concerned with discovery as a means of the justification of rationalist theses. Accordingly, this study seeks to identify and measure the conservativeness or innovativeness of the anatomical content of *The Purple Island* using a method that recognizes this fact of the material determinacy of discovery, and the fact that the figurative conceptualization of anatomical parts had to change over time in response to discoveries which affected conceptions of their functions and structures, and hence their identity and relations in the contexts of analogy and correspondence. The traditional justifications would be in this way often adapted to the context of discovery rather than simply abandoned by it, as we shall see in the cases of Andrea Cesalpino, and later, Harvey.

Fletcher may have been aware of Erasistratus's theory, for although it had been partially eclipsed by that of Galen, there were still those who adhered to aspects of the pre-Galenic theory.[78] But if the poet of *The Purple Island* was aware of this theory, he chose not to follow it over the embryological origin of the veins, which Fletcher maintains is the liver, not the heart, and he resisted a further aspect of it, which had been proven by Galen to be incorrect. Erasistratus had argued that while the veins contain blood for the nourishment of the body, the arteries contain only *pneuma* or spirit.[79] It is of Erasistratus that Singer in his positivist and idealist history of anatomy writes, "he was only hindered from reaching the idea of the circulation of the blood by [t]his pneumatic theory."[80] In contrast to this theory, the marginal note to stanza 11 of Canto 2 of *The Purple Island* clearly states: "An arterie is a vessel long, round, hollow, formed for conveyance of that more spritely bloud, which is elaborate in the heart."

So if it was not Erasistratus who influenced *The Purple Island*'s presentation of the anatomy of the cardiac valves and the movement of blood through the heart, perhaps it was Colombo, who followed not Erasistratus but Galen in holding that the arteries (except the "veiny artery") contain blood, but developed a theory of perpetual movement from the vena cava through the heart and lungs into the aorta, resembling that of Erasistratus, and, indeed, that of Fletcher, but for two very basic differences. Fletcher's concurrence with the advocates of pulmonary transit, including Banister, is on the operation of the cardiac valves and the direction of flow in the pulmonary vessels; it does not extend to the substance that flows in the "Veiny arterie" or pulmonary vein, for *The Purple Island*'s "ayer-tribute" moving through the "Veiny arterie" conforms to Galen (*P.I.* 4.23.6 & margin. x). Nor, consequently, does the poem agree with Colombo or Valverde about the location of the preparation or generation of vital spirit, which Colombo (and Valverde inconsistently) proposes is the lungs rather than the left ventricle of the heart.[81] Banister describes and seems to concur with Colombo's thesis that an office of the lungs "is (sayth he) the preparation, yea almost the generation of vitall spirites, which after are perfected in the hart."[82] This was in fact Colombo's major thesis, to which he simply gave weight by adopting the corollary that the blood he found in the pulmonary vein had come across the lungs from the pulmonary artery.[83]

Neither of these differences necessarily entails that Fletcher was not influenced by the theory of pulmonary transit, insofar as it involved an account of, or at the very least implied, the physiological competence of the cardiac valves. Guinther and Plater accepted pulmonary transit, and the competence of all but the mitral valve, and yet maintained the theory of the generation of vital spirits or *pneuma zoticon* in the heart. For advocates of pulmonary transit inclined to do this, a particular problem arose. This was to explain how the lungs might be furnished with this spirit and arterial blood. Ignoring the competence of the mitral valve, Bauhin, and others, "allowed a two-way passage of blood through the pulmonary vein: of natural blood inwards to be made vital, and of vital blood outwards for the use of the lungs."[84] *The Purple Island,* for which this problem also arises, fails to address it, and maintains the competence of the mitral valve. Advocates of pulmonary transit who also maintain that the pulmonary vein contains only air, such as Paré and Carlo Ruini, are obviously inconsistent in doing so.[85] In fact, Ruini contradicts himself over whether blood and air can be found in the pulmonary vein. Paré's physiology is internally contradictory, maintaining both the Galenic pulmonary physiology and the nonexistence of perforations in the septum. Together, these two ideas preclude the flow of blood from the right to the left ventricle of the heart.

But despite the anomalies and contradictions in these accounts, it does not necessarily follow, as Whitteridge suggests it does, that the competence of the cardiac valves precludes any compromise with Galen, and could not be accommodated to the Galenic physiology of the heart and lungs.[86] But for its failure to address the problem of how the lungs are supplied with vital spirit, *The Purple Island*'s physiology is an entirely consistent and noncontradictory, if incorrect, physiological model, and we may summarize it as a model whereby venous blood from the right ventricle of the heart moves along the pulmonary artery in one direction only, to nourish the lung; and whereby air travels in one direction only from the lungs to the left ventricle of the heart via the pulmonary vein; and whereby the existence of the intraventricular pores allows blood to pass in one direction only between the ventricles of the heart. The model is indebted to Galenic physiology of the heart and lungs, and to accounts of the competence of the cardiac valves written probably by the advocates of pulmonary transit rather than by Erasistratus or his Renaissance adherents. But because *The Purple Island* is emphatic about the competence of the valves, one suspects that provenance probably does not lie in those works by Guinther, Plater, and Bauhin that deny or ignore the competence of the mitral valve. Again a probable provenance is Banister's *Historie of Man,* which succinctly summarizes the competence of the cardiac valves, saying, "two of them are made to carie into the hart, whilest ye hart is dilated: but the other ii. to beare forth in the time of constriction." If Fletcher did take the physiological structure of the valves from Banister, however, he replaced the figurality in which Banister provides his explanation of their function—Banister describes them as "litle wickets."[87]

The question of who among the advocates of pulmonary transit may have directly contributed to the anatomy in *The Purple Island* is by no means easy to answer, but circumstantial biographical and bibliographical evidence can lend weight to a hypothesis. Among the medical books bequeathed to Cambridge University Library by Thomas Lorkyn, in 1591, was a copy of the 1562 edition of Colombo's *De re anatomica,* published at Paris. If Fletcher had access to the second folio edition of the *Fabrica* in Lorkyn's bequest, for which there is inconclusive circumstantial evidence discussed in the previous chapter, there is no reason to suppose he would not also have been able to read *De re anatomica.*[88] It remains to be determined whether there is any textual evidence to support a claim that he did.

Colombo's anatomy can be understood in relation to that of Vesalius. In his myology in particular, Vesalius correlated structure with function and understood the function of each particular muscle in terms of the way it operates in the functional context of the other parts to which it is connected. As this study has remarked, Vesalius's myological plates present anatomical structure in action. But in representing the "building" of anatomy (principally the skeleton and musculature) as the instrument and edifice of the soul for which it is built and which gives it life, Vesalius's anatomy was largely concerned with structure. French argues that "Vesalius' woodcuts could not present knowledge of a proper philosophical kind, which included knowledge of function," and of more besides myological structure from the perspective of its function. Vesalius's work is teleological, but "in correcting Galen's [morphological] errors, Vesalius was obliged to keep very close to Galen's exposition." Consequently, "Vesalius' treatment is almost entirely a *historia,* an account of the 'fabric' of the body. . . ." Vesalius followed the anatomical *accessus,* "the traditional rote of observables, used by anatomists since Mondino," of which function or use was only one among nine criteria, all of which are listed in chapter 2 of this study.[89] Whereas at the end of the last chapter of the last book of the *Fabrica,* Vesalius had spent five pages explaining that "the anatomist's task was also to consider function, which could only be obtained from vivisections,"[90] Colombo's aim was that "of restoring the practice of the early Alexandrian anatomists, Herophilus and most particularly Erasistratus," the key feature being "to put *vivisection* centre-stage in the search to discover more about the *actions* of the parts," their use and purpose.[91] While insisting like Vesalius on the need for an observational epistemology to supplant deference to the authority of Aristotle and Galen, Colombo's anatomy, although not deeply philosophical, was if anything more telelogical than Vesalius's had been. Although it is arguable that Colombo's anatomy may have imbibed a kind of empiricism from the human anatomy of the partially skeptical Alexandrian anatomists, little about Alexandrian anatomy was in fact knowable.[92]

Colombo's description of the pulmonary transit of the blood and of the action of the cardiac valves occurs in Book VII of *De re anatomica,* the section detailing the heart and arteries.[93] In his description of the "membranes" composing the structure of the valves, he says that "there are three at the opening of the vena cava," like the "three little folding doores, with three forks" in *The Purple Island,* "three again at the pulmonary artery, three at the aorta," the "three little folding doors" of the semilunar valves in Fletcher's *Purple Island,* "and two at the pulmonary vein," the mitral valve, "which hath two folds" in Fletcher's poem. Colombo acknowledges also that those "which are at the vena cava and pulmonary vein are different in shape from the membranes of the aorta and pulmonary artery."[94] Moreover, he adds: "Know that just as their shape is different, so also is their employment. The valves of the aorta and pulmonary artery from their internal position are moved outwards to be useful for the emission of blood [from the heart]; the valves of the other two vessels [vena cava and pulmonary vein] on the contrary are set from outside inwards, so that they seem to be made for the containing of the enclosed blood [in the heart]."[95] In *The Historie of Man,* Banister's rendering of this passage confuses the positions and operations of the semilunar valves with those of the mitral and tricuspid, saying: "The wickettes therfore . . . of the hollow veyne and veniall arterie, are sited from within forth, as seruyng to the emission of bloud: but the wickettes of the other ii. vessels, contrariwise, from without forth, or inward, that so they might seeme unto us to be made, for the conteining of included bloud."[96] Banister partially rectifies the situation further down the page, where he says, "you shall finde the bloud which is now gone into the right ventricle, not able any more to go backe agayne into the hollow veyne"; so both Colombo's and Banister's descriptions of the anatomy of the cardiac valves resemble that in *The Purple Island* insofar as they derive their physiology from structure, and they detail them precisely according to the *occasiones* of the anatomical *accessus.*

It is therefore possible that, even if Fletcher derived the anatomy of the valves, and even their operation, from elsewhere, possibly a Galenic text, he took their physiological competence from *De re anatomica* or *The Historie of Man,*[97] a physiology moreover that if not identical is entirely compatible with that of the cardiac valves in Canto 4 of Fletcher's poem. Servetus, not understanding the action of the valves of the heart, had "retained the notion of fuliginous vapors regurgitating through the pulmonary vein"; and Guidi accepted that "most of the blood must go . . . through the pulmonary artery, the lungs and the pulmonary vein to reach the left ventricle," yet retained the regurgitation of fuliginous vapours. But Colombo's, Banister's, and Fletcher's accounts of the mitral valve, and indeed of all the cardiac valves, do not permit such regurgitation of vapors, blood, or spirit.[98] This compatibility, together with the use of the term "membranes" in a single case in *The Purple Island* (4.24 margin. y), to describe the folds of the semilunar valve of the aorta, suggests a not improbable provenance in either Colombo or Banister. There are further considerations, with regard to cardiovascular and more general anatomy and physiology, which might either support or counter such a provenance. For instance, making such provenance less singularly probable is the fact that the term "membranes" is used to refer to the structure of the cardiac valves by other anatomists too; but because these anatomists' understanding of the function of the membranes resembles and usually follows that of Colombo,[99] this does not question the conceptual consistency between them.

Although the publication of *De re anatomica* in 1559 cannot be said to have been especially recent in the first decade or so of the seventeenth century, when Fletcher came to write

Canto 4 of his poetic allegory of human anatomy, the effect of Colombo's theory of pulmonary transit on later anatomists like Valverde, Banister, Ruini, Paré, Guinther, and Plater, has been noted in this chapter, and the theory was to become generally accepted, rather than superseded, in the seventeenth century. In accordance with our preliminary guidelines for the comparative studies undertaken in this chapter, where probable provenance is ascertained, the possibility that the anatomy in *The Purple Island*'s recent sources is in fact conceptually metaphorical in the way that early sixteenth-century and Vesalian anatomy has been found to be, must be investigated. For were this in fact to be the case, our task would be to develop an argument capable of discerning how *The Purple Island* figurative anatomy compares with that of Fletcher's sources, and how the *difference* in the *Purple Island* figurality modifies the figurative conceptual character of the anatomy of its sources.

Examples have been given of how the *Purple Island* figurality of cardiovascular anatomy diverges from that in *The Historie of Man,* but Colombo made "the practical basis" of his anatomical work a conceptualization of the body "as being nourished, being vivified, and being given motion, by three life-giving rivers"—the veins, the arteries, and the nerves and spinal medulla, respectively.[100] The liver, heart, and brain, in this theory, constitute, respectively, the three fountainheads from which flow these rivers. Other similar figurative conceptions existed, such as that in Raynalde's *The Birth of Man-Kinde,* which draws on Paracelsian analogies between the human body and the earth, and imagines the three sources as mines in which occurs "a transmutation of . . . matter" by "dayly and long coagitation": the "bloud" mine, which is the liver; the heart which draws the blood from the "great master vaine" (vena cava) and "doth ingender vehement and lively spirit"; and the "brayne, of whom all the sinewes take their originall."[101] The "mettals of the body" that are produced by these mines are "vaine blood," by the liver, "arteriall blood," by the heart, and "lively spirits," by the brain.

This study of *The Purple Island* has considered mainly its figurative representation of the arteries and veins not as mines but as "rivers" or "streams" that emerge from the "fountains" or "springs" of the liver and heart (*P.I.* 2.9 margin.; 2.10.1; 2.11.1). But the poem consistently groups these systems of transmission with that of the nerves and their source, the brain. The poem's division of the "thousand brooks," that in "azure chanels glide" (*P.I.* 2.9.2, 3), is into "Three diff'ring streams from fountains different" (*P.I.* 2.10.1), the third of which—the nerves—are described as "a spermaticall part rising from the brain and the pith of the backbone," that is, the spinal medulla (*P.I.* 2.12 margin.). Of these three "streams," "the first gives nourishment, / Next life, last sense and arbitrarie [that is, voluntary] moving" (*P.I.* 2.13.2–3). This attribution of functions Fletcher does not of course only share with Colombo, but also Galen, Vesalius, and most others; but the figure of three rivers issuing from three fountainheads is a conceptually mobilizing "invention" of Colombo,[102] and if Fletcher did not ingeniously "invent" the same notion, its use in *The Purple Island* supports a provenance for at least this figural conceptualization in *De re anatomica,* which Fletcher seems in this respect not to modify at all. Despite the provenance of much of Banister's anatomy in *De re anatomica,* I would suggest it is unlikely that Fletcher would have derived such a figure from the anatomy in *The Historie of Man.*

Unlike *The Purple Island* and Vesalius's *Fabrica,* however, and as a consequence of its emphasis upon function or physiology, *De re anatomica* deals with each of the major organs "*together with their vessels*" (as Fabrici and Harvey were later to do) and does not represent at all in its scheme the three venters of Galen, Mondino, and Vesalius, which, of course,

Fletcher's division reiterates.[103] Banister deals with the heart, for example, with the arteries, and the brain with the nerves, thereby not having a separate book on the nerves, veins, and arteries, but otherwise retains the Vesalian order or method, as was said in the previous chapter. Although Fletcher may have derived from Colombo a significant image, that image is perhaps less conceptually effectual in *The Purple Island* than in *De re anatomica,* or at least, its effectiveness is not consistent with that in its source.

Fletcher's poem is in some of these respects perhaps closer in conception to Valverde's *Historia de la composicion del cuerpo humano,* which was the first anatomical work to announce the pulmonary transit of the blood, and attributed its discovery to the author's "preceptor" Colombo.[104] This work adheres to the Vesalian procedure of dissection, except in dealing with the arteries and veins in one book, while simultaneously "paying . . . lip-service to the concept of the rivers," as its use of the metaphor of "aquaducts" for the arteries, veins and nerves, according to Cunningham, demonstrates.[105] Valverde's use of a "mixed metaphor," draws on concepts of the body's constructedness and naturalness, the two basic conceptualizations of the body that underlie the conceptual metaphors of anatomy in the schools and in medicine, and which characterize the anatomical figurality in Fletcher's poem. Together with the incomplete conceptual shift that is evident in Valverde's book's division of the body into three venters, Valverde's use of figurality suggests a close resemblance to the anatomy of the heart and vascular system in *The Purple Island.* But, besides the fact that no copy of Valverde's book, or of its Italian translation, was held by any Cambridge library until 1634,[106] Fletcher could not have derived from Valverde his descriptions of the anatomy and physiology of the cardiac valves, for Valverde only alludes to Colombo's realization of their competence, and describes the function of only the mitral valve.[107] It is more probable that Fletcher derived this description and the image of three systems of rivers from Colombo, though possibly indirectly, but only partially integrated them into an order of dissection and a physiological system which owe a great deal to Galen and Vesalius, which he derived from elsewhere, not improbably Vesalius's *Fabrica* or Banister's *Historie of Man,* as was proposed in the previous chapter.

Establishing with more certainty the specific provenances of those aspects of the *Purple Island* cardiovascular physiology and anatomy discussed heretofore would require further and closer textual comparisons; but of the general character of this anatomy and physiology it is possible to be fairly sure. Broadly speaking, the order of the dissection adheres to Vesalius and Banister. The hepatic physiological origin of the veins and the hepatic embryological origin of the blood, as well as the embryological priority of the liver over the heart, are all Galenic. It is not the case, in *The Purple Island,* however, that the liver is superior in nobility or principality to the heart, as Nicolson claims: even in Canto 2, the most Galenic in respect of this physiology, the heart is represented as a "greater fountain" than the liver (*P.I.* 2.11.2); and of the three "Metropolies," "regions" or venters, "the lowest," in which "the liver is sovereigne," "is the widest, but meanest" (*P.I.* 2.14.2 & margin. l). In addition, the differentiation between veins and arteries is clearly Galenic, as is the imagery of the body soaking up or absorbing both the venous blood's nutriment and natural spirit, and the vital spirit of arterial blood. But the fact of the one-directional, possibly circular movement of blood in these vessels remains unattributed. The character of the poem's cardiac anatomy and physiology, which appears in Canto 4, is more mixed in origin. The concoction of blood in the right ventricle, and of this rarefied blood together with *pneuma* in the left ventricle, is Galenic, except for the poem's use of the terms "distillation" and "confection" to describe one or both of these processes. The one-directional flow of blood to the lungs and that of air from the lungs to the left ventricle is similarly an ideational combination of Galenism and pulmonary transit. Fletcher derived the competent physiology of the cardiac valves most probably from the theory of pulmonary transit, though it is probable that he drew from Colombo exclusively the figurative theory of three life-giving rivers arising from three fountainheads; but his anatomy and physiology of the interventricular septum of the heart is thoroughly Galenic.

Standing out among the features of the poem's cardiovascular anatomy as unresolved as to attribution and probable provenance are physiological phenomena: the movement of blood in the veins and particularly the arteries, and the use of the terms "distillation" and "confection" to describe the ventricular concoctions. The former of these features is all the more puzzling in that, as Whitteridge remarks: "Although Colombus [the most progressive of anatomists thus far attributed with being a probable source of the *Purple Island* anatomy] discovered the pulmonary transit of the blood, there is no reason to suppose that he queried the notion of blood ebbing and flowing in other parts of the body." He did not, for instance, "question the validity of the notion of the two-way movement of blood and chyle in the branches of the portal vein."[108] And Banister's heavy indebtedness to Colombo did not induce him to infer the movement of the blood in the veins and arteries in one direction only. It is said above that Fabrici presents a model of at least a predominance of movement *from* the liver and heart *to* the rest of the body, and it might be supposed, given the function of the veins in *The Purple Island* to nourish the parts with unrefined blood and natural spirit, that Fletcher's text adheres to some such model as this. But Fabrici and all Galenists generally supposed the ebb and flow movement of blood in the arteries to be, if anything, more pronounced than that in the veins, as was said above. It is notionally possible that Fletcher inferred, from the action of the semilunar valve of the aortic orifice, the one-directional movement of blood in the aorta and the arteries throughout the body.[109] But, in order to avoid the attribution to Fletcher of a natural-philosophical ingenuity or invention in measure beyond that of his anatomist contemporaries, in its capacity to advance, in a conceptually coherent and groundbreaking way, on the conflicting theories of his forebears, we should explore the possibility that Fletcher derived the *Purple Island* physiology of the arteries, and possibly that of aspects of the heart and cardiac valves, from a single and distinct source. A likely source may be one in which it is possible to trace the terms "distillation" and "confection" as well as the *Purple Island* vascular physiology.

The term "distillation" is alchemical, and has philosophical connotations, whereas "confection" could denote any of several meanings: "[m]aking or preparation by mixture of ingredients; compounding; composition, . . . making up, manufacture"; "[a] preparation made by mixing," potentially "[a] medicinal preparation"; or, in the sixteenth century, "[t]he performance of the sacrifice of the mass."[110] Colombo's revival of Erasistratus's anatomical projects and concepts of the body did not involve developed philosophical or, indeed, theological dimensions.[111] The writer of *De re anatomica* leaves the discussion of the difficulties of his theory of the generation of vital spirits in the lungs (rather than the heart) "to the great philosophers"; and apportions to himself the task, "as far as in me is possible," to "describe correctly the parts of the body and their uses."[112] Because the meaning of the heart in *The Purple Island* may depend for its conception upon notions of an alchemical and philosophical character, it is necessary, notwithstanding the obvious Galenism of much of this analogy and character, to consider the possibility of another influence.

Towards the end of the sixteenth century, besides a shift among anatomists in Protestant countries away from Aristotle and toward Plato and Neoplatonism, even in some cases reviving Galen, there was also, particularly at Padua, a shift from the Galenism of Vesalius and his followers to a revived Aristotelianism.[113] The Aristotelian revival at Padua and elsewhere was in no insignificant part due to Fabrici, though it may be traced in the work of the Aristotelian pupil of Colombo, and later Professor of Anatomy at Pisa, Andrea Cesalpino.[114] The Aristotelian philosophical emphasis on the action, use, and purpose of structure in anatomy can be seen in the *Questiones Medicae,* where Cesalpino presented his theory of the "perpetual movement" of blood through the heart and lungs: "The passages of the heart are so arranged by nature that from the vena cava a flow takes place into the right ventricle, whence the way is open into the lung [through the pulmonary artery]. From the lung moreover there is another entrance into the left ventricle of the heart [via the pulmonary vein], from which then a way is open into the aorta artery, certain membranes being so placed at the mouths of the vessels that may prevent return. Thus there is a sort of perpetual movement from the vena cava through the heart and lungs into the aorta artery. . . ."[115] This perpetual movement is secured by the competent operation of the cardiac valves, which, in Cesalpino's *Questionum peripateticarum libri quinque,* are described as "little membranes whose function is to secure that the [cardiac] orifices letting in do not lead out and that those leading out do not let in."[116] Although the use of the term "membranes" is not exclusive to Cesalpino among anatomists, this, together with the competence of the valves, and the notion of the perpetual movement of blood might suggest the writings of Cesalpino as a source for *The Purple Island.* But even if Fletcher could have read *Questionum peripateticarum* at Cambridge, which is unlikely, he could not have found there a description of the cardiac valves as detailed as his own, or as that of Colombo or even that of Banister.[117]

Notwithstanding the improbability of a provenance in Cesalpino's works for the unresolved features of the cardiac and vascular anatomy and physiology in *The Purple Island,* it will shed some light on Fletcher's use of the terms "Circling" and "distillation" to examine the ambiguity, or rather multivalence, of the term "*circulatio*" in the writings of Cesalpino. Walter Pagel rightly comments that, "Caesalpino . . . regards the cooling down of the blood as the purpose of its passage through the lungs—rather than the nutrition of the latter [as for Galenists] or an actual mixture of blood with air [as for Banister and the followers of Colombo and Valverde]."[118] A question which arises in discussions of Cesalpino's use of the term "*circulatio,*" though, is whether it refers to the cyclic process of heating and evaporation of blood in the heart and cooling down and condensation in the lungs, a process akin to that "known in chemistry and alchemy as *distillation* [*distillatio*]";[119] or to the perpetual physiological movement of "very hot blood from the right ventricle through the pulmonary artery and . . . to the left auricle by the anastomoses of the artery with the pulmonary vein."[120] Providing a solution to this difficulty has the repercussion of suggesting whether or not Cesalpino should be understood not merely as advocating a version of pulmonary transit, but as anticipating Harvey in the notion of the blood's continuous and circulatory flow; and this is because Cesalpino added that, "in each part of the body such an upward and downward movement of the blood" occurs, "the former being due to arterial influx and the latter to venous backflow."[121]

Opinions have differed greatly over this. Foster felt forced to "admit that Cesalpinus had not only clearly grasped the pulmonary circulation, but had also laid hold of the systemic circulation. . . ."[122] By contrast, Whitteridge thinks that, "Cesalpinus is not applying the word *circulation* to the movement of the blood in the vessels, but to the action of cooling the hot blood from the heart"; and, therefore, that his use of "circulation" is alchemical rather than physiological.[123] While conceding that Cesalpino "did know of the pulmonary transit of the blood and, perhaps, understood its significance," she alludes, in support of her argument, to the instances in his writings where Galenism directly contradicts the competence of the cardiac valves: "In the heart is the fountain of the blood which is distributed into the four veins, namely the vena cava, the aorta, the pulmonary vein and the pulmonary artery, which irrigate the whole body like the four rivers that flow out from Paradise."[124] It may be added to this debate, however, that self-contradiction, neither logically nor historically (as is evident in the cases of Ruini and Paré), precludes adherence (albeit inconsistent) to a specific meaning in anatomy; and Cesalpino's Galenism does not erase the ambiguous innovation of his theory of "circulation" as perpetual movement.[125]

This view is deftly articulated by Pagel, whose own account of the passage from *Questionum peripateticarum,* cited above, is that it conveys "the chemical sense of 'distillation' . . . namely the cooling down and condensation of an overheated and vaporous fluid"; but that this chemical or alchemical process "is continuously repeated and cyclical, for it is the same fluid which is subjected to it in an ever recurring cycle, owing to the anatomical arrangements of valves and vessels."[126] This physiological signification is borne out by the use of physiological structure, in *Questionum peripateticarum,* as evidence for "circulation": "That such 'circulation' does operate from the left ventricle through the lung is borne out by the following anatomical facts: There are two vessels terminating in the right and two terminating in the left heart; one that must be afferent and the other one efferent—because of the arrangement of the valves."[127] It is genuinely plausible to conclude, as Pagel does, that "Cesalpino's claim for priority in conceiving of a continuous motion of the blood in the right direction is therefore not negatived by his use of the term 'Circulatio' also in the chemical sense."[128] Cesalpino was Aristotelian and adhered to alchemical principles, and when the natural philosophy of Aristotle or alchemy is used to justify a mode of procedure, the knowledge to which it gives rise, even if new, may yet be validated according to the precepts and projects of Aristotelian philosophy or alchemy. Such is the case of Cesalpino, in whose anatomical project the "context of discovery" is distinct from the "context of justification," and since physiological circulation is already interpreted or justified in his work in terms of alchemical circulation, it is misleading of Foster to say, however regretfully on his part, that Cesalpino had laid hold of the systemic circulation before Harvey. This is a fine example of there not being a timeless standard by which to judge the validity of claims to scientific knowledge.

Cesalpino's considerable theoretical innovation has been diminished by some historians of anatomy and physiology, for whom, as we shall see later in this chapter, Harvey's advance on his predecessor, by devising experiments which should test the validity of the hypothesis of systemic circulation, and by discerning correctly its mechanism in the muscular systole and diastole of the heart, occupies an unprecedented threshold in the history of scientific method. In Harvey's case, at least for these historians, the contexts of discovery and of justification more nearly coincide than in the case of Cesalpino. But without independent justification for believing that Fletcher would only include in his allegory anatomical ideas that are supported by experiment, it is misleading to dismiss Cesalpino as an influence on Fletcher, as Langdale does, on the grounds that the anatomist "had no proofs for his hypotheses."[129] But there are other grounds, including their lack of a description of the anatomy of the cardiac valves, for arguing that Cesalpino's writings had little or no effect on *The Purple Island.* Even though the description of the perpetual movement of the blood through the heart (cited above) may seem to a modern reader to imply, or even require, the extension of this movement to the veins and arteries, it is not clear that such a claim is made by Cesalpino. It is true that, unlike his predecessor Colombo, Cesalpino did not confine his innovation of physiological "circulation" to the pulmonary artery and vein. However, his suggestion that "when we are awake a large supply of blood and of spirits is conveyed to the arteries," so that the centrifugal movement of blood predominates over the centripetal, and that "[w]hen we are asleep . . . the same heat is carried back to the heart not by the arteries but by the veins," so that the centripetal predominates over the centrifugal, is nowhere alluded to in *The Purple Island,* which represents a continuous centrifugal movement in the arteries from the heart, and in the veins from the liver.[130] While Fletcher's use of the phrase "Circling about" in respect of the whole vascular system might suggest alchemical connotations of distillation and circulation, and might appear superficially to be an allusion to Cesalpino, it is doubtful whether there is any authority in Cesalpino for Fletcher's insistence that the veins and arteries "creeping forward slide, but never turn again" (*P.I.* 2.9.7), and it is necessary, therefore, to look elsewhere for a more probable source.

Not adhering to the theory that blood flows in the pulmonary vein, Fletcher could hardly have regarded the function of the lungs to be to cool the hot blood therein, as Cesalpino does. But had the alchemical signification of "circulation" in Cesalpino's work found its way into Fletcher's use of the term "distillation," it seems likely that the application of the shared meaning of the two terms would have been removed to the processes of concoction in the left ventricle of the heart, where a process of "distillation" is found in *The Purple Island* (4.21.1–5, cited above). This is because the traditional Galenic function of the passage of air from the lungs to the left ventricle was not only to provide *pneuma,* but as Banister (cited above) states, to cool the heart's innate heat.[131] This explanation in no way directly accounts for the occurrence, in Fletcher's poem, of a similar process of "distillation" in the right ventricle, where there cannot be, and is not said to be, any cooling and condensing influence of inspired air (*P.I.* 4.18.7, cited above).[132] But, by Cesalpino's account, which may be partly consistent in this respect with that of Fletcher, the blood leaving the right ventricle in the pulmonary artery has been heated in that ventricle and requires cooling in the lungs. (The process of heating in *The Purple Island,* which was described above, separates the thinner blood from the "grosser waves," sending the former through the porosities in the interventricular septum and the latter through the pulmonary artery, into the lungs for their nourishment.) The fact that Cesalpino's work is only very rarely cited by other anatomists seems further to exclude it from the probable sources of Fletcher's poem, however.[133] Yet, notwithstanding the improbability of Cesalpino's actual influence upon *The Purple Island,* concoction of blood in Fletcher's poem may be an alchemical, as well as a physiological and possibly even sacramental process.

The provenance of this alchemical process may be in the work of Paracelsus and his followers, who cherished the prevailing Christian Neoplatonic ideas about the microcosm and the "mystical neo-Platonic universe," and developed theories of inspired or revelatory knowledge in medicine. As a reaction against "Aristotelian scholarship" and scholastic logic, these ideas in the sixteenth and early seventeenth century were "enjoying a new popularity due to the attention focussed on [them] by the Florentine Platonists of the late fifteenth century," such as Marsilio Ficino and Pico della Mirandola, whose thoughts on the microcosm were discussed in chapter 1.[134] Alchemy in late antiquity had been strongly loaded with Gnosticism and Neoplatonism, and continued macrocosm-microcosm speculation in the Renaissance and seventeenth century, but, as "a special approach to the general neo-Platonic universe," alchemy combined spiritual apprehension with experimentalism, and was "generally unthinkable without the laboratory and man's own attempt to observe and carry out experiments on the metals" and other matter.[135]

Like *The Purple Island* (1.39–48), the work of Paracelsians on the universe commonly began with the Creation. The leading French Paracelsist of the sixteenth century was Joseph Duchesne, whose views influenced those of Robert Fludd, the "foremost English alchemical theorist of the first half of the seventeenth century." Duchesne "based his cosmology on the Biblical story of Creation," wherein God is represented initially as "separating Light from Darkness,"[136] as in Fletcher's rendition:

> First stepp'd the Light, and spread his chearfull rayes
> Through all the Chaos; darknesse headlong fell,
> Frighted with suddain beams, and new-born dayes;
>
> (*P.I.* 1.40.1–3)

For all Paracelsians, the Creation was an act of separation, of distillation; for Paracelsus himself "the great generating substance from which all other more special mysteries proceed" and from which all other substances are separated, is the *Mysterium Magnum,* of which *The Purple Island,* however, makes no mention. As separated from this prime matter, Paracelsus and Duchesne continued to use the four elements, as does the account of the Creation in *The Purple Island* (1.41.1), though Paracelsus postulated the existence of "true spiritual elements" of which the "material substances" which Classical philosophers had identified were to be considered "but crude approximations." But unlike Fletcher, "he dropped the concept of a conjunction of qualities and considered each element to stand for a power in its greatest intensity."[137] Thus, whereas for Paracelsus "[f]ire is simply hot, not a combination of the qualities hot and dry," for Fletcher "Warm air with mutuall love hot Fire infolds" (*P.I.* 1.42.3), air being hot and moist, fire hot and dry.

In addition, Paracelsus and Duchesne discarded the four humors, which Fletcher partly retains (*P.I.* 3.18.2; 3.15.2–3 & margin. m), and introduced a chemical theory of three principles, or spiritual substances, of which there is no mention in *The Purple Island:* namely sulfur (the principle of combustibility, structure, and substance), mercury (the principle of vaporous qualities), and salt (the principle of solidity and color). These principles form the *tria prima* in the cosmology of Duchesne who interprets God's separation of waters from waters to refer to the separation of these three principles from the prime matter.[138] Undoubtedly biblically inspired, like the Paracelsian version, Creation in *The Purple Island* is nonetheless more indebted to the Classical tradition of warring elements bound by love, which stretches back to Empedocles.

Nevertheless, *The Purple Island* is in some measure influenced by the Paracelsian flowering of the time, not least by the aggregation by Paracelsus and others of the geocosmic and macrosmic correspondences for Man, but also in the conception and figuration of the functions of organs. On the subject of the concoction of vital spirit, in its aspect as distillation, there is no explicit mention in Fletcher's anatomy of the Paracelsian *archei,* which like internal alchemists in the different organs, separate or distill pure from impure parts, for example, of the food or blood.[139] But by Paracelsians, "mountain streams were explained in terms of earthly distillation," just as in *The Purple Island* the left ventricle of the "fountain" of the heart "Stills" "thinne aire" and the "right sides rent," and "Sends" them "both together" "Through many a dale . . . and many a mountain" (*P.I.* 2.11.2; 4.21.5, 2, 7, 4; 2.11.4).[140] *The Purple Island* also describes a "distillation" or "separation" of urine from blood in the kidneys, which are explicitly described as "mountains" (*P.I.* 3.21.1 & margin. t):

[s] The third bad water, bubbling from this fountain, Is wheyish cold, which with good liquours meint, Is drawn into the double *Nephros* mountain; Which suck the best for growth, and nourishment:	s The watery humour with some good bloud (which is spent for the nourishment of those parts) is drawn by the kidneys.
[t] The worst, as through a little pap, distilling To divers pipes, the pale cold humour swilling, Runs down to th'Urine-lake, his banks thrice daily filling.	t The Ureters receive the water separated from the bloud, distilled from little fleshie substances in the kidneys, like to teats.

(*P.I.* 3.20)

There can be little doubt of this geocosmic figurative description of the function of the kidneys having been influenced by Paracelsian or iatrochemical works; but what of the possible Paracelsianism of cardiovascular anatomy and physiology in *The Purple Island?*

Duchesne "went on to suggest that as there is a continual circulation of the elements in the macrocosm, so also there is a circulation of the blood in man as well," though even less than Cesalpino's theory of circulation does Duchesne's approximate to "the circulation of the blood as we now understand it."[141] It is more emphatically "the circulation of a liquid in a chemical distillation" and not a physiological circulation. So, although alchemical "circulation" seems to have affected the *Purple Island* conception of concoction as distillation, possibly through the work either of Robert Fludd or of Paracelsus, who was very well represented in Thomas Lorkyn's bequest to Cambridge University Library,[142] the theory can have no bearing on our account of the conceptual character and provenance of that poem's detailed anatomical description of the cardiac valves, or of pulmonary transit of blood from the right ventricle of the heart to the lungs and of inspired air from the lungs to the left auricle. For these, except the passage of air in the pulmonary vein, provenance, though uncertain, is probably to be located in either Colombo's *De re anatomica* or Banister's *Historie of Man.* But for *The Purple Island*'s representation of one-directional movement in the veins and arteries, we must proceed once again to look elsewhere, this time at another anatomist's work, which was in some measure Aristotelian though also "modern," in order to discover if Fletcher may have found in this work some aspect, or perhaps the integration of all, of the movement of blood in the arteries, the competence of the cardiac valves, and the alchemical connotations of "circulation" or "distillation," which he was to represent in Canto 2, stanza 9, and Canto 4, stanzas 18 to 24, of *The Purple Island.*

Of the cardiovascular work of William Harvey, that which was prior to, and led to, his theory of the circulation of the blood, presented in *Exercitatio anatomica de motu cordis et sanguinis in animalibus,* in 1628, is relevant.[143] It is by no means clear from the arguments of Nicolson and Baldwin that cardiac and vascular anatomy in *The Purple Island* is in conception entirely removed from Harvey's theory.[144] It has been argued in the previous chapter and the first section of this chapter that the method and much of the anatomical content of Cantos 2 to 5 of Fletcher's poem is Vesalian in character; yet despite the fact that there is hardly an inconsistency or contradiction within the cardiovascular physiology that is presented by *The Purple Island,* for this is almost entirely coherent, a significant and substantial part of the physiology of the heart, veins, and arteries, especially much of that which is presented in Canto 4, clearly has a different character from the Vesalian anatomy, is probably from a different ideational context, and is possibly of different provenance. But a Harveian source for Fletcher's physiology is a near intractable problem for several different reasons.

It is not clear that the description in Fletcher's poem is not in composition prior to the discovery by Harvey of systemic circulation, and its subsequent publication. That Fletcher may have incorporated details of systemic circulation from *De motu cordis* close to the publication of his anatomical poem is not implausible, but the date of any late revision of *The Purple Island* is uncertain, and there are considerations of a conceptual kind that make this unlikely. The conceptual character of the *Purple Island* anatomy and physiology is insufficiently similar to that of *De motu cordis* to warrant either the view that Fletcher can be supposed to have read *De motu cordis,* or—and Langdale is of this view—that the hypothesis, and possibly the *demonstratio,* of systemic circulation were communicated to Fletcher by Harvey. *The Purple Island* makes no mention of the muscular dilation (diastole) and contraction (systole) of the heart. The observation of, experimentation with, and correct understanding of systole and diastole, and the simultaneity of arterial pulse and cardiac systole, which only Colombo had observed and understood prior to Harvey, constituted the evidential ballast with which Harvey proceeded to undertake experiments in blood flow through the heart, which led more directly to circulation theory.[145]

Colombo's observations and understanding in this respect might be thought to argue against *De re anatomica* as a source for *The Purple Island.* But the competence of the cardiac valves, for which Fletcher is here supposed to be indebted to Colombo (if he is indebted to him for anything, besides, that is, the figurative concept of three life-giving rivers), does not logically depend upon the muscular action of the heart itself. Whereas Harvey could probably not have arrived at the theory of systemic circulation without knowledge of the systole and diastole of the heart, Fletcher could have derived the physiological competence and structure of the cardiac valves from Colombo without paying much attention to his description of systole and diastole. As seems likely if Harvey followed a disputational-commentarial method, consistent with a natural-philosophical research method, rather than a modern scientific one, Harvey developed a thesis of forceful systole as a kind of commentary or development upon Colombo's observations, yet "[t]he defence of his new thesis of forceful systole had led him into impossibility and threatened the very credibility of the thesis itself. If so much blood passed out of the heart, where did it all go?"[146]

Another suggestion as to the emergence of the theory of systemic circulation is based on Harvey's testimony, as twice recorded by Robert Boyle, that it was a consideration of the valves of the veins which first hinted to him of the circulation. Because the valves gave free passage to the blood towards the heart, but presented an obstacle to it the contrary way, as Fabrici had observed, Harvey "was invited to imagine that so provident a cause as nature had not so placed so many valves without design; and no design seemed more probable, than that . . . [the blood] should be sent through the arteries and return through the veins."[147] Although the reliability of Boyle's record of Harvey's testimony has not gone unchallenged,[148] the movement of blood in the veins is by contrast centrifugal—flowing from the liver to the parts—in *The Purple Island,* and Fletcher's poem does not so much as mention the valves in the veins. Consequently, the character of the *Purple Island* anatomy and physiology is insufficiently similar to that upon which the theory of systemic circulation is based in *De motu cordis,* for Fletcher to have incorporated that theory into *The Purple Island.*

There is another way into this problem of possible Harveian provenance, however. Sawday claims, on the basis of one topical intertextual reference, that "Fletcher was still revising parts of his poem" in 1616.[149] This is the year in which Harvey, as Lumleian Lecturer since August 20, 1615, delivered his first Lumleian lectures on the whole of anatomy at the London College of Physicians.[150] That Fletcher should have revised the already commenced, if not almost complete, *Purple Island,* in order to incorporate aspects of Harveian physiology, at some point after 1615 and prior to the poem's publication, would be consistent with his late revision of *Locustae, Vel Pietas Jesuitica,* and *The Locusts, or Apollyonists* before their publication in 1627; though these were unusual in being the only major poetic works Fletcher acknowledged as his own, indicating, perhaps, that he may also have taken more pains over their composition.[151] Whether it was to incorporate Harveian or Colombian or another anatomy will be difficult to determine with certainty, but a late revision of *The Purple Island* is also suggested by the inconsistencies between the ideational systems of the cardiovascular anatomy and physiology in Fletcher's poem.

It is important for a consideration of the *Prelectiones anatomiae universalis* (as Harvey himself called these lectures, though they were not published until 1886) as a record of anatomical and physiological ideas, that may have been communicated by some means to Fletcher, to ascertain their composition and content.[152] It is initially incumbent, however, to cast some doubt on the lectures presented in London as a direct source for *The Purple Island*'s cardiac and vascular physiology, in order more clearly to ascertain the relative probability of this as opposed to other possible scenarios.

It is generally believed that Harvey prepared the *Prelectiones* as the rough notes for anatomical lectures delivered as Lumleian Lecturer at a dissection carried out on April 16, 17, and 18, 1616, although the title page, upon which this view is based has been interpreted to signify differently.[153] It is not clear in what capacity Fletcher would have attended the anatomy theater of the College of Physicians, or indeed that of the Barber Surgeons, along with the physicians of the College and the surgeons of the Barber Surgeons Company.[154] However, according to Langdale, "[s]erious medical students were present in the anatomical theaters, but they were outnumbered by the gentlemen and ladies on holiday."[155] Given Fletcher's successive residences in Cambridge, Derbyshire, and Norfolk, such attendance could only have been occasional: on December 16, 1614, he returned to Cambridge from an unknown location, which may have been London, and he resided at King's College until March 25, 1615, when he left for Risley, Derbyshire, to occupy the chaplaincy of Risley Hall.[156] As Langdale admits, *The Purple Island* is the product of sustained rather than occasional exposure to anatomy;[157] but such attendance at Harvey's lectures as Fletcher was capable of on his sojourns to London may have been sufficient for him to have absorbed at least some anatomical principles, which he was then able to integrate into a more conventional, Vesalian, and Galenic method of exposition and basic anatomical character. Although Fletcher was frequently in London between 1602 and 1610, during the holidays, there is no evidence that he visited London between 1616 and 1633, the year *The Purple Island* was published.[158] Between 1616 and 1621 the lack of evidence cannot be regarded as conclusive, but if after the first of November 1621, he had taken leave of his duties as rector of All Saints Church, at Hilgay, in Norfolk, at any time in this period (perhaps in order to make arrangements with his London publishers), the register at Hilgay, in which Fletcher made the entries, would, as Langdale maintains, betray the fact.[159] But it does not.

Given that Harvey's anatomical lectures were not published until the nineteenth century, a biographical context for the communication of anatomical ideas therein should provide a means whereby either the transmission of the lectures in manuscript or their verbal communication becomes possible. Any communication, if outside the context of the public lectures, would in all likelihood have been partial as regards the content of those lectures, but the character of the anatomy in Cantos 2 to 5 of *The Purple Island* is such that whether the communication of the lectures was verbal or written, it did not need to have been in their entirety, and may have been only a summary of part or the whole. My research into a putative biographical connection between Harvey and Fletcher, which is outlined in Appendix B, draws the conclusion that although plausible, communication between Harvey and Fletcher remains a conjecture. By way of qualification, however, it may be added that whereas communication is perhaps more plausible between 1615 and 1621, during Fletcher's residence in Derbyshire with the Willughbys of Risley Hall, it may be comparatively less likely after the first of November 1621, when Fletcher took up the position of rector of All Saints Church, at Hilgay, for still far, though not quite so far, from the London environs in which his more illustrious contemporary could then be found,[160] Hilgay was also far from the great midlands and Kentish estates of the Will[o]ughbys through whom it may or may not be most conceivable that Fletcher may have had occasion to meet Harvey.

The *Prelectiones* may represent Harvey's ideas at the least unlikely time of contact of one sort or another between Fletcher and himself; though, to adduce the *Prelectiones* so, is only to be undertaken advisedly. As the basis for his notes for the lectures, Harvey used the textbook of Bauhin, *Theatrum anatomicum,* initially the first edition of 1605, and later the second edition of 1621, but he frequently disagrees with his source and with other anatomists by entering symbols and marginal comments.[161] Although it might rather be thought that the *Prelectiones* and *The Purple Island* share a common source, given the speculative or conjectural nature of communication between Harvey and Fletcher, there were no copies of any of Bauhin's anatomical works at any Cambridge library during the period in which Fletcher might be thought to have been reading anatomy textbooks at Cambridge while composing Cantos 2 to 5 of *The Purple Island.* While certainly not precluding the possibility that Fletcher read the work during this period or indeed later, when revising *The Purple Island,* the fact cannot count in favor of a provenance in Bauhin.[162] Further, independent evidence against a provenance in Bauhin lies in critical features of *The Purple Island's* cardiovascular anatomy clearly not having hailed from Bauhin, particularly perhaps the competence of the mitral valve.

The *Purple Island* cardiovascular anatomy is not influenced by systemic circulation, but there are strong reasons for postulating that physiological ideas therein are consistent with ideas represented in the *Prelectiones,* and that these may have been communicated to Fletcher, and incorporated into *The Purple Island.* Despite the impediments to a specific biographical thesis with which to explain this consistency and possible provenance, there is value in understanding aspects of the *Purple Island* anatomy and physiology in relation to that in the *Prelectiones* and in studying the content, method, and form of Harvey's lectures. Quite apart from the issue of sources for Fletcher's poem is the consideration of the congruity of anatomy with figurality in the early seventeenth century. The present part of this study was introduced, at the beginning of chapter 7, with the proposition that our theory of metaphorical and allegorical meaning in *The Purple Island* allows that the proper meanings of recent anatomy could be among the significant connotations of the Isle of Man, but that the identification of recent anatomy in the poetic stanzas depends upon the figurality with which it is presented being consistent with its essential conceptual character. As we know, the *Purple Island* figurality is highly literary, yet in it the conceptual metaphors and the analogies of the microcosm coalesce in a consistent and coherent image, the threshold figure of the Isle of Man, and the figurality unfolds an etymology of anatomical nomenclature that was not yet in the early seventeenth century standardized. But even though the combination of anatomy with figurality in *The Purple Island* need not necessarily be intellectually conservative, we have not yet fully discerned whether *in fact* it is conservative or not. With the exception of Banister's *Historie of Man,* which is at any rate a compendious synthesis mainly of Vesalius and Colombo, we have not examined sufficiently closely any anatomy text written in the late sixteenth or early seventeenth century, in order to discern whether the soteriological allegory of Fletcher's poetic anatomy would remain congruous, coherent, and consistent if the anatomy were predominantly, or even significantly, more recent than that of the middle of the sixteenth century, that is, more recent than the anatomy of Vesalius and Colombo, notwithstanding the endurance of their influence and legacy.

A comparison of aspects of the anatomy of the heart, veins, and arteries in *The Purple Island* with that in the *Prelectiones* provides us with an opportunity to test the assumption of critics like Nicolson and Baldwin, that had the poem's method of inquiry been more anatomically specialized, the anatomy more faithful to observation, and the physiology more mechanistic and modern, then the poem's allegory could not have been as systematic as it is. As has been said in the introduction to this part of the study, the full extent to which the anatomy in the poem is conceptualized in terms of its allegorical figuration, or whether it is intellectually conservative or innovative, and whether or not this character is a result of or even relates to its figuration, can finally only be answered fully and with precision if we can be reasonably sure of the conceptual character of the *Purple Island* anatomy and its provenance. But the less directly intertextual, more general, comparative question raised here, of whether the allegory *could in principle* (rather than *would*) remain congruous, coherent, and consistent if the anatomy were significantly (consistent with that) of the late sixteenth or early seventeenth century, may be answered whether or not the question of whether Fletcher's *Purple Island* is actually indebted to the Harveian anatomy and physiology of the *Prelectiones* is answerable on the evidence that exists. Furthermore, Harvey's lecture notes throw light upon *The Purple Island*'s rhetorical, methodical, and formal approximation to an anatomical demonstration. But that is a subject for the final chapter of this study.

This section aims to demonstrate that there are strong reasons for postulating that physiological ideas in *The Purple Island* are consistent with ideas represented in the *Prelectiones,* whether or not these were communicated to Fletcher, but there are impediments of two kinds to understanding the status of physiological ideas in Harvey's *Prelectiones:* one of these is chronological; the other is interpretative. The manuscript text of the *Prelectiones* bears many additions, "written between the lines, in the margins and on originally blank pages, containing as they do references to autopsies and quotations from new books or from new editions of books appearing after 1616."[163] The date of significant aspects of the content cannot be ascertained with any certainty. But, it is possible to be reasonably sure that, when he originally wrote the *Prelectiones,* in 1616, Harvey had no knowledge of the systemic circulation of the blood. If the *Prelectiones* should be found to have influenced the anatomy in Fletcher's poem, this at least is consistent with the view that the *Purple Island* cardiovascular anatomy is not influenced by systemic circulation. The passage on folio 80ᵛ which outlines the physiology of "the perpetual movement of the blood in a circle," has been isolated by Whitteridge as "a subsequent addition to the text"; and she defends this contention on the basis of the "character of the handwriting," which "differs considerably from that of the rest of this MS. and resembles more closely that of the MS. of the *De motu locali animalium* which is dated 1627."[164] It is Whitteridge's conjecture that "this passage was added to the *Prelectiones* somewhere about 1627 or 1628, but while this supposition remains plausible," she concedes, "it is incapable of proof."[165]

In support of the conjecture, Whitteridge charts the differences of conception between the *Prelectiones* (excluding the passage on circulation) and *De motu cordis,* disclosing that although, in 1616, Harvey accepted pulmonary circulation because of the simultaneity of the systole of the heart and the pulse of the aorta and pulmonary artery, and the action of the cardiac valves of these vessels in preventing the return of blood to the heart, the *Prelectiones,* like *The Purple Island,* omit any mention of the valves of the veins, which were to be crucial in determining or hinting at circulation for Harvey.[166] This omission may in itself perhaps be thought to render the *Prelectiones* a more likely source than, say, Bauhin's *De corporis humani fabrica Libri IIII* of 1590, or any of his subsequent anatomical works that include a description of the valves in the veins,[167] but Fabrici's demonstration of the venous valves, in Padua sixteen years before Bauhin's description was published, did not acquire great significance until their function in systemic circulation was realized, and might have been omitted by Fletcher. Further to the state of Harvey's ideas in the *Prelectiones,* the description of veins of the leg in the unpublished *De musculis* (MS Sloane 486), which was probably written in 1619, is pre-circulatory, in holding, as does *The Purple Island,* that veins carry blood to the parts of the body.[168] There are, in addition, in the *Prelectiones* several remarks which are "quite incompatible with the hypothesis of the circulation of the blood throughout the whole body."[169]

A determination of when Harvey's experimentation and observation led him to discover systemic circulation would indicate the latest possible date for communication of pre-circulatory Harveian ideas about cardiovascular physiology that may or may not be represented in *The Purple Island.* It obviously does not follow from Whitteridge's interpretation that the passage on folio 80ᵛ of the *Prelectiones* was added somewhere about 1627 or 1628 that Harvey did not arrive at the theory of systemic circulation before this time. Harvey refers, both in the dedicatory epistle to the President (Dr. Argent) and Fellows of the College that prefaces *De motu cordis,* and in the main text of that work, to his "Anatomy Lectures" in relation to his "opinion concerning the motion and use of the heart and Circulation of the blood."[170] In the prefatory reference, Harvey claims to have "confirm'd" this opinion "by ocular demonstrations for nine years and more."[171] After considering the absence of any need to doubt the truth of this statement and the familiarity with Harvey's experiments that Dr. Argent may be supposed to have experienced, Whitteridge proposes that, although the date of Harvey's discovery seems to be "attached by him to some time about 1619," "this does not necessarily mean any more than that he then began to think that the evidence which he had accumulated seemed to point towards the hypothesis that the blood circulated throughout the whole of the body."[172] French is broadly in agreement. He argues that Harvey's reference to the demonstrations refers only to the "forceful systole and nature of the pulse" rather than its "use" which is the circulation. Besides, whatever Harvey may have meant to imply in 1628, French argues, his first announcement was to do only with the systole and pulse.[173] Whitteridge proceeds to state that the process of "observation, vivisection, demonstration and argument" culminated in the confirmation of the hypothesis, as Harvey says, "some years before" *De motu cordis* and she "hazard[s] a cautious guess that 1625 is approximately the right date."[174]

Consequently, the reason there is no contemporary account of the impact of Harvey's lectures on his audience from 1616 until after the publication of *De motu cordis*—a fact that is often remarked upon—is because the public demonstrations to which Harvey refers in *De motu cordis,* are of the preliminary stages of this process, as recorded in the *Prelectiones* in 1616, and not the experiments that led to the culmination of the process in systemic circulation.[175] The absence from any of the additions to the *Prelectiones* made during the period 1616–26 of any indication that Harvey was in possession of the theory of circulation, and his failure to amend any of the statements therein which are inconsistent with this theory, may be accounted for by the probability that the public anatomies were broadly designed to teach basic anatomy, and not to engage in controversial subjects.[176] The observations, experiments, and arguments regarding, for example, the simultaneity of cardiac systole and the pulse of the aorta and pulmonary artery, which are recorded in the *Prelectiones,* "were sufficiently advanced [by 1616] for [Harvey] to be certain that what he had to say on the subject was true,"[177] or that it was significantly, even consensually, regarded as such.

It is appropriate in Harvey's case to reiterate the view of this study that there are no ahistorical criteria of judgment by which the truth of scientific statements may be tested. Generally with regard to an anatomical doctrine, and specifically in the case of pulmonary circulation in the context of the *Prelectiones,* there were many reasons why a consensus might be reached—economic, religious, and political—and among them there were those concerned with education, such as "formal structures of argument and expression" (using, for example, the anatomical *accessus*), rhetoric, and observation, as well as social and intellectual groupings, age, training, and occupation, among others. The precise function of systole and diastole was notoriously elusive, and, as French argues, one could not "see" the motion of the blood, so this network of mechanisms and conditions "lay between the uncommitted and his acceptance or rejection of the circulation."[178]

For our study of Fletcher, the net effect of the protracted process whereby the theory of systemic circulation was reached by Harvey himself, the circulation then demonstrated to colleagues, and the theory gradually accepted by them, is to introduce 1625 as approximately the latest possible date for communication between Harvey and Fletcher. But given the nature of the 1625 date as a cautious estimate, and the significance of 1619, when the unpublished *De musculis* was probably written, it might be safer to assume that communication probably occurred (if it occurred at all) much earlier than 1625, and thus quite possibly in the period of Fletcher's residence in Derbyshire until 1621.

It may be found on the evidence of the *Prelectiones* that, for some of the period between 1616 and 1625(?), Harvey's ideas on the following are interestingly similar to those articulated in *The Purple Island:* the action of the cardiac valves of the aorta and pulmonary artery and the movement of blood in the arteries; the existence of pores in the interventricular septum; (to some degree) the action of the mitral valve and also that of the tricuspid valve; the movement of blood in the vena cava, and in the other veins. As a proportion of the conceptually significant features of cardiovascular anatomy and physiology in *The Purple Island* discussed thus far in this chapter, I do not think that any other single author or work has a greater tally of similarities than does Harvey in the *Prelectiones*. Harvey does not describe the physiological structure of the cardiac valves, however, and so Fletcher could not have derived his detailed descriptions from Harvey's anatomical lectures. It has been argued that the provenance of *The Purple Island*'s description of the cardiac valves, and of pulmonary transit of blood from the right ventricle of the heart to the lungs is of uncertain provenance, but may find its origin in either Colombo's *De re anatomica* or Banister's *Historie of Man*. Although these represent greater consistency with the cardiovascular anatomy of *The Purple Island* than any other works discussed thus far, Colombo, and Banister after him, deny the existence of pores in the interventricular septum, and are comparatively silent on the movement of blood generally in the arteries and veins.

The competent operation of those valves in the aortic and pulmonary orifices is assumed in Harvey's observations and experiments on systole and diastole, referred to above, and in his acceptance of pulmonary transit.[179] Harvey concluded from these observations that the pulse comes from "the thrusting forth of the blood by the heart." He extends this conclusion to the pulse throughout the body: "Action of the heart: . . . contracted, scups it over. The whole of the body responds to the [aorta] artery as my breath in a glove."[180] If the valves of the aortic and pulmonary orifice prevent a flow backward into the heart, then the blood in the arteries *throughout the body* must, as they do in *The Purple Island,* "creeping forward slide, but never turn again" (*P.I.* 2.9.7), though it might be added that "creeping forward slide" hardly conveys the propulsion of arterial blood as effected by the forceful systole of the heart in Harvey's *Prelectiones*. Nevertheless, Harvey was making a point about the muscular cardiac action whereby the blood in the arteries is moved in one direction, not necessarily about the velocity at which it moves throughout the body, and the description in *The Purple Island,* for which it is doubtful there is any authority in Cesalpino, and for which there is none elsewhere that I know of, except perhaps in the work of Galen (though such a reading of Galen, as is explained above, would be unorthodox, even exceptional), may find no more likely provenance than Harvey's direct or indirect communication.

Unlike in Fletcher's poem, there is ambivalence in the *Prelectiones* about the interventricular septum.[181] Harvey marks the belief in its porosity with an "X," which is used to mark his disagreements with Bauhin or other anatomists, and yet he cites as evidence in favor of the pores Bauhin's positive observation in the heart of an ox.[182] It has been suggested that Harvey's opinion was that of many advocates of pulmonary circulation, namely that, although most of the blood went through the lungs, nevertheless a little seeped through the pores.[183] If so, Harvey's opinion, in 1616, was not as dissimilar to the opinion of Fletcher as it was to become when Harvey developed his theory of systemic circulation.

Also differing from the physiology of systemic circulation is the operation of the tricuspid valve and the movement of blood in the veins, in the *Prelectiones,* where Harvey's account is considerably more involved, and more enigmatic, than that of Fletcher. He describes how during systole of the heart, unlike the other veins which do pulsate, the vena cava and pulmonary vein do not pulsate, but "suffer an attraction" or "suffer to be dragged down."[184] Harvey explains that the reason for this "contraction" in the veins that do not pulsate is "because the valves set in a contrary direction break off the pulse both in the heart and in the other veins."[185] The explanation presents difficulties for interpretation—"break off the pulse" may mean "prevent" the pulse, or it may mean "slow it down"—but Harvey seems to suggest that whereas the veins that do pulsate have valves that "break off" or impede the pulse to some degree, the pulse is impeded in the pulmonary vein and the vena cava, which do not have valves, by the competent operation of the mitral and tricuspid valves "in the heart." His suggestion is that the closing of these cardiac valves prevents or slows down a flow of blood from the left and right ventricles into the pulmonary vein and vena cava, which thereby fail to pulsate. Harvey goes on to say, that "for this reason, the veins have very many valves opposed to the heart, while the arteries have none except in the exit from the heart in contrary fashion. Hence these veins are capable of pulsating and these indeed are not."[186] Given Harvey's conclusion that the pulse in the arteries is caused by "the thrusting forth of the blood by the heart," those veins which do pulsate, which are, in this view, all but the vena cava and the pulmonary vein, logically must also receive a "thrust" of blood from the heart.

Whitteridge's judgment on this enigmatic passage is as follows: "If Harvey thinks that [the valves in the veins'] use is to break off the pulse in the veins, as the valves of the heart break off the pulse, then it follows by implication that he thinks that their use is to delay the flow of blood in the veins and this is what Fabricius had believed."[187] The reader may recall that Fabrici's theory of the use of the valves in the veins presumes at least a predominance of flow *from* the liver and heart *to* the rest of the body. And this is in contradistinction to the function of the valves of the veins in systemic circulation, which is to prevent a *backflow* of blood in the veins and thereby maintain its *forward* movement *toward,* not *from,* the heart. In the Galenic model of Fabrici, to which Harvey seems to adhere in the *Prelectiones,* blood being thrust into the veins from the heart is compelled by the nature of the anatomical connections to enter them via the vena cava, or, in smaller quantities, the pulmonary vein. So the blood from the liver that is thrust from the heart passes through the vena cava or pulmonary vein, neither of which pulsate, in order to reach the other, pulsating, veins through which the blood reaches the other parts and the extremities. Although the vena cava and pulmonary vein are said to have neither pulse nor valves, this is because the attractive power of the heart, rather than valves, slows the blood down, though the tricuspid, and mitral "valves of the heart" contribute to this outcome. What is achieved by a combination of the attractive power of the heart and the physiological structure of the tricuspid and mitral valves in the vena cava and the pulmonary vein is achieved by the physiological structure of the "very many valves" in the veins.[188] On this view, neither the mitral valve nor the tricuspid valve nor the valves in the pulsating veins actually prevent the flow of blood from the heart to the extremities, they temporarily intercept or slow it down, but the two cardiac valves in this description "break off" the pulse, that is, prevent it, whereas the valves in the veins "break off" the pulse in the sense of impeding rather than preventing it.

It would appear that Harvey attributes less physiological significance to the anatomical structure of the cardiac valves than we might retrospectively expect from such an observant and innovative anatomist, and indeed less than is attributed by Fletcher in Canto 4 of *The Purple Island.* Because Harvey notes that the valves in the veins and in the atrioventricular orifices of the heart are set "in a contrary direction" to the pulse, that is, the venous, tricuspid and mitral valves are afferent with respect to the heart, he evidently did not think that their structure implies that they prevent blood flowing in that direction, that is, from the heart to the extremities. In this, at least in respect of the mitral valve, he appears to concur not with Fletcher, but with those like Bauhin who ignored the valve's competence and allowed a two-way passage of blood through the pulmonary vein: of natural blood inwards to be made vital, and of vital blood outwards for the use of the lungs. Yet when in the passage cited above on "attraction," Harvey pays attention to the physiological structures of the semilunar valves of the aortic and pulmonary orifices, he states that the semilunar valves, where the arteries "exit from the heart," are placed "in contrary fashion" to the "very many valves" in the veins, that is, they are efferent with respect to the heart. If we adopt Harvey's reasoning on the tricuspid, mitral, and venous valves, it ought to follow that the semilunar valves "break off" the pulse of blood flowing back *toward* the heart in the arteries, that is, not actually prevent the flow of blood from the extremities, but temporarily intercept it or slow it down, as the tricuspid, mitral, and venous valves are thought to slow down blood moving *from* the heart. But Harvey does not draw this inference; rather, as was said earlier, he says that because of the forcefulness of cardiac systole the whole of the body responds to the aorta as breath in a glove, and there is no suggestion that the blood flows back along the arteries to the heart at all. In this respect at least, Harvey's *Prelectiones* is alone among those anatomical works examined in this chapter in concurring strictly with the movement of blood in the arteries in *The Purple Island.*

Moreover, Harvey's account of the movement of blood through the heart and lungs assumes the competence of the semilunar valve at the orifice of the pulmonary artery, in preventing a flow backward into the heart. Even more significantly, and in contradistinction to the account of "contraction" in the veins that do not pulsate, discussed above, Harvey's account of pulmonary transit assumes the competence of the mitral valve in the left atrioventricular orifice, and bears a close resemblance to the account by Colombo, whose theory of pulmonary transit Harvey had read and accepted by 1616 and refers to in the *Prelectiones.*[189] Harvey's account also bears similarities to the account by Cesalpino, which, however, he appears not to have read.[190] Although there is no evidence to suppose that Colombo's observations of the cardiac valves and of pulmonary transit led him to question the notion of blood ebbing and flowing in the other parts of the body,[191] Cesalpino did realize that the competence of the tricuspid valve, like that of the mitral valve, implies the centripetal flow of blood in all the veins towards the heart.[192]

In the *Prelectiones,* though, Harvey seems to have achieved a semantically obscure and awkward compromise between two ideational systems. On the one hand, there is the competence of the cardiac valves, especially the mitral and semilunar valves, in the context of forceful systole, arterial pulse, and pulmonary transit; on the other hand, there is the centrifugal flow of blood in the veins from the liver and the heart to the body. In the first context the mitral valve is competent and the tricuspid valve can be assumed competent in admitting blood into the right ventricle of the heart from the liver through the vena cava, as does that in *The Purple Island.* But in the second context the mitral valve and the tricuspid valve serve to "break off," rather than prevent, the pulse of blood thrust out of the heart into the pulmonary vein and vena cava and thence into the veins throughout the body. Harvey's adherence to a broadly Galenic venous system, combined with his interest in the mechanics of systole and diastole as the means whereby blood is propelled both into the arteries and (erroneously) into the veins, not only gives rise to opacity approaching contradiction, but also changes the movement of blood in the veins and arteries from the predominantly but not exclusively centrifugal flow from the liver and the heart to the extremities, to an exclusively centrifugal movement of blood in both the arteries and the veins, such as we find in *The Purple Island.* And it is in concurring with this that *The Purple Island* is unusual among anatomies of any kind—scholastic, Hellenist, humanist, Galenist, Aristotelian, and even very recent anatomy, as well as literary, poetic, and allegorical.

The Purple Island achieves greater clarity than the *Prelectiones*. By sidestepping the question of how blood might physiologically be moved through the veins, Fletcher's poem obviates the need to introduce a mechanism like diastole, systole, and pulsation. It does so by directing blood in a "Nile-like inundation" from the liver "to all the Isle" (*P.I.* 3.11.6, 7). That venous blood flows from the liver as though downhill like water in the geocosm gives the impression that there is no need for blood entering the right ventricle of the heart to be thrust out again, in order to reach the extremities of the body, and therefore no need of systole, and no need of the tricuspid valve's function to "break off" the pulse. The blood in the vena cava which does not enter the right ventricle of the heart needs merely to flow as though according to gravity along the veins to the other parts, thereby leaving the competence of the tricuspid valve unchallenged. Thus all four of the cardiac valves are always in every way effective in Fletcher's cardiac physiology—even the mitral valve—a luxury perhaps that, due to the need to demonstrate some way by which *pneuma zoticon* might reach and vitalize the lungs, many anatomists felt they could not afford. The ambiguity of function of the mitral and tricuspid valves in Harvey's lectures is perhaps less damaging than Bauhin's ignorance of the mitral valve to an argument for provenance, and although the explanatory force of fluvial distribution in the geocosm is considerably different from the muscular mechanism of cardiac systole and diastole, the physiological outcome in each case is similar.

But for their difference in physiological interest and complexity, and consequently (in this case) clarity, the *Prelectiones* and *The Purple Island* agree on the (at least partial) competence of the tricuspid and mitral valves, the total competence of the semilunar valves, and the continuous centrifugal flow of blood in both the arteries and the veins: a combination of physiological ideas that to my knowledge would be unique to either Harvey or Fletcher, if it were not for the fact that they share them. This might be thought to imply that the *Purple Island* cardiovascular physiology is only internally consistent because of the ingenuity of its geocosmic figuration, and so the figurality of the poem is detrimental to recent (possibly Harveian) anatomy after all: but perhaps that would be an overly retrospective evaluation, relying heavily as it does on the absolute discontinuity that would emerge between the Galenic venous system and the competence of the cardiac valves in pulmonary transit, a discontinuity that would loom more largely into view when Harvey's theory of systemic circulation became consensually accepted as true, after 1628.

For a similar reason, it may be historically and bibliographically accurate to say that while the *Purple Island* cardiovascular physiology appears divided from the point of view of provenance, its description of the competent function of the tricuspid valve and of the other cardiac valves need not be indebted to an ideational system and source distinct from the Galenism to which its description of the one-directional flow of blood in the veins may be partly indebted. Not least because that flow is exclusively rather than predominantly centrifugal in *The Purple Island,* it does not seem improbable that, from among the varieties of compromise reached by anatomists in the period between the publication of *De re anatomica* in 1559 and that of *De motu cordis* in 1628, the author of *The Purple Island* derived these aspects of the poem's cardiovascular anatomy from Harvey, whose anatomical lectures demonstrate the fullest satisfactory, if inconsistent, integration of the exclusively centrifugal movement of blood in the arteries and veins with the (albeit inconsistent) competence of the cardiac valves. It remains true to say, however, that the determination of provenance is uncertain even where conceptual character is known.

William Harvey and the Character of Anatomy in the Early Seventeenth Century

The question of provenance aside, the conceptual consistency of aspects of cardiovascular anatomy and physiology in *The Purple Island* with anatomy found in work by Colombo, Valverde, Banister, Cesalpino, and Bauhin, and in Harvey's *Prelectiones,* has consequences for my interpretation of figurality in *The Purple Island,* and for our understanding of the differentiation between poetry and late sixteenth- and early seventeenth-century cardiovascular anatomy. It has been said earlier that if the view were correct that because the principal and subsidiary subjects of the Isle of Man metaphor are substantially modified in conception by their involvement in analogy, the degree of integration of anatomical description in allegory ought to ensure that the meanings of the heart are intellectually enclosed within the superimposed circles of correspondence, we probably could not have discerned the original character of the anatomy upon which Fletcher drew for his poetic allegory. But it has been possible to see how the poem's process of discovery or invention of significant connotations of the Isle of Man extends the threshold figure systematically and clearly, giving coherent structure to similarities between the principal and subsidiary subjects. Often these similarities are present, if only in germinal form, in the conceptual metaphors and analogies of textbook anatomy. So although the cardiovascular anatomy in Fletcher's poem has been assigned only an uncertain provenance, much can be said to be conceptually consistent with comparable, if less figurative, accounts in anatomy textbooks by the aforementioned authors, particularly Colombo, Banister, and Harvey. Nicolson's assumption that, when anatomy is involved in analogy, its conceptual character is not only derived from that involvement, but is consequently conservative, can be seen to be as susceptible to a self-fulfilling hermeneutic circularity, as is Langdale's attempted extraction of the poem's anatomical content from allegory in order to demonstrate its relatively recent character, a character that purports to be confirmed by its independence of the allegory from which it is purportedly extracted.

The Purple Island does not only exceed the poetic exploration of human anatomy in the literary works compared in part 2 in a more precise application of anatomical descriptive criteria to the body, or in a generally more precise, less schematic anatomy, and more systemic, less formulaic physiology; *The Purple Island* also connotes more recent and perhaps consequently more intellectually innovative and challenging anatomy; challenging not to the overall figurative framework of metaphor and allegory, but challenging in the sense that conventional figures, analogies, and metaphors may not always apply to new anatomical discoveries, and so new figurations must be "discovered" or invented to provide an appropriate context of interpretation or justification.

Examples have been selected in this chapter not only to demonstrate the consistency of the *Purple Island* anatomy with that of the aforementioned anatomists, but therein to illustrate how the anatomy in the poetic stanzas is identifiable not because it is conceptually independent of analogy or allegory for what might be regarded as its essential conceptual character, and may therefore be isolated or extracted for analysis, but rather because the figuration of anatomy is consistent with its essential character. The innovativeness of the anatomy has been identified, and Nicolson's assumption has been found to be unjustified not only in the light of *The Purple Island*'s descriptions of the eye and inner ear, but also its figurations of the movement of the blood and the operation of the cardiac valves. Although imagined in a way that is consistent with the figuration of the body as an isle—"thousand brooks / In azure chanels glide on silver sand"—the fluvial figuration of the movement of blood is consistent with the anatomical criteria of functions or movements, for these brooks "creeping forward slide, but never turn again" (*P.I.* 2.9.2–3, 7). Although fanciful in a way that is consistent with the concept of *ingenium* as invention or *fantasia,* something close to imagination or wit, *The Purple Island*'s "band" of "Three *Tritons*" (*P.I.* 4.22.4–5) bearing tridents is shaped by the requirement that they accommodate the detailed structure and function of the tricuspid valve in terms of the descriptive criteria of the anatomical *accessus.* The cusps of the semilunar valve between the right ventricle and the pulmonary artery are figured as "carefull guards" (*P.I.* 4.23.3), in a way that is consistent with the ideological configuration of the kingdom, the order of which is constructed around the overriding principle that it must be defended.

These fanciful, inventive, and in the latter case even conceivably xenophobic, geopolitical figurations do not detract from the physiological function of the valves or the one-directional movement of blood through the vessels. In *The Purple Island*'s pursuit of the allegorical ideal of continuous superimposition, the anatomical rote of observables is applied to the subsidiary subject, the Isle, and the description of the anatomy of the heart is integrated systematically in the poem's allegory, in ways that are consistent with the poem's adherence to a descriptive anatomy that may be found in textbooks or lectures, particularly those written by Colombo, Banister, and Harvey. For this reason it has been possible to suggest not only where Fletcher consistently maintains the anatomical figurality of a probable or possible source, such as Banister's *Historie of Man,* in attributing the number of folds of which each cardiac valve consists, but also where Fletcher may be conjectured as having augmented or departed from the figurality of his probable or possible sources, as when Fletcher declines to call the valves "wickets." Discerning the conceptual character and identifying probable provenance of anatomy is clearly not the same as claiming that the anatomy is essentially free of allegory: if we were to throw off the cloak of allegory in *The Purple Island,* there would be no anatomy to extract, except perhaps for some of that which is in the margin.

The precondition of the fidelity of the anatomical exposition in Cantos 2 to 5 of *The Purple Island* to recent anatomy is the special construction of a set of associated implications of the Isle consisting of many implications that are shaped by the requirement that in the poetic stanzas they accommodate the detailed structure and function of anatomy in terms of the descriptive criteria of the *accessus.* Now that we know that a significant portion of this anatomy was recent at the time when Fletcher was composing *The Purple Island,* or at least Cantos 2 and 4, this fidelity can be explained partly by what we already know about the work's figurative design, which was the subject of part 1 of this study. Although it is based on analogy, broadly conceived within the traditions of the microcosm and microchristus, the *Purple Island* allegory unfolds a creative metaphor rather than an analogy, which not only extends the analogies upon which it is based, but explores the ordinarily used part of the defining concept of the metaphor and uses the ordinarily unused part of that domain (the Isle), in order thereby to "discover" and give coherent structure to similarities between the

Isle and recently observed anatomical phenomena, thereby giving expression to unconventional, previously unseen similarities between the human body and the Isle. Nicolson's assumption that the integration of anatomical description in analogy ought to ensure that it is intellectually enclosed within circles of correspondence is by no means entirely misguided in principle. If it were not for the *Purple Island*'s metaphorical interaction and "discovery" of previously unseen and newly highlighted similarities between its principal and subsidiary subjects, its anatomy would have remained thus enclosed and intellectually conservative. In order to comprehend this in the contexts of hermeneutics and the explanation and interpretation of a figurative work, we require distinctions between the ways by which anatomical content is allegorized; distinctions such as have been formulated in chapters 1, 2, and 4, that rest on the differences in the kind of figure, its relative conventionality and innovation, the extent of its ingenuity or invention, and the degree of metaphorical interaction and of definition of one subject in terms of another.

By way of example, the description of the heart's form and posture in *The Purple Island* derive from the office of principality or sovereignty that the heart shares with its analogues, Jesus Christ (the Son of God), the sun, and the Prince, as discussed above in the introduction to this chapter. Each of the analogues relates to its respective sphere of influence through *convenientia* and corresponds to the others through *aemulatio*. The King or Prince, or a synecdoche for the Prince such as the crown, may metaphorically stand for or give rise to the form of the heart, such that the vena cava and two coronary arteries are said to "Infold" the base or root of the heart "like a crown," and the heart is described as "crown'd with triple wreath, & cloath'd in scarlet pall" (*P.I.* 4.17.5,4,7). But even though the interpretation of this involves the working out of the significant connotations of a metaphor and an allegorical personification, conventionally there are only a given number of preexisting, isolated ways in which this may be done and these are determined by the analogical context. This is metaphor as comparison, as a contracted simile, otherwise the simple-analogy case of metaphor. It is possible that an innovative metaphor may "discover" previously hidden similarities, and the *Purple Island* allegory reproduces many such antecedently existing similarities and "discovers" many which were formerly hidden. But as a creative metaphor that selects, emphasizes, suppresses, organizes, and extends features of the principal subject, the Isle of Man is not merely a matter of language or conventional thought; it is capable of highlighting previously unexplored parts of the domain of the principal and the subsidiary subjects, "discovering" wholly new sets of similarities between the previously unconnected subjects of recent anatomy and a geopolitical realm which is not only Britain but a fanciful contrivance of *ingegno* or *inventio*.

The creative metaphor thereby involves multiple potential points of similarity which open out so that the subject is the predicate in more ways than one; simultaneously the metaphor exhibits a tension between the literal meanings of the newly combined subjects. The unfamiliarity of the combination of subjects ensures that the aspects which the subjects do not have in common are not masked, as they are in a conventional simile or analogy. The disparities in the metaphor are operative and their role opens the significance of the metaphor so that they act to bring out the resultant meanings not actually given in the antecedently existing similarities or even in the meanings or systems of associated commonplaces of the subjects the metaphor brings together. The unproblematic comprehension of an analogy is replaced by the cognitive processing of the similarities between, for example, tricuspid valves and trident-bearing tritons, and in order that the reader is able to "conceit" or imagine the world the work intends to create, he or she is required to focus on the consistency with which number, size, shape, operations, and so on, are preserved in the description, constituting the points of similarity, and ensuring recognition of the anatomy.

In an allegorical structure of conventional analogy, even where the anatomy is masked and unrecognizable, the conceit of the reader in the form of conventional wisdom or erudition would step in to support the explicit comparisons being made. This obviously has a role in reader responses to *The Purple Island,* but because the allegory of *The Purple Island* is structured by a system of entailment metaphors, the most specific of which is the Isle of Man, and only based upon, rather than structured by, at least predominantly, a system of conventional *aemulatio* or analogies, the integrity of the principal subject is maintained and not masked, even as it undergoes a metaphorical interaction. Even the cardiovascular anatomy of Harvey's *Prelectiones* remains unmasked as the cardiac valves become tritons, guards, and nymphs.

So although it may not be entirely misguided to assume that the integration of anatomical description in analogy implies that it is intellectually enclosed within correspondence, nevertheless, as it is, *The Purple Island* connotes more recent and perhaps consequently more intellectually innovative and challenging anatomy than any of the literary works with which it is compared in this study because (or rather, it is enabled so to do because) it exceeds them in the range of previously *unseen* similarities it draws between the principal and subsidiary subjects of the Isle of Man metaphor, and exceeds them in the systematicity of its exploration of the significant connotations or satellite conceptual metaphors to which the metaphorical interaction gives rise. In short, the figurative design, ingenuity, and anatomy of *The Purple Island* imply that its author was equally intellectually engaged by the *inventio* or discoveries of anatomy as he was creatively engaged in the composition of an intricate allegory attributed with *ingenium* and *inventio* by its readers. The character of the connotations unraveled in the allegory is an effect of the work as poetry, insofar as the poetry foregrounds the metaphorical interaction of some of the most recent anatomy of Man with the Isle's associated implications, many of which are specially constructed to accommodate the anatomical *accessus. The Purple Island* is a rare example, even in the early modern period, of the body's ingenious incision or dissection, involving the invention or discovery of recent anatomy as subject matter, with its eloquent figuring forth in poetry, the contemporaneousness of the anatomy being directly dependent upon the creativeness of the figurative design.

On the evidence of this chapter, then, the Isle of Man's innovative connotations are not confined to those relating to the historically incipient and ideologically emergent concepts of individualism and nationalism: they extend to those which, as it were, embody recent anatomy in a figurative, allegorical context which provides anatomical discoveries with a context of justification and interpretation. In theory of metaphor of the kind this study has employed, there is, strictly speaking, only one kind of metaphor that by putting aside, hiding, or downplaying some aspects of the defined concept and by virtue of giving coherent structure to a newly highlighted range of aspects, actually *creates similarities of a new kind* between the ranges of aspects of the defined and defining concepts. This is novel metaphor. The metaphorical interaction of associated implications of Man and the realm that are ideologically emergent, recently discovered, or historically incipient, gives rise to some similarities that are new. But as was said earlier, in Chapter 2, this is not to claim that the threshold figure of the Isle of Man is novel per se.

The knowledge of very recent anatomy would have constituted a remote set of connotations of the human body for poetry, but, unlike in novel metaphor, the contextual effect of this anatomy probably goes no further than actualization of the *potential* range of commonplaces or connotations belonging to the system of associated commonplaces of the focal word or subsidiary subject, the Isle. Beardsley adds to the potential range of connotations "the range of *properties* which do not yet belong to the range of connotations of my language."[193] According to this theory, "remote properties (those not understood as standard parts of the clusters of connotations commonly accepted as relevant to a term) shift when selected and attributed to the referent of another term in the metaphor."[194] "The metaphor transforms a *property* (actual or attributed) into a *sense*." But as Ricoeur argues, "to speak of properties *of things* (or *of objects*) which were not yet meant is to concede that the novel meaning is not *drawn from* anywhere, in language at least (property is a thing-implication, not a word-implication)."[195] Thus it is that novel metaphor, according to Ricoeur and Hausman, is a momentaneous creation of language, a semantic innovation which has no status in language, as already established, neither as designation nor as connotation, and consequently brings about the qualitative growth of language.[196] Karsten Harries agrees that, "in poetry at least, metaphor joins dissimilars not so much to let us perceive in them some previously hidden similarity but to create something altogether new."[197]

It is true that, in *The Purple Island*, the construal of something new "is forced on the hearer or reader because of [the] tension" between the antecedent meanings of the two subjects and the logical absurdity of the metaphorical statement (tricuspid valves are not literally tritons with tridents);[198] yet as was said in chapter 2, the *Purple Island* case is more like the modeling of anatomy on architecture in Leonardo's anatomical drawings, though on a smaller scale. In *The Purple Island* there is an interpretation that reaches beyond the traditionally held correspondence between Man and geopolitical island kingdoms. The analogy becomes a metaphor in which the correspondence not only attains systematicity and considerable conceptual content, but is extended to encompass and provide a justifying context for recent anatomical investigations. The specially constructed set of associated implications of the subsidiary subject of the Isle of Man are shaped to accommodate the knowledge derived from these investigations, and novel similarities between islands and Man emerge from the metaphorical interaction.

Langdale's claim that the anatomical character of the heart and vascular system in *The Purple Island* is influenced by Harvey's theory of systemic circulation, may seem to imply that Fletcher met a much greater challenge to his allegorical figurative design and soteriological purposes than the descriptive anatomy of Vesalius or Banister, or even Colombo or Bauhin or Cesalpino. But the cardiovascular anatomy and physiology in Fletcher's poem is not consistent with systemic circulation, except perhaps incidentally, and is very unlikely to have been derived in any measure whatsoever from an account of it. Nevertheless, the previous section's analysis demonstrates that *The Purple Island* is indebted to accounts of the operation of the tricuspid valve written between the publication of Colombo's *De re anatomica* in 1559 and the composition and presentation of Harvey's *Prelectiones* in 1615–16. A difficulty that might be thought to have latently beset anatomy in this period was of acknowledging the competence of the tricuspid valve in allowing blood from the liver to enter the right ventricle of the heart from the vena cava and the auricle, preventing its return by the same route, but simultaneously permitting the valve to provide no obstruction to, or perhaps merely to "break off," the flow of blood from the ventricle into the vena cava during cardiac systole. As was said in the previous section, the difficulty might be thought to suggest that the *Purple Island* cardiovascular physiology is only internally consistent because of the ingenuity of its geocosmic figuration of the movement of blood in the vena cava as a fluvial inundation of the parts with blood and *pneuma physicon* or *spiritus naturalis,* that is, natural spirit. Such an interpretation would suggest, as Nicolson did, that the figurality of the poem is detrimental to the anatomy, in this case not improbably the anatomy in the *Prelectiones* of Harvey, which account for seemingly mutually exclusive competences of the tricuspid valve only by almost becoming self-contradictory. No doubt there is tension in Harvey's account, but it is only with hindsight that one can analytically discern contradiction between distinct and incompatible ideational systems.

In order better to understand the conceptual, linguistic, figural, and discursive differentiation between the *Purple Island* anatomy and the philosophical and medical anatomy of the early seventeenth century, in the present section, we will look at how the *Purple Island* figurality relates consistently or otherwise to the figurality of Harveian anatomy, which is here understood not specifically as a source of the *Purple Island* cardiovascular anatomy, but as representative of the most progressive anatomy not only among probable sources of the *Purple Island,* but generally in England in the period until the absolute discontinuity emerged between the Galenic venous system and the competence of the cardiac valves in pulmonary transit, that is, when Harvey's theory of systemic circulation became consensually accepted as true (which is not the same as saying, 1628, when Harvey published *De motu cordis*). The establishment of the theory of systemic circulation and the demise of the Galenic model of the distribution of blood, humors and *spiritus naturalis,* coincided more or less with the emergence of what has been referred to as the Cartesian construction of the mechanical body, and of other versions of the mechanistic model of bodily functions, movements, uses, and purposes, upon which much subsequent anatomy was to be based.

The dominance of the machine model of human anatomy coincided historically with attempts by the natural philosophers of the Royal Society to exclude theology, questions of the soul, parable, allegory, metaphor, including conventional conceptual metaphor, and much analogy, including Aristotelian analogy, such as we find in Benedetti, from the discourse of natural philosophy and medicine. This was no historical accident, for the seventeenth-century machine model accompanies this desire for linguistic purification, and in broader terms, both may be seen as part of the beginning of the bifurcation of the ways and traditions of poetry and science. At this time it was believed that purifying "scientific" language of figurality and analyzing the operations of the body in terms of automata or moving machines would grant anatomists the capacity to describe in an epistemologically transparent way that which exists independently of language and of the conceptualizations according to which it is described.

It was concluded in chapter 6 that a determination of the precise contours in *The Purple Island,* of the conceptual, linguistic, and discursive differentiation between anatomy and poetry, and consequently of how far they are congruous or synthesized, is possible only by resisting the collapse of this differentiation into that between the stanzas and the marginalia in the work's formal disposition. Rather, it was proposed, we must examine the conceptual character of the anatomy and its relation to the poem's figurality, and in order to judge how *The Purple Island* in this regard relates to anatomy in the early seventeenth century, whether in both there is a "refusal" to acknowledge the difference in the forms of discourse appropriate to their respective disciplines, we must examine seventeenth-century anatomy. This section will begin by examining the proposition that, if the theory of systemic circulation is essentially mechanistic, as is thought by some historians,[199] and the mechanistic conception of nature is a nonmetaphorical model, as was thought by its seventeenth-century advocates, then *The Purple Island*'s allegorical figuration of aspects of cardiovascular anatomy that were to contribute to Harvey's hypothesis, such as forceful systole and the exclusively centrifugal movement of blood in the arteries, might be regarded either as detrimental to the original anatomy, or as superficial, as indeed Langdale thought they were, though not for precisely the same reason: he did not think that the anatomy that could be extracted was only Harveian anatomy, but also Vesalian and other Renaissance anatomy. As either detrimental to the original anatomy or as superficial, Fletcher's allegory might be thought to have "refused" to acknowledge the difference in the forms of discourse appropriate to the disciplines of poetry and anatomy. But the question of whether in his work Harvey "refused" to acknowledge such a distinction must also be answered.

It seemed to seventeenth-century advocates of the notion of the human body as an incomparable moving machine, that this mechanistic conception was inherently inimical to allegory and other forms of figurality. But it did so because of a number of related assumptions, among which are, for example, the objectivist assumption that metaphor is misleading because it is semantically false, and the confusion that the mechanistic conception of nature is bereft of metaphor, that the mechanistic model for understanding nature is nonmetaphorical. Regardless of whether one holds out for the possibility of objectively verifiable, true linguistic statements about natural phenomena, a philosophical issue that lies beyond the scope of this study, the linguistic structure of the project to understand human anatomy mechanistically as formulated and proposed by René Descartes (1594–1650) is metaphorical: "the operations of the body have to be analysed *in terms of* the 'many different automata or moving machines the industry of man can devise'" (my italics).[200] The project thus conceived has a mediated not an unmediated or transparently mediated relation to the (always putative) ontology of the body. The project is moreover ideological, in that the metaphorical mediation is occluded from the conception of the model, and the body is imagined as being in reality, in itself, mechanistic, when understood objectively.

The machine metaphor is like the visual feature of anatomical illustrations that was examined in chapter 6 as having arisen from techniques to overcome representational challenges and technical questions, and as having optimistically gestured towards an epistemologically transparent form of representation. The similarities newly and coherently highlighted by the machine metaphor and the "truths" purportedly revealed by the anatomical incision or dissection were located increasingly within the confines of the body itself, as though the truths and connotations were neither still replete with nor produced by an imaginative rationality, and so in each case the "discovery" was frequently at least as rhetorical as objectively real, that is, an elegant or eloquent figure which is also artificial or extravagant in gesturing at something which is not. That the mechanistically inclined among the linguistic purifiers of the Royal Society were antipathetic to allegory and other forms of figure only because their distinction between the machine model and other figurative modes and types was ideological, can be seen from an effect of the seventeenth-century rise to dominance of the machine model, which was to erase or suppress the signs of its historical production as one among other metaphorical modes of comprehension which, as chapter 1 has argued, were all rationalized in relation to teleology and rendered coherent by sharing philosophically and theologically significant entailments.

This study has developed a theory of allegorical, analogical, and metaphorical meaning to interpret the figurality of *The Purple Island,* to reconstruct and analyze its rationality, and to explore its congruity. Analysis of the metaphorical structuring of the anatomy of the human body in terms of vegetation and "building," in sixteenth-century anatomy textbooks, was crucial to developing this theory of the imaginative rationality of *The Purple Island,* and revealed that analogies between animal (including human) anatomy and that of plants or the structural organization of buildings were found to give rise to metaphorical expressions which were conceptually shaping, structuring, and defining of that anatomy. The conceptual metaphors of vegetation and "building" were found jointly to provide a coherent understanding of the concept of anatomy as a whole, as both natural and constructed, and human anatomy was found to be systematically conceptualized in terms of a host of metaphorical concepts, such that conceptual metaphor underlies surface metaphorical expressions and characterizes anatomical knowledge, eschewing the distinction between "literal," "scientific" language and purportedly ornamental figurative language, particularly with regard to the conceptual content of utterances of either kind.

Contrary to the views of the mechanical philosophers and linguistic purifiers of the seventeenth century is the view that conceptual metaphor, including the metaphor of the human body as an automaton or a complex set or unit of mechanisms, is the means whereby human beings get a handle on concepts. This has been discussed in chapter 1, where it is also proposed that truth, generally speaking, is not objective but relative to a conceptual system in large part defined by metaphor, by virtue of which metaphor can be true or false. Natural-philosophical *inventio* consists not only of discovery, but also of the interpretation of what is found via metaphor, models, and other forms of figurality, the understanding of which contextualizes discovery in the conceptual system, which may in turn alter in response to the emergence into dominance of new metaphors and models. Analysis of the figurative design of *The Purple Island* in this study has revealed not only that the anatomy therein is figured in vegetative, terrestrial, and architectural terms, and that it is rationally consistent with descriptions in anatomy textbooks, but also that it provides often in allegorical form a clear indication of the function of structure, and the *telos* or final cause of structure and function in some purpose of the soul, which is embedded within and implicit to the figurative language. Although this view has been found to be valid in respect of the sixteenth-century anatomy of the schools and medical profession, what has not been analyzed are the implications and entailments of the metaphoricity of the anatomy of the early seventeenth century, and this is to be done in the present section, most immediately with regard to the machine model in anatomy.

The reduction of the body "to a mere mechanical contrivance" in Cartesian philosophy, following the principle that the operations of the body have to be analyzed in terms of the "many different automata or moving machines *the industry of man can devise*" (my italics), was preceded historically by other less reductive and less totalizing mechanical conceptions.[201] These can be found at least as far back as Galen; they did not accompany a desire for linguistic purification and they were consistent with the more basic conceptualization of the body as constructed and created by God, though not as a building for the soul but nevertheless as a product of divine ingenuity. But whereas for Galen, "the demonstration of divine art is the main aim" because "everything in the body . . . is understandable only from the point of view of its instrumentality to the soul," for Descartes, by contrast, "the body functions according to mechanical laws. It does not need a soul."[202] So when in the *Discourse on Method,* Descartes says that we shall consider the body "as a machine which having been made by the hands of God is disposed incomparably better and has in itself more attainable movements than any of those that can be invented by men"; "[t]he idea of divine workmanship in analogy to human automata merely serves to explain the possibility of such a marvellous construction. From a biological point of view, Descartes" identification of the body with a machine is a metaphor by which he tries to give a concept of an organized natural object regulated merely by matter and motion."[203]

As a metaphorical conceptualization of human anatomy, Cartesian mechanism has metaphorical entailments that it shares with the conceptualization of anatomy as "building," and among these is the entailment that on account of machines being built for a purpose, the machine body needs also to have a purpose. But unlike the case of anatomy as "building," this entailment assumes significance "if applied to the concept of man who has a rational soul totally distinct from his body," because "[f]or the purpose of man's conscious and purposeful life, the body can indeed be considered as a machine that will run according to the manipulations of the machinist."[204] This is to ascribe purpose not to the body-in-itself, which has no purpose of its own, but to its use by the mind, which is distinct from the final cause or purpose in Aristotelian teleology, for that manifests itself as structure understood physiologically. This is why Descartes answers the question of the purpose of animals by supposing them not to have a soul at all (not even a vegetative or a motile or a sensitive soul).[205] This shift in physiological model led to that which Foucault describes as "the *disciplines:* an *anatomo-politics of the human body,*" which, beginning in the seventeenth century, "centered on the body as a machine: its disciplining, the optimization of its capabilities, the extortion of its forces, the parallel increase of its usefulness and its docility, its integration into systems of efficient and economic controls."[206]

As with Renaissance anatomy, it has been found in Fletcher's poem that the metaphorical concepts of anatomy as vegetation and as "building" share philosophically and theologically significant entailments: the concepts of creation or construction; death and mortality or ruin; and regeneration, rebuilding, and renewal. Because the first of these sets of entailment concepts is the topic of events 1 to 3 of the main story line of *The Purple Island,* as analyzed and delineated at the beginning of chapter 3, and because the second set of concepts concerns events 4 and 5, and the third set occupies the thematic focus of events 6, 7, and 8, the character of the *Purple Island* anatomy also derives from Christian theology, and particularly from the moral and natural theology of Creation and the Fall, and the soteriological doctrine of the incarnation and salvation. In providing moral and theological argument from design, *The Purple Island* conceptualizes the intricate structure and movements of the human body as an ingenious artifice or God's wondrous art, but there is no suggestion in Fletcher's work that the body should be understood as a machine, and certainly no presentiment of a concept of the body as functional without a soul.

Arising from these observations is the question of the extent to which early seventeenth-century anatomy and that of Harvey in particular can be seen to have been mechanistic in conception. The question is important for our assessment of the congruity between anatomy and theologically rationalized figurality in the early seventeenth century, and our understanding of the relationship between the *Purple Island* anatomy and that taught in the universities at Bologna, Padua, and elsewhere in the years during and immediately preceding the poem's composition. *The Purple Island* demonstrates that the conceptual character and imaginative rationality of sixteenth-century anatomy need not be in conflict with a soteriological perspective on Man, and that this perspective permits the coherence and systematicity of conceptual metaphors and analogies, through concepts about the body's creation, mortality, and salvation. The poem also demonstrates that the idea about the human body, that God who made it never will forsake it, is dependent upon the incarnation, death, and resurrection of Christ. But if the idea of divine workmanship in relation to human automata merely serves to explain the possibility of such a marvelous construction, which otherwise functions as an organized natural object regulated merely by matter and motion, there is no need to insist upon the necessity either of an enforming soul or a savior to explain its construction. For these reasons the seventeenth-century machine model of human anatomy marks the commencement of a polarization of religion versus the investigation of Nature.

Intent on charting the progressive ways by which Harvey's methodological innovation and the discovery to which it would ultimately lead, contributed to the refinement of scientific method and consequently to other discoveries, positivist historians of science have characterized Harvey's innovation as moving beyond mere observational anatomy in devising and repeatedly conducting experiments to test hypotheses, such as the one that the pulsation of the arteries is caused by the forceful systole of the heart, therein demonstrating the mechanism of cardiac systole and diastole.[207] Additionally, Harvey's quantitative observations on the amount of blood pumped by the systolic action of the heart are often thought to have led directly, as Harvey himself might seem to claim, to his realization that the liver could not possibly produce so much blood from the chyle, and, therefore, to his hypothesis that the same blood must continually circulate in the body. On this view: "It was clearly the meditation on the amount of blood leaving the heart and the allied investigations which moved him to think of the circulation."[208] Harvey's supposed hypothetico-deductive method is thought to comprise the contexts of project formulation and justification in which he conducted his anatomical investigations and quantitative experiments, and for this reason Harvey occupies an unprecedented threshold in the history of scientific investigation.[209]

More recently, however, Roger French has argued that Harvey followed a disputational-commentarial method, consistent with a natural-philosophical research method as opposed to a modern scientific one, and as was suggested in the previous section in this chapter, Harvey developed a thesis of forceful systole as a kind of commentary or development upon Colombo's observations, and this provided the contexts of project formulation and justification. French argues: "Harvey did not come to the notion that the blood circulated by making inductions from a mass of evidence and neither was he making quantitative experiments to discover how much blood passed from veins to arteries. Somewhere deep in the modern historiography of Harvey scholarship is the notion that Harvey's method was important because it was quantitative, like modern science. Undoubtedly Harvey's estimate of the amount of blood ejected from the heart helped to convince some of his readers of his case. But it was hardly scientific. As a 'quantitative method' it was far less precise than those of Sanctorius and van Helmont; it was not an experiment or even an observation, but a mode of argument, or rhetoric."[210] Accordingly, to draw from Harvey's innovation an interpretation of his work as exemplary of "the new experimental and mechanical approach to nature of seventeenth-century science" is retrospectively to elide both the disputational-commentarial method by which anatomy was conducted in the early seventeenth century, and Harvey's deeply informing Aristotelianism.[211]

It is not the object of this chapter to analyze thoroughly the metaphoricity of the machine model of nature for two reasons. The first is that the mechanics of Harvey's physiology of the heart and blood have been overestimated by positivist historians of science. In the passage on folio 80ᵛ of *Prelectiones* which outlines the physiology of the perpetual movement of the blood in a circle, referring to the systole of the left auricle and ventricle, Harvey does say that "the blood is perpetually carried across through the lungs into the aorta as by two clacks of a water bellows to raise water."[212] But the technological simile is not enlarged. Similarly, the well-known analogy between the human heart and the pump does not occur in *De motu cordis* and was used for the first time only in the Second Letter to Riolan, written some time before the First Letter, probably before, but not much before, 1640, and the analogy may have been suggested to Harvey "by the sight of the fire-engine of new design at work in London in 1637 on the occasion of the fire in the house of Harvey's friend, Thomas Howard, Earl of Arundel."[213]

The second reason that this chapter does not thoroughly analyze the metaphoricity of the machine model of nature is that, as this example suggests, the machine model emerged only towards the middle of the seventeenth century in the form that was to become dominant. In the first half of the century, figurative insights such as this in Harvey's work were isolated and not conceptually mobilizing, at least, that is, outside of the work of the early mechanist philosophers, like Descartes, but such instances of mechanistic imaginative rationality were incremental, and over the course of the century we find what David Edge calls "the slow, steady working out of the affective implications" of the machine metaphor: "As the metaphor begins to 'bite,' cognitively, it brings with it attitudes appropriate to [the subsidiary subject's] literal referent."[214]

It is my view, therefore, that the model's absence from the *Purple Island* anatomy is not an indication of the poem's intellectual conservatism. This view is clearly distinguishable from that of Healy, for whom the intellectual struggle in Fletcher's poem is situated in "the apparent strains within the poem which seem to have emerged from the fracture between ideas of correspondence with more mechanistic accounts of the body's function." Healy assigns to the conjunction of poetic stanzas and marginal scholia the burden of this struggle, and puts his finger on the means of modification required for the poem's appropriation when he says that Fletcher embraces ambiguity or the "instability of poetic metaphor." In summary, Healy understands Fletcher's poem as appropriating "contemporary anatomical ideas" and (what he understands as) methods of empirical investigation and presentation, in order to validate ideologically "an intellectually reactionary project."[215]

By comparison with the sites in the text that stage the purportedly reactionary political project, however, the "strains" of the struggle between the emergent "mechanistic" conceptualizations of anatomy and its implications for the demands of correspondence gain little of Healy's attention.[216] There is one obvious reason for this. Healy is unconcerned to pursue the accuracy of *The Purple Island*'s anatomical content in relation to recent anatomy, as established initially through Langdale's efforts to provide a case for the progressive character of the work, which is of course denied by Nicolson and Baldwin.[217] Healy dismisses the project of Langdale and others as "based on what can be termed largely teleological concerns" about poetry's "accuracy in depicting scientific explorations" and ideas, and about the ideas and explorations in so far as "they helped to inform later (positive) developments."[218] Moreover, Healy's use of the concept of "appropriation" ought theoretically to allow the poem's description of purportedly empirical anatomy and mechanical physiology to contradict that which is based upon correspondence and teleology, without threatening (or subverting) the overall conservative framework of correspondence required of allegory. This is because appropriation as a strategy is capable of resolving ideologically the contradictions between the appropriating power and the social or cultural elements or practices that are appropriated, which means resolving them in the interest of the appropriating power. Consequently, unlike Baldwin and Nicolson, who also think of the poem as conservative, Healy has no difficulty in admitting that *The Purple Island* "exceeds any other comparable English text of the period in detailing a description of the body's anatomy according to Vesalian designs."[219]

In one sense, Healy is clearly right. When positivist or idealist concerns to extract anatomical content govern readings of texts other than the properly scientific, in the modern sense, they do an injustice to the modifications—figurative, linguistic, generic, discursive, narrative, ideological, or psychological—which characterize those texts. Such texts gain their distinctiveness from, among other things, the fact and character of this modification, which is also capable of revealing the material, intertextual, and cultural relations of texts and their rationalization of their subject matter. But the fact that Healy declines to detail a defense of the Vesalian character he claims for much of the anatomical content, and that he does not even comment on the debate over its Harveian influence, has adverse consequences for his arguments. In my view, it is only by focusing on the "sites of struggle" in the text, when one is aware of the accuracy or otherwise of the anatomy being described, in relation to contemporary anatomical knowledge and language, that one can assess the validity of the application of the term "appropriation" to the struggle or negotiation occurring within the text.

Since the "strains" of the struggle between the purportedly emergent mechanistic conceptualizations of anatomy and its implications for correspondence gain little of Healy's attention, it is incumbent upon the present study to present an objection to the proposition advanced above, that the absence of mechanism from the *Purple Island* anatomy is not an indication of the poem's intellectual conservatism. An objection might refer to the correct observations of diastole and systole, recorded in the *Prelectiones,* which would lead indirectly, yet ultimately, to the mechanistic and quantitative methods of later seventeenth-century anatomy and physiology.[220] Although *The Purple Island* represents the heart as a muscle, it makes no mention of the diastole and systole of the heart, as the previous section remarked. If one leaves aside the possibility that Fletcher may have followed Hippocrates in regarding the heart to be a very strong muscle, and leaves aside also the fact that what may or may not have been communicated to Fletcher by Harvey is unknown, an argument can be made that Fletcher follows anatomically, but not physiologically, Harvey's preliminary observation (with regard to his discovery of systemic circulation) that the hardness of the heart during systole proceeds from the same kind of tension as that of other muscles during contraction, and must therefore be considered as like that of any other muscle.[221] Fletcher may appear to have been careful to omit the most mechanistic aspect of the action of the heart and the movement of the blood in Harvey's physiological account prior to and including *De motu cordis.* Moreover, Fletcher does not merely omit diastole and systole. He relies instead on the explanatory power of fluvial motion, as the previous section remarked, and on macrocosmic and christological correspondences, in which the heart is said to be "In motion still (a motion fixt, not roving) / Most like to heav'n in his most constant moving" (*P.I.* 4.16.5–6). In this way, an argument may be constructed that Fletcher perhaps appropriates mechanism (insofar as it existed) for the dominance of the theories of the microcosm and microchristus, by appearing to have reconciled mechanistic and microcosmic theories in describing the heart as a muscle.

A defense of the proposition that the absence of mechanism from the *Purple Island* anatomy is not indicative of the poem's intellectual conservatism might likewise refer to Harvey, on the grounds that he was the most progressive anatomist not only among Fletcher's probable sources, but generally in England in the period until the absolute discontinuity emerged between the Galenic venous system and the competence of the cardiac valves, that is, when Harvey's theory of systemic circulation became consensually accepted as true, the Cartesian and other constructions of the mechanical body emerged into dominance, and the natural philosophers of the Royal Society sought to exclude theology, questions of the soul, and various figures from the discourse of natural philosophy. Firstly, there is the general point, made above, that the mechanics of Harvey's physiology of the heart and blood have been overestimated by positivist historians of science, whose accounts have proved influential. More specifically, what the marginal comment in Canto 4 of *The Purple Island* actually says is: "The flesh of the heart is proper and peculiar to it self, not like other muscles" (*P.I.* 4.17 margin. n). Because this does not even sound like a response to Harvey's observation that the hardness of the heart during systole proceeds from the same kind of tension as that of other muscles during contraction, and must therefore be considered as like that of any other muscle, the comment is unlikely to have been devised to give the poem the appearance of having reconciled mechanistic and microcosmic theories in describing the heart as a muscle. Secondly, the systole and diastole of the heart constituted for Harvey the simple action or motion. In the reorganization of the anatomical *accessus* by Fabrici, whom Harvey followed in this, simple action was less important than use or purpose, which could be induced from observation and experiment and which in this case was the nature of the pulse and the one-directional movement of blood in the arteries (the latter of which Fletcher does describe).[222] Thirdly, sixteenth- and early seventeenth-century anatomists had "no real understanding" of what was actually involved in the movement of muscle, and even for Harvey, "the action of the heart occurred because of a pulsative faculty or power of the soul," just as it did for Galen, though not for the Hippocratic anatomists; and furthermore, in the *Prelectiones,* which are more directly relevant than *De motu cordis* for our consideration of the conservativeness or innovation of the *Purple Island* anatomy, the muscularity of the heart's action and structure does not yet replace the Galenic notion of its attractive power.[223]

If the arguments on either side of this debate seem inconclusive, the remainder of this chapter aims to contribute a fuller argument along the lines that *The Purple Island*'s (use of) anatomy is not best interpreted *either* as conservative, after Nicolson, Baldwin, and Healy, *or* progressive, after Langdale. Rather, this dichotomy presents a misleading set of alternatives for a writer subject to both innovative and conservative impulses and involved in a more complex negotiation of anatomical, rhetorical, poetic, philosophical, and theological ideas and discourses, which are not clearly differentiated from each other even in his anatomical sources.[224] As to whether this supports Healy's view that in anatomy and literature in the early seventeenth century, there is a "refusal" to acknowledge the difference in the forms of discourse appropriate to their respective disciplines, the remainder of this chapter will endeavor to gauge the extent of the imbrication of figurative or poetic and anatomical discourses, and the extent of the availability of a discursive or conceptual differentiation which anatomists and poets might be thought to have refused to acknowledge.

On the subject of whether, and if so how, Harvey moved beyond simple observational method, philosophically he upheld observation rather than the application of principles derived from mechanics. Harvey's Aristotelianism is suggested in some measure by his postulation of the existence of the fabric of Nature herself. Keynes argues: "Since Harvey was so largely using ancient authority, his arguments tend to be deductive from first principles, but he also used induction from observed facts."[225] Aristotelian philosophy, as Wear argues, could be observational, and "could therefore be a proper philosophical framework for an anatomist."[226] Harvey insists many times in all his writings that "the only way to achieve any knowledge of biological phenomena is to interrogate Nature." In *De motu cordis* he professes to learn and teach anatomy not "from books or from the maxims of philosophers, but from dissections and from the fabric of Nature herself."[227] And in *De generatione animalium* he says that although it is a new and difficult way to find out the nature of things by things themselves, rather than by reading books or taking our knowledge from the opinions of philosophers, yet the former is "a much more open way to the hidden secrets of natural philosophy and one which leads less into error."[228]

Because his arguments tend often to be deductive from first principles, however, Porter summarizes Harvey's Aristotelianism, saying: "With Aristotle [Harvey] shared a teleological view of the body and the belief that its workings depended upon the distinctive soul."[229] The view "that all parts of the body had been constructed with a purpose, and that nature did nothing in vain," was "anathema to the protagonists of the new science," who ascribed purpose not to the body-in-itself, which has no purpose of its own, but to its use by the mind; yet this Aristotelian, teleological view had a formative role in hinting to Harvey the idea of the circulation.[230] The actions or uses of the parts, in which Harvey as an Aristotelian was more interested than in their morphology, are directly related by him to their designs and final causes, which are "the very reasons for their existence."[231] So, although Harvey's method of anatomical division involved *historia,* or the descriptions of a part's location and posture or continuity with other parts, his "philosophical division of the parts" is in accordance with "the end for which [the parts] were designed."[232] In this Harvey's anatomy accords with the philosophical anatomy of his teacher at Padua, Fabrici, whose criteria also begin with *historia,* after which is the inquiry into an "action" and "use," which is elicited from structure; this is followed by "utility," which is also sometimes translated as "use" or "usefulness," and is a general account of why a part exists and why it is as it is in the Animal, a general or universal category of final cause or purpose, discerned by induction from the description and action of the particular animal.[233] Finally, in Fabrici's criteria, there is "demonstration" of the truth of the account given.

Unsurprisingly, the teleological view of the body is closely related in Harvey's anatomy to the belief in the relationship between the body and the soul. Whereas the parts of the body, which are designed by Nature as "an instrument" for some action or operation, are various, because "the operations and actions of Nature are manifold and diverse"; similarly, the whole body is made in order to serve as "the instrument of the soul, or rather the whole man," and this "requires the performance of many functions."[234] In Luke Wilson's analysis of these relations of subordinate parts to the whole in the *Prelectiones,* "the soul is understood to be the body's formal and final cause. . . . In this sense it is to be identified with Harvey's 'Nature.'"[235] The teleological meaning and nature of the body's anatomy is thus grasped by Harvey through physiology, and "what is body from a morphological point of view is, from a teleological or physiological perspective, soul and body as a unitary actuality"—"the whole man."[236] Thus, Harvey's "philosophical division of the parts is in accordance with the instrument of the soul which it represents, in accordance with the divisions of the soul or in accordance with the [soul's] faculties, and in accordance with its final cause."[237]

As far as the *Prelectiones* are concerned, despite Harvey's allegiance to Aristotle, Fabrici's tutelage, and Colombo's example, the division of the body partly follows a three-venter order, the method of necessity or discipline, which is in part followed by *The Purple Island:* "*Venter Inferior*" (the lower belly), followed by "*De Ventre Superiore*" (of the upper belly), and finally the head.[238] This order is differentiated from the order of "philosophical anatomy," which is primarily concerned to reveal action, function, and purpose, as well as microcosm-macrocosm correspondences; but it is probable that the three-venter arrangement was chosen by Harvey to serve the specific purposes of the Lumleian lecture, while philosophical anatomy was to determine the physiological identity of the subjects of the later *De motu cordis* and *De generatione.*[239] Besides which, French is right to argue: "Although Harvey announced that the lectures were to be *anatomia popularis,* public or teaching anatomy, yet his interest throughout is philosophical."[240] Like Harvey's *Prelectiones,* though, the poetic texts of Davies of Hereford and Fletcher divide the body in accordance with the divisions and faculties of the soul; so that, for example, both *Microcosmos* and Canto 4 of *The Purple Island* anatomize the thorax in which the heart is the seat of the soul's irascible or "*Vitall*" faculty, and of its "*Affections*" and "passions untam'd."[241] Chapter 5 of this study explored the theological skeptical context in which the didactic poems of Davies of Hereford indicate that if the senses are affected badly, they wrong "th'*Intelligence,*" which consequently misleads the "*Will*" which moves the "*Members,*" the "instrumentall agents of the *Minde.*"[242] Over teleology and the instrumentality of the body, there are concepts and vocabulary that are shared by Galenic, Aristotelian, and Harveian anatomy, as well as the Christian poetic anatomies of Davies of Hereford and Fletcher.[243]

There are other ways in which Aristotelian teleology may be distinct, however. If even the "least parts" are said to be exquisitely formed and located by "wondrous art" for their physiological functions (*P.I.* 5.42.3) in Harvey's anatomy, as in that of Galen, de Laguna, and Vesalius, then Harvey's anatomy would be close in this respect to that of Davies of Hereford and that of Fletcher.[244] This teleology derives from Galen whose own Aristotelian influence was combined with that of Plato: "Galen's world, for example in *On the Use of the Parts,* was a providential created world in which the subject-matter that he claims will be of interest to philosophers is close to a natural theology which reveals the success of a divine nature (which is not far removed from a Platonic demiurge) in handling recalcitrant matter."[245] Galen's anatomy "assumes that the human body is the product of design in the same way that Plato does," though not necessarily the same design.[246] This is why Vesalius and Berengario, among other sixteenth-century anatomists, find no obstacle in Galen to the notion that accuracy of observation is a kind of piety. And this is also why anatomy in the sixteenth century could be seen to be *revelatory* of structure, function, cause, creation, mortality, salvation, and ultimately God, who is the body's efficient cause.

But the notion that is found in Plato's *Timaeus,* of the human body as "the product of [the] intelligent design" of the Creator, may be contrasted with the Aristotelian notion of "The Animal [including the human animal], teleologically arranged."[247] Even in the early *Prelectiones,* which are concerned primarily with human anatomy, Harvey's Aristotelianism aims at providing a general, universal account of a function, a physiological operation, an account which pertains to "The Animal," and in this respect it differs from *The Purple Island* (subtitled *The Isle of Man*).[248] Moreover, Aristotelian natural philosophers and anatomists of the Renaissance and early modern period have little to say about God, revealed knowledge, or God's relation to the world, other than in the terms framed by Aristotle himself, because they aimed to read Aristotle "aristotelically."[249] Fletcher's difference from Harvey on this point may constitute further evidence of his influence by Galenist and Vesalian anatomy. The Christian, as distinct from Aristotelian, schemes of *Mirum in modum, Microcosmos* and *The Purple Island* try to recover the "rare existence" of God the ingenious artist, in the exquisite design of the parts, and, by contrast to Aristotelianism, they eschew any identification of Nature and the soul.[250]

Within the Christian context of the sixteenth and early seventeenth century, however, distinctions do not hold consistently, and even the Aristotelian Fabrici "speaks often in praise of God the Creator in his writings" and concentrates on "finding intelligence, purpose and design in Nature."[251] And although Fabrici's immediate concern is actually with "the nature-of-the-thing, not the providential and wise actions of 'nature,' "[252] his pupil Harvey's teleology invokes a concept of the "efficient cause and divinity of nature which works at all times with consummate art, and providence and wisdom, and ever for a certain purpose, and to some good end."[253] So although Harvey was philosophically Aristotelian, he shared with Galen this appreciation of the role of deity in Creation, and demonstrated enormous admiration for Galen's work, particularly in the *Prelectiones*.[254] Harvey's faith commitment is not known, and "among his contemporaries and predecessors Harvey was unusual in not giving a religious purpose to anatomy," yet this does not prevent him from attributing in Christian fashion the cause in Nature to "the Divine Architect" and omnipotent Creator,[255] despite what his Aristotelian teleology may lead us to expect, and despite the fact that it has been said that Aristotle's God does "not lead . . . very far toward the production of a God available for religious purposes."[256] Notwithstanding Harvey's attribution of cause, there is a marked absence from his natural philosophy of natural-theological arguments,[257] that is, of theological justifications of anatomy in concepts of the creation and redemption of the human body. By contrast, it is argued in chapter 3 of this book that the soteriological purpose or illocutionary act of *The Purple Island* is to provide a verbally constituted form of ocular demonstration of the creative ingenuity and saving love of God.

Despite their different purposes and emphases, there are many similarities between the anatomies of *Mirum in Modum, Microcosmos, The Purple Island*, and the *Prelectiones*, which do not stop at teleology. It has been argued that after Harvey, the names of the spirits, the three kinds of *pneuma*, "survived," "but the names were henceforward devoid of any really essential meaning."[258] This is because an absolute discontinuity emerged between the Galenic venous system and pulmonary and systemic circulation, as was mentioned above. For later physiological investigations, Harvey's "virtue of restraint," in declining to add to debates on the life, *pneuma* or spirit, and the innately hot complexion of the heart, which he regarded as "remote problems," was exemplary; and for consequent physiological discoveries mechanical explanation became a model.[259] But, again, the *Prelectiones*, and arguably *De motu cordis*, betray that, to draw from their influence an interpretation of Harvey's work itself as exemplary of "the new experimental and mechanical approach to nature" is to disregard the adherence of his language to notions of vital spirit and innate heat, to anatomical conceptual metaphors, and to the figurality and philosophy of theories of correspondence.[260]

Views of vital spirit, heat, and the microcosm were not uniform, however, and Harvey's require elucidation. The *Prelectiones* express a conviction in an interpretation of an opinion of Galen. Harvey professes to agree with the "Prince of Physicians" who "holds rather (not as the vulgar do that spirits are distinct and separate from the humours and the parts . . .) but that spirit and blood are one thing, like whey and cream in milk," and that spirit exists "in continual generation" from blood, as flames from fire, or light from a candle.[261] In chapters 8 and 15 of *De motu cordis,* Harvey follows Aristotle's proposition that the heart is the source of innate heat in all living things, and is therefore the *principium vitae* (though not in an embryological sense, for the blood appears first and is the primary source of life).[262] Likewise, in the *Prelectiones,* the heart is "the store-room of the spirits" and the "fountain" of heat.[263] Harvey argues that, "following the definition of Aristotle the very word blood connotes the presence of heat as an accident in a substance which is essentially cold, just as . . . light is the result of the act of something which gives light."[264] Spirit and heat are not dismissed; yet, subsumed in the blood itself and accidental to it, respectively, they became less immediately significant problems for research into the physiology of the movement of the blood.

There are instances, however, even in *De motu cordis,* where heat and spirit are proposed, speculatively, to be intrinsic to the purpose of cardiovascular physiology, and, in the case of the discovery of systemic circulation, they help to construct Harvey's thesis and may provide the context of justification: "The blood, having given its life to the parts of the body [where it was 'refrigerated, coagulated, and made as it were barren'], came back to the heart to regain its original perfection and did so 'by the natural, powerful, fiery heat, a sort of stove of life, [wherein] it is reliquefied and becomes impregnated with spirits.' "[265] By these vital spirits in the heart, the blood is perfected, as it, and consequently the whole body, "is defended from corruption."[266] The heart thereby acts as the material substratum necessary for the appropriate effects of heat and life to be obtained in animals, and its inherent natural function is (that of) the "soul."

It should be borne in mind that Harvey was to make it clear that his essentially Aristotelian considerations about the heart and the circulation's heating and vivifying purpose were "separate from his anatomical demonstration of the circulation."[267] And in a letter to Caspar Hofmann, written on May 20, 1636, Harvey retracted what confidence he may once have had in these Aristotelian speculations about the spirituous purpose of circulation, and he came to regard them as tentative.[268] He refuses to commit himself on whether or not the heart does add anything beyond the motive force required for systemic circulation, to which the justification of his discovery was to be reduced by later, mechanistic natural philosophers.[269] However, in *De generatione,* teleology and the instrumentality of the body to the soul is undiminished, though the blood, rather than the heart, becomes the substratum necessary for the function of the soul to preserve the life of the body.[270]

We may get a clearer sense of Harvey's thinking at the time when Fletcher was composing *The Purple Island* by referring to the *Prelectiones. The Purple Island* represents the heart as "the seat of sure and active loving," and of "passions" (*P.I.* 4.16.7; 4.25 margin. z), and similarly, the *Prelectiones* situates therein not only heat and vital spirits, but "emotion, [and] the passions."[271] Besides, although Harvey was to regard his Aristotelian considerations about the heart and the circulation's heating and vivifying purpose as separate from his anatomical demonstration of the circulation, he did not dismiss the speculations as unimportant in their own right or in the context of the natural philosophy to which anatomy was residually still subalternated. Rather, it was Harvey's considered opinion by 1628, that the perfection of the blood by virtue of its revivification through heat and vital spirits "must be inquired into . . . and decided upon other grounds" than those of observation and experimentation.[272] This did not preclude heat and vital spirit from anatomy, and so speculation was not omitted from *De motu cordis,* where the purpose ascribed by Harvey to the physiology of the heart, blood, and spirits, in Chapter 8, takes on a politically sexual and gendered character that is not overt in Fletcher's *Purple Island* for physiological and probably other reasons.

The respective choices of metaphor by Harvey and Fletcher for the purpose of the concoction of vital spirits in the heart demonstrate the antithetical as well as analogous relation between sexual generation and two kinds of spiritual regeneration—the regeneration of *pneuma zoticon* or vital spirit and the regeneration of the will or the vital faculty of the soul in the heart indwelt by Christ or the Holy Spirit. In Harvey's figurative modification, "the warmer more perfect vaporous spirituous . . . blood" in the arteries is masculine in relation to the body, which is "cherished, quickened" by it; and through contact with the body, the blood is "made as it were barren" and, "so to speak, effete."[273] In relation to the heart, to which it returns in the veins to be "impregnated with spirits," venous blood is feminine.[274] The context of justification hereby created is philosophically Aristotelian. Oppositional terms—cold/hot, solid/fluid, matter/spirit, imperfection/perfection, corruption/vivification, impotence/power, passivity/agency—many of which are structural to Aristotle's anatomy of sexual difference and to his physiology of sexual reproduction, politically engender the process whereby the individual life is perpetuated, not by generation but by circulation. But for Fletcher, in contrast, the perfection of the blood, and indirectly the preservation of the parts, takes on a different character.

As was indicated above, Fletcher's use of the term "confection," for which I have found no provenance in anatomy, could denote making or preparation by mixture of ingredients, compounding; composition, making up, manufacture; a preparation made by mixing, potentially a medicinal preparation; or, in the sixteenth century, the performance of the Sacrifice of the Mass. There is reason in the plurisignification of the *Purple Island* anatomy for thinking that all these meanings are connoted when the left ventricle "makes a strange confection" of "aire" and "bloud," and "Sends it to every part, both heat and life inspiring" (*P.I.* 4.21.3, 2, 7 & margin. t). It does not matter that it is not a part of systemic circulation; this process has a kind of medicinal effect whereby the individual life is perpetuated through vital spirit. The process may be construed thereby as soteriological, in its meaning of the promotion and preservation of health, or the restoration and perfection of human nature. But we know also that the soteriology of *The Purple Island,* in respect of its figurative design and purposed effect, is not only this but also, theologically, the doctrine of salvation.

We have already encountered the post-Reformation, Protestant culture of northern Europe, in which great emphasis was placed on the nobility of the human body, as *imago dei* and as God's dwelling place; and we have seen how in this culture Neoplatonism, Paracelsianism, and related esotericism thrived. But let us put aside Fletcher's Protestantism momentarily and consider briefly how the Roman Catholic doctrines of the sacrifice of the Mass might apply through the concept of confection to the concoction of vital spirit in the heart. The christological significance of the heart in Fletcher's poem, discussed particularly in chapter 3, is such that, at the Creation, the hand of God set the picture of his own state in the heart, a picture that may yet be visible because of the incarnation of the "Sunne"/"Sonne" who inheres therein. As a "picture" of "that loftie one," the human heart is Christ's "glorious cell," full of "saintly spirits" confected by the heart and distributed as the "spritely bloud" in the arteries (*P.I.* 4.9.3,5,6,7; 2.11 margin. b). Were the *Purple Island*'s use of confection to be metaphorical, a system of associated implications characteriztic of the subsidiary subject would be drawn from the system of associated commonplaces. We would be led by this system to construct a corresponding system of implications about the principal subject. If the situation and perpetually regenerating efficacy of the heart in the individual body is understood to correspond to that of *corpus christi* in *ecclesia* (the collective body of members), the effect of the arterial blood and vital spirit in the individual body, which is to cherish, quicken, and defend from corruption the parts, would correspond to the Eucharistic "confection" of bread and wine and "the propitiatory effects of [Christ's] passion [which] flow to men through the Mass, for the remission of sins and for the welfare of the living and dead."[275] How might the confection of blood and sacrament effectuate these causes in ways that correspond?

Prior to the doctrine of transubstantiation, the earlier Roman Catholic doctrine was formulated along Aristotelian lines in which the incorporeal body of Christ inheres in the articles or species of bread and wine, turning them into the accidents of which his body becomes the essence. There is reason to think that this Eucharistic doctrine may be seen on an analogy with the Aristotelian concept of the particularized soul, which as form and cause of the whole body actualizes and animates it, perhaps through the heart or blood, the material substrata proposed by Harvey as necessary to the function of the soul.[276] The analogy is made explicit by Crooke, when he says, in *Microcosmographia,* that "the Soule of man *Is wholly in the whole, and wholly in every particular part*" of the body, a formula which parallels exactly the Catholic one of the real presence of the body of Christ in the Eucharist.[277] Unfolding the significant connotations of the metaphorical use of confection, it may follow from the analogy between Christ's body and the soul, and between the accidents of bread and wine and the human body, that the material substratum necessary for the dispensation of life and heat in the body corresponds to the accidents of bread and wine which are necessary for the appropriate soteriological effects of regenerate life to be obtained in the bodies and souls of the faithful, living and dead. Both the reproductive figurality of the perfection of the blood in Harvey's Aristotelian anatomy and the sacrifice or confection of the Mass are gendered, for *ecclesia* is of course the bride of Christ.

But in this sense as the performance or sacrifice of the Mass, confection is contrary to Reformation doctrine, and contrary to "The Order for the Administration of the Lord's Supper or Holy Communion," in The Book of Common Prayer.[278] The 1552 Order of the Communion is trenchant in this respect, stating that "we be unworthy through our manifold sins, to offer unto thee any sacrifice" and replacing the relative sacrifice of the Mass with the "reasonable, holy, and lively sacrifice" of "ourselves, our souls and bodies," supplemented in the 1549 edition with "our bounden duty and service," and "our prayers and supplications."[279] But it is conceivable that confection is consistent with that for which petition is made in the invocation of God "to bless and sanctify these thy gifts and creatures of bread and wine," that "whosoever shall be partakers of this holy communion may worthily receive ["may be partakers of"] the most precious body and blood of thy Son Jesus Christ, and be fulfilled with thy grace and heavenly benediction, and made one body with thy Son Jesu Christ, that he may dwell in them, and they in him."[280]

Perhaps the heart's concoction of rarefied blood and *pneuma* into vital spirit (which Harvey regarded as one with blood), which sent to the extremities cherishes and quickens the parts of the body, is understood in Fletcher's allegory in terms of the way in which the "holy mysteries" and efficacy of Christ's "full, perfect, and sufficient sacrifice, oblation, and satisfaction for the sins of the whole world" are dispensed, "with the spiritual food of [his] most precious body and blood," to each of the penitent elect or faithful, through the "perpetual memory" and signification of the ("creatures" of) bread and wine, and meditated upon in their hearts, that they "may be very members incorporate in [Christ's] mystical body, which is the blessed company of all faithful people."[281] As was remarked in the introduction to this study, penitence, or the preparation for the regeneration of will in the heart, in Reformed contexts, is understood on an analogy with the concoction or regeneration of venous blood therein (*P.I.* 6.57–77), and the purple color of venous blood is the color of penitence in ecclesiastical contexts. Although Fletcher's Eucharistic theology does not adhere to Roman Catholic doctrines of the Sacrifice of the Mass, nevertheless his figuration of concoction as confection draws significance for human anatomy from the poem's main story-line of divine history and its eschatology; for the Eucharist "dost assure" the partakers of the body and blood of Christ that they "be also heirs through hope of [the] everlasting kingdom," thereby, in faith, auguring eternal life in the *parousia* beyond the general resurrection.[282]

Whereas for Harvey the circulation of the blood throughout the body is justified by its reliquefication and impregnation with spirits in the heart and the consequent revivification of the parts, for Fletcher the concoction or confection of *pneuma* and refined venous blood into arterial blood in which the vital spirit inheres is justified on an analogy between what happens in the microcosm and what happens in *ecclesia,* in both of which the Son of God is believed to reside, in microchristus and the Eucharist, in order that the heart, and so the whole human being, may turn to heaven from hell and be redeemed. The context of justification for Harvey may be Aristotelian; but for Fletcher it is soteriological.

In *The Purple Island,* the sacramental process of confection in the left ventricle of the heart bears also the alchemical significance of "distillation"; as does the concoction of venous blood in the right ventricle, and possibly the description of the blood as "Circling about" the body. In the right ventricle, the "pension of liquours" "first by *Hepars* streams . . . hither brought" are "here distill'd with art, beyond or words or thought," and "The grosser waves of these life-streams" from the liver are separated and sent "down to the lungs," while the "labour'd extracts" pass through the interventricular septum (*P.I.* 4.18.5,6,7; 4.19.1,4 & margin. q). The left ventricle takes in air from the pulmonary vein and "Of this thinne aire, and of that right sides rent, / (Compound together) makes a strange confection" and "Stills them with equall never-quenched firing" (*P.I.* 4.21.2–3,5). Though the ventricular distillation in the right ventricle separates the "laboured" and "thinner part" from the "lesse laboured, and thicker" parts of the blood (*P.I.* 4.19 margin. q), thereby separating pure from impure parts as do the *archei* in Paracelsian physiology, it is conceivable that both ventricular distillations signify in an alchemical sense.

It remains to be determined whether this alchemical significance or connotation may be held in common with the perfection of the blood in Harvey's accounts. There is probably no question here of provenance, for the *Prelectiones* do not sufficiently elaborate upon the heart's purpose in this relation. The question in regard to the *Prelectiones* would be whether or not they adhere implicitly to a doctrine of separation or distillation of the blood in the heart. The account that is in part cited above is from *De motu cordis,* and the only Paracelsianism which might reasonably be accorded to the whole passage is its qualified use of the term "balsam," to describe that with which, besides "spirits," the venous blood "might be said" to be "impregnated."[283] The restraint of the passage ensures that no wholesale borrowing of the concept from Paracelsian medicine occurs here, but rather the suggestion of something "which brings it about that the human body will not decay."[284] The "constancy of human balsam" which constitutes the "chain of Venus" (as in John Donne's "Love's

Alchymie"), forms an aspect of the reproductive continuity which is the subsidiary subject of Harvey's figurality.[285] But the principal subject of the "treasury of life" for *De motu cordis,* as for *The Purple Island,* is the heart, not the womb.

Besides this, the passage from *De motu cordis* visualizes circulation "as a cyclic process of blood regeneration," in which "blood that went out from the heart to the tissues in a hot and vaporous condition returns to it cooled down and condensed."[286] Whether or not it is significant for understanding Harvey's theory of the circulation physiologically, to realize that this may be akin in some measure to the process known in alchemy as "*circulatio*" or "*distillatio,*" has been strongly contested.[287] But it occurs to me that, regardless of whether or not it is important for the development of experimentally demonstrable anatomical knowledge or the later formulation and justification of anatomical projects, it is historically significant that Harvey's philosophical context of justification and his thesis construction should be Aristotelian and possibly alchemical.

In *The Purple Island,* on this matter, the image of the left ventricle's "never-quenched firing," which "Stills" labored venous blood and *pneuma* "in one vessel" (*P.I.* 4.21.5, 4) (that is, heats and vaporizes the mixture, before cooling and condensing it on entry into the aorta) bears a resemblance to the alchemical "circulation" of Cesalpino, to Harvey's philosophical speculations on "circulation," and even to Descartes's theory of the blood's rarefication and distillation in the heart. But just as Fletcher could hardly regard the function of the lungs to be to cool the hot blood passing through, as Cesalpino did,[288] neither could he, having presented a centrifugal venous system, regard the heart's distillation to be a regeneration of blood returning to the heart in the veins. The blood may be described as "Circling about" the body in Fletcher's poem, but it cannot be regarded as consistent with systemic circulation without considerable damage to the consistency of the poem's model of cardiovascular anatomy and physiology. So notwithstanding the resemblances, distillation in *The Purple Island* is circular only in the sense of returning overheated and vaporous blood to its original condition as a liquid.[289]

At the commencement of the passage in chapter 8 of *De motu cordis,* Harvey suggests a purpose for the circulation of the blood by drawing an analogy with Aristotle's description of the "circular" movements of air and rain and those of the "superior bodies" of the cosmos. Both of these phenomena are themselves presented by both Aristotle and Harvey as analogues of each other.[290] Whereas "some iatrochemists postulated a chemically inspired theory of the circulation of the blood prior to the work of Harvey," only Cesalpino implied also a physiological circulation.[291] But this metaphor of the movements of air and rain and of the cosmic bodies "allows Harvey, in a few paragraphs, to indicate his notions of the physiological significance of the blood for the life of the whole organism."[292] In the same passage, Harvey also relates how the concept of circulation came into his mind. So the agency of cosmological, geocosmic, and alchemical circular movements in the formulation of Harvey's anatomical projects, that is, in suggesting to him the hypothesis of the circular motion of the blood, has been much contested.[293]

Both Aristotle's and Harvey's analogies were elaborated by alchemists and iatrochemists, many of whom, Debus says, compared the "motion of the blood" to "the great macrocosmic circulations of the heavenly bodies and the rains."[294] For the Paracelsians, alchemists, and iatrochemists, as for Harvey, in this analogical configuration of microcosm, geocosm, and macrocosm the "heart . . . is . . . the sun of the microcosm, even as the sun in his turn might well be designated as the heart of the world."[295] The alchemical theorist Robert Fludd, who like Harvey was a member of the College of Physicians, and who carried out anatomical experiments and gave public anatomy lectures to the College at least once, is reputed to have observed Harvey's dissections of the heart and vascular system which ultimately led to his theory of systemic circulation.[296] It was Fludd, in what has been perhaps unjustifiably called "an ironic twist," who was the first to agree in print with Harvey's theory of systemic circulation, saying that it was "in consonance with his own philosophy," in a work on the pulse published in 1629, in *Medicina Catholica,* at about the same time that the mechanist philosopher Gassendi had prepared his *Examen philosophiae Roberti Fluddi,* to be published in 1630, rejecting Harvey's theory.[297] Systemic circulation continued to be accepted by those like John Heydon, writing in the Paracelsian, alchemical, or Rosicrucian traditions, even after it had been reinterpreted as a mechanistic phenomenon as opposed to a vitalistic one.[298]

The cases of Gassendi and of Descartes represent noteworthy examples of how the metaphor of the body as a machine, just like any other metaphor for anatomy when used alone, could impede the apprehension of a discovery by overdetermining the formulation of hypotheses and the selection of the evidence of observation. Perhaps surprisingly, their case is not dissimilar to that of Davies of Hereford's *Microcosmos,* as discussed in Chapter 5 of this study. In the adherence to a narrow metaphorical scope in describing the body in terms of automata, or in terms of architecture and carpentry, as in *Microcosmos,* generalization from particulars may characterize the inductive application of the concept of a machine or a building from one feature of the body to another, but, rather than being properly inductive, the character of the body is already determined either as an organized natural object regulated merely by matter and motion, or, in Davies of Hereford's case, as an effect, the principal cause of which is in its divine architect. In both cases, in practice, reasoning is deductive, and in the absence of evidence from dissection and sensory observation, the anatomy is overdetermined by a generalized and unnuanced conceptualization. It was said in chapter 5 that *Microcosmos* seems to assume that propositions regarding human anatomy could be judged as true or false according to whether they are consistent with the criterion that the human building is created by God as the instrument and abode of the soul. Similarly, according to Keynes, "Descartes was unable to entertain the idea of the heart acting as an involuntary muscle, believing that all muscles were controlled by the will." "Descartes was unable to support his views by any convincing experiments, but was satisfied with his mechanistic explanation" of the heart as a kind of furnace, its movement an expansion caused by heat.[299] In Davies of Hereford and Descartes, reasoning reduces itself to a self-confirmatory circularity, because both the philosopher and the poet fail to realize all hypotheses may in principle be tested against the ontology of the body-in-itself by observation and experiment, and, given the absence of an epistemologically transparent and neutral nomenclature to relate this putative ontology, that a different metaphor for the body could replace that of "building," and so more metaphorical traffic may be preferable to less in the contexts of project formulation and justification. There is not very much difference in this respect between the *inventio* or *ingegno* of anatomical discovery and that of poetic discovery.

Whatever the degree of agency of Harvey's Aristotelian speculation in his formulation of the hypothesis of systemic circulation, it remains the case that, in the context of justification, that is, in the formulation of his thesis, (re)generation is the purpose of each analogous phenomenon of the analogical configuration he speculatively proposes, just as it is the purpose of the circulation conceived along politically sexual and gendered terms. In the narrative of Harvey's geocosmic analogue, "the moist earth, warmed by the sun evaporates; the vapours drawn upwards are condensed, and descending in the form of rain moisten the earth again; and by this arrangement are generations of living things produced."[300] This analogy was adopted by Philip Jacob Sachs von Lewenheimb, in his *Oceanus Macro-Microcosmicus,* where it is transmogrified so that the circular motion of the blood from and back to the heart is analogous to that of water from and back to the sea: "The sea lets water rise by evaporation and return in the form of rain whereby the rivers and subterranean waters are nourished and these finally return the same water to the sea."[301]

This geocosmic-microcosmic analogy must surely have existed prior to its particular development in von Lewenheimb's work, which was not published until 1664, for it seems to have been combined by Fletcher with the analogy in *De motu cordis* in an untitled poem of three ten-line stanzas, written to conclude a chapter of *A Father's Testament,* a prose work which was published posthumously in 1670, but written much earlier.[302] In this poem, the mystical "Vast Ocean of light, whose rayes surround / The Universe," is a symbol of God, and is related by analogy and antithesis (within hierarchy) to both "Our Seas" and "Our Sun" (1.1–2; 2.1,3). Lines in stanza 2 describing "Our Seas" alternate with those describing "Our Sun"; and between each of these descriptions, linguistic and metrical parallels convey the effect that the two phenomena are comparable, hinting at a single corporeal analogue, which, as we shall see, is the heart. Seas are described in lines 1–2 and 5–6; the sun in lines 3–4 and 7–10. The adjectival phrase in parenthesis in line 1 "(a drop of thine)" parallels that in line 3 "(a spark of thine)," and both are combined in the first stanza as derivatives of the "Vast Ocean of light," "Who lend'st the Sun his sparkling drop" (1.1,3). The action of making "drunk the thirsty plains" in line 2 of the second stanza parallels that of reviving the dead in line 4; and the adverbial phrases describing these actions—"all the earth" in line 2 and "all the world" in line 4—are also comparable in syntax and vocabulary. As water is distilled "through earth pipes," that is, streams and rivers, so the sun's rays peep "through jet clouds" and his light is "through ayers spread" (2.5,7,8). This parallelism culminates in line 9, where the sun's "gleams" of light are described as "life-giving streams."

Together with their actions, both of the derivative analogues—seas and sun—come to interact with and signify the metaphorical term—the human heart—which is engaged (albeit indirectly) to describe them in the two central lines of the middle stanza. At the beginning of that stanza, the "Seas . . . with arms dispread / Through all the earth make drunk the thirsty plains" (2.1–2), in language reminiscent of *The Purple Island*'s description of how the veins and arteries "Emptie themselves into th'all-drinking main" (2.9.6). The "Seas" in the later poem are analogous to the heart (though this is not clear until line 5), and they "spread" their inundation to the body's extremities. The two central lines proceed to deploy the terminology of the *Purple Island* cardiovascular anatomy and physiology, with the effect that at the very center (the heart) of the stanza, and indeed of the poem, the reader loses sight of the determination of which is the principal and which the subsidiary subject of the extended metaphor: "Seas (through earth pipes distill'd) in Cisterns shed, / And power their liver springs in river veins." (2.5–6). In *De motu cordis,* "Harvey points to the fact that the heart alone of all the parts of the body contains blood for the general use of the parts. . . . Therefore, as Aristotle had said, the heart is to be considered as a storage cistern," a tank or underground reservoir for storing rain water.[303] When this poem was composed, had Fletcher read *De motu cordis,* or just his Aristotle? Unlike in *The Purple Island* the precise conceptual character of anatomy is difficult to ascertain because the late poem explores anatomy and physiology systemically but schematically rather than systematically, and without applying the *occasiones* of the anatomical *accessus* anything like as fully as the earlier poem, either to the principal or the subsidiary subjects. Whatever the answer and the nature of the anatomy, these few short lines of Fletcher's late three-stanza poem achieve the simultaneous signification of superimposed analogues, as does *The Purple Island* on a more poetically ambitious scale.

In the second stanza of the poem, the location of the verb, "shed," at the end of line 5 (cited above), and of the adjectival phrase in parenthesis in the same line, ensure that on a close reading interpretation is deeply entangled in syntax. One way of paraphrasing the line might be: "Seas," which are "in Cisterns," "shed" water (in the geocosm) or blood (in the microcosm) "through earth pipes" or arteries, by first distilling them. This renders line 5 close in meaning to Canto 4, stanza 21, lines 1–5 of *The Purple Island* (cited above), wherein the heart distills refined blood and *pneuma* into the aorta; meanwhile, the use of the adjective "through" suggests a one-directional movement in the arteries, as in *The Purple Island.* On this reading, the cistern-heart contains the sea of blood, as in the analogy of von Lewenheimb. Another feature of that analogy is the operation of the sun's heat on the water or blood in the heart; and in Fletcher's little poem, the sun is responsible for the evaporation of "Earths foggs" (2.10). Additionally, like the heart or sun of the microcosm in Harvey's physiology, the "Sun" in Fletcher's little poem "revives the dead" by the action of its heat (2.4). "Seas" and "Sun" therefore coalesce as symbols of the liquid "drop" and fiery "spark" of the "Vast Ocean of light," to be found as microchristus within the human heart, whose function in the body combines those of the sea and the sun in the geocosm, and, more particularly, the function of the cyclical distillation of water as it is heated by the sun and condenses as rain.

Galen had contended that the liver, not the heart, fulfilled the role of storing blood, and if, in stanza 2 of Fletcher's little poem, the prepositional phrase, "in Cisterns," refers not to whence the seas are "through earth pipes distill'd," but whereto they are so "shed" or moved, then these subterranean reservoirs may be analogous to the liver in the microcosm, which receives blood from the heart through the abdominal aorta and hepatic artery. The liver and springs interact metaphorically in line 6 of this stanza as the source(s) of venous blood and of rivers; but the movement of blood in the liver's outlets is "powered" by the "Seas," by, in fact, the heart, and not by the same force of gravity as governs the flow of blood in a "Nile-like inundation" from the liver along the vena cava and the other veins "to all the Isle," including the heart, in *The Purple Island* (*P.I.* 3.11.6, 7). There venous blood flows from the liver as though downhill like water in the geocosm giving the impression that there is no need of cardiac systole to thrust the blood entering the right ventricle of the heart out again into the vena cava, in order to reach the extremities of the body. Though not overtly mechanistic, blood is by contrast "powered" by the "Seas" of the heart in Fletcher's late little poem, which, in this respect, is an example of how the figuration of anatomy may change in response to the material determinacy of anatomical discovery and the changing contexts of justification. Perhaps the heart's attractive faculty or power, the Galenic notion that had not yet been replaced in *De motu cordis* by the muscularity of the heart's action, draws the venous blood, though not as in the *Prelectiones* where the vena cava and pulmonary vein are said to have neither pulse nor valves because the attractive power of the heart in combination with the tricuspid and mitral valves is said to "break off" the pulse, but rather as in systemic circulation where the direction of flow is from the "liver springs" through the vena cava towards the heart, as that of rivers is towards the sea. There might even be a suggestion that the impact of the blood distilled by the forceful systole of the heart into the arteries forces that in the veins to "spring" and flow. Benefiting from the Aristotelian speculation surrounding Harvey's major physiological discovery, Fletcher's late little poem yet lacks either a decisively muscular or a mechanistic character of the way "liver springs," or veins, or, indeed, arteries are "powered," and leaves in doubt whether or not the anatomy may be characterized by an uneasy synthesis. As such the poem may be seen perhaps as an index or microcosm of the earlier larger poem, which also omits to figure cardiac systole.

Nevertheless, there are strong indications that the little poem is informed by accounts of physiological observations of aspects of the theory of systemic circulation which could not have found their way into the earlier *Purple Island,* and by alchemical and teleological speculations, which advance on the analogy between the heart and the sun in the earlier, allegorical poem. As in *The Purple Island,* God is sought in the late poem obliquely through the knowledge that "his" creation is a metonym and analogue for, and emulation of, its creator, and moreover, that it is redeemed. But in the broader context of the whole late poem, the speaker's admission of failure to find that "Vast Ocean of light"—"How miss I then?" ("S.G." 3.1)—in spite of "knowing" God to be present through his creation and through the incarnation of the Redeemer, suggests that the author may have been dissatisfied with anatomical knowledge, natural philosophy, and natural theology. The poem turns, like the rector of Hilgay perhaps, to rusticity and religiosity, to meek and homely ways, uncomplicated desires, and the significance of the cross, where "Love stands nail'd with love" ("S.G." 3.10).

Although the figurative design of Fletcher's late three-stanza poem achieves the simultaneous signification of superimposed subjects, in which the reader loses sight of the determination of which is the principal and which the subsidiary subject of the extended metaphor, the figurality, like that of *The Purple Island,* is based on, though not structured by, analogy. The analogical correspondences on which the figurality of the poem is based include the macrocosm/microcosm correlations in which the heart "is the chiefest part of the body for it occupies the chiefest place like the middle point in a circle," an Aristotelianism shared by the *Prelectiones,* chapters 8 and 15 of *De motu cordis,* and *De generatione.*[304] Harvey was candid about his Aristotelianism which, as we have seen, informed the philosophical contexts of the project formulation and justification in which Harvey conducted his anatomical disputations, investigations, and experiments.[305]

Further consideration of Harvey's Aristotelianism also addresses itself directly to the question of the congruity of anatomy and figurality in early seventeenth-century anatomy. Parts of *De motu cordis* are even marked by a sociopolitical conceptualization of the heart's centrality and necessity, which resembles that in *The Purple Island,* where it is not only christological and geopolitical, but also Aristotelian. In *De motu cordis:* "The heart, like a prince in a kingdom, in whose hands lie the chief and highest authority, rules over all; it is the original and foundation from which all power is derived, on which all power depends in the animal body."[306] According to Temkin, this should not be dismissed as "the euphuistic style of the body physician Harvey," for the statement has "a definite biological meaning": "Harvey compares the heart with the ruler of the state, because the heart originates first, because it does not depend on other organs, notably brain and liver, but has the organs of its own movements in itself—'as if it were some internal animal'—so that all the rest of the body depends upon it."[307] In *De motu cordis,* according to Pagel, Harvey feels he has proved the main tenets of Aristotle's natural philosophy: "the sovereignty of the heart, its independence of the brain, in sense and motion, . . . its containing blood, life, sensation, motion, before either the brain or the liver were in being."[308] While it is true that for Harvey the physiological circulation of the blood "is the sole and only end of movement and pulse of the heart,"[309] this "end" is understood as central, like the sun, in a constellation of functions that cohere as the soul or purpose of the whole body.[310]

Relating to Harvey's Aristotelianism, a twofold concept of order, one hierarchical, of the principality of organs, and the other sequential, of their embryological development, informs the imagery of "building" in the *Prelectiones.* In another case of how the figuration of anatomy changes in response to the material determinacy of discovery and the changing context of justification, Harvey departs from the Galenic and Vesalian conception of osteology and myology as the foundations or framework of the fabric of the human edifice, for he discovered that Nature or God does not have a singular care, like an architect, first in edifying his house, to be advised of the manner of his foundation and framing of the timberwork, as Banister had said. Harvey retains the anatomical aim to build up the body, but his embryological experiments forced him in *De generatione* to abandon the view that Nature's constructive, compositive, and embryological method begins with the bones, as discussed in chapter 3. Embryologically, he discovered, the blood appears first, and it is not true that at the first constitution of the parts the bones spring from the seminal excrement, rather "the Veines are the first woof, and the first visible foundation of the body, and . . . all the other parts are superadded to them."[311] Consequently, for Harvey, the hierarchical and sequential orders coincide because "all principal parts are the first origins of those which are less principal, for they are more at the beginning because they are more nearly first origins than the other." The origin of the nerves, for example, is the brain.[312]

Harvey was not so Aristotelian that he believed, like Aristotle, that the nerves arise from the heart. Like Galen, he locates voluntary motion, "the senses, [and] the animal spirits from which come understanding and reason" in the brain.[313] For Harvey the heart's preeminence rests in its necessity ("it is the first beginning and foundation of the rest"), whereas the head is "more worthy," for it is "more worthy to exist well than simply to exist."[314] Precisely the same hierarchical distinction is made in *The Purple Island* (2.14 margin. l). The fealty of the parts of the body in *The Purple Island* is owed to the ruler of the venter, region, or metropolis to which they belong, but at the higher level this fealty is divided between the heart and the brain. How the anatomical description of the heart represents its "ends, place, office, state" as corresponding "so nearly neare" to that of the "Prince" of the geopolitical realm (*P.I.* 4.3.4; 1.45), has been expounded heretofore in this study. But in Canto 6, "The Islands Prince, of frame more then celestiall, / Is rightly call'd th'all-seeing *Intellect,*" and is identified with "The understanding," "this Islands highest Soveraigne," and consequently located in the brain (*P.I.* 6.28.1–2; 6.51.2 & margin. *). In this case the descriptions of the monarch's "Sun-like face" and "frame" personify the eye of the intellect, rather than the innate heat and revivifying power of the heart (*P.I.* 6.28.4; 1.45.6). Moreover, the christological significance of the heart slides over to the "most divine aspect" and "so bright majestie" of the "*Intellect*" (*P.I.* 6.28.4, 7, 2). Although the language of the *Prelectiones* is less illustrious in its praise, it is not unlike *The Purple Island* in according the brain political sovereignty for moral-anatomical reasons, if not for physiological necessity: "so is it deemed the prince of all parts," whose influence is transmitted through the nervous system by the "sensory and motor spirits" (the two kinds or functions of animal spirits), which "shed radiance like the sun and cause actions" in "everything that is under their influence."[315] Despite the similarities, differences emerge once again between the Christian scheme of *The Purple Island* and Harvey's Aristotelianism. Since Harvey's Aristotelianism aims at providing a general, universal account which pertains to "The Animal," Harvey adds a rejoinder to his acknowledgment of the brain's preeminence: "However, there is no disputing with the heart because its sway is wider, for the heart is seen even in those creatures that want a brain."[316]

Like sixteenth-century anatomy, and like the *Purple Island* anatomy, Harvey's descriptive passages partake not only of the imaginative rationality that gives rise to the metaphorical conceptualization of human anatomy as "building," but also that which gives rise to its figuration in terms of vegetative growth and the human-built environment. In the *Prelectiones,* as in earlier anatomy, the nervous and vascular systems, arising from the brain and the heart respectively, are rendered in terms of vegetation. The fourth group of nerves are described as arising at the beginning of the spinal medulla "from a double root" and sending out "a branch like a vine tendril." The mesenteric veins are described as "looking like the roots of a tree"; and the resemblance of simile is exchanged for the affinity of metaphor where "the portal vein is the tree and . . . the trunk has roots . . . and branches."[317] This extended metaphoricity is supported and rationalized in the *Prelectiones* by a general Aristotelian analogy drawn between plants and animals, along similar lines to that in Benedetti's *Anatomice,* which is discussed in chapter 1. In the *Prelectiones* the analogy is useful even where a differentiation is being made: in animals "there must be a place for concoction and something like the earth which performs the action, for plants take in [from the earth] concocted nourishment. Animals carry the earth inside them."[318]

The systematicity of this general analogy in the *Prelectiones* finds its rationale in Harvey's adherence to Aristotelian concepts of the soul, wherein the instruments of the actions of the most basic aspect of the animal soul—the "nutritive or vegetative soul"—include the whole trunk, including the heart, lungs, and blood vessels.[319] This is in contrast to Galen's allocation of only the organs concerned with the nutritive, auctive, and generative faculties (together with the attractive, retentive, concoctive, assimilative, and expulsive faculties) to the most basic aspect of the soul—the natural faculty, Nature or *physis.* Consequently, imagery of vegetative growth is applied very widely in the anatomy and physiology in Harvey's *Prelectiones.* But at the epicenter of the vegetative soul for Aristotle and Harvey lies the heart, which in embryological development betrays its affinities to growing plants: "the ventricles . . . are made from the drop of blood which is found in the egg and the heart is fashioned along with the remaining parts, as sprouts come forth in an ear of corn, all together from something which is too small to be seen."[320]

The body's constructedness permeates Harvey's anatomical language outside of the subjects of embryology and generation as well as within them. What is written in the *Prelectiones* of Galen's three venters or "bellies," to which the poems of Davies of Hereford and Fletcher partly adhere, is called "popular anatomy"; but it is not thought to be without reason. Nor, for that matter, is the representation of the upper belly ("the head") as "the sacristy," for it is "the seat of the brain, the habitation of the soul." Otherwise it is "the citadel," or "safest tower," for "Nature made no part more greatly defended." The middle belly, "the chest, the abode of the heart" is imagined as "the inner room, the shrine," though it has a pericardium "to strengthen and defend" it "like a rampart," as does the House of Alma and the Isle of Man, though they both exceed the *Prelectiones* in the linguistic and conceptual elaboration of defense and (in Fletcher) surveillance. The lower belly, "the abdomen, the abode of the liver" is "the kitchen, the offices" (as in Alma's House, the Isle of Man, and

Benedetti's *Anatomice*).[321] Thus the body's analogues range extensively in Harvey's *Prelectiones,* as they do in *The Purple Island,* from a building to a city, in which anatomy and physiology take on social and political functions such as the defense or feeding of the members of the population. The order of the body conceptualized as a city is hierarchical in the *Prelectiones,* and the brain's preeminence is inseparable from its fortification: "As Nature made no part more greatly defended, so is it deemed the prince of all parts."[322] A comparable association, albeit with the reverse direction of explanation, is made in *The Purple Island:* the "Foure severall walls, beside the common guard, / For more defence the citie round embrace," "For here the Prince his court & standing palace places" (*P.I.* 5.11.1–2, 7).

However much it is true to say that the *Purple Island* allegory extends the range of similarities between anatomy and the built and social environments of the geopolitical realm by exploring the ordinarily used part of the defining concept of the Isle of Man metaphor and by using the ordinarily unused part of that domain, on the evidence of Harvey's *Prelectiones,* the differentiation between poetry and early seventeenth-century anatomy is conceptually, figuratively, and linguistically less than perfect. It would seem that what was said in chapter 6 in relation to Vesalian anatomy remains true of Harveian anatomy, and not only the anatomy of Harvey, but also that of Banister and Colombo: that *The Purple Island*'s discovery or invention of significant connotations of the Isle of Man, by extending the threshold figure systematically and clearly, and giving coherent structure to similarities between the principal and subsidiary subjects, subtly develops the inherent figuration of the (albeit not yet standardized) vocabulary used to describe the body, as either constructed or natural, shading it into the built, natural, and social features of the geopolitical realm of the island and the cosmos above it. The conclusion drawn from the analyses conducted in the first section of this chapter, that the anatomical character of the heart and vascular system in *The Purple Island* may be influenced by Harvey's *Prelectiones,* may no longer seem to imply that Fletcher met a much greater challenge to his figurative design and soteriological purposes than the descriptive anatomy of Vesalius, because some of the similarities which the poetic stanzas "discover" in *The Purple Island* are present, if only in germinal form, in the conceptual metaphors and analogies by means of which anatomical discoveries are interpreted and justified in Harvey's *Prelectiones.*

In the early seventeenth century, teleological, qualitative, and vitalistic conceptions facilitated comparisons not only between humans and other animals, but between humans and other (purportedly) purposive constructed and vegetative things. In Harvey's works we find adherence to anatomical conceptual metaphors, and to the figurality and philosophy of theories of correspondence. Considering the qualitative and vitalistic view of the body that Harvey, Aristotle, Galen, and Vesalius share, as opposed to one that is mechanical and quantitative, and given the inherent potential for figurality in Harveian and Renaissance Galenist and Vesalian anatomy, and the potential for the extension of that figurality into allegory, it is not appropriate to oppose Harveian anatomy and physiology, as modern, to the antiquity of Aristotle or Galen, in conformity to a positivist historiography.[323] The differences between Harvey and either the ancients or the anatomists of the sixteenth century should at least be distinguished from the differential between the degrees of effect of *The Purple Island*'s allegory on its anatomy. Allegorical complexity and personification in Cantos 2 to 5 of *The Purple Island* have been found to increase in relation to the degree to which a part is imbued with *pneuma* or spirit, or is markedly physiologically active; they have not been found to increase in inverse proportion to the relative modernity of the anatomy. Even understood mechanistically, as Temkin says, human biology "is defined by convictions entering from outside. Therefore, the language too comes from outside; it is metaphorical. The use of metaphors in human biology is not an aberration from which even great men have failed to escape."[324]

In order to discern whether *The Purple Island* unfolds the entailments of the microcosm's analogies and the significant connotations of the allegorical threshold figure to the detriment of the anatomy of its most recent sources, this chapter has focused on what might have been the difficulty for the poem to integrate what appears to be its most recent anatomy—cardiovascular anatomy and physiology—into what some of its critics have regarded as an intellectually conservative framework of correspondences. The chapter set out to build upon or qualify the conclusions drawn from the evidence gathered and discussed in the previous chapter, with regard to the figuration of Vesalian anatomy in Fletcher's poem. The present chapter thus set out to explore the possibility that the anatomy of even *The Purple Island*'s most recent sources was in fact conceptually metaphorical in the way that early to mid-sixteenth-century anatomy was, and that the relevant *Purple Island* anatomy is consistent with the essential character of its most recent anatomical sources. In both cases the answer is a qualified "yes."

The principle has already been established that the combination of anatomy with figurality in *The Purple Island* need not necessarily be intellectually conservative, but this chapter has shown that the anatomy is *in fact* not conservative. Even where provenance cannot be ascertained with anything approaching certainty, conceptual character often can be discerned, and at the very least the comparative question of whether the allegory *could in principle* remain congruous, coherent, and consistent if the anatomy were significantly more recent than that of Vesalius, has been answered.

Although the contemporary attribution of ingenuity to its author denotes *The Purple Island*'s innovative and imaginatively resourceful extension of correspondences already assumed to exist, it also suggests that the innovative metaphoricity of *The Purple Island* is incommensurable with the conservatism of didactic poetry. The poetic allegory unfolds a creative metaphor, not a metaphor as comparison, or a contracted simile or analogy. By exploring the ordinarily used part of the defining concept of the creative metaphor of the Isle of Man, and by using the ordinarily unused part of that domain, *The Purple Island* is able to connote more recent anatomy than any of the literary works with which it has been compared in this study because it exceeds them in the range of previously unseen similarities it draws between the principal and subsidiary subjects of the Isle of Man metaphor. The disparities in the creative metaphor of the Isle of Man (as an index or microcosm of all the satellite entailment metaphors of Cantos 2 to 5 of the poem) are not masked; they are operative and their role opens significance so that they act to bring out the resultant meanings not actually given in the antecedently existing similarities or even in the meanings or systems of associated commonplaces of the subjects the metaphor brings together. The contextual effect actualizes a potential range of commonplaces or connotations belonging to the system of associated commonplaces of the focal word or subsidiary subject, and "discovers" a remote range of properties which did not yet belong to the range of connotations of the language of poetic anatomy, giving coherent structure to "previously hidden" similarities between the previously unconnected subjects of recently observed anatomy and a geopolitical realm which is not only Britain but a fanciful contrivance of *ingegno* or *inventio.*

The intellectual curiosity, which *The Purple Island* implies in its author, did not distinguish the ingenious discoveries of anatomical *inventio* from the conceits of poetic *inventio.* In the absence of an epistemologically transparent and neutral nomenclature to describe the putative ontology of the body, an imaginative rationality was exercised in the natural-philosophical contexts of project formulation and justification, and so there is not very much difference in this respect between the *inventio* or *ingegno* of anatomical discovery and that of poetic discovery. In both poetic and anatomical contexts, the ingenuity of invention is such that the figuration of anatomy changes in response to the material determinacy of discovery and the changing context of justification. In *The Purple Island* the contemporaneousness of the anatomy is directly dependent upon the *ingenium* of the figurative design.

In the metaphorical generation of similarities, the anatomy in the poetic stanzas is identifiable not because it is conceptually independent of analogy or allegory for what might be regarded as its essential conceptual character, and may therefore be isolated or extracted for analysis, but rather because the figuration of anatomy is consistent with its essential character. Put another way, *The Purple Island*'s pursuit of the allegorical ideal of continuous superimposition ensures that the anatomical rote of observables is applied to the subsidiary subject, the Isle, and the description of the anatomy of the heart is integrated systematically in the poem's allegory. For these reasons it has been possible to suggest not only where Fletcher consistently maintains the anatomical figurality of a probable or possible source, but also where he may have augmented or departed from the similarities that may be present, if only in germinal form, in the conceptual metaphors and analogies of these sources.

The character of the figuration of anatomy, either in textbooks or poetry, is of significance not only because it is rationally consistent and provides often a clear indication of the function of structure and the *telos* or final cause of structure and function in some purpose of the soul, but also because it gestures toward the ways in which "discovery" is interpreted and justified; it provides a germ of the context of justification and rationalization, of hypothesis as well as thesis construction.[325] Thus in the context of the circular motion of the blood, in chapter 8 of *De motu cordis,* the description of the heart as the sun of the microcosm, even as the sun is designated as the heart of the world, introduces the analogical complex of microcosm, geocosm, and macrocosm to the demonstration of cardiac systole, arterial pulse, and systemic circulation, such that the cyclic process is understood as one of blood regeneration, in which blood that went out from the heart to the tissues in a hot and vaporous condition, in which *pneuma zoticon* inheres, returns to it cooled down and condensed, more or less as in *distillatio.* This faintly alchemical process corresponds to the great geocosmic and macrocosmic circulations of the heavenly bodies and the rains, and allows Harvey to indicate his thesis on the physiological significance of the blood for the life of the whole organism. In the following chapter, our attention will turn to another aspect of the analogical mode of thought in Harvey's work—its epistemology. We shall see that rhetorically, this gives rise to constitutive, analogical relations, which include that between the identification of corporeal identity as the teleological physiology of the body, and anatomical theory that examines or anatomizes itself to give rise to a physiological account of the body.

Our task in this chapter to develop an argument capable of discerning how *The Purple*

Island's figurative anatomy compares with that of Fletcher's most recent anatomical sources and with contemporaneous anatomical works, and how the difference in the *Purple Island* figurality modifies the figurative conceptual character of the anatomy of either its sources or contemporaneous works, has a direct bearing upon our interpretation of how *The Purple Island*'s illocutionary act and the soteriological purpose of the anatomy that Thirsil demonstrates differ from the philosophical purposes and justifications, and the physiological uses and purposes, attributed to the anatomy of Fletcher's sources or contemporaneous works. Whereas the physiology of the heart, blood, and spirits, in chapter 8 of *De motu cordis,* for example, was said to take on a politically sexual and gendered character that is consistent with Aristotle's anatomy of sexual difference and his physiology of generation; the heart's concoction of rarefied blood and *pneuma* into vital spirit and sent to the extremities to cherish and quicken the parts, in Canto 4 of *The Purple Island,* is understood in terms of the way in which the holy mysteries of Christ's sufficient sacrifice for the sins of the world are dispensed in the Eucharist to the penitent elect, that they may be incorporate members in Christ's mystical body. Fletcher's figuration of concoction as confection thereby draws significance for human anatomy from the poem's main story-line of divine history and its eschatology.

In the poem's critical history, its account of the heart and blood has given rise to divergent and contradictory readings because, broadly speaking, the influence of Galenic anatomy, probably through Renaissance Galenism, on, for example, the poem's account of the embryological origin of the veins in the liver, the physiological origin of the blood in the liver, and the embryological priority of the liver over the heart, has seemed retrospectively, from a positivist historical perspective, to contradict Fletcher's description, for example, of the competence of the cardiac valves, and of the streams of the vascular system as "Circling about" and as "creeping forward" but "never turning again," a description which has suggested to some the influence of Harvey's theory of systemic circulation. The attribution of provenance in this chapter remains in all cases uncertain, though a consideration of the apparently varying weight of conceptual and figural influence of a variety of possible sources has made it possible to arrive at conclusions regarding the probability of provenance and, less tentatively, the conceptual character of the anatomy.

One thing that was considered necessary to our inquiry in this chapter was a clear account of the cardiovascular anatomy and physiology in *The Purple Island.* The period from the discovery of pulmonary transit up to about 1628 was one during which it was possible to ignore, to refuse to acknowledge, willfully or unconsciously, the contradictions that were emerging in anatomy between, for example, the competence of the tricuspid valve and the predominantly centrifugal flow of blood and natural spirit from the liver and heart to the extremities, because these contradictions were as yet germinal and appear irreconcilable only in retrospect from beyond the discovery of systemic circulation. In this climate, the *Purple Island* cardiovascular anatomy constitutes an uncommonly self-consistent and noncontradictory physiological model whereby venous blood mixed with natural spirit or *pneuma physicon,* moving along the inferior vena cava from the liver, enters the right auricle or atrium of the heart and either passes into the superior vena cava, whence it flows centrifugally to the parts of the body to nourish them, or enters the right ventricle of the heart via the tricuspid valve of the right atrioventricular orifice, and is heated, distilled, or concocted, the separated refined portion of the blood passing through the minute foramina or intraventricular pores of the septum, from the right to the left ventricle, and the unrefined, cruder blood passing from the right ventricle, via the semilunar valve of the pulmonary orifice, along the pulmonary artery to nourish the lung. Meanwhile air travels from the lungs via the pulmonary vein to the left auricle of the heart, where it is prepared into *pneuma,* and onto the left ventricle via the mitral valve of the left atrioventricular orifice. In the ventricle this *pneuma* meets the refined portion of the venous blood from the right ventricle, which has passed through the interventricular septum, and together the blood and *pneuma* are concocted or distilled making arterial blood in which the vital spirit or *pneuma zoticon* inheres. This spirituous blood is sent out of the heart through the semilunar valve of the aorta centrifugally to the parts of the body to give them life. The only inconsistency in this account is its failure to address the problem of how the lungs are supplied with vital spirit.

Although otherwise entirely self-consistent and noncontradictory, on close inspection this cardiovascular anatomy has been found to be consistent with ideational systems the divergence of which seems to become more apparent when the anatomy is placed in relation to the Hellenist, humanist, and scholastic anatomy of the early to mid-sixteenth century (examined in previous chapters and, with a view to determining provenance, in chapter 7) on the one hand, and the theory of systemic circulation on the other hand. The conceptual divergence might, but does not necessarily, suggest more than one distinct source, a possibility that is consistent with at least one stage of revision by Fletcher of Canto 2, and particularly Canto 4, of *The Purple Island.* The concoction of the blood entering the right ventricle is Galenic or Galenist, except that the process is also described as a chemical or alchemical distillation with art "beyond or words or thought." That in the left ventricle is similarly Galenic but for its description as distillation and the remark that it "makes a strange confection." But Galenism was seen to recede further when the actions of the cardiac valves were considered. Two of the four great vessels leaving the heart, which in Galenist physiology were likened to the four great rivers of Paradise, permit no flow from the heart, the flow of blood (or air) through the valves at the venous and arterial orifices of the heart being consistently one-directional, into the heart through the tricuspid and mitral valves, and out of the heart through the semilunar valves.

Although they are physiologically consistent within the anatomy of Man in Fletcher's allegory, the physiological anatomy of the cardiac valves and that of the veins and arteries is described separately, in Cantos 4 and 2 respectively. This, and the conclusion that most of the anatomy of the blood vessels in Canto 2 is Galenic or Renaissance Galenist in origin, prompted me initially to explore a provenance for the cardiac valves independently. The competence of the valves in the orifices of the aorta and the pulmonary artery had been recognized by a number of anatomists, including Mondino, Berengario, and Vesalius, but the source for Fletcher's account of the anatomical structure, operation, and physiological competence of these and the other cardiac valves is probably to be found among the advocates of pulmonary transit in the second half of the sixteenth century, and a probable provenance is either Colombo's *De re anatomica* or Banister's *Historie of Man.* Circumstantial biographical and bibliographical evidence may be thought to lend weight to either hypothesis. Many advocates of pulmonary circulation or transit, including Colombo and Banister, however, rejected the existence of the porosities in the septum, and Fletcher on this matter follows Galenic theory. Fletcher's concurrence with the advocates of pulmonary transit is on the operation of the cardiac valves and the direction of flow in the pulmonary vessels; it does not extend either to the substance that flows in the pulmonary vein or to the location of the preparation or generation of vital spirit, which Colombo, Banister following him, and Valverde inconsistently, propose is principally the lungs rather than the left ventricle of the heart.

There is further, persuasive, evidence for a provenance of some of the *Purple Island* vascular and neural anatomy in *De re anatomica,* and this lies in the practical basis of Colombo's anatomical work, which has been described as a conceptualization of the body as being nourished, vivified, and given motion by three life-giving rivers—the veins, the arteries, and the nerves and spinal medulla, respectively. Just as in *The Purple Island,* the liver, heart, and brain constitute the three fountainheads from which flow these rivers. Although in the light of the geocosmic, terrestrial figuration of the Isle of Man, toward the creation of which the work intends, it is not improbable that Fletcher ingeniously invented this notion independently of Colombo, its use in *The Purple Island* credibly supports a provenance for at least this figural conceptualization in *De re anatomica.* But if Fletcher did derive this anatomical conception from Colombo, he chose rather to retain the order or method of anatomy from Vesalius's *Fabrica* than to adopt the method in *De re anatomica,* which deals with each of the major organs together with their vessels and omits altogether from its scheme the three venters. Evidence of this kind supports a view that the cardiovascular anatomy and physiology is not only consistent with divergent ideational systems but might suggest a provenance in more than one distinct source.

None of the advocates of pulmonary circulation drew from their observations hypotheses regarding the physiological consequences for the flow of blood in the aorta and its branches, which *The Purple Island*'s diegetic narrator, Thirsil, describes in Canto 2, and so none shares *The Purple Island*'s coherent and consistent combination of the competence of the cardiac valves and the one-directional, centrifugal movement of blood in the arteries. With a view to shedding light on this combination and the use of the terms "distillation" and "confection" to describe the ventricular concoctions, this chapter discussed the work of Cesalpino, only to find that Fletcher could not have found there a description of the cardiac valves as detailed as his own, or as that of Colombo or Banister. However improbable the derivation of cardiovascular anatomy from the work of the relatively unknown Cesalpino, it was thought that Fletcher's use of the phrase "Circling about," in respect of the whole vascular system, might suggest an alchemical use of the term "*circulatio,*" the cyclic process known in chemistry and alchemy as distillation, namely the cooling down and condensation of an overheated and vaporous fluid, the process which is also said to occur in the ventricles of the heart in Canto 4.

But Fletcher did not regard the function of the lungs to be to cool hot blood, as Cesalpino does. Nor does *The Purple Island* share Cesalpino's meaning of the physiologically ambivalent circular movement of very hot blood centrifugally in the arteries predominantly during the day, and centripetally back to the heart in the veins predominantly when we are asleep, a movement wherein the chemical or alchemical process is continuously repeated and cyclical. The possibility that Fletcher derived concepts of *distillatio* and *circulatio* instead from the work of Paracelsians was also considered. But although the geocosmic figuration of the function of the kidneys was judged to have been influenced by Paracelsian or iatrochemical works, they seem to have had no influence on the conceptual character of *The Purple Island*'s detailed anatomical description of the cardiac valves, or on the "pulmonary transit" of blood from the right ventricle of the heart to the lungs and of inspired air from the lungs to the left auricle, or on the centrifugal flow of blood in the arteries.

Harvey's anatomical lectures to the London College of Physicians are conceivably the most plausible unitary source not of all of the conceptually significant features of cardiovascular anatomy and physiology in *The Purple Island,* but of the greatest proportion, including the exclusively centrifugal flow of blood in the veins and arteries, described in Canto 2 of Fletcher's poem, and (ambivalently) the existence of pores in the interventricular septum, the competence (though not the physiological structure) of the cardiac valves of the aorta and pulmonary artery, (to some degree) the action of the mitral valve and the tricuspid valve, and the physiology of concoction in the ventricles of the heart, all described in Canto 4. But partly because Harvey's lectures were not published until the nineteenth century, a proposition that a provenance may be found in Harvey requires qualification both in terms of the probability of Fletcher's attendance at the lectures or of communication between Harvey and Fletcher, which remains a conjecture, and in terms of the conceptual consistency of the anatomy. Regarding the biographical context of the possibility of communication, my conclusions are highly provisional, maintaining merely that whereas communication is perhaps more plausible between 1615 and 1621, during Fletcher's residence in Derbyshire with the Willughbys of Risley Hall, it may be comparatively less likely after November 1621, when Fletcher took up the position of rector at Hilgay. In addition, Fletcher's attendance at Harvey's lectures appears extremely unlikely after 1621.

Despite impediments of a chronological and an interpretative nature to understanding the state of Harvey's ideas as represented in the Lumleian lectures, and some residual uncertainty, the *Prelectiones* may represent Harvey's ideas between 1616 and about 1625, that is, a period that overlaps with what may be the least unlikely time of contact of one sort or another between Fletcher and Harvey. In addition there are strong reasons for postulating that physiological ideas in the lectures are consistent with ideas represented in *The Purple Island.* Even without firm corroboration from biographical research, I do not think a greater tally of similarities can be attached to any other single author or work than to Harvey in the *Prelectiones,* and the *Prelectiones* are alone among those anatomical works examined in this chapter in concurring strictly with the movement of blood in the arteries in *The Purple Island,* a movement that, as far as I know, sixteenth-century anatomists, excepting perhaps Cesalpino, thought of as an ebb and flow. Agreement on the operation of the mitral and tricuspid valves, however, is ambivalent. In respect of the mitral valve, Harvey assumes its competence in his account of pulmonary transit, yet appears elsewhere not to concur with Fletcher, but with those like Bauhin who ignored the valve's competence and allowed a two-way passage of blood through the pulmonary vein: of natural blood inwards to be made vital, and of vital blood outwards for the use of the lungs. On balance, though, Fletcher could indeed have drawn his cardiovascular physiology in large part from Harvey's lectures, and if he did he perhaps replaced Harvey's account of venous blood flow in the pulmonary vein with his own account of air flow in the opposite direction, in order to maintain the valve's competence, albeit thereby excluding any means whereby the lungs might receive vital blood and *pneuma zoticon* for their life.

Curiously, it is probably Harvey's interest in the mechanics of systole and diastole, an interest he shared with his predecessor Colombo, combined with his adherence to a broadly Galenic venous system, which gives rise to his view in the *Prelectiones* that blood is propelled by cardiac systole both into the arteries and into the veins, in which it flows in an exclusively centrifugal movement to the extremities, as it does in *The Purple Island.* Although the explanatory force of fluvial distribution in the geocosm in *The Purple Island* is considerably different from the muscular mechanism of cardiac systole, the physiological outcome in each case is similar. This chapter has considered the view that the *Purple Island* cardiovascular physiology is only internally consistent because of the ingenuity of its geocosmic figuration, which, by enabling the allegorical anatomy to avoid the mechanics of systole and diastole, is consequently detrimental to the anatomy of the work's most recent probable sources. This is a view, however, that accentuates the differentiation between the anatomy that is Galenic or Renaissance Galenist, Hellenist, humanist, or scholastic in conception, such as the physiological origin of the blood in the liver, and the anatomy that has been attributed to Colombo or Harvey, such as the competence of the cardiac valves. But the historical divergence of the anatomy which emphasized function, like that of Colombo and Harvey, from the anatomy that emphasized structure, like that of Vesalius, did not actually occur until the absolute discontinuity emerged between the Galenic venous system and the competence of the cardiac valves in pulmonary and systemic circulation, after 1628. For this reason, despite the perspective gained by positivist historiography, it is not improbable that the *Purple Island* description of the competent function of the tricuspid valve and of the other cardiac valves need not be indebted to an ideational system and source that is distinct from a Galenist description of the flow of blood in the vena cava and the other veins. Indeed the apparently transitional character of the exclusively centrifugal flow in the veins in Harvey's *Prelectiones* and in *The Purple Island,* being neither entirely Galenic nor consistent with systemic circulation, suggests that, from among the varieties of compromise reached by anatomists in the period between the publication of *De re anatomica* in 1559 and *De motu cordis* in 1628, Fletcher not improbably derived cardiovascular anatomy from Harvey, or, if not, from some other recently published source.

In this respect, the ingenuity of *The Purple Island* extends to its inventive integration of contemporary anatomy, by which it distances itself from its didactic and allegorical poetic and prose precursors and sources, which were discussed in chapters 4 and 5, by surpassing them in anatomical detail, systematicity, and contemporaneity. The poignancy with which we may now see that *The Purple Island* was acutely vulnerable to imminent historical change is perhaps enhanced by our knowledge of the contemporaneity of the anatomy at the time of the work's composition together with our realization that at the time of its publication in 1633, the theory of systemic circulation had been in print for five years.

In evaluating *The Purple Island*'s conservatism or innovativeness, we ought not to confuse emergent contradictions between, say, the competence of the tricuspid valve and the predominantly centrifugal flow of blood and natural spirit from the liver and heart to the extremities, with contradictions between teleological, vitalistic models and mechanistic models, for they are of a different kind, the first demonstrating the early fissures between physiological models or systems, the second being of the order of metaphorical models. But this chapter has also attempted to respond in its analysis to the degrees of divergence between metaphorical models, and to avoid traditional historical conclusions about the retrospectively apparent incompatibility between the idea that the action of the heart occurred because of a pulsative faculty or power of the soul and the idea that the hardness of the heart during systole proceeded from the same kind of tension as that of other muscles during contraction. Notwithstanding the evidence that the two ideas were not perceived by Harvey as mutually incompatible, and that the latter did not emerge from a mechanistic metaphorical model, many have characterized the latter idea as mechanistic, partly because the conflict of models followed swiftly on the heels of the publication of *De motu cordis,* and mechanistic explanations were embraced by those who were also to champion Harvey's purportedly experimental method.

If the theory of systemic circulation had been found in this chapter to be essentially mechanistic, and the mechanistic conception of nature found to be a nonmetaphorical model, then *The Purple Island*'s allegorical figuration of aspects of cardiovascular anatomy that were to contribute to Harvey's hypothesis of systemic circulation, such as forceful systole and the exclusively centrifugal movement of blood in the arteries, might have been regarded either as detrimental to the original anatomy, or as superficial. *The Purple Island* is composed of anatomy that has been found to be consistent with Galen, Renaissance Galenism, Vesalius, Colombo, and Harvey, yet the differences between the conceptual characters of these anatomies are not as great as might be suggested by an interpretation of Harvey's experimental method as hypothetico-deductive and his physiology as quantitative, inductive, or mechanistic, shorn of correspondences traditionally assumed implicitly to exist.

But even the difference between seventeenth-century mechanistic anatomy and the vitalism, teleology, and rationalism of Aristotelian and Galenic anatomy in the Renaissance is ideologically rather than actually insuperable, for the machine model was discovered by natural-philosophical *inventio,* and was an elegant or eloquent metaphor which happens also to have been artificial or extravagant in gesturing at something which is not, objectively at least. Like any other conceptual metaphor, when adhered to exclusively, it could impede the apprehension of a discovery by overdetermining the formulation of hypotheses and the selection of the evidence of observation. In time, however, the "truth" of the machine metaphor or model became more generally apparent as the conceptual systems, in respect of which statements could be regarded as true or false, altered in conjunction with the emergence into dominance of the new metaphor or model, and the polarization of religion versus the investigation of Nature commenced.

But the difference between the anatomy of Harvey and that of Colombo, or even Vesalius, is even less than that between sixteenth-century anatomy and the mechanistic anatomy of the later seventeenth century. In the positivist historiography of anatomy, physiology and natural-philosophical and scientific method, Harvey's supposed hypothetico-deductive method is thought to comprise the contexts of project formulation and justification in which he conducted his anatomical investigations and quantitative experiments, and for this reason the discovery of systemic circulation represents a philosophical watershed and occupies an unprecedented threshold in the history of scientific investigation; but the discovery was not seen to do so by Harvey, whose *De generatione* of 1651, due to its Aristotelian redolence, was largely ignored in England by a philosophical community among whom the emergence into dominance of the mechanistic and quantitative model of physiology had occurred in the wake of the publication of *De motu cordis* in 1628.[326]

Harvey's arguments tend often to be deductive from first principles, and although he also used induction from observed facts, his philosophical method or division of the parts is in accordance with the end for which the parts were designed. For Harvey, the body was constructed by Nature for a purpose; it "was not a machine, but was moved by vital forces,"[327] as the adherence of his language to notions of vital spirit and innate heat demonstrates. What is body from a morphological point of view is, from a teleological or physiological perspective, soul and body as a unitary actuality. From this perspective, Thirsil's allegorical presentation of the *telos* of the blood in the veins and arteries in the function of rivers is not very different from Harvey's presentation of the physiology of the heart, blood, and spirits in terms of sexual generation. Both convey the idea of Nature's purpose for and design of anatomical structure. Although he forged ahead experimentally in the context of discovery, Harvey followed a disputational-commentarial method by which anatomy was conducted in the early seventeenth century, and in the formulation of hypotheses he sought to give a new birth to Aristotle's anatomical project which would provide for Harvey's results an equivocal justification.[328] French argues: "Harvey's programme of 'research' which he called into operation when faced with a particular anatomical problem . . . was derived ultimately from Aristotle."[329] Though, as Cunningham maintains, Harvey's deeply informing Aristotelianism does not mean that the ancients, and particularly Aristotle, were "more modern than we thought."[330]

There are differences of emphasis between Aristotelian teleology and the Galenic concept of design which was shared by many Christian anatomists in the Renaissance, like Vesalius. On the one hand there is the Aristotelian notion of the Animal teleologically arranged, while on the other is the human body as the product of the intelligent design of the Creator. Whereas Aristotelian anatomists like Harvey identify Nature with the soul, have little to say in the way of natural-theological argument about God, revealed knowledge, or God's relation to the world, other than in the terms framed by Aristotle himself, and may not ascribe a religious purpose to anatomy, Vesalius and other sixteenth-century anatomists treat accuracy of observation as a kind of piety, the knowledge to which it gives rise as *revelatory* even of God, whose "rare existence" as the body's ingenious artist and efficient cause may be found in the exquisite design of the parts. On most of these points of difference Fletcher follows Vesalius and the Galenists, and further consequences of Fletcher's choice, for the verbally constituted form of ocular anatomical "demonstration" of the creative ingenuity and saving love of God by Thirsil in Cantos 2 to 5 of *The Purple Island,* will become apparent in the following chapter. Within the Christian context of the sixteenth and early seventeenth century, however, distinctions do not hold consistently, and there are concepts of teleology and the instrumentality of the body that are shared by Galenic, Aristotelian, Vesalian, and Harveian anatomy, as well as by the Christian poetic anatomies of Davies of Hereford and Fletcher. Harvey and his teacher, the Aristotelian Fabrici, were also influenced by Galen, and acknowledge a Christian God as the Creator, efficient cause, and divinity of Nature, in their writings.

Harvey turns to the authority of Aristotle "in search of some statement which he can cite in support of his theory" and to venture for it a purpose,[331] even perhaps reading Aristotle "aristotelically," rather than from within the Christian tradition of commentary that stems from Thomas Aquinas's commentaries on Aristotle's *Physics, Metaphysics, Nicomachean Ethics,* and *De Anima.* Meanwhile Fletcher draws significance and purpose for the anatomy presented by Thirsil in Cantos 2 to 5 of *The Purple Island* from the narrative of divine providence in history, from Creation, mortality, and salvation, thereby to fashion the presentation of anatomy with a soteriological epistemology and hermeneutics of ingenuity and eloquence. Fletcher and Harvey may have been situated at opposite ends of a spectrum philosophically and theologically, as practitioners of literary and anatomical *ingegno* and *inventio,* but as they consolidated and justified their knowledge of the meaning and purpose of the body, they were facing each other.

In the first half of the seventeenth century, insights into the mechanistic operations of the human body were often isolated, as they are in Harvey's work, but over the course of the century the affective implications of the machine metaphor were slowly, but steadily worked out, bringing attitudes appropriate to a mechanistic imaginative rationality. Mechanistic accounts of the body's function do not occur in the *Purple Island* anatomy, but the "fracture" between them and ideas of correspondence which Healy perceives as having given rise to the apparent strains within the poem, did not exist in the early seventeenth century. So the absence of mechanism is not an indication of the poem's intellectual conservatism. Of all the anatomy to which Fletcher was indebted in composing *The Purple Island,* it is because Harvey's anatomy was not consistent with mechanism that Healy's interpretation of *The Purple Island* as appropriating contemporary anatomical ideas and (what he understands as) methods of empirical investigation and presentation, in order to validate ideologically an intellectually reactionary project, seems to me to be unjust. *The Purple Island*'s (use of) anatomy is not best interpreted either as conservative or progressive, because Fletcher was involved in a more complex negotiation of imperfectly differentiated anatomical, rhetorical, poetic, philosophical, and theological ideas and discourses, in which he was subject to innovative and conservative impulses.

The imperfection of these early seventeenth-century differentiations does not lend support to Healy's view that in anatomy and literature in the early seventeenth century, there is a "refusal" to acknowledge the difference in the forms of discourse appropriate to their respective disciplines. There seems instead to have been a deep imbrication of poetic and anatomical discourses. In previous chapters this imbrication has been described in terms of the conceptual metaphors of vegetation and "building," analogies of the microcosm, the unfolding of the etymological figurality of anatomical nomenclature, the rationalization of conceptual metaphors in their provision of the *telos* of anatomical structure or function in some function and faculty of the soul, and the coherence of these metaphors in the entailments which they share, that the body is created, mortal, and redeemed. In this chapter our understanding of the discursive imbrication has been extended to the early seventeenth-century use of terms like ingenuity, invention, and discovery in anatomical and natural-philosophical contexts, terms the poetic and rhetorical meanings of which we have had cause to examine in part 2 of this book, in relation to the contemporary attribution of ingenuity to *The Purple Island* and its author. The following chapter returns to the poetic and theoretical development of ingenuity in *The Purple Island* and the work of Davies of Hereford, which was discussed in chapter 5, and examines this in relation to the demonstration of anatomy in the anatomy theater.

To argue, as Healy does, that *The Purple Island*'s inclusion of recent anatomy creates the impression of scientific verification for the poem's intellectually reactionary project is to misconstrue the conceptual character of early seventeenth-century anatomy and the intellectual character of the natural philosophy and theology in the *Purple Island* allegory. Harvey's Aristotelian, and possibly alchemical, speculations in *De motu cordis* about the process that may be akin in some measure to *distillatio,* and which propose the innate heat of the heart and the spiritual perfection of the blood as intrinsic to the purpose of systemic circulation, thereby contributing to the construction of Harvey's thesis and providing the context of justification, were perhaps neither immediately relevant to his demonstration of systemic circulation nor essential to the physiological understanding of that demonstration. As we know, Harvey was to withdraw what confidence he may once have had in any purpose but the motive force supplied by cardiac systole. But when, in 1628, Harvey said that the perfection of the blood by virtue of its revivification through heat and vital spirit must be inquired into and decided upon other grounds than those of observation and experimentation, Harvey did not see, nor did he refuse to see, anything clearer or sharper than an imperfect differentiation between dissection and vivisection on the one hand and the broader subject of anatomy, which was subalternated to natural philosophy, on the other. Harvey's hypothesis and project formulation and his justifications of discovery were steeped in Aristotelian rationalism, and remained so after the publication of *De motu cordis,* and, more significantly for this study, after the period in which Fletcher might realistically be conjectured to have derived anatomical knowledge from Harvey. This surely is of historical importance for our interpretation of the comparative or intertextual relationships between *The Purple Island* and contemporaneous anatomy, even if the same rationalist speculations were not to be important for the development of experimentally demonstrable anatomical knowledge or for the later formulation and justification of anatomical projects.

Contrary to the judgment of Healy, it is not apposite to describe *The Purple Island* as appropriating modern anatomical sources. Rather, the mode of representational exchange which occurs between the poem and the anatomy in textbooks and in lectures is better described as acquisition through metaphor or synecdoche: a mimetic strategy or exchange whereby the poem draws out homologies between dissection by the anatomist and dissection by words alone, or between what Harvey called observation and experimentation, on the one hand, and anatomy and natural philosophy, on the other.[332] These homologies were not even latent, but explicit and expounded by Corti and other anatomists in the sixteenth century as that between architecture, to which anatomy is analogous, and building, to which the art of dissection corresponds, though by the seventeenth century, the precise articulation of the distinction which Corti provides would have been regarded as old-fashioned scholastic Galenism: "Natural philosophy was not only a body of opinion or a corpus of knowledge, set out by statute and modified by teachers. It was also a set of techniques that in some sense *demonstrated* that knowledge."[333]

The Purple Island's acquisition isolates text and method from the illustrated textbook and, as the following chapter will expound in more detail, isolates the verbal account from anatomical demonstration in the theater, but it compensates for the limitations of its linguistic form with suprarealism, mimesis, and Renaissance theories of poetry as a speaking picture, which provide a kind of justification for such an acquisition, as the ingenuity and eloquence of the conceit of the Isle of Man and its allegorical extension by Fletcher the *concettist* provide the means. These extracted elements then stand together as synecdoche for the whole demonstration or presentation of which they were originally a part. The poem does not perform this acquisition by epistemological elision or by sleight of poetic hand, but exposes with aesthetic interest, though not with polemical technique,[334] the ingenious device to the view of the reader. This is done by juxtaposing poetic stanzas and anatomical prose marginalia, thereby formally exaggerating the contrast between poetry and anatomy, which in the early seventeenth century were differentiated not by a disjunction or a concealed discontinuity, but by a graded continuity of language, figurality, and conception.

9

Lectures and Demonstrations in the Anatomy Theater

FROM ONE STANDPOINT THE PROCESS OF UNDERSTANDING THE FIGURE OF THE ISLE OF Man in *The Purple Island* has been seen in this study to be the key for the understanding of the whole of Cantos 2 to 5. The poetic foregrounding of the creative metaphor in the literary work enhances the contextual effect or interaction. As an interactive, creative metaphor, rather than a compressed simile reducible to the antecedently existing analogical and conceptual-metaphorical strands upon which it is based, and which it unifies, the figure of the Isle of Man is abundantly meaningful, opening out degrees of similarity between the associated implications of its principal and subsidiary subjects. The significant connotations created thereby systematically unfold allegorically throughout Cantos 2 to 5, combining to fulfill the intentional direction of the work toward the creation or signification of a geological and geographical location, an island, with its built environment and human society, a dynastic realm or proto-nation which is an imaginary Britain. This realm is simultaneously composed of the anatomy, the structure and functions, of the human body as a microcosm, a little world.

From one other standpoint, it has been the understanding of Cantos 2 to 5 as a whole that has given the key to the interpretation of the metaphor, for this develops the work's intentional direction toward the world it creates and its reflective direction toward a self. The entire anatomical cantos combine with the "conceit" of the reader to create a context in which the metaphor's significant connotations are generated and by which that generation is constrained, thereby deepening and extending meaning in accordance with a notion of the intelligibility of the world of its readers, revealing an obscure unity of the world conceived as a kind of divine and witty poem, the metaphors of which could be brought to light by a sufficiently insightful or ingenious poet.

This study has approached *The Purple Island* not as an object but as a work without clearly defined boundaries between what is inside and what outside, and with a desire to erode the text/context, literature/historical background dichotomy in the interests of a more nuanced conception of the work of literature as a culturally emergent event, albeit one that is graphically embodied in the extent of its signifiers, and materially in bookhood. The digressions to which this approach has led in this study, extensive as they are, at least *en masse,* continue in this chapter with a study of the mode of relationship, negotiation, or exchange between *The Purple Island* and the techniques, conventions, conceptual and demonstrative content, duration and method of public anatomical demonstrations and lectures, as well as the form and function of the anatomy therein displayed or demonstrated, and the iconography and style in the design of anatomy theaters.

Throughout much of this study, in order to indicate the extent of the poem's indebtedness to the figurative and metaphorical language of sixteenth-century scholastic, Hellenist, and humanist anatomy, and the philosophical and medical anatomy of the sixteenth and early seventeenth century, one aim has been to relate this microcosmic geopolitical realm, toward which the intentional direction of the work refers, to figurations of the body as created, constructed, or designed, as natural, as a microcosm, as mortal, and as redeemed, in anatomy textbooks. This analysis of the concepts and figurality of anatomy textbooks has revealed that human anatomy, like that in *The Purple Island,* was conceptually structured in terms of vegetation, geological terrain, and "building," in the way that abstract or less clearly defined or delineated concepts are often structured in terms of concrete, more clearly delineated ones, in our conceptual systems. In Renaissance and early modern anatomy, metaphorical expressions that conceptually shape, structure and define anatomy often arise from analogies between human or animal anatomy and either that of plants or the construction (building), carpentry, masonry, and general structure of buildings. As linguistic metaphors of this kind trigger cognitive processing in the mind of the reader, each of the conceptual metaphors of vegetation and "building" can be found to structure part of the concept of human anatomy, and together provide a coherent understanding of the concept as a whole, as both natural and constructed. Moreover, human anatomy was systematically conceptualized in terms of a host of these metaphorical concepts, which are related in terms of entailment relationships, so that conceptual metaphor underlies surface metaphorical expressions and characterizes anatomical knowledge.

In the anatomy of the schools and in medicine there was also a coherence between the analogies of the body as a microcosm of the terrestrial globe or the macrocosm and the conceptual metaphors of anatomy, that is, between anatomy as vegetation and the analogies of the geocosm and macrocosm, and between anatomy as "building" and the analogy of the body politic. The Creation, the Fall, and the hope or promise of salvation being coherent within Christian doctrine, the body's creation, mortality, and salvation provide the conceptual metaphors of anatomy with coherence. Neither the special sensory epistemology of Renaissance anatomy nor its metaphorical conceptual character and imaginative rationality were in conflict with a soteriological perspective on Man. Moreover, these beliefs about the human body also elevated investigation with the hands as an integral part of the "holy art" of medicine, thereby providing moral and theological justification for the study of anatomy using dissection, in the context of anthropological problems, like the revulsion generated by contact with cadavers, the cultural association of public display and bodily disintegration with shame and dishonor, the taint of the infamy either of the condemned criminal or the executioner, and more specifically religious concerns about the delaying of burial and a belief in the nobility and dignity of the human body and the sacredness of the human corpse.

Providing the rationale for the metaphorical conceptualization of morphology and function in anatomy, beyond the basic conceptualizations of the human body as both natural and constructed, was the understanding of the purpose of functions, their reason for being, or what was called their final cause. The most basic aspect of the animal soul, which humans share with other animals and with plants, and which in humans and some animals rules in the whole of the trunk of the body, according to Aristotle, or in the abdomen, according to Galen, is the Aristotelian vegetative soul or the Galenic *physis.* This provided the reason for the metaphorical structure of the concepts of the nutritive, auctive, and reproductive parts and functions of bodies, because this faculty of the soul's functions is appetitive or desirous, giving reason for the functions of generation, nutrition, and growth. In the Galenic division, the vital soul, or vital faculty of the mortal soul, whose function is agitative, irascible, spiritual, or passionate, provided the reason for the figuration of those parts concerned with the concoction and distribution of vital spirit or *pneuma zoticon,* whereas for Aristotle this was a part of the function of the vegetative soul. The immortal, animal (from *anima,* soul) soul, whose only functions are rational or reasonable for Aristotle, also provided the justification for the figuration of the sensitive and motive functions for Galen. Thus the human body was defined according to its instrumentality and accompaniment or inhabitation by the rational, vital, and vegetative or concupiscent faculties of the soul, and by *physis,* or Nature, and this rationality provided a significant part of the justification of projects of anatomy and dissection in the Renaissance and early modern period.

This study's reconstruction and analysis of the rationality of the use of figurality in *The Purple Island* has demonstrated that the poetry shares figurative and rational coherence and consistency with descriptions of the human body in anatomy textbooks of this period. The contemplation of the creative ingenuity of God as the divine Creator or architect, for example, establishes the ways in which the Christian allegory maintains the systematicity that obtains between the metaphors of anatomy as vegetation and as "building," the analogies of the microcosm as like a landscape or a built environment, and the significant connotations of the Isle of Man. This systematicity even extends to those connotations which arise from the metaphorical interaction of newly emergent ideology, incipient nationhood, and recently discovered anatomy. In the figurative design of *The Purple Island,* however, the conceptual metaphors of anatomy and the analogies of the microcosm coalesce in a way that they do not in textbook anatomy. Being an interactive, creative metaphor, the Isle of Man, the allegorical threshold figure of the work, unifies in a single image the coherent, but in anatomy inconsistent, metaphorical concepts of vegetation and "building," with the analogues of the microcosm, thereby imposing an order on anatomy and fulfilling the principles of coherence *and* consistency.

The single image of the Isle of Man unfolded provides in allegorical form the *telos,* the reason for Renaissance and early modern anatomy's figurality, in some purpose of the soul, which is embedded within and implicit to the figurative language of the allegory. For example, one of the functions of fluvial distribution in *The Purple Island* is that it "gives to all the Isle their food and vegetation" (*P.I.* 3.11.7 & margin. i), and the rational consistency of this figurality is that Fletcher actually provides in allegorical form the reason for the vegetative figurality of the veins in textbooks, which is that they perform the same nourishing function. As chapter 6 has argued, some apparent distortions of composition, magnitude, shape, position, and connexion, in the poetic stanzas, are caused by the purpose that the reader's exploration of the world it is the work's intention to create should be characterized by an appreciation of function as more important than other criteria, and by a teleological understanding.

The Renaissance critical term to which the invention of such metaphors as the Isle of Man related and the most commonly occurring critical attribution made by Fletcher's contemporary and subsequent seventeenth-century readers was ingenuity. In the sixteenth and seventeenth centuries, and as a term applied to *The Purple Island,* ingenuity connoted intellect, wit, literary or poetic talent, inspiration, cleverness, and skill in contriving or making something, such as the poem's figurative design and artifice of allegory. Ingenuity meant the capacity for invention of subject matter and of conceits and creative metaphors. Invention was a part of dialectic, and denoted also imagination or discovery. The ingenious discoveries of anatomical *inventio,* in some measure indicative of the cultural figure of the anatomist as an intrepid explorer of undiscovered territory, have been associated in this study with the conceits of poetic *inventio,* not only in the intellectual curiosity and imagination of Fletcher, but also in the imaginative rationality exercised in the natural-philosophical contexts of project formulation and justification. It has been argued in this study that there is not very much difference in some respects between the *inventio* or *ingegno* of anatomical discovery and that of poetic discovery. In both scenarios, there is a sense of revealing that which existed antecedently but had hitherto been hidden, and a sense of inventing something new, a hypothesis, a theory, a conceit, a relationship between words or objects. In both, the ingenuity of invention is such that the figuration of anatomy changes in response to the historical, material determinacy of discovery and the changing context of justification. As chapter 8 concluded, this is very apparent in *The Purple Island,* where the contemporaneousness of the anatomy is directly dependent upon the ingenuity of the figurative design.

In *The Purple Island,* Fletcher surpasses the ingenuity of all allegorizing poets to whose work his has been compared, not only in the extended range of analogues of anatomy (combined in the subsidiary subject, the geopolitical island realm), but also in the extent of the similarities the poem generates between the geopolitical realm and the anatomical microcosm, and the comprehensiveness, systematicity, and coherence with which the allegory unfolds these similarities, and thus the extent to which its intention ingeniously to contrive a world is achieved. The question of the innovative character of the tropology of *The Purple Island,* that is, to what extent the allegorical threshold figure and the figurative design of *The Purple Island* might be thought to have retrenched and corroborated what the work's contemporary readers might already have known or might have expected to be didactically taught, and to what extent it might be thought to have expanded their vision, is a question to which this study has returned frequently, and will do so again in this chapter, in respect of the public anatomical lesson and demonstration.

This study's conclusions about this question thus far have been related to what the study has described of the poetic achievement of *The Purple Island*'s allegory of human anatomy. A specially constructed set of associated implications of the subsidiary subject interact with the conceptual content, conventional associations, and nomenclatural etymology of sixteenth- and early seventeenth-century anatomy, and although both the principal and subsidiary subjects of this allegory or extended metaphor are modified by their involvement in the unfolding of this host of metaphorical entailments and significant connotations, the Isle, which is the defining domain or subsidiary subject of the metaphor, is not allowed to take priority of meaning. This is for several reasons.

Although the figurality of anatomy in *The Purple Island* is clearly extended beyond what occurs in anatomy textbooks, the differentiation between poetry and anatomy in the Renaissance and even in the early seventeenth century was conceptually, figuratively, and linguistically less than perfect, and *The Purple Island* implies of its author a particular interest in the subtle development of the inherent figurality of the etymology of anatomical nomenclature, which trailed in the wake of the standardization of anatomical vocabulary. His interest has been explained in this study in terms of the way in which etymology facilitates allegory and the metaphorical extension of meaning by eroding the totalization of anatomical concepts which is the goal of the standardization of nomenclature. In addition, in pursuit of the allegorical ideal of continuous superimposition, the poem consistently applies nearly all the descriptive criteria of the anatomical rote of observables, the *accessus,* to the description of the geopolitical realm. The anatomical *accessus* is the unifying feature of the formal design of Cantos 2 to 5, since it underlies the descriptions in both the component parts of the juxtaposition of poetic stanzas and prose marginalia. No other comparable literary work of the period as systematically integrates human anatomy and its analogues and subsidiary subject, or more precisely applies the *occasiones* of the anatomical *accessus*. The result is not only a generally more precise, less schematic anatomy, and more systemic, less formulaic physiology. Anatomical structure and function in *The Purple Island* are meticulously imbricated with social structures and processes, and may not be extracted from the bed of allegory, but must be identified by means of figuration. Although the anatomy in Cantos 2 to 5 has been assigned only an uncertain provenance, much has been identified as conceptually consistent with recently discovered anatomy, and innovative significant connotations have been apprehended as created by the interaction of this anatomy with the specially constructed associated implications of the Isle, its inhabitants, society, and activities.

As was remarked in the introduction to this study, it may follow from the idea of literary criticism that the construction or interpretation of a work of literature may be said more probable than another, or most probable only when it accounts for the greatest number of facts provided by the text, including potential connotations. But this study is a literary criticism that treats meaning as culturally dispersed and possibly conflicted, and therein owes its allegiance to interdisciplinary historical interpretation and analysis of anatomy rather than to formal criticism. Although for this reason, the study concentrates on the anatomical cantos of Fletcher's poem, the context thereby provided, in which the Isle of Man's significant connotations are generated and by which that generation is restrained, is to be expanded in some measure in the present chapter to Cantos 6 to 12, which tell us significantly more than the earlier, anatomical cantos about the Isle's inhabitants.

The crucially significant events recounted in *The Purple Island*'s main story-line of divine providence (which are the Creation, the Fall, the incarnation, death, and resurrection of Jesus Christ, the apocalypse, and the general resurrection) affect Thirsil's anatomical exposition with eschatological delay or deferral of fulfillment. Additionally they provide the perspective through which anatomy is interpreted in terms of the significance of the body's creation, mortality, and salvation, and they inform *The Purple Island*'s poetics and the effect that the work implies was desired or intended. As discussed in Chapter 3, while the concepts of creation, mortality, and salvation thematically characterize the principal events of the main story line of Thirsil's hypodiegesis in *The Purple Island,* they also provide the conceptual metaphors of anatomy with coherence. I have said very little so far in this study about the psychomachia, the internecine struggle for the human body and mind and the island realm, each of which is represented in a partially recovered state since the incarnation, death, and resurrection of Jesus Christ; yet this contestation, together with the gallery of allegorical portraits or personifications of the Isle's inhabitants, the moral and theological virtues and vices, and the events of the apocalypse—the second intervention by Christ the King, followed by victory for the virtues, their regeneration or resurrection, and the defeat and damnation of vice—comprise six cantos out of the twelve of which the poem is composed.

This study has classified the generic character of *The Purple Island* as Christian poetic allegory and literary anatomy, with epic aspirations and pastoral elements. On the question of the generic unity of the work, it has been said that Grundy and Kastor, among other critics of Fletcher, consider the pastoralism of the diegesis, and the pastoral mode of Thirsil's narration within that story, to be perfunctory, superficial, and essentially external to the anatomical exposition, and to the procession of virtues and vices into battle, and their subsequent engagement in conflict. Yet the disunity of the work seems less if one considers that the character of *The Purple Island*'s diegetic narrator, Thirsil, combines the Classical, pagan, and sensuous tradition of pastoral, which Fletcher Christianizes after Spenser, with the Christian, even puritan, identification of the pastor as the under-shepherd to the Master-and-Owner Shepherd or Pastor. Thirsil invokes this "Great Prince of shepherds" with the rhythm of a prayer for the inspiration to figure forth the anatomy and epic spiritual struggle of Man as a secondary Scripture that reveals the Creator, in poetry as a supreme source of information concerning divine mysteries, or God's "sacred skill" in Man. Moreover, as Linda Flasch Redman has argued, the religious pastoral allegory "was a natural expression of the religious poets and of preachers as well," particularly puritan preachers, in whose writings the "imagery of the humble shepherd, the pilgrim (or traveller), and of the warrior were frequent."[1]

Besides the generic character of *The Purple Island,* disunity has been located by critics elsewhere in the work as well. In developing a theory of allegorical, analogical, and metaphorical meaning to interpret the figurality of *The Purple Island,* to reconstruct its rationality, to explore its congruity, and to assess the validity of the arguments of its critics, this study has been concerned, like several others, with the work's combination of anatomical exposition and allegory, and hence with the status of figurality, rhetoric, and knowledge therein. Because the narrative, with its emphasis on soteriology and eschatology, has seemed, to many of *The Purple Island*'s critics, to constitute a rupture with the anatomy of Cantos 2 to 5, it has been necessary to understand the narrative's effect on the conceptual character, coherence, systematicity, and rationality of anatomical figurality in the figurative design. This was the specific aim of chapter 3, but throughout this study analyses of figurality and anatomy in *The Purple Island* have led in digressive ways to analyses of related passages of anatomy textbooks, as well as to works which are literary, theological, religious, natural-philosophical, and medical, in order to conduct this reconstruction and analysis of the work's imaginative rationality.

This exploration has been either intertextual or analytical of the modes of negotiation or exchange within a more general cultural context. But the aspect of Renaissance anatomy that has featured prominently in recent histories and literary and cultural studies, especially those which approach the subject of anatomy from a cultural-historical or social-historical or anthropological perspective, is the public demonstration of anatomy through dissection, or of physiology through vivisection, in either temporary or permanent anatomy theaters. As this is an aspect of our subject with which this study has not yet dealt in specific relation to *The Purple Island,* the concern of the present chapter is to assess the effect of the narrative of providential history on the representation of lectures and demonstrations in the anatomy theater, to discern the congruity or incongruity between public anatomy and the literary genres in which *The Purple Island* is classified, and to determine whether Fletcher's work creatively develops the metaphor of Man as an island in ways that relate directly to the demonstration of anatomy in the anatomy theater, with the focus firmly on human anatomy because of the character of the *Purple Island* anatomy. Analysis in this chapter is intended to enhance and deepen our understanding of *The Purple Island* as a culturally emergent event, albeit a physically embodied one, but one without clearly defined boundaries between what is inside and what was happening outside, that is, between anatomy as knowledge and rhetoric, and dissection and vivisection as practice and performance.

It is the aim of this chapter by examining the historical account of the conceptual content, method, and form of the presentation of anatomy in the theater, the context, function, and purpose of the lecture or lesson and demonstration, and the iconography and design of anatomy theaters, to analyze and detail the significance therein for our understanding of the provenance, conceptual character, method, form, and narrative of the presentation of the *Purple Island* anatomy by Thirsil, and the soteriological purpose of his anatomical exposition in the context of the whole work. This study is yet fully to explore the question of whether in its conceptual, figurative, rhetorical, illocutionary, and perlocutionary aspects, the *Purple Island* anatomy is not significantly different from the presentation of anatomy by lecture and by demonstration in the theater, whether that too was grasped by an imaginative rationality, through conceptual metaphor, analogy, symbolism, and emblematics, which are of teleological, cosmological, metaphysical, moral, and theological origin or justification in *The Purple Island.*

If there were to have been a correspondence of this kind then we might conclude that the perspective of empirical verifiability and philosophical or scientific objectivity was simply not available to Fletcher or anyone else in the early seventeenth century. It is the conclusion toward which the analyses of anatomy in this study have been steadily moving, that only emerging in the early seventeenth century in germinal form, without the coherence it was to assume later in the century, was the objectivist philosophical view that the body could provide its own justification for anatomical dissection (a purpose in and of itself), so that it would no longer signify as an element or part of a whole context which it reflects, but could be an ontologically discrete object present to discourse only across the distance of representation, and could be described by an epistemologically transparent and neutral nomenclature, which in effect, however, was the objectifying application of ideas derived from mechanics. If the coherence of this philosophical position was as yet still out of reach in the late sixteenth- and early seventeenth-century public anatomical demonstration and lecture, as the previous chapter has demonstrated it was for Harvey in his *Prelectiones,* and even his *De motu cordis,* it need not imply that anatomical opinions could not have been tested or demonstrated by dissection, but that anatomy, either in the theater or the textbook, sought to demonstrate much besides that which could be decided on the basis of observation, dissection, vivisection, and experimentation, which Harvey distinguished within the broader subjects of anatomy and natural philosophy to which they belong.

Prior to undertaking this projected analysis in this chapter, it is incumbent upon me to assess the evidence for or against Fletcher having visited an anatomy theater to attend a public anatomical lesson and demonstration. While it may contribute a reason for doubting a provenance for any of the *Purple Island* anatomy in an anatomy theater or a dissection, either public or private, a negative answer to this question is unlikely to be anything other than provisional and inconclusive; but a categorically positive answer would enhance the plausibility of a claim that the influence of the theater and its lectures and demonstrations can be felt on the presentation of anatomy in *The Purple Island* as delivered verbally by Thirsil in Cantos 2 to 5.

Royal licenses, university statutes, and other legally binding documents in the Renaissance and early modern period generally testify to the widespread existence of regulations governing and strictly circumscribing the conditions under which anatomical dissection could be practiced. One of these conditions is frequently that dissection must be performed publicly.[2] The appointment of regulatory officers to ensure the selection of suitable cadavers and to prevent physicians and students from exhuming the dead or from stealing the publicly displayed cadavers of criminals, and the regulations to limit the number of bodies provided for dissection in any given period of time, suggest that universities and colleges were obliged to dissect publicly a significant proportion of the total number of bodies acquired legitimately.[3] Public anatomy lectures and demonstrations were in most cases the only way in which a person, who was neither a physician nor a student specializing in medicine, could attend a dissection. They were also quite often the only legitimate way available to physicians and students of medicine.

Langdale is of an opinion that Fletcher was present at anatomical dissections and considers that the site of such demonstrations attended by Fletcher was Gonville and Caius College, Cambridge. A charter of foundation, secured for the college, in 1557, followed by a Royal License, in 1564, both secured by John Caius, a former student of Vesalius, allowed the Masters and Fellows to receive annually for dissection the bodies of two criminals executed in the town, castle, or the county of Cambridge.[4] "The cadavers were finally to be interred in St. Michael's, the whole college reverently attending";[5] but what Langdale does not notice is that the records of St. Michael's list only one instance of burial of a dissected subject, and that in 1601.[6] Although the bodies of convicted criminals may otherwise have remained unburied, it would seem from university statutes and licenses such as that secured for Gonville and Caius College that the use of cadavers in dissection, if acquired legitimately, entailed their being granted a decent burial afterwards, as a part of the series of prudent strategies implemented to limit the circumstances under which anatomical demonstration could be legitimately allowed (these were discussed in chpater 3).[7] There are no identifiable records of burials of anatomical subjects at the University Church of St. Mary the Great during Fletcher's time at Cambridge, and references to anatomical dissection at Cambridge in the hisotrical record of the period are decidedly few, those being limited to recording the dissection at Gonville and Caius College, in 1601. It may be telling that, in 1587, while celebrating the foundation at the London College of Physicians of the Lumleian Lectureship in anatomy, Holinshed's *Chronicles* lamented "hitherto . . . the wante and lacke so hurtfull; sith that onelie in ech universities by the foundation of the ordinarie and publike lessons, there is one of physicke, but none of surgerie."[8]

In the event that public demonstrations of anatomy and surgery did occur at Cambridge while Fletcher was there, it is Langdale's suggestion that the demonstrator may have been Thomas Grymstone (Grimston) or John Gostlyn, two Fellows of Caius. Grymstone quit his fellowship in 1595, five years prior to Fletcher's arrival in August 1600 at King's College,[9] but he did undertake a public anatomy in 1601, "two years after Harvey had left Cambridge," the cadaver being that interred at St Michael's in that year.[10] Fletcher's anatomical interest may have encouraged him to attend, but although it seems that the physician Grymstone may possibly have conducted the dissection himself, it is not known whether Grymstone's dissection followed Vesalius' method. There is no evidence of Gostlyn ever having undertaken anatomical demonstrations at the university, and in spite of an abundance of statutes and "provisions made for the holding of anatomies both in the University, . . . and in Caius College in particular, there is little evidence to show that such dissections actually took place."[11] "A grace issued in 1627 recounted the statutory requirements for anatomical study and noted that 'for some years past in order to spare a small expense, the practice of making anatomies has utterly ceased in our very distinguished faculty of medicine and surgery.'"[12] So, although it is conceivable that Fletcher attended occasional demonstrations, including perhaps that of Grymstone in 1601, there was apparently no opportunity for repeated and prolonged exposure at Cambridge from 1600 to Lady Day 1616, when Fletcher's name was removed from the King's College rolls.[13]

It appears unlikely that the *Purple Island* anatomy is substantially related discursively or conceptually to its author's attendance at demonstrations in Cambridge. Whether he attended elsewhere is a question that has partly already been asked in this study in relation to the provenance of the *Purple Island* cardiovascular anatomy. Some of the anatomy in Cantos 2 to 5 of *The Purple Island* has been found to be conceptually consistent with that in textbooks by Vesalius, Banister and Colombo, but some cardiovascular anatomy is consistent with that in anatomical lectures presented to the London College of Physicians by Harvey, lectures that would have been accompanied by a demonstration of the anatomy described with reference to a dissected cadaver. The terms of the foundation of the lectureship stipulate that the subjects for the anatomies were to be the bodies of executed criminals.[14]

Chapter 8 not only concluded that Harvey's anatomical lectures are conceivably the most plausible unitary source of the greatest proportion of the conceptually significantly "recent" features of cardiovascular anatomy and physiology in *The Purple Island,* but also that Fletcher's attendance at the lectures remains conjectural, and appears extremely unlikely after 1621. Fletcher presumably did visit his mother in London prior to this date, even though the evidence is inconclusive, and he may have attended the anatomy theater along with the physicians of the College and the surgeons of the Barber Surgeons Company, as did other gentlemen and ladies on holiday. Fletcher's successive residences in Cambridge, Derbyshire, and Norfolk mean that such attendance could only have been occasional, but *The Purple Island* is the product of sustained exposure to the anatomy of textbooks rather more than to that of the theater, and such attendance at Harvey's lectures as Fletcher was capable of on his sojourns to London may nevertheless have been sufficient for him to have absorbed not only some anatomical principles, but perhaps also something of the techniques, conventions, character, duration, and method of public anatomical demonstrations and lectures, as well as the form and function of the anatomy therein displayed, and the design of anatomy theaters.

As with determining the provenance and conceptual character of the *Purple Island* anatomy, in the previous chapter, the uncertainty of whether Fletcher actually attended lectures and demonstrations at the London College of Physicians, or anywhere else, does not entail the impossibility of making a judgment as to the conceptual character of the anatomy insofar as it may be described as probably or improbably deriving from experience of, or at least ideas about, the lessons and demonstrations that took place in anatomy theaters. As was remarked in the introduction to this part of the study, where evidence, or the lack of it, does not allow biographical connections, there may yet be reason to postulate relationships that depend instead on considerations of conceptual, linguistic, semiotic, and figurative similarities between *The Purple Island* and lectures and demonstrations.

Didacticism, Disputation, and Illustration

Although the extradiegetic narrator of *The Purple Island* "speaks" to the implied reader through the medium of linguistic text arranged in poetic stanzas, the story he narrates includes, in Cantos 2 to 5, the diegetic narrator Thirsil's verbal delivery of the anatomical exposition to the diegetic narratees, "a lovely crue / Of Nymphs & shepherd-boyes" (*P.I.* 2.1.6–7).

Thirsil is not described by the extradiegetic narrator as actually performing a dissection, a physical demonstration of anatomy, but this does not obviate the usefulness of asking whether he is nonetheless delivering a public anatomy lesson. As Chapter 7 has demonstrated, the *Purple Island* method or order of dissection combines the method of Nature with the method of discipline. The latter which is also called the method or order of necessity or preservation or continuance, applies to the sequence of anatomy from the abdomen to the head, and corresponds to the tripartite division according to the functions of the animal, vital, and natural faculties, used by Mondino. This was the order anatomists tended to observe in the public anatomies of the schools or universities, when they desired to show all, or the most part of the whole of anatomy, in one and the same body. The order was adopted because the "lower belly" or abdomen is more subject to putrefaction than the rest, by reason it was thought, of its cold and moist temperature and the excrements contained therein.

Although, as was said in chapter 7, *The Purple Island* follows Vesalius's *Fabrica* in departing from the sequence of Mondino by placing the muscles, bones, and other parts of the body before the abdominal viscera, thoracic viscera, and the brain, in *The Purple Island* the method of discipline, insofar as it is adopted, is also consistent with the fact that in the story time of the diegesis, the anatomization is concluded within three days—the same period of time often, though by no means always, allotted to a practical dissection in an anatomy theater.[15] Three days was the period over which Harvey delivered the Lumleian lectures at the London College of Physicians, in 1616,[16] wherein the division of the body follows in large part the "popular" three-venter order. In some universities, meanwhile, a four-day practical dissection was statutory, while the ordinances of the Tomlins Readership, the first anatomical lectureship at a British university, endowed to the University of Oxford in 1624, stated that every spring after Lenten Assizes, four anatomical lectures, each of two hours, were to occur over two days.[17] However, the sequence of the dissection prescribed in the ordinances conforms in other respects to the method of discipline: on the first or second day after execution, the "Scituation, Nature and office of the parts commonly called Naturall videlicet Liver Spleene Stomake Guttes &c." are to be demonstrated, followed in the second lecture by the rest of the natural parts, in the third lecture by the vital parts, and in the fourth lecture by the animal parts.[18]

Variation in the number of days allotted to a practical dissection does not in the least detract from *The Purple Island*'s diegetic mimesis of the popular three-day public anatomy, or, indeed, from the correspondence of the anatomical exposition's story-time of three days to text-time, which can be measured by the division of the anatomical portion of the poem into cantos, each comprising a number of stanzas ranging from thirty-three (Canto 4) to seventy one (Canto 5), and each representing half a day in the diegesis, except for Canto 5 which covers a whole day. Thus, the anatomical exposition commences, in Canto 2 in the afternoon of the first day of the diegesis, with the bones, cartilage, ligaments, "flesh," veins, arteries, and nerves, proceeding to the skin, cuticle, fleshy panicle, abdominal muscles, and peritoneum, and the physiological anatomy of the processes of ingestion, digestion ("concoction"), and excretion of feces and urine. The demonstration of the abdomen recommences in Canto 3 in the morning of the second day, discontinuing around noon, and the demonstration of the thorax, the "middle Province" (*P.I.* 4.2.5), commences towards evening, in Canto 4. Canto 5 describes the anatomy of the "highest region" (*P.I.* 5.2.1), the head, which proceeds throughout a whole day, ending in the evening. Aside from the question of provenance in the lectures and demonstrations in the anatomy theater, the story-time and text-time of Cantos 2 to 5 of *The Purple Island* are designed so that the narrative and division of the anatomical cantos mimics a public anatomy lesson.

Among a small number of features of the anatomical descriptions in *The Purple Island* which, in his biography of Fletcher, Langdale identifies as indicating a provenance for at least some of the anatomical content in lectures and demonstrations in the anatomy theater, is the relative modernity of items of anatomical knowledge. Should Langdale be correct in his identification of this anatomy, it might be one obvious way in which the temporal and textual organization of the story of Cantos 2 to 5 gains significance. Langdale claims, for instance, that the only persons who could inform Fletcher of the aqueduct of Sylvius and the vermiform appendix were the progressive physiologists, not found in libraries, but in anatomical theaters. The aqueduct of Sylvius is probably incorrectly identified by Langdale, however, and it was suggested in chapter 7 that the relevant description was of the anatomical means of the passage of animal spirit from the third ventricle under the testes of the brain into the fourth ventricle, which is common to the cerebellum and the dorsal medulla. On the other hand, although *The Purple Island* does not name the appendix, its description is unmistakable and conforms in several ways to modern accounts.[19] But self-contradictorily, Langdale correctly states elsewhere in his biography that the appendix was first described by Berengario da Carpi; moreover, it "is clearly drawn, named, and described in the *Fabrica* of Vesalius,"[20] which the present study has described as one of Fletcher's likeliest textual sources for much of the anatomy. Neither of these items of *The Purple Island*'s anatomical content either strengthens the case for a provenance of some of the anatomy in the anatomy theater, or lends significance to the narrative and poetic organization of the story and the text.

So although the influence of the theater and its lectures and demonstrations can be felt on *The Purple Island's* presentation of anatomy as delivered verbally by Thirsil in Cantos 2 to 5, in an order that is in part prescribed by the method of discipline, over a period of three days, there has been found no reason except the possible indebtedness to Harvey's lectures, to suppose that the poem's anatomical content is derived from lectures and demonstrations in the theater; and a further reason for doubting such a provenance might be the lack of any evidence aside from the anatomical content that *The Purple Island* shares with the *Prelectiones,* that Fletcher visited an anatomy theater or was present at a dissection, either public or private (he may have received communication of the content of Harvey's account of cardiovascular anatomy independently of the lectures in the College of Physicians).

Whatever conclusions we may draw from this must be inconclusive and should allow for the probability that with regard to the progressiveness of the anatomy in Fletcher's poem, Langdale's premise may anyway be mistaken. The public anatomy even in the early seventeenth century may not have been the forum in which recently discovered anatomy was demonstrated. Although caution is advised in adducing the *Prelectiones* thus, Harvey's observations, experiments, and arguments regarding the simultaneity of cardiac systole and the pulse of the aorta and pulmonary artery were only recorded in the *Prelectiones* because in some part they were indebted to Colombo, and in further part they were sufficiently advanced by 1616 for Harvey to be certain that what he had to say on the subject was true, or that it was regarded as such either consensually or by a significant part of the medical community, such as Harvey's colleagues at the London College of Physicians. This is not all that should be said on this matter, however, and we will return to this point in due course.

So what was the principal function of the anatomy lesson and demonstration in the theater and how may this be related to *The Purple Island?* In chapter 5, my study of encyclopedic poetry and expository and fictional anatomy came to the conclusion that although works thus classified do not constitute a firmly knit family, each adheres to at least some of four interrelated concepts: erudition and self-knowledge, opening or discovery, division or dissection, and analysis or justification. These I have referred to as the conventional, literary associations or commonplaces of anatomy and along with ingenuity or invention and sensory epistemology or experience, each will be found in this chapter to have been significant in the anatomy theater. It was argued in chapter 1 that the public dissections took place for didactic and educational purposes, for the communication of erudition, and it was concluded therein that in the early to mid-sixteenth century, a desire to revive the anatomical projects of the ancients, combined with adherence to Empedocles' theory of elements in the world, to the Hippocratic dogma of qualities and the theory of humors in the body, and to the microcosm-macrocosm analogy between the parts of the human body and those of the cosmos, motivated the study of Man as an impetus to dissect not merely in order to demonstrate that which was already known or believed to be true, but to "discover" also in the sense of finding that which had been unknown. But a distinction can and should be made between two kinds of anatomical dissection, as this chapter is actually less concerned than the previous two chapters with diachronic genealogies or trajectories of discovery; it is concerned instead with the relationship of the *Purple Island* account of anatomy, its anatomical speaking picture, with pedagogical and other reasons for the lectures or lessons and demonstrations of anatomy in the anatomy theater, rather than with the confirmation of anatomical knowledge or discovery of new knowledge for which dissection was conducted privately, by small groups of students and physicians, or individual physicians or artists.[21]

Historically, there was the tendency of demonstration, discussed in chapter 1, deductively to confirm authority and to justify belief or faith therein. This was because it incorporated an epistemological process of inference from authoritative text to cadaver, wherein inferences are most likely to have been made about anatomy from the authoritative texts, rather than about the truth or falsity of authoritative texts from the human body. Much early anatomical dissection, down even into the sixteenth century, had as its immediate object to prove that what Galen or Avicenna or Mondino had said about a given part of the body was correct. About demonstrations, William Heckscher explains: "Their character was that of the so-called *quodlibeta,* the sophisticated public disputes that, from the thirteenth century onward, had become as it were the show windows through which the nonacademic outsider could observe and enjoy the goings-on of the universities."[22] According to Carlino, the public anatomy lessons at Padua and Rome were "carried out in accordance with what . . . is identified as the quodlibetarian model, in which the roles pertaining to practice are kept distinct from those pertaining to theory."[23] By this is meant either the separation of debate from demonstration or the delegation of roles in the demonstration: as discussed in chapter 1, the dissection itself was delegated to the *anatomista, sector, incisor, prosector, resector,* or, in sixteenth-century England, the steward, who was usually a barber or surgeon or even a lay person; the indication of the line of incision in the corpse and the exposed anatomy was delegated to the *ostensor* (demonstrator); and the theoretical component of the lesson—the delivery of a lecture from Mondino, one or other of Galen's works on anatomy, or from Avicenna, *ex cathedra*—belonged to the *praelector/lector* (teacher or professor) of anatomy. The anatomical *quodlibeta* "from a strictly scientific point of view . . . led practically nowhere," in Heckscher's opinion: "The crowning feature of the formal anatomy down to 1543 was the verbal disputation, often carried on without reference to the corpse on display," because attention was focused on the anatomical text recited or lectured upon by the presiding *praelector.*[24] This is not to accord the disputations the function only of confirmation of a consensus; "as a product of the schools," natural philosophy was "adversarial." But in the Middle Ages, even outside the *quodlibeta,* natural philosophy "did not have what we would recognise as a 'research' component, with the expectation that new knowledge could be regularly generated."[25]

Giving epistemological priority to the skill of the hands and to demonstration to the senses, Vesalius reclaimed the practice of the *chirurgus, anatomista, sector,* or *incisor,* the menial who dissects, and that of the *ostensor,* the demonstrator, for physicians, even for the Regius Professors of anatomy, who in the anatomical demonstration had formerly confined themselves to the role of *praelector.* For Vesalius anatomical investigation with the hands was an integral part of the very holy and pious art of medicine, as we have seen, and the result of his innovations was a new relation to the body, which, according to Wilson, was "presentative or performative," rather than "ostensive," because the cadaver acquired a measure of authority over the presenter, the anatomist, and there was an epistemological shift from the text to the body.[26] But, in Carlino's view: "In spite of the innovative suggestions made by Vesalius and of the publication of numerous anatomy texts intended for university teaching, this model [the *quodlibeta*] survived for most of the sixteenth century, and Mondino's *Anatomy* continued to function as the textbook that accompanied and guided dissection in Padua."[27]

During the sixteenth century, the public anatomical demonstration changed, and this involved at many institutions the lapse of the traditional disputation; though at Bologna, despite having been dropped by the doctors during the fifteenth century, by a decree of 1586 the disputation was ensured survival as an integral part of proceedings.[28] Nevertheless, the function of the demonstration continued to be didactic and the methods adopted by anatomists tended still to be disputational and commentarial, as we know from the case of Harvey discussed in chapter 8. In its intentional direction toward a world substantiated with erudition, and its reflective direction toward the reader whom it aspires to teach, *The Purple Island* reflects this didacticism, as it does that of the *Devine Weekes* of Du Bartas, but it differs from the conservatism of the *Devine Weekes,* and possibly that of the anatomy lesson, in its invention, especially in respect of the detail, sytematicity, and contemporaneity of its anatomical subject matter, which consists of the recent products of anatomical *inventio,* and in respect of the witty, cleverly constructed allegorical representation and fanciful, mimetic invention of a strange, yet familiar location that is both a geopolitical island realm and a human body. If *The Purple Island*'s readers approached the work with an expectation based on experience of public anatomical demonstrations and lectures, these "ingenious" features of the work may have challenged rather than corroborated that expectation.

If we consider that the crowning feature of the formal anatomy was the verbal disputation, during which attention was focused on the anatomical text recited or lectured upon by the presiding *praelector,* what in the context of the lesson might have been the function of the anatomy exposed to view when reference was made to the corpse on display? Was it to illustrate the text or one or other opinion represented in the dispute, and if so, what would *to illustrate* mean in this context? It is with the epistemology of the public anatomy and the representational status of the dissected cadaver, which permits the demonstration of anatomy to the audience, that this section of the chapter is primarily concerned. In relation to that concern and to the mimesis of the *The Purple Island,* it is our immediate aim to interpret the character of the relationship between the demonstration of anatomy during the anatomy lesson or lecture in the theater and the illustration of anatomy in the textbook or in fugitive sheets.

Carlino argues that fifteenth-century public anatomies at Bologna and Padua "could only take place for didactic purposes, . . . and that they were intended to 'illustrate' the books read by medical students and commented on by the teachers, thus providing a visual aid to facilitate the memorization of these texts."[29] One other feature of the *Purple Island* anatomy identified by Langdale as indicating a provenance in the anatomy theater may relate to this notion of anatomical demonstration as like book illustration. It comprises aspects of the descriptions of the appearance of anatomical parts, which, Langdale claims, suggest direct observation.[30] Phrases such as "the nerves of hearing plainly appeare" (*P.I.* 5.46 margin. *) create a sense of the immediacy and visual display of the proceedings, and thus may aid memory. Since they occur in the marginalia they may be thought to correspond to the lesson or lecture given *ex cathedra* by the *lector* or *praelector.* They may enhance the impression that the marginal text, like the *lector,* is commenting directly on the anatomical speaking picture of the poetry which (it was argued in chapter 5) has acquired materiality from the mortal corpse and (it was argued in chapter 6) mimics the illustrations of anatomy textbooks.

As regards *The Purple Island*'s formal relationship to the public anatomy lesson, Sawday argues that the poetic stanzas of *The Purple Island* become the equivalent not only of the anatomical illustration, but also of the dissected body in public anatomical demonstration, which I have said is revealed by the combined roles of the *ostensor* and *chirurgus* or *anatomista.* On this view the marginal notes become not only the text or main written explanation of the figure, but the observations or lesson of the anatomist, or more traditionally the *lector/praelector,* on the human anatomy displayed in the theater.[31] In this way, as was mentioned at the end of chapter 6, Sawday argues that in *The Purple Island,* the poetic body, the visual image, is "harmonized with the revealed knowledge of the human form pursued in the anatomy theatre," which in his view is represented by the marginalia.[32]

Let us explore Sawday's proposition by referring to the argument that Fletcher employs the representational status of anatomical illustration and the expository function of marginal notes. As regards the purpose of this mimesis (to reiterate the argument presented by both Healy and Sawday), Fletcher deploys the marginal notes to act as an exposition of his anatomical speaking picture, in the way that the explanatory labels of keying mechanisms elucidate anatomical illustration, and in this the commentary's accuracy depends on how well it describes what is depicted in the poem. Chapter 6 of this study concluded in support of this view that because their purpose is to elucidate, the marginalia do not constitute an autonomous descriptive discourse, but reproduce the allegorical personification and social conceptualization of anatomy found in the stanzas, which, as a demonstration of what is said in the margin, extend the figurality inherent in the etymological nomenclature. Sawday's proposition implies that the relative statuses of the verbal lecture and the demonstration of a dissected human body replicate the relative statuses in the illustrated anatomy textbook of the text and the illustration which it accompanies and explains. My analysis in chapter 6 suggests that the respective roles in *The Purple Island*'s formal juxtaposition of poetry and marginal prose do correspond to the layout and functional relationship of illustration and accompanying labels in Vesalius's *Fabrica.* On the basis of this study's analysis of the relative roles and functions, the implication of Sawday's contention would be that the public lecture and the explanatory label both correspond to the explanatory, elucidatory, clarifying, nomenclatural, and arguably supplementary function of the marginal notes of Cantos 2 to 5 of *The Purple Island.*

It may be that Sawday's argument for what appears to be a kind of mimesis requires qualification, for in the fifteenth and most of the sixteenth centuries, the observations in the lesson of the anatomist were usually not observations of the dissected cadaver at all, and certainly they were not merely supplementary, explanatory, or elucidatory, but more than the dissected corpse itself they were authoritative (as are the *Purple Island* marginalia occasionally), and they would have supplied nomenclature (as do the *Purple Island* marginalia often). On the other hand, it is arguable whether the marginal notes of Cantos 2 to 5 of *The Purple Island* are merely supplementary. Their expansiveness raises the status of notation and, as was argued in chapter 6, they achieve in the work of Fletcher a status comparable with that of the poetry. Combined with expansiveness the experience of reading the notes as almost a sequentially continuous commentary rather than a series of notes suggests that they constitute a discourse parallel to but not autonomous from the discourse of poetry. The consequent gain in authority, combined with the consistency of their deployment of anatomical nomenclature, might suggest that the marginalia do correspond to the lesson delivered by the physician or Regius Professor of anatomy. But in addition, Sawday's account seems not to fully recognize the important role played by the diegetic narrator, Thirsil, in actually pronouncing on the anatomy rather than physically demonstrating it.

In support of Sawday's argument I can say that Thirsil's narration or verbal delivery of the anatomical exposition to his audience of nymphs and "shepherd-swains" (*P.I.* 3.2.5) takes on something of the aspect of the physical demonstration, indication, or illustration of anatomy in the public anatomy lesson, that is, the functions of the *ostensor* and the *chirurgus,* but these functions were not to provide a verbal exposition or disputation. It may be that in putting an extensive, albeit poetic, disquisition on anatomy into the mouth of a character and narrator within the story or diegesis of Cantos 2 to 5, rather than that of the extradiegetic narrator (the narrator who is on a level called the extradiegetic level, immediately superior to the framing story of *The Purple Island,* which he narrates but does not himself appear in) Fletcher was responding to Vesalius's innovative suggestions for the formal, public anatomy: the appropriation of the roles of the *chirurgus* and the *ostensor* by the physician or Regius Professor of anatomy, who would no longer merely provide the lecture or lesson *ex cathedra.* Such a view is supported not so much by the public anatomy as it had been conducted until the seventeenth century, but by the probability that one of the most important sources for much of the anatomy in Cantos 2 to 5 of *The Purple Island* was Vesalius's *Fabrica,* in the Preface of which the author outlines his suggestions which he also rigorously defends. According to this interpretation, the marginalia of Cantos 2 to 5 perhaps correspond to the text commented on or read in the theater, which guides or stabilizes, yet is illustrated by and perhaps less important than, the demonstration, which is narrated by Thirsil.

If, as Park and Carlino argue, fourteenth- and fifteenth-century dissections were rather like an extension of illustration,[33] the question arises, in what epistemological sense is the indication by the *ostensor* of anatomy in the dissected cadaver, or the combined roles of the *ostensor* and *chirurgus* or *anatomista,* illustrative in the sense that the plates and figures of anatomy textbooks illustrate the anatomy therein? It must also be asked whether the public anatomy persisted in this epistemological mode in the sixteenth and early seventeenth century.

It matters less whether it did so in Vesalian demonstrations because Fletcher was neither geographically nor historically positioned to take advantage, and we must assume that any Vesalian influence would have been via the *Fabrica* or indirectly through Vesalius's influence on the direction of the development of the public anatomy. Yet, as Carlino has said (cited above), in spite of the innovative suggestions made by Vesalius and of the publication of numerous anatomy texts intended for university teaching, the *quodlibeta,* albeit without the disputation everywhere but at Bologna, continued to be the model for most of the sixteenth century, and works by either Galen, Avicenna, or Mondino continued to guide dissection.

Let us examine further how *The Purple Island,* in the context of its mimetic relationship to anatomical illustration, may be related to knowledge-demonstrating or knowledge-illustrating methods that might have been operative in demonstrations in the anatomy theater. Amid those descriptions which combine with the effect of creating an anatomical speaking picture not unlike those displayed as fugitive sheets or in anatomy textbooks, is the rhetorical communication of the apprehension of discrete anatomical parts by means of such phrases as "the nerves of hearing plainly appeare." This is intended imaginarily to convey the impression of an ongoing ambivalently verbal and persistently visceral anatomical dissection undertaken by Thirsil, or perhaps by the implied author in the course of the composition or the reading of the poem. The insistence on the observable may correspond both to a special sensory epistemology in anatomy and to the increase in observational and representational accuracy and precision in illustration. The phrase-gestures remind us of how Berengario used *demonstratio* "in a way that we might call "physical logic." Although the practice of *demonstratio* derived from the simple *ostensio* of the public anatomies, the word "demonstration" carried over something of the force with which it was used in dialectic."[34] The gestures in *The Purple Island* resemble the poem's rhetoric of exposing "these dissected wreathes," but their appearance in the marginalia suggests that they help to clarify or stabilize the connotations of the demonstration undertaken by Thirsil. If we relate them to the concept of ingenuity they seem to imply that the ingenuity of the poem has overcome the incommensurability of the signified with the putative ontology of the body-in-itself, and in some sense has recreated a substantially present body with the immediacy of the anatomy theater or private dissection, so that it might confirm or demonstrate the existence of the part about which it eloquently speaks (in the way the *ostensor* might indicate what the *lector* speaks about).

In order better to compare the representational statuses and epistemological functions of anatomical illustration and public anatomical demonstration, let us be reminded of the conclusions reached hitherto in this study with regard to the development of techniques and accuracy in the field of textbook illustration, and with regard to *The Purple Island*'s formal mimesis of anatomical illustration with accompanying text. There emerged in the sixteenth century a belief in the possibility of discovering a body possessed of its own meanings, which in turn gave rise to optimism about the formation and standardization of a nomenclature and set of visual techniques objectively to convey such meanings. In germinal form this optimism motivated early modern anatomists to uncover the body's truths hidden in its inner recesses and "wreathes," and to convey the knowledge gained thereby in text, lecture, demonstration, and illustration. The positioning of the body within a three-dimensional matrix may have been the key to anatomical understanding, but for the artists who illustrated textbooks technical questions arose, such as whether the anatomy should be presented from a particular viewpoint or a combination of viewpoints. Pictorial techniques developed among artists were realized in medical illustration, and the advantages and disadvantages of wood or copper plate for the transmission of three-dimensional space onto a flat surface were explored.

In defense of his anatomical drawings, Leonardo gives voice to a distinctive opinion of the relative merits of observation in public or private dissection and illustration. The comparison he makes is between perusing an illustration and watching "an anatomist at work," not dissecting for oneself, which of course Leonardo did. Nevertheless, Leonardo's advocacy of illustration rests on its potential to respond to the technical question of whether to present the anatomy at a particular stage of dissection or as a synthetic demonstration by incorporating the observations of successive dissections.[35] During the sixteenth and seventeenth centuries, ways of reading an illustration as presenting particular, individual or typical, exemplary anatomy, that could only be a result of successive dissections, were learned and became conventional, enhancing the effects of the naturalistic techniques that had been developed.

One implication of this might be that although private dissection by small groups of students and physicians or artists probably had the epistemological advantage of being a means whereby anatomical knowledge could have been confirmed or new knowledge discovered, the combined roles of the *ostensor* and *chirurgus* or *anatomista* in public demonstration may well have shared a pedagogical, didactic, and illustrative function with textbook illustration or anatomical fugitive sheets. But of more immediate importance, according to Leonardo, a single illustration is capable of demonstrating that which can only be demonstrated by successive dissections, and for that reason illustration might be regarded as a more effective pedagogical, didactic tool. Harsh verdicts were passed on the "scientific" and educational value of public anatomy lessons, referring to them as a relic of scholastic procedures and conclusions and inconvenient for scientific investigation and education,[36] particularly in the eighteenth and nineteenth centuries, after the emergence into dominance of the metaphor of the human body as a machine, when medicine was becoming increasingly professional and anatomy an increasingly specialized, less accessible body of knowledge with its particular use to medicine, and with an increasingly standardized, specialist nomenclature.

A somewhat different proposition relating to this correspondence between illustration and demonstration might be that, although the public anatomy was not traditionally the forum in which recently discovered anatomy was demonstrated, its illustrative function may well have responded to innovations in representational accuracy in the field of artistic illustration. It is important to recognize that when we consider how Harvey must have thought that the simultaneity of cardiac systole and the pulse of the aorta and pulmonary artery were or would be regarded as true by his colleagues at the London College of Physicians, and therefore could be demonstrated publicly, he included among those observations and experiments, if only perhaps by implication, even relatively recently observed aspects, such as the exclusively centrifugal movement of blood in the arteries, in which it seems he had advanced upon the observations of Colombo. Although we must acknowledge that it may be true that the function of the lectures and demonstrations at the London College of Physicians differed from that at the universities, because the College was not an educational facility responsible for the teaching of students, the suggestion is that the pedagogical, didactic function of public anatomical lectures and demonstrations may have been shifting subtly in the early seventeenth century so as to take greater account of the confirmation of anatomical knowledge or discovery of new knowledge in dissections conducted privately.

Support for this view is to be found in the presence of recently discovered anatomical knowledge in the *Purple Island* allegory, but only if the poetry's mimesis of the anatomy lesson and demonstration extends beyond the Lumleian lectures and extends to this proposed general shift in the anatomy lesson's function, and we cannot know either of these with any certainty. Support is not, however, to be found reliably in the rhetorical communication of the apprehension of discrete anatomical parts by means of phrases like "plainly appeare," for this is reminiscent of the way in which figures were sometimes drawn in such a way as to suggest they were drawn "from life" when in fact they were not, and in particular the phrase-gestures may remind us of the pictorial device of incision, which in the plates of female reproductive anatomy in Estienne's *De dissectione,* discussed in chapter 6, functions as a means whereby the illustration presents the "scientifically nugatory"[37] anatomy as derived from dissection and reproduced in book illustration by a method that is epistemologically transparent in respect of its object and empirically verifiable. The device's design is to suggest that the object depicted by the illustration is so faithfully represented that it is even perhaps substantially present to its representation, as the cadaver is present to its dissection. The artists' tools apparently substitute for those of the anatomist, and the graphic incision becomes a synecdoche for the general practice of dissection.

This rhetoric may or may not support a view that Fletcher attended anatomical demonstrations, but as with the presence of recently discovered anatomical knowledge in the *Purple Island* allegory, the rhetoric, taken alone, may do so only if such a verdict could be reached independently, because the rhetorical appeal to the imaginative "senses" of the reader does not endorse a view that the anatomy described has a source in the anatomy theater, and it cannot support what Langdale claims with regard to provenance. But whether or not the rhetorical form relates to actual dissections, each phrase places the anatomy it describes before the mind's eye or conceit of the reader, perhaps for mnemonic purposes, doing so not with extensive and detailed pictorial description of the appearance of a part, but merely by indicating that at this particular stage of dissection the said part does appear. In this the phrase-gestures resemble such phrases from practical advice on dissection in anatomy textbooks or from manuals of dissection such as Achillini's *Annotationes anatomicae,* which were used by physicians and students.[38]

It was argued in chapter 6 that the incision in Estienne's plates of female reproductive anatomy tries to justify science and inspire in the implied reader faith in its *telos,* in belief in the possibility if not inevitability of discovering a body possessed of its own meanings and of forming a nomenclature and set of visual techniques objectively to convey those meanings. As I tried to show in chapter 6, the knowledge that one derives from the anatomy revealed in these plates by the incision is not independent of the pictorial situation in which the figure is placed, and this context reflects the conceptual metaphoricity with which anatomical knowledge located within the confines of the body itself is replete. The plates are designed to bring together the intellectual pleasure of knowledge and aesthetic enjoyment, and their meaning is theatrical; it is about appearances and erotic fetishes which reinforce the conventional view that pregnancy is the fulfillment for women of the architectural or anatomical design in the mind of God.

In chapter 6, without determination of the conceptual character and probable provenance of much of the *Purple Island* anatomy, and without further examination of the possibility of a provenance of at least some of this anatomy or of the modes and method of presentation in the theater itself, however, it would have been premature to conclude from a perspective of empirical verifiability and philosophical or "scientific" objectivity, that *The Purple Island* is rhetorically disingenuous about the representational status of its anatomical "pictures." The evidence of recent anatomy suggests that it is not entirely disingenuous, for it is far from arbitrary and false. But it seems to be at least partly true that the *Purple Island* anatomy, like that of the plates of *De dissectione,* is theatrical, and designed to bring together the intellectual pleasure of knowledge and aesthetic enjoyment. This is consistent with the opinion expressed by Redman, that "Fletcher admired above all others those poets who inculcated virtue while 'charming the senses,'" and with Leah Jones's view that Fletcher's theory of poetry consisted in part of a concept of "sweetness."[39]

Leaving aside as conclusively unanswerable the question of whether such phrases as resemble those from practical advice on dissection in anatomy textbooks, or from manuals of dissection, relate Cantos 2 to 5 of *The Purple Island* directly to Fletcher's attendance at lessons and demonstrations in the anatomy theater, or indeed his unlikely presence at private dissections, I will now return to the consequences of the influence of the mastery of technical questions arising from the illustration of anatomy for the understanding of the illustrative function of the demonstration of anatomy in the anatomy theater. It was remarked a little earlier in this chapter, that the design of the pictorial device of the incision in Estienne's plates and the phrases in Fletcher's poem that seem to be borrowed from dissection manuals, is to suggest that the object depicted is not only faithfully represented, but perhaps substantially present to its representation, as the cadaver is present to its dissection. This is an epistemological claim, and therefore I will look more closely at the epistemology of the demonstration's response both to innovations in accuracy in the field of artistic illustration and to the confirmation of anatomical knowledge or discovery of new knowledge in dissections conducted privately or by groups of students and physicians.

If the public anatomy secured a new pedagogical function to demonstrate recently discovered anatomy, is it not plausible that an epistemological or representational distinction would have been perceived between frequently elaborate and artistic anatomical illustration and the physically present human corpse in public anatomical demonstration? In principle one can make a theoretical distinction between the representational status of the anatomical figure in book illustration or fugitive sheets, even taken perhaps from the life, and the substantial presence of the cadaver in the theater, however illustrative or demonstrative its function. Although one cannot distinguish on the basis of an association of dissection with newly acquired knowledge, a distinction might be that however much one learns and deduces from illustration about even recently discovered human or animal anatomy, only the theater, besides private dissection, could provide the necessary epistemological conditions for unmediated, direct personal observation, *autopsia* (personal experience), by which the ancient empiricist physicians had proceeded to the administration of therapeutic measures. Does it not follow, in this respect at least, that in its form, or more specifically, in its poetic stanzas, *The Purple Island* could not replicate the empirically verifiable representation of the body which can be found in the anatomy theater?

It has already been said that in the Preface to the *Fabrica,* Vesalius remarks on how greatly pictures assist pedagogically in anatomy, placing it more exactly before the eyes than even the most precise language is capable of doing: he thought visibly and he drew with charcoal on the dissection table; the illustrated fugitive sheets of his *Tabulae sex* were on a more generous page than any of their predecessors; and in the publication of the *Fabrica* he presided over technically astute illustrations of comprehensive visual description. The illustrations of Vesalius's *Fabrica* present in many respects empirically verifiable anatomy while combining this with teleology, artistic beauty, and Classical sculptural and *vanitas* presentation.

Even though Vesalius thought that the observational detail and artistic beauty of the illustrations should be privileged over the text on which they are based, yet he implies in his *Fabrica* that while possible in the anatomy theater, personal engagement is not possible by looking at anatomical illustrations, at least not in the same measure. There is even in Vesalius's advocacy of illustrations a barely concealed barb of condemnation of those physicians who do not venture to be present at anatomical demonstrations perhaps for the reasons outlined in chapter 3 of this study, and above in this chapter, as recalcitrance, fear, revulsion, infamy, shame, dishonor, and the violation of the nobility and dignity of the human body and the sacredness of the human corpse: "Furthermore, the illustrations of the human parts will greatly delight those for whom there is not always a supply of human bodies for dissection; or, if there is, those who have such a fastidious nature, little worthy of a physician, that even if they are enthusiastic about the most pleasant knowledge of man attesting the wisdom of the Great Creator—if anything does—yet they cannot bring themselves even occasionally to be present at dissection."[40] Vesalius advocated attendance at dissection for unmediated personal observation. In doing so, however, it should be remembered that Vesalius was alluding primarily to nothing more than demonstration's pedagogical function. Even if it is implied that what is demonstrated may be a "discovery" for the "fastidious" physicians whom Vesalius addresses, it does not seem that Vesalius is advocating the public anatomy for investigation or for a project of confirmation or discovery, for which private dissection would have been undertaken.

We should not look to Vesalius, or any Renaissance rationalist anatomist, for objectivism or empirical epistemology. It may have been for inductive reasons but it was not for the observation of particular differences between bodies that Vesalius advocated attendance at dissection. The presentation of the anatomy in the illustrations of the *Fabrica* is of "a putatively normative description of human structure in general, encompassing and systematizing an ever-expanding body of discrete anatomical data."[41] Similarly analytical, systematizing principles extended to the presentation of anatomy in the theater. Vesalius desired "that the body employed for public dissection be as normal as possible . . . so that you may compare other bodies to it."[42] This gave rise in anatomy to a demand for cadavers with specific characteristics that not only assisted clear demonstration, but which avoided overt signs of their particular history: "it is good practice to get a body with all its flesh and of a solid and firm age, medium height, comely, uncorrupted and sturdy in all its parts. It should not have died from disease or wounding, but should have been hanged, strangled or submerged in water."[43] In Vesalian and post-Vesalian anatomy, the individuality of the corpse is erased in the public anatomical dissection by its "embodiment of a *norm* to be demonstrated."[44] Even though dissection provided an immediate means of learning anatomy by approaching the human body directly, and constituted an investigative tool, a technique for unveiling the body's secrets without the mediation of texts, the public anatomy, for which most bodies yielded by the authorities to the anatomists were destined, was still determined largely by prior judgments as to its illustrative, pedagogic, and didactic function, and therefore by prior knowledge, even when it was conducted by Vesalius, though its receptivity to knowledge recently discovered through dissection varied considerably and before the early seventeenth century was probably greatest in those lessons and demonstrations conducted in the Vesalian tradition, where the physician was the *anatomista,* and was able and prepared to refute the authority of texts.

In England, at the end of the period characterized by the "[a]rtistic conception of the anatomic norm," which extended from Vesalius to Casserio (i.e., 1543–1627),[45] Harvey's Aristotelianism combines the approach of usually Protestant natural philosophers of the late sixteenth century, from particulars to universals or "the great Universal in the mind," with the usually Roman Catholic philosophical method of proceeding from universals to particulars; but to the historian Roger French, Harvey's view seems to be closer to those philosophers who argued that the intellect knows primarily singulars, and generates knowledge of universals from particulars.[46] This approach suggests that Harvey's practice was inductive; moreover, it was also to insist on divulging the particulars of the body actually being dissected, as Luke Wilson says with reference to the *Prelectiones,* the lectures presented by Harvey as Lumleian lecturer to the College of Physicians: "the notion that the anatomist ought to surrender to the particular body a full measure of authority was familiar to and even axiomatic for Harvey, who practiced an anatomy that opened itself to the significance of the differences that occurred among particulars rather than only the similarities on which induction is based."[47] A conflation of the divisions inherent in the body-as-referent and those of the method of anatomy as knowledge and discourse is said by Wilson to give rise to "Harvey's tendency to make the anatomist's agency of division qualitatively similar to a 'natural' or bodily agency that would resist or oppose the anatomist's endeavours."[48]

It seems to be true that in Harvey's case the anatomized and displayed cadaver in the theater did not have a role that was merely illustrative, and the individuality of the corpse was not erased by its embodiment of a norm to be demonstrated, and yet Harvey's method in the *Prelectiones* was neither especially inductive nor empirical. In Wilson's view, the location of somatic *plenum,* "that is to say, corporeal 'depth' and with it corporeal identity in epistemological resistance" to inquiry, is physiology "encountered under the aspect of *telos.*"[49] If so, this is the conceptual character or putative ontology of the body in Harvey's *Prelectiones.* So despite the apparent empiricism of the deference to the actual particular body, and the induction of generalization from particulars, the agency of division is derived to some extent from the location of the body's identity in a physiology understood not so much empirically or inductively as teleologically and deductively.

This location of *plenum* is understood by Wilson as "one of the ways in which the body is constituted" as "a matrix for exchanges involving, but transcending, simple corporeal elements."[50] The putative ontology of the body for Harvey was not therefore independent of what Wilson identifies as Harvey's "analogical mode" of thought and of epistemology, which, Wilson claims, gives rise in the *Prelectiones* to recursive logic as one way by which the body as this matrix transcends corporeality. The empirical or "natural" corporeal body is only one term among several in constitutive, circular, "analogical relations."[51]

Because the notion that the anatomist ought to surrender to the particular body a full measure of authority was axiomatic for Harvey, anatomical authority itself became an object of scrutiny within the anatomy theater, subjected to Nature's own divisions inherent in the body. But this did not take the form merely of giving priority of an epistemological kind to the body and revising the anatomical account accordingly. Rather, the aforesaid analogical relations and recursive logic involve "the implicit analogy between the body that is anatomized and the anatomical theory that anatomizes itself—conceives of itself as a body of knowledge that examines its own parts." The "slippage" from the anatomist's agency of division, the ideational structure of anatomy, to the "natural," organic or "bodily agency" of "inherent" divisions generates ambiguities, as Wilson explains: "the body is confused with the science of its study, not in the sense that the science provides an internally coherent account of the body that substitutes for any recourse to an empirical origin, but rather in the opposite sense, that it reapplies to itself its own account of the body, as though somehow it had lost a sense of the difference between them."[52]

So although there is deference to the particular body in Harvey's approach, the body is not an ontologically discrete object present to discourse across the distance of epistemologically transparent representation. Rather, the analogical mode of thought and of epistemology is at work, and this analogy between, even conflation of, modulations of division in the text and the categories of division in the body "seems quickly to gather momentum in which the 'divisions' of the two tend toward a nearly explicit identification," such that, for example, "the idea of the bodily division of the parts slides into the notion of the differences of opinion that cause controversy among the authorities." In this case, rather than giving epistemological priority to the body and revising the anatomical account accordingly, Harvey's analogical mode gives rise to a view characterized by similarities, such that "[t]he body, specifically in its diversity, is then the proximate agent of the formation of the diversity of opinion that structures the science of anatomy," itself a corpus or body of knowledge.[53] More than being revised, the authority of anatomical opinions is justified. On this view, Harvey's anatomy in the *Prelectiones* does not present itself as empirical, inductive, or epistemologically transparent.

Harvey's analogical mode can be related to that of *The Purple Island* and the literary works examined in chapter 5, which share a teleological and soteriological epistemology and hermeneutics. These works include Davies of Hereford's *Mirum in modum* and *Microcosmos,* and the relevance of their mode of "figuring forth" and their suprarealist mimesis for this comparison lies in their epistemological and representational claim that the poetic anatomy reproduces or recreates the actuality, life, or truth of the body-in-itself as the signs of poetic work. Anatomy, teleology, and language for Davies of Hereford and for Fletcher were understood more in Christian than in Aristotelian terms, but by adopting stances of humility before the omniscience of God or the "grace" of Nature, respectively, *The Purple Island* and *Microcosmos* approach Harvey's notion that the anatomist ought to surrender to the body a full measure of authority.[54] In addition, the poetic anatomies of Fletcher and Davies of Hereford imply that incising a body, that is, opening and dividing it for anatomical investigation or demonstration, is imaginarily reconciled in poetry to (re)creating that body in the structure or composition, extent or size, form, internal relations, quality, effects, and life of a poem, and the number and disposition of its parts. This is a mimetic, rhetorical reproduction of the body-as-referent as the text, which in turn acquires a rhetorical corporeality, ingeniously appearing as a body of words to replicate, or speak the truth of, the body it depicts.

In this way each of these poetic texts purports to combine incisive inquiry with the eloquent articulation or dilation of meaning found in the body understood as a text, because they convey in figurative form the *telos* of structure and function, and in part because they combine the meaning of pictorial accuracy with eloquence in the notion of poetry as a speaking picture, which thereby pierces and possesses the sight of the soul. Rather than the body merely illustrating the text that is read or commented on in the lesson, the poetic anatomies of Davies of Hereford and Fletcher aim to give life to a body of words purportedly by recreating in description the body's reason for being. In so far as they do this, by mimesis as categorized under *descriptio,* they reproduce or reconstitute the somatic *plenum,* the fullness, actuality, life, or truth of the body-in-itself, in the anatomico-poetic account of the body, that is, in the signs that constitute its description. The poetic anatomies thereby rhetorically overcome the incommensurability of the signified with the putative ontology of the body-as-referent in a way that is not very dissimilar to the *Prelectiones* of Harvey.

Harvey's lectures are rhetorical in this respect, because their mode of thought and of epistemology is analogical, and as was argued in chapter 5, the reconciliation of incisive inquiry and eloquent figurality is possible in an imaginary way for Davies of Hereford and Fletcher because of the purported character of the body's particulars, which is metaphorical, and that of its causes, which is inferred metonymically from the postulate that the body is an effect of the soul and of God's act of creation. Yet in the works of Davies of Hereford, in particular, knowledge is purportedly acquired metonymically, through reasoned inference of cause from figurative effect, and by synecdoche, through deductive generalization from metaphorically figured particulars. This model of reason and ingenuity has been found to be vulnerable to skepticism and to the development of the special sensory epistemology of anatomy. In professing the limits of poetic ingenuity, the poetic voice of *Microcosmos* does not actually acknowledge the epistemological resistance of Nature, but how it is aesthetically impossible to represent and theologically inscrutable; and in deferring the task of dissection "unto *Anatomists,*" the poetic voice of that poem does so dismissively, as though their concerns with philosophy of language, epistemology, and method were insignificant as compared with the poet's concerns with aesthetics, theology, and philosophy.[55] Like his skepticism, Davies of Hereford's teleology and epistemology are pious, and on this point the epistemological commitments of the poetic works of Davies of Hereford depart from those of Harvey's *Prelectiones.*

By contrast, the analogical and metaphorical correspondences in *The Purple Island* do not determine the body by a single conceptualization; and the poem's more thorough application of anatomical descriptive criteria, and its integration of anatomy that appears to derive from recent sources, perhaps from Banister's *Historie of Man,* of 1578, and possibly even Harvey's own Lumleian lectures, ensures that the knowledge specific to anatomy is not subsumed in that poem by that of architecture, carpentry (as in *Microcosmos*), or the geography or politics of the island-realm.

Wilson's analysis of the *Prelectiones* proceeds to a convergence of the preceding theoretical notions in the lectures' identification of the account or "*logos*" of the body as a whole with the soul. The strands of argumentation may be outlined as follows. Given, firstly, that the physiological uses of the parts are, for Harvey, directly related to their causes; and, secondly, that the whole body is made "in order to serve as the instrument of the soul," which "is understood to be the body's formal and final cause, the determinant and purpose of its structure";[56] it follows that the lectures, which are an account of the physiological anatomy of the whole body, a summation of each of its parts' various reasons for being, approximate to the soul. This understanding of anatomical discourse is distinguishable from the model of discourse that was to characterize much of later seventeenth-century scientific investigation, based as it was to be not upon seeking the answers to questions of *why* a function in the body happens or a part was made, but upon the objectifying application of ideas derived from mechanics to the body's processes, in order to explain *how* they happen.

Wilson does not seem to recognize that Harvey's analogical mode, as he describes it, departs from Aristotelian philosophy in respect of this rhetorical approximation of the lectures to the soul. As was outlined in chapter 5, even though the Aristotelian formal and final cause of the body, which is the soul or form, is theoretically separate from the body and exists, as it were, in the account, it cannot exist without a body, and so the soul cannot survive the death of the cadaver in order to comprise, or persist in, the anatomical account any more than it can truly be thought to persist in the poetic anatomies.[57] The Aristotelian commentators Costelloe and Muirhead write: "Inasmuch as the soul is dependent upon the body for its existence and activity, it must come into existence and perish with it. Only incorporeal spirit can precede and outlast the bodily life. But this, according to Aristotle, is to be found only in the reason and in that part of it which is without taint of the lower activities of the soul—namely, the Active Nous."[58] The immortality of Active *Nous* or Reason, which "has no material organ," which "cannot be entangled in the life of the body," and "is wholly unaffected by the body," does not constitute a survival of personal identity or *form,* and some commentators have found even the question of the relationship between incorporeal, yet incorporated, individual human reason and Active Reason a vexing one in Aristotle.[59]

For this reason the relationship of Harvey's epistemology to the model of reasoning in Davies of Hereford's *Microcosmos,* which has "Witt's-wonder-working *Ingeny,*" with which *The Purple Island* is also attributed, and which is said to be "[f]ixt to" the "soule's facultie[s],"[60] although not easy to define, is closer than we might assume. The rhetorical figure which Harvey's *Prelectiones* employ to accomplish the approximation of the account to the soul is that which Wilson identifies as the analogical mode of thought and epistemology. As is argued above, this mode (re)constitutes or "reproduces" the body in the signs of its account, by locating somatic *plenum,* and the agency for the divisions and the account, simultaneously, in the empirical or "natural" corporeal body and in a physiology which is understood as teleological, and which becomes apparent only in its account, in the anatomical theory that anatomizes itself to give rise to a physiological account of the body.

Let us draw to a provisional conclusion the investigation undertaken thus far in the present chapter into the relationship, negotiation, or exchange between *The Purple Island* and the techniques, conventions, duration, and method of the public demonstration or presentation of anatomy through dissection, as well as the conceptual and demonstrative content, form, and function or purpose of the anatomy therein displayed. Although it is conceivable that Fletcher attended occasional demonstrations at Cambridge, there was apparently no opportunity there for repeated and prolonged exposure, but in addition or instead Fletcher may have attended Harvey's lectures on his sojourns to London. However inconclusive our knowledge may be about this, insofar as it is adopted in *The Purple Island,* the method of discipline is consistent with the fact that in the story-time, the dissection is concluded within three days, and this corresponds to text-time, such that the narrative and division of the anatomical cantos mimic a public anatomy lesson.

There are no items of *The Purple Island*'s anatomical content which strengthen a case for a provenance in the anatomy theater, except those in Cantos 2 and 4 which might derive from Harvey's Lumleian lectures, but the public anatomy was not anyway traditionally the forum in which recently discovered anatomy was demonstrated. It was designed traditionally for didactic and educational purposes, and followed the quodlibetarian model, in which the roles pertaining to practice are kept distinct from those pertaining to theory, and this model, albeit without the disputation everywhere except at Bologna, continued to be the model for most of the sixteenth century. In this respect a distinction is drawn between the public anatomy and anatomical dissection undertaken privately for the confirmation or discovery of knowledge. *The Purple Island* was found to reflect the didacticism of the traditional public anatomical demonstration, but may differ from it in *inventio,* not in the sense that the poem undertakes an actual anatomical investigation involving dissection, but in the "discovery" of its anatomical subject matter, which consists of the recent products of anatomical *inventio.*

Traditionally, public anatomies were intended to illustrate the works by either Galen, Avicenna, or Mondino, read by medical students and commented on by the teachers, and so this section of the present chapter has examined the possibility that in addition to its mimesis of the anatomical illustration and accompanying explanatory text, the formal disposition of the text in Cantos 2 to 5 of *The Purple Island* may mimic the form and purposes of the lesson and demonstration of anatomy in the theater. Thirsil's diegetic narration or verbal delivery of the anatomical exposition takes on the functions of the *ostensor* and the *anatomista* or *chirurgus,* to demonstrate, indicate, and illustrate anatomy in the dissected cadaver, but it was argued that in putting the poetic disquisition on anatomy into the mouth of Thirsil, who is the diegetic narrator, rather than that of the extradiegetic narrator, Fletcher was responding to Vesalius's innovative suggestions for the formal, public anatomy. On Vesalius's model, Thirsil not only has the roles of the *chirurgus* and the *ostensor,* but also that of the physician. *The Purple Island*'s mimesis of the formal anatomy is therefore manifest in the levels of narration as well as in the formal disposition of the text.

If this analysis brought us to a conclusion regarding *The Purple Island*'s narrative and formal mimesis of the public anatomy, it was yet necessary, in order to make a fuller comparison with Fletcher's poem, to establish the epistemological sense in which the combined roles of the *ostensor* and *chirurgus* or *anatomista* were illustrative in the way that the plates and figures of anatomy textbooks illustrate the anatomy therein. Certain aspects of the anatomical exposition of Cantos 2 to 5 remind us of how Berengario's use of *demonstratio,* which derived from the simpler *ostensio* of the public anatomies, involved something of the force of logic in dialectic. On the basis of the discussion in chapter 5, it was possible to infer that the rhetorical effect of this in *The Purple Island* is to imply that the ingenuity of the poem has created a substantially present body with the immediacy of the anatomy theater, so that it might confirm or demonstrate the existence of the part about which it speaks. But those aspects of the marginalia which produce this effect might as easily be modeled on practical advice on dissection in anatomy textbooks or manuals of dissection, and so need not refer the reader to the anatomy theater. Nevertheless, the public anatomical demonstration of a dissected body may well have shared the rhetorical confirmatory function of *demonstratio* with textbook illustration, and with the use of phrase-gestures in the marginalia of Cantos 2 to 5 of Fletcher's poem. These phrases refer the reader to the poetic stanzas because these stanzas have what might arguably be called epistemological priority over the marginal notation, which in the work's formal mimesis corresponds to the text upon which the lecture and demonstration are based.

The combined roles of the *ostensor* and *anatomista* in public anatomical demonstration have been found to have shared a pedagogical, didactic, and illustrative function with textbook illustration and anatomical fugitive sheets, and with *The Purple Island*'s exposition of anatomy, and yet illustration was regarded by some as a more effective pedagogical and didactic medium than public demonstration. However, this section has also argued that the public anatomy's illustrative function may have responded to innovations in accuracy in the field of artistic illustration, shifting subtly in the early seventeenth century so as to take greater account of the discovery of new knowledge in dissections conducted privately. In Vesalian public anatomies, the most innovative of the sixteenth century and in those after Vesalius which took their cue from his recommendations, however, the individuality of the corpse was erased by its embodiment of a norm to be demonstrated. So the public anatomy continued to be determined largely by prior judgments as to its illustrative, pedagogical, and didactic function, and therefore by prior knowledge, even at its most innovative. In this the demonstration of anatomy in the theater was not unlike illustration in anatomy textbooks. Yet in demonstrating relatively recently discovered anatomy, as well as in the mimetic function of its form and levels of narration, *The Purple Island* may be compared to those lessons and demonstrations conducted in the Vesalian tradition, where the physician was the *anatomista.* Combined with the poem's mimesis of the Vesalian method or order of anatomy, and its extensive use of Vesalian anatomy, these findings provide further evidence of a provenance for much of the conceptual character and presentation of the anatomy in Cantos 2 to 5, in Vesalius's *Fabrica,* possibly the copy of the second folio edition which was bequeathed by Lorkyn to Cambridge University Library, in 1591.

The previous chapter concluded by suggesting that *The Purple Island*'s negotiation of sixteenth- and seventeenth-century anatomy in lectures, like its negotiation of that in textbooks, might be described as an acquisition through metaphor and synecdoche: a mimetic strategy or mode of representational exchange whereupon the poem draws out homologies between dissection by the anatomist and dissection by words alone, or between what Harvey called observation and experimentation, on the one hand, and anatomy and natural philosophy, on the other. As was said earlier, these homologies were expounded as analogous to that between architecture, to which anatomy corresponds, and building, to which the art of dissection corresponds. *The Purple Island*'s acquisition of the anatomy of the public lecture and demonstration has been found so far to separate from the dissection each of the following: the text, verbal account, method, duration, and pedagogical, didactic, and illustrative function. This negotiation privileges architecture over building, anatomy over the practice of dissection. The extracted elements then stand together as a synecdoche for the whole demonstration and lesson of which they were originally parts.

As was said in chapter 7, in drawing on a particular source of anatomy, Fletcher may have accepted Banister's invitation to the implied reader of his *Historie,* to lay the stones of his gathering to the building of a cleverly wrought tower of health, but it is inferred that *The Purple Island*'s building is architecturally conceived and designed by the work's (implied) author with a vision and a purpose to preserve or renew the spiritual and physical health of his Protestant nation and its inhabitants, or at least a few friends, to restore and perfect their human nature, and to contribute to their salvation. Whether or not this soteriological aim may be considered as consistent with the actual dissection in the theater will go some way to answering the question of whether *The Purple Island* elides the significance of that part of the lesson which it omits.

By the seventeenth century, as was said in the previous chapter, the implicit Galenism of privileging anatomy as architecture over dissection as building had been superseded by the view that anatomy was not only a body of opinion or a corpus of knowledge, set out by statute and modified by teachers, but that it was also a set of techniques that in some sense demonstrated that knowledge. A probable seventeenth-century source of the *Purple Island* anatomy has been identified in Harvey's Lumleian lectures, and therein Harvey's practice, unlike that of Vesalius, was to insist on divulging the particulars of the body actually being dissected. Harvey's method was no less teleological than that of Vesalius, however, and may have been more so. The agency of division is derived to some extent from the location of bodily identity in a physiology understood not so much empirically or inductively as teleologically and deductively. This is because the somatic *plenum* or putative ontology of the body arises from a mode of thought and of epistemology in which the empirical or natural corporeal body is understood to exist as only one term among several in constitutive, analogical relations.

Notwithstanding these conclusions, it was proposed in this section that it may have been the case that however much one learns and deduces from illustration about even recently discovered anatomy, only the theater or private dissection could provide the necessary epistemological conditions for unmediated, direct personal observation, or *autopsia.* Does it then follow that in its mimesis of the public anatomy through metaphor and synecdoche, *The Purple Island* could not provide the epistemological conditions for unmediated, direct personal observation, or perhaps even those required for genuine *demonstratio,* and therefore could not replicate the empirically verifiable representation of the body which can be found in the anatomy theater? Vesalius implies that this kind of personal engagement is not possible by looking at anatomical illustrations, at least not in the same measure as in the theater.

The Purple Island attempts to compensate for the limitations of its linguistic form with the dominant Renaissance conceptualization of the relationship between painting and poetry, in such a way as to combine in poetry the meaning of pictorial imitative accuracy with figurality or poetic imagery. The poetry thereby in its own way represents an attempt to achieve what Vesalius, or indeed Leonardo, tried to achieve in pictures by providing a shifting, even structurally distorting, perspective that synthesizes the observations of a series of dissections and points of view, and not only outlines structure and morphology but gives metaphorical expression and emphasis to function and teleology. In poetry as an "image," a speaking picture, which strikes, pierces, and possesses the sight of the soul, Fletcher sought to fulfill the didactic, pedagogical, demonstrative and, most of all, the illustrative function of dissection in the anatomy theater.

Healy's argument is that the formal arrangement of Cantos 2 to 5 elides the epistemological differences between anatomical and poetic modes of inquiry, because "the poetic description [is presented] as though it were an empirically verifiable representation of the body which supposedly can be found replicated in each anatomy performed . . . in the anatomy theatre of the Royal College of Physicians."[61] As I argued in chapter 6 in relation to the *Purple Island* mimesis of the illustrated anatomy textbook, the charge of epistemological elision only follows if the poetic allegory is, as Healy sees it, a far cry from anatomy's "usual terms of reference" which "are to the actual human [body] dissected in the anatomy theatre,"[62] and if Fletcher might for that reason be considered as having appropriated anatomy for his own intellectual agenda. We now have reason to believe that although they created a representation that was not without observation or a sensory epistemology, neither anatomical illustration nor demonstration created a representation that was empirically verifiable in all respects, not only because the function of the anatomy lesson in the theater was generally didactic and pedagogical, but more importantly because the function of the dissected cadaver was not confirmatory but to illustrate either the text that was being recited or commented on or some other source of prior knowledge. This remained so throughout the sixteenth and early seventeenth century, even though the relationship of authority between the cadaver, the *anatomista,* the *lector,* and the text shifted during this period.

The function of the dissection of the body in the anatomy theater was not investigative; it did not serve the purposes of discovery or research, or the inductive, quantitative, and hypothetico-deductive methods of modern science. Consequently, it is critically more accurate to describe Cantos 2 to 5 of *The Purple Island* as a mimetic representation, rather than an appropriation, of the lectures and demonstrations of the anatomy theater, the means of which can be theorized rhetorically as via metaphor and synecdoche. In this as in other respects, *The Purple Island*'s negotiation of anatomy is better theorized in ways relating to aesthetics, rhetoric, poetics, and suprarealist theories of language, than the exchanges that predominate in early capitalist market relations.

The Soteriology of Ritual, Drama, and Carnival

With the aim of showing that the dissected body was not presented anatomically from the perspective of empirical verifiability or philosophical or scientific objectivity, or on the model of a machine or automaton, since there is reason to think that none of these perspectives, however inconsistently associated, was available to Fletcher in the early seventeenth century, this section will describe some of the contemporary natural-philosophical and theological interpretations and justifications of dissection in the anatomy theater. More specifically, it is to be concluded on the basis of further investigation in this section of the chapter, whether in Harvey's lectures, or indeed in lessons in the anatomy theater more generally, the analogical relations in which the corporeal body was constituted comprised any part of the analogical complex to which sixteenth- and early seventeenth-century anatomists have been found to have adhered, a complex which included Empedocles' theory of elements in the world, the Hippocratic dogma of qualities and the theory of humors in the body, and the microcosm-macrocosm analogy between the parts of the human body and those of the cosmos. These analogies were found in chapters 1 and 8 of this study to have motivated the study of Man and added to the reasons to dissect, thereby constituting a part of the contexts of project formulation and the justification of discoveries, but the present section of this chapter examines whether they may also have constituted reasons publicly to demonstrate that which was already known or believed to be true about the body.

From the perspective now gained by this study's analysis of *The Purple Island*'s mimesis of illustration in anatomy textbooks and of the public anatomy lesson and demonstration in the anatomy theater, a central question posed for this chapter must continue to be how it is possible to think of the function of the opened cadaver in anatomical demonstration as like an illustration in an anatomy textbook, which sought to illustrate or demonstrate the anatomy described linguistically by an accompanying text, but also didactically to transmit knowledge untranslatable into words, or pedagogically to perform a mnemonic or synthesizing function, or to bring together the intellectual pleasure of knowledge and aesthetic enjoyment. The knowledge conveyed by demonstration might have been either scientifically nugatory or empirically verifiable in some respects, but whether it may also have sought theatrically to demonstrate divine order and teleology, or the astrological conception of the cosmos, or the corruptibility of all natural things, or cosmic and human redemption, are questions for the present section to answer.

It is necessary also to determine whether Fletcher's work creatively develops the metaphor of Man as an island in ways that relate directly to the demonstration of anatomy in the anatomy theater; to turn our attention to the context in which lectures or lessons and demonstrations took place; to look at the iconography and style in the design of either temporary or permanent anatomy theaters; to determine whether anatomy in the theater, like that in *The Purple Island,* was grasped by an imaginative rationality, through conceptual metaphor, analogy, symbolism, and emblematics; to assess the effect of the narrative of providential history on the representation of lectures and demonstrations in the anatomy theater; to discern the congruity or incongruity between public anatomy and the literary genres by which *The Purple Island* is classified. In these ways this study is to explore more fully the question of whether in its conceptual, figurative, rhetorical, illocutionary, and perlocutionary aspects the *Purple Island* anatomy is not significantly different from the presentation of anatomy by lecture and by demonstration in the theater.

There was a change which throughout Europe emerged initially during the sixteenth century, due to new objectives for anatomical lessons and demonstrations in the anatomy theater. One aspect of this change was that these lectures and demonstrations began to be held in the interests of the honorific requirements of the institutions involved, bringing cultural dignity and establishment both to them, in the first place, and to the professional field of anatomy.[63] At many institutions this coincided with the lapse of the traditional disputation that was of the character of the *quodlibeta,* the sophisticated public disputes of the medieval universities, and in most cases the disputation was discontinued in the sixteenth century, the public anatomy at Bologna being the notable exception. But that part of the change which corresponds most closely with the new objectives that interest this study was an emergent emphasis on the ritual and drama of the demonstration of anatomy as an extension of the means of bringing cultural dignity and justification. Drama should be understood in this context in a strong sense, as well as in a weaker sense of ceremony, and it is with the latter that we shall be concerned initially, and only later with the former and with the objectives of this ritual or drama.

The medieval and Renaissance disputation was ceremonial even as it was keenly competitive;[64] though this has been qualified. Giovanni Ferrari's comparison of the fifteenth-century demonstrations, as suggested by the stipulations adhering thereto, and the sixteenth-century demonstrations, at Bologna, concludes that whereas the proceeding of the former "constituted a lesson, albeit of a rather particular type," those of the latter constituted "a public ceremony."[65] The organization, ritual, and ceremony of public anatomical demonstration was related to other prudent strategies implemented to legitimize anatomical demonstration in the context of recalcitrance, fear, or revulsion generated by contact with cadavers, and religious concerns about the violation of the nobility, dignity, and sacredness of the human body.[66] Though not reducible to a strategy of this kind, I would suggest that ceremony in this context corresponds to the desire on the part of the physicians and their institutions to dissociate from the infamy of the executioner and compensate for the notion of dissection as a further, needless punishment bringing an increase of ignominy and shame to the criminal, who may well have been persuaded to repentance, and therein to accept the sentence of death, on the assurance that even the sins that had attracted a death sentence could be forgiven by God. For this reason the ceremony of public anatomies is propitiatory in the same way as the Masses celebrated for the redemption of the souls of the dissected victims of execution.

Arguably related to these strategies was the practice of holding the annual public dissections in January and February, especially during the Carnival period, not only in Bologna, where the customary practice for more than a century entered the statutes in 1602,[67] but in several other Italian universities; and also possibly related is the practice of holding the public anatomies around the festive Christmas season in Holland.[68] There can be little doubt of these practices being widespread, though in Italy it is not always possible to be sure that the event took place specifically in the Carnival season, since, for example, the statutes of the University of Pisa specify "winter-time" as the period during which the dissection should be conducted, and the Paduan statutes "stipulate that it should be held 'after studies have begun and before the end of February.'" Nevertheless, even in Rome, where the statutes gloss over the period during which the dissection should be conducted, "throughout the sixteenth century, dissections were held primarily between early January and late February, with rare exceptions."[69]

The practice extended also to England, to the London College of Physicians, the Annals for 1569–1570 providing the earliest evidence of anatomical lectures for physicians, and the College's anatomy thenceforward being held each year in the winter, though in 1602, it was decided that the annual College anatomy should thereafter be held at the end of the Hilary Term and not during the earlier part of the winter. Additionally, the terms of the foundation of the Lumleian Lectureship, established in 1582, envisage the lectureship, which commenced May 6, 1584, as consisting of a public anatomy in winter, though, as is remarked above, in 1616, Harvey's Lumleian lectures were delivered on April 16, 17, and 18.[70] Notwithstanding individual variations and exceptions, generally speaking, according to Heckscher, public anatomies in all Europe were celebrated only once a year, and they always "marked the height of the winter season," while for the rest of the year the anatomy theaters were "devotional and moralizing museums," a point to which we shall return.[71]

Carlino speculates that the choice of the period of Carnival as the proper time for the dissection had a twofold justification: "first of all, it was the coldest time of the year, and thus the period when it would be possible to best preserve the cadaver; second, since it was a time when the university was on vacation, *lectors* and students, free of their courses, could attend an event that was exceptional from a didactic and sensational point of view."[72] But as Carlino admits, the scheduling of the anatomy lesson during this period "would give it a cultural as well as a practical connection to the Carnival," which is "the only clearly defined time of the year during which certain behavior (generally considered transgressive) is permitted by virtue of an implicit social pact." Therefore, he surmises: "If the dissection was considered macabre and sacrilegious in certain respects and was looked upon as barely permissible—an act that avoided prohibition only by being circumscribed by certain regulations and by a rigorous ceremonial—then to perform it during Carnival implied that it too, along with other transgressive practices, was temporarily channeled into the sphere of the licit by its ritualization. Even if dissection was to be considered a transgressive and profaning act, it was to be tolerated because it took place at a time when every form of subversion and inversion was concealed under the guise of performance."[73]

By contrast with the public anatomies throughout Western Europe, it is in May, the month of regeneration, rather than January or February, the months of the Carnival, in which the story narrated by the extradiegetic narrator of *The Purple Island* takes place. The significance of this difference may be that it suggests that Cantos 2 to 5 are less dependent on the actual public anatomy for their character, though there is one other association of the month that will be explained in due course.

In the meantime, let us explore, in its relation to lectures and demonstrations in the anatomy theater, the strong sense or denotation of drama, which is of course related to the performances of the Carnival. Although for the most part they were not complied with during the sixteenth century, Vesalius's innovations in the formal public anatomy had contributed to the institution of a new relation to the body, which, as was remarked above, has been described as "presentative or performative," rather than "ostensive," and in which the cadaver acquired a measure of authority over the anatomist, as a result of an epistemological shift from the text to the body. Wilson describes the effect of these changes in the following way: "The static spectacle of the authority of the text has been superceded by the dynamic performance of an identifiable performer who, both in his own corporeal particularity and in his physical contiguity with the cadaver, finds himself fundamentally not unlike the body whose structure he performs."[74] Some of these remarks will become clearer in the course of this section of the chapter. Initially, however, let us substantiate the link with the dramatic theater through its conventional practices, accoutrements, and associations.

It is consistent with the characterization of anatomical demonstration as drama, in the strong sense, that Heckscher remarks on how, in 1497, tickets were sold for a public anatomy at Padua, and this practice became widespread in anatomy theaters elsewhere in Europe.[75] In the London College of Physicians, "an iron framework was put up round the dissecting table so that the body could be decently hidden by curtains before the demonstration began"; a practice which relates to the conventional association of opening or disclosure with anatomy, and could not fail to add to the "mysterie" and suspense of the occasion.[76] In reference to the dramatic aspect of the traditional public anatomy, prior to Vesalius's innovations, and even afterwards despite them, Carlino says that at Rome, "it seems that the person of greatest prominence in the demonstration must have been the *lector,* who as reader coordinated the entire didactic/theatrical operation."[77] In addition, the anatomy lesson's division into different phases (which as we shall see may be construed as sequences of dramatic events in a story); "the performance of music to entertain the audience"; the introduction of rules "to regulate the behaviour of those attending and the care taken over the 'production'"; all these things suggest the demonstration's similarity to theatrical performance.[78]

In architectural construction, some anatomy theaters, like that at Padua, responded primarily to functional demands, while others, such as the new theater at Bologna, built between 1637 and 1649, responded primarily to "spectacular" demands, as did public ritual or ceremony, and of course dramatic theater.[79] Even so, when designing the anatomy theater at Padua, in his *Anatomice,* Benedetti declared that he was modeling the theater with ranks of seats on the Arena of Verona and Rome's Colosseum; and Estienne's chapter, in *De dissectione,* on the ideal Vitruvian-style anatomy theater compares anatomy to any other public show, a dissected human body to "anything that is exhibited in a theatre in order to be viewed," and he takes into account the necessity of good vantage points and sight lines.[80] Heckscher proposes the view that formal anatomies represented "important chapters in the historical development of the stage," and Ferrari credits the anatomy theater with full status as a "theatrical site."[81]

The new objectives for anatomical demonstration which throughout Europe emerged during the sixteenth century included the investigation and theatrical display of Man and Nature as Creation,[82] a display which arguably shames the dissected criminal less than it celebrates the divine ingenuity of the construction of human anatomy. According to Cunningham, the anatomy theater became "a place for *showing*—exposing to gaze—the high point of God's creation. In this sense it was a religious drama which was being enacted in public dissection."[83] The drawing of curtains, the uncovering and opening of the cadaver, and the coordination of its display by the *lector* or the anatomist, in the anatomy theater, thereby relate not only to the commonplace or literary association of anatomy with disclosure, but also to the religious revelatory rhetoric we have found in Renaissance anatomy.

Chapter 3 of this study has explained how the rational inference of function and cause from natural knowledge of structure, in the special sensory epistemology of anatomy, was thought to reveal God, and how this, together with the process whereby this might occur, related anatomical knowledge and epistemology to the knowledge purportedly revealed by God and the process of revelation. Although Augustine among others expressed metaphysical anxiety about the potential of anatomy to isolate judgments which record simple observations of fact from other types of judgment, and although Aquinas and most of his successors carried the distinction between natural or rational and revealed knowledge not only to the two ways of knowing, but also to the respective areas of knowledge to which these ways were able to conduct us, neither distinction was consistently upheld, and anatomy was defended against metaphysical and epistemological criticism. It was presented as a body of theory and set of investigative procedures that were revelatory of the truth of naturally hidden structure, function, and material cause, and of the truth of Greek anatomical ideas, which had been hidden because there were barriers in Arabic or scholastic "error" and mistranslation of original Greek medical and anatomical texts, which prevented the assimilation and appropriation of that truth. But anatomy was also presented as revelatory of the efficient cause in God, the formal cause in the conjoint state of body and soul, and the final cause in the instrumentality and inhabitation of the body whereby it might serve the functions of the soul. Public anatomy lessons were demonstrative or illustrative of this revelation, in the sense of revealed knowledge rather than the process of revelation, which was retrieval of Greek texts or the practice of private dissection, though, as we shall see, the mystery of the performance could take on associations of the process of revelation too.

As was argued in chapter 6, and reiterated in the section immediately above, anatomical illustrations usually sought to achieve more than just specific understanding of the structure of the human frame, and in this the representational status of the dissected cadaver in the anatomy lesson and demonstration was no different, partly because its function was traditionally to illustrate. Like the didactic allegory of *The Purple Island,* illustrations spoke of the divine order of the universe as it could be seen represented or imitated in the body as a microcosm or microchristus. They represented human anatomy not as a discrete ontological entity, but as possessing an illustrative function within analogical conceptual systems. Having said this, we know there was an increase in observational and representational accuracy and precision, in both illustration and public anatomical demonstration, and there was an emergent belief in the possibility of discovering a body possessed of its own meanings, and an optimism about the possibility of objectively conveying such meanings, which in germinal form motivated early modern anatomists to uncover the body's truths hidden in its inner recesses, and to convey the knowledge gained thereby in text and illustration. But despite the retrospective tendency to associate objectivism with secular science, this complex of technical accomplishment, epistemological development, belief, optimism, motivation, and aims contributed to the association of anatomical knowledge with the distinctive truths of revelation, which are hidden and must be uncovered, and convey "news" in what is revealed. As chapter 6 has demonstrated with particular reference to the illustrations of Vesalius's *Fabrica* and Estienne's *De dissectione,* these epistemological and ideological shifts bear no necessary correlation, either historically or theoretically, to a diminution of the capacity of illustrations to register on levels other than the strictly corporeal. As in Harvey's *Prelectiones,* perhaps, the empirical or "natural" corporeal body is only one term among several in constitutive, circular, "analogical relations," which convey the createdness, design, and teleology of the anatomy displayed in the illustration and the theater.

Although Cunningham acknowledges that it was a religious drama which was being enacted in public dissection, he fails to see in the anatomy demonstration the scope for dramatic conflict between what have come to be seen since the work of William Schupbach as the two contrasting interpretations of anatomy that were adopted within Renaissance and early modern learned culture. It is with this and related conflicts and tensions as emerging or as being accentuated within the anatomy theater, and as registered and negotiated by *The Purple Island,* that the remainder of this section is concerned.

Schupbach argued succinctly that among anatomists the invocation of the Greek proverb, *nosce teipsum* (know thyself), inscribed over the oracle of Apollo at Delphi, to justify the study of anatomy acquired two aspects. Though we have encountered this injunction in the context of the comparative literary studies undertaken in chapter 5, as one of the principles around which figurality in some of those works was organized, we have not yet attended to the distinction of which Schupbach has made us aware.

It is the optimistic interpretation which informs Cunningham's idea that the anatomy theater was a place for showing the pinnacle of God's Creation, and this interpretation was codified in a formula in which *cognitio sui* or knowledge of oneself "led to the second and more valuable use of anatomy, *cognitio Dei* or knowledge of God."[84] This aspect of anatomy was related to Marsilio Ficino's focus on the central position of Man between visible creation and the things which can only be grasped by the intellect, and to the dignity of Man as proposed by Pico della Mirandola and by later Neoplatonists, who continued the Platonic tradition of Man's correspondence to the All; Man as a *minor mundus,* every creature, in the sense that "he" contains all things. At the same time this tradition retained something of the medieval emphasis upon God rather than cosmos or Nature as Creator of (as well as containing) all created things in the figurative macrocosm, including Man, the last thing created, the end or goal of the Creation process, and hence created in God's own image, the highest likeness of God and the summation of all other Creation.

Combined with this Platonist tradition in the optimistic interpretation of human anatomy, because only imperfectly differentiated from it, was the literal microcosmic view of the human body as a little world, which combines total inclusiveness and miniature proportion. In chapter 1 of this study it was argued that the microcosm-macrocosm analogy between the parts of the human body and those of the cosmos should be acknowledged as having been a motivating concept in the study of Man. *The Purple Island* displays adherence to the Platonic and the literal microcosm theories, though the body as microchristus, rather than microcosmos, frequently carries the weight of meaning; yet additionally, if the macrocosm is conceived of as including the geocosm, *The Purple Island*'s adherence is firm and the analogy is meticulously explored. The microcosm as central to the optimistic interpretation was something anatomists sought to demonstrate dramatically in the anatomy theater, and was adopted particularly by those anatomists who took an exclusive view of anatomy.[85] Neill characterizes it as expressive of the powerful drive towards humanist triumphalism.[86]

Whether temporary or permanent, anatomy theaters were traditionally amphitheaters; if temporary, they were constructed from wooden scaffoldings (oval, semicircular, or circular) placed inside buildings that already existed.[87] In Italy, these purpose-built temporary theaters were frequently erected in churches, which are of course consecrated buildings and purportedly extend the sacredness and solemnity of their purpose to the activities within. The 1540 series of lectures by Corti and demonstrations by Vesalius, at Bologna, purportedly took place in the churches of San Salvatore and San Francesco, for at that time the university had no appropriate structure.[88]

For reason of their location public dissections were conducted with solemnity and ceremony. But the proceedings of the Bologna demonstrations by Vesalius were certainly combative, particularly in the twenty-second demonstration, when Vesalius, having heard his views on venesection refuted by Corti in his twenty-fourth lecture, asked Corti to accompany him to the anatomical demonstration.[89] Perhaps in this and other cases, its occurrence during the Carnival period opened the public anatomy's representations to Carnivalesque possibilities of interpretation and response, and vistas of possibility with regard to the "performance" of ordinarily inappropriate social behavior. Ferrari suggests that the students arranged the "clash" between Corti and Vesalius, as an expression of their sense of the Carnivalesque.[90]

The demonstrations could also be vivacious and on one occasion, in the fifteenth demonstration, Vesalius, dissecting the sixth pair of nerves of the brain going to the chest, and the seventh to the tongue, became "confused, upset and bewildered owing to the noise and disorder that the students then made."[91] Many of the rules introduced to regulate the behavior of those attending, were designed to offset precisely the lack of solemnity and decency the Carnival introduced, the participation of maskers being only the most obvious example.[92] The "Ordinances of the Surgeons of Amsterdam (1605, 1625) stipulated that the audience refrain from laughing and talking and that anyone caught in the act of helping himself to organs," in what was perhaps a grotesque development of Vesalius's insistence on touching and seeing for oneself, "was fined the prohibitively high sum of six guilders."[93] This is only one aspect, or rather a manifestation, of the scope for dramatic conflict between the dignified, optimistic interpretation of anatomy and the conditions and ideas with which it had to struggle, in order to ennoble the body.

In Protestant countries in particular, wherein, as French has remarked, "the Renaissance classicism of the middle of the [sixteenth] century gave way to something approaching natural theology,"[94] the emphasis upon the need to understand the body as "the temple and simulacrum of God" probably facilitated the sense of the appropriateness of the ecclesiastical locations for demonstrations, creating spaces in which that which contained reflected that which was contained, just as the geocosm, the macrocosm, the body politic, *ecclesia,* and Cosmic Christ each reflect the microcosm, or microchristus. The intended effect of such a design might be deduced from Melanchthon's advice on beginning his anatomy of the inner parts in his *Commentarius de Anima:* "At this point truly you think yourself introduced into a temple and a kind of shrine; on this account you ought not to simply look at the material with particular reverence, but to take into consideration the plan and diligence of the

Maker."[95] The sentiment echoes entirely that in *The Purple Island,* for which the ingenious artist or artificer implied by the "wondrous art" of human anatomy is not only, nor even primarily, Nature, but God. As with Crooke's enthusiasm for "cleer arguments" leading from design to the architect of the built environment and landscape of the microcosm, referred to in chapter 3, these writers strove to indicate how dissection might reveal the dignity of the human body as created by God.

When permanent theaters for anatomical demonstration were built,[96] they frequently incorporated elements of the interpretative framework offered by humanist and religious concepts of Man, and the cosmological and sacred correspondences of anatomy, in their architectural design and iconography. The ceiling of the new anatomy theater at Bologna was adorned with "symbolic representations of fourteen constellations gravitating around the figure of Phoebus, the god of medicine";[97] thereby constructing an analogical and reflective correspondence between heaven and the body of Man on earth, which reflected the astrological conception of the cosmos in the "Zodiacal Men" in illustrations, with the heart as the sun of the microcosm, surrounded by organs corresponding to the zodiacal constellations. Reference in *The Purple Island* to the astrological conception of the human microcosm refrains from positively affirming analogical identity between organs and planets, and, as was said in chapter 1, insists on the use of simile; but the correspondences between the heart or "*Kerdia,*" Christ, the Prince and the sun, in form, function, posture, end or purpose, place or position, and office or nobility, have been found to be saturated with significance for our understanding of the poetics, allegory, anatomy, figurality, philosophy and theology of the work. The "four degrees of cedar seats, rising one above the other," in the elliptical anatomy theater of the Barber-Surgeons in London, were reported in the eighteenth century to be adorned with the figures of the seven liberal sciences, the twelve signs of the zodiac, and a bust of King Charles I. This theater was designed by Inigo Jones, and built in 1636, adjoining the north aspect of the Barber-Surgeons' Hall, adjacent to the Roman and medieval city wall and the churchyard of St. Giles Cripplegate, in Monkwell Square.[98] It was the only part of the Barber-Surgeons' buildings to survive the Great Fire of 1666, after which the Hall was rebuilt and the theater remained in use until the separation in 1745 of the Barbers and the Surgeons, who constructed a new octagonal anatomy theater at their new site.[99]

Increasingly, in the late sixteenth and early seventeenth century, the anatomy theater was the architectural construction purpose-built for the dramatic public display and solemn appreciation of the creation and design of the human fabric, the microcosm of the entire Creation. But the term "theatre" carried this positive interpretation of anatomy over to its use in the titles and descriptions of anatomy textbooks. According to French, the term "theatre" "was a natural one to use for a textbook on human anatomy, already associated with a physical theatre."[100] But acquiring a literary connotation from expository anatomy or encyclopedic prose, "theatre" came to connote a comprehensive survey or "collective display" of the whole of a subject or created work of Nature, such as the entire field of human and/or animal anatomy, in a confined or compendious form. In this way the textual "theatre" also connoted the way that in seventeenth-century Holland, the anatomy theater itself reflected in miniature the geocosm and macrocosm, for in it "the whole world was represented and mastered" through its function as a public museum, "in which a collection of 'curiosities,' that is to say, objects of natural history and of natural science, was exhibited."[101] The vast collections of curiosities held by anatomy theaters in their capacity as "devotional and moralizing museums" (to use Heckscher's phrase, cited above in this section) share the same erudite and aggregative method of the textual "theatres" of anatomy proposed by Casserio and Bauhin. Casserio writes, in *Pentæstheseion, hoc est de quinque sensibus liber:* "Accordingly I will presently have published, for your benefit and the furthering of the study of anatomy, my Theater (birdseye view), which I have now on hand, a work containing the structure, actions, and uses of all the parts of the body, together with a wealth of observations and a great variety of new ideas and illustrations from life."[102]

Perhaps the most important example historically of the use of "theater" in the title of an anatomy textbook (because it was popular and widely used) is Kaspar Bauhin's *Theatrum anatomicum,* which Whitteridge describes as "systematic," giving "adequate consideration to the ancient authorities" without going into "too much detail over the controversies." She describes its author's great merit as "his ability to treat his subjects in an orderly and methodical manner," and this perhaps is reflected in his particular "theatre of anatomy."[103] Due to its author's Protestantism, *Theatrum anatomicum* displays something which in certain significant respects is "very different from the anatomy of the earlier sixteenth century,"[104] though it shares its didacticism. The purpose of this Protestant anatomy is more in keeping with the soteriological illocutionary act of *The Purple Island:* "There are vestiges only of the Hellenism of Benedetti, the aggregative method of Zerbi, the high-scholastic commentary of Berengario, and the attack and defence of Galen by the Vesalians and their opponents. Bauhin, at the end of the century, is displaying the human body in a new glory. It is a glory at the root of the reasons for doing anatomy. . . ."[105] These reasons were the "dignity and glory of the human body" as the "miracle of miracles" which "houses the soul," and "the consequent excellence, use and necessity of anatomy" in discovering and displaying, as in a theater, "the instruments of the soul" and so revealing the soul and the self, as a necessary "preliminary stage in coming to know God." The judgment of the ancients has been largely supplanted in Bauhin and other late sixteenth- and early seventeenth-century Protestant anatomists by Augustinian theology, Neoplatonic philosophy, and even hermeticism,[106] all of which we know to have influenced Fletcher's own conception of the significance of anatomy, and of its linguistic and poetic-pictorial representation.

In the Renaissance anatomy theater, there was effectively, however, a dramatic tension or conflict between the purpose to demonstrate divine Creation and providence, on the one hand, and, on the other, in an age prior to the development of effective preservatives, the slow yet steady putrefaction of the corpse displayed on the dissection table. The presentation was determined by this imperative that issues from the body, which therefore posed a practical problem,[107] but the conflict also had symbolic value, and for this reason it was dramatic.

The body had been opened and divided specifically to display the construction of Man as the pinnacle of Creation, yet that body's progressive decay and dissection rendered it increasingly unlike the *imago dei* it was supposed to represent. As the dissection which revealed the body's createdness simultaneously dismantled it, the anatomical account sought progressively to reconstruct it from dismembered parts. Harvey's anatomical lectures, for example, cumulatively demonstrate the form and final cause of "the whole of anatomy" (the soul), and despite the obvious differences between Harveian and Vesalian anatomy, the anatomical exposition of *The Purple Island* which follows the entire Vesalian sequence or method, is designed to replicate the way Nature or God fashions human bodies, and similarly seeks progressively to reconstruct the whole body as it divides it. As the body is divided so Thirsil attends to the building up of a cleverly wrought tower of health. Contributing to this is the accretion of teleological meaning when the signifiers of the poetic text accumulate linearly in the process of reading, thereby cognitively generating denotations and figurative connotations, which the reader "conceits" or apprehends, so that the *telos* gradually becomes more apparent as the allegory unfolds. As the ensuing signifiers accumulate in the process of reading and interpretation, they individually and cumulatively pierce or divide the body *and* expose to the apprehension its nature, truth, and life as a whole.

The association of death with the anatomy theater may have found its material cause in the steadily putrefying corpse, but it achieved symbolic status as anatomy's second aspect or interpretation, which few if any anatomists accepted exclusively, but wherein "know thyself" meant "recognize the mortality of man." Schupbach himself thought that *vanitas* iconography was conventional but inessential to anatomy: "it was the pessimistic images—skull, skeleton, corpse—that were chosen, because they alone illustrated 'know thyself' in the physical sense that linked the proverb with anatomy."[108] But death has more shaping and informing significance in the anatomy theater than either Cunningham or Schupbach allows. For one thing, death was difficult to avoid or suppress, and that the events of the anatomical demonstration "had a certain rhythm and a liturgy which was not without decorum and style" was partly in order to ameliorate or effect a symbolic distance from the sights, sounds, and smells associated contiguously with death, which were, incidentally, unavoidable in hospitals.[109] Moreover the pervasiveness of these morbid perspectives in the early modern period can be appreciated when we realize that even in the living, some parts of the body, such as hair and fat, that we now regard as integral to our human organism, even if relatively less essential than other parts, were regarded by many as "comming neare the condition of an excrement, rather than a part," and thus as particularly subject to corruption.[110] On the symbolic level, moral-didactic commentaries on the public anatomies suggest their service as *memento mori,* declaring the transience and vanity of Man's life on earth. A poem by Barlaeus, "On the Anatomical House Which Can Be Visited in Amsterdam," hails the human "*Fabrica*" in generally Protestant and Neoplatonic terms as "the seat of the soul, . . . the venerable tabernacle of the mind," an "insignificant receptacle [which] contains divinity concealed."[111] But equally it meditates morbidly on the "house we found for Death," on "a life's span [that] has withered," and on the fact that "once the light has been extinguished, man disintegrates in putrefaction."[112]

So, unlike the textual "theatres" of Casserio and Bauhin, the Protestant anatomy theater's illustrative demonstration did not seek to display its subject only in a new glory. There are several elements of cultural significance to the pessimistic interpretation of this dramatic conflict between glory and vanity. Frequently, for example, the expository literary anatomy took the pessimistic line on knowing oneself, and often, as in Ward's *The Anatomy of Warre,* it dealt exhaustively with its subject in order somberly to reveal it for what it is. Ward sought thoroughly to delineate war, in his words, "for the information of the judgement, the clearing of the understanding, and the satisfaction of the conscience of all those, who will peruse this Treatise" during the time of England's "*Civill Warre,*" in order "therefore that we may see clearly, as in a glasse, the true nature of this heavy plague of *Warre.* . . ."[113] Insofar as the literary anatomy sought somberly to reveal its subject for what it is, a kind of analytical and discerning or judicious realism, perhaps an emergent realist mimesis, as opposed to idealism, was intimately related to disclosure, discovery, the laying open of the subject for the reader.

When anatomy sought to reveal its subject for what it is in the theater, death was exhibited partly as metaphysical, as "nothing but the result of the true operation of the laws of nature."[114] Dollimore relates a conflict between idealist and realist mimesis in the dramatic theater to a debate in the early seventeenth century between providentialism and the advocates of the decay thesis, the most noted of whom was Godfrey Goodman whose *The Fall of Man; or, The Corruption of Nature* was published in 1616: "In it he argued that the material world was subject to progressive and irreversible deterioration. There were two interrelated processes whereby this occurred, privation and the conflict of opposites."[115] The second of these is rooted in Empedocles' notion of *discordia concors,* and relates also to Hippocrates' theory of the four forces or qualities of the human body, and the theory of humors, which, as chapter 1 of this study outlined, were originally associated with the elements, but which subsequently lost their immediately cosmic significance and developed an exclusively corporeal significance. It is under the guise of *discordia concors* that we encountered the pessimistic aspect of anatomy in the comparative literary studies of chapter 5, as a further principle, besides *nosce teipsum,* around which the figurality in some of those works was organized. It may be argued that whereas *nosce teipsum* was associated predominantly with the optimistic, humanist interpretation of anatomy, *discordia concors* was often a way of representing philosophically a more pessimistic view of Man. But it was not only Man, for as Dollimore argues, "[t]he conflict exists in everything comprised of the four elements, from microcosm through to macrocosm, within individuals and between species. Even the universe is subject to it."[116] Since, as this study has shown, the elemental microcosm is related consistently with the conceptual metaphors of anatomy, the natural explanation of human mortality was also that, just as it is natural for vegetation to wither, for buildings to crumble with age, even for civilizations to pass away, so it is natural for Man to die, whose complexion derives from the elements.

From the perspective of the clash between providentialism and the decay thesis, the two aspects or interpretations of anatomy in the sixteenth and early seventeenth centuries are doubtless contrastive and in conflict. According to Dollimore, "decay theory threatens what was perhaps the most powerful tenet of providentialism," "the idea of purpose and order teleologically encoded both in the universe generally and in the identity of things in particular. Instead it posits a universe where future, purpose, and identity disintegrate in perpetual strife."[117] So prevalent and relentless is the world's disintegration that, despite Goodman's intention "in relating these miseries," to make God's goodness "appear," Dollimore says, "[u]nfortunately for Goodman it did not appear; the disjunction between divine perfection and secular chaos was too great"; and Goodman's main opponent, George Hakewell, "opposed the belief in decay because 'it makes men murmure and repine against God under the name of Fortune and Destinie.'" On the whole, Hakewell's view prevailed, with the concomitant opinion that things "can bee improved by industry."[118] As Dollimore concludes, "those following him take up the ideas of uniformity in change and nature's encoded order (the 'constancy of nature'), aligning them now with an optimistic protestantism, itself on the side of the new science, the 'moderns' against the 'ancients.'"[119]

But the two aspects or interpretations of anatomy are not in fact mutually exclusive or as conflicting as is suggested by understanding them from this perspective. The extradiegetic narrator of *The Purple Island* opens Canto 1 with a description of the world as "more aged by new youths accrewing," and as growing "worse by age, & older by renewing" (*P.I.* 1.1.5, 7).[120] Bleak though this is, when *Thirsil* resumes the theme in Canto 2, he qualifies it with reference to the incarnation, death, and resurrection of Christ, the consequence of which is deemed to be that the Isle of Man was "pull'd from that horrid main, / Which bears the fearfull looks and name of death" (*P.I.* 2.1–2). This does not indicate a difference of perspective between the narrative's two narrators, but rather the implied author's theological awareness of the convergence in the present of the effects of the Fall and the partial redemption of Man to "A baser state then what was first assignd" at the Creation (*P.I.* 2.1–2, 5). Martin Luther himself was among the "most melancholy of chronologists," and expounded a view of history as "degenerative cycles," and Thomas Browne's *Urne-Buriall* suggests that his own and immediately preceding or ensuing generations, "in the setting part of time, are providentially taken off from such imaginations" as the extension of their "Memories by Monuments," implying that the perspective from which "the remaining particle of futurity . . . maketh Pyramids pillars of snow, and all that's past a moment," was consistent with a view of being "naturally constituted unto thoughts of the next world."[121]

We have encountered the pessimistic aspect of anatomy in chapter 3, in the discussion of the theme of mortality in textbook anatomy of the sixteenth century, which Cunningham neglects due to the lack of significance he attributes to anatomy's *vanitas* iconography. We discovered there that Renaissance anatomy, in which Nature was not the eternal creator it was for Aristotle, provided moral and theological argument not only from design, but from the demise to which Man as the chemical microcosm was inevitably subject. In this manner, the anatomy theater echoed illustration in disseminating the moral that the great Creator of things has fashioned our body not only that it may live but that it may, subject to corruption, live properly. This moral-didacticism resonated with that in the dramatic theater and with the realist mimesis that was also emerging in the dramatic theater and the expository literary anatomy.

The pessimistic revelation in the anatomy theater or in the comprehensive survey or literary anatomy, therefore, might equally be about the moral and religious perspective on death as the wages of sin, an idea which was associated with the anatomy theater particularly because the corpses dissected in public anatomical demonstrations "were invariably those of criminals."[122] A clear example of a literary anatomy that fits this category, which does not perceive irreparable separation between a pessimistic interpretation of human nature as vanity and an optimistic view of Man as *imago dei,* is Samuel Purchas's *Microcosmus; or, The Historie of Man,* an exegetical anatomy, the second chapter of which commences "The Anathomie of the Text" (by which is meant the work's threshold text), which like that of a sermon is biblical, and is suitably pessimistic: "Verily, every man, at his best state, is altogether vanitie. Selah."[123] Purchas proceeds to unfold the meaning of "knowledge of Mans selfe," but not in Neoplatonic fashion "as if Selfe-conceit, and reknowledging the Soules heauenly Originall, by Conscience of Mans Nobilitie, were the only way to vertue." Purchas declares that rather than "KNOW THY SELFE . . . written in golden letters on the front of the Delphian Temple," he meant "another of knowledge of Mans selfe, *that I may know* (sings *David* here) *how fraile I am*"; that is, the knowledge that "all Our Actions dance the sober *Measures of Mortalitie;* Man *liuing,* as *continually dying.*"[124]

Notwithstanding this pessimism, for Purchas, whose analogies appear to be influenced by Paracelsus and by Neoplatonism, the body is "a Microcosme, & created after the rest, as an Epitome of the whole Universe, and truest Mappe of the World, a summarie and compendious other World"; but simultaneously he insists that his reader should not forget that the matter of the body "was not derived from the Sunne, starres, or any part of the Æthereall superiour World; no, not from the higher and nobler Elements, but from this lowest and basest of all, the Earth (the very dregges and settled Lees, or Sinke . . . of the whole Creation)."[125] Having attended a lecture by Harvey at the College of Physicians in the spring of 1618,[126] Purchas nevertheless wrote the chapters on the anatomy of the human body in order to relate the wonders of Man's creation, initially exploring the two main conceptual metaphors of anatomy as vegetation and as building and extending them to the analogies of the microcosm with the body politic, the Church, and the natural universe; and then, with copious references to Scripture, drawing moral lessons wherever possible, the chapters proceed systematically to detail anatomically a conceptualization of the human body as an "Ædifice, the House of the Soule, the Temple of Deitie, the severall Roomes, Offices, Walls, Partitions, Architecture," and so on.[127] When, however, in the succeeding chapters, he turns his attention to Man's degeneration, no human age, occupation, affection, passion, virtue, or member of the body escapes the puritanical execration of all as vanity. Redemption is said to occur only in Christ.[128]

As well as combining elements of the cosmological and sacred correspondences in their architectural design, permanent anatomy theaters frequently incorporated solemn features from the ecclesiastical location of the temporary theaters, and thus juxtaposed the iconography of Creation and mortality, because from a theological perspective they are not mutually exclusive.[129] The challenge of death was a reminder of the Fall in the permanent anatomy theater of Leiden, built by Peter Paauw in 1597, which in the manner of a museum mounted exhibits during all but the brief annual season of demonstrations, including two skeletons representing Adam and Eve, flanking the Tree of Knowledge of Good and Evil.[130] The paintings, etchings, and engravings on display in and around the Leiden theater represented biblical, historical, and allegorical subjects, and the *vanitas* genre dominated.[131] This preoccupation with mortality is perhaps more pronounced at Leiden than elsewhere in Holland. Neither at Delft nor at Amsterdam did the anatomy theaters exhibit so consistently in the *vanitas* tradition, though as a museum the Amsterdam theater was by no means without such connotations.[132]

Death was symbolized, then, by the iconography of *vanitas* exhibits, but was most appreciable by turning one's attention once more to the semiotics of the centrally placed dissected criminal cadaver, "a perfect emblematic display of fallen man's mortal condition."[133] It was said in chapter 6 that the poetics and the formal design of *The Purple Island,* which imitates the layout of text and illustration in anatomy textbooks, may also be seen as emblematic, with a poetic picture and a clarifying, explanatory, and elucidatory text. Emblematics extended to the illustrative function of the anatomical demonstration, and in one particular case, the emblematic interpretation seems especially appropriate. Heckscher cites another poem by Barlaeus, written not later than 1646, inscribed on the anatomical table of the Amsterdam anatomy theater in such a way that the criminal cadaver thereon would have visually illustrated not only the verbally communicated anatomical text or lesson, but also Barlaeus's poem, which begins: "Here lies spread out Man and offers to all the World spectacles of his pitiful state. To miserable mortals he exposes his naked limbs, and since this happens on account of the Fall of Sin, he lacks shame."[134]

We can see in the *vanitas* and *memento mori* iconography and the eloquent representational status of the cadaver, anatomy's relation to dramatic tragedy. On the metaphysical idea of the brevity and inevitable dissolution of human grandeur, Sidney unfolds that tragedy "teacheth the uncertainty of this world, and upon how weak foundations gilden roofs are builded." Among the conventional, literary associations or commonplaces of anatomy, opening connotes disclosure, handling, or discovery of a thing or subject, and implies that what is laid open may be fictional or historical. In tragedy it may be either of these and it was frequently both, thereby often dramatizing the faultlines between realist and idealist mimesis.[135] In anatomical dissection the laying open could also register in the domain of psychology or morality, as Berengario acknowledges: "Some men are said to be born with a shaggy [bristly with hair] heart; they are bolder and stronger than other men."[136] Epistemologically, for Sidney among others, the dramatic theater had attracted the conventional moral associations of the anatomy theater, such that, for example, tragedy was characterized by its opening of "the greatest wounds" and the exposure of moral "ulcers that are covered with tissue."[137] Therein anatomy's special sensory epistemology, which extended to autopsy, and perhaps surgery, was put in service of moral discernment. This is one small part of the widely acknowledged phenomenon in early modern English literature and culture which expresses mental experiences and phenomena using the language of the body.[138] More specifically, for example, David Hillman has observed that the rise of anatomy may have suggested a more systematic method or practice whereby moral or psychological knowledge might be extracted from within the human body.[139] In addition, if one considers the associated conventional or literary commonplaces of anatomy, provided in chapter 5 of this study, the associated implications of opening may be thought of as relating to the notion of a reiterated incision of the subject, a division or dissection into several parts, phases, or acts, to expose to judicious or discerning view the seat of the disease, if pathological, or the villainy, if moral. We will have cause to look more closely at this phenomenon of expressing the mental or spiritual in corporeal terms in the final section of this chapter.

Like tragedy, and like the *Purple Island* allegory, the narrative, dramatic semiotics of public anatomical demonstration encompasses the analogy between the natural human body and the organically conceived body politic, partly because of the moral and social status of the cadaver. The analogy holds that the purportedly villainous member of the body politic may be analyzed judiciously, and, like the pathological anatomical part of a living person, treated surgically and therapeutically, and thereby reintegrated or healed for the sake of the health of the whole organism, the body of the polity, and that of its citizens. But as was outlined in chapter 2, when some versions of the body politic draw an analogy between the ruler and the physician, in distancing the ruler from the sick or decaying body to which he or she administers, they evince absolutist and authoritarian political sympathies, which often advocate severe punitive measures and justify violent state oppression as the natural clemency of the sovereign on the grounds of a medical analogy with amputation or venesection, a sharp remedy applied to a small and infected part, purportedly to save the whole from dissolution and destruction. Capital punishment, in which public anatomical dissection was infamously implicated, was often justified by analogy with amputation to restore the health of the body, and for some the "health" of the political body was more important than justice.

Conversely, in revenge tragedy, morally "abused" yet disenfranchised without recourse to legal redress, and, according to "certain extreme notions of personal honour," morally vindicated,[140] the protagonist, who is often a malcontent with deepening political awareness, vows to commit "a kind of wild justice."[141] The justification of this is sometimes in a kind of parody of the anatomical or medical analogy of capital punishment, as, for instance, when *The Revenger's Tragedy* opens with Vindice's moral "anatomical" description of the "Duke" who has poisoned and killed his betrothed:

O, that marrowless age
Would stuff the hollow bones with damned desires,
And 'stead of heat kindle infernal fires
Within the spendthrift veins of a dry duke,
A parched and juiceless luxur.

(1.1.5–9)

Vindice is a character written with some independence of the morality-play didacticism according to which he should personify revenge, and by him, it has been said, "providentialism is obliquely but conclusively discredited."[142] Vindice embarks on an inexorable descent into moral duplicity, in which revenge becomes a "sweet, delectable, rare, happy, ravishing!" (3.5.1), and anatomical judgment and retributive punishment become a further poisoning and torture (3.5.141–216). Vindice's relish in vengeful murder is the dramatized obverse of Keats's relish in thoughts of dying—"for many a time / I have been half in love with easeful Death, / Call'd him soft names in many a mused rhyme."[143] The "awareness of death" ceases to be a reminder of "upon how weak foundations gilden roofs are builded," that nothing one has, not even one's life, is truly one's own to possess, and all may be taken at any time; the skull beneath the skin appears to become morally reassuring or sensuously comforting when faced with the futility of hope for either justice or beauty uncontaminated by either sinfulness or time.

As Michael Neill argues, anatomy lessons were "at least as much a piece of drama, a species of didactic tragedy, as . . . a scientific event,"[144] though the high-moral, "tragic" seriousness of the strife between mortality and providence could be unsettled by parody, as in *The Revenger's Tragedy,*[145] or by the Carnivalesque encroachment of a popular sense of what was comic about the life and death of the body and its dissection. Baldasar Heseler wrote, on witnessing the fifteenth of the series of anatomical demonstrations undertaken by Vesalius, at Bologna, that the anatomist "took off and showed us half of the cranium which he previously had severed from the substance of the brain by section. Then the Italians cried: O what a beautiful cap."[146]

Death is accommodated to providence not without tension, and sometimes in the theater and the Carnival, trifles were made of the terror that death instills. If the lecture and demonstration in the anatomy theater comprised a mimetic spectacle or performance like that in the dramatic theater, it understandably harbored, dramatized, and intensified the same fault-lines appearing in the idealist demonstration of providential order and the tensions between this order and realist mimesis. As Dollimore expounds, "drama foregrounds, perhaps more acutely than any other literary genre, the problematic relations between the two realms" of the mundane and the divine.[147] But although the subversive power of the corruptible body, the worldly processes of birth and death, and the perceived ubiquity of degeneration may seem to conflict with providentialism, they do not do so theologically with the trust that even when all that is created tends toward disintegration, the human and the cosmos are redeemed.

There is a subtle equivocation in the doctrine of salvation, over the manifestation in the temporal, earthly dimension of human history of that which in the domain of heaven is eternal. Redemption is thought to take place in the future, at the apocalypse, but also to have taken place already in Christ's historical, "full, perfect, and sufficient sacrifice, oblation, and satisfaction for the sins of the whole world."[148] Whereas Neill characterizes the tension in the symbolic drama and ritual meaning of the anatomy lesson and demonstration as between the didacticism of tragedy, with its analysis of Man and its unveiling of death, and the powerful drive towards humanist triumphalism,[149] I would rather think that optimism about the glory of Man has more to do with faith in the assurance of things hoped for.

From the perspective of the doctrine of salvation through the incarnation, death, and resurrection of Jesus Christ, the two aspects or interpretations of anatomy—that which pessimistically concentrates on human mortality and is influenced by realist mimesis and the decay thesis, and that which optimistically focuses on the dignity of Man as the microcosm and is influenced by Neoplatonism—remain somewhat contrasting, even as they are coherent. But the contrast and the potential for conflict is what makes for dramatic tension. The anatomical demonstration was a drama conducted in the shadow of the Fall and the light of salvation; a religious and moral conflict played out in the theater between death as the wages of sin, symbolized by the criminal cadaver, and redemption as revealed by the natural-theological and soteriological aspects of the lesson, and represented in the iconography of resurrection.[150]

Thus demonstrations, like illustrations, reflected the design and teleology of the universe in the design and teleology of human anatomy, particularly osteology and myology, but also of the theater itself, and they expounded the corruptibility of all natural things by using *vanitas* iconography and the emblematics of the criminal cadaver. This should not surprise us if, as Heckscher claims, this period "was stirred more deeply by the *arcana* of death and life, generation and degeneration, birth and rebirth, than any era before or after."[151] Moreover demonstrations and illustrations reconciled Creation and mortality in the cosmic redemption they displayed in the redemption of Man and the propitiatory prayers and sacrifice of the Masses performed for the sake of the redemption of the souls of the condemned. As Sawday has argued, aside from the character of the anatomy lesson, the moral design of the Renaissance anatomy theater itself, and the emblematic status of the body within it, expressed the significance of human anatomy and mortality for providentialist schemes of human salvation.[152] But on the whole, the significance of salvation and resurrection in anatomy has received insufficient attention even from historians and literary critics who have emphasized either the relationship to mortality through the *vanitas* tradition, or that to Creation through Classical concepts of design and teleology. As we have seen, the soteriological perspective not only permitted the coherence and systematicity of vegetative and architectural conceptual metaphors with the analogies of the microcosm, but was also reconciled to the imaginative rationality and special sensory epistemology of Renaissance anatomy.

By modeling its diegetic narration on the method of practical or theoretical anatomical demonstration, as Sawday argues, *The Purple Island* adopts the moral design, "in order to explicate the incarnation" as "[c]entral to the body's significance in the divine plan."[153] For this reason, the dissected human body displayed to the audience at a public anatomy, or to the reader of *The Purple Island,* acquires another symbolic meaning, and this too should perhaps be regarded partly as an aspect of *The Purple Island*'s mimesis of the demonstration of anatomy in the theater. It was explained in chapter 3 of this study that Fletcher drew on theory of poetry as a speaking picture, emblematics, a suprarealist view of language, and a possibly Augustinian theory of the relationship between language and incarnation; he sought to restore the knot of analogical correspondence between the Isle of Man and God; and he adopted a method of dissection that both purports to imitate Nature's method of creation and partially follows the order of demonstration to an audience in the theater. In all these ways, Fletcher designed his work to have a restorative and saving effect on its readers. The kernel of this efficacy purportedly lay in *The Purple Island*'s representation of the idea about human anatomy, that God "who first mad'st" "never wilt forsake it" (*P.I.* 3.4.5), and the demonstration that belief in this idea is entirely dependent upon the incarnation, death, and resurrection of Jesus Christ. Because of the parallels drawn between this article of faith and the necessity of the death of the criminal or poor foreigner, in order that the natural opacity and epistemological resistance of living bodies is overcome and anatomy may be demonstrated in the theater,[154] the body becomes seen as microchristus in actual or poetic dissection, and faith in the resurrection may be both demonstrated to and fortified in the audience or reader. The emblematic representation of the body as microchristus in *The Purple Island* and in the anatomy theater eloquently demonstrates not only the ingenuity of its Creator, but also the sacrificial and saving love of its Redeemer.

If the "picture," the illustrative display, in the anatomy theater was like the crucified anatomical Christ or crucifixion figure in Berengario's *Commentaria* and the 1522 edition of his *Isagogae breves,* or that of Christ or microchristus in Remmelin's *Catoptrum microcosmicum,* which we encountered in chapter 3, the *anatomista* would have inherited the moral ambivalence of wielding the instruments of the passion, as both injurious to the body of Christ and salutary in the efficacy of Christ's death and resurrection. Sawday's argument regarding anatomical illustration, referred to in chapter 3, that the anatomist's direct or symbolic implication in the exercise of judicial power, and unavoidable proximity to the infamy of the executioner, sat uncomfortably with his desire for the cultural inclusion and acceptability of his profession, seems to be relevant to the public anatomy too. As was suggested in chapter 3, the iconographic associations with Creation, the Fall, and redemption in Christ were made perhaps in order to acknowledge, rather than vainly try to avoid, this involvement which gave rise to the main source of cadavers for dissection. Yet by association with Christ it was implied that the criminal was willing to die, either in the cause of anatomical knowledge, or as profoundly repentant and of the view that acceptance of a severe yet just death penalty might be followed by forgiveness by God and admission into God's presence in Paradise. That the dissection should then grant the criminal a vicarious resurrection perhaps in the illustrated pages of anatomy textbooks, where they are monumentalized and transformed through their representation as artifacts or Classical statuary, is also implied, as is the justification of anatomy's direct or symbolic involvement in the rites of judicial power, in the contribution it makes to medicine or to self-understanding.[155] Allaying anxieties about the provenance of bodies for dissection, about the taint of the infamy of the condemned criminal or the executioner, or about the delaying of burial, or about the borderline shameful, illicit, and sacrilegious violation of the integrity of the human body, this justification specifically echoes a legitimization made elsewhere in medical literature, despite the lack of direct medical application for the anatomical projects of the sixteenth and early seventeenth centuries.[156]

Just as solemnity and serious-mindedness were perhaps encouraged in the theater by the sacred correspondence between Church or *ecclesia* and Man, and between both these analogues and the design in the mind of God, so the officiation of the Professor of Anatomy or the *anatomista,* with respect to the microcosmic temple of the soul on the table before him, could easily be seen as "a religious observance," a ritual in which natural philosophy takes on revelatory or even sacramental connotations.[157] Late sixteenth- and seventeenth-century theaters in Protestant regions, and particularly in Holland, were not only accommodated by churches or chapels; "[i]n one or two instances it is even possible to demonstrate that the anatomical table took the place of the altar."[158] The ritualization of the anatomical demonstration as analogous, or as in some more tenuous way related, to the "Administration of the Lord's Supper or Holy Communion,"[159] or, in Roman Catholic contexts, the Sacrifice of the Mass, traces in outline at least, if not in semantic content, an association of the criminal cadaver with the body of Christ, as having died and risen, and as present bodily or "spiritually" in the sacrament of the Eucharist.[160]

If the demonstration was understood as in some sense like a sacrament, the proposition emerges that perhaps the death and dissection of the criminal cadaver either reflects or vicariously participates in the sacrifice of Christ, perhaps as a distantly "relative sacrifice," or in some symbolic sense memorializes it,[161] so that others may benefit through a quasi-sacramental medical efficacy. Perhaps this is why contact with the cadaver, or with extracted parts thereof, may have been regarded in some contexts as therapeutic.[162] In Catholic contexts, dissection perhaps was imagined as having performed something like a confection on the bodies of criminals;[163] and in Reformed contexts the ritual may have been presented so as to give the impression that the recipients, those attending the lesson and demonstration, had in some sense been granted "remission of [their] sins."[164] On a political and perhaps even an ecclesiastical level, the criminal was in the role of a scapegoat, or a victim of severe punitive measures, which, as argued above, were justified on the grounds of a medical analogy with amputation or venesection, a sharp remedy applied to a small and infected part, purportedly to restore the whole and each of its members to medical, moral, and social health, perhaps as "one body with . . . Jesu Christ, that he may dwell in them, and they in him."[165] The purpose of anatomical demonstration was thereby not unlike the illocutionary act of *The Purple Island,* to communicate erudition and knowledge, and dispense the balm of social and religious salvation.[166]

Beyond the effects of its dependence on criminal execution, there was much in Renaissance anatomy that gave rise to the idea of representing the dissected body as redeemed or redemptive, and this has been discussed in chapter 3, specifically, and at various places throughout this study. In the changing system of conceptual frameworks, meanings, representations, and relations that Sawday calls the "culture of dissection," the poet-anatomist as well as the *anatomista* could have been a guide through "the bewildering mazes of the human form";[167] but (it should be added), at least as far as Fletcher is concerned, he is only so, on his own admission, by the grace and light of God, to whom alone is ascribed full knowledge of human anatomy. That *The Purple Island* more than explicates the incarnation is supported by the confidence of "A.C." that, although anatomical knowledge has been "much defac'd" by the Fall (*P.I.* 2.3.3), the anatomico-poetic process of dissecting the body-as-microchristus in "divine / And higher aires,"[168] is capable of tracing evidence of the ingenuity and love of the divine artist, thereby recovering God's art and creative intent for fallen humanity. This is one reason why it is in May, the month of regeneration, rebirth, and (by analogy, contiguity, antithesis, or some other association) resurrection, rather than January or February, the months of the Carnival, in which the story narrated by the extradiegetic narrator of *The Purple Island* takes place. An untitled poem of three ten-line stanzas, written to conclude a chapter of *A Father's Testament,* has Christ as the Divine Lover say, "I make the foul *December* flowry *May*" (II.9).[169]

The drama with its conflict as well as its ritual dignity, its narrative division and progression, and its performative display or presentation, could have been integral to the understanding of the public anatomy as late as the early seventeenth century. In his examination of the "constitution" or construction of the body not only in the text of Harvey's *Prelectiones,* but in the demonstration or (what he calls) the "performance" of the anatomy in the theater of the College of Physicians, Luke Wilson's analysis shows that it was not, moreover, only iconography, architecture, emblematics, revelatory opening, ritual, and dramatic division and conflict, that in the theater conveyed the messages that the human body was created, mortal, and redeemed. The epistemology and teleology of the demonstration, which are discussed in the earlier section of the present chapter, were themselves soteriological, and intimately related to the "performance" in the theater, in that the acquisition of knowledge was itself dramatic and soteriological.

In the context of the discussion in the present section, above, about the dramatic tension or conflict between the purpose to demonstrate divine creation and providence, on the one hand, and the steady putrefaction of the displayed corpse, on the other, it was suggested that the presentation was determined by this imperative that issues from the body. The tension was outlined as the divergence between the integrity of the whole body of Man, which was opened to demonstrate the *imago dei* as the pinnacle of Creation, and the dissection which, while it progressively reveals this image, progressively renders the body unlike that which the anatomical account sought to convey. Paradoxically, as it were, the *anatomista,* in the Vesalian tradition, or the *lector* or *praelector,* in the traditional formal anatomy, attempts to reconstruct the body from its dismembered parts in the account of the body in the lecture. As was said in chapter 1 and has been maintained at several places, in respect of several discourses, this study seeks to argue that it is consistent with the understanding of anatomy in early modern culture, that division, or indeed disintegration or death, is a means to, and simultaneous with, unification or reintegration through the creative and analytic, even redemptive, establishment of a relationship of part to whole. But there was undoubtedly a tension.

It is the immediate aim of the present discussion to define precisely the character of the reconstitution of the body in the theater, by examining further the extent to which *The Purple Island*'s teleological and soteriological epistemology and hermeneutics is relevant to the demonstration of anatomy in Harvey's Lumleian lectures, and to some extent more generally in the public anatomy. As was said heretofore in this chapter, Harvey's presentation of the body in his *Prelectiones,* according to Wilson, consists of an analogical mode the consequence of which is the loss of distinction between the body-as-referent and the signifying system of anatomical theory, which, as an adjacent similitude of the body-as-referent, anatomizes itself. As such Harvey's epistemological mode has been related in this study to the "figuring forth" and mimesis of *The Purple Island* and Davies of Hereford's *Mirum in modum* and *Microcosmos,* which share a teleological and soteriological epistemology and hermeneutics. In these poetic works, incising or anatomizing a body is imaginarily reconciled to a mimetic, rhetorical reproduction of the body-as-referent as the signs of the poetic text, which in turn acquire a rhetorical corporeality, a somatic fullness or *plenum,* ingeniously appearing as bodies of words. Three reasons for this have been discussed.

The poetic anatomies convey in figurative and expository form the *telos* of structure and function, the actuality, life, or truth of the body-in-itself, thereby purportedly giving life to a body of words. As was argued in chapter 6, it may be inferred from the character of the anatomical descriptive criteria upon which the poetic stanzas place emphasis that the aim could be that the reader, by exploring the little world of Man's body, should appreciate function as giving rise to an understanding of why a part is as it is. Secondly, these poetic works combine in the notion of poetry as a speaking picture the meaning of pictorial accuracy with eloquence which thereby pierces and possesses the sight of the soul. This they do by modeling their theories of representation and their epistemology on the concept of the ingenuity of the rational soul. Thirdly, without the suprarealist view of language, which informed the idea that the discovery, invention, and creation of metaphor related to that of the body, poets would not have regarded metaphor as having life and true meaning of its own, and not only in relation to empirical fact, and Fletcher would not have explored the connotations of the Isle of Man figure, which for him existed in reality, some antecedently, some as etymons of anatomical nomenclature, and some in the working out of novel implications. Since, on the suprarealist view of language, words were perceived to be as real and as powerful as the things named, to have a potency as solidly meaningful as physical fact, and were sensed in terms of a nearly physical presence, the poetic works also claim, through mimesis and ingenuity, to reproduce or reconstitute the somatic *plenum,* the actuality of the body-in-itself, in the anatomico-poetic account.

Although these poetic works thereby claim to overcome the incommensurability of the signified with the putative ontology of the body-as-referent, they do so only by assuming that the signs of the text acquire a rhetorical corporeality, which is mortal and corruptible, like that of the body-in-itself, as would have been especially appreciable in the anatomy theater, in the sensory oppression of the steadily putrefying dissected criminal cadaver, and in its semiotics as an emblematic display of fallen Man's mortal condition. As discussed in chapter 5, the claim for the poetic works is that because the body, as signified by the poem, is composed of poetic eloquence, which acquires corporeality from the mortal body, it is subject to imminent dissolution. The sense of this is especially strong in the works of Davies of Hereford, for, unlike Sidney's poetics, for example, they polarize immortal wisdom or ingenuity and mortal conceit or eloquence, this firmly privileging the former as grasping the spirit, while the latter is the letter, the expression in the flesh or matter of words. Of the poetic works, however, *The Purple Island* makes more explicit reference to the condition of the breathlessness of the actual body or "pattern," in order that the "picture" or account of the whole of anatomy "breathes" and may be demonstrated from the "dissected wreathes" (*P.I.* 4.9.3, 2, 1, cited in chapter 3). Fletcher's poem has the clearer, if imperfect, sense of the dramatic conflict in which the presentation of the account of the anatomy of *imago dei* was determined by the imperative that issues from the actual body, whose progressive decay and dissection rendered it increasingly unlike that which it was supposed to represent, and which the anatomical account sought to reconstruct.

The way that Harvey's lectures overcome this incommensurability of the signified of the anatomical account or lectures and the body in the theater is to imply that because the lectures are an account of the physiological anatomy of the whole body, a summation of each of its parts' various reasons for being, or *telos,* and because (in Aristotelian teleology and rationalist anatomy) the purpose of the function of the whole body is to serve the soul, the lectures simultaneously approximate to the particularized soul, which is the life and form of the body.

In both these ways—in the poetic texts by acquiring the mortal corporeality of the body and in the anatomical lectures by approximating to the form of the body—graphic or verbal signs become identified with living bodies. But there is a further dimension to this interaction between text and body, which has already been discussed in chapter 5 in relation to the poetry. The relationship between the anatomical account, as the soul-as-form of the body, in the *Prelectiones,* and the cadaver in the theater of the College of Physicians is not merely analogical, but hierarchical, as is that between the soul's ingenuity, which creates the fore-conceit of the poem in the mind of the poet, and either the body-as-referent or the poem's acquired corporeality of mortal eloquence, in *Microcosmos, Mirum in modum* and *The Purple Island.* Notwithstanding the differences in their appreciation of the dramatic conflict, the designed effect of hierarchization in the poetry of Davies of Hereford and Fletcher is that ingenuity, which belongs to the immortal rational soul, will stem the tide of corruptibility by mimetically perfecting and resurrecting it in its ideal and immortal form. In the anatomy theater too, the effect is rhetorically and ingeniously to resurrect the body.

In order better to understand the anatomical context, and specifically that of the anatomy theater, in relation to which the rhetorical resurrection of the body-as-referent in the poetic works might contemporaneously have been understood, let us look more closely at the implications or ramifications of this reconstitution or resurrection, and at how it is imagined to occur, comparatively, in Harvey's lectures and in the poetic works of Davies of Hereford and Fletcher, focusing firstly on the lectures.

It was argued in the previous chapter that in Harvey's vitalism and teleology, what is body from a morphological point of view is, from a teleological or physiological perspective, soul and body as a unitary actuality. Further to this, Wilson suggests that the Aristotelianism of the *Prelectiones* includes the proposition in *De anima,* "that the body, in what we would call its morphology, subsists in potentiality only, until by virtue of its association with the *psyche* it is enabled to step into actuality teleologically defined (*entelechy*)." This actuality is inherent in a *telos* that manifests an intention in physiological structure.[170] Anatomical form is not, therefore, ontologically distinct; it is *actually* the manifestation of function as physiological structure. And, for Aristotle and Harvey, the "*stable form*" of the whole body, understood in terms of function, is the soul (as distinguished from the Active *Nous* or Reason).[171] As Pagel explains, the Aristotelian conception of the vital principle, the *anima,* was the "'prote entelecheia,' i.e., the 'first grade of actuality,' the 'form' of a natural body (e.g., the eye) that has life (e.g., vision) potentially within it."[172] As the soul, therefore, brings the potentiality of morphology into the actuality of physiology teleologically defined, so the account of the whole body brings the body from potentiality into actuality. Because the "*stable form*" of the body, that is its soul, is precisely what distinguishes it from a corpse, it is not unreasonable to conclude, as Wilson does, that "the anatomy in some sense bring[s] the body [in the theater] (back) into existence."[173] This is analogically rhetorical because as was said above, the Aristotelian soul or form cannot exist without a body, and so cannot survive the death of the body in order to resurrect it, but it is theoretically separate from the body and is approximated by the account. Moreover, because of the recursivity of the demonstration, alluded to above in this chapter, both the cadaver and the corpus of anatomical knowledge are reconstituted as whole, saved, as it were, from degeneration, either material or into unproductive dispute.[174]

By comparison, in Davies of Hereford's poetic theory, as was described in chapter 5 of this study, "*Ingeny*" partakes of the efficacy of the incarnation of Jesus Christ, and resurrects the body-as-signified from the corruptibility it has rhetorically acquired from the mortal "*Corpes,*" and from that of the eloquence of which it is composed.[175] But like the *Prelectiones,* the poetic theory, which is relevant also to *The Purple Island,* makes the apparently soteriological representational and epistemological claim not only to recompose the body-as-signified from the corruptibility of the morphology of the dissected cadaver, but in addition to perform a mimetic perfecting of the real, a homage to divine creation.[176] In this way, invention, which could also apply to the faculty of mimetic representation of Nature, that is, the faculty of "imagination," brings back to life the mortal, deceased body-as-referent, which is, as it were, on the dissection table, insofar as the poetry mimics the anatomy lesson. From this there emerge structural and semantic parallels in which the body, in what we would call its morphology, in Harvey's Aristotelianism, is equivalent in the poetry to the mortal "*Corpes,*" the dissected, deceased "pattern," and to the corporeality of "crude *Conceit*"; and physiology teleologically defined, in the lectures, is equivalent to ingenious erudition or knowledge of the body, in the poetry.

The works of Davies of Hereford and Fletcher broadly share with Harvey's lectures a concept of anatomical form as subsisting in potentiality and corruptibility, until it steps into actuality by virtue of its association with the soul as either ingenuity or as the form of the body. The parallels and differences are most clearly perceptible in what is quintessentially the place, both in the body-in-itself and in the (poetic) account of the body, where the soul's regenerative or revivifying effects are discerned. This place is (the account of) the heart, and it illuminates how *The Purple Island*'s illocutionary act, and the soteriological purpose of the anatomy that Thirsil demonstrates, differ from the philosophical purposes and justifications, and physiological uses, attributed to the heart by Harvey.

In Harvey's speculation and in the work of the poets, the soul acts through the heart as the cause of the appropriate effects to be obtained in physical life. But thereafter differences emerge. In the philosophical speculation of chapter 8 of *De motu cordis* (discussed in chapter 8 of this study) it is specifically physiology defined teleologically that is the body's preservative, and this takes on a politically sexual and gendered character that is consistent with Aristotle's anatomy of sexual difference and physiology of generation. In the work of the poets, physiology defined teleologically is important but it is not the body's preservative: ingenuity purportedly partakes of the efficacy of the incarnation of Jesus Christ, which is thought to effectuate human salvation, and so, as we discovered in the previous chapter, *The Purple Island*'s account of the heart's concoction of rarefied blood and *pneuma* into vital spirit, sending it to the extremities to cherish and quicken the parts, renders it in terms of the way in which the efficacy of Christ's sufficient sacrifice for the sins of the world is dispensed in the Eucharist to the penitent elect, so that they may be incorporate members in Christ's mystical body, the Church. These holy mysteries are demonstrated in the sacramentalism and ritual of the body as microchristus in the theater, and the parallel between the necessity of the heart in vivification and the necessity of Christ for resurrection would have been felt as palpable and seen, as it were, in the iconography and drama of the scene.

As is argued above, the analogical mode of Harvey's lectures (re)constitutes the body by locating somatic *plenum,* and the agency for the account, simultaneously, in the natural, corporeal body-in-itself and in a teleological physiology which becomes apparent progressively in its account in the lectures. This cumulative demonstration of the form and final cause of the whole of anatomy, as the particularized soul, was compared earlier in this section with the way in which *The Purple Island*'s anatomical exposition by Thirsil accumulates progressively and in narrative stages or phases of story-time and text-time, and also linearly in the process of reading, replicating the way Nature or God fashions the body, and cognitively generating denotations and teleological metaphorical connotations, which the reader apprehends as the allegory unfolds. With regard to the cumulative or temporal process whereby the body's reanimation occurs in the *Prelectiones,* Wilson contends that his "argument by which account as function of part is linked to account as soul in relation to body, furthermore, connects anatomical procedure and the notion of the body as agent": "Harvey's physiological anatomy means that the soul manifests itself, increasingly as the dissection progresses, in the body, as an agency providing an account of (the function of) that body. Thus the body Harvey dissects is in this specialized sense understood to be alive, and the dissection resuscitates it in reconstituting it as an intentional structure as it demonstrates the functions of its parts."[177] This demonstrates more clearly what has been shown hitherto in shards of a broken pane—that the anatomical demonstration and lesson in the theater was a narrative dramatic performance. As Wilson argues, it is in the temporality of the enactment of the anatomy lesson and demonstration, that "the scene of anatomy emerges as a drama in which the visual constrains in various ways the oral presentation that accompanies it, and in which anatomist and cadaver encounter one another in a context of authority, performance, and the constitution of self."[178]

Not only Harvey's Lumleian lectures, but the anatomy lesson generally, as a drama, was the performance of a narrative, and as such, it comprised a text, a narrator, and a story.[179] The text was the lecture, though in the traditional formal anatomy there were effectively, if not actually, two texts. These were the authoritative anatomy textbook which guided the dissection and the commentary on it given by the *lector* or *praelector,* though the latter involved also the performance of a text (the lecture, commentary, or the authoritative textbook) in verbal delivery. Insofar as it contributes to the work's anatomical speaking picture, *The Purple Island*'s narrative comprises modes that are graphically (rather than orally) linguistic and rhetorically visual, in story-time and text-time. The implied author and the extradiegetic and diegetic narrators, the latter of whom provides for his audience of nymphs and shepherd-boys an anatomy lesson that mimics the Vesalian formal anatomy, together manipulate "visibility" and text, like the *anatomista* or *lector* or *praelector,* in an attempt to disclose or open their subject and provide a comprehensive treatment or display of the human subject. The picture of poetry not only lives and speaks, but spawns images and linguistic connotations, which the reader unfolds, dilating meaning in accordance with the intelligibility of the world toward which the work is intentionally directed. The effects of ingenuity and eloquence in *The Purple Island* thereby extend to the hermeneutic context, to the relationship between the work as embodied graphically in the extent of its signifiers and the intentional direction of that work, as an emergent event, toward a world the realization of which is initiated by the interpretation of the work's signification, but the "inward conceit" or apprehension of which accompanies the expansion of the reader's self-understanding in front of the world. That world is the one which is created by the story.

In narrative fiction, including *The Purple Island,* the reader has no unmediated access to the story of events described by the narrative text; he or she must construct it from the text itself, because the text is directly available to the reader. Access by the audience to the "events" of a public anatomy, however, was granted, and the function of the text(s) in relation to the events, including the disclosure visually of human anatomy, and the encounter and confrontation between the anatomist and the cadaver, have been discussed in this chapter in various places. The sequential organization of the dissection into parts and phases told the story of human creation, of the efficient cause of human anatomy in God, the formal cause in the conjoint state of body and soul, and the final cause in the instrumentality and inhabitation of the body whereby it might serve the functions of the soul. But the story was also that of human mortality and the resolution of conflict in salvation; the ceremony and ritual comprised an emblematic and sacramental theatricality or (in *The Purple Island*) poetics. Unlike the traditional public anatomy perhaps, the lesson in the story of *The Purple Island* is not wholly didactic, for the hermeneutic circle is initiated by the implied invitation to the reader to complete the process of semantic generation by using their own "conceit" or imagination. Interpretation, constrained by the context of the whole work, holds in tension a retrieval of what is to some extent immanent in the work as denotation, connotation, implication, or merely suggestion, and the individual and historical variation implied by the perlocutionary act of the work.

From one perspective the difference between Aristotelian philosophical anatomy, on the one hand, and Harveian Aristotelian philosophical anatomy and poetic anatomy, on the other hand, looms largely. In the Christian, as distinct from pagan Aristotelian, schemes of *Mirum in modum, Microcosmos, The Purple Island,* and the *Prelectiones,* the immortal *rational* soul is also particularized, individualized; whereas in Aristotle, the particularized rational soul which enforms, uses instrumentally and inhabits the whole body, is mortal. However, it was the conclusion of the analysis in chapter 5 of the soteriological effect of poetry on human anatomy-in-itself, including the anatomy of the reader, that the relation of ingenuity to the materiality of the body may be modeled on a combination of the Pauline and Thomistic eschatological doctrines of the Resurrection of the Dead, the latter of which was itself influenced by Aristotelian ideas of the relationship of body to soul.

In Harvey's lectures on the whole of anatomy, the body is brought back to life in some sense through Harvey's Aristotelianism combined with his Christian notion of the immortality of the individual rational soul, but there is no apparent claim such as the poetic works rhetorically make, that by partaking of the efficacy of Christ's incarnation, ingenuity dispenses the balm of salvation to the body and soul of the reader and informs or enforms the mortal body-as-referent, resurrecting it as the poetic text, in its ideal and immortal form, not as another body which is subject to death. Despite the rhetoric of Harvey's analogical mode of epistemology and knowledge, which suggests that the Aristotelian final and formal cause of the body, that is, the soul, can exist without a body, and is theoretically separate from it, there is no suggestion that it can reconstitute the body as immortal. This leads to the last difference between the accounts, which lies in the attribution of responsibility for the ingenious actualisation of wondrously made structure to Nature as efficient cause, in the case of Harvey, and to God, in the poetic works. Harvey's lectures occasion a struggle against nature as unyielding and cadaverous, in order to demonstrate Nature as living purpose, but unlike Nature, God is purportedly able to resurrect the body in its immortal form.

The reason for the similarities between the Aristotelian reconstitution of the body-as-account-of-body in Harvey's lectures and the resurrection of the body in *Mirum in modum, Microcosmos,* and *The Purple Island,* enabled by suprarealism and mimesis, and supported by eschatology and Christology, is, as chapter 5 argued, that Davies of Hereford's theory of ingenuity as the agent of resurrection is not a Neoplatonic transcendent theory. Nor is the anatomy of *The Purple Island* entirely within a Platonic-Galenic-Vesalian tradition. Neither the works of Davies of Hereford nor *The Purple Island* can be regarded as having attempted to resurrect the body-as-referent without Aristotelianism, as well as the Christianity whereby ingenuity has the power to invest eloquence with immortality. The similarity that has been found between the accounts is not to suggest that either of these poets derived their theorization of the effects of ingenuity in anatomical poetry from the anatomical demonstrations of Harvey, or indeed any other anatomist, though Fletcher may have attended Harvey's Lumleian lectures. It is rather to argue that the works of Harvey, Fletcher, and Davies of Hereford share a broadly similar set of ideas about accurate, or truthful, and ingenious, or teleological, linguistic representation as reconstituting or reanimating the object it describes.

A similarity, as well as a distinction, can also be discerned in the different ways in which the *Prelectiones* and the poetic works equivocate over the body's reanimation or resurrection. Chapter 5 of this study concluded by saying that there was equivocation in *Microcosmos* and *The Purple Island* over whether their sapience had sufficiently salted their eloquence to resurrect the referent of the text into a living body of knowledge that preserves the body of the reader. By way of comparison, let us examine the equivocality in the poetic works and in Harvey's *Prelectiones.* Wilson suggests that the "performance" in the anatomy theater be divided "into a ritual dismantling (dissection) and a reconstitution (anatomy) of the cadaver, acts probative respectively of the body's inanimation and its figurative reanimation."[180] There is some correlation here with what the Galenist Matteo Corti's lectures on *Anatomia Mundini* say on the distinction between anatomy and dissection, that the latter was contained in the former, just as house building is contained in architecture.[181] In the performance in the anatomy theatre of the College of Physicians, while "the body-as-account-of-body is understood . . . as entailing the reanimation of the cadaver as an intentional structure offering an account of itself," a tension remains with the inanimate condition of the cadaver, the emblematic display of fallen Man's mortal condition.

But Wilson's analysis suggests that the equivocation goes further, as Harvey's agency of division tends to be made similar to a "natural" or bodily agency, and anatomizes the collective authority of textual anatomical knowledge, with which the anatomist is identifiable by way of synecdoche. "In gaining the spectators" complete attention, Harvey himself becomes, like the cadaver, a body 'subject to the view'—a body rather than simply an oral presence, because the performance *is* a performance and values enactment over information."[182] The recursivity of the demonstration therefore threatens the certainty of "the absolute difference between anatomist and cadaver," not only because it reanimates the cadaver, but also because it objectifies and "cadaverizes the living body" of the anatomist, which differs in its epistemological accessibility from the cadaver only in degree.[183]

It was remarked in chapter 5 of this study that Katherine Rowe has observed a similar recursive logic in Crooke's *Microcosmographia,* in which the hand that actually cuts and orders is revealed by anatomy as performing a morally and spiritually discriminating act, and is thereby endowed with efficacy as the instrument of the rational soul. This hand embodies the rational design it produces, so analogies were drawn between the internal structures that facilitate grasp and are revealed by dissection, and the hand of the anatomist which facilitates the demonstration. An illustrative example of this recursivity, which is also an expression of the anatomical application of the concept of *nosce teipsum,* and of the "truth" of the body's having been created appears in Valverde's *Anatomia del corpo humano.* Therein a dissected anatomist "looks heavenward as his fingers resect the ribs of a Vesalian Apollo Belvedere or perhaps himself," reflected in the cadaver lying prone before him.[184]

Wilson suggests that in the *Prelectiones,* the body-as-account-of-body should be understood "neither simply as entailing the reanimation of the cadaver as an intentional structure offering an account of itself, nor simply as bringing about a cadaverization or deanimation of subjects made objects," but as provoking the anxieties which each of these suppositions occasions.[185] The tension achieved in this equivocation, in Wilson's analysis, is as between the genres of tragedy, which Neill has argued is public dissection's exact literary counterpart, and romance. Whereas "anatomy *as dissection* is, like tragedy, a cathartic ritual" (at least, that is, as Aristotle understood tragedy, and as many in the Renaissance, including Sidney, following him, did); in contrast, anatomy "*as anatomy*" "results in a fantasy in which the body is felt to be alive."[186] Whereas the former is dismantling, pessimistic and a reminder of death, the latter is "anatomy as romance; its trope is the reanimation of the dead," which, while occasioning "a morbid anxiety," "produces a certain elation in the apparent reversibility of mortality": "The anatomy is the reversal in fantasy both of the dissection and of the execution that preceded it, a retroactive pardoning of the guilty that repairs the damage that the punisher in his punishment, and the anatomist in his dissection, has inadvertently worked against himself."[187] As optimistic and a reminder of redemption, the reconstitution of the body in anatomy as romance augments the "religious observance." Moreover, it further allays anxieties about the provenance of bodies for dissection, the taint of the infamy of the condemned criminal or the executioner, the delaying of burial, and the possibly shameful or illicit violation of the human body.

To compare the equivocation in the *Prelectiones* with that over the resurrection of the body in the poetic works is to reveal interesting differences, but also general similarities. As was argued in chapter 5 of this study, *Microcosmos* acknowledges a failure to "comprize" the "grace" of Nature, in both aesthetic and theological senses, thereby implying a failure to regenerate the body without the will of God. There is a discernible wish to avoid the intellectual pride that led to the Fall, including the fall of human reason, and so this is an equivocation that sees human frailty and vanity in need of grace. The anatomy of the body in *The Purple Island* is placed as an event in the narrative context of divine history, and the resurrection is thereby deferred along the sequence of events of an eschatological narrative; which is to say that for similar reasons of the Fall, the intelligibility of the body's true reason for being, like the true reason for one's actions, is located solely in God's omniscience until after the general resurrection.[188] So the equivocation in Fletcher's poem is an effect of eschatological delay.

The dimension of eschatological narrative relating to the identity of the body has been discussed in chapter 5, where it was argued that as a defeat of death and, therefore, of corporeality, the resurrection must necessarily transform the mortal, earthly, natural, physical body into an immortal, spiritual body; yet paradoxically, the immortal must remain identical with its mortal precursor, for the identity of the individual to be continuous. It was also argued that continuity of identity after death cannot be asserted without accounting in some recuperative way for the discontinuity between death, which the soul but not the body survives, and the resurrection of the body at the Last Judgment. The theories of resurrection, to which the poetic anatomies herein discussed are indebted, simultaneously provide for the continuity and the transformation of the identity of the body after death. Whereas the Pauline doctrine of the resurrection proposes that the body in the sense of the "I," the person, will survive death through the creative act of God, the Thomistic doctrine combines Aristotelianism and Christian eschatology, professing that the soul as a substantial form is immortal and survives the death of the body, as Harvey's thought has been found to imply, though the full person does not exist until body (matter or morphology) is actualized again by virtue of its association with the *psyche* or the soul, and restored to its form teleologically defined, at the end of time. Although God is responsible for this actualization or restoration, it is purported to be possible because it is soul that accounts for the "whatness" of body, and thus, any matter which soul informs at the end of time will be its body. Though these theories may offer means whereby identity (as a metaphysical category) could be imagined as surviving death, the resurrection was also thought theologically to involve a moral and qualitative change, a glorification and a shedding of sinful corruptibility. It is to this moral change as an equivocation and source of conflict in the narrative of *The Purple Island* that we shall turn in the final section.

The Island-Body and the Grotesque Flesh

The anatomy of the human body in Cantos 2 to 5 of *The Purple Island* acquires moral significance in relation to "The flesh," which is personified in the later cantos as "*Caro*" (*P.I.* 7.14.1 & margin. *). This antithetical relation corresponds to that between Creation and mortality; it corresponds also to the work's equivocation over its resurrection of the body-as-referent as the signs of the poetic text, and to the way in which this qualified rhetorical resurrection reveals itself to be a prolepsis that foreshadows the Resurrection of the Dead while speaking of its own soteriological and eschatological significance. This correspondence exists because the antithesis characterizes the sequence of sub-events that comprise event 7 of the main story line as an epic conflict for full possession of embodied identity between virtue, which combined with grace may lead to salvation, and vice, which lacking repentance may lead to eternal death.[189]

The redemptive efficacy of Christ's death and resurrection (event 6) applies across story time, breaking "ope the jayl" of mortality and original sin (*P.I.* 7.12.5). But this is a partial regeneration to a situation described in chapter 2 of this study as a place of only relative safety within the castle's "weak" defensive walls, where the "King" of the Isle of Man (Intellect), and "Queen" "*Voletta*" (the will), with their daughter Eclecta, who is the elect, *ecclesia,* the mystical body or bride of Christ (though not a perpetual extension of the incarnate Godhead), "might trust and seek an higher providence" (*P.I.* 6.3.6; 6.57.3, 5 & margin. *; 7.12.6, 7).[190]

Insofar as the poem's soteriological exposition in Cantos 6 to 12 is of Man as body and soul, the latter is comprised principally of the intellect (the reasonable faculty of the principal faculty of the animal faculty or soul) resident in the head, and secondarily, of the will which was thought to reside in the vital faculty of the mortal soul.[191] That the "will" was located in the heart and thorax, *The Purple Island*'s allegorical portrait of the repentance or contrition of Queen Voletta for her sin of "pride," and her subsequent conversion, can confirm (*P.I.* 6.60.1), for the contrition is figured inwardly as cardiac (distillatory) physiology, a medicinal cordial, and a phlebotomy (*P.I.* 6.62–64):[192]

> Thereto [Repentance] addes a water soveraigne,
> Of wondrous force, and skilfull composition:
> For first she pricks the heart in tender vein,
> Then from those precious drops, and deep contrition,
>   With lips confession, and with pickled cries,
>   Still'd in a broken spirit, sad vapours rise,
> Exhal'd by sacred fires, and drop through melting eyes.

(*P.I.* 6.63)

As the distillation of blood and *pneuma* in the left ventricle of the heart revivifies the purple venous blood, turning it into arterial blood, which is evacuated through the semilunar valves of the aortic orifice to preserve the body in life and health; so the "cordiall drops, these spirit-healing balms," of penitence, which as Sawday remarks is linked to the color purple in ecclesiastical contexts, the color worn by the priest or minister on Good Friday,[193] "Cure all [Voletta's] sinfull bruises" (*P.I.* 6.64.1–2), regenerating the sinful but repentant Christian. Partly because affect and passion in the "weak" materialism articulated here presuppose embodiment, within a theistic context in which subjectivity is in relationship with God, and bodily references assume an affective context or consequence, the poem's soteriological exposition in Cantos 6 to 12 is not of Man as comprising a conflict between body and soul;[194] the poetic exposition, which owes some debt to Book II, Canto 11, of Spenser's *Faerie Queene,*[195] is of fallen Man in thrall to the flesh and to sin, for "spread round about this little hold," which encloses Intellect and a repentant Voletta, are "armies infinite encamped" (*P.I.* 7.13.1, 2), in the vanguard of which is "Th'enraged Dragon and his Serpents bold" (*P.I.* 7.13.3), and "*Caro,*" "cursed damme of sinne" (*P.I.* 7.14.1).[196]

Of the representation of the army of vices which encircle the refuge of Intellect, Voletta, and Eclecta, Cable observes: "We are given . . . not merely a linear parade of the fleshly vices, but a genealogy of Caro's children, every relation of which is telling."[197] Offering a kind of negative image of the genealogy of virtues, which is described in Cantos 9 and 10 and springs from the union of "*Spiritto*" (*P.I.* 9.6) and "*Urania,*" or "Heaven" (*P.I.* 9.6), this gallery of religious vices derives from the "fruits of the flesh" in Galatians 5:19–21, which the marginal at Canto 7, stanza 16, cites. The derivation is faithful but for the following minor alterations: a minor change of order;[198] "Uncleanness" becomes "Sodomie," which is personified by "*Acatharus*" (*P.I.* 7.21.1 & margin. *);[199] "revellings" become "*Gluttonie,*" the "gutling brother" to "*Methos*" or "Drunkennesse" (*P.I.* 7.80.1; 7.73.3 & margin. *); the addition of "*Hypocrisie*" (*P.I.* 7.35.1) among the sons of "*Caro*" and "*Asebie,*" or "Irreligion" (*P.I.* 7.14.1 & margin. *; 7.16.5 & margin. *); the addition of "*Theeverie*" and "*Detraction*" (*P.I.* 7.84.4) in subordinate relation to Methos and Gluttonie; and the placement of several new vices in subordinate relation to the sons of Caro and Asebie.[200] All these religious vices are represented either as having coupled sexually with Caro, or as the offspring of such couplings from which the genealogy thereby stems. These couplings are with "*Anagnus,*" or "Unchastitie," Asebie, "*Adicus,*" or "Unrighteous[n]ess," and "*Acrates,*" or "Intemperance" (*P.I.* 7.16.4, 6, 7 & margin. *). Thereby, as the marginal at Canto 7, stanza 16, explains, Caro's children "may be ranked into foure companies," composed of the intricately interrelated nature of vices. There are in addition ranks of moral vices described in Canto 8, that are not among these companies or the genealogy of Caro.

According to Cable, "Fletcher's panoramic tapestry of the inhabitants of the mortal mind shows a moral and psychological perceptiveness as acute as Spenser's and in some ways more relentless."[201] Moreover, the allegorical virtues and vices are emphatically embodied and figured forth individually, according to Headley, with "an air of life, which in that particular renders them superior to those of Spenser and Sackville."[202] Comparing the personification of "Gluttonie," for example, in *The Faerie Queene* (*F.Q.* I.4 .21–23) and in *The Purple Island* (*P.I.* 7.80–84), Grosart finds the latter "is the more living," and that it shows "the re-action of the over-burdened be-putrified 'body' on the inner intellect"; the soul "weigh'd down with muddie chains" (*P.I.* 7.81.1).[203] Cable observes: "The Vices bristle with anarchic energy, the rude compelling sensuality of the grotesque. The Virtues, if less sensual, are equally mortal, and apprehensible as such."[204] In this, *The Purple Island* is consistent with the theological ethics of Thomas Cranmer and other Reformation theologians, who held that "our own imperfection is so great, through the corruption of original sin, that all is imperfect that is within us, faith, charity, hope, dread, thoughts, words and works."[205]

In the vanguard of the objects and offspring of her couplings, and of the ranks of the other, moral vices, is Caro, but what morally, theologically, and politically does she represent in *The Purple Island,* and what does her rebellion imply of the relationship between the soul (Intellect and Voletta) and the body?[206] In Paul, "bodily existence does not in itself denote something either good or bad. Rather the body [*sōma*] is the concrete sphere of existence in which man's relationship with God is realized."[207] The implied author of *The Purple Island* seems to concur with Paul in this respect, because the vices coupled with or born of Caro strive to control Man, as the microcosm and *imago dei.* Demonstrating the "dynamic relationship" in which "perceived problems in . . . national and economic bodies may well impact on the schema of the individual body," and vice versa, the vices strive simultaneously to control Britain, as the dynastic realm or proto-nation, with Eclecta as the mystical body of Christ or the elect of the Church of England.[208] The vices do so by engaging in "intestine" conflict with the virtues (*P.I.* 6.33.3; 11.15.7), a conflict that is internal to the island realm in the political sense of "civil strife" (*P.I.* 6.20.5), and borrows an anatomical sense from the meaning of intestine as a noun.[209] This is because, with the Intellect, Voletta, Eclecta, and the "*Pemptarchie,*" or "five senses" (*P.I.* 6.42.2 & margin. *), the vices and virtues are "this purple Islands nation, / A people never seen, yet still in sight; / Our daily guests, and natives, yet unknown" (*P.I.* 6.6.3–5).[210]

In this manner, *The Purple Island* may be unusual in not wholly subscribing to the particular ideology of the functionalist organicism of the body politic analogy, as defined by Jonathan Gil Harris, that "social illness has its origins exclusively in external factors."[211] Out of the embodied identity which Paul distinguishes as "I," using the term "*sōma,*" as distinct from *sarx* (Rom. 7:14–18), Caro and her tribe seek to create what Paul refers to as a "body of sin" (Rom. 6:6) or the "sinful flesh [*sarx*]" (Rom. 8:3).[212] In so doing, they represent not the body as opposed to the incorporeal soul, but "man himself, in so far as he gives himself up to his own aims in opposition to God's."[213] The measure of their success is that Thirsil describes them as "Our servants born, but now commanders grown" (*P.I.* 6.6.6).[214]

Caro, as the flesh, represents a broader notion than the pleasure of the flesh, and therefore more than what might be regarded as the most sensual of vices, which along with pride (Voletta's sin) and curiosity, or "avarice for knowledge," was one of the three kinds of vice identified by Augustine, and Roman Catholic and Reformation theologians, as comprising the sin of Adam and Eve which reached "into all of sin's dimensions."[215] Among the vices in the *Purple Island* allegory, "*Anagnus,*" or "Unchastitie," and "*Acrates,*" or "Intemperance," and the entirety of the issue of their unions with Caro, personify the pleasure of the flesh, strictly speaking, although Caro herself either embodies, occasions or gives rise (birth) to and leads in the field "*Pharmacus,*" or "Witchcraft and curious arts" (*P.I.* 7.31.1 & margin. *), "*Chaunus,*" or "Arrogancie" (*P.I.* 8.36.2 & margin. *), "*Philotimus,*" or "Ambition" (*P.I.* 8.38.1 & margin. *), and a host of vices that extends well beyond the religious vices in Galatians 5:19–21, to include the moral vices, and details and extends all the dimensions of the sin of Adam and Eve. If *sōma* or the body as microchristus is the concrete sphere of existence in which Man's relationship with God is realized, Caro, or *sarx,* is the sphere of sin, of pride, curiosity, and the pleasure of the flesh, in which that relationship is blighted by vice, and her appearance in the narrative of *The Purple Island* is as a consequence of the Fall and as an impediment to salvation and to the resurrection.

With regard to the anthropological as opposed to sociopolitical aspect of this intestine conflict, the way in which Fletcher's prose theological work, *The Way to Blessednes,* describes the seventh chapter of Paul's epistle to the Romans is revealing as a less figurative, non-allegorical version of the same spiritual conflict: "It is no small light to us, when we see the combate between the *Spirit,* and the flesh doctrinally expressed. *Gal. 5. 17.* But when we behold it acted personally in the Apostle (Rom. 7.) and view him wrastling with that loathsome, and cumbersome enemie, even the body of death, sometime talking, sometime giving the foyle, crying out for helpe, and breaking forth into thanks for his victorie, it doth not onely cleare our understanding, but powerfully workes on our affections, shewes our enemie, and kindles us against him; first teacheth us, that fight we must; then instructs us how to fight[,] how to overcome, and how to use the conquest."[216]

The way in which *The Way to Blessednes* describes the Book of Psalms is also revealing, for it proposes David as the type of the "new, and inward man," whose spiritual warfare is understood in relation to that of Paul, and whose emergence from the interaction of the creative metaphor of the Isle of Man in *The Purple Island* this study has referred to in previous chapters. Fletcher's work aims to regenerate Man through the emergence of new significant connotations in the imagination (conceit) of the reader, who (ideally) reflects the redeemed Isle of Man and strives to emulate the Redeemer about whom he or she reads in the work. It may be recalled that among the associated implications of the subsidiary subject of the Isle of Man figure are concepts of the geographical or national isolation, insularity, perhaps xenophobia, of the island or dynastic realm (specifically Britain), concepts which relate to a historically incipient nationalism. When these concepts are applied to the principal subject, the metaphorical interaction selects and emphasizes features of similarity as characteristic of the proto-nation and the human being, creating corresponding associated implications, that a human being is alone, individual, possibly alienated, self-determining, perhaps self-fashioning, and by implication even (spiritually and religiously) paranoid. It has been said that this particular set of significant connotations is innovative and relates to an ideologically emergent concept of individualism, which is perhaps not reducible to extensions or discoveries of antecedently associated yet hidden connotations of the figure of Man as an island, and therefore must be distinguished from Classical and medieval occurrences of the metaphor.

This study has argued that the characteristic feature of the innovation of the Isle of Man figure is the way in which the ingenuity and effort it expends in trying to shore up and preserve correspondences and analogies, and the entrenchment of its reaction to "what e"re is in the continent" (*P.I.* 1.53.3), especially Roman Catholicism, help to concretize and give succinct yet infinitely expandable expression to the quasi-Paracelsian and alchemical notion of natural philosophy as individual revelation, and to the emergent ideologies of English (or British) Protestant individualism and nationalism, which ironically introduced a language which would in the history of the discourse of science, religion, and nationhood help to bring about a collapse of what the implied author of *The Purple Island* endeavored to preserve.

As so often in cases of Protestant innovation in the early modern period, Fletcher sought to restore and regenerate rather than create entirely anew, and this soteriological aim is reflected in his choice of type of this "new, and inward man," in *The Way to Blessednes,* which describes the Book of Psalms as a kind of divine anatomical speaking picture of this Man: "And this is the reason why the booke of *Psalmes* is by all acknowledged the most profitable, and usefull, because we here behold lively drawne, by *Gods* owne finger, not onely the face, and hands (the outward profession, and actions) but the very heart of that man, who was according to *Gods* owne heart: and so may take there a full view of every limb, and be acquainted with every joynt, and feature of the new, and inward man: wee shall see that heart ever in action, continually drawing from that living fountain the word (to which it cleaues unseparably) the waters of life, and having throughly concocted them in the affection, returning them againe to every part for growth and practice."[217]

Whereas Psalms is an anatomy of "every limb, . . . every joynt, and feature of the new, and inward man," which David exemplifies, so *The Purple Island* represents the same of Man in the purportedly universal and abstract as opposed to exemplary mode. Whereas the Isle of Man, as the threshold metaphor or conceit of *The Purple Island,* is systematically unfolded in the course of Thirsil's poetic anatomy by a host of metaphorical entailments, which include the representation of concoction in the heart as partaking of the soteriology of the incarnation and as sacramental in its efficacy among the members; so David is perceived to be the figure at the center of the Psalms which are interpreted to anatomize and unfold the type of his character and relationship with God, and David's divine inspiration is articulated figuratively in the physiological terms of the concoction and distribution of *pneuma zoticon.* The "word" of God is represented in metaphorical terms as the "living fountain" of the liver and the "waters of life" (grace) as the blood which the liver produces and sends through the inferior vena cava to the heart, where it is "concocted." The heart being the locus of the affective faculty of the mortal soul, it stands for "affection." Grace is then spread like *pneuma zoticon* "to every part for growth and practice" in the faith.

Here, as where Voletta's repentance is also described in terms of cardiac physiology, the use of metaphor produces a system of entailments, the accruing of which sews additional thread between anatomical and ethico-spiritual schemes in the work of Fletcher. The direction of metaphorical structuring is the reverse of that which predominates in the anatomical cantos of *The Purple Island.* It creates or develops a corporeal spirituality rooted in the understanding of bodily existence and experience, which thereby augments with concrete metaphoricity the philosophical and theological context in which the figurality of anatomy itself is rationalized, justified, and finds coherence.

The reversal of metaphorical structuring in *The Way to Blessednes* also helps us to understand the soteriological illocutionary act of *The Purple Island* as conceived in anatomical terms. Since that poem purports to become a body, and its diegetic narrator implores for and purports to draw upon the inspiration of the Trinity, of the "dread Spirit," "who of late . . . / This nest of hellish fogges and *Stygian* night / With thy bright orient Sunne hast fair renew'd" (*P.I.* 6.26.1, 3–5), *The Purple Island*'s illocutionary act also becomes (grace permitting) like concoction in the heart of the body-as-signified and in the heart of its readers, a perpetuation of the cordial balm of salvation, and possibly a sharp remedy like phlebotomy, dispensed by the incarnation in the inspired poetic word, to dispel "Earths foggs."[218] The soteriological illocutionary act of *The Purple Island* is thereby, through ingenuity and eloquence, to replicate the operation of the Holy Spirit in *ecclesia,* understood in Fletcher's *Joy in Tribulation* metaphorically as the operation of the soul through the heart and brain in the human body: "[A]s the soule . . . dwels in the body, so doth the holy Ghost dwel in the faithfull, quickening, sustaining, & leading them on to the rest of their soules, and Lieging with us. Look as in the body the hart and Head produce life, sense, and motion by the vitall, and animall spirits; which being sent by them, and diffused into every particular member, quicken, and move it; So the Spirit of Christ flowing from him into his mysticall bodie, fils every part with life, sense, and motion."[219]

What makes *The Way to Blessednes* all the more relevant to this study of the *Purple Island* anatomy is that when the former describes what the anatomy in Psalms reveals, Fletcher might almost be providing a paraphrase of Cantos 6 to 12 of *The Purple Island:* "In a word, the Booke of *Psalmes* is a field, where we may be most skilfully trained in all our Christian warfare: . . . where, of the one side wee shall learne to know all our spirituall enemies, their colours, weapons, strategems, their manner of charging; of the other side our Captain, succours, armours offensive, and defensive, our discipline, conflict, victory, and triumph, and all these not onely by relation, but action in our fellow-souldier, and an old experienced Commander."[220] According to Fletcher, Psalms and the seventh chapter of Paul's epistle to the Romans, like *The Purple Island,* Cantos 6–12, are attempts to "paint" in an expository and narrative speaking picture the same interior, spiritual warfare, in David, in "the Apostle," and in Man, respectively, that might less effectively be expressed otherwise. It is therefore in a moral, as well as an ontological, metaphysical way, that, as was argued in chapter 5, the poetic anatomy of Cantos 2 to 5 of *The Purple Island,* rather than accomplishing a resurrection of the body-as-referent, is qualitatively changed into a verbally constituted form of ocular evidence for believing in the divine promise of bodily resurrection, and hence of reintegration, "Reunion," "Re-efformation," and "Re-collection" of the whole person,[221] when the body will become morally perfected, glorified, impassible, and fully intelligible and revealed to the redeemed. The epistemology of the *Purple Island* anatomy may be soteriological, but inference about Man and God can be made only without certainty, by faith, until the moral conflict between vice and virtue, *sarx* and *sōma* or microchristus, death and providence, has ceased at the general resurrection. But before we look at the outcome of this struggle and the body that is resurrected in Canto 12 of Fletcher's poetic allegory, it is desirable that we should explore analytically two facets of the antithetical relationship between body and flesh, both in the anatomy theater and in Fletcher's allegorical poem.

The first of these facets is the antithesis between teleological and moral physiology and anatomy, on the one hand, and the sensuality, sexual curiosity, and concupiscence of the flesh, on the other. Curiosity was one of three kinds of vice identified by theologians as comprising the sin of Adam and Eve which reached into all of sin's dimensions, and Aquinas perceived curiosity as "a vice related to the lust of the eyes." The "morbid curiosity,"[222] and prurience or appetite for autoptic sexual or sensual knowledge, to which the public anatomical demonstration gave rise or occasion for expression can be illustrated in the series of demonstrations given by Vesalius at Bologna, in 1540. Because at the end of his twelfth demonstration, Vesalius had promised that the next demonstration will be of "the muscles moving the head and the anatomy of the penis," and the Italian students knew that in this demonstration Vesalius would dissect the penis of a dog, Heseler observes that at the end of the lecture by Corti, just prior to Vesalius's thirteenth demonstration, "the Italians did not want to wait any longer. They ran like mad to the dissection." In the event, they would have been disappointed, because Heseler writes that by the end of the thirteenth demonstration, "the anatomy of all the muscles was complete except the muscles of the penis, which [Vesalius] said he should demonstrate in a fresh body, that we shall get tomorrow." Gratification was deferred even longer, however, as Heseler recounts in his entry for the fourteenth demonstration: "Because our next subject was hanged this morning (there were two of them) during Curtius' lecture, and they were still hanging, as I myself saw, the students in the meantime had killed a pregnant bitch, which was the fourth dog. Hence, this morning D. Vesalius demonstrated the anatomy of the uterus and of the embryos contained in it, the puppies." The anatomy of the penis was not demonstrated until the seventeenth demonstration, and there is a moment of autoptic curiosity gratified, yet unworthy perhaps of the wild excitement to which the prospect had given rise, when Heseler writes: "Afterwards I went up and took the dissected penis in my hands. I saw that the fistula spermatis was rather spongy. And I felt also that the testicles were soft and light."[223]

Inevitably found in combination and in tension in the anatomy theater, sensual, sexual curiosity, and the erudite account of the teleology of the structure and function of the whole body are also to be found in the consequences of event 5 in the anatomical speaking picture and the main story line of the hypodiegetic level of the narrative of *The Purple Island,* even though, as has been said, Thirsil's demonstration omits sexual anatomy. In its place, immediately prior to his decision to "Flie then those parts, which best are undescri'd," Thirsil contrasts the proper attitude of "some few," who "in these hid parts would see / Their Makers glory, and their justest shame," with that of "the most [who] would turn to luxurie" (*P.I.* 3.25.5, 1–2, 3). In the prose panegyric of Daniel "Featly," published in 1633 as prefatory to *The Purple Island,* the assertion that "He that would learn *Theologie,* must first studie *Autologie,*"[224] implies that the key to knowledge of God and redemption consists of knowledge of the anatomy of Man, which as we know was associated with the injunction, *nosce teipsum.* The prefatory epistle of "Featly" to the readers of *The Purple Island* claims that "[t]his Poem will make thee understand that way" "to God by our selves," but when "Featly" says this way is "a blinde and dirty way; it hath many windings, and is easie to be lost," although he is referring to the forbiddingly intricate forms, connections, and functions of internal anatomy in the ingenious design of God, he is also warning the implied reader about the necessity of acknowledging sin and the danger of becoming ensnared by sin, and about the necessity of constant vigilance for moral self-amendment, regeneration, and personal and national salvation.[225]

"Featly" is only echoing the poet whose work he praises, for through his diegetic narrator, Thirsil, Fletcher attempts morally to distinguish the "curious search" of "what e"er is in the continent" (*P.I.* 1.53.6, 3), which led the "happy Isle" out of the "bounds of happinesse" and "blessed ignorance" (*P.I.* 1.54.2; 1.53.2; 1.51.1), from the "curious labour" with which "we" "Prie in the nature of this King and Queen" (*P.I.* 6.65.2, 1, 3), and from his poetic aim "To frame this curious Isle, whose framing yet / Was never throughly known to any humane wit" (*P.I.* 3.3.6–7). For Fletcher, anatomy unavoidably harbors or occasions an empty sensual curiosity and impulse to know the secrets of the flesh, together with a pious studiousness to know God's "wondrous art." A moral delicacy about providing the occasion in a vernacular work of poetic anatomy, from which impulsive rather than studious curiosity might arise, partly explains Fletcher's choice to omit a detailed anatomy of the sexual organs.[226]

So how do the flesh and this curiosity to know the secrets of the flesh combine with or frustrate the pious studiousness to know God's "wondrous art" in *The Purple Island?* Fletcher introduced the flesh in *The Purple Island*'s anatomical cantos in (non-Aristotelian) rationalist anatomical, Hellenistic rather than Judeo-Christian terms, as a "curious work . . . rais'd" upon a "base" of bones, cartilage, and ligaments (*P.I.* 2.7.1).[227] But an example in the anatomical cantos, which gives figurative, allegorical expression to the corporeal ethical imperatives suggested by "Featly" as necessary to avoid the "sinful flesh [*sarx*]," is the anatomy of "*Lingua*" or "The Tongue" (*P.I.* 5.56.1; 5.58 margin. c). Although Lingua's words "sometimes heav'ns inspire," "Her voice tears open hell with horrid blasphemies" (*P.I.* 5.56.6; 5.57.7). Jesus's advice in the Gospel According to St. Matthew is to cut out the eye and the hand, that is, literally, "the lustful look and the malicious act," for, he says, "it is profitable for thee that one of thy members should perish, and not *that* thy whole body should be cast into hell."[228] Instead the "great Creatour" in *The Purple Island* "Curb'd [Lingua] with iron bit," "a ligament, ordinarily called the bridle" (*P.I.* 5.58.4 and margin. c), and "close within [the mouth] he sets twice sixteen guarders" ("thirtie two teeth"), while "Without the gate he plac'd two other warders" ("the lips"), all "bold libertie restraining" (*P.I.* 5.59.1, 3 & margin. d; 5.58.7).[229]

Herein *The Purple Island* construes the tongue's extension outside the body as a metonymy for linguistic extension, such that the tongue acquires an unruly subjectivity that strains its status as part of a whole when the whole (speaking) subject (*sōma*) declines to own its malevolent, blasphemous language. What Carla Mazzio describes as "the grammatically convenient misogyny that enables the officious organ of taste for Fletcher to be male ('tongue') and the 'strange' organ of speech to be female ('lingua')" certainly draws a politically sexual distinction which contrasts Lingua with the "tongue" in its capacity to "taste" and "deliver . . . up the meat (well chewed) to the instruments of swallowing," as "*Gustus,*" the "Groom with wondrous volubilitie" (*P.I.* 5.54.3 & margin. z; 2.31.1 & margin. f). This is a distinction which, according to Mazzio, "emphasizes a logic of self-estrangement commonly encoded in textual, iconographic, and dramatic representations of male speech in the early modern period."[230]

It was argued in chapter 1 of this study that Man as represented by Renaissance anatomy is presumed to represent male and female, and to do so in the form of the male except where sexual differences are expressly the object of inquiry, but whether the anatomy is male except where sexual differences are expressly the object of inquiry in *The Purple Island* is unclear. What is clear, however, is that the body conceived of as the flesh (*Caro*) is female, and yet "*Profanenesse*" and "*Blasphemie*" (*P.I.* 7.45.2, 3), the vices of which Lingua is reputedly guilty, are both represented as male. It might also be remembered that in Canto 5, the occasion for Thirsil's digression into the myth of how "*Orpheus* wanne his lost *Eurydice*" (*P.I.* 5.61.1) is his description of Lingua, the tongue, and, as was argued in chapter 4, from this it becomes clear that what the prefatory verse of "A.C." claims for Fletcher's poem derives from what Fletcher in the voice of Thirsil claims for the tongue: "She heav'n to earth in musick often brings, / And earth to heaven" (*P.I.* 5.60.5–6). The context of the sexual identification of Lingua in the whole work, including its eloquence and soteriological illocutionary act, not only renders an interpretation of misogyny more uncertain and problematic than Mazzio allows; it also makes it more difficult to separate what is virtuous from what is vicious.

Less severe also than customary fantasies of taking revenge on the tongue, "to articulate in the strongest possible way the way in which words (quite literally) hurt,"[231] is *The Purple Island*'s form of punishment. The difference between the punitive, dismembering measures advised in the Sermon on the Mount and this preemptive constraint placed upon the (feminized) tongue of *sōma* as Man, by a "foreseeing" deity (*P.I.* 5.58.1), in the form of inbuilt mechanisms and structures of self-restraint, is that the latter is anatomically manifest in the functions of a body which remains whole, even if it does suggest the notorious "scold's bridle."[232]

The conception of restraint manifest in ethical structure is traditional in anatomy and goes back to Aristotle. It is related to the way in which anatomical form is not ontologically distinct, but actually the manifestation of function as physiological structure. But equally traditional is the conception of the body's physiologically necessary appetite. Davies of Hereford expounds a conventional view that while the belly or abdomen's appetite is "Naturall," it is divided into an "inclination voide of Sense or Soule" and that with a capacity for action and desire.[233] The latter belongs to the vegetative soul, which plants, beasts, and humans share, and which, although located in the abdomen, is "bodilesse" and without bounds,[234] and therefore simultaneously suffused throughout the body as the appetite of all the "*members,*" "[w]ho wanting *nutriment* doe sucke the *vaines.*"[235] Since every organ shares this "naturall" and vegetative hunger for blood and chyle, the application of the belly's likeness to vegetative growth extends throughout the human frame in the works of Davies of Hereford, in anatomy textbooks, and in *The Purple Island,* as we have seen. Anatomical structure in Aristotle's *Physica* is consequently inherently both desirous and ethical: "The narrowness and windings of these passages [of the bowels] serve to moderate the appetite" that is inherent in physiological structure.[236] The marginalia and poetic stanzas of *The Purple Island* concur in presenting "the entrails" as "knotty and full of windings, lest the meat too suddenly passing through the body should" "whet the appetite, still cloy'd, & still desiring," making it "too subject to appetite and greedinesse" (*P.I.* 2.37.7 & margin. l). In fact *The Purple Island*'s anatomical description of every organ which serves the functions of the concupiscent or appetitive faculty of the soul (in the Platonic division) is morally constructed along these lines, an obvious example being that already given in chapter 1—the uterus, which is delivered from its own dangerous animality when its appetite is providentially reconfigured as having its cause in the divinely ordained purpose of procreation.

The example of the entrails demonstrates not only the manifestation of appetite and ethical restraint in physiological structure, but also that self-knowledge and self-examination, and therefore curiosity, extend "naturally" from a consideration of anatomical structure and function to the moral sphere of desire, intention, and action. For desire and intention in *The Purple Island,* we may also look to the representation of the heart in the anatomical cantos, which the doctrine of the partially regenerating effect of the incarnation informs particularly. In the geopolitical analogue, the heart is where the Prince is situated (though he is also in the brain) and Christ's "thrones fair picture" dwells therein (*P.I.* 4.9.2). The heart is Christ's "glorious cell," full of "saintly spirits" which the heart confects (*P.I.* 4.9.3, 5, 6, 7), and with which the "spritely bloud" of the arteries is said to be "full" (*P.I.* 2.11 margin. h). *Sōma* is therefore a "picture" both of "that loftie one" and, physiologically, of the prayed-for effect on *ecclesia* of the administration of the Lord's Supper or Holy Communion (*P.I.* 4.21.3; 4.19.4). But following upon the metrical heels of the dissection of the heart (*P.I.* 4.15–24) are two stanzas which describe both how "Oft very hell in this strait room resideth" (*P.I.* 4.25.4), and how "that great Light, by whom all heaven shines / With borrow'd beams, oft leaves his loftie skies / And to this lowly seat himself confines" (*P.I.* 4.26.1–3). This is a deftly constructed split vision, in which simultaneously and "oft," the most vitiating effects of the Fall reside alongside sacramentalism in the physiology of the heart and the vital soul, which is driven into conflict through the force of the startling juxtaposition of "hell" and "heaven," and the divergent perspectives on Man as virtuous *sōma* (body) or microchristus and as vicious *sarx* (flesh).

Although regarded as theoretically, and often practically, distinguishable as objects of ethical inquiry by the prevailing skeptical and Christian opinion in the early modern period, intention and action were not inquired into by rationalist anatomists except insofar as the latter was the expression of the function of the body as a whole, a structural potential the purpose of which is to serve the soul.[237] It is arguable though that for this reason the actions of religious contemplation, eloquence, and rhetoric in the illustrative plates of the *Fabrica* that we have examined are presented as exemplary actions, the perfect or final expression of normative structure in a religiously informed humanism. Additionally, the analysis in chapter 6 of the plates of female sexual anatomy in Estienne's *De dissectione* argued that the architecture of the pictorial context and the windows into the figures' abdominal interiors seem to inform the peruser that pregnancy is the end, aim, or final cause of the function or action of the female sexual organs, the purpose for which they are designed and constructed, and the fulfillment for women of the architectural design in the mind of God. Among the organs performing the functions of the vegetative or concupiscent soul, particularly the sexual organs are recuperated for a role in providence in *The Purple Island.* The way that they are so is to be distinguished in tone, if not entirely in kind, from Harvey's statement in the *Prelectiones* that Man is "By the string [penis] tied to eternity."[238]

Since it is in the form of Paul's conception that the connotations of the terms body (*sōma*) and flesh (*sarx*) break upon the shore of *The Purple Island,* we turn to Pauline ethics for insight into the representation of the relation between action and inner intention in Fletcher's poem. For Paul, "[h]uman existence—even in the sphere of the *pneuma* (spirit)—is a bodily, somatic existence"; "the *sōma* is not merely an outer form but the whole person."[239] Accordingly Paul writes in 1 Corinthians 6:12–20, that in fornication, bodies which are "members of Christ" become one with sin, become "members of an harlot," "joined to an harlot [in] one body." In *The Purple Island,* the same idea informs the description, or *contreblason,* of Caro, whose fleshly appetite in excess of its recuperation for reproduction is suggested by "Her empty breasts [which] hang like lank hollow bagges" (*P.I.* 7.14.3–5),[240] but more particularly it informs the description of vices, such as "*Porneius,*" or "Fornication" (*P.I.* 7.19.1 and margin. *), whose "flesh and bones were gnawn with hot desire" (*P.I.* 7.20.4), his "leaden eyes" and "joylesse look" (*P.I.* 7.19.2, 3) signifying both metonymically and causally the psychological degeneration precipitated by his particular sin.

Maus argues that the point of distinctions between "an unexpressed interior and a theatricalized exterior" in the early modern period "is normally to privilege whatever is classified as interior," because "[p]ersons and things inwardly *are* [and] persons and things outwardly only *seem.*"[241] However, in Pauline ethics, bodily acts "affect not only the individual act of sin but the whole person to his innermost being,"[242] thereby perhaps enabling the practice of induction, or "artificial conjecture," "from the superficial to the deep, from the effect to the cause, from seeming [or rather, in this case, action] to being."[243] Given Fletcher's Pauline moral theology, his relative lack of skepticism, and the theistic allegorical context in which the representation of moral intention and action occurs, and therefore where the problem of human inwardness is posed, in *The Purple Island,* personification generally presumes to assume a standard of what would constitute certainty, "an effortless transcendence of the boundaries that frustrate human knowledge."[244] Therefore, making a virtue of a complete lack of concealment or deceit, there is "*Erythre*," or "Modestie," who is described as possessed of an "often-blushing face," which "Sweetly her in-born shame-fac't thoughts commended; / The faces change prov'd th'hearts unchanged grace" (*P.I.* 10.41.2, 3–4 & margin. *).

At the opposite extreme, demonstrating the deceitful non-convergence of an unexpressed interior and a theatricalized exterior is Caro, who "seems (skin-deep) most fair by witching gin / To weaker sight" (*P.I.* 7.14.3–4). Yet what Caro conceals, what she *actually* personifies, is exposed to the purged eye of theistic allegory, which draws on the idea of God's anatomical judgment upon Man's heart, and constitutes the standard of certainty (*P.I.* 7.14.4).[245] Therefore the faculties, virtues, and vices bear some similarities to the dramatized virtues and vices of "the morality play tradition from medieval England,"[246] but are generally unlike Vindice who in *The Revenger's Tragedy* is a character with some subjective independence of the morality-play didacticism according to which he should personify revenge.

Necessarily, then, given this ethics, in Fletcher's poem, Intellect is himself participant in the Fall of whole Man, and takes refuge "in a Tower where thousand enemies / Assault the fort" (*P.I.* 6.38.2–3). Therein he is confined to "his base bonds, and loathsome prison thrall, / Whom thousand foes besiege, fenc'd with frail yeelding wall" (*P.I.* 7.8.6–7). While this tower in the geocosm may be the "vaulted Tower" with "glitt'ring turrets" (*P.I.* 5.51.1; 5.10.2) in the "Citie" of the brain and its facial "Suburbs" (*P.I.* 5..6.3, 5 & margin. c), because Man is not only *imago dei* but also fallen, it is also the "wofull bondage and captivitie" of a "sinfull and unhappy estate."[247] Here, in this penitential retreat, "Prest down in captive chains, and pent in earthly mould," Intellect and Voletta, his queen, no longer possess the spiritual sight to discern what is good and true: "every natures difference . . . / . . . now they scarcely see with pain" (*P.I.* 6.67.4–5) because they are "now with blackest clouds . . . thick enroll'd" (*P.I.* 6.65.6).[248] Their imprisonment is a metaphysical, psychological, moral, and physical condition and its etiology is sin. Cable explains it as follows: "Stripped of the delusive metaphor of the perfectly ordered body politic, Prince Intellect on earth no longer appears an omniscient, omnipotent ruler, but rather a soul striving desperately for survival in a body newly abased by the exposure of its gross physicality. . . . The human body, so glorious a piece of work in a world of material phenomena and empirical truth, becomes in the world of moral and spiritual values a tent, a rag whose inadequacies make the Intellect no ruler but a fugitive."[249]

Cable resists the division by Kastor and others of the work into three parts—"the external, physical anatomy of Man the Island [Cantos 2 to 5]; the internal, moral-mental-spiritual anatomy of Man-Island [Cantos 6 to 10]; and the war of virtues and vices in the Man-Island [Cantos 11 and 12]"[250]— or into two parts, with a break at the end of Canto 5 between representation of the physical and representation of the nonphysical aspects of Man.[251] Cable is accurate in her comment that, "[c]ontrary to general critical opinion, the body of *The Purple Island* does not get 'left behind' after the fifth canto; it is very much present throughout the poem. . . ." As she insists, "criticism's facile willingness to separate *The Purple Island*'s body from its poetic and spiritual soul is one of the errors the poem seeks to deter."[252] Notwithstanding her resistance against such criticism, Cable does exaggerate the distinction between the moral-spiritual-aesthetic world (of the poetic stanzas throughout, but particularly of Cantos 6 to 12) and the purportedly "empirical," natural-philosophical realm of material phenomena (in the marginalia of Cantos 2 to 5, particularly, but also to less extent the poetic stanzas of those cantos).[253] In the allegorical narrative in Cantos 6 to 12, clearly regeneration of the human body is not so much an Aristotelian actualization of morphology in its account as physiological structure, but a shedding of sinful corruptibility; but this is not so much because of a body/soul relationship, the dichotomization of which Cable imagines the poem seeking to resist, but because of a moral and theological struggle for possession of embodied Man between *mortal* virtues and the fruits of the flesh. In this struggle, action and intention are inscribed by what Maus has described as "the way the structure of internal experience is thought necessarily to imply observation by a divinity," in the early modern period.[254]

Cable does astutely notice the shift in representation of the anatomical body politic from Canto 5, the last of the anatomical cantos, and Canto 6, but again she dichotomizes the perspectives or focalizations unnecessarily sharply.[255] It is recognized throughout *The Purple Island* that the body politic, like the body, is simultaneously both "perfectly ordered" and transient because fallen, and the desirability of the regeneration of the body and soul is true also of the providential salvation of England under James I, because the bemudded progeny of Caro struggle also to take control of the geopolitical realm and *ecclesia* or *sōma Christou* (the body of Christ), which is personified by Eclecta. This is why in the anatomical cantos the body which is also Britain is not only an armed fortress, but home to the intelligence-center of a government (or the sensory apparatuses of a human individual), which is wary, suspicious, and continuously vigilant not only against "what e"re is in the continent" (*P.I.* 1.53.3), the supposedly encroaching forces of Roman Catholicism, but also of a perceived internal threat from insurgent members of its own flesh, "the body politic's endogenous pathologies." So although it is true that *The Purple Island* conspires in the functional utility of accusations of treachery in consolidating a reified sacral authority, and imagines the source of treachery as in some measure invasive of the island-realm's boundaries and vulnerable apertures, it does not elaborate the external threat in order to negotiate and erase, repudiate, or contain domestic conflict or discontent, as Harris has shown was and is so frequently the case elsewhere.[256] Because of the acknowledged need for penitence, the internal threat is not displaced but is assessed as posing an even greater risk to Prince Intellect, whom the poetry does not lead us gradually to identify with the body he rules, as Cable suggests,[257] but rather, who is "present" and "President" "In every quarter of this blessed Isle," "all in all, and all in every part," thereby assuming the property of the individuated or particularized Aristotelian or scholastic rational soul (*P.I.* 6.32.2, 1, 6):

> He guards all entrance, and by divers spies
> Searches into his foes and friends designes:
> For most he fears his subjects wavering mindes.
> This Tower then onely falls, when treason undermines.
>
> (*P.I.* 6.38.1, 4–7)

As the discussion of *The Purple Island*'s soteriological purpose has indicated in this study, the relationship in which disturbances are perceived simultaneously in the human body and the polity not only corresponds to the body politic analogy, but relates to the diegetic narrator's expressed purpose for his work to have a soteriological effect in the sense of the promotion and preservation of health, or the restoration and perfection of human nature. As Margaret Healy has argued: "Medical explanations can be 'exopathic' (disease an external force of some kind by which the body is invaded), or 'endogenous' (disease as an internal disorder or derangement, a state of being out of step with the environment). Because the body has an inside and an outside separated by a protective carapace, there is always an interplay between endogenous and exopathic explanations, and ideas about disease are part of a wider system of beliefs regarding contagion, pollution, sin and death."[258] By exploring emergent ideological formations and incipient state apparatuses, and drawing them into metaphorical interaction with the anatomy of Man, Fletcher then systematically figures Man not as escaping in the spiritual substance of the Christian Neoplatonic immortal, rational soul as a criterion for personal identity, but rather as a nervous and defensive human individual, who is as vigilant against her or his own sinful, fleshly appetite and impulses, as he or she is against the "other," that was Roman Catholicism. The body is the field in which the relationship with God flourishes or flounders, and this conflict or equivocation extends from the conception of anatomical structure and physiological function to the co-extension of action and inner intention in the rational mind or intellect. For reason of this vulnerability of even the rational soul to fleshly appetite, the human body cannot be fully reconstituted by a rationalist Aristotelian actualization of morphology in its account as physiological structure, or resurrected in the signs of the poetic text in an immortal, incorruptible form by rational ingenuity.

The inevitable tension between the curiosity and sensuality of the flesh, on the one hand, and the body conceived on teleological principles by rationalist anatomy, on the other, not only relates the *Purple Island* anatomy to the equivocation in the *Prelectiones* between the morphology of the body and its actualization as physiologically understood structure, but also to the practice of holding annual public dissections, though not the Lumleian lectures, in January and February, during the Carnival period. Carlino suggests that if it is true that the word "Carnival" originates in the Latin *caro* (flesh), and "if several Carnival practices are indeed linked to flesh in its multiple aspects (food, sex, violence), . . . dissection, practiced during the Carnival period, fits rather appropriately into those activities connected with sensuality characteristic of that time of year."[259] Ferrari claims that the conceptual categories of Carnival, its "concretely sensuous ritual-pageant 'thoughts' experienced and played out in the form of life itself," are also dramatized in the anatomy theater.[260] But if these suggestions are the case, the conflict that arose in the dramatic presentation of anatomy in the theater, between the body as created and as mortal, would have been augmented. The etymology of "Carnival," as Schoenfeldt notes, however, "derives from the farewell to the flesh (the literal meaning of *carne-vale*) that is celebrated by a brief but intense indulgence in carnal desire," which should be understood in terms of its social function in the "calendrical process of self and communal regulation."[261] In relation to the struggle with the flesh for possession of Man as embodied, this conflict must therefore be understood as thus qualified, and can be analyzed as twofold.

Firstly, there is that which we have been focusing on so far in this discussion, the dichotomy and struggle between human anatomy, as ethically designed and created as teleological by God to serve the rational soul, and the flesh as concupiscent, rebellious, and sensuous. The teleological understanding of the body in classically inspired rationalist anatomy, that the parts are designed to perform their functions, which are integrated into the function of the whole body, the purpose of which is to serve the rational soul (as distinguished from Aristotle's Active Reason), is potentially subverted as the conceptual categories of Carnival undermine the dominance of the immortal, rational mind over the concupiscent and nutritive functions of the mortal soul, the pleasure of the flesh and sexual curiosity. This is the moral confrontation with which we have been concerned.

Secondly, the struggle that affects the resurrection of the body-as-referent in *The Purple Island* is also between the body as created and redeemed from corruption in a stable, closed form, and the mortal, corruptible *vanitas* of the flesh that is grotesquely open to sensation. This struggle, which is related to the first, moral reason for the augmentation of the conflict that arose in the presentation of anatomy in the theater, would have involved the idea that, in the "grotesque body," "there are many points of contact with the ambivalent image that anatomy offered of the body: 'open', 'revealing its proper substance'. . . ."[262] Yet, in the drama of the demonstration of anatomy in the theater of the College of Physicians, the body which was progressively reanimated was not a Carnivalesque or grotesque body, which is open, full of potential but without clear boundaries. It was rather an actualized, teleological body with "stable form" and corporeal integrity. But it is also argued above, after Wilson, that in the *Prelectiones,* the body-as-account-of-body should be understood neither simply as entailing the reanimation of the cadaver as an intentional structure offering an account of itself, nor simply as bringing about a cadaverization or deanimation of the anatomist, but as provoking the anxieties which each of these suppositions occasions. It is said that the tension achieved in this equivocation is an aesthetic tension, as between the genres of romance and tragedy. Whereas anatomy as dissection, like tragedy, is decomposing, pessimistic, and a reminder of death, anatomy as anatomy, like romance, is recomposing and produces elation in the apparent reversibility of mortality. This struggle or equivocation can be analyzed as consisting of two constituent, interrelated struggles, one of which is moral and aesthetic, the other metaphysical, ontological, and theological. In the paragraphs immediately following, we shall see how the moral and aesthetic equivocation relates to the moral confrontation between the body and the flesh in Cantos 6 to 12 of *The Purple Island.*

The aesthetic, moral confrontation has two distinguishable though related manifestations. The first constitutes a further way in which Thirsil's diegetic narration, like an illustration in an anatomy textbook, or like the theatrical performance of anatomy in the demonstration and lesson, is designed to bring together the intellectual pleasure of knowledge and aesthetic enjoyment. Moreover, this manifestation of the moral and aesthetic confrontation figures the body, which is open, permeable, and vulnerable, as female,[263] and like the plates of female sexual anatomy in *De dissectione* discussed in chapter 6, the manifestation is erotic, and draws upon the Neo-Petrarchan tradition of anatomical love poetry, that was associated with the style of the School of Fontainbleau artists, the intellectual and social context in which Estienne placed his work.[264]

When Valerie Traub suggests that critics of early modern culture and literature should reconceptualize the boundaries of the person as "a permeable membrane, through which cultural-psychic osmosis occurs," or as apertures or "mutually interfacing and constitutive *spaces*" of "interchange and circulation," "rather than inflexible demarcations functioning to restrain movement and exchange," she proposes: "Without slipping into biologism, we can see that the body provides an analogy for the criss-crossings of cultural exchange. In particular the *erotic* body metaphorically figures precisely the permeability that is constitutive of the early modern subject, its apertures palpably embodying social relations as linguistic and bodily discourse."[265] In the following paragraphs, we shall see how *The Purple Island* envisages the subject, Man, as constituted without clear boundaries, by erotic and psychic openness to the "other," even as the poem strategically seeks an authoritative discourse with which to defend and insulate the Isle of Man not only from the "other" of Roman Catholicism, but also the "otherness" and purportedly corrupting influence of "his" own flesh.

Bodily integrity is personified in the *Purple Island* allegory by "*Parthenia,*" or "Chastitie in the single" (*P.I.* 10.27.2 & margin. *), "a warlike Maid, / . . . all in steel, and gilded arms" (*P.I.* 10.27.1–2), who has been compared with several literary Amazons who find their progenitress in Camilla of Virgil's *Aeneid.*[266] Consistent not only with the epic battle they are to engage in, but also with the pageantry of Carnival, all the virtues and vices in the *Purple Island* allegory are described in terms of their arms, behavior, dress or armor, and shield with device and motto.[267] The details of each of these bear a metaphorical, metonymic, or symbolic relationship to the virtue or vice the personification represents, and Parthenia's device, "Th'Arabian bird" (*P.I.* 10.28.4), symbolizes her "*chaste single state,*" "Phoenix-like alone in narrow cells" (*P.I.* 10.28.7; 3.29.4). Her "Love" is self-regenerating and heavenly (*P.I.* 10.28.5–6; 11.32.4). She is an allegorical embodiment of the ideological and ideal configuration of the human body, the kingdom and Eclecta, or *ecclesia,* the order of each of which is constructed around the overriding principle that they must be defended from the desires of the flesh.[268] As such Parthenia stands in moral-religious defiance of physicians' warnings, outlined in chapter 1 of this study, that any woman who is unaccustomed to the benefit of marriage though is "apt thereunto," and is led thereby to "containe her selfe," or any woman who refrains from marital sexual activity through her own choice, for reasons of religious vocation, is regarded as liable to retain seed in her uterus, which may lead to the suffocation of the Mother.[269]

Parthenia's defeat of Porneius, son of Caro and Anagnus, or Unchastity, in Canto 11, stanzas 19–21, is followed by a confrontation with "*Aselges,*" "Lasciviousness," brother to Porneius (*P.I.* 7.23; 11.22–23), whom she fights and unsaddles. Thereafter, Parthenia continues to prevail against all the vices of the flesh (*P.I.* 11.24–25), until "The subtil Dragon," who is Satan (*P.I.* 7.11; 11.26.1), "with hellish charms" makes "false Delight" seem "true" (*P.I.* 11.26.6, 7). This "craftie swain" "whets [Parthenia's] wrath with many a guilefull word / Till she lesse carefull did fit time afford," when he had deceived or distracted her sufficiently, for him to inflict a wound deep "in her side" (*P.I.* 11.26.5; 11.27.5–6; 11.29.1).

The "dolefull sight" of the wounded "Conquerour" (*P.I.* 11.28.2, 3), with her "life[-blood]" "From that wide spring" in her side flowing out "in long rivers" and "purple streams drowning the silver plains," is a grotesque image of the body's vascular anatomy and physiology turned inside out, open and revealing its proper substance (*P.I.* 11.29.2, 3, 4). The narration decelerates considerably in pace, lingering disbelievingly upon Parthenia's mortal injury, which it cannot reconcile to a body and realm providentially ordered by the "powerfull heav'ns" (*P.I.* 11.28.1). The image of her wounding is the undoing of *sōma* itself; a kind of anatomization which follows upon her having been not only (naturalistically) unbalanced and undermined by her own wrath, but arguably (in the context of the allegory) sorely tempted by "false Delight" to sin, if not actually guilty of commission.[270] In her invocation of "Love" she fears this may lead to her defilement and shame:

> Let not this Temple of thy spotlesse love
> Be with foul hand and beastly rage defil'd:
> But when my spirit shall his camp remove,

And to his home return, too long exil'd;
  Do thou protect it from the ravenous spoil
  Of ranc'rous enemies, that hourely toil
Thy humble votarie with loathsome spot to foil.

(*P.I.* 11.33)

Parthenia is consumed by a dread of the shame and dishonor associated with the ignominious public display, violation, and possible dismemberment of her corpse, the sacredness of which she implicitly assumes will survive her death.

The reason why this happens in Thirsil's diegesis is related to the narrator's displacement of erotic desire in the anatomical cantos from the sexual parts—"Forbear, my maiden song, to blazon wide / What th'Isle and Natures self doth ever strive to hide" (*P.I.* 3.25.6–7)—to the face and head, where "beautie . . . keeps her chief residence" (*P.I.* 5.6.7).[271] Not uncommon as a strategy in Renaissance erotic expository and narrative poetry, the displacement from the "obsessively unspoken" part to the "obsessively spoken part" takes on an anatomical character.[272] Whereas the genital organs are described as "the vilest part," despite their "wondrous art" (*P.I.* 2.26.7, 6), an aesthetic and epistemophilic desire drives the dissection of the head (*P.I.* 5.10.3–7): "The gazers eye still more and more entices / Of the inner rooms to get a fuller sight; / Whose beautie much more winnes his ravisht heart, / That now he onely thinks the outward part / To be a worthie cov'ring of so fair an art."[273] When describing the "curious work" of the flesh (*P.I.* 2.7.1), by the pressure of this desire, the narrator is drawn to describe bodily surfaces aesthetically and erotically as lightly clothing the interior: "All fair is hung with coverings slight and thinne; / Which partly hide it all, yet all is partly seen" (*P.I.* 2.7.6–7).[274]

When Thirsil turns to blazon the face and upper body of Parthenia, Fletcher's writing displays the "astonishing" sensuousness with which he has been associated primarily on account of some of his other work.[275] Divine "Love" (*P.I.* 10.31.1), who "Upon her brows lies his bent Ebon bow," becomes ambivalently, yet progressively, like "wanton *Cupid,*"[276] as the blazon makes a division of Parthenia's eyes, eyebrows, cheek, lips, teeth, voice and "daintie breasts," with "little rising heads" that "may seem to sight / To be enwombed both of pleasure and delight" (*P.I.* 10.31.5; 3.22.4; 10.32–37; 10.37.1, 3, 7).[277] In the contexts of Cantos 2 to 5 of the work, and of the grotesque anatomical image produced by her wounding, Parthenia's eroticization perhaps functions to manage anxieties originating in the development of the special sensory epistemology in anatomy, which, not only in Harvey's case but elsewhere too, attempts to reconcile the pursuit of knowledge with the materiality and inherent divisions of the body-in-itself, an aim that may lead to the cadaverization or abjection of the anatomist and to the morbid anxieties resultant thereupon.[278] The comparable psychic safety of Neo-Petrarchan poetry turns anxiety to arousal, but only temporarily. At the culminating suggestion regarding Parthenia's breasts, the narrator anxiously equivocates and, as though to erase the unfolding implications, abruptly ceases to divide her body, and turns to the admiration of her virtue, restoring her integrity as *sōma,* and as "chastest" Love's "Temple" and "humble votarie" (*P.I.* 11.32.4; 11.33.1, 7), whose beauty is "As when a taper shines in glassie frame," the bright emanation of "th'inward Sunne," "the highest Love" of the Redeemer "thron'd in her heart" (*P.I.* 10.38.6, 2, 4, 3).[279]

Through this scopophilic sublimation, the diegetic narrator proffers a Neoplatonic moral justification for blazoning Parthenia's body and eroticizing her flesh. Avoiding the imputation of erotic love to her, this justification is that Divine Love is manifest in her "mortal flesh."[280] But the form of the body and the kind of love which was expressed initially in the Neo-Petrarchan appreciation of Parthenia, were not only the Neoplatonic notions of the Ideal Form or "Celestial Love"; Thirsil is not "ravish'd at the sight of Intellectual Beauty, because blinde to the sensible."[281] Much as Man may strive to shore up a commitment to virtue, and to aspire to the condition of being *sōma,* incorporate in *sōma Christou* and no longer in the flesh (Rom. 7:5; 8:8), in the sense that it matters not to her or him if he or she lives or dies,[282] Thirsil implies that Man gazes at the flesh, enticed by its pleasures, but equally horrified by the prospect of living by it, as though there were a necessary relation between the flesh and dependence on the flesh, and therefore no possibility of removing from a consideration of the sins of the flesh a desire to commit them.[283]

This predicament, and the danger of death and damnation it purportedly poses, finds expression in a further element of the denigrating descriptive blazon of Caro. The reason that Caro's shield bears the device of "a Mermaid," who "sung and playd," and "Whose humane beauties "lure the wandring sight," is also because desire or the pleasure of the flesh renders her enticingly beautiful (*P.I.* 7.15.3, 4, 7). But in sharp contrast to Parthenia's idealized beauty from within, Caro "seems (skin-deep) most fair by witching gin / To weaker sight; but to a purged eye / Looks like (nay worse then) hells infernall hagges" (*P.I.* 7.14.3–5).[284] It is apposite, therefore, that after blazoning the hair, forehead, teeth, and eyes of "*Erythre,*" Parthenia's page (*P.I.* 10.42.1–2), Thirsil refrains, unconvincingly exclaiming, "But neck and breast no man might bare descrie, / So sweetly modest was this bashfull girle" (*P.I.* 10.42.3–4), only to abandon his self-restraint and proceed to describe her mammae, or breasts, and her mammillae, or nipples, in terms evoking the prototypical prohibition and disobedience: "But that sweet paradise (ah!) could we see, / On these white mountlets daintier apples be, / Then those we bought so deare on *Edens* tempting tree" (*P.I.* 10.42.5–7). The structural problematic that characterizes the laudatory poetic description of Parthenia—the combination of verbal enticement to desire and its displacement or sublimation—finds in the blazon of Erythre a theological, moral rationalization of the fallen human failure ever to succeed either in removing shame and guilt from the sensuous celebration of Creation, or in being released from the paradoxically providential "fire of generation" that burns "in *Venus* courts without cessation" (*P.I.* 3.23.5, 6), spurring Man in "his" entirety as flesh, and in repetition of the first disobedience, to know and possess what so ingeniously consumes "him" with desire and curiosity—"Sure heav'n with curious pencil, at thy birth, / In thy rare face her own full picture drew" (*P.I.* 10.30.4–5).[285]

The role of Repentance in Queen Voletta's regeneration is played in Parthenia's ordeal by "*Continence,*" brother to "*Encrates*" ("Temperance") and "*Agnia*" ("Chastitie in the married"), Parthenia's sister (*P.I.* 11.36; 10.21; 10.24; 10.27.1). "Love" intercedes (*P.I.* 11.34), rousing Encrates, Agnia, and Continence, who pours a "precious liquor" into the wound of Parthenia (*P.I.* 11.37), with "decidedly sacramental overtones."[286] The "precious liquor" echoes the 'precious drops' of contrite Voletta's blood, and both draw upon the soteriological efficacy of Christ's bloody sacrifice on the cross. Continence (which means self-restraint) is then described as "binding up the gaping orifice" (*P.I.* 11.37.5), thereby rendering Parthenia's body intact and enclosed again. For Crooke, the "orderly" procedure of dissection and its artificial preservation of the body's integrity distinguish it from the "accidental investigations of warfare."[287] Following on from the grotesque "dissection" by "false Delight" of

Parthenia's body, open and revealing its vascular anatomy, the sacramental effectiveness of her recovery, as an optimistic reminder of redemption, corresponds to the reversal in fantasy of the dissection and of the execution that preceded it, in the "religious observance" of the anatomy theater. The ministration of the sacrament by Continence becomes a channeling of the shameful or illicit violation of Parthenia's body into the sphere of the licit by its ritualization, and a retroactive pardoning of the guilt, not of "false Delight," but of the diegetic narrator, who has blazoned Parthenia's body and eroticized her flesh, and then punished her for a sin of which she is at most guilty only in the sense of having been tempted.

The second manifestation of the conflict between the openness of the grotesque body and the stable form of the teleological, classical body is similarly aesthetic, but also cognitive, and concerns figurality. That the grotesque body was conceived, in contrast to the classical body, as not separated from the world by clearly defined boundaries is consistent with the theory of the body as a microcosm, with "inseparable links to all parts of the creation," and particularly with the character of the conceptual metaphor of anatomy as vegetation.[288] As was remarked above, Davies of Hereford's poems articulate the connection, only implicit in *The Purple Island,* between the appetite of organs and the conceptual metaphoricity of anatomy and its rationalization in terms of the functions of the soul. In the imaginatively rationalist conception of the human body that characterizes textbook and demonstrated anatomy from at least as early as the late fifteenth century to as late as Harveian anatomy in the seventeenth century, there is a discernible tension, but also a coherence, between that which is consistent with the conception of anatomy as vegetation (primarily the functions of nutrition, growth, and reproduction), and might be allied to the Carnivalesque, and that which is consistent with the conception of anatomy as "building" (principally the skeleton and the musculature), which is emphatically teleological. But the grotesque and classical bodies are not only "mutually constitutive."[289] They are coherent because, as Schoenfeldt has argued with clarity and subtlety, the cogency of humoral terminology not only linked the body's corporeal flux to the conception of the grotesque body; humoral theory encouraged constitutional solubility, not Carnivalesque liquefaction; Galenic medicine and ethics rendered the obstructed, impermeable body a source of pathogens and the disordered, undisciplined self a site of subjugation, as does *The Purple Island.*[290] Coherence also arises because each conceptual metaphor tries to grasp different aspects of the concept of human anatomy, and because the vegetative metaphor and thus appetite are themselves teleological; the vegetative metaphor arises from and is justified by Aristotelian analogy and the vegetative soul, *Physis,* or Nature. Both conceptual metaphors of anatomy also find coherence in concepts of Creation, mortality, and salvation.

The body in *The Purple Island* is characterized by the same tension and coherence for all the reasons of its indebtedness to textbook and theatrical anatomy, and because of its creative, poetic development of the figurative conceptualizations found therein. It might even be said that *The Purple Island* metaphorically acquires the generative grotesque body as the figurality of its threshold conceit.[291] With its potentially infinite generation of connotations, many of which the poem unfolds to the reader, the creative metaphor of the Isle of Man is like the grotesque body, which is open-ended, unfinished, outgrows itself, and transgresses its own limits.[292] The work's tissue of metaphoricity combines conceptually with the grotesque, feminized flesh to lay siege to phallogocentrism, to Intellect in his tower, who thereby decisively ceases to "stand proudly erect, towering above the squishy world of hope and fear, displaying the clean-cut, virile autonomy that is [his] birthright."[293] The creative metaphor extends the concept of Man to include a collection of curious objects not only from within Man's anatomy but which are found in the geocosm, macrocosm, and body politic.

Notwithstanding this plurisignification, interpretation is not only realized, but restricted or constrained, by the work as a whole, as chapter 2 has shown. Furthermore, semantic generation and perspective are also stabilized by the marginalia, and, on the level of semantic content, the vegetative and generative figurality is at length limited by the figuration of the body as an island which is purportedly providentially separated from the rest of the world ("the continent") by clearly defined, and defended, boundaries or shores, like the classical body.[294] This explains how, as is claimed in an early part of this chapter, the meanings generated by the work, which provide in allegorical form the *telos* or reason for the figurality in the purposes of the soul, gradually make the soul more apparent as the allegory unfolds. In this respect, possibly the most suitable visual counterpart of *The Purple Island* is the second plate of Book III of Valverde's *Historia de la composicion del cuerpo humano,* referred to in chapter 4 of this study. The combination of the relatively unprotected abdomen of the two figures sporting Roman military dress, which is dissected while the thorax of each figure remains protected from the tools of the anatomist and artist by a breastplate, not only serves to give powerful expression to the concept of the chest as "most safe and armed," but figuratively embodies and embellishes the simultaneous destruction and preservation of the classical male body, taken to its most defensive extreme.[295]

The struggle or equivocation between dissection, as like tragedy, and anatomy, as like romance, can be analyzed as consisting of two constituent, interrelated struggles, which were introduced above. The first of these we have now analyzed in both its manifestations as one which is moral and aesthetic; the second reason why the concretely sensuous ritual-pageant "thoughts" in the anatomy theater would have augmented the conflict that arose in the dramatic presentation of anatomy is ontological or metaphysical and theological. Described by Mikhail Bakhtin as when ambivalence dominates the conception of every state "as inevitably subjected to change, to alternation," the Carnivalesque perceives everything, including Man, though especially the body, as tied to worldly processes.[296] The Carnivalesque, grotesque body was not therefore identical to the dead, dying, or decaying body, but it was tied to the processes of birth and death, devouring and being devoured, growth and multiplication—the processes of the flesh, that in its Scriptural rendition was identified by Purchas as "the Corruption of Nature."[297]

According to Kocher, Man was "deeply participant in the two orders of existence": "In so far as he was a body he was subject to the natural laws of motion and change, generation and corruption, growth, procreation, and death. As soul he escaped those laws and received the higher injunction to know, to love, and to obey God and cherish his fellow man."[298] But in the Old Testament and frequently in Hellenism, "flesh denotes man as a whole" in the sense that "man is flesh in his essence," for flesh describes the transitoriness of life.[299] Some of the natural-philosophical and medical strands involved in expository or fictional anatomies and the anatomy of the schools and colleges, in the early modern period, do not maintain a perfectly clear differentiation between the matter and mortality of the body, on the one hand, and on the other, the immaterial substance of the soul, which was of course in part mortal and in other part, a rational part, immortal. These intellectual strands seem rather to have thought about a graded continuum of created substances separated both by the degree of their rarefaction and by that of their participation either in time or eternity. This gave rise to the world of correspondences, of analogies and antitheses, between the substances of the microcosm and the cosmos, and those of the body and the soul within the microcosm itself, such that cardiovascular physiology was thought by Fletcher, in *The Way to Blessednes,* to be a reasonable means of figuring out the properties of divine inspiration.

When either the Judeo-Christian idea of *sōma* as the whole person, or the Pythagorean and Paracelsian notion of the body's "[i]ntrinsique Balme," the "astrum" or the "anatomy of the world," or the Aristotelian conception of the soul, was uppermost in an early modern representation of the body and soul(s) of Man, the immortal, rational soul or spirit could become something particularized or individualized and in no way "absolutely free from bodies,"[300] it could become implicated detrimentally in the development of the lower souls, as in John Donne's *An Anatomy of the World* and, here, through the lens of Aristotelianism, in *The Second Anniversarie:*

> Thinke further on thy selfe, my soule, and thinke
> How thou at first wast made but in a sinke;
> Thinke that it argued some infermitee,
> That those two soules, which then thou foundst in mee,
> Thou fedst upon, and drewst into thee, both
> My second soule of sence, and first of growth.[301]

In Donne's poetry one finds intimations of the philosophical view that even the rational soul may be material in substance and mortal in nature, suspicions of which already existed among physicians,[302] and debates about which would emerge later in the seventeenth and early eighteenth centuries, conducted in tracts, treatises, and epistolary discourses.[303]

Although there is no implication, in the poetic anatomies of Davies of Hereford or of Fletcher, that the body that undergoes moral alteration, as it is resurrected as the signs of the poetic text, is by inference a specifically criminal body, *The Purple Island*'s mimesis of the anatomy theater implies that its microcosmic Man partakes of the emblematics of the criminal cadaver as fallen, but the lessons learned about death, not only as natural and inevitable but as the wages of sin, are intended as generalizable to all human beings. The service of *memento mori,* and indeed of mortality, was, according to Vesalius, the lesson that the great Creator of things has fashioned our body not only that it may live but that it may live properly. In *The Purple Island,* though she does not signify death directly, the allegorical personification of the flesh, Caro, is a *memento mori,* a reminder of death as the wages of sin. Her description is characterized by her consumption by age and disease, which is both metonymic and symbolic: "And *Iris* ulcer'd skin is patcht with leprous ragges" (*P.I.* 7.14.7). Underneath Caro's device, "a Mermaid," "Whose humane beauties "lure the wandring sight," the "word" or motto is "*Heare, gaze, & die*" (*P.I.* 7.15.3, 4, 7), so Caro is not so much a personification and a *memento mori* of the transitoriness of the flesh and of its sinful pleasures, though this does inform her image, but of Paul's explicitly moral edict: "if ye live after the flesh, ye shall die" (Rom. 8:13).[304] If Intellect were to "neglect his Masters law" (*P.I.* 6.36.1) and live "according to the flesh" (Rom. 8:13), and if he "with those traitours [the vices] 'gainst his Lord rebells" (*P.I.* 6.36.2), he "sows to the flesh [and] will from the flesh reap corruption" (Gal. 6:8), "night, death, despair and horrour" (*P.I.* 6.36.4).

At length in the apocalyptic battle, a "pois'nous gorge" is vomited by the "Dragon," who is Satan (*P.I.* 7.11; 12.26). This consists of Hamartia, a "foul deformed wight" who personifies "Sinne" (*P.I.* 12.27.2 and margin. *), "*Despair*"(*P.I.* 12.32.1), and "two brethren" "in common name of death agreeing" (*P.I.* 12.36.1, 2). In addition to her "loathly" genesis from the Devil's "stinking panch" (*P.I.* 12.26.5, 6), Hamartia as original sin is "The foulest bastard" of "that first woman," Eve (*P.I.* 12.29.2, 1), her parents giving rise to her curious form: "A woman seem'd she in her upper part" (*P.I.* 12.27.5); "The rest (though hid) in serpents form arayd" (*P.I.* 12.28.1).[305] As "chaffer fit for fools their precious souls to sell" (*P.I.* 12.31.7), she also personifies the sinful pleasures occasioned by the flesh,[306] and her personification of both inherited original sin *and* sins of commission and omission gives rise to the "double sting" at the end of "her knotty tail" (*P.I.* 12.28.5, 3). The theological logic or rationale of the organization and order of Satan's putrid progeny is that the inheritance of original sin made Man subject to the first death who as Death the Mower is "arm'd with a sithe" (*P.I.* 12.36.3), hence Hamartia's "dead life, her carrion smell" (*P.I.* 12.35.7; 12.31.5); but as Despair's motto testifies, the "*Reward of sinne is death*" in a second sense of damnation or eternal death as the punishment of vice (*P.I.* 12.39). Since "nought," no virtue or righteousness, "could help the wound" inflicted by Hamartia's sting ("but bloud of heav'nly King") (*P.I.* 12.28.7), Despair has "looks fast nail'd to *Sinne*" (*P.I.* 12.32.3); yet despair is itself a sin which abandons faith in salvation and therefore abrogates responsibility for what is necessary for justification in Protestant theology.[307]

Comparable in several respects with the description of Caro is that of Hamartia, because she personifies the sins occasioned by the flesh. She shares Caro's "deceiving art": "such lying glosse" she could "impart" to her "upper part" (*P.I.* 12.27.7, 6, 5), that "fair and lovely [she] seems to fools dimme eyes" (*P.I.* 12.31.1). Through being willfully misled towards misinterpretation,[308] therefore, one confuses dependence on the flesh, particularly in the form of sensuous, sexual curiosity and pleasure, with the aesthetic celebration of Creation. And this is said to happen when the Isle of Man is a "nest of hellish fogges and *Stygian* night," before "the bright orient Sunne hast fair renew'd, / And with unwonted day hast it endu'd" (*P.I.* 6.26.4, 5–6); before the Redeemer's cordial balm of salvation has concocted spirit in Man's "mistie breast," and shed God's "sparkling light" of inspiration, or revelation, to "guide" Thirsil's "erring footsteps" and "weak hand" through "this fogge" that conceals the "art" of microchristus (*P.I.* 3.4.3, 4, 6). Thirsil's rhetorical, mimetic resurrection of the body-as-referent is forestalled because of this foggy misinterpretation.

It is *The Purple Island*'s illocutionary act to discover or reveal, and so recreate, invent or renew, the "picture" of Christ in the heart of its readers, and in the heart of the body-as-signified, through the work's incarnational language and inspired poetic words. Therefore the "purged eye" of theistic allegory and anatomy purportedly exposes what Hamartia conceals. When "her true form true light descries" (*P.I.* 12.31.3), it is at once grotesque—"Her loathsome face," "Her shapelesse shape," "foul," "monstrous," and "deformed" (*P.I.* 12.31.4, 5; 12.27.2, 3; 12.29.2, 3; 12.30.2)[309]—and expressive of death as the wages of sin—with "blancht skinne," she is "The devils dung" (*P.I.* 12.31.4, 6). But Hamartia's description is not simply either of these.

In respect of her association with sinful transgression, her apparent beauty and her "snakie hair," with "long adder chains, / In thousand knots and wreaths infolded round" (*P.I.* 12.31.4; 12.30.4–5), Hamartia bears comparison with Medusa.[310] Uncannily her "viperous locks" (*P.I.* 12.30.1) remind us also of the "dissected wreathes" of the anatomized body (*P.I.* 4.9.1), and of entrails "which are knotty and full of windings," "with hundred turnings giring" (*P.I.* 2.37 margin. l; 2.37.5). Stranger still, Hamartia's description reflects the teleological body created with inbuilt mechanisms to restrain appetite. In the abdomen, the entrails "are seven-fold longer than the Isle, / Yet all are folded in a little pile" (*P.I.* 2.39.5–6); Hamartia's "viperous locks hung loose about her eares; / Yet with a monstrous snake she them restrains, / Which like a border on her head she wears" (*P.I.* 12.30.1–2). This reflection of images of entrails resonates yet again with the second plate of Book III of Valverde's *Historia de la composicion del cuerpo humano,* for in each breast-plated figure therein a series of incisions allows the peritoneum to be folded back to reveal the stomach and the entrails—the duodenum, jejunum, ileum, cecum, and colon.

For Sawday, "the familiar tale of the Medusa," who in myth and emblem typically signifies horror, "is immediately appropriate to any description of the potentially transgressive gaze of the subject who studies his or her own bodily interior."[311] But in the moralized context of *The Purple Island,* Hamartia, with "dreaded" power to inflict sin (*P.I.* 12.28.6), is the "true" image of the body in so far as it has been given up to aims in opposition to God's; she is the "body of sin" (Rom. 6:6), exposed, open, and revealing its proper substance. As such she is a salutary reminder of the need for greater moral vigilance along the "blinde and dirty way" that, like the entrails, "hath many windings, and is easie to be lost."

According to Sawday: "the cultural history of the body suggests that a body which escapes its boundary—demanding attention by allowing its tokens of interiority to emerge on the surface—tends to be constructed, within the widely different forms of representation, as female, whatever the biological sex of the subject in question."[312] Hamartia is at her most threatening when "in her anger lightly she unbound" her "snakie hair," "And darting farre away would sure and deadly wound" (*P.I.* 12.30.6–7). She evokes a horror which is at the height of its intensity when her "shapeless shape" grotesquely escapes its own already ill-defined boundaries. It is a horror perhaps of matter without form and gendered female,[313] or of women's genitals,[314] the sight of which may tug at the entrails of desire and consume with curiosity; certainly it is a horror of the "fire" that burns "in *Venus* courts without cessation" (*P.I.* 3.23.5, 6), turning Man in "his" entirety to a body of sin, or the sinful flesh. The consequences for the Intellect are vividly imagined as desire turns to torture in the belly of hell:

> Down to the deeps ten thousand fiends him draw,
> Deeps, where night, death, despair and horrour dwells;
>   And in worst ills, still worse expecting fears:
>   Where fell despite for spite his bowels tears,
> And still increasing grief, and torment never wears.
>
> (*P.I.* 6.36.3–7).

If on the other hand, Intellect is cognizant of his own flesh as a kind of *memento mori* and a mortal threat, and if he lives "as knowing he may die," he "Bathes his crown'd head in soft eternitie" (*P.I.* 6.34.2, 4).[315] Therefore the extent of the difficulty that the grotesque perception of Man posed for notions of design, providence, and teleology is also, paradoxically, the extent to which reminders of mortality were readily available spurs to faith in salvation. While Man is still with sin, and still mortal, the limitation of the soteriological perlocutionary act of *The Purple Island* and of *Joy in Tribulation,* as Fletcher readily acknowledges, is that spiritual "*wounds*" "*may be suppled,*" but not entirely "*cured,*" the latter being the exclusive prerogative of God, the remembrance of whom is David's "*maine cordiall.*"[316] The "new, and inward man" which Fletcher, in *The Way to Blessednes,* says is anatomized by Psalms, is the type of "*The new birth,*" "whereby you *are borne, not of bloud, nor of the will of the flesh, nor of the will of man, but of God.*"[317] In this way, Fletcher's prose theological works sharpen our insight also into the *Purple Island* anatomy in respect of redemption and sin.

With the conquest and destruction of the Isle of Man all but complete, and seeing "the field swimme in her Champions bloud" (*P.I.* 12.43.3), Eclecta frames a speech as a reply to the male speaker in the opulently, tenderly sensuous Canticles, or Song of Solomon. Christian mystical tradition as early as Origen took an allegorical approach to The Song of Solomon, "celebrating Christ's love for his church or for the believing soul."[318] Eclecta's speech though differs in character from the biblical text. As the fallen, yet chosen, believing soul of the Reformed Church Militant, and as the allegorical personification of that elected Church, Eclecta addresses her "dearest Lord, [her] hearts sole Soveraigne" (*P.I.* 12.47.1) in supplication and penitence—"Take, take my contrite heart, thy sacrifice, / Washt in her eyes that swimmes and sinkes in woes" (*P.I.* 12.53.1–2). The emphasis in *The Purple Island*'s representation is on what has been lost of the relationship between Christ and his Bride, of the restoration of which there can be no assurance in righteousness. "Sweet Spouse," says Eclecta, "see not my sinnes, but through my plaints and tears" (*P.I.* 12.51.7). She becomes the bride of Christ only in the *parousia* (*P.I.* 12.69–86); until then, although she addresses Christ as her "Dear Spouse," she is besieged by Christ's "foes," "Hells troops [her] soul beleaguer" (*P.I.* 12.53.5; 12.49.1; 12.51.5).[319] Whereas the female voice in Solomon's Song says, "I found him whom my soul loveth: I held him, and would not let him go" (3.4), Eclecta raises her "suppliant hands," saying:

> These to be lillies thou hast often told me;
> Which if but once again may ever hold thee,
> Will never let thee loose, will never more unfold thee.
>
> (*P.I.* 12.48.5–7)

The Reformers broadly agreed on the doctrine of justification, consistent with which is Eclecta's apprehension of her pardon through saving "Faith" in her spouse's "former word and love," demonstrated by his "birth, life, death," though "Experience" and "Hope" assist her, hopes that "some comforts borrow / From firmest truth" (*P.I.* 12.50.1, 4; 12.46.3, 5–6). But for her "dawn of joy," her May of regeneration or resurrection, she acknowledges entire dependence upon "God's favour freely imputed" to her,[320] and implores her "dearest Lord," "deign" to intercede from "thy heav'ns, where thou dost safely reigne" (*P.I.* 12.46.7; 12.47.1, 7, 3). For Eclecta as the fallen yet penitent soul and Church, the "Reviving Cordiall" of her "dying sprite, / The best Elixar for souls drooping pain," is her "Sweet Spouse" Jesus Christ (*P.I.* 12.51.1–2, 7). Making a firm distinction between her weak flesh and *sōma Christou,* and acknowledging the necessity of the transformation, the regeneration, of the former, she pleads, "Harbour my fleshly bark safe in thy wounded side" (*P.I.* 12.52.7).[321]

In the text of *The Purple Island* there are several sites of particularly intense struggle between, on the one hand, Man as virtuous *sōma* (body), or microchristus, manifesting moral purpose in teleological and ethical structure, and redeemed from corruption in a stable, closed form, and, on the other hand, fallen Man as vicious *sarx* (flesh), mortal, corruptible, morbidly curious, concupiscent, and grotesquely open to sensation. Among these sites is the moral ambiguity of Parthenia, but more particularly the difficulty with which the narrator is posed, of removing from the sins of the flesh the desire to commit them. There are even instabilities in the dichotomies which the diegetic narrator diligently sets up at these sites, whereby it may be implied of the author, that he provided the broader theological context of the work in which the fallen yet redeemed and partially recovered nature of humanity interposes between, interrogates and to some degree collapses the distinctions between the opposed terms, in order to expose to view the paradoxes of this "strange, though native coast" which is the Isle of Man, and in which both hell and heaven often dwell (*P.I.* 1.34.6). These instabilities, which include the uncanny correspondence of the "viperous locks" of Hamartia, who personifies the body of sin, with the inherently appetitive yet ethical anatomical structure of the intestines, find their most resonant echo in the fictive approximation to the possession of what is hoped for in heaven.

At the institution of the *parousia, The Purple Island*'s representation of the relationship between Eclecta and her heavenly spouse as the restoration of intimacy and innocence lost in the Garden of Eden is a parabolic or typological reading of The Song of Solomon.[322] Because in the Kingdom of God, the vitiating effects of the Fall on Man—principally death and sin—are defeated, the "Bridegroom" and the "Bride," "Their hearts in love," "No change, no grief, no age can . . . befall" (*P.I.* 12.76.3, 4, 5); "all dayes are but one" (*P.I.* 12.76.7) because eternity, while encompassing the temporal dimension of which humanity had become aware at the Fall, yet has neither beginning nor end, and is that existence with reference to which the relation of succession has no application. There is no sense any longer that melancholy "dwells with Beauty—Beauty that must die";

> And Joy, whose hand is ever at his lips
> Bidding adieu; and aching Pleasure nigh,
> Turning to Poison while the bee-mouth sips:[323]

Appetite, desire, and pleasure need no longer be restrained either by definition in antithesis to their opposite or, moreover, morally, because there is no possibility of redeemed Man becoming ensnared in the sins of the flesh. The Bridegroom and the Bride are both blazoned in heaven (*P.I.* 12.78–86), "Upon his lightning brow Love proudly sitting" (*P.I.* 12.78.1), yet even though "his thoughts" are "sweetly feasted" on her breast, "Love" is mild and "gentle"; Cupid "His bow unbent disclaims all tyrannie" (*P.I.* 12.86.6; 12.79.3, 4). In the "heav'nly hall" (*P.I.* 12.76.6), are "sweet delights, which know nor end, nor measure," "eating" without regulation, "Pleasure full grown, yet ever freshly breeding," the soul of joy ever pregnant "yet still conceiving" (*P.I.* 12.75.1, 2, 4, 6). No amount of "wastefull spending can empair [the] treasure" (*P.I.* 12.75.3) in this heavenly Carnival. Bathing "his crown'd head in soft eternitie," Prince Intellect had savored only provisionally such "joyes," "delight," and "sweets" (*P.I.* 6.35.2, 3, 4; 6.34.7), but once "his long imprisonment" ends in "never-ending libertie," he is bid to "spend in joy his never spending dayes" (*P.I.* 12.68.1, 2, 7).

The diegetic narrator's efforts to morally dichotomize Cantos 6 to 12 appear to be simultaneously justified and undermined by the authority of the narrative's ending in eternal glory, which must necessarily represent both continuity and transformation of earthly existence, beauty, love, and pleasure. Oriented throughout towards this figurative, imaginative projection of the eternal reward in the future, as the restoration of a sensuously innocent mythic past before the Fall, the narrative of *The Purple Island* is characterized, in all its allegorical events and expositions or explanations of Man, by a belated anticipation of sufficient bodily and soulful pleasure free of guilt and shame.[324]

The question of the creative and innovative character of the tropology of *The Purple Island*—of the extent to which the unfolding of the figure of the Isle of Man retrenches and corroborates or expands what Fletcher's seventeenth-century readers might already have known or expected—has exerted a pressure on this chapter as it has on all the previous chapters of this study. As an interactive, creative metaphor that is abundantly meaningful, the Isle of Man's significant connotations extend beyond textbook anatomy, whether textual or pictorial, to a complex set of associations surrounding the demonstration of anatomy in the anatomy theater. By turning our attention to the context in which anatomy lectures or lessons and demonstrations took place, and by looking at the iconography and style in the design of temporary and permanent anatomy theaters, this chapter has related some of the interpretations and justifications of dissection to those in *The Purple Island,* exploring the question of whether in its locutionary, illocutionary, and perlocutionary aspects the *Purple Island* anatomy is not significantly different from the presentation of anatomy in the theater.

This chapter has described ways in which the anatomical speaking picture in *The Purple Island,* and the opened cadaver in the anatomy theater, function like an illustration in an anatomy textbook: by seeking to illustrate or demonstrate didactically the knowledge described both linguistically, in either the marginal text of Fletcher's poem or the anatomy textbook prescribed for the lesson, and verbally either by Thirsil to his audience of nymphs and shepherd-swains, or by the Regius Professor of anatomy or the anatomist himself in the theater; by either actually or rhetorically providing for autoptic experience untranslatable into words; by performing educative mnemonic and synthesizing functions; and by bringing together the intellectual pleasure of knowledge and aesthetic enjoyment. The function of the public demonstration was at least partly to illustrate anatomy thus, but its justification, which became its objective during the sixteenth century, was natural-philosophical and theological.

In this period, the first in the history of Christendom in which anatomical dissection had been revived and the glories of the instrument and vehicle of the immortal soul were being displayed and demonstrated publicly, the anatomy theater became a place for theatrically presenting a comprehensive survey or collective display of the whole of the anatomy of Man as the pinnacle of God's Creation. This optimistic interpretation of anatomy combined the Platonic and the literal microcosmic views of Man or the human body as the cosmos confined in a compendious form—a little world. These ideas found their way into the *Purple Island* mimesis in the reflection of that which contains in that which is contained, in the reflection of the geocosm, the macrocosm, the body politic, *ecclesia* and Cosmic Christ in the microcosm or microchristus. Anatomy in illustration, the theater, and *The Purple Island,* was grasped by an imaginative rationality, through conceptual metaphor, analogy, symbolism, emblematics, and the narrative of providential history. As Sawday argues, the poetic body—the visual image—which speaks in the eloquent language of gesture via metaphors, similes, and the devices of figurative language, is harmonized with the revealed knowledge of the human body pursued in the anatomy theater.

This harmony is not least because the differentiation is neither conceptually, philosophically, linguistically, nor rhetorically perfect. The natural knowledge which the public anatomy communicated, the sensory epistemology and technical accomplishment of anatomy, and the beliefs, optimism, motivation, and aims of anatomists, were related not only to the commonplace association of anatomy with disclosure of what is hidden, but to the distinct truths and processes of revelation. In the demonstration of anatomy in the theater and in illustrations in anatomy textbooks, the dissected body was presented neither from the perspective of empirical verifiability or philosophical or scientific objectivity, nor on the model of a machine or automaton, since none of these perspectives was substantially or consistently available in the early seventeenth century. Additionally, for Thirsil, and by implication *The Purple Island*'s author, the body is not a natural fact with an ontology independent of the ways in which it is painted, sculpted, blazoned, and described (though nor is it reducible to these) for its sensations are not only culturally mediated by concepts like enjoyment and suffering, sin, and the pleasure of the flesh; they are structured in some sense by such concepts.[325]

The anatomical demonstration and lesson in the theater was a narrative dramatic performance, with a text, a narrator, and a story with actors. The body was opened and divided to display the construction of Man, yet that body's progressive dissection and decay simultaneously rendered it increasingly unlike the *imago dei* it was supposed to represent. So the anatomical account, including that of Thirsil, sought progressively and cumulatively from dismembered parts to build up a cleverly wrought tower of health, with teleological meaning which gradually became apparent as the allegory or lesson unfolded. The decomposition of the body achieved symbolic status as anatomy's second aspect or interpretation, wherein "know thyself" meant recognize the transience and vanity of Man's life on earth, and "his" mortality as the wages of sin. This meaning was combined with a kind of analytical and judicious realism which discovered or lay open the body and revealed or disclosed it for what it was. But the division of the body was also thereby reconciled in the theater to the exposition and apprehension of its nature, truth, and life as a whole. Everywhere we have looked in this study, the understanding of anatomy in early modern culture was that division, disintegration, and death were a means to, and coherent with, a progressive, yet also retroactive, integration through the creative, analytic, and redemptive establishment of a relationship of part to whole. The theatrical demonstration reflected the theological perspective which *The Purple Island* implies of its author, from which the two contrastive aspects or interpretations of anatomy converge, as do the vitiating effects of the Fall and the partial redemption of Man. Cantos 7 to 12 of *The Purple Island* poetically allegorize the drama conducted in the theater in the shadow of the Fall and the light of salvation.

Fletcher designed his work to have a restorative and saving effect on its readers, and in this too, aside from its relationship to the traditions of encyclopedic poetry and literary anatomy that shared this aim, *The Purple Island* has a mimetic relationship to the ritual demonstration and performance of anatomy in the theater, the mystery of which could take on associations of revelation, a spiritual assurance of salvation, a religious observance, and the administration of a sacrament, in which the parallel between the necessity of the heart in vivification and the necessity of Christ for salvation would have been felt as palpable and seen as central to the drama of the scene. Dramatic tragedy, the *Purple Island* allegory, and the narrative, dramatic semiotics of public anatomical demonstration encompassed the analogy between the human body and the political state, whose judicial execution and, by extension, dissection of the criminal represented a sharp remedy applied to an infected part, purportedly to restore each of its members to medical, moral, and social health, perhaps as one body with Jesus Christ. The geocosmic analogue for the redemption, which the sacrament was believed to augur and instill, is the season of the earth's regeneration, which appropriately sees Thirsil "crown'd" as May "Lord" when his song is ended (*P.I.* 12.89.7).

The epistemology and hermeneutics of the narrative, emblematic, and poetic presentation of anatomy in *The Purple Island* are soteriological. Similarly, the epistemology of the ritual, dramatic conflict, narrative division and progression, and performative presentation of the public anatomy, even as late as the early seventeenth century, was broadly soteriological. In *The Purple Island* and in Harvey's *Prelectiones,* texts of graphic or verbal signs become identified with bodies and souls, and there emerge structural and semantic parallels in which the body, in what we would call its morphology, or the mortal "*Corpes*" or "pattern," and the corporeality of "crude *Conceit,*" are brought into actuality by the soul as the stable form of the body, by rhetoric and by ingenuity. The effect of this is that the anatomy in some sense brings the body in either the theater or the speaking picture back into existence, increasingly as the dissection progresses. As far as *The Purple Island* is concerned, it is arguable that the claim of the attribution of ingenuity is that by partaking of the efficacy of Christ's incarnation, poetic eloquence dispenses the balm of salvation to the body and soul of the reader, and enforms the mortal body-as-referent, resurrecting it as the poetic text, not as another body which is subject to death, but in its ideal and immortal form.

But my analysis has found, in chapter 5 and in the present chapter, that claims to bring the body to life through the faculties of the soul are invariably qualified and equivocal. In Harvey's Lumleian lectures, the pessimistic *memento mori* and tragedy of the cathartic ritual of anatomy as dissection is counterpoised by the romance of anatomy as anatomy or "building," an optimistic reminder of redemption and a fantasy in which the body is felt to be alive. Neither the aesthetic nor the religious concerns of these genres is incongruous with the literary genres by which *The Purple Island* is classified. Since in Fletcher's poetic allegory and anatomy, the intelligibility of the body is located solely in God's omniscience until after the general resurrection, the eschatological narrative delay creates an equivocation which affects the body's identity as a metaphysical category and as a moral and qualitative category. The final section of this chapter explored this equivocation between sinful corruptibility and bodily glorification.

The responses of historians of science and natural philosophy to *The Purple Island* have demonstrated a lack of interest particularly in the development of the work's reference in Cantos 7 to 12, its intentional direction toward a world in which allegorically personified vices and virtues engage in epic warfare for full possession of embodied human identity as either *sōma* or *sarx.* The conflicts therein are among the specially constructed set of associated implications of the subsidiary subject of the Isle of Man metaphor, but they correspond to the conflict in the theater between mortality and Creation or resurrection, and to the fact that public anatomical dissection was frequently performed during the Carnival, along with other potentially transgressive and profaning practices. The conflict in Cantos 7 to 12 of Fletcher's poetic allegory is not one in which the body is opposed to the incorporeal soul, but one in which Man himself is either a member incorporate in Christ's mystical body or a body of sin, the sinful flesh, in which the relationship with God is blighted by vice. The flesh, or Caro, is not a personification merely of the sensual vices, the pleasure of the flesh, but represents a broader notion which includes pride and curiosity, or avarice for knowledge, these three comprising the sin of Adam and Eve which reached into all of sin's dimensions and caused Man to fall from grace.

Dissection in the Renaissance was temporarily channeled into the sphere of the licit by its ritualization, and not merely because it was propitiatory in the same way as the Masses celebrated for the redemption of the souls of the dissected victims of execution. "The almost fashionable success of anatomy cannot be attributed solely to scientific curiosity. It is not hard to understand; it corresponds to an attraction to certain ill-defined things at the outer limits of life and death, sexuality and pain."[326] And through ceremony and performance, anatomy's fleshly and deathly fascination was channeled into contemplation. It may be implied from its soteriological illocutionary act that *The Purple Island*'s negotiation of Carnival or of this fascination is in part a moral one to convince his readership of the value and worth of a practice that borders on the shameful, sinful, illicit, sacrilegous, and infamous.

In Cantos 7 to 12 *The Purple Island* also develops the reflective direction toward a self which by implication comprises internally this mental, moral, and spiritual warfare. In previous chapters, we have discussed the concept of Man which emerges from the interaction of the allegory's threshold figure, a quasi-Paracelsian, English (or British) Protestant individual. But our appreciation of the "new, and inward man" which *The Purple Island* articulates, is described with vivid compression in *The Way to Blessednes* as typified by David and by Paul.

In this chapter we have explored how the widely acknowledged phenomenon in early modern English literature and culture which expresses mental experiences and phenomena using the language of the body, relates to Cantos 6 to 12 of *The Purple Island,* where the "literal" discursive frame or context is ethical, religious, psychological, and spiritual. In this context the metaphorical interaction of specially constructed implications of the subsidiary subject with human anatomy and physiology, reveals previously hidden, if antecedently existing, similarities, and produces significant connotations which combine to construct a corporeal spirituality rooted in the understanding of bodily existence and experience. This corporeal spirituality becomes a foundational root-metaphor, like the Isle of Man, an "axial metaphor" which "allows for contextual stability without absolute certainty," because its entailments and ramifications may be endlessly unfolded without a context.[327] In the context of Cantos 6 to 12, therefore, the figurality not only further imbricates the concept of Man, as physical, moral, mental, and spiritual, and the concept of the geopolitical realm, but augments with concrete metaphoricity the philosophical and theological context in which the figurality of anatomy itself is rationalized, justified, and finds coherence.

It becomes possible to postulate the soteriological illocutionary act of *The Purple Island* as the replication, through ingenuity and eloquence, of the operation of the Holy Spirit in *ecclesia,* understood in Fletcher's *Joy in Tribulation* metaphorically as the operation of the soul through the heart and brain in the human body. However, the implication of the ambivalent description of the tongue, in *The Purple Island,* that perception and expression are not merely cognitive and spiritual but unavoidably sensory and sensuous, extends not only to the difficulty of separating what is virtuous *sōma* (body) or microchristus from what is vicious *sarx* (flesh), but also to the corruptible eloquence (what Davies of Hereford calls "crude *Conceit*") of the work itself, and its soteriological illocutionary act.

The world created by the diegetic narration of Thirsil, enhanced by the epic warfare for full possession of embodied human identity as either *sōma* or *sarx,* is characterized by facets to the conflict. The first of these was identified as the antithesis—found in the anatomy theater—between pious knowledge of the teleological and moral physiology and anatomy of the wondrous art of Man's body, on the one hand, and the sensuality, sexual curiosity, and concupiscence of the rebellious flesh, on the other. But the difficulty of separating pious studiousness from morbid curiosity is related to the simultaneous manifestation of appetite and restraint in ethical, physiological structure. A self-reflexive consideration of the moral sphere of desire, intention, and action extends "naturally" from an examination of anatomical structure, function, and appetite, making it difficult to separate not only the rational soul from the body it contemplates, but contemplation itself from the appetite it struggles not to entertain. This is far removed from the postulation of ahistorical mechanical ontology and the search conceived in earnest in the later seventeenth century for a type of linguistic origin, a moment of Edenic purity of expression, bereft of metaphor, neutral and transparent, to speak the truth about that ontology.

Part of the explanation of *The Purple Island*'s perspective on the ethics of the anatomy of Man is that for Paul, human existence—even in the sphere of the *pneuma* (spirit)—is a bodily, somatic existence. Fletcher's Pauline moral theology, his relative lack of philosophical skepticism, and the theistic allegorical context in which the representation of moral intention and action occurs in *The Purple Island,* result in a kind of allegorical personification which presumes to assume a standard of what would constitute certainty with regard to the problem of human inwardness. Although this standard implies observation by a divinity, in a way this is only an extrapolation from the Pauline view, that *sōma* is not merely an outer form but the whole person, and is the field in which the relationship with God flourishes or flounders. The descriptions of Caro and Hamartia ensure that the work does not refuse to recognize the possibility of deceit or "dissembling"; but the "purged eye" of theistic allegory moves representation away from the theatrical and Carnivalesque preoccupation with the potential disparity between what persons inwardly are and outwardly seem. Fletcher systematically figures Man, not as escaping in the spiritual substance of the Christian Neoplatonic or Cartesian immortal, rational soul as a criterion for personal identity to which the subject alone has "privileged access,"[328] but rather as a nervous and defensive Protestant individual, who is vigilant not only against Roman Catholicism, but against her or his own hidden sinful, fleshly appetite and impulses, which are revealed to God and to a "purged eye."

The inevitable tension between the curiosity and sensuality of the flesh, on the one hand, and the body conceived on teleological principles by rationalist anatomy, on the other, has been found to relate to the equivocation in the *Prelectiones* between the morphology of the body and its actualization as physiologically understood structure. The vulnerability of even the rational soul to fleshly appetite suggests that the human body cannot be fully actualized or resurrected in the signs of the poetic text in an immortal, incorruptible form by rational ingenuity alone, and that God as the efficient cause of resurrection is necessary.

A second facet to the conflict for embodied identity, which affects the resurrection or reconstitution of the body-as-referent, was identified as between the body as redeemed from corruption in a stable, closed form and the mortal, corruptible *vanitas* of the flesh that is grotesquely open to sensation. This struggle or equivocation was analyzed as consisting of two constituent, interrelated struggles, one of which is moral and aesthetic, the other metaphysical, ontological, and theological.

The aesthetic, moral confrontation has two distinguishable though related manifestations. The first is that in which *The Purple Island* envisages the subject, Man, as constituted without clear boundaries, by erotic and psychic openness to the "other," even as the poem strategically seeks an authoritative discourse with which to defend and insulate the Isle of Man from Caro. *The Purple Island* seeks overtly to bring together the intellectual pleasure of knowledge and aesthetic enjoyment through representations characterized by this manifestation of the conflict, principal of which is Parthenia, an allegorical embodiment of the ideological and ideal configuration of the human body, the kingdom and *ecclesia,* constructed around the overriding principle that they must be defended from the desires of the flesh. But her representation suggests that much as Man may strive to shore up a commitment to virtue, and to aspire to the condition of being incorporate in *sōma Christou,* there may be a necessary relation between being in the flesh and dependence on the flesh, and therefore perhaps no possibility of removing from a consideration of the sins of the flesh a desire to commit them, or from the sensuous celebration of Creation the shame and guilt associated theologically with the (albeit providential) "fire of generation" that spurs Man to know and possess what ingeniously consumes "him" with desire and curiosity. Through ceremony, ritual, sacramentalism, and performance in the theater, anatomy's fleshly fascination was channeled into contemplation, and this process finds allegorical expression in the ministration of the sacrament by Continence in *The Purple Island,* which becomes a retroactive pardoning of the guilt of the diegetic narrator, who has erotically blazoned and then punitively violated Parthenia's body for a sin of which she is at most guilty only in the sense of having been tempted.

The second manifestation of the aesthetic conflict between the openness of the grotesque body and the stable form of the teleological, classical body was found to be cognitive, and to concern figurality. Consistency has been found between the metaphorical conception of anatomy as vegetation, the functions of nutrition, growth, and reproduction, and the Carnivalesque or grotesque conception of the body as the flesh. Equally consistency has been found between the metaphorical conception of anatomy as "building," the skeleton and the musculature, and the teleological conception of the body. Between these two systems or sets of consistent concepts arises tension and coherence because each conceptual metaphor tries to grasp different aspects of the concept of human anatomy, because both metaphorical concepts arise from and are justified by Aristotelian analogy and by different faculties of the soul, and because both find coherence in concepts of Creation, mortality, and salvation. Tension is expressed in *The Purple Island*'s allegory in which the grotesque, feminized flesh lays siege to Intellect in his tower. But coherence is manifest in the way in which figurality gives rise to semantic generation, yet the semantic content is at length limited by the figuration of the body as an island, providentially separated from the rest of the world by borders, like the classical body, and progressively revealing the soul and providing the key to interpretation.

The second constituent struggle of the second facet of the conflict for embodied identity in Cantos 7 to 12 of *The Purple Island* and in the anatomy theater was found to be ontological or metaphysical and theological. The Carnivalesque, grotesque body was tied to the processes of birth and death, devouring and being devoured, growth and multiplication—the processes of the flesh in its Scriptural rendition as "the Corruption of Nature." Even the apparently antithetical properties of mortality and immortality seem in some early modern intellectual strands to have been separated by a graded continuum of created substances and degrees of rarefaction and participation in time and eternity. The Pauline strand affects the representation of Caro, not so much as the personification of the transitory flesh, but of the moral edict: "if ye live after the flesh, ye shall die" eternally, body and soul. Similarly, Hamartia, the "true" image of the "body of sin," is an uncanny and grotesque reflection of the manifestation of moral restraint in ethical anatomical structure, exposed, open and revealing its proper substance, as it would have in the anatomy theater. As such her function is soteriological; she is a salutary reminder of the need for moral vigilance along the "blinde and dirty way" that Fletcher proposes that his implied reader undertakes.

The purported congruity of mortality with Creation and providence, in the early modern period, is evident in the way in which the extent of the difficulty that the grotesque perception of Man posed was also the extent to which reminders of mortality were readily available spurs to the assurance of things hoped for. But the dichotomy within this congruity persists only until the Last Judgment, after which appetite, desire and pleasure need no longer be restrained because there is no possibility of becoming ensnared in the sins of the flesh and no metaphysical precondition that life or Creation, and each and every quality or value, is defined in terms of its opposite.

This study has been a work of literary criticism, and as such holds that an interpretation may be said to be more probable than another when it accounts for the greatest number of facts provided by the text of the whole work, including potential connotations. Clearly the concentration of this study on the anatomical cantos makes it a kind of literary criticism that treats meaning as culturally dispersed, rather than formally contained in a work, and as such it aspires to be a work of interdisciplinary historical analysis. Notwithstanding this qualification, the digression into the underlying moral and religious conflict of the presentation of anatomy in the theater, and the traditional social context of the public anatomy in the Carnival, has permitted a reading that accounts for more of the facts provided by the text than would otherwise have been possible, and so deepens and enriches our understanding of the meaning of Thirsil's anatomical exposition in Cantos 2 to 5, and its intellectual, cognitive, and sensuous grasp of the reality of the body.

Conclusion

IN THIS STUDY OF *THE PURPLE ISLAND* I HAVE ENDEAVORED TO PROVIDE A CRITICAL ACcount of the work's figurative design, narrative structure, formal disposition, generic character, intertextuality, literary context, reception, and critical history. In so doing, and in determining the conceptual character of the work's anatomical content and method, as well as the provenance of these in textbooks and the anatomy theater in the sixteenth and early seventeenth centuries, I have addressed the interpretative difficulty of general questions of the ontological, epistemological, linguistic, conceptual, and discursive differentiations of anatomy from poetry; of the imaginative and rational congruity and consistency between them; and of conceptualizations of the modes of exchange.

In this interdisciplinary historical and critical analysis and interpretation of *The Purple Island,* with its comparative, intertextual, and contextual study of specific bibliographical, biographical, conceptual, formal, linguistic, and narrative connections between Fletcher's work and other works of literature, natural philosophy, and theology, it is hoped that a contribution is made to the sum of existing scholarship and knowledge relating to early modern literary studies and the history and philosophy of human anatomy, particularly with regard to our understanding of the intellectual, cognitive, sensory, and sensual apprehension of the anatomy of the human body among English authors of the early seventeenth century. The following paragraphs will reiterate the conclusions of this study that I think are distinctive and most concrete, and will venture to suggest a few further inferences that may be drawn from these conclusions.

At the outset of the "scientific revolution," the conceptualization of the body in the anatomy of the schools, colleges, and medical profession, in textbooks and in the anatomy theater, was characterized jointly by a sense of the body as having been created or constructed as teleological like a building for its inhabitation, accompaniment, and operative instrumental use by the rational soul, and of its being natural like vegetation, and therefore subject to corruption, in disease and in death. Compositional methods or sequences of anatomical description were thought by anatomists to emulate the method of construction followed by Nature or by God in Creation, and the body was opened and divided in the anatomy theater in order to display this construction whereby the soul could be known. Although the body's progressive dissection and decay rendered it simultaneously less like the abode of the soul, the anatomist or *lector* sought cumulatively to build the *imago dei* from dismembered parts in the anatomical account.

In the absence of an epistemologically transparent and neutral nomenclature to describe the putative ontology of the body, an imaginative rationality was exercised in the natural-philosophical contexts of project formulation, hypothesis, interpretation, justification, and public demonstration. Human anatomy was not only *compared* to "building" and vegetation, but was systematically structured in terms of metaphorical concepts derived thereof, which were related in terms of entailment relationships and were coherent with the analogies of the body as a microcosm, so that conceptual metaphor and analogy characterized anatomical knowledge and underlay and informed surface metaphorical expressions, iconography, symbolism, and emblematics in text, illustration, and demonstration.

The role of conceptual metaphor in the rationalization of structure and function in terms of the faculties of the soul, Nature, or *physis,* in the rationalist tradition in medicine and natural philosophy from Aristotle through Galen to Harvey, has been overlooked not only by historians of anatomy, natural philosophy, and medicine, but by pioneers of the cognitive or conceptual theory of metaphor, such as Lakoff and Johnson, who categorize Plato and Aristotle within the objectivist tradition of Western philosophy as suspicious of the semantics, semiotics, and cognitive content of metaphor.

Relationships between the figurative modes of reference, knowledge, and discourse specific to anatomy and to poetry were inherited as deeply structural in the early modern period, though not as monolithic, predetermining frameworks. Rather they were continuously negotiated and reconceptualized in relation to the material determinacy of history, as, for example, in the early to mid-sixteenth century, when the anatomical projects of Aristotle, Galen, and Erasistratus were revived with consequences especially for the context of anatomical discovery, and again in the late sixteenth and early seventeenth century, when Christian Neoplatonism and alchemical physiology affected particularly the context of justification.

During the sixteenth and early seventeenth centuries, imaginative rationality combined, and rarely clashed, with a sensory epistemology in anatomy. It was only later, in the second half of the seventeenth century, after the establishment of the theory of systemic circulation and the demise of the Galenic model of the distribution of blood, humors, and *spiritus naturalis,* that a consensus began to form that objectivist suspicions of metaphor had rational validity. Analogy and metaphor became regarded as conservative modes of apprehension in natural philosophy, and the positivist and objectivist search was conceived in earnest for what until then had been only a dream—a neutral and epistemologically transparent visual technique and verbal nomenclature of Edenic purity, far removed from the figurative design of *The Purple Island.* That this was thought by many to have been discovered in the machine model of human anatomy, which rationalized the analysis of the operations and functions of the body in terms of the objectifying application of ideas derived from the construction of automata or moving machines, testifies to the perennial utility of figurative language in human biology, and, in this case, to the ideology of claims to have eradicated it from the (mediated) relationship of knowledge to the (putative) ontology of the body.

The author of a literary anatomy was not without constraint in negotiating the inheritance of relationships between figurative modes of reference, knowledge, and discourse specific to anatomy and to poetry, but was ineluctably caught up in the material determinacy of history. The conceptual metaphoricity of the rationalist tradition of anatomy is demonstrably congruous and consistent, rationally, conceptually, and linguistically, with the poetic figurality of *The Purple Island,* the implied author of which was not greatly influenced by skepticism and, more pertinently, was not an objectivist. Consequently the anatomy in the poetic stanzas of *The Purple Island* is not independent of the figurality with which it is presented, and provides numerous examples of the way in which the descriptive vocabulary and the figurative conceptualization of anatomical parts had to change in response to the material determinacy of anatomical discoveries.

Because the figuration is in each case consistent with the anatomy's essential conceptual character, the anatomy in *The Purple Island* is often identifiable, though not always in terms of provenance. It is at various places Galenic in conceptual character, such as in the account of the embryological origin of the veins in the liver; it is Vesalian in method as in much of its character; and, in the account of the anatomy of the heart, veins, and arteries, the movement of blood and the concoction of vital spirit, it is consistent with anatomy found in writings by Colombo, Valverde, Banister, Cesalpino, Bauhin, and Harvey, among whom, however, several were advocates of pulmonary transit, a theory which Fletcher does not seem entirely to have held.

Despite the contradictions between these authors and between sources which this study has only uncertainly identified, and often even within a single author's account, the *Purple Island* cardiovascular anatomy constitutes an uncommonly self-consistent and almost non-contradictory physiological model. Perhaps most interestingly, though, Fletcher's account seems to imply that unlike all these anatomical authors with the exception of Harvey, he drew a hypothesis regarding the physiological consequences of the flow of blood through the semilunar cardiac valve into the aorta and its branches, for which Harvey's anatomical lectures to the London College of Physicians are conceivably the most plausible source, as they may be for other conceptually significant features of cardiovascular physiology; though what evidence has been found to support the conjecture either that Fletcher attended the lectures or that their content was communicated to him in some other way, though suggestive, is weak.

Even though, in response to the material determinacy of anatomical discoveries, the descriptive vocabulary and the figurative conceptualization of anatomical parts had to change in accordance with their newly understood properties and functions, what particularly emerges from this study for Fletcher's poem is not the difficulties for successfully negotiating an allegorical artifice of correspondence in the early seventeenth century, but the facility with which even comparatively recent anatomical discoveries could be allegorized without being detrimental to the anatomy and without being superficial. It is the systematicity of the allegory in Fletcher's poetic anatomy that is responsible for the identifiableness of the anatomy. Theoretically, this bears no direct relation either to how allegorical the poetry is, or to the conservativeness or innovativeness of the anatomical content, because the allegory can be seen inventively to have adjusted to the anatomy, or to have been determined by it.

The historiography that tends retrospectively to see the anatomy which emphasized function, like that of Colombo and Harvey, as divergent from that which emphasized structure, like that of Vesalius, fails to recognize that the divergence of physiology did not occur until the absolute discontinuity emerged between the Galenic venous system and the competence of the cardiac valves in pulmonary and systemic circulation, after 1628, and therefore after the composition of *The Purple Island* in or around 1608, but for revisions and additions made probably until about 1621 and possibly thereafter. Given this transitional context, it is not improbable that the *Purple Island* description of the competent function of the valves may be indebted to an ideational system in a source that is Galenist in its description of the flow of blood in the vena cava and the other veins, probably one that had not long been in print.

The rhetorical concept that governs *The Purple Island*'s relationship to its subject matter is mimesis, a mode of *descriptio*, though the mode of representational exchange which occurs between the poem and its anatomical sources might also be conceived as acquisition through metaphor and synecdoche. By this means the work isolates the text, method, and pedagogical and didactic function from the illustrated textbook, and these and the lecture or lesson, together with its division and duration, from the anatomy theater. The mode of exchange makes these extracted components stand for the whole of which they were parts. Although the evidence for Fletcher's attendance at anatomical demonstrations either at Cambridge or London is not strong, there is evidence in *The Purple Island* of the influence of the techniques and conventions of the public demonstration or presentation of anatomy through dissection. Additionally, in the levels of narration and in the formal disposition of the text, there is evidence of mimesis of those lessons and demonstrations conducted in the Vesalian fashion outlined in the Preface to his *Fabrica.*

Through a combination of incisive accuracy, or ingenuity, and figurality, or eloquence, the nature, truth, or heart of the body's meaning is purportedly exposed or demonstrated to the implied reader of *The Purple Island,* whose inward conceit or imagination is pierced and possessed, or so it is intended. To this end, the figurality of *The Purple Island* was designed to be distinct from that in anatomy textbooks, but in a way that is in part modeled formally and imaginatively on the human body as encountered either in the aesthetically pleasurable and technically accomplished illustrated pages of anatomy textbooks or in the demonstrations which accompanied lessons in the anatomy theater, both of which shared not only pedagogical, didactic, and illustrative functions with the presentation of anatomy in *The Purple Island,* but also the rhetorical confirmatory function of *demonstratio* with the use of certain indicative phrase-gestures in Fletcher's work. These are attempts by the implied author of *The Purple Island* to fulfill the illustrative function of dissection and to compensate for the limitations of linguistic forms of representation by drawing upon the Renaissance understanding of the relationship between painting and poetry.

The conceptual metaphors of anatomy and the analogies of the microcosm coalesce in the allegorical threshold figure and the figurative design of *The Purple Island,* providing what is not only a coherent understanding of anatomy, but also a consistent, figuratively complex image, emblem, or speaking picture of anatomy, which explores the shared entailments and significant metaphorical connotations of islands, bodies, and buildings to an extent that is unprecedented in anatomy textbooks in its systematicity, coherence, and persistence. Although this may occasionally distort some of the descriptive criteria of the anatomical *accessus* or rote of observables, by providing vantage points on particular organs or members that best display their reason for being what they are and where they are in the microcosmic world, its purpose is vividly to impress upon the reader an appreciation of the importance of function in teleological or rationalist anatomy.

The speaking picture of anatomy is created from a metaphorical interaction between a set of associated implications of the geopolitical island realm (the focal term or defining concept of Man) and the conceptual content, significant connotations or associations, and nomenclatural etymology of items of both conventional and recently discovered anatomical knowledge. This interaction in the foreground of the poetic work not only highlights antecedent conventional analogies and similarities but also discovers those which were previously unseen; it even invents or creates a potentially infinite number of novel similarities as a contextual effect of the clash of recently discovered anatomy with the connotations belonging to the system of associated commonplaces of the focal word. Consequently, there is a dilation of meaning whereby signs give birth to others, and an implied invitation to the reader to complete this semantic process of generation that is nevertheless constrained by the context of the whole work, such that interpretation holds in tension a retrieval of what is to some extent immanent in the work as connotation, implication, or merely suggestion, and the individual and historical variation implied by the perlocutionary act—the reception of the work.

Without *The Purple Island*'s discovery or invention of previously unseen and novel similarities, the remote connotations of recent anatomy could not have been represented in the work. The anatomy would have remained either intellectually conservative or shadowed and in either case enclosed within the circles of analogical correspondence, wherein there are only a given number of preexisting, isolated ways in which comparison can be made. Instead even the cardiovascular anatomy of Harvey's Lumleian lectures (prior to the development of his theory of systemic circulation) remains unmasked as the disparities in the creative metaphors of the allegory open significance to bring out the resultant novel meanings. The contemporaneousness of the anatomy in the work is directly dependent upon the adaptable creativeness of the figurative design.

In the ways summarized immediately above, *The Purple Island* has been examined in this study from the explanatory perspective of the allegorical threshold figure—the Isle of Man—as the key for understanding the whole work. But it has also been approached from the interpretative perspective of the whole work, and the world it intends to create, which give the key to the interpretation of metaphor and provide in allegorical form the teleological reason for anatomical figurality. In the whole work, the analogy between Man and an island becomes a metaphor in which the correspondence is extended to provide a justifying context for recent anatomical discoveries. Consequently, we have been able to see how the work's figurality is also distinct in that it coherently, systematically, and persistently deepens and extends the unifying single image or idea of the Isle of Man, more or less continuously superimposing the two metaphorical subjects one upon the other, and eloquently exploring and developing the threshold figure's significant connotations or entailment metaphors in allegory.

Throughout the anatomical cantos, personification is regulated in relation to anatomical connections and functions, and the anatomical *accessus* is applied consistently in both prose marginalia and poetic stanzas, where the inherent figurality of anatomical etymology, the entailments of the microcosm-geocosm analogy, and the significant connotations of the allegorical threshold figure are unfolded. They are unfolded in a way that is neither ignorant of contemporaneous anatomy nor detrimental to such anatomy, as has often been thought. Rather the allegory shades inherent nomenclatural figurality into the built, natural, and social features and dimensions of the body as a geopolitical island realm, and specially constructs and shapes the associated implications of this realm according to the requirement that they accommodate the descriptive criteria of the anatomical *accessus*. It is the strain of this extension of the poetic of *concettismo* into an ideological artifice, rather than anticipation of the mid-seventeenth-century program of linguistic reform in natural philosophy, that gives rise to faultlines in the coherence of the allegory.

Like some comparable poetic anatomies by Davies of Hereford and Du Bartas, *The Purple Island*'s mimesis of human anatomy rhetorically substantiates the accomplishment of the work's intentional direction toward a microcosmic and proto-national world by making a number of interrelated claims, all of which are partly mimetic in character. The poetic anatomy makes a rhetorical claim that the narrator dissects the human body of which the poem speaks; an epistemological claim that the body-in-itself is dissected thereby; a representational claim that the eloquent words of poetic anatomy, acquiring materiality, corporeality, and mortality, invoke, reproduce, or reconstitute the body-in-itself as the signs of the speaking picture, thereby overcoming the incommensurability of the signified with the body-as-referent; and a soteriological claim, or perhaps a hope, which is also hermeneutic, that poetic ingenuity, partaking of the redemptive efficacy of the incarnation and the immortality of the rational soul to which it belongs, will stem the tide of corruption by mimetically perfecting and resurrecting the body (that of the reader and generally the body-as-referent) in its ideal form.

The relationship between poetic ingenuity and the epistemology of anatomy in the early seventeenth century is not easy to define, but closer than might be assumed. Harvey's analogical mode of thought and of epistemology reconstitutes the body in the signs of its account, by locating the agency for the division simultaneously in the empirical or natural corporeal body and in a physiology which is understood as teleological, and which becomes apparent only in its account. In *The Purple Island* and in Harvey's *Prelectiones,* texts of graphic or verbal signs become identified with bodies and souls, and there emerge structural parallels in which the body, in its morphology or mortality, and corporeal eloquence are brought into actuality by the soul as the stable form of the body, by either poetic or "scientific" ingenuity.

The works examined in this study are for the most part either literary or anatomical (the latter categorized as natural-philosophical). Only a few works discussing human anatomy are theological, but the study has demonstrated that anatomy was partly understood in the conceptual framework and contexts of project formulation and justification provided by natural and moral theology, eschatology, soteriology, and the doctrines of the incarnation and the Eucharist. Despite inconsistent religious, theological (but probably not ecclesiastical), and anthropological anxieties about anatomical dissection, it was defended as a piety by virtually all but the most Aristotelian of anatomists, who had little to say in the way of natural-theological argument about God. Moreover, the natural knowledge of anatomy, its sensory epistemology and the technical accomplishment of illustration were related not only to the commonplace association of anatomy with disclosure of what is hidden, but were seen as an analogy of revelation, though again not by the Aristotelians. Anatomy was seen particularly by Neoplatonists and those influenced by the idea of the spiritual apprehension of "cosmic anatomy" in Paracelsian medicine (if not by its objection to dissection) as revelatory in a religious sense and involving not only reason but inspiration.

Besides the theological complexion of these epistemologies, theology affected the demonstration of anatomy in the theater. The struggle to display the anatomy of Man, via the decaying and divided corpse, as the pinnacle of the ingenious art of the Creator provided theological justification for the use of the hands in the holy art of medicine to dissect, and became a narrative dramatic performance with a structure of symbolic and actual conflict. Anatomy attracted a positive, optimistic interpretation which combined Platonic and literal microcosmic views of Man under the moral injunction—"know thyself." But the decomposition also achieved symbolic status as anatomy's negative, pessimistic interpretation, wherein "know thyself" usually meant recognize the mortality of Man as the wages of sin. But just as death and disintegration were thought to lead to a redemptive reintegration into the body of Christ so division of the body dramatically led to analytic comprehension of its nature, truth, and purpose as a whole. The epistemology of this ritual performance was broadly soteriological, though its ending might have hung in the balance between the pessimistic *memento mori* of tragedy and the optimistic romance or eschatological epic of redemption.

The anatomical demonstration was performed in the shadow of the Fall and the light of salvation, and its mystery could take on associations of revelation, a meditation on mortality, a spiritual assurance of salvation, a religious observance, and the administration of a sacrament. This had a resounding impact on the intellectually curious, sensuous, but rationalist and religious writer Phineas Fletcher, whose literary father, Spenser, had already provided an anatomical description in the House of Alma in *The Faerie Queene.* Although the anatomy of Cantos 2 to 5 is transformed through Christology, eschatological narrative delay, and metaphorical interaction with the dynastic realm or proto-nation, neither conceptually nor aesthetically nor even generically does the Christian allegorical narrative constitute an absolute rupture with the expository anatomy. The broad themes of creation, mortality, and salvation maintain the coherence and systematicity that obtains between the metaphors in anatomy, the analogies of the microcosm, and the significant connotations of the Isle of Man figure, and extend even to those connotations that are innovative or previously unseen. The theological perspective which *The Purple Island* implies of its author converges the two contrastive interpretations of anatomy in the narrative that encompasses the vitiating effects of the Fall and the redemption of Man through the incarnation, death, and resurrection of Christ.

In *The Purple Island,* as in the anatomy theater, the created yet mortal microcosm is also microchristus, indwelt by the incarnate Christ, and this affects the work's soteriological illocutionary act, which is to preserve the spiritual and physical health of the Protestant nation of Britain and its inhabitants, restoring or perfecting their human nature, and contributing to their salvation. The poetry's suprarealist analogical and metaphorical language, punning on "Sunne" and "Sonne," evokes, or rather *invokes,* the incarnation, and attempts ingeniously to participate in the communication of that event's ostensibly cosmic redemptive efficacy; though in *The Purple Island* this is actually contentiously narrowly national. When the diegetic narrator traces the love of the Redeemer and the lineaments of providence and salvation in the anatomical properties and physiological function of the human heart, which participate in all the allegory's assemblage of superimposed analogous systems, he does so simultaneously in the property and function of the sovereign of Britain, rather than the whole world. It is as an anatomical foreshadowing of the redemption and resurrection of the body and the English Church after death at the Last Judgment that Fletcher's poetic anatomy is modeled formally and imaginatively on the human body, and rhetorically claims to reconstitute the body as a body not subject to corruption. The anatomical speaking picture is designed thereby to provide a poetically constituted form of ocular evidence for believing in the divine promise of the resurrection and the implied reader's salvation.

This theological context of justification had certain implications for the apprehension of the anatomy of the human body with the senses, which were understood as fallen and yet indispensable not only to bodily, somatic human existence, but also to the sensory epistemology of anatomy, which is consequently mediated and structured by concepts like enjoyment and suffering, life and mortality, virtue and the pleasure of the flesh, sexuality and pain. The eschatological narrative of *The Purple Island* creates an equivocation which affects the redemption of the body metaphysically and morally; the moral equivocation being that between sinful corruptibility and bodily glorification. This takes allegorical form in the epic warfare between personified vices and virtues for full possession of embodied human identity and somatic existence as either *sōma* or *sarx* (personified as Caro). This moral equivocation and allegorical conflict correspond to the dramatic conflict in the anatomy theater between mortality and Creation or resurrection, which was performed routinely, though not without exception, during the Carnival, along with other potentially transgressive and profaning practices. This equivocation or conflict was broken down in this study into the moral antithesis between pious knowledge of the teleological and moral physiology and anatomy of the wondrous art of Man's body and the sensuality, sexual curiosity, and concupiscence of the rebellious flesh; and the moral, aesthetic, metaphysical, and theological struggle between the body, as redeemed from corruption in a stable, closed form like that of a building, and the mortal, corruptible *vanitas* of Nature and the flesh that is grotesquely open to sensation and tied to the vegetative functions of growth, nutrition, and reproduction. *The Purple Island* figures Man, not as essentially the spiritual substance of the Christian Neoplatonic or Cartesian immortal, rational soul, but rather as a nervous, even paranoid, Protestant individual, who in "his" endeavors to know "himself" as created and mortal, must remain vigilant not only against Roman Catholicism, but against hidden sinful appetite and impulses which arise as though of necessity by being in the flesh.

In the anatomy theater, and in *The Purple Island,* the opened, dissected cadaver that reveals the manifestation of moral restraint in ethical anatomical structure is also an emblem of the body of sin, the wages of which are death. Through ritualization and performance, and through the consistency and coherence of anatomical figurality, anatomy's fleshly and morbid fascination is channeled into the licit as contemplation of creation and death with the prospect of the assurance of salvation, but only by the grace of God, who having "first mad'st" the body, "onely know'st it right, / And hid'st that art from all the world beside" (*P.I.* 3.4.5, 1–2).

The critical reputation of Phineas Fletcher, which rests upon *The Purple Island,* has been damaged by those who from the eighteenth to the twentieth century have been unwilling to accommodate the aesthetic criteria of their critical evaluation to such sustained treatment of human anatomy. Frequently criticism has been particularly sensitive to the use of the supposedly denotatively limited vocabulary of anatomy and its "unusual ugliness," tastelessness, or incongruity as subject matter for poetry. The reason has often been rationally flawed, unavoidably historically situated, and even anachronistic. Nevertheless it has been consequential, leading critics like Nicolson to dismiss the poetry as negligible and evaluate the work on the grounds of its interest for intellectual history alone.

In this study, I have tried to make an argument that the figurative design of *The Purple Island,* its ingenuity and the anatomy which it represents are entwined and integrated allegorically in the work, but also cognitively, discursively, and linguistically in the culture, and so judgments of aesthetic and of historical value ought not to be separated as they have been by literary critics and by positivist historians of anatomy and physiology. The ontological and epistemological justification of this argument has found support not only in some recent cultural-materialist history of the human subject but also in conceptual theory of metaphor and cognition. This study also concludes that the emergent political, anthropological, and literary implications of the Isle of Man both confirm *and* challenge the firmest interdisciplinary convictions of the Renaissance about the intelligibility of the world and Man, and the poem's participation in the vogue of *ingegno* and *concettismo* is indicative of its innovativeness and its conservatism. Both these conclusions are to some extent supported by the extant written responses of *The Purple Island*'s earliest readers, whose attribution of ingenuity recognizes the poem's intellectually intriguing and fanciful, yet erudite and "truthful," (re)invention or reconstitution of anatomical subject matter, and the talented, imaginative, witty contrivance of the poem's innovative conceit and figurative design, which is skillfully, cleverly constructed and sustained throughout the "conservative" allegorical artifice.

Contrary to what is suggested by critical history, *The Purple Island* is neither strange nor incongruous in any historically or philosophically defensible sense, but rather had an intellectually and aesthetically sympathetic readership, as is also shown by comparison with literary works of the period which resemble *The Purple Island* in figurative design, formal disposition, or choice of subject matter. The *Purple Island* allegory aggregates, deepens, and extends conventional analogical knowledge through the invention of a creative metaphor for Man, but in the very effort of doing so, the work gives rise to significant connotations that relate to concepts that were historically incipient, ideologically emergent, and anatomically recently discovered by Colombo and Harvey. It is curious, yet typical of Protestant innovation, that in Fletcher's other religious writings, *The Way to Blessednes* and *Joy in Tribulation,* Man is the "new, and inward man," yet like the Hebrew David or like the apostle Paul.

These innovative connotations, however, were to help bring about a collapse of such correspondences as *The Purple Island* endeavors to preserve as belonging properly to the ontological nature of the world and of Man. But the demands of poetry and of anatomy in the early seventeenth century ought not to be seen as contradictory. *The Purple Island*'s allegorical figuration of anatomy ought not to be disparaged as either ignorant of or as appropriating and concealing, and therefore detrimental to, the essentially mechanistic character of aspects of cardiovascular anatomy that were later to contribute to Harvey's hypothesis of systemic circulation. There is a certain historical anachronism in this charge and a retrospective realignment both of Harvey's program of anatomical research, which was derived ultimately from Aristotle, and of his conception of the body, which unlike that of Descartes, was not regulated merely by matter and motion, but was constructed like a building as an abode and instrument of the soul. On the evidence of Harvey's *Prelectiones,* and even his *De motu cordis,* the differentiation between anatomy and poetry in the early seventeenth century was ontologically, epistemologically, conceptually, linguistically, and discursively less than perfect. Teleological, qualitative, and vitalistic conceptions of anatomy facilitated precisely the analogies and comparisons between similarly purposive constructed and natural objects that we find highly developed in *The Purple Island.*

As he wrote, in about 1608 to 1621, Fletcher was challenged by the discoveries of anatomical *inventio* to exercise his poetic ingenuity in the task of discovering a conceit the allegorical extension of which would allow the figurative reconceptualization of anatomical parts in accordance both with their newly understood properties and functions and with a proto-nationalist and religious context of justification that, although not inconsistent with the presentation, interpretation, and justification of anatomy in textbook illustration and the anatomy theater, would express his conservative political and intellectual instinct to innovate in order to preserve. Although the discovery of systemic circulation did not represent a philosophical watershed for Harvey, there emerged into dominance thereafter the mechanistic and quantitative model of physiology. The epistemological, ontological, and linguistic shift to which this was central retrospectively seems to render *The Purple Island* more acutely vulnerable to imminent historical change than Fletcher could possibly have imagined.

The critical neglect or maligning of a significant and sophisticated work of the English literary Renaissance is here contested as unjust. The anatomical cantos of *The Purple Island* exceed its literary sources and all anatomical passages of literary works of the period to which it is compared in this study, in innovation, ingenuity and invention, in figurative and poetic organization, coherence, complexity, extensiveness, and systematicity, and in anatomical comprehensiveness, systematicity, precision, mimesis, and modernity. Although the poetic anatomies of Davies of Hereford compare well in some respects, *The Purple Island* stands apart from comparable work of the period, or possibly of any period, in simultaneously being highly literary and displaying in the conceptual content and method of its anatomy deep and current erudition. The form which Fletcher adopted and developed for Cantos 2 to 5 in particular of his work, in order to accentuate the differences between poetry and anatomy while exploring new ways of presenting the correspondences, is especially suited to this congruity. Although they are not differentiated as former critics have suggested, while the work's poetic stanzas tend to grant the reader a shifting perspective on the anatomized microcosmic world towards which its creative intention is directed, the marginalia tend to stabilize that world without reducing it of its figurative etymology and meaning. In addition, *The Purple Island* may be almost alone among works of the period in the extent to which it develops a highly sophisticated soteriological epistemology and hermeneutics of the "scientific" and poetic concepts of ingenuity and eloquence, which Fletcher's religious prose shows us is modeled on the operation of the Holy Spirit in *ecclesia,* which is in turn modeled on the operation of the soul through the heart and brain in the human body.

On examining comparable fictional and expository anatomies, including encyclopedic poetry, for this study, I found that when they organize or conceptualize subjects other than the morphology and function of the body using a method often referred to as anatomical, they generally adhere to at least some of a number of interrelated concepts—erudition and self-knowledge, ingenuity and invention, opening or disclosure, sensory epistemology or experience, division or dissection, and analysis or justification—and these I refer to inclusively as the conventional, literary associations and commonplaces of anatomy. In the context of the widely acknowledged phenonmenon in early modern English literature which expresses mental experiences using the language of the body, perhaps the most intriguing yet undeveloped topic of this study is the phenomenon in early modern vernacular literature of using not only the language of anatomy (and not only Galenic humoral theory and Paracelsian medicine), but also the conceptual character, methods, descriptive criteria, and commonplace associations of nonmechanistic anatomy and dissection, in the soul's description, that is, where the literal discursive frame or context and the experience or phenomenon expressed are mental, psychological, spiritual, or ethical, rather than physical. In Fletcher's seldom-discussed religious prose, the metaphors of corporeal spirituality augment with concrete metaphoricity the philosophical and theological context which in turn provides coherence, rationalization and justification for the anatomy in Cantos 2 to 5 of *The Purple Island.* But this observation is insufficiently developed here to warrant the claim that has been made for the conceptual metaphors of anatomy, that they belong to or give rise to a host of metaphorical concepts that are related in terms of entailment relationships.

Appendix A
The First Edition of *The Purple Island*

THE EDITION OF 1633 IS ENTITLED *THE PURPLE ISLAND; OR, THE ISLE OF MAN: TOGETHER With Piscatorie Eclogs and Other Poeticall Miscellanies,* by "P.F." It was printed by the Printers to the University of Cambridge, who issued this edition twice: the issues are cataloged as 11082 and 11082.5 in *STC.*[1] *STC* 11082.5 has the same imprint as *STC* 11082, but is on "large paper," and contains three additional engravings: (a) a coat of arms, on the first title page *verso;* (b) an emblem with English verse, "on either the [fir]st blank or the blank conjugate w[ith] the subt[itle]-p[age]" (*STC,* I, 491.);[2] (c) an emblem with Latin verse, on a separate leaf. Variant copies of both issues exist. The Harvard University Library copy of *STC* 11082.5, which was photographed for microfilm for the series *Early English Books 1475–1640,*[3] possesses two identical subtitle pages; it also lacks emblem (c),[4] as does a British Library copy of *STC* 11082.5 (Shelfmark: G.11463.), in which emblem (b) appears on the blank facing the first title page. *STC* acknowledges that in some copies of *STC* 11082.5, emblem (c) is not present (*STC,* I, 491). In all copies that I have seen of either issue the pagination of *Poeticall Miscellanies,* which is separately paginated, skips from 96 to 101. Since none of the text of the work is affected I take this to be merely a printer's error.

A poem by Francis Quarles, "To my dear friend, the Spencer of this age," is printed on the unpaginated leaf following p. 130 of all copies that I have seen of either issue, except the copy of *STC* 11082 which was photographed for microfilm for the series *Early English Books 1475 -1640* (Reel 836) and a copy of *STC* 11082 held by the British Library (Shelfmark: 239.i.23.[1.]), which in all probability is the copy which was photographed for the microfilm. This poem is supplied in manuscript in this "variant" copy, and its absence in print is explained by the British Library's online catalog as an imperfection of the copy. This copy is bound with copies of *Locustae, vel pietas iesuitica* (Cambridge, 1627) and *The Locusts, or Apollyonists* (Cambridge, 1627).

Copies of the first edition of *The Purple Island* which conform in all particulars to *STC*'s descriptions are the Bodleian Library's copy of 11082 (Shelfmark: Douce F.228), which is reprinted in facsimile in The English Experience series, number 313 (Amsterdam: Theatrum Orbis Terrarum; New York: Da Capo Press, 1971), and the British Library's presentation copy of 11082.5 to Edward Benlowes (Shelfmark: C.344.g.33).

In both issues of the first edition the title page is followed by the author's Dedication to Edward Benlowes, which precedes the epistle "To the Readers," by Daniel "Featly" (Daniel Fairclough), six commendatory verses and a prefatory poem. These are, in the order they appear: "On the Excellent Morall Poem, Entituled the *Isle Of Man,*" by "Edward {Benlowes / Benevolus}";[5] an anonymous poem in Latin;[6] "To the Learned Authour, Sonne and Brother to Two Judicious Poets, Himself the Third, Not Second to Either," by "W[illiam] Benlowes"; "To the Ingenious Composer of this Pastorall, the Spencer of this Age," by

"Fran[cis] Quarles"; "To the Unknown Mr. *P.F.* Upon Survay of his *Isle Of Man,*" by "Lod. Roberts [Lewis Roberts]"; "On the Most Accurate Poem, Inscribed *The Purple Island,*" by "A.C." (who might be either Abraham Cowley or Anthony Champney, though the former is more probable);[7] and an untitled prefatory poem by "Fran[cis] Quarles," which begins, "Mans Bodie's like a house."

A copy of the first edition in the National Library of Wales (NLW) (Shelfmark: w.b.8262) is identified by the NLW's microfiche catalog, and by the *Online Computer Library Centre, Inc. (OCLC) Firstsearch* catalog database, as a copy of *STC* 11082. Importantly, this copy is identical to other copies of *STC* 11082 in that it is not printed on "large paper," and in that it lacks the engravings added in *STC* 11082.5. However, the NLW copy is unlike *STC* 11082 and 11082.5 in that it contains, of the prefatory material, only the dedication to Edward Benlowes, the epistle "To the Readers," and the first commendatory poem. This copy has not lost the relevant leaves from the current binding, which, however, is not original, but predates the copy's library accession date (1909/19).[8] Inside the back cover is affixed a description from the catalog of sale claiming the copy contains the absent verses. This is late nineteenth century on the authority of Mr Charles Parry (Assistant Librarian, Department of Printed Books, NLW), although handwritten on the affixed description slip is "Cat. No 395. Feb. 1919," which may or may not refer either to the catalog of sale or the library accession date. Since the current binding in any case predates this catalog of sale, in respect of the catalog's reference to the absent verses the description must refer either mistakenly to the copy in question, or to another copy. In either case, two explanations are plausible. NLW w.b.8262 may be a copy of a separate issue, lacking all but the first commendatory poem, but this is unlikely, especially in view of the fact that *STC* makes no mention of such an issue. Alternatively it is a copy of *STC* 11082 from which the relevant leaves were lost prior to its current binding. The likelihood is that the latter explanation is true.

Appendix B
The Biographical Relation of Phineas Fletcher to John Banister and to William Harvey through the Willughbys of Wollaton and Risley Halls

John Banister (1533–1610) and William Harvey (1578–1657) were both contemporaries of Phineas Fletcher (1582–1650), and both had connections with Cambridge and with the Willughby family by whom Fletcher was patronised. So aside from engaging with the texts of Banister's *Historie of Man* and Harvey's *Prelectiones* for the order and state of anatomical ideas within, which is done in chapters 7 and 8, I considered it necessary in the course of this study to attempt to establish whether there was a possibility of communication of ideas to the poet by Banister or by Harvey. It was thought that in Banister's case the biographical context should provide, at the very least, a means whereby it becomes possible for Fletcher more easily to have acquired access to and to have read Banister's *Historie of Man,* which was published in 1578, and that this means should be further to that arising from Cambridge University's acquisition in 1588/89 of what may have been a copy of Banister's anatomical work, and Banister's own donation of a copy of his work to King's College, in 1596, as discussed in chapter 7. In respect of Harvey's *Prelectiones,* however, because they were published only much later, it was thought that the biographical context, if forthcoming, should provide a means whereby either the transmission of the lectures in manuscript, or their verbal communication becomes possible. The character of the anatomy in Cantos 2 to 5 of *The Purple Island* is such that whether the communication of the lectures was verbal or written it did not need to have been in their entirety, and may have been only a summary of part or the whole. But it needs obviously to be further to Fletcher's attendance at Harvey's Lumleian lectures at the College of Physicians, for which no documentary evidence exists.

Possible connections have been suggested by Langdale between Fletcher and Harvey via the Will[o]ughbys among whom was Fletcher's patron, Sir Henry Willughby (d. 1653); via Lewis Roberts; and via Sir Francis Bacon, whose physician Harvey was; as well as directly between the Fletcher and Harvey families. On March 25, 1614/15, two weeks after the performance at King's College, Cambridge of his piscatory drama *Sicelides,* and sorely disappointed by the premature departure of King James, Prince Charles, and their train, for whose recognition the play had been written, Fletcher left Cambridge for Risley, Derbyshire, never to return as a fellow.[1] In Risley, he occupied the chaplaincy at Risley Hall, the seat of Sir Henry Willughby since the decease of his grandfather, George Willughby, in 1613.[2] The Will[o]ughbys were an ancient family with seats at Wollaton, Nottinghamshire, Risley, Derbyshire, Bore Place (Boreplace), Kent, and elsewhere; they were "an important part of an extended network of gentry and noble families bound together by service, land ownership, and political patronage."[3] Sir Henry was thereafter Fletcher's patron, and had been from 1612 according to Langdale's reasonable extrapolation from the few facts available for the period of Fletcher's biography from September 1612 to April 1614.[4]

Investigating the biographical relation between Fletcher and the Will[o]ughbys, Langdale argues that Sir Henry's sudden appearance in the records of Fletcher's life, as his patron, in 1615, belies a probable association between them from even earlier than 1612, and he conjectures that the Kentish baronet Percivall Willoughby (d. 1643), Member of Parliament, and knighted by James I, and who was related to the Risley Willughbys through his wife, Bridget (Bridgett) Willughby of Wollaton Hall (d. 1629?), was one of the possible connections between Sir Henry and Phineas, on account of the proximity of Sir Percivall's Kent home, Bore Place, to that of Maximilian Brook, one of the elder of the comrades of Giles Fletcher (Phineas's father).[5]

Bridget Willughby was the eldest daughter of Elizabeth Lyttleton and Sir Francis Willughby (1543–1596), who built Wollaton Hall, and Sir Francis "acted as patron for Dr. John Banister . . . anatomist and surgeon who lived [or, at least, frequently visited] at Wollaton in the 1570s and attended Lady Willughby as her personal physician. In 1578, Banister dedicated his *Historie of Man* to Sir Francis,"[6] in addition to dedicating it to the "Fraternitie of Chirurgians in London," as was stated in chapter 7. Banister's decease probably prior to Fletcher's patronage by the Risley Willughbys makes it doubtful their paths would have crossed in the corridors of Wollaton Hall. But my research on the papers of the Will[o]ughby family of Wollaton, Risley, and Bore Place (The Middleton Collection, Department of Manuscripts and Special Collections, Hallward Library, Nottingham) and on histories of the Will[o]ughby family,[7] has revealed that when Fletcher was chaplain at Risley Hall, the prestigious library at Wollaton Hall probably held a copy of Banister's *Historie,* presumably presented by the author. Other sixteenth-century works on anatomy and surgery, which include a copy of the 1559 edition of Thomas Gemini's *Compendiosa totius anatomiae delineatio,* as well as other medical books, were acquired for the library at Wollaton Hall, though I have not been able to determine whether or not Fletcher may have had access to the books in this library.[8] At the very least the Willughby family connection plausibly facilitated the opportunity for a copy of *The Historie of Man* to have found its way to Fletcher.

Langdale's biography of Fletcher provides a set of connections between Fletcher and Harvey, which in various ways takes in the Will[o]ughbys and Sir Francis Bacon, whose physician Harvey was. Sir Henry Willughby's cousin, Peregrine Willoughby of Eresby,[9] "had been on cordial terms with Anthony Bacon, Dr. Giles Fletcher's friend and employer."[10] Peregrine's cordial relationship with Anthony Bacon indirectly connects him with the elder Giles Fletcher, whose own friendship and employment connects him with Anthony's brother, Francis Bacon. The slightness of these connections changes complexion in the light of a related set of circumstances that link Phineas and Sir Roger Townshend through the latter's friendship and patronage. Sir Roger's uncle was Francis Bacon, who is also drawn more closely to Phineas's brother, Giles, as his patron, in which capacity the Lord Chancellor presented Giles to the rectory of Helmingham in Suffolk, in 1617, and was active in securing an exchange of livings for him in 1618, both of which occasions put Bacon in communication with Giles in temporal proximity to Harvey's early anatomical demonstrations.[11] In addition, I have been able to establish that, in 1621, Francis Bacon was deeply involved in a cause depending between Dorothy Tamworth, Lady Wharton (who was Sir Francis Willughby's second wife, and who after his death married Lord Wharton), and Sir Percivall Willoughby. This not only engages Bacon more intimately in Will[o]ughby affairs but also may closely draw in Harvey.[12] But the only documented connection between the Will[o]ughbys and William Harvey is that the obstetrician Percival Willoughby (1596–1685), fifth son of Sir Percivall Willughby, M.P., and Bridget Willughby, of Wollaton Hall, and who was thereby related to the Risley Willughbys, was visited at Derby as a friend by Harvey.[13] If the obstetrician's admission as an extra licentiate of the Royal College of Physicians, on February 20, 1640/41, rather than an earlier family connection, marked the occasion for the commencement of his friendship with Harvey, seven or eight years after the publication of *The Purple Island,* then Percival would have been unable at any earlier date to introduce the anatomist to the poet. It seems unlikely, in any case, that his attendance upon Bacon as his physician alone would have occasioned Harvey's introduction to Phineas Fletcher; and Harvey's derisive disparagement of his patient's philosophical writing—"he writes Philosophy like a Lord Chancellour"—should countenance against hasty speculation about anything more than a professional relationship.[14]

In addition to the aforesaid possibilities linking Fletcher biographically to Harvey, Langdale suggests that Fletcher may have been directly acquainted with the Harveys, who resided in Folkstone, Kent, twenty miles from Cranbrook, where the Fletchers dwelt until they, like the Harveys, moved to London around 1585–86.[15] Of equal significance for Langdale is Phineas's father's intimate association, as the remembrancer of the city and the ambassador for the Merchant Adventurers, until 1611, with the commercial contexts inhabited by Harvey's five brothers, who were all to become prosperous London merchants.[16] The only evidence for a familial acquaintance, however, is slight, and resides in the little encomium, entitled "To the Unknown Mr P. F. Upon Survay of his *Isle of Man,*" which was published among the commendatory verses that prefaced the first edition of *The Purple Island.* The Welsh author of this encomium, Lewis Roberts, appended "Lod. Roberts" to the encomium, and was the apprentice of Thomas Harvey, eldest brother of William.[17] Langdale plausibly argues that the stance of unfamiliarity with Fletcher, adopted by the encomium's voice, is in keeping with the anonymity with which *The Purple Island* was published, and may even be ironic.[18] Although Izaak Walton, a mutual friend of Fletcher and Roberts in Langdale's account, may have introduced them, Langdale argues that, "there is more reason for thinking that one of the Harveys was the go-between"; and that Fletcher was brought into contact with the Harveys through the circumstances of habitation, and later, through them, with their employee, Roberts. Although it is reasonable for Langdale to claim that Roberts "must have been bound to Fletcher in some fashion, because otherwise his rhymed couplets would never have found their way to these island shores," the circumstances of their acquaintance and its extension to William Harvey remain thus far conjectural.[19]

It is necessary to regard Langdale's speculation of a Cambridge and/or London acquaintance between Fletcher and Harvey, between 1602, when it is likely that Harvey returned from Padua to London, and 1610, as useful only in so far as it may account for the beginning or growth of their acquaintance.[20] The series of experiments that would lead to the hypothesis and proof of systemic circulation had not even begun at this time, and so an early acquaintance may be unlikely to have provided the context in which the communication to Fletcher of the physiological ideas of the operation of the cardiac valves and the alchemical connotations of "circulation" or "distillation" took place, and even less likely to have occasioned Harvey's communication of the movement of blood in the arteries. The earliest evidence of Harvey's ruminations on these ideas exists in his *Prelectiones,* and so, it would be unsafe to suggest a date much earlier than 1616 for the possibility of communication of any of them. As chapter 8 shows, Langdale was of an incorrect opinion as to the state of Harvey's ideas before 1616.

In the light of the necessity for a later date than Langdale proposes for the communication of Harvey's physiological ideas to Fletcher, it is worth considering that if Phineas had not been introduced to Harvey before he took up the chaplaincy at Risley, his opportunity for being so while ensconced within the Willughby family seat, and so close to Wollaton Hall, might have increased, though this surmise depends of course on the preexistence of a familial acquaintance between Harvey and the Will[o]ughbys of Wollaton, of which there is only uncertain evidence in Harvey's friendship with Sir Percivall's physician son. Communication may be comparatively less likely after the first of November 1621, when Fletcher took up the position of rector of All Saints Church, at Hilgay, for though not quite so far as Derbyshire from either London where Harvey lived and worked, or Bore Place, the Kentish estate of the Willoughbys, Hilgay was also far from Wollaton and Risley Halls. By 1628, Fletcher had been ensconced in the fens of Norfolk for seven years, and it is improbable (yet, not impossible, not least because Sir Henry Willughby was still Fletcher's patron) that Harvey communicated complex physiological ideas to him. It is more likely, for reasons not only of geography, but also of the conceptual character of the *Purple Island* physiology, that Harvey's ideas, such as they were, were made known to Fletcher, if at all, between 1615 and 1621, during Fletcher's residence in Derbyshire.

Notes

Introduction

1. James Howell (1594?–1666), "*To* E. Benlowes *Esq;* upon the Receipt of a Table of Exquisit Latin Poems," dated "*Fleet, 25. Aug., 1645,*" *Epistolæ Ho-Elianæ: Familiar Letters, Domestic and Forren; Divided into Four Books, Partly Historical, Political, Philosophical, upon Emergent Occasions,* 4 vols. (bound together), 5th ed. (London: Thomas Guy, 1678), II, Letter 66. The letter is also cited in Alexander B. Grosart, *The Poems of Phineas Fletcher, B.D., Rector of Hilgay, Norfolk: For the First Time Collected and Edited,* with memoir, essay, and notes, 4 vols., Fuller Worthies' Library (Blackburn: C. Tipelady, 1869), IV, cvii–cviii. Whether the copy of *The Purple Island* which Howell received was in manuscript or a copy of one of the two issues of the first edition is unclear. The first edition of *The Purple Island* appeared as *The Purple Island; or, The Isle of Man: Together With Piscatorie Eclogs and Other Poeticall Miscellanies,* by "P.F." (Cambridge: University of Cambridge, 1633). The printers issued this edition twice: the issues are cataloged as 11082 and 11082.5 in *A Short-title Catalogue of Books Printed in England, Scotland, & Ireland and of English Books Printed Abroad 1475–1640,* first compiled by A. W. Pollard and G. R. Redgrave, 2nd ed., rev. and enlarged, begun by W. A. Jackson and F. S. Ferguson, completed by Katharine F. Pantzer, 3 vols. (London: Bibliographical Society, 1976–1991), I (1986), 491 (further references to this text use the abbreviated short-title form: *STC*). Further details of the first edition of *The Purple Island* are given in Appendix A of this study. The "*Scarlet* Island which lies near the *Baltic* Sea" I take to be Delos, which *The Purple Island* represents as a precedent for the figurative concept of the "*Isle of Man.*"

2. Herbert Grierson, *The First Half of the Seventeenth Century,* Periods of European Literature (Edinburgh: Willliam Blackwood and Sons, 1906), p. 142.

3. Giles and Phineas Fletcher, *Poetical Works,* ed. by Frederick S. Boas, 2 vols. (Cambridge: Cambridge University Press, 1908–9; repr. 1970), II. This is the most recent scholarly edition which contains *The Purple Island,* and all further references to *The Purple Island* are to this edition unless otherwise stated, and are included in the text, prefixed by the initials of the short title: *P.I.* References are given in the following order: canto number, stanza number, line number(s); e.g., (*P.I.* 1.1.1–2). References to the marginal text are given as follows: canto number, stanza number, "margin." (= marginalia) and marginal letter or symbol; e.g., (*P. I.* 2.5 margin. a). Other editions of *The Purple Island* are cited as and when relevant in the notes hereafter.

4. Grosart, *The Poems of Phineas Fletcher,* IV, cclxiv. Joan Grundy, *The Spenserian Poets: A Study in Elizabethan and Jacobean Poetry* (London: Edward Arnold, 1969), p. 201, suggests that the color purple in the title of Fletcher's poem corresponds to a baroque disposition for "gaudy colour effects."

5. Grundy, *The Spenserian Poets,* p. 182. Fletcher's anti-Catholicism is most virulent in his *Locustae vel pietas Iesuitica,* with *The Apollyonists* (Cambridge, 1627). A good modern edition is Phineas Fletcher, *Locustae vel pietas Iesuitica,* ed. with introduction, translation, and commentary by Estelle Haan, Supplementa Humanistica Lovaniensia, IX (Leuven: Leuven University Press, 1996).

6. Edmund Gosse, *The Jacobean Poets,* 2nd impression (London: John Murray, 1899), p. 148.

7. See, for example, Frank S. Kastor, *Giles and Phineas Fletcher,* Twayne's English Authors Series, 225 (Boston: Twayne, 1978), pp. 113–15.

8. George Parfitt, *English Poetry of the Seventeenth Century,* 2nd edn., Longman Literature in English Series (London: Longman, 1992), p. 193, says, "Phineas Fletcher operates in a . . . self-consciously witty manner and his poem displays itself as artifact in a distinctly mannered way" (p. 190). He adds that Fletcher creates "pseudo-epic worlds in which art is nearly all, with the result that the sense that epic deals with vital issues is weakened" (p. 194). Also on Fletcher's style, see Grundy, *The Spenserian Poets,* pp. 181–84, 196, 201–2. On Fletcher's derivativeness and dependence on Spenser, see Grundy, *The Spenserian Poets,* Ch. 9; Abram Barnett Langdale, *Phineas Fletcher: Man of Letters, Science and Divinity,* Columbia University Studies in English and Comparative Literature, 125

(New York: Columbia University Press, 1937), p. 131; Parfitt, *English Poetry of the Seventeenth Century*, p. 193. On the derivativeness of the "Spenserian poets" generally, see Grundy, *The Spenserian Poets*, pp. 8, 42–43.

9. George MacDonald, *England's Antiphon* ([London]: Macmillan, [1868]), p. 156.

10. For questions of form and linguistic character, see Kastor, *Giles and Phineas Fletcher*, pp. 114–16; Parfitt, *English Poetry of the Seventeenth Century*, p. 191. For questions of genre, see Kastor, *Giles and Phineas Fletcher*, pp. 112–15; Parfitt, *English Poetry of the Seventeenth Century*, p. 191. For questions of overall conception or design, see Gosse, *The Jacobean Poets*, p. 148; Kastor, *Giles and Phineas Fletcher*, pp. 107, 112–16; Parfitt, *English Poetry of the Seventeenth Century*, p. 191. On questions of design in the work of "Spenserian poets" generally, see Grundy, *The Spenserian Poets*, p. 62.

11. See, for example, Gosse, *The Jacobean Poets*, p. 146, and Grundy, *The Spenserian Poets*, pp. 181–82, although others are discussed in the chapters of this study.

12. MacDonald, *England's Antiphon*, p. 156.

13. Gosse, *The Jacobean Poets*, p. 146.

14. See , for example, Gary Taylor, *Reinventing Shakespeare: A Cultural History from the Restoration to the Present* (London: Hogarth Press, 1989), p. 406.

15. Leonard Barkan, *Nature's Work of Art: The Human Body as Image of the World* (New Haven: Yale University Press, 1975); Philippe Ariès, *L'Homme devant la mort* (Paris: Éditions du Seuil, 1977), trans. by Helen Weaver and published in English as *The Hour of Our Death* (London: Allen Lane, 1981).

16. See, for example, Ellen Spolsky, *Gaps in Nature: Literary Interpretation and the Modular Mind* (Albany: SUNY Press, 1993); *Satisfying Skepticism: Embodied Knowledge in the Early Modern World* (Aldershot: Ashgate, 2001).

17. Paul de Man, "The Epistemology of Metaphor," *Critical Inquiry* 5 (1978): 13–30. De Man reads "semiotically" (his term) Locke's philosophical endeavor in *An Essay Concerning Human Understanding* to establish how it is possible to formulate propositions which truthfully convey true ideas, some of which are simple, some of which may correspond to substances, and some of which may correspond to complex ideas or "mixed modes" such as Man (although Locke abandons the latter enterprise, concluding that to consider mixed modes to be real is to mistake words for things) (pp. 14–20). De Man also deconstructs claims that the names of complex entities refer truthfully to the essence or ground of their properties (p. 19).

18. Catherine Belsey, "Literature, History, Politics," in *New Historicism and Renaissance Drama*, ed. and introduced by Richard Wilson and Richard Dutton (London: Longman, 1992), pp. 33–44 (pp. 36–37), describes how a suspicion found in some of Jacques Derrida's work of the difference between the signifier and the signified, and a subsumption of the order of the signified under "presence," "which is understood indiscriminately as concept, intention or referent, so that meaning, being and truth are collapsed together," has led "vulgar deconstructionists to take it that there is no such thing as meaning, and in consequence, since meaningless language is literally unthinkable, that words mean whatever you want them to mean."

19. Langdale, *Phineas Fletcher.*

20. See, for example, Augustus Grote Pohlman, "*The Purple Island* by Phenias Fletcher: A Seventeenth Century Layman's Poetical Conception of the Human Body," *Johns Hopkins Hospital Bulletin* [18], no. 197 (1907): 317–21 (p. 320, col. 2); Samuel W. Lambert, "Phineas Fletcher and Francis Quarles and *The Purple Island* of Phineas Fletcher," *Annals of Medical History*, repr. from n.s., 3 (1931), 568–78 (p. 576, col. 2); Anon., "A Poet's Anatomy," *British Medical Journal*, 9 May 1925, p. 890 (col. 1).

21. R. G. Baldwin, "Phineas Fletcher: His Modern Readers and His Renaissance Ideas," *Philological Quarterly* 40 (1961): 461–75 (p. 468).

22. Baldwin, "Phineas Fletcher," p. 471.

23. Ibid., p. 470.

24. Ibid., pp. 467, 470.

25. Grosart, *The Poems of Phineas Fletcher*, IV, xxi; Kastor, *Giles and Phineas Fletcher*, p. 79.

26. Jonathan Sawday, "Bodies by Art Fashioned: Anatomy, Anatomists, and English Poetry 1570–1680" (doctoral thesis, University of London, 1988), p. 134.

27. In this regard, Hayden White's structuralist critiques of historiography were seminal. See, for example, White, "The Historical Text as Literary Artifact," in *The Writing of History: Literary Form and Historical Understanding*, ed. by Robert H. Canary and Henry Kozicki (Madison: University of Wisconsin Press, 1978), pp. 41–62. In the same volume, Lionel Gossman, "History and Literature: Reproduction or Signification," pp. 3–39, examines the Classical association of "rhetoric" with writing history, a view to which the Renaissance largely conformed, and the historical emergence in the eighteenth and nineteenth centuries of the notion that the historian's

objects are situated on the other side of evidence, whether problematically or not depending on one's method. Gossman argues that in this period, the historical discipline became associated with positivism, which aimed to bring history as close as possible epistemologically and methodologically to the natural sciences. Gossman argues that for reasons of rhetoric or presentation, the tense and perspective of historical narration, the nature of the linguistic sign and the character of historical "truth," the relationship between fictional and historical narratives is a close one. In literary studies, Stephen Greenblatt's self-confessedly thwarted attempts to speak with dead authors of Renaissance literature contributed to a now burgeoning critical literature which examines the historicist approach to interpreting texts. See Stephen Greenblatt, *Renaissance Self-Fashioning: From More to Shakespeare* (Chicago: University of Chicago Press, 1980). This tendency has been described by Alan Liu, "The Power of Formalism: The New Historicism," *English Literary History* 56 (1989): 721–71 (p. 746), as having become so sceptical of "a regulated method of reaching the historical other from the domain of the text," that "it at last studies itself in the anxious pose of reaching the other." Cited in H. Aram Veeser, "Introduction: The New Historicism," in *The New Historicism Reader*, ed. by H. Aram Veeser (London: Routledge, 1994), pp. 1–32 (p. 29, note 7).

28. Jonathan Sawday, *The Body Emblazoned: Dissection and the Human Body in Renaissance Culture* (London: Routledge, 1995), pp. 170–82.

29. Walter Pagel, "The Vindication of 'Rubbish,'" *Middlesex Hospital Journal* 45 (1945): 42–45 (repr. in Walter Pagel, *Religion and Neoplatonism in Renaissance Medicine*, ed. Marianne Winder, Collected Studies Series, CS226 (London: Variorum Reprints, 1985), pp. 1–14 (p. 8)).

30. Andrew Cunningham, *The Anatomical Renaissance: The Resurrection of the Anatomical Projects of the Ancients* (Aldershot: Scolar Press, 1997).

31. Roger French, *Dissection and Vivisection in the European Renaissance* (Aldershot: Ashgate, 1999), p. 1.

32. French, *Dissection and Vivisection in the European Renaissance*, p. 2.

33. The main work consulted in the field of conceptualist theory of metaphor is George Lakoff and Mark Johnson, *Metaphors We Live By* (Chicago: University of Chicago Press, 1980). Others are cited where relevant.

34. Ivan Illich, "A Plea for Body History," *Michigan Quarterly Review* 26 (1987): 342–48 (p. 345)

35. Owsei Temkin, "Metaphors of Human Biology," in *Science and Civilization*, ed. by Robert C. Stauffer (Madison: University of Wisconsin Press, 1949), pp. 169–94; repr. in Temkin, *The Double Face of Janus and Other Essays in the History of Medicine* (Baltimore: Johns Hopkins University Press, 1977), pp. 271–83 (p. 283).

36. See, for example, Judith Butler, *Bodies that Matter: On the Discursive Limits of 'Sex'* (New York: Routledge, 1993), p. 8.

37. Belsey, "Literature, History, Politics," p. 44; Veeser, "Introduction: The New Historicism," p. 10.

38. See Michael C. Schoenfeldt, *Bodies and Selves in Early Modern England: Physiology and Inwardness in Spenser, Shakespeare, Herbert, and Milton*, Cambridge Studies in Renaissance Literature and Culture, 34 (Cambridge: Cambridge University Press, 1999), Ch. 5, pp. 131–68.

39. Thomas McKeown, *The Role of Medicine: Dream, Mirage or Nemesis*, 2nd edn (Oxford: Basil Blackwell, 1989), p. xvi. McKeown argues that in modern medicine this approach "has led to indifference to the external influences and personal behaviour which are the predominant determinants of health"; "the approach to biology and medicine established in the seventeenth century was an engineering one based on a physical model. . . . Physics, chemistry, and biology are considered to be sciences basic to medicine" (p. 7). On scientific paradigms and the cyclical history of science, in which the "scientific revolutions" like that of the seventeenth century establish new paradigms, see Thomas S. Kuhn, *The Structure of Scientific Revolutions* (Chicago: University of Chicago Press, 1962).

40. Illich, "A Plea for Body History."

41. On this power claimed by artists and writers, see François Rigolot, "The Rhetoric of Presence: Art, Literature, and Illusion," in *The Cambridge History of Literary Criticism*, ed. by H. B. Nisbet and Claude Rawson, 9 vols. [7 vols. published] (Cambridge: Cambridge University Press, 1989–), III, ed. by Glyn P. Norton (1999), 161–67 (p. 161).

42. Thomas Healy, "Sound Physic: Phineas Fletcher's *The Purple Island* and the Poetry of Purgation," *Renaissance Studies* 5 (1991): 341–52; Sawday, "Bodies by Art Fashioned," Ch. 3: "'A Strange Though Native Coast': *The Purple Island* of Phineas Fletcher," pp. 107–61; Sawday, *The Body Emblazoned*, pp. 170–82.

43. Veeser, "Introduction: The New Historicism," pp. 2, 8. Of considerable significance in the development of the idea that a wide range of material practices embody ideology was Louis Althusser's reinterpretation of Karl Marx's classical model of society as infrastructure (economic base) and superstructure (ideology and the political and legal apparatuses) in "Ideology and Ideological State Apparatuses (Notes Towards an Investigation)" (January–April 1969), in *Lenin and Philosophy and Other Essays*, trans. from the French by Ben Brewster (London: NLB,

1971), pp. 121–73 (first publ. in *La Pensée,* 1970). Rather than revising the metaphorical notion of the state as an edifice which endows the base with an index of effectivity, the determination in the last instance of what happens in the upper floors by what happens in the economic base, Althusser develops the metaphorical entailments of Marx's model, extending state power beyond the familiar ground of the repressive state apparatuses and of political practice to other private and public material practices. More influential for new historicism and cultural materialism, however, was Foucault's refusal "to find a central and determining locus of power in the model of production," his decentralisation of the operations of power, his equation of power with language and knowledge, and his insistence on distinguishing knowledge from what is true, the signified from the referent and from intention. Belsey, "Literature, History, Politics," pp. 37–38.

44. Influential in my formulation of this critical commitment, expanded in the chapters of the study, is Joseph Margolis, *Art and Philosophy* (Sussex: Harvester Press, 1980). Annette Barnes, *On Interpretation* (Oxford: Basil Blackwell, 1988), discusses Margolis's position in ch. 4 (pp. 47–48). According to Richard Wilson, "Introduction: Historicising New Historicism," in *New Historicism and Renaissance Drama,* pp. 1–16 (p. 11), new historicism broke the base/superstructure, text/context dichotomy "by insisting that works of art are themselves events, which intervene materially in history, rather than mirrors of reality." Wilson, however, warns that "the danger of this iconolatry" and of the corresponding textualization of history, "was that history would be aestheticised into a spectacle or game," without material determinacy.

45. Stephen Greenblatt, *Shakespearean Negotiations: The Circulation of Social Energy in Renaissance England* (Oxford: Clarendon Press, 1988), especially pp. 8–11.

46. Veeser, "Introduction: The New Historicism," p. 3; Greenblatt, *Shakespearean Negotiations,* p. 8. According to Wilson, "Introduction: Historicising New Historicism," p. 9, Foucault's equation of language with power, was "the 'discourse theory' that allowed New Historicism to equate the transactions of language with those of capitalism, by conflating verbal with market forces, as Greenblatt did in his book *Shakespearean Negotiations.*" However, Veeser indicates that "for Greenblatt the 'circulation of social energy' meant something like rhetorical *energia* or force. *Economy* and its metaphors mean something very much more specifically capitalist for later N[ew] H[istoricist]s."

47. On the subject of intention I partly follow Barnes, *On Interpretation,* ch. 4 (pp. 48–50), who argues that "if an interpretation accords with a successfully executed artistic intention, then it is true." But this does not mean that, either, only such interpretations are true or that truth is either necessary or sufficient for an interpretation to be acceptable. As culturally emergent, a work of literature like any work of art is not identical with a physical object where only extensional predicates are needed to talk about it; in art *intensional* predicates are needed, "for artworks exist and can be identified as artworks only within a cultural context." Not all these culturally significant properties however need be intentional by the author, for emergent properties are "interpretatively accessible" and what can "defensibly be attributed to the work depends in part on the culturally determined myths or schemes of imagination that interpretations bring to bear on the work." Because these schemes are plural, it is logically possible that "plural, non-converging, even incompatible" interpretations can be generated. Furthermore, "the appeal to authorial intent as a way to privilege certain interpretations is rejected" on the grounds that the author has no privileged voice over whether intentions have been fulfilled.

48. Illich, "A Plea for Body History," p. 345.

49. Ibid., p. 345.

50. This is the espoused focus of Routledge's four-volume collection entitled *The Body,* Critical Concepts Sociology (London: Routledge, 2003).

51. Barkan, *Nature's Work of Art,* p. 62.

52. Steven Shapin, "The Philosopher and the Chicken: On the Dietetics of Disembodied Knowledge," in *Science Incarnate: Historical Embodiments of Natural Knowledge,* ed. by Christopher Lawrence and Steven Shapin (Chicago: University of Chicago Press, 1998), pp. 21–50 (p. 23). Mari Sorri and Jerry H. Gill, *A Post-Modern Epistemology: Language, Truth and Body* (Lewiston: Edwin Mellen, 1989), p. 1, claim, with some justice, that the theme of the "disembodied character of human knowledge" "has remained pivotal throughout Western epistemological discussion all the way from Plato to Ayer and beyond."

53. Steven Shapin and Christopher Lawrence, "Introduction: The Body of Knowledge," in *Science Incarnate,* pp. 1–15 (p. 1).

54. Paul Ricoeur, "Metaphor and the Main Problem of Hermeneutics," *New Literary History* 6 (1974): 95–110 (p. 104).

55. Belsey, "Literature, History, Politics," p. 43.

56. Ricoeur, "Metaphor and the Main Problem of Hermeneutics," pp. 95, 110.

57. On the classification of literary texts as either "declarative," "imperative" or "interrogative," see Catherine Belsey, *Critical Practice*, New Accents Series (London and New York: Routledge, 1980; repr. 1992), especially ch. 4. In "Literature, History, Politics," p. 42, Belsey admits that the classification is "excessively formulaic, implying that texts can unilaterally determine their reception by the reader."

Chapter 1. Conceptual Metaphors of the "*Isle of Man*"

1. Linda J. Flasch Redman, "*The Purple Island:* A Critical Study" (unpublished doctoral dissertation, Oklahoma State University, 1974), p. 2, persuasively argues that although "the failure on Fletcher's part to call his poem an allegory appears to confuse, even contradict, interpretative efforts, yet one must understand the work as allegory." This study takes the view of Maureen Quilligan, that "allegory is, in fact, a class, a genre—a legitimate critical category of a prescriptive status similar to that of the generic term 'epic.'" Allegory is not only *allegoresis*, from which allegory arose, a form of criticism (which began with the Greeks) "whereby meanings not originally intended were extracted from texts," and which was justified by allegorical critics not "by faithful reference to their texts, for the disparity between literal meaning and allegorical interpretation only proved the sacredness of the process and the care the author had used to hide his meaning from the vulgar." Maureen Quilligan, *The Language of Allegory: Defining the Genre* (Ithaca: Cornell University Press, 1979), pp. 14, 30. Northrop Frye generalizes the applicability of allegory or *allegoresis* to all commentary, the "attaching of ideas to the structure of poetic imagery": "The instant that any critic permits himself to make a genuine comment about a poem . . . he has begun to allegorize." Northrop Frye, *Anatomy of Criticism: Four Essays* (Princeton: Princeton University Press, 1971; repr. 1973), p. 89. As a narrative genre, however, which arose in the fourth century A.D., in the work of the Christian Latin poet Prudentius, allegory does not only denote reader competence, but, according to Quilligan, "codifies the rules for reader expectations" and "has its source in a culture's attitude toward language, and in that attitude, as embodied in the language itself, allegory finds the limits of its possibility." With regard to its particular "generic status," Quilligan argues: "Allegory [is] . . . much like satire, which is unarguably a genre in its own right, but which shares with other works in other genres a quality we can legitimately term 'satirical.' Just as some works are satires and others are satirical so some works are allegories while others are merely allegorical." Quilligan thereby makes a distinction between literary genres that are labeled by content (e.g., elegy) and those which are labeled by form (e.g., sonnet). *The Language of Allegory*, pp. 16, 15, 18. *The Purple Island* is an allegory in formal genre and its content is allegorical throughout, with epic and pastoral elements which are discussed as their relevance arises in this study.

2. Kastor repeatedly emphasizes the criterion of unity in his aesthetic evaluation of the work of Phineas Fletcher. See Kastor, *Giles and Phineas Fletcher*, pp. 95, 107, 113–16. For a study of the mutual influence between Giles Fletcher's *Christ's Victorie, and Triumph* and his brother's *The Purple Island*, and of the latter in the context of the "Spenserian poets" to whom both brothers have been regarded as belonging, see Grundy, *The Spenserian Poets*, ch. 9, pp. 181–203; also pp. 8, 42.

3. Kastor, *Giles and Phineas Fletcher*, p. 114.

4. Ibid., p. 114.

5. Edwin Honig, *Dark Conceit: The Making of Allegory* (Oxford: Oxford University Press, 1966), p. 72.

6. Ibid., p. 72; Quilligan, *The Language of Allegory*, p. 58.

7. As an agent in the story, the Isle of Man does not qualify as a character, that is, as a non (or pre-) verbal abstraction, which by no means human in the literal sense, is partly modeled on the reader's conception of people and in this sense person-like. It will become increasingly clear over this and the next eight chapters, however, that the island is by no means passive as a location, but is rather animated, its occupants or inhabitants personifications of its own innate virtues and capacities, and its destiny the subject of the entire poem. As such the Isle of Man must qualify as an agent or, perhaps as in structuralist terms, an actant or acteur, which "can include not only human beings (i.e., 'characters') but also inanimate objects (e.g., a magic ring) and abstract concepts (e.g., destiny)." Shlomith Rimmon-Kenan, *Narrative Fiction: Contemporary Poetics*, New Accents Series (London: Routledge, 1983; repr. 1994), p. 34; also see pp. 32–35.

8. Further retrospective or prospective intratextual references in this study are given in the text in the abbreviated form in which they appear at the head of the relevant part or chapter. References to parts (e.g., part 1) should be regarded as inclusive of all chapters therein. Chapter references (e.g., 1) should be regarded as inclusive of all sections therein, unless such sections as are referred to are specified.

9. Thirsil is a character in the *diegesis* (story) narrated by the *extradiegetic narrator,* that is, the narrator who is on a level called the *extradiegetic level,* immediately superior to the framing story of *The Purple Island,* which he narrates but does not himself appear in. Within that story, that is on the *diegetic level* of narration, the character Thirsil himself narrates a story, and so is a *diegetic narrator.* The story which Thirsil tells forms a second-degree narrative, hence a *hypodiegetic level.* In terms of this organization of subordination relations in narrative levels, narration is always at a higher narrative level than the story it narrates. See Rimmon-Kenan, *Narrative Fiction,* pp. 91–92. Pohlman, "*The Purple Island* by Phenias Fletcher," identifies Thirsil as Fletcher the author (p. 317). I think it is more precise to say that in much of his literary work, Fletcher adopted a narratorial voice with which to speak in his poetry to a female narratee on the diegetic level of the narrative. Just as his narrators corresponded in some way to Fletcher the author, so the narratees to whom they speak corresponded to a real or implied female reader who was presumably also Fletcher's lover, or the object of his affection or desire. As Fletcher's allegiance changed during the course of his literary output, so he adopted a differently named narratorial voice. The narrator Thirsil is associated with the narratee "Fusca." Both appear in work written between about 1608 and 1615, but Fusca is not the diegetic narratee of Thirsil's story in *The Purple Island.* This is "a lovely crue / Of Nymphs & shepherd-boyes" (*P.I.* 2.1.6–7). Langdale, *Phineas Fletcher,* pp. 48–50, suggests that Fusca may have corresponded to the actual Lady Elizabeth Cheyney Colepeper.

10. Giles and Phineas Fletcher, *Poetical Works,* II. References to *The Purple Island* are to this edition throughout the study unless otherwise stated, and are included in the text in the form disclosed in the Introduction, note 3.

11. On the character of allegory to develop a commentary on its own intrinsic threshold text, see Quilligan, *The Language of Allegory,* pp. 58, 61.

12. William Camden, *Britain; or, A Chorographicall Description of the Most Flourishing Kingdomes, England, Scotland, and Ireland, and the Ilands Adioyning, out of the Depth of Antiqvitie; Beavtified with Mappes of the Severall Shires of England,* trans. by Philémon Holland, revised, amended, and enlarged with sundry additions by William Camden (London: George Bishop and John Norton, 1610), p. 154. Camden's work first appeared in Latin as *Brittania siue florentissimorum regnorum, Angliae, Scotiae, Hiberniae chorographica descriptio* ([London (?)]: [Eliot's Court Press] for R. Newbery, 1586). Other Latin editions appeared in 1587, 1590, 1594, 1600, and 1607. The first English translation appeared in 1610 (and all references are to this edition). Another edition of this English translation appeared in 1637, of which there are four variants, all published that year.

13. Besides this, Camden's "chorographicall description" postulates unity through etymological recourse to "the originals of names," and hence the singularizing authority of "antiqvitie," after Plato's *Cratylus.* Camden, *Britain,* "The Author to the Reader." *The Purple Island,* as we shall see in this and later chapters, explores etymologically the nomenclature of anatomy, also to unifying effect, in this case superimposing features of the island realm onto those of the human body.

14. This argument employs Benedict Anderson's distinctions between the dynastic realm and the nation, in *Imagined Communities: Reflections on the Origin and Spread of Nationalism,* revised ed. (London: Verso, 1991; repr. 1996); see chs. 1. "Introduction," and 2. "Cultural Roots"; especially p. 9.

15. Michael Drayton (1563–1631), *Poly-Olbion; or, A Chorographicall Description of Tracts, Rivers, Mountaines, Forests, and Other Parts of this Renowned Isle of Great Britaine, With Intermixture of the Most Remarquable Stories, Antiquities, Wonders, Rarityes, Pleasures, and Commodities of the Same* (London: H.L., 1613). This, the First Part of *Poly-Olbion,* consisting of eighteen songs and accompanying notes by John Selden of the Inner Temple explaining the historical and antiquarian allusions, was first published in 1612. It was reissued (with a new general title substituted for the printed title of 1613) in 1622 with *The Second Part; or, A Continvance of Poly-Olbion from the Eighteenth Song: Containing All the Tracts, Riuers, Mountaines, and Forrests: Intermixed With the Most Remarkable Stories, Antiquities, Wonders, Rarities, Pleasures, and Commodities of the East, and Northerne Parts of this Isle, Lying Betwixt the Two Famous Riuers of Thames, and Tweed* (London: Augustus Mathewes, 1622). (The Second Part has a separate title page.) For details of the issues of *Poly-Olbion* see the bibliography in volume V of *The Works of Michael Drayton,* ed. J. William Hebel, 6 vols. (Oxford: Blackwell for the Shakespeare Head Press, 1961). *Poly-Olbion* appears in volume IV of this edition, to which are all further references. Drayton's chief source for *Poly-Olbion* was Camden's *Britannia,* and although Drayton's plan for the work was known by 1598, it was not ready for publication until 1612, which delay enabled him to incorporate some of the new material introduced into the 1607 edition of Camden's *Britannia* and into Philemon Holland's translation of that work first published in 1610.

16. Charles Singer, *The Evolution of Anatomy: A Short History of Anatomical and Physiological Discovery to Harvey; Being the Substance of the FitzPatrick Lectures Delivered at the Royal College of Physicians of London in the Years 1923 and 1924* (London: Kegan Paul, Trench, Trübner & Co., 1925), pp. 115–16.

17. Thomas Laqueur, *Making Sex: Body and Gender From the Greeks to Freud* (Cambridge, Mass.: Harvard University Press, 1992), p. 40.

18. Alessandro Benedetti (c. 1450–1512), *Anatomice, siue historia corporis humani; eiusdem collectiones medicinales seu aforismi* (Venice, 1502), bk. I, ch. 2, translated in L. R. Lind, *Studies in Pre-Vesalian Anatomy: Biography, Translations, Documents* (Philadelphia: The American Philosophical Society, 1975), p. 84 (all references to Benedetti's *Anatomice* henceforth in the text and notes are to this translation): "the male genitals are external, the female genitals or vagina are located within the body; if they fall forward the male groin appears." Andreas de Laguna (1499–1560), *Anatomica methodus, seu de sectione humani corporis contemplatio* (Anatomical Procedure, or a Survey of the Dissection of the Human Body) (Paris, 1535), translated in Lind, *Studies in Pre-Vesalian Anatomy,* pp. 277–78 (all references to de Laguna's *Anatomica methodus* henceforth in the text and notes are to this translation): "If the testicles were not enclosed, since men are warmer than women, the testicles would grow warm with an excessive heat, and since they are the warmest members of the body, nature caused them to hang down so that their extraordinary heat might be fanned by external cold." "In women, however, since they are colder than men, nature placed the testes near the matrix." De Laguna derives his opinion about the relation of heat to sexual difference from Aristotle's *De generatione animalium.*

19. De Laguna, *Anatomica methodus:* "Although the female testicles are inside her body they are endowed by nature, however, with little semen; and it is much colder than in men." See Robert Martensen, "The Transformation of Eve: Women's Bodies, Medicine and Culture in Early Modern England," in *Sexual Knowledge, Sexual Science: The History of Attitudes to Sexuality,* ed. by Roy Porter and Mikulas Teich (Cambridge: Cambridge University Press, 1995), pp. 107–33 (pp. 108, 113).

20. Laqueur, *Making Sex,* pp. 8, 11.

21. Ibid., p. 109.

22. Susanne Scholz, *Body Narratives: Writing the Nation and Fashioning the Subject in Early Modern England* (Basingstoke: Macmillan, 2000), pp. 3–4. For a description of the utero-centric model, see Martensen, "The Transformation of Eve."

23. De Laguna, *Anatomica methodus,* p. 280.

24. Laqueur, *Making Sex,* pp. 109–10.

25. De Laguna, *Anatomica methodus,* p. 280.

26. *The Birth of Mankinde; Otherwise Named the Womans Booke,* English translation by Thomas Raynalde (fl. 1539–1552?) (London: T. Adams, 1604), fol. 29. The first English version of this book appeared as *The Byrth of Mankynde Newly Translated out of Laten into Englysshe,* trans. Richard Jonas (London, 1540). It was translated from Eucharius Roesslin (Rösslin, d. 1526), *De swangern frawen und hebammen Roszgarten* (Strassburg, 1513; or 1526), often referred to as *Rosengarten* (Rose garden), of which at least forty editions were published. Jonas's translation was followed by a translation by Thomas Raynalde, *The Byrth of mankynde . . .* (London, 1545), following which there were twelve editions of Raynalde's translation between 1552 and 1654. For bibliographical details of all fourteen editions see K. F. Russell, "A Checklist of Medical Books Published in English Before 1600," *Bulletin of the History of Medicine* 21 (1947): 922–58.

27. John Sadler, *The Sicke Woman's Private Looking-Glasse; Wherein Methodically Are Handled All Uterine Effects, or Diseases Arising from ye Wombe Enabling Women to Informe the Physitian about ye Cause of their Griefe* (London: Ph[ilip] Stephens and Ch[arles] Meredith, 1636; repr. Amsterdam: Theatrum Orbis Terrarum; Norwood, N.J.: Walter J. Johnson, 1977), ch. 6: "Of the suffocation of the Mother" (pp. 61–62, 66).

28. Martensen, "The Transformation of Eve," pp. 108–09.

29. While some physicians, including Jean Fernel, argued that one of the components of menstrual blood is inherently venomous, Sadler objected in *The Sicke Woman's Private Looking-Glasse,* pp. 9–11 (pp. 49–50 cited in the text). He argued that it is the retentive faculty of the uterus, which makes blood venomous. Galen provides the anatomical etiology for this pathological activity, in a pathological development of the womb's natural retentiveness. Of all the retentive organs, the uterus was regarded by Galen and by Renaissance anatomists as exemplary. But it was not essentially different from the male testicles, either in its coldness or its retentiveness. For the opinion of Jean Fernel, see Evelyn Berriot-Salvadore, "The Discourse of Medicine and Science," in *A History of Women in the West,* 5 vols, ed. by Georges Duby, Michelle Perrot, and Pauline Schmitt Pantel (Cambridge: Belknap Press of Harvard University Press, 1992–94), III, *Renaissance and Enlightenment Paradoxes,* ed. by Natalie Zemon Davis and Arlette Farge (1993), 348–88. Writing on a condition that as a further kind of "Melancholy" was imperfectly differentiated from "the Suffocation of the Matrix" or Mother—"Furor Uterinus"—James Ferrand argues that the etiology of ravenous feminine desire consists in "venemous and malignant" vapors retained by the uterus. James Ferrand, Ερωτομανια; *or, A Treatise Discoursing of the Essence, Causes, Symptomes, Prognosticks,*

and Cure of Love or Erotique Melancholy, trans. by E. Chilmed (Oxford, 1640), pp. 120, 99; originally published in French as *La maladie d'amour, ou mélancholie érotique* (Paris, 1623). Also see Robert Burton, *The Anatomy of Melancholy,* ed. by Thomas C. Faulkner, Nicolas K. Kiessling, and Rhonda L. Blair, and commentary by J. B. Bamborough and Martin Dodsworth, 6 vols. (Oxford: Clarendon Press, 1989–2000; repr. 1992–97), III, 231: "Foelix Platter *in the first booke of his observations,* tels a story of an ancient Gentleman in Alsatia, that married a young wife . . . but she because of this inhibition of Venus, fell into a horrible fury, and desired every one that came to see her, by words, lookes, and gestures to have to doe with her, &c." (In this quotation I have altered that which in the original is in roman font to italics, and vice versa.)

30. Berriot-Salvadore, "The Discourse of Medicine and Science," p. 373.

31. Edward Jordan, *A Briefe Discourse of a Disease Called the Suffocation of the Mother. Written uppon Occasion Which Has Beene of Late Taken thereby, to Suspect Possession of an Evill Spirit, or Some Such Like Supernaturall Power. Wherein in Declared that Divers Strange Actions and Passions of the Body of Man, Which in the Common Opinion, are Imputed to the Diuell, Have their True Naturall Causes, and Do Accompanie this Disease* (London, 1603), repr. in Michael MacDonald, ed., *Witchcraft and Hysteria in Elizabethan London: Edward Jorden and the Mary Glover Case* (London: Tavistock / Routledge, 1991), fol. 17ᵛ [19ᵛ]. The surgeon Ambroise Paré (1510–90) ascribes to "too violent and immoderate evacuations of the flowers" the prognosis of "the ascention of the wombe," which he advises is "diligently to bee distinguished from the strangulation thereof" on account of a different and less severe symptomatology. *The Workes of that Famous Chirurgion Ambrose Parey,* translated from Latin and compared with the French by Thomas Johnson (London: Thomas Cotes and R. Young, 1634), Bk. XXIV, p. 940; published in French as *Les Oeuvres de M. Ambroise Paré* (Paris, 1575; repr. 1579); and *Les Oeuvres d'Ambroise Paré* (Paris, 1585). The English translation by Thomas Johnson was republished with *Three Tractates out of Adrianus Spigelius of the Veines, Arteries, and Nerves, with Large Figures* (London: R. Cotes & W. Dugard, 1649). I have also made reference to another edition: *The Works of that Famous Chirurgeon Ambrose Parey,* translated out of Latin, and compared with the French, by Th[omas] Johnson; together with *Three Tractates Concerning the Veines, Arteries, and Nerves: Exemplified with Large Anatomical Figures: Translated out of Adrianus Spigelius* (London: John Clark, 1678).

32. Paré, *Workes* (1634), Bk. XXIV, p. 939.

33. Berriot-Salvadore, "The Discourse of Medicine and Science," p. 362. It is not uncommon for a medical text to advise hysterical patients to masturbate or to have their genitals manipulated by a midwife in order to emit seed from the uterus. See Paré, *Workes* (1634), Bk. XXIV, p. 945. Other medical advice warns against masturbation for precisely the reason that it deposits seed within the uterus, which may subsequently corrupt on account of not being emitted.

34. MacDonald, *Witchcraft and Hysteria in Elizabethan London,* fol. 22ᵛ; Paré, Workes (1634), Bk. XXIV, p. 942; Berriot-Salvadore, "The Discourse of Medicine and Science," p. 362.

35. Paré, *Workes* (1634), Bk. XXIV, p. 945. The persistently reiterated cure of hysteria was: "If [the illness is] from the retention of seed; a good husband will administer cure"; "if she be married, let her forthwith use copulation, and bee strongly encountered by her husband, for there is no remedy more present than this." Sadler, The Sicke Woman's Private Looking-Glasse, p. 74; Paré, Workes (1634), Bk. XXIV, p. 945. Similarly, Ferrand's therapeutic advice, in cases of "furor uterinus," is "speedy Marriage." Ερωτομανια, p. 97. In cases of hysteria, the physician's administrations for the restoration of the female body for use would seem to have induced abjection. Paré's account is the fullest in detail and the least discrete: the patient is placed on her back, her garments loosed; her hair, especially her pubic hair, is pulled; her legs and arms are bound and tied "with painfull ligatures"; her body is rubbed with salt and vinegar "until it be very sore and red"; a pessary is inserted into the uterus; cupping glasses are applied to the belly, thighs, and vulva; "glysters" are injected into the anus and vagina; and finally she is copulated with by her husband (pp. 943–45). Each medical procedure is anatomically and medically rationalized. For example, pubic hair is pulled so that "the patient may not onely be brought to her selfe again, but also that the sharpe and maligne vapour ascending upwards, may be drawne downwards" (p. 943). For discussions of the seventeenth-century anatomical and medical discourses, which posit women as "natural" and subject to reproductive or scientific "use," see Berriot-Salvadore, "The Discourse of Medicine and Science"; Martensen, "The Transformation of Eve"; Sawday, *The Body Emblazoned,* ch. 8: "Royal Science," pp. 230–70.

36. Raynalde, trans., *The Birth of Mankinde,* fol. 29.

37. Sadler, *The Sicke Woman's Private Looking-Glasse,* p. 3. Raynalde, trans., *The Birth of Mankinde,* fol. 41, argues that the effect of the womb's appetite is to persuade women to forget the pains and labors of childbirth, and become pregnant again.

38. Paré, *Workes* (1634), Bk. XXIV, p. 939.

39. Ibid., p. 925.

40. Sadler, *The Sicke Woman's Private Looking-Glasse,* p. 86, lists these and other etiologies.

41. Paré, *Workes* (1634), Bk. XXIV, p. 925. Françoise Mauriceau agreed "that the main factor in the formation of a mole was too frequent intercourse." James V. Ricci, *The Genealogy of Gynaecology* (Philadelphia: Blakiston Company, 1950), pp. 297–98.

42. Sadler, *The Sicke Woman's Private Looking-Glasse,* p. 86.

43. Paré, *Workes* (1634), Bk XXIV, p. 939

44. Although Jorden's *A Brief Discourse of a Disease called the Suffocation of the Mother* was one of the first medical accounts of hysteria in English, its partially unorthodox etiology fell on the deaf ears of a medical profession seemingly steeped in a utero-centric model of female physiology and pathology. Despite its representation in some histories of medicine as a precursor of the neural etiologies proposed by Thomas Willis and Thomas Sydenham, Jordan's etiology does less than reject the malady as a disturbance of the uterus. MacDonald, ed., *Witchcraft and Hysteria in Elizabethan London,* p. xlvii, remarks that "Jorden's *Brief Discourse* had little lasting influence on the diagnosis of either witchcraft or hysteria. . . . Nor was his influence over the subsequent history of hysteria as great as Veith's superb survey implies. For although he added mental disturbance to the aetiological factors of the disease, he did not renounce traditional uterine explanations. And those who did, Thomas Willis and Thomas Sydenham, offered completely different explanations for the nervous and psychological basis of the illness." While the detachment of Charles Lepois's "animal" and neural etiology from the utero-centric model of female physiology and pathology was absolute, it made little impact on the development of the concept of hysteria in England during the first half of the century. In *Selectiorum observationum et consiliorum,* Lepois reasons: "Since in hysterical suffocation the entire body is seized with convulsions and stiffens, the principle of the nerves [which extend throughout the body and are responsible for movement and sensation] is necessarily diseased." Charles Green Cumston, *An Introduction to the History of Medicine from the Time of the Pharaohs to the End of the XVIIIth Century* (London: Kegan Paul, Trench, Trübner & Co., 1926), p. 299. According to Ilza Veith, *Hysteria: The History of a Disease* (Northvale, N.J.: Jason Aronson, 1993; 1st edn. 1965), pp. 128–29, Charles Lepois was the first European physician to argue unequivocably for a brain etiology of hysteria.

45. Stephen Blancaard's *Anatomica Practica Rationalis* (1688) records that in the autopsy of a patient who had died of "hysterical passion," the uterus and ovaries were found to be markedly congested. The anatomist's prior diagnostic convictions encouraged him to invest pathological organic abnormality with etiological significance.

46. Charles Lepois / Carolus Piso (1563–1633) observes in his *Selectorium observationum et consiliorum de praetervisis hactenus morbis affectibusque praeter naturam ab aqua seu serosa colluvie et diluvie ortis, liber singularis, &c* (1618): "The symptoms ordinarily called hysterical are almost all common to both men and women." Cumston, *Introduction to the History of Medicine,* p. 299. Similarly Michael Ettmüller, *Etmullerus Abridg'd: or, A Compleat System of the Theory and Practice of Physic, Being a Description of All Diseases Incident to Men, Women and Children,* 3rd ed. (London: Andrew Bell, 1699), bk. IV, section 3, ch. 3, p. 600, notes: "Tho this Disease is accounted peculiar to Women, and to proceed from the Womb, yet it is found by experience to be common also to Men, and to be of the same Nature with their hypocondriac and scorbutical Distempers. . . ."

47. Raynalde, trans., *The Birth of Mankinde,* fol. 48.

48. Even Lepois's description of the disease, while unconventional in etiology, retains as principal symptoms the paralysis of movement and sensation, and a deep torpor, from which "nothing will arouse the patient." Lepois's nervous etiology arises from an attempt to explain this nervous symptomatology. Cumston, *Introduction to the History of Medicine,* p. 301. Michael Ettmüller, *Opera Omnia,* cited in Ricci, *The Genealogy of Gynaecology,* p. 308, describes three ways in which a physician may distinguish between a "hysterical sleep" and death: "by using a Looking glass, or Feather, or blowing Snush up the Nose."

49. Paré, *Workes* (1634), bk. XXIV, p. 942. The account cited is also told by Sadler, *The Sicke Woman's Private Looking-Glasse,* pp. 62–63. Lazarus Riverius, in *The Practice of Physick* (London: Peter Cole, 1658), p. 421, writes that "some have been accounted dead, and laid out for Burial, and yet afterward Revived." Indeed Lepois records the case of a young girl, "who was about to be buried and would have been had she not awakened in time." Cumston, *Introduction to the History of Medicine,* p. 301.

50. Laqueur, *Making Sex,* p. 110: "The new anatomy, and more specifically the widespread distribution of anatomical illustrations . . . well beyond the bounds of the learned community to midwives, barber surgeons, and laypeople, showed that not only was the uterus kept more or less in place by very broad ligaments but that the space between it and the throat was full of other organs and divided by thick membranes."

51. Roy Porter, *Mind-Forg'd Manacles: A History of Madness in England from the Restoration to the Regency* (Harmondsworth: Penguin, 1990), p. 48.

52. Benedetti, *Anatomice,* p. 100.

53. Paré, *Workes* (1634), bk. XXIV, pp. 940, 939.

54. Realdo Colombo (*c.* 1515–59), *De re anatomica* (Venice, 1559)

55. Gabriele (Gabriello) Falloppio (c. 1523–62), *Observationes anatomicae* (Venice, 1562), [another edn.] (Helmstadt, 1588)

56. Katharine Park, "The Rediscovery of the Clitoris," in *The Body in Parts: Fantasies of Corporeality in Early Modern Europe,* ed. by David Hillman and Carla Mazzio (London: Routledge, 1997), pp. 170–93 (p. 178).

57. De Laguna, *Anatomica methodus,* p. 280. Unfortunately, if de Laguna was in possession of a view that differed fundamentally from the "one-sex" view of Aristotle and Galen, he did not divulge it.

58. Park, "The Rediscovery of the Clitoris," p. 173.

59. Lyndal Roper, *Oedipus and the Devil* (London and New York: Routledge, 1994), p. 16.

60. Sawday, *The Body Emblazoned,* p. 213.

61. *The Blank-Paged Bible: The Holy Scriptures of the Old and New Testaments* [*Authorised Version*]*; With Copious References to Parallel and Illustrative Passages: and the Alternate Pages Ruled for MS. Notes* (London: Samuel Bagster and Sons, [n.d.]). All further references to the Bible are to the Authorised Version, unless otherwise stated.

62. David Lee Miller, "Spenser's Poetics: The Poem's Two Bodies," *Publications of the Modern Language Association* 101 (1986): 170–85 (p. 178).

63. For a discussion of Fletcher's sources, including *The Faerie Queene,* see Chapter 4.

64. Miller, "Spenser's Poetics: The Poem's Two Bodies," p. 178.

65. Edmund Spenser (1552?–99), *The Faerie Queene,* ed. A. C. Hamilton, assisted by R. J. Manning, Longman Annotated English Poets (London: Longman, 1977). All references to *The Faerie Queene* are to this edition, and are cited in the text in the manner in the text where the reference to this note appears, that is: Book number, Canto number, stanza number, line number(s), all of which is prefixed by the initials of the short title: *F.Q.*

66. Margaret Healy, *Fictions of Disease in Early Modern England* (Basingstoke: Palgrave, 2001), p. 3.

67. See, for example, Barkan, *Nature's Work of Art,* pp. 62–66. Barkan says that "Renaissance England represents in a way the heyday of the anthropomorphic image of the commonwealth" (p. 75).

68. *The Republic of Plato,* trans. with an introduction and notes by Francis MacDonald Cornford (Oxford: Clarendon Press, 1941; repr. 1966), ch. 13 (bk. IV, 434D–41C), pp. 126–35.

69. *Aristotle's Politics,* trans. Benjamin Jowett (Oxford: Clarendon Press, 1931), bk. I, ch. 2, pp. 26–30. See Paul Archambault, "The Analogy of the 'Body' in Renaissance Political Thought," *Bibliothèque D'Humanisme et Renaissance* 29 (1967): 21–53; especially pp. 22–23.

70. See Archambault, "The Analogy of the 'Body' in Renaissance Political Thought," p. 52.

71. An influential medieval political theorist who assigned to the sovereign the analogue of the head is John of Salisbury (c. 1110–1180), whose model of the social organism in his *Policraticus: Courtiers' Trifles and Philosophers' Footprints,* 8 bks. (1159), is based on an apocryphal text of Plutarch, the *Institutio Traiani.* John Fortescue (1400?–1476), in his *De laudibus legum Anglie,* argues on the premise of this analogy the necessity of one ruler. Thomas Starkey's *Dialogue between Cardinal Pole and Thomas Lupset,* written c. 1536, but not published until 1871, draws an analogy between the Prince and the heart, understood on the Aristotelian view. See Archambault, "The Analogy of the 'Body' in Renaissance Political Thought," pp. 26–27, 35, 40; p. 28 for Cambrensis. Archambault says of Starkey's *Dialogue:* "No other work of the period allows the image [of the body politic] so much importance, and dwells with such insistence upon its various forms. . . . In Starkey's work, rather than being merely an instrument of thought, the analogy becomes an entire system of expression." "The Analogy of the 'Body' in Renaissance Political Thought," p. 43.

72. Barkan, *Nature's Work of Art,* p. 62.

73. Ibid.

74. *The Works of Aristotle,* trans. into English under the editorship of J. A. Smith and W. D. Ross, 12 vols. [IV–V ed. by J. A. Smith; I–III, VI–XII ed. by W. D. Ross] (Oxford: Clarendon Press, 1908–52; repr. 1937–63), XI: *De poetica,* trans. by Ingram Bywater (1946; repr. 1959), ch. 21, 1457b. Also see F. E. Sparshott, "'As,' or The Limits of Metaphor," *New Literary History* 6 (1974): 75–94, especially pp. 81–82.

75. Redman, "*The Purple Island,*" pp. 2–3, argues, typically in this case, that allegory implies that "writers often focused on two planes of reality at once—that which reproduced what they perceived visibly and that which conveyed what they perceived imaginatively (spiritually)." Although, as Redman argues, "definitions all stress the dualistic statement of allegorical works" (p. 4), Quilligan's redefinition of the mode and genre of allegory entails a "fundamental shift in emphasis away from our traditional insistence on allegory's distinction between word said

and meaning meant, to the simultaneity of the process of signifying multiple meaning." *The Language of Allegory,* p. 26. Such a definition does not preclude the intermittence of one or other signified being uppermost in the signification of the narrative, but it does suggest that allegory is characterized by polysemy or multiplicity of analogical or metaphorical meaning as distinct from "different 'levels' of meaning." The latter conception of allegorical meaning is popular and "signals our tendency to think about allegory in terms of a vertically organized spatial hierarchy, where the gaps between the levels of meaning, rather than the relationships across the gaps, holds for us the definitive allure of the form." *The Language of Allegory,* p. 27. My use of the idea of intermittent submergence and emergence of signifieds is intended to imply simultaneity of signification, albeit with shifts of emphasis upon one or other signified, as well as to allude to the linearity of the narrative succession of signification. As Quilligan writes, the "vertical conceptualization of allegory and its emphasis upon disjunct 'levels' is absolutely wrong as a matter of practical fact. All reading proceeds linearly, in a word-by-word fashion, but allegory often institutionalizes this fact by the journey or quest form of the plot. . . . It would be more precise to say therefore that allegory works horizontally, rather than vertically, so that meaning accretes serially, interconnecting and crisscrossing the verbal surface long before one can accurately speak of moving to another level 'beyond' the literal." *The Language of Allegory,* p. 28.

76. Kastor, *Giles and Phineas Fletcher,* p. 115.

77. Parfitt, *English Poetry of the Seventeenth-Century,* p. 191. The view held by Parfitt and Kastor, that the poem is discursively incongruous, also looks unfavorably upon its generic hybridity: see Kastor, *Giles and Phineas Fletcher,* pp. 112, 115; Parfitt, *English Poetry of the Seventeenth-Century,* p. 191.

78. Gosse, *The Jacobean Poets,* p. 146.

79. This minority includes Henry Headley: "Nor is Fletcher wholly undeserving of praise for the intelligibility with which he has struggled through his difficulties, for his uncommon command of words, and facility of metre." Henry Headley, ed., *Selected Beauties of Ancient English Poetry,* 2 vols. (London: T. Cadell, 1787), I, liii. Headley's critical remarks are reproduced in *The Purple Island, a Poem: by Phineas Fletcher,* with the critical remarks of the late Henry Headley, A.B. and a biographical sketch by William Jaques (London: Burton and Briggs, 1816), pp. v–xi.

80. Sawday, "Bodies by Art Fashioned," p. 108.

81. Parfitt seems to make this qualification: *English Poetry of the Seventeenth-Century,* p. 191.

82. Langdale, *Phineas Fletcher,* p. 167.

83. Ibid., p. 169.

84. Ibid., ch. 11. Langdale's claims for the character and derivation of the poem's anatomical content are debated in more detail in the third part of this study.

85. David M. Knight, *Natural Science Books in English 1600–1900* (London: Portman Books, 1989; first published 1972), p. 1. Before the 1830s, "scientific men called themselves 'natural philosophers', or 'natural historians'; and indeed the word 'science' meant any organized body of knowledge, down to about the same date."

86. Peter Dear, "The Church and the New Philosophy," in *Science, Culture and Popular Belief in Renaissance Europe,* ed. by Stephen Pumphrey, Paolo Rossi, and Maurice Slawinski (Manchester: Manchester University Press, 1991), pp. 119–39 (p. 123); Bruce Janacek, "Catholic Natural Philosophy: Alchemy and the Revivification of Sir Kenelm Digby," in *Rethinking the Scientific Revolution,* ed. by Margaret J. Osler (Cambridge: Cambridge University Press, 2000), pp. 89–118 (pp. 94, 96).

87. Langdale, *Phineas Fletcher,* p. 170.

88. George Lakoff, "The Meanings of Literal," *Metaphor and Symbolic Activity,* 1 (1986): 291–96 (p. 294). Lakoff analyses four allegedly conjunctive assumptions characterizing the term "literal," showing that they are actually disjunctive and restricts the meaning to that given in the text. This definition of "literal" is the one to which this study initially adheres, and may be assumed where the use in the text of the term "literal" is not qualified so as to signify a divergence in its meaning.

89. Wayne Booth, "Metaphor as Rhetoric," *Critical Inquiry* 5 (1978): 49–72 (p. 53).

90. Langdale, *Phineas Fletcher,* p. 176.

91. Baldwin, "Phineas Fletcher."

92. See, for example, Michel Foucault, *The Order of Things: An Archaeology of the Human Sciences,* Series: World of Man: A Library of Theory and Research in the Human Sciences (London: Tavistock Publications, 1970), pp. 23–25; originally published in French under the title *Les Mots et les choses* by Éditions Gallimard, 1966. Foucault argues that the principle of sympathy has the power of assimilating, of rendering things identical, and is compensated for by antipathy. Sympathy and antipathy together explain how things can resemble others without being swallowed up, without the world being reduced to a point of identity.

93. See, for example, Barkan, *Nature's Work of Art,* p. 15.

94. Galen's adherence to the four qualities theory of Hippocrates (c. 400 B.C.) was an aspect of his view that "in medicine as in geometry there were a certain number of self-evident maxims upon which reason, conforming to the rules of logic, might build up a scientific structure." In *The Healing Art,* Galen "devotes a number of chapters to such subjects as signs of a hot and dry heart, signs of a hot liver, and signs of a cold lung," thereby introducing the four qualities into medical prognostics. Lynn Thorndike, *A History of Magic and Experimental Science During the First Thirteen Centuries of Our Era,* 8 vols., vols. I–II (London: Macmillan, 1923–58), vols III–VIII (New York: Columbia University Press, 1947, 1960), I (1923), 154–55, 176. On Galen's hugely influential adherence to the Hippocratic view that the four elements and four qualities compose all natural objects, see Thorndike, *A History of Magic and Experimental Science,* I, 139.

95. Barkan, *Nature's Work of Art,* pp. 19–20.

96. Ibid., p. 9. *Timaeus* was "the one book by Plato that continued to be available throughout the rest of Greek and Roman Antiquity and also throughout the so-called Dark and Middle Ages in the Latin West, at least in part." Cunningham, *The Anatomical Renaissance,* p. 11. Joseph A. Mazzeo, "Universal Analogy and the Culture of the Renaissance," *Journal of the History of Ideas* 15 (1954): 299–304, refers to the view of the cosmos as "a living universe, pervaded by vitality and rationality," as hylozoism, which, he argues, was accepted from Thales to Kepler (p. 299). I have used the modern edition of *Timaeus,* in *The Dialogues of Plato,* trans. into English with analyses and introductions by Benjamin Jowett, 4 vols., 4th ed. (Oxford: Clarendon Press, 1953), III, 631–780.

97. Langdale, *Phineas Fletcher,* p. 170; also see pp. 147–48, 169–70. The difference between the Renaissance Platonic idea of the human as an index to the whole cosmos and the idea of the microcosm may be theorized in terms of the distinction that Foucault makes between the sixteenth-century concepts of "*aemulatio*" and "*analogy,*" in Foucault, *The Order of Things,* pp. 19–23. "*Aemulatio*" or emulation is described as the similitude of reflection from a distance, that is, without the need of proximity in the great chain of being (p. 19). Analogy superimposes *aemulatio* and *convenientia,* which is similitude as the visible effect of propinquity or proximity in the great chain of being (pp. 18–19, 21). Analogy therefore involves resemblances across space—in the case of the microcosm, between the human body and the cosmos—in the context of adjacencies or joints (p. 21). So that, for example, the sun in the center of the cosmos is analogous to the heart which is positioned in the center of the human body. This is consistent with Aristotle's definition of analogy, which is discussed in the text above and below.

98. Mazzeo, "Universal Analogy and the Culture of the Renaissance," p. 302.

99. See Foucault, *The Order of Things,* p. 22.

100. Langdale, *Phineas Fletcher,* p. 170.

101. See Barkan, *Nature's Work of Art*, pp. 10, 11.

102. Ibid., p. 12.

103. Ibid., p. 3.

104. Ibid., pp. 3, 28.

105. Mazzeo, "Universal Analogy and the Culture of the Renaissance," p. 303. Furthermore, Mazzeo says of Renaissance poets: "Their criterion of the truth or falsity of a statement seems to have been its degree of richness in essentially 'poetic' elements." Mazzeo argues, however, "philosophers and poets of the Renaissance not only often shared a common vocabulary but the same *forma mentis,*" which was "richly metaphorical and symbolic" (pp. 300, 303). With this last statement I am in agreement, but on a different premise, that analogy is logical and conceptual.

106. Langdale, *Phineas Fletcher,* p. 170. "Mundinus" is the Latinized name for Mondino dei Luzzi (*c.*1270–1326), professor of anatomy and surgery in the medical faculty at Bologna University from the 1290s until his death, and author of *Anatomia Mundini,* which dates from 1316 in preserved form in manuscript and print, and which for two and a half centuries was the most used textbook in North Italy and even north of the Alps as well. See, for example, Ruben Eriksson, *Andreas Vesalius' First Public Anatomy at Bologna 1540: An Eyewitness Report by Baldasar Heseler Medicinae Scolaris together with his notes on Matthaeus Curtius' Lectures on Anatomia Mundini,* ed., with an introduction, translation into English, and notes (Uppsala: Almquist and Wiksell, 1959), pp. 34–36.

107. Langdale, *Phineas Fletcher,* p. 170.

108. Ibid., p. 170.

109. Barkan, *Nature's Work of Art,* p. 14.

110. According to Barkan, the unity of the cosmos for Plato consists in the bodily nature of its constituents, as the unity of the human body consists in that of its constituent parts. Although both cosmos and human body consist of parts, unification in the cosmos is so perfect that the parts are hard to distinguish, whereas in the human body the parts are distinguishable. *Nature's Work of Art,* pp. 11–12.

111. Foucault, *The Order of Things*, p. 18.

112. See Don Parry Norford, "Microcosm and Macrocosm in Seventeenth-Century Literature," *Journal of the History of Ideas* 38 (1977): 409–28 (p. 413).

113. Baldwin, "Phineas Fletcher," p. 467.

114. It is argued by many historians of medicine that the practice of anatomical dissection after the Greeks began again with Mondino, although this is a controversial point. According to Charles Singer, Mondino "was the first known to have dissected the human body systematically in public. He thus introduced practical anatomy as an academic subject." Charles Singer, "Beginnings of Academic Practical Anatomy," introduction to Ludwig Choulant, *History and Bibliography of Anatomic Illustration*, trans. and annotated by Mortimer Frank (London: Hafner Publishing Co., 1962), pp. 21-A–21-R (p. 21-E); originally published as *Geschichte und Bibliographie der anatomischen Abbildung nach ihrer Beziehung auf anatomische Wissenschaft und bildende Kunst* (Leipzig: Rudolph Weigel, 1852). On this matter Eriksson, *Andreas Vesalius' First Public Anatomy at Bologna 1540*, p. 36, writes: "It was thanks to Mundinus that anatomies of human corpses were made part of the study of medicine."

115. Baldwin, "Phineas Fletcher," p. 467.

116. F. W. Dillistone, *Christianity and Symbolism* (London: Collins, 1955), p. 45. Dillistone is making reference to R. G. Collingwood, *The Idea of Nature* (Oxford: Oxford University Press, 1945).

117. "Nature," with an uppercase initial letter, refers in the text to the meaning understood by the ancient Greek, medieval, Renaissance, or early modern natural philosophers, whichever is implied by the local discursive context of the use of the term. It does not correspond usually to the Post-Enlightenment, Romantic, or modern scientific uses of the same term, which are implied by the use in the text of "nature" with a lowercase initial letter. In Renaissance and early modern texts, however, the term Nature is used with the historical meaning appropriate to it, but with a lowercase initial letter. The meaning may be regarded as consistent, however, insofar as consistency is possible over historical change, with that connoted by its use in the text outside of quotations, where that is intended to convey the Renaissance or early modern conception.

118. Barkan, *Nature's Work of Art*, p. 30.

119. French, *Dissection and Vivisection in the European Renaissance*, p. 12.

120. Ibid.

121. Ibid., pp. 12–15.

122. See Lind, *Studies in Pre-Vesalian Anatomy*, pp. 5, 6. "The very word "scholasticism" could be argued to be the invention of humanist writers, anxious to discredit the movement which it represented," according to Alister E. McGrath, *Reformation Thought: An Introduction*, 2nd ed. (Oxford: Blackwell, 1993; repr. 1994), p. 67.

123. See, for example, Cunningham, *The Anatomical Renaissance*, p. 207; Foucault, *The Order of Things*, p. 30, who states that the microcosm was revived during the Middle Ages.

124. Thorndike, *A History of Magic and Experimental Science*, III (1960), 240–47.

125. McGrath, *Reformation Thought*, p. 68.

126. Lind, *Studies in Pre-Vesalian Anatomy*, p. 39: "[M]en did not dissect in order to discover new knowledge but simply to verify the old and to demonstrate [the] accuracy [of Galen, Avicenna, or Mundinus] to their observers, students or laymen." In the Statutes of the medical faculty at the University of Padua as late as 1570, "the teachers were warned not to present views opposed to those in the prescribed texts. . . ." Eriksson, *Andreas Vesalius' First Public Anatomy at Bologna 1540*, p. 41. Avicenna lived in Persia in the eleventh century. Eriksson argues that statutes like the above-mentioned ones of the medical faculty of the University of Padua "contributed towards the retention of *Anatomia Mundini* as a text-book in anatomy for such a long period of time at Bologna, Padua and other schools of medicine in northern Italy" (p. 41). There have been harsh assessments, usually by historians of early modern anatomy, of the period from Galen to the early sixteenth century, such as: "Truth and science came to mean simply that which was written, and inquiry became mere interpretation." M. Foster, *Lectures on the History of Physiology During the Sixteenth, Seventeenth and Eighteenth Centuries* (Cambridge: Cambridge University Press, 1901), p. 4.

127. Andrea Carlino, *Books of the Body: Anatomical Ritual and Renaissance Learning*, trans. by John Tedeschi and Anne C. Tedeschi (Chicago: University of Chicago Press, 1999), p. 2 (originally published as *La fabbrica del corpo: Libri e dissezione nel Rinascimento*, at Turin, in 1994)

128. Eriksson, *Andreas Vesalius' First Public Anatomy at Bologna 1540*, p. 36; Glenn Harcourt, "Andreas Vesalius and the Anatomy of Antique Sculpture," *Representations* 17 (1987): 28–61 (p. 36). The "*chirurgus*" was also known as the "*incisor*," "*prosector*," or "*resector*," and in sixteenth-century England, as the "steward." See William Heckscher, *Rembrandt's Anatomy of Dr. Nicolaas Tulp: An Iconological Study* (New York: New York University Press, 1958), p. 48. Also see Carlino, *Books of the Body*, p. 78. Harcourt says that this pedagogical

procedure required of the professor of anatomy either a lecture "or even simply the recitation of a text—most often Galen or the fourteenth-century *Anathomia* [1316] of Mondino da Luzzi" (p. 36). Carlino, *Books of the Body,* pp. 85–86 (p. 86), has revealed that the functions of the persons involved in the anatomical demonstration varied between places. The arrangement outlined in the text was that adhered to at Rome and other places, while at Padua the *lector*'s role was merely to read the *Anatomia Mondini,* while the responsibility of the more prominent figure of the "public reader" was "to clarify and to comment on the passages read and then [like the *ostensor* elsewhere] to demonstrate and affirm his commentary on the cadaver." Doing a lot of human dissection and animal vivisection himself, Berengario da Carpi cast himself in the role of "operator" (equivalent to the *anatomista*). French, *Dissection and Vivisection in the European Renaissance,* p. 96. Notwithstanding the innovations of Berengario, and later Vesalius, Carlino, *Books of the Body,* p. 2, argues: "During the entire sixteenth century such writings [as were used in public anatomies] consisted of Galen's surviving works and of a series of *compendia*—manuals and treatises inspired by them—of which the most popular was Mondino dei Liuzzi's *Anatomia,* written in 1316."

129. There is an illustration in the edition of Mondino's *Anatomia* which is published in Johannes de Ketham (d. *c.* 1490), *Fasciculo di medicina* (Venice, 1493), which according to Singer's introductory essay in Choulant, *History and Bibliography of Anatomic Illustration,* p. 21-G, is "supposed to represent a successor of Mundinus in the 'chair' of (practical) anatomy at Bologna." The Ketham work is a compilation of texts (including Mondino's *Anatomia*) and illustrations, first published in Venice in 1491 in a Latin edition, *Fasciculus medicinae.* That the collection was done by Ketham, however, has been called into doubt. It is reproduced as *The Fasciculo Di Medicina, Venice 1493,* intr. by Charles Singer, 2 parts, in *Monumenta Medica,* 2 vols. (Florence: R. Lier and Co., 1925), II (all further references are to this edition unless otherwise stated). See Choulant, *History and Bibliography of Anatomic Illustration,* pp. 75–97.

130. See, for example, Foucault, *The Order of Things,* pp. 30–31, where it is argued that by the sixteenth century the microcosm had a fundamental role in the field of knowledge.

131. Mazzeo, "Universal Analogy and the Culture of the Renaissance," pp. 301–2, outlines this interpretation, saying that Nominalism "had rejected 'faith metaphysics' and made the world of experience the proper object of philosophy" (p. 302). But without a proper method for the study of this new philosophical object, Neoplatonism revived to provide universal analogies and cosmic affinities, microcosm-macrocosm and "varying versions of *coincidentia oppositorum,*" as a basis for a new theory of knowledge.

132. Norford, "Microcosm and Macrocosm in Seventeenth-Century Literature," pp. 409, 410, 412 (p. 409 cited); Barkan, *Nature's Work of Art,* pp. 32–33. For Marsilio Ficino (1433–99), see his epistle "Five Questions Concerning the Mind," trans. with an introduction by Josephine L. Burroughs, in *The Renaissance Philosophy of Man,* ed. by Ernst Cassirer, Paul Oskar Kristeller, and John Herman Randall Jr. (Chicago: University of Chicago Press, 1948), pp. 183–212. Ficino's notion of the *appetitus naturalis* is, according to Burroughs, dependent upon a concept of the essence of a species such as Man, arising directly from that species and "ultimately related to God"; it is consequently a relation between particular things and God as the source of order, and as boundless truth and goodness to which the *appetitus naturalis* of Man is directed (pp. 188–89). For Giovanni Pico della Mirandola (1463–94), see his *Oration on the Dignity of Man,* translated by Elizabeth Livermore Forbes, with an introduction by Paul Oskar Kristeller, in *The Renaissance Philosophy of Man,* pp. 213–54. The *Oration,* which was never published during Pico's lifetime, was written as an introductory speech for a projected disputation on his nine hundred theses, published in December 1486, which was suspended by Pope Innocent VIII.

133. Norford, "Microcosm and Macrocosm in Seventeenth-Century Literature," p. 411.

134. Cunningham, *The Anatomical Renaissance,* p. 41.

135. Norford, "Microcosm and Macrocosm in Seventeenth-Century Literature," pp. 409, 415, 410.

136. Ibid., pp. 425–26.

137. Helkiah Crooke, *Microcosmographia: A Desription of the Body of Man* (London: W. Jaggard, 1615), fol. 4.

138. Foucault, *The Order of Things,* pp. 17, 18, 19, 21.

139. Crooke, *Microcosmographia,* fol. 3.

140. Ibid., fol. 7. Sir Walter Ralegh (Raleigh) (1552–1618) also evokes this understanding of Man as a chemical microcosm: "In this also is the little world of man compared, and made more like the Vniversall (man being the measure of all things; *Homo est mensura omnium rerum,* saith *Aristotle* and *Pythagorus*) that the foure complexions resemble the foure elements." *The History of the World* (London: [W. Stansby] for Walter Burre, 1614), bk. I, ch. 2, section 5, p. 31.

141. Barkan, *Nature's Work of Art,* p. 2. According to Temkin: "In the classical and explicit form which [the microcom theory] reached in the late Middle Ages and Renaissance it consisted of a point by point concordance between the world and animal life." "Metaphors of Human Biology," p. 279.

142. Crooke, *Microcosmographia,* fol. 3.

143. Lind, *Studies in Pre-Vesalian Anatomy,* p. 6. John Hale, *The Civilization of Europe in the Renaissance* (London: HarperCollins, 1993), p. 546, says: "Anatomical dissections, carried out by the humbler surgeons under the direction of a presiding physician, from being visual demonstrations of information included in classical texts, became more questioning." A public anatomical theater was founded by Alessandro Benedetti in the University of Padua.

144. See Lind, *Studies in Pre-Vesalian Anatomy,* pp. 5–6. An example of the continuation of scholastic disputation in sixteenth-century anatomy is in Alessandro Achillini (1463–1512), *Annotationes anatomicae* (Bologna, 1520), translated in Lind, *Studies in Pre-Vesalian Anatomy,* p. 47, where Achillini judiciously favors Galen in a debate about the function of the portal vein, yet simultaneously manages a reconciliation with two Arabic medical authors, Almansor and Averroes (all references to Achillini's *Annotationes anatomicae* in the text and notes hereafter are to this translation): "Almansor in his first book says that the portal vein does not carry blood; so does Averroes, *Colliget,* I, ch. 27, but in chapter 7 he had said a branch of the portal vein comes to nourish the upper part of the stomach in the right-hand region of the back. Therefore, either this vein carries blood or the stomach is nourished from outside but not with blood. Then Galen insisted that the spleen in some manner makes blood; therefore, the vein will carry impure blood from the spleen." Lind, *Studies in Pre-Vesalian Anatomy,* p. 47.

145. Lind, *Studies in Pre-Vesalian Anatomy,* p. 6. On the intellectual paternity chosen by medical teachers in the schools of Western Europe and the tradition they claimed and fostered, see French, *Dissection and Vivisection in the European Renaissance,* pp. 74–78.

146. French, *Dissection and Vivisection in the European Renaissance,* pp. 74–86 (p. 85); Gabriele de Zerbi (b. 1445), *Liver Anathomie Corporis Humani et singulorum membrorum illius* (Venice, 1502). From 1494, Zerbi was professor of theoretical medicine in the university at Padua.

147. The revival of Greek and Greek medicine among the professors of medicine at the University of Ferrara in the fifteenth and sixteenth centuries illustrates this point. See Vivian Nutton, "The Rise of Medical Humanism: Ferrara, 1464–1555," *Renaissance Studies* 11 (1997): 2–19 (p. 9). On Alessandro Benedetti, see Lind, *Studies in Pre-Vesalian Anatomy,* p. 6.

148. See Vivian Nutton, "Greek Science in the Sixteenth-Century Renaissance," in: J. V. Field and Frank A. J. L. James, eds., *Renaissance and Revolution: Humanists, Scholars, Craftsmen and Natural Philosophers in Early Modern Europe* (Cambridge: Cambridge University Press, 1993), pp. 15–28 (p. 19). For example, on the question of the identity of the *mirach, siphac* and *zirbus* of the Latin translations of the Arabic of Avicenna and others, see French, *Dissection and Vivisection in the European Renaissance,* pp. 49, 54.

149. An early sixteenth-century example of such investigative dissection is recorded in de Laguna, *Anatomica methodus,* p. 277: "Many, however, have doubted whether the wine distils in this way into the bladder; this doubt has surely not been justified by dissection since all these ducts become occluded owing to the cold air to which they are exposed so that these valves and foramina completely escape detection by the senses. In order to confirm by experience what I had conceived in my imagination, I placed a rather long probe through the ureters and dilated the closed mouths without much trouble so that no further occasion for doubt remained." Later, in *Anatomica methodus,* p. 284, de Laguna writes: "For I know that there are two instruments by which we always search out the truth: the one is reason, the other is experience." These became watchwords for Renaissance and early modern anatomists.

150. Andreas Vesalius (1514–64), *De humani corporis fabrica librorum epitome* (Basle: Joannes Oporinus, 1543). References to this edition will hereafter be in the short-title form: Vesalius, *Epitome.* I have cited a scholarly edition in English translation: *The Epitome of Andreas Vesalius,* trans. from the Latin with a preface and introduction by L. R. Lind (Cambridge: The M.I.T. Press, 1969), pp. xxxiv, xxxiii. All further references to the 1969 edition use the short title *The Epitome of Andreas Vesalius.* The seven books on the fabric of the human body were published as Andreas Vesalius, *De humani corporis fabrica libri septem* (Basle: Joannes Oporinus, 1543) (all references to this text are to Vesalius, *Fabrica,* or merely the *Fabrica*). Hale, *The Civilization of Europe in the Renaissance,* p. 546, remarks upon this combination of anatomical dissection, medicine, and microcosmic theory: "While the human body was there to be undressed, palpated, bled, cut into, poulticed and medicated, it was also, as Galen had implied, invisibly linked to the rest of the natural world by its chemical composition and the physical principles that theoretically connected the jerk in a muscle to a comet's trail across the heavens."

151. "The Preface of Andreas Vesalius to His Books *De Humani corporis fabrica* Addressed to the Divine Charles, Great and Invincible Emporer" is translated by C.D. O'Malley in an appendix to his *Andreas Vesalius of Brussels 1514–1564* (Berkeley: University of California Press, 1964), p. 324 of which is here cited. Vesalius's acknowledgment of the reasonableness of the microcosm theory of the ancients militates against Charles Singer's

view, that the *Fabrica* and Nicholas Copernicus's *On the Revolution of the Celestial Spheres* (1543) together "destroyed for ever the favourite mediæval theory of Macrocosm and Microcosm." Singer, *The Evolution of Anatomy,* p. 122. The "scientific" basis of the theory of the microcosm may or may not have been destroyed, but the theory continued to exercise considerable influence on ways of thinking about the relationship of Man to the universe.

152. Crooke, *Microcosmographia,* fol. 12. Also see Nicholas Culpeper's Dedication to Samuel Hyland in Johann Veslingus, *The Anatomy of the Body of Man . . . Published in Latin by Joh[ann] Veslingus, Reader of the Publick Anatomy in the Most Famous University of Padua . . . and Englished by Nich[olas] Culpeper* (London: Peter Cole, 1653), sig. A: "the knowledg of mans self being of all Natural knowledg the most profound and most to be desired, for he that knows himself aright cannot but know all the world, because he is an Epitome of it." Also see Veslingus's address to the reader, where he calls Man "a delicate Epitome of the whol world" (sig. C).

153. Benedetti, *Anatomice,* bk. I, ch. 4, p. 86. Benedetti's general conception derives in some measure from the tradition extending from Empedocles' notion of the "*discordia concors.*"

154. Ibid., p. 85. Benedetti's scheme includes the accretion of correspondences to the elements in Empedocles' theory. These include, most notably, Hippocrates' theory of the four forces or qualities of the human body, and the theory of humours. In a passage on what he calls the doctrine of the Peripatetikes (followers of Aristotle), Crooke, *Microcosmographia,* fol. 7, outlines the chemical or elemental microcosm-macrocosm analogy in anatomy. To the four elements of earth, air, fire, and water, he adds that of heaven; and he draws correspondences between these and melancholy, blood, choler, phlegm, and the "aethereal" spirits, respectively. The spirits will be discussed in the text of this and the next chapter in due course.

155. "Vesalius on the Human Brain: Being a Translation of a section of his Fabrica of 1543," in Charles Singer, *Vesalius on the Human Brain: Introduction, Translation of Text, Translation of Descriptions of Figures, Notes to the Translations, Figures* (London: Oxford University Press, 1952), p. 18. Lind, in *The Epitome of Andreas Vesalius,* p. 51, argues: "The Vesalian discussion of blood and bile is an effort to bring to the prevailing humoral theory of physiology and pathology some degree of anatomical specificity." The evidence of microcosmic theory in Vesalius's conception of the complexion of the parts of the body does not necessarily relate to whether or not the *Fabrica* subscribes to astrological theory. Charles Singer supposes that the two theories are related, and argues that "the detailed display of the structure of the human body by Vesalius and his followers . . . made the astrological point of view less tenable." Charles Singer, "Historical Relations of Religion and Science," in *Science, Religion and Reality,* ed. by Joseph Needham (London: Sheldon Press; New York: Macmillan, 1925), pp. 85–148 (p. 130).

156. This point is maintained also by Healy, *Fictions of Disease in Early Modern England,* p. 6, who argues that "the age tended to combine these ostensibly competing theories [the Galenic humoral, microcosmic paradigm and the Vesalian or Paracelsian models] in idiosyncratic manners: certainly, the Galenic bodily model . . . together with its body politic analogies, was not eclipsed by the major seventeenth-century challenges to its authority."

157. See, for a further example, de Laguna, *Anatomica methodus,* p. 273. A notable exception to the rule is Achillini, *Annotationes anatomicae,* p. 39, which, according to Lind, avoids entirely "any of the theories on humors, temperaments, complexions, influences, or virtues which vitiated the medical thinking of his time."

158. Crooke, *Microcosmographia,* fol. 14. Within the same range of metaphorically consistent expressions as "Index" are "abstract or modell, or brief Storie of the Vniversall, in whom God concluded the creation, and worke of the world." Ralegh, *The History of the World,* bk. I, ch. 2, section 5, p. 30.

159. The distinction between identity and similarity is recognized by Marjorie Hope Nicolson as of significance for the history of ideas of correspondence: "What once seemed 'identicals' have become in our modern world only 'similars'. Metaphor, based upon accepted truth, inscribed by God in the nature of the universe, has given way to simile." Marjorie Hope Nicolson, *The Breaking of the Circle: Studies in the Effect of the 'New Science' on Seventeenth-Century Poetry,* rev. ed. (London and New York: Columbia University Press, 1965), p. 7. Astrological correspondences to the microcosm could take the form either of analogies between the "Zodiake" and the various parts, or between the heavenly planets and either those parts or "the seven Ages of man." Crooke, *Microcosmographia,* fol. 7; Ralegh, *The History of the World,* bk. I, ch. 2, section 5, p. 31.

160. It was widely accepted in the sixteenth and early seventeenth centuries that an analogy existed between the heart in the body and the sun in the macrocosm, which was a part of a wider correspondence between the "middle venter" or thorax of "Man" and the "heavenly part" of the universe, that is, the sun, stars, and planets. See, for example, Crooke, *Microcosmographia,* fols 6–7.

161. The pun on "Sunne, Sonne" is explicit in Canto 1, stanza 45, line 7. According to Allan D. Galloway, *The Cosmic Christ* (London: Nisbet, 1951), the doctrine of cosmic redemption emerged among the early prophets of the Old Testament, was extended by Amos as an idea of judgment not only upon Man but the whole environment,

with a corresponding redemptive aspect in the covenant, and found its full development in the apocalyptic writings of Daniel and the Apocrypha (pp. 10–12).

162. Galloway, *The Cosmic Christ,* p. x.

163. Ibid., p. 50.

164. See ibid., p. 53, which argues that Paul's general intention "was to assert that the work of Christ was of universal or cosmic significance." Galloway contends that, not only for Paul, but for the New Testament generally, "there can be no doubt of the general currency of the notion that the Christ came not only to individuals, but to the whole universe, and that His work is significant for the whole of creation" (p. 55). A further examination of this analogy occurs in Chapter 8 of this study, in relation to the anatomy and physiology of the heart as represented in *The Purple Island.*

165. Nicolson, *The Breaking of the Circle* (ch. 1: "'A Little World made cunningly,'" especially p. 28).

166. Crooke, *Microcosmographia,* fol. 7.

167. Ibid., fol. 7.

168. Nicolson, *The Breaking of the Circle,* p. 22. Barkan identifies in classical natural philosophy four approaches to cosmology arising from microcosmic speculation, all but the second of which include the relation of the microcosm to the terrestrial globe: the chemical geocosm, which consists of the combination of the four elements, fire, air, earth, and water; the astral geocosm, which presupposes a heavenly cosmos that is both spiritual and physical; a numerial geocosm, which is in a mathematical relation to the human body; a natural geocosm, consisting of objects apprehended by the senses. *Nature's Work of Art,* p. 14. It is the aspects of the latter in particular which *The Purple Island* unfolds in the course of its allegory. The natural geocosm appears in Ralegh, *The History of the World,* bk. I, ch. 2, section 5, p. 30, where, as in *The Purple Island,* "the bones" of Man's body are compared to "the hard rockes and stones" of the earth, but Fletcher's poem exceeds Ralegh's *History,* and many similar evocations of the natural geocosm, in its consistency, as this chapter goes on to show.

169. Philippus Aureolus Theophrastus (Philip Bombastus von Hohensheim / Paracelsus) was born in Einsiedeln (in modern Switzerland) around 1493 and died around 1542. On Paracelsus and analogy, see, for example, Temkin, "Metaphors of Human Biology," pp. 279–80. In the twelfth century Robert Grosseteste's *Quod Homo sit Minor Mundus* draws analogies between the flesh and the earth, the blood and water, the breath and air and between vital heat and fire. The microcosm thereby internalizes the elements, but the scheme lacks the variety of Paracelsus's analogizing. For the emergence of Grosseteste's analogical scheme out of theories of Empedocles, Hippocrates, and Polybus, see Barkan, *Nature's Work of Art,* pp. 15–20.

170. Barkan, *Nature's Work of Art,* p. 15.

171. Cunningham, *The Anatomical Renaissance,* pp. 246, 245; also see pp. 241–47. See Barkan, *Nature's Work of Art,* p. 38. For the concept of "sympathies" in sixteenth- and early seventeenth-century thought, see Foucault, *The Order of Things,* pp. 23–25.

172. Barkan, *Nature's Work of Art,* p. 38.

173. Foucault, *The Order of Things,* p. 30.

174. Ibid., p. 31. Foucault's concept of the microcosm is in fact the earth, everything contained within which finds its macrocosmic reflection in the heavens. But the same principle of limitation applies if everything within the geocosm must be reflected within the microcosm of Man.

175. Ibid., p. 30.

176. All but the last of the criteria listed are provided in James Deese, "Mind and Metaphor: A Commentary," *New Literary History* 6 (1974): 211–17 (p. 213).

177. Deese, "Mind and Metaphor: A Commentary," p. 213.

178. David Edge, "Technological Metaphor and Social Control," *New Literary History* 6, no. 1 (1974): 135–47 (p. 135).

179. Deese argues that scientific metaphors "must not be ambiguous": "an explication is the working part of the scientific metaphor so far as science as an abstract body of knowledge independent of human concern is at issue." Ibid., p. 213.

180. Monroe C. Beardsley, *Aesthetics: Problems in the Philosophy of Criticism* (New York: Harcourt, Brace & World, 1958), p. 434.

181. See Karsten Harries, "Metaphor and Transcendence," *Critical Inquiry* 5 (1978): 73–90 (p. 87).

182. Lind, *Studies in Pre-Vesalian Anatomy,* pp. 5, 6.

183. Ibid., p. 71.

184. Nutton, "Greek Science in the Sixteenth-Century Renaissance," p. 21.

185. The doctrine of humors in the body is frequently ascribed to Aristotle or to the "Peripatetickes," the followers of Aristotelian philosophy. See, for example, Crooke, *Microcosmographia,* fol. 7.

186. See Lind, *Studies in Pre-Vesalian Anatomy,* pp. 5–6, 71, 77; Benedetti, *Anatomice,* bk. I, ch. 5, p. 86. Nutton, "Greek Science in the Sixteenth-Century Renaissance," says that very few manuscripts of Greek medicine in Latin or Greek were available to any but "a few cognoscenti or collectors" before 1525, when as a direct result of the coming onto the market in late 1524 of the greatest private library of Greek medical texts, and the largest single collection of Greek scientific, philosophical, and medical writings in Western Europe, owned by Niccolò Leoniceno, "the Nestor of medical humanism," there was an outburst of Greek texts, including a series of medical authors in Greek from the Aldine Press in Venice in the 1520s and 1530s (pp. 23, 22, 20). As compared with literary and historical Greek texts, it was a relatively late date at which Greek scientific texts became available in print, whether in original Greek or Latin translations newly made from the Greek. Before 1500 only three scientific authors had been printed in Greek: Aristotle and Theophrastus, both in 1497, and Dioscorides in 1498 (p. 20). Two works by Galen were printed in the Greek in 1500, two years prior to the publication of Benedetti's *Anatomice;* but it was not until 1525 that the first collected edition of Galen in Greek appeared, from the Aldine Press (p. 17). "By the middle of the sixteenth century [the humanists] had recovered practically all the medical classics that we now possess." Singer, *The Evolution of Anatomy,* p. 103. Benedetti does however say that, "if ever I lack Latin words I have preferred to use Greek terms, rejecting the barbarian [Arabic] names." *Anatomice,* bk. I, ch. 2, p. 84. Julius Pollux (134–92 AD), *Onomasticon,* written in the time of Galen, replaced Arabic and Hebraic terms with Greek ones, and became influential after its publication in 1502, and is "the source of a good deal of our modern anatomical terminology." Singer, *The Evolution of Anatomy,* p. 107. Also see Singer, "Beginnings of Academic Practical Anatomy," in Choulant, *History and Bibliography of Anatomic Illustration,* p. 21–L; *The Epitome of Andreas Vesalius,* p. 8.

187. Lind, *Studies in Pre-Vesalian Anatomy,* p. 6.

188. See Nutton, "The Rise of Medical Humanism: Ferrara, 1464–1555," p. 18.

189. Benedetti died in 1512, but his concerns for the retrieval of the Greek sources of anatomy was shared by those who met the efforts of the Aldine scholars in Italy with phrases such as "new birth," "bringing the Greeks out of darkness," and "the opening up of pure springs." It was thought by such scholars in the early sixteenth century, that "[s]uch medical errors as there were might be put right with new manuscripts to hand, or by emendation." "The arguments in medical letters, commentaries, and treatises also focussed on linguistic problems: the meaning of words and their changes of meaning over the passage from Greek through the Arabic to the Latin of the Middle Ages." New humanist Latin translations were "on the whole, concerned with providing a more accurate version of [Greek] standard texts. Questions are mainly of identification and translation of words and phrases, rather than of wholesale interpretation, and are often resolvable by philological means." Only later in the wake of Vesalius's *Fabrica* and his and other anatomists' resurrection of the projects of the ancients, not only their medical terminology, did "[t]he deficiencies of a philological explanation for apparent errors of the ancients become clearer." Nutton, "Greek Science in the Sixteenth-Century Renaissance," pp. 19, 23–24.

190. Benedetti, *Anatomice,* bk. II, ch. 23, pp. 98–99; *The Works of Aristotle,* ed. by Smith and Ross, IV: *Historia animalium,* trans. by D'Arcy Wentworth Thompson (1910; repr. 1962), Bks. V–VII.

191. Benedetti, *Anatomice,* Bk. II, Ch. 23, p. 99; Aristotle, *Historia animalium,* VI, 10, 564b 25–30.

192. Benedetti, *Anatomice,* Bk. II, Ch. 23, p. 99.

193. Ibid.

194. Max Black, "Metaphor," *Proceedings of the Aristotelian Society,* n.s. 55 (1955): 273–94 (p. 291) (repr. in Max Black, *Models and Metaphors: Studies in Language and Philosophy* (Ithaca: Cornell University Press, 1962), pp. 25–47 (all further references are to the earlier publication).

195. Ricoeur, "Metaphor and the Main Problem of Hermeneutics," p. 99. Ricoeur is referring to Black, "Metaphor," p. 276. Black's terminology is by other authors substituted by the grammatical terms "subject" and "predicate" or "modifier" (Beardsley, in his *Aesthetics,* uses the terms "subject" and "modifier"), or the words "tenor" and "vehicle." The latter pairing was first suggested by I. A. Richards, although Black appends to his essay "*A note on terminology*" which exposes the imprecision of Richards' application of these terms. See Black, "Metaphor," p. 294, note 23.

196. Lakoff and Johnson, *Metaphors We Live By,* p. 5. Lakoff and Johnson's work, both together and individually, continues to exercise a profound influence on the fields of cognitive linguistics and philosophy of language. See, for example, the articles in Raymond W. Gibbs Jr. and Gerard J. Steen, eds., *Metaphor in Cognitive Linguistics: Selected Papers From the Fifth International Cognitive Linguistics Conference, Amsterdam, July 1997* (Amster-

dam: John Benjamins, 1999). It is, according to Gerard Steen, the understanding of metaphor as a highly revealing instance of the human capacity for making sense, which makes it comparable to models in science. Gerard Steen, *Understanding Metaphor in Literature: An Empirical Approach* (London: Longman, 1994), pp. 3–4.

197. See Steen, *Understanding Metaphor in Literature,* Section 1.3.1, pp. 10–13 (p. 11). Healy, *Fictions of Disease,* extensively uses the metaphorical approach to cognition that has emerged in specific form from the work of Lakoff and Johnson, in her analysis of the "dynamic relationship" between social and political instability and writings about disease, in the early modern period.

198. Lakoff and Johnson, *Metaphors We Live By,* pp. 6, 4.

199. Ibid., pp. 115–16, 186–87, 198, 202.

200. Ibid., p. 217.

201. The principle of fallibilism is enunciated by the American philosopher Charles Pierce, and is described by Umberto Eco, "Testing, testing . . . ," *Guardian Review,* 4 September 2004, p. 7, as the principle "according to which science progresses by continually correcting itself, falsifying its hypotheses by trial and error, admitting its own mistakes—and by considering that an experiment that doesn"t work out is not a failure but is worth as much as a successful one because it proves that a certain line of research was mistaken and it is necessary either to change direction or even to start over from scratch."

202. Lakoff and Johnson, *Metaphors We Live By,* pp. 187, 186, 196. Lakoff and Johnson acknowledge that theories of ontological relativity, which postulate the objective given-ness of meaning and truth relative to a given language, are nonetheless objectivist (p. 203).

203. Ibid., pp. 208, 10.

204. Ibid., p. 116; also see pp. 117–25.

205. Ibid., ch. 23, pp. 156–58, ch. 24, pp. 159–84, p. 197 (pp. 159, 179).

206. Mark Johnson, *The Body in the Mind: The Bodily Basis of Meaning, Imagination, and Reason* (Chicago: University of Chicago Press, 1987), p. xiii.

207. Lakoff and Johnson, *Metaphors We Live By,* ch. 13, pp. 61–68.

208. Ibid., pp. 56–57, 59, ch. 4, pp. 14–21 (p. 14).

209. Ibid., ch. 6, pp. 25–32; ch. 12, pp. 56–60.

210. Ibid., p. 59, ch. 14, pp. 69–76. Also see Johnson, *The Body in the Mind,* pp. xiv–xv, where these ideas about metaphor are integrated into a more general philosophical theory.

211. Barkan, *Nature's Work of Art,* p. 62.

212. Lakoff and Johnson, *Metaphors We Live By,* p. 58.

213. Ibid., ch. 13, pp. 61–68, p. 61 cited.

214. Ibid., p. 29; also see p. 69.

215. Barkan, *Nature's Work of Art,* pp. 3, 28.

216. See Lakoff and Johnson, *Metaphors We Live By,* p. 213.

217. Ibid., pp. 209, 213.

218. Ibid., p. 109.

219. This objection also counts against what Lakoff and Johnson refer to as the "*strong* homonymy view," which would argue that the two uses of the term "plant" "are entirely different and have nothing to do with each other," since one refers to an living organism in the uterus and the other to a living organism on the ground. Ibid., pp. 107, 111.

220. See, for example, Veslingus, *The Anatomy of the Body of Man,* p. 32, which is discussed below.

221. Lakoff and Johnson, *Metaphors We Live By,* ch. 2, pp. 7–9 (p. 7).

222. Steen, *Understanding Metaphor in Literature,* p. 7.

223. The more recently developed cognitive view of metaphor differs from that of Lakoff and Johnson, in eschewing some of their philosophical assumptions about doing cognitive science, and in distinguishing itself from their analytic method which assumes language structure as the sole basis for understanding language use. See Steen, *Understanding Metaphor in Literature,* pp. 5–6, 9.

224. Ibid., p. 20.

225. Ibid., pp. 17, 18. For this reason, Steen proposes a modification of Lakoff's definition of "literal," as referred to above in the text. His proposition is that literal language or nonmetaphorical literality is "directly meaningful language—not language that is *potentially* understandable, even partly, in terms of something else" (p. 24). Steen thereby excludes from his definition those linguistic metaphors that are conceptual metaphors and those that are dead or highly conventional. Steen, furthermore, argues that since the work of Lakoff and Johnson, research

on the cognitive character of metaphor has revealed that there are three aspects: metaphor as conceptual structure; metaphor as linguistic structure; metaphor as an ongoing process of cognition or interpretation. Steen says that although the first and second aspects affect the third, the precise interaction remains unclear (pp. 25–26).

226. Eriksson, *Andreas Vesalius' First Public Anatomy at Bologna 1540,* p. 16. Baldasar Heseler entered the University of Wittenberg where he studied theology under Martin Luther before moving to the University of Leipzig in 1533, and then to the University of Bologna possibly in 1539, where as a student of medicine in the following year he attended Matteo Corti's lectures on *Anatomia Mundini* and Vesalius's accompanying anatomical demonstrations.

227. For further discussion of these and other considerations of the historical context of anatomical representations, see Martin Kemp, "'The mark of truth': Looking and Learning in Some Anatomical Illustrations from the Renaissance and Eighteenth Century," in W. F. Bynum and Roy Porter, eds., *Medicine and the Five Senses* (Cambridge: Cambridge University Press, 1993), pp. 85–121 (p. 89).

228. Steen, *Understanding Metaphor in Literature,* p. 11.

229. Lakoff and Johnson, *Metaphors We Live By,* pp. 8–9.

230. Ibid., p. 9.

231. Ibid., p. 9; for internal systematicity, see pp. 17–18.

232. Veslingus, *The Anatomy of the Body of Man,* p. 32.

233. Jacopo Berengario da Carpi (1460–1530), *A Short Introduction to Anatomy (Isagogae breves),* trans. with an introduction and historical notes by L. R. Lind and with anatomical notes by Paul G. Roofe (Chicago: University of Chicago Press, 1959), p. 73. (All further references in the text and notes are to this edition unless otherwise stated.) *Isagogae breves* (the short title by which the above text will be referred to henceforth) has been translated as "Little Anatomical Handbook" by Harcourt, "Andreas Vesalius and the Anatomy of Antique Sculpture," p. 35. It was first published in Latin as *Isagogae breves, perlucide ac uberimme, anatomia humani corporis* (Bologna: Benedictus Hectoris, 1522). Further editions were printed in Venice, 1523, 1533, and 1535; Cologne, 1529; and Strasbourg, 1530, edited by Heinrich Sybold. The Venice edition of 1535 was chiefly a reprint of the earliest edition with only minor alterations; while the Strasbourg edition of 1530 has the same figures and arrangements as the 1523 edition. Lind's translation is a collation of the 1523 and 1535 edtions. In 1660 Henry Jackson produced a literal English translation based on either the 1522 or 1535 edition, and probably the latter: Μικροκοσμογραφια*; or, A Description of the Little World or Body of Man; Being a Brief and Practical Anatomy of the Body of Man; not only Shewing a Methodical Description of the Parts, but also the Manner of Anatomizing from Part to Part . . .* (London: Livewell Chatman, 1660). This was reissued with a different title page, as Μικροκοσμογραφια*; or, A Description of the Body of Man; Being a Practical Anatomy, Shewing the Manner of Anatomizing from Part to Part . . .* (London: Livewell Chatman, 1664). Both of these issues are discussed by S. V. Larkey and Linda Tum Suden in "Jackson's English Translation of Berengarius of Carpi's 'Isagogae Breves', 1660 and 1664," *Isis* 21 (1934): 57–70.

234. Thomas Vicary (1490–1561), *The Anatomie of the Bodie of Man: This Edition of 1548, As Reissued by the Surgeons of St. Bartholomew's in 1577 With a Life of Vicary,* ed. by Frederick J. Furnivall and Percy Furnivall (London: Published for the Early English Text Society by N. Trübner, 1888), p. 77. (All further references are to this edition.) First published in 1548 as *A Profitable Treatise of the Anatomie of Man's Body,* this was the "first printed book actually devoted to Anatomy in English." Singer, *The Evolution of Anatomy,* p. 170. It was reprinted several times. In 1586 it was published as *The Englishman's Treasure, with the True Anatomy of Man's Body.* It is a largely unoriginal work, copied from a fourteenth-century English manuscript based on the works of the medieval writers on surgery, Lanfranchi and Mondeville, and its anatomy is predominantly Galenic.

235. Veslingus, *The Anatomy of the Body of Man,* p. 32.

236. Berengario, *Isagogae breves,* p. 15. Berengario writes (pp. 80–82): "The entire uterus with its testicles and seminal vessels is similar to the members of generation in men, but the male members are completed outside, since they are thrust out on account of their heat. The members of women are diminished and retained within the body because of their lesser heat." "The uterus is, as it were, a converse [analogous] instrument. For the neck of the uterus is like a penis, and its receptacle with testicles and vessels is like the scrotum. The scrotum has a cavity within itself when turned in, and when turned outward the testes and the seminal vessels lie outside like the receptacle of the uterus. The testes and vessels of men, however, are larger."

237. Lakoff and Johnson, *Metaphors We Live By,* pp. 10–13.

238. The terms member and part were used by Renaissance and early modern anatomists according to how their ancient counterparts used them, that is, usually interchangeably. According to the overview provided by Crooke, *Microcosmographia,* fol. 38, Hippocrates used the terms in this way; meanwhile, "*Aristotle* calleth those

only *Members* which are compounded of parts of divers natures, as the Head, the Feate, and the Hands; and those that are similar he calleth properly *Parts*" (for similar and dissimilar parts, see chapter 2, note 112); whereas Galen, albeit influenced by Aristotle, in some cases uses the terms interchangeably. Crooke himself accepts as a definition of both terms "a body cohearing or cleaving to the whole, and joyned to it in common life, framed for his use and function." In general and not necessarily in cases of citation, the use of the terms in the text of this study is interchangeable, although member has been found to carry political connotations lacking in the term part, and the use of both terms implies the definition here provided by Crooke.

239. Lakoff and Johnson argue that many concepts are understood partly in terms of inherent properties, and partly metaphorically. *Metaphors We Live By,* p. 119.

240. Choulant, *History and Bibliography of Anatomic Illustration,* pp. 73–75, provides a brief history of the presentation of the female reproductive organs in anatomical illustration from the ninth-century drawings of a flask-shaped uterus to Ketham's *Fasciculo di medicina* of 1493.

241. Vittorio Putti, *Berengario da Carpi* (Bologna: L. Cappelli, 1937), cited in Berengario, *Isagogae breves,* p. 18. Also see Lind's Introduction to Berengario, *Isagogae breves,* and the text of *Isagogae breves,* where Berengario writes of the uterus: "It has a single cavity or cell, which somewhat near its fundus is divided into two parts, as if there were two uteruses terminated at one neck." (p. 82).

242. See Lind, *Studies in Pre-Vesalian Anatomy,* p. 14. Charles Estienne (1504–64) divulges that he himself has seen no more than three cells in any uterus. *De dissectione partium corporis humani libri tres* (Paris: Simon de Colines, 1545) was translated into French and published as *La Dissectione des parties du corps humain* (Paris, 1546).

243. K. B. Roberts and J. D. W. Tomlinson, *The Fabric of the Body: European Traditions of Anatomical Illustration* (Oxford: Clarendon Press, 1992), pp. 122, 15, offer the explanation that these structures, described by Mondino as "two thick strong ligaments," are "the uterine horns of cattle and other species, erroneously transferred to women." A "horned" uterus or vagina appears in *The Fasciculo Di Medicina,* II, and Singer suggests that the "horns" might be "the cut edge of the vagina, altered and misunderstood during the passage of time" (p. 23). In either case, the implication is that medieval and Renaissance drawings of the female reproductive system were either seldom done from the object or incorporated what the artist expected to see.

244. Singer, *The Evolution of Anatomy,* p. 130.

245. Ibid. The correct position of the fetus in utero is depicted by Leonardo, but of course his drawings remained unpublished. See Choulant, *History and Bibliography of Anatomic Illustration,* p. 105.

246. Lakoff and Johnson, *Metaphors We Live By,* p. 110.

247. Raynalde, trans., *The Birth of Mankinde,* fol. 24.

248. Scholz, *Body Narratives,* p. 2, says: "On a very abstract level, [the early modern] epistemological shift could be described as the tendency to isolate worldly phenomena and internalize their meaning instead of attributing signification from the point of an overarching structure of correspondences held in place, ultimately, by God. This shift in the epistemological framework of things effected all kinds of alterations on various levels of early modern cultures. One of the outcomes of this shift, and one which is not usually perceived by those that have analysed it as completed before the eighteenth century, is the development of a medical discourse of the human body. Certainly, the human body had been the focus of 'medical' attention before that, but it had not been the object of a discourse, by which I mean a set of texts, practices, institutions, forms of knowledge, ways of thinking, feeling, acting, and social power-relations, all of which constituted the body as an object of a scientific gaze quite separate from its 'possessor's' location in the world and society."

249. Gabriele Falloppio, author of *Observationes anatomicae,* "gave the scientific names that they now bear to the *Vagina* and *Placenta.*" Giulio Cesare Aranzi(o) (Julius Caesar Arantius, 1530–1589) "gave the first adequate printed account of the gravid uterus," and "a noteworthy description of the Anatomy of the foetus, by far the best up to his time," in *De humano fœtv libellus* (On the human fetus) (Basil, 1579). Adriaan van der Spieghel (Adrian Spigelius / Spigel, 1578–1625) demonstrated great improvement in anatomical terminology. Singer, *The Evolution of Anatomy,* pp. 143, 144, 163.

250. Sawday, *The Body Emblazoned,* p. 24.

251. *The Fasciculo Di Medicina,* II, 22–23. Roberts and Tomlinson, *The Fabric of the Body,* p. 45, suggest alternatively that the first illustration drawn directly from dissection may well be a visceral figure in Lorenz Friesz (Fries / Phryesen / Phries / Frisius), *Spiegel der Artzny* (Looking-glass of Medicine) (Strasburg, 1518). The superscription to this plate indicates that it was based on an actual dissection. The fugitive sheets of the illustrations of Friesz's work were published in Strasburg, by Johann Schott, in 1517, and the work itself was reissued by Grüninger in 1519.

252. Lakoff and Johnson, *Metaphors We Live By,* p. 9; for internal systematicity, see pp. 17–18.

253. Lind, *Studies in Pre-Vesalian Anatomy,* p. 5. In Berengario da Carpi, *Commentaria cum amplissimis additionibus super Anatomia Mundini* (Bologna, 1521), "Berengario uses Zerbi (often to refute him) an extremely high number of times." Lind, *Studies in Pre-Vesalian Anatomy,* p. 10.

254. For Galen's insistence on trusting no authority without testing it and determining by observation and the experience of his own senses how much it agrees with actual phenomena, see Thorndike, *A History of Magic and Experimental Science,* I, 152–53. An example of Galen's influence on Berengario in this regard can be found in the description in *Isagogae breves* of "the muscles of the neck and head" (p. 104). Initially he refers his readers "to Galen, Avicenna, and to other authors' because "we cannot hold a complete investigation of these unless the anatomy of their nerves, veins, and aforesaid arteries is disposed of." He then urges that these muscles "should be examined with care," and, referring to Galen, *On the Use of the Parts,* XII, 8, he repeats Galen's advice that an anatomist should be "trained carefully beforehand if he wishes to follow with certainty the things which are said here." Finally, Berengario reiterates Galen's declaration "that speech alone is not sufficient in anatomy, but touch and sight are needed."

255. Roger French, *William Harvey's Natural Philosophy* (Cambridge: Cambridge University Press, 1994), p. 20. According to French, Berengario "deliberately excluded from anatomy what could not be seen . . . ," and used *demonstratio* "in a way that we might call 'physical logic.' Although the practice derived from the simple *ostensio* of the public anatomies, the word "demonstration" carried over something of the force with which it was used in dialectic" (pp. 19, 20).

256. Lind, *Studies in Pre-Vesalian Anatomy,* p. 10. Also see Berengario, *Commentaria . . . super Anatomia Mundini* (Bologna, 1521), fols. 153^{v}, 412^{v}, 443^{r}; Lind's Introduction to Berengario, *Isagogae breves,* p. 7, and the Dedication "To the Most Renowned and Noble Lord Alberto Pio, Count and Most Meritorious Master of Carpi," where Berengario writes: "The authors [of other anatomical texts] seem to have borrowed fables from other volumes instead of writing genuine anatomy" (p. 35).

257. Cited by Lind, *Studies in Pre-Vesalian Anatomy,* p. 10.

258. Berengario, *Isagogae breves,* p. 15.

259. Benedetti, *Anatomice,* bk. II, ch. 24, p. 100; citation inserted in square brackets also from p. 100.

260. Eriksson, *Andreas Vesalius' First Public Anatomy at Bologna 1540,* p. 209.

261. A modern embryological description helps to clarify the parts and connections with which early modern anatomy is here concerned, and provides an almost entirely contrasting conceptualization: "The *chorion* is developed from the outer layer of the blastocyst [also known as the blastodermic vesicle, this is a "modified blastula (hollow ball of cells)," which forms in the very early stages of mammalian development], together with a layer of mesoderm [cells which form a layer in the blastoderm, that part of the blastocyst which develops into the embryo], and surrounds the embryo. It develops villi ["minute finger-like projections"], which erode the mucosa ["lining membrane"] of the uterus, forming pits into which they grow. The villi extending into the decidua basalis [that part of the "mucosa, or lining membrane, of the uterus" "against which the embryo rests and which forms the maternal part of the placenta"] become numerous, long, and highly branched. This portion of the chorion is called *chorion frondosum* and becomes the fetal part of the placenta. The villi on the other portions of the chorion do not usually continue to grow with the embryo and may even disappear, leaving this surface of the chorion smooth (*chorion laeve*). Through the *umbilical cord,* which attaches the fetus to the placenta, two umbilical arteries carry fetal blood from the fetus to the capillaries of the fetal placenta, and the umbilical vein returns this blood to the fetus." Diana Clifford Kimber, Carolyn E. Gray, and Caroline E. Stackpole, *Textbook of Anatomy and Physiology,* 11th edn. (New York: Macmillan, 1942; repr. 1943), pp. 710–11; citations in square brackets are from pp. 706, 491, 709.

262. Achillini, *Annotationes anatomicae,* p. 49. Nowadays, the term embryo is used to refer to the unborn offspring during that period of development from the beginning of the third week after conception to the end of the second month, and the term fetus to the same during the period from the beginning of the third month to the time of birth. See, for example, Kimber, Gray, and Stackpole, *Textbook of Anatomy and Physiology,* p. 678. I cannot be sure that Benedetti, Berengario, or Achillini's use of the terms is entirely in accordance with their modern denotations in this regard, but those instances of their use that I have cited are so.

263. Benedetti, *Anatomice,* bk. II, ch. 24, p. 100.

264. Veslingus, *The Anatomy of the Body of Man,* p. 32; Adriaan van der Spieghel, *De formato fœtv liber singularis aeneis figvris exornatus* (Padua: Johann Bap de Martinis and Liuius Pasquatus, 1627), [another edition] (Frankfurt: Matthaeus Merian, 1631), *Tab.* IIII. Figures of the dissected individual parts of the female reproductive system in Veslingus, *The Anatomy of the Body of Man,* also appear like the reproductive parts of plants. See, for example, ch. 8, figures II, III. Of Giulio Aranzi Singer, *The Evolution of Anatomy,* p. 144, says, that in his

De humano foetu liber (On the human foetus) (1564), [3rd ed.] (Venice, 1595), he "finally dispelled the idea of a human cotyledonous placenta. Many had believed in this before his time." Although Singer mentions the "cotyledonous human placenta figures in the magnificent drawings of Leonardo," he makes no reference to that in the much later *De formato fœtv* of van der Spieghel.

265. Thomas Bartholin (1616–1680), *Anatome ex omnium veterum recentiorumque observationibus* (Lyons: Hackiana, 1673), "*De infimo ventre*," *Tabula* LI, also presents the fetus as emerging from a flower-like uterine structure, which in this plate represents the placenta, and *Tabula* LII presents the womb as floral in very much the same way as does the plate in van der Spieghel, *De formato fœtv. Tabula* LIII in this series in Bartholin's work presents the blood vessels of the placenta as a many-divided structure akin to the highly divided leaf structure of fennel. However, it is difficult to avoid the impression that the actual structure of these blood vessels, when all placental tissues have been stripped away, suggests just such a comparison. See, for example, a photograph reproduced in Geraldine Lux Flanagan, "Going to Work on an Egg," *Observer Life*, 2 June 1996, p. 20, which is accompanied by the following text: "A tree of life. In the womb the baby was sustained by nourishment coming through these branching blood vessels of the placenta."

266. Benedetti, *Anatomice*, bk. II, ch. 23, pp. 98–99.

267. This metaphorical concept, which was attributed to Aristotle, had a long period of currency, finding its way into popular medical works as late as the eighteenth century. See, for example, *Aristotle's Masterpiece; or, The Secrets of Generation Display"d in All the Parts Thereof* (London: F.L., 1690), ch. 2, especially p. 11, which says that the embryo "may be called the rough draught, or Embrio, and well likened to Seed which is Sown in the ground, which through kindly heat and moisture grows up by degrees into a perfect form, either in Plant or Grain." All further references are to this edition. Several English-language editions of *Aristotle's Masterpiece* were published in the late seventeenth century, in 1684, 1690, 1694, and 1697. Popular medical works relied heavily upon metaphorical conceptualization and analogy for their pedagogical aims, without the strict observance of anatomical description demanded for the academic textbooks. *Aristotle's Masterpiece*, for example, calls the uterus "the Field of mans Generation" (p. 110); and it argues that "[i]f the Woman be stricken in years, and the time of breeding children be past . . . those Women must despair of further Generation: for as the learned in this Art frequently observe, where there is neither Buds nor Blossoms there can be no fruit" (p. 79).

268. The characterisation of the metaphorical concept of male seed as belonging to a system of internally coherent metaphorical concepts of aspects of animal and human reproduction centering anatomically on the uterus, is comparable with Lakoff and Johnson's concept of "external systematicity." *Metaphors We Live By*, p. 18.

269. Benedetti, *Anatomice*, bk. II, ch. 24, p. 100.

270. Sawday, *The Body Emblazoned*, p. 25, describes the figure of the anatomist in Renaissance culture in these terms.

271. Ibid.

272. The *implied reader* is the embodiment of the projected recipient of the ideas, beliefs, and ideologies in a work, who must, for example, be in a position to comprehend the work and must be the holder of certain assumptions or knowledges in order to facilitate interpretation. See Wolfgang Iser, *The Implied Reader* (Baltimore: Johns Hopkins University Press, 1974).

273. Daniel "Featly," "To the Readers," in Giles and Phineas Fletcher, *Poetical Works*, II, 4. Although commonly known as "Featly," the author of this prose panegyric was Dr. Daniel Fairclough. Langdale, *Phineas Fletcher*, p. 86, note 6, comments: "Fairclough, born in Oxfordshire in 1582, and thus a coeval of Fletcher, entered Corpus Christi College at Oxford and later went to France as chaplain to the English ambassador, Thomas Edmunds. After a variety of ecclesiastical employment, he became the third provost of Chelsea College. He retired to Kennington in 1625, and died at about the time of the Restoration. What acquaintance he had with the poet is unknown. They may have met in London, where the provost was famous for his anti-papist disputations. At least the two had a common enthusiasm."

274. See Sawday, *The Body Emblazoned*, pp. 175–76.

275. According to Bernard Schultz, *Art and Anatomy in Renaissance Italy*, Studies in the Fine Arts: Art Theory, 12 (Ann Arbor: UMI Research, 1985), p. 76, "early in the *Trattati* [*di architettura ingegneria e arte militare*], Francesco di Giorgio equates the rivers of Mt. Athos to the veining in an antique sculpture by Dinocrates." The *Trattati* were composed in the last two decades of the quattrocento, "a time concurrent with Leonardo's initial architectural and anatomical investigations in Milan" (p. 75) (see above). Moreover, Francesco di Giorgio and Leonardo were colleagues, and Leonardo is known to have entered notes into one of the original versions of Francesco's treatise, the MS Codex Ashburnham 361, in about 1490, which Schultz argues demonstrate Leonardo's "literal" interpretation of the correspondence between architecture and anatomy (pp. 75–76). Ralegh,

The History of the World, bk. 1, ch. 2, section 5, p. 30, combines the vocabulary of both the vegetative and fluvial figurations of the veins of Man: "His bloud, which disperseth it selfe by the branches of veines through all the bodie, may be resembled to those waters, which are carried by brookes and riuers over all the earth. . . ."

276. Cunningham, *The Anatomical Renaissance,* pp. 14, 15.

277. My use of the terms "structure," "particular form," "goal," and "function" are here and elsewhere in this context indebted to Cunningham's elegant and precise deployment of them in relation to Aristotle and Galen in *The Anatomical Renaissance,* ch. 1.

278. See French, *Dissection and Vivisection in the European Renaissance,* p. 9, for an explanation of the four causes that are involved in all natural change in Aristotle's philosophy.

279. *The Works of Aristotle,* ed. by J. A. Smith and W. D. Ross, V: *De partibus animalium,* trans. by William Ogle (1912; repr. 1958), bk. III, ch. 1, 661b.

280. Eriksson, *Andreas Vesalius' First Public Anatomy at Bologna 1540,* p. 49.

281. Cunningham, *The Anatomical Renaissance,* p. 24.

282. Ibid., p. 16.

283. "Aristotle, by contrast [with Plato], claimed that the perfect, eternal and stable could indeed be seen within the imperfect and transitory things of this world." "Aristotle found form, intelligibility, definiteness, where Plato had never found it: in the limited, recurrent but orderly processes of nature itself." Cunningham, *The Anatomical Renaissance,* p. 15. For Plato, and for the late sixteenth-century Platonist anatomists, the soul is not the form of the body, rather the body is a temporary shelter for the soul. See, for example, French, *Dissection and Vivisection in the European Renaissance,* pp. 215–16.

284. Siegfried Wibbing, "*Sōma,* body," in *The New International Dictionary of New Testament Theology,* 3 vols., ed. by Colin Brown (Exeter: Paternoster Press, 1975), I, 232–38 (p. 232). Wibbing refers to Aristotle, *De Anima,* 2.1.

285. B. F. C. Costelloe and J. H. Muirhead, *Aristotle and the Earlier Peripatetics: Being a Translation from Zeller's "Philosophy of the Greeks,"* 2 vols. (London: Longman's, Green, and Co., 1897), II, 6, say that Aristotelian form cannot exist without a body.

286. Costelloe and Muirhead, *Aristotle and the Earlier Peripatetics,* II, 3–4, 95.

287. Cunningham, *The Anatomical Renaissance,* p. 16. In the early seventeenth century, Crooke reiterated this motivation for studying anatomy, saying that "by the dissection of the body, and by Anatomy, wee shall easily attaine unto this knowledge [of our selves]," because the body is the soul's organ and instrument, through which her functions and offices are discharged. Crooke, *Microcosmographia,* fol. 12.

288. De Laguna, *Anatomica methodus,* p. 264.

289. See R. J. Hankinson, *The Sceptics* (London: Routledge, 1995), pp. 225–36.

290. Richard H. Popkin, *The History of Scepticism from Erasmus to Spinoza* (Berkeley: University of California Press, 1979), pp. xviii–xix; the quotation in brackets within the former quotation is from pp. xviii, xiv. In the Hellenistic period, there was Pyrrhonian skepticism, which began with Pyrrho of Elis (c. 360–275 BC), and Academic skepticism, which was formulated in the Platonic Academy in the third century BC. As a skeptical theoretical formulation, however, Pyrrhonism is attributed to Aenesidemus (c. 100–40 BC). The sole surviving texts of Pyrrhonian skepticism are those of Sextus Empiricus. In 1562, Henri Estienne published a Latin edition of *Pyrrhonean Hypotyposes,* and a Latin edition of the entire works of Sextus was published in 1569, and republished in 1601. The 1569 Latin edition was published by Gentian Hervet. In 1621, a Greek text of the entire works was published by the Chouet brothers, P. and J. A further source for skeptical ideas was the corpus of works by Cicero. Meanwhile, Academic skepticism was known through Diogenes Laertius's *Lives of Eminent Philosophers.*

291. Sextus Empiricus, *Selections from the Major Writings on Scepticism, Man, & God,* ed. Philip P. Hallie, trans. by Sanford G. Etheridge (Indianapolis: Hackett, 1985), pp. 32–33. The antitheses generally take various forms, appearances opposed to appearances, judgments to judgments, or appearances to judgments. Antithesis is meant not in the sense of denial and affirmation, but rather as "conflicting."

292. Popkin, *The History of Scepticism from Erasmus to Spinoza,* pp. xiv, 3. The cited work by Sextus Empiricus is *Pyrrhonean Hypotyposes* (compiled by Sextus about AD 200), II, ch. 4.

293. De Laguna, *Anatomica methodus,* p. 264. De Laguna also characterizes the "methodics" (or Methodists), another major branch of Hellenistic medical tradition, as "so addicted to their most detailed methods that sometimes they took no account whatever of the results of clear demonstration or of the evidence of the senses" (p. 264). Although in some respects holding even less theoretical commitments than the Empiricists, they categorized all pathological conditions as either "relaxed" or "constricted." See Hankinson, *The Sceptics,* pp. 234–36. Regarding methodical physicians or "methodists," Celsus explains in his *De medicina,* that they thought that "knowledge

acquired from anatomizing bore no relation whatsoever to the knowledge required by the physician for healing the body: the body works on a level quite inaccessible to the anatomist." Cunningham, *The Anatomical Renaissance,* p. 24.

294. Hankinson, *The Sceptics,* pp. 228, 236.

295. See French, *Dissection and Vivisection in the European Renaissance,* pp. 169–71.

296. De Laguna, *Anatomica methodus,* p. 264.

297. Hankinson, *The Sceptics,* pp. 231, 236.

298. French, *Dissection and Vivisection in the European Renaissance,* p. 1

299. De Laguna, *Anatomica methodus,* p. 264; Cunningham, *The Anatomical Renaissance,* p. 24. Celsus, *De medicina* was written in the first century A.D., but was unknown thereafter until rediscovered in the mid-fifteenth century and reprinted in Florence in 1478. In his introductory essay to Choulant, *History and Bibliography of Anatomic Illustration,* p. 21–L, Singer calls *De medicina* the single most significant medical text recovered in the fifteenth century.

300. Cunningham, *The Anatomical Renaissance,* pp. 18–19. Aristotle supposed that some animal species other than humans may possess a rational soul, whereas others do not even possess a motile soul—they lack the faculty of movement—and even less lack even a sensitive soul, and thus the capacity for sensation.

301. Emily Michael and Fred. S. Michael, "Two Early Modern Concepts of Mind: Reflecting Substance vs. Thinking Substance," *Journal of the History of Philosophy* 27 (1989): 29–48 (p. 36).

302. Cunningham, *The Anatomical Renaissance,* pp. 20–21; also see pp. 18–19. Gail Kern Paster, "The Body and Its Passions," *Shakespeare Studies,* 29, ed. by Leeds Barroll (Madison: Fairleigh Dickinson University Press, 2001), 44–50 (p. 44), argues that the organization of the natural world, in early modern Europe, was analogical: "a network of analogy and mutual functionality enlivened all animate life in a hierarchical continuum of ensoulment (*empsychos*) ascending from vegetables, to imperfect animals such as sponges, to perfect animals such as birds and mammals, and finally to the human being, uniquely endowed with an intellective soul. The human body, though set apart from animals, vegetables, and minerals by the existence of the rational soul, was nevertheless joined indissolubly to the rest of ensouled nature in the universal scale."

303. See Cunningham, *The Anatomical Renaissance,* p. 18.

304. It may be seen as a consequence of the Aristotelian conceptualization of the relationship between soul and body that, as Lakoff and Johnson argue in *Metaphors We Live By* (p. 135), "[t]he word or grammatical device that indicates ACCOMPANIMENT also indicates INSTRUMENTALITY."

305. This is also an objection to the "abstraction view" and counts against what Lakoff and Johnson call the "*weak* homonymy view," which would argue that the two uses of the term "plant" signify two "distinct and independent concepts . . . but allows that their meanings may be similar in some respects and that the concepts are related by virtue of this similarity." The weak homonymy view claims that in such cases there is similarity rather than understanding one concept in terms of another; there is, in other words, no direction of metaphorical structuring. *Metaphors We Live By,* pp. 107, 108–9, 111–13 (p. 107).

306. Eriksson, *Andreas Vesalius' First Public Anatomy at Bologna 1540,* p. 123. Matteo Corti (Matthaeus Curtius) was born circa 1475 at Pavia and died in 1542 or 1544.

307. Cunningham, *The Anatomical Renaissance,* p. 206.

308. De Laguna, in his *Anatomica methodus,* is further at pains when confronted with an aspect of human anatomy that suggests no purpose whatsoever: "Thus it happens that the emphasis of the jejune intestine brought about for no purpose nevertheless may follow that which is brought about by necessity and produce such an effect" (p. 270). Matteo Corti argued in his Lectures at the University of Bologna in 1540 that "Nature created no organ without purpose; for even if they do not perform all the operations, yet they support them." Eriksson, *Andreas Vesalius' First Public Anatomy at Bologna 1540,* p. 49.

309. Paré, *Workes* (1634), bk III, p. 80 (emphases mine). Galenic anatomy is also indebted to Plato in ways that are discussed in due course in the text. Among these ways is one that is immediately relevant, however. For Plato, in *Timaeus,* the body exists for the soul, its *telos* (purpose, design) is the service of the soul. See, for example, Cunningham, *The Anatomical Renaissance,* pp. 12–13.

310. Cunningham, *The Anatomical Renaissance,* pp. 29–30, p. 30 cited. Also see Temkin, "Metaphors of Human Biology," p. 275. Paré, *Workes* (1634), p. 80, says that in *De usu partium* (On the use of the parts), Galen "shewes why, and for what use they are made." In his First Lecture in the series of Lectures accompanied by Vesalius's anatomical demonstrations at the University of Bologna, in 1540, Corti refers to Galen's *On the Use of the Parts* and Hippocrates's treatise *De alimentis,* to support not only the argument of design, that "all the parts of the body are given to maintain, to perform and to fill the functions necessary for preserving the whole," but also

the teleological implication that "from the action or service of the organs you will learn the arrangement and the nature of the parts themselves." Eriksson, *Andreas Vesalius' First Public Anatomy at Bologna 1540,* p. 49.

311. Eriksson, *Andreas Vesalius' First Public Anatomy at Bologna 1540,* p. 103.

312. See, for example, Cunningham, *The Anatomical Renaissance,* pp. 28–29; Singer, *Vesalius on the Human Brain,* p. 1.

313. According to Cunningham, *The Anatomical Renaissance,* p. 28, Galen argued that "since sensation and voluntary motion are peculiar to animals, whilst growth and nutrition are common to plants as well, we may look on sensation and voluntary motion as the work of the *soul,* and growth and nutrition as the work of *Nature.*"

314. A detailed division of the faculties of the soul along the lines of Galen is provided by Paré, *Workes* (1634), Bk. III, p. 83. The animal faculty of the soul is divided into the "principal," "sensitive" and "motive" faculties. The principal faculty is further divided into the "imaginative," "reasonable," and "memorative" faculties; the sensitive into "seeing," "hearing," "smelling," "tasting," and "touching"; the motive into the "progressive," and the "apprehensive" faculties. The vital soul is divided into the "dilative" and "contractive" faculties, which are together termed the "pulsificke" faculty of the heart and arteries. Finally, the natural soul is divided into the "nutritive," "auctive," and "generative" faculties, which three perform with the help of five other faculties: the "attractive," "retentive," "concoctive," "assimilative," and "expulsive."

315. Once again, Paré, *Workes* (1634), Bk. III, pp. 83–84, provides a detailed division of the bodily functions for which each faculty of the soul is responsible. The animal faculty is in the head but is allied also with the extremities, and is divided into the "principle," "sensitive," and "motive" functions. The principle functions are the "imaginative," "reasonable," and "memorative"; the sensitive functions are the "visive," "auditive," "smelling," "tasting," and the "tactive" or "touching"; the motive is divided into the movement of each of the three "bellies," the abdomen, the thorax, and the head. The vital faculty in the throrax is divided into the functions of "respiration" and "vital action," the latter being the movement of the heart and arteries. The natural faculty, meanwhile, is divided the "nutritive," "auctive" and "generative" functions, and the subdivisions of these are of so great a number that I must refer the reader to the work in question.

316. In his first lecture of the series accompanied by demonstrations by Vesalius, at the University of Bologna in 1540, Corti expresses why he thinks Galen and Aristotle differ in their opinions on "the form and the purpose" of the organs: "For as a good philosopher Aristotle in respect of Nature considered the efficient and material cause, and in the same manner also Galen considered them. Hence, in respect to the material and effective cause of the functions of the organs there is no ambiguity between them concerning anatomy." Their differences, therefore, are either a matter of the final cause, that is, the reason for the material cause of the functions, the reason for the substance, shape, position, number, and so on, of the organs being made to perform those functions, or a matter of the efficient cause, which is the soul or Nature. Eriksson, *Andreas Vesalius' First Public Anatomy at Bologna 1540,* p. 53; French, *Dissection and Vivisection in the European Renaissance,* p. 9. It should be noted that "efficient" is not used consistently in anatomy textbooks to signify the "agent who effected" a change or function, as it does here; sometimes it is used to mean the effective cause.

317. Cunningham, *The Anatomical Renaissance,* pp. 28–29; also see pp. 18–19.

318. Lind, in his General Interpretation of chapter 4 of *The Epitome of Andreas Vesalius,* p. 62, says: "The pneuma physicon, or natural spirit, is that [spirit] associated with the lowest or vegetative side of man: it is a vapor formed in the liver and carried with the humors along the veins." Also see, for example, Singer, *Vesalius on the Human Brain,* p. xvii; Paré, *Works* (1678), Bk. III, p. 52. Some, however, such as Crooke, *Microcosmographia,* fol. 44, questioned the liver's provision of the natural faculty of the soul to the body, given that all the other parts and members are imbued with this faculty anyway. In either case, it is the natural faculty of the soul that may be seen as responsible for the conceptualization of the parts in terms of vegetation.

319. Cunningham, *The Anatomical Renaissance,* p. 21.

320. Costelloe and Muirhead, *Aristotle and the Earlier Peripatetics,* II, 93–95 (p. 95).

321. Berengario, *Isagogae Breves,* pp. 94–95: "Its [= the heart's] roots are attached to the highest part of itself. These are solid and hard, as if cartilaginous, so that above these its continuous agile motion may take place." Cunningham, *The Anatomical Renaissance,* pp. 20–21.

322. This is what Lakoff and Johnson, *Metaphors We Live By,* p. 18, call "external systematicity" among various metaphors with the same defining concept. This external systematicity in anatomy may be rationalized in another way. During pregnancy, the vessels of the umbilicus "descend" like the roots of plants not only into the uterine wall, understood analogously as the earth or soil, but beyond that into the maternal vascular system. But this does not explain the density of vegetative figurality in passages describing the liver and heart, nor the consistency of that figurality with Aristotelian and Galenic physiology.

323. Paré, *Works* (1678), bk. III, p. 76.

324. Seventy seven copper plates of Giulio Casserio of Piacenza (Iulius Casserius Placentinus, 1561–1616), together with twenty others, drawn and engraved by the same artists who made those of Casserius (Odoardo Fialetti [1573–1638], the artist, and Francesco Valegio [Valesius], the engraver), were published under the title *Ivlii Casserii Placentini tabulae anatomicae LXXIIX* with Adriaan van der Spieghel, *De humani corporis fabrica libri decem* (Venice: [n. pub.], 1627), [another edition] (Frankfurt: Matthaeus Merian, 1631/32). For the history of these plates see Choulant, *History and Bibliography of Anatomic Illustration,* p. 225; Thornton and Reeves, *Medical Book Illustration,* pp. 66–68. The figure referred to in the text is figure 2, *Tabula VIII* of book 8, fol. 71^{r}. Another, figure 4, *Tabula IX,* book 8, presents also on a model of vegetation the vessels arising from the spleen. Bartholin, *Anatome ex omnium veterum recentiorumque observationibus,* "De infimo ventre," *Tabula* XXIV, also presents the liver and its major connecting vessels on the model of vegetative growth, and "De venis," *Tabula* I, presents the many divisions of the vena cava and portal vein as leaf-like.

325. Juan Valverde da Hamusco (c. 1525–c. 1587), *Historia de la composicion del cuerpo humano* (Rome: Antonio Salamanca and Antonio Lafrery, 1556), bk. VI, Tab. 1. The plate is reproduced in the Italian translation of this the Spanish original of Valverde's text: *Anatomia del corpo humano* (Rome: Antonio Salamanca and Antonio Lafrery, 1560), bk. VI, Tauola 1. The plate is reproduced again in Valverde's *Vivae imagines partium corporis humani aereis formis expressae* (Antwerp: Christophe Plantin, 1566), bk. VI, Tabula I.

326. *The Epitome of Andreas Vesalius,* p. 47. The direction of the venous distribution of the veins in Vesalius's description is counter to what a modern reader may expect, given modern knowledge of the circulation of the blood, and consequently it requires some explanation. It is determined by two considerations. Firstly, it follows the division of greater vessels into lesser ones, as the branching of a tree arises out of its trunk. Indeed, describing the juncture of the inferior vena cava and the two common iliacs, Vesalius says: "The most outstanding of these ["clusters" of "offshoots" from the vena cava] are those which arise from the vena cava where it divides into two equal trunks above the union of the os sacrum and the lunar vertebrae." *The Epitome of Andreas Vesalius,* p. 48. Secondly, the direction of distribution is guided by Vesalius's understanding of the function of the veins, which is also that of all of his contemporaries, that the veins distribute blood and humors from the liver to the parts. This physiology, which is Galenic, is discussed in greater depth in part 3. Vegetative terminology is extensively used in the passages of *The Epitome of Andreas Vesalius,* pp. 46–47, describing the venous system. The most commonly occurring terms are "offshoot," "branch," "twig" and "shoot."

327. Girolamo Fabrici of Acquapendente (Hieronymus Fabricius ab Aquapendente, 1533–1619), *De venarum ostiolis* (Padua: Lorenzo Pasquati, 1603; repr. in *Anatomices et chirurgiae . . . tractatus quatuor: Quorum, I, De formato fœtv, II, De locutione et ejus instrumentis, III, De loquela brutorum, IV, De venarum ostiolis . . .* (Frankfurt: Jacques de Zetter, 1624)), plate II; Cunningham, *The Anatomical Renaissance,* pp. 178, 181.

328. See, for example, *The Epitome of Andreas Vesalius,* pp. 57–66.

329. Vesalius was introduced to the works of Galen when he studied medicine and anatomy in the University of Paris, under Jacobus Sylvius (Jacques Dubois, 1478–1555) and Johannes Guinterius from Andernach, "both exponents of the Galenical school of medicine." Eriksson, *Andreas Vesalius' First Public Anatomy at Bologna 1540,* p. 20. Foster, *Lectures on the History of Physiology During the Sixteenth, Seventeenth and Eighteenth Centuries,* pp. 5–7, argues, rather harshly, that instruction in anatomy was to Jacobus Sylvius "reading a chapter of Galen," and he adds with more justice, "though he did make use of dissections, these were used as mere concrete illustrations to render easy the comprehension of what he was teaching, not as tests by which the truth of what he was stating might be tried." Cunningham, *The Anatomical Renaissance,* pp. 133–36, says that Sylvius, in his address to the reader in his *Isagoge,* describes the work as "commentaries on anatomising gathered from the diligent observation (ex diligenti observatione) of many human bodies dissected by me in public and in private." In bk. VII, ch. 6, of the *Fabrica,* however, Vesalius says: "In saying these things I admit that innumerable tenets of Galen contradict me." Furthermore, and by way of self-justification: "These and many others of the kind [tenets of Galen] are, I am convinced, learned rather from copious discussion than from dissection, for these things are so obvious that it is superfluous to warn against them those who dissect the brain themselves or watch others dissecting, and do not trust merely in books." Singer, *Vesalius on the Human Brain,* pp. 34–35.The citation in the text of my chapter at this point is of the notes taken by Heseler on Vesalius's third demonstration of anatomy to students at the University of Bologna in 1540. Eriksson, *Andreas Vesalius First Public Anatomy at Bologna 1540,* p. 109. Significantly, Heseler's notes record that when Vesalius "sometimes" pronounced Galen's partial ignorance on the anatomy of the muscles, sinews, and tendons of the hands and wrists, "Curtius, Boccaferrius, Erigius and many others went away" (p. 109). It is unclear, however, whether the departure of these professors of anatomy and medicine was occasioned by Vesalius's refutation of Galen or his manner of refuting him. In the twenty-second demonstration

(p. 273), Vesalius's vehemence in defense of his position in the debate about venesection in pleurisy caused Corti to accuse him of being angry. Alternatively, the departing professors may have been simply responding to a professional divide: the anatomy being demonstrated is said to be relevant to "surgeons" rather than "physicians." Heseler notes that "only few physicians used to practice it. However, it was good to have seen it once." (p. 109). The latter reason, suggested by Heseler's notes, illustrates the superiority assumed by physicians over surgeons and over even an "excellent" "anatomista" such as Vesalius is acknowledged to be by Corti himself (p. 273).

330. Cunningham, *The Anatomical Renaissance,* p. 113; and for Vesalius's revival of Galenic anatomy, see Chapter 4, pp. 91–137.

331. French, *Dissection and Vivisection in the European Renaissance,* p. 169; Cunningham, *The Anatomical Renaissance,* pp. 29–30; Eriksson, *Andreas Vesalius' First Public Anatomy at Bologna 1540,* p. 49.

332. See French, *Dissection and Vivisection in the European Renaissance,* pp. 165–68, 176.

333. See, for example, O'Malley, *Andreas Vesalius of Brussels 1514–1564,* p. 320.

334. Singer, *The Evolution of Anatomy,* pp. 115, 115–16, 116.

335. Ibid., p. 116. Also see Singer, "Historical Relations of Religion and Science," p. 128.

336. Lind, *Studies in Pre-Vesalian Anatomy,* p. 17. Lind notes that the incomplete index of Vesalius's *Fabrica* "lists at least 265 references to Galen, most of them involving corrections" (p. 17). But, in his introduction to *Andreas Vesalius' First Public Anatomy at Bologna 1540,* p. 23, Eriksson comments that a number of mistakes are corrected in the second edition of the *Fabrica,* published in 1555, "but the many occasions on which he accepted Galen's conceptions of anatomy show that the duties at the Imperial Court [to which Vesalius was at that time affiliated] did not allow Vesalius opportunities to prosecute systematically his studies of human anatomy." These perceived shortcomings in the work of Vesalius have led some, like Singer, to conclude: "When all is said, the most permanent achievement of Vesalius was his triumphant enlistment of graphic art in the service of biological structure in general and of anatomy in particular." Singer, *Vesalius on the Human Brain,* p. xxvi. It is important to be aware, however, that the tendency of such conclusions is to employ the illustrations of the *Fabrica* and the *Epitome* as an index of the continuity of a tradition of descriptive anatomy begun by the Renaissance. There are good reasons for arguing also for a sharp discontinuity and much work has been done recently on establishing this perspective. See, for example, Harcourt, "Andreas Vesalius and the Anatomy of Antique Sculpture," p. 28. I wish to add that among the immediately relevant reasons to argue for discontinuity is the pervasiveness of the conceptual metaphor of vegetation in Renaissance and early modern anatomy, but other reasons regarding the general theological character of anatomical investigation in the period are discussed in the text forthwith.

337. Vesalius's *Fabrica* and *Epitome* were published in the period which Nutton has called the period of Hellenicizing, from about 1525 to 1557, during which Greek editions of the medical and anatomical works of Galen, Aristotle, Hippocrates, and others of the ancients, were published in abundance, mostly from the Aldine Press in Venice. Also during, before and after this period humanist translators made a concerted effort to produce good, accurate Latin translations of the Greek medical and anatomical texts, but their concern, as that of the Aldine scholars, was with linguistic problems, and this concern persisted beyond and in direct relation to Vesalius. John Caius (1510–1573), who lodged with Vesalius while he was at Padua, is best known perhaps for his "philological approach to saving the credibility of Galen" in the face of the accumulating evidence of dissection produced by Vesalius and those influenced by him. Caius was convinced that "Vesalius's attack on Galen was the result of Vesalius's poor knowledge of Greek, not of anatomy." Subsequent to Vesalius and partly as a result of his approach to anatomy, "[d]iscussions of Galenic medical theory rarely involved discussions of textual minutiae, or could be controverted by the discovery of a new reading," and they were consequently conducted on the basis of more readily available and more accessible (to those who did not read Greek) Latin versions, rather than the Greek. Nutton, "Greek Science in the Sixteenth-Century Renaissance," pp. 17, 19, 26, 27.

338. Lind, *Studies in Pre-Vesalian Anatomy,* p. 17. Eriksson, *Andreas Vesalius' First Public Anatomy at Bologna 1540,* p. 34: "Vesalius himself expressed [his reliance on observational method] according to Heseler: 'tangatis uos ipsi uestris manibus, et his credite', feel with your hands and trust them" (p. 34). It is Eriksson's opinion that "Vesalius' methods and demands on unprejudiced investigations at a time when medieval belief in authority prevailed not only in theology but also in the sphere of medicine constitute his lasting contribution to anatomy."

339. Eriksson, *Andreas Vesalius' First Public Anatomy at Bologna 1540,* p. 33. Singer, *Vesalius on the Human Brain,* p. xvii, also says: "It is particularly to be observed that [Vesalius] usually displays the great negative virtue of refusing to see things with the eyes of another, and thus often avoids seeing the non-existent."

340. For the same reason, the metaphorical expressions or statements in anatomical texts are not to be regarded as semiotic or conceptual symbols, which are replaceable, because in some sense arbitrary. See Carl R. Hausman,

Metaphor and Art: Interactionism and Reference in the Verbal and Nonverbal Arts (Cambridge: Cambridge University Press, 1989), pp. 13–14.

341. *The Epitome of Andreas Vesalius,* p. 47.

342. Steen, *Understanding Metaphor in Literature,* p. 11.

343. The analogical structure behind metaphorical expressions in anatomy textbooks is a fact about the language of anatomy as a generalized code, as *langue,* and not necessarily a fact about the language use of every individual contribution to the field of anatomy, which may be regarded as the *parole* of the anatomical language system.

344. *The Epitome of Andreas Vesalius,* pp. 70–71.

345. Ibid., p. 70. The identification and modern appellations of the nerves of which Vesalius writes are in part taken from Lind, *The Epitome of Andreas Vesalius,* p. 70. But in further part they must remain insecure, for they are ascertained by comparison of the description in the *Epitome* with the following editions of *Gray's Anatomy:* Henry Gray, *Anatomy Descriptive and Surgical,* ed. by T. Pickering Pick and Robert Howden, 15th ed., first published in 1901 (London: Chancellor Press, 1985), p. 709, fig. 394; Roger Warwick and Peter L. Williams, eds., *Gray's Anatomy,* 35th edn (Edinburgh: Longman, 1973), p. 1008, fig. 7.181.

346. *The Epitome of Andreas Vesalius,* p. 71. For the brachial plexus, see Gray, *Anatomy Descriptive and Surgical,* pp. 745–47 and fig. 410. Identifying the nerves which Vesalius refers to is rendered difficult by the variable nature of the plexus: "The nerves which form the plexus are all similar in size, and their mode of communication is subject to considerable variation, so that no one plan can be given as applying to every case" (p. 745).

347. Lakoff and Johnson, *Metaphors We Live By,* p. 109. This constitutes a further objection to the "Abstraction View."

348. For Lakoff and Johnson's view on the figurative elaboration or extension of metaphorical concepts in ways (such as the use of unused parts and extensions of the used part of the defining domain) that remain consistent with the ordinary metaphorical structuring of the concept, see ibid., pp. 13, 53.

349. *The Epitome of Andreas Vesalius,* p. 66.

350. Ibid., p. 49. A folded plate of Bartholin, *Anatome ex omnium veterum recentiorumque observationibus,* p. 294, presents the arteries and veins connecting with the vagina as forming an interlacing pattern reminiscent of the spread of the branches of trees.

351. Matteo Corti is recorded by Baldasar Heseler as professing the opinion that the pudenda are internal parts in the third of his lectures given at Bologna in 1540 and accompanied by anatomical demonstrations by Vesalius. Eriksson, *Andreas Vesalius' First Public Anatomy at Bologna 1540,* p. 63.

352. *The Epitome of Andreas Vesalius,* p. 48.

353. Eriksson, *Andreas Vesalius' First Public Anatomy at Bologna 1540,* pp. 207–9.

354. This idea may derive from Book III of Vitruvius's *Ten Books on Architecture,* where it is argued that "it was from the members of the body that they (Greeks) derived the fundamental ideas of the measures which are obviously necessary in all works." Schultz, *Art and Anatomy in Renaissance Italy,* p. 74.

355. Schultz, *Art and Anatomy in Renaissance Italy,* p. 74.

356. An historical instance of the overlap of the arts of anatomy and architecture is the work of the Spanish architect and sculptor Juan de Arphe Y Villafañe. He studied anatomy at Salamanca and Toledo, and his *Varia commensuracion para la escultura y arquitectura* (Seville, 1585) consists of four books on the following subjects: geometry and gnomonics; the proportions of the human body; quadrupeds and birds; architecture. See Choulant, *History and Bibliography of Anatomic Illustration,* p. 218.

357. Cited in Schultz, *Art and Anatomy in Renaissance Italy,* p. 74. Schultz adds that Book 4 of Vitruvius's *Ten Books on Architecture* compares Doric and Ionic columns with male and female human bodies, respectively (p. 74). Moreover: "During the Renaissance, book IX, chapter 7 of Alberti's *De re aedificatoria* completed in 1452 and first printed in Florence in 1485, repeated the proportional correspondence between columns and the 'members of the human body.'"

358. Schultz, *Art and Anatomy in Renaissance Italy,* pp. 74–76 (pp. 74, 76); also see pp. 69, 73, 227 note 45. In those of Leonardo da Vinci's (1452–1519) anatomical drawings which belong to the period he worked in Milan in the 1480s, when his serious inquiry into human anatomy began, it has been noticed by Bernard Schultz and others, that two-way sections of anatomical parts are in some measure modeled on architectural techniques of "representing a building in terms of the ground plan." Given this form of modeling and the parallel development realised between the structure of the neck and studies of columns in other of Leonardo's early anatomical studies, it has been speculated that perhaps Leonardo's novel use of cross-sectional anatomy, which would become a standard feature of modern anatomical illustrations, was suggested or inspired by architectural drawings and by his

observations on the fastenings and horizontal sections of a column (p. 73). A later drawing of the female viscera, datable to about 1508, "utilizes a full range of architectural drawing methods to present the internal anatomy, displaying, for example, the esophagus in full relief while sectioning the uterus below" (p. 94).

359. Schultz, *Art and Anatomy in Renaissance Italy,* p. 224, note 4: "On [MS] Windsor 19070 *verso* Leonardo notes that he has dissected 'more than ten human bodies,' and from the later period in France, Cardinal Louis of Aragon recorded that the artist dissected more than thirty bodies." Giorgio Vasari, *Lives of the Most Eminent Painters, Sculptors and Architects,* 10 vols., trans. De Vere (London: Macmillan, 1912–15), IX, 103–4, claims that "in order to be entirely perfect, innumerable times [Michelangelo] made anatomical studies, dissecting men's bodies in order to see the principles of their construction and the concatenation of the bones, muscles, veins, and nerves, the various movements and all the postures of the human body; and not of men only, but also of animals." Cited in Schultz, *Art and Anatomy in Renaissance Italy,* p. 67. Schultz concludes that on comparison with Leonardo, however, "Michelangelo's primary path of inquiry was toward the explication of 'the movements and gestures of human beings'" (p. 95). In other words his figures respond to the anatomical knowledge necessary for art, whereas, by contrast, for Leonardo "the minutiae of anatomical structure has indeed become an active concern" by the early sixteenth century (p. 95).

360. Schultz, *Art and Anatomy in Renaissance Italy,* pp. 88, 95; also see p. 94. Leonardo's early anatomical studies in Milan during the 1480s, by contrast to those of the Florentine period, commonly display his "'filling in' of anatomical details based not upon observation but rather drawn from traditional anatomical assumptions" (p. 71).

361. Opinions vary as to the association of Leonardo with the humanist physician Marcantonio della Torre, according to Schultz, "ranging from a belief that Leonardo's drawings were to illustrate a [projected] treatise by Marcantonio" to the opinion that "the results of this association did not significantly alter Leonardo's approach to anatomical study." *Art and Anatomy in Renaissance Italy,* p. 97. Schultz's view is that "a more careful consideration of Vasari does not reveal an intent toward a publishing venture based on this professional collaboration. . . . This, however, is not to say that Leonardo did not project an anatomical treatise of his own, for the Windsor anatomical pages reveal the developing complexity of just such an intent" (p. 97).

362. Berengario, *Isagogae breves,* pp. 37–41.

363. When the term "building" occurs in single inverted commas in the text, other than when it is a citation, its signification should be regarded as ambivalent in this sense; but it should be assumed that throughout the study the term anatomy may either signify the object or knowledge of and discourse on it, or, ambivalently, both. The reasons for and effects of this ambivalence will be explored throughout the study.

364. Crooke, *Microcosmographia,* fol. 37.

365. Thus Corti uses the distinction to assert the traditional supremacy of the physician and natural philosopher over the *anatomista.* Eriksson, *Andreas Vesalius' First Public Anatomy at Bologna 1540,* pp. 45, 47.

366. In the Preface to his *Fabrica,* Vesalius locates the root of "the too-great separation that has taken place between those disciplines that complement one another for the fullest comprehension of a single art," "[e]specially after the devastation of the Goths when all the Sciences, formerly so flourishing and fittingly practiced, had decayed, [and] the more fashionable physicians, first in Italy in imitation of the old Romans, despising the use of the hands, began to relegate to their slaves those things which had to be done manually for their patients and to stand over them *like architects.*" O'Malley, *Andreas Vesalius of Brussels 1514–1564,* pp. 317, 318 (emphasis mine). In "a celebrated and righteously vitriolic passage" of this preface, fol. 3^r (pp. 319–20), Vesalius deplores the pedagogical procedure described in the text, which was "mandated by statute for the conduct of Vesalius' Paduan chair." Harcourt, "Andreas Vesalius and the Anatomy of Antique Sculpture," 57 note 22, 36. Harcourt adds that "this regulation was contravened by Vesalius from the outset" (p. 57 note 22). The passage of the *Fabrica* is translated by C. D. O'Malley.

367. O'Malley, *Andreas Vesalius of Brussels 1514–1564,* p. 317. Vesalius herein claims for the use of the hands in dissection the advocacy of Hippocrates and Galen (pp. 317–18).

368. Berengario, *Isagogae breves,* pp. 37–41.

369. Ibid., pp. 37–41, 50.

370. Eustachio did, however, publish his *Opuscula Anatomica* in Venice in 1563, as well as a "small book on the kidney" with copper plates. See French, *Dissection and Vivisection in the European Renaissance,* pp. 183–85; Roberts and Tomlinson, *The Fabric of the Body,* pp. 189–92.

371. Choulant, *History and Bibliography of Anatomic Illustration,* p. 200. In the eighteenth century these prints were found with the heirs of Pinis, the family Rossi (de Rubeis) by the papal physician Lancisi, who published them with his own commentaries in Rome in 1714. Choulant also provides further details regarding the distribution and publication of these plates.

372. Roberts and Tomlinson, *The Fabric of the Body*, p. 192.

373. Ibid., p. 192; Choulant, *History and Bibliography of Anatomic Illustration*, p. 200. One of the reasons for the dominance of Vesalius in the historical trajectory of Renaissance to early modern anatomy, is perhaps provided by Kemp, "'The Mark of Truth,'" p. 85: "[I]t has been claimed that the Vesalian revolution in anatomy was predicated upon the new modes of visual representation in the Renaissance—that the revolutionary procedures for the systematic imitation of real forms from specific viewpoints were themselves ultimately responsible for developing new modes of looking amongst investigators and for breaking the mould of scholastic commentaries, which were dominantly shaped by verbal formulae." Had Eustachio's "maps" of human anatomy been published when they were produced this trajectory may now look very different; and historians of anatomical illustration would have less reason to look to the development in art of the representation of visual space from one-point perspective for the models of anatomical illustration.

374. Roberts and Tomlinson, *The Fabric of the Body*, p. 192.

375. Choulant, *History and Bibliography of Anatomic Illustration*, p. 200.

376. Schultz, *Art and Anatomy in Renaissance Italy*, p. 71. Schultz goes on to assert: "It was precisely this correspondence between the 'seat of the soul,' the *sensus communis*, [in the brain] and the anatomy of physical movement that allowed Leonardo [da Vinci] to assert that, 'A picture or representation of human figures ought to be done in such a way that the spectator may easily recognize, by means of their attitudes, the purpose in their minds.'"

377. Singer, *Vesalius on the Human Brain*, p. 1.

378. O'Malley, *Andreas Vesalius of Brussels 1514–1564*, p. 324. In the statement cited in the text,Vesalius appeals to the Platonic and Platonist concept of the body as the temporary shelter of the soul, and the Aristotelian notion of the body as the soul's form and instrument. On instrumentality, Schultz, *Art and Anatomy in Renaissance Italy*, p. 71, says: "This pneumatic physiological construct, steeped in medical tradition, addressed a continuum between mind and muscle as related cause to effect."

379. See Temkin, "Metaphors of Human Biology," p. 275.

380. Lakoff and Johnson explain as a feature of some literary or imaginative metaphors the use of the conventionally unused parts of the defining domain of a conventional metaphorical concept. *Metaphors We Live By*, p. 53.

381. Singer, *Vesalius on the Human Brain*, p. 32.

382. *The Concise Home Doctor: Encyclopedia of Good Health*, written by a distinguished group of general practitioners and specialists in medicine, surgery, psychology, hygiene, physical culture, and veterinary practice, with plates in full colours and nearly 4,000 illustrations, 2 vols. (London: Educational Book Company, [1932(?), 1933(?)]), I, 656, col. 2. According to Singer, *The Evolution of Anatomy*, pp. 109, 117, Sylvius, at the University of Paris, introduced the term corpus callosum, although it was introduced to the printed page by Vesalius.

383. *The Concise Home Doctor*, I, 406, col. 1.

384. Singer, *Vesalius on the Human Brain*, p. 32.

385. Ibid.

386. On metaphorical coherence, see Lakoff and Johnson, *Metaphors We Live By*, ch. 16, pp. 87–96, and ch. 17, pp. 97–105 (p. 89).

387. For Lakoff and Johnson on metaphor and truth, see *Metaphors We Live By*, ch. 23, pp. 156–58; ch. 24, pp. 159–84; especially pp. 159, 165–66, 169–72.

388. See Foucault, *The Order of Things*, p. 22, on the role of analogy in comparative anatomy.

389. Eriksson, *Andreas Vesalius' First Public Anatomy at Bologna 1540*, pp. 57–59.

390. Also see Vesalius's description of the elements of the skull (cited above in the text). In commencing his treatise on water, Leonardo also followed the anatomical/geological comparison, saying: "By the ancients man has been called the world in miniature; and certainly this name is well bestowed, because inasmuch as man is composed of earth, water, air and fire, his body resembles that of the earth; and as man has in him bones, the supports and framework of his flesh, the world has its rocks, the supports of the earth." The passage goes on to qualify the analogy and then rationalize the difference thus: "The body of the earth lacks sinews, and this is because the sinews are made expressly for movements and, the world being perpetually stable, no movement takes place, and, no movement taking place, muscles are not necessary. But in all other parts they are much alike." *Manuscript A*, cited in Schultz, *Art and Anatomy in Renaissance Italy*, p. 76. Schulz argues that "the microcosmic analogy of the human figure," especially the more "literal" equation of "human anatomy with the geological structure of the earth," together with "nearly contemporary developments in architectural drawing," both contributed to "the means of perspective employed by Leonardo in his early anatomical studies" (pp. 77, 76). Moreover, the "macrocosmic/ microcosmic analogy . . . governed the whole of [Leonardo's] approach to anatomical study, beginning with the

literal adaptation of architectural drawing techniques through the current and more complex teleological methodology, uniting his observations of the physical world with the physical existence of being" (p. 100).

391. Crooke, *Microcosmographia,* fol. 6.

392. Berengario, *Isagogae Breves,* p. 93. The "spirit" to which Berengario refers is "the thin blood [carried from the heart by the aorta] which by some is called the vital spirit" (p. 63). According to Heseler, Corti cites Aristotle's *Historia Animalium,* VIII, as a source of the idea that "man is a little world, because his parts are as disjointed as in the great universe above." Eriksson, *Andreas Vesalius' First Public Anatomy at Bologna 1540,* p. 61. One can see in such provenance the development of the microcosm theory from Empedocles' notion of the "*discordia concors*" of the four elements. Also see Crooke, *Microcosmographia,* fols. 6–7.

393. Veslingus, *The Anatomy of the Body of Man,* p. 66. Also see Ralegh, *The History of the World,* bk. I, ch. 2, section 5, p. 30, which compares "our eies" "to the light of the Sunne and Moone."

394. Crooke, *Microcosmographia,* fols 6–7.

395. The microcosm was, according to Foucault, in *The Order of Things,* pp. 31–32, a surface effect of the analogical mode that made knowledge possible in the sixteenth century.

396. Paré, *Works* (1678), bk. IV, p. 95.

397. Cunningham, *The Anatomical Renaissance,* p. 12, is my source for this knowledge, but his remarks upon *Timaeus* do not recognize, or find useful for his own concerns, how implicit in Plato's account is the analogy of anatomy with the constructed environment.

398. See, for example, Temkin, "Metaphors of Human Biology," pp. 271–72.

399. Galen followed a different sequence of dissection when it came to understanding what the different parts do with respect to the whole: "the account Plato gave in the *Timaeus* of the human body being divided according to three grand functions [those of the animal, vital, and natural faculties of the soul]. All of Galen's understanding of how the body functions [as opposed to how it is constructed] is based on this." Cunningham, *The Anatomical Renaissance,* p. 28.

400. Colombo, *De re anatomica.* According to Schultz, *Art and Anatomy in Renaissance Italy,* p. 103, A. Parronchi's "Michelangelo e Realdo Colombo," *Opere giovanili di Michelangelo* (Florence, 1975), II, 199–200, "demonstrates that the preface to the first chapter [of *De re anatomica*] could also suffice as the introduction to an architectural treatise." Schultz adds: "Parronchi is correct in asserting the influence of Vitruvius and especially Michelangelo's 'organic' architectural theory, but, as we have seen, this tradition was far more widespread, enveloping both medical literature with Galen and art theory, becoming a continuing consideration in the architectural writings following Alberti" (p. 103).

401. Cunningham, *The Anatomical Renaissance,* p. 27. Cunningham is citing Galen, *On Anatomical Procedures.* In this order, following the dissection of the bones and muscles is that of the vessels (veins, arteries, nerves) and then the viscera (organs in the trunk and head). Cunningham suggests: "That Galen took such a view of the relation of the skeleton to the flesh and organs may owe something to his experience of the building techniques of his time—the use of load-bearing pillars and beams, the space between them being filled in with curtain walls—and the fact that his own father was an architect by profession." None of this detracts from Galen's philosophical motivation for conceptualising anatomy in this way, however. A detailed exposition and functional and pedagogical justification of this metaphorical concept determining the order of anatomical dissection occurs in the First Book ("Of Bones, the frame of the Body") of John Banister (1533–1610), *The Historie of Man, Sucked from the Sappe of the Most Approued Anathomistes, in this Present Age, Compiled in Most Compendious Forme, and Now Published in English, for the Vtilitie of All Godly Chirurgians, within this Realme . . .* (London: John Daye, 1578; repr. Amsterdam: Theatrum Orbis Terrarum; New York: Da Capo Press, 1969), sig. C_i^r (fol. 1^r) (N.B. the text has two forms of pagination neither of which is continuous throughout the whole text, but which overlap): "As the good and expert *Architect,* hath a singular care, first in edifying his house, to be well aduised of the manner of his foundation, and heedy framyng of his Timberworke, before he lay on Thack, Tile, Slate, Lime, or Plaster: so it behoueth us, in searchying the true Secresie of this Misterious Science, and body of Man, first, before the partes supported, distinctly, and deliberatly, with diligent care, to consider the partes supporting. And the reason is good. For if you doe not first exactly examine, and to a sufficient fulnes feede your appetite, with the diligent peruse of this History of Bones: you shall finde the rest of this booke, not onely in readyng unsauery, but also to your sensible understanding uery obscure." (The term which in this citation appears in italics is in Roman font in the original text, where the remainder of the text is in gothic.)

402. De Man, "The Epistemology of Metaphor," pp. 19, 22. The evidence mounted in the course of this study suggests that most language (including much so-called literal language) is more or less actively metaphorical, at least linguistically if not cognitively, because it includes idioms, polysemy, dead metaphors, trivial metaphors,

highly conventional metaphors, conventional metaphors, creative or novel extensions of conventional metaphors, and novel metaphors. This metaphoricity is often, though not always, conceptual, producing of and by itself the entities it signifies which may have no equivalence in nature or in ideas existing independently, because the truth or falsity of the cognitive content is somewhat dependent on conceptual systems which are themselves largely metaphorical. The extensive implications of de Man's argument for understanding and for language arise in part, however, because he includes in his definition of metaphoricity occurrences of catachresis (p. 19), with which this study is not concerned. He also enrolls etymology to the purpose of eroding the totalization of properties even of simple ideas (pp. 17–18); and I will return to this particular aspect of his argument later in the next chapter.

403. Terence Hawkes, *Metaphor,* Critical Idiom Series, 25 (London: Routledge, 1972; repr. with revised bibliography 1989), p. 20.

404. Ibid., p. 19.

405. Ibid., p. 91.

Chapter 2. The Creative Metaphor of the "*Isle of Man*"

1. De Laguna, *Anatomica methodus,* p. 273. Temkin, "Metaphors of Human Biology," p. 272, says that Galen "used the simile of the food supply of a city for explaining the function and name of the veins of the portal system, which carried chyle to the liver just as many routes carried food to the city's bakeries."

2. Kimber, Gray, and Stackpole, *Textbook of Anatomy and Physiology,* p. 477.

3. The humors are called "carriers" in an uncited part of the passage in de Laguna's *Anatomica methodus,* p. 273.

4. De Laguna, *Anatomica methodus,* p. 273. On the origin of diseases in imbalance of the four qualities or the four humours, see Schoenfeldt, *Bodies and Selves in Early Modern England,* pp. 13–15, 20–33.

5. Hausman stresses: "The generation of multiple meanings—the open-endedness, or what Ricoeur calls 'plurisignification'—of metaphor must not be taken for ambiguity. Ambiguity is dependent on a condition of undecidability about which of several determinate meanings is appropriate" (p. 14). This definition of ambiguity is indebted to Beardsley, *Aesthetics:* "I reserve the term 'ambiguity' for linguistic expressions that are doubtful in meaning because they could have either, but not both, of two possible meanings and provide no ground for a decision between them. Where there is suggestion or connotation, no choice is called for; several things are meant at once" (p. 126). Metaphor's generation of multiple meanings is also said to distinguish it from semiotic or conceptual symbols whose meaning is fixed. Hausman, *Metaphor and Art,* p. 14.

6. See Hausman, *Metaphor and Art,* for a summary of this view of metaphor ("Introduction," pp. 1–11; pp. 5, 1 cited), and of the theories of Black and Beardsley, (pp. 32–39). The work of Black, Beardsley and Ricoeur on metaphor is in the fields of philosophical aesthetics, poetics, and rhetoric; whereas Lakoff and Johnson's work is in philosophy of language and cognitive linguistics. There is also a historical difference for much of the work of the former group is written before 1980, whereas that of Lakoff and Johnson and others in what Steen calls the "cognitive turn" in metaphorology is largely post-1980. Steen, *Understanding Metaphor in Literature,* pp. 3–5.

7. Black, "Metaphor," p. 286. As a source for an "interaction view" of metaphor, Black is here drawing upon as well as disagreeing with I. A. Richards, *The Philosophy of Rhetoric* (Oxford: Oxford University Press, 1936), especially Chs. 5 and 6.

8. For Black's discussion of the "substitution view" of metaphor, see especially pp. 278–79.

9. Black, "Metaphor," pp. 280–81.

10. Lakoff and Johnson, *Metaphors We Live By,* pp. 193, 235.

11. George Puttenham (*c.* 1529–90/91), *The Arte of English Poesie 1589,* A Scolar Press Facsimile (Menston: Scolar Press, 1968), p. 128.

12. Hawkes, *Metaphor,* pp. 13–14.

13. Puttenham, *The Arte of English Poesie,* pp. 115, 114. Sir Philip Sidney (1554–86) also acknowledges that, in philosophy and history, "Poesy" "trimmeth . . . garments with guards [ornamental borders]," but this view carries less weight for Sidney than for Puttenham. Sir Philip Sidney, *The Defence of Poesy,* in *Selected Writings: Astrophil and Stella, The Defence of Poesy and Miscellaneous Poems,* ed. with an introduction and notes by Richard Dutton (Manchester: Carcanet, 1987), p. 136. This edition is a conflation of two independent texts or variant manuscripts by Sidney, both published in 1595: *An Apology for Poetry* and *The Defence of Poesy.* Another use of "figure" to signify "ornamentation" is by Thomas Campion, *Obseruations in the Art of English Poesie . . .*

Wherein It Is Demonstratiuely Prooued, and By Example Confirmed, that the English Toong Will Receiue Eight Seuerall Kinds of Numbers, Proper To It Selfe, Which Are All in this Booke Set Forth, and Were Neuer Before this Time By Any Man Attempted (London: Richard Field, 1602); repr. The English Experience (Amsterdam: Theatrum Orbis Terrarum; New York: Da Capo, 1972), p. 4. Campion describes rhyme as "*figura verbi*" (verbal ornamentation): "a continual repetition of that Rhetoricall figure which we tearme *similiter desimentia,* and that being but *figura verbi,* ought . . . sparingly to be vsed, least it should offend the eare with tedions [*sic*] affectation." But Campion also says, "[t]he eare is a rationall sence," alluding principally to its judgment of proportion in poetry, but hinting also at the conceptual dimension of figurality (which here includes verbal ornamentation), and to the instrumentality of the sense for the rational soul (pp. 4–5).

14. See Hawkes, *Metaphor,* p. 13.

15. See Quilligan, *The Language of Allegory,* p. 46: "*Allegoresis,* elder cousin to narrative allegory, begins by saying that texts are, superficially, lies; they must be interpreted, or 'allegorized' into telling the truth." De Man, "The Epistemology of Metaphor," p. 29, concedes that, although it is "absurd to ask whether a code [in the structuralist sense] is true or false," it is "impossible to bracket this question when tropes are involved—and this always seems to be the case."

16. Hawkes, *Metaphor,* p. 24. Hawkes argues that Peter Ramus (1515–72) was "[o]ne of the most formative although indirect influences on the nature of metaphor at this time" (p. 22). Rosemond Tuve, *Elizabethan and Metaphysical Imagery: Renaissance Poetic and Twentieth-Century Critics* (Chicago and London: University of Chicago Press, 1947; 4th impression 1961), Part II, argues for logical functions of poetic imagery, and that the Ramistic reorganization of logic pressed writers toward a more conscious attempt to relate poetry to dialectic (p. 252).

17. Ullrich Langer, "Invention," in *The Cambridge History of Literary Criticism,* ed. by Nisbet and Rawson, III: *The Renaissance,* ed. by Norton, 136–44 (p. 136). Also see Mazzeo, "Metaphysical Poetics and the Poetry of Correspondence," p. 223.

18. Langer, "Invention," p. 136: "The noun *inventio* corresponds to the verb *invenire* [to find, to discover, to come upon]. In rhetorical treatises, at the place of *invenire* we often find *reperire* [to find, to discover] or *excogitare* [to think of, to find by reflection]."

19. Christopher Braider, "The Paradoxical Sisterhood: 'ut pictura poesis,'" in *The Cambridge History of Literary Criticism,* ed. by Nisbet and Rawson, III, 168–75 (p. 171).

20. W. Tatarkiewicz, *History of Aesthetics,* 3 vols. (The Hague: Mouton; 1970), III, 167, cited by Jonathan Dollimore, *Radical Tragedy: Religion, Ideology and Power in the Drama of Shakespeare and his Contemporaries,* 2nd ed. (London: Harvester Wheatsheaf, 1989), p. 81.

21. Langer, "Invention," p. 138. For some theorists, Sidney included, "it is only at the level of her 'intention', as *natura naturans* rather than *natura naturata,* that nature can be said to be 'perfect' and therefore worthy of imitation." Braider, "The Paradoxical Sisterhood: 'ut pictura poesis,'" p. 175. According to Langer, "[p]oetic invention allows, for [Philip] Sidney, a mimetic perfecting of the real, a homage to divine creation" (p. 140); whereas George Puttenham's *Arte of English Poesie* tends to posit a "more tempered link between invention and imagination" (p. 140). This is also true of late sixteenth-century Italian theorists and sixteenth-century French theorists, examples of both of which could have influenced the distinctly continental baroque style of *The Purple Island.* For Sidney, see *The Defence of Poesy,* p. 107.

22. William Benlowes, "To the Learned Authour, Sonne and Brother to Two Judicious Poets, Himself the Third, Not Second to Either," Giles and Phineas Fletcher, *Poetical Works,* II, 7. Langdale, *Phineas Fletcher,* pp. 90–91, says that William Benlowes was "evidently a kinsman of the heir to Brent Hall [Edward Benlowes (1603?-1676)] and by him introduced to Fletcher. Little is known about this second Benlowes."

23. See Dollimore, *Radical Tragedy* (1989), ch. 4: "Renaissance Literary Theory: Two Concepts of Mimesis," pp. 70–82.

24. Steen says, in *Understanding Metaphor in Literature,* pp. 18–19, that when people solve riddles, "nonliteral analogical processing takes the form of an active search for mappable relations between the domains of [the riddle or metaphor]." Steen distinguishes such metaphors from cases of individual polysemy and from scientific models.

25. Hausman, *Metaphor and Art,* p. 2.

26. Black, "Metaphor," 282–83.

27. Analogy is one of the four types of transference involved in metaphor, as discussed in chapters 21–25 of Aristotle's *Poetics.* See Hawkes, *Metaphor,* p. 7; Hausman, pp. 22–23. Hausman explains that Aristotle's classification serves as the basis for later distinctions among figures of speech, such as synecdoche, metonymy, and anal-

ogy. According to Hawkes, Aristotle's account of metaphor in his *Rhetoric*, III, regards it "as a decorative addition to language." (p. 8). Clarity is said to come from ordinary language. Metaphor adds charm and distinction, but not cognitive content which could not otherwise be provided by a paraphrase. Aristotle does, however, admit of surprising resemblances in metaphor, and recognizes its creative and educative aspects or uses (pp. 9–10). Hausman also admits that Aristotle "does indicate that metaphorical figures of speech are, at bottom, comparisons" (p. 23). Cicero says that metaphor is a "contracted" simile, giving decorous effect, and Quintilian's view, which had considerable influence in the Renaissance, is that metaphor adorns, is "ornament." For Cicero, see Hawkes, *Metaphor*, p. 11.

28. Hausman, *Metaphor and Art*, p. 17.

29. Black, "Metaphor," 283; Beardsley, *Aesthetics*, p. 137; Hausman, *Metaphor and Art*, p. 28.

30. Beardsley, *Aesthetics*, p. 137.

31. Lakoff and Johnson, *Metaphors We Live By*, p. 153–54. Aristotle claims that "a good metaphor implies an intuitive perception of the similarity in dissimilars," thereby assuming the antecedent existence of the similarities perceived therein. *De Poetica*, ch. 22, 1459a. On this passage from *De Poetica*, Rufus Wood, *Metaphor and Belief in "The Faerie Queene"* (Basingstoke: Macmillan, 1997), p. 2, erroneously collapses the categories of discovery of antecedently existing similarities and creation of new similarities: "If, as Aristotle here says, to metaphorize well is a gift of genius, through which hidden relationships can be intuitively grasped, then metaphor becomes instrumental in the processes of invention and discovery." Wood's error is compounded by running definitions of metaphor into those of simile, without acknowledging that there might even be a distinction (p. 9). Wood is correct, however, in indicating that Aristotle's theory suggests that "[m]etaphor is not a mere ornament of language, it is involved in the redefinition of reality" (p. 2).

32. It should be conceded that some similes have a stronger impact of transformed meaning than others, and even than some metaphors. Certain to qualify as such a simile is "O My Luve's like a red, red rose." Robert Burns, "A Red, Red Rose" (1796), in *The Norton Anthology of English Literature*, ed. by M. H. Abrams, 2 vols., 5th ed. (London: Norton, 1986), II, 107. This simile and others which have a strong impact of transformed meaning frequently have "metaphorical roots"; i.e., *a* is like *b*, when and only when *b* is understood metaphorically. Hausman, *Metaphor and Art*, p. 17. It would be more precise to say that such similes have *active* metaphorical roots, for, given that our conventional conceptual system is arguably largely defined metaphorically, similes as well as metaphors have (more or less) metaphorical roots, though these roots may be relatively actively or inertly metaphorical.

33. Black, "Metaphor," 284.

34. Beardsley, *Aesthetics*, pp. 137–38.

35. Sparshott, "'As,' or The Limits of Metaphor," pp. 82–83.

36. Beardsley, *Aesthetics*, p. 435.

37. Ibid., pp. 137–38. Also see Black, "Metaphor," 284.

38. Joseph Anthony Mazzeo, "Metaphysical Poetry and the Poetic of Correspondence," *Journal of the History of Ideas* 14 (1953): 221–34 (pp. 222–25), argues that there was a disparity between the seventeenth-century poetic use of metaphor and seventeenth-century theory of metaphor, which still maintained that it was a modification of a literal statement or a kind of ornamentation. Theory tended to abide by a disciplinary divide between grammar, or plain language, and rhetoric, understood as ornate language, and this, rather than poetic practice, according to Mazzeo, determined the definitions of poetry and poetic metaphor. For the idea that in this period, metaphor was used to deepen and extend meaning in the way described in my text, see, for example, Hawkes, *Metaphor*, pp. 18, 21. When Foucault argues that the question in the Renaissance was how to know a sign designated what it signified, and that only after the Renaissance did it became possible to ask how a sign could designate what it signified, he provides for this dimension of deepening meaning beyond what a sign ordinarily designates, by reference to what he calls the "primal text" beneath the surface of language, from which linguistic meaning grows and to which it must always return. This primal text is generally the Bible. *The Order of Things*, pp. 42–44.

39. See Hawkes, *Metaphor*, pp. 21, 24.

40. Black's article lists these claims on pp. 291–92 of "Metaphor," but reference is also made in the text and notes of this chapter to Black's fuller expositions of the respective points elsewhere in his text. Black discusses the pertinence of the seven claims or principles on page 292.

41. Ricoeur, "Metaphor and the Main Problem of Hermeneutics," p. 99. On the "contextual effect," Ricoeur says: "If the metaphorical sense is more than the actualization of one of the potential meanings of a polysemic word (and all our words in common discourse are polysemic), it is necessary that this metaphorical use is only *contextual;* by that I mean a sense which emerges as the result of a certain contextual action."

42. In the threshold figure of *The Purple Island* is a compressed analogy, but one in which there is a principal and a subsidiary subject, for throughout the first five cantos the metaphorical term or focal word is almost invariably "Island" or (even more often) "Isle" or (occasionally) some metonym or synecdoche of the part for the whole, such as "coast" or "land." Following is a list of references to each of the terms referred to by the text, as they occur in the first two cantos of *The Purple Island:* "Island": 1.34.2; 1.48.7; 1.60.2; 2.13.1; 2.39.1; 2.42.7. "Isle": 1.38.1; 1.43.4; 1.46.1; 1.48.7; 1.53.1; 1.54.2; 1.57.1; 7, 2.2.1; 2.5.5; 2.9.2; 2.10.4; 2.14.1; 2.16.2. "Coast": 1.34.6. "Land": 1.51.6–7; 2.9.1. Rarely is the principal subject referred to directly, and even then the reference takes a metaphorical form. The term "Man" occurs only once in the first two cantos of *The Purple Island,* as "*Isle of Man*" (1.48.7).

43. Black, "Metaphor," p. 293. According to Hausman, Richards refers to "'the interanimation of words' to cover the way the words depend on one another when they function in metaphorical sentences. The degree of interdependence varies." "[T]his interaction is the condition for a meaning not carried by any of the terms as they function independently of the metaphorical context." But Richards also "wants a ground for metaphor, and this seems to lie in something in common that is referred to by the terms." Even so, he "repudiates the idea that the terms of metaphors are related only through common meanings or resemblances." Hausman, *Metaphor and Art,* p. 31.

44. Ricoeur, "Metaphor and the Main Problem of Hermeneutics," p. 102; Ricoeur cites Black, *Models and Metaphors,* p. 46 ("Metaphor," p. 293).

45. Black, "Metaphor," p. 293; Ricoeur, "Metaphor and the Main Problem of Hermeneutics," p. 101. Further to the inadequacy of the literal paraphrase, see Black, "Metaphor," p. 283. Beardsley provides the criterion, that a discourse can be regarded as a paraphrase of another discourse, if both discourses have "identical cognitive content." *Aesthetics,* p. 433; also see pp. 432–37. He discusses how "a set of literal statements that are equivalent to the metaphor" "do not seem to exhaust its meaning completely," adding "nor does it seem as if, in practice at least, we can ever exhaust that meaning, or be certain we have exhausted it" (pp. 431, 432). Moreover, "it is significant that when we try to explicate metaphorical attributions, that is, give partial paraphrases, we fall back, more often than not, upon other metaphors, or upon figurative language of other sorts." This, however, does not prove to be a fatal difficulty, an argument to show the impossibility of literal paraphrase, as Beardsley shows (p. 434). But, although "[i]t does not seem that there is no conclusive argument against the possibility, in principle, of paraphrase . . . there are excellent reasons for supposing it to be usually impossible in practice" (pp. 435–36). Also see p. 452.

46. Black, "Metaphor," 284–85. Also see Hausman, *Metaphor and Art,* pp. 9–10. Sam Glucksberg and Boaz Keysar, in "Understanding Metaphorical Comparisons: Beyond Similarity," *Psychological Review* 97 (1990): 3–18, also argue against the view that metaphors are implicit similes, and develop a theory that the category of the defined term is assigned to the category of the defining term, with the resultant effect that: "The similarity that is thereby perceived among category members is thus a product of that categorization, not an antecedent of it. That is, the categorization produces the similarity, not the other way around" (p. 11). According to Steen, this is "a well-founded psychological explanation of the . . . view in metaphorology that some metaphors create similarities rather than formulate pre-existing ones." He adds that, "for conventional metaphors . . . there may be an *ad hoc* category which is already there in the mind," but for less familiar and novel metaphors, analogical processing may still have to work out how, at the level of categorization, the defining and defined terms are similar, and "similarity and analogy may play a crucial role in creating the *ad hoc* category." *Understanding Metaphor in Literature,* Section 1.3.2, pp. 13–15 (pp. 14, 15).

47. George Lakoff, "A Figure of Thought," *Metaphor And Symbolic Activity* 1 (1986): 215–25 (p. 223) cited in Steen, *Understanding Metaphor in Literature,* p. 15.

48. Black, "Metaphor," p. 291.

49. Ibid., pp. 287, 291.

50. Ibid., p. 291.

51. Ibid., pp. 288, 291.

52. Hausman, *Metaphor and Art,* p. 7.

53. Kastor, *Giles and Phineas Fletcher,* p. 55 note 4.

54. Black, "Metaphor," pp. 291–92; also see p. 289.

55. Beardsley, *Aesthetics,* p. 142. Black and Beardsley are concerned in this respect neither with utilitarian metaphors, which barely produce a change of meaning at all, nor with "dead" or inert metaphors, such as the leg of a table, which transfers the concept of supportive function from the concept of the corresponding part of animals, to that for which there is no other literal term, and does so with probably no processing of a metaphorical concept. According to Ricoeur, a dead metaphor is a metaphor that is "adopted by an influential part of the speech

community, . . . become[s] a standard meaning and [is] added to the polysemy of the literal entities, contributing in that way to the *history* of language as *langue,* code, or *system.*" "Metaphor and the Main Problem of Hermeneutics," pp. 99–100.

56. On the use of metaphor to criticize metaphorically constituted world-views or visions, see Booth, "Metaphor as Rhetoric," especially pp. 63–64.

57. Deese, "Mind and Metaphor: A Commentary," p. 214.

58. Beardsley, *Aesthetics,* p. 150.

59. Ricoeur, "Metaphor and the Main Problem of Hermeneutics," p. 99. The same "clash" is what Beardsley calls the "*Logical Absurdity*" of a metaphorical statement, when read literally, and allows, for Black and Beardsley, the recognition of metaphor and the consequent exposition of the extension or generation of meanings from interaction. Beardsley, *Aesthetics,* pp. 138–40, 161–62; Black, "Metaphor," pp. 273–75.

60. See Ted Cohen, "Metaphor and the Cultivation of Intimacy," *Critical Inquiry* 5 (1978): 3–12 (p. 7).

61. Hausman, *Metaphor and Art,* p. 8; also see pp. 26, 32.

62. Marjorie Donker and George M. Muldrow, *Dictionary of Literary-Rhetorical Conventions of the English Renaissance* (Westport, Conn. and London: Greenwood, 1982), p. 52.

63. Donker and Muldrow, *Dictionary of Literary-Rhetorical Conventions of the English Renaissance,* p. 52: "In this way, the general meaning of *thought* or *idea* and an emphasis on intellectual activity carry over into the literary use of *conceit* to designate a similitude based on unusual relationships." Also see Norford, "Microcosm and Macrocosm in Seventeenth-Century Literature," p. 427.

64. Hausman, *Metaphor and Art,* p. 32; also see pp. 29–30.

65. Donker and Muldrow, *Dictionary of Literary-Rhetorical Conventions of the English Renaissance,* p. 55. Bruno uses the term *concettismo* for conceit. Norford, "Microcosm and Macrocosm in Seventeenth-Century Literature," p. 427.

66. Ibid., p. 55.

67. Ibid., p. 52.

68. See Quilligan, *The Language of Allegory,* p. 64; also see pp. 28–29.

69. Ibid., p. 67. In the context of the criticism of allegory, "literal" is often used to refer to the "sustained or continuous fictional activity or plot." Understood in terms of the words of a text, however, allegorical action "can provide the basis for a continued, coherent allegorical interpretation." Quilligan prefers to refer to the allegorical meaning as the "literal" or "letteral" meaning, and to the literal or "lifelike" meaning of the action or plot as the "metaphorical level" of "imaginary action" (p. 67): "This dispensable 'plot' is *only* metaphor [i.e., literally understood modifier] compared to the literal truth inherent in the words of the text, and the allegorist is always at pains lest his text be read as mere metaphor of any sort" (p. 68). Quilligan is herein using "metaphor" in a reductionist way to signify something like "ornament." Wood, *Metaphor and Belief in "The Faerie Queene,"* p. 11, sees how allegory, by masking some aspects of the meaning of a subject, seeks to illuminate or discover other aspects, thereby accentuating the tension between literal and metaphorical meanings.

70. Hausman, *Metaphor and Art,* p. 17.

71. Black, "Metaphor," 289; Beardsley, *Aesthetics,* p. 143.

72. Lakoff and Johnson, *Metaphors We Live By,* p. 53.

73. Ibid., pp. 147–55 (p. 151).

74. Ricoeur, "Metaphor and the Main Problem of Hermeneutics," p. 99, note 1.

75. Black, "Metaphor," p. 292.

76. Hausman, *Metaphor and Art,* p. 40.

77. Ibid., p. 40.

78. Ricoeur, "Metaphor and the Main Problem of Hermeneutics," pp. 99 (note 1), 101. For his "Controversion Theory" of metaphor, see Beardsley, *Aesthetics,* pp. 138–44, 161–62. Beardsley refers to Black, "Metaphor," saying that Black's "'system of associated commonplaces' . . . is the set of beliefs that determine the range of connotation of the metaphorical modifier" (p. 161). Beardsley states, on the distinction between primary meanings and secondary meanings, or connotations: "The distinction is sometimes made in terms of rules: there are 'rules' of designation but not of connotation. This is puzzling too, but 'rules' can presumably refer only to dictionary definitions—what other rules are there in general use?—and, if so, the distinction is misleading. For no one would say that a dictionary can tell *all* the meaning of a word in any particular context, and so the meaning cannot be reduced to what is reported in a dictionary." *Aesthetics,* p. 149.

79. Ricoeur, "Metaphor and the Main Problem of Hermeneutics," p. 102.

80. Quilligan, *The Language of Allegory,* pp. 26, 33, 46.

81. Hawkes, *Metaphor,* p. 71.

82. Black, "Metaphor," pp. 289, 291. Arguably, Black's view is shared by Beardsley, for whom (Hausman argues): "The modifier . . . must be given a 'metaphorical twist.' The interpreter of a metaphor, then, must look for connotations or marginal meanings that can be attributed without contradiction or falsity of application to what the central meaning of the other term conveys." Hausman, *Metaphor and Art,* p. 33. With regard to the qualitative growth of meaning, however, Beardsley seems at best ambivalent, and this is aptly represented by the following. He claims that his "Controversion Theory explains one of most puzzling and important features of metaphor, its capacity to create new contextual meaning." But merely a few lines later, he qualifies this statement, adding that "metaphor does not create the connotations, but it brings them to life," implying, as Hausman succintly puts it, that the connotations were "antecedent and actual, or potential and ready to be actualized." Beardsley, *Aesthetics,* p. 143; Hausman, *Metaphor and Art,* p. 33.

83. Black, "Metaphor," p. 289.

84. It may be important to add that the associated commonplaces and metaphorical and analogical concepts of each of the subjects may be regarded as pertaining in some way to, without being *true* of, the subject; and the requirement that they are true, or even coherent, may be disposed of in certain contexts by the overriding requirement of the effectiveness of the metaphor. The purpose of such metaphors as Booth, "Metaphor as Rhetoric," p. 53–54, discusses is to communicate clearly, to persuade. Their meaning is consequently stabilized by the context and largely paraphrasable, yet because more is communicated than the words literally mean or even what might be regarded as cognitive content, they are less effective when paraphrased.

85. Lakoff and Johnson's view is that cognition of conceptual metaphor is relatively independent of the way it is expressed grammatically in language.

86. See Steen, *Understanding Metaphor in Literature,* pp. 7–8.

87. This idea is somewhat indebted to Terence Hawkes's discussion in *Metaphor,* pp. 70–79.

88. See Hausman, *Metaphor and Art,* p. 31. On the contextual effect, see note 41, in this chapter.

89. Black, "Metaphor," 290. Also see Harries, "Metaphor and Transcendence," p. 83.

90. See Steen, *Understanding Metaphor in Literature,* p. 26 and Ch. 2.

91. Healy, "Sound Physic," p. 345, argues that *The Purple Island* employs "a guise of medical and empirical impartiality" in its description of an ideologically configured kingdom, the order of which is said to be "constructed around one overriding principle," which is that "it must be defended." Sawday, *The Body Emblazoned,* p. 176, describes the display of the body as a dismantled structure in *The Purple Island* as "a schizophrenic collection of fragments behind which the poet-narrator hides, terrified of the hostile exterior world."

92. See Harries, "Metaphor and Transcendence," pp. 74–76.

93. Hausman, *Metaphor and Art,* p. 14.

94. Paul Valéry, "Pure Poetry," in *The Art of Poetry,* trans. Denise Folliot (New York: Random House, 1961), p. 192, cited in Harries, "Metaphor and Transcendence," p. 78.

95. Ricoeur, "Metaphor and the Main Problem of Hermeneutics," p. 97. In Ricoeur's work, this idea develops upon Beardsley's related idea in *Aesthetics.* Ricoeur defines a "work" as "the closed sequence of discourse which may be considered as a text" (p. 96). He goes on to ask a further question: "to what extent may the hermeneutical problem of text interpretation be considered as a large-scale expansion of the problems condensed in the explication of a local metaphor in a given text?" (p. 97). Also see Cohen, "Metaphor and the Cultivation of Intimacy," especially p. 7; Hausman, *Metaphor and Art,* pp. 6–7.

96. Ricoeur, "Metaphor and the Main Problem of Hermeneutics," p. 98.

97. Ibid., p. 98. Ricoeur suggests that this "extralinguistic reality" may be "the world" or "a world" (p. 98).

98. Ibid., p. 100.

99. Roland Barthes, "Introduction to the Structural Analysis of Narratives" (1966), in *A Barthes Reader,* ed. and with an introduction by Susan Sontag (London: Jonathan Cape, 1982), pp. 251–95. Barthes's "functions," or "grammatical units," can be combined and recombined in infinite ways, yet the two basic principles of combination are said to be temporal succession (consecutive) and causality (consequential). Indices are the same as functions, in so far as they are units which are consequential or consecutive, but they have a further significance.

100. Indices either provide the reader with information with which it becomes possible to understand the story, or they are indices proper, which signify the "actions" of the story. Barthes, and other structuralist analysts of narrative, radically reduce the minutiae of significance and meaning that readers may draw from the actions of characters to a limited number of categories of act. In narrative theory, the concept of story has been subject to definitions which propose that it is made of separable components which may be analyzed in isolation and in re-

lation to each other. "Such a view," according to Rimmon-Kenan, *Narrative Fiction,* pp. 6–7, "justifies attempts to disengage a *form* from the substance of the narrated content, a specific narrative form" (also see p. 20).

101. Ricoeur, "Metaphor and the Main Problem of Hermeneutics," p. 105.

102. Ibid., p. 103.

103. Ibid., p. 109.

104. Ibid., p. 96. Also see Hausman, *Metaphor and Art,* p. 29.

105. Deese, "Mind and Metaphor: A Commentary," p. 211. Deese argues: "the interpretation of an analogy, the ability to produce or interpret a metaphor is limited only by the individual mind's knowledge and the accessibility of that knowledge to conscious search. This means, among other things, that the semantic structure of a language can never be characterized in any complete and exhaustive way." (p. 211). Deese adds: "It is characteristic of the poetic attitude to embrace a perceived object simultaneously from different sides. What those sides are depends upon context. The solution of the riddle provided by a given metaphor depends upon the context in which it occurs. Particular syntactic contexts influence the solution of metaphors, but context is more general. Sometimes . . . an entire work provides a context. . . ." (p. 215). I am arguing that *The Purple Island* is just such a work. Also see Hausman, *Metaphor and Art,* p. 3; Ricoeur, "Metaphor and the Main Problem of Hermeneutics," p. 104.

106. Beardsley, *Aesthetics,* p. 144. Hawkes considers the contextual effect in a slightly different though also important way: "In any context, any metaphor will be modified by such considerations as rhythmic quality, the syncopation between that and the overt 'meaning' of the words, the effect of rhyme or of the absence of rhyme, the position of the metaphor in the poem, in the line, even, in terms of its relative 'modernity' or archaic qualities, in time itself. The trouble with much modern linguistic analysis of metaphor is that it does not (and perhaps cannot) take into account the *full* context involved. . . ." *Metaphor,* p. 79.

107. This unfolding is to be distinguished from that which Quilligan argues is characteristic of the genre of narrative allegory: "allegorical narrative unfolds as a series of punning commentaries, related to one another on the most literal of verbal levels—the sounds of words." *The Language of Allegory,* p. 22. This chapter in due course reveals *The Purple Island*'s concern with polysemy, etymology, and homonymy, but of these, only etymology is of central concern to the unfolding of metaphorical implications and entailments.

108. Hausman, *Metaphor and Art,* pp. 17–18. The Isle of Man is not, however, from any perspective a symbol. Symbols may be repeated across different contexts and retain semantic stability, as may "frozen" or dead metaphors. See, for example, Hausman, *Metaphor and Art,* p. 15.

109. George Gascoigne (1534?-77) hints at both senses of allegorical figurality as able to "discover" meaning (that is, to disclose what is antecedently existing or, if novel, "meanings that are not arbitrary or are constrained by something independent of them") and to hide meaning in "shadowes," to mask it in a system of figures which downplay and hide aspects of the ordinary meanings of words. Gascoigne, *Certayne Notes of Instruction Concerning the Making of Verse or Ryme in English, Written at the Request of Master Edouardo Donati* (1575), in *The Complete Works of George Gascoigne,* ed. by John W. Cunliffe, Cambridge English Classics, 2 vols. (New York: Greenwood Press, 1969) (originally published in 1907 by Cambridge University Press), I, 465–73 (p. 466); Hausman, *Metaphor and Art,* pp. 39–40.

110. Ricoeur, "Metaphor and the Main Problem of Hermeneutics," p. 110; also see p. 109.

111. It should not be concluded from the creativity of the Isle of Man figure or its dependence upon a poetic context for our understanding of it, that the significance of the figure is "not a matter of the meaning of the words, but a feature of the contexts of their use, of their 'pragmatics,'" for, as Ricoeur argues, "the 'metaphorical twist' (to speak like Monroe Beardsley) is something which happens to words; the shift of meaning which requires the whole contribution of the context affects the word; it is the word that has a 'metaphorical use,' or of a nonliteral meaning, or a novel 'emergent meaning' in specific contexts." Cohen, "Metaphor and the Cultivation of Intimacy," p. 6; Ricoeur, "Metaphor and the Main Problem of Hermeneutics," p. 96.

112. According to Roger French, "A Note on the Anatomical Accessus of the Middle Ages," *Medical History* 23 (1979): 461–68 (p. 463), "*De sectis* was the first book to be studied in the post-classical Alexandrian medical curriculum, and so, apart from its complexity, it posed special problems for the inexperienced reader, and as a result it attracted much commentary." An edition of John of Alexandria's commentary was published in 1515. It divided anatomy into anatomy of the living (vivisection) and of the dead. In vivisection, says John, we must distinguish between two criteria, and in the anatomy of the dead between six criteria or "*occasiones.*" These are discussed in the text in relation to the list provided therein, but a fuller account of their provenance in other forms of *accessus,* and in Aristotle's *Categories,* and their development in the Middle Ages, is given by French. In his "Introduction" to Berengario, *Isagogae breves,* pp. 15–16, Lind writes: "Vesalius [in his *Epitome*] abandons the

old custom followed by Mundinus and Berengario of patiently listing for each organ the main facts about its substance, size, shape, location, connections, and complexion." But, in the Dedicatory Letter to Prince Philip which prefaces the *Epitome,* Vesalius criticizes "the mob of physicians that they perform their duty so carelessly in distinguishing the parts of the human body that not even enumeration is made use of in learning them. For when, beyond the function and use of each part, its location, form, size, color, the nature of its substance, the principle of its connection with the other parts, and many things of this sort in the medical examination of the parts may never be sufficiently perceived, how many can be found who know even the number of the bones, cartilages, ligaments, muscles, and veins, arteries, and nerves running in numerous succession throughout the entire body and of the viscera which are found in the cavities of the body?" *The Epitome of Andreas Vesalius,* pp. xxxiv–xxxv. Vesalius evidently encouraged the application of traditional criteria of description to anatomy, but for the sake of brevity he found it expedient in the purview of anatomy in the *Epitome* to incorporate traditional criteria without necessarily announcing them or dwelling on them in its descriptions of the position, operation or use, and appearance of each of the parts, and the adjacency of other members. These descriptive criteria are by no means the only criteria by which the anatomy of the human body is divided and described. Vesalius says, for example: "All the parts of the human body are either similar, or simple to the senses, such as the bones, cartilages, ligaments, fibres, membranes, flesh, and fat, or dissimilar, instrumental agencies, such as the veins, arteries, nerves, muscles, individual fingers, and the remaining organs of the entire body. These latter parts are made all the more instrumental to the degree that they (eg., the hand and head) are composed of many similar parts and also of many functional mechanisms." *The Epitome of Andreas Vesalius,* p. 1. Lind remarks that these "first few lines [of the *Epitome*] contain an idea which is now commonplace in elementary anatomy, viz: organs are composed of tissue. Probably it was the first great generalization in the analysis and synthesis of the main anatomical facts." *The Epitome of Andreas Vesalius,* p. 1. *The Purple Island* utilizes the definition of parts as "similar" (*P.I.* 2.1.5 margin. a), but because this definition is necessarily specific to anatomy and does not translate well into descriptions of the features of the island realm, it is confined to the marginalia of the text. Some anatomists use a distinction between simple and compound parts instead of, or as well as, that between similar and dissimilar (or instrumental). The two distinctions are synonymous, and are used interchangeably, for example, by Paré, *Works* (1678), bk. III, p. 51. Paré further divides the dissimilar, "Compound," "instrumental," or "organical" parts into "a fourfold order": "They are said to be of the first order, which are first and immediately composed of the simple, as only the authors of some one action, of which kinds are the muscles and vessels. They are of a second, which consist of these first simple, and others besides, as the fingers. They are counted of the third rank, which are composed of parts of the second order, and some besides, as the hand taken in general. The fourth order is the most compound, as the whole body, the organ and instrument of the Soul" (pp. 51–52). A further way of dividing parts is positional. Berengario, *Isagogae breves,* p. 185, says that the terms "Common," "proper," and "more proper" "indicate the position of the parts of the body in their relation to the observer or dissector, and their relation to one another. The common parts are those which lie at the outer edge of the lower [or any other] cavity where it impinges upon the middle cavity at one end or at the extremities at the other end."

113. Substance is not among John of Alexandria's "six *occasiones,*" but it is one of Aristotle's *Categories.* See French, "A Note on the Anatomical Accessus of the Middle Ages," p. 464. French (p. 467) says: "As a surgeon, Paré did not have a regular medical education, and it seems probable that when he came to write his short text on anatomy he sought advice from a physician. The result is that he dutifully runs through all the standard, contemporary, and introductory 'theory of medicine' from the 'orders' of the *Ars parva* to the *occasiones* in a slightly naïve way that well reveals the educational source of these ideas." Indeed the *occasiones* appear most frequently in teaching texts. Paré, *Works* (1678), bk. III, p. 52, distinguishes between "substance" and "composition" as two separate things "to be considered." Crooke, *Microcosmographia,* fol. 28, explicitly associates substance with the metaphor of the body part as a building, as "the mansion of the particular and determinate faculty."

114. Crooke, *Microcosmographia,* fol. 28, condenses several criteria, including form, into "Conformation," which he describes as including natural proportion, constitution, symmetry. Crooke is very concerned to relate all of the descriptive criteria to the metaphorical conceptualization of anatomy as "building" or architecture.

115. Berengario, *Isagogae breves,* pp. 37–40. Paré, *Works* (1678), bk. III, p. 52, says that by "connexion," "we also understand the original and insertion" of a part.

116. Berengario, *Isagogae breves,* pp. 37–40; Paré, *Works* (1678), bk. III, p. 52; L. R. Lind, *Studies in Pre-Vesalian Anatomy,* p. 43; Eriksson, *Andreas Vesalius' First Public Anatomy at Bologna 1540,* p. 87. Veslingus's description of the "temperature of the Liver" as "hot and moist, that so it may the better concoct the blood," is a good example of how complexion enters anatomical description and is related to function. Veslingus, *The Anatomy of the Body of Man,* p. 17. The concept of the mingling of elements in the body to produce a hot, cold, moist, or

dry complexion, derives from Hippocrates. For the emergence of this idea from the "chemical geocosm" and "microcosm" of Empedocles, see Barkan, *Nature's Work of Art,* pp. 15–20. One of Aristotle's *Categories* is quality, of which the anatomical criteria complexion is a sub-category. According to French, "A Note on the Anatomical Accessus of the Middle Ages," p. 464, "Aristotle's 'quality' also includes 'shape.'" He also indicates (p. 463) that the *occasiones* of John of Alexandria do not include complexion, but a further subcategory of Aristotle's quality, which is the "character of the part (whether bony, fibrous, or cartilaginous)." He suggests (p. 464): "Perhaps an additional *rationale* was that the anatomical 'qualities' are reducible to the elementary qualities of the 'similar' parts (e.g., bone, 'fibre', and cartilage) of which the organs are composed." For more overtly religious anatomists like Crooke, complexion, or "Temper," is "the immediate receiver of the soule." Crooke, *Microcosmographia,* fol. 28. The distinction between complexion and composition is not one that is consistently maintained in medical and natural philosophical works of the Middle Ages and Renaissance.

117. When dividing anatomy into that of the living (vivisection) and that of the dead, John of Alexandria distinguished two criteria which may be obtained by vivisection: "Operation and Necessity, that is, some organs have an operative Faculty and others do not, but are necessary adjuncts of the Faculty of the other." French, "A Note on the Anatomical Accessus of the Middle Ages," p. 463. French says that later commentators take these two topics or criteria from vivisection as an extension of John's anatomical *occasiones,* thereby having necessity and operation, both as equivalent to "action" in Aristotle's *Categories.* Matteo Corti's delineation of the descriptive criteria separates "utilities" or uses, from "functions," and "functions" from functions that are in direct relation to organs in proximity to that which is being described. Eriksson, *Andreas Vesalius' First Public Anatomy at Bologna 1540,* pp. 73–75. Corti's distinction gestures towards but bears only superficial relation to that of Fabrici of Acquapendente between action and use (see note 118, below).

118. Cunningham, *The Anatomical Renaissance,* p. 176. Fabrici's revival of the philosophical anatomy of Aristotle, at Padua in the late sixteenth century, involved his development of a four-part presentation of anatomy. The summary in my text, and this supplementary note, are heavily indebted to Cunningham, *The Anatomical Renaissance,* pp. 175–77. Fabrici's criteria begin with "*Historia.*" Next is the inquiry into an "action." Third is "use," which is a general account of why a part exists and why it is as it is in the Animal, a general or universal category. "Use" thereby induces from the description and action of the particular animal an account of the general. Finally, there is a "demonstration" of the truth of the account given either by showing how the particulars are instances or instantiations of the general, or by referring to variations from the norm, which are nonetheless dealt with teleologically.

119. The view of Redman, "*The Purple Island,*" p. 17, is: "The consistency readers expect in long poems of the seventeenth century is often too extensive. The allegory is generally not systematic; the kind of consistency one might justly expect is nothing more than that anticipated in *any* long poem—the metaphors are various in kind and material, yet all subserve a consistent tone and purpose." The following in the text refutes this view, though only with regard to *The Purple Island.*

120. Giovambattista Giraldi Cinthio, *Discorsi . . . intorno al comporre de i Romanzi, della Comedie, e delle Tragedie, e di altre Maniere di Poesie* (Venice, 1554), cited in Langer, "Invention," p. 137. Invention could also apply to "the faculty of mimetic representation" of nature, that is, the faculty of "imagination" (p. 140).

121. In critical vocabulary the metaphor appears as late as 1787 in Headley, ed., *Selected Beauties of Ancient English Poetry,* I, p. xi, as "the 'disjecta membra Poetæ.'"

122. An ideally convergent interpretation is logically possible if it discloses what is descriptively or perceptibly present, but possibly hidden, in a literary work. For the view that not only description, but interpretation and even value are inherent in a work of art, and are objectively perceptible to the critic, see Colin Lyas, "The Evaluation of Art," in *Philosophical Aesthetics: An Introduction,* ed. by Oswald Hanfling (Oxford: Blackwell in association with The Open University, 1992), pp. 349–80. I suggest, however, that the elusiveness of such an interpretation is historical and ontological. Such an interpretation cannot by implication be superseded by an even more ideally convergent interpretation that may arise at a later date. Given that the precondition of criticism is that interpretation may be improved upon, an ideally convergent interpretation must by definition remain forever elusive. The elusiveness is also ontological. My discussion thus far of metaphor, particularly of creative metaphor, supports a view that although the context of the work as a whole may rule out some connotations of a metaphor in the work, it remains that in literature some things are suggested, others are gently hinted, and whatever methods of literary interpretation we use, we can never establish all of them decisively as "in" or "out" of the work. We lack criteria for determining putatively internal properties to be or actually not to be internal; and consequently we lack the requisite theory of identification for a literary work to qualify as an object. Lyas's argument seems to me to fall foul of the ontological and historical problems with deciding properties inherent to the literary work.

Also see Margolis, *Art and Philosophy,* pp. 111–13, who as a consequence of this elusiveness attempts to distinguish logically between interpretation and description of a literary work. He suggests that, as distinct from interpretation, description "implies a stable, public, relatively well-defined object available for inspection" (p. 111). He adds that "the effort of the describer calls for no special notice; differences in description are to be reconciled by a further examination of the object or the points of view from which it is described. 'Interpreting', on the other hand, suggests a touch of virtuosity, an element of performance, a shift from a stable object whose properties are enumerable (however complex they may be) to an object whose properties pose something of a puzzle or challenge. . . ." (p. 111). It occurs to me, however, that the definition of the literary work as an object fails to be useful insofar as its description as such bears so little relation to what is important to the critical endeavor. My own position is, however, in agreement with Margolis's observation that "once we construe artworks as culturally emergent, we cannot apply the usual criteria for demarcating what is 'in' and what is not 'in' an object" (p. 122).

123. Plato, *Timaeus,* refers to the abdomen, thorax, and head as the "three natural regions of the body." Cunningham, *The Anatomical Renaissance,* p. 12.

124. Paré, *Works* (1678), bk. III, p. 51.

125. Cunningham, *The Anatomical Renaissance,* pp. 28–29.

126. See, for example, Singer, *Vesalius on the Human Brain,* p. 1. A detailed division of the faculties of the soul along the lines of Galen is provided by Paré, *Workes* (1634), bk. III, p. 83 (quoted in the notes to chapter 1).

127. The principal organs are among the "organical parts." Paré, *Works* (1678), bk III, p. 52. The natural, vital and animal members are further divided according to various differentiating criteria. In the eighteenth of the series of lectures presented by Corti at Bologna in 1540, for example, the organs of the lower venter are divided into those of digestion, such as the stomach, liver, intestines, and meseraic veins, those which purge or purify the blood, the gallbladder, spleen, and bladder, and those which regulate the expulsion of the superfluous humors, such as "the muscle in the anus around the rectum [the sphincter]," and the "neck of the urinary bladder."

128. See, for example, Singer, *Vesalius on the Human Brain,* p. 1. Cunningham, *The Anatomical Renaissance,* p. 22, says that for Plato the diaphragm separates the higher part of the mortal soul (which for Galen governs the vital members) from the lower part (which for Galen governs the natural members); whereas the diaphragm was of no particular interest to Aristotle because most animals do not possess one. Therefore, he says, "we can take the diaphragm and concern with it as a 'marker' to distinguish the anatomical tradition which looked back to Plato from that which looked back to Aristotle. It is a 'marker' in two senses: concern with it indicates a concern with the Platonic view of the body, and it is seen as a physical marker which sets off one set of internal organs from another, both functionally and conceptually." These Aristotelian and Platonic/Galenic divisions were not the only divisions employed in anatomy textbooks. The distinction between similar and dissimilar parts was another, as was that between simple and compound parts (see note 112, above). Furthermore, Paré makes a distinction between the parts necessary for life, which are the heart, brain, and liver, those necessary for better and more commodious living (but not for life per se), such as the arms, eyes, nose, ears, hands, and those necessary only for propagation, the womb, testicles, and so on. Paré, *Workes* (1634), bk. III, p. 85. Paré, *Workes* (1634), bk. III, p. 85, and *Works* (1678), bk. III, p. 55, further suggests another form of division, according to those parts which contain and those which are contained.

129. See, for example, Paré, *Workes* (1634), bk. III, p. 85. Paré recommends the common and vulgar way of division to be used in dissection because otherwise the animal parts, owing to their distribution to the extremities in the performance of their motive and sensitive functions, are mixed with vital and natural parts.

130. See, for example, Eriksson, *Andreas Vesalius' First Public Anatomy at Bologna 1540,* p. 205; Paré, *Workes* (1634), bk. III, pp. 83–84 (cited above in chapter 1, note 315).

131. The theory of *pneuma* is, according to Singer, based on Galen's views, which in this respect "rest ultimately on belief in a world-spirit or pneuma. This came naturally to Galen as part of a Pagan philosophic theory of his age, but it cannot be made to fit the Christian philosophy of the time of Vesalius." Singer, *Vesalius on the Human Brain,* p. xvii. Paré, *Works* (1678), bk. III, p. 52, says that the principal organs are so called, "not only, because they are necessary for life . . . but because from each of these three, some force, power and faculty, or also matter necessary for the whole body, flow over all the body, when no such thing proceeds from the rest of the parts. For from the Liver a matter fit for nourishment is distributed by the veins through all the body; from the Heart the vital force, diffused by the arteries, imports life to the whole body, which gives them sense and motion." Crooke, *Microcosmographia,* fol. 44, concurs with the opinion of Paré: "That is called Principal part, which alloweth or affordeth to the whole, either some faculty, or at least some matter." Crooke's qualification in this definition arises from an uncertainty as to whether the liver provides the body with the natural faculty of the soul (see chapter 1, note 318).

132. In sixteenth-century anatomy a point of contention was "whether the spiritus naturales are engendered out of the spiritus vitales transferred to the liver, as the spiritus animales are engendered out of the spiritus vitales transferred to the rete mirabile of the head." In a discussion of this problem, Corti is reported by Heseler as saying that in *De morbis curandis,* bk. XII, ch. 5, "Galen seems to doubt if there is a spiritus naturalis distinguished from the spiritus vitalis. . . . Nature created the rete mirabile in the brain to produce the spiritus animales, so that the spiritus vitales had to pass through those narrow arteries and after having remained there some time, be turned into spiritus animales. If now Nature in a similar way had wanted to produce in the liver the spiritus naturales out of the spiritus vitales, then she would also have created such narrow ducts there as in the rete mirabile of the brain. But nature did not lead to the liver such arteries as it did to the brain in the rete mirabile, consequently spiritus naturalis is not produced out of spiritus vitalis, as is the spiritus animalis, because spiritus vitalis cannot longer be retained in the liver to be changed in[to] spiritus naturalis as it is changed into spiritus animalis in rete mirabile. . . . One could, however, well fancy that out of the subtle vapours of the vital blood and of these spiritus also the spiritus naturales should be produced. But these things certainly cannot be seen by dissection." Eriksson, *Andreas Vesalius' First Public Anatomy at Bologna 1540,* pp. 155–57. (Please note: the spelling of "naturalis/-es," "vitalis/-es," and "animalis/-es" is variable in the text of Heseler's lecture notes in accordance with number, i.e., singular or plural.)

133. Singer, *Vesalius on the Human Brain,* p. 2.

134. See, for example, Berengario, *Isagogae Breves,* p. 111, on Galen's demonstration.

135. Benedetti, *Anatomice,* bk. II, ch. 2, p. 89. In Aristotelian fashion, Benedetti explains the embryological priority of the heart (which he says is followed in development by the brain and liver) arguing that "it contains the principle of life and sense" (p. 83). But in book I, chapter 8 "On the Nerves," Benedetti challenges Aristotle's theory of nerves arising from the heart, and locates the faculty of sensation in the brain.

136. De Laguna, *Anatomica methodus,* p. 283.

137. In de Laguna's *Anatomica methodus,* the Greek opinion that the mind and the reasoning part of the soul were located in the diaphragm or transverse septum is dismissed as "a stupid confusion," an erroneous deduction from the observation that "when [the diaphragm] is injured men are quickly seized with delirium paraphrosyne [Greek for madness]." Ibid., p. 281. De Laguna notes that "Galen, however, very often calls [the diaphragm] phren, as does Plato, signifying 'mind,'" because of the effect of its injury on the mind. *The Purple Island* alludes to this sympathetic connection between the diaphragm which is the "border-citie" "*Diazome,*" and the head, or "*Cephals* towers" (*P.I.* 4.11.2, 1; 4.13.6):

> But if this town [Diazome] be fir'd with burnings nigh,
> With selfsame flames high *Cephals* towers fry;
> Such is their feeling love, and loving sympathie.
>
> (*P.I.* 4.13.5–7)

138. Fletcher is not alone in using this figuration. In Windsor 19054 *recto,* Leonardo considers the flow of urine into the bladder, and records that Nature has fashioned the organ in accord with his observations from his fourth book on conduits. Schultz, *Art and Anatomy in Renaissance Italy,* pp. 93–94.

139. Paré, *Works* (1678), bk. III, p. 52, says that of the "organical parts" "there be three, by whose power the body is governed; which for that cause they call regent and principal, because they govern all the rest; they are the Liver, Heart, and Brain." Crooke, *Microcosmographia,* fol. 43, discusses and dismisses the view that the stomach is principal because necessary to the whole body, saying this "kinde of necessity is but relative or collaterall, by which some parts owe observance to the principall, as necessary servitors unto them."

140. Temkin, "Metaphors of Human Biology," p. 272, says that although "the comparison with a social organism was not Galen's main biological metaphor," "he found the concept of social justice valuable." The concept of hierarchy is however more integral to the organization of the body in Galenic anatomy.

141. De Laguna, *Anatomica methodus,* p. 287.

142. Veslingus, *The Anatomy of the Body of Man,* pp. 39–40, uses metaphor: "The Heart itself, which is the Prince of all the Bowells."

143. Crooke, *Microcosmographia,* fols. 45, 47. The doctrine of consent did not imply that the faculties of each of the principal organs is dependent on the others for its "Idea or forme" (final cause).

144. See, for example, Temkin, "Metaphors of Human Biology," pp. 271–72.

145. Benedetti, *Anatomice,* bk. II, ch. 13, p. 94. In *Anatomica methodus,* pp. 273–74, de Laguna writes: "Plato has clearly indicated that the part of the soul which desires and which feels pleasure and sorrow is located in this

organ [the liver], for if I remember correctly these are his words on the subject in both Greek and Latin: 'As for the liver, which I have just mentioned, the third part of the soul is based within it; the liver lies deservedly between the umbilicus and the transverse septum [diaphragm]. It has almost no thought, mind or reason at all but shares the senses of pleasure and sorrow and of the desires which pertain to those senses.'" Lind comments that this passage is Plato, *Timaeus* 70D–71B, considerably condensed. It was also maintained by some anatomists as late as the seventeenth century that embryologically the liver has precedence over the heart and brain. See, for example, Crooke, *Microcosmographia,* fol. 46.

146. De Laguna, *Anatomica methodus,* p. 281. According to French, *Dissection and Vivisection in the European Renaissance,* p. 225, André du Laurens thought not dissimilarly to de Laguna, that "Aristotle's general principle (what du Laurens called his 'universal nature') was sound but that he was ignorant of many obvious things, like . . . the origin of the veins and nerves." According to Cunningham, *The Anatomical Renaissance,* p. 21, Aristotle did not "attribute much in the way of function to the brain," and because of his view that the mind or rational soul is "solely a Form," he concluded that "there are no organs associated with it to be investigated." Aristotle, however, also argues that the heart is the center not only of the vegetative soul but of "all the other aspects of the animal soul as well: the sensitive, the motile, the rational." He held further, according to Thorndike, *A History of Magic and Experimental Science,* I, 146, that "the brain was of secondary importance [to the heart], being the coldest part of the body, devoid of blood, and having for its chief or only function to cool the heart."

147. Galen's public demonstration by dissection, performed in the presence of various philosophers and medical men, of the the connection between the brain and voice and respiration and the commentaries which he immediately afterwards dictated on this point were so convincing, he argued fifteen years later, that noone had ventured openly to dispute them. Thorndike, *A History of Magic and Experimental Science,* I, 146.

148. One such early modern anatomist to insist upon the priority or the heart as far as necessity to the whole body is concerned, was Crooke, in his *Microcosmographia,* fol. 46.

149. Quilligan, *The Language of Allegory,* p. 42, says that: "Relying on the process of making inanimate nouns animate, [allegorical personification] requires a curious treatment of language as language. The violation of grammatical categories necessary for personification emphasizes the very operation of language and, having become self-conscious about the grammar, it is only logical for the poet (and reader) to become sensitive to other surface verbal structures."

150. Jonathan Gil Harris, *Foreign Bodies and the Body Politic: Discourses of Social Pathology in Early Modern England* (Cambridge: Cambridge University Press, 1998), p. 2, notes that in Tudor and Stuart versions of the body politic analogy, the liver was often compared to the High Treasurer.

151. "The function of the liver is to turn the chyle [the end-product of the "digestion" in the stomach of ingested food] into blood," or "chyme," which it distributes via the venous system to all the body. Eriksson, *Andreas Vesalius' First Public Anatomy at Bologna 1540,* p. 151 (citation in square brackets is of p. 153).

152. De Laguna, *Anatomica methodus,* p. 271.

153. Heseler's notes of the eleventh lecture in the lecture series given by Corti at the University of Bologna in 1540 explain that the spleen "was created by Nature for receiving the melancholic humour from the liver, so that the blood should be purged from the superfluous melancholic humour." This humor is further described as "cold and thick." In addition the notes explain the function of the gallbladder to draw the other of the "superfluous humidities," the yellow bile, from the liver, and describe this as "less thick [than the melancholic humor], warm and subtle." Eriksson, *Andreas Vesalius' First Public Anatomy at Bologna 1540,* p. 141; also see p. 153.

154. Because of this function, Veslingus, *The Anatomy of the Body of Man,* p. 18, argues: "The magnitude of the *Spleen* is bigger in melancholy Men than in others."

155. Berengario says, in a passage of his *Isagogae breves,* "Concerning the Cyst containing the Gall or Bile, which is called the Gall Bladder": "The other duct according to some goes to the pylorus of the stomach to aid digestion with its gall. If this duct is large, it makes a man unhappy because of its continuous emission of gall" (p. 62). Although not engaging in allegorical personification, or even drawing a political analogy, this description assumes the continuity of bodily humors with emotions and character traits and dispositions.

156. According to Rimmon-Kenan, *Narrative Fiction,* p. 44, "time in narrative fiction can be defined as the relations of chronology between story and text." Story-time can be conceived of as the linear succession of possibly causally linked events in chronological order. Strictly speaking text has no temporality. However, the text does acquire a kind of temporality, and it does so metonymically through the spatiality of the text, through the linear distribution of the words over poetic lines, stanzas, cantos, chapters, or pages.

157. While employing the analogy of the body in his *Republic,* Plato's "ideal" or "just" state distinguishes itself from a state in which secession of members and dissolution of the social contract were possible. *The Republic*

of Plato, Part II (books II–IV, 445B); especially ch. 6 (367E-372A), pp. 40–140 (p. 52). For Aristotle, the state is an association, among the principles defining which are a combination of self-interest and the striving for that which is defined as the social good. Like Plato, but to a lesser extent, Aristotle is more concerned with unity than diversity or, indeed, than internal conflict. *Aristotle's Politics,* bks. I, II, VII. In Seneca, however, "the diversity implied in the analogy to the body begins to have as much imaginative life as the unity." And in the Middle Ages, the notions of *res publica* and *corpus mysticum* become "expressions of that essentially medieval tension between the passionate vision of comprehensive cosmic unity and the overwhelming evidence of fragmentation and diversity within the world." The *corpus mysticum* compares exactly with the medieval relation of microcosm and macrocosm, for in both cases an abstract conceptual order is seen as responsible for the order within the human, the world or the state. Barkan, *Nature's Work of Art,* pp. 66, 69–70. Cantos 7 to 12 of *The Purple Island* develop ideas of the unity of the mystical body or the body of Christ and describe a conflict within the human body and body politic, in relation to the Reformation doctrine of the Fall.

158. For Lakoff and Johnson on metaphor and truth, see *Metaphors We Live By,* ch. 23, pp. 156–58, ch. 24, pp. 159–84; especially pp. 159, 165–66, 169–72.

159. See, for example, Barkan, *Nature's Work of Art,* p. 62; Archambault, "The Analogy of the 'Body' in Renaissance Political Thought."

160. Archambault, "The Analogy of the 'Body' in Renaissance Political Thought," p. 32.

161. The four Royal Proclamations referred to are: 71 "A Proclamation for suppressing of persons riotously assembled for the laying open of Inclosures" (Westminster, 30 May 1607); 72 "A Proclamation signifying his Majesties pleasure as well for suppressing of riotous Assemblies about Inclosures, as for reformation of Depopulations" (Greenwich, 28 June 1607); 74 "A Proclamation signifying his Majesties gracious parden for the Offendours about Inclosures" (Windsor Castle, 24 July 1607); 75 "A Proclamation concerning Starch" (Salisbury, 23 August 1607). J. F. Larkin and P. L. Hughes, eds., *Stuart Royal Proclamations,* 2 vols. (Oxford: Clarendon Press, 1973), I. Cited in the text are Proclamations 72, pp. 155, 156, and 75, p. 164.

162. Archambault, "The Analogy of the 'Body' in Renaissance Political Thought," p. 21.

163. Ibid., p. 22.

164. Larkin and Hughes, *Stuart Royal Proclamations,* Proclamation 72, p. 156.

165. Ibid.

166. Ibid. In the *De ortu et authoritate* [*sic*] *imperii Romani* (Basle, 1559), written around 1446, Aeneas Sylvius (1405–64, later Pope Pius II) justified capital punishment by analogy with amputation to restore the health of the body. For Sylvius the concept of the "health" of the political body is more important than justice, an important departure from Plato, for whom justice in the state and the individual life was paramount, and in contrast to Aristotle's conception wherein the well-being of all members of the polity is the end and purpose of the political body. See Archambault, "The Analogy of the 'Body' in Renaissance Political Thought," pp. 32–34.

167. William Shakespeare's (1564–1616) *Coriolanus* has been related to the Midlands Rising of 1607 by, among others: Dollimore, *Radical Tragedy* (1989), pp. 218–30; C. C. Huffman, *Coriolanus in Context* (Cranbury: Associated University Presses, 1971), pp. 147–49; A. Patterson, *Shakespeare and the Popular Voice* (Oxford: Basil Blackwell, 1989), pp. 120–45 (pp. 136–41). References in the text are to *The Arden Edition of the Works of William Shakespeare: Coriolanus,* ed. by Philip Brockbank (London and New York: Routledge, 1988). I have also consulted William Shakespeare, *The Complete Works,* compact edition ed. by Stanley Wells and Gary Taylor (Oxford: Clarendon Press, 1999), in which the second of the references in the text is act 1, scene 5, lines 2–5.

168. The *De ortu et authoritate* [*sic*] *imperii Romani* of Aeneas Sylvius (see note 166, above) is a case in point. Another Renaissance political theorist with absolutist sympathies, and who drew the analogy of the Prince and the physician of a sick body, is Sir Thomas Elyot (1490–1546), in *The Boke Named the Governour* (London, 1580). In this work, the Prince's knowledge of the parts of the countries under his governance, and of the customs and manners of the people of those countries, is made analogous to the physician's knowledge of the diet, concoction, quietness, exercise, and sleep of his patient. See Archambault, "The Analogy of the 'Body' in Renaissance Political Thought," pp. 37–39.

169. In Shakespeare, *The Complete Works,* the reference is 3.1.155–59.

170. Ibid., 3.1.23.

171. Ibid., 1.1.195–98.

172. Larkin and Hughes, *Stuart Royal Proclamations,* p. 155, note 2.

173. In Shakespeare, *The Complete Works,* these three references are: 1.1.148; 1.1.149; 1.1.146.

174. Ibid., 1.1.148–51.

175. This is regarded as an aspect of the operation of ideology, whether one holds a cognitive or materialist

view of ideology. See Jonathan Dollimore and Alan Sinfield, eds., *Political Shakespeare: New Essays in Cultural Materialism,* 2nd edn., with additional chapters (Manchester: Manchester University Press, 1996; c. 1994), pp. 6–7, for various meanings of ideology; Dollimore, *Radical Tragedy* (1989), pp. 9–10, for cognitive and materialist views of ideology in relation to Renaissance culture. Belsey's view of ideology, in *Critical Practice,* p. 57, is both materialist and poststructuralist, and her analysis is directly relevant to the effect which Menenius tries by his relation of the belly fable to achieve in the agitated plebeians: "ideology is both a real and an imaginary relation to the world—real in that it is the way in which people really live their relationship to the social relations which govern their conditions of existence [a materialist view], but imaginary in that it discourages a full understanding of these conditions of existence and the ways in which people are socially constituted within them [a cognitive view]."

176. Veslingus, *The Anatomy of the Body of Man,* p. 33 (emphases mine).

177. De Laguna, *Anatomica methodus,* p. 269.

178. Larkin and Hughes, *Stuart Royal Proclamations,* Proclamations 74, p. 161; 71, p. 153; 75, p. 163. On the point about enclosures of common land, Larkin and Hughes comment: "1 James I, c. 25, it is true, had re-enacted the last tillage statute of Elizabeth, but 4 James I, c. 11 had authorised private enclosure of perhaps one-third of the common fields in Herefordshire" (p. 153, note 2). Patterson indicates that: "While enclosures were the immediate cause of the Rising, the documents of its aftermath indicate that food shortages were also a major issue," and adds that these shortages were caused by making starch, grain hoarding, "engrossing," export of grain, unfair pricing, quality control and unemployment. Moreover: "We can . . . infer that when Shakespeare began work on *Coriolanus,* probably in 1609, the issue of the Rising had merged in the public consciousness with the problem of dearth (in both its senses)." *Shakespeare and the Popular Voice,* pp. 137 (and note 48), 138. Jonathan Dollimore alludes to "a report from Warwickshire during the dearth of 1608 (probably the year of [*Coriolanus*'s] production) which confirmed that the shortage of corn was caused partly by hoarding—something which was widely recognised anyway." Dollimore, *Radical Tragedy* (1989), p. 228.

179. Patterson, *Shakespeare and the Popular Voice,* p. 11.

180. In Shakespeare, *The Complete Works,* these two references are: 1.1.91–92; 1.1.77–78. As Patterson argues, the plebeians have a "considerable degree of political self-consciousness." "Shakespeare, having identified the undeniably physical content of the citizens" motivation, does not leave them . . . clapping their hands spasmodically upon their stomachs. On the contrary, they proceed to discuss the political economy, not without wit, and not without considerable social perspective." *Shakespeare and the Popular Voice,* pp. 127, 133.

181. This particular instance exemplifies Catherine Belsey's generalization: "Analysis reveals that at any given moment the categories and laws of the symbolic order [including language] are full of contradictions, ambiguities and inconsistencies which function as a source of possible change. The role of ideology is to suppress these contradictions in the interests of the preservation of the existing social formation, but their presence ensures that it is always possible, with whatever difficulty, to identify them, to recognize ideology for what it is, and to take an active part in transforming it by producing new meanings." *Critical Practice,* pp. 45–46.

182. John of Salisbury, *Policraticus,* limits monarchical authority as against sacerdotal powers of the clergy, who are represented in the body politic as analogous with the soul; although insofar as the polity is also the *corpus mysticum,* the human soul is analogous to God. See Barkan, *Nature's Work of Art,* pp. 72–73; Archambault, "The Analogy of the 'Body' in Renaissance Political Thought," pp. 27–28. In Barkan's view: "In the late Middle Ages, the characteristically English approach to the organic theory of the State stressed the importance of the whole realm to the government of the commonwealth at least as much as that of the king" (p. 75). Some French theorists of the period shared this view: Claude de Seyssel (1450–1520), *La Monarchie de France,* written between 1515 and 1520, argues for tempered, balanced or political monarchy and against absolute monarchy, as does John Fortescue, in his *De laudibus legum Anglie,* which posits the will of the people as like the heart in the body, in that both contain the blood and therefore the source of life respectively of the state and human body. In the sixteenth century, Thomas Starkey, *Dialogue between Cardinal Pole and Thomas Lupset,* imputes the cause of decay in the polity to the physician or ruler. Many sixteenth-century political treatises deploy the analogy to argue for monarchy while not being apologies for absolutism. But, says Barkan: "The English Renaissance picture of world order was above all a defense of the hierarchical status quo, and the human body was a useful analogy because, like society, it was clearly a strange and arbitrary combination of functions which all seemed to work perfectly together." Archambault, "The Analogy of the 'Body' in Renaissance Political Thought," pp. 44, 35, 39, 48; Barkan, *Nature's Work of Art,* p. 77, and see p. 75 on John Fortescue.

183. See Curtis C. Breight, *Surveillance, Militarism and Drama in the Elizabethan Era,* Language, Discourse, Society series (Basingstoke: Macmillan, 1996).

184. In early modern anatomy the nerves were regarded as possessing two innate and distinguishable faculties, the physiological functions of which were often said to require slightly different anatomical structures: "Thus nature with the most brilliant foresight in all things used soft nerves for sense perception and hard nerves for motion in putting together the bodily parts. . . ." De Laguna, *Anatomica methodus,* p. 288. In the twelfth of the series of lectures given to students at the University of Bologna, in 1540, Corti has the following to say on the distribution and anatomical substance or composition of the nerves: "to some organs [Nature] has led and allotted large nerves to serve both motion and sensation alike, e.g. in the muscles, which are instruments for voluntary motion, and to certain organs She has sent very small nerves rolled up and intertwined only in their membranes to make them feel what might hurt them, as we see here in the membrane of the liver and in other viscera. For the motion requires the largest nerve, the faculty of sensation a large one, but a little degree of sensation the smallest one. For these very small nerves perceive immediately the matter injuring these organs, which starts their expulsive faculty, driving out the noxious elements, and therefore Nature clothed these principal and nobler organs with such membranes or coats interwoven with such subtle nerves. The remaining less noble organs, which are receptacles for the superfluous humours and excrements, do not need such nerves, to wit, the intestines, the spleen, the two bladders, etc., because these humours by themselves infer a certain disagreeable feeling and incitement of the expelling faculty in the organs, as the gall-bladder has the yellow bile, which it sends to the intestines to expel the unnecessary matter in them." Eriksson, *Andreas Vesalius' First Public Anatomy at Bologna 1540,* p. 157; also see p. 275. In the twenty-first of the series of anatomical demonstrations accompanying Corti's lectures, Vesalius attests to the probability that "hard nerves" are in fact that which in early modern anatomy are also referred to (as they are in modern anatomy) as "tendons." Eriksson, *Andreas Vesalius' First Public Anatomy at Bologna 1540,* pp. 265–67. According to Lind, Vesalius at times confuses ligaments and nerves, or regarded the terms as synonymous, but in the text of his *Epitome,* ligaments, sinews and what we now call nerves are distinct and the latter are "the organs by means of which the animal spirit is led from the brain." *The Epitome of Andreas Vesalius,* p. 14.

185. Nicolson, *The Breaking of the Circle,* p. 40.

186. The omitted descriptive criterion of anatomical textbooks is passions, or illnesses. However, the consequences for human health of the body's political and ecclesiastical incorporation, and of its redemption in Christian eschatology, do become apparent in later cantos.

187. Edmund Gosse, *The Jacobean Poets* (London: John Murray, 1894; 2nd impression 1899), pp. 144–51.

188. Ibid., p. 146.

189. Ezekial Sanford, ed., *The Works of the British Poets, With Lives of the Authors,* 50 vols. (XVIII–XXI, XXIII, XXV–L, ed. by R. Walsh) (Philadelphia: Mitchell, Ames, and White, 1819), V, 92.

190. Gosse, *The Jacobean Poets,* p. 146.

191. Ibid., p. 148.

192. Sanford, ed., *The Works of the British Poets,* V, 92.

193. Sanford claims that "Phineas, with a livelier fancy [than his brother Giles], had a worse taste," and Gosse's remarks on the "unusual ugliness and aridity" of Fletcher's chosen theme are cited above. Sanford, ed., *The Works of the British Poets,* V, 92.

194. Ferdinand de Saussure's *Course in General Linguistics* refutes the notion that language is the transparent medium that objectivists purport it to be, or at least, that it may be. According to Belsey, *Critical Practice,* p. 38, he argues that "far from providing a set of labels for entities which exist independently in the world, language precedes the existence of independent entities, making the world intelligible by differentiating between concepts." Furthermore, the "inseparability of the signifier and the signified . . . creates the illusion of the transparency of language."

195. In his introduction to *Andreas Vesalius' First Public Anatomy at Bologna 1540,* p. 29, Eriksson says of Vesalius's contribution to the establishment of a standardized anatomical nomenclature: "Both in *Tabulae anatomicae* and *Fabrica* Vesalius devoted much space to terminology, in that he gives variants of the anatomic terms in the illustrations." Moreover, "the terminology of *Fabrica* (1543) is the result of a systematic scrutiny of the contemporary anatomic vocabulary, during which Vesalius rejected non-Latin terms and revived older, good Latin names, and did not hesitate to create new ones when he did not find suitable terms in the ancient writers" (p. 29). F. J. Cole, *A History of Comparative Anatomy from Aristotle to the Eighteenth Century* (London: Macmillan, 1944), p. 57, claims that Vesalius introduces or stabilizes "corpus callosum, a translation of Galen's term for this structure; [and] incus, the small anvil." In addition, Lind says, Vesalius employed terminology derived from "the more ancient pre-Galenic anatomists who, in his opinion, derived their nomenclature from direct anatomical examination; the terminology of such observers as Galen (who, he contends, saw only the superficial aspect of the human body) is subject to criticism." *The Epitome of Andreas Vesalius,* p. 90.

196. As a "storehouse from which the Humanists replaced the current Arabic terms," the section on anatomical terms in Pollux, *Onomasticon*, written in the time of Galen, was, according to Lind, of great significance after its publication in 1502. *The Epitome of Andreas Vesalius*, p. 8. Celsus, *De Medicina* was written in the first century AD, but was unknown thereafter until rediscovered in the mid-fifteenth century and reprinted in Florence in 1478. Of Celsus, Singer, *The Evolution of Anatomy*, p. 104, says: "His excellent Latin created a new standard for medical writing, and many of his anatomical terms came into use. These replaced Arabic and Latino-Arabic words." Unlike Benedetti, Vesalius adopted a pragmatic approach to the employment of anatomical terms and defended "the use of non-Latin terms for such conceptions as were unknown to the antique Latin writers, such as Plinius, Celsus, . . . he shows that even in his first printed work [a paraphrase of Rhazes' ninth book] he appreciated the significance of medical terminology." Eriksson says that "when describing an organ he generally gave the name used by Greek writers (Graecis, i.e. Hippocrates and Galen), and that in Latin literature, and other terms in common use, and finally his own term." Eriksson, *Andreas Vesalius' First Public Anatomy at Bologna 1540*, p. 29.

197. Eriksson, *Andreas Vesalius' First Public Anatomy at Bologna 1540*, p. 29. In Singer's essay, "Beginnings of Academic Practical Anatomy," which prefaces Choulant, *History and Bibliography of Anatomic Illustration*, Singer remarks that "the same name is often used by [Mondino] for several parts. Thus . . . by *anchae* he may mean the hips, or the pelvic skeleton, or the acetabulum, or again, the corpora quadrigemina" (p. 21-F).

198. In Singer's essay, "Beginnings of Academic Practical Anatomy," which prefaces Choulant, *History and Bibliography of Anatomic Illustration*, Singer observes that in *Anatomia Mundini*, "there are many parts for each of which Mundinus has several names. . . . Thus the sacrum is variously termed *alchatin, allanir*, and *alhavius*" (p. 21-F). De Laguna's *Anatomica methodus* employs as referentially interchangeable terms *hypogastrium* (Greek), *abdomen* (Latin), and *mirach* (Arabic) (p. 272). But a glance at Berengario's *Isagogae breves*, pp. 41, 185, demonstrates that while *mirach* refers to the anterior and posterior parts of the abdomen, the Greek *hypogastrium* is qualified, because the anterior and lateral parts are referred to as the *epigastrion*. Berengario refers to the posterior abdomen as the *imum dorsi*. Meanwhile in Latin terminology, *abdomen* is above and *sumen* is below; although frequently only *abdomen* is used for both areas.

199. Raynalde, trans., *The Birth of Mankinde*, fol. 24. Gordon Williams, *A Dictionary of Sexual Language and Imagery in Shakespearean and Stuart Literature*, 3 vols. (London and Atlantic Highlands, N.J.: Athlone Press, 1994), III, 1464–65, says that "womb and vagina tend to blur in sexual folklore."

200. Vicary, *The Anatomie of the Bodie of Man*, p. 77.

201. Veslingus, *The Anatomy of the Body of Man*, p. 58.

202. John Davies of Hereford (1565?–1618), *Mirum in modum: a Glimpse of Gods Glorie and the Soules Shape* (London, 1602), in *The Complete Works of John Davies of Hereford (15..-1618)*, with memorial-introduction, notes and illustrations, glossarial index, and portrait and facsimile, &c., ed. by Alexander B. Grosart, 2 vols., Chertsey Worthies' Library (St George's, Blackburn: Edinburgh University Press, 1878; repr. Anglistica & Americana Series, 2 [Hildesheim: Georg Olms Verlagsbuchhandlung, 1968]), I, p. 7, col.1, ll. 7, 20–21. (All further references are to this edition and the reference system used here and hereafter is that employed by the "Notes and Illustrations" of the cited text, where "l." stands for "line" and "ll." for "lines.") "*Mater*" is of course the Latin term for mother.

203. Banister, *The Historie of Man* declines to provide an anatomical description of female reproductive anatomy, and is decidedly circumspect when it comes to that of the male: "asmuch as of the Male may commodiously be spoke[n] (for more I thought not good to translate into English) shall now specially be declared." (sig. Bb$_i^r$ [fol. 85^r]). Banister writes in English and, according to the prefatory letter by William Clowes "Chirurgeon," he "wrightst not for the learned sort (I know) that were vayne, But hopest to helpe the meaner folke." The moral issue for Banister is that access to sexual anatomical knowledge would be widened had he not decided to omit its description from his work. Echoes of this perceived problem can be found in other popular anatomical and medical works in the vernacular.

204. See Roberts and Tomlinson, *The Fabric of the Body*, p. 222.

205. De Man, "The Epistemology of Metaphor," pp. 17–18, uses etymology to the purpose of eroding the totalization of properties of substances and simple ideas.

206. The *Epitome* lacks highly technical terms and is very descriptive. For this reason, Lind assumes that it was written for students, although according to his Dedicatory Letter, Vesalius says the work is for physicians and may be treated as an index to the *Fabrica*. *The Epitome of Andreas Vesalius*, p. 8.

207. Henri Estienne, *Dictionarium medicum; vel, expositiones medicinalium ad verbum excerptae ex Hippocrate, aretæo, . . . cum Latina interpretatione . . . lexica duo in Hippocratem . . . prefixa sunt, unum erotiani . . . alterum Galeni* (Geneva: H. Stephanus, 1564); Robley Dunglison, *A Dictionary of Medical Science, Containing*

a Full Explanation of the Various Subjects and Terms of Anatomy, Physiology, [. . .] 1st ed. (Philadelphia, 1833). Several subsequent editions of Dunglison's work appeared culminating with *A Dictionary of Medical Science . . .*, 23rd edn., thoroughly rev. by T. L. Stedman (London: J. & A. Churchill, 1904) (all further references are to this edition).

208. Richard S. Snell, *Clinical Anatomy for Medical Students*, 3rd edn. (Boston & Toronto: Little, Brown & Co., 1986), p. 807, says: "The ventricles of the brain consist of the two lateral ventricles [the first and second], the third, and the fourth ventricle. The lateral ventricles are in communication with the third ventricle through the *intraventricular foramina* . . . the third ventricle communicates with the fourth ventricle by the cerebral aqueduct."

209. The septum lucidum is also known as the septum pellucidum. See Snell, *Clinical Anatomy for Medical Students*, p. 799, figs. 11–44.

210. See Foucault, *The Order of Things*, p. 35.

211. Richard Rorty, "'He had both a bubbling wit and the ability to make fun of himself,'" *Times Higher Education Supplement*, 12 November 2004, p. 16, says on the subject of Jacques Derrida's critique of the insistence among philosophers on the stability of meaning, that is, of reference: "To ask that words be used literally is simply to ask that certain inferential relationships be held stable for the duration of the argument."

Chapter 3. Creation, Mortality, and Salvation

1. Baldwin, "Phineas Fletcher," pp. 471–72.

2. Redman, "*The Purple Island*," p. 12.

3. Plot is the order of events as they appear in the text of a narrative, that is, in the sequence they are told by the narrator(s). Rimmon-Kenan, *Narrative Fiction*, p. 3, defines story as "the narrated events, abstracted from their disposition in the text and reconstructed in their chronological order, together with the participants in these events." According to Chris Baldick, ed., *The Concise Oxford Dictionary of Literary Terms* (Oxford: Oxford University Press, 1990), p. 212, this distinction between story and plot is derived from Russian Formalism and its opposed terms "*fabula*" and "*sjuzet*." "The story, then, is the hypothetical 'raw material' of events which we reconstruct from the finished product of the plot." The events in the story are arranged sequentially "as we assume them to have occured in their likely order, duration, and frequency": "it includes events preceding and otherwise omitted from the perceived action, and its sequence will differ from that of the plot if the action begins 'in media res' or otherwise involves an anachrony." So, "the plot is the selected version of events as presented to the reader or audience in a certain order or duration, whereas the story is the full sequence of events as we imagine them to have taken place in their "natural" order and duration" (p. 170). The distinction acknowledges how the events or situations in a narrative or dramatic work are "selected and arranged both to emphasise relationships—usually of cause and effect—between incidents and to elicit a particular kind of interest in the reader or audience, such as surprise or suspense" (p. 170).

4. For the possibility that an "influence upon Fletcher's handling of moral and religious thought in *The Purple Island* may have been Dante, who provided in his *Divine Comedy* figurative Christian concepts which later Christian writers repeated," see Redman, "*The Purple Island*," p. 10.

5. An analepsis is "a narration of a story-event at a point in the text after later events have been told." A homodiegetic analepsis provides information about a character, event or story-line at that point in the text. And an internal homodiegetic analepsis refers within the scope in story-time of the whole work. Rimmon-Kenan, *Narrative Fiction*, pp. 46–47 (p. 46).

6. Sometimes an analeptic sequence of events between 1 and 6 is narrated within another internal homodiegetic analepsis, consisting of a later sequence of events between 1 and 6 (see, for example, *P.I.* 2.3).

7. Rimmon-Kenan, *Narrative Fiction*, pp. 125–29, argues that a narrative text defers knowledge and intelligibility in two ways. The first of these is what is called "delay," which may be future-oriented, if the knowledge or intelligibility delayed belongs to events in the future of the story, or past-oriented, if the knowledge or intelligibility belongs to the past. Whereas the future-oriented delay causes the reader to ask, "What next?," the past-oriented delay causes the reader to ask, "What happened?" (as one does, for example, in reading a detective novel) (p. 125).

8. See Ibid., p. 127.

9. Lakoff and Johnson, *Metaphors We Live By*, pp. 92–94.

10. On how an abstract entailment of two or more coherent metaphorical concepts may not be concrete

enough to form an image of the defined concept, thereby requiring more concrete defining concepts, see Lakoff and Johnson, *Metaphors We Live By,* pp. 93–94.

11. Singer, *The Evolution of Anatomy,* p. 115.

12. Thorndike, *A History of Magic and Experimental Science,* I, 149 (the citation includes a citation from Galen's *On the Use of the Parts*).

13. Singer, *Vesalius on the Human Brain,* 31–2. A mullion is a "vertical bar dividing the lights" or "perpendicular divisions" of a window. *The Oxford English Dictionary* . . . , 12 vols. and supplement (Oxford: Clarendon Press, 1933; repr. 1970) (hereafter cited as *OED*), VI, "M," p. 752, col. 1.

14. Singer, *Vesalius on the Human Brain,* p. 15.

15. Ibid., p. 13.

16. Eriksson, *Andreas Vesalius' First Public Anatomy at Bologna 1540,* p. 123.

17. Dillistone, *Christianity and Symbolism,* p. 45.

18. De Laguna, *Anatomica methodus,* p. 271.

19. Heseler's notes of this lecture include the following passage: "*De usu partium,* V: c.14 and 15, Galen explains this very beautifully. There he tells why not more nor fewer muscles have been created by Nature. . . . Consequently it was necessary, that Nature made four muscles at each side, namely two oblique ones ascending and two descending to get a perfect pressure and expulsion in the intestines, as it appears in the bladder full of water." Eriksson, *Andreas Vesalius' First Public Anatomy at Bologna 1540,* p. 79.

20. Temkin, "Metaphors of Human Biology," p. 275, says, "the Greek biologists of the Aristotelian tradition did not forget that an instrument has an artist who designs it and a master who uses it. The artist [of the body] to them was divine nature and the master, the soul. Whether in part or as a whole, the human body was conceived as an instrument of the soul, formed so as to suit the requirements of the soul."

21. Temkin, "Metaphors of Human Biology," p. 275, says it was Galen's conviction that "the divine craftsman, the demiurge, has created man as a sublime work of art. This metaphor of the work of art must be taken literally, against the background of Greek civilization. Just as the workman did not count in the highest ranks of ancient society, so the demiurge did not necessarily hold the highest place among the gods." Temkin says that not only this metaphor, but that of anatomy as society, "were not always mere figures of speech, but integrating concepts used by biologists in guiding their thought" (p. 279).

22. Benedetti, *Anatomice,* bk. II, ch. 23, p. 99. The citation in square brackets is from the same passage in *Anatomice.*

23. Harcourt, "Andreas Vesalius and the Anatomy of Antique Sculpture," p. 29. Indeed the eighth member of the series of visceral figures that accompany Book 5 of the *Fabrica* "is unmistakably based on one of the most famous antique fragments to come to light during the Renaissance, the so-called Belvedere Torso, first recorded in the Roman collection of Cardinal Prospero Colonna in the early 1430s" (p. 30; also see p. 55, note 7 for a bibliography of contributions to this argument). Harcourt argues: "Even a superficial survey of the remainder of the series is sufficient to show that this is by no means an isolated instance. Of the ten such figures, only the ninth is given as though actually modelled on a body laid out for dissection. . . . And even here the characteristic truncation of head, arms, and legs appears rather as fractured stone than severed flesh. As for the rest, they all display a similar conceit: they exist for the eye not as actual bodies dissected but as fragments of sculpture somehow miraculously able to sustain the act of real anatomical demonstration" (p. 30).

24. Harcourt argues that representation has a central role in the constitution of early modern anatomical knowledge, not in the sense that anatomy as a descriptive science is constituted by the artistic activity of the Renaissance to which the figures in the *Fabrica* rightfully belong (as has been argued by some historians of art and anatomy), but rather in the sense that the demonstration of anatomy on models conceived as sculptural fragments serves anatomical functions, or rather meets the concerns of early modern anatomists. Harcourt, "Andreas Vesalius and the Anatomy of Antique Sculpture," pp. 33–34, 52–53 (where these concerns are summarized); also see p. 44. Harcourt's interpretation of these concerns does not, however, involve the concept of the body's creation.

25. See, for example, *The Purple Island,* 2.26.6–7.

26. Cunningham, *The Anatomical Renaissance,* p. 202; Banister, *The Historie of Man,* sig. C_i^r (fol. 1^r). For Plato, *Timaeus,* "the body is the product of intelligent design, the intelligent design of God the Creator." Cunningham, *The Anatomical Renaissance,* p. 13.

27. Cunningham, *The Anatomical Renaissance,* p. 206; also see p. 267.

28. Crooke, *Microcosmographia,* fol. 14. Crooke's point about learning about the Creator from study of the created is essentially Aristotelian. He expands the point to say that "all the knowledge of God that can be had, must

be derived not *à priori*, but *à posteriori*, not from any cause or matter preceding, but from the effects and thinges subsequent."

29. Ibid., fol. 14.

30. The citation is from Vicary, *The Anatomie of the Bodie of Man*, p. 11. In similar vein to the citation of Vicary, the last paragraph of Berengario's *Isagogae breves*, p. 178, says: "These are the words which for the present I have given to my students in common anatomy. And now, may He who is three in one be praised. He whom most humbly I pray to direct me toward greater things, Amen." Heseler's notes on Vesalius's anatomical demonstrations at the University of Bologna in 1540 end with an even more entirely conventional statement and with even greater brevity: "Here ends the anatomy of three human subjects and six dogs, to the praise and glory of the Supreme God, and to the utility of our neighbour. Amen." Eriksson, *Andreas Vesalius' First Public Anatomy at Bologna 1540*, p. 293.

31. Benedetti, *Anatomice*, The Dedication to Maximilian Caesar Augustus, in Lind, *Studies in Pre-Vesalian Anatomy*, p. 81.

32. Paré, *Workes* (1634), bk. III, p. 97.

33. Crooke, *Microcosmographia*, ch. 6.

34. The citations which distinguish the religious conviction of Crooke from the less directly religious style of Vesalius, are taken from Veslingus, *The Anatomy of the Body of Man*, p. 30.

35. Crooke, *Microcosmographia*, fol. 15.

36. French, "A Note on the Anatomical Accessus of the Middle Ages," p. 467, note 25, speculates that when Paré "came to write his short text on anatomy he sought advice from a physician."

37. Kaspar (Caspar, Casper, Kasper, Gaspard) Bauhin (1560–1624), *Theatrum anatomicum novis figuris aençis illustratum et in lucem emissum opera* . . . (Frankfurt, 1605), [another edition, unillustrated] (Frankfurt, 1621). According to Roberts and Tomlinson, *The Fabric of the Body*, p. 222, "Bauhin was responsible for introducing a more rational approach to anatomical nomenclature." He studied with and succeeded Felix Platter as professor of medicine at Basle.

38. Singer, *Vesalius on the Human Brain*, p. 18.

39. *The Epitome of Andreas Vesalius*, p. xxxiv.

40. The first of these two appellations is taken from the first broadsheet of Andreas Vesalius, *Tabulae anatomicae sex* (Venice, 1538), the second from *The Epitome of Andreas Vesalius*, p. 83. Also see Roberts and Tomlinson, *The Fabric of the Body*, p. 49; Singer, *The Evolution of Anatomy*, p. 116.

41. Paul H. Kocher, *Science and Religion in Elizabethan England* (New York: Octagon Books, 1969; c. 1953 by Henry E. Huntington Library & Art Gallery), surveys medical works generally (as opposed to anatomical works specifically), published or available in England, and finds that: "The most devout among contemporary medical works were likely to lean upon religion not merely in their prefaces but pervasively throughout their texts, sometimes culminating in a closing prayer for divine aid" (p. 263). "The average Elizabethan work on medicine or surgery is much less intent on religion. It normally makes a pious reference or two in the preface and dedicatory epistle, but when it buckles down to the business of medical analysis is likely to forget the supernatural, except for a possible passing allusion now and then" (p. 264). "Then there is also a smaller class of treatises in which religion does not appear at all, or so infinitesimally as to be negligible" (p. 265).

42. See French, *Dissection and Vivisection in the European Renaissance*, pp. 58, 66, 73, 83, 89.

43. Cunningham, *The Anatomical Renaissance*, p. 38.

44. C. A. Patrides, *Premises and Motifs in Renaissance Thought and Literature* (Princeton: Princeton University Press, 1982), pp. 83, 84. On the prevalence of this motif in the Reformation and Renaissance, Patrides says it was adopted among the Reformers by Calvin, and its adherents among English poets include "not only Milton but Chapman, Benlowes, and Wither; . . . among English divines, William Perkins, Thomas Adams, Joseph Hall, Donne, Henry King, and Richard Baxter. . . ." (p. 86). Aquinas had interpreted God's words in Genesis regarding Man as the image of God, "as referring only to the spiritual, not the physical side of man's nature." "But," adds French, *Dissection and Vivisection in the European Renaissance*, p. 130, "the anatomists also came to see the body also in some sense as the image of God, and to dissect it was thus a religious performance."

45. The citations are of the notes taken by Baldasar Heseler from Corti's second lecture of the series of lectures accompanied by Vesalius's anatomical demonstrations, in Bologna in 1540. Eriksson, *Andreas Vesalius' First Public Anatomy at Bologna 1540*, pp. 55–57. There were various causes or reasons given of the upright form. In his *Anatomice*, Benedetti gives voice to the widely held opinion that "man is the warmest" among animals, saying that blood is very abundant in Man, causing much heat and consequently erect posture. Benedetti, *Anatomice*,

bk. I, ch. 2, pp. 83–84; Eriksson, *Andreas Vesalius' First Public Anatomy at Bologna 1540,* p. 57. Crooke, *Microcosmographia,* fol. 5, also provides this cause, arguing that Man is hotter than all other creatures. This hot complexion is an efficient cause and requires the material condition or cause that "mans body . . . is soft, pliable and temperate, readie to follow the Workeman in every thing, and to every purpose."

46. Crooke, *Microcosmographia,* fol. 5.

47. We have already considered how for Vesalius the hand was integral to the discipline of anatomy and its reintegration with other aspects of medicine. Additionally, Cunningham, *The Anatomical Renaissance,* p. 123, says: "When Galen was writing his book *On the use of the parts,* it was the hand with which he began his account, singing a paean of praise to it. Indeed it was these particular muscles [of the forearm and hand] which, in his view, presented the unanswerable argument that there was design in Nature, and that man was the hight point of that design. Vesalius too started his Bologna teaching with a demonstration of these same muscles. . . . [Vesalius] was making a statement about the existence of design in Nature and about the status of man." Actually, Eriksson, *Andreas Vesalius' First Public Anatomy at Bologna 1540,* p. 109, describes Vesalius's demonstration of the tendons of the hands and wrists as the third (not the first) demonstration. Finally, for Aristotle, the hand is the "instrument of instruments," enabling "Man" to use other instruments. Cunningham, *The Anatomical Renaissance,* p. 123. Cunningham argues that all three connotations of the hand are suggested by Vesalius's self-portrait which was drawn by Jan Stephan Van Calcar (Jan van Kalkar, 1499–1546) and cut on wood for the first edition of the *Fabrica,* bearing the date 1542, and appears also in the *Epitome,* and in later editions of both works. See M. H. Spielmann, *The Iconography of Andreas Vesalius (André Vésale) Anatomist and Physician 1514–1564: Paintings, Pictures, Engravings, Illustrations, Sculpture, Medals,* with notes, critical, literary, and bibliographical, Wellcome Historical Medical Museum London, Research Studies in Medical History, 3 (London: John Bale, Sons & Danielsson, 1925), p. 1. A further connotation of the hand, or at least of its connected structures, for Vesalius is suggested by Spielmann's examination of the above-mentioned portrait of Vesalius: "The muscles of the fore-arm and of the lower portion of the arm have been dissected in such a manner as to enable Vesalius to demonstrate one of Galen's anatomical errors: an open scroll set conspicuously on the table is audaciously inscribed with a text drawing attention to the corresponding text in the *Fabrica.* Thus in the face of all Europe he bearded the oracle in the very forefront of his book" (pp. 2–3).

48. The citations are of Aristotle, *De partibus animalium,* bk. II, ch. 10, 656a, and of Heseler's notes of Corti's second lecture, in Eriksson, *Andreas Vesalius' First Public Anatomy at Bologna 1540,* p. 57. Corti agrees with Aristotle that: "It is because man by his intelligence belongs to God, that man has upright stature. . . ."

49. Crooke, *Microcosmographia,* fol. 5.

50. Benedetti, *Anatomice,* p. 83.

51. Thomas Milles, *The Treasurie of Auncient and Moderne Times* (1613–19), I, 18–19, cited in Patrides, *Premises and Motifs,* p. 84. Also see Crooke, *Microcosmographia,* fol. 5: "man had an upright frame & proportion, that he might behold and meditate on heavenly things." Crooke, *Microcosmographia,* fol. 5, provides as an "efficient cause" of the upright frame, that the soul, "as mindfull of her owne Originall, lifteth her building up on high."

52. Sir Thomas Browne (1605–82), *Pseudodoxia Epidemica; or, Enquiries into Very Many Received Tenents and Commonly Presumed Truths,* 6th edn., corrected and enlarged by the author (London: J.R., 1672), pp. 210, 212–13 (p. 213) (all further references are to this edition). *Pseudodoxia Epidemica* was first published in London in 1646, and printed by "T.H." Galen's definition of erectness of posture, as provided by Browne, is that it occurs when the "spine and thigh-bone are carried in right lines" (p. 210). Browne counters the confirmation "by several testimonies," and "the *Greek* Etymology of *Man,*" that the principal cause of this posture is "for to behold and look up toward heaven" (pp. 212, 210). Patrides, *Premises and Motifs in Renaissance Thought and Literature,* p. 87, comments that the Greek "man" derives from the Greek for "upward" and "leap."

53. For further opposition, see Patrides, *Premises and Motifs in Renaissance Thought and Literature,* p. 89.

54. Not only spatial verticality but also spatial proximity determined the position of an anatomical member in the hierarchy. Writing in *Isagogae breves,* p. 127, of the "temporal muscles," whose function is to close the lower jaw, and which are attached to the temples "between the cranium and the ossa paris," Berengario says: "Because of their nobility nature located them in the cave of the temples between the bones that guard them on all sides." He adds: "The muscles of the eyes are also very noble because they are near the brain, but they do not have as many nerves as these temporal muscles." The point about the nerves responds to the idea that they furnish the parts with animal spirit, and do so unequally, so that those parts with more nerves adhering to or connecting with them are regarded as nobler. Not only the abundance of animal but also of natural and vital spirits in an organ determines its nobility. Thus, according to Corti, "the liver is situated in the right side as being nobler or stronger owing to the

abundance there of natural warmth thanks to the hepatic veins among which it is situated." Proximity to organs abundant in one or other spirit lends nobility to other members: "Hence also the kidney ought to be situated near the liver as attracting more strongly the wine [urine]. And thus the kidney on the right side is placed higher because of the liver. And the attraction becomes better and stronger when direct than if it takes place from the opposite side, namely from one side to the other. . . . Consequently, the right kidney is warmer and has stronger attractive force than the left one because it attracts directly and quicker since it is nearer the liver than the left one." Eriksson, *Andreas Vesalius' First Public Anatomy at Bologna 1540*, p. 167. The opinion that the right kidney is located higher in the abdomen than the left in humans was the result of the weight of Galenic authority. As Hale, *The Civilization of Europe in the Renaissance*, p. 549, rightly remarks, "even Leonardo da Vinci and Vesalius described the right kidney in humans as being higher than the left, as it was in animals and in Galen."

55. The discussion of the colon occurs in *Anatomice*, bk. II, ch. 7, p. 92: "Not without reason is this intestine included among the lower organs although it seems to hold a higher position. Likewise it is rightly very close to the liver [a principal organ] in order to receive through a duct from the latter the bile from the gall bladder. The biting effect of the bile causes the excrement to be expelled; hence that part to which the bile duct is attached is yellow and bitter to taste."

56. In the tenth of the series of lectures given by Corti at Bologna in 1540, Corti is reported by Heseler to have rejected Mondino's view that the position of the stomach between the spleen below and the liver above is due to the yielding of the more noble organ, the stomach, to the less noble, the colon. Rather Corti's explanation is that "the stomach was situated in the middle between the inferior and the middle venter . . . to have its digestional activity assisted [by heat] from all sides." Eriksson, *Andreas Vesalius' First Public Anatomy at Bologna 1540*, p. 129. Corti's explanation of the situation of the stomach is functional, rather than hierarchical (like Mondino's explanation), yet it depends upon the necessity of digestion for the life of the Animal.

57. *The Epitome of Andreas Vesalius*, p. 51.

58. Kocher, *Science and Religion in Elizabethan England*, p. 26.

59. Sawday, *The Body Emblazoned*, pp. 175–76

60. Crooke, *Microcosmographia*, fols. 6–7.

61. Sawday, *The Body Emblazoned*, p. 176

62. See, for example, Temkin, "Metaphors of Human Biology," p. 279: "Neither for Galen nor for the mechanists of the seventeenth century did the animal body stand alone as a divine work of art. A contemptible part like the foot moved Galen to a comparison with the sun and to an enraptured hymn upon the divinity of nature [*On the Use of the Parts* (Bk. III, ch. 10)]." The relative position of the foot in the body and its political analogue is the subject of a series of metonymic ruminations on the semiotics of feet by Peter Stallybrass: "Footnotes," in *The Body in Parts*, ed. by Hillman and Mazzio, pp. 312–25.

63. Singer, *Vesalius on the Human Brain*, p. 18.

64. Eriksson, *Andreas Vesalius' First Public Anatomy at Bologna 1540*, p. 61.

65. Ibid., p. 181.

66. Cunningham, *The Anatomical Renaissance*, p. 22; also see pp. 30, 38, 41–42. Temkin, "Metaphors of Human Biology," p. 275, writes of Nature, the demiurge, in Galen's thought: "And indeed, his abilities were rather limited. He did not create the world out of nothing; he had to use matter with all its imperfections."

67. See Foucault, *The Order of Things*, p. 25.

68. *The Epitome of Andreas Vesalius*, p. 75. The phrase *inevitable fatum* (or *vatum*) appears on the title page of Johann Eichmann (Dryander, 1500–1560), *Anatomia capitis humani* (Marburg, 1536), a monograph of the anatomy of the head. It was also "cut into the woodblocks for his anatomy book issued the next year": *Anatomiae, hoc est corporis humani dissectionis pars prior* (Marburg, 1537). Roberts and Tomlinson, *The Fabric of the Body*, p. 86.

69. John L. Thornton and Carole Reeves, *Medical Book Illustration: A Short History* (Cambridge: Oleander Press, 1983), p. 61, claim that the artist of the foreground views, as of the anatomical figures, was Jan Stephan van Calcar, although the background views "were drawn by another artist, possibly Domenico Campagnola." The plates of the *Fabrica* and those of the *Epitome* are reproduced in J. B. de S. M. Saunders and C. D. O'Malley, eds., *The Illustrations from the Works of Andreas Vesalius of Brussels* (New York: Dover Publications, 1950).

70. G. S. T. Cavanagh, "A New View of the Vesalian Landscape," *Medical History* 27 (1983): 77–79 (p. 77).

71. This illustration is in the first book of Vesalius's *Fabrica*, and is reproduced in *Andreae Vesalii icones anatomicae* ([Munich]: [Bremen], [1934]), as Figure 97.

72. The illustration is Figure 6 in Vesalius, *Epitome*. The citations in the text are of Michael Neill, *Issues of Death: Mortality and Identity in English Renaissance Tragedy* (Oxford: Clarendon Press, 1997), p. 106, and the introduction by Lind to *The Epitome of Andreas Vesalius*, p. xxiii.

73. The verses upon the pedestal are from Silius Italicus, *Punica* XII, 243–44.

74. Figure 2 of the *Tabulae libri secundi* of the *Fabrica* is reproduced in *Andreae Vesalii icones anatomicae.*

75. Many have concluded from this derivation that the workshop of Titian was responsible for the plates throughout the *Fabrica,* and that this component of the book was basically a Venetian project. Jan Stephan van Calcar, it is widely agreed, is the artist responsible for drawing these and the plates of the *Epitome,* although "the hands, and manner, of two or three different engravers have been identified in them." Spielmann, *The Iconography of Andreas Vesalius,* p. 11. Thornton and Reeves, *Medical Book Illustration* p. 58, identify Francesco Marcolini da Forlì (c. 1505–c. 1560), of Venice, as the woodcutter. Less recently, from an anatomical comparison between an initial preparatory sketch for the figure that derives from Titian's *Allocution,* Martin Kemp has argued for an extremely close relationship between the artist(s) of the *Fabrica* and its author. Martin Kemp, "A Drawing for the *Fabrica;* and Some Thoughts upon the Vesalius Muscle-men," *Medical History* 14 (1970): 277–88. "Thus, both the preparation of detailed anatomical studies and their subsequent transformation into finished blocks must almost of necessity have taken place not at Venice but at Padua, to which Vesalius himself was confined for the bulk of every year by the regulations governing his university chair." The *Fabrica* was produced while Vesalius held the post of *explicator chirurgiae* (lecturer in surgery) at the University of Padua. Harcourt, "Andreas Vesalius and the Anatomy of Antique Sculpture," p. 32; also see p. 31. It is only fair to say that in 1969, Lind argued that the plates of the *Epitome* were "closely supervised throughout by Vesalius himself." *The Epitome of Andreas Vesalius,* p. xxii. According to Harcourt, "Andreas Vesalius and the Anatomy of Antique Sculpture," p. 44, the second figure in the myological series is not alone in reproducing the *adlocutio;* the ninth in the same series also does so. But in the text here, I am referring to the former of these two figures.

76. A military interpretation of the gesture of the figure does not in my opinion preclude a religious interpretion of the figure's heavenward gaze. Many of the figures in the *Fabrica* gesture with their eyes heavenward, with one hand toward heaven, and/or with one or both hands open in order to draw the gaze of the peruser of these illustrations to the anatomical detail. See, for example, figure 5 of *Tabulae libri secundi,* reproduced in *Andreae Vesalii icones anatomicae,* a male "muscleman" who is open-mouthed as if imploring, his face turned toward heaven, whose left hand bears some remnant of the *adlocutio* gesture, lifted upwards from the elbow toward heaven, and who holds his right hand open. Figures 3 and 4 of the same book of the *Fabrica* display both hands openly to the viewer.

77. Singer, *The Evolution of Anatomy,* pp. 115–16, argues that Vesalius's anatomy is "essentially *living* anatomy," the figures "are posed as in the living body, and given a background such as that to which they were accustomed during life"; and he offers the explanation that "[t]he parts are not the subject of morphological treatment, but of investigation as contributing to the existence of that complex vital unit we call a Man."

78. Neil, *Issues of Death,* pp. 106, 108, fig. 15, notes that when reproduced in Banister, *The Historie of Man,* this plate contained an additional lily, traditional emblem of *vanitas,* over which the skeletal figure hunches. Figure 13 of *Tabulae libri secundi* of the *Fabrica,* reproduced in *Andreae Vesalii icones anatomicae,* also stages a particularly striking convergence of the conceptual metaphors of anatomy in the concept of ruin. A deeply dissected male figure stands before a town, his dismembered tibia, fibula, bones of the foot and connecting muscles lie on the ground in the fashion of a severed tree branch. Directly beside the dismembered lower leg is in fact a tree stump, and furthermore the human limb rests against a crumbled stone. The practice in the illustrative plates of anatomy textbooks of placing anatomically demonstrative figures in landscapes of buildings, which often are ruined, and vegetation, often weathered and dying, extends to published books throughout the sixteenth and early seventeenth centuries. This is in many cases because the plates of the *Fabrica* are reproduced. See, for example, van der Spieghel, *De humani corporis fabrica,* which contains five of the Vesalian plates under the title *Ivlii Casserii tabulae anatomicae LXXIIX.*

79. Harcourt, "Andreas Vesalius and the Anatomy of Antique Sculpture," p. 45, argues that the action of the three osteological figures in bk. I of the *Fabrica* is rooted in the iconography of *vanitas.* He says that while the first, spade in hand, signifies mortality, this becomes a subject for meditation in the second, and finally mourning in the third. My interpretation is in agreement with this except that I consider the third of the series to represent prayer or contemplation rather than mourning, thereby shifting the emphasis from the loss entailed by mortality to its didactic lesson and association with redemption.

80. Felix Plater (b.1536), *De corporis humani structura et usu libri III, tabulis methodice explicati, iconibus accurate illustrati* (Basle: Ambrosius Froben, 1583), bk. III, *Tabulae* I, II, III. The iconography of the bow and arrow is polysemous, meaning possibly the speed and irresistability with which either death or love strikes, and being the armory of Cupid in relation to the latter case. But also arrows are the symbol of infection, especially in cases of plague or pestilence.

81. Guido Guidi (also Vido Vidio, Vidi Vidii or Vidus Vidius, d.1569), ed., *Ars medicinalis,* 3 vols. (Venice: Juntas, 1611), III, *De anatome,* bk. I, *Tabula* 1. Many of the illustrations of *De anatome* are Vesalian in origin, although their rendition is altered. The editor of this work is the son of an anatomist of the same name, who authored *Chirurgia è Graeco in Latinum conuersa* (Paris, 1544), the woodcuts of which have been attributed to François Jollat, by Choulant, *History and Bibliography of Anatomic Illustration.*

82. Cunningham, *The Anatomical Renaissance,* p. 128. For a contrasting argument, see Neill, *Issues of Death,* pp. 106–13, though Neill takes only cursory notice of Vesalius's original sketch, which is so important for Cunningham's interpretation. Neill argues that whereas "the anatomist [in the scene depicted in the title page] projects himself as the active hero of [the] drama [of Vesalius's "revolutionary practice" of performing the dissection himself], he has lost the central, controlling position occupied by his predecessors as they intoned the ancient texts *ex cathedra.* This place has been usurped by the figure at which the lecturer, in a gesture repeated by the subjects of numerous later Anatomy Lessons . . . , points with his left index finger. This rhetorical stress upon the prominence of the skeleton—added, interestingly enough, after the artist had submitted his original sketch—has been glossed by anatomy historians simply as a reference to Vesalius' doctrine that 'the study of anatomy begins with the bones, to which constant reference must be made during the process of dissection'. But certain details resist this reductive reading—the skeleton's unaccountably heroic scale, and its seated posture, which seems to contradict Vesalius' instruction in the *Fabrica* that it should be displayed in a standing position 'supported by iron rod.'" (Neill is quoting from Saunders and O'Malley, eds., *The Illustrations from the Works of Andreas Vesalius of Brussels,* p. 42, which reproduces the title pages of the first and second editions of the *Fabrica,* as well as Vesalius's original sketch, and O'Malley, *Andreas Vesalius of Brussels 1514–1564,* p. 143.) Neill's glancing acknowledgment of the omission of the skeleton from Vesalius's original sketch is revealing because it would seem to compromise, although not fatally flaw, his argument of the skeleton's prominence. Vesalius presided closely over the production of the first edition, including the title page, but the question that requires an answer is whether Vesalius was aware of the alteration to the title page of the second edition, which conspicuously incorporates the connotations of Death the Mower from the *memento mori* tradition, and if aware, whether he had power to amend it had he chosen to do so.

83. See Kemp, "A Drawing for the *Fabrica.*"

84. Lind offers this explanation in *The Epitome of Andreas Vesalius,* p. xxii.

85. Neill, *Issues of Death,* pp. 105, 104. Also see Harcourt, "Andreas Vesalius and the Anatomy of Antique Sculpture," pp. 31, 45, for a discussion of the osteological figures of the *Fabrica* which draw upon this tradition. Neill, *Issues of Death,* p. 106, argues of the osteological and myological plates of the *Fabrica:* "In plate after plate skeletons and cadavers adopt the attitudes of figures from the *memento mori* tradition—above all from the *Totentanz.*" Choulant, *History and Bibliography of Anatomic Illustration,* p. 47, and Roberts and Tomlinson, *The Fabric of the Body,* p. 21, also argue for the influence of the *Danses Macâbres* (Dances of Death) on the tradition of anatomical illustration. In the *Danses Macâbres* "skeletons appear in a great variety of movements, with their bones still connected by ligaments, often even with the skin and internal organs still visible." Moreover the artists responsible for these figures were illustrating more accurately the thoracic cage and forearm bones, than those who drew the properly anatomical illustrations of much of the same period. They began in the fifteenth century but reached the climax of their sway in the early sixteenth century. Also of possible influence upon the development of anatomical figures are the lemurs, or spirits of the dead, which in the later Middle Ages represent death. They are described by Roberts and Tomlinson, *The Fabric of the Body,* p. 32, as resembling "an extremely emaciated, decaying corpse, with a sunken and sometimes black abdomen, often shown with a mid-line incision through which the rapidly-decomposing intestines had been removed." The title page of Pieter Pauw (Paaw, 1564–1617), *Primitiae anatomicae: de humani corporis ossibus* (First Fruits of Anatomy: On the Bones of the Human Body) (Leiden: Justi à Colster, 1615), "includes a picture of an orchestra of lemur-Deaths playing trumpets; a skeleton uses bones as drumsticks." Roberts and Tomlinson, *The Fabric of the Body,* p. 236.

86. Singer, *The Evolution of Anatomy,* pp. 134–35: "The work of Vesalius was often reprinted. After the first publication, the *Fabrica* appeared in at least twenty-five editions between the years 1543 and 1782, being issued from the presses at Augsburg, Basel, Cologne, Ingdstadt, Leyden, London, Nuremberg, Paris, and Venice [and Lyons]. The figures of Vesalius were copied and plagiarized from the beginning." The *Fabrica*'s series of osteological and myological plates, for example, are recut in copper engraving with some alterations in a work by the Spanish anatomist Juan Valverde da Hamusco: *Historia de la composicion del cuerpo humano.* They are reproduced also in the Italian translation of the Spanish original of Valverde's text: *Anatomia del corpo humano.* Roberts and Tomlinson, *The Fabric of the Body,* p. 208, offer this explanation for "the centrality of [Vesalius's] work in the hundred years following 1543": "The illustrations in Vesalius' works were the ones most frequently copied and modified, for few other publications of this kind were as desirable and available. His predecessors' work was

obviously less complete, and the figures of his contemporary, Estienne, were much less satisfactory and less effective." In addition, the plates of others remained unprinted or printed only in part.

87. See Cunningham, *The Anatomical Renaissance,* p. 7.

88. Thorndike, *A History of Magic and Experimental Science,* I, 151.

89. On the liminal state of the figures demonstrating "living anatomy" in anatomical illustrations, see Sawday, *The Body Emblazoned,* pp. 113–14.

90. Thorndike, *A History of Magic and Experimental Science,* II (1947), 150.

91. Sawday, "Bodies by Art Fashioned," p. 148.

92. Sawday, *The Body Emblazoned,* p. 178.

93. See Braider, "The Paradoxical Sisterhood: 'ut pictura poesis,'" p. 168.

94. Ibid., p. 168.

95. Ibid., p. 170. Braider explains that this misreading was "favoured by a corrupt punctuation of Horace's text tying the key verb *erit* ["will be"] to the initial phrase *ut pictura poesis* rather than to the actual subject of the main clause."

96. Ibid., p. 170.

97. Francis Quarles (1592–1644), *Emblemes* (Cambridge: R.D., 1643), in *The Complete Works in Prose and Verse of Francis Quarles,* ed. with memorial-introduction, notes, and illustrations, portrait, emblems, facsimiles, &c. by Alexander B. Grosart, 3 vols. (Edinburgh: Edinburgh University Press, i.e.: T. & A. Constable, 1880), III, Verse, p. 45.

98. See Peter M. Daly, "George Wither's Use of Emblem Terminology," in *Aspects of Renaissance and Baroque Symbol Theory 1500–1700,* ed. by Peter M. Daly and John Manning, AMS Studies in the Emblem, 14 (New York: AMS Press, 1999), pp. 27–38 (pp. 28, 29). It is notable that Wither's use of "figure," as either a noun or a verb, is "to denote a pictorial sign of any kind," though for some kinds (including personification) more than others (p. 31).

99. See Braider, "The Paradoxical Sisterhood: 'ut pictura poesis,'" p. 169. The idea that art and poetry are imitation is derived from Aristotle: see, for example, *De Poetica,* ch. 4, 1448b. In Renaissance and early modern poetics, however, the idea spawned a number of derivatives describing poetry as "painting, limning, drawing, sketching, colouring, depicting, or portraying." Braider, "The Paradoxical Sisterhood: 'ut pictura poesis,'" p. 168.

100. See, for example, Foucault, *The Order of Things,* pp. 27, 34.

101. Mazzeo, "Universal Analogy and the Culture of the Renaissance," p. 302.

102. Ibid. Similarly Foucault, *The Order of Things,* p. 26, attempts to articulate the rationale of signatures, saying that buried similitudes, which would remain indefinitely dormant, must be indicated by visible marks to testify that the abundance of resemblances "has long been prepared by the order of the world." And: "The world of similarity can only be a world of signs," which need to be unearthed and deciphered.

103. Quarles, *Emblemes,* p. 45.

104. My use of the term "suprarealist" to signify the attitude to language described, is intended to be consistent with its use by Quilligan, *The Language of Allegory,* ch. 3 (especially pp. 156–57). Quilligan explains: "Briefly, what I am calling the suprarealist attitude toward language existed generally throughout late antiquity into the Middle Ages and up through the Renaissance. At the end of the seventeenth century the attitude gave way before a vast epistemological shift; neoclassical decorum not only decreed a ban on wordplay, the philosophy of the Enlightenment did not concern itself directly with language. In a sense, words lost the battle to 'things' and language disappeared as a potent force for shaping man's sense of his cosmos" (p. 157).

105. Foucault, *The Order of Things,* pp. 27–29, 34–35.

106. Walker Percy, *The Message in the Bottle* (New York: Farrar, Straus and Giroux, 1976), p. 261, cited in Sorri and Gill, *A Post-Modern Epistemology,* pp. 170–71.

107. Quilligan, *The Language of Allegory,* p. 156. Language was studied by grammarians as sharing the same characteristics as the created world, as even having a "building" stucture—syntax—like the bodies of animals and humans. Foucault, *The Order of Things,* pp. 27–29. Foucault describes how Ramus's grammar divides its subject into etymology (intrinsic properties of words) and syntax, and argues that the representative content of language has "no role to play here" (p. 35). Grammar has the same epistemological arrangement as science of Nature or esoteric disciplines (p. 35). Foucault's "archaeology," a methodological attempt to articulate his subject from within as well as to analyse it, relies heavily upon a small selection of sixteenth-century natural philosophers, medical theorists and other writers, which omits too much humanism and Greek revivalism to be considered a representative sample of concepts of language. Hence, my next paragraph in the text.

108. Vesalius, The Dedicatory Letter to Prince Philip, *The Epitome of Andreas Vesalius,* p. xxxiii.

109. Quarles, *Emblemes,* p. 45.

110. Sawday, *The Body Emblazoned,* p. 177; "Bodies by Art Fashioned," p. 148.

111. Cunningham, *The Anatomical Renaissance,* p. xi.

112. On the crucified anatomical figure in Berengario, see Choulant, *History and Bibliography of Anatomic Illustration,* p. 138. This was redrawn for Johann Eichmann (Dryander), ed., *Anatomia Mundini* (Marburg, 1541). French, *Dissection and Vivisection in the European Renaissance,* p. 3, maintains what might be called the artistic-influence view of this figure in Berengario: "Artists studied anatomy in order to give added force to their paintings of the human figure, generally in action, and when illustrating anatomical texts often placed the human figure, and parts of it, in surroundings they might have used in a painting. This image of death is in a context symbolic for the Christian reader, and also represents a not uncommon method of execution at the time." Another, I would suggest partial, explanation might be the universal or ideal human nature of Jesus Christ, on which see James Ussher, *Immanuel; or the Mystery of the Incarnation of the Son of God* (London: John Parker, 1638), p. 11: "we must consider, that as he took upon him not an humane *Person,* but an humane *Nature:* so it was not requisite hee should take upon him any *Personall* infirmities, such as are, madnesse, blindnesse, lamenesse, and particular kinds of diseases which are incident to some only, and not to all men in generall; but those alone which do accompany the whole Nature of mankinde, such as are hungring, thirsting, wearinesse, griefe, paine and mortality." On the provision of cadavers from the scaffold for dissection, a disputed historical issue as far as the extent and significance of the practice is concerned, see Carlino, *Books of the Body,* esp. pp. 92–98; Samuel Y. Edgerton Jr, *Pictures and Punishment: Art and Criminal Prosecution During the Florentine Renaissance* (Ithaca: Cornell University Press, 1985), pp. 157–64; Heckscher, *Rembrandt's Anatomy of Dr. Nicolaas Tulp,* chs. 13, 15; Katharine Park, "The Criminal and the Saintly Body: Autopsy and Dissection in Renaissance Italy," *Renaissance Quarterly* 47 (1994): 1–33; Sawday, *The Body Emblazoned,* pp. 113–14; Giovanni Ferrari, "Public Anatomy Lessons and the Carnival: The Anatomy Theatre of Bologna," *Past and Present* 117 (1987): 50–106 (pp. 59–61, 87–88).

113. Johann (Hans) Remmelin (1583–1630), *Catoptrum microcosmicum suis aere incisis visionibus splendens cum historia et pinace de nova prodit* (Ulm: Johann Görlin, 1639), figure 2, p. 15 (all further references are to this edition). The first authorized edition of this work was published by David Franck, in Ulm, in 1619; although there was an unauthorized edition published in 1613, also probably in Ulm. Several editions and translations of this work were published, the latest in 1720, and an English translation is discussed in chapter 6 of this study. In the early editions the plates were engraved by Lucas Kilian (1579–1637) of Augsburg.

114. The English rendition of this biblical text which is closest to the Latin used in *Catoptrum microcosmicum* is that of the *Authorised Version,* p. 509, col. 2. *The Holy Bible Revised Standard Version Containing the Old and New Testaments Translated from the Original Tongues, Being the Version Set Forth* A.D. *1611, Revised* A.D. *1881–1885 and* A.D. *1901, Compared With the Most Ancient Authorities and Revised* A.D. *1952* (London: Thomas Nelson and Sons, 1952), renders the verse as follows: "Man is like a breath, / his days are like a passing shadow." Close to this in conception is the rendering by *The Revised English Bible With the Apocrypha* (Cambridge: Cambridge University Press, 1996): "They [i.e., "human beings" and "frail mortals" (v. 3)] are no more than a puff of wind, / their days like a fleeting shadow." *Nascentes morimur* comes originally from a first century AD poem on astrology by Manilius.

115. Choulant, *History and Bibliography of Anatomic Illustration,* p. 232.

116. Ibid.; Thornton and Reeves, *Medical Book Illustration,* p. 66.

117. Hildegard of Bingen claimed that humans have been weaker since the deluge; the sun, moon, planets and other stars were "somewhat turbulent from excessive heat" after Adam's fall but prior to the deluge. This heat meant that humans were possessed of great strength to endure it. Thorndike, *A History of Magic and Experimental Science,* II, 136.

118. Thirsil's faith echoes that of the speaker of Psalm 16, v. 10: "For thou wilt not leave my soul in hell; / Neither wilt thou suffer thine Holy One to see corruption." *The Holy Bible Revised Standard Version* translates "Holy One" as "godly one," and *The Revised English Bible* as "your faithful servant," both suggesting perhaps an inclusive as opposed to exclusive meaning of the appellation.

119. W. L. Walker, *The Spirit and the Incarnation in the Light of Scripture, Science, and Practical Need,* 3rd ed. (Edinburgh: T. & T. Clark, 1907), p. 110.

120. French, *Dissection and Vivisection in the European Renaissance,* pp. 217, 219. French argues that anatomy at the Protestant university of Wittenberg "not only displayed the not unusual connections between the body and God's wisdom and goodness, but it also carried a number of other messages," among which is "the injunction that as God's dwelling-place the body should be kept in order as one would a church" (p. 217). According also to French (p. 219), although some Protestant anatomies make no mention of *imago dei,* the concept was

more important to Protestant anatomy than to Roman Catholic anatomy. Walker, *The Spirit and the Incarnation,* p. 98, argues: "According to Paul (I Cor. i. 30) and the New Testament generally, God has given us in Christ these three things [of which the second is here of relevance]: (1) 'Wisdom,' that is Revelation or Knowledge; (2) 'Righteousness,' that is Reconciliation or Acceptance with Himself; and (3) 'Sanctification,' implying the indwelling and grace of the Holy Spirit; the whole culminating in complete and eternal 'Redemption' as the sons and heirs of God."

121. Galloway, *The Cosmic Christ,* p. 47. Galloway concedes, however, that in Paul there is no systematic doctrine of cosmic redemption; rather the idea served a homiletic purpose (p. 48).

122. Ibid., p. 51.

123. Ibid., especially pp. 21–56 (p. 29 cited).

124. The time of the year at which the diegesis of *The Purple Island* is narrated is given by the first two lines: "The warmer Sun the golden Bull outran, / And with the Twins made haste to inne and play" (*P.I.* 1.1.1–2). The "Bull" signifies the constellation of Taurus, the "Twins" that of Gemini. So the time specified is when the sun moves from the zodiacal sign of Taurus (20 April–20 May) to that of Gemini (21 May—20 June). On the hypodiegetic level of narration, see Chapter 1, note 9, of this study.

125. Galloway, *The Cosmic Christ,* p. 47.

126. See Ibid., pp. 45–47, who regards this Pauline notion as Greek, "the pagan counterpart to the Jewish myth of the Fall."

127. Galloway, *The Cosmic Christ,* p. 29.

128. Sawday, *The Body Emblazoned,* p. 178.

129. Augustine, *De anima et ejus origine libri quatuor* (4.2.3), cited in Harcourt, "Andreas Vesalius and the Anatomy of Antique Sculpture," p. 37.

130. Harcourt, "Andreas Vesalius and the Anatomy of Antique Sculpture," p. 37; also see p. 57, note 26. Regardless of whether Augustine objected also on moral grounds to either vivisection or dissection of human subjects, he articulates a distinctly theological position, from which, for example, the practice of obtaining knowledge of human anatomy on the basis of comparative studies of animals is theologically flawed by its failure to recognize the fundamental difference between a human being and a beast, that the former has a spirit or immortal soul.

131. Galloway, *The Cosmic Christ,* p. 133, calls this "the experimental method," the systematic application of which presupposes that "I must confine myself to those judgements which will submit to experimental verification." Vesalius approached such a method, which, however, need not necessarily be described as experimental, and in Vesalius's case was descriptive or observational. Galloway's view, which approaches that of Augustine, is: "Within the total *Gestalt* of any given moment of experience, judgements of the kind in which we record simple sense-data are held together with judgements in which we record moral, aesthetic and religious factors in the situation" (p. 135).

132. Dear, "The Church and the New Philosophy," p. 121.

133. Stewart R. Sutherland, "The Concept of Revelation," in *Religion, Reason and the Self,* Essays in Honour of Hywel D. Lewis, ed. by Stewart R. Sutherland and T. A. Roberts (Cardiff: University of Wales Press, 1989), pp. 35–45 (pp. 39, 37).

134. Thomas Aquinas, *Summa contra Gentiles,* IV, Ch. 1, cited by John Baillie, *The Idea of Revelation in Recent Thought* (London: Oxford University Press; 1956), p. 4. Baillie makes the distinction between natural or rational knowledge which is "the unaided exercise of . . . powers of thinking," and the way of revelation, which is "direct communication from God."

135. Sutherland, "The Concept of Revelation," p. 43.

136. Ibid., p. 39.

137. Dear, "The Church and the New Philosophy," p. 120.

138. Sutherland, "The Concept of Revelation," pp. 40–43, outlines in more detail these distinguishing criteria of the truths of revelation.

139. Ibid., p. 42, claims that this statement of the criterion must be analytically rather than synthetically true: "If X is revealed, X is true. . . . This form shows *both* that if something is revealed and recognized as such then it may not properly be doubted, *and* that if it is doubted, then *ipso facto* its status as revealed is under question."

140. Ibid., p. 43.

141. Upon reading Sutherland, "The Concept of Revelation," I became aware of the need to make clear that in discussing the anatomical use of the concept of revelation, I am not exploring the use of a term which is different and unrelated to revelation in theology, but one which semantically has much in common (see pp. 35–36).

142. This illustration occurs in Berengario da Carpi, *Isagogae breves, perlucide ac uberrime, anatomia humani corporis* (Bologna: Benedictus Hectoris, 1523), fol. 6.

143. The claim that medical practice and theory was based on error and mistranslation of Greek texts by the Arabs and their medieval commentators, had of course been voiced earlier than when its effects fed directly into the practice of dissection. Initially the authority of the Greeks and the quest for philological accuracy were preferred by humanists to either the need for the demonstration of the senses or the needs of medical practice. See, for example, Nutton, "The Rise of Medical Humanism: Ferrara, 1464–1555," pp. 6–8.

144. An imaginative understanding of this process may be adduced from Giorgio Vasari, *Le vite de' più eccellenti pittori, scultori, ed architettori* (Lives of the Most Eminent Painters, Sculptors and Architects), as cited in Schultz, *Art and Anatomy in Renaissance Italy,* p. 67: Antonio della Torre "was one of the first [as I have heard tell] that began to illustrate the problems of medicine with the doctrine of Galen, and to throw true light on anatomy, which up to that time had been wrapped in the thick and gross darkness of ignorance."

145. Ficino, "Five Questions Concerning the Mind," pp. 186–87 (p. 186); Baillie, *The Idea of Revelation in Recent Thought,* p. 5.

146. Baillie, *The Idea of Revelation in Recent Thought,* p. 4, incorrectly says that in the seventeenth and eighteenth centuries, all agreed that the processes were entirely different.

147. Within the context of religion, there are two sorts of truths which have been thought to belong particularly to the context of revelation: those about the nature of God; and those about the human, or (more personally) the individual's, condition or state. See Sutherland, "The Concept of Revelation," pp. 43–44.

148. Cunningham, *The Anatomical Renaissance,* p. 43: "By 1268 the teachers of arts and medicine had formed a (single) 'college' and were examining students in medicine, and by about 1295 students of arts and medicine from outside Bologna had formed a (single) 'university of arts and medicine.'" By the end of the thirteenth century, the students of arts and medicine at Bologna were taught in a single "university of arts and medicine," and most were "students both of philosophy—and that now included Natural Philosophy—and of medicine at the same time and in the same faculty and studying for a degree in both subjects combined."

149. Brasavola, Professor of Medicine at the University of Ferrara, graduated in medicine in 1521, becoming distinguished for his work in theoretical and practical botany and pharmacology, but taught (during his career) logic, natural philosophy, medical theory and *practica,* and even published a theological work, *Imitatio Christi.* See Nutton, "The Rise of Medical Humanism: Ferrara, 1464–1555," pp. 11–16. Perhaps the most extreme example of multidisciplinary distinction is Kaspar (Caspar) Bartholin (1585–1629), who was a student of Fabrici at Padua, and of Kaspar Bauhin, and later became Professor of Philosophy at Basel, of Anatomy at Naples, of Greek at Montpellier, and then successively of Eloquence, of Medicine, and finally of Theology at Copenhagen. See Singer, *The Evolution of Anatomy,* p. 167.

150. Crooke, *Microcosmographia,* fol. 15. Culpeper's Dedication to Samuel Hyland, which prefaces Veslingus, *The Anatomy of the Body of Man,* sig. A_2, also claims that in the text and illustrations that follow, "you may see man turned the wrong side outwards, and all his internal parts laid open to your view, whereby your Spirit may be even ravished in the consideration of the Wisdome and Power of that God who hath made man in such an absolute form, so accurate in all his delineaments; if such be the Creature, what is the Creator?"

151. This is a model of revelation in which the metaphor of "depth" is central. Sutherland, "The Concept of Revelation," explains that according to this model: "One is depicted as trying to perceive the structures or substructures of this world, rather than trying to understand this world better by gaining knowledge of another ultimate reality" (pp. 36–37). Early modern examples of the view of the world as a revelation of God, a theophony, are many, but particularly developed is that of Jacob Boehme.

152. See, for example, Foucault, *The Order of Things,* pp. 33, 34.

153. Baldwin, "Phineas Fletcher," p. 470.

154. Ibid.

155. See French, *Dissection and Vivisection in the European Renaissance,* p. 220.

156. Walter Pagel, "Religious Motives in the Medical Biology of the XVIIth Century," *Bulletin of the Institute of the History of Medicine* 3 (1935): 97–128, 213–31, 265–312; repr. in Walter Pagel, *Religion and Neoplatonism in Renaissance Medicine,* ed. by Marianne Winder, Collected Studies Series, 226 (London: Variorum Reprints, 1985) pp. 97–128, 213–31, 265–312 (p. 99).

157. Mazzeo, "Universal Analogy and the Culture of the Renaissance," p. 303.

158. Pagel, "Religious Motives in the Medical Biology of the XVIIth Century," p. 103.

159. Ibid.

160. Ibid., pp. 311, 99. Knight, *Natural Science Books in English 1600–1900,* pp. 19–20, provides evidence in the publications of editions of relevant works, that Paracelsian medicine and theory did not "enjoy any popularity in sixteenth-century England." He goes on, in further detail: "In 1605 Thomas Tymme translated passages

from a celebrated Paracelsian, Duchesne; in 1585 R. Bostocke had published an apology for Paracelsian theory, but his books went through only one edition, as did a shorter work by I.W. which appeared in the following year. In the 1650s a great number of such works, including those of Hermes, Helmont, Croll, Paracelsus, and Lemnius appeared" (p. 20). Knight also remarks (p. 21): "It is by no means easy to separate the Paracelsians from the alchemists, but in fact . . . the former wished to be understood in their writings, while the latter wrote only for adepts."

161. Pagel, "Religious Motives in the Medical Biology of the XVIIth Century," pp. 119–20. Pagel argues that also important in regard to this anti-intellectualist tendency of seventeenth-century science is the "scientia experimentalis" (experimental science) of Roger Bacon, which neglected "the 'argumenta,' because these alone do not give any certainty, and even if they are very strong, they provide it only in connection with the 'experientia.'" Bacon admitted, however, the necessity of reason in order that "mere empiricism" may be a "science" (p. 116).

162. Cunningham, *The Anatomical Renaissance,* pp. 241–42.

163. Pagel, "Religious Motives in the Medical Biology of the XVIIth Century," p. 98.

164. Ibid., p. 99.

165. Ibid., pp. 119–20.

166. As translated by Harris L. Coulter, *A Divided Legacy: A History of the Schism in Medical Thought,* 3 vols (Washington: Wehawken Book Co., 1973–77), I, 407, from the German in Karl Sudhoff, *Theophrastus von Hohenheim, genannt Paracelsus: Sämtliche Werke,* 14 vols. (Munich: Oldenbourg, 1929–33), VI, 332–33, and cited by Cunningham, *The Anatomical Renaissance,* p. 245.

167. In Vesalius's ninth demonstration of the series accompanying Corti's Lectures, Vesalius is reported by Heseler to refuse to show the "small arteries in the concavity of the liver," when asked to do so by Corti, and in his defense Vesalius says, "everyone can see them for himself." In the twenty-sixth and last demonstration, Vesalius is asked by some students whether the arteries" movements follow those of the heart or if they have a movement different from that of the heart. "Vesalius answered: I do not want to give my opinion, please do feel yourselves with your own hands and trust them." Eriksson, *Andreas Vesalius' First Public Anatomy at Bologna 1540,* pp. 165, 293.

168. Cunningham, *The Anatomical Renaissance,* pp. 243–44.

169. Temkin, "Metaphors of Human Biology," p. 280.

170. Cunningham, *The Anatomical Renaissance,* pp. 245–46 (p. 246).

171. Sutherland, "The Concept of Revelation," p. 44. Also see Walker, *The Spirit and the Incarnation,* p. 98 (cited in note 120, above), which argues that the first of the things God has given us in Christ is wisdom, that is revelation or knowledge. Also see p. 108.

172. Pagel, "Religious Motives in the Medical Biology of the XVIIth Century," pp. 98–99.

173. Ibid., p. 99.

174. For other expressions of the same sentiment, see *The Purple Island* 1.33; 4.14.6–7.

175. John Hick, *The Metaphor of God Incarnate* (London: SCM, 1993), Ch. 10, especially pp. 108–9, discusses this view of the incarnation, which he calls a Christology of inspiration, in which the principal activity of God towards and in his human creation is in inspiration. On such a view, "we see in the life of Jesus a special instance of the fusion of divine grace / inspiration and creaturely freedom" (p. 109). Hick's discussion works toward consolidating his own conclusion that "in the case of divine incarnation the initial idea has proved to be devoid of literal meaning and accordingly identified as metaphor" (p. 104). The relation of such a view to the ways in which the incarnation was contemplated in the early modern period is beyond the scope of this study.

176. See Ariès, *The Hour of Our Death,* pp. 353–61; Ferrari, "Public Anatomy Lessons and the Carnival," p. 101.

177. Park, "The Criminal and the Saintly Body," pp. 4, 19–20 (p. 19); Sawday, *The Body Emblazoned,* pp. 54–66.

178. O'Malley, *Andreas Vesalius of Brussels 1514–1564,* pp. 321–22. See, for example, Crooke, *Microcosmographia,* fol. 18.

179. Crooke, *Microcosmographia,* fol. 18.

180. Ibid., fol. 18.

181. Ibid., fol. 12.

182. Mary Niven Alston, "The Attitude of the Church towards Dissection before 1500," *Bulletin of the History of Medicine* 16 (1944): 221–38. At the School of Salerno (Salernum), principal anatomical and medical center between c. 985 A.D. and c. 1225 A.D., dissection of human bodies was regarded by Nicholas and Richard of Salerno as inhuman or abhorrent, although anatomical dissections on pigs were carried out. Cunningham, *The Anatomical Renaissance,* p. 37. At the beginning of the eleventh century in Salerno medical courses began to be regularly organized, and by the beginning of the twelfth century, regulations for the first state examinations in med-

icine were instituted, with anatomy (of the pig) as a required subject. In the middle of the thirteenth century, Frederick II made it a rule that the students at Salerno should be present at at least one human dissection each year. *The Catholic Encyclopedia: an International Work of Reference on the Constitution, Doctrine, and History of the Catholic Church,* ed. by Charles G. Herbermann and others, 15 vols. and index (New York: The Encyclopedia Press, 1913), I, 458, col. 2. Alston (pp. 225–26) infers from the earliest mention of anatómies in a decree of Frederick II in 1240, that dissection was an established practice. The first autopsy mentioned dates from 1286, and in 1302 a judicial postmortem was ordered at Bologna in a question of poisoning (p. 226). *The Catholic Encyclopedia,* I, 458, col. 2, states that before the end of the thirteenth century anatomical demonstrations had been conducted at Paris by Hermondaville.

183. Alston, "The Attitude of the Church towards Dissection before 1500," p. 226. Also on the relationship between autopsy and dissection, see Park, "The Criminal and the Saintly Body," pp. 6–8 et passim.

184. Alston, "The Attitude of the Church towards Dissection before 1500," p. 227. In Charles Singer's essay, "Beginnings of Academic Practical Anatomy," which prefaces Choulant, *History and Bibliography of Anatomic Illustration,* Singer claims that "by the second decade of the fourteenth century public dissection had become well recognised at Bologna," and suggests that it began around 1300 (p. 21–B). He further claims that "practical anatomical study at Bologna" began "in the decade between 1266 and 1275" with William of Saliceto (*c.* 1215–1280) (p. 21–C). Eriksson writes: "According to one view, Bartolommeo da Varignano was the first to dissect a human corpse at Bologna. This dissection was performed in 1302, but it should rather be regarded as an obduction in the service of justice. To my mind this suggests that private anatomies had been performed earlier, already in the latter half of the thirteenth century." *Andreas Vesalius' First Public Anatomy at Bologna 1540,* pp. 35–36.

185. Alston, "The Attitude of the Church towards Dissection before 1500," p. 222. The bull was "soon incorporated into the canonical collection *Extravagantes communes* and thus became part of the Church's law regulating burial practices." Elizabeth A. R. Brown, "Death and the Human Body in the Later Middle Ages: The Legislation of Boniface VIII on the Division of the Corpse," *Viator* 12 (1981): 221–70 (p. 271).

186. Brown, "Death and the Human Body in the Later Middle Ages," p. 221; also see p. 227. "This custom, Boniface proclaimed, was not only abominable in the sight of God but also abhorrent to the human mind . . . [and] violated standards of common decency" (p. 221). There were, according to Brown, sentiments expressed, for example by Louis IX, in the thirteenth century, that what God has kept whole should not be destroyed (pp. 231–32). Brown, "Death and the Human Body in the Later Middle Ages," pp. 227–28. After 1270–71, a further related funerary practice became more widespread: "the separate burial of both the entrails and the heart," the former usually close to the site of death, owing to rapid putrefaction, but the latter (like the bones) at a place selected with special care (p. 233).

187. Singer, *The Evolution of Anatomy,* p. 85. Singer cites Guido de Vigevano who produced in France an anatomical text in 1345, "illustrated by figures which show the actual process of dissection," and wherein it is explained that the Church prohibits dissection. Yet, de Vigevano also claims to have performed dissections "many and many a time on human bodies." Alston, "The Attitude of the Church towards Dissection before 1500," p. 225. It is possibly in relation to a gloss on the bull by Cardinal Lemoine which absolutely forbade evisceration, that Henry of Mondeville wrote that entrails could be removed from the corpse for purposes of embalming only if special permission were granted by the Church. Brown, "Death and the Human Body in the Later Middle Ages," p. 250, suggests that the gloss by Cardinal Lemoine was probably written before Boniface's death, thus suggesting the pope's consent (also see pp. 247, 250, note 102). Embalming continued to be practiced in Italy despite the gloss by Cardinal Lemoine. Mondino, in his *Anatomia,* probably in response to the original papal bull, says he refrains from boiling bones, "owing to the sin involved in this." Alston, "The Attitude of the Church towards Dissection before 1500," p. 224. (The citation is from *Anatomia Mundini.*)

188. Alston, "The Attitude of the Church towards Dissection before 1500," p. 225. By way of argument against the view that the papal bull had a significant effect on anatomy, Park, "The Criminal and the Saintly Body," argues that "the debates over bodily integrity that so exercised northern canonists and theologians seem to have found few echoes among Italian writers" (p. 11).

189. Alston, "The Attitude of the Church towards Dissection before 1500," p. 229.

190. Singer, *The Evolution of Anatomy,* p. 121: "A Paduan surgeon (Bertapaglia) tells of having witnessed dissection in 1439 and 1440 and Montagnana . . . in 1444 had seen fourteen post-mortems there." Singer also notes that in fourteenth- and fifteenth-century Padua, "even the skeleton could be adequately studied." In his introductory essay to Choulant, *History and Bibliography of Anatomic Illustration,* p. 21–J, he remarks that dissection received official recognition in the University Statutes at Padua in 1429. A record of the practice in nearby Venice suggests that there it had begun much earlier. "In Venice where there was no medical school the authorities, in

1308, ordained that one dissection every year should be made for the benefit of physicians of the city." *The Catholic Encyclopedia,* I, 459, col. 1.

191. Alston, "The Attitude of the Church towards Dissection before 1500," pp. 229, 231. *The Catholic Encyclopedia,* I, 459, col. 1, argues that "the declaration that the development of anatomy was interfered with by ecclesiastical authorities is founded on nothing more substantial than a misunderstanding of the purport" of the papal bull *Detestande Feritatis.*

192. Brown, "Death and the Human Body in the Later Middle Ages," p. 266.

193. Park, "The Criminal and the Saintly Body," p. 10. Brown, "Death and the Human Body in the Later Middle Ages," p. 248, suggests that "a congeries" of motivations inflamed Boniface's attitude to the custom of dispersed burial, none of which seems to reflect a general abhorrence at the custom or the Church's united opposition to it (also see p. 249). In about 1410 Pietro d'Argellata embalmed Pope Alexander V, a process which like dissection and autopsy involves considerable mutilation of the corpse, and could not have been undertaken if the opening of the body was forbidden by the Church. Alston, "The Attitude of the Church towards Dissection before 1500," p. 228. This lack of evidence has led some, like Eriksson, to argue: "It is, however, an undeniable fact that the dissection of human bodies was frowned upon by the Church, even if the ecclesiastic authorities did not actually forbid such dissection in the interests of science." Eriksson, *Andreas Vesalius' First Public Anatomy at Bologna 1540,* p. 36.

194. Alston, "The Attitude of the Church towards Dissection before 1500," pp. 228, 232. Alston lists sources where mention of dissection is lacking where it might be supposed to be mentioned were it an offense to the Church (p. 223). She does note an apparent slight variation between the Church's attitude in France and that in Italy. While there is no testimony that anatomy was hindered in Italy at any time, in France there is no evidence of wholesale permission, and the severe limits on numbers of cadavers made available for dissection suggests to Alston that the authorities may have been conservative but probably not obstructive of the practice (p. 232).

195. Bologna did not experience the freedom from Rome's ecclesiastical authority enjoyed by Venice or by Padua, a domain of the Venetian Republic; yet within a decade of the issuance of *Detestande feritatis,* the medical faculty of the university of Bologna was able to stage possibly the first systematic series of public demonstrations of human anatomy, upon which Mondino founded his *Anatomia.* Guy de Chauliac attended these demonstrations and while he was surgeon to three popes he encouraged, even insisted upon, the practice for the benefit of surgery. *The Catholic Encyclopedia,* I, 459, cols. 1, 2. Furthermore, pontifical regulations on dissection for postmortem examinations empowered the *Medici Chirurghi* (college of surgeons) to perform public anatomies at Venice in 1368, and annually thereafter at the Hospital of Saints Peter and Paul at Castello and in other places. The Pontifical regulations empowered the college of surgeons in this way, initially, "by an act of the *Consiglio Maggiore* for a period of twenty-two days continuously, beginning on May 27, 1368." "Benedetti records at least three such anatomies performed by him, on the heart, on the brain, and on the veins, muscles, bones, and nerves: *Anatomice,* III, Preface, p. 36; IV, pp. 49–50; V, p. 68." Lind, *Studies in Pre-Vesalian Anatomy,* p. 82, note 7.

196. Schultz, *Art and Anatomy in Renaissance Italy,* p. 61. The brief appears to Schultz to have been published only once, by Dr Ludwig Friedrich von Froriep in 1811, but the original document is preserved in the archives at the University of Tübingen (p. 62). The brief states that "the Rector, doctors and scholars of the university" had "proposed in our presence that in order to become learned and expert in the art of medicine," they wished "to receive bodies or cadavers of several malfactors [*sic*]," "once their death has been accomplished," "and to have those bodies cut up or dismembered according to the canons and practice of doctors." The representatives of the University of Tübingen are reported also to have declared that "this is not at all allowed . . . without the dispensation of the apostolic chair or without special license" (p. 221, note 119). Schultz comments: "Following approval from the 'apostolic chair,' [the brief] does indeed allow for the dissection of bodies of criminals put to death by the state," on the condition that the university "cause the condemned of this sort to be buried just as such are usually buried" (p. 63). This condition may have been intended to counteract Boniface VIII's denial of Christian burial to those bodies that have undergone the practices described in the bull. At least its effect would have been to overturn the penalty. Eriksson's comments on Sixtus's brief differ considerably from those of Schultz, but are the result of less assiduous research: "The aversion of the Church towards the dissection of human corpses led at times to application to the Papal Court for permission to perform dissections." Eriksson, *Andreas Vesalius' First Public Anatomy at Bologna 1540,* p. 36, note 1.

197. Schultz, *Art and Anatomy in Renaissance Italy,* pp. 61, 63.

198. Singer, *The Evolution of Anatomy,* pp. 86; also see p. 121.

199. Schultz, *Art and Anatomy in Renaissance Italy,* p. 63.

200. Eriksson, *Andreas Vesalius' First Public Anatomy at Bologna 1540,* p. 36. By comparison, in 1485, the

University Statutes of Ferrara stipulate an annual anatomy, but according to Nutton, "The Rise of Medical Humanism: Ferrara, 1464–1555," p. 17, it is not obvious that these anatomies took place regularly, or even at all.

201. Carlino, *Books of the Body,* p. 3. Jonathan Sawday, "The Fate of Marsyas: Dissecting the Renaissance Body," in *Renaissance Bodies: The Human Figure in English Culture c. 1540–1660,* ed. by Lucy Gent and Nigel Llewellyn, Critical Views Series (London: Reaktion, 1990; repr. 1995), pp. 111–35 and pp. 260–64 (p. 126 et passim), characterizes this anthropological obstacle to the public display and dismemberment of the corpse as a taboo. Park, "The Criminal and the Saintly Body," however, argues against what she describes as the persistent misconception that there was in medieval and Renaissance Europe a deep-seated "taboo" connected with corpses and the disclosure of the body. In the ensuing discussion in the main text, I hope to have qualified my argument in respect of the opposing positions.

202. Florike Egmond, "Execution, Dissection, Pain and Infamy—A Morphological Investigation," in *Bodily Extremities: Preoccupations with the Human Body in Early Modern European Culture,* ed. by Florike Egmond and Robert Zwijnenberg (Aldershot: Ashgate, 2003), pp. 92–127; Park, "The Criminal and the Saintly Body," p. 13.

203. Carlino, *Books of the Body,* pp. 3, 93. On grave robbing and other clandestine ways of securing corpses for dissection, which became more widespread in the sixteenth century, despite prohibitive laws and heavy fines, see, for example, Park, "The Criminal and the Saintly Body," pp. 17–18.

204. Carlino, *Books of the Body,* p. 92. For references to contributions to the historical debate surrounding the provision of cadavers from the scaffold for dissection, see note 112, above, in this chapter.

205. Park, "The Criminal and the Saintly Body," pp. 12, 16. Also see Ferrari, "Public Anatomy Lessons and the Carnival," pp. 54, 66, 87.

206. Carlino, *Books of the Body,* p. 93.

207. Ibid., pp. 78, 109, 113 (pp. 109, 113); Sawday, *The Body Emblazoned,* p. 113; Sawday, "The Fate of Marsyas," pp. 116–17; Ferrari, "Public Anatomy Lessons and the Carnival," p. 51.

208. Carlino, *Books of the Body,* p. 109.

209. Edgerton, Jr., *Pictures and Punishment,* pp. 134–35.

210. Sawday, *The Body Emblazoned,* pp. 113–14. See Mark S. R. Jenner, "Body, Image, Text in Early Modern Europe," *Social History of Medicine* 12 (1999): 143–54, on the consequences of the provenance of Sawday's sources for his argument with regard to infamy and anatomy. For Sawday's arguments, also see Sawday, "The Fate of Marsyas" (pp. 116–17, 120).

211. Park, "The Criminal and the Saintly Body," p. 4; Carlino, *Books of the Body,* pp. 78, 109. Park argues that the procedures carried out on a corpse to gain information about the physical state of a particular individual appear to have grown out of the practice of embalming, for which evisceration is necessary. This practice, which was done to saints whose organs purportedly possessed unusual structures which were construed as evidence of their holiness, gave rise to casual observations and interpretations of cause of death.

212. Carlino, *Books of the Body,* p. 115.

213. Park, "The Criminal and the Saintly Body," p. 4; Carlino, *Books of the Body,* pp. 109, 116, 117.

214. Foster, *Lectures on the History of Physiology During the Sixteenth, Seventeenth and Eighteenth Centuries,* pp. 4–5.

215. Cunningham, *The Anatomical Renaissance,* pp. 196, 197.

216. Ibid., pp. 198, 201–2. The view described by Cunningham is held by Singer, "Historical Relations of Religion and Science." Singer's understanding of the relationship of religion and science, one to another, is as of "the two sides of man's nature," and he claims that through history "only under certain circumstances do the two sides of man's nature represented by these departments come to affect each other" (p. 87). It is also Singer's view, however, that sixteenth-century anatomy (and he writes specifically of Vesalius) represents just such an effect of religion upon science. Even in such a case, Singer still views the two disciplines as essentially separate: "I shall merely seek to indicate how the two activities have diverged, approached, united, receded, or have run parallel to each other, or again, how they have crossed each other's paths and reinforced or injured each other during the passage of time" (p. 88).

217. Pagel, "Religious Motives in the Medical Biology of the XVIIth Century," p. 99; also see p. 103.

218. Carlino, *Books of the Body,* pp. 115, 117.

219. Roberts and Tomlinson, *The Fabric of the Body,* pp. 214–22, argue that the Counter-Reformation Church looked with suspicion on those anatomists and medical doctors who challenged received opinions.

220. Roberts and Tomlinson, *The Fabric of the Body,* p. 214: "autopsies were performed in 1550 by [Realdo] Colombo on a cardinal—Valverde was present—and, in 1556, on Ignatius Loyola."

221. *The Catholic Encyclopedia,* I, 459, col. 2.

222. Singer, *The Evolution of Anatomy,* p. 122. Eustachio was also physician to the Pope. *The Catholic Encyclopedia,* I, p. 460, col. 1.

223. Vivian Nutton, "Wittenberg Anatomy," in *Medicine and the Reformation,* ed. by Ole Peter Grell and Andrew Cunningham (London: Routledge, 1993), pp. 11–32. Also see Cunningham, *The Anatomical Renaissance,* p. 230; French, *Dissection and Vivisection in the European Renaissance,* p. 217.

224. Galloway, *The Cosmic Christ,* p. 29 (the citation within the citation is from Isaiah 11:9).

225. Soteriology means primarily, theologically, the "doctrine of salvation"; rarely, the promotion and preservation of health; as either soteriology or "Soterialogy," "the restoration and perfection of human nature." *OED,* X, "S," p. 457, col. 3. The argument in the text relates to Redman's point that "[t]he goal of a [puritan] writer (or a preacher) is, then, to lead readers to spiritual maturity through the allegorical narrative"; but it does not wholly share her view of the pedagogical function of the marginal notes in this respect. Redman, "*The Purple Island,*" pp. 12–13 (p. 12).

226. Quilligan, *The Language of Allegory,* pp. 160–61, argues that during the Christian Middle Ages, "Christ's special position as the second person of the Trinity, the Word of God, gave to man's language a special sanction. Thus in *De Trinitate* Augustine argues for the analogy between human and divine speech. Defining a 'word' as a 'locution of the heart' which is also vision, Augustine explains that it is possible to understand the meaning of a word 'not only before it is uttered aloud, but even before the images of its sounds are rehearsed in thought.' This 'unspoken word' is an enigma, Augustine cautions, yet one in which the human understanding may see 'a certain likeness of the Word of which it is written; "In the beginning was the Word, and the Word was with God, and the Word was God."' The act of human speech furthermore becomes an analogue for the incarnation when Augustine considers how 'the word in its outward sounding is a sign of the word that is inwardly luminous.' 'We may compare the manner in which our own word is made as it were a bodily utterance . . . with that in which the Word of God was made flesh. . . . Even as our word is made utterance yet not changed into utterance, so the Word of God was made flesh, but most assuredly not changed into flesh.'"

227. Baldwin, "Phineas Fletcher," pp. 469–70. According to Leah Jones, the general theory of literature in the seventeenth century was didactic, and stressed "the delightful teaching of mankind as the divine purpose of poetry," to which theory Phineas Fletcher's poetry firmly adheres by striving "to expound the nature, struggle, and rescue of the saved." *The Divine Science: The Aesthetic of Some Representative Seventeenth-Century English Poets* (New York: Columbia University Press, 1940), pp. 15, 121. On didacticism and *The Purple Island,* see also Sawday, "Bodies by Art Fashioned," pp. 117–18, 144–45; *The Body Emblazoned,* pp. 141–46.

228. Sawday, "Bodies by Art Fashioned," p. 144.

229. Sawday, *The Body Emblazoned,* p. 144.

230. Giles Fletcher, *Christs Victorie, and Triumph in Heaven, and Earth, over and after Death* (Cambridge, 1610)

231. Baldwin, "Phineas Fletcher," p. 470.

232. Mazzeo, "Metaphysical Poetry and the Poetic of Correspondence," p. 232.

233. "Diegetic" is used to describe a level or frame in the structure of the narration in *The Purple Island.* Thirsil is a character in the diegesis (story) narrated by the extradiegetic narrator, that is, the narrator of the framing story at the highest level who does not himself appear in that or any other story in the poem. The character Man, that is, the Isle of Man figure, is expounded in Thirsil's narration, which constitutes another frame inside that of the extradiegetic narration, and which is called the diegetic narration because it takes place in the diegesis told by the extradiegetic narrator. The story told by the diegetic narrator forms a second-degree narrative, hence a hypodiegetic level. See Chapter 1, note 9, of this study, and, for example, Rimmon-Kenan, *Narrative Fiction,* pp. 91–92.

234. John Banister, *The Historie of Man,* bk. I, sig. C_i^r (fol. 1^r). The passage is cited in full, above, in chapter 1, note 401.

235. Summers, *Michelangelo and the Language of Art,* p. 405, cited in Schultz, *Art and Anatomy in Renaissance Italy,* p. 106.

236. Benedetti, *Anatomice,* bk. I, chs. 5, 6; p. 86 cited. According to French, *Dissection and Vivisection in the European Renaissance,* p. 188, there were two contrasting methods of anatomy, called the resolutive and the compositive. "The method of resolution was to start with the whole body and work down to its ultimate components, the "similar" parts." The compositive method was synthetic, "a verbal presentation that proceeded as nature had proceeded in making the body, from first principles and components ["the smallest parts—the similars"] to the whole" (citation in square brackets is of p. 224). It is to the compositive method that Banister alludes. Crooke, *Microcosmographia,* fol. 14, suggests that perhaps the most visible and palpable sign of God's power in the creation of the human body is embryological, "in that of so small a quantitie of seede, the parts whereof seeme to be all homogenie or of one kinde; and of a few droppes of blood, he hath framed so many and so divers particles. . . ."

237. Hick, *The Metaphor of God Incarnate,* p. 102, says that the incarnation as defined by the definitive Chalcedonian settlement presents an obvious puzzle in the contradictions between the properties of each of the two natures, divine and human, which purportedly are incorporated in the one person of Christ. These contradictions or dissimilarities and the doctrine of the incarnation as promulgated in early seventeenth-century England are beyond the scope of this study.

238. Ussher, *Immanuel,* p. 2.

239. The term "unfolded" is used by Ussher in the title of his work—"unfolded by . . . Archbishop of Armagh"—to describe how he intends to undertake the "untying of this knot" (Ibid., p. 6) of the incarnation.

240. Ibid., p. 4.

241. Ibid., p. 6.

242. Hick, *The Metaphor of God Incarnate,* p. 101.

Chapter 4. Hermeneutics and Intertextuality

1. Baldwin, "Phineas Fletcher," p. 471. See also Nicolson, *The Breaking of the Circle,* especially p. 40.

2. Nicolson, *The Breaking of the Circle,* p. 40.

3. Ibid., p. 42. Also see Anon., "A Poet's Anatomy," p. 890, which doubts Fletcher's practical knowledge of anatomy, and recommends *The Purple Island* not as a textbook of anatomy but for its "melodious verse, wealth of language, and picturesque imagery" (cols 1, 2).

4. Baldwin, "Phineas Fletcher," p. 463.

5. Ibid., pp. 468–69.

6. See Terry Eagleton, *Literary Theory: An Introduction,* pp. 54–90 (p. 74): "Literary texts do not exist on bookshelves: they are processes of signification materialized only in the practice of reading. For literature to happen, the reader is quite as vital as the author."

7. According to Ricoeur, "Metaphor and the Main Problem of Hermeneutics," p. 98: "It is easy to show that all texts are discourse, since they proceed from the smallest unity of discourse, the sentence."

8. Ibid., p. 97. Ricoeur is referring to the theory of speech acts expounded in John L. Austin, *How To Do Things With Words,* The William James Lectures delivered at Harvard University in 1955 (Oxford: Clarendon Press, 1962; repr. 1975), especially Lectures VIII and IX.

9. Paul de Man, "Semiology and Rhetoric," in *Textual Strategies: Perspectives in Post-Structuralist Criticism,* ed. with an introduction by Josué V. Harrari (London: Methuen, 1980), pp. 121–40 (p. 125) (first publ., Ithaca: Cornell University Press, 1979).

10. Ibid., p. 127.

11. Ibid.

12. Ibid., p. 125.

13. Ibid., p. 129. De Man adds: "Although it would perhaps be somewhat more remote from common usage, I would not hesitate to equate the rhetorical, figural potentiality of language with literature itself" (p. 130).

14. Ibid., p. 127. De Man acknowledges his indebtedness to Charles Sanders Pierce's concept of the interpretant.

15. See Jonathan Culler, *The Pursuit of Signs: Semiotics, Literature, Deconstruction* (London: Routledge, 1981; repr. 1992), p. 191, where this point is raised.

16. See, for example, Culler, *The Pursuit of Signs,* p. 209.

17. Hausman, *Metaphor and Art,* pp. 9–11 (p. 9). Also see Ricoeur, "Metaphor and the Main Problem of Hermeneutics," p. 95 et passim.

18. For a list of this prefatory material and its occurrence in copies of the first edition, see Appendix A. Redman, "*The Purple Island,*" p. 2, likewise, albeit briefly, turns to the "verses affixed to the published poem" as "testimony to the acceptance of the poem."

19. "A.C.," "On the Most Accurate Poem, Inscribed *The Purple Island,*" in Giles and Phineas Fletcher, *Poetical Works,* II, 10. I have reversed the appearance of the text from the original, in which capitalized words and proper names are in plain type, and the remainder of the text is italicized. The text of further citations of this panegyric is reversed in the same way. In the British Library copy of the first issue of the first edition of *The Purple Island* (*STC* 11082), (Shelfmark: 239.i.23.[l.]), the attribution of this brief panegyric to "A.C." has been altered by hand to suggest that Abraham Cowley (1618–67) was its author (for details of this and other copies of each of the two issues of the first edition, see Appendix A). See Grosart, *The Poems of Phineas Fletcher,* IV, 32, where it

is claimed that "Thompson" is responsible for the amendation by hand of British Library 239.i.23.(l.); also see Langdale, *Phineas Fletcher,* p. 86, note 6. The candidature of Anthony Champney (1569?–1643?) for the authorship of the panegyric is proposed by Lambert, "Phineas Fletcher and Francis Quarles and *The Purple Island* of Phineas Fletcher," p. 576, col. 2. But the proposal is doubtful for two reasons. Firstly, Champney was a Catholic divine, and the anti-Catholic invective of Fletcher's *Locustae vel pietas iesuitica* (Cambridge, 1627) is unlikely to have endeared him to the author of *The Purple Island.* Other contributors of commendatory poems, such as Francis Quarles and Edward Benlowes, share very much more closely Fletcher's Protestant spirituality. Secondly, during the composition of *The Purple Island* and its preparation for publication, Champney held the following posts which may be considered to have kept him from England for substantial periods of time. He was fellow of the Sorbonne in Paris and superior of Arras College; from April 25, 1619, he was vice-president of the English college of Douay; later, he was confessor to the English Benedictine nuns at Brussels, a post that he held for three years, surrendering it on September 23, 1628 in consequence of a complaint made by the Benedictine monks that he was one of the thirteen priests who had signed the protestation of allegiance to Queen Elizabeth. After leaving Brussels he exercised his former employments at Douay until he was sent to England (at an unknown date), where he was chosen a canon of the chapter, and afterwards, in 1637, dean, on the death of Edward Bennet, by which time anyway *The Purple Island* had been published. On the life and occupations of Anthony Champney, see *Dictionary of National Biography* (hereafter cited as *DNB*) 63 vols., plus supplement, 3 vols.; second supplement, 3 vols.; indexes and epitomes (London: Smith, Elder, & Co., 1885–1900; supplement [1901]; second supplement [1912]), ed. by Sidney Lee [XXVII—LXIII, supplement, second supplement, indexes, and epitomes] and Leslie Stephen [I–XXI] [XXII–XXVI ed. by Sidney Lee and Leslie Stephen], X (1887), 35, col. 1.

20. Langdale, *Phineas Fletcher,* p. 171.

21. The anonymous Latin poem appears in Giles and Phineas Fletcher, *Poetical Works,* II, 6 (therein the poem is italicised). These verses are omitted from an edition of 1783, entitled *The Purple Island; or, The Isle of Man: An Allegorical Poem By Phineas Fletcher, Esteemed the Spenser of his Age,* a new edition, corrected and revised, with additional notes by the editor (London: Frys & Couchman, 1783). The editor's introduction to the second part of this edition is signed "P. B.," which may signify Peregrine Branwhite.

22. For this English translation of the first verse of the commendatory Latin poem, and for numerous helpful suggestions about its interpretation, I am indebted to Dr. A. Doug Lee, Senior Lecturer, Department of Classics, University of Nottingham. I also wish to thank Mr. Geoffrey Eatough, of the Department of Classics, University of Wales, Lampeter, for further helpful comments regarding translation and interpretation.

23. P. G. W. Glare, ed., *Oxford Latin Dictionary* (Oxford: Clarendon Press, 1968), p. 906, cols 1–2.

24. Francis Quarles, "To the Ingenious Composer of this *Pastorall,* the Spencer of this Age," Giles and Phineas Fletcher, *Poetical Works,* II, 8.

25. *OED,* V, "I," p. 285, col. 1.

26. Ibid.

27. Langer, "Invention," p. 136.

28. Ibid.

29. Ibid., p. 141.

30. Ibid., p. 136, says that *inventio* is the finding of apt subject matter to achieve the aim of the work. According to Alessandro Lionardi, invention comes from quickness of wit among other things (p. 138).

31. Francis Quarles, "To my deare friend, the Spencer of this age," Giles and Phineas Fletcher, *Poetical Works,* II, 284. Grundy observes that: "In complimentary verses Fletcher is praised three times as the "ingenious" author of *The Purple Island.*" Grundy, *The Spenserian Poets,* p. 193.

32. William Winstanley, *The Lives of the Most Famous English Poets; or, The Honour of Parnassus* (London: H. Clark, 1687), p. 159. Phineas's father was Giles Fletcher (c. 1548–1611) , author of *Of the Russe Commonwealth; or, Maner of Government by the Russe Emperour . . .* (London, 1591); *De literis antiquae Brittaniae,* ed. by Phineas Fletcher (Cambridge, 1633), and others. His brother was Giles, author of *Christ's Victorie, and Triumph.* Winstanley is incorrect in providing for Phineas two brothers: "*George Fletcher,* the Author of a Poem, entitled, *Christs Victory and Triumph over and after Death;* and *Giles Fletcher,* who wrote a worthy Poem, entitled, *Christs Victory.*" Winstanley, *The Lives Of the Most Famous English Poets,* p. 159. The misnaming of the author of *Christ's Victorie, and Triumph* as George had already occurred before Winstanley in Edward Phillips, *Theatrum Poetarum* (1675), p. 53. Grosart remarks that from the outset the scanty facts of the Fletcher brothers's lives have been "blundered." *The Poems of Phineas Fletcher,* I, xcvi. The only thoroughly researched biography of Phineas Fletcher is Langdale, *Phineas Fletcher.*

33. Grundy, *The Spenserian Poets,* p. 193. On the baroque qualities of Fletcher's poetry, see pp. 181–82, 184, 196, 201–02.

34. See, for example, Mazzeo, "Metaphysical Poetry and the Poetic of Correspondence," p. 227.

35. Donker and Muldrow, *Dictionary of Literary-Rhetorical Conventions of the English Renaissance,* p. 55. Also see, for example, Mazzeo, "Metaphysical Poetics and the Poetry of Correspondence," pp. 224–25. Mazzeo suggests the critical criteria available to critics and theorists made it difficult for them to avoid falling into accounts of metaphor as ornament, attempts to describe the "furore" or "gusto" of poetic creation, and exclusively intellectual accounts of literary creation, such as perhaps, Ramus's logical definition of metaphor, which Mazzeo claims always ended up "finding a kind of intellectual truth in the conceit" (p. 225). Mazzeo's point here is that Renaissance theorists did not really have a concept of the *differentia specifica* of poetic metaphor, except as ornament. He says that even theorists of the conceit fell back on ornamental theories of metaphor, which were not intrinsic to the arts (p. 228). This, however, did not prevent poets fashioning conceits, and should not prevent the application of modern theory of metaphor to this practice. Mazzeo's article was written when originativist or interactive theories of poetic metaphor were coming to the fore, from Black and Beardsley, and I think his insistence on the specificity of poetic metaphor, especially its cognitive processing, responds to these theories. Subsequent to the development of cognitive theories of metaphor, commenced by Lakoff and Johnson, a much broader scope of metaphorical occurrences have been seen as conceptual. This shift has tended to downplay rather than accentuate the differences that do exist between some literary or poetic metaphors and nonliterary metaphors, a tendency that in particular affects detrimentally the formulation of the specific properties of creative and novel metaphors. Culler, *The Pursuit of Signs,* pp. 202–6, divides the traditions of metaphorology up to the twentieth century into the *via philosophica* and *via rhetorica.* The first of these, Culler, argues grants to metaphor cognitive respectability, but in its thorough problematizing of the distinction between literal and metaphorical language it tends toward a view that all language is metaphorical and fails to explain how we respond to metaphor as distinct from "literal" language. The *via rhetorica* Culler summarizes as regarding metaphor as constituting the space between what is said and what is meant, a more concise and vivid version than a literal paraphrase. The latter view of metaphor is contested throughout Chapters 1 and 2 of this study, while a more nuanced version than Culler allows of the former view, combined with a certain *differentia specifica* of literary metaphor, most evident in its foregrounded character and its propensity for creative exploration and novelty, is advocated.

36. Donker and Muldrow, *Dictionary of Literary-Rhetorical Conventions of the English Renaissance,* p. 53.

37. Langer, "Invention," p. 140.

38. Dollimore, *Radical Tragedy* (1989), p. 74. Dollimore quotes Sidney, *An Apology for Poetry,* ed. by Geoffrey Shepherd (Manchester: Manchester University Press, 1973), p. 102.

39. Sidney, *The Defence of Poetry,* pp. 107–8, 138.

40. Dollimore, *Radical Tragedy* (1989), p. 75. Dollimore quotes Sidney, *An Apology for Poetry,* ed. by Shepherd, pp. 100, 103.

41. Mazzeo, "Metaphysical Poetry and the Poetic of Correspondence," p. 228.

42. For the thesis of cosmic decay, see Dollimore, *Radical Tragedy* (1989), Ch. 5: "The Disintegration of Providentialist Belief," pp. 83–108 (pp. 92–103).

43. Drayton, *Poly-Olbion,* "From the Author of the *Illustrations.*"

44. See Sidney, *The Defence of Poetry,* p. 105, where conceit is used in this sense.

45. See, for example, Joseph Beaumont, *Psyche; or, Loves Mysterie in XX Cantos: Displaying the Intercourse Betwixt Christ, and the Soule* (London: John Dawson, 1648), "The Author to the Reader," wherein Beaumont writes, "I will venture to cast my self upon thy Ingenuitie."

46. J. B. Leishman, *The Monarch of Wit: An Analytical and Comparative Study of the Poetry of John Donne,* 6th edn (London: Hutchinson, 1962; repr. 1967), p. 123.

47. Ibid., p. 123.

48. See, for example, Mazzeo, "Metaphysical Poetry and the Poetic of Correspondence," p. 233.

49. See Ibid., p. 234.

50. Nicolson, *The Breaking of the Circle,* pp. 4–6 (p. 5). Also see Leishman, *The Monarch of Wit,* pp. 208–9.

51. Leishman, *The Monarch of Wit,* p. 88.

52. For a discussion of some stylistic features of Fletcher's poetry in relation to baroque and metaphysical styles, see Grundy, *The Spenserian Poets,* p. 196.

53. Grosart, *The Poems of Phineas Fletcher,* I, cclxiii–cclxiv. For Henry Headley, whose "indiscriminating generalization on the 'anatomical' cantos" Grosart is criticizing directly, see Headley, ed., *Selected Beauties of*

Ancient English Poetry, I, liii (cited below in the text). Headley's critical remarks are reproduced in *The Purple Island, a Poem: by Phineas Fletcher,* pp. v–xi.

54. Sanford, *The Works of the British Poets,* V, 93.

55. Benedetti, *Anatomice,* Bk. II, Ch. 1, p. 89.

56. Grosart, *The Poems of Phineas Fletcher,* I, ccli.

57. Kastor, *Giles and Phineas Fletcher,* p. 139. Izaak Walton quoted seventeen lines from *The Purple Island* in his *Compleat Angler* (London, 1653), and this provides the occasion for a revealing editorial comment in a mid-eighteenth-century edition of the work. In *The Compleat Angler; or, Contemplative Man's Recreation, Being a Discourse on Rivers, Fish-Ponds, and Fishing, in Two Parts, the First Written by Mr. Izaak Walton, the Second by Charles Cotton, Esq; to Which Are now Prefixed, the Lives of the Authors; Illustrated with Cuts . . . and Notes . . .* (London: Thomas Hope, 1760), pp. 232–34, note *, the editor says of Phineas Fletcher: "It would be great injustice to the memory of this excellent person, whose name is now hardly known, to pass him without notice." The note then proceeds to provide a brief biography and to print some stanzas from *The Purple Island*—"a more entire quotation from that work" than that which Quarles supplies in his text. See note 77 below.

58. R. Anderson, ed., *A Complete Edition of the Poets of Great Britain* (Edinburgh, 1793), IV, 378–80, cited in Kastor, *Giles and Phineas Fletcher,* p. 140. See also Sanford, *The Works of the British Poets,* V, 92.

59. Headley, ed., *Selected Beauties of Ancient English Poetry,* I, liii.

60. For the attribution of accuracy, see "A.C.," "On the Most Accurate Poem, Inscribed *The Purple Island,*" Giles and Phineas Fletcher, *Poetical Works,* II, 10.

61. Donker and Muldrow, *Dictionary of Literary-Rhetorical Conventions of the English Renaissance,* p. 54.

62. See Mazzeo, "Metaphysical Poetry and the Poetic of Correspondence," pp. 230–33.

63. Donker and Muldrow, *Dictionary of Literary-Rhetorical Conventions of the English Renaissance,* p. 54.

64. Ibid.

65. Mazzeo, "Universal Analogy and the Culture of the Renaissance," p. 304. Of this quality of some metaphors, Deese, "Mind and Metaphor," p. 214, writes: "Our minds run in particular ways, ways determined by some deep, sometimes universal, and sometimes culture-bound, characteristics. There must be a fine tuning of a metaphor to the structure of particular human minds in their cultural settings before it will work. Thus, the metaphor is neither objective nor subjective but a relation between the objective and the subjective."

66. Gascoigne, *Certayne Notes of Instruction Concerning the Making of Verse or Ryme in English,* p. 465.

67. Donker and Muldrow, *Dictionary of Literary-Rhetorical Conventions of the English Renaissance,* p. 54.

68. In the philosophy of Nicholas of Cusa, this exemplification of the infinity of God in the finite mind of Man was a case of *coincidentia oppositorum,* a coincidence of the maximum with the minimum. See Ernst Cassirer, *The Individual and the Cosmos in Renaissance Philosophy,* trans. Mario Domandi (Oxford: Clarendon Press, 1963), chapter 1, especially pp. 9–40; Norford, "Microcosm and Macrocosm in Seventeenth-Century Literature," pp. 412–14.

69. Barkan, *Nature's Work of Art,* p. 33.

70. On this function of metaphor see Harries, "Metaphor and Transcendence," p. 88.

71. Edward Benlowes, "On the Excellent Morall Poem, Entitled the *Isle of Man,*" in Giles and Phineas Fletcher, *Poetical Works,* II, 5; Quarles, "To the Ingenious Composer of this Pastorall, the Spencer of this age," in Giles and Phineas Fletcher, *Poetical Works,* II, 8.

72. "Lod. Roberts," "To the Unknown Mr. P. F. Upon Survay of his *Isle Of Man,*" in Giles and Phineas Fletcher, *Poetical Works,* II, 9.

73. "A.C.," "On the Most Accurate Poem, Inscribed *The Purple Island,*" in Giles and Phineas Fletcher, *Poetical Works,* II, 10.

74. Langer, "Invention," p. 141. Langer refers to an oft cited phrase from Terence: "all that is said has been said before."

75. For similar functions of metaphor, see Cohen, "Metaphor and the Cultivation of Intimacy," p. 11.

76. Langdale, *Phineas Fletcher,* pp. 92–93, says: "The longing for an immortality in print, so common with literary artists, was lacking in Fletcher—even in the Cantabrigian poet of 1610. He wrote to provide courteous little offerings to his friends, to gain the respect of his university associates, and to better his academic position. Since the audience which he addressed was small, all these ends could be achieved by the circulation of his manuscripts, and hence the printing press had little allurement." Regarding the eventual publication of many of his works, Langdale adds that "in the neighbourhood of 1627, the patron [Edward Benlowes] seems to have induced the poet to publish; and by 1633, the title pages of Fletcher's various works had established when and where they were published" (p. 94). Published in this period were: *Locustae vel Pietas Iesuitica; The Locusts or Apollyonists*

(Cambridge, 1627); *Brittain's Ida; Written by that Renowned Poët, Edmond Spencer* (London, 1628); *Sicelides, or Piscatory, As It Hath Been Acted in King's Colledge in Cambridge* (London, 1631); *The Way to Blessednes; a Treatise . . . on the First Psalm* (London: James Boler, 1632); *Joy in Tribulation; or, Consolation For Afflicted Spirits* (London: James Boler, 1632); *Sylva Poetica* (Cambridge, 1633); and Phineas's father's *De literis antiquae Brittaniae*, ed. by Phineas Fletcher, which was published in one octavo edition with *Sylva Poetica* (details given above). A copy of *The Purple Island* held by the British Library (C.34.g.33) is presented to Edward Benlowes in the autograph of Fletcher. See Langdale, *Phineas Fletcher*, p. 94. Also see Appendix A.

77. Langdale, *Phineas Fletcher*, pp. 85–86, 89–91 (pp. 90, 91 cited). Walton quoted seventeen lines from *The Purple Island* in his *Compleat Angler* (1653), but intriguingly, almost one half of the lines of the *Compleat Angler*'s rendition contain important variations from the text published in 1633. *The Compleat Angler* (1760) has an editorial note, saying that it is *The Purple Island* "from whence the passage in the text, with a little variation is taken" (p. 233 note *). Langdale surmises (p. 52): "The author appears to have written at least two versions of the work, although only one is now extant." He adds (p. 86): "If Walton was familiar with the poet, it is not surprising that he came into possession of a variant manuscript of *The Purple Island*, from which he quoted in his discourse of fish and fishing." Quarles and Benlowes were both natives of Essex and exchanged verses. Benlowes was a gentleman commoner of St John's College, Cambridge, matriculating 8 April 1620. On leaving the university he traveled on the continent before returning later that decade. See *DNB*, XLVII (1896), 93, col. 2; *DNB*, IV (1885), 226, col. 1.

78. Daniel 'Featly', "To the Readers," in Giles and Phineas Fletcher, *Poetical Works*, II, 4.

79. Francis Quarles, *Divine Fancies: Digested into Epigrammes, Meditations, and Observations* (London: M.F., 1632), in *The Complete Works*, ed. by Grosart, II: Verse, p. 196.

80. On the function of the conceit to elicit wonder, see Donker and Muldrow, *Dictionary of Literary-Rhetorical Conventions of the English Renaissance*, pp. 52–53, 55.

81. Quarles, *Divine Fancies*, p. 196.

82. Sawday, "Bodies by Art Fashioned," p. 137.

83. The *implied author* is a phrase coined by Wayne Booth in *The Rhetoric of Fiction* (Chicago: University of Chicago Press, 1961), and employed since to refer to a construct inferred and assembled by the reader from the text, yet "[m]ore than just a textual stance," in fact almost "an anthropomorphic entity, often designated as 'the author's second self.'" Rimmon-Kenan, *Narrative Fiction*, pp. 86–87 (p. 86). The implied author has no voice, no direct means of communication, as has the narrator. Rather the implied author instructs the reader silently through the design of the whole work (p. 87).

84. Richard Waswo, "Theories of Language," in *The Cambridge History of Literary Criticism*, ed. by Nisbet and Rawson, III, ed. by Norton, 25–35 (p. 27). See Puttenham, *The Arte of English Poesie*, Bk. II, Ch. 12.

85. Nicolson, *The Breaking of the Circle*, p. 109: "One of the most melancholy of chronologists, Luther read into the lesson taught by the past an unbroken history of what may be called 'degenerative cycles' of man and of the world, which was created in 4000 B.C. Each 'great age' of two thousand years had begun well; during each period steady degeneration occurred in both man and the world. . . . Living in the last age of the world, Luther was persuaded that a just God would not permit that world to fulfill the course He had once ordained." Nicolson offers a pertinent interpretation of *The Purple Island*, Canto 1, stanza 36, ll. 5–6—"The aged world, though now it falling shows, / And hastes to set, yet still in dying grows"—saying that the globe "ironically was growing to its greatest extent [through geographical exploration] just before its death" (p. 110).

86. Reading a text semiotically rather than semantically is advised and performed by de Man, "The Epistemology of Metaphor," in order to realize how the signs of a text undermine its explicit statements or conclusions (p. 16).

87. Kastor, *Giles and Phineas Fletcher*, p. 112. MacDonald, *England's Antiphon*, p. 868, writes: "Of all the strange poems in existence, surely this is the strangest." Also see Gosse, *The Jacobean Poets*, p. 148; Parfitt, *English Poetry of the Seventeenth Century*, p. 191.

88. Gosse, *The Jacobean Poets*, p. 148.

89. For the influence of the *Devine Weekes* of du Bartas on Cantos 1 to 5 of *The Purple Island*, see Baldwin, "Phineas Fletcher," p. 471 (brief, but pertinent to this chapter); Philip Alan Tennant Gardner, "The Banquet of the Word: Biblical Authority and Interpretation in Spenser and the Fletchers" (unpublished doctoral thesis, University of Toronto, 1985), pp. 421–22; Langdale, *Phineas Fletcher*, p. 115. For the influence of "The House of Alma," in Cantos 9 to 12 of Book II of Spenser's *Faerie Queene*, on Cantos 1 to 5 of *The Purple Island*, see Redman, "*The Purple Island*," pp. 6–8; Gardner, "The Banquet of the Word," p. 422; Kastor, *Giles and Phineas Fletcher*, p. 155, note 4; Langdale, *Phineas Fletcher*, pp. 115, 117, 145–46. Sawday also consults Crooke's *Microcosmographia*,

which he relates to both the "House of Alma" passage in *The Faerie Queene* and *The Purple Island.* Sawday, "Bodies by Art Fashioned," pp. 109–15. Kastor's is alone among this collection of studies in arguing that the "House of Alma's" influence upon Cantos 1 to 5 of *The Purple Island* is unlikely. He adds: "Islands figure in much of Phineas['s] poetry: piscatory Isles, 'this Isle' England (*Apollyonists*), Sicily, etc." (p. 55, note 4) For a list of borrowings of specific lines from Fletcher's models and sources, see Langdale, *Phineas Fletcher,* Appendix B.

90. Alfred Horatio Upham, *The French Influence in English Literature from the Accession of Elizabeth to the Restoration* (New York: Columbia University Press, 1911), pp. 170, 198–201, "Appendix B." Also see Langdale, *Phineas Fletcher,* pp. 115–17.

91. Langdale, *Phineas Fletcher,* p. 115. *The Faerie Queene,* Books I–III were published in 1589/90, William Ponsonby having secured a license for their publication on 1 December 1589; and a second edition appeared in 1596. Books IV–VI were completed in 1594, were brought to London by Spenser at the close of 1595, and licensed for publication by Ponsonby on 20 January 1595/96.

92. Langdale, *Phineas Fletcher,* p. 115.

93. See, for example: Langdale, *Phineas Fletcher,* p. 146; Sawday, "Bodies by Art Fashioned," p. 111. Kastor's is again the unconventional voice: "the influence of Vergil and Du Bartas appears to have been even stronger than Spenser's—and, has yet to be explored fully." *Giles and Phineas Fletcher,* p. 145.

94. Langdale, *Phineas Fletcher,* pp. 146–47 and Appendix B.

95. Baldwin, "Phineas Fletcher," p. 471; Upham, *The French Influence in English Literature,* p. 199.

96. Langdale, *Phineas Fletcher,* p. 171; Francis Quarles, "*Man's Bodie*'s like a *house,*" Giles and Phineas Fletcher, *Poetical Works,* II, 11.

97. *Du Bartas His Devine Weekes and Workes,* trans. by Joshua Sylvester (London: H. Lownes, 1611), pp. 158, 161. All further references are to this edition, unless otherwise stated, and occur in the text prefixed by the initials of the abbreviated title: *D.W.*

98. See Louis B. Wright, *Middle-Class Culture in Elizabethan England* (Ithaca: Cornell University Press, 1965), p. 597.

99. Compare with *The Purple Island* 3.13.6.

100. See also *Devine Weekes,* p. 164.

101. For an argument for the more general influence of Cantos 9 to 12 of Book II of Spenser's *Faerie Queene* on Cantos 1 to 5 of *The Purple Island,* see Langdale, *Phineas Fletcher,* pp. 145–46. Redman, "*The Purple Island,*" p. 6, correctly, I think, argues that *The Purple Island* is based "generally on Book II, cantos ix and xi," of Spenser's work, though the influence of Canto 11 is not generally felt on *The Purple Island* until Canto 7, yet is stronger thereafter.

102. The phrase in quotation marks is used by Spenser to describe "the Faery Queene," his poem, in a letter "To the Right noble, and Valorous, Sir Walter Raleigh knight," "expounding his whole intention in the course of this worke." Spenser, *The Faerie Queene,* Appendix 1, p. 737.

103. The sole use in *The Faerie Queene,* Book II, Canto 9, of the conceptualization of the body in terms of natural vegetation or terrain, is in stanza 46, ll. 1–2.

104. See, for example, the *Devine Weekes,* p. 163 (cited above); *The Purple Island* 3.11–12 (3.11.1 & margin. i; 3.12.5–7).

105. James Nohrnberg, *The Analogy of "The Faerie Queene"* (Princeton: Princeton University Press, 1976), pp. 326–27.

106. Ibid., p. 326. Nohrnberg adds that the landscapes and the temperament confronting them often "obey a principle of allegorical chiasmus, as we might call it, whereby the external features of the environment reflect the internal ones in inverse form. A man in the environment here often stands for the environment in man" (p. 327).

107. See, for example, Robin Headlam Wells, *Spenser's "Faerie Queene" and the Cult of Elizabeth* (London: Totowa, N.J.: Barnes & Noble, 1983), "Preface," "To Sound Her Praises: Introduction," pp. 1–28 et passim; Quilligan, *The Language of Allegory,* pp. 101–2. In the *Commedia,* all four levels of allegory are operable at any one time, whereas signification in *The Faerie Queene* and *The Purple Island* is more flexible.

108. The terms "survey" or "suruew" derive from Old French *surveeir, souveeir,* from medieval Latin *supervidere* (oversee, supervise). "The sense of determine the form, extent, and position of (land) is first recorded in 1550." Robert K. Barnhart, ed., *Dictionary of Etymology* (Edinburgh: Chambers, 1988), p. 1097, col. 1.

109. In the sense of close or intense scrutiny over potential sources of threat, *The Purple Island* here anticipates the emergence of the word "surveillance" in 1802, which is borrowed from the French *surveillance* (oversight, a watch), from *surveiller* (oversee, watch). Barnhart, ed., *Dictionary of Etymology,* p. 1097, col. 1

110. Paré, *Works* (1678), Bk. IV, p. 95.

111. Valverde, *Historia de la composicion del cuerpo humano,* Bk. III, Tab. 2 (Figures V and VIIII); *Anatomia del corpo humano, Lib.* III, *Tauola* 2 (Figures V and VIIII); *Vivae imagines partium corporis humani aereis formis expressae, Lib.* III, *Tabula* II (Figures V and VIIII).

112. On mannerism, style, and the combination of abstractly or figuratively related details in an imaginary framework, see Arnold Hauser, *The Social History of Art: Renaissance, Mannerism and Baroque,* 4 vols. (London: Routledge & Kegan Paul, 1968), II, 88–96 (p. 93). On the "fertile" cognitive figurality of *maniera* or style, see Schultz, *Art and Anatomy in Renaissance Italy,* p. 91. As a historical and critical category of art that separates the High Renaissance from the baroque, however, "mannerism" is difficult to define, as Henri Zerner observes in "Observations on the Use of the Concept of Mannerism," in *The Meaning of Mannerism,* ed. by Franklin W. Robinson and Stephen G. Nichols, Jr (Hanover, N.H.: University Press of New England, 1972), pp. 105–21.

113. Healy, "Sound Physic," p. 345.

114. For the extension in *The Purple Island* of the figuration of anatomy as fortification from the abdomen to the thorax, see Canto 4, stanzas 4 and 24; and for its extension to the head, see Canto 5, stanzas 27 (quoted in chapter 2) and 31.

115. John Heydon, *A New Method of Rosie Crucian Physick: Wherein is Shewed the Cause; and Therewith Their Experienced Medicines for the Cure of All Diseases* (London: Thomas Lock, 1658), pp. 2, 3.

116. Ibid., p. 2.

117. See, for example, Simon Shepherd, *Spenser* (London: Harvester Wheatsheaf, 1989), pp. 12–13.

118. After Henry VII had ascended the throne, he called his son Arthur to reestablish an imaginary lineage and to institute a proto-nationalistic departure from the foreign houses that had been on the throne in England since the invasion of the Normans. The Tudor House represented the sovereignty of the first English royal lineage since 1066.

119. Miller, "Spenser's Poetics: The Poem's Two Bodies," p. 170.

120. Ibid., p. 178.

121. Ibid., p. 178.

122. See chapter 1, entitled "Prince of Peace from Heaven Blest," in Wells, *Spenser's "Faerie Queene" and the Cult of Elizabeth,* pp. 29–51.

123. Ibid..

124. Spenser, *The Faerie Queene,* p. 250, note to Book II, Canto 9, stanza 21, ll. 1–2. Also see Caroll Camden, "The Architecture of Spenser's 'House of Alma'" *Modern Language Notes* 58 (1943): 262–65 (p. 263). Camden refers to William Austin, *Haec Homo; Wherein the Excellency of the Creation of Women is Described, By Way of an Essay* (1636; repr. 1637, 1638, 1639), which Camden argues "antedates by some seven years the issuing of Digby's *Observations,* which has heretofore been considered the earliest commentary on Spenser's passage" (p. 263). Austin's solution to Spenser's geometrical proportions of the human body is twofold: mystical and literal. As to the latter, Austin interprets the geometry as in the image of God, arguing that the form of the human body "must be excellent because God gave his own form to it" (p. 263).

125. See Nohrnberg, *The Analogy of "The Faerie Queene,"* p. 293.

126. The figure of the body as a castle or fortress occurs in the allegorical poem *King Hart,* the authorship of which is uncertain, but it may be by Gavin Douglas (c. 1475–1522). According to Denton Fox, "The Scottish Chaucerians," in *Chaucer and Chaucerians: Critical Studies in Middle English Literature,* ed. by D. S. Brewer, Nelson's University Paperbacks (London: Thomas Nelson and Sons, 1966), pp. 164–200 (p. 188, note 2), the only poems that can safely be attributed to Douglas are his translation of the *Aeneid, The Palace of Honour* and "the short and trivial *Conscience.*" "The allegorical poem, *King Hart,* has usually been thought to be by Douglas, but now it seems very unlikely that it is really his." There is no early printed edition of *King Hart,* and the unique copy of the poem is found in the Maitland Folio, a large manuscript collection of Scottish poetry of the fifteenth and sixteenth centuries, now in the possession of the Pepysian Library, Magdalene College, Cambridge (MS 2553). However, it has since appeared in various printed editions, including Priscilla J. Bawcutt, *The Shorter Poems of Gavin Douglas,* The Scottish Text Society, 4th Series, 3 (Edinburgh and London: William Blackwood & Sons, 1967), pp. 139–70. Bawcutt says: "The conception of the body as a castle was widespread in medieval didactic literature. The idea appears to derive ultimately from ancient images of the body as a microcosm, an island, a city, and above all as a house; it was further popularised by its use in Scriptural exegesis, especially of such verses as Proverbs iv, 23; Luke xii, 39; Matthew xxi, 2, and xxiv, 43" (p. lxi).

127. Spenser, *The Faerie Queene,* p. 252, note to Book II, Canto 9, stanza 27.

128. Ibid.

129. See Nohrnberg, *The Analogy of "The Faerie Queene,"* p. 326.

130. See Wells, *Spenser's "Faerie Queene" and the Cult of Elizabeth,* "Preface." In Renaissance poetics it was generally believed that the highest poetic kind was the epic, and most critics agreed that the function of epic was essentially epideictic, that is, to display the virtues of some great man or woman as a pattern for emulation. The purpose of epideictic poetry is not elaborately to flatter. Rather, it is to set up a pattern of moral goodness, in such a way as to reform bad rulers and improve the good (pp. 1–7).

131. Wells, *Spenser's "Faerie Queene" and the Cult of Elizabeth,* "Preface" and pp. 1–28.

132. Benedetti, *Anatomice,* p. 90.

133. De Laguna, *Anatomica methodus,* p. 267. Vapors from the cooking of food in the stomach are emitted in plentiful amount in humans because the process of digestion or concoction is slower and weaker than in other animals, hence its figuration as cooking. In the seventh in the series of lectures given at Bologna in 1540, Corti explains: "For we see that all animals quickly and easily digest even raw nourishment in great quantity. Man, however, digests with difficulty and slowly even well-cooked food in small quantities." Eriksson, *Andreas Vesalius' First Public Anatomy at Bologna 1540,* p. 99.

134. Berengario, *Isagogae breves,* p. 138. It is not clear that the opinion that the formation of hairs is the congelation of superfluous vapors emitted through the pores of the skin, was held by all or even a majority of anatomists in the sixteenth century. In the seventh in the series of lectures given at Bologna in 1540, for example, Corti omits to mention this aspect of their function and substance; but I have not found any anatomists unequivocably opposed to the theory. Eriksson, *Andreas Vesalius' First Public Anatomy at Bologna 1540,* p. 99. On diet as medicine and as self-fashioning in the early modern period, see Schoenfeldt, *Bodies and Selves in Early Modern England,* pp. 20–33 *et passim.*

135. Achillini, *Annotationes anatomicae,* p. 62. On this general economy of fluids in the body and the sexually differentiated means of fluid elimination, see Martensen, "The Transformation of Eve."

136. *The Epitome of Andreas Vesalius,* p. 40.

137. Berengario, *Isagogae breves,* p. 91; de Laguna, *Anatomica methodus,* p. 281. On the diaphragm as a marker in so far as concern with it indicates a concern with the Platonic and hence Galenic view of the body, and in the sense that it sets off one set of internal organs (the abdominal organs) from another (those in the thoracic cavity), both functionally and conceptually, see chapter 2, note 128.

138. De Laguna, *Anatomica methodus,* p. 267; Berengario, *Isagogae breves,* p. 138.

139. Berengario, *Isagogae breves,* p. 139; also see p. 148.

140. Spenser, *The Faerie Queene,* p. 253, note to Book II, Canto 9, stanza 31.

141. Compare, for example, *The Faerie Queene* II.9.31.6–8 and II.9.32.6 with *The Purple Island* 2.36.2–7; *The Faerie Queene* II.9.36.6 with *The Purple Island* 2.36.7. For further verbal similarities between *The Faerie Queene* Book II, Canto 9 and *The Purple Island* Cantos 1, 2, and 5, compare, for example, the following: *F.Q.* II.9.22.1–2 and *P.I.* 1.44.1–4; *F.Q.* II.9.26.2 and *P.I.* 2.30.1; *F.Q.* II.9.46.3 and *P.I.* 5.22.5; *F.Q* II.9.46.7–8 and *P.I.* 5.27.3–4; *F.Q.* II.9.47.4 and *P.I.* 5.4.3.

142. For Fletcher's open declarations of his indebtedness to "the influencing genius of Spenser" throughout his work, see Redman, "*Phineas Fletcher,*" pp. 6–7. Of "*Bartas*" (*P.I.* 1.14 margin. *) the extradiegetic narrator of *The Purple Island* says:

> And that **French* Muses eagle eye and wing
> Hath soar'd to heav'n, and there hath learn'd the art
> To frame Angelick strains, and canzons sing
> Too high and deep for every shallow heart.
>
> (*P.I.* 1.14.1–4)

143. For Fletcher's indebtedness to Spenser see, for example, Langdale, *Phineas Fletcher,* pp. 131–41. Langdale identifies one hundred and three borrowings from Spenser in Fletcher's works: "A study of the complete group leads to these conclusions: the borrowings are more frequent in *Brittain's Ida,* the *Piscatorie Eclogs, The Apollyonists,* and *The Purple Island,* but they are most frequent in the latter or allegorical cantos of *The Purple Island,* which are drenched in a flood tide of verbal influence" (p. 140). Langdale adds (p. 141): "Many of Thirsil's borrowings are so close that they are almost transcriptions; of these he must have been fully conscious, for no one knew his Spenser better than the poet of the Cam[;] . . . on the other hand, his pages are crowded with phrases so turned because of involuntary Spenserian imitation, phrases which I have not included among the 103 parallels." And he concludes (pp. 131–32): "A comparison of the work of the two poets has shown that Spenser's influence upon his follower was even greater than previous students have indicated. Actually he dictated the style, the imagery, the philosophy, the vocabulary, the subject matter, the character portrayal, the prosody, and the wording, of

parts of almost every Fletcherian poem." Also see Grundy, *The Spenserian Poets,* pp. 7–8, who identifies qualities of the group of poets whom she identifies as Spenserian, and among whom she identifies Phineas Fletcher and his brother Giles. The relevant qualities include conservatism, patriotism, concern to represent the English countryside, a combination of traditionalism and robust independence and strength, a derivativeness which "set them free to accomplish something new" (p. 8), expansiveness, real gifts of feeling and imagination, an emphasis on the distinction of the poet's office.

144. The complexion of the first of the honorable sages in *The Faerie Queen* II.9.52.2–7 is similar to that of Phantastes in *The Purple Island* 6.46.2–5; there are also verbal similarities between *F.Q.* II.9.50.3–9 and *P.I.* 6.47.2–4. The description of the sage who keeps things past in memory resembles that of Eumnestes (see especially *F.Q.* II.9.55.1–8 and *P.I.* 6.50.1–4; also see *F.Q.* II.9.56.1–7 and *P.I.* 6.49.1–4; there is also a partial repetition of the phrase "both their infancies" (*F.Q.* II.9.57.3) in *P.I.* 6.49.7, although with a different application). In addition the "litle boy" attending on this sage in *The Faerie Queene* II.9.58.4 resembles the page in *The Purple Island* 6.50; and there are verbal similarities between *F.Q.* II.9.58.5 and *P.I.* 6.50.6.

145. See Chapter 1, note 314.

146. Grundy, *The Spenserian Poets,* p. 8.

Chapter 5. The Literary Context

1. Raymond Williams analyzed the complex historical process in terms of "residual, dominant and emergent elements which coexist at any cultural moment," and Jonathan Dollimore, *Radical Tragedy: Religion, Ideology and Power in the Drama of Shakespeare and His Contemporaries* (Brighton: Harvester Wheatsheaf, 1984), passim (p. 7), explores English Renaissance drama and culture with these criteria very much in mind.

2. Langdale refers to the following: Girolamo Fracastoro, *Syphilis, sive morbi Gallicis* (Verona, 1530); David Kynloch, *De hominis procreatione, anatome, ac morbis internis priores libri duo heroico carmine donati* (Paris, 1596); another edn in *Delitiae poetarum Scotorum,* ed. by Arthur Johnson (Amsterdam, 1637); Robert Grove, *Carmen de sanguinis circuitu, a G. Harvæo primum invento; adjecta sunt miscellanea quædam* (London, 1685). *Phineas Fletcher,* p. 171. See also Baldwin, "Phineas Fletcher"; Nicolson, *The Breaking of the Circle,* pp. 40–42; Sawday, "Bodies by Art Fashioned," Chapter 3, pp. 107–61; *The Body Emblazoned,* pp. 170–82.

3. Sawday, "Bodies by Art Fashioned," p. 117; also see pp. 115–19. Sawday vastly expands the intertextual link between *The Purple Island* and Kynloch's text, made by Langdale in *Phineas Fletcher,* p. 171.

4. Wright, *Middle-Class Culture in Elizabethan England,* pp. 550, 551. Wright observes: "Caxton . . . translated from the French and published *The myrrour of the worlde* [1481], an encyclopedia compiled from various medieval authorities, incorrectly attributed to Vincent of Beauvais. It provided in compendious form a survey of medieval scientific knowledge, beginning with a discussion of the nature of God, which precedes the account of the creation, and ending with an exposition of the system of the universe" (p. 552). The most influential of all medieval encyclopedias was, however, Bartholomaeus Anglicus, *De proprietatibus rerum* (1398), first printed, in Trevisa's translation, by Wynkyn de Worde in 1495. "It was the basis of an Elizabethan encyclopedia, compiled by Stephen Batman as *Batman uppon Bartholome, his Booke De Proprietatibus Rerum* (1582). This was a storehouse of curious learning from which Shakespeare and his contemporaries garnered much of their scientific knowledge," such as "the properties of the soul, man's body with its parts and qualities, diseases with their causes and cures" (p. 552).

5. Sawday, "Bodies by Art Fashioned," p. 123; see pp. 119–23.

6. Frye, *Anatomy of Criticism,* passim (pp. 311–12). The Menippean satire or anatomy is prose fiction, but developed out of verse satire "through the practice of adding prose interludes" (p. 309).

7. Ibid., p. 312.

8. Ibid., p. 310, defines satire as "a structural principle or attitude, what we have called a *mythos,*" which is "a combination of fantasy and morality," and "as the name of a form," which "confined to literature . . . can be either entirely fantastic or entirely moral."

9. Ibid., p. 311. Frye identifies Rabelais as an author who exemplifies this tendency.

10. Wright, *Middle-Class Culture in Elizabethan England,* p. 551.

11. Frye, *Anatomy of Criticism,* pp. 310–11.

12. Jeremy Shakerley, *The Anatomy of Urania Practica; or, A Short Mathematicall Discourse; Laying Open the Errors and Impertinencies Delivered in a Treatise Lately Published by Mr. Vincent Wing, and Mr. William*

Leybourne, under the Title of "Urania Practica" (London. Thomas Brudenell, 1649). Wing and Leyborne (Leybourn) responded with the publication of *Ens Fictum Shakerlœi; or, The Annihilation of Mr. Jeremie Shakerley, his In-artificial Anatomy of Urania Practica* (London: Robert Leybourn, 1649), wherein, it is promised, Shakerley's "falacies or ignorance, are demonstratively detected, his malice in its groundlesse colours display'd," as in an anatomy.

13. John Norden (*c.* 1547–1625), *A Store-Hovse of Varieties, Briefly Discoursing the Change and Alteration of Things in this World* (London: [n. pub.], 1601). This work is "the first part" of *Vicissitudo rerum*, which was first published in 1600, and of which the second part was not written. According to the description appended to the microfilm copy, *Early English Books, 1475–1640*, reel 1283, the work is taken point by point from Book I of Robert Ashley's *Of the Interchangeable Course and Varietie of Things in the Whole World* (London, 1594), translated from Le Roy, and is "largely similar to Ashley's work in expression of thought, identity of allusions, etc."

14. Burton divides and subdivides his subject according to "partitions," "sections," "members," and "subsections."

15. Foucault, *The Order of Things*, pp. 38–40 (p. 40).

16. Ibid., p. 40.

17. Kocher, *Science and Religion in Elizabethan England*, p. 21.

18. On demonstration in the development of method in Renaissance medical schools and its relationship to other fields of Renaissance scholarship, see Neal W. Gilbert, *Renaissance Concepts of Method* (New York: Columbia University Press, 1960), especially pp. 98–102. Gilbert argues that, "at the start of the fifteenth century, although methodology of a fairly complex sort was being discussed by . . . medical theoreticians of science [particularly in northern Italy, and especially at Bologna and Padua, in relation to Aristotle, Galen, the Arab Haly Abbas, and Averroes, "whose views they combined in various ways to produce theories of demonstration or of teaching"], *methodus* had not yet become a technical term." Direct translations from Galen in the late fifteenth and sixteenth centuries changed this (p. 101).

19. Richard Ward, *The Anatomy of Warre; or, Warre with the Wofull, Fruits, and Effects thereof, Laid out to the Life: Wherein from Scripture, and Experience, These Things are Clearly Handled; . . .* (London: John Dalham and Rich[ard] Lownds, [1642])

20. Pierre Du Moulin, *The Anatomy of Arminianisme; or, The Opening of the Controversies of These Times (Formerly Handled in the Low-Countries) Concerning the Doctrine of Prouidence, of Predestination, of the Death of Christ, of Nature and Grace, . . .* (London, 1620)

21. Henry Denne, *The Levellers Designe Discovered; or, The Anatomie of the Late Unhappie Mutinie, Presented unto the Souldiery of the Army Vnder the Command of His Excellency the Lord Fairfax, for Prevention of the Like in Others* (London, 1649).

22. Fletcher, *The Way to Blessednes*, "The Preface to every Christian Reader."

23. Ephraim Huit, *The Anatomy of Conscience; or, The Svmme of Paul's Regeneracy; Wherein are Handled the Places of Conscience, Worship, and Scandall; With Divers Rules of Christian Practise; Very Profitable for the Weake Christian* (London, William Sheffard 1626).

24. Barnhart, ed., *Dictionary of Etymology*, p. 32, col. 2, traces anatomy, which is extant from before 1398, through the Late Latin *anatomia* to the Greek etymon *anatomé* (dissection), which is related to *anatémnein* (to cut up, dissect). After 1398 it was used to refer to the knowledge or "science" of bodily structure, as discovered by dissection. *OED*, I, "A," p. 310, col. 2. *OED*, I, "A," p. 310, col. 1, gives as its principal definition of anatomy, from 1541: "The process, subjects, and products of dissection." It defines "anatomist," in the first instance, from 1594, as "[o]ne who practices, or is skilled in, the art of dissecting bodies"; and in the second instance, from 1587, as: "A dissector of anything . . ." (p. 309, col. 3). From 1541, to anatomize primarily meant: "To dissect or cut up," especially but not exclusively, to dissect "a human body, or an animal . . ." (p. 309, col. 3).

25. Eriksson, *Andreas Vesalius' First Public Anatomy at Bologna 1540*, p. 47; Paré, *Works* (1678), Bk. III, p. 51. In the opening lecture of the Bologna series, Corti explains that anatomy proper is the explanation of the internal, rather than the external parts, and "[a]ccordingly, anatomy is the same as dissection"; furthermore, "anatomy embraces the art of dissection, both performed actually and by description." For the definition of anatomy as dissection in regard of expository and fictional anatomy, see Frye, *Anatomy of Criticism*, p. 311. This meaning of anatomy is also provided by *OED*, I, "A," p. 310, col. 2: "The dissection or dividing of anything material or immaterial, for the purpose of examining its parts."

26. Sawday, *The Body Emblazoned*, p. 3.

27. John Lyly, *Euphues: The Anatomy of Wit; Verie Pleasant for All Gentlemen to Reade, and Most Necessarie to Remember; Wherein are Contained the Delights that Wit Followeth in his Youth, by the Pleasantnesse of Loue; and the Happinesse He Reapeth in Age, by the Perfectnesse of Wisedome* (London: G. Eld, 1617). Lyly's *Euphues*

was first published as Part 1 of his *Euphues and his England; Containing his Voyage and Adventures* . . . (London, 1580–81), and was first published independently in 1581.

28. *The Works of Aristotle,* ed. by J. A. Smith and W. D. Ross, I: *Categoriae,* trans. by E. M. Edghill (1928; repr. 1963).

29. Gilbert, *Renaissance Concepts of Method,* p. 110. Method in this case is as defined by the English Ramist, George Downham (d. 1634), as applying "not to the finding of arts nor to the arts themselves, but to their 'common form, by which the precepts are disposed.'" George Downham, *Commentarii in P. Rami regiiprofessoris dialecticam* . . . (Frankfurt, 1605), p. 752, cited in Gilbert, *Renaissance Concepts of Method,* p. 65.

30. *The Body Emblazoned* is reviewed by Jenner, "Body, Image, Text in Early Modern Europe," p. 144, who argues that Sawday's work "is *not* concerned with 'all forms of cultural endeavour,' but with 'high' literature, philosophy, and painting. . . . Sawday does not discuss ballads, almanacs, proverbs, court depositions, or any traces of oral tradition such as folklore, let alone the myriad practices which come within anthropological definitions of culture." This, at least, is a valid consideration for the application of the appellation in question.

31. Paré, *Works* (1678), bk. III, p. 51.

32. See Corti's distinction between anatomy and the art of dissection in the opening lecture of the series he presented at Bologna in 1540. Eriksson, *Andreas Vesalius' First Public Anatomy at Bologna 1540,* p. 47. Corti's comments on anatomy *as* dissection (see note 25, above) should be read in the context of this distinction; for an explanation of which see chapter 1. For the definition of anatomy as analysis in regard to expository and fictional anatomy, see Frye, *Anatomy of Criticism,* p. 311; Sidney, *The Defence of Poesy,* p. 110. *The Norton Anthology to English Literature* , I, 509, note 8, glosses Sidney's use of "anatomies" as "analyses." *OED,* I, "A," p. 310, col. 2, provides the following as its tenth definition of anatomy, as extant from 1569: "The dissection or dividing of anything material or immaterial, for the purpose of examining its parts; detailed examination, analysis." From 1553–87, to anatomize is also (p. 310, col. 1): "To lay open minutely; to analyse." But foremost among definitions of anatomy, extant from 1541, is this (p. 310, col. 1): "The artificial separation of the different parts of the human body or animal (or more generally of any organized body), in order to discover their position, structure, and economy." Thus dissection or division is also a recuperation of each part to its situation and function in relation to the whole.

33. *OED,* I, "A," p. 310, col. 2, adds to the meaning provided above in note 32: "detailed examination, analysis."

34. Katherine Rowe, "'God's handy worke': Divine Complicity and the Anatomist's Touch," in *The Body in Parts,* ed. by Hillman and Mazzio, pp. 284–309 (pp. 290–96).

35. Rowe, "'God's handy worke'" pp. 296, 291, 292. Rowe notes that in Crooke, this "orderly" procedure of dissection is to be distinguished from the "accidental investigations of warfare," in that it preserves the body's integrity.

36. Ibid., p. 288.

37. Ibid., pp. 303, 288.

38. Giles and Phineas Fletcher, *Poetical Works,* II, 4.

39. Lyly, *Euphues: The Anatomy of Wit,* sigs. A2r–A2v. In this passage, Lyly contrasts the veins of the heart with the muscles of the heel, for reason of the former's nobility (on account of function and proximity to the noble organ, the heart) and the latter's comparative subordination in the hierarchy of organs and members. In "Footnotes," Stallybrass argues that in early modern organic representations of the body politic, the poor, in being poor, have no feet; they are feet. However, implicit the analogy of the body politic and the metaphors it produces is a notion of interdependence: "Without the great toe, the body limps, the body politic totters" (p. 316). The final vignette in Stallybrass's series of "footnotes" or metonymic ruminations on the semiotics of feet, asks, why of all craftsmen and professionals were shoemakers most likely to be politically and religiously radical? His answer acknowledges individual agency within the determining logic of the body politic.

40. Wright, *Middle-Class Culture in Elizabethan England,* pp. 555–56.

41. Frances A. Yates, *The French Academies of the Sixteenth Century,* Studies of the Warburg Institute, 15 (London: Warburg Institute, University of London, 1947), pp. 2, 95–104.

42. Wright, *Middle-Class Culture in Elizabethan England,* p. 556. Wright (pp. 554–55) argues: "Moralistic purposefulness often garbed the popular literature of natural science. The allegorical interpretations and the moral lessons of the medieval bestiaries and lapidaries continually reappeared in Elizabethan treatises and accentuated their vogue with a public always ready to see in the simplest phenomena the manifestations of the supernatural. Compilers of encyclopedias assured their readers that a study of the physical world and of man in his mental and physical aspects revealed profitable lessons of God's will. Thus what we should today call psychology, physiology, and natural history found a pious excuse, in addition to the newer appeal of an aroused curiosity."

43. Wright, *Middle-Class Culture in Elizabethan England,* p. 557.
44. Sawday, "Bodies by Art Fashioned," pp. 117, 116.
45. See Frye, *Anatomy of Criticism,* p. 315.
46. Charles D. Murphy, "John Davies's Versification of Sidney's Prose," *Philological Quarterly* 21 (1942): 410–14 (p. 411). Sir Philip Sidney and Arthur Golding, trans., *A Woorke Concerning the Trewnesse of the Christian Religion* (London, 1587).
47. Murphy, "John Davies's Versification of Sidney's Prose," p. 410; "The passages versified by Davies are largely from Sidney's portion of the translation, the early chapters on God" (p. 411).
48. Henoch Clapham (fl. 1585–1614), *Ælohim-Triune, Displayed by His Workes Physicall and Meta-Physicall, in a Poeme of Diuerse Forme, Adapted to the Hebrue Text, the Frame of Diuinitie, and Catholike Exposition; Togither With Necessarie Marginall Notes For Relieuing of the Young Student; the First Part* (London: Iohn Harrison, 1601), sig. B^r.
49. Clapham, *Ælohim-Triune,* sig. B^r.
50. For the prevalence of biblical texts as the threshold texts of allegories, see Quilligan, *The Language of Allegory,* pp. 51–64.
51. Wright, *Middle-Class Culture in Elizabethan England,* pp. 556–58. Joshua Sylvester's translation of Du Bartas's encyclopedic religious poem provided Puritan and other readers for the next half century with information about their universe. "So popular was it that new editions continued to appear throughout the first half of the seventeenth century" (p. 558). Other than the 1611 edition, to which references in the text are, unless otherwise stated, I have consulted the following edition: *Du Bartas His Divine Weekes and Workes With A Compleate Collecti[on] Of All the Other Delight-full Workes,* trans. and written by that famous *Philomusus* Iosvah Sylvester (London: Robert Young, 1641). References to this edition also occur hereafter in the text, prefixed by the initials of the abbreviated title, and the date of publication in parentheses, thus: *D.W.* (1641). Also see Upham, *The French Influence in English Literature,* Ch. 4 (p. 148): Du Bartas's "pages were often filled with lavish displays of encyclopedic knowledge, frequently arranged in the popular medieval catalogue form."
52. Upham, *The French Influence in English Literature,* p. 158.
53. Sawday, "Bodies by Art Fashioned," p. 130.
54. John Davies of Hereford, *Mirum in modum* and *Microcosmos: The Discovery of the Little World, With the Government Thereof* (Oxford, 1603). Further editions of *Microcosmos* were published in 1605 and 1611. Like *Mirum in modum, Microcosmos* is also in *The Complete Works of John Davies of Hereford,* I.
55. Langdale, *Phineas Fletcher,* pp. 92–93. See chapter 4, note 76, of this study, for the circumstances surrounding the publication of *The Purple Island.*
56. Wright, *Middle-Class Culture in Elizabethan England,* p. 561.
57. *Mirum in modum,* in *The Complete Works,* ed. by Grosart, I, p. 6, col. 2, l. 7.
58. Ibid., p. 11, col. 2, l. 6 (from bottom).
59. *OED,* II, "C," p. 517, col. 2
60. *Microcosmos,* in *The Complete Works,* ed. by Grosart, I, p. 30, col. 2, ll. 15–16. (All further references are to this edition and the reference system used here and hereafter is that employed by the "Notes and Illustrations" of the cited text.)
61. *Microcosmos,* p. 31, col. 2, ll. 20, 22, 19.
62. Norden, *A Store-Hovse of Varieties,* stanza 47 (sig. C_1^v), stanza 48 (sig. C_2^r).
63. Ibid., stanza 50 (sig. C_2^r).
64. Ibid., stanza 47 (sig. C_1^v).
65. Ibid., stanza 93 (sig. D_3^v), stanza 51 (sig. C_2^r).
66. Ibid., stanza 48 (sig. C_2^r); also see sig. A_2^r, stanza 45 (sig. C_1^v).
67. John Hagthorpe (bap. 1585, d. in or after 1630?), *Visiones Rervm: The Visions of Things; or, Foure Poems, Wherein the Authour Expresseth His Invention By Way of Dreame* (London: Bernard Alsop, 1623), "*Principium & Mutabilitas Rerum;* or, The Beginning and Mutabilite of All Things," stanza 52, ll. 1, 2. *Visiones Rervm* is divided into "Foure Poems," thematically organized under the following titles: "1. *Principium & Mutabilitas Rerum;* or, The Beginning and Mutabilitie of All Things"; "2. *Cursus & Ordo rerum;* or, Art and Nature"; 3. "*Opineo & Ratio rerum;* or, Wealth and Pouertie"; 4. "*Malum & [S]inis rerum;* or, Sinne and Vertue, Concluding With the Last Iudgement and End of All Things." All further references to the poetic stanzas or notes of *Visiones Rervm* are included in the text, prefixed by the initials of the short title: *V.R.* References are given in the following order: poem number, stanza number, line number(s), thus, for example: II.3.4. References to the notation are given as follows: poem number, stanza number, "note" and notational symbol or number.

68. John Hagthorpe, *Divine Meditations, and Elegies* (London: Bernard Alsop, 1622).

69. Ibid., p. 84.

70. *Microcosmos*, p. 31, col. 2, l. 1 (from bottom)–p. 32, col. 1, l. 2.

71. Hagthorpe, *Divine Meditations, and Elegies*, p. 82.

72. Ibid.

73. For the relationship in Platonic and Galenic theories between the three faculties of the soul, their bodily residence in the principal members of the three venters of the body, and the three spirits that arise from these members, see chapter 2.

74. Hagthorpe, *Divine Meditations, and Elegies*, p. 82.

75. Ibid.

76. Ibid., p. 84.

77. *Mirum in modum*, p. 22, col. 2, ll. 18–17 (from bottom).

78. Norden, *A Store-Hovse of Varieties*, stanza 67 (sig. C_4ʳ).

79. *Microcosmos*, p. 85, col. 2, ll. 8–11.

80. Sawday, "Bodies By Art Fashioned," pp. 122–23.

81. Hagthorpe, *Divine Meditations, and Elegies*, p. 83.

82. Ibid., p. 83.

83. Ibid., pp. 83–84.

84. Sawday, "Bodies by Art Fashioned," p. 144.

85. See Upham, *The French Influence in English Literature*, p. 187.

86. *Mirum in modum*, p. 6, col. 2, l. 1 (from bottom)–p. 7, col. 1, l. 9.

87. "Ro. Vn.," *A New Anatomie, Wherein the Body of Man is Compared: 1 To a Household, 2 To a Cittie, With Divers Approved Medicines, Wittie, and Pleasant To Be Read.* (1605); another issue with a new title: *The Little World; or, A Lively Description of All the Partes and Properties of Man; For Invention Wittie, For Iudgement Learned, and For Practise Necessarie* (London: W. Iones, 1612), pp. 1, 2. All references are to the 1612 issue. Sawday, *The Body Emblazoned*, p. 295, note 3, says that "'Ro. Un.' has been identified as Robert Underwood." An intradiegetic narrator is a narrator who is both above the story, like an extradiegetic narrator, and also a diegetic character in the first or diegetic level of the narrative told by the extradiegetic narrator (see Chapter 1, note 9). The intradiegetic (or second-degree) narrator is therefore a character in the story he or she narrates. See Rimmon-Kenan, *Narrative Fiction*, p. 94.

88. "Ro. Vn.," *The Little World*, p. 3.

89. Ibid., p. 1.

90. Ibid., pp. 24, 42.

91. Ibid., pp. 26–27.

92. Ibid., pp. 9, 27, 43.

93. Ibid., pp. 33, 43.

94. Ibid., p. 44.

95. Ibid., p. 37.

96. Ibid., p. 42.

97. Hagthorpe, *Visiones Rerum*, "Dedication to Prince Charles," sig. A_2ʳ: "the Sunne (most excellent Prince) the Treasurer of Light and Time"; "the Prince, who resembles the Sunne in his spheare."

98. "Ro. Vn.," *The Little World*, p. 39.

99. Richard Bernard (bap. 1568, d. 1641), *The Isle of Man; or, The Legall Proceeding in Man-Shire Against Sinne; Wherein, By Way of a Continued Allegorie, the Chiefe Malefactors Disturbing Both Church and Common-Wealth, Are Detected and Attached; With their Arraignment, and Iudiciall Triall, According to the Lawes of England; the Spiritual Use Thereof, With an Apologie for the Manner of Handling, Most Necessary To Be First Read, For Direction in the Right Use of the Allegory Thorowout, Is Added in the End*, 5th edn, much enlarged (London: Edward Blackmore, 1628), sig. A_3ʳ. The division of "*Soules-towne*," however, bears little resemblance either to professional anatomy or to the literary anatomies of Fletcher or Davies of Hereford in terms of what it reveals. This town is divided into and by "Sense-street," "Thought-street," "Word-street," and "Deed-Street" (pp. 10–11).

100. *Microcosmos*, p. 27, col. 2, ll. 27–28.

101. Ibid., ll. 30–32.

102. Ibid., p. 28, col. 1, ll. 24–25, 33, 28.

103. Ibid., p. 29, col. 1, l. 24; p. 28, col. 1, l. 13 (from bottom); p. 28, col. 1, ll. 6–5 (from bottom); p. 28, col. 2, l. 3; p. 28, col. 1, ll. 2–1 (from bottom).

104. Ibid., p. 28, col. 1, note 3.

105. Norden, *A Store-Hovse of Varieties,* stanza 95 (sig. D_3^v).

106. "Ro. Vn.," *The Little World,* p. 42.

107. Norden, *A Store-Hovse of Varieties,* stanza 95 (sig. D_3^v), describes among the "parts" of the body politic: "The *king,* the *subject,* and the magistrate, / *Noble* and *base, rich, poore.*"

108. "Ro. Vn.," *The Little World,* pp. 3, 13.

109. Ibid., pp. 13, 14.

110. The identification of Alma with "the rational soul or Mind . . . which controls the body" is made by Florio in 1611 (*F.Q.* 2.9.18. note to l. 1).

111. The order of these expositions and disquisitions in *Microcosmos* may be ascertained by the following references: three kinds of appetite, p. 23, col. 2, l. 8 (from bottom)–p. 27, col. 2, l. 9; heart and thorax, p. 27, col. 2, l. 10–p. 29, col. 2, l. 7; pleasures and affections, p. 29, col. 2, l. 8–p. 83, col. 1, l. 6 (from bottom), including the digression on policy, p. 45, col. 1, l. 10–p. 63, col. 1, l. 20; soul's composition, location, origin, and end, p. 83, col. 2, l. 14–p. 88, col. 2, l. 1 (from bottom).

112. *Mirum in Modum,* p. 6, col. 1, ll. 7–6 (from bottom). The order of these divisions, dissections and explorations in *Mirum in Modum* may be ascertained by the following references: division of faculties of soul, p. 6, col. 1, l. 7 (from bottom)–p. 6, col. 2, l. 9 (from bottom); p. 7, col. 2, l. 24; skull and brain, p. 6, col. 2, l. 8 (from bottom)–p. 7, col. 2, l. 20; subdivisions of animal faculty of soul, p. 7, col. 2, l. 21–p. 30, col. 2, l. 1 (from bottom).

113. Sawday, "Bodies By Art Fashioned," p. 119.

114. *Microcosmos,* p. 45, col. 1, l. 10–p. 63, col. 1, l. 20.

115. According to *OED,* I, "A," p. 310, col. 2, the application of "anatomy" to the act of "dividing of anything material or immaterial," for the purpose of detailed examination or moral analysis was extant from the late sixteenth century.

116. *Microcosmos,* p. 61, col. 2, ll. 1–2.

117. Sawday, "Bodies By Art Fashioned," p. 120.

118. Ibid., p. 121.

119. *Microcosmos,* p. 28, col. 2, ll. 1–2.

120. Ibid., p. 27, col. 2, ll. 17 (from bottom), 13 (from bottom); p. 28, col. 1, l. 22, 2–3.

121. Ibid., p. 28, col. 2, ll. 10–18.

122. Ibid., col. 1, ll. 7, 3–5.

123. An ambiguous description in *Microcosmos* (p. 29, col. 1, ll. 18–12 [from bottom]) may signify the mitral valve between the left auricle and ventricle and the tricuspid valve between the right auricle and ventricle.

124. Ibid., ll. 5–13.

125. *Microcosmos,* p. 28, col. 1, ll. 24–41. The physiology of the heart and lungs in *Microcosmos* is precirculatory in both the senses of being prior in character to the pulmonary or lesser circulation, proposed originally by Michael Servetus and by Realdo Colombo, although independently of each other, and prior in character and conception to William Harvey's theory of systemic circulation.

126. Ibid., p. 29, col. 1, ll. 20–19 (from bottom).

127. "Ro. Vn.," *The Little World,* pp. 1, 2.

128. Ibid., p. 2.

129. Pace is a narratological concept which describes the relationship between story time and text time. Story time is conceived of as the linear succession of events conventionally and conveniently arranged in ideal chronological order. Strictly, the text has no temporality but rather spatiality, though it acquires temporality metonymically through the process of reading. Complete correspondence between the linearity of text time and the unilinearity (one story line) or multilinearity (more than one story line) of story time is very rare. This is either because of discrepancies between the order of events in the story and the order of events as narrated in the text, because the number of times an event appears in the story differs from the number of times it is narrated or mentioned in the text, or because of a lack of correspondence between the duration of the story and the duration of the text. Unlike with order and frequency, it is very difficult to describe in parallel the duration of the text and that of the story, because there is no way of measuring text duration (time of reading varies). But a different kind of correspondence is suggested between duration in story and the length of text devoted to it (a temporal-spatial correspondence). If the norm is constant pace, two modifications are possible: acceleration, which involves ellipsis or omission, and deceleration, which may involve a descriptive pause, such as we find in *The Little World* when the narrator embarks upon his more detailed anatomical descriptions. See Rimmon-Kenan, *Narrative Fiction,* pp. 16–17, 44–58.

130. Rimmon-Kenan, *Narrative Fiction*, p. 56, makes the claim that usually deceleration in the text corresponds to significance in the story.

131. "Ro. Vn.," *The Little World*, pp. 8, 9.

132. Ibid., p. 9.

133. *Mirum in modum*, p. 7, col. 1, ll. 20–18 (from bottom); p. 7, col. 1, l. 33.

134. *Microcosmos*, p. 31, col. 2, ll. 33, 35–36.

135. Sawday, "Bodies by Art Fashioned," p. 120.

136. Sawday, *The Body Emblazoned*, p. 88. On the skepticism of Sir John Davies's poetry of the body, see pp. 88–89. On that of John Donne's poetry, see, for example, John Carey, *John Donne: Life, Mind and Art*, new edn (London: Faber and Faber, 1990), pp. 217–46, notes pp. 292–94.

137. Sawday, "Bodies by Art Fashioned," p. 120–21.

138. See, for example, H. E. G. Rope, "John Davies of Hereford, Catholic and Rhymer," *Anglo-Welsh Review* 11 (1961): 20–36 (p. 34); Upham, *The French Influence in English Literature*, pp. 187–93.

139. Ruth L. Anderson, "A French Source for John Davies of Hereford's System of Psychology," *Philological Quarterly* 6 (1927): 57–66 (p. 57). It should be added that La Primaudaye often quotes Du Bartas. See Yates, *The French Academies of the Sixteenth Century*, p. 131, note 1; p. 134, note 5. Anderson concludes: "The connection between *Mirum in modum* and the *Microcosmos*, on the one hand, and the translation of *The French Academie*, on the other, is so immediate in arrangement, substance, and language that we may be sure that we have found in La Primaudaye's work an actual source for the two poems of John Davies of Hereford" (p. 66). On Davies of Hereford's indebtedness to other French encyclopedic works, see Charles D. Murphy, "John Davies's Versification of Sidney's Prose," *Philological Quarterly* 21 (1942): 410–14.

140. Sir Francis Bacon, *The Advancement of Learning*, ed. by G. W. Kitchin, introduction by Arthur Johnston (London: J. M. Dent & Sons, 1973), p. 113. See Sawday, "Bodies by Art Fashioned," p. 121.

141. Sawday, "Bodies By Art Fashioned," p. 121.

142. "Ro. Vn.," *The Little World*, p. 5.

143. Ibid., p. 6.

144. Ibid., p. 8.

145. Ibid., pp. 13–14.

146. Ibid., p. 24.

147. *Microcosmos*, p. 83, col. 1, ll. 2–1 (from bottom).

148. Ibid., p. 83, col. 2, ll. 5–9.

149. Sawday, "Bodies By Art Fashioned," p. 122.

150. Ibid., p. 121.

151. French, *Dissection and Vivisection in the European Renaissance*, p. 219.

152. *Mirum in modum*, p. 6, col. 1, l. 6 (from bottom), l. 4 (from bottom), l. 2 (from bottom); p. 6, col. 2, l. 5; p. 13, col. 1, l. 18 (from bottom). For the location of the animal faculty and "*Reas'n*" in the "*Braine*" in the "*Scull*," and the vital faculty and "*Ire*" in the "*Hart*," see *Mirum in modum*, p. 6, col. 2, ll. 4–8. p. 13, col. 1, ll. 23 (from bottom), l. 18 (from bottom). For Paré's division of the faculties and functions of the animal faculty of the soul, see chapter 1, notes 314 and 315. Anderson, "A French Source for John Davies of Hereford's System of Psychology," p. 63, argues, after a more thorough examination of the ideas and language of *Mirum in modum*, that the "identical classification and description of the powers arising from the animal virtue" she has not found elsewhere, except in La Primaudaye's *The French Academie*.

153. For Paré's division of the faculties and functions of the natural soul, see chapter 1, notes 314 and 315.

154. *Mirum in modum*, p. 6, col. 1, l. 6 (from bottom); p. 13, col. 1, l. 17 (from bottom); p. 6, col. 2, l. 10,8.

155. Ibid., p. 7, col. 2, ll. 6–7, 1.

156. *Microcosmos*, p. 83, col. 2, ll. 1–4.

157. *OED*, I, "B," p. 637, col. 1, provides this meaning as extant from 1440. The continuity of this meaning is apparent in R. E. Allen, ed., *The Concise Oxford Dictionary of Current English*, first ed. by H. W. Fowler and F. G. Fowler, 8th edn. (Oxford: Clarendon Press, 1990), p. 93, col. 1.

158. *Mirum in modum*, p. 15, col. 1, l. 11 (from bottom).

159. Sawday, "Bodies By Art Fashioned," p. 122.

160. Hankinson, *The Sceptics*, p. 24; Popkin, *The History of Scepticism from Erasmus to Spinoza*, pp. xiv, 3.

161. For Aristotle, form is separate from body; form moves, the body is moved; form is the efficient and final cause, the final aim or end of the body, which is disclosed in movement. This motive or efficient force is called "Entelechy." Costelloe and Muirhead, *Aristotle and the Earlier Peripatetics*, II, pp. 1–2, 6–9.

162. Harcourt, "Andreas Vesalius and the Anatomy of Antique Sculpture," 46; French, *William Harvey's Natural Philosophy,* p. 25.

163. De Laguna, *Anatomica methodus,* p. 265.

164. For example, Achillini, *Annotationes anatomicae,* p. 49: "Avicenna, however, said in his anatomy that two veins come from the left emulgent vein, and twice I saw them thus in 1502 and 1503. Galen is doubtful about the matter in his *On the Use of the Parts,* XIV, 7; his solution is that the right-hand vessel is warmer and hence carries warmer material."

165. Lind, *Studies in Pre-Vesalian Anatomy,* p. 78. Lind is citing *Anatomice,* III.

166. French, *William Harvey's Natural Philosophy,* p. 19.

167. Ibid., p. 20. Also see French, *Dissection and Vivisection in the European Renaissance,* pp. 91–102 (p. 102).

168. Niccolò Massa, *Introductorius anatomiae siue dissectionis corporis humani* (Venice, 1536), cited in Lind, *Studies in Pre-Vesalian Anatomy,* p. 11. French, *William Harvey's Natural Philosophy,* p. 23, observes: "Massa complained that the other anatomists of the time (he must mean primarily Zerbi and Berengario) were too much concerned with authorities and too little with the actual processes of dissection and observation; this is why his own text is abundantly practical."

169. French, *William Harvey's Natural Philosophy,* pp. 24, 23, 25. French argues: "Where Berengario had angled for some link between 'physical logic' and a more formal 'demonstration' to persuade his audience, Massa uses *sensata demonstratio* in a more confident appeal to their senses" (p. 25). Of the "traditional rote of observables in the dissected body," the anatomical *accessus,* "Massa omits 'complexion' because, although fundamental in the natural philosophy of the body, it cannot be seen by the senses" (p. 24).

170. In the ninth demonstration of this series which Vesalius undertook to accompany Corti's lectures, Vesalius is reported to have said to his audience: "There is no truth in what others say about the five lobes of the liver. They are only flaps that cover the stomach, *as you see.*" Eriksson, *Andreas Vesalius' First Public Anatomy at Bologna 1540,* p. 165 (emphases mine).

171. Singer, *Vesalius on the Human* Brain, p. 35. Choulant, *History and Bibliography of Anatomic Illustration,* pp. 174–75, note 1, comments upon this "sudden leap in Vesalius' power of seeing things correctly."

172. French, *Dissection and Vivisection in the European Renaissance,* p. 169.

173. The osteological and myological figures of Vesalius's *Fabrica* are the best known of all its illustrations. The anatomy they present was of the utmost priority to Vesalius. In the Preface to his *Fabrica,* Vesalius writes that the separation of the art of surgery, of "the hand," from that of the physician arose from a preoccupation in medicine with internal complaints. This, he says, and the entrusting of the use of the hands entirely to the barbers, gave rise to a knowledge among physicians only of the viscera and a neglect of the structure of the bones, muscles, nerves, and veins and arteries. However, the physicians' neglect of dissection ensured that "true knowledge of the viscera [was also] lost to the physicians." And finally, the practice of dissection "soon died away," because the physicians "did not undertake it," and "those to whom the manual skills had been entrusted were so unlearned that they did not understand the writings of the professors of dissection." O'Malley, *Andreas Vesalius of Brussels 1514–1564,* p. 319. The osteological series of figures accompanies Book II of the *Fabrica,* while those figures demonstrating the muscles illustrate Book III. In the *Fabrica,* fol. 3^{v}, Vesalius says: "The books [of the *Fabrica*] present in sufficient detail the number, site, shape, size, substance, connection to other parts, use and function of each part of the human body." Cited by Harcourt, "Andreas Vesalius and the Anatomy of Antique Sculpture," p. 38.

174. Harcourt, "Andreas Vesalius and the Anatomy of Antique Sculpture," p. 46.

175. Ibid., pp. 45–46.

176. This analysis of the relationship between structure and function in the myological figures of the *Fabrica* relates them theoretically to Michelangelo's definition of the upright form of Man, as described by David Summers in *Michelangelo and the Language of Art* (Princeton: Princeton University Press, 1981), p. 87, cited in Schultz, *Art and Anatomy in Renaissance Italy,* p. 84: "Thus the tradition that began with Leonardo's definition, at once hypothetical and tragic, that uprightness was a continual equilibrium of opposing movements, that life was at most a harmony of opposites, at least a struggle against gravity, is negated. For Michelangelo life no longer necessitates movement, but movement signifies life, the soul; ideal movement is identified with eternal essence, the forms of time thus defeating time."

177. A passage from the *Epitome* demonstrates this point: "A second muscle of those which move the scapula arises from the occiput and, following along the length of the neck to the spine of the eighth thoracic vertebra, originates from the apexes of the vertebral spines as well; it inserts on the spine of the scapula, the acromion, and a portion of the clavicle. All of that part of it lying in the neck pulls the scapula upward; that part below the neck

in the posterior portion of the thorax and which resembles a monk's hood, draws the scapula downward." *The Epitome of Andreas Vesalius*, p. 19. Lind comments on the *Epitome:* "While the outline of Vesalius' text corresponds to the 'systematic' anatomy, the chapters themselves are written as functional anatomy (or, to borrow a term noticed recently, 'dynamic morphology'). Vesalius' conception of muscles as contractile masses associated with a joint and acting across it in co-ordination is more in accord with our ideas of synergy than is the treatment of them as single structures. The reasoning which derives a muscle's action by considering it as an isolated contractile element tending to shorten the distance between its points of attachment is incorrect physiologically (for it is rare, except in pathologic states of major paralysis, for a muscle to act alone). . . . The modern student has difficulty in finding in any of these great reference texts [standard encyclopedias of anatomy] as clear a concept of integration as was available to the Renaissance student who owned a Vesalian *Epitome;* only with effort do we lead the student from the morass of isolated osteologic and myologic textual statement to the terra firma of synergy." *The Epitome of Andreas Vesalius*, pp. 31–32. Lind recognizes the model emergent in kinesiology in the nineteen thirties and forties, in which muscles are considered in functional groups rather than as separate units, and stress is laid on the muscular attachments and movements at joints, as "only a rediscovery of a principle derivable from Vesalius' text and stated more than two centuries earlier with conciseness and clarity by Mondino: '[W]e must first gain an idea of the whole, and then of the parts.'" But this assertion of a rediscovery elides one crucial dimension to the presentation of structure and function in Vesalius. The tendency therein to understand structure and function in the operational context of joints is, as Mondino's statement suggests, the desire to understand the purpose of the function of the parts, and to assimilate that into the purpose of the whole, which is to be the soul's instrument.

178. French, *William Harvey's Natural Philosophy*, p. 36. Also see pp. 37–38 for Jacobus Sylvius's piety in respect of Galen and Hippocrates, and his attack on Vesalius.

179. Ibid., pp. 36, 38.

180. Choulant, *History and Bibliography of Anatomic Illustration*, p. 200. Also see Roberts and Tomlinson, *The Fabric of the Body*, pp. 189–90. Singer, *The Evolution of Anatomy*, p. 135, claims that although the plates of Eustachio (d. 1574) "are less beautiful than those of Vesalius," and "present dead and not living Anatomy"; nevertheless, they are "more accurate, and they contain such a multitude of discoveries that for originalty Eustachius has only Leonardo and Vesalius as superiors among modern anatomists."

181. French, *William Harvey's Natural Philosophy*, p. 39.

182. O'Malley, *Andreas Vesalius of Brussels 1514–1564*, p. 318. According to Foster, *Lectures on the History of Physiology During the Sixteenth, Seventeenth and Eighteenth Centuries*, p. 7, Vesalius "tells us how he and a fellow student were wont to try their knowledge by a test which has been often used since, the recognition of the individual bones by touch alone, with the eyes shut."

183. Singer, *Vesalius on the Human Brain*, p. 4.

184. Park, "The Criminal and the Saintly Body," p. 16. Interestingly, both Berengario and Vesalius are most effusively in awe of the divine ingenuity of the creation of the human body when examining the head and brain. A likely explanation must lie in the principality accorded that organ by Galen.

185. Charles Taylor, *Sources of the Self: The Making of the Modern Identity* (Cambridge: Cambridge University Press, 1989), p. 213. Knight, *Natural Science Books in English 1600–1900*, p. 11, discusses Bacon's *Advancement of Learning* in this regard. The *Advancement of Learning* "first appeared in 1605; in 1623 it was translated into Latin, in an expanded form . . . and this was translated back into English in 1640. . . . [P]art of its object was to clear the ground from lumber, but one nevertheless finds in it Bacon's precepts, which appeared in more detail in his *Novum Organum.*" Knight goes on to describe how the *Advancement of Learning* discusses induction and warns of premature theorizing and the futility of logic-chopping purportedly in the manner of scholastic philosophy. We should remember that the anatomical *accessus* evolved out of several forms, including a literary or philosophical *accessus*, Aristotle's *Categories*, and the Aristotelian apparatus of fourfold causality or causality *accessus*.

186. Knight, *Natural Science Books in English 1600–1900*, p. 11. Devon L. Hodges, *Renaissance Fictions of Anatomy* (Amherst: University of Massachusetts Press, 1985), p. 104, draws attention to Bacon's claim in *Novum Organum*, that in order for him to make any progress through induction, the humiliation of the human spirit has been required.

187. Francis Bacon, *Novum Organum*, I, 73, in *Novum Organum With Other Parts of The Great Instauration*, trans. and ed. by Peter Urbach and John Gibson, Paul Carus Student Editions: 3 (Chicago: Open Court, 1994), p. 83.

188. Cunningham, *The Anatomical Renaissance*, p. 267.

189. Ibid., pp. 227, 218. Vesalius "would have been aware of, and perhaps even quite familiar with, Erasmian thinking about religious reform," because of his father's membership of the court of Charles V (Charles of Ghent,

1500–1566), Holy Roman Emporer, whose young courtiers were admirers of Erasmus of Rotterdam. The period of Vesalius's life from 1530, which he spent at the new humanist bilingual college at Louvain, and that which he spent at Paris, once a bastion of orthodoxy, but by 1533 possessing Jean Calvin in the faculty of Law and Michael Servetus in Medicine (pp. 223–25), may also be important in this relation. But his later employment as physician in the service of Charles V, who tried in the latter part of his life to unite his empire under Catholicism, suggests that Vesalius was not a Lutheran. Cunningham admits it quite possible that "he may have kept his anatomizing separate from his religious commitment, whatever it was" (pp. 224, 225), but his conclusion is that, irrespective of the inconclusive knowledge of Vesalius's religious commitment, "[a]natomy—indeed, Vesalius's anatomy—became a Lutheran Protestant endeavour, not in the medical schools but in the far larger arena of the philosophy course of universities, the basic course taken by all students at university" (p. 230).

190. While in Spain with Philip II, and unable to carry out anatomical investigations, Vesalius reportedly said: "I still live in hope that at some time or other, by some good fortune I may once more be able to study that true bible, as we count it, of the human body and of the nature of man." Foster, *Lectures on the History of Physiology During the Sixteenth, Seventeenth and Eighteenth Centuries,* p. 18, interprets Vesalius as saying that "[t]he true bible to read is nature itself, things as they are, not the printed pages of Galen or another; science comes by observation, not by authority." Perhaps because of his unyielding criticism of the role of the Church in the history of anatomy, Foster ignores the other implication of the metaphor of the body as a bible, that God is its author.

191. "Ro. Vn.," *The Little World,* p. 4, 41.

192. Ibid., pp. 1, 4–5.

193. Francis Clark, S.J., *Eucharistic Sacrifice and the Reformation,* 2nd edn. with a foreword by Cardinal John Heenan (Oxford: Basil Blackwell, 1967), p. 156. In saying this I am not unaware of the opinion of Rope, "John Davies of Hereford, Catholic and Rhymer," pp. 20, 36, that there is "no reasonable doubt that [Davies of Hereford] was a Catholic," "truly a faithful Catholic."

194. Hagthorpe, *Divine Meditations and Elegies,* p. 83.

195. Ibid., p. 83.

196. Veslingus, *The Anatomy of the Body of Man,* the address "To the Reader" by Nicholas Culpeper, sig. B_2.

197. *Mirum in modum,* p. 6, col. 1, l. 1 (from bottom)–col. 2, l. 1; *Microcosmos,* p. 23, col. 2, l. 8 (from bottom).

198. *Microcosmos,* p. 23, col. 2, ll. 8–5 (from bottom).

199. *Mirum in modum,* p. 5, col. 2, ll. 23–24; *Microcosmos,* p. 23, col. 1, note 3. For philosophical discussion of the senses, also see *Mirum in Modum,* p. 5, col. 2, l. 7–p. 6, col. 1, l. 10; p. 7, col. 2, ll. 24–5 (from bottom); p. 12, col. 1, ll. 19–22.

200. See Pagel, "Religious Motives in the Medical Biology of the XVIIth Century," pp. 121–24.

201. At some point prior to his death in 1618, Sir Walter Ralegh wrote an essay, entitled *Sceptick; or, Speculations,* which was first published in *Sir Walter Raleigh's Sceptick; or, Speculations, and Observations of the Magnificency and Opulency of Cities* (London: W. Bentley, 1651). After this edition, *Sceptick* was frequently reprinted as part of a volume entitled *Remains.* See Ernest A. Strathmann, *Sir Walter Ralegh: A Study in Elizabethan Skepticism* (New York: Columbia University Press, 1951), pp. 222–27.

202. Thomas Hobbes, *Leviathan; or, The Matter, Forme, and Power of a Common-Wealth Ecclesiasticall and Civill* (London: Andrew Crooke, 1651), p. 4, remarks on the view in "the Philosophy-Schools," and "grounded upon certain Texts of *Aristotle,*" that the cause of understanding lies in the following: "the thing Understood, sendeth forth an *intelligible species,* that is, an *intelligible being seen;* which coming into the Understanding, makes us Understand." It is not clear, though, that *The Purple Island* account of the conveyance of perception of the object to the understanding conforms to this scholastic model.

203. Bernard, *The Isle of Man,* pp. 51, 54.

204. Ibid., p. 55.

205. Ibid., p. 50.

206. Hankinson, *The Sceptics,* p. 25; Empiricus, *Selections from the Major Writings on Scepticism, Man, & God,* pp. 38–39 (italics mine).

207. See, for example, Hankinson, *The Sceptics,* pp. 17–18, 25; Charles Larmore, "Scepticism," in *The Cambridge History of Seventeenth-Century Philosophy,* 2 vols., ed. by David Garber and Michael Ayers, with the assistance of Roger Ariew and Alan Gabbey (Cambridge: Cambridge University Press, 1998), II, 1145–92 (p. 1148).

208. Ralegh, *Skeptick,* p. 12.

209. See Ibid.

210. By contrast, for Fulke Greville (1554–1628), the "faculties of apprehension," although "perfect" at "the soules creation," are now not merely distinct from the corruption of the world which they perceive, but share in

its "Debility, misprision, imperfection." Greville, *A Treatise of Human Learning,* stanza 18, cited by Dollimore, *Radical Tragedy* (1989), pp. 80–81.

211. *Mirum in modum,* p. 7, col. 1, ll. 11, 9; p. 24, col. 2, ll. 1–3.

212. French, *Dissection and Vivisection in the European Renaissance,* p. 223.

213. Ibid.

214. *Mirum in modum,* p. 24, col. 2, ll. 21, 24, 23, 25, 26.

215. Ibid., p. 12, col. 1, ll. 19–20.

216. *Microcosmos,* p. 23, col. 1, note 3; *Mirum in modum,* p. 5, col. 1, l. 31; *Microcosmos,* p. 87, col. 2, ll. 7–4 (from bottom).

217. *Mirum in modum,* p. 8, col. 2, ll. 18–19.

218. *Microcosmos,* p. 44, col. 2, ll. 4, 19–21.

219. "Ro. Vn.," *The Little World,* p. 41.

220. Ibid.

221. *Microcosmos,* p. 79, col. 2, ll. 24–26.

222. Ibid., ll. 27–29.

223. Butler, *Bodies that Matter,* p. 8.

224. *Microcosmos,* p. 83, col. 2, l. 13; *Mirum in modum,* p. 10, col. 1, l. 17 and note to this reference on leaf following p. 30; *Microcosmos,* p. 81, col. 2, l. 4 (from bottom); p. 82, col. 1, ll. 13, 12. Sidney, *The Defence of Poesy,* p. 126, also identifies the most ancient human learning as poetry.

225. *Microcosmos,* p. 82, col. 1, ll. 15–17.

226. Ibid., p. 81, col. 2, ll. 3 (from bottom), 1 (from bottom).

227. Donker and Muldrow, *Dictionary of Literary-Rhetorical Conventions of the English Renaissance,* p. 78.

228. *Microcosmos,* p. 64, col. 1, ll. 8–9.

229. Ibid., p. 64, col. 1, ll. 17–25.

230. Davies of Hereford utilizes the concept of poetry as a speaking picture in his work entitled *The Holy Roode; or, Christs Crosse; Containing Christ Crucified, Described in Speaking-Picture* (London: N. Butter, 1609). This work is collected in *The Complete Works,* ed. by Grosart, I.

231. The term "Unfolding" is used in *Microcosmos,* to describe the process whereby "the *Poets* from the *World's* first *Age*" acquired knowledge of "*Heav'n & Earth, Sun, Moone, & Starres.*" *Microcosmos,* p. 81, col. 2, l. 4 (from bottom); p. 82, col. 1, l. 15 (from bottom).

232. De Man, "Semiology and Rhetoric," pp. 127–28.

233. Donker and Muldrow, *Dictionary of Literary-Rhetorical Conventions of the English Renaissance,* pp. 78–79. According to Donker and Muldrow, Henry Peacham's *The Garden of Eloquence, Conteyning the Figures of Grammar and Rhetorick, From Whence May Bee Gathered All Manner of Flowers, Coulors, . . . Formes and Fashions of Speech* (London, 1593) claims that the principal instruments by which eloquence allures men's minds and affections are figures (p. 79).

234. De Man, "Semiology and Rhetoric," p. 28. After Charles Sander Pierce, de Man characterizes this process as "pure rhetoric, as distinguished from pure grammar, which postulates the possibility of unproblematic dyadic meaning, and pure logic, which postulates the possibility of the universal truth of meanings" (p. 128). Donker and Muldrow, *Dictionary of Literary-Rhetorical Conventions of the English Renaissance,* p. 79.

235. According to Donker and Muldrow, *Dictionary of Literary-Rhetorical Conventions of the English Renaissance,* p. 80, Peacham's *The Garden of Eloquence* categorizes *allegoria* (allegory) under tropes of sentences, rather than of words, as an extended metaphor or trope of words.

236. Donker and Muldrow, *Dictionary of Literary-Rhetorical Conventions of the English Renaissance,* p. 84.

237. See Dollimore, *Radical Tragedy* (1989), ch. 4: "Renaissance Literary Theory: Two Concepts of Mimesis," p. 71 *et passim.*

238. Donker and Muldrow, *Dictionary of Literary-Rhetorical Conventions of the English Renaissance,* p. 84.

239. *Mirum in modum,* p. 7, col. 2, l. 21. See also: *Microcosmos,* p. 29, col. 2, l. 8.

240. Drayton, *Poly-Olbion,* p. 422.

241. Sidney, *The Defence of Poesy,* p. 108.

242. Ibid., pp. 113–14.

243. Dollimore, *Radical Tragedy* (1989), pp. 76–77.

244. Eriksson, *Andreas Vesalius' First Public Anatomy at Bologna 1540,* p. 47.

245. For Paré, there is a clear subordination of textual anatomy to "the sight of the eye, and touching and handling each part with the hand, yet neverthelesse," he adds, "the labour of describing Anatomy is not unprofitable.

For by reading, such as have often exercised themselves in the dissecting of men's bodies may refresh and helpe their memories, and such as have not, may make plaine and easie the way to the understanding of dissections." *Workes* (1634), III, p. 83.

246. Achillini, *Annotationes Anatomicae* contains instructions of the following kind: "Separate the tendon of the transverse muscles from the siphac [peritoneum] and demonstrate each of them" (p. 44).

247. See Rebecca Ferguson, "'Intestine Wars': Body and Text in *An Epistle to Dr Arbuthnot* and *The Dunciad*," in *Pope: New Contexts*, ed. by David Fairer (London and New York: Harvester Wheatsheaf, 1990), pp. 137–52 (p. 139), where it is argued that in *An Essay on Criticism*, "literary 'Judgement' is attuned to 'Nature', a body of organic and spiritual perfection with nature as the *nous*, 'th'informing Soul', in/forming in the literal sense of giving it shape, 'feeding' from within without taking visible form itself."

248. Sidney, *The Defence of Poesy*, pp. 138–39 (p. 139). For Sidney, as for others, eloquence, unlike ingenuity, is style rather than meaning (pp. 143–44).

249. Nicholas Deeble, "To the Booke as it is dedicated unto his most excellent Majestie," a preliminary poem to *Microcosmos*, p. 8; *Mirum in modum*, p. 5, col. 1, l. 6.

250. Thomas Stanley's exposition of the Epicurean school of philosophy, in *The History of Philosophy*, 2nd ed. (London, 1687), p. 871, contains "a fascinating discussion in which the myriad potential combinations of the letters of the alphabet to form words (both comprehensible and incomprehensible) is compared with the sheer profusion of the processes of generation and corruption in bodies." In this discussion "atoms are selective in how they combine themselves to form compound beings and species. . . . Likewise, letters cannot be merely jumbled anyhow to produce words 'innumerable, inexpressible, and incomprehensible', but by selection and combination we form 'words, accommodated to Pronunciation and Reason', suitable to be spoken or written. Words (properly defined) are selectively bred, and this is what makes them legitimate." Ferguson, "'Intestine Wars'" pp. 137–38. Ferguson examines the metaphors of writing as a physical body and of literary creation as sexual generation in Stanley and in Pope, focusing on two key concerns among others: "the nature of literary engendering"; "creativity and organic growth" (p. 138).

251. Ricoeur, "Metaphor and the Main Problem of Hermeneutics," pp. 100–101.

252. Sidney, *The Defence of Poesy*, p. 114.

253. Eagleton, *Literary Theory*, pp. 86, 87.

254. Ibid., p. 87.

255. *Microcosmos*, p. 83, col. 2, ll. 13, 10, 12 & footnote 1.

256. Ferguson, "'Intestine Wars,'" p. 138. Ferguson is making a generalizable remark in specific relation to Thomas Stanley, *The History of Philosophy* (see note 250, above).

257. *Mirum in modum*, p. 5, col. 1, ll. 6–8.

258. George Gascoigne, *Certayne Notes of Instruction Concerning the Making of Verse or Ryme in English*, p. 465, advises that invention must have in it "*aliquid salis*" or "something salty," "[f]or it is not inough to roll in pleasant woordes." Gascoigne explains: "By this *aliquid salis*, I meane some good and fine devise, shewing the quicke capacitie of a writer: and where I say some *good and fine invention*, I meane that I would have it both fine and good." See Gerald Hammond, ed., *Elizabethan Poetry: Lyrical and Narrative*, A Casebook (London: Macmillan, 1984), pp. 21, 22, note 1, which reproduces Gascoigne's text in extract and provides a translation of his terms in the notes. In *An Essay on Criticism*, Pope's speaking voice is disquieted by the "Length of *Fame* (our *second* Life)" which is vastly reduced from that in the "Golden Age," "When *Patriarch-Wits* surviv'd a *thousand Years*" (ll. 480, 478, 479). However, the concern seems to arise in part from the impermanent nature of the English language, "a common complaint in Pope's time"; though this is not Davies of Hereford's concern. *The Poems of Alexander Pope*, a one-volume edition of the Twickenham text with selected annotations, ed. by John Butt (London: Routledge, 1963; repr. 1989), p. 159.

259. *OED*, X, "S," p. 457, col. 3: Soteriology may be used to mean the promotion and preservation of health, or the restoration and perfection of human nature. See the meanings of soteriology in note 225 in chapter 3. The term "*Corpes*," or "corps," is used regularly in both of Davies of Hereford's poems, to signify either the anatomized or elementally complex body (*Mirum in modum*, p. 22, col. 2, l. 17 [from bottom]; *Microcosmos*, p. 35, col. 2, l. 18 [from bottom]), or the body in hierarchical relation to the soul, e.g.: "For looke how much the *Minde* the *Corpes* excels" (*Microcosmos*, p. 75, col. 2, l. 22; see also *Mirum in modum*, p. 24, col. 2, l. 29).

260. Francis Quarles, *Hadassa; or, The History of Qveene Ester: With Meditations Thereupon, Diuine and Morall* (London: for Richard Moore, 1621), in *The Complete Works*, ed. by Grosart, p. 42.

261. Quarles, *Hadassa*, p. 42.

262. See Sidney, *The Defence of Poesy*, p. 114.

263. Miller, "Spenser's Poetics: The Poem's Two Bodies," p. 173.

264. Edmund Spenser, "An Hyme in Honour of Beautie," *Fowre Hymnes*, in *The Works of Edmund Spenser*, The Globe Edition, ed. from the original editions and manuscripts by R. Morris, with a memoir by John W. Hales (London: Macmillan, 1869; repr. 1929), p. 597, ll. 117, 132–33.

265. Miller, "Spenser's Poetics: The Poem's Two Bodies," pp. 173–74.

266. On the significance for allegory of the correspondence of human to divine speech, and thus the embodiment of God, see Quilligan, *The Language of Allegory*, pp. 156–61.

267. MacDonald, *England's Antiphon*, pp. 94–96 (p. 96).

268. Sidney, *The Defence of Poesy*, pp. 110–111 (p. 110).

269. "Ro. Vn.," *The Little World*, pp. 42, 41.

270. For difficulties experienced by ecclesiastical writers in trying to persuade people of this congruity, see Ariès, *The Hour of Our Death*, p. 32.

271. "Ro. Vn.," *The Little World*, p. 41.

272. The *Authorised Version* consistently uses the terms "natural body" and "spiritual body" (1 Cor. 15:44–47); whereas the *Revised Standard Version* uses "physical body" and "spiritual body," as does *The Revised English Bible With the Apocrypha. The New English Bible: New Testament*, Library Edition (Cambridge: Oxford University Press and Cambridge University Press, 1961) uses "animal body" and "spiritual body," the former of which could have caused confusion in an early modern reader familiar with the etymological derivation of "animal" from the Latin *anima* (soul), and the use of the adjective "animal" to describe faculties and properties of the rational soul.

273. Wibbing, "*Sōma*, body," p. 236.

274. Ibid.

275. Ibid.

276. Caroline Walker Bynum, *Fragmentation and Redemption: Essays on Gender and the Human Body in Medieval Religion* (New York: Zone Books, 1991), p. 228.

277. Ibid.

278. Costelloe and Muirhead, *Aristotle and the Earlier Peripatetics*, II, 131, 1–9.

279. Ibid., 120–25, 130–34 (p. 131). Also see pp. 95–96, 120, where it is argued that the *Nous* is "not born into existence like the other parts of the soul" at the generation of the body, and is not "affected by the death of the body."

280. Pomponazzi, *De immortalitate animae* (Bologna, 1516), cited in Michael and Michael, "Two Early Modern Concepts of Mind," p. 33. A number of prominent philosophers in the Renaissance, especially at Padua, questioned Aristotle's commitment to the personal immortality of the human soul, because although "the intellect operates without the use of any bodily organ," and so "can be said to 'participate' in the immortality of purely spiritual beings like God and angels," nevertheless it cannot exercise its powers of abstracting concepts without the body, and so "the human soul is dependent on the body and consequently mortal" (pp. 32, 36, 37).

281. Richard Crashaw, *Steps to the Temple: Sacred Poems* (London, 1646), in *The Poems English, Latin and Greek of Richard Crashaw*, ed. by L. C. Martin, 2nd ed. (Oxford: Clarendon Press, 1957), p. 86.

282. *The Pocket Oxford Dictionary of Current English*, Compiled by F. G. Fowler and H. W. Fowler, 4th edn., revised by H. G. Le Mesurier and E. McIntosh (Oxford: Clarendon Press, 1946; repr. 1947), p. 347, col. 2; *Mirum in modum*, p. 26, col. 1, ll. 25–27. Aesthetic and theological meanings of "grace" are first recorded in 1340 and about 1330, respectively. Barnhart, ed., *Dictionary of Etymology*, p. 444, col. 1.

283. *Microcosmos*, p. 37, col. 2, l. 21; p. 26, col. 2, ll. 13–12 (from bottom). For further references to intellectual pride as the cause of the fall, see *Mirum in modum*, p. 12, col. 1, ll. 3–10; p. 27, col. 2, l. 4 (from bottom);

284. MacDonald, *England's Antiphon*, p. 94.

285. Hagthorpe, *Divine Meditations, and Elegies*, p. 3.

286. Ibid., p. 3.

287. Ibid., p. 4.

288. For reasons also of the Fall, Pope's *An Essay on Criticism* finds that "wit," "the most interior and originary faculty of creation," is degenerate. Ferguson, "'Intestine Wars,'" p. 139.

289. See Quilligan, *The Language of Allegory*, p. 19, who argues that, historically, "the late rise of narrative allegory is due to the effect Christian theology had on notions of classical rhetoric; by adding historical dimension to the classical Greek *logos*, Christianity gave to classical rhetorical figures (*paranomasia*, *prosopopoeia*) a capacity for massive narrative extension." However, according to Wells, *Spenser's "Faerie Queene" and the Cult of Elizabeth*, p. 7, "Virgil [in the *Aeniad*] links events widely separated in time in order to show that they form steps

in an historical progression culminating in the reign of Augustus." Wells continues: "If the patterns of recurrence and prefiguration we find in the *Aeniad* bear similarities to those in the Bible, this is because they are the product of a theory of history which sees events not simply as sequence, but as significant elements of a divine plan. Such a providential view of history is not, of course, unique to Christendom."

290. Norden, *A Store-Hovse of Varieties,* stanza 100 (sig. D_4^r).

291. Given the ambiguity in *Mirum in modum* of the simultaneous expression of concern about the judgments of posterity, and hope that ingenuity will stem the tide of corruptibility in the poetic body of eloquence, *The Purple Island*'s lack of accomplishment of a rhetorical resurrection may reflect it's author's lack of a longing for an immortality in print (see chapter 4, note 76).

292. "Prolepsis" is a narratological term to describe a certain narratorial manipulation of the relationship between text-time and story-time. If story-time is conceived of conventionally and conveniently as a linear succession of events, various discrepancies may occur between this linearity of story-order and the order in which the events are narrated. A narration of a story-event at a point in the text before earlier events have been mentioned has traditionally been referred to as "foreshadowing" or "anticipation." In order to avoid the psychological and cinematic-visual connotations of these terms, Genette coined a specifically literary nomenclature for such discrepancies, referring to "anticipation" as "prolepsis." See Rimmon-Kenan, *Narrative Fiction,* pp. 44–46. The qualified rhetorical resurrection of the body in *The Purple Island* occurs in Cantos 2 to 5, and is followed by a narration of further events prior to eschatological events which are narrated in Canto 12. Not only for this reason is it a prolepsis, but also because in the divine history to which the events of the *Purple Island* narrative correspond, the Resurrection of the Dead may be considered as taking place at the end of time, and only foreshadowed in time, although there was also the view that the Last Judgment was not only imminent but even eternally now.

293. Sawday, "Bodies By Art Fashioned," p. 123.

294. Kastor, *Giles and Phineas Fletcher,* p. 112; Grundy, *The Spenserian Poets,* pp. 101, 102. On *The Purple Island*'s pastoralism, also see Leah Jones, *The Divine Science,* pp. 117–22; Langdale, *Phineas Fletcher,* pp. 160–62.

295. Kastor, *Giles and Phineas Fletcher,* pp. 107, 115, 107. On the "epic aspirations" of *The Purple Island,* see Baldwin, "Phineas Fletcher," pp. 469–70; Jones, *The Divine Science,* pp. 121–23; Parfitt, *English Poetry of the Seventeenth Century,* p. 194.

296. Kastor, *Giles and Phineas Fletcher,* pp. 107, 115; see also pp. 113, 116. On the unity of *The Purple Island,* in relation to subjects, genre, and narrative, see Gardner, "The Banquet of the Word," p. 391; Grosart, *The Poems of Phineas Fletcher,* pp. cclxxvii-cclxxviii, footnote; MacDonald, *England's Antiphon,* p. 156; Parfitt, *English Poetry of the Seventeenth Century,* p. 191.

297. Clapham, *Ælohim-Triune,* title page.

298. The subtitle of *The Isle of Man*—"*The Legall Proceeding in Man-Shire Against Sinne*"—provides the focus for the actions and anatomy described in the account, although this focus is not discussed in this study. Furthermore, *The Isle of Man* shares with *Mirum in modum* and *Microcosmos* the perspective on its material of the ancient injunction *nosce teipsum,* and in addition "the Apostles charge": "*Quisqué explorat seipsum.*" Bernard, *The Isle of Man,* sig. A_2^v.

299. Cunningham, *The Anatomical Renaissance,* p. 201.

Chapter 6. The Form of Stanza and Margin

1. Sawday, "Bodies by Art Fashioned," pp. 115–16

2. The arrangement and form of referencing altered in subsequent editions, such as that of 1783, *The Purple Island, or The Isle of Man,* ed. by [Peregrine Branwhite]. This edition moves the marginalia to the foot of the page, while retaining the first edition's letter reference system. The later edition of 1816, *The Purple Island, a Poem: by Phineas Fletcher,* additionally replaces the letters of the reference system with symbols, such as asterisks.

3. Langdale, *Phineas Fletcher,* p. 176.

4. "Ro. Vn.," *The Little World,* p. 1.

5. Ibid., p. 8.

6. Lana Cable, "Such Nothing is Terrestriall: Philosophy of Mind on Phineas Fletcher's *Purple Island,*" *Journal of the History of the Behavioural Sciences* 19 (1983): 136–52 (p. 139)

7. Ibid., p. 140

8. Ibid.

9. Ibid., p. 141
10. Ibid., pp. 140–41
11. Sawday, "Bodies by Art Fashioned," p. 144.
12. Ibid., p. 116.
13. For the explanatory function of the notes of *Microcosmos*, see p. 30, col. 2, ll. 1–2: "And yet the *body's* but the *Instrument* / Whereon the [1]*soule* doth play what she doth please." The note (1) reads: "The Soule worketh by motion, and the Body by Action."
14. *Microcosmos*, p. 28, col. 2, ll. 23–21 (from bottom).
15. Ibid., p. 28, col. 2, note 3.
16. For this function in the notes of *Visiones Rervm*, see p. 2, where, in the note to stanza 5, "Cronos" is merely translated to "Time."
17. On historiography of natural philosophy and natural history in the seventeenth century, and on some specific interrelationships between anatomy and natural history, see Harold J. Cook, "The Cutting Edge of a Revolution? Medicine and Natural History Near the Shores of the North Sea," in *Renaissance and Revolution: Humanists, Scholars, Craftsmen and Natural Philosophers in Early Modern Europe*, ed. by J. V. Field and Frank A. J. L. James (Cambridge: Cambridge University Press, 1993; repr. 1997), pp. 45–61. Cook argues that the historiographical distinction between the end of natural history in description and that of natural philosophy in causal analysis is "somewhat doubtful" (p. 49).
18. Sawday, "Bodies by Art Fashioned," p. 145. In the [Branwhite] edition of 1783, and that of 1816 which contains the critical remarks of Headley, the relocation of the notes to the foot of the page seems to reduce their potential to be an "autonomous discourse," and to accentuate their supplementary relation to the poetic stanzas, in which they explicate the stanzas's anatomical content.
19. Cable, "Such Nothing is Terrestriall," p. 140
20. Ibid., p. 141
21. Ibid.
22. Sawday, "Bodies by Art Fashioned," p. 147. By way of explanation, in an endnote, Sawday states: "In the seventeenth century 'whey' was not a term used exclusively in the dairy. It also refered to the watery fluid exuded by serous membranes (OED)." (p. 160).
23. *The Purple Island*, 2.9 margin. f (my emphases): "The whole body *is as it were* watered with great plenty of rivers, veins, arteries, and nerves."
24. The English term "Cavern" is "borrowed from Old French *cavern* [which means] cave, [a] learned borrowing from Latin *caverna* [which means] cave, cavity." *Dictionary of Etymology*, p. 152, col. 2.
25. In relation to the allegorical use of wordplay and etymology, Quilligan, *The Language of Allegory*, p. 35, note 16, remarks: "In the *Interpretation of Dreams* . . . Freud explains that wordplay—which includes simple puns—is important to the process of condensation in dreams which can be seen 'at its clearest when it handles words and names'. . . . Etymology also has an important function in the dreaming human subconscious: 'Language has a whole number of words at its command which originally had a pictorial and concrete significance, but are used today in a colorless and abstract sense. All that the dream need do is to give these words their former full meaning or to go back to an earlier phase of their development'. . . . This is all an allegory need do as well. The importance of Freud's discussion of wordplay in dreams ought not to be understood as proof that allegorical poets were writing down real, or realistically dreamlike dreams (even though many allegories are 'dream visions'), but to suggest that wordplay has once again become a potent device by which man may investigate the truths of his psyche. Freud's sensitivity to the linguistic structure of dreams prepared the way for our increased sensitivity to language, which has in part made itself felt in the renewed appeal of older allegory as well as the reappearance of allegorical narrative and *allegoresis*." It is Quilligan's contention that "allegory is (and always has been) the most self-reflexive and critically self-conscious of narrative genres, and . . . its purpose is always to make its reader correspondingly self-conscious" (p. 24). "Other genres appeal to readers as human beings; allegory appeals to readers as readers of a system of signs, but this may be only to say that allegory appeals to readers in terms of their most distinguishing human characteristic, as readers of, and therefore as creatures finally shaped by, their language" (p. 24).
26. Singer, *Vesalius on the Human Brain*, pp. 31–32.
27. Sawday, "Bodies by Art Fashioned," p. 149.
28. Sir Thomas Browne, *Selected Writings*, ed. by Sir Geoffrey Keynes (London: Faber and Faber, 1968), p. 248, cited in Sawday, *The Body Emblazoned*, p. 234. Sawday comments that the utterance is perhaps an unlikely one to have come from the pen of "the great discoverer of endless similitudes." Even so, Alexander Ross

undertook in his *Arcana microcosmi,* 2nd edn. (London, 1652), to "refute 'Doctor Browne's VULGAR ERRORS, the Lord Bacon's NATURAL HISTORY and Doctor Harvey's DE GENERATIONE.'" *The Body Emblazoned,* p. 234.

29. Quilligan, *The Language of Allegory,* p. 175.

30. According to Hodges, *Renaissance Fictions of Anatomy,* p. 104, Bacon argued that, in order to find the truth of nature, "one must paradoxically accept the estrangement of human perception and language from nature." This estrangement of intellect from nature "marks the shift from the order of microcosm and macrocosm to the order of taxonomy." Bacon also opposed *allegoresis* in the reading of ancient texts. See Quilligan, *The Language of Allegory,* pp. 173–75.

31. Quilligan, *The Language of Allegory,* p. 175. Also see Knight, *Natural Science Books in English 1600–1900,* p. 11.

32. Hobbes, *Leviathan,* p. 20. "Of Reason, and Science" is the title of Chapter 5 of *Leviathan,* and the objection cited in the text is the sixth such objection to what Hobbes calls "Absurd assertions" (p. 20). In Chapter 4, "Of Speech," Hobbes counts among "Abuses of Speech" the use of words "metaphorically; that is, in other sense than that they are ordained for; [which] thereby deceive others" (p. 13).

33. Ibid., p. 20; *Leviathan,* ed. by Richard Tuck (Cambridge: Cambridge University Press, 1991), p. 52, cited by Sawday, *The Body Emblazoned,* p. 235.

34. Joseph Glanvill, *The Vanity of Dogmatizing* (London, 1661), p. 160, cited by Sawday, *The Body Emblazoned,* p. 235.

35. J. B. Van Helmont, *Oriatrike or Physicke Refined,* trans. by John Chandler (London, 1662), p. 323, cited by Sawday, *The Body Emblazoned,* p. 235.

36. Sawday, *The Body Emblazoned,* p. 235.

37. Knight, *Natural Science Books in English 1600–1900,* p. 12.

38. R. F. Jones, "Science and Language in England of the Mid-Seventeenth Century," in *Seventeenth-Century Prose: Modern Essays in Criticism,* ed. by Stanley E. Fish (New York: Oxford University Press, 1971), pp. 94–111 (p. 101), cited in Quilligan, *The Language of Allegory,* p. 175. According to Sawday, *The Body Emblazoned,* p. 235, "Glanvill's post-Restoration ideal of a reformed scientific language—nakedly simple—was to be the much-vaunted linguistic touchstone of the Royal Society." Knight, *Natural Science Books in English 1600–1900,* p. 12, claims that "down to the founding of the Royal Society Bacon was the most important and influential philosopher of science for those whose first language was English."

39. Quilligan, *The Language of Allegory,* p. 175.

40. Ibid., pp. 177–78.

41. Cable, "Such Nothing is Terrestriall," p. 139

42. Quilligan, *The Language of Allegory,* p. 177.

43. On *The Purple Island* 5.47, Quilligan (p. 177) writes in derogatory manner: "This quaintly charming versification of physics should remind us that Dante's 'world of circles in the skie,' analogous to other circles elsewhere in the cosmos, had circumscribed an entirely different *episteme.*" Quilligan thereby interposes (erroneously I think) a complete discontinuity between the epistemic modes of the Middle Ages and those of the early seventeenth century.

44. I have derived concepts applied to literary works in this paragraph of the text, from the discussion in Alan Sinfield, *Faultlines: Cultural Materialism and the Politics of Dissident Reading* (Oxford: Clarendon Press, 1992), pp. 82–83, of the function of the dominant ideology to obscure the actual power structure of a state, giving rise to a whole series of dislocations between the dominant ideology and both the politically hegemonic and the economically dominant classes.

45. Sawday, *The Body Emblazoned,* p. 179.

46. For Fulke Greville on the relationship between word and object, or referent, his rejection of intellectual speculation, his attack on rhetoric, his view that linguistic structures "constantly carry the danger of obscuring reality or, worse, actually becoming a substitute reality," his distinction between images of life and images of wit in literature, and his contradictory but simultaneous investment of poetry with "an idealist function," see Dollimore, *Radical Tragedy* (1989), pp. 78–79.

47. The composition of *The Purple Island* is discussed in chapter 7 of this study.

48. Stephen Greenblatt, "Shakespeare and the Exorcists," in *Shakespeare and the Question of Theory,* ed. by Patricia Parker and Geoffrey Hartman (London: Methuen, 1985), pp. 163–87 (p. 164).

49. The formulation of this position with respect to the relation that obtains between early modern and modern understandings of the body's religious or scientific meanings is indebted to Stephen Greenblatt's articulation

of the similar relation which in his view obtains between the psychological phenomena of Renaissance texts and psychoanalysis. See Greenblatt, "Psychoanalysis and Renaissance Culture," in *Literary Theory / Renaissance Texts*, ed. by Patricia Parker and D. Quint (Baltimore: Johns Hopkins University Press, 1986), pp. 210–24. However, because *The Purple Island*'s influence upon other literature and the history of ideas is in this study uncharted, it may be inappropriate to apply to it further permutations of Greenblatt's theory for Renaissance texts. Whether, for example, the post-Hobbesian proscription of poetic language in the languages of science, and the divergence of discourses it helped to secure, were in any way resultant upon experiments in textual form, such as that of the form of *The Purple Island,* would require a thorough study of textual influence.

50. Baldwin, "Phineas Fletcher," p. 468.

51. Healy, "Sound Physic," p. 346. Quotation in brackets is of Sawday, *The Body Emblazoned,* p. 178.

52. Sawday, "Bodies By Art Fashioned," p. 116.

53. See, for example, *Microcosmos,* p. 47, col. 2, note 3; *Summa Totalis,* in *The Complete Works,* ed. by Grosart, I, p. 17, col. 2, note 1.

54. The notes in Hagthorpe, *Visiones Rervm,* also occasionally perform this function by quoting and citing an authoritative source. See, for example, I. 57 note 6.

55. Michael Spaher and Johann Remmelin, *A Survey of the Microcosme; or, The Anatomy of the Bodies of Man and Woman; wherein Skin, Veins, Nerves, Muscles, Viscera, Bones, and Ligaments thereof Are Accurately Delineated, and So Disposed by Pasting, as that All the Parts of the Said Bodies, Both Internal and External, Are Exactly Represented in Their Proper Site,* corrected by Clopton Havers (London: James Moxon, [1695]).

56. Figures 1 and 3 of *A Survey of the Microcosme,* in addition to figure 2 ("*Visio Secunda,*" the Christlike figure), are based on the figures of the same numbers in *Catoptrum microcosmicum.* Other than the omissions listed in the text, the later figures are also unlike their orginal counterparts in that they are colored. In addition to these figures, *A Survey of the Microcosme* borrows from Estienne, *De dissectione,* a figure on the first blank page verso after the title page. There are two plates in the copy of *Catoptrum microcosmicum* to which I refer (British Library shelfmark: Tab. 583. f. 45.) which do not appear in *A Survey of the Microcosme.* One is a plate of the abdominal and thoracic viscera of a male figure, on an unpaginated leaf. The anatomy is much more rudimentary than in the other illustrations, suggesting perhaps that it does not belong to this work. A further unpaginated leaf contains a female figure, which is annotated in French (as opposed to the Latin of the other illustrations), and for this reason and the fact that the flaps which open the abdomen and thorax reveal only a blank page beneath, it is likely that the page containing this plate also does not belong to the original work, and was added at a later date.

57. Thomas Banks (1735–1805), *Anatomical Crucifixion,* 1801, plaster cast on wooden cross, is in the Royal Academy of Arts, London. A photographic reproduction appears in Martin Kemp and Marina Wallace, *Spectacular Bodies: The Art and Science of the Human Body from Leonardo to Now* (London: Hayward Gallery Publishing, 2000), cat. 41, p. 86.

58. L. Price Amerson Junior, *L'Ecorché,* Cahier no. 3, l'Ecole des Beaux Arts de Rouen, p. 33, cited in Kemp and Wallace, *Spectacular Bodies,* p. 87. The initiators of the project were Thomas Banks himself, Benjamin West, then President of the Academy, and the painter Richard Cosway, "who was a keen student of anatomical books."

59. See, for example, *The Purple Island* 3.10 margin. h; 5.24 margin. x.

60. Sawday, *The Body Emblazoned,* p. 178.

61. Clapham, *Ælohim-Triune,* sig. B^{r}, notes in the margin (a), "The sixt dayes worke," next to the adverbial phrase in the poetic text, "Before that a*Adam* breath'd or being had."

62. "Ro. Vn.," *The Little World,* pp. 5, 10.

63. Healy, "Sound Physic," p. 346.

64. Ibid.; Sawday, *The Body Emblazoned,* p. 178.

65. See Quilligan, *The Language of Allegory,* p. 61, for this understanding of the formal relationship of allegory to *allegoresis,* or allegorical interpretation.

66. See Quilligan, *The Language of Allegory,* p. 28. Quilligan develops this idea of the reader of allegory accumulating awareness, out of which "comes a consciousness not only of how he is reading, but of his human response to the narrative, and finally his relation to the only 'other' which allegory aims to lead him to, a sense of the sacred" (p. 29).

67. See Ibid., (p. 31 cited in the text) for this idea of allegory as both text and commentary upon its own "threshold text" or "figure."

68. Healy, "Sound Physic," p. 346.

69. Both terms, "symbiosis" and "synthesis," are used by Sawday: the first, for example, in "Bodies by Art Fashioned," p. 108; the second, for example, in *The Body Emblazoned,* p. 179.

70. See Kemp, "'The mark of truth,'" p. 89, whose discussion of the sets of conventions pertaining to the presentation of anatomy in illustration includes the relationship of illustration to the main text as well as to the descriptive notes keyed to the illustration.

71. Sawday, "Bodies by Art Fashioned," p. 145. Insofar as the statement in the text is true, it cannot follow that, as Redman, "*The Purple Island*," pp. 12–13, argues, Fletcher complies with the goal of a Puritan writer, "to lead readers to spiritual maturity through the allegorical narrative," by "supplying his readers with the copious marginalia at the beginning of the poem [Cantos 2 to 5] (for the reader is just beginning his travel through knowledge)," and proceeding to reduce this instruction "to only occasional definitions in later cantos, as the reader progresses in understanding." Rather the marginalia in the later cantos serve a slightly different clarifying, authoritative function, and do not express a mimetic relationship of the kind expressed by the formal disposition of Cantos 2 to 5, where the notes are much more extensive.

72. Choulant, *History and Bibliography of Anatomic Illustration,* p. 30. According to Roberts and Tomlinson, *The Fabric of the Body,* p. 190, anatomy in the period of the sixteenth century from "Berengario's pioneering *Commentaria*" to the copperplates of Eustachio, "may be seen to be in process of maturing as an observational science; illustrations, correctable illustrations, were major factors in achieving this maturity."

73. Heckscher, *Rembrandt's Anatomy of Dr. Nicolaas Tulp,* p. 62. As an explanation, Heckscher suggests that book illustration's "unavoidable association with the illuminated manuscript page of the medieval codex militated against its employment for classical and related texts. Any subject matter hallowed by ancient authority, such as anatomy, had to be kept from being contaminated by illustrative images."

74. The graphic incunabula of anatomy are pictorial representations of dissections in the form of plates "either used individually for teaching, or sold as sets between covers. The dual publication of [these] atlases as either separate plates or as collections was mainly due to economic factors, but the 'collections' were often broken up and the plates (often framed and glazed) were displayed for teaching purposes in dissection rooms." Thornton and Reeves, *Medical Book Illustration,* p. 44. Also see Singer, *The Evolution of Anatomy,* p. 97, according to which the *Commentaria . . . super Anatomia Mundini* "is the earliest anatomical treatise that can properly be described as having figures illustrating the text. . . . Some are not devoid of beauty, and are prepared for the use of artists rather than anatomists."

75. See Choulant, *History and Bibliography of Anatomic Illustration,* p. 22; Kemp, "'The mark of truth,'" p. 89. Roberts and Tomlinson, *The Fabric of the Body,* p. 54, state that many illustrations prior to the visceral figure in Friesz's *Spiegel der Artzny* were symbolic or conceptual in form; and that this visceral figure (See chapter 1 and the notes thereof for a discussion of which) "marks the transition in published anatomical illustrations . . . to the more factual." According to Choulant, pp. 73–74, the earliest known drawings of the uterus represent it schematically as flask-shaped. Schematic forms of representation are used by Vesalius in three plates of his *Tabulae sex.* See Harcourt, "Andreas Vesalius and the Anatomy of Antique Sculpture," pp. 40–41, who argues that these or other such plates were used by Vesalius as aids to students in his lectures.

76. Vesalius, The Dedicatory Letter to Prince Philip, *The Epitome of Andreas Vesalius,* p. xxxiii.

77. MS Windsor, Royal Library, 19013v, cited in Kemp, "'The mark of truth,'" p. 94. Leonardo's achievements in the field of anatomical illustration are exceptional, partly for the very reason (given in chapter 1) that they are in large part omitted from this study. Kemp argues: "The extraordinary inventiveness and unlimited ambition of Leonardo's illustrations only became possible in a context in which practical considerations remained in the background" (p. 90).

78. *The Birth of Mankinde* (1604), fol. 32.

79. Gray, *Anatomy Descriptive and Surgical,* p. 1001. The following are descriptions taken from Gray of the "convoluted plexus (*plexus pampiniformis*)" of which the chief mass of the spermatic cord is composed, and which is formed by the spermatic veins and the tributaries of the epididymis; the ramifications of the artery of the vas deferens and its anastomoses with the spermatic artery; and the divisions of the spermatic artery, its anastomoses with the cremasteric artery, and its "tortuous" descent into the testicle. The structure of the spermatic cord is said to be composed of "arteries, veins, lymphatics, nerves, and the excretory duct of the testicle." The arteries of the cord are: "the spermatic, from the aorta; the artery of the vas deferens, from the superior vesical; and the cremasteric, from the deep epigastric" . The spermatic artery is "a branch of the abdominal aorta" (p. 1000). It "descends to the testicle, and, becoming tortuous, divides into several branches, two or three of which accompany the vas deferens and supply the epididymis, anastomosing with the artery of the vas deferens" (pp. 1000–001). The cremasteric artery is "a branch of the deep epigastric artery. It accompanies the spermatic cord and supplies the Cremaster muscle and other coverings of the cord, anastomosing with the spermatic artery." The artery of the vas deferens is "a branch of the superior vesical . . . which accompanies the vas deferens, ramifying upon the coats of

that duct, and anastomosing with the spermatic artery near the testis." The spermatic veins "emerge from the back of the testis, and receive tributaries from the epididymis: they unite and form a convoluted plexus (*plexus pampiniformis*) which forms the chief mass of the cord" (p. 1001).

80. Kemp, "'The mark of truth,'" p. 85.

81. O'Malley, *Andreas Vesalius of Brussels 1514–1564,* p. 323.

82. Cunningham, *The Anatomical Renaissance,* p. 91, remarks that a student at Vesalius's first anatomy, which took place from the sixth to the twenty-fourth of December 1537, when Vesalius was Demonstrator of Anatomy and Lecturer in Surgery at the University of Padua, said that Vesalius declared his intention to publish his illustrations of the distribution of the veins, arteries, and nerves. Cunningham adds: "Already, in his first surgical lectures and his first anatomical demonstration, Vesalius was thinking visually" (p. 91). The student's notes which Cunningham has consulted, contain sketches, copied down from diagrams produced by Vesalius in the course of the anatomical demonstrations.

83. Cunningham, *The Anatomical Renaissance,* p. 111.

84. See Singer, *The Evolution of Anatomy,* p. 114.

85. The letter is published as the Preface to both the 1543 and 1545 editions of the *Fabrica.* The citation in the text is as it appears in the 1543 edition, and is cited by Choulant, *History and Bibliography of Anatomic Illustration,* p. 172. In the introductory paragraph entitled *Lectori,* on the title page of the *Epitome,* Vesalius "addressed the reader with the astonishing statement that it did not matter one way or the other whether one started a problem with text or with *tabulae* (i.e. illustrations)." Heckscher, *Rembrandt's Anatomy of Dr. Nicolaas Tulp,* p. 63.

86. See Kemp, "'The mark of truth,'" p. 89, where he discusses the two conventions described in the text, and the division of labour and space between illustration and text in anatomy textbooks. Heckscher, *Rembrandt's Anatomy of Dr. Nicolaas Tulp,* p. 62, says: "The terms 'illustrare' and 'elucidare' were originally used to denote verbal commentaries on authoritative texts. Illustrations of books by pictures that aspired to the glories of a scholarly commentary or glossary were, for a long time to come, anathema to the devotees of the word."

87. Sidney, *The Defence of Poesy,* pp. 113–14.

88. Singer, *The Evolution of Anatomy,* pp. 134–35, says that among the more shameless of plagiarizers of the plates of Vesalius was Paré, "who reproduced them without acknowledgement in 1551 and afterwards." Despite the use of illustrations, the anatomy of Book III of Paré's *Works* (1634) ("Treating of the Anatomy of Mans body") relies heavily on verbal rather than visual description, even when Paré (p. 80) proposes not only to subject the parts "to the eye in the way of knowing them, but also to the minde in the faithfull understanding them." Paré relates this distinction to that between the anatomy of Galen's *On Anatomical Procedures* and that of his *On the Use of the Parts.* Galen, himself, was skeptical of the utility of illustration.

89. Crooke, *Microcosmographia,* fols. 17–18, is rather conservative on this matter: "But I am not altogether of that opinion [as was Galen] to thinke these pictures vaine and ydle, because I see everie day represented unto mine eyes by these Pictures manie things, and those oftentimes of great consequence, which were unknowne to the skilfullest artists of former ages."

90. According to Harcourt, "Andreas Vesalius and the Anatomy of Antique Sculpture," p. 53, Vesalius's impact was the introduction of visual representation as necessary for anatomy. With regard to the later seventeenth century, Thornton and Reeves, *Medical Book Illustration,* p. 75, remark that, despite original research promoted by Harvey and others, the development of the microscope, and the publication of outstanding books in other branches of science, "there were comparatively few books on medical subjects containing noteworthy illustrations. Many of the anatomical texts were illustrated with plates copied from earlier writings, and most of the shorter communications were in the form of papers submitted to the Royal Society, founded in 1662, and published in its *Philosophical Transactions.*" Similar societies existed in Italy, Germany, and France.

91. The letter is published as the Preface to both the 1543 and 1545 editions of the *Fabrica.* The citations in the text are as they appear in the 1543 edition, and are cited by Choulant, *History and Bibliography of Anatomic Illustration,* p. 172. Thinking visibly, Vesalius developed techniques of anatomical illustration that solved the representational problems of the body's volumetric proportions and its complexity of structure, and achieved a precision that arguably was not possible in words.

92. For a discussion of the artistic concept of *rilievo,* see Michael Baxandall, *Painting and Experience in Fifteenth Century Italy: A Primer in the Social History of Pictorial Style* (Oxford: Oxford University Press, 1972; repr. 1983), p. 121. Leon Battista Alberti used *rilievo* to translate the Latin word *prominentia* (projection), and explained that it is the appearance of a form modeled in the round.

93. Kemp, "'The mark of truth,'" p. 89. According to Kemp, Leonardo became aware of the usefulness of combining illustration and text. "Early in his career, he seems to have assumed that drawings could themselves

communicate all he had seen and all he wished to say, not only on questions of form but also on matters of function and cause." "As his contacts with the Galenic tradition encouraged him to explain functionally every detail of created form, so he became aware that illustrations reigned supreme for description, while the text remained best adapted to explaining how something worked" (p. 94). Kemp argues (p. 89) that further considerations of visual representation of human anatomy should include wherever possible and relevant, its context of publication. According to Kemp, this context is composed of the intellectual position of the book's author; its patronage and that of the illustrator; its intended and actual production and audience; and the broad values in society relating to the roles of books, medicine, and observational science. At various points in this study, in the context of discussions about illustrations, items on Kemp's list of considerations are included among the interpretative criteria employed. Their employment is not, however, systematic, because other criteria become more immediately useful. Some of these criteria become evident in this chapter's discussion of illustrations of Estienne, *De dissectione.*

94. Sawday, *The Body Emblazoned,* pp. 133–34.

95. Francis Barker, *The Tremulous Private Body: Essays on Subjection* (London: Methuen, 1984), p. 23.

96. Ibid., p. 80. In the passage from which the quotation in the text is taken, Barker is discussing Rembrandt van Rijn's painting entitled *The Anatomy Lesson of Dr Nicolaas Tulp* (1632). The consequences of the relationship between subjectivity, representation and the body are more widely applicable, in this case, to pictorial representations of anatomical dissection. However, the bifurcation of text and image of the body in anatomical and medical texts occurs as early as the late fifteenth century, and in no sense directly presages the mediation of rational intellect and will between the text and body, as represented in Rembrandt's painting, which has been interpreted by Sawday, *The Body Emblazoned,* p. 153, in Cartesian terms.

97. "The body was one possible authority, but its organization was mysteriously complex and, because of the biological tendency towards decay, continually eluding the practical dissector. The text was another possible source of authority, but how could bodies be represented in texts when, in their material reality, they seemed to crumble and dissolve before the eyes of the onlooker?" Sawday, *The Body Emblazoned,* p. 133.

98. Sawday, *The Body Emblazoned,* p. 86.

99. See Kemp, "'The mark of truth,'" pp. 88–89.

100. Schultz, *Art and Anatomy,* p. 72. According to Baxandall, *Painting and Experience in Fifteenth Century Italy,* pp. 139–41, 124, *designo* was associated with linear, as opposed to tonal, representation of the object; with profiles, contours, lines, edges, and perspective. (For *scorci,* see pp. 143–45.) For Cennino Cennini, the foundations of the art of painting were *disegno* and *colorire,* which goes with brush, tones, the representation of surfaces. Annibale and Agostino Carracci's *Accademia Degli Incamminati* (first called the *Accademia delli Desiderosi*) gave lessons principally in drawing and painting, thereby recognizing the distinction between these two arms of the handiwork of the artists. Giorgio Vasari, meanwhile, established the *Accademia del Disegno* under the patronage of Cosimo de' Medici. Antonio Manetti defined *prospettiva* as the systematic diminution or enlargement, as it appears to human eyes, of objects that are respectively remote or close at hand. On the academy of the Caracci, see A. W. A. Boschloo, *Annibale Carracci in Bologna: Visible Reality in Art after the Council of Trent,* trans. by R. R. Symonds, Kunsthistorische Studiën van het Nederlands Instituut te Rome Deel III, Band I (The Hague: Government Publishing Office; Maarssen: Gary Schwartz, 1974), pp. 39–40.

101. Windsor 19061r, cited in Schulz, *Art and Anatomy,* p. 99.

102. Kemp, "'The mark of truth,'" p. 97.

103. Schultz, *Art and Anatomy in Renaissance Italy,* p. 83; also see p. 106.

104. Giorgio Vasari, *Le vite de' più eccellenti pittori, scultori, ed architettori,* ed. by Milanesi, 9 vols. (Florence: G. C. Sansoni, 1878–85), IV, 11, cited in Schultz, *Art and Anatomy,* p. 68.

105. Leon Battista Alberti explains the significance for artists of the movements of the body: "Movements of the soul are recognized in movements of the body. . . . There are movements of the soul, called affections—grief, joy, fear, desire and others. There are movements of the body: growing, shrinking, ailing, bettering, moving from place to place. We painters, wanting to show movements of the mind with movements of the body's parts, use only the movements from place to place." Leon Battista Alberti, *Leon Battista Alberti On Painting and On Sculpture: The Latin Texts of "De Pictura" and "De statua,"* ed. with translations, introduction, and notes by Cecil Grayson (London: Phaidon, 1972), cited in Baxandall, *Painting and Experience in Fifteenth Century Italy,* p. 60; also see pp. 61–71 for possible origins and contemporary conventions for understanding muscular movement and gesture in fifteenth-century painting.

106. Sawday, *The Body Emblazoned,* p. 86.

107. Kemp, "'The mark of truth,'" p. 89.

108. Ibid., p. 101. According to Choulant, *History and Bibliography of Anatomic Illustration,* p. 224, Casserio

claims in *De vocis auditusque organis historia anatomica* (Ferrara, 1601), that he will make public "pictures finished and complete in every part, which will, I venture to say, excel in nicety, clearness, and finally in workmanship and pains all that have hitherto been published." The painter and etcher of the plates in this work is Josias Maurer, but it is not known whether he is responsible for the plates published later under the title *Ivlii Casserii Placentini . . . Tabulae anatomicae LXXIIX.* According to Choulant, p. 228, Casserius's plates "mark a new epoch in the history of anatomic representation, owing to the correctness of their anatomic drawing, their tasteful arrangement, and the beauty of their technical execution. And this all the more, since they cover the whole field of anatomy."

109. Kemp, "'The mark of truth,'" p. 94. For Berengario's devotion to *anatomia sensibilis,* see, for example, *Isagogae breves,* p. 104, where with regard to the muscles of the neck and head, he declares that speech alone is not sufficient, but touch and sight are needed. Lind, *Studies in Pre-Vesalian Anatomy,* p. 7, regards Berengario's "first use of adequate anatomical illustrations, not mere schemata but drawn from the cadaver by a good artist," as not less significant than his contributions to the study of "the brain, the sphenoid sinuses, the vermiform appendix, the tympanum, the thymus gland, the arytenoid cartilages, the action of the cardiac valves, the papillae of the kidneys, the chyliferous ducts and . . . the abdominal muscles."

110. Kemp, "'The mark of truth,'" p. 94.

111. Berengario, *Isagogae breves* (1522), fols 63^v–64^r, cited in Kemp, "'The mark of truth,'" p. 94. For the same passage in *Isagogae breves,* see p. 160.

112. Kemp, "'The mark of truth,'" pp. 97–98. It is significant that, as Lind, *Studies in Pre-Vesalian Anatomy,* pp. 16–17, says: "While the pre-Vesalians [i.e., anatomists] had many friends among physicians, philosophers, and politicians they seem to have been personally acquainted with few if any artists." An exception would be Marcantonio della Torre, lecturer at Pavia, and possibly one of the first humanist Galenists in the field of medicine, who collaborated with Leonardo da Vinci. Also see Schultz, *Art and Anatomy in Renaissance Italy,* p. 67.

113. Kemp, "'The mark of truth,'" p. 87.

114. See chapter 1 and the notes thereof for a discussion of these illustrations.

115. Sawday, *The Body Emblazoned,* p. 86. According to Sawday, there was "a widely held view that the universe as a structure was organized anthropocentrically, and the artist's chief endeavour was to replicate that structure in the world of representation" (p. 87).

116. Kemp, "'The mark of truth,'" pp. 85–87.

117. Healy, "Sound Physic," p. 347.

118. Ibid.

119. Sawday, *The Body Emblazoned,* p. 16.

120. Barker, *The Tremulous Private Body,* pp. 12–13.

121. The student notebook is MS 150 in Aberdeen University Library, and is discussed by Roger French in *Dissection and Vivisection in the European Renaissance,* pp. 218–19.

122. French, *Dissection and Vivisection in the European Renaissance,* p. 218.

123. Sawday, *The Body Emblazoned,* p. 16. Sawday adds: "To consider the body in isolation was not merely difficult but, strictly speaking, impossible, since the body's primary function, it was held, was to act as a vessel of containment for the more significant feature of the soul." Sawday holds the view that "Plato's depiction of a perpetual struggle between body and soul, transmitted via Neoplatonism, patristic authority, and medieval theology, is the dominant model for understanding the relationship between body and soul within western culture prior to the fifteenth century" (p. 17). Furthermore, he holds that, "with the advent of the preliminary results of scientific investigation in the later sixteenth century, the interior war became even more fierce. . . . Science gave an added impetus to the urge to peer into the recesses of the body. But Calvinistic theology, with its seemingly obsessive desire to chart the inner state of each individual's spiritual well-being, was to argue with a conviction equal to that of the scientist that the division between the realm of the body and the realm of the soul was now the concern of every thinking person" (p. 17). Although Sawday may be justified in his assessment of culture at large, as a description of Renaissance and early modern anatomy, his overview neglects the significance of Aristotelianism and its hylomorphic and instrumental conception of the relationship between body and soul.

124. French in *Dissection and Vivisection in the European Renaissance,* pp. 219–26 (p. 219).

125. Sawday, *The Body Emblazoned,* p. 179. Sawday's anatomical interest in *The Purple Island* can be distinguished from Langdale's, in that Sawday is concerned not to retrieve the anatomy from its bed of allegory, but to understand the relation between anatomy and allegory, in this poem in particular, and in the "culture of dissection" out of which he claims it grew.

126. *The Fasciculo Di Medicina,* II, p. 15. In his remarks on this figure, Singer says (p. 17): "This exquisite

figure, though bearing a general resemblance to that on p. 9 of the 1491 edition [*Fasciculus medicinae*], has been entirely redrawn and differs in many details."

127. Hodges, *Renaissance Fictions of Anatomy*, pp. 96–97.

128. Ibid., p. 103. For Hodges's analysis of Bacon's use of aphorisms, see pp. 102–3.

129. Ibid., p. 103.

130. Kemp, "'The mark of truth,'" pp. 94–97.

131. *The Birth of Mankinde* (1604), fol. 3. "Anathomy" is here used as a noun to denote "[a] body . . . anatomized or dissected." *OED*, I, "A," p. 310, col. 1.

132. French, *William Harvey's Natural Philosophy*, p. 33.

133. Choulant, *History and Bibliography of Anatomic Illustration*, p. 152, argues: "The preparations for this work seem to have been made long before its first appearance for several plates bear the dates 1530 (p. 154), 1531 (p. 155), 1532 (pp. 150, 151)." For the sequence of events in the printing and publishing of Estienne's book, see Roberts and Tomlinson, *The Fabric of the Body*, p. 170. Also see Thornton and Reeves, *Medical Book Illustration*, p. 54.

134. Choulant, *History and Bibliography of Anatomic Illustration*, pp. 152–53.

135. See Ibid., p. 29; Kemp, "'The mark of truth,'" pp. 101–3.

136. Barker, *The Tremulous Private Body*, pp. 63–65.

137. Andrea Carlino, *Paper Bodies: A Catalogue of Anatomical Fugitive Sheets 1538–1687*, trans. by Noga Arikha, *Medical History*, Supplement No. 19 (London: Wellcome Institute for the History of Medicine, 1999), p. 26. C. E. Kellett, "A Note on Rosso and the Illustrations to Charles Estienne's *De dissectione*," *Journal of the History of Medicine* 12 (1957): 325–36 (pp. 325, 330), indicates his opinion that Perino del Vaga completed eighteen of the twenty drawings, and that the remaining two were done by Rosso Fiorentino. Kellett proposes that Rosso Fiorentino borrowed the drawings from Perino del Vaga (p. 330). Choulant, *History and Bibliography of Anatomic Illustration*, p. 152, says of *De dissectione*: "The author speaks with praise of the assistance rendered him by the surgeon Etienne Riviere, who is named in the title, and who assisted him both in his dissection and in the drawing of the illustrations." For the anatomical portions inset, see, for example, those figures in *De dissectione* on pages 260, 267, 271, 275, 276, 279, 281.

138. Carlino, *Paper Bodies*, p. 23.

139. Ibid.

140. Ibid., pp. 23–26.

141. Laqueur, *Making Sex*, pp. 130–34 (pp. 133, 132).

142. Valerie Traub, "Gendering Mortality in Early Modern Anatomies," in *Feminist Readings of Early Modern Culture: Emerging Subjects*, ed. by Valerie Traub, M. Lindsay Kaplan, and Dympna Callaghan (Cambridge: Cambridge University Press, 1996), pp. 44–92 (pp. 46, 48, 80–82). Although aspects of the history and philosophy of anatomy are a little imprecise or insecure in Traub's article, in a sophisticated theoretical manner Traub develops gender and eroticism as a rubric for understanding anxieties purportedly implicit in the emergence of dissection, and the construction and stabilization of gendered differentiations of embodiment. For the development of "a coherent scheme of dualities"—"men are to women as culture is to nature, as mind is to body, as subject is to object, as domination is to subordination"—from "the patriarchal division of the 'civilized' world into public and private spheres that institutionalized the division into the political world of men and the domestic world of women," see Ruth Bleier, *Science and Gender: A Critique of Biology and Its Theories of Women*, Athene series (New York: Pergamon Press, 1984), pp. 164–65.

143. Traub, "Gendering Mortality in Early Modern Anatomies," pp. 81–82. There is strong suggestion of the popular association of anatomy with prurience, in *Rare Verities: The Cabinet of Venus Unlocked, and Her Secrets Laid Open; Being a Translation of Part of Sinibaldus His Geneanthropeia, and a Collection of Some Things out of Other Latin Authors, Never before in English* (London: P. Briggs, 1658). This work borrows the sentiment from Ovid, that even though women frequently cover their "love" with modesty, they are "ever grateful" to men for "stolen pleasure" (p. 16). The term "stolen" in this context teeters uncertainly between the theft of pleasure and the theft of knowledge, or the English translation of "Latin Authors." The two senses are combined in the author's admission "To the Amorous Readers": "I am content to be the pick-lock of Venus her Cabinet, to let you with more ease enter and rifle and despoile her inestimable treasure." There are, significantly, a number of cabinets and chests in the series of plates that present female sexual anatomy, in *De dissectione*. Linda Williams, *Hard Core: Power, Pleasure, and the "Frenzy of the Visible"* (Berkeley and Los Angeles: University of California Press, 1989; repr. London: Pandora, 1990), argues that hard-core pornography develops more from the *scientia sexualis* and "body knowledge" emerging in the early modern period, than from "ancient traditions of erotic art" (p. 35).

144. Traub, "Gendering Mortality in Early Modern Anatomies," p. 81. Bleier, *Science and Gender,* p. 180, argues: "The important elements in this erotic art of the science of sexuality have to do with the multiplication and intensification of pleasures connected to the production of the 'truth' about sex. But, we may ask, pleasure for whom? From my point of view, this is a constructed male pleasure based largely (at least for many heterosexual men) in the realities and fantasies they enjoy within the context of their representations and constructions of women's sexuality: a fabricated woman and her fabricated sexuality, which has been, contradictorily, represented either as passive and masochistic or untamed fury." In the latter alternatives, we may see, as we did in chapter 1, the symbolic significance of hysteria-related illnesses, or particularly "Furor Uterinus," for the anatomy of female sexuality. It is conceivable that one may draw a developmental line from the theatrical, visual demonstration which occurs in the plates of *De dissectione,* to the significance and displacements of "the visual evidence of the mechanical 'truth' of bodily pleasure caught in involuntary spasm; the ultimate and uncontrollable—ultimate because uncontrollable—confession of sexual pleasure in the climax of orgasm," in fetishist and hard core pornography. Williams, *Hard Core,* p. 101.

145. On some of these characteristics of the viewer's experience of pornography, see Edward Marriott, "Men and Porn," *Guardian Weekend,* 8 November 2003, pp. 44–47, 89.

146. Traub, "Gendering Mortality in Early Modern Anatomies," p. 81.

147. Estienne, *De dissectione,* p. 267.

148. The left arm of the female figure on p. 270 of Estienne, *De dissectione,* rests upon a plaque, on which is written the key with anatomical descriptions of the parts in view, which I have loosely translated in the text: "A. Propria & prima matricis tunica, à peritonaeo revulsa. B. Venulae communes, quibus chorion intexitur: ab exteriore matricis parte, intró demissae. C. Chorij substantia, & contextus."

149. For details of the first and second issues of the first edition of *The Purple Island,* see Appendix A.

150. This emblematic plate is reproduced on the blank conjugate with the first page (i.e., p. 223) of *Poeticall Miscellanies,* in Giles and Phineas Fletcher, *Poetical Works,* II.

151. Healy, "Sound Physic," p. 342 (my emphasis). Although Healy attributes this interpretation of the discursive imbrication of the early seventeenth century to Sawday's "Bodies by Art Fashioned," as an ostensibly disinterested description of that work, Sawday actually argues for the congruity of anatomical and poetic discourses in sixteenth and early seventeenth-century culture. Upon this misreading, Healy's interpretation of *The Purple Island* partly rests. My italics emphasize the single most important word in the passage quoted from Healy for indicating his unacknowledged, and therefore inadvertently revealed, disagreement with Sawday's argument.

152. Sawday, *The Body Emblazoned,* p. 179.

153. Ibid.

Chapter 7. Questions of Provenance

1. Nicolson, *The Breaking of the Circle,* p. 40.

2. Baldwin, "Phineas Fletcher," p. 468.

3. Carlino, *Books of the Body,* p. 66 (the quotation is from the text to fig. 29).

4. Baldwin, "Phineas Fletcher," pp. 467–68, 473. Nicolson, *The Breaking of the Circle,* p. 138.

5. For definitions of these phrases, see H. Floris Cohen, *The Scientific Revolution: A Historiographical Inquiry* (Chicago and London: University of Chicago Press, 1994), p. 154.

6. Scholz, *Body Narratives,* p. 3.

7. Sanford V. Larkey, "The Vesalian Compendium of Geminus and Nicholas Udall's Translation: Their Relation to Vesalius, Caius, Vicary and De Mondeville," *Library,* 4th series, 13, no. 4 (1933): 367–94 (p. 367).

8. Thornton and Reeves, *Medical Book Illustration,* p. 61. According to Larkey, "The Vesalian Compendium of Geminus and Nicholas Udall's Translation," p. 369, the copper engravings are "among the first to appear in England." Larkey adds (p. 368): "Geminus [*fl.* 1540–1562] was typically Renaissance in his versatility; a printer and engraver, he also appears on the 'Accounts of Payments of Edward VI' as a surgeon to the King."

9. Larkey, "The Vesalian Compendium of Geminus and Nicholas Udall's Translation," pp. 369, 370.

10. Thomas Gemini, *Compendiosa totius anatomiae delineatio, œre exarata,* trans. and modified by Nicholas Udall (London, 1553); Larkey, "The Vesalian Compendium of Geminus and Nicholas Udall's Translation," pp. 376, 378. The 1553 edition of *Compendiosa totius* was reissued with a few minor additions in 1559. Thomas Vicary was one of the early Masters of the Barbers and Surgeons Company. His *Profitable Treatise,* later published

as *The Anatomie of the Bodie of Man,* went through some ten later editions, though, according to Larkey (p. 367), there is no known copy of the first edition, "the earliest extant version being that published by the 'Chirurgions of Saint Bartholomewes Hospitall' in 1577" (see pp. 379–86 for Larkey's conjecture, referred to in the text).

11. Sawday, *The Body Emblazoned,* p. 300, note 82.

12. Langdale, *Phineas Fletcher,* p. 206; Sawday, "Bodies by Art Fashioned," p. 142; *The Body Emblazoned,* pp. 176–77.

13. Sawday, "Bodies by Art Fashioned," p. 142. For the method of Nature, also see chapter 3 of this study.

14. De Laguna, in his *Anatomica methodus,* refers to this order of anatomy as "the method of nature" or the method which "follows nature," the implication being that this order follows that whereby Nature creates the body (p. 265).

15. Eriksson, *Andreas Vesalius" First Public Anatomy at Bologna 1540,* p. 49. The method of Nature relates also to the order or division according to "Composition," described by Paré, *Works* (1678), bk. III, p. 50, as deriving from Aristotle, and "being very commodious for the teaching of Arts." According to Paré, it is "the order and beginning taken from the least and most simple, to the more compound."

16. On the sequence of dissecting that Vesalius adopted in the Bologna demonstrations, "and which conforms to his repeated exhortations about the right sequence to pursue," see Cunningham, *The Anatomical Renaissance,* p. 113.

17. Larkey, "The Vesalian Compendium of Geminus and Nicholas Udall's Translation," p. 368.

18. Sawday's description of the anatomical content of the cantos, in "Bodies by Art Fashioned," p. 142, is not entirely accurate, for Canto 2 begins the description of the abdomen. It covers the anatomy of the skin, cuticle, fleshy panicle, abdominal muscles, and peritoneum (*P.I.* 2.15–23), and the physiological anatomy of the processes of ingestion, digestion ("concoction") and excretion of feces and urine (*P.I.* 2.23–46). In the view of Andrew Cunningham, Vesalius chose his order of dissection "because this is precisely the dissecting sequence that Galen recommends." Cunningham, *The Anatomical Renaissance,* p. 114.

19. Crooke, *Microcosmographia,* fol. 19; Paré, *Works* (1678), bk. III, p. 54. Paré further divides the "lower belly" into those which "contain" and those which "are contained" (p. 55). For Crooke, the order of "Preservation or continuance" is among those methods that may be employed when there is a scarcity of cadavers, and one needs to enquire into everything in the one body. He adds, especially "if we would keepe a body long, the dissection must be begun at those parts which are most subject to corruption; & therefore, first cut up the lower belly, then the Chest, after the Head, and lastly the ioynts." In de Laguna, *Anatomica methodus,* the "method of discipline" is said to follow "necessity," and begins the dissection "from the thicker intestines lest their smoky excrements which putrefy in short order interfere by means of their invisible heaviness of odor with our contemplation of the noble members. . . ." (p. 265) De Laguna follows the method of discipline, yet he makes it clear why he does so: "Compelled by utility and necessity, however, I employ the method of the discipline and have arranged my material in reverse order, placing that which is posterior before that which is prior. . . ." (p. 265). Many other anatomists adopt a similar expediency: the method Berengario follows in *Isagogae Breves* of dividing the body for dissection into three cavities plus the extremities is that of Mondino. The division is distinguished by animal (i.e., spiritual, from *anima,* spirit, soul), vital, and natural members (brain, heart, liver). Veslingus explains that he has divided his anatomy textbook into chapters according to the method whereby "publickly it is demonstrated in the Anatomies, and the Series of every Chapter is in the same manner as it is shewed to the eye in publick Dissections." *The Anatomy of the Body of Man,* sig. C. Thus, the anatomy of the skeleton and muscles in chapters 1 and 2 is followed by that of the "Mesenterium, Sweet-Bread [pancreas], Liver and Spleen," in chapter 4, which is followed by the rest of the nutritive members in chapter 5, the anatomy "Of the Instruments of Generation in Man," in chapter 6, and in "Women" in chapter 7. Then follows embryology and the anatomy of the fetus or "Fruit in the Womb," in chapter 8. Consistent with the method of necessity, the anatomy continues with the thorax, the "Heart and Lunges" (chapter 10), and eventually onto "the Brain and Cerebellum" (chapter 14), and "the Eyes" (chapter 15). The method of discipline is to be identified with the method Galen employs when understanding what the parts do with respect to the whole. See Cunningham, *The Anatomical Renaissance,* p. 28.

20. See Larkey, "The Vesalian Compendium of Geminus and Nicholas Udall's Translation," pp. 368, 375, who says that Vicary's *Profitable Treatise* is taken from an earlier English manuscript, which has disappeared and was "based largely on Henri de Mondeville, a surgeon of the thirteenth century" (pp. 374–75).

21. Larkey, "The Vesalian Compendium of Geminus and Nicholas Udall's Translation," pp. 378, 385–86

22. Ibid., p. 376. The copy referred to in the text is said by Larkey to be in The British Library.

23. Ibid., pp. 386, 390

24. Ibid., p. 370

25. *The Epitome of Andreas Vesalius,* p. 8

26. Banister, *The Historie of Man,* sig. A_{iiii}^{r}. The five Vesalian plates included in *The Historie of Man* are the following: the anterior view of the skeleton, leaning on a spade, like Death the Sexton, which appears prior to the first book, on the page opposite to sig. C_{i}^{r}; the posterior view of the skeleton, mourning or praying, with a lily (of funereal significance) added to the picture, on sig. m_{ii}^{r} (fol. 38^{r}); an anterior view of a muscle-man, entitled "The forepart of the Muscles," on [sig. n_{iii}^{v} (fol. 43^{v})]; a posterior view of a muscle-man, entitled "The backe part of the Muscles," on sig. s_{iii}^{r} (fol. 63^{r}); "A Table of the Instrumentes seruyng to Anathomicall dissection," at the end of book IX, between sig. Hh_{iiii}^{v} and sig. I_{i}^{r}.

27. Ibid., sig. A_{iiii}^{v}. In the quoted passage, Banister refers to Leonhart Fuchs (1501–66), probably his *Institutionum Medicinae, ad Hippocratis, Galeni, aliorumque veterum scripta recte intelligenda mire utiles Libri Quinque* (Lyon, 1555), which served as an introduction to the big medical texts of ancients, and in which he adopted some Vesalian doctrines. See French, *Dissection and Vivisection in the European Renaissance,* pp. 179, 253–54. The reference to Realdo Colombo, *De re anatomica* will become more significant in the following chapter.

28. Banister, *The Historie of Man,* sig. C_{i}^{r} (fol. 1^{r}); sig. M_{ii}^{v} (fol. 38^{v}); sig. N_{i}^{r} (fol. 41^{r}); sig. N_{iiii}^{r} (fol. 44^{r}); [sig. S_{iii}^{v} (fol. 63^{v})]; sig. Bb_{i}^{r} (fol. 85^{r}); sig. Cc_{i}^{r} (fol. 89^{r}); sig. Ee_{ii}^{r} (fol. 98^{r}); sig. Hh_{iii}^{v} (fol. 111^{v}).

29. Ibid., sig. C_{i}^{r} (fol. 1^{r}). The passage quoted in the text is quoted in full in note 401, in chapter 1. (The term which in this citation appears in italics is in Roman font in the original text, where the remainder of the text is in gothic.)

30. Ibid., sig. C_{i}^{r} (fol. 1^{r}). The passage quoted in the text is quoted in full in note 401, in chapter 1.

31. Ibid., sig. A_{iiii}^{v}.

32. Ibid., sig. Bb_{i}^{r} (fol. 85^{r})

33. Ibid., fol. 89^{r} (sig. CC_{i}^{r})

34. On the date of the award of Fletcher's fellowship, see J. and J. A. Venn, *Alumni Cantabrigiensis: A Biographical List of all Known Students, Graduates and Holders of Office at the University of Cambridge, from the Earliest Times to 1900,* 2 parts: Part I, Vols. I–IV; Part II, Vols. I–VI (Cambridge: Cambridge University Press, 1922–54), Part I, Vol. II, p. 150, col. 1. Also see Langdale, *Phineas Fletcher,* p. 40. King's College still holds the copy of *The Historie of Man* which Banister donated (M.73.13 formerly F.12.8). The copy is referred to in early author catalogues, classified catalogues, and lists of donors. See King's College Lib[rary] MS. 1, fols. 23^{r}, 56^{r}, 145^{r}, 170^{r}, 177^{r}; Lib. Ms. 2, fol. 49; Lib. MS. 13. E. S. Leedham-Green, *Books in Cambridge Inventories: Book-Lists from Vice-Chancellor's Court Probate Inventories in the Tudor and Stuart Periods,* 2 vols. (Cambridge: Cambridge University Press, 1986): Volume I of this work, entitled "The Inventories," is a transcription of all the book lists included in inventories presented for probate in the Court of the Vice-Chancellor, strictly of the Chancellor. Volume II, entitled "Catalogue," is an author / title catalog of the books, nearly twenty thousand of them, with a subject catalog, on sixteenth-century lines, of the three thousand odd authors and/or headings under which the books are listed. See the Preface in Vol. I for further explanatory details. The anonymous inventory of 1588/89 was appraised by experts, Reignold Bridges, himself a bookseller, and "Hiero Bowricell" (I, 402). Item 245 appears on folio 3 of the inventory, is valued as 1 s. 6 d, and referred to as "banisters anatomy parchment" (II, 72). Whether this suggests a copy of *The Historie of Man* bound in limp vellum (rather than printed on parchment) is unclear, but Leedham-Green identifies it as the published work of 1578 (*STC* 1359). I have not succeeded in tracing the item in any of the Cambridge libraries. According to the *Catologue of Books Printed on the Continent of Europe, 1501–1600, in Cambridge Libraries,* compiled by H. M. Adams (Cambridge: Cambridge University Press, 1967), 2 vols. (the form of the appearance of authors' names and other headings in which, Leedham-Green, *Books in Cambridge Inventories,* borrows for consistency and purposes of cross-reference), there are five copies of the first edition of the *Fabrica* in Cambridge libraries, although not King's College Library—the University Library has two copies, and Clare College, Peterhouse and the School of Anatomy each have one copy (Vol. II, 322, col. 1). One of the University Library's copies has a bookplate, which Dr. J. J. Hall, Under-Librarian, Rare Books Department, informs me indicates that it is from the library of John Moore, Bishop of Ely (d. 1715), and presented to the University Library by King George I in 1715. Dr. Hall cannot tell when the other copy held by the University Library was acquired, but thinks it likely that it was after Fletcher's time. The copy in Peterhouse Library was bequeathed by Perne, in 1589 (see Leedham-Green, *Books in Cambridge Inventories,* II, 781). I do not know at what date a copy of the *Fabrica* of 1543 was acquired by Clare College, but it does not appear in Leedham-Green's author / title catalog of books in book-lists derived from inventories presented during the Tudor and Stuart periods, and so was probably acquired by Clare College later. The School of Anatomy did not exist until long after Fletcher's time. So, of these copies of the first edition of the *Fabrica* the accession of only one—that held by Peterhouse—occurred prior to or during Fletcher's periods as scholar and Fellow of King's. There is no reason to

suppose, however, that as a Fellow of King's, Fletcher would have had access to the Library of Peterhouse, and consequently to the copy bequeathed by Perne. Dr. David McKitterick, Librarian of Trinity College Library, informs me that as a Fellow of King's, Fletcher would not normally have access to other college libraries, though he might have negotiated his way in. Langdale, *Phineas Fletcher,* makes no suggestion, however, that any of Fletcher's Cambridge friends were connected with Peterhouse. A 16o edition of the *Fabrica* was issued from Lyons, by Jean de Tournes, in 1552, and a copy of this was acquired by the University Library, although Dr. Hall is unable to say when, except that again it is likely to have been after Fletcher's time.

Of the second folio edition of the *Fabrica,* published in 1555, there are single copies in Cambridge University Library, Gonville and Caius, Emmanuel, and St John's (Vol. II, p. 322, col. 1). Dr. Hall informs me that the University Library copy bears the signature or monogram of Thomas Lorkyn, who matriculated as a pensioner from Pembroke College in 1549, was elected to a fellowship at Queen's in 1551, proceeded B.A. and was elected to a fellowship at Peterhouse in 1553, graduating M.A. in 1555, was licensed to practice medicine in 1559, proceeded M.D. in 1560, and was Regius Professor of Physic from 1564 to 1591 (see Leedham-Green, *Books in Cambridge Inventories,* I, 492). The University Library's copy of the second folio edition of the *Fabrica* was indeed a part of Lorkyn's bequest to the Library in his will, which was made and proved in 1591, although "there is reason to believe that the books [in the inventory in his will, among which was the *Fabrica*] were not received, or at any rate not finally arranged, until the end of 1594." But this was still in advance of Fletcher's matriculation at King's. J. C. T. Oates, *Cambridge University Library: A History,* 2 vols. (Cambridge: Cambridge University Press, 1986), I: *From the Beginnings to the Copyright Act of Queen Anne,* 143. According to correspondence which I have received from Mr. Mark Statham, Librarian of Gonville and Caius College, the copy belonging to that library was acquired from Thomas Grymstone, Fellow of the College, who died in March 1607/08, and I have been able to verify this. His will was published in August 1608, in which he specified that a number of his books, including the Vesalius, be given to the library, where it resides to this day. Although this date falls within the period of Fletcher's fellowship at King's, I can find no reason to suppose that Fletcher would have had access to this copy, the same conditions and restrictions applying as with Peterhouse Library. Nothing in Grymstone's will suggests otherwise. I have been told in correspondence from Dr. H. C. Carron, Librarian of Emmanuel College, that the copy in the Library of that College is from the collection of William Sancroft (1617–93), Master of Emmanuel (1662–64), Archbishop of Canterbury (1677–1691), whose bequest of the bulk of his private library to the College, in two batches in 1693 and 1694, is obviously too late to be of relevance to Fletcher. The copy in St John's College Libary was given by, or bought with money given by, John Collins (d. 1634) (St. John's B.A. 1595–96, M.D. 1608, F.R.C.P. 1613), Fellow of St. John's (1598) and later Regius Professor of Physic 1626–34), according to correspondence which I have received from Mr Jonathan Harrison, Special Collections Librarian, at St. John's College. The donation label pasted inside the copy indicates its bequest in the year of Collins's death, which is of course too late to have rendered it useful to Fletcher. Another folio edition of the *Fabrica* was published in 1568 at Venice, and single copies of this were acquired by Gonville and Caius College, Pembroke College, and Sidney Sussex College. I have been unable determine the accession dates of any of these copies, despite the efforts of Mr Mark Statham at Gonville and Caius, but none of these copies appears in Leedham-Green's author / title catalogs of books, and so they were probably acquired after the Tudor and Stuart periods. For the information which has been recorded in this note, and more besides, I am grateful to Dr. H. Carron, Dr. J. J. Hall, Mr. Jonathan Harrison, Dr. David McKitterick, and Mr. Mark Statham.

35. Leedham-Green, *Books in Cambridge Inventories,* I, 493. Nearly half of Lorkyn's library of 570 books consisted of medical books, that is, those which his will names as "phisicke bookes" and "those that appertaine . . . to phisicke." Oates, *Cambridge University Library,* I, 142.

36. Rotherham's library room was the upper room of the eastern range built in 1469–71(?) to complete the quadrangle of the Schools. Oates, *Cambridge University Library,* I, 41–42. Oates (p. 142) adds: "It was thus Lorkyn's intention that his books should be used in the teaching of medicine. This is a development for which the Library's regulations of 1583 make no provision, since only Masters of Arts, Bachelors of Law or Physic, and Doctors were admitted as readers in their own right. Thus if by 'Students of Phisicke' Lorkyn meant undergraduates pursuing the six-year course that would turn them into Bachelors, he must have envisaged them visiting the Library in the company of their professor as his 'Auditours' while he displayed to them copies of the texts on which he was lecturing."

37. For this view, I am indebted to Dr. J. J. Hall, Under-Librarian of the Rare Books Department of Cambridge University Library.

38. Leedham-Green, *Books in Cambridge Inventories,* I, 493; Oates, *Cambridge University Library,* I, 142.

39. William B. Hunter Jr, ed., *The English Spenserians: The Poetry of Giles Fletcher, George Wither, Michael*

Drayton, Phineas Fletcher and Henry More (Salt Lake City: The University of Utah Press, 1977), p. 107. For a study of the mutual influence between Giles Fletcher's *Christ's Victorie, and Triumph* and his brother's *Purple Island,* and of the latter in the context of the "Spenserian poets" generally, see Grundy, *The Spenserian Poets,* ch. 9, pp. 181–203; also pp. 8, 42.

40. Kastor, *Giles and Phineas Fletcher,* p. 106; also see pp. 110, 113.

41. Langdale, *Phineas Fletcher,* p. 48; Kastor, *Giles and Phineas Fletcher,* p. 92.

42. Kastor, *Giles and Phineas Fletcher,* p. 106; also see p. 121.

43. Langdale, *Phineas Fletcher,* p. 52.

44. Crooke, *Microcosmographia,* fol. 19. Andrés de Laguna calls the order of dignity, the order of nobility. See French, *Dissection and Vivisection in the European Renaissance,* p. 225.

45. French, *William Harvey's Natural Philosophy,* pp. 21–22.

46. Langdale, *Phineas Fletcher,* pp. 182; 183. *The Purple Island*'s descriptions of the "epiploon" (gastrocolic omentum) and "mesenterium" (mesentery) occur in Canto 2, stanza 45 and the marginals x and y. In addition, Langdale, *Phineas Fletcher,* p. 184, argues that Fletcher's "word for the functioning of the kidneys, namely distillation [*P.I.* 3.20.5], implies a better knowledge of the nephritic glands than was owned by Galen, Mundinus, or Vesalius, each of whom believed that the urine was literally filtered off." However, the term "distillation" may signify alchemically as well as physiologically, and is discussed in relation to concoction in the heart, in the following chapter.

47. Langdale, *Phineas Fletcher,* pp. 193–95.

48. Ibid., p. 194.

49. Banister, *The Historie of Man,* fol. 102^{v} (sig. $Ff_{ii}{}^{v}$)

50. Johann Kepler, *Ad vitellionem, paralipomena* (Frankfurt, 1604).

51. In *De re anatomica,* Colombo "states that he was the first to describe . . . the stapes," but Ingrassius's "discovery was made known in 1548 to Falloppius, who inquiring of his friends at Rome about it was assured by them that neither Colombo nor any one else had ever mentioned it." Foster, *Lectures on the History of Physiology During the Sixteenth, Seventeenth and Eighteenth Centuries,* p. 31. Also see Cole, *A History of Comparative Anatomy From Aristotle to the Eighteenth Century,* p. 57; Langdale, *Phineas Fletcher,* p. 195.

52. Healy, "Sound Physic," p. 343.

53. Langdale, *Phineas Fletcher,* p. 193, claims that Fletcher omits the special functions traditionally assigned to the three cerebral ventricles—sensation, imagination, memory—which would have adapted well to the fanciful arrangement of the poem, but which were omitted because they were discredited anatomically. Baldwin, "Phineas Fletcher," pp. 473–74, says that Langdale's statement is demonstrably false. In fact, while Fletcher omits these functions from his anatomy of the brain in Canto 5, he describes a version of them in Canto 6. Rather than assigning them to the cerebral ventricles, however, he locates common sense and fancy in the central and anterior parts of the brain, respectively, implying that Eumnestes, memory, is located in the posterior part. For the intertextual relationship between this account in *The Purple Island* and that in Spenser's *Faerie Queene,* see chapter 4. For accounts in anatomy textbooks, see, for example, de Laguna, *Anatomica methodus,* p. 287, which locates common sense in the anterior and memory in the posterior of the brain. Berengario, *Isagogae breves,* pp. 143–44, at first places reason in the third ventricle, which is situated in the middle of the cerebrum, common sense in the anterior of the cerebrum (the lateral ventricles), and memory behind the third ventricle (possibly in the fourth ventricle). Berengario then disagrees with this assignment and places cognition, reason, and memory in the first two (the lateral) ventricles, and acknowledges that the middle (third) ventricle does have some relation to common sense. In the seventeenth century, Veslingus, *The Anatomy of the Body of Man,* pp. 59–60, places memory in the cerebellum (still, therefore, in the posterior region of the brain, but no longer in a ventricle). The differentiation of brain function in regions of the brain can of course be scientific: see *The Concise Home Doctor,* I, 226, col. 3.

54. De Laguna, *Anatomica methodus,* pp. 286–87. Ever the rational anatomist, de Laguna argues that "it is consistent with reason to suppose that the faculty from which motion and sense perception proceed [i.e., the rational faculty] would not excel in heat but would be as temperate as possible, since qualities which are excessive do not effect or promote sense perception but destroy it the fullest extent."

55. Berengario, *Isagogae breves,* p. 137 (pp. 146–47).

56. Achillini, *Annotationes Anatomicae,* pp. 60–61. Achillini refers to the statement by Averroes, in his *Colliget,* I, Ch. 16, that "vertebrae of the nape of the neck cover the net just as the cranium covers the brain"; in response to which he asks, "when the bone is broken is the marvellous net in the direction of the colatorium?" And he claims that this is where he saw it in 1503.

57. Berengario, *Isagogae breves,* pp. 147, 210. Lind writes: "It is true that Berengario does not say he has seen

it and he does say 'according to common opinion' [in the title of the relevant section of *Isagogae breves,* i.e., "Concerning the Marvelous Net According to Common Opinion, and Something About the Nape of the Neck" (p. 146)], and 'as the authors commonly say' [p. 146], but he doesn't deny its existence." I think that Lind's desire for Berengario to have more emphatically broken from the ranks of textual tradition into the realm of observation and the senses than he actually did, leads him to underestimate the actual significance of Berengario's declaration not to have seen this organ, which was believed to play a vitally important function. Berengario repeats his negative declaration on p. 147 and effectively denies its existence: "Yet I have never seen this net, and I believe that nature does not accomplish by many means that which she can accomplish by few means." The Circle of Willis is named such because it was studied by Willis. It is an "arterial circle [though by no means round in shape], an anastomosis in the subarachnoid space at the base of the brain," and is "formed by the anterior and posterior cerebral arteries and the communicating arteries of Willis." An anastomosis is a communication between two or more vessels; a "union of ramifications of leaf veins; union of blood vessels arising from a common trunk; union of nerves; . . . formation of a network or anastomotic meshwork." The subarachnoid space is "a cavity filled with cerebrospinal fluid between [the] arachnoid and pia mater." *The Catholic Encyclopedia,* I, 460, col. 1; *Henderson's Dictionary of Biological Terms,* 9th edn. (London: Longman, 1979), pp. 21, 465, 417–18; Dunglison, *A Dictionary of Medical Science,* p. 250, cols. 1–2.

58. Berengario, *Isagogae breves,* p. 137 (pp. 141, 147). Also see Larkey and Suden, "Jackson's English Translation of Berengarius of Carpi's 'Isagogae Breves', 1660 and 1664," p. 69.

59. Berengario, *Isagogae breves,* pp. 141, 147.

60. Ibid., pp. 147, 209. Berengario mentions the "vermis [worm] of the cerebellum" in this respect on more than one occasion (also see pp. 141–42).

61. Ibid., pp. 137, 141–42.

62. *Stedman's Medical Dictionary,* 26th edn., illustrated in color (London: Williams & Wilkins, 1995), p. 1384, col. 2; Dunglison, *A Dictionary of Medical Science,* p. 242, col. 2. W. G. Hayle and J. P. Margham, *Dictionary of Biology* (London and Glasgow: HarperCollins, 1988), p. 128, describes the choroid plexus as "a non-nervous EPITHELIUM which projects into and forms part of the roof of the brain of vertebrates. It secretes the CEREBROSPINAL FLUID." See Snell, *Clinical Anatomy for Medical Students,* pp. 800–801, for a description of the production of cerebrospinal fluid by the choroid plexuses within the lateral, third, and fourth ventricles of the brain.

63. Eriksson, *Andreas Vesalius' First Public Anatomy at Bologna 1540,* p. 25.

64. Ibid., p. 221.

65. Ibid.

66. Ibid., pp. 289–91 (my emphases). Corti, who delivered the accompanying lectures, by contrast, confirmed the existence of the *rete mirabile* with very little comment, as though its location and function were uncontentious (p. 155).

67. See Ibid., pp. 25–26, whose opinion is that because Vesalius appears to have demonstrated the *rete mirabile* in the sheep's brain, he inferred its existence in the human brain. I think this claim may be made only advisedly, as in the fifteenth demonstration Vesalius is recorded as saying that the creation of animal spirit occurs in the "network of arteries" which extend to the choroid plexuses, and in the twenty-fifth demonstration Vesalius was locating the part in question higher up in the cranium than it was traditionally thought to be, and therefore identifying the choroid plexuses as the locus of the creation of animal spirit.

68. See Singer, *The Evolution of Anatomy,* p. 97, which claims that Berengario "denies the existence of the *rete mirabile* below the brain, thereby contradicting not only Mondino but also Galen." Berengario is also said to advance on Mondino in "recognizing the general form of the [cerebral] ventricles, the composition of the choroid plexus out of the arteries and veins, the pineal gland and the relations of the fourth ventricle."

69. Eriksson, *Andreas Vesalius' First Public Anatomy at Bologna 1540,* pp. 25–26, argues that, "being aware that both in his writings and during his lectures he had accepted the rete mirabile in human beings too, [Vesalius] was compelled to give an explanation: so as not to be judged unskilled in dissection he always had at his demonstration, in addition to a human brain, the brain of a sheep, and in that way he was able to show the rete mirabile that he did not find in human brains." This explanation seems not to account for Vesalius's reticence with regard to the function of the traditionally located *rete mirabile* and his insistence on its effective (that is, functional) location higher in the cranium than was thought before Berengario.

70. Singer, *Vesalius on the Human Brain,* p. xx. Elsewhere Singer argues that Vesalius "does not wholly reject" the existence of the *rete mirabile,* by which I can only presume that he means that Vesalius moves the physiological function associated with the *rete mirabile* to another structure as opposed to rejecting the function entirely. Singer, *The Evolution of Anatomy,* p. 132. Singer draws from his observation that Vesalius says in the *Fabrica* that

his figure of the *rete mirabile* is made to fit the description of Galen, the conclusion that this "shows how deeply steeped [Vesalius] was in the Galenic physiology."

71. *The Epitome of Andreas Vesalius,* pp. 61–66.

72. Ibid., p. 69; also see Lind's summary remarks on p. 76.

73. Langdale, *Phineas Fletcher,* p. 193. No critic subsequently, including those who deny the *Purple Island* anatomy's modernity, has cared to comment on whether the *rete mirabile* is omitted or not.

74. *Dorland's Illustrated Medical Dictionary,* 28th edn. (London: W. B. Saunders, 1994), p. 1309, col. 1.

75. I have been unable to determine whether it is entirely consistent with this identification of the "knot of veins and arteries" as the choroid plexus that "Hereby two little hillocks joyntly rise," described in the marginal note (i) as "two little bunches like teats, the instruments of smelling" (*P.I.* 5.15.5 & margin. I). These may be the basal ganglia, "masses of grey nervous tissue, that is tissue containing nerve cells, situated at the lower part of each hemisphere of the brain just above its peduncle or stalk." These structures are "concerned with the appreciation of sensation, and with movement," although *The Purple Island* calls them "the instruments of smelling" (*P.I.* 5.15 margin. i). *The Concise Home Doctor,* I, 153, col. 2 and figure.

76. Banister, *The Historie of Man,* fol. 96^{r} (sig. Dd$_{iiii}$^{r}); fol. 100^{r} (sig. Ee$_{iiii}$^{r}). The anterior or lateral ventricles are referred to by Banister as the "upper ventricles" on account of his acceptance of Realdo Colombo's preference to call them thus, or "the superiour: for asmuch as they are contained or situated aboue the rest." Banister follows Colombo, suggesting that this "large and notable plicature, or weavyng," which "may worthely, for the marueilous workemanshyp thereof, be called the marveilous nette," is formed of the branches of "foure Arteries": "two Arteries named *Carwtides,*" and two which arise from either the "*Soporariæ*" (thus called the "distributions of the *Soporiferous* Arteries") or "from the arteries *Axillares,*" "the whiche Arteries haue their Veynes to them ioyned in fellowshyp" (fol. 96r [sig. Dd$_{iiii}$^{r}]; fol. 95^{v} [sig. Dd$_{iii}$^{v}]; fol. 96^{v} [sig. Dd$_{iiii}$^{v}]). A "plicature" is a "Folding, mode of folding; a fold." *OED,* VII, "P," p. 999, col. 3.

77. Ibid., fol. 100^{v} (sig. Ee$_{iiii}$^{v}).

78. Singer, *Vesalius on the Human Brain,* p. 32.

79. Banister, *The Historie of Man,* fol. 100^{v} (sig. Ee$_{iiii}$^{v}).

80. Ibid., fol. 98r (sig. Ee$_{ii}$^{r}).

81. Ibid., fol. 98r-v (sig. Ee$_{ii}$$^{r-v}$).

82. The acqueduct of Sylvius, or "cerebral aquaduct," is the "narrow cavity of the midbrain . . . which connects the third and fourth ventricles." Snell, *Clinical Anatomy for Medical Students,* p. 805.

83. Langdale refers to Foster, *Lectures on the History of Physiology During the Sixteenth, Seventeenth and Eighteenth Centuries,* p. 147. Langdale, *Phineas Fletcher,* p. 193 and note 57.

84. *The Epitome of Andreas Vesalius,* p. 69.

85. Banister, *The Historie of Man,* fol. 100^{v} (sig. Ee$_{iiii}$^{v}).

86. Ibid., fol. 98^{v} (sig. Ee$_{ii}$^{v}).

87. A tangible alternative whereby Fletcher may have gained access to and read a copy of Banister's *Historie of Man* is explored in Appendix B.

88. Banister, *The Historie of Man,* fol. 1r (sig. C$_{i}$^{r}). References to "God" include the following: "But besides, unto man, for the apt acte of generation, the omnipotent maker hath giuen a member (called in Latin *Coles, penis, mentula, virga,* or *Pudendum virile,* in English the yard) most fit for the effusion of seede into the wombe. . . ." fol. 87^{v} (sig. Bb$_{iii}$^{v})

Chapter 8. The Heart, Veins, and Arteries

1. In early sixteenth-century anatomy, the shape of the heart was derived symbolically from its complexion. The explanations of this derivation take two forms: the Platonic is based on the correspondence of complexion to the elements, and the heart follows fire, which "tends naturally to take the form of the pyramid"; the version more readily accepted in the sixteenth century is possibly medieval in origin, and holds that the heart is shaped thus "because a pyramid signifies heat." Achillini, *Annotationes Anatomicae,* p. 53. The view that form or shape was derived from complexion was countered by Galenists, like Berengario da Carpi—"It has the form of a pyramid, humped however, not principally because it is hot, following the form of fire, but because it is mixed, perfect, and animate and so possesses a form suitable for its operation"—and Matteo Corti, who says that if heat were the reason for the heart's shape, "the liver ought to have such a figure, as it, too, is warm in man"; but while concluding

that "the pyramidic shape of the heart will be evident from its operation," Corti finds the explanatory power of heat too attractive to dismiss. Berengario, *Isagogae breves,* p. 94; *Andreas Vesalius' First Public Anatomy at Bologna 1540,* pp. 231, 231–33. Although not explicitly complexional in origin, the pyramidal form of the heart in *The Purple Island* is implicitly related to its sovereignty, as is its dispensation of innate heat. That the heart is said to be "of a figure pyramicall" is hermeneutically indeterminate (*P.I.* 4.17 margin), for from the fourteenth century, "figure" signifies "a shape" or "form," and "a literary device." Barnhart, ed., *Dictionary of Etymology,* p. 381, col. 2.

2. Sawday, *The Body Emblazoned,* p. 180.

3. Langdale, *Phineas Fletcher,* pp. 189–92.

4. Langdale contends that, "All signs point to the conclusion that the poet's understanding of the blood system was much closer to Harvey's than we have heretofore suspected." Ibid., p. 192.

5. Gweneth Whitteridge, *William Harvey and the Circulation of the Blood* (London: Macdonald; New York: American Elsevier Inc., 1971), p. 44; Eriksson, ed., *Andreas Vesalius' First Public Anatomy at Bologna 1540,* pp. 233–35.

6. See Cunningham, *The Anatomical Renaissance,* Ch. 6 "Fabricius: The Revival of Aristotelian Anatomy," pp. 167–87 (pp. 172–75). A perusal of the titles (translated into English) of some of the books by Fabrici and by Harvey allows one to see how their shared physiological conceptualization of anatomy affected their approach to the body. By Fabrici: *On Vision; or, On the Eye the Organ of Sight* (1600); *On Speech and its Instruments* (1603); *On Respiration and its Instruments* (1615). By Harvey (supplying the original Latin title for the book referred to later in the text): *Exercitationes de generatione animalium* (Disputations touching the generation of animals) (London, 1651). Hereafter reference to this work, either specifically in this edition, or, in more general discussion of the work, in no specific edition, will be made to *De generatione,* unless otherwise specified.

7. Cunningham, *The Anatomical Renaissance,* p. 175.

8. Langdale writes emphatically, with reference to Canto 2, stanza 9, line 7: "Surely the author of those words had some intimation of the doctrine of complete circulation." *Phineas Fletcher,* p. 192.

9. Whitteridge, *William Harvey and the Circulation of the Blood,* p. 43.

10. *OED,* VI, "M," p. 46, col. 3, gives the date 1555 as the earliest for the use of main ("mayne") for mainland, and 1579–80 as the earliest for the use of main ("maine") for main sea. (Some words that are in upper case in the cited work, I have rendered in lower case in the text, except for the initial letter.)

11. Whitteridge, *William Harvey and the Circulation of the Blood,* pp. 43–44.

12. According to Nicolson, *The Breaking of the Circle,* p. 138, note 6, Fletcher "still followed the most conventional belief that the blood in the arteries differed from that in the veins." Berengario, *Isagogae breves,* pp. 108–9 provides a basic set of distinctions between the veins, which have a "simple, soft tunic," and the arteries which "have two tunics and are hard." The arteries are also not supported by any solid body. He adds that the blood in the veins is "more apt to descend than to ascend because it is heavy," whereas that in the arteries is "light" for they contain more spirit (*pneuma zoticon* or vital spirit).

13. For a clear and concise account of the function of the liver as producer of the blood in Galenic physiology, see Whitteridge, *William Harvey and the Circulation of the Blood,* pp. 41–42.

14. For the hepatic origin of the veins in the anatomy of Corti and of Fabrici, see, respectively, Eriksson, ed., *Andreas Vesalius' First Public Anatomy at Bologna 1540,* p. 243; Whitteridge, *William Harvey and the Circulation of the Blood,* p. 25. For Vesalius's observations on the cardiac origin of the veins in the anatomy of Aristotle, and for this view in the earlier anatomy of Harvey, see, respectively, Eriksson, ed., *Andreas Vesalius' First Public Anatomy at Bologna 1540,* p. 245; *The Anatomical Lectures of William Harvey: Prelectiones Anatomie Universalis, De Musculis,* edited, with an introduction, translation, and notes by Gweneth Whitteridge (Edinburgh and London: Published for the Royal College of Physicians, London, by E. & S. Livingstone Ltd., 1964), pp. 253–55. For Harvey's final opinion that the blood is "[w]ithout all question . . . to be counted *Author* of the *Liver,* rather that the *Liver* the *Author* of the *Blood,*" and further that, contrary to the opinion of Aristotle, the heart is "born after the Blood," and that even the veins embryologically precede the heart, being the "first visible foundation of the body," see Harvey, *De generatione,* Ex.xix, p. 115, cited in Whitteridge, ed., *The Anatomical Lectures of William Harvey,* pp. 254, note 1.

15. For Harvey's observations in support of the opinion that the heart is embryologically prior to the liver, see Whitteridge, ed., *The Anatomical Lectures of William Harvey,* pp. 253–55.

16. Whitteridge, *William Harvey and the Circulation of the Blood,* p. 41. Also see Whitteridge, ed., *The Anatomical Lectures of William Harvey,* p. xxxix.

17. Whitteridge, *William Harvey and the Circulation of the Blood,* p. 43.

18. Foster, *Lectures on the History of Physiology During the Sixteenth, Seventeenth and Eighteenth Centuries,* pp. 36–37. Also see Whitteridge, ed., *The Anatomical Lectures of William Harvey,* p. xli.

19. Alessandro Benedetti first used the term "valve" to signify these structures in the heart. Singer, *The Evolution of Anatomy,* p. 104. Cunningham explains that Fabrici uses the term "little door" because, as an Aristotelian physiological anatomist, he chose to name the part from its use or function, which was to slow down, not to prevent, the progress of the blood. Cunningham, *The Anatomical Renaissance,* p. 180. According to Geoffrey Keynes, *The Life of William Harvey* (Oxford: Clarendon Press, 1966), p. 29, Fabrici's claim to be the first observer of the *ostiola* in 1574 was not justified: "They had been demonstrated in the *azygos* and renal veins by Giambattista Canano about 1536 and shown to Vesalius at Ferrara when he visited there in 1545; but Canano understood their function no better than Fabricius and did not publish his observations. Claims have also been made for Charles Estienne and Jacque Dubois (Sylvius), but their observations were too meagre to be seriously considered. More serious are the claims made for the theologian, Pietro Paolo Sarpi (or Father Paul), who is stated by his biographer [Friar Fulgenzio Micanzio] to have discovered the valves in veins by dissection and to have demonstrated them to Fabricius." Father Paul's investigations suggest further the imbrication between anatomy and theology, not only in discourse but also in practice.

20. Fabrici, *De venarum ostiolis* (1574), cited in Foster, *Lectures on the History of Physiology During the Sixteenth, Seventeenth and Eighteenth Centuries,* p. 37. Also see Singer, *The Evolution of Anatomy,* p. 158; Whitteridge, ed., *The Anatomical Lectures of William Harvey,* p. xli.

21. Whitteridge, *William Harvey and the Circulation of the Blood,* p. 23. Also see Cunningham, *The Anatomical Renaissance,* p. 178.

22. Banister, *The Historie of Man,* fol. 89^r (sig. CC_i^r).

23. *The Epitome of Andreas Vesalius,* p. 59.

24. For a clear and concise account of the function of the heart and its adjoining vessels in Galenic physiology, see Whitteridge, *William Harvey and the Circulation of the Blood,* pp. 41–43.

25. Whitteridge, *William Harvey and the Circulation of the Blood,* p. 42.

26. Eriksson, *Andreas Vesalius' First Public Anatomy at Bologna 1540,* p. 237.

27. Whitteridge, *William Harvey and the Circulation of the Blood,* p. 42. Alternatively, it was thought that "the thick [venous] blood leaving the liver was . . . unsuitable nourishment for the light and frothy lungs," and so the rendering of it fine and thin "had the double advantage of making it suitable for the lungs and allowing it more easily to pass through the pores of the interventricular septum." This was a view to which Vesalius and Fabrici adhered, but not Fletcher. Whitteridge, *William Harvey and the Circulation of the Blood,* p. 47. For Fabrici, see Foster, *Lectures on the History of Physiology During the Sixteenth, Seventeenth and Eighteenth Centuries,* p. 40.

28. Banister, *The Historie of Man,* fol. 89^r (sig. CC_i^r).

29. Ibid., fol. 89^r (sig. CC_i^r).

30. Whitteridge, *William Harvey and the Circulation of the Blood,* p. 42.

31. Andrew Wear, "Introduction," in William Harvey, *The Circulation of the Blood and Other Writings,* trans. by Kenneth J. Franklin with an introduction by Andrew Wear (London: J. M. Dent & Sons, 1990) p. ix.

32. Whitteridge, *William Harvey and the Circulation of the Blood,* p. 42.

33. Wear, "Introduction," in Harvey, *The Circulation of the Blood,* p. viii.

34. Snell, *Clinical Anatomy for Medical Students,* p. 103. "The *tricuspid valve* guards the atrioventricular orifice. . . . It consists of three cusps formed by a fold of endocardium with some fibrous tissue enclosed" (p. 104).

35. Cunningham, *The Anatomical Renaissance,* p. 178.

36. Whitteridge, ed., *The Anatomical Lectures of William Harvey,* p. xli.

37. Berengario, *Isagogae Breves,* p. 96.

38. Ibid., p. 96.

39. Cole, *A History of Comparative Anatomy from Aristotle to the Eighteenth Century,* p. 57. "The *mitral valve* guards the atrioventricular orifice. . . . It consists of two cusps, one anterior and one posterior, which have a similar structure to those of the tricuspid valve." Snell, *Clinical Anatomy for Medical Students,* p. 107.

40. Langdale, *Phineas Fletcher,* pp. 191–92. Both the pulmonary valve and the aortic valve are semilunar. "The *pulmonary valve* guards the pulmonary orifice. . . . It consists of three semilunar cusps formed by folds of endocardium with some fibrous tissue enclosed." "The *aortic valve* guards the aortic orifice and is precisely similar in structure to the pulmonary valve." Snell, *Clinical Anatomy for Medical Students,* pp. 104, 107.

41. Adams, *Catalogue of Books Printed on the Continent of Europe, 1501–1600, in Cambridge Libraries,* lists no copies of *De venarum ostiolis.*

42. Whitteridge, *William Harvey and the Circulation of the Blood,* pp. 42–3.

43. Ibid., p. 44. The four rivers are the Nile, Ganges, Tigris, and Euphrates.

44. Singer, *The Evolution of Anatomy,* pp. 183–84.

45. Banister, *The Historie of Man,* fol. 94[r] (sig. Dd_{iii}[r]).

46. Ibid., fol. 91[r] (sig. Cc_{iii}[r]).

47. Ibid.

48. Ibid., fol. 94[r] (sig. Dd_{iii}[r]).

49. Singer, *The Evolution of Anatomy,* p. 179. Also see Whitteridge, *William Harvey and the Circulation of the Blood,* p. 45.

50. Singer, *The Evolution of Anatomy,* p. 177. Also see Whitteridge, ed., *The Anatomical Lectures of William Harvey,* p. xl.

51. The accounts of pulmonary circulation by Valverde, Colombo, Cesalpino, Bauhin, and Banister are dealt with below in the text, insofar as they relate to the physiology of the heart and adjoining vessels in *The Purple Island.* The accounts by Michael Servetus (1511–53), Paré, Guidi, Carlo Ruini (Ruinus), Johann Guinther of Andernach (1487–1574), Plater, Giulio Cesare Aranzi, and the positions of these anatomists on the theory, are referred to in relation to the former anatomists whose accounts are regarded in this study as more comparable with that in *The Purple Island,* and/or as possibly constituting source material for that poem. For an overview of the formation of, defense of, questioning of, and opposition to, the theory of pulmonary circulation in the period up to Harvey, see Whitteridge, *William Harvey and the Circulation of the Blood,* ch. 2, esp. pp. 48–68. Also on this subject, see Singer, *The Evolution of Anatomy,* p. 177; French, *Dissection and Vivisection in the European Renaissance,* pp. 247–48.

52. Colombo, *De re anatomica,* p. 177, cited in Whitteridge, *William Harvey and the Circulation of the Blood,* p. 49.

53. For Colombo, see *De re anatomica,* p. 177; Wear, "Introduction," in Harvey, *The Circulation of the Blood and Other Writings,* p. ix. For Paré, see his *Works* (1678), bk. IV, p. 97. For Guido Guidi, see his *De anatomia corporis humani libri VII* (Frankfurt, 1626), bk. VI, ch. 5, p. 301: although observationally forced to conclude that "no pores can be demonstrated going through [the septum] from side to side," Guidi concedes that, "you may say that some portion of [blood] is refined in the right ventricle and passes into the left *per foramina obscura,* through some hidden pores." Whitteridge, *William Harvey and the Circulation of the Blood,* p. 60. For Ruini, see his *Dell'anatomia et dell'infirmita del cavallo* (Bologna, 1598); Whitteridge, *William Harvey and the Circulation of the Blood,* p. 60.

54. Berengario, *Isagogae breves,* p. 95.

55. Whitteridge, *William Harvey and the Circulation of the Blood,* p. 45.

56. Keynes, *The Life of William Harvey,* p. 170.

57. A modern description of the *trabeculae carneae* is provided by Snell, *Clinical Anatomy for Medical Students,* p. 104: "The walls of the right ventricle [of the heart] are much thicker than those of the right atrium and show a number of internal projecting ridges formed of muscle bundles. The projecting ridges give the ventricular wall a sponge-like appearance and are known as *trabeculae carneae.*"

58. Singer, *The Evolution of Anatomy,* pp. 132, 177; Vesalius, *Fabrica,* bk. 6, ch. 2, p. 589, cited in Whitteridge, *William Harvey and the Circulation of the Blood,* p. 46; Roy Porter, *The Greatest Benefit to Mankind: A Medical History of Humanity from Antiquity to the Present* (London: HarperCollins, 1997), p. 212. In the second folio edition of the *Fabrica,* Vesalius promises to give "elsewhere" an alternative explanation of how blood is received into the left ventricle from the right; but, according to Whitteridge, this promise "does not appear to have been fulfilled." Vesalius, *Fabrica* (1555), cited in Whitteridge, *William Harvey and the Circulation of the Blood,* p. 46; also see p. 48 for Whitteridge's comment. Vesalius's skepticism with regard to the intraventricular pores may be seen in relation to his otherwise "purely Galenic" explanations "for the role and function of the various blood vessels connecting the heart and the lung," and of the blood vessels "as comprised of two systems, the arteries containing *aer* or spirit in the blood to vivify the body, the veins carrying blood to nourish it [from the liver]." Cunningham, *The Anatomical Renaissance,* p. 137.

59. Michael Servetus, *Christianismi restitutio: Totius ecclesiæ apostolicæ ad sua limina vocatio, in integrum restituta cognitione dei, fidei Christi, justificationis nostræ* (Vienna, 1553). See Cunningham, *The Anatomical Renaissance,* pp. 247–55; Foster, *Lectures on the History of Physiology During the Sixteenth, Seventeenth and Eighteenth Centuries,* p. 23; French, *Dissection and Vivisection in the European Renaissance,* p. 247; Whitteridge, ed., *The Anatomical Lectures of William Harvey,* p. xl. The thirteenth-century Arabic writer, Ibn an-Nafis, in his "Commentary on the Anatomy in the Canon of Ibn Sina," had arrived at a similar view about the pulmonary transit of blood, but the "Commentary" has not been translated into Latin "and no intermediary source is known."

Owsei Temkin, "Was Servetus Influenced by Ibn an-Nafis?," in *The Double Face of Janus and Other Essays in the History of Medicine,* pp. 284–86 (p. 284). Temkin's analysis of Servetus's ideas also suggests to him that Servetus's "conception is *different* from that of the Arab," thereby supporting the view that "Servetus developed his theory independently from Ibn an-Nafis" (pp. 285, 284). Moreover, as Wear writes, Ibn an-Nafis's ideas had no influence on other anatomists. Wear, "Introduction," in Harvey, *The Circulation of the Blood and Other Writings,* p. ix.

60. Colombo, *De re anatomica,* p. 177, cited in Whitteridge, *William Harvey and the Circulation of the Blood,* p. 49. Also see Foster, *Lectures on the History of Physiology During the Sixteenth, Seventeenth and Eighteenth Centuries,* p. 28; Wear, "Introduction," in Harvey, *The Circulation of the Blood,* p. ix. Banister, *The Historie of Man,* fol. 94^r (sig. $Dd_{iii}{}^r$), merely reiterates Colombo's opinion regarding the interventricular septum.

61. Whitteridge, *William Harvey and the Circulation of the Blood,* p. 56.

62. The problem of how blood got from the right ventricle of the heart to the left exercised the anatomists of the second half of the sixteenth century. See Whitteridge, *William Harvey and the Circulation of the Blood,* ch. 2.

63. *Microcosmos,* p. 28, col. 2, ll. 14–6 (from bottom); p. 29, col. 1, ll. 18–16 from bottom; Sawday, "Bodies By Art Fashioned," p. 121.

64. Wear, "Introduction," in Harvey, *The Circulation of the Blood,* p. ix.

65. Whitteridge, *William Harvey and the Circulation of the Blood,* pp. 64, 56.

66. Ibid., p. 45.

67. Banister, *The Historie of Man,* fol. 89^r (sig. $Cc_i{}^r$); fol. 91^r (sig. $Cc_{iii}{}^r$); fol. 94^r (sig. $Dd_{ii}{}^r$)

68. Gweneth Whitteridge, "Gaspard Bauhin," *Dictionary of Scientific Biography* 1 (1970): 522–25 (p. 523, col. 2); *William Harvey and the Circulation of the Blood,* p. 63. Bauhin, in his first complete anatomy textbook, *De corporis humani fabrica: Libri IIII. Methodo anatomico in praelectionibus publicis proposita* . . . (Basel, 1590), "pointed out that . . . the venous blood could more easily go to the lungs from the right ventricle through the pulmonary artery, there be refined and mixed with air, and return to the left ventricle through the pulmonary vein." All Bauhin's anatomical writings were brought together, revised and enlarged, and published in *Theatrum anatomicum novis figuris aenis ilustratum et in lucem emissum opera et sumptibus Theodori de Bry relictae viduae et filiorum Joann. Theod. et Joann. Israel de Bry* (Frankfurt, 1605); [a new and enlarged edn.] (1621).

69. Whitteridge, "Gaspard Bauhin," p. 523, col. 2. According to Adams's *Catologue of Books Printed on the Continent of Europe, 1501–1600, in Cambridge Libraries,* the only anatomical work by Bauhin in any Cambridge library is *Appendix ad Theatrum anatomicum . . . sive Explicatio characterum omnium* (Frankfurt, 1600), which was frequently bound with the *Theatrum anatomicum* of 1605. Two copies of this appear, one in each of St. John's and the University Library (see Vol. I, 102). Leedham-Green's author / title catalogs of books in book lists derived from inventories presented for probate during the Tudor and Stuart periods, records that the first copy of Crooke's *Microcosmographia* to be bequeathed to the University appears in an inventory made in 1659. Leedham-Green, *Books in Cambridge Inventories,* II, 250.

70. According to French, *Dissection and Vivisection in the European Renaissance,* p. 248, "Leonardo Botali claimed that blood moved from the right to left ventricle through a passage between the auricles, discovered by himself in dissection. He called it the *vena nutrix* and found it sometimes rather tortuous and valve-like in man, but always open in oxen, pigs and dogs. Botali was a pupil of Fallopio, but apparently less skilled than his teacher. He seems not to have understood Colombo completely and he did not know that Galen had already described the structure." French adds (p. 252, note 45): "The structure may perhaps now be called a persistant foramen ovale." Banister, *The Historie of Man,* fol. 89^r (sig. $Cc_i{}^r$), alludes to Botali's theory, "prouyng that bloud is brought into the left Ventricle an other way, that is, by a propper Veyne that is found somewhat aboue the coronall Veyne, neare adioyning to the right auricle, thence with direct tract marchyng to the left"; though Banister does not account Botali among "the best approued" anatomists.

71. Singer, *The Evolution of Anatomy,* p. 33. Erisistratus lived at the beginning of the third century B.C. According to *The Catholic Encyclopedia,* I, 458, col. 1, he "discovered the heart valves and called them, from their forms, sigmoid and tricuspid." On Erasistratus as a rationalist or dogmatist anatomist, see Cunningham, *The Anatomical Renaissance,* p. 24.

72. Charles Singer, *A Short History of Anatomy and Physiology from the Greeks to Harvey* (New York: Dover Publications, 1957) (an unabridged republication of the first edition with a new Preface by the author, originally published as *The Evolution of Anatomy*), p. 32.

73. Ibid.

74. Cunningham, *The Anatomical Renaissance,* p. 198.

75. See French, *William Harvey's Natural Philosophy,* p. 32. Hankinson, *The Sceptics,* p. 226, says that

Herophilus accepted causes on the basis of a hypothesis: "Herophilus offers causal explanations not as being certain, or even probable, but simply as offering a rationally satisfying heuristic reconstruction of the physical world." Hankinson maintains that "Herophilus clearly advocates . . . as a *modus operandi* a procedure which he admits may not capture the way things really are"; and that Herophilus thought that our "epistemological limitations demand that our ambitions be modest—and practical medicine requires no more."

76. See, for example, Cohen, *The Scientific Revolution,* p. 153.

77. Cohen, *The Scientific Revolution,* p. 154.

78. See Whitteridge, *William Harvey and the Circulation of the Blood,* p. 137.

79. Keynes, *The Life of William Harvey,* p. 169, says that Erasistratus shared this belief with Praxagoras of Cos (c. 320 BC), who was the first to distinguish between the veins and arteries.

80. Singer, *The Evolution of Anatomy,* p. 33.

81. For Colombo's theory of the generation of the vital spirits in the lungs, and their further elaboration or perfection in the left ventricle of the heart "as if by the application of the tips of the fingers," see Colombo, *De re anatomica,* bk. VII, p. 223. For a translation of the relevant passage, see Whitteridge, *William Harvey and the Circulation of the Blood,* pp. 53–54. Valverde states that, the heart's left ventricle generates the spirits whose "substance" is "the vaporous exhalation of the fine blood," "together with some part of the air which we take in continuously through the mouth and nose while breathing." Valverde, *De animi et corporis sanitate tuenda* (Paris, 1552), p. 145; *Historia de la composicion del cuerpo humano,* bk. IV, ch. 1, p. 102; both texts cited in Whitteridge, *William Harvey and the Circulation of the Blood,* p. 56. Contradictorily, Valverde also states that, "[t]he principal office of the lungs is to receive the air and prepare it so that from it can be made the vital spirits (if indeed they are made)." Valverde, *Historia de la composicion del cuerpo humano,* bk. IV, ch. 7, p. 104v, cited in Whitteridge, *William Harvey and the Circulation of the Blood,* p. 56. Valverde's parenthetical remark evinces a skepticism he may have shared with Colombo regarding the real existence of the vital spirits (see below for a discussion of this).

82. Banister, *The Historie of Man,* fol. 91^r (sig. $Cc_{iii}{}^r$); also see fol. 92^v (sig. $Cc_{iiii}{}^r$).

83. French, *William Harvey's Natural Philosophy,* p. 82.

84. Jean Guinther (Guinter) of Andernacht (Johannes Guinterius, Andernacus) (1505–74), *De medicina veteri et nova* (Basel, 1571); Plater, *De corporis humani structura et usu.* According to Whitteridge, Plater does not refer to either Valverde or Colombo, and she argues that it is conceivable "that it was from Guinther that Plater derived his knowledge" of the pulmonary transit. Whitteridge, *William Harvey and the Circulation of the Blood,* pp. 62–63; 102.

85. For Paré and Ruini, see Whitteridge, *William Harvey and the Circulation of the Blood,* pp. 58–59; 60–1, respectively.

86. See Whitteridge, *William Harvey and the Circulation of the Blood,* pp. 58–59.

87. Banister, *The Historie of Man,* fol. 94^v (sig. $Dd_{ii}{}^v$).

88. There are also single copies of the first edition (1559) of Colombo's *De re anatomica* in both Cambridge University Library and St. John's College Library. The University Library's copy was presented to the library by King George I in 1715 (see chapter 7, note 34). According to correspondence which I have received from Mr Jonathan Harrison, Special Collections Librarian at St. John's College Library, the copy which that library holds was given by, or bought with money given by, John Collins (see chapter 7, note 34). The donation label pasted inside the copy indicates its bequest in 1634, which is of course the year after *The Purple Island* was first published. Cambridge University Library also holds single copies of editions of Colombo's *De re anatomica* published at Paris in 1572, and at Frankfurt in 1593. These were also presented to the library by King George I in 1715. A copy of the Paris, 1572 edition is also held by Trinity College Library. Dr. David McKitterick informs me that this copy belonged to John Nid, who bought it for a shilling in 1653. Nid's bequest was made to the Library in 1659. St. John's College also holds a copy of *De re anatomica* published at Frankfurt in 1590, though it appears in a catalog which the Library has dating from around the middle of the seventeenth-century (MS. U.1), but cannot be found in the 1634 shelf list (MS. U.3), strongly suggesting that it was acquired after 1634 and before the production of the mid-seventeenth century catalog. Again, this places it out of reach of Fletcher during his residence at Cambridge. For this information I am gratefully indebted to Mr Jonathan Harrison. A copy of this same edition is held by Emmanuel College Library, and this was acquired between 1693 and 1694, from the private collection of William Sancroft (see chapter 7, note 34). A further copy of the same edition is held by Sidney Sussex College Library, but I have been unable to discover when this was acquired. It does not appear in Leedham-Green's author / title catalogs of books in book lists derived from inventories presented during the Tudor and Stuart periods, and so was probably acquired by the College later. Curiously, Leedham-Green, *Books in Cambridge Inventories,* II, 235, lists a copy of an early edition of *De re anatomica* in the bequest of Hatcher. Given the date of the bequest

in 1587, this must have been a copy of either the first edition (1559) or either of the Paris editions of 1562 or 1572. Unfortunately Leedham-Green is unable to identify this copy with any existing copy in Cambridge libraries, and without positive identification I am unable to make any judgment as to whether or not Fletcher may have had access to it. There is, however, a copy of the Paris, 1572 edition in Gonville and Caius College Library, for which Mr. Mark Statham, Librarian, is unable to specify a date of acquisition. If this were to be the copy in Leedham-Green, however, it would not particularly strengthen the chances of Fletcher having read it, for the reasons given in chapter 7, note 34, regarding Fletcher's access to other colleges.

89. French, *William Harvey's Natural Philosophy,* pp. 32, 26.

90. Ibid., p. 26. These last five pages are not of course all that Vesalius has to say in the *Fabrica* about function. Lind, *Studies in Pre-Vesalian Anatomy,* p. 17, even suggests that there is a lot of physiology rather than only anatomy, although this should be read in the context of his study of pre-Vesalian anatomy.

91. Cunningham, *The Anatomical Renaissance,* p. 147. Also see French, *William Harvey's Natural Philosophy,* pp. 83–84.

92. See French, *William Harvey's Natural Philosophy,* p. 32.

93. Whitteridge, *William Harvey and the Circulation of the Blood,* p. 49.

94. Colombo, *De re anatomica,* bk. VII, p. 179, cited in Whitteridge, *William Harvey and the Circulation of the Blood,* pp. 51–52.

95. Ibid., p. 52. The second sentence of this quotation, in *De re anatomica,* reads: "The valves of the vena cava and pulmonary vein," but that this is an error probably, according to Whitteridge, due to the printer is evident from the textual context. Whitteridge, *William Harvey and the Circulation of the Blood,* p. 76, endnote 3.

96. Banister, *The Historie of Man,* fol. 94^{v} (sig. $Dd_{ii}{}^{v}$).

97. A number of anatomists, including and perhaps most importantly Galen (given his immense influence), had described the "operation" of some or all of the cardiac valves in relation to their structure, but had failed to infer from function the implications for the movement of blood in the vessels at whose orifices they occur; and had thus failed to appreciate fully the "competence" of the valves.

98. For Servetus and Guidi, see Whitteridge, *William Harvey and the Circulation of the Blood,* pp. 53, 60, respectively. Like Colombo, Ruini, in his *Dell'anatomia,* maintains that all the valves of the heart "prevent regurgitation." Whitteridge, *William Harvey and the Circulation of the Blood,* p. 60. Similarly, Guinther, in his *De medicina veteri et nova,* writes that "when the valves of the heart are closed, nothing can pass through them" (Comment.1, Dial.5, p. 179) (The relevant passage is translated in Whitteridge, *William Harvey and the Circulation of the Blood,* p. 62). Plater, in his *De corporis humani structura et usu,* also "states all the essential features of [Colombo's] theory, including the competence of the valves of the heart," but, unlike Guinther, he adds that fuliginous vapours are given off in the lungs during the preparation of blood and air there. Whitterige, *William Harvey and the Circulation of the Blood,* p. 62. On the non-competence of the mitral valve in relation to the regurgitation of blood and vital spirit, see above, in the text at p. 325.

99. The term "membranes" is used by Guinther and by Paré to refer to the structures of the cardiac valves, whose accounts of the valves and of pulmonary transit are probably based directly on Colombo's writing. For Guinther, see his *De medicina* (Comment.1, Dial.5, p. 179); Whitteridge, *William Harvey and the Circulation of the Blood,* p. 61. For Paré, see *Works* (1678), bk. IV, pp. 96–97.

100. Cunningham, *The Anatomical Renaissance,* p. 156; also see p. 155.

101. Raynalde, *The Birth of Man-kinde; Otherwise Named the Womans Booke* (London: James Boler, 1626), fols. 34–35. Raynalde's work provides a fourth mine, "the stones" (testes) which produce seed.

102. Cunningham *The Anatomical Renaissance,* p. 165, note 33, says that, "Galen does not use this image of 'rivers' flowing from a fountain-head." In part, this is because Galen specifically says "that the heart is *not* the source of the arterial system in the sense in which a river has a source." See David J. Furley and J. S. Wilkie, *Galen on Respiration and the Arteries* (Princeton: Princeton University Press, 1984), pp. 30–31, cited in Cunningham, *The Anatomical Renaissance,* p. 165, note 33. Galen does, however, use the image of channels irrigating a garden. See, for example, Peter Brain, *Galen on Bloodletting: A Study of the Origins, Development and Validity of his Opinions, With a Translation of the Three Works* (Cambridge: Cambridge University Press, 1986), pp. 154–55, cited in Cunningham, *The Anatomical Renaissance,* p. 165, note 33.

103. Cunningham, *The Anatomical Renaissance,* p. 153.

104. The pulmonary transit of the blood is described by Valverde in his *Historia de la composicion del cuerpo humano,* bk. 6, ch. 14, p. 131v. Whitteridge, *William Harvey and the Circulation of the Blood,* p. 48; Cunningham, *The Anatomical Renaissance,* p. 162.

105. Cunningham, *The Anatomical Renaissance,* p. 162.

106. None of the Cambridge libraries possesses a copy of the first edition (in Spanish) of Valverde, *Historia de la composicion del cuerpo humano,* but St. John's College Library and the University Library have copies of early editions of the Italian translation, *Anatomia del corpo humano.* St John's copy of the first edition (1559) was donated along with a copy of Valverde's *Vivae imagines partium corporis humani aereis formis expressae* (Antwerp, 1579), by John Collins, in 1634 (see chapter 7, note 34). The University Library possesses a copy of *Anatomia del corpo humano* published at Rome in 1586, which together with two copies of *Vivae imagines partium corporis humani aereis formis expressae,* one being the first edition of 1566, the other being the 1579 edition, were acquired from the library of John Moore, Bishop of Ely, and presented to the University Library by King George I in 1715 (see chapter 7, note 34). These findings make it unlikely that Fletcher saw a copy of the work in which Valverde announced the pulmonary transit of the blood.

107. For Valverde's description of the mitral valve see his *Historia de la composicion del cuerpo humano,* bk. VI, ch. 14, p. 131v.

108. Whitteridge, *William Harvey and the Circulation of the Blood,* p. 102.

109. Carlo Ruini of Bologna wrote in 1598 "a very admirable book on the Anatomy of the Horse, in which he showed that he had grasped, in a striking manner, the actions of the valves of the heart." Foster, *Lectures on the History of Physiology During the Sixteenth, Seventeenth and Eighteenth Centuries,* p. 53. Aranzio, in his *Anatomical Observations* (Venice, 1587), "gave a good description of the lesser circulation . . . [and] described the little nodules of cartilage in the semi-lunar valves to which his name is now attached. Vidus Vidius [(Guido Guidi)] . . . had already described them in . . . [*De anatome corporis humani*] which however, was not printed till 1611," in *Ars Medicinalis.* Singer, *The Evolution of Anatomy,* p. 144.

110. *OED,* II, "C," p. 796, col. 3.

111. For a differentiation between (the perception of) Galen as the physician, Erasistratus as the anatomist, and Aristotle as the philosopher; for the philosophical character of Aristotle's anatomy; and for the body in the anatomy of Galen as the creation of God and Nature, see Cunningham, *The Anatomical Renaissance,* pp. 14–30, 38. Cunningham maintains "that Aristotle saw and was talking about a different concept of the body to that seen and talked about by Galen, and similarly for their Modern followers in the Renaissance." Notwithstanding, Cunningham admits, Galen took aspects of rationalism and teleology from Aristotle. Cunningham, *The Anatomical Renaissance,* pp. 7, 30. In fact, the differentiation, which Cunningham draws, becomes less clear in *The Purple Island,* which combines aspects of the anatomy derived from both Galenic and Aristotelian, and possibly Erasistratian, projects of inquiry; as do Harvey's *Anatomical Lectures,* or *Prelectiones.*

112. Colombo, *De re anatomica,* p. 177; cited in Whitteridge, *William Harvey and the Circulation of the Blood,* p. 55. Whitteridge argues that, "Columbus's account of the pulmonary transit of the blood reveals the anatomist and physiologist reasoning from observed phenomena," and that, "[t]he only compromise that Columbus made with authority concerned the generation of vital spirits." She adds that, "from his remark about leaving difficulties of this kind to philosophers, it may, perhaps, be asked how much he believed in their real existence." Whitteridge, *William Harvey and the Circulation of the Blood,* p. 55. But, equally, Colombo's remark may indicate his acknowledgment of the limitations of his empirical approach to anatomy; limitations which the more philosophical approach of Aristotelian anatomy did not share.

113. On the shift among anatomists in Protestant countries away from Aristotle and toward Plato, see French, *Dissection and Vivisection in the European Renaissance,* pp. 215–29. On the shift also among anatomists in Protestant countries away from Aristotle and toward Galenic medical anatomy, see French, *William Harvey's Natural Philosophy,* pp. 40–45. On the shift, particularly in Padua, from the Galenism of Vesalius and his followers to a revived Aristotelianism, see, for example, Wear, "Introduction," in Harvey, *The Circulation of the Blood,* p. viii; French, *Dissection and Vivisection in the European Renaissance,* pp. 230–52; French, *William Harvey's Natural Philosophy,* pp. 51–70.

114. On the revival of Aristotelian anatomy, see, for example, Cunningham, *The Anatomical Renaissance,* ch. 6; Singer, *The Evolution of Anatomy,* chs. 8–9.

115. Andrea Cesalpino (1519–1603), *Questiones medicae* (Venice, 1593), cited in Foster, *Lectures on the History of Physiology During the Sixteenth, Seventeenth and Eighteenth Centuries,* p. 34.

116. Andrea Cesalpino, *Questionum peripateticarum libri quinque* (Florence, 1571), cited in Foster, *Lectures on the History of Physiology During the Sixteenth, Seventeenth and Eighteenth Centuries,* p. 33.

117. Cambridge University Library holds two copies of the edition of *Questionum peripateticarum libri quinque,* published at Venice in 1593; one was presented to the Library by King George I in 1715 (see chapter 7, note 34); the other has no obvious indication of its date of acquisition, but Dr. J. J. Hall, Under-Librarian, Rare Books Department, has conveyed to me that he thinks it likely that it was acquired after Fletcher's time. Emmanuel

College Library also holds a copy of the same edition, and according to the Librarian of Emmanuel College, Dr. H. C. Carron, this comes from the collection of William Sancroft (see chapter 7, note 34). It was therefore added to the library between 1693 and 1694.

118. Walter Pagel, "William Harvey and the Purpose of Circulation," *Isis* 42 (1951): 22–38 (p. 25).

119. See Ibid., pp. 23, 37. Whitteridge's objection, that the movement of blood in Cesalpino's theory of "circulation" is not out and around, like Harvey's physiology of the circulation of the blood, but up and down, and consequently, not circular, is inappropriate. Whitteridge, *William Harvey and the Circulation of the Blood,* p. 127. Firstly, this motion is circular, not because it describes the geometrical figure of a circle, but because the blood reverts to its point of departure—the heart. Secondly, Harvey's physiology of the circulation of the blood involves not only a movement "out and around," but also "up and down," as in blood rising to the head through the innominate and common carotid arteries, and returning to the heart in the innominate and jugular veins.

120. Cesalpino, *Questionum peripateticarum,* cited in Pagel, "William Harvey and the Purpose of Circulation," p. 25.

121. Pagel, "William Harvey and the Purpose of Circulation," p. 26.

122. Foster, *Lectures on the History of Physiology During the Sixteenth, Seventeenth and Eighteenth Centuries,* p. 35.

123. Whitteridge, *William Harvey and the Circulation of the Blood,* p. 67. In support of her view of "circulation" in Cesalpino, Whitteridge quotes his *Speculum artis medicae Hippocraticum* (Frankfurt, 1605), bk. VI, ch. 9, p. 442 (first publ. at Rome, in 1601), in which it is explained that, as a result of the cooling of the blood first in the pulmonary artery, then in the pulmonary vein, "a kind of circulation" occurs, whereby "the blood is perfected into the nature of spirit." Whitteridge, *William Harvey and the Circulation of the Blood,* p. 67.

124. Cesalpino, *Speculum artis medicae Hippocraticum,* p. 1, cited in Whitteridge, *William Harvey and the Circulation of the Blood,* p. 66.

125. Singer dismisses Cesalpino from the "evolution" of anatomical discovery, saying, in reference to the same passage from *Speculum artis medicae Hippocraticum* (p. 1) as cited by Whitteridge, that "Cesalpino had given a hint of the greater circulation, but had withdrawn it." Singer, *The Evolution of Anatomy,* p. 177.

126. Pagel, "William Harvey and the Purpose of Circulation," p. 25.

127. Cesalpino, *Questionum peripateticarum,* cited in Pagel, "William Harvey and the Purpose of Circulation," p. 25.

128. Pagel, "William Harvey and the Purpose of Circulation," p. 25.

129. Langdale, *Phineas Fletcher,* p. 190. Nicolson's forceful argument against the Harveian provenance of Fletcher's physiology of the heart and blood rests partly on a similar premise: "The originality of Harvey's discovery was less in the theory that blood circulates, which others had suggested, than in demonstration of the process by which circulation is effected." Nicolson, *The Breaking of the Circle,* p. 138, note 6.

130. Cesalpino, *Questiones medicae,* bk. II, Quest. 17, cited in Foster, *Lectures on the History of Physiology During the Sixteenth, Seventeenth and Eighteenth Centuries,* p. 34. Also see Pagel, "William Harvey and the Purpose of Circulation," p. 26.

131. See Paré, *Works* (1678), bk. IV, p. 94.

132. The most recently published work on the subject of respiration, Fabrici's *On Respiration and its Instruments* (1615), "assumes that the lungs were made to fan and refrigerate the heart." Whitteridge, *William Harvey and the Circulation of the Blood,* p. 136.

133. The fact that few contemporary anatomists refer to the writings of Cesalpino suggests to Whitteridge that, his writings "belonged to the domain of philosophical speculation rather than to the practical world of anatomical study and so were neglected by those with more immediately concrete and physical interests." Whitteridge, *William Harvey and the Circulation of the Blood,* p. 65. This need not have excluded Cesalpino from Fletcher's breadth of reading, however.

134. A. G. Debus, *The English Paracelsians* (London: Oldbourne, 1965), p. 18.

135. Ibid., p. 22.

136. Ibid., p. 90 (also see p. 25); for Duchesne, also see pp. 67, 87, 90–92. Two of the most important theoretical works of Joseph Duchesne (Quercetanus/Quersitanus) were translated by Thomas Tymme and published in England in 1605 as *The Practise of Chymicall, and Hermeticall Physicke, For the Preseruation of Health* (London, 1605). Debus describes Robert Fludd (bap. 1574, d. 1637) as "the English culmination of all the occult strains of alchemical, Paracelsian, kabbalistic, and neo-Platonic thought" (p. 105).

137. Debus, *The English Paracelsians,* pp. 25, 27.

138. Ibid., pp. 26–27, 30, 94–95, 90.

139. Ibid., p. 30.

140. Ibid., p. 29.

141. Ibid., p. 95.

142. Leedham-Green, *Books in Cambridge Inventories,* I, 494, says: "Lorkin's medical books seem to show a greater interest in the works of Paracelsus than Sayle suggests, and it is certain that no pains were spared to acquire the latest continental publications." For the list of works by Paracelsus in the inventory of books in Lorkyn's bequest, see Leedham-Green, II, 596.

143. William Harvey (1578–1657), *Exercitatio anatomica de motu cordis et sanguinis in animalibus* (Frankfurt am Main, 1628). Hereafter, reference to this work, either specifically in this edition, or, in more general discussion of the work, in no specific edition, will be made to *De motu cordis,* unless otherwise specified.

144. Langdale's contention, that "the poet's understanding of the blood system was much closer to Harvey's than we have heretofore suspected," is refuted by Nicolson in *The Breaking of the Circle,* p. 138 note 6; and Baldwin finds Nicolson's refutation "convincing," in "Phineas Fletcher," p. 473.

145. For Colombo on the diastole and systole of the heart, see Colombo, *De re anatomica,* bk. VII, p. 179, cited in Whitteridge, *William Harvey and the Circulation of the Blood,* p. 52. For Harvey's observation of and experimentation with the diastole and systole of the heart and the arterial pulse in the context of the formation of circulation theory, see, for example, Singer, *The Evolution of Anatomy,* pp. 177–79; Whitteridge, ed., *The Anatomical Lectures of William Harvey,* pp. xliii–xliv.

146. French, *William Harvey's Natural Philosophy,* p. 92.

147. Richard A. Hunter and Ida Macalpine, "William Harvey and Robert Boyle," *Notes and Records of the Royal Society* 13 (1958): 115–27 (p. 125), cited in Whitteridge, ed., *The Anatomical Lectures of William Harvey,* p. xlvii.

148. Hunter and Macalpine write: "[I]t is reasonable to assume that Boyle had no certain memory of his contacts with Harvey when he wrote it—three years before his own death and thirty-one years after that of Harvey. Indeed, the evidence of his own writings shows that Boyle's statement that he had only one 'Discourse' with Harvey cannot be accepted unless with the qualification 'as far as he remembered at the time of writing.'" "William Harvey and Robert Boyle," p. 115; also see p. 119, where Boyle is quoted as saying, in *Some Considerations Touching the Usefulnesse of Experimental Naturall Philosophy* (Oxford, 1663), that the consideration of the valves of the heart and veins "first hinted the circulation to our famous *Harvey,*" "as he himself told me." Foster argues, that "[i]t was not that the true action of these [valves] led to the true view of the motion of the blood, but that the true view of the motion of the blood led to the true understanding of their use." Foster, *Lectures on the History of Physiology During the Sixteenth, Seventeenth and Eighteenth Centuries,* p. 46. Either way, it is clear that Fabrici's *De venarum ostiolis* influenced his student Harvey, "who borrowed figures from it and based much of his argument concerning the circulation of the blood on the action of the valves." Singer, *The Evolution of Anatomy,* p. 156.

149. Sawday, "Bodies By Art Fashioned," p. 159, note 30.

150. The Lumleian Lectureship was established July 3, 1582, by Richard Caldwell, a Fellow of the College of Physicians, and Lord Lumley, to provide lectures for surgeons. The terms of the foundation envisage the lectureship, which commenced May 6, 1584, as consisting of twice-weekly lectures on surgery and a public anatomy in winter, for a period of six years. The subjects for these anatomies were the bodies of executed criminals. In the first year, the lecturer was to dissect the whole human body, and it is for this demonstration that the *Prelectiones* are generally regarded to have been written. The College had obtained a royal charter in 1565, giving it the right to dissect the bodies of one, two, three, or four criminals hanged in the City of London, in Middlesex or in any county within a radius of sixteen miles. William Munk indicates that the first mention of anatomical lectures for physicians is to be found in the College Annals for 1569–1570. William Munk, *The Roll of the Royal College of Physicians of London,* 2nd ed., revised and enlarged (London: The College of Physicians, 1878). The College's anatomy was thenceforward held each year in the winter, and one of the fellows of the College was selected to give it. In some years, the College's and the Lumleian full-scale anatomy seem to have coalesced, and, in 1599, it was decided that the College's anatomy should be given in alternate years by a Fellow of the College and by the Lumleian lecturer. In 1602, it was decided that the annual College anatomy should thereafter be held at the end of the Hilary Term and not during the earlier part of the winter. Harvey gave this lecture in the spring of 1618. For further history of the Lumleian Lectureship, and of its duties, prior to and during Harvey's tenure, and for that of the London College of Physicians's annual public anatomy, see William Harvey, *Lectures on the Whole of Anatomy,* an annotated translation of *Prelectiones Anatomiae Universalis,* by C. D. O'Malley, F. N. L. Poynter, and K. F. Russell (Berkeley: University of California Press, 1961), pp. 1–7; Keynes, *The Life of William Harvey,* pp. 84–90; Whitteridge, ed., *The Anatomical Lectures of William Harvey,* pp. xxvi–xxvii; Whitteridge, *William Harvey and the Circulation of the Blood,* pp. 83–87.

151. Langdale, *Phineas Fletcher,* p. 94.

152. The Royal College of Physicians published a facsimile reproduction together with a transcription of these lectures as *Prelectiones Anatomiae Universalis* (London: Churchill, 1886). Two more recent editions have been consulted: Harvey, *Lectures on the Whole of Anatomy;* Whitteridge, ed., *The Anatomical Lectures of William Harvey.*

153. See, for example, Harvey, *Lectures on the Whole of Anatomy,* pp. 6–9; Geoffrey Keynes, *A Bibliography of the Writings of Dr William Harvey 1578–1657,* 3rd ed., revised by Gweneth Whitteridge and Christine English (Winchester: St Paul's Bibliographies, 1989), p. 16 (first published, Cambridge: the Syndics of the Cambridge University Press, 1928); Whitteridge, ed., *The Anatomical Lectures of William Harvey,* pp. 2–3; Whitteridge, *William Harvey and the Circulation of the Blood,* p. 87. In Keynes, *A Bibliography of the Writings of Dr William Harvey 1578–1657,* p. 16, Whitteridge says: "On the title-page, Harvey had written: *Prelectiones anatomiae universalis per me Gulielmum Harveium medicum Londoniensem Anatomie et Chirurgie Professorem Anno Domini 1616 anno aetatis 37 praelectae Aprilis 16, 17, 18,* . . . ; which being interpreted reads: Lectures on the whole of Anatomy by me William Harvey Doctor of London, Professor of Anatomy and Surgery, Anno Domini 1616, aged 37, delivered on April 16, 17, 18, [on the male and female body]." According to Whitteridge, *William Harvey and the Circulation of the Blood,* p. 87, Harvey dated the manuscript (MS Sloane 230, MS Sloane 486) 1616, and "says that he delivered the lectures either at the age of thirty-seven, or during his thirty-seventh year, *aetatis suae 37.* As he was born in 1578, this implies 1614 or 1615, a manifest impossibility having regard to the date of his appointment as Lumleian lecturer." Neither is it possible that the notes were first used on an earlier occasion, for "the College's anatomies in those years were not entrusted to Harvey." Whitteridge's entirely plausible solution, in the absence of further evidence, is that, although he had "just passed his thirty-eighth birthday when he delivered [the lectures], . . . [he] had doubtless been preparing them for more than two weeks" (Harvey's date of birth was April 1).

154. For the composition of the audience at William Harvey's Lumleian Lectures, in the anatomy theater of the London College of Physicians, see Luke Wilson, "William Harvey's *Prelectiones:* The Performance of the Body in the Renaissance Theater of Anatomy," *Representations* 17 (1987): 62–95 (p. 90). Keynes, *The Life of William Harvey,* pp. 49–50, says: "The Barber-Surgeons' Hall and the College of Physicians were the official centres for the study of human anatomy, which was forbidden anywhere else in London. A heavy fine could be inflicted for disobeying this rule. No other form of anatomical work was countenanced or would have aroused much interest in most of Harvey's acquaintances."

155. Langdale, *Phineas Fletcher,* p. 201. Admission to the anatomical demonstrations in the anatomy theaters of Leiden, Delft, and Amsterdam, and to the museums, libraries, and botanical gardens attached to or within them, was also in principle open to everyone, although subject to regulations. Rupp, "Matters of Life and Death," p. 264.

156. Were Harvey's lectures to have been given in 1614 or 1615, which is unlikely, it is relevant that Fletcher was for short periods of a week or less in Cambridge, in April, June, and August, 1614, and otherwise in an unknown location, which might have been London, where his mother lived at that time. See Langdale, *Phineas Fletcher,* pp. 68–70. It has been known that Fletcher also spent some considerable portion of his life in Norfolk long before he went to Hilgay in 1621. The date and cause of this earlier residence has been sugested by Langdale, and these are discussed in Appendix B. For our present purpose Langdale locates Fletcher in Norfolk until 1614.

157. Langdale, *Phineas Fletcher,* p. 203.

158. Ibid., pp. 185–86. Harvey was instructed to present the anatomy lectures at the College of Physicians in 1618, in addition to 1616, and "his turn would come up again in 1620, 1622, 1624, 1626, and so on. We have, indeed, definite evidence that he was lecturing in March 1623/4, when Sir Simonds D'Ewes was a member of his audience." Keynes, *The Life of William Harvey,* p. 86. Whitteridge estimates that, while it is difficult to ascertain from records of the College, "there is some reason to suggest" that Harvey held a general public anatomy "every year between 1620 and 1627, with the exception of 1621 and 1625." Whitteridge, *William Harvey and the Circulation of the Blood,* p. 107.

159. Langdale, *Phineas Fletcher,* p. 96. Of Fletcher's writings, *Brittain's Ida* (an edited version of *Venus and Anchises*) was published in London, in 1628, by Thomas Walkley, and *The Way to Blessednes* and *Joy in Tribulation* were both published in London, by James Boler, in 1632. The Hilgay Parish Registers (1583–1674) are held at Norfolk County Record Office, Archive Centre, Norwich.

160. Keynes, *The Life of William Harvey,* assumes that Harvey returned to London not earlier than June 1602. He was examined for a Candidateship at the London College of Physicians on August 7, 1604, and approved and elected. He was eventually elected Fellow of the College on May 16, 1607, and formally admitted on June 5. Having gained permission to practice in London in May 1603, he presumably exercised that privilege. Furthermore

he was appointed Physician to St. Bartholomew's Hospital "in reversion," on February 25, 1608/09, and was admitted to the office of the Physician on October 14, 1609. The appointment as Lumleian Lecturer in 1615 and further appointments, for example, to the position of an Elect on December 3, 1627, and as Treasurer of the College of Physicians in 1628 and 1629, ensured that he was based firmly in London until the latter date, when the King commanded him to attend the Duke of Lennox in his travels abroad.

161. There are instances in the *Prelectiones* where Harvey copies Bauhin verbatim; his own remarks and opinions are "frequently, but not always, prefixed by the initials WH," and his disagreements with Bauhin or other anatomists are marked with an X. Sometimes Harvey's disagreement may only be registered "a little further on in the text," and so apparent self-contradiction occurs, which is actually Harvey contradicting the views of Bauhin's textbook. Whitteridge, *William Harvey and the Circulation of the Blood,* p. 90.

162. According to Adams, *Catalogue of Books Printed on the Continent of Europe, 1501–1600, in Cambridge Libraries,* there are two copies of Bauhin, *Appendix ad Theatrum anatomicum,* each bound with the *Theatrum anatomicum* of 1605, in Cambridge libraries. Dr. J. J. Hall, Under-Librarian, Rare Books Department, informs me that the copy in the University Library has a bookplate which shows that it is from the library of John Moore, Bishop of Ely, and was presented to the University Library by King George I in 1715 (see chapter 7, note 34). The copy in St. John's College Libary was given by, or bought with money given by, John Collins (see chapter 7, note 34), according to correspondence which I have received from Mr Jonathan Harrison, Special Collections Librarian, at St. John's College. In 1631 and mainly in 1634, Collins gave the library all his medical books and £100 to buy more. Once again a book label indicates provenance. In both cases, therefore, acquisition by a Cambridge library occurred after the period during which Fletcher might be thought to have been reading anatomy textbooks at Cambridge while composing Cantos 2 to 5 of *The Purple Island.*

163. Whitteridge, *William Harvey and the Circulation of the Blood,* pp. 87–88.

164. Whitteridge, ed., *The Anatomical Lectures of William Harvey,* pp. 273, 272. In Keynes, *A Bibliography of the Writings of Dr William Harvey 1578–1657,* p. 6, Whitteridge states that the MS of *De motu locali animalium* is "of considerable interest and importance because it shows Harvey in possession of a technique by which he could perform all the animal experiments necessary to prove the hypothesis of the [circulation of the] blood."

165. Whitteridge, ed., *The Anatomical Lectures of William Harvey,* p. li. In support, perhaps, of Whitteridge's supposition, there are "numerous and successive interpolations" in the text of the *Prelectiones,* "of which the latest that can be dated refers to the 1626 edition of Jean Riolan's *Anthropographia et osteologia.* Wilson, "William Harvey's *Prelectiones,*" p. 75. Also see Whitteridge, *William Harvey and the Circulation of the Blood,* p. 88.

166. Whitteridge, ed., *The Anatomical Lectures of William Harvey,* pp. xliv–xlvii. Also see Harvey, *Lectures on the Whole of Anatomy,* pp. 15–16.

167. Whitteridge, "Gaspard Bauhin," p. 523, col. 2. In *De corporis humani fabrica, Libri IIII,* Bauhin, who had been at Padua, "describes [the valves] in terms which may well derive directly from the teaching of Fabricius." Bauhin's interpretation of the use of the valves in the veins is similar to that of Fabrici, and although "not entirely in accordance with Galen's teaching . . . does not contradict it." Whitteridge, *William Harvey and the Circulation of the Blood,* pp. 21, 22.

168. Whitteridge, ed., *The Anatomical Lectures of William Harvey,* pp. xlvii-xlviii. Whitteridge argues elsewhere that the "anatomical description of the muscles [in the manuscript] corresponds well enough with the scope of the second Lumleian lecture, the "Muscular" lecture, and if Harvey followed the terms of his appointment as Lumleian lecturer then he gave this special lecture on the anatomy of the arms and legs for the first time in 1619." Keynes, *A Bibliography of the Writings of Dr William Harvey 1578–1657,* p. 19.

169. Whitteridge, ed., *The Anatomical Lectures of William Harvey,* p. xlix. Despite Whitteridge's observations and plausible conclusion, historians repeat the error of ascribing to the Lumleian lectures of 1616 the significance of staging the communication of the theory of systemic circulation, or at least, of its "first indications." See, for example, D. Geraint James, ed., *Circulation of the Blood* (Tunbridge Wells: Pitman Medical Publishing Co., 1978), p. 6; Lisa Jardine, *Ingenious Pursuits: Building the Scientific Revolution* (London: Little, Brown & Co., 1999), p. 111.

170. The references occur in *De Motu Cordis,* pp. xi, 17. On these references, see Harvey, *Lectures on the Whole of Anatomy,* p. 11; Whitteridge, *William Harvey and the Circulation of the Blood,* ch. 5, pp. 104–12.

171. Harvey, *De Motu Cordis,* p. ix, cited in Whitteridge, *William Harvey and the Circulation of the Blood,* p. 104. Also cited in Harvey, *Lectures on the Whole of Anatomy,* p. 11.

172. Whitteridge, *William Harvey and the Circulation of the Blood,* p. 105. Conversely, O'Malley et al. argue that, "it is most unlikely that [Harvey] meant anything but precisely what he wrote [in the dedicatory epistle of *De motu cordis*]; that is, more than nine but not as long as ten years: a statement that accords completely with Jan-

uary 1618, as the date of that memorable occasion," on which the systemic circulation of the blood was first demonstrated. Harvey, *Lectures on the Whole of Anatomy,* p. 11. Wear opts for the slightly later date: "From Harvey's own recollection of events in *De Motu Cordis* it appears that he had the idea of the circulation of the blood around 1619." "Introduction," in Harvey, *The Circulation of the Blood,* p. ix.

173. French, *William Harvey's Natural Philosophy,* p. 86.

174. Whitteridge, *William Harvey and the Circulation of the Blood,* p. 105.

175. Harvey, *Lectures on the Whole of Anatomy,* p. 17; Whitteridge, *William Harvey and the Circulation of the Blood,* pp. 107–8.

176. Whitteridge, *William Harvey and the Circulation of the Blood,* pp. 107–9. Whitteridge suggests that, Harvey "continued to teach in accordance with the best authorities of his day at a time when he was beginning to realise, and ultimately to know, that some of these opinions were false." Whitteridge, *William Harvey and the Circulation of the Blood,* p. 110.

177. Whitteridge, *William Harvey and the Circulation of the Blood,* p. 110.

178. French, *William Harvey's Natural Philosophy,* pp. 1–2, Ch. 1, pp. 3–17 et passim (p. 1). With regard to "formal structures of argument and expression," French, p. 16, note 32, adds: "It would be a mistake to assume that the Renaissance swept away the scholasticism of the thirteenth century. Although Renaissance commentaries sometimes ignore the rigid syllogistic form of some of the older commentaries and concentrate on the ancient authors, yet many scholastic features often remained." On the application of the anatomical *accessus* and the disputational—commentarial mode of exposition in Harvey's *Prelectiones,* see French, chapter 4.

179. For Harvey's observations and experiments on systole and diastole, see Whitteridge, ed., *The Anatomical Lectures of William Harvey,* pp. 257, 267–69. For further discussion, see Whitteridge, *William Harvey and the Circulation of the Blood,* pp. 95–97. For Harvey's acceptance of pulmonary transit, see Whitteridge, ed., *The Anatomical Lectures of William Harvey,* p. 271; Whitteridge, *William Harvey and the Circulation of the Blood,* p. 98.

180. Whitteridge, ed., *The Anatomical Lectures of William Harvey,* pp. 269, 273.

181. See Whitteridge, *William Harvey and the Circulation of the Blood,* p. 101.

182. Whitteridge, ed., *The Anatomical Lectures of William Harvey,* p. 261.

183. Advocates of pulmonary transit, sharing this belief, were Servetus, Valverde, and Guidi. As Whitteridge observes, only Colombo denied the transventricular passage of blood through these invisible pores, though he was followed in this by Banister. Whitteridge, *William Harvey and the Circulation of the Blood,* p. 101.

184. This had been observed in the vena cava by Coiter, and had been thought to denote "the force of the attractive power of the heart by which the blood was sucked into the ventricles." Whitteridge, *William Harvey and the Circulation of the Blood,* p. 99. Such an explanation, however, is inconsistent with the understanding of systole and diastole reached by Colombo and Harvey, for the "contraction" of the vena cava and pulmonary vein occurs during systole, when blood is thrust out of the heart. The ambiguity of how this "contraction" during systole is described in the *Prelectiones* is, according to Whitteridge, an inherent difficulty of translation. The same terms "*attrahitur/attrahi*" may be translated to suggest Harvey's adherence to the traditional idea of the attractive power of the heart, or to indicate his "observation of a physiological fact." Whitteridge, *William Harvey and the Circulation of the Blood,* pp. 99–100.

185. Whitteridge, ed., *The Anatomical Lectures of William Harvey,* p. 273.

186. Ibid.

187. Ibid., pp. 100–101.

188. In this interpretation, Harvey's account of the attractive faculty of the heart alters entirely its function from that which was previously attributed to it. It had been thought that this function was to suck blood into the ventricles during what was thought to be the active filling of the heart, but which was demonstrated by Colombo and Harvey to be its contraction or systole. The observable "contraction" of the vena cava, which occurs during systole, was therefore thought to be an effect of or response to the heart's attractive power.

189. Of the cardiac valves, the mitral (as well as the tricuspid) valve's competence is ambiguously represented in the *Prelectiones.* See Whitteridge, ed., *The Anatomical Lectures of William Harvey,* p. 261; Whitteridge, *William Harvey and the Circulation of the Blood,* p. 102.

190. Whitteridge, ed., *The Anatomical Lectures of William Harvey,* p. 271.

191. Whitteridge, *William Harvey and the Circulation of the Blood,* p. 102.

192. Cesalpino, *Questiones Medicae,* bk. II, quest. 17, cited above in the text of this chapter.

193. Beardsley, *Aesthetics,* p. 138, cited in Ricoeur, "Metaphor and the Main Problem of Hermeneutics," pp. 102–3. In addition, Ricoeur cites Beardsley as claiming "the metaphors would not only actualize a potential connotation, but establish it as a staple one," so that "the metaphor transforms a property actual or attributed into

a sense" (p. 103). Ricoeur allows Black's qualification to his fourth claim: "Metaphor and the Main Problem of Hermeneutics," p. 102. Ricoeur cites Black, *Models and Metaphors,* p. 43.

194. Hausman, *Metaphor and Art,* p. 33.

195. Ricoeur, "Metaphor and the Main Problem of Hermeneutics," p. 103; Hausman, *Metaphor and Art,* p. 34. Hausman argues that this transformation is "simply a rearrangement of antecedent properties": "What is new, then, is the 'sense,' and presumably this depends on the 'new status' of the relevant, 'eligible' properties. And if novelty is present, it must be found in the way that the new status of properties shows them to be transformed and to yield something unprecedented." But, as Hausman maintains, Beardsley does not explain "why a sense should be new, although it may have a status (ontological?) different from that of the properties" (p. 34). In *Aesthetics,* Beardsley uses for "sense" "designation" (p. 149). The same concept, as he recognizes, "is usually in literary criticism . . . called 'denotation,' but this latter word is needed for the relation between words and things they refer to" (p. 151). Also see Hausman, who argues that Beardsley "falls back on an appeal to antecedent 'marginal' meanings." *Metaphor and Art,* p. 28.

196. Hausman, *Metaphor and Art,* p. 30; also see pp. 24–25, 30, for a discussion of what Hausman calls his "originativist view" of metaphor, "which sees the original and specific formulation of creative metaphors as indispensable," and which he opposes to "the substitution, or reductionist, view" of metaphor (p. 30). Hausman argues that the issue between these views "turns on the question whether language can grow qualitatively. Can language evolve in the sense that meanings and references not present implicitly or explicitly in the language at one time are introduced into it for the first time? Or is language pre-formed so that the first spoken words imply all the significance found in each language in the present and in the future?" (p. 30).

197. Harries, "Metaphor and Transcendence," 73. Harries recognizes that metaphors that so join dissimilars are perhaps more common in modern poetry, for in Renaissance and early modern poetry, "Scripture and Nature, God's two books, furnish authoritative texts, rich with figures from which the poet can draw in his attempt to embellish or to exhibit the deeper significance of what he is celebrating." And, moreover: "For the modern poet there is no authoritative text, no commonly shared understanding of the spritual significance of things that can give necessity to his metaphors" (p. 80).

198. Hausman, *Metaphor and Art,* p. 32. According to Hausman, however, Black rejects "Beardsley's notion that tension in the form of falsity or logical absurdity is a necessary condition of a metaphor." But Black grants: "An expression or statement may be interpreted metaphorically not when it shows falsity or incoherence but instead when it shows 'the banality of that [literal] reading's truth, its pointlessness, or its lack of congruence with the surrounding text and non-verbal setting.'" Hausman comments: "But banality, pointlessness, and lack of congruence in context all serve at least as negative conditions. If such conditions are not instances of tension as *tension* is ordinarily used, they are instances of the absence of 'fit' in the context" (p. 35). Hausman's citation is of Black, "More About Metaphor," in Andrew Ortony, ed., *Metaphor and Thought* (Cambridge: Cambridge University Press, 1979), pp. 19–43; p. 36 cited. This is a slightly modified version of an earlier essay entitled "More about Metaphor," *Dialectica* 31, fosc. 3–4 (1977): 431–57. It seems to me that on a literal reading metaphors are either false (logically absurd in Beardsley's terms), non-sensicle, or trivial (banal in Black's terms), and only the latter, an example of which is "no man is an island" (linguistically the obverse of the *Purple Island* figure, with some implications opposed to that figure), is free of tension.

199. See, for example, Foster, *Lectures on the History of Physiology During the Sixteenth, Seventeenth and Eighteenth Centuries,* p. 47: "Harvey's argument is essentially a physical mechanical argument; the problem which he puts before himself to solve is essentially a mechanical physical problem; the solution of that problem at which he arrived is essentially a mechanical solution of the phenomena of the circulation."

200. Sawday, *The Body Emblazoned,* p. 28. The quotation within the quotation is from René Descartes, *Discourse on the Method of Properly Conducting One's Reason and of Seeking the Truth in the Sciences,* first publ. with the *Essays* (Leiden, 1637), trans. F. E. Sutcliffe (Harmondsworth: Penguin Books, 1968), p. 73.

201. I think that with regard to the development of the machine model of human anatomy, Sawday, *The Body Emblazoned,* p. 28, is right to emphasize that "one needs to stress the slow process of paradigmatic change."

202. Temkin, "Metaphors of Human Biology," p. 276: "The metaphor of the divine work of art does not disappear when the organism begins to be likened to a machine." Thorndike, *A History of Magic and Experimental Science During the First Thirteen Centuries of Our Era,* I, 167, says that in order to illustrate "his theory that the gods prepare the sperms of plants and animals, and set them going as it were, and afterwards leave them to themselves, Galen compares them to wonder-workers—who were perhaps not magicians but men similar to our sidewalk fakirs who exhibit mechanical toys—who start things moving and then go away themselves while what they have prepared moves on artificially for a time."

203. Temkin, "Metaphors of Human Biology," p. 276.

204. Ibid.

205. On Aristotle's attribution of nutritive or vegetative souls to plants, animals and humans, of sensitive souls to some animals and to humans, of motile souls to some but fewer animals and to humans, and of rational souls only to humans (although there seems to be some doubt in Aristotle as to whether some animals may also possess a rational soul), see Cunningham, *The Anatomical Renaissance,* pp. 14–21. Descartes "rejected the existence of the vegetable and sensitive soul, reduced the body's three kinds of spirit (natural, vital, and animal) to only the last of these, and altogether denied thought and passion in animals." Paster, "The Body and Its Passions," p. 44.

206. Michel Foucault, *The History of Sexuality: An Introduction,* trans. from the French by Robert Hurley (London: Allen Lane, 1979; repr. Penguin Books, 1990), p. 139.

207. On Harvey's experiments and others on the movement of the heart, in the *Prelectiones,* see Whitteridge, *William Harvey and the Circulation of the Blood,* pp. 95–97. For an example of the positivist historical interpretation of Harvey's experiments, see Foster, *Lectures on the History of Physiology During the Sixteenth, Seventeenth and Eighteenth Centuries,* p. 53. On Harvey's innovation, Whitteridge writes that, by his achievement, Harvey "had shown that mere observation was not enough; that observation had to be followed by the formulation of an hypothesis, and the validity of this hypothesis tested by experiments expressly designed for that purpose, and that the process had to be repeated over and over again until there was no room for doubt that the true explanation had been found." Whitteridge, *William Harvey and the Circulation of the Blood,* p. 4. My argument against positivist history of anatomy does not discount the significance placed by anatomists on the tradition of anatomy as progressive. Colombo, for example, saw the *res anatomica* (that is, the tradition and his book *De re anatomica*) "as progressive in an accumulative sense within a tradition." Colombo "felt that he was making additions to [this tradition] by his vivisectional experiments," among other things. But Colombo saw this tradition as extending from Galen, or even beyond Galen, from Erasistratus, Herophilus, and others. This accumulation of knowledge and technique within a tradition is not identical with the positivist or idealist notion of internal developement within the scientific project.

208. Wear, "Introduction," in Harvey, *The Circulation of the Blood,* p. xi. Wear's remark responds to the English rendition of the Latin of page 45 of the edition of Harvey's work which he introduces. Wear argues that this rendition conveys the "literal sense," and shows "how, in a very long sentence, Harvey ruminates about the factors that led him to the circulation." "Introduction," p. xi. It is on pp. 57–58 of *De motu cordis,* that Harvey lists all that he "had often and seriously considered with myself" regarding his experiments and observations in cardiovascular physiology, prior to the formulation of the hypothesis of systemic circulation; in which the abundance of blood passing through the heart in how short a time leads directly to the question of whether food or chyle could furnish this or not, and then to the formation of the hypothesis. See Whitteridge, *William Harvey and the Circulation of the Blood,* pp. 125–26.

209. See, for example, Singer, *The Evolution of Anatomy,* p. 160. Besides her alchemical interpretation of Cesalpino's use of the term "circulation, Whitteridge also finds reason to distinguish Harvey's methods from those of his contemporaries, aligning the former with those of modern biologists: "So all scientific discovery is made. First comes the uneasy feeling that something is wrong with accepted theory, that there is some dissonance which breaks the harmony. Then follows the search for more and new facts. A temporary hypothesis is made; experiments designed to test it; and this not once but many times. So little by little the facts, demonstrable to the senses in the seventeenth-century phrase, are accumulated and then, and only then, comes the postulating of an hypothesis which seems to account for every known fact and which is dictated by the facts themselves. After that follow more experiments designed to test its truth. This process is sometimes called the 'hypothetico-deductive' method. It is the method used by the modern biologist. It was the method used by Harvey." Whitteridge, *William Harvey and the Circulation of the Blood,* p. 126.

210. French, *William Harvey's Natural Philosophy,* p. 92.

211. Wear, "Introduction," in Harvey, *The Circulation of the Blood,* p. xii. Also see Keynes, *The Life of William Harvey,* p. 161; Porter, *The Greatest Benefit to Mankind,* pp. 215–16.

212. Whitteridge, ed., *The Anatomical Lectures of William Harvey,* p. 273.

213. Keynes, *A Bibliography of the Writings of Dr William Harvey 1578–1657,* p. 10. Harvey's two letters or anatomical essays to Jean Riolan were published in the form of a small book: *Exercitationes duae anatomicae de circulatione sanguinis* (Cambridge, 1649; Rotterdam, 1649).

214. Edge, "Technological Metaphor and Social Control," p. 141.

215. Healy, "Sound Physic," p. 342, 345–47, 343.

216. For Healy's interpretation of *The Purple Island*'s imagery of defensive fortification as a mechanism for

the poem's appropriation of empirically founded anatomy for a politically reactionary agenda, which is anti-Catholic, parochial and xenophobic, see "Sound Physic," pp. 345–48. For *The Purple Island*'s anti-Catholicism, also see Sawday, *The Body Emblazoned,* pp. 175–77; Gardner, "The Banquet of the Word," p. 395. Sawday introduces a psychological dimension to the poem's elaborate language of defense, a discussion of which is beyond the scope of this study. Gardner discriminates between *The Purple Island*'s anti-Catholicism, which is politically militant and theologically Calvinistic, and its biblical exegesis, which "appears, like his brother's, to be consistent with the Elizabethan settlement as continued by James." Parfitt also reads Fletcher politically as conservative, for reasons of *The Purple Island*'s interest in "the decorative and rhetorical possibilities of the material" of epic, and its lack of engagement with the society of the author. Parfitt, *English Poetry of the Seventeenth Century,* p. 194.

217. For Sawday's perspective on this type of criticism, see "Bodies by Art Fashioned," pp. 108, 142–43.

218. Healy, "Sound Physic," p. 343.

219. Ibid., p. 341.

220. For Colombo on the diastole and systole of the heart, see Colombo, *De re anatomica,* bk. VII, p. 179, cited in Whitteridge, *William Harvey and the Circulation of the Blood,* p. 52. For Harvey's observation of and experimentation with the diastole and systole of the heart and the arterial pulse, in the context of the formation of circulation theory, see, for example: Singer, *The Evolution of Anatomy,* pp. 177–79; Whitteridge, ed., *The Anatomical Lectures of William Harvey,* pp. xliii–xliv.

221. Singer, *The Evolution of Anatomy,* p. 177.

222. See French, *William Harvey's Natural Philosophy,* p. 86.

223. For Galen, because non-voluntary movements in the body served the needs of the body without the intervention of the will, "they were held to be associated with the natural spirit and were therefore known as natural movements." Thereafter, the "movement of the heart was generally categorised as natural, for although it was a muscle it was not controlled by the will." Whitteridge comments: "When it is realised that the sixteenth and early seventeenth-century anatomists had no real understanding of what was actually involved in the movement of muscle, it is obvious that the statement that the heart moves like a muscle is little more than an empty phrase." Whitteridge, *William Harvey and the Circulation of the Blood,* pp. 68, 69. It was not until Walter Needham, in 1667, that the muscular structure as Harvey had described it, was emphasized so as to be "alone responsible for the pumping of the blood and incompatible with its ebullition or vaporisation." Pagel, "William Harvey and the Purpose of Circulation," p. 25. On the idea of the attractive power of the heart in the *Prelectiones,* see Whitteridge, *William Harvey and the Circulation of the Blood,* pp. 99–100.

224. A similar point is raised by Alan Sinfield, in "Power and Ideology: An Outline Theory and Sidney's *Arcadia,*" *English Literary History,* Summer 1985, pp. 259–77, and commented upon by Jonathan Dollimore, in relation to the concept of "containment": "in a materialist critique of the notion of containment Sinfield describes a way of approaching texts which shows them not so much as *either* conservative *or* subversive, but as sites of struggle." Dollimore, "Introduction to the Second Edition," *Radical Tragedy* (1989), p. xxii.

225. Keynes, *The Life of William Harvey,* p. 164.

226. Wear, "Introduction," in Harvey, *The Circulation of the Blood,* p. xiii.

227. Harvey, *De motu cordis,* p. 7, cited in Keynes, *A Bibliography of the Writings of Dr William Harvey 1578–1657,* p. 13.

228. Keynes, *A Bibliography of the Writings of Dr William Harvey 1578–1657,* p. 13.

229. Porter, *The Greatest Benefit to Mankind,* p. 216.

230. Wear, "Introduction," in Harvey, *The Circulation of the Blood,* p. xii.

231. Whitteridge, ed., *The Anatomical Lectures of William Harvey,* p. 23, cited in Wilson, "William Harvey's *Prelectiones,*" p. 80. The point about morphology and use is Wilson's, and is extended here from Wilson's study of the *Prelectiones* to Harvey's formulation of the theory of systemic circulation.

232. Whitteridge, ed., *The Anatomical Lectures of William Harvey,* pp. 13, 5, 9. It should be added that, although Aristotle's emphasis upon understanding anatomical structure in terms of function is greater than that in Galen, or that of Vesalius which was heavily indebted to Galen, it does not distinguish his anatomy in kind from Galen's. Galen took from Aristotle the rationalist approach to anatomy, that argues function from structure and structure from function, and the teleological belief that "every structure has its particular form in order to fulfil the goal of its function." Cunningham, *The Anatomical Renaissance,* p. 30.

233. For this description of the philosophical anatomy of Fabrici, I am heavily indebted to Cunningham, *The Anatomical Renaissance,* pp. 175–77, and French, *William Harvey's Natural Philosophy,* p. 67. Also see chapter 2 of this study, especially note 118. On the distinction between "use" and "utility," see French, *William Harvey's Natural Philosophy,* p. 74.

234. Whitteridge, ed., *The Anatomical Lectures of William Harvey,* p. 5. Also see Wilson, "William Harvey's *Prelectiones.*"

235. Wilson, "William Harvey's *Prelectiones,*" pp. 78–79. See Pagel, "William Harvey and the Purpose of Circulation," p. 29: "Soul [for Aristotle and for Harvey] was something functional, the plan of form and function."

236. Wilson, "William Harvey's *Prelectiones,*" p. 80.

237. Whitteridge, ed., *The Anatomical Lectures of William Harvey,* p. 13. In the *Prelectiones,* Harvey offers his own definition of a part, or that with which he agrees, as "that which by any means whatever completes a whole in so far as it is integrate and composite" (p. 9).

238. Ibid., pp. 72, 215, 307.

239. Ibid., p. 5. French, *William Harvey's Natural Philosophy,* p. 3, deduces from Harvey's discussion in the *Prelectiones* of theories and kinds to which anatomy could be reduced, that there are in fact four types: public, teaching or "popular" anatomy, involving principally *historia;* philosophical anatomy; medical anatomy, which is interested in the application of anatomical knowledge of rarities and morbid conditions to medical practice; and mechanical anatomy, which is skill in dissection.

240. French, *William Harvey's Natural Philosophy,* p. 4.

241. *Microcosmos,* p. 70, col. 2, l. 21; *P.I.* 4.25.3.

242. *Mirum in modum,* p. 5, col. 2, ll. 5 (from bottom), 8 (from bottom), 3 (from bottom), 2 (from bottom).

243. Cunningham, *The Anatomical Renaissance,* pp. 41–42, argues: ". . . when our Christian anatomists . . . discussed the operations of the body as being expressions of the soul (or of the souls, or of different aspects of the one soul), their vocabulary and concepts were common both to the Christianity in whose defence and for whose maintenance and strengthening their anatomizing was ultimately being undertaken, and to the Greek philosophers and physicians whose works they were using to undertake it."

244. *Mirum in modum,* p. 24, col. 2, l. 23.

245. French, *William Harvey's Natural Philosophy,* p. 68.

246. Cunningham, *The Anatomical Renaissance,* p. 29.

247. Ibid., pp. 13, 22.

248. For the generalizing tendency of Aristotelian anatomy, see Cunningham, *The Anatomical Renaissance,* ch. 1 and, for the Aristotelianism of Harvey's teacher at Padua, Fabrici, ch. 6.

249. Cunningham argues that the aim of these Paduan philosophers of the sixteenth century was to read Aristotle "aristotelically," or in his own terms, but that the positions at which they consequently arrived "were found highly provocative outside the Veneto. For if you spoke like Aristotle then, whether you wished it or not, you ran the risk of being understood to deny the immortality of the soul, the creation of the universe (for Aristotle's universe was eternal), and the creative activity of God. This was potentially 'atheistic' in the sixteenth-century sense of denying that there is divine punishment or an afterlife (rather than that there is no God)." Cunningham, *The Anatomical Renaissance,* p. 264. The risk of being understood to deny the immortality of the soul arose because for Aristotle and sixteenth-century Aristotelians like Jacobus Zabarella, the soul is the act, form, and perfection of the body, which cannot exist without the body, and can be understood only through its operations in the body. For Zabarella and Aristotelian philosophy at Padua, see Cunningham, *The Anatomical Renaissance,* pp. 170–73. For Aristotle, however, as for Aristotelians, there was an immortal as well as a mortal part of the soul. Immortality pertains only to the incorporeal spirit in the Active Nous. See Costelloe and Muirhead, *Aristotle and the Earlier Peripatetics,* II, 130–31.

250. *Mirum in modum,* p. 24, col. 2, l. 24. See, for example *P.I.* 2.26.6–7.

251. Cunningham, *The Anatomical Renaissance,* p. 265.

252. French, *William Harvey's Natural Philosophy,* p. 68. Also see p. 69: "When Harvey used the Latin term *natura* he meant what other Englishmen meant in the seventeenth century: very largely 'nature' in the sense we use it today, as 'the natural world' or 'mother nature' as a personalised organising and productive principle. . . . But Aristotle's *physis* was the nature-of-a-thing, not a generalised or personalised principle. It was for Aristotle the nature of the *animal* that derived what was good and avoided what was evil for that animal. In this way Harvey is interpreting Aristotle in a way consistent with seventeenth-century medicine."

253. Harvey, *Exercitationes de generatione animalium* (London, 1651), cited in Pagel, "William Harvey and the Purpose of Circulation," p. 31.

254. Harvey's adherence to Aristotle did not preclude a respect for Galen, whose work he was reluctant to refute. According to O'Malley et al., in Harvey, *Lectures on the Whole of Anatomy,* p. 12: ". . . so reluctant was Harvey to refute some of Galen's errors in human anatomy that he suggests men have changed in 2,000 years." The unlikely notion of such anatomical or physiological alteration in the human body occurring between the time of Galen

and early modern England achieves the effect of locating authority simultaneously in the practice of dissection and the texts of Galen. It might be relevant that as a Fellow of the College of Physicians, Harvey would have been discouraged from advancing ideas contrary to Galen. See Keynes, *The Life of William Harvey,* pp. 91, 95. The same notion was given expression by Davies of Hereford, in *Microcosmos.* After delineating the physiological anatomy of the ventricles of the heart, the poetic voice exclaims: "But now (alas) most harts all hollow are, / That *Bloud* and *Spirits* therein confuséd lie, / So as no *Art* can one from other spie" (p. 29, col. 1, ll. 2–4). The notion can also be found in Banister, *The Historie of Man,* sig. B_{ii}^{r}, where he offers an explanation for the contradictions between the Ancients Galen and Hippocrates and modern authors: ". . . neither are the same in all our bodies in these dayes, as, in elder time the auncient Anathomistes most commo[n]ly observed. . . ."

255. French, *William Harvey's Natural Philosophy,* p. 4; Harvey, *De generatione,* cited in Pagel, "William Harvey and the Purpose of Circulation," p. 31; Keynes, *The Life of William Harvey,* pp. 330, 422. Keynes says, p. 422, that although "Harvey was reticent concerning his religious beliefs," "[p]assages in *De generatione animalium,* 1651, express a profound consciousness of a Divine Providence watching over man's affairs and forming the laws of Nature." Chapter 3 has argued the importance placed by anatomists on understanding human anatomy in relation to the Creator. French, p. 4, comments: "It was common for anatomists to place at the top of their lists of the parts or the functions of anatomy the purpose of knowing oneself in order to revere the Creator."

256. The judgment regarding Aristotle's God was made by Alfred North Whitehead, in the Lowell Lectures, at Harvard University, and is quoted by Walter Lippmann, in his *A Preface to Morals* (London: George Allen & Unwin, 1931), p. 26.

257. French, *William Harvey's Natural Philosophy,* p. 70: "Harvey did not share Galen's abundant wonder at the skill of a creator, or at least he did not express it when arguing about the action of the heart and the motion of the blood. Nor did he express the feeling shown by others [for example, the Protestant natural historians] of being in a theatre of creation, able to read the nature of the Creator from the objects of creation. Unlike most other anatomists, he did not put in first place a religious justification of anatomy. The reason is no doubt that Harvey adopted what he saw as Aristotle's method in the books on animals. Aristotle's world was eternal, and did not reflect a creator; his natural philosophy was search for causes within the operations of the nature of things."

258. Foster, *Lectures on the History of Physiology during the Sixteenth, Seventeenth and Eighteenth Centuries,* p. 47.

259. Singer, *The Evolution of Anatomy,* p. 175. See also Foster, *Lectures on the History of Physiology during the Sixteenth, Seventeenth and Eighteenth Centuries,* p. 47.

260. Wear, "Introduction," in Harvey, *The Circulation of the Blood,* p. xii. Also see Porter, *The Greatest Benefit to Mankind,* pp. 215–16.

261. Whitteridge, ed., *The Anatomical Lectures of William Harvey,* p. 293.

262. See, for example, Whitteridge, *William Harvey and the Circulation of the Blood,* p. 143.

263. Whitteridge, ed., *The Anatomical Lectures of William Harvey,* p. 275.

264. Ibid., p. 293.

265. Wear, "Introduction," in Harvey, *The Circulation of the Blood,* p. xiii. The citation is from chapter 8 of *De motu cordis.* The citation in square brackets is taken from the 1653 translation of *De motu cordis,* entitled *The Anatomical Exercises of Dr. William Harvey, Professor of Physick and Physician to the Kings Majesty, Concerning the Motion of the Heart and Blood,* with a preface by Zachariah Wood (London, 1653), cited in Porter, *The Greatest Benefit to Mankind,* p. 215. On the "perfection" of the blood as the purpose of circulation, see Pagel, "William Harvey and the Purpose of Circulation," p. 26. It is the view of Whitteridge, and of others, that the philosophical notions of the purpose of circulation "have no fundamental part to play in *De motu cordis* which remains pre-eminently the exposition and proof of a purely physiological hypothesis." But, as Whitteridge admits, such notions "do underlie *De motu cordis*"; and the "proper understanding of *De motu cordis* [and of Harvey's physiology elsewhere] can only be had from the knowledge of the whole of Harvey's thought." Whitteridge, *William Harvey and the Circulation of the Blood,* p. 144. For Pagel, "William Harvey and the Purpose of Circulation," by contrast, Harvey's adherence to the doctrines of Aristotle on innate heat is "to be recognised as a true motive and stimulus in scientific 'inventio'" (p. 26).

266. *The Anatomical Exercises of Dr. William Harvey,* cited in Porter, *The Greatest Benefit to Mankind,* p. 216.

267. Wear, "Introduction," in Harvey, *The Circulation of the Blood,* p. xiii.

268. In a letter that Harvey wrote on 20 May 1636 in reply to a letter from Caspar Hofmann, written on the previous day, Harvey denies any conviction in this philosophical speculation as to the purpose of circulation: "But about concoction and about the causes of this movement and circulation, particularly about its final cause, I have never spoken, but have omitted it altogether, deliberately"; he adds "I do not deny that in cap. 8 I insert in pass-

ing for the sake of an illustration," the teleological explanation (which is cited in the text), "[b]ut whether this be so or not, I openly declare that I have not yet proved, or even said much about." Whitteridge, *William Harvey and the Circulation of the Blood,* pp. 165, 165–66. Harvey's restraint concerning the purpose of circulation has been noted (see above in the text); but in *De motu cordis,* Harvey offers the explanation in question as "so in all likelihood." Harvey, *De motu cordis,* ch. 8, cited in Pagel, "William Harvey and the Purpose of Circulation," p. 23. Whatever degree of assent Harvey may be thought to have given to his speculation, it seems certain that he recognized that it was "incapable of demonstration in an autopsy or a vivisection, and so incapable of proof." Whitteridge, *William Harvey and the Circulation of the Blood,* p. 168. Also see Keynes, *The Life of William Harvey,* pp. 232–36.

269. See Whitteridge, *William Harvey and the Circulation of the Blood,* pp. 165–68.

270. In *De generatione,* Harvey "deviated from the one-sided emphasis laid by Aristotle on the heart as the principal organ and shifted it on to the blood." Therein, Harvey argues that "there dwells in the blood some power which acts beyond the power of elements [by which blood is constituted], its chief function being nutrition and preservation of the parts of the body. It is a nature and indeed a 'soul' inherent *in* the blood." Pagel concludes, therefore, that if "blood and 'soul' are identical, the latter, however, being neither the product of the elemental constitution of the blood nor of something 'ethereal' ('heat,' 'spirit,' or 'astral body') added to it, 'soul' can mean nothing but the natural function inherent in blood which acts as the material substratum necessary for the appropriate effects to be obtained in physical life. Only as far as the blood displays a certain function ('virtues and powers') can it be called 'spirituous' or 'celestial.'" Pagel, "William Harvey and the Purpose of Circulation," pp. 29, 29–30, 30. In the latter of these quotations, Pagel is citing *De generatione* (Amsterdam, 1651), Ex. 71, pp. 475, 480. Keynes, *The Life of William Harvey,* p. 350, says: "In Exercise 52 of *De generatione* [Harvey] made the unequivocal statement that the blood 'is the builder and preserver of the body and principal part wherein the soul hath her Session.'"

271. Whitteridge, ed., *The Anatomical Lectures of William Harvey,* p. 33.

272. Harvey, *De motu cordis,* cited in Foster, *Lectures on the History of Physiology during the Sixteenth, Seventeenth and Eighteenth Centuries,* p. 47.

273. Harvey, *De motu cordis,* ch. 8, cited in Pagel, "William Harvey and the Purpose of Circulation," p. 23; *The Anatomical Exercises of Dr. William Harvey,* cited in Porter, *The Greatest Benefit to Mankind,* p. 215. "Barren" and "effete" are the respective translations of the same Latin term in *De motu cordis,* by the two works cited.

274. This translation is provided by Pagel, "William Harvey and the Purpose of Circulation," p. 23; and Wear, "Introduction," in Harvey, *The Circulation of the Blood,* p. xiii.

275. Thomas Aquinas, *Summa Theologica,* pars III, qu. 83, cited by Clark, *Eucharistic Sacrifice and the Reformation,* pp. 77–78.

276. The particularized soul is the form of the body and is to be distinguished in Aristotelian philosophy from the Active *Nous* or Reason, which "stands aloof from [the body] and takes not part in its activity," and is "real . . . only in the act of thinking." Costelloe and Muirhead, *Aristotle and the Earlier Peripatetics,* II, 100, 96; also see p. 131.

277. Crooke, *Microcosmographia,* fol. 4.

278. *The Book of the Common Prayer and Administration of the Sacraments, and other Rites and Ceremonies of the Church: after the Use of the Church of England* (London, 1549), which contains "The Order for the Administration of the Lord's Supper or Holy Communion," was revised in 1552, 1559, 1604, and 1662. Given the uncertainty of the dates of composition of *The Purple Island,* it is not known whether the 1559 or 1604 editions would have been current when Fletcher composed Canto 4, though arguments made in this study as regards a possible Harveian provenance for the conceptual character of some of the cardiovascular anatomy in that canto would suggest that it was composed after 1604. If so, Fletcher's adherence to that edition might be expected given his own Calvinistic leanings and *The Purple Island*'s allegorical identification of James I/VI, "Our late most learned Soveraigne," with the "Angel full of heav'nly might, / . . . / From Northern coast," who with "his remonstrance and comment on the Apocal[ypse]" is said "Through all the earth his glorious beams [to] dispread, / And open laies the Beasts and Dragons shame" (*P.I.* 12.55.1,3,4–5 & margin. *). But as Alfred Barry argues in *The Teacher's Prayer Book: Being The Book of Common Prayer, with Introductions, Analyses, Notes, and a Commentary upon the Psalter* (London: Eyre & Spottiswoode, Queen's Printers [n.d.]), p. v, the changes in the revised edition of 1604 were small, and confined to "the addition of some prayers and Occasional Thanksgivings, the restriction of Private Baptism to a 'lawful minister,' and the addition to the Catechism of the Exposition of the Sacraments." Accordingly, I have quoted from *The Two Books of Common Prayer, Set Forth by Authority of Parliament in the Reign of King Edward the Sixth: Compared with Each Other,* ed. by Edward Cardwell, 2nd ed. (Oxford: Oxford University Press, 1841), which contains the editions of 1549 and 1552.

279. Cardwell, ed., *The Two Books of Common Prayer,* pp. 309, 299. It is an influential view that theologians who defended the Catholic position on the sacrifice of the mass, from the beginning of the Protestant revolt up to at least the middle years of the sixteenth century, agreed that "the Mass is not an absolute but a relative sacrifice, drawing its efficacy from the sacrifice of Calvary"; though there were "different theories about the nature of that mysterious relationship which unites Christ's unbloody sacrifice of the altar with his unique redemptive sacrifice of the cross." Clark, *Eucharistic Sacrifice and the Reformation,* p. 97.

280. Cardwell, ed., *The Two Books of Common Prayer,* pp. 297–99, 308. The quoted passages are from the 1549 edition, while the quotation in square brackets within a quotation is taken from the 1552 edition.

281. Ibid., pp. 307, 310. The quotations are of the 1552 edition.

282. Ibid., p. 310. The quotations are of the 1552 edition.

283. Harvey, *De motu cordis,* ch. 8, cited in Pagel, "William Harvey and the Purpose of Circulation," p. 23.

284. W. A. Murray, "Donne and Paracelsus: An Essay in Interpretation," in John R. Roberts, ed., *Essential Articles for the Study of John Donne's Poetry* (Hassocks, Sussex: Harvester Press, 1975), pp. 122–28 (p. 124).

285. Murray, "Donne and Paracelsus," p. 124. In the Paracelsian natural philosophy of Peter Severinus, "the semen is identical with the 'mummy,' the vital balsam which matures and is elaborated in the testicles, but acts in conformity with the vital spirit (vital 'sulphur') that is present in all organs and tissues." Pagel, "William Harvey and the Purpose of Circulation," p. 35. This identification enhances the sexual and reproductive figuration in *De motu cordis.*

286. Pagel, "William Harvey and the Purpose of Circulation," p. 23.

287. Pagel argues that, although the Paracelsian speculations of Peter Severinus, and others, "are in no way connected with it," "the discovery of the blood circulation . . . invited mystics to 'support' and embellish it with speculation around the analogy between the greater and the lesser world." He claims, however, that in Harvey's work the term "circulation," "on the physiological level . . . suffers from an ambiguity which is also found in Caesalpino's ideas and later on in those of Harvey's supporters and followers." Ibid., pp. 37, 38. Also see Nicolson, *The Breaking of the Circle,* pp. 132–33, on the significance of Harvey's capitalization in his announcement that he began "to think whether there might not be a MOTION, AS IT WERE, IN A CIRCLE." Whitteridge stoutly resists the suggestion that the physiology in *De motu cordis* is mixed conceptually with alchemy. She argues that, although Chapter 8 draws an analogy between systemic circulation and the falling and evaporation of rain, which process "had already been used by alchemists and some esoteric philosophers to describe the process of distillation" in alchemy; notwithstanding this, Harvey had forged the "hypothetico-deductive" method of the modern biologist, although, "having found incontrovertible experimental proof of his hypothesis," he turns "to the past in search of some statement which he can cite in support of his theory." Whitteridge, *William Harvey and the Circulation of the Blood,* pp. 127, 126. Appropriately, the analogy Harvey turns to in Chapter 8 is Aristotelian in origin, as is explained in the text below. For French, *William Harvey's Natural Philosophy,* p. 92, note 41, it is by no means clear that Harvey had developed any such method as Whitteridge suggests: "The reading back of characteristics of modern science into Harvey's words makes every scholar see Harvey in his own image. The true significance of Harvey is thus seen variously in his comparative anatomy, his hypothetico-deductive method, his zero-level statements and first-level hypotheses, his contribution (even) to a modern philosopher's attempt to analyse modern scientific method. None of this is helpful in the present attempt to describe how William Harvey's natural philosophy led him to his discovery and how other people's natural philosophy made them accept or reject that discovery." I am in agreement with French, considering such retrospective projections as those of which he speaks, to fail fully to overcome the temptation to see the history of science extending back into the Renaissance as progressive in a positivist and idealist way.

288. Although he did not read Cesalpino's work, Harvey concurs in attributing to inspiration "a cooling effect preventing the blood from being overheated and blown up like boiling milk or honey, whereby the lungs may become overdistended and the animal suffocate." Pagel, "William Harvey and the Purpose of Circulation," p. 23. In the *Prelectiones,* "Harvey did not find the idea of blood effervescing in the heart [which could be used to "explain" systole and diastole] inconsistent with his own idea of forceful systole. He even thought the boiling motion of the blood and the forceful systole might have a ventilating function." French, *William Harvey's Natural Philosophy,* p. 89.

289. For René Descartes's theory of the movement of the heart and the pulse of the arteries, and of the rarefication of the blood, see René Descartes, "Treatise on Man," in *The World and Treatise on Man,* first publ. separately, the latter part as *L'Homme de René Descartes* [Paris, 1664]), in *The Philosophical Writings of Descartes,* trans. by John Cottingham, Robert Stoothoff, and Dugald Murdoch, 3 vols. (Cambridge: Cambridge University Press, 1992), I, 99–108 (pp. 100, 104, 107); Descartes, *Discourse on the Method of Rightly Conducting One's Reason*

and Seeking the Truth in the Sciences (first publ. with the *Essays* [Leiden, 1637]), in *The Philosophical Writings of Descartes,* I, 109–51 (pp. 134–39); Descartes, *Description of the Human Body,* first publ. as *La Description du corps humain* (Paris, 1664), in *The Philosophical Writings of Descartes,* I, 313–24 (pp. 316–19). For secondary reading on these theories of Descartes, see Pagel, "William Harvey and the Purpose of Circulation," p. 24; Whitteridge, *William Harvey and the Circulation of the Blood,* pp. 151–59.

290. Harvey, *De motu cordis,* Ch. 8, cited in Pagel, "William Harvey and the Purpose of Circulation," p. 23.

291. Debus, *The English Paracelsians,* p. 32.

292. Temkin, "Metaphors of Human Biology," p. 282.

293. Pagel argues that, because Harvey relates how the concept of the circulation of the blood came into his mind in the passage which visualizes that circulation "as the microcosmic copy of a general cosmological pattern and principle" derived from Aristotle, the discovery "no longer remains a discovery of scientific detail but obtains a position in a view of the world which is based on two main tenets of Aristotle: (I) the excellence of the circular motion and (2) the parallel of the macrocosm and microcosm." Pagel, "William Harvey and the Purpose of Circulation," p. 28; also see p. 26. Whitteridge argues that: "Each chapter [of *De motu cordis*] is concerned with the understanding of how various phenomena happen, not why they do. Only here, in attempting to relate Aristotle's cosmic analogy to the circulation of the blood does he for the first time suggest a reason," such as the excellence of circular motion. In her view it is therefore unlikely "that it was speculation on the purpose of the blood's movement that led to the hypothesis of its circular motion"; and she denies that "Harvey was much influenced by the philosophy of circles current among the more esoteric thinkers of his day, and that he thought first of the circle and then of the facts." In support of this interpretation, she cites pages 57 and 58 of chapter 8 of *De motu cordis,* where Harvey "makes it abundantly plain that the facts which he had accumulated" compelled him to "bethink myself if [the blood] might not have a circular motion." Whitteridge, *William Harvey and the Circulation of the Blood,* pp. 128, 125, 125–26.

294. Debus, *The English Paracelsians,* p. 31.

295. Harvey, *De motu cordis,* ch. 8, cited in Pagel, "William Harvey and the Purpose of Circulation," p. 23.

296. See William H. Huffman, *Robert Fludd and the End of the Renaissance* (London: Routledge, 1988), p. 22. Fludd's public anatomy lecture to the College of Physicians took place on June 27, 1620. Huffman, p. 207. On the history of the attitude of the College, which was overwhelmingly Galenic in its medical practice, towards Paracelsian medicine, see Keynes, *The Life of William Harvey,* p. 64. According to Keynes, pp. 134–35, on first presenting himself to the College of Physicians on November 8, 1605, Fludd was found insufficient in "both Galenical and spagyrical medicines" and enjoined to abstain from practice. On Fludd's possible influence on a comment made by Harvey in his *Prelectiones,* see Keynes, *The Life of William Harvey,* p. 104.

297. Huffman, *Robert Fludd and the End of the Renaissance,* p. 64. Robert Fludd, *Pulsus seu nova et arcana pulsum historia* is the fifth volume of *Medicina Catholica; seu, mysticum artis medicandi sacrarium: in tomos diuisum duos; in qvibvs metaphysica et physica tam sanitatis tuendæ, quam morborum propulsandorum ratio pertractatur* (Frankfurt: William Fitzer, 1629).

298. See Heydon, *A New Method of Rosie Crucian Physick,* p. 4, which demonstrates knowledge of the competence of the cardiac valves.

299. Keynes, *The Life of William Harvey,* pp. 321–22. Keynes adds (p. 322) that Descartes "could allow that some movements, such as scratching, could be reflex, that is, initiated without conscious direction."

300. Harvey, *De motu cordis,* ch. 8, cited in Pagel, "William Harvey and the Purpose of Circulation," p. 23.

301. Philip Jacob Sachs von Lewenheimb, *Oceanus Macro-Microcosmicus* (Breslau, 1664), cited in Pagel, "William Harvey and the Purpose of Circulation," p. 23.

302. Phineas Fletcher, *A Father's Testament,* written long since for the benefit of the particular relations of the authour, Phin[eas] Fletcher, sometime minister of the gospel at Hilgay in Norfolk, and now made publick at the desire of friends (London: Henry Mortlock, 1670), pp. 91–92. The little poem is reprinted in Giles and Phineas Fletcher, *Poetical Works,* II, 322–23, where it is given by Boas the title "The Search after God" in brackets to indicate that it was originally untitled; and in *The Penguin Book of Renaissance Verse 1509–1659,* selected and with an introduction by David Norbrook, ed. by H. R. Woudhuysen (London: Penguin Books, 1993), pp. 559–60. Further references to the poem occur in the text in parentheses, in the following order: stanza number, line number.

303. Whitteridge, *William Harvey and the Circulation of the Blood,* p. 143. None of the blood contained by the heart is regarded as for its own use, because "[t]he blood for its own use is conveyed to it by the coronary veins and arteries." John Smith, *The Portrait of Old Age: Wherein is Contained a Sacred Anatomy Both of Soul, and Body, and a Perfect Account of the Infirmities of Age Incident to Them Both; Being a Paraphrase upon the Six Former Verses of the 12. Chapter of Ecclesiastes,* 2nd edn., corrected (London: J. Macock, 1666), uses both the metaphors

of cistern and fountain for the heart in an allegorical passage in which he determines to fit Harvey's discoveries into his exegesis of the words of Solomon. Also see Keynes, *The Life of William Harvey,* pp. 166–68, where the passage is cited.

304. Whitteridge, ed., *The Anatomical Lectures of William Harvey,* p. 245.

305. In *De generatione,* Ex. 11, p. 87, Harvey writes: "The authority of Aristotle has always such weight with me that I never think of differing from him inconsiderately." Cited in Pagel, "William Harvey and the Purpose of Circulation," p. 28.

306. Harvey, *De motu cordis,* cited in Pagel, "William Harvey and the Purpose of Circulation," p. 28. According to Pagel, "[p]assages in which the heart is extolled as the 'king' and 'royal fortress' of the body, etc., are obviously derived from Aristotle." Pagel, "William Harvey and the Purpose of Circulation," p. 26, note 19.

307. Temkin, "Metaphors of Human Biology," p. 281.Temkin notes that "Harvey here uses the Aristotelian metaphor of the heart as the 'acropolis' and 'the supreme power.'"

308. Pagel, "William Harvey and the Purpose of Circulation," p. 28. "In Caesalpino's, as well as Harvey's, case adherence to the doctrines of Aristotle is, therefore, not to be 'excused' as the spirit of the time or as a cloak used in order to stave off ecclesiastical persecution, but to be recognised as a true motive and stimulus in scientific 'inventio.'" (p. 26). Also see Pagel, "The Vindication of 'Rubbish,'" p. 5.

309. Singer, *The Evolution of Anatomy,* p. 182.

310. Wear, "Introduction," in Harvey, *The Circulation of the Blood,* p. xii.

311. Harvey, *De generatione animalium,* Ex. 19, p. 11, cited in Whitteridge, ed., *The Anatomical Lectures of William Harvey,* p. 254.

312. Whitteridge, ed., *The Anatomical Lectures of William Harvey,* p. 255.

313. Ibid., p. 31.

314. Wear, "Introduction," in Harvey, *The Circulation of the Blood,* p. xiii; Whitteridge, ed., *The Anatomical Lectures of William Harvey,* p. 33. Elsewhere in the *Prelectiones,* Harvey states that, "[t]he pre-eminence of a part follows its usefulness, that is usefulness to the greater number." Whitteridge, ed., *The Anatomical Lectures of William Harvey,* p. 25.

315. Whitteridge, ed., *The Anatomical Lectures of William Harvey,* pp. 307, 329.

316. Ibid., p. 307. Harvey inscribed "WH" between the second and third words of this sentence in the manuscript of the *Prelectiones,* an inscription, according to Whitteridge, which stresses a point with which Harvey agrees or introduces an original remark or comment.

317. Ibid., pp. 333, 91.

318. Ibid., p. 99.

319. Cunningham, *The Anatomical Renaissance,* pp. 18–21. On analogy in Aristotle as "the bond which unites different genera," which "transcends generic differences," see Costelloe and Muirhead, *Aristotle and the Earlier Peripatetics,* pp. 25–28 (p. 25). The correspondence is based on anatomical functions, so that in the analogy between animals and plants, mouths in animals correspond to roots in plants (pp. 26–27). Moreover, Aristotelian analogy is accommodated to the general scale of being, such that "one life unfolds itself from the same fundamental forms in continually ascending degrees of perfection" (pp. 27–28).

320. Whitteridge, ed., *The Anatomical Lectures of William Harvey,* p. 251.

321. Ibid., pp. 31, 307, 33, 247.

322. Ibid., p. 307.

323. Wear, "Introduction," in Harvey, *The Circulation of the Blood,* p. viii. Harvey's firmness about the incorrectness of the Galenic physiology of the heart and vascular system appears only in and after *De motu cordis;* e.g., in his two anatomical essays to Jean Riolan, published as *Exercitationes duae anatomicae de circulatione sanguinis.* Wear, "Introduction," in Harvey, *The Circulation of the Blood,* p. xiv. It seems unlikely, however, that, as suggested by Wear, Harvey did not realize that "the circulation of the blood had the potential to bring down the whole of traditional medicine."

324. Temkin, "Metaphors of Human Biology," p. 283.

325. Healy, *Fictions of Disease in Early Modern England,* p. 13, writes that tropes "facilitate understanding and reasoning, initiate hypotheses, and enable us to have a cognitive hold on the more problematic, intangible experiences of our everyday existence."

326. See Wear, "Introduction," in Harvey, *The Circulation of the Blood,* p. xv: "The discovery of the circulation was a fact of observation for Harvey, and not the basis of a new philosophy."

327. Porter, *The Greatest Benefit to Mankind,* p. 215.

328. For the argument that the Renaissance or early modern anatomical projects were attempts to give new birth to the Ancient projects, not to reject them, see Cunningham, *The Anatomical Renaissance*, p. 7.

329. French, *William Harvey's Natural Philosophy*, p. 66.

330. Cunningham, *The Anatomical Renaissance*, p. 5.

331. Whitteridge, *William Harvey and the Circulation of the Blood*, p. 126.

332. For an explanation of the concepts of acquisition through metaphor and synecdoche, see Greenblatt, *Shakespearean Negotiations*, pp. 10–11.

333. French, *William Harvey's Natural Philosophy*, p. 14.

334. An interesting parallel for this exposure of device may be found in Barker's description of seventeenth-century stage representation, which he describes as "an artisanal world in which the device is naked not out of polemical technique but as its normal condition." By contrast, and historically subsequently, "a discursive situation before production has quite "disappeared" . . . at the level of representation, into the conventions of that bourgeois naturalism which has nothing to do with nature, and everything to do with naturalizing the suppression of the signs of the artifact's production." Barker, *The Tremulous Private Body*, p. 18.

Chapter 9. Lectures and Demonstrations in the Anatomy Theater

1. Redman, "*Phineas Fletcher*," p. 8. Redman adds: "The philosophical outlook [of pastoral allegory] which is said to have derived from Spenser is one shared by Fletcher's fellows at Cambridge as well—it is not restricted to tradition in secular literature alone." Also see pp. 14–15.

2. See, for example, Carlino, *Books of the Body*, pp. 77–78, on the 1531 statutes at Rome.

3. Relating to the phenomenon of anatomists or students stealing the publicly displayed cadavers of criminals, Vesalius is said to have conducted midnight raids on gibbets to gather skeletal pieces for the construction of a whole human skeleton. See Sawday, *The Body Emblazoned*, pp. 195–96. On the setting of limits on numbers of dissections and cadavers legitimately made available to the University of Bologna, see Ferrari, "Public Anatomy Lessons and the Carnival," pp. 53–54.

4. See Keynes, *The Life of William Harvey*, p. 10; Langdale, *Phineas Fletcher*, p. 204; Whitteridge, *William Harvey and the Circulation of the Blood*, p. 5; Frances Valadez, "Anatomical Studies at Oxford and Cambridge," in *Mecidine in Seventeenth Century England*, ed. by Allen G. Debus (Berkeley, Los Angeles and London: University of California Press, 1974) pp. 393–420 (p. 396); John Caius, *The Annals of Gonville and Caius College*, ed. by John Venn (Cambridge: Printed for the Cambridge Antiquarian Society, 1904), p. 90; John Venn, comp., *Biographical History of Gonville and Caius College*, III: *Containing the Biographies of the Successive Masters, the History of the Various Endowments and Benefactions, and Transcripts of Many early Deeds and Charters* (Cambridge: Cambridge University Press, 1901), pp. 38, 48; Gonville and Caius College Library MSS. 615/768 (p. 177), 617/549 (pp. 43, 45), 617 bis/770 (p. 7), 707/693 (pp. 10, 12) (English versions of the annals). On his death, the practice of dispersed burial was performed on the body of John Caius, which Keynes, *The Life of William Harvey*, p. 56, calls "a rather macabre form of piety." For a discussion of dispersed burial in relation to ecclesiastical policy on dissection and its historical development, see chapter 3 of this study.

5. Langdale, *Phineas Fletcher*, p. 204.

6. Valadez, "Anatomical Studies at Oxford and Cambridge," p. 396; John Venn, ed., *The Register of Baptisms, Marriages and Burials in St. Michael's Parish, Cambridge (1538–1837)* (Cambridge: Printed for the Cambridge Antiquarian Society, 1898), pp. vii, 180.

7. Egmond, "Execution, Dissection, Pain and Infamy," pp. 109–11 (p. 111, note 45), assumes the opposite, that it would not have been common practice for "the fragmented bodily remains of a (publicly) dissected person" to be "handed over for burial afterwards," because "so many of the subjects were convicted criminals whose dead bodies would otherwise have remained unburied." However, it seems that, as part of the series of prudent strategies mentioned in the text, special provision was made in the case of those cadavers provided for dissection. For example, see Schultz, *Art and Anatomy in Renaissance Italy*, on Pope Sixtus IV's brief to the University of Tübingen, in 1482, which permitted dissection following the approved authority of the Church, on the condition that the university "cause the condemned of this sort to be buried just as such are usually buried" (p. 63). Also see Venn, comp., *Biographical History of Gonville and Caius College*, III, 47–8.

8. Cambridgeshire Record Office, MS. P30/1/1 (Parish Register of Saint Mary the Great, 1559–1859) (the

register records its earliest burial of an anatomical body in 1566, but no further such records occur before Fletcher left Cambridge in 1615); R. Holinshed, *The First and Second [and Third] Volumes of Chronicles . . . now newlie augmented and continued . . . to the yeare 1586 by John Hooker* (London, 1587), III, 1349, cited in Harvey, *Lectures on the Whole of Anatomy,* p. 4. The historical record provided by manuscript and printed annals, histories, parish registers and other documents at Gonville and Caius College Library and Archive, the University Library, and Cambridgeshire Record Office suggests no evidence of further anatomical demonstrations at Cambridge which Fletcher may have attended.

9. Langdale, *Phineas Fletcher,* pp. 205–6. Keynes, *The Life of William Harvey,* p. 11, says that Grymstone "probably performed the dissection of the bodies granted to the College"; but adds: "Whether this demonstration was carried out regularly during Harvey's years at Cambridge there is no means of knowing, but many years later in his anatomical lectures in London he referred to the small liver and spleen that he had seen in a corpse in Cambridge, and he is not known to have returned to the University after going down at the end of 1599." This may well have been a postmortem autopsy rather than a public dissection, as Keynes (p. 17) seems to acknowledge. For further details of the life of Thomas Grymstone and his bequest of a copy of the second folio edition of the *Fabrica* to Gonville and Caius College Library, see chapter 7, note 34, of this study. Also see Valadez, "Anatomical Studies at Oxford and Cambridge," pp. 393–7, Caius, *The Annals of Gonville and Caius College,* pp. 209, 213; Venn, comp., *Biographical Hisotry of Gonville and Caius College,* I, 115.

10. Keynes, *The Life of William Harvey,* p. 11; Whitteridge, *William Harvey and the Circulation of the Blood,* p. 8. Harvey left Cambridge on 30 October 1599. It is not known when he arrived at Padua, "for the matriculation registers for this period no longer exist in the University Archives. The first reference to his presence in Padua occurs in August 1600, when he was elected Councillor for the English Nation in the Jurist university." Whitteridge surmises, however, that, "he arrived in Padua in all probability for the beginning of the Michaelmas Term, November 1599" (p. 9).

11. Whitteridge, *William Harvey and the Circulation of the Blood,* p. 8. Also see Valadez, "Anatomical Studies at Oxford and Cambridge," p. 397. Keynes, *The Life of William Harvey,* p. 16, admits that the teaching of anatomy at Gonville and Caius College was "never properly organized." On a related but different issue, Keynes, pp. 14–15, points out: "There do not appear to have been any formal medical lectures at Cambridge, though some informal discourses were occasionally given."

12. Valadez, "Anatomical Studies at Oxford and Cambridge," p. 406.

13. Lady Day is 25 March. Fletcher was married to Elizabeth Vincent on August 23, 1615; and although a fellowship at Cambridge lapsed automatically with matrimony, "the person involved continued to enjoy his privileges throughout one year of grace." Langdale, *Phineas Fletcher,* pp. 74–75, 71.

14. When the Lumleian lectureship was founded, it was stipulated that the subjects for the anatomies were to be the bodies of executed malefactors. The Sessions Records for Newgate prison, preserved in the Record Office of the City of London, show that on Friday, April 12, 1616 "two men were condemned to be hanged, and on one or other, or both, of their bodies, Harvey could well have given his lectures on the following Tuesday, Wednesday and Thursday." Whitteridge, ed., *The Anatomical Lectures of William Harvey,* p. xxvi. Also see Whitteridge, *William Harvey and the Circulation of the Blood,* pp. 83, 87. Also see chapter 8, note 150, of this study.

15. The fact was first observed by Sawday, "Bodies by Art Fashioned," p. 142.

16. Harvey delivered the lectures on April 16, 17, and 18, 1616. See chapter 8, note 153, of this study.

17. French, *William Harvey's Natural Philosophy,* p. 21, says that a four-day practical dissection was statutory. The ordinances of the Tomlins Readership, the first anatomical lectureship at a British university, endowed to the University of Oxford in 1624, stated that every spring after Lenten Assizes, four anatomical lectures, each of two hours, were to occur over two days. The first Tomlins reader was Thomas Clayton, who had studied at Oxford and Montpellier. Clayton selected Bernard Wright as the first surgeon for the readership, to dissect and demonstrate according to the reader's direction. See Valadez, "Anatomical Studies at Oxford and Cambridge," pp. 397–98. The Tomlins reader was also expected to give a lecture on osteology in the autumn of each year. Annual records of performances of anatomies were not kept, however, though Valadez finds some evidence for their occurrence in student diaries (p. 399). The Laudian Code, which recodified Oxford's statutes in 1633–36, reiterated the Tomlins statutes (p. 400); and in 1636, Oxford obtained royal approval of the university charter, known as the Great Charter of Charles I, which granted several new privileges, including the legal authority to claim the bodies of executed felons for dissection. The legality was disputed, however, and never confirmed by Parliament (p. 401).

18. Valadez, "Anatomical Studies at Oxford and Cambridge," p. 398.

19. For a modern description of the vermiform appendix, see, for example, *The Concise Home Doctor,* I, p. 86, col. 3—p. 87, col. 1.

20. Langdale, *Phineas Fletcher,* pp. 182–83; Baldwin, "Phineas Fletcher," p. 468. On Berengario da Carpi as the first to describe the appendix, see Lind, *Studies in Pre-Vesalian Anatomy,* p. 7; Singer, *The Evolution of Anatomy,* p. 97.

21. Ferrari, "Public Anatomy Lessons and the Carnival," p. 90, enlarges upon this distinction historically, saying that during the seventeenth century, when most medical research undertaken was in the fields of physiology, or microscopic or pathological anatomy, and the anatomy of particular systems, there was far stronger emphasis on dissection conducted privately or in scientific academies, than on the public anatomy.

22. Heckscher, *Rembrandt's Anatomy of Dr. Nicolaas Tulp,* pp. 45–46.

23. Carlino, *Books of the Body,* p. 86.

24. Heckscher, *Rembrandt's Anatomy of Dr. Nicolaas Tulp,* p. 46.

25. French, *William Harvey's Natural Philosophy,* pp. 58, 66.

26. Wilson, "William Harvey's *Prelectiones,*" p. 69.

27. Carlino, *Books of the Body,* p. 86.

28. Ferrari, "Public Anatomy Lessons and the Carnival," pp. 70–71. Ferrari remarks upon how during the sixteenth century, the disputes of the doctors were said to have descended to undignified and shameful terms (p. 71). Nevertheless, this special feature of the Bolognese theater, the dispute, attracted foreign scholars (p. 76).

29. Carlino, *Books of the Body,* p. 2.

30. Langdale, *Phineas Fletcher,* pp. 207–8.

31. Sawday, "Bodies by Art Fashioned," p. 145.

32. Sawday, *The Body Emblazoned,* p. 179.

33. Park, "The Criminal and the Saintly Body," p. 14.

34. French, *William Harvey's Natural Philosophy,* pp. 19, 20.

35. *The Literary Works of Leonardo da Vinci,* compiled and ed. by J. P. Richter, 2 vols., 2nd edn. (Oxford: Oxford University Press, 1939), II, no. 796, p. 85: "And you, who say that it would be better to watch an anatomist at work than to see these drawings, you would be right, if it were possible to observe all the things which are demonstrated in such drawings in a single figure, in which you, with all your cleverness, will not see nor obtain knowledge of more than some few veins."

36. Ferrari, "Public Anatomy Lessons and the Carnival," pp. 89–93.

37. Laqueur, *Making Sex,* p. 130.

38. Achillini's *Annotationes Anatomicae* is written as a manual for dissection and demonstration, including imperative instructions such as: "Separate the tendon of the transverse muscles from the siphac [peritoneum] and demonstrate each of them." It contains also indicative statements resembling those of *The Purple Island,* such as: "When you lift the panniculi you can see the lung" (pp. 44, 53).

39. Redman, "*The Purple Island,*" p. 13; Jones, *The Divine Science,* pp. 117–18.

40. O'Malley, *Andreas Vesalius of Brussels 1514–1564,* p. 323.

41. Harcourt, "Andreas Vesalius and the Anatomy of Antique Sculpture," p. 39.

42. Vesalius, *Fabrica,* Bk. V, p. 19, cited in Harcourt, "Andreas Vesalius and the Anatomy of Antique Sculpture," p. 28.

43. T. Garzoni, *La piazza universale di tutte le professioni del mondo* (Venice, 1587), p. 298, cited by Piero Camporesi, *The Anatomy of the Senses: Natural Symbols in Medieval and Early Modern Italy,* trans. by Allan Cameron (Cambridge: Polity Press, 1994), p. 131.

44. Harcourt, "Andreas Vesalius and the Anatomy of Antique Sculpture," p. 43.

45. Choulant, *History and Bibliography of Anatomic Illustration,* p. 25.

46. French, *William Harvey's Natural Philosophy,* pp. 47, 53 (and note 12).

47. Wilson, "William Harvey's *Prelectiones,*" p. 80. As early as 1578, the surgeon John Banister was keen to acknowledge that: "It may fall out to be no new saying, that almost in all bodies, some varietie is to be sene." But he attaches no particular importance to this observation. *The Historie of Man,* sig. B_{iiii}^{r}.

48. Wilson, "William Harvey's *Prelectiones,*" p. 78.

49. Ibid., pp. 77, 76. See also p. 75, where Wilson interprets two epigraphs that appear on the makeshift title page of the *Prelectiones,* as initiating "an association of *plenum,* or fullness constitutive of body, with an attribution of the origin and organization of matter to divine teleology."

50. Ibid., p. 76.

51. Ibid., pp. 76, 77.

52. Ibid., pp. 66–67, 76, 77, 78, 79.

53. Ibid., pp. 77, 79.

54. In *Microcosmos,* see the narrative of "a *Child,* confin'de t'a *Dungeon* deepe" (*Microcosmos,* p. 64, col. 1, ll. 8–25, cited and discussed in chapter 5); in *The Purple Island,* see the diegetic narrator's invocation of God as guide through his anatomy (*P.I.* 3.4, cited and discussed in chapter 3 and the introduction to chapter 7).

55. *Microcosmos,* p. 83, col. 2, ll. 5–9 (cited and discussed in chapter 5).

56. Wilson, "William Harvey's *Prelectiones,*" pp. 79.

57. Costelloe and Muirhead, *Aristotle and the Earlier Peripatetics,* II, 131, 1–9.

58. Ibid., II, 120–25, 130–34 (p. 131). This argument understands soul, in this Aristotelian context of Harvey's *Prelectiones,* as the particularized soul which is the form of the body. This is to be distinguished from the Active *Nous* or Reason, which is real only in the act of thinking and is "the universal as distinguished from the individual element in man." It may not be clear in Aristotle whether and to what extent the soul-as-form is bound to the individual body or to anatomy defined teleologically in terms of the final cause of the species. In other words, there may be no *personal* as opposed to *species* identity in Aristotelian metaphysics of either the body or the lower or higher souls. Ibid., II, 131.

59. Ibid., II, 99–102, 94.

60. *Microcosmos,* p. 83, col. 2, ll. 13, 12 (cited and discussed in chapter 5).

61. Healy, "Sound Physic," p. 347.

62. Ibid., p. 347.

63. See, for example, Ferrari, "Public Anatomy Lessons and the Carnival," p. 75, on the new anatomy theater at Bologna, built 1637–49.

64. French, *William Harvey's Natural Philosophy,* p. 59; Sawday, "The Fate of Marsyas," p. 114.

65. Ferrari, "Public Anatomy Lessons and the Carnival," p. 54. On the regulations relating to dissections in the University of Bologna statutes (relating to the university of arts) from 1405, see pp. 53–54.

66. See, for example, Ferrari, "Public Anatomy Lessons and the Carnival," pp. 59–61, who argues that the "general feeling of discomfort about dissection" was offset by the organization of anatomy sessions, the slow growth of demand for cadavers, and recourse to bodies from hospitals (p. 61).

67. At Bologna, by the sixteenth century, it had become common practice for the students to have holidays during the carnival period, and although the statutes drawn up in the fifteenth century did not provide explicitly for the public dissections to take place during this period, they often did so nonetheless during the sixteenth century. In 1579, the senate decided that the lessons would begin at the beginning of January, that is, at the start of the carnival holiday period. The decree of 1579 was disregarded on several occasions in order to enable other public dissections to be held at other times in the year. "None the less . . . the anatomy course came increasingly to revolve around a single annual public dissection performed by the full anatomy professor" in January; and in 1602, regulations "finally determined that there would only be one public anatomy course, and that it would take place during the carnival holiday period, beginning on the feast of Saint Anthony Abbot (January 17)." Ferrari, "Public Anatomy Lessons and the Carnival," pp. 66, 68–71 (pp. 69, 71). Also see Carlino, *Books of the Body,* p. 80.

68. Ferrari, "Public Anatomy Lessons and the Carnival," p. 97. Jan C. C. Rupp, "Matters of Life and Death: The Social and Cultural Conditions of the Rise of Anatomical Theatres, with Special Reference to Seventeenth Century Holland," *History of Science* 28 (1990): 263–87 (p. 264), indicates that in Holland, "[t]he anatomical dissections of the corpse of a criminal were annual public manifestations around Christmas."

69. Carlino, *Books of the Body,* pp. 80–81.

70. See Chapter 8, note 153, of this study.

71. Heckscher, *Rembrandt's Anatomy of Dr. Nicolaas Tulp,* pp. 32, 98.

72. Carlino, *Books of the Body,* p. 81.

73. Ibid., p. 81.

74. Wilson, "William Harvey's *Prelectiones,*" pp. 69–71.

75. Heckscher, *Rembrandt's Anatomy of Dr. Nicolaas Tulp,* p. 32; Rupp, "Matters of Life and Death," pp. 267–68. Carlino, *Books of the Body,* p. 78, says that the 1531 University of Rome statutes stipulate that the medical students, and all surgeons and barbers attending the public anatomies "were required to pay a 'tax,' which was basically an admission fee to a spectacle."

76. Keynes, *The Life of William Harvey,* p. 88, 87.

77. Carlino, *Books of the Body,* p. 85.

78. Ferrari, "Public Anatomy Lessons and the Carnival," pp. 82–84; Neill, *Issues of Death,* p. 117. On the extreme care of staging the anatomy demonstration and the decorum of the proceedings, also see Heckscher, *Rembrandt's Anatomy of Dr. Nicolaas Tulp,* p. 28.

79. Ferrari, "Public Anatomy Lessons and the Carnival," pp. 76, 82; Neill, *Issues of Death*, p. 115.

80. Rupp, "Matters of Life and Death," p. 267; Ferrari, "Public Anatomy Lessons and the Carnival," pp. 84–85 (p. 85). Ferrari discusses the architectural character of Estienne's design (p. 85).

81. Heckscher, *Rembrandt's Anatomy of Dr. Nicolaas Tulp*, p. 28; Ferrari, "Public Anatomy Lessons and the Carnival," p. 84.

82. See Ferrari, "Public Anatomy Lessons and the Carnival," p. 55.

83. Cunningham, *The Anatomical Renaissance*, p. 208.

84. William Schupbach, *The Paradox of Rembrandt's "Anatomy of Dr. Tulp"* (London: Wellcome Institute for the History of Medicine, 1982), pp. 31–34 (p. 33).

85. See Schupbach, *The Paradox of Rembrandt's "Anatomy of Dr. Tulp,"* pp. 32, 36.

86. Neill, *Issues of Death*, pp. 120–22

87. Ferrari, "Public Anatomy Lessons and the Carnival," p. 62; Heckscher, *Rembrandt's Anatomy of Dr. Nicolaas Tulp*, p. 29. Anatomy theaters were built at the following cities: Padua, by Benedetti (temporary) (see above in the text of this section); also Padua, by Fabrici, in 1584 (temporary wooden structure erected firstly in the courtyard and later inside the central university block); again Padua, by Fabrici, in 1594 (permanent); Rome and Verona by the early sixteenth century; at Montpellier, by Rondelet, in 1556; London, 1583, for the London College of Physicians, then, in 1614, at the College's new premises at Amen Corner; Basel, by Kaspar Bauhin, in 1589; Leiden, by Peter Paauw (1564–1619), in 1597 (permanent); Delft, in 1614 (permanent); Amsterdam, in 1619 (permanent); London, by Inigo Jones, in 1636 (permanent), for the Barber-Surgeons' Company; and Bologna, in 1637–49 (permanent). Leiden even had two anatomical theaters for some time, one connected with the university, the other with the guild of surgeons. Ferrari, "Public Anatomy Lessons and the Carnival," pp. 72, 61, 78; Keynes, *The Life of William Harvey*, pp. 25, 87; Rupp, "Matters of Life and Death," pp. 263, 270, 281, note 9; Singer, *The Evolution of Anatomy*, pp. 166–67; Carlino, *Books of the Body*, p. 83.

88. Cunningham, *The Anatomical Renaissance*, p. 101; Sawday, "The Fate of Marsyas," p. 114.

89. Eriksson, *Andreas Vesalius First Public Anatomy at Bologna 1540*, p. 273. The Venesection debate began in Vesalius's eighteenth demonstration, followed by Corti's twenty-first lecture, and proceeded until the confrontation in the twenty-second demonstration

90. See Ferrari, "Public Anatomy Lessons and the Carnival," pp. 64–66, 98.

91. Eriksson, *Andreas Vesalius First Public Anatomy at Bologna 1540*, p. 221.

92. See Ferrari, "Public Anatomy Lessons and the Carnival," pp. 52, 70, 99. A law against the wearing of masks by those attending the demonstration in Bologna was introduced in 1594, only to be reversed in 1616 (pp. 98–99). In his recommendations for the building of anatomy theaters and the provision of public anatomical demonstrations, Benedetti, in *Anatomice*, advises that warders be provided for the regulation of the unruly incoming rabble. Rupp, "Matters of Life and Death," p. 267.

93. Heckscher, *Rembrandt's Anatomy of Dr. Nicolaas Tulp*, p. 28. Rupp, "Matters of Life and Death," p. 268, explains that "during the dissection dissected parts were passed around so that everyone, the spectators on the upper rows of the theatre included, could have a look at them; these parts were very much sought-after [*sic*] for constructing a skeleton or for stuffing."

94. French, *William Harvey's Natural Philosophy*, p. 43.

95. Melanchthon, *Commentarius de Anima*, fol. 44^{v}-45^{r}, cited in Cunningham, *The Anatomical Renaissance*, p. 231.

96. For a list of permanent anatomy theaters, see note 87, above, in this chapter.

97. Ferrari, "Public Anatomy Lessons and the Carnival," p. 80.

98. Museum of London Archeology Service (MoLAS), *Barber-Surgeons' Hall Garden, Monkwell Square, London EC2 . . . An Archaeological Research Evaluation* (London: Museum of London, 1998), pp. 10, 4, 9. The work cited here is a report commissioned by Professor Dennis Hill, a member of the Worshipful Company of Barbers, "following a research exercise designed to locate the site of the post-medieval Anatomy Theatre associated with the Barber-Surgeon's Hall." During February 1998 archeologists from the Museum of London Archaeology Service (MoLAS) undertook an archeological research evaluation on the site in order to discover whether the foundation of the anatomy theater, built in 1636, survived in any form beneath the ground. The research evaluation was intended to discover the position, extent and nature of any surviving remains of the post-medieval theater. Although in no way involved with the project, I had hoped that some discovery might have shed light upon the cultural significance of the design of the theater. Unfortunately nothing was found which could be related to the theater (see pp. 4, 6, 33). The following prints and designs exist in the Guildhall Library, London, cataloged under the reference

number PR. 45/BAR. One print (*c.* 1750) shows an interior view and a plan, drawn by Isaac Ware, of what was formerly identified as Hick's Hall, but is in fact the anatomy theater built in 1636. Both the interior view and the plan show the four elliptical levels of raked seating, whereas the view also shows the pitched roof, described in the eighteenth-century report as "an elliptical cupola," and the plan shows two staircases at the front of the theater and an entrance for the surgeon or physician with the cadaver to demonstrate the anatomy. A second set of six prints from later in the eighteenth century includes drawings of both the exterior anterior and exterior posterior aspects.

99. MoLAS, *Barber-Surgeons' Hall Garden,* p. 10. The theater at Monkwell Square is illustrated on John Rocque's map of 1746 (figure 6) as a circle; but by 1784, "the original Theatre had fallen into disrepair. It was pulled down to make way for housing and had disappeared from the maps by the time Horwood published his version in 1799" (p. 10).

100. French, *Dissection and Vivisection in the European Renaissance,* p. 221.

101. Rupp, "Matters of Life and Death," pp. 269, 264 (also see p. 272); French, *William Harvey's Natural Philosophy,* p. 66. According to Rupp, the Dutch anatomy theaters typically also established a public library of scientific books in the same building as the theater, and a public *hortus botanicus* adjacent to the theater (p. 264). French, p. 66, says: "Fabricius" programme of 'research' was directed towards 'an account of the structure of the whole animal', *Totius animalis fabricae theatrum.* By *theatrum* he meant, like the anatomist Bauhin, a collective display. It was a display for which Fabrici constructed a physical theatre, the first permanent anatomy theatre, designed to allow as many people as possible to view the dissection at the same time."

102. Giulio Casserio, *Philosophi, medici et anatomici, in vtraque medicina patavii versantis Pentæstheseion, hoc est de quinque sensibus liber* (Venice: Nicolavs Misserinus, 1609), cited in Choulant, *History and Bibliography of Anatomic Illustration,* p. 224. Casserio died before completion of this project. See Thornton and Reeves, *Medical Book Illustration,* p. 66.

103. Whitteridge, "Gaspard Bauhin," p. 523, col. 2; p. 524, col. 1.

104. French, *Dissection and Vivisection in the European Renaissance,* p. 221.

105. Ibid.

106. Ibid., pp. 222–23.

107. See Wilson, "William Harvey's *Prelectiones,*" p. 83. Also see Eriksson, *Andreas Vesalius First Public Anatomy at Bologna 1540,* p. 177, where Vesalius is recorded as saying, "For tomorrow we shall have another body—I believe they will hang another man upon which I shall demonstrate to you the anatomy of all the veins, arteries and nerves. For this subject is now too dryed and wrinkled."

108. Schupbach, *The Paradox of Rembrandt's "Anatomy of Dr. Tulp,"* pp. 33–34, 36. Ferrari, "Public Anatomy Lessons and the Carnival," p. 95, summarizes "the two contrasting interpretations of anatomy that were adopted by anatomists and thinkers" as follows: "it was viewed as the discipline that disclosed the magnificent hand of God in the perfection of our corporeal machinery, while at the same time declaring the transience and nullity of man's life on earth." Ferrari suggests further that the "moralized anatomy" which drew on the *vanitas* tradition "never became a major pictorial current in Italy as it did, for example, in Holland" (p. 95). He draws brief attention to the inessential *vanitas* iconography of the frontispiece to Vesalius's *Fabrica,* but makes no mention of the osteological plates of Book I of the *Fabrica* (for my discussion of these, see Chapter 3). Also see, for example, Heckscher, *Rembrandt's Anatomy of Dr. Nicolaas Tulp,* p. 31

109. Camporesi, *The Anatomy of the Senses,* pp. 130–31 (p. 131).

110. Paré, *Workes* (1634), Bk. III, p. 90.

111. Barlaeus, "On the Anatomical House Which Can Be Visited in Amsterdam," *Poematum Miscellaneorum / Elegiarum libri,* 2 vols.3 (Amsterdam, 1645–46), cited in Heckscher, *Rembrandt's Anatomy of Dr. Nicolaas Tulp,* p. 114.

112. Barlaeus, "On the Anatomical House Which Can Be Visited in Amsterdam," cited in Heckscher, *Rembrandt's Anatomy of Dr. Nicolaas Tulp,* p. 114.

113. Ward, *The Anatomy of Warre,* p. 1.

114. Heckscher, *Rembrandt's Anatomy of Dr. Nicolaas Tulp,* p. 108; also see p. 98.

115. Dollimore, *Radical Tragedy* (1989), pp. 99–100 (p. 100).

116. Ibid., p. 100.

117. Ibid., p. 102.

118. Godfrey Goodman, *The Fall of Man; or, The Corruption of Nature, Proved by the Light of Our Natural Reason* (London, 1616), p. 65; George Hakewell, *Apologie; or, Declaration of the Power and Providence of God* (1627), Preface, p. 22, cited by Dollimore, *Radical Tragedy* (1989), pp. 100, 101.

119. Dollimore, *Radical Tragedy* (1989), p. 101.

120. Nicolson, *The Breaking of the Circle*, pp. 110–11, discusses one other allusion in *The Purple Island* to the decay thesis or cycles of degeneration. The thirty-sixth stanza of the first canto describes how "Vain men" "plough the seas," "Adding new worlds to th'old," with the result that "The earths vast limits dayly more unbinde!" Moreover, "The aged world, though now it falling shows, / And hastes to set, yet still in dying grows." Fletcher completes the stanza with a moral didactic assertion: "Whole lives are spent to win, what one deaths houre must lose" (*P.I.* 1.36.1, 3, 4, 5–6, 7). Nicolson remarks that it was often thought that the globe "ironically was growing to its greatest extent [through discovery] just before its death" (p. 110). As with poetic discovery, Fletcher and his contemporaries tended to equate discovery of that which was antecedently existing, but hidden, with creation or generation of something new.

121. Nicolson, *The Breaking of the Circle*, p. 109; Thomas Browne, *Hydriotaphia*, v, cited in Nicolson, *The Breaking of the Circle*, p. 112.

122. Heckscher, *Rembrandt's Anatomy of Dr. Nicolaas Tulp*, pp. 45; also see p. 98. The doctrine of death as the wages of sin is traced by Heckscher from Paul to the *decreta* of the Council of Trent (June 17, 1546), which confirmed the doctrine.

123. Samuel Purchas (1577–1626), *Purchas his Pilgrim: Microcosmus; or, The Historie of Man: Relating the Wonders of his Generation, Vanities in his Degeneration, Necessity of his Regeneration; Meditated on the Words of David. Psal. 39.5. Verily, Every Man at his Best State is Altogether Vanitie* (London: W.S., 1619), fols. 11–16. A modern facsimile of this work is published under the same title (Amsterdam: Theatrum Orbis Terrarum; New York: Da Capo, 1969).

124. Ibid., fols. 6–7.

125. Ibid., fols. 25–26, 21. For the Paracelsianism of the analogies of the microcosm, see fols. 30–31.

126. *The Purchas Handbook: Studies of the Life, Times and Writings of Samuel Purchas 1577–1626*, with bibliographies of his books and of works about him, 2 vols., ed. by L. E. Pennington, Series: Works issued by the Hakluyt Society, 2nd series, nos. 185–86 (London: Hakluyt Society, 1997), I, 349. In chapter 9 of *Microcosmus*, Purchas mentions having "heard that worthie, wittie, learned, industrious Physician, Doctor *Harvey*, reading unto Curiositie of this Subject," that is, of the ears. Purchas began to write *Microcosmus* after the death of his mother, Anne Purchas, aged seventy-five, on March 13, 1619.

127. Purchas, *Microcosmus*, Chs. 3–10 (fol. 31).

128. Ibid., Chs 11–85.

129. On the solemn features suggested by Benedetti for the first permanent anatomy theater at Padua, see Ferrari, "Public Anatomy Lessons and the Carnival," p. 61.

130. Heckscher, *Rembrandt's Anatomy of Dr. Nicolaas Tulp*, p. 98. Heckscher adds: "Other skeletons hold banners proclaiming traditional death warnings: PVLVIS ET UMBRA SUMUS. NOSCE TE IPSUM. MORS ULTIMA LINEA RERUM, etc. A reclining putto with skull and hourglass presides over the whole." He also claims: "There is evidence that the displays in the *theatra anatomica* when these were used as museums were not intended for medical instruction; serious osteology was taught in Leiden in the auditorium medicum in the University Building" (p. 98). The implication is that the theater is designed as a spectacular lesson in mortality and redemption.

131. Rupp, "Matters of Life and Death," p. 272. Rupp adds: "The Vanitas-genre [*sic*] in the strict sense was in Holland a typical Leiden-genre, Leiden being the production centre of Vanity. As a genre it was a continuation of an Italian humanistic tradition (compare the illustrations in *De fabrica* of Vesalius)" (p. 273). In conclusion, Rupp claims: "The anatomical theatre in Leiden was as an art-museum a museum of Vanitas art" (p. 273).

132. Rupp, "Matters of Life and Death," pp. 273–79 (p. 278).

133. Neill, *Issues of Death*, p. 131.

134. Heckscher, *Rembrandt's Anatomy of Dr. Nicolaas Tulp*, p. 113.

135. See Dollimore, *Radical Tragedy* (1989), ch. 4, pp. 70–82.

136. Berengario, *Isagogae breves*, p. 93. This idea is explicable in terms of the general theoretical economy of fluids and vapors in the body, which is mentioned in chapter 4 of this study. Therein vapors are superfluous byproducts of digestion and heat-producing physiological processes, and among other means to excrete the vaporous matter is the growth of hair. Since the heart is the "sun" of the microcosm, which is responsible for the body's heat and illumination with spirit (pp. 93, 138), and, according to Benedetti, swells in the fetus with "the boiling spirit within it," and since the heart is the seat of the vital faculty of the mortal soul, whose function is agitative, irascible, spiritual, or passionate, the appearance of hairs or bristles on the heart was thought to indicate an excess of heat in the very organ that was the source of the body's heat and the seat of that which in the moral or psychological

domain is the equivalent of heat, such as, passion, courage, irascibility, or impetuousness. Benedetti, *Anatomice*, p. 99. Interestingly, for the context of dramatic theater, "heat" denotes both the enticement to venery, to sin, and to honor or revenge, following indignation, in *The Revenger's Tragedy*, published anonymously at London, in 1606/07, and ascribed first to Cyril Tourneur (*c.* 1570–1626) and then to Thomas Middleton (1580–1627). *Four Revenge Tragedies*, ed. with an introduction and notes by Katharine Eisaman Maus, Oxford English Drama (Oxford and New York: Oxford University Press, 1995; reissued 1998). Also see *Three Jacobean Tragedies*, ed. with an introduction by Gâmini Salgâdo (Harmondsworth: Penguin, 1965; repr. 1978)

137. Sidney, *The Defence of Poesy*, p. 124. Also see Dympna Callaghan, *Woman and Gender in Renaissance Tragedy: A Study of "King Lear," "Othello," "The Duchess of Malfi" and "The White Devil"* (London: Harvester Wheatsheaf, 1989), p. 50; Neill, *Issues of Death*, p. 121.

138. Katharine Eisaman Maus, *Inwardness and Theatre in the English Renaissance* (Chicago: University of Chicago Press, 1995), and Schoenfeldt, *Bodies and Selves in Early Modern England*, demonstrate that phenomena like emotion and desire were explained as rising from an economy of internal organs, fluids, and spirits, and Schoenfeldt, in particular, demonstrates that this phenomenon persisted even in the late seventeenth century, by referring to John Milton's development of moral, spiritual, and physiological strategies for the regulation of appetite and sexual desire. The causes of this phenomenon are seen as Galenic humoral theory and Paracelsian medicine, as distinct from anatomy, but this may be because there is an influential, but questionable, view in historicist literary studies that the rise of anatomical science led inexorably to a mechanistic conception of human anatomy among anatomists in the seventeenth century, and contributed to a diminution of this phenomenon.

139. David Hillman, "Visceral Knowledge: Shakespeare, Skepticism, and the Interior of the Early Modern Body," in *The Body in Parts*, ed. by Hillman and Mazzio, pp. 81–105.

140. *The Revenger's Tragedy*, Act 1, scene 1, l. 13, in *Four Revenge Tragedies*, further references are to this edition, and are provided in the text in the following order: Act number, scene number, line number(s); e.g., (1.1.13); Gâmini Salgâdo, "Introduction," *Three Jacobean Tragedies*, p. 15.

141. The quotation is from Francis Bacon's essay "Of Revenge," in *The Essays or Counsels Civil and Moral of Francis Bacon, Lord Verulam, Viscount St. Albans*, ed. with introduction and notes, by A. S. Gaye (Oxford: Clarendon Press, 1911), pp. 27–29 (p. 27). Also see Salgâdo, "Introduction," *Three Jacobean Tragedies*, pp. 14–17, on contemporary attitudes to revenge, and its legal and religious contexts.

142. Dollimore, *Radical Tragedy* (1989), p. 140.

143. John Keats, "Ode to a Nightingale" (VI. ll. 1–3), *The Poetical Works of John Keats*, ed. with an introduction and textual notes by H. Buxton Forman (London: Oxford University Press, 1915), p. 232. Also see *Romanticism: An Anthology*, 2nd edn., ed. by Duncan Wu (Oxford: Blackwell, 1998), pp. 1058–60 (p. 1059), and the editor's introduction to "John Keats (1795–1821)," pp. 1009–12 (p. 1011).

144. Neill, *Issues of Death*, p. 119.

145. Dollimore, *Radical Tragedy* (1989), pp. 139–43.

146. Eriksson, ed., *Andreas Vesalius' First Public Anatomy at Bologna 1540*, p. 219.

147. Dollimore, *Radical Tragedy* (1989), p. 73.

148. Cardwell, ed., *The Two Books of Common Prayer*, p. 307.

149. Neill, *Issues of Death*, pp. 120–22

150. Schupbach, *The Paradox of Rembrandt's "Anatomy of Dr. Tulp*," pp. 37–38, draws attention to the optimistic and pessimistic iconography of Rembrandt Harmensz. Van Rijn's *The Anatomy Lesson of Dr Nicolaes Tulp* (1632): "there are two demonstrative hatted figures and Rembrandt's praelector [Tulp] taught (following Laurentius) that anatomy had two uses: *cognitio sui*, which was expressed through the phrase 'know thyself', and *cognitio Dei*, which was expressed in some form of words equivalent to *est Deus in nobis*. If we read the picture from left to right, the second hatted figure (Dr. Tulp) illustrates the second lesson of anatomy, by using the flexor-tendons to demonstrate *est Deus in nobis*. It remains for the first hatted figure (Frans van Loenen) to demonstrate the first lesson of anatomy, 'know thyself', which he does by pointing to the corpse." Rupp, "Matters of Life and Death," p. 272, argues that the iconography of *vanitas* compositions included a group of symbols of resurrection, such as "ears of corn and sprigs of laurel or ivy, often twisted around the skull." Referring to *The Paradox of Rembrandt's "Anatomy of Dr. Tulp*," Rupp argues: "Schupbach, who clearly demonstrates the presence of Vanitas-style [*sic*] in the painted anatomical lessons in the first period of the Amsterdam theatre, sees a paradox in Rembrandt's painting of Dr Tulp between a pessimistic (Vanitas) emblem and an optimistic emblem. But the Vanitas-genre [*sic*] could, as we have already mentioned, very well include symbols of resurrection" (p. 278).

151. Heckscher, *Rembrandt's Anatomy of Dr. Nicolaas Tulp*, p. 107.

152. Sawday, *The Body Emblazoned*, pp. 178; 179–80. In connection with the emblematic status of the body

in (and of) *The Purple Island,* I wrote in chapter 3 of the speaking picture of the poetry, and in Chapter 6 of its mimesis of the juxtaposition of text and illustration in anatomy textbooks.

153. Sawday, *The Body Emblazoned,* pp. 178; 177.

154. See Wilson, "William Harvey's *Prelectiones,*" pp. 72–73.

155. Traub, "Gendering Mortality in Early Modern Anatomies," pp. 51–52, has argued that the series of visceral figures of Book V of Vesalius's *Fabrica,* act as a bulwark against the anxiety of the deadening, objectifying force of the anatomical "narrative."

156. Several recent studies concur in this interpretation of Renaissance anatomy. See, for example, Cunningham, *The Anatomical Renaissance;* French, *Dissection and Vivisection in the European Renaissance.*

157. French, *William Harvey's Natural Philosophy,* p. 43.

158. Heckscher, *Rembrandt's Anatomy of Dr. Nicolaas Tulp,* p. 29. Heckscher assumes that the "curious predilection for chapels and church choirs [. . .] was restricted to countries of the Reformation" (p. 31). Certain aspects of the convention were so, but not the broader convention of conducting public anatomy lessons in churches.

159. Cardwell, ed., *The Two Books of Common Prayer,* pp. 266–316 (p. 266).

160. Ibid., p. 273. The quoted passage is from the 1549 edition.

161. It is an influential view that theologians who defended the Roman Catholic position from the beginning of the Protestant revolt up to at least the middle years of the sixteenth century, agreed "that the Mass is not an absolute but a relative sacrifice, drawing its efficacy from the sacrifice of Calvary"; although there were "different theories about the nature of that mysterious relationship which unites Christ's unbloody sacrifice of the altar with his unique redemptive sacrifice of the cross." For the Reformers, Lutheran, Zwinglian, and Calvinist, the atonement was the substitutionary punishment of Christ on the gibbet in the place of guilty mankind, for the sins of whom "he paid once for ever," thus averting from the elect God's anger. Since, on account of Christ's sacrifice God "no longer imputed to [the elect] their sins," the sacrifice "was essentially *past,*" and the Church could not "perpetuate its reality" or "mediate its efficacy to men" through sacramental rites. Clark, *Eucharistic Sacrifice and the Reformation,* pp. 97, 109, 110 (also see pp. 144–45, 153, 154).

162. See Ferrari, "Public Anatomy Lessons and the Carnival," pp. 100–01.

163. For the effects of Christ's Eucharistic intercession as described by Thomas de Vio Cardinal Cajetan and other Counter-Reformation theologians, see, for example, Cardinal Thomas de Vio Cajetan, *De missae sacrificio et ritu adversus Lutheranos* (1531), ch. 6, cited in Clark, *Eucharistic Sacrifice and the Reformation,* p. 90.

164. Cardwell, ed., *The Two Books of Common Prayer,* p. 298. The quoted passage is from the 1549 edition.

165. Ibid., p. 308. The quoted passage is from the 1549 edition.

166. In this my view differs from that of Sawday, *The Body Emblazoned,* p. 118, who argues that although the convention of "self-demonstration" in anatomical illustration draws upon Christ's own division of his symbolic body at the Last Supper, the object of anatomical demonstration is not redemption but knowledge. For a related discussion of the sacramentalism of dissection, see Edgerton, *Pictures and Punishment,* p. 213. For a discussion of redemption in relation to the sacrament of the Eucharist, and of its Christological and eschatological aspects, see Neville Clark, *An Approach to the Theology of the Sacraments,* Studies in Biblical Theology, 17 (London: SCM, 1956), ch. VII: Towards a Theology of the Sacraments (pp. 71–85).

167. Sawday, *The Body Emblazoned,* p. 177.

168. "A.C.," "On the Most Accurate Poem, Inscribed *The Purple Island,*" Giles and Phineas Fletcher, *Poetical Works,* II, 10.

169. Fletcher, *A Father's Testament,* pp. 126–27, repr. in Giles and Phineas Fletcher, *Poetical Works,* II, 326–27, where it is given by Boas the title "The Divine Lover" in brackets to indicate that it was originally untitled.

170. Wilson, "William Harvey's *Prelectiones,*" pp. 80, 81.

171. Cunningham, *The Anatomical Renaissance,* p. 16. See Costelloe and Muirhead, *Aristotle and the Earlier Peripatetics,* II, 96, 100, 131.

172. Pagel, "William Harvey and the Purpose of Circulation," p. 29. Pagel is referring to Aristotle, *De Anima* 2.1.412a and 1.2.405b.

173. Cunningham, *The Anatomical Renaissance,* p. 16; Wilson, "William Harvey's *Prelectiones,*" p. 81.

174. Wilson, "William Harvey's *Prelectiones,*" p. 84.

175. *Microcosmos,* p. 83, col. 2, l. 13; *Mirum in modum,* p. 5, col. 1, ll. 6–8.

176. Langer, "Invention," pp. 137, 140.

177. Wilson, "William Harvey's *Prelectiones,*" p. 81.

178. Ibid., pp. 82, 83.

179. It is arguable that narratives are composed essentially of three elements: story, text, and narration. Though it is not reducible to the events of which it is composed, there can be no narrative without events. These events must also be organized into a sequence, a story, though some of these events may occur simultaneously. The sequential arrangement of events in a narrative text, which is called the plot, and the story that is told in that narrative need not, and usually are not, the same (on the distinction between story and plot, see chapter 3, note 3, of this study). For the events of a narrative to be arranged sequentially or simultaneously a medium in which they are so arranged is required. The extent of media to which the definition of text applies need not be prematurely closed, but in this context text is written and/or spoken words. In order that a text tells a story, a teller, or narrator, is required: a voice, and of course a writer. Each of these elements is indispensable to the construction of a narrative. Although "text is the only [aspect] directly available to the reader," "unless [the text] told a story it would not be a narrative, and without being narrated or written, a story would not be a text." Rimmon-Kenan, *Narrative Fiction*, p. 4.

180. Wilson, "William Harvey's *Prelectiones*," p. 89.

181. Eriksson, ed., *Andreas Vesalius' First Public Anatomy at Bologna 1540*, p. 45. See chapter 1 of this study for an explanation of Corti's distinction between anatomy and dissection; also see the text and notes of chapter 5, for further related comments on the relationship between the two.

182. Wilson, "William Harvey's *Prelectiones*," pp. 87–88 (p. 87).

183. Ibid., pp. 88, 89; also see pp. 83–84.

184. Valverde, *Anatomia del corpo humano, Tauola delle fig. del Lib. IIII* (this work is the Italian translation of the Spanish original); Laqueur, *Making Sex*, p. 77, fig. 12. On the so-called Belvedere Torso, and the series of visceral figures of Book V of Vesalius's *Fabrica*, in which the anatomy is presented as if through the dissection of antique sculpture, see chapter 3, note 23, of this study.

185. Wilson, "William Harvey's *Prelectiones*," p. 89.

186. Ibid., p. 89; Neill, *Issues of Death*, pp. 119–22. As Clifford Leech acknowledges, "what underlies Aristotelian *catharsis* has been of major importance in tragic writing, even when Aristotle has not been much or at all in the dramatist's mind." *The Dramatist's Experience: With Other Essays in Literary Theory* (London: Chatto & Windus, 1970), ch. 7: "*Catharsis* in English Renaissance Drama," p. 124. Also see Clifford Leech, *Tragedy*, The Critical Idiom: 1 (London: Methuen, 1969; repr. 1970), ch. 4: "Cleansing? or Sacrifice?," pp. 47–55. In the passage in his *Defence of Poesy* (p. 124), where Sidney compares tragedy to anatomy, he also describes how tragedy "with stirring the affects [feelings] of admiration and commiseration, teach the uncertainty of this world."

187. Wilson, "William Harvey's *Prelectiones*," p. 89.

188. According to Donne, in a sermon "Preached to the Nobility [undated, probably fairly early]," "a great part" of glorification consists in "Denudation," "that manifestation of all to all; as, in this world, a great part of our inglorious servitude is in those disguises, and palliations, those colours, and pretences of *publique good*, with which men of power and authority apparell their oppressions of the poore; In this are we the more miserable, that we cannot see *their ends*, that there is none of this denudation, this laying open of our selves to one another, which shall accompany that state of glory, where we shall see one anothers *bodies*, and *soules*, *actions* and *thoughts*." *John Donne's Sermons on the Psalms and Gospels*, ed. by Evelyn M. Simpson (Berkeley: University of California Press, 1963), p. 144.

189. For the outline of the main story-line of the narrative of *The Purple Island*, see the introduction to chapter 3.

190. Eclecta represents the Reformation idea of the Church as the chosen, the elect, rather than the Roman Catholic idea of the Church, whose role was understood by the "theology of *mediation* and *participation*." Francis Clark summarizes this theology, saying, "the Church, transforming the world by the divine life of grace which flows to her from Christ her head, is a perpetual extension of Christ's Incarnation." Apostolic succession and mystical sacramentalism are integral parts of this theology: "Endowed with Christ's priesthood, the Church through her ministers has the function of mediating to all men the fruits of Christ's all-sufficient work of salvation. This is the 'work', the *opus operatum*, of the sacramental system." Clark, *Eucharistic Sacrifice and the Reformation*, pp. 105, 103.

191. Aristotle's division of the souls of Man and of the Animal does not involve a location of the will. See Cunningham, *The Anatomical Renaissance*, pp. 14–22.

192. Voletta is attended by a "faithfull Counseller, *Synteresis*," who is "Conscience" (*P.I.* 6.61.2 and margin.), so the marginal note informs the reader; and by "a sad-fair maid, *Repentance*," who "in her arms her fainting Lady bears, / Washing her often stains with ever-falling tears" (*P.I.* 6.62.5, 6–7). According to William Whately, *The New Birth; or, A Treatise of Regeneration, Delivered in Certaine Sermons* (London: Felix Kyngston, 1619), sig. A_4^v,

the "principall graces of the new man" lie in the following: (1) "Vnderstanding," consisting of "Knowledge" and "Faith" in God and the Word of God; (2) "Conscience," which contains "Peaceablenesse" and "Wakefulnesse"; (3) "Will," which consists of "Being carried after God" and "Subjection to the will of God"; and (4) "The inferiour powers" of "Memory," "Imagination" and "Affections." (In the original the cited text appears in italics.) Repentance is used interchangeably with contrition in early seventeenth-century religious discourse, and according to the influential preacher Lancelot Andrewes, in a sermon given on Ash Wednesday in 1619, contrition is a turn "wherein we look backward to our sins wherein we have turned to God, and with beholding them our very heart breaks." Contrition is the first and preparatory stage of regeneration, and involves "reflecting and sorrowing for [what] is past"; the heart "sentencing itself for evil done heretofore." *The Liturgical Sermons of Lancelot Andrewes,* introduced and ed. by Marianne Dorman (Edinburgh, Cambridge, and Durham: Pentland Press, 1992), I: *Nativity, Lenten and Passion,* 127. Whately, *The New Birth,* p. 30, describes how in the first stage of the "*working of Regeneration,*" "the Spirit of God working in, and with the Law (but tempered with the Gospell) becomes a spirit of contrition, causing a man to see and feele his extreame sinfulnesse and wretchednesse, in so much that he is even wounded at the very heart therewith. . . ."

193. Sawday, "Bodies by Art Fashioned," pp. 134–35: "But the full range of religious association with the colour is hinted at later in the poem, and finally developed in the closing stanzas. In the sixth canto, the sleeping conscience of the island ('Synteresis'), is compared to flowers which in winter disappear. One of these flowers, the violet, is taken as a metaphor for the eventual awakening of conscience as it flowers once more in the spring. . . . [(*P.I.* 6.71–71)]" Sawday's analysis of the "code" of purple is not quite secure, here, for the "flowres that . . . / Retire into their mother *Tellus* wombe," and "in the Spring in troups new mustered / Peep out again from their unfrozen tombe," are a metaphor for the dimming and resurgence not of conscience but of the perceptual powers of Intellect and Voletta to "descrie" "every natures difference" (*P.I.* 6.68.1–4; 6.67.4). It is not that conscience sleeps, but that temperance or restraint and knowledge are no longer possessed of Voletta, who is "The will" (*P.I.* 6.57.5 margin. *), and Intellect, respectively. Conscience is Voletta's constant and alert attendant, but the will is often "deceived" (*P.I.* 6.61.4; and see 6.61.1), and in need of regeneration. Thus contrition, like winter, is a necessary preparation for regeneration of will (as of venous blood) in the heart (*P.I.* 6.57–77). The regeneration analogy drawn here is between the geocosm and the soul and body of the microcosm.

194. On what I have categorized as "weak materialism," see Paster, "The Body and Its Passions," p. 45.

195. Redman, "*The Purple Island,*" p. 6, argues that *The Purple Island* is based "generally on Book II, cantos ix and xi," of Spenser's work. The influence of *The Faerie Queene,* Book II, Canto 11, which is prefaced by the verse beginning, "The enimies of Temperaunce / Besiege her dwelling place," is generally felt on *The Purple Island* beginning at Canto 7. *The Works of Edmund Spenser,* ed. by R. Morris, p. 139.

196. Purchas, *Microcosmus,* fol. 8, remarks that "the Corruption of Nature (which the Scripture termeth *Flesh*) is now become another nature, having her owne God and Father, the author of Corruption"; but *The Purple Island* does not represent Caro as the offspring of the "Dragon," or Satan (*P.I.* 7.11.2), whose progeny include "*Cosmos,*" or "The World or Mammon" (*P.I.* 8.4.2 & margin. *), "*Despair*" (*P.I.* 12.32.1), and "two brethren" "in common name of death agreeing," the first of whom is Death the Mower (*P.I.* 12.36.1, 2, 3), the second damnation or eternal death (*P.I.* 12.39). See Redman, "*The Purple Island,*" p. 16: "Available to [Fletcher] for depicting the spiritual warfare in which man is involved was all the traditional psychomachia—the prolonged dragon fight from medieval allegory, the personification of the virtues and vices from religious drama, the familiar Puritan epic heard from lecterns and pulpits daily."

197. Cable, "Such Nothing is Terrestriall," p. 149.

198. In *The Purple Island,* "*Hæreticus*" or "Heresie" (*P.I.* 7.34.1 & margin. *) is moved from before "envyings," where it appears in the *Authorised Version,* to the group of sons to Caro and "*Asebie*" or "Irreligion" (*P.I.* 7.16.5 & margin. *), the others of which are "*Idololatros*" or "Idolatrie" (*P.I.* 7.28.3 & margin. *), "*Pharmacus*" or "Witchcraft and curious arts" (*P.I.* 7.31.1 & margin. *), and "*Hypocrisie*" (*P.I.* 7.35.1).

199. In the iteration of uncleanness, there are variations between versions of the Bible. In the *Revised Standard Version,* "impurity" is preferred; in the *Revised English Bible,* "indecency."

200. The vices in subordinate relation to the sons of Caro and Asebie are "*Ignorance,*" "their nurse" (*P.I.* 7.40.1), with her daughters "Errour" and "*Superstition*" (*P.I.* 7.41.5; 7.43.1). Errour "multiplies / Her num'rous race in endlesse progenies: / For but one truth there is, ten thousand thousand lies" (*P.I.* 7.41.5–7). Her "brood" is described as "o're-spread[ing] her round with sinne and bloud," "envie, malice, [and] mischiefs," which "flie about & spread their seed," breeding "hate, pride, schisme, warres & seditions" (*P.I.* 7.4.1, 2, 5, 6). These vices, with the exception of Ignorance, Errour, and Superstition, are not highly personified; indeed "*Envie*" is personified more fully elsewhere (*P.I.* 7.66.1), as a son to Caro and "*Adicus*" or "Unrighteous[n]ess" (*P.I.* 7.16.6 & margin. *).

201. Cable, "Such Nothing is Terrestriall," p. 149.

202. Henry Headley, "Supplement," in Headley, ed., *Selected Beauties of Ancient English Poetry,* II, 180. Headley's critical remarks on *The Purple Island* are reproduced in *The Purple Island, A Poem: By Phineas Fletcher,* pp. viii–ix. Gosse, *The Jacobean Poets,* p. 149, claims that "the allegorical figures [Fletcher] creates in so great abundance are distinct and coherent, with, as a rule, more of Sackville than of Spenser in the evolution of their types." Most studies have examined the influence of Spenser, however. For studies of the influence of Books I and II of Spenser's *Faerie Queene* on Cantos 6 to 12 of *The Purple Island,* see Gardner, "The Banquet of the Word," p. 401; Kastor, *Giles and Phineas Fletcher,* p. 155 note 4; Langdale, *Phineas Fletcher,* pp. 145–46. Less commonly, links are made between the vices and virtues of *The Purple Island* and Aristotle's *Nicomachean Ethics,* and Joseph Hall's *Characters of the Vertues and Vices* (1608). See Grundy, *The Spenserian Poets,* p. 189.

203. Grosart, ed., *The Poems of Phineas Fletcher,* pp. ccxxvii–ccxxxi.

204. Cable, "Such Nothing is Terrestriall," p. 149.

205. Thomas Cranmer, *Homilies,* 1849 edn., p. 33, cited in Clark, *Eucharistic Sacrifice and the Reformation,* p. 157.

206. The vices that are described in Canto 7, which are on the whole religious vices, are all described as having coupled with or as being the offspring of Caro. Those described in Canto 8, which are on the whole moral vices, are not so described, but all fall in rank behind her in the order the vices assume as they advance into battle with the virtues.

207. Wibbing, "*Sōma,* body," p. 236; also see p. 233. Brown, ed., *The New International Dictionary of New Testament Theology,* I, 229, which says that the basic idea of *sōma* in the Old and New Testaments is understood by some modern theologians as "the body in the sense of the whole person." *Sōma* has a further significance in the New Testament, however. It denotes the physical aspect to a person, expressed as a possession of that person. Nevertheless, it is argued that the meaning in the sense of a whole person persists in and is developed in the writings attributed to Paul.

208. Healy, *Fictions of Disease in Early Modern England,* p. 3.

209. According to Barnhart, ed., *Dictionary of Etymology,* p. 539, col. 1, intestine as a noun, extant from before 1424, was "borrowed through Middle French *intestin,* or directly from Latin *intestina,* neuter plural of *intestinus,*" the adjective meaning internal, and was "probably altered (by influence of *clandestinus* hidden) from [the] earlier *intustinos,* from *intus,*" which means "within."

210. Principal among the faculties, Intellect is described by the diegetic narrator Thirsil ambiguously as incorporeal but bodily in a nonspatial manner: "His strangest body is not bodily, / But matter without matter; never fill'd / Nor filling" (*P.I.* 6.30.1–3). Cable, "Such Nothing is Terrestriall," p. 145, comments: "Thirsil the sensemaker throws himself into describing the indescribable, and his 'too bold vent'ring gleam' [*P.I.* 6.29.5] is flung back at him by a string of paradoxes." Yet, as Cable adds, although "the substance and dimension of Intellect are no sooner affirmed than denied by the paradoxes, still their cumulative effect is to persuade us that something has indeed been described." "Like Intellect himself," argues Cable (p. 150), "the [other] inhabitants of the Purple Island . . . cannot be apprehended directly, but they can be understood through our experience of sensory poetic language."

211. Harris, *Foreign Bodies and the Body Politic,* p. 4.

212. See Wibbing, "*Sōma,* body," p. 236. For Paul, "[t]he desires of the body and the desires of the flesh are synonymous." On occasion, in Paul, *sarx* (flesh) may denote the human body (2 Cor. 12:7; Gal. 4:13, 14), or "the material of which the body (*sōma*) is composed" (Col. 1:22; 2:11). H. Seebass and A. C. Thiselton, "Flesh," in Brown, ed., *The New International Dictionary of New Testament Theology,* I, 671–82 (p. 676). But the vices listed in Galatians 5:19–21, upon which *The Purple Island* draws faithfully, are the works of the flesh, rather than of the body, because whereas the body may be a member of Christ, the flesh "not only occasions sin but also becomes entangled in it" (pp. 674–75).

213. Seebass and Thiselton, "Flesh," p. 676.

214. See John Donne, *LXXX Sermons, preached by that learned and reverend Divine John Donne Dr. in Divinity late Deane of the Cathedrall Church of S. Pauls London* (London: R. Royston and R. Marriot, 1640), p. 145, where, in a sermon on 1 Corinthians 15:26, "Preached at White-hall, March 8. 1621," Donne says, "for thy body hath thee, and not thou it, if thy body tyrannize over thee. If thou canst not withdraw thine eye from an object of tentation, or withhold thy hand from subscribing against thy conscience, nor turne thine eare from a popular, and seditious Libell, what hast thou towards a man?"

215. Nohrnberg, *The Analogy of "The Faerie Queene,"* pp. 335–43 (pp. 337, 335). The sin of Adam and Eve had been delineated by theologians as corresponding to the three temptations of Jesus in the wilderness.

216. Fletcher, *The Way to Blessednes,* "The Preface to every *Christian Reader.*" Roman font in the original is rendered in the text as italicized, and italicized text in the original is rendered in roman font, in this and all other citations of "The Preface to every Christian Reader."

217. Fletcher, *The Way to Blessednes,* "The Preface to every Christian Reader."

218. The quotation is from line 10 of the second stanza of the little poem by Fletcher which is reprinted in Giles and Phineas Fletcher, *Poetical Works,* II, 322–23, where by Boas it is given the title "The Search after God." The poem is discussed above, in chapter 8 of this study, and the action of dispelling earth's fogs is that of the sun and the heart. In *The Purple Island,* as is cited in chapters 3 and 7 of this study, Thirsil also invokes God: "Shed in my mistie breast thy sparkling light, / And in this fogge my erring footsteps guide" (*P.I.* 3.4.3–4). Also in *A Father's Testament,* pp. 126–27, repr. in *Poetical Works,* II, 326–27, the poem that is given by Boas the title "The Divine Lover," has Christ say, "my ray / Thy mists, and hellish foggs shall pierce" (II, 6–7).

219. Fletcher, *Joy in Tribulation,* p. 202.

220. Fletcher, *The Way to Blessednes,* "The Preface to every Christian Reader."

221. Donne, *LXXX Sermons,* p. 206. The terms are quoted from Sermon XXI, which is said to have been preached "at S. Pauls, in the Evening, upon Easter-day. 1626."

222. Thomas Aquinas, *Summa Theologica,* Pt. II, 2nd Pt., q. 167, art. 1–2, and Augustine, *Confessions,* X. 30, cited in Nohrnberg, *The Analogy of "The Faerie Queene,"* pp. 336, 338. Augustine says that, "through the eyes especially prevails curiosity." He explains that, "there is present in the soul a kind of empty longing and curiosity which aims not at taking pleasure in the flesh, but at acquiring experience through the flesh, and this empty curiosity is dignified by the names of learning and science." Augustine, *Ennar. in Psalmos,* VIII. 8, and *Confessions,* X. 30, cited in Nohrnberg, *The Analogy of "The Faerie Queene,"* pp. 336, 338.

223. Eriksson, *Andreas Vesalius First Public Anatomy at Bologna 1540,* pp. 195–201, 203, 209, 229. Ferrari, "Public Anatomy Lessons and the Carnival," p. 98, says that huge crowds attended Vesalius's dissection of the genital organs.

224. Giles and Phineas Fletcher, *Poetical Works,* p. 4.

225. Ibid. Michael Schoenfeldt, "Fables of the Belly in Early Modern England," in *The Body in Parts,* ed. by Hillman and Mazzio, pp. 242–61 (p. 258), chooses to interpret the remarks of "the Conformist divine Daniel Featley" as referring to the "alimentary path," the "'blinde and dirty way,' through the interior spaces of the consuming subject," which Schoenfeldt says "is a voyage that a surprising number of early modern writers felt obliged to make." Significantly, Schoenfeldt is simultaneously aware of the physical, corporeal anatomy and physiology of this voyage, and its ethical character and meaning, as a "route to knowledge of self and God," and as "pertinent to the comprehension and expression of the inward self."

226. In the dedication of the first edition of *The Purple Island* "To my most worthy and learned friend, Edward Benlowes *Esquire,*" Fletcher provides for his anonymity the explanation that the poems collected in that volume may thereby "flie under [the] *Shadow*" of his patron, "to cover them from the piercing eye of this very curious (yet more censorious) age." "The contrary of curiosity in Aquinas is studiousness, and studiousness is described as a part of the virtue of temperance; it is classified as a kind of restraint." Thomas Aquinas, *Summa Theologica,* Pt. II, 2nd Pt., q. 167, art. 1–2, cited in Nohrnberg, *The Analogy of "The Faerie Queene,"* p. 336. In the "Turret" of "The Castle of Alma," Guyon restrains his curiosity, but he is "encouraged to indulge his studiousness." Nohrnberg, *The Analogy of "The Faerie Queene,"* p. 346. Similarly, in *The Purple Island,* when Intellect has "view'd with searching eye" the "shapes deriving" from the "busy brain" of Phantastes, "He straight commits them to his Treasurie," kept by studious Eumnestes, "Father of memorie" (*P.I.* 6.48.5.1; 6.47.1; 6.48.6, 7).

227. For Aristotle, "[t]he soul resides primarily in the Pneuma, which is the cause of vital heat, and which in turn has its chief seat in the heart." The heart's function "consists partly in preparing the blood," the concoction of *pneuma zoticon,* and "partly in producing sensation and motion." "The organ of touch is the heart: the medium through which impressions are transmitted to the heart is the flesh." Costelloe and Muirhead, *Aristotle and the Earlier Peripatetics,* II, 40–42, 66. Fletcher did not follow Aristotle, as chapter 2 of this study has made clear, for to do so would have involved, firstly, ignoring Galen's demonstration that the reversive nerves in the neck arise from the brain, and secondly, courting the problem of how "the sensations communicated by particular senses can have their seat in the head, while the seat of the sensitive life itself is in the heart, and all sensation belongs to one and the same part of the soul," which Aristotle fails to explain. "If [Aristotle's] view is that the pictorial image is generated in the organs of sense, while its reference to the object takes place in the heart, the question still remains, how can sensation originate in organs in which the sensitive soul does not reside?" Costelloe and Muirhead, *Aristotle and the Earlier Peripatetics,* II, 66.

228. Matthew 5:29–30; Brown, ed., *The New International Dictionary of New Testament Theology,* I, 231, argues

that Jesus' advice, at this point in his Sermon on the Mount, bears "no trace" of the Aristotelian "idea of perfection as the complete number of the members in their entirety."

229. *The Purple Island*'s account of the tongue bears a resemblance to that of Helkiah Crooke, in *Microcosmographia*, fols. 628–29, cited in Harris, *Foreign Bodies and the Body Politic*, p. 109.

230. On the anatomical, linguistic, and moral ambivalence with respect to the representation of the tongue in early modern literature, see Carla Mazzio, "Sins of the Tongue," in *The Body in Parts*, ed. by Hillman and Mazzio, pp. 52–79. Mazzio argues that the tongue functions as a metonym of language in a process of "somatization of speech" (p. 55). However, it might be that the relation between language and organ elsewhere could more usefully be interpreted as metaphor, wherein the tongue might be regarded as condensing, rather than, as Mazzio suggests, displacing, anxieties about reference itself, the slipperiness of language, the movement of speech away from the individual body, and the movement of signs away from any singularly discernible, naturalized context (see p. 54). Mazzio argues that the tongue and the mouth became the repository of these collective anxieties about meaning and authority in an increasingly textualized culture, in which words are decontextualized from the body. Mazzio also notices how expressions to the effect that the tongue wanders through the world perform as synecdochical, representing the whole person, the effect of whose utterance is so widely felt. An effect of the tongue's suitability "for the articulation of collapsing distinctions, be they linguistic, sociopolitical, geographic, or cosmic," is that "[f]antasies of the tongue's mobility were often explicitly linked to disturbances of social and political order" (p. 57).

231. Mazzio, "Sins of the Tongue," p. 63.

232. The comparative lack of severity of *The Purple Island*'s form of punishment of the unruly tongue should not lead us to ignore the possibility that, as Harris, *Foreign Bodies and the Body Politic*, p. 110, argues: "A memory of the scold's bridle appears to lurk in Phineas Fletcher's account of the malevolent female tongue and her 'bridling' in *The Purple Island*."

233. Davies of Hereford, *Microcosmos*, p. 23, col. 2, l. 2 (from bottom); p. 24, col. 1, l. 2; p. 24, col. 1, ll. 6–23.

234. Ibid., p. 86, col. 1, ll. 7–14.

235. Ibid., p. 24, col. 1, l. 16.

236. Costelloe and Muirhead, *Aristotle and the Earlier Peripatetics*, II, 46.

237. Anatomists did not search into what Katharine Eisaman Maus refers to as "the difference between an unexpressed interior and a theatricalized exterior." Her *Inwardness and Theater in the English Renaissance* explores "the afflictions and satisfactions that attend upon [this difference]: the epistemological anxieties that gap generates, the social practices that are devised to manage it, and the sociopolitical purposes it serves" (p. 2).

238. Whitteridge, ed., *The Anatomical Lectures of William Harvey*, p. 175. The wondrous mystery in reproductive and post-natal physiology of how the blood of "*Hepar* rivers" (*P.I.* 4.6.4) turns in the female breasts "a lily white" (*P.I.* 4.6.7), "by some strange hidden might" (*P.I.* 4.6.6), is related as evidence of "the providence of God," in *The Purple Island* (*P.I.* 4.6 margin.c).

239. Wibbing, "*Sōma*, body," p. 234.

240. The description is reminiscent of Clément Marot's *blason*, or rather *contreblason*, *Laid tetin* (Ugly breast), which, according to Alison Saunders, *The Sixteenth-Century Blason Poétique* (Bern, Frankfurt am Main, and Las Vegas: Peter Lang, 1981), p. 56, reiterated and developed "Villon's evocation of the sagging, shrivelled breast of the *Belle heaulmiere*." The literary form of the blazon is "not very far removed" from the emblem (p. 39; also see pp. 168–77), and the pattern in the anatomical *blason* is an illustrative woodcut followed by a descriptive and analytical poem (p. 138). On the interpretation of the *contreblasons*, Saunders distinguishes between those which draw on serious, didactic moralizing traditions, which in cases like that of the *contreblasons* written by La Hueterie sought to denigrate the body as the unduly idolized subject of the *blasons*, and those by Marot, Peletier, and Rus, which "describe the ugly female body for mere amusement" (p. 72; and see pp. 141–56). Fletcher need not have read the collected *Sensuivent Les blason anatomiques du corps femenin, ensemble les contreblasons de nouveau composez, & additionez, avec les figures* . . . , as the first collected edition was entitled, in 1543, not because, as Saunders observes (p. 69), a wealth of potential source material for the *contreblasons* existed within the native French tradition in the form of anti-feminine poems, one category of which concentrates on the transience of female beauty (because Fletcher did not read French) but because by the early seventeenth century, the tradition had become very widely disseminated, and similar moralizing depictions of the body existed in English and other traditions which treated with morbid relish the subject of the decay of the body which is temporal. During the later sixteenth and early seventeenth centuries, the amatory and anatomical preoccupations of the *blason anatomique* evolved and diversified, and "by the slightly later period fashions had . . . evolved away from printed illustration." As in *The Purple Island*, the poetry itself became the picture: "Metaphor and myth and linguistic adornment replace woodcut illustration and decorative borders" (p. 277).

241. Maus, *Inwardness and Theater in the English Renaissance,* pp. 2, 4, 5.

242. Wibbing, "*Sōma,* body," p. 235.

243. Maus, *Inwardness and Theater in the English Renaissance,* p. 5. Maus is referring to the method expounded by the physician John Cotta.

244. For Fletcher's relative lack of skepticism, see chapter 5 of this study. Maus, *Inwardness and Theater in the English Renaissance,* pp. 6–12 (pp. 7, 6) discusses the relationship between the early modern distinction between interior and exterior and "the [skeptical] 'problem of other minds' [which] presents itself to thinkers and writers not so much as a question of whether those minds exist as a question of how to know what they are thinking," and the "possibility of some secret motive, some unexposed residue [which] can never be wholly discounted, even when the gesture of self-revelation seems most generous and complete." Maus (pp. 10, 12) argues that: "Even as Renaissance theism generates misgivings about human access to truth, it provides a context for thinking about what certainty would mean and to whom it could be ascribed. The Christian God exemplifies not only mysterious inwardness, but an effortless transcendence of the boundaries that frustrate human knowledge." "On the one hand, the theistic context in which the problem of human inwardness is posed provides a standard of what would constitute certainty. On the other hand, faith itself encourages a kind of mistrust: for what is most true about human beings in such a system is simultaneously least verifiable."

245. See John Donne, *LXXX Sermons,* p. 140, where, in a sermon "Preached at White-hall, March 3. 1619," Donne describes God's judgment as "when the records of all thoughts shall be laid open, . . . when the heart shall be laid open."

246. For the morality play tradition as, at least "a parallel expression," if not a possible source, of the general religious expression in *The Purple Island,* see Redman, "*The Purple Island,*" p. 11.

247. Whately, *The New Birth,* p. 30, explains that the contrite sinner is "wounded at the very heart" with "his extreame sinfulnesse and wretchedness," and "his sinfull and unhappy estate becomes a wofull bondage and captivitie unto him."

248. The relation in Paul between the flesh as Man and dependence on the flesh as sin, is such that even the *Nous,* the Active Reason of Aristotelianism, "is unable to find out what is good, since the true meaning of life remains hidden from it by sin." But, as Paul explains in Colossians, the believer holds fast to the head, who is Christ, in whose body they are a member (Col. 1:24; 2:19). "Thus, living in the flesh, [the believer] is not in subjection to these powers [of the flesh], but in his flesh he already shares in the life of Christ." Here, the meaning of the flesh almost converges with that of *sōma.* Seebass and Thiselton, "Flesh," p. 676. The authors refer to Colossians 1:24; and 2 Corinthians 4:11; 10:3.

249. Cable, "Such Nothing is Terrestriall," p. 148. Cable is referring to *The Purple Island* 6.38–39.

250. Kastor, *Giles and Phineas Fletcher,* p. 107.

251. Ibid., p. 113, argues: "The physiological or anatomical allegory, physical man as a purple island (II–V), is, however, something quite different [from the rest of the poem]. This use makes the poem read like science turned into literature and forms an almost separate (and separable) section."

252. Cable, "Such Nothing is Terrestriall," pp. 148, 138.

253. See the examination in chapter 6 of this study, of Cable's analysis of the differentiation between the poetic stanzas and prose marginalia in Cantos 2 to 5 of *The Purple Island.*

254. Maus, *Inwardness and Theater in the English Renaissance,* p. 11, note 33. Maus says further (p. 10): "In fact, for some writers the presence of an omniscient spectator seems so fundamental to the structure of human subjectivity that the fact of that subjectivity becomes part of the proof of God's existence."

255. Focalization in narrative theory is a concept which was developed by Genette, and is to be distinguished from narration. It is discussed clearly and at some length by Rimmon-Kenan, *Narrative Fiction,* pp. 71–81. It is the way in which the story is presented in the narrative text through some "prism," "angle of vision," which is verbalized by the narrator, though not necessarily hers or his. In addition, "its purely visual sense has to be broadened to include cognitive, emotive and ideological orientation" (p. 71). Rimmon-Kenan therefore divides focalization into the perceptual facet, the psychological facet, and the ideological facet (pp. 77–81). Genette's introduction of the term focalization dispelled confusion between perspective and narration, which occurs when "point of view" is used. Rimmon-Kenan usefully suggests that the answer to the question "who sees?" at any point in a text, concerns focalization; whereas the answer to the question "who speaks?" concerns narration (p. 72). Focalization may be further divided however, for it has both a subject and object: the focalizer and the focalized (p. 74). So, the focalized is the anatomy of the body politic in the focalization discussed in the text, above. The focalizer may be internal or external to the diegesis or story; the focalized may be seen from within or from without (pp. 75–76). Both the focalizer and the focalized may remain fixed throughout a story, or alternate, or be multiple (p. 76).

256. Harris, *Foreign Bodies and the Body Politic,* pp. 13, 9,

257. Cable, "Such Nothing is Terrestriall," p. 146.

258. Healy, *Fictions of Disease in Early Modern England,* p. 10.

259. Carlino, *Books of the Body,* p. 81. It is Peter Burke who has suggested, in his *Popular Culture in Early Modern Europe* (London, 1978), that the word "Carnival" originates in the Latin *caro.*

260. Mikhail Bakhtin, *Problems of Dostoevsky's Poetics,* ed. and trans. by Caryl Emerson, Theory and History of Literature Series, 8 (Manchester and Minneapolis: Manchester University Press, 1984), p. 123, cited in Ferrari, "Public Anatomy Lessons and the Carnival," p. 103.

261. Schoenfeldt, *Bodies and Selves in Early Modern England,* p. 14.

262. Ferrari, "Public Anatomy Lessons and the Carnival," p. 104.

263. See Traub, "Gendering Mortality in Early Modern Anatomies," p. 54, on the association of the male body with the Classical body, and with the powers of agency, rationality, and impermeability, and of the female body with the grotesque body, and with traits of permeability, and with need of enclosure or containment.

264. On the Italian tradition of anatomical love poetry, and particularly its association with the court of Francis I, see Saunders, *The Sixteenth-Century Blason Poétique,* pp. 91–112 (p. 100). It is also conceivable that the influence on *The Purple Island* may have come from the poets of the *Pléiade* or by their successors (see p. 266). Clément Marot, author of the challenge that gave rise to the early *blasons anatomiques* of the sixteenth century (see note 240, in the present chapter), according to Sawday, was one of the so-called *Grands Rhétoriqueurs* also associated with the courts of Francis I and his sister Margerite d'Angoulême (later Navarre).

265. Valerie Traub, *Desire and Anxiety: Circulations of Sexuality in Shakespearean Drama* (London: Routledge, 1992), pp. 14, 15.

266. For the influence of Virgil (70–19 B.C.) on *The Purple Island,* see Langdale, *Phineas Fletcher,* Ch. 8: "Foreign Economy," especially p. 114.

267. A device in a picture with a motto or "word" was also known as an *imprese,* which appears to have originated earlier than the emblem, which consists often of a motto, picture, and epigram. "Whereas, too, the emblem was understood to be expressive of some general moral truth, an impresa was always specific and particular. It was designed to be used by a single individual to express his or her situation, aim, or state of mind, often for a unique occasion, such as a tournament . . . , pageant, or masque." Alan R. Young, assisted by Beert Verstraete, *The English Tournament Imprese* (New York: AMS Press, 1988), p. 1. This is why Fletcher chooses the *imprese* over the emblem to express the "subjective" truths of the flesh and the many vices, though he chooses the same, for the sake of consistency, for the virtues, whose mottos and pictures could be thought to be expressive of general moral truths. However, discussing the emergence of the emblem as a distinct literary form in the first half of the sixteenth century, John Manning, *The Emblem* (London: Reaktion Books, 2002), p. 15, says: "Being a new invention, however, the emblem was free to define itself. . . . The taxonomical question becomes further muddied when we look at closely related linguistic forms in various vernacular traditions: *impresa* in Italian and *empresa* in Spanish in many cases refer to what we would commonly accept as emblems."

268. Parthenia's significance in the battle for the island and body, between the works of the flesh and the virtues, is reflected by her prominence in text-time. In total, Parthenia is the principle subject of twenty of the fifty stanzas in Canto 11. The vices, or fruits of the flesh, in *The Purple Island* include, but are not reducible to, those with explicit applicability to Roman Catholicism as represented ideologically from an English Protestant position. These consist principally of the daughters of "*Ignorance*" (*P.I.* 7.40.1). These are "Errour" and "Owl-ey'd *Superstition*" (*P.I.* 7.41.5; 7.43.1). Ignorance is nurse to the progeny of Caro and Asebie, or Irreligion, and consequently these offspring carry some connotations applicable to an ideologically represented Roman Catholicism. They are "*Idololatros,*" or "Idolatrie," Pharmacus, or Witchcraft and curious arts, "*Hæreticus,*" or "Heresie," and "*Hypocrisie*" (*P.I.* 7.28.3 & margin. *; 7.34.1 & margin. *; 7.35.1).

269. MacDonald, *Witchcraft and Hysteria in Elizabethan London,* fol. 22^{v}; Paré, Workes (1634), Bk. XXIV, p. 942.

270. The diegetic narrator's explicit praise of Parthenia, and avoidance of blame, is not all that the reader is able to infer, from the events of the story, about her, or about the relationship between vice and virtue in the poem.

271. The verb "to blazon" "first appears in English in the sixteenth century," and is used broadly for "to describe" in a variety of heraldic, colorful or ornamental ways, and for "to publish vauntingly or boastfully." The verb relates also to the English verb "to blaze," "to proclaim . . . to publish . . . and, by extension, to defame or celebrate, to depict, to portray." To blazon is, however, French in origin, from *blasonner,* and although used so commonly there "that it signified little more than 'to describe,' . . . its usage was rooted in two specific descriptive traditions, one heraldic and the other poetic." Nancy J. Vickers, "'This heraldry in Lucrece' face,'" in *The*

Female Body in Western Culture: Contemporary Perspectives, ed. by Susan Rubin Suleiman (Cambridge: Harvard University Press, 1986), pp. 209–22 (p. 213). Saunders, *The Sixteenth-Century Blason Poétique*, pp. 10, 17–19, agrees that "*Blason*," either as a title or when not used as a title to a poem, "was wide-embracing in its range of meaning at this period," but that the early *blason* was heraldic, and was explanatory and analytical of illustrated arms. However, "[i]n the context of the anatomical *blason*, the word is generally taken to have the sense of praise (when used as the title to a poem), with the word '*contreblason*' being coined to designate the opposite meaning of blame" (p. 11). Saunders (p. 12) also adds the *blason* could be a poem of analysis, definition, interpretation, and simple description, and as such the blazon could have lent its meaning to the description of anatomical parts in Cantos 2 to 5 of *The Purple Island*, though it should be added that the term is only used at this point where the narrator Thirsil refrains to "blazon" the sexual organs, and this perhaps reflects the amatory as well as anatomical character of the *blasons anatomiques*. To "blazon" is used by the diegetic narrator of *The Purple Island* to mean, to "limme forth" (*P.I.* 3.32.2), to poetically paint, depict, describe, define, analyze, praise, celebrate, or vaunt. It is also relevant to my more general remarks about the literary expository or fictional anatomy and encyclopedic poetry, that Saunders identifies the "medieval encyclopedias" and the "poetic compendium, in which all the various aspects of a given phenomenon are broken up, described and analysed and classified," as providing "a marked parallel with," or as having had a discernible "direct influence upon," the sixteenth-century *blason poétique* (pp. 81–82). Sawday associates the poetic form of the blazon very closely with anatomy itself, in chapter 7: "The Realm of *Anatomia:* Dissecting People," in *The Body Emblazoned*, pp. 183–229. For the psychoanalytic concept of displacement, see J. Laplanche and J.-B. Pontalis, *The Language of Psycho-analysis*, with an introduction by Daniel Lagache, trans. by Donald Nicholson-Smith (London: Karnac Books and the Institute of Psychoanalysis, 1988), p. 121: "The fact that an idea's emphasis, interest or intensity is liable to be detached from it and to pass onto other ideas, which were originally of little intensity but which are related to the first idea by a chain of associations."

272. Vickers, "'This heraldry in Lucrece' face,'" p. 220, describes "a deep ambivalence" in the "heraldry in Lucrece' face," in William Shakespeare's *The Rape of Lucrece:* "The monstrous becomes the other side of the beautiful; the obsessively spoken part—the face—the other side of the obsessively unspoken but violated part—the genitalia; . . ." *The Rape of Lucrece* appears in Shakespeare, *The Complete Works*, pp. 237–55, and the line which is quoted in the title of Vickers's essay is line 64 (p. 239).

273. For Melanie Klein's view on the child's urge for knowledge, the epistemophilic impulse, see her "Early Stages of the Oedipus Conflict" (1928), in *The Selected Melanie Klein*, ed. by Juliet Mitchell (London: Penguin, 1991), pp. 69–83 (pp. 69, 72, 75, 77).

274. This is reminiscent of the "frisson of desire" with which Vesalius has been described conducting midnight raids on gibbets as a way of gathering skeletal pieces for the construction of a whole human skeleton. See Sawday, *The Body Emblazoned*, pp. 195–96. Sawday also describes the erotics of dissection in early modern poetry, in chapter 3: "The Body in the Theatre of Desire," and chapter 7: "The Realm of *Anatomia:* Dissecting People," in *The Body Emblazoned*, pp. 39–53, 183–229. For the view that the flap anatomy, such as we find in Remmelin's *Catoptrum microcosmicum*, constitutes anatomy as pornography, in which the reader is coerced into an experience of sin and transgression, see Traub, "Gendering Mortality in Early Modern Anatomies," p. 82.

275. Grundy, *The Spenserian Poets*, pp. 181–85 (p. 183) surveys *The Purple Island, Venus and Anchises*, the latter of which she regards as "pornographic art, a youthful day-dream of quintessential love-making," and the Sion manuscript epithalamion, which she describes as "desire climbing with a tongue of fire."

276. The conceit of the eyebrows as Cupid's bow is also to be found in Maurice Scève's *Sourcil*, which is influenced by Italian anatomical conceits. See Saunders, *The Sixteenth-Century Blason Poétique*, p. 102.

277. Saunders, *The Sixteenth-Century Blason Poétique*, pp. 57–58, describes how beauty featured above maternity as fulfillment in some poetic treatments of the breast.

278. For a similar argument relating to the eroticization of female anatomical figures in textbook illustrations, see Traub, "Gendering Mortality in Early Modern Anatomies," pp. 44–48, 80–82.

279. On the part-whole relationships in the blazon tradition, and the tension or balance between dismemberment and incorporation or integrity, see Saunders, *The Sixteenth-Century Blason Poétique*, pp. 117–18, 136. For a contrastive view, see Nancy J. Vickers, "Members Only: Marot's Anatomical Blazons," in *The Body in Parts*, ed. by Hillman and Mazzio, pp. 3–21 (pp. 8, 4, 9, 16), who argues that in blazoning, "few attempts are made to recover the body's unity." Vickers understands anatomical blazons as "the extension, the exaggeration to the point of violation of the descriptive mode itself: they isolated each unit of conventional body-part enumeration; they attempted a microdescription that ultimately failed to be descriptive; and they diminished synecdoche" by addressing a part rather than a whole woman. The blazon's fetishistic overdetermination of individual parts is "most

graphically enacted in the radically fragmentary illustrations that preceded their poems," though the texts are also characterised by the "tropic disfiguration" which privileges a single signifier, whether the breast, thigh and so on.

280. See 2 Corinthians 4:11; Seebass and Thiselton, "Flesh," p. 679; Giovanni Pico della Mirandola, *A Platonick Discourse on Love, Written . . . in Explication of a Sonnet by Hieronimo Benivieni,* trans. by Thomas Stanley (London, 1651), pp. 239–40: "Vulgar Love is the Appetite of sensible Beauty, through corporeal sight. The cause of this Beauty is the visible Heaven by its moving Power. As our motive faculty consists in Muscles and Nervs, (the Instruments of its Operation,) so the motive faculty of Heaven is fitted with a Body proper for circular sempiternal motion; Through which Body the Soul (as a Painter with his Pencil changeth this inferiour matter into various forms) hath her causal being from the moving power of the Heavens, her formal from colour, enlightened by the visible sun as Ideas by the invisible. . . ." What the description of Parthenia's beautiful body attempts to do is comparable with the accomplishment of the *blasons* of Maurice Scève, which, according to Saunders, *The Sixteenth-Century Blason Poétique,* p. 123, elevate their subject "beyond the particular, towards a more universal significance; worship of individual beauty is transcended, in the manner advocated to Socrates by the wise woman, Diotima [in Plato's *Symposium*], towards a religious adoration of an ideal of spiritual beauty, of which physical beauty is no more than a minor outward manifestation." For "scopophilia," see, for example, Laura Mulvey, "Visual Pleasure and Narrative Cinema," *Screen* 16, no. 3 (1975): 6–18, though Mulvey's thesis has been much contested. For sublimation as a process in which the sexual instinct is "diverted towards a new, non-sexual aim," and in which "its objects are socially valued ones," arguably like Neoplatonic philosophy and Neo-Petrarchan poetry, see Laplanche and Pontalis, *The Language of Psycho-analysis,* p. 431.

281. Pico della Mirandola, *A Platonick Discourse on Love,* pp. 242–43. Saunders, *The Sixteenth-Century Blason Poétique,* p. 53, counters the argument that "the model for the whole corpus of *blasons anatomiques* [of Clément Marot] should be sought in the neo-petrarchan love poetry of such Italian poets as Nocturno, Cei and, above all, Olympo de Sassoferrato." She argues, on the contrary, that the blazon's "analytical descriptive aspect" "derived from a purely French heraldic tradition" (p. 52; also see pp. 91–112). The influence of both the anatomical *blasons* and the Neo-Petrarchan poets is likely in Fletcher's case. In addition, as in the passage of *The Purple Island* discussed in the text, Neoplatonic and Neo-Petrarchan influences sometimes converged, for example, at Lyons and the court in France in the sixteenth century, where the *blason* became a prominent poetic form (p. 122).

282. Seebass and Thiselton, "Flesh," p. 676. Paul's letters offer the view that believers simultaneously remain "in the flesh" in that they are living; but they live in Christ. Pico della Mirandola, *A Platonick Discourse on Love,* offers a mystical interpretation of some of Paul's remarks on this matter of no longer being in the flesh: "Though Celestial Love liveth eternally in the Intellect of every Soul, yet onely those few make use of it, who declining the Care of the Body, can with S. Paul say, *Whether in the Body or out of the Body they know not.* To which state a Man sometimes arrives; but continues there but a while, as we see in Extasies."

283. According to Seebass and Thiselton, "Flesh," p. 673, "Judaism in its various forms connected man's carnality with his sin, but without interpreting the flesh as the actual cause of sin." The connection is defined such that dependence on the flesh is sin (Isa. 31:3). But equally, "to strive after evil is inherent in man" (Gen. 8:21). Given these Old Testament strands, Seebass and Thiselton argue that "[t]here must be a relation between the flesh and dependence on the flesh."

284. In this Fletcher's opinion is not consistent with the "Neoplatonic" sentiment expressed in the textual notes on the sonnet by Hieronimo Benivieni, in Pico della Mirandola's *A Platonick Discourse on Love,* pp. 256–57: "Plotinus himself averres, that there was never any beautiful person wicked, that this Gracefulnesse in the Body is a certain signe of Perfection in the Soul." The contrast between the beautiful body parts of Parthenia and the (actually, if not apparently) degenerated body parts of Caro corresponds perhaps to the idealization of female body parts in the *blasons anatomiques* and the denigration of the same in the *contreblasons.* Vickers, "Members Only," pp. 5, 13, 17, argues: "From the beginning, blazon and counterblazon, fascination and repulsion, desire and disgust went hand in hand." Even though they remained outside the ostensible terms of the debate inspired by Marot's challenge, La Hueterie's *contreblasons* enter into the "competition between poets within the genre," because "when the *blazons anatomique* finally appeared independently [they had formerly been published as supplements to other texts], their format mirrored Marot's proposal; the anthology began with the assembled poems in praise (*laudatio*) and concluded with de la Hueterie's poems in blame (*vituperatio*)." Whereas the blazons and counterblazons limit their praise and blame to external anatomy, their bifurcation of the female body into a site of both an admiration and a denigration, has a corresponding representation in anatomical and medical texts. Vickers (p. 7) argues, the "association of the body part with divine 'glory' . . . was a common trope of almost all anatomical treatises of the sixteenth century," and she relates this trope to *laudatio.* But she does not remark upon the denigration of the female sexual organs in French medical texts by Jean Fernel and others. On idealization as a mental process

"by means of which the object's qualities and value are elevated to the point of perfection," disavowal is implemented, sexual difference and sexuality are repressed in favor of aim-inhibited instincts, and "[i]dentification with the idealised object contributes to the formation and elaboration of the individual subject's so-called ideal agencies (ideal ego, ego ideal)," or the object as the Other begins to take the place of the ego ideal, see Laplanche and Pontalis, *The Language of Psycho-analysis,* p. 202.

285. On the relationship between desire, possession and knowledge in Neoplatonism, see Pico della Mirandola, *A Platonick Discourse on Love,* p. 228.

286. Gardner, *The Banquet of the Word,* p. 412.

287. Rowe, "'God's handy worke,'" p. 291.

288. Scholz, *Body Narratives,* p. 1. On the conception of the grotesque body, which is expressed in the text, see Peter Stallybrass, "Patriarchal Territories: The Body Enclosed," in *Rewriting the Renaissance: The Discourses of Sexual Difference in Early Modern Europe,* ed. by Margeret Ferguson, Maureen Quilligan, and Nancy Vickers, Women in Culture and Society Series (Chicago: University of Chicago Press, 1986), pp. 123–42 (p. 124)

289. Traub, "Gendering Mortality in Early Modern Anatomies," pp. 54–55.

290. Schoenfeldt, *Bodies and Selves in Early Modern England,* pp. 12, 15.

291. For an explanation of the concept of acquisition through metaphor, see Greenblatt, *Shakespearean Negotiations,* pp. 10–11. Also see the conclusions to the previous chapter of this study and to chapter 9.

292. On this conception of the grotesque body, see Stallybrass, "Patriarchal Territories," p. 124.

293. Rorty, "'He had both a bubbling wit and the ability to make fun of himself,'" p. 16, on Jacques Derrida's critique of phallogocentrism: "Most of the canonically great philosophers have shared an almost manic desire to achieve an impossible purity ["a fundamental immobility and a reassuring certitude . . . beyond the reach of play"]. . . . Such attempts to imitate God were described by Heidegger as denials of human finitude. Derrida, picking up on Freud, characterised them as expressions of phallogocentrism—of the conviction that the human intellect must stand proudly erect, towering above the squishy world of hope and fear, displaying the clean-cut, virile autonomy that is its birthright." "In a witty and brilliant essay called 'White mythology", Derrida describes what happens when philosophers obsessed with purity turn their attention to language. They typically try to cleanse discourse of any trace of metaphor. Derrida thought this ludicrous. His essay shows that the Western philosophical tradition itself was a tissue of imaginative metaphors, and none the worse for that." On the matter of how sovereignty in the Isle of Man is Intellect's birthright, it is helpful to trace his ancestry to the original inhabitants of the Isle. The biblical events between Adam and Eve's seclusion in Eden and the Fall (events 4 and 5) are ambiguously reflected in the incidents affecting the original inhabitants of the geopolitical realm, who "claim'd their birth from that eternal Light," "heav'ns dread King" (*P.I.* 6.7.5; 6.14.4). These "forefathers" (*P.I.* 6.21.7) of the "nation," which later comes to be ruled by Intellect and Voletta, are described variously as "Saints" (*P.I.* 6.10.1), "Graces" (*P.I.* 6.12.1), and "Heroes, who in better times / This happy Island first inhabited / In joy and peace" (*P.I.* 6.7.1–3). They depart the isle, as Adam and Eve are exiled from Eden, and Mercy's subsequent interposition with "heav'ns dread King" (*P.I.* 6.14.4) has the same effect as the incarnation (event 6), "unloosing" the "chains of death" (*P.I.* 6.16.4). For these reasons the story-strand of the geopolitical realm is placed in analogy or parallel to, and draws significance from, the biblical story of the Fall. But inconsistency emerges from an unresolved ambiguity or gap in the representation of the Fall. Rather than being punished for the "rebellious crimes" of the sins of the flesh, pride and curiosity (*P.I.* 6.7.3), as Adam and Eve are in their exile, the original inhabitants of the isle complain "at their fathers throne" (*P.I.* 6.12.1) to heaven's high court (*P.I.* 6.12), and subsequently the earth is punished by "*Dicaea,*" or Justice (*P.I.* 6.13), implying that the inhabitants are in no way culpable for the sins committed by the "fiend-like companie" (*P.I.* 6.10.2), to whom they unwillingly leave the isle. This is why there is no indication that the displaced exiles are corrupted along with the corruption of the body and geopolitical realm which they leave behind. The inhabitants' "faces bore their parents image bright" (*P.I.* 6.7.7), but when they are displaced, "The goodly temples which those Heroes plac't, / By this foul rout were utterly defac't, / And all their fences strong, and all their bulwarks raz'd" (*P.I.* 6.10.5–7). Thus the ideal aspect of *sōma,* made in God's image, and only glimpsed by the diegetic narrator-dissector of the anatomical cantos, is defaced. Nevertheless, the virtues, described in Cantos 9 and 10, and the new "nation," including Intellect and Voletta, sent by Mercy to inhabit the isle are descendents of the original inhabitants, and consequently claim their birthright. Furthermore, the vices described in Cantos 7, 8, and 12 are related to the "false, foul, fiend-like companie" which sent the original inhabitants into exile (*P.I.* 6.10.12), not only by their structural opposition to the forces of moral and theological virtue, but also in their vicious character, and their generation and multiplication as a result of the Fall. The polarity of vice and virtue in Cantos 7 to 12 of *The Purple Island* therefore originates in the earlier conflict.

294. Ann Jefferson, "Bodymatters: Self and Others in Bakhtin, Sartre and Barthes," in *Bakhtin and Cultural*

Theory, ed. by Ken Hirschkop and David Shepherd (Manchester: Manchester University Press, 1989; repr. 1989), pp. 152–77, describes the "classical body" in Bakhtin as "a completed entity sealed off both from the world which is its context and from other bodies" (p. 165). Some of the psychological implications of this are explored in Jefferson's article.

295. In relation to this argument, see Traub's discussion of the illustrations in works of anatomy by Valverde, in "Gendering Mortality in Early Modern Anatomies," pp. 53–54.

296. Ferrari, "Public Anatomy Lessons and the Carnival," pp. 102–3 (p. 102).

297. Purchas, *Microcosmus,* fol 8.

298. Kocher, *Science and Religion in Elizabethan England,* p. 225.

299. Seebass and Thiselton, "Flesh," pp. 672–73, 671. See Isaiah 40:6–8 (citation in brackets is of Isaiah 40:3): "The voice ["of him that crieth in the wilderness"] said, Cry. And he said, What shall I cry? All flesh is grass, and all the goodliness thereof is as the flower of the field: the grass withereth, the flower fadeth: because the spirit of the LORD bloweth upon it: surely the people is grass. The grass withereth, the flower fadeth: but the word of our God shall stand for ever." In the synoptic gospels the transitoriness of the flesh and the world is also contrasted with the enduring significance of the Law: see Matthew 5:18; Luke 16:17.

300. The quotation is from Pico della Mirandola, *A Platonick Discourse on Love,* p. 216, and is descriptive of the Platonists" notion of an invisible, incorporeal, and immortal intellectual nature, which is called Angelical by the divines, or theologians (also see p. 224).

301. John Donne, *The Second Anniversarie: Of the Progres of the Soule; Wherein, by Occasion of the Religious Death of Mistris Elizabeth Drury, the Incommodities of the Soule in this Life and her Exaltation in the Next, Are Contemplated,* in *The Epithalamions, Anniversaries and Epicedes of John Donne,* ed. by W. Milgate (Oxford: Clarendon Press, 1978), p. 45; John Donne, *An Anatomy of the World; Wherein, by Occasion of the Untimely Death of Mistris Elizabeth Drury the Frailty and the Decay of this Whole World is Represented,* a facsimile of the 1st edn: London, 1611 (Cambridge: Cambridge University Press, 1951), sig. A_6^r. See Thomas Willard, "Donne's Anatomy Lesson: Vesalian or Paracelsian?," *John Donne Journal* 3 (1984): 34–61 (pp. 36, 42). Donne owned a copy of Paracelsus's *Chirurgia magna,* a translation of *Die Große Wundarznei,* "one of the most accessible Paracelsian books" (p. 37); but his position with regard to Paracelsian medicine and "anatomy" is rightly said by Willard to shift between skepticism and purposeful deployment for logical poetic ends (p. 43). Donne is said by Willard to prefer the "[i]ntrinsique Balme" of Paracelsus to the "vital spirits" of Galenic anatomy because the former is also a restorative remedy, and so may preserve and bring about the new world, now that the old has died and corrupted (p. 44). Paracelsus treated the microcosm as a spiritual body, connected to the physical body through the blood and studied through signatures, Paracelsus's alternative to Galen's *semeiotike* or symptomatology (pp. 41–42). This spiritual body was the astrum, or astral body, or "sidereal body," which received the impressions of the stars; and as a study of the parallelism between the greater and lesser worlds, Paracelsian "anatomy" disclosed in the cosmos to the Paracelsian physician that which was hidden in Man (pp. 42, 54, 39, 41). The "anatomy of the world" or "whole world" is a technical term in Paracelsian medicine, wherein the *corpus physicum* of Man is seen in the *anatomia mundi* (p. 38). *Anatomia mundi* could therefore be an anatomy either of the microcosm or the macrocosm, or a procedure touching on both (p. 42).

302. Kocher, *Science and Religion in Elizabethan England,* p. 226, writes that from the perspective of theologians, "physicians must [it was argued] recognize soul as a spiritual substance, radically different from and superior to body, not merely a phase or consequence of it. That is, they must avoid materialism." Medically inspired monism should not be confused with Cartesian and Hobbesian mechanism, for the latter's view of matter was as "completely inert," all vital processes being epiphenomena of the interaction of particles of matter, whereas in the medical tradition, influenced as it was by Aristotelianism, "matter itself, or some forms of it, was inherently active, self-motive, and endowed with seminal or organizing powers." John Henry, "The Matter of Souls: Medical Theory and Theology in Seventeenth-Century England," in *The Medical Revolution of the Seventeenth Century,* ed. by Roger French and Andrew Wear (Cambridge: Cambridge University Press, 1989), pp. 87–113 (pp. 90–91).

303. Within the period concerned, there are three debates on the question of the soul's materiality and mortality, one in the seventeenth century, and two in the early eighteenth century. The earliest began with the publication of Richard Overton's *Mans Mortalitie: or, A Treatise wherein 'tis Proved . . . that Whole Man—as a Rationall Creature—is a Compound Wholly Mortall, Contrary to that Common Distinction of Soule and Body* (Amsterdam, 1644), which provoked, within one year, the publication of at least two counterarguments, one of which was entitled *The Prerogative of Man: or, The Immortality of Humane Soules Asserted against the Vain Cavils of a Late Worthless Pamphlet, Entituled, Mans Mortality* (Oxford, 1645). The issue was taken up again with the publication of William Coward's *Second Thoughts Concerning Human Soul,* in 1702. This provoked responses, which in turn

provoked more. But argument and counterargument was at its most rapid and prolific following the publication of Henry Dodwell's *Epistolary Discourse,* in 1706. In the four years following, no fewer than eleven writers contributed around thirty separate publications to the debate

304. See Seebass and Thiselton, "Flesh," p. 681.

305. On the female body as the "mortal site of primal sin and worldly knowledge," see Traub, "Gendering Mortality in Early Modern Anatomies," p. 84.

306. *OED,* II, "C," p. 245, col. 1, provides as an obsolete meaning of "Chaffer," which was still used in the seventeenth century: "That which is bought and sold; wares, merchandise, goods for barter or sale." This would seem to be the most appropriate denotation in the context of *The Purple Island*'s description of Hamartia.

307. For the Reformation doctrine of justification, see Clark, *Eucharistic Sacrifice and the Reformation,* pp. 106–7.

308. *OED,* IV, "G," p. 230, col. 2, provides for "Gloss," "a comment, explanation, interpretation. Often used in a sinister sense: A sophistical or disingenuous interpretation." The meaning of "Superficial lustre" is not irrelevant here, either. *OED,* IV, "G," p. 230, col. 3.

309. On some of these qualities of the grotesque body, see Traub, "Gendering Mortality in Early Modern Anatomies," p. 80.

310. John Freccero, "Medusa: The Letter and the Spirit," *Yearbook of Italian Studies* (1972): 1–18 (p. 7): "Whatever the horror the Medusa represents to the male imagination, it is in some sense a female horror. In mythology, the Medusa is said to be powerless against women, for it was her feminine *beauty* that constituted the mortal threat to her admirers."

311. Sawday, *The Body Emblazoned,* p. 8. Sawday argues that the myth of Medusa suggests the possibility of containing the effects of the transgressive gaze, in which "the male protagonist confronts his female opponent and renders her powerless through a dispersal of her body parts. Or rather, her killing and healing power—her head and her blood—are appropriated by male archetypes: the wandering hero [Perseus] and the surgeon or healer," Aesculapius, mythical founder of surgery, who is said to have drawn from Medusa's veins blood with the power either to raise the dead or destroy whoever drank it (p. 9).

312. Ibid., p. 9.

313. On the gendering of matter, and on gender as matter, in early modern anatomy, see Traub, "Gendering Mortality in Early Modern Anatomies," pp. 83–84.

314. The association of Medusa with the female genitalia draws on Sigmund Freud, "Medusa's Head," in *The Standard Edition of the Complete Psychological Works,* 24 vols., translated from the German under the general editorship of James Strachey in collaboration with Anna Freud, assisted by Alix Strachey and Alan Tyson, editorial assistant Angela Richards (London: Hogarth Press, 1953–74), XVIII (1964), 273–74. Introducing his interpretation of the mythological theme of the decapitated head of Medusa, Freud states categorically that: "To decapitate = to castrate. The terror of Medusa is thus a terror of castration that is linked to the sight of something. Numerous analyses have made us familiar with the occasion for this: it occurs when a boy, who has hitherto been unwilling to believe the threat of castration, catches sight of the female genitals, probably those of an adult, surrounded by hair, and essentially those of his mother." Along psychoanalytic lines, it can be argued that whereas in the Phallic Stage, it was structurally necessary for the boy to consider the mother as possessing the Phallus, thereby ensuring his own imaginary possession of it, the sight of the female genitals, and more specifically, of his mother's lack of a penis, heralds the onset of the Castration Complex. The Phallic Stage is pre-Oedipal, or lasts until the decline of the Oedipus Stage. According to Jacques Lacan, it is characterised, firstly, by "imaginary dominance of the phallic attribute," in which the mother is "phallic," and, secondly, by masturbatory "*jouissance.*" Jacques Lacan, "The Signification of the Phallus," in *Ecrits: A Selection,* trans. Alan Sheridan (London: Tavistock, 1977; repr. London: Routledge, 1992), p. 282. Also see Jacqueline Rose, *Sexuality in the Field of Vision* (London: Verso, 1991), p. 66.

315. Paul says in various letters that the believer, who is *sōma,* is no longer in the flesh (Rom. 7:5; 8:8f), in the sense that it matters not to them if they live or die; although they remain "in the flesh" in that they are living; but they live in Christ. Seebass and Thiselton, "Flesh," p. 676.

316. Fletcher, *Joy in Tribulation,* Dedication to "Sir Walter Roberts, Knight and Baronet" and "his *gracious Lady.*"

317. Ibid. In reference to 2 Corinthians 5:21, the speaking voice of *Joy in Tribulation* pleads with its reader, "*Oh let not thy sorrow,* (*no not for sinne*) *drowne thy rejoycing in that Saviour, who is become sinne for thee, that thou* mightest bee the righteousnesse of God in him." As was remarked in chapter 3 of the present study, defeated beneath the foot of the Christ figure in Remmelin's *Catoptrum microcosmicum* are mortality and sin, represented by the skull and the serpent.

318. James M. Reese, "Song of Solomon," in *The Oxford Companion to the Bible,* ed. by Bruce M. Metzer and Michael D. Coogan (Oxford: Oxford University Press, 1993), pp. 708–10 (p. 709). According to Reese, allegorical interpretation of The Song of Solomon begins with the Aramaic translations called Tagrums, that read it as signifying the Lord's love for Israel, a love which became marriage in the prophetic tradition.

319. It may be that Christ's "foes" and "Hells troops," in *The Purple Island,* develop upon the "watchmen that went about the city," of which the female voice in Song of Solomon says, they "found me, they smote me, they wounded me; the keepers of the walls took away my veil from me" (5.7)

320. Clark, *Eucharistic Sacrifice and the Reformation,* p. 107.

321. Eclecta's use of the term "bark" in this context should not be confused with the Neoplatonic notion of either the physical, elemental body or the eternal body (*vehiculum coeleste*) as a vehicle for the soul. See Pico della Mirandola, *A Platonick Discourse on Love,* p. 224. Rather, Eclecta's figure is discursively consistent with her metaphorical identification of the Church as her spouse's "ark," which is regarded as a type of the Church. This metaphor is extended, particularly in Canto 12, stanzas 52–53.

322. On parabolic or typological readings of The Song of Solomon, see, for example, Reese, "Song of Solomon," p. 709.

323. John Keats, "Ode on Melancholy" (III. ll. 1–4), *The Poetical Works,* p. 248.

324. Heather Leach, "Crossing the Line: Stories on Mapping the Maze," *Times Higher Education Supplement,* 5 November 2004, pp. 18–19, discusses the effects of the linearity of narrative, to which the eschatological narrative is particularly prone (p. 18, cols. 2, 1): "The linear form is organised sequentially—there is only one authorised way to move though the work. . . it is closed and there is no going back. So long as the aim of the story is towards an imagined reward—even if that reward is not forthcoming, these metaphors of paths and journeys are potentially, if not actually, benign. But once the trope is applied to the whole of life; and in particular, to secular life—life with neither heaven nor hell—then linearity becomes a threat," as one discovers, for example, "that the story of ageing itself is already written, inscribed into culture." Further to the debate about the relationship between narrative, life, history, and subjectivity, see, for example, Barbara Hardy, "An Approach through Narrative" (1968), in *Towards a Poetics of Fiction,* ed. by Mark Spilka, Essays from *Novel: A Forum on Fiction,* 1967–1976 (Bloomington: Indiana University Press, 1977), pp. 31–40, who argues that "narrative, like lyric or dance, is not to be regarded as an aesthetic invention used by artists to control, manipulate, and order experience, but as a primary act of mind transferred to art from life"; "we dream in narrative, daydream in narrative, remember, anticipate, hope, despair, believe, doubt, plan, revise, criticize, construct, gossip, learn, hate, and love by narrative" (pp. 31–33); Louis Mink, "History and Fiction as Modes of Comprehension," *New Literary History* 1 (1970): 541–58, who argues, "[w]e do not dream or remember in narrative, I think, but tell stories which weave together the separate images of recollection"; "[s]tories are not lived but told. Life has no beginnings, middles, or ends; there are meetings, but the start of an affair belongs to the story we tell ourselves later, and there are hopes, plans, battles, and ideas, but only in retrospective stories are hopes unfulfilled, plans miscarried, battles decisive, and ideas seminal." Also see White, "The Historical Text as Literary Artifact," who argues that "no given set of events attested by the historical record constitutes a story manifestly finished and complete. . . . We do not live stories, even if we give our lives meaning by retrospectively casting them in the form of stories." (p. 43). On the significance of death for the intelligibility and story of a life, see Walter Benjamin, "The Storyteller: Reflections on the Works of Nikolai Leskov," in *Illuminations,* trans. by Harry Zohn, ed. by Hannah Arendt (London: Fontana/Collins, 1973), pp. 83–109: "It is characteristic that not only a man's knowledge or wisdom, but above all his real life—and this is the stuff that stories are made of—first assumes transmissable form at the moment of his death. Just as a sequence of images is set in motion inside a man as his life comes to an end—unfolding the views of himself under which he has encountered himself without being aware of it—suddenly in his expressions and looks the unforgettable emerges and imparts to everything that concerned him that authority which even the poorest wretch in dying possesses for the living around him. This authority is at the very source of the story." Furthermore: "A man . . . who died at thirty-five will appear to *remembrance* at every point in his life as a man who dies at the age of thirty-five" (pp. 94, 100).

325. See Scholz, *Body Narratives,* pp. 1–2: "Historicizing the body means deconstructing the implicit nature-culture dualism that traditionally prevails in historical description, which assumes the human body to be the unchangeable substratum on which culture and history imprint their traces." "[T]he conception of the body as perceived and thus *written* differently in different historical or cultural settings still retains the idea of a natural, prediscursive body. 'Historicity' [in Scholz's work] also implies that the body in different historical periods was 'produced' differently, by normative discourses that did not get imprinted on its surface but which, as Judith But-

ler [in *Bodies that Matter*] insists, structure the very shape in which the body materializes, and then proclaim the result to be 'natural.'"

326. Ariès, *The Hour of Our Death*, p. 369.

327. Sorri and Gill, *A Post-Modern Epistemology*, p. 196.

328. For a critique of the related philosophical viewpoints that "minds constitute a world other than 'the physical world,'" that "there exist ways of discovering the contents of this other world which are counterparts to our ways of discovering the contents of the physical world," but which constitute "a species of non-sensuous perception," and that "minds possess powers of apprehending their own states and operations superior to those they possess of apprehending facts of the external world," including those of other minds, indeed that "a mind cannot help being constantly aware of all the supposed occupants of its private stage," thus granting a subject "Privileged Access" to her or his own mind, see Gilbert Ryle, *The Concept of Mind* (London: Hutchinson, 1949; repr. 1959), ch. 6: "Self-Knowledge," pp. 154–98 (p. 154).

Appendix A

1. *STC*, I, 491. (Further references in this appendix to this work are in the text.)

2. For a discussion of this emblem, see Chapter 6 of the main body of this study.

3. *Early English Books 1475–1640* (Ann Arbor: University Microfilms International), Reel 1924.

4. This lack in the microfilm copy of *STC* 11082.5 is noticed in *Early English Books 1475–1640 Selected from Pollard and Redgrave's Short-title Catologue: A Guide to Units 53–61 of the Microfilm Collection, Reels 1783–1979* (Ann Arbor: University Microfilms International, 1988), p. 283.

5. For the significance of this anagrammatic transposition of the name of Phineas's dearest friend, see Chapter 6 of the main body of this study.

6. For a partial translation and discussion of this Latin poem, see chapter 4 of the main body of this study.

7. For conjecture as to the authorship of this brief encomium, see Chapter 4, note 19, in the main body of this study.

8. According to a NLW stamp, with additional handwritten remarks, the copy was presented to the library by Sir John Williams, and is dated the second of January 1919. A card affixed to the page attached to the front cover identifies in Welsh the book as a gift from the library of Sir John Williams in 1909.

Appendix B

1. Langdale, *Phineas Fletcher*, pp. 69–70.

2. Ibid., p. 68. Sir Henry Willughby was born at Risley and married Elizabeth Knollys, niece of the Earl of Banbury. He was made a baronet in 1611, and was buried at Risley on 22 November 1653.

3. Alice T. Friedman, *House and Household in Elizabethan England: Wollaton Hall and the Willoughby Family* (Chicago and London: University of Chicago Press, 1989), pp. 13–14 (p. 14).

4. Langdale, *Phineas Fletcher*, pp. 64–68 (p. 64): "It has been know that Fletcher spent some conderable portion of his life in Norfolk long before he went to Hilgay in 1621. The date and cause of this earlier residence have been unsolved mysteries. The commons books [at King's Collge, Cambridge] provide the year, and Fletcher's own address to Henry Willughby offers the reason: 'for my long continuous entertainment in your house (such as I never saw any gentleman give unto their minister) or that I first initiated by weake Ministerie in your Familie and Hamlet' (*The Way to Blessednes*, dedicatory address). There is but one 'long continuous' period in the poet's life before April, 1616 (at which time he 'initiated' his 'weake Ministerie' at Risley), when we are unable to locate him precisely, and that is the ninteen months from September, 1612, to April, 1614. Thus late in 1612 or shortly thereafter, Fletcher was honored with Willoughby's hospitality and patronage."

5. Ibid., p. 67. Sir Percival Willoughby was the first son of Thomas Willoughby (d. 1596) and Catherine Hart (daughter of Sir Percivall Hart [d. 1580]), and grandson of Robert Willoughby of Bore Place, Kent (d. 1555), and Dorothy Willoughby, who was divorced for adultery and died in 1553 (?).

6. Friedman, *House and Household in Elizabethan England*, p. 31 (also see p. 45). The dedication of *The*

Historie of Man to Sir Francis Willoughby reads: "To the Right Worshipfull, Syr Frauncis Willoughby, Knight," "I hope, my labour shalbe thankfully taken, of all honest Chirurgians, considering the barren draught, that Chirurgerie throughout the Realme of England, in this present age, endureth: and which can neuer be quenched, by the fruitefull water that floweth from the fountains of Anathomie. Of which, I haue now endeuoured to turne one cocke."

7. On the history of the Will[o]ughby family, see the work of Cassandra Willoughby (1670–1735), Duchess of Chandos, entitled "The Account of the Willughby's of Wollaton Taken out of the Pedigree, Old Letters, and Old Books in My Brother Sir Thomas Willoughby's Study, Dec. A.D. 1702," and "The Continuation of the Account of the Willoughby's of Willoughby and Eresby in Lincolnshire and of the Willughby's of Willughby and Wollaton in Nottinghamshire and of Middleton in Warwickshire taken out of the Pedigrees, Old Letters, Books of Accounts, and other Manuscripts, which still remain in the Library at Wollaton," which are held in the Middleton Manuscripts Collection, in two volumes: LM26 and LM27. A large part of the work has been published, in two parts: the first volume, in part, as *Report on the Manuscripts of Lord Middleton Preserved at Wollaton Hall, Nottingham,* ed. by W. Stevenson (London: Historical Manuscripts Commission, 1911); the second volume as *The Continuation of the Willoughby Family,* by Cassandra, Duchess of Chandos, being Vol. II of the manuscript edited by A. C. Wood (Eton, Windsor: Shakespeare Head Press for the University of Nottingham, 1958). Also on the history of the Will[o]ughby family, see Friedman, *House and Household in Elizabethan England;* Pamela Marshall, *Wollaton Hall and the Willoughby Family* (Nottingham: Nottingham Civic Society, 1999); Elizabeth May, *Wollaton Hall: Family House and Natural History Museum* (Derby: J. H. Hall & Sons, 1989).

8. See Middleton Manuscripts Collection Mi I 17/1 "Catalogue of the Library [at Wollaton Hall] in the handwriting of the Natural Philospher [Francis Willoughby (1635–1672)]," fols. 6, 27. A partial list of these books is provided by Friedman, *House and Household in Elizabethan England,* pp. 31–32. On Sir Percival and Bridget Willoughby, see Marshall, *Wollaton Hall and the Willoughby Family,* especially Ch. 4.

9. The Eresby Willoughbys were related to the Wollaton Willughbys through Sir Percivall Willoughby M.P., his marriage to Bridget Willughby. Sir Percivall's grandfather, Robert Willoughby of Bore Place, Kent, was the first son of Sir Thomas Willoughby of Eresby and Bridget Read, daughter of Sir Robert Read of Bore Place, and Margaret Alphew. Sir Thomas was the fourth son of Christopher Willoughby of Eresby and Margaret Jenney, daughter of Sir William Jenney.

10. Langdale, *Phineas Fletcher,* p. 67.

11. Ibid., pp. 67, 77–79, 185.

12. An inserted slip in Mi LM26, p. 158, with a ms. note written in Cassandra Willoughby's hand, cites from the Journal of the House of Lords, 4 April 1621, the "2nd Article of the Charge against the Lord Chancelor Bacon, that he took as a Bribe from the Lady Wharton 310 l. in a Cause depending between her and Sir Percivall Willoughby." Stevenson, ed., *Report on the Manuscripts of Lord Middleton,* p. 583, n. 1, notes this.

13. See Langdale, *Phineas Fletcher,* p. 186. Keynes, *The Life of William Harvey,* p. 286, says that in September 1642, Harvey "is known to have ridden over [from Nottingham, where he is assumed in his capacity as Physician in Ordinary to have been with the King] to Derby to visit his medical friend Percival Willoughby, who had been in practice there for many years." Percival Willoughby was born at Wollaton Hall. He matriculated from Magdalen College on 23 March 1620/21, and graduated B.A. on 6 July 1621. However, due to his father's financial problems, he gave up his studies and was "articled for seven years to Feamer [Feames] van Otten, after which he was to have joined his uncle [Robert Willoughby, seventh son of Thomas Willoughby (d. 1596) and Catherine Hart, and brother to Sir Percivall Willoughby, M.P., and a medical man]; but Van Otten dying in 1624, Willoughby soon after commenced practice for himself, and in 1631 he settled in Derby, where he married Elizabeth, daughter of Sir Francis Coke of Trusley, by whom he had two or three sons and two daughters [one of whom was called Anne, who married Thomas Hart (Hurt)]." On 20 February 1640/41, Percival was admitted an extra licentiate of the Royal College of Physicians. In 1655 he removed to London, returning in 1660 to Derby and resuming practice as a physician, "enjoying a high reputation throughout the neighbouring counties for his skill in obstetric operations." He "committed to writing the conclusions at which he had arrived after long years of study and observation, revising and transcribing the manuscripts in English and in Latin," but "he seems to have hesitated at the last at their publication." 'The earliest copy of his work is a closely witten quarto in the British Library Sloane MS. 529. The second is an amplification of this, "while the third and greatly enlarged edition consisted of two exquisitely written copies in Latin and in English." This latter was published as *Observations in Midwifery; As Also the Countrey Midwifes Opusculum or Vade Mecum,* by Percivall Willughby, Gentleman, ed. from the original MS. by Henry Blenkinsop (Warwick: H. T. Cooke and Son, at the Shakespeare Printing Press, 1863). A Dutch translation had also been printed as an octavo at Leiden in 1764. *DNB,* LXII (1900), 57, col. 1. Also see Stevenson, ed.,

Report on the Manuscripts of Lord Middleton Preserved at Wollaton Hall, Nottingham, pp. 609–10; Munk, *The Roll of the Royal College of Physicians of London,* p. 231; Marshall, *Wollaton Hall and the Willoughby Family,* p. 93.

14. John Aubrey, *Brief Lives,* cited in Porter, *The Greatest Benefit to Mankind,* p. 216. Also see Wear, "Introduction," in Harvey, *The Circulation of the Blood,* p. xv.

15. Langdale, *Phineas Fletcher,* pp. 11–12, 187–88. According to Keynes, *The Life of William Harvey,* p. 49, when Harvey's mother died on 8 November 1605, Harvey's parents had been living until this time in Folkstone. William and probably Thomas (1584/85–1623), the eldest of William's brothers, were living in London from about 1602 and 1600, respectively.

16. Langdale, *Phineas Fletcher,* pp. 186, 188.

17. See Keynes, *The Life of William Harvey,* p. 131.

18. Langdale, *Phineas Fletcher,* pp. 186–87.

19. Ibid., pp. 188, 186.

20. Ibid., pp. 185–86. According to Langdale, Fletcher frequented London during his holidays from King's College, Cambridge, between 1602 and 1610. Harvey left Padua some time after April 1602, and, according to Whitteridge, it is likely he was settled in London by early 1603, intending to practice medicine. Whitteridge, *William Harvey and the Circulation of the Blood,* p. 81.

Bibliography

Manuscripts and Prints

Cambridge University Library, Manuscripts, University Wills, Bundle No. 6. 1608–1612 (including the will of "Thomas Grimestone" [Grymstone]).

Cambridge Record Office, MS P30/1/1 (Parish Register of Saint Mary the Great, 1559–1851).

Guildhall Library, PR. 45 / BAR (interior view and plan by Isaac Ware of the anatomy theater of Barber-Surgeons' Hall, c. 1750; exterior anterior and exterior posterior aspects of the anatomy theater of Barber-Surgeons' Hall [eighteenth century]).

Gonville and Caius College, Library. Cambridge, MS. 615/768; MS. 617/549; MS. 617 bis/770; MS 707/693 (English versions of the annals).

King's College, Cambridge, Archive Centre, Lib[rary] MS. 1 (early author catalogue, classified catalogue, and list of donors, begun in 1612); Lib. MS. 2 (donors book of the early seventeenth century with continuations to about 1710); Lib. MS. 13 "Register of Donors to King's College Library 1527–1645," compiled by F.L. Clarke.

Middleton Manuscripts Collection, Hallward Library, Nottingham. The family and estate papers of the Willoughby family, Lords Middleton, from the twelfth to the twentieth centuries.

Norfolk Record Office, Archive Centre, Norwich, Hilgay Parish Registers (1583–1674).

Printed Books, Articles, and Theses

Abrams, M. H., ed. *The Norton Anthology of English Literature.* 5th edn. 2 vols. London: Norton, 1986.

Achillini, Alessandro. *Annotationes anatomicae.* Bologna: 1520. Trans. in Lind, *Studies in Pre-Vesalian Anatomy.*

Adams, H. M., comp. *Catalogue of Books Printed on the Continent of Europe, 1501–1600, in Cambridge Libraries.* 2 vols. Cambridge: Cambridge University Press, 1967.

Aers, David. "A Whisper in the Ear of Early Modernists; or, Reflections on Literary Critics Writing the 'History of the Subject.'" In *Culture and History 1350–1600: Essays on English Communities, Identities and Writing,* ed. David Aers, pp. 177–202. London: Harvester Wheatsheaf, 1992.

Allen, D. C. "John Donne's Knowledge of Renaissance Medicine." In *Essential Articles for the Study of John Donne's Poetry,* ed. Roberts, pp. 93–106. Originally published in *Journal of English and Germanic Philology* 42 (1943), 322–42.

Allen, R. E., ed. *The Concise Oxford Dictionary of Current English,* first edn. by H. W. Fowler and F. G. Fowler. 8th edn. Oxford: Clarendon Press, 1990.

Alston, Mary Niven. "The Attitude of the Church towards Dissection before 1500." *Bulletin of the History of Medicine* 16 (1944): 221–38.

Althusser, Louis. "Ideology and Ideological State Apparatuses (Notes Towards an Investigation)"

(January–April 1969). In *Lenin and Philosophy and Other Essays,* trans. Ben Brewster, pp. 121–73. London: NLB, 1971.

Anderson, Benedict. *Imagined Communities: Reflections on the Origin and Spread of Nationalism.* Rev. ed. London: Verso, 1991; repr. 1996.

Anderson, Ruth L. "A French Source for John Davies of Hereford's System of Psychology." *Philological Quarterly* 6 (1927): 57–66.

Andrewes, Lancelot. *The Liturgical Sermons.* Ed. Marianne Dorman. Vol. I, *Nativity, Lenten and Passion.* Edinburgh: Pentland Press, 1992.

Archambault, Paul. "The Analogy of the 'Body' in Renaissance Political Thought." *Bibliothèque D'Humanisme et Renaissance* 29 (1967): 21–53.

Ariès, Philippe. *The Hour of Our Death.* Trans. Helen Weaver. London: Allen Lane, 1981.

Aristotle. *Generation of Animals.* Ed. T. E. Page and others., trans. A. L. Peck. Loeb Classical Library. London: William Heinemann, 1942; rev. and repr. 1963.

———. *Politics.* Trans. Benjamin Jowett. Oxford: Clarendon Press, 1931.

———. *The Works.* Ed. J. A. Smith and W. D. Ross. 12 vols. (IV–V ed. J. A. Smith; I–III, VI–XII ed. W. D. Ross). Oxford: Clarendon Press, 1908–52; repr. 1937–63. I, *Categoriae,* trans. E. M. Edghill (1928; repr. 1963); IV, *Historia Animalium,* trans. D'Arcy Wentworth Thompson (1910; repr. 1962); V, *De partibus animalium,* trans. William Ogle (1912; repr. 1958); XI, *De poetica,* trans. Ingram Bywater (1946; repr. 1959).

Aristotle's Masterpiece; or, The Secrets of Generation Display'd in All the Parts Thereof. London: F.L., 1690.

Atherton, Ian, ed. *Norwich Cathedral: Church, City and Diocese, 1096–1996.* London: Hambledon Press, 1996.

Austin, John L. *How to Do Things with Words.* The William James Lectures delivered at Harvard University in 1955. Oxford: Clarendon Press, 1962; repr. 1975.

Bacon, Sir Francis. *The Advancement of Learning.* Ed. G. W. Kitchin, intr. Arthur Johnston. London: J. M. Dent & Sons, 1973.

———. Sir Francis. *The Essays or Counsels Civil and Moral of Francis Bacon, Lord Verulam, Viscount St. Albans.* Ed. A. S. Gaye. Oxford: Clarendon Press, 1911.

———. Sir Francis. *Novum Organum,* in *Novum Organum With Other Parts of The Great Instauration.* Trans. and ed. Peter Urbach and John Gibson, Paul Carus Student Editions, 3. Chicago: Open Court, 1994.

Bagnall-Baker, Thomas. *Anatomico-Theology; or, A Critical Dissection of Various Scriptures Explanatory of the Doctrines, Precepts, Hopes, and Prospects of the Christian Religion.* London: B. Wertheim, Aldine Chambers, 1843.

Baillie, John. *The Idea of Revelation in Recent Thought.* London: Oxford University Press; New York: Columbia University Press, 1956.

Baldick, Chris, ed. *The Concise Oxford Dictionary of Literary Terms.* Oxford: Oxford University Press, 1990.

Baldwin, R. G. "Phineas Fletcher: His Modern Readers and His Renaissance Ideas," *Philological Quarterly* 40 (1961): 461–75.

Banister, John. *The Historie of Man, Sucked from the Sappe of the Most Approued Anathomistes, in this Present Age, Compiled in Most Compendious Forme, and Now Published in English, For the Vtilitie of All Godly Chirurgians, within this Realme. . . .* 1578. Reprint, Amsterdam: Theatrum Orbis Terrarum; New York: Da Capo Press, 1969.

Barkan, Leonard. *Nature's Work of Art: The Human Body as Image of the World.* New Haven: Yale University Press, 1975.

Barker, Francis. *The Tremulous Private Body: Essays on Subjection.* London: Methuen, 1984.

Barnes, Annette. *On Interpretation.* Oxford: Basil Blackwell, 1988.

Barnhart, Robert K., ed., *Dictionary of Etymology.* Edinburgh: Chambers, 1988.

Barry, Alfred, ed. *The Teacher's Prayer Book: Being The Book of Common Prayer, with Introductions, Analyses, Notes, and a Commentary upon the Psalter.* London: Eyre & Spottiswoode, Queen's Printers, n.d.

Barthes, Roland. "Introduction to the Structural Analysis of Narratives." 1966. In *A Barthes Reader,* ed. Susan Sontag, pp. 251–95. London: Jonathan Cape, 1982.

Bartholin, Thomas. *Anatome ex omnium veterum recentiorumque observationibus.* Lyons: Hackiana, 1673.

Bawcutt, Priscilla J. *The Shorter Poems of Gavin Douglas,* Scottish Text Society, 4th Series, 3. Edinburgh and London: William Blackwood & Sons, 1967.

Baxandall, Michael. *Painting and Experience in Fifteenth Century Italy: A Primer in the Social History of Pictorial Style.* Oxford: Oxford University Press, 1972; repr. 1983.

Beardsley, Monroe C. *Aesthetics: Problems in the Philosophy of Criticism.* New York: Harcourt, Brace & World, 1958.

Beaumont, Joseph. *Psyche: or, Loves Mysterie in XX Cantos: Displaying the Intercourse betwixt Christ, and the Soule.* London: John Dawson, 1648.

Beckwith, Sarah. *Christ's Body: Identity, Culture and Society in Late Medieval Writings.* London: Routledge, 1993.

BeDuhn, Jason David. "A Regimen for Salvation: Medical Models in Manichaean Asceticism." *Semeia* 58 (1992): 109–34.

Belsey, Catherine. *Critical Practice,* New Accents Series. London and New York: Routledge, 1980; repr. 1992.

———. "Literature, History, Politics." In *New Historicism and Renaissance Drama,* ed. Wilson and Dutton, pp. 33–44.

Benedetti, Alessandro. *Anatomice, siue historia corporis humani; eiusdem colleciones medicinales seu aforismi.* 1502. Trans. in Lind, *Studies in Pre-Vesalian Anatomy.*

Benjamin, Walter. *Illuminations.* Trans. Harry Zohn, ed. Hannah Arendt. London: Fontana/Collins, 1973.

Berengario da Carpi, Jacopo. *Isagogae breves, perlucide ac uberimme, anatomia humani corporis.* Bologna: Benedictus Hectoris, 1522.

———. *Isagogae breves, perlucide ac uberimme, anatomia humani corporis.* Bologna: Benedictus Hectoris, 1523.

———. *A Short Introduction to Anatomy (Isagogae breves).* Trans. L. R. Lind, with anatomical notes by Paul G. Roofe. Chicago: University of Chicago Press, 1959.

Bernard, Richard. *The Isle of Man; or, The Legall Proceeding in Man-Shire Against Sinne; Wherein, By Way of a Continued Allegorie, the Chiefe Malefactors Disturbing Both Church and Common-Wealth, Are Detected and Attached; With their Arraignment, and Iudiciall Triall, According to the Lawes of England; the Spirituall Use Thereof, With an Apologie for the Manner of Handling, Most Necessary to Be First Read, For Direction in the Right Use of the Allegory Thorowout, Is Added in the End,* 5th edn., much enlarged. London: Edward Blackmore, 1628.

Berriot-Salvadore, Evelyn. "The Discourse of Medicine and Science." In *A History of Women in the West,* 5 vols., ed. Georges Duby, Michelle Perrot, and Pauline Schmitt Pantel. Cambridge: Belknap Press of Harvard University Press, 1992–94. Vol. III, *Renaissance and Enlightenment Paradoxes,* ed. by Natalie Zemon Davis and Arlette Farge (1993), 348–88.

The Bible Authorised Version. Ed. John Stirling. London: British & Foreign Bible Society, 1954.

The Blank-Paged Bible: The Holy Scriptures of the Old and New Testaments [Authorised Version]; with Copious References to Parallel and Illustrative Passages; and the Alternate Pages Ruled for MS. Notes. London: Samuel Bagster & Sons, n.d.

The Holy Bible Revised Standard Version Containing the Old and New Testaments Translated from the Original Tongues, Being the Version Set Forth A.D. 1611, Revised A.D. 1881–1885 and A.D. 1901, Compared With the Most Ancient Authorities and Revised A.D. 1952. London: Thomas Nelson & Sons, 1952.

The New English Bible: New Testament, Library Edition. Cambridge: Oxford University Press and Cambridge University Press, 1961.

The Revised English Bible with the Apocrypha. Cambridge: Cambridge University Press, 1996.

Black, Max. "Metaphor." *Proceedings of the Aristotelian Society,* n.s. 55 (1955): 273–94.

———. *Models and Metaphors: Studies in Language and Philosophy.* Ithaca: Cornell University Press, 1962.

———. "More About Metaphor." In *Metaphor and Thought,* ed. Andrew Ortony, pp. 19–43. Cambridge: Cambridge University Press, 1979. This is a slightly modified version of an earlier essay entitled "More about Metaphor," *Dialectica* 31, fosc. 3–4 (1977): 431–57.

Bleier, Ruth. *Science and Gender: A Critique of Biology and Its Theories of Women.* Athene series. New York: Pergamon Press, 1984.

Booth, Wayne. "Metaphor as Rhetoric." *Critical Inquiry* 5 (1978): 49–72.

———. *The Rhetoric of Fiction.* Chicago: University of Chicago Press, 1961.

Boschloo, A. W. A. *Annibale Carracci in Bologna: Visible Reality in Art after the Council of Trent.* Trans. R. R. Symonds, Kunsthistorische Studiën van het Nederlands Instituut te Rome Deel III, Band I. The Hague: Government Publishing Office; Maarssen: Gary Schwartz, 1974.

Braider, Christopher. "The Paradoxical Sisterhood: 'ut pictura poesis.'" In *The Cambridge History of Literary Criticism,* ed. Nisbet and Rawson, vol. III, ed. by Norton, 168–75.

Breight, Curtis C. *Surveillance, Militarism and Drama in the Elizabethan Era.* Language, Discourse, Society series. Basingstoke: Macmillan, 1996.

Bridson, Gavin D. R., and James J. White, comp. *Plant, Animal & Anatomical Illustration in Art & Science: A Bibliographical Guide from the 16th Century to the Present Day.* Winchester: St. Paul's Bibliographies in association with Hunt Institute for Botanical Documentation, 1990.

Brockman, B. A., ed. *"Coriolanus": A Casebook.* Basingstoke: Macmillan, 1969.

Brown, Colin, ed. *The New International Dictionary of New Testament Theology.* 3 vols. Exeter: Paternoster Press, 1975.

Brown, Elizabeth A. R. "Death and the Human Body in the Later Middle Ages: The Legislation of Boniface VIII on the Division of the Corpse." *Viator* 12 (1981): 221–70.

Browne, Thomas. *Pseudodoxia Epidemica; or, Enquiries into Very Many Received Tenents and Commonly Presumed Truths,* 6th edn, corrected and enlarged by the author. London: J.R., 1672.

———. *Religio Medici.* 7th edn., corrected and amended, with annotations upon all the obscure passages therein; also observations by Sir Kenelm Digby. London: Andrew Crook, 1672.

Burton, Robert. *The Anatomy of Melancholy.* Ed. Thomas C. Faulkner, Nicolas K. Kiessling, and Rhonda L. Blair, and commentary by J. B. Bamborough and Martin Dodsworth. 6 vols. Oxford: Clarendon Press, 1989–2000.

Butler, Judith. *Bodies that Matter: On the Discursive Limits of "Sex."* New York: Routledge, 1993.

Bynum, Caroline Walker. "The Body of Christ in the Later Middle Ages: A Reply to Leo Steinberg." *Renaissance Quarterly* 39 (1986): 399–439.

Bynum, Caroline Walker. *Fragmentation and Redemption: Essays on Gender and the Human Body in Medieval Religion.* New York: Zone Books, 1991.

Cable, Lana. "Such Nothing is Terrestriall: Philosophy of Mind on Phineas Fletcher's *Purple Island.*" *Journal of the History of the Behavioural Sciences* 19 (1983): 136–52.

Caius, John. *The Annals of Gonville and Caius College,* ed. by John Venn. Cambridge: Printed for the Cambridge Antiquarian Society, 1904.

Callaghan, Dympna. *Woman and Gender in Renaissance Tragedy: A Study of "King Lear," "Othello," "The Duchess of Malfi" and "The White Devil."* London: Harvester Wheatsheaf, 1989.

Camden, Caroll. "The Architecture of Spenser's 'House of Alma.'" *Modern Language Notes* 58 (1943): 262–65.

Camden, William. *Britain; or, A Chorographicall Description of the Most Flourishing Kingdomes, England, Scotland, and Ireland, and the Ilands Adioyning, out of the Depth of Antiqvitie; Beavtified with Mappes of the Severall Shires of England,* trans. by Philémon Holland, revised, amended, and enlarged with sundry additions by William Camden. London: George Bishop and John Norton, 1610.

Campion, Thomas. *Obseruations in the Art of English Poesie . . . Wherein It Is Demonstratiuely Prooued, and By Example Confirmed, that the English Toong Will Receiue Eight Seuerall Kinds of Numbers, Proper to It Selfe, which Are All in this Booke Set Forth, and Were Neuer before this Time by Any Man Attempted.* 1602. Reprint, The English Experience, Amsterdam: Theatrum Orbis Terrarum; New York: Da Capo, 1972.

Camporesi, Piero. *The Anatomy of the Senses: Natural Symbols in Medieval and Early Modern Italy.* Trans. Allan Cameron. Cambridge: Polity Press, 1994.

———. "The Consecrated Host: A Wondrous Excess." *Zone* 3, *Fragments for a History of the Human Body,* Part 1, ed. Michel Feher, with Ramona Naddaff, and Nadia Tazi, 220–37. New York: Zone, 1989.

———. *The Incorruptible Flesh: Bodily Mutation and Mortification in Religion and Folklore.* Trans. Tania Croft-Murray, Latin texts trans. Helen Elsom. Cambridge: Cambridge University Press, 1988.

Canary, Robert H., and Henry Kozicki, eds. *The Writing of History: Literary Form and Historical Understanding.* Madison: University of Wisconsin Press, 1978.

Cardwell, Edward, ed. *The Two Books of Common Prayer, Set Forth by Authority of Parliament in the Reign of King Edward the Sixth: Compared with Each Other.* 2nd ed. Oxford: Oxford University Press, 1841.

Carey, John. *John Donne: Life, Mind and Art.* New edn. London: Faber & Faber, 1981.

Carlino, Andrea. *Books of the Body: Anatomical Ritual and Renaissance Learning.* Trans. John Tedeschi and Anne C. Tedeschi. Chicago: University of Chicago Press, 1999.

———. *Paper Bodies: A Catalogue of Anatomical Fugitive Sheets 1538–1687.* Trans. Noga Arikha, *Medical History* Supplement No. 19. London: Wellcome Institute for the History of Medicine, 1999.

Cassam, Quassim. *Self and World.* Oxford: Clarendon Press; New York: Oxford University Press, 1997.

Casserio, Giulio. *Philosophi, medici et anatomici, in vtraque medicina patavii versantis Pentæstheseion, hoc est de quinque sensibus liber. . . .* Venice: Nicolavs Misserinus, 1609.

———. *Tabulae anatomicae LXXIIX,* in Van der Spieghel, *De humani corporis fabrica libri decem* (1627).

———. *Tabulae anatomicae LXXIIX,* in Van der Spieghel, *De humani corporis fabrica libri decem* (1632).

Cassirer, Ernst. *The Individual and the Cosmos in Renaissance Philosophy.* Trans. Mario Domandi. Oxford: Clarendon Press, 1963.

Cassirer, Ernst, Paul Oskar Kristeller, and John Herman Randall Jr., eds. *The Renaissance Philosophy of Man*. Chicago: University of Chicago Press, 1948.

Cavanagh, G. S. T. "A New View of the Vesalian Landscape." *Medical History* 27 (1983): 77–79.

Choulant, Ludwig. *History and Bibliography of Anatomic Illustration*. Trans. and annotated by Mortimer Frank. London: Hafner Publishing, 1962. Originally published as *Geschichte und Bibliographie der anatomischen Abbildung nach ihrer Beziehung auf anatomische Wissenschaft und bildende Kunst* Leipzig: Rudolph Weigel, 1852.

Clapham, Henoch. *Ælohim-Triune, Displayed by His Workes Physicall and Meta-Physicall, in a Poeme of Diuerse Forme, Adapted to the Hebrue Text, the Frame of Diuinitie, and Catholike Exposition; Togither With Necessarie Marginall Notes For Relieuing of the Young Student; the First Part*. London: Iohn Harrison, 1601.

Clapham, Henoch. *Three Partes of Salomon His Song of Songs, Expounded. . . .* London: Valentine Sims, 1603.

Clark, Francis, S.J. *Eucharistic Sacrifice and the Reformation*. 2nd ed. Oxford: Basil Blackwell, 1967.

Clark, Neville. *An Approach to the Theology of the Sacraments*. Studies in Biblical Theology, 17. London: SCM, 1956.

Coakley, ed. *Religion and the Body*, Cambridge Studies in Religious Traditions, 8. Cambridge: Cambridge University Press, 1997.

Cohen, H. Floris. *The Scientific Revolution: A Historiographical Inquiry*. Chicago: University of Chicago Press, 1994.

Cohen, Ted. "Metaphor and the Cultivation of Intimacy." *Critical Inquiry* 5 (1978): 3–12.

Cole, F. J. *A History of Comparative Anatomy from Aristotle to the Eighteenth Century*. London: Macmillan, 1944.

The Concise Home Doctor: Encyclopedia of Good Health, written by a distinguished group of general practitioners and specialists in medicine, surgery, psychology, hygiene, physical culture, and veterinary practice, with plates in full colours and nearly 4,000 illustrations. 2 vols. London: Educational Book Company, [1932?, 1933?].

Cook, Harold J. "The Cutting Edge of a Revolution? Medicine and Natural History Near the Shores of the North Sea." In *Renaissance and Revolution: Humanists, Scholars, Craftsmen and Natural Philosophers in Early Modern Europe*, ed. J. V. Field and Frank A. J. L. James, pp. 45–61. Cambridge: Cambridge University Press, 1993; repr. 1997.

———. "Institutional Structures and Personal Belief in the London College of Physicians." In *Religio Medici: Medicine and Religion in Seventeenth-Century England*, ed. Ole Peter Grell and Andrew Cunningham, pp. 91–114. Aldershot: Scolar Press, 1996.

Costelloe, B. F. C., and J. H. Muirhead. *Aristotle and the Earlier Peripatetics: Being a Translation from Zeller's "Philosophy of the Greeks."* 2 vols. London: Longman's, Green, and Co., 1897.

Crashaw, Richard. *The Poems English, Latin and Greek of Richard Crashaw*. Ed. L. C. Martin. 2nd ed. Oxford: Clarendon Press, 1957.

Crooke, Helkiah. *Microcosmographia: A Description of the Body of Man*. London: W. Jaggard, 1615.

Cuddon, J. A. *A Dictionary of Literary Terms and Literary Theory*. 3rd ed. Oxford: Basil Blackwell, 1991.

Culler, Jonathan. *The Pursuit of Signs: Semiotics, Literature, Deconstruction*. London: Routledge, 1981; repr. 1992.

Cumston, G. C. *An Introduction to the History of Medicine*. London: Kegan Paul, Trench, Trübner & Co., 1926.

Cunningham, Andrew. *The Anatomical Renaissance: The Resurrection of the Anatomical Projects of the Ancients*. Aldershot: Scolar Press, 1997.

Da Vinci, Leonardo. *The Literary Works.* Ed. and comp. J. P. Richter. 2nd edn. 2 vols. Oxford: Oxford University Press, 1939.

Daly, Peter M. "George Wither's Use of Emblem Terminology." In *Aspects of Renaissance and Baroque Symbol Theory 1500–1700,* ed. Peter M. Daly and John Manning, pp. 27–38. AMS Studies in the Emblem, 14. New York: AMS Press, 1999.

Davies of Hereford, John. *The Complete Works.* Ed. Alexander B. Grosart. 2 vols. Chertsey Worthies' Library. St George's, Blackburn: Edinburgh University Press, 1878. Reprint, Anglistica & Americana, 2. Hildesheim: Georg Olms Verlagsbuchhandlung, 1968.

De Laguna, Andreas. *Anatomica methodus, seu de sectione humani corporis contemplatio.* 1535. Trans. in Lind, *Studies in Pre-Vesalian Anatomy.*

De Man, Paul. "The Epistemology of Metaphor." *Critical Inquiry* 5 (1978): 13–30.

———. "Semiology and Rhetoric." In *Textual Strategies: Perspectives in Post-Structuralist Criticism,* ed. and intr. Josué V. Harrari, pp. 121–40. London: Methuen, 1980. Originally published, Ithaca: Cornell University Press, 1979.

Dear, Peter. "The Church and the New Philosophy." In *Science, Culture and Popular Belief in Renaissance Europe,* ed. Stephen Pumphrey, Paolo Rossi, and Maurice Slawinski, pp. 119–39. Manchester: Manchester University Press, 1991.

Debus, Allen G. *The English Paracelsians.* London: Oldbourne, 1965.

Deese, James. "Mind and Metaphor: A Commentary." *New Literary History* 6 (1974): 211–17.

Derrida, Jacques. *Margins of Philosophy.* Trans. Alan Bass. Chicago: University of Chicago Press, 1982.

Descartes, René. *The Philosophical Writings.* Trans. John Cottingham, Robert Stoothoff, and Dugald Murdoch. 3 vols.. Cambridge: Cambridge University Press, 1992.

Dictionary of National Biography (*DNB*). 63 vols. plus supplement, 3 vols; second supplement, 3 vols; indexes and epitomes. Ed. by Sidney Lee (XXVII–LXIII, supplement, second supplement, indexes, and epitomes) and Leslie Stephen (I–XXI) (XXII–XXVI ed. by Sidney Lee and Leslie Stephen). London: Smith, Elder, & Co., 1885–1900; supplement (1901); second supplement (1912).

Dillistone, F. W. *Christianity and Symbolism.* London: Collins, 1955.

Dollimore, Jonathan. *Radical Tragedy: Religion, Ideology and Power in the Drama of Shakespeare and his Contemporaries.* Brighton: Harvester Wheatsheaf, 1984.

———. *Radical Tragedy: Religion, Ideology and Power in the Drama of Shakespeare and his Contemporaries.* 2nd ed. London and New York: Harvester Wheatsheaf, 1989.

Dollimore, Jonathan, and Alan Sinfield, eds. *Political Shakespeare: New Essays in Cultural Materialism.* 2nd ed. with additional chapters. Manchester: Manchester University Press, 1996; c. 1994.

Donker, Marjorie, and George M. Muldrow. *Dictionary of Literary-Rhetorical Conventions of the English Renaissance.* Westport, Conn.: Greenwood, 1982.

Donne, John. *An Anatomy of the World: Wherein, by Occasion of the Untimely Death of Mistris Elizabeth Drury the Frailty and the Decay of this Whole World is Represented.* 1611. Reprint, Cambridge: Cambridge University Press, 1951.

———. *Deaths Duel, 1632.* Reprint, Menston: Scolar Press, 1969.

———. *Deaths Duell; or, A Consolation to the Soule, Against the Dying Life, and Living Death of the Body; Delivered in a Sermon at White-Hall, before the Kings Majestie, in the Beginning of Lent, 1630, by that Late Learned and Reverend Divine, John Donne, Dr. in Divinity, and Deane of S. Pauls, London; Being His Last Sermon, and Called by His Maiesties Houshold the Doctors Owne Funeral Sermon.* London: Benjamin Fisher, 1633.

———. *The Divine Poems.* Ed. Helen Gardner. 2nd ed. Oxford: Clarendon Press, 1978.

———. *The Epithalamions, Anniverseries and Epicedes.* Ed. W. Milgate. Oxford: Clarendon Press, 1978.

———. *Five Sermons upon Special Occasions, 1626.* Reprint, Menston: Scolar Press, 1970.

———. *The Poems.* Ed. Herbert J. C. Grierson. 2 vols. Oxford: Oxford University Press, 1912; repr. 1958.

———. *The Second Anniversarie: Of the Progres of the Soule; Wherein, by Occasion of the Religious Death of Mistris Elizabeth Drury, the Incommodities of the Soule in this Life and her Exaltation in the Next, Are Contemplated,* in *The Epithalamions, Anniversaries and Epicedes of John Donne.* Ed. W. Milgate. Oxford: Clarendon Press, 1978.

———. *The Sermons.* Ed. G. R. Potter and E. M. Simpson. 10 vols. Berkeley: University of California Press; London: Cambridge University Press, 1953–62.

———. *Sermons on the Psalms and Gospels, With a Selection of Prayers and Meditations.* Ed. Evelyn M. Simpson. Berkeley: University of California Press, 1963.

———. *LXXX Sermons Preached by that Learned and Reverend Divine John Donne, Dr in Divinity, Late Deane of the Cathedrall Church of S. Pauls London.* London: R. Royston and R. Marriot, 1640.

———. *X Sermons Preached by that Late Learned and Rev[erend] Divine John Donne, Doctor in Divinity, Once Dean of the Cathedral Church of Saint Paul's,* chosen from the whole body of Donne's sermons by Geoffrey Keynes. London: Nonesuch Press, 1923.

Dorland's Illustrated Medical Dictionary. London: W. B. Saunders, 1994.

Downing, Crystal Nelson. "The *Charmes Backe to Reverse:* Deconstructing Architectures in Books II and III of *The Faerie Queene.*" *Comitatus* 13 (1982): 64–83.

Drayton, Michael. *The Works.* Ed. J. William Hebel. 6 vols. Oxford: Blackwell for the Shakespeare Head Press, 1961.

Dunglison, Robley. *A Dictionary of Medical Science.* . . . 23rd edn., thoroughly rev. by T. L. Stedman. London: J. & A. Churchill, 1904.

Eagleton, Terry. *Literary Theory: An Introduction.* Oxford: Basil Blackwell, 1983; repr. 1989.

Earle, John. *Microcosmography; or, A Piece of the World Discovered: In Essays and Characters.* 1628; London: White & Cochrane [et al.], 1811.

Early English Books 1475–1640 Selected from Pollard and Redgrave's Short-Title Catalogue: A Guide to Units 53–61 of the Microfilm Collection, Reels 1783–1979. Ann Arbor: University Microfilms International, 1988.

Eco, Umberto. "Testing, testing. . . ." *Guardian Review,* 4 September 2004, p. 7

Edge, David. "Technological Metaphor and Social Control." *New Literary History* 6 (1974): 135–47.

Edgerton, Jr., Samuel Y. *Pictures and Punishment: Art and Criminal Prosecution During the Florentine Renaissance.* Ithaca: Cornell University Press, 1985.

Egmond, Florike. "Execution, Dissection, Pain and Infamy—A Morphological Investigation." In *Bodily Extremities: Preoccupations with the Human Body in Early Modern European Culture,* ed. Florike Egmond and Robert Zwijnenberg, pp. 92–127. Aldershot: Ashgate, 2003.

Elmer, Peter. "Medicine, Religion and the Puritan Revolution." In *The Medical Revolution of the Seventeenth Century,* pp. 10–45, ed. French and Wear.

Empiricus, Sextus. *Selections from the Major Writings on Scepticism, Man, & God.* Ed. Philip P. Hallie, trans. Sanford G. Etheridge. Indianapolis: Hackett, 1985.

Eriksson, Ruben. *Andreas Vesalius' First Public Anatomy at Bologna 1540: An Eyewitness Report by Baldasar Heseler Medicinae Scolaris together with his notes on Matthaeus Curtius' Lectures on Anatomia Mundini.* Ed. with an introduction, translation into English, and notes. Uppsala: Almquist and Wiksell, 1959.

Estienne, Charles. *De dissectione partium corporis humani libri tres.* Paris: Simon de Colines, 1545.

Estienne, Henri. *Dictionarium medicum; vel, expositiones medicinalium ad verbum excerptae ex Hippocrate, aretæo, . . . cum Latina interpretatione . . . lexica duo in Hippocratem . . . prefixa sunt, unum erotiani . . . alterum Galeni. . . .* Geneva: Henri Estienne, 1564.

Ettmüller, Michael. *Etmullerus Abridg'd: or, A Compleat System of the Theory and Practice of Physic, Being a Description of All Diseases Incident to Men, Women and Children.* 3rd edn. London: Andrew Bell, 1699.

Fabrici of Acquapendente, Girolamo. *De venarum ostiolis.* 1603. Reprint in *Anatomices et chirurgiae . . . tractatus quatuor: Quorum, I, De formato fœtv, II, De locutione et ejus instrumentis, III, De loquela brutorum, IV, De venarum ostiolis. . . .* Frankfurt: Jacques de Zetter, 1624.

Feher, Michel, with Ramona Naddaff, and Nadia Tazi, eds. *Zone* 3–5, *Fragments for a History of the Human Body,* 3 Parts. New York: Zone, 1989.

Felperin, Howard. *The Uses of the Canon: Elizabethan Literature and Contemporary Theory.* Oxford: Clarendon Press, 1992.

Ferguson, Rebecca. "'Intestine Wars': Body and Text in *An Epistle to Dr Arbuthnot* and *The Dunciad.*" In *Pope: New Contexts,* ed. David Fairer, pp. 137–52. London and New York: Harvester Wheatsheaf, 1990.

Ferrand, James. Ερωτομανια*; or, A Treatise Discoursing of the Essence, Causes, Symptomes, Prognosticks, and Cure of Love or Erotique Melancholy.* Trans. E. Chilmed. Oxford, 1640. Originally published in French as *La maladie d'amour, ou mélancholie érotique* (Paris, 1623).

Ferrari, Giovanni. "Public Anatomy Lessons and the Carnival: The Anatomy Theatre of Bologna." *Past and Present* 117 (1987): 50–106.

Ficino, Marsilio. "Five Questions Concerning the Mind." Trans. Josephine L. Burroughs, in *The Renaissance Philosophy of Man,* pp. 183–212, ed. Cassirer, Kristeller, and Randall.

Flanagan, Geraldine Lux. "Going to Work on an Egg." *Observer Life,* 2 June 1996, pp. 18–20.

Fletcher, Giles, and Phineas. *Poetical Works.* Ed. Frederick S. Boas. 2 vols. Cambridge: Cambridge University Press, 1908–09.

Fletcher, Phineas. *A Father's Testament,* written long since for the benefit of the particular relations of the authour, Phin[eas] Fletcher, sometime minister of the gospel at Hillgay in Norfolk, and now made publick at the desire of friends. London: Henry Mortlock, 1670.

———. *Joy in Tribulation; or, Consolation for Afflicted Spirits.* London: James Boler, 1632.

———. *Locustae vel pietas Iesuitica.* Ed. with introduction, translation, and commentary by Estelle Haan. Supplementa Humanistica Lovaniensia, 9. Leuven: Leuven University Press, 1996.

———. *The Poems . . . for the First Time Collected and Edited,* with memoir, essay, and notes by Alexander B. Grosart. 4 vols. Fuller Worthies' Library. Blackburn: C. Tipelady, 1869.

———. *The Purple Island; or, The Isle of Man: Together With Piscatorie Eclogs and Other Poeticall Miscellanies.* Cambridge: University of Cambridge, 1633. [*STC* 11082]

———. *The Purple Island; or, The Isle of Man: Together With Piscatorie Eclogs and Other Poeticall Miscellanies.* Cambridge: University of Cambridge, 1633. [*STC* 11082.5]

———. *The Purple Island; or, The Isle of Man: An Allegorical Poem By Phineas Fletcher, Esteemed the Spenser of His Age. . . .* New edn., corrected and rev., with additional notes by the editor [Peregrine Branwhite?]. London: Frys & Couchman, 1783.

———. *The Purple Island, a Poem: By Phineas Fletcher.* With the critical remarks of the late Henry Headley, A.B. and a biographical sketch by William Jacques. London: Burton & Briggs, 1816.

———. *Venus & Anchises (Brittain's Ida) and Other Poems by Phineas Fletcher.* Ed. from a Sion College Ms. for the Royal Society of Literature by Ethel Seaton, with a Preface by F. S. Boas. London: Oxford University Press, 1926.

———. *The Way to Blessednes; a Treatise . . . on the First Psalm.* London: James Boler, 1632.

Fludd, Robert. *Medicina Catholica; seu, mysticum artis medicandi sacrarium: in tomos diuisum duos; in qvibvs metaphysica et physica tam sanitatis tuendæ, quam morborum propulsandorum ratio pertractatur.* Frankfurt: William Fitzer, 1629.

Foster, M. *Lectures on the History of Physiology during the Sixteenth, Seventeenth and Eighteenth Centuries.* Cambridge: Cambridge University Press, 1901.

Foucault, Michel. *The History of Sexuality: An Introduction.* Trans. Robert Hurley. London: Allen Lane, 1979; London: Penguin, 1990.

———. *The Order of Things: An Archaeology of the Human Sciences.* World of Man series: A Library of Theory and Research in the Human Sciences. London: Tavistock Publications, 1970. Originally published as *Les Mots et les choses* by Éditions Gallimard, 1966.

Fox, Denton. "The Scottish Chaucerians." In *Chaucer and Chaucerians: Critical Studies in Middle English Literature,* ed. D. S. Brewer, pp. 164–200. Nelson's University Paperbacks. London: Thomas Nelson and Sons, 1966.

Freccero, John. "Medusa: The Letter and the Spirit." *Yearbook of Italian Studies* (1972): pp.1–18

French, Roger. *Dissection and Vivisection in the European Renaissance.* Aldershot: Ashgate, 1999.

———. "A Note on the Anatomical Accessus of the Middle Ages." *Medical History* 23 (1979): 461–68.

———. *William Harvey's Natural Philosophy.* Cambridge: Cambridge University Press, 1994.

French, Roger, and Andrew Wear, eds. *The Medical Revolution of the Seventeenth Century.* Cambridge: Cambridge University Press, 1989.

Freud, Sigmund. "Medusa's Head." In *The Standard Edition of the Complete Psychological Works.* 24 vols. Trans. from the German under the general editorship of James Strachey in collaboration with Anna Freud, assisted by Alix Strachey and Alan Tyson, editorial assistant Angela Richards. London: Hogarth Press, 1953–74. XVIII (1964), 273–74.

———. *Studies on Hysteria,* trans. James and Alix Strachey, ed. James and Alix Strachey, assisted by Angela Richards, this volume ed. Angela Richards. Vol. III of *The Penguin Freud Library* (formerly *The Pelican Freud Library*), ed. Angela Richards and Albert Dickson. Harmondsworth: Penguin, 1974; repr. 1991.

———. *On Sexuality,* trans. from the German under the general editorship of James Strachey, this volume comp. and ed. Angela Richards. Vol. VII of *The Penguin Freud Library* (formerly *The Pelican Freud Library*), ed. Angela Richards and Albert Dickson. Harmondsworth: Penguin, 1977; repr. 1987.

Friedman, Alice T. *House and Household in Elizabethan England: Wollaton Hall and the Willoughby Family.* Chicago: University of Chicago Press, 1989.

Frye, Northrop. *Anatomy of Criticism: Four Essays.* Princeton: Princeton University Press, 1971; repr. 1973.

Galloway, Allan D. *The Cosmic Christ.* London: Nisbet, 1951.

Gardner, Philip Alan Tennant. "The Banquet of the Word: Biblical Authority and Interpretation in Spenser and the Fletchers." PhD thesis, University of Toronto, 1985.

Gascoigne, George. *Certayne Notes of Instruction Concerning the Making of Verse or Ryme in English, Written at the Request of Master Edouardo Donati.* 1575. In *The Complete Works of George Gascoigne.* Ed. John W. Cunliffe. 2 vols. Cambridge English Classics. New York: Greenwood Press, 1969. I, 465–73. Originally published in 1907 by Cambridge University Press.

Gibbs, Raymond W., Jr., and Gerard J. Steen, eds. *Metaphor in Cognitive Linguistics: Selected Papers From the Fifth International Cognitive Linguistics Conference, Amsterdam, July 1997.* Amsterdam: John Benjamins, 1999.

Gilbert, Neal W. *Renaissance Concepts of Method.* New York: Columbia University Press, 1960.

Glare, P. G. W., ed. *Oxford Latin Dictionary.* Oxford: Clarendon Press, 1968.

Glucksberg, Sam, and Boaz Keysar. "Understanding Metaphorical Comparisons: Beyond Similarity." *Psychological Review* 97 (1990): 3–18.

Gosse, Edmund. *The Jacobean Poets.* London: John Murray, 1899.

Gossman, Lionel. "History and Literature: Reproduction or Signification." In *The Writing of History,* ed. Canary and Kozicki, pp. 3–39

Gray, Henry. *Anatomy Descriptive and Surgical.* Ed. T. Pickering Pick and Robert Howden. 15th ed. 1901. Reprint, London: Chancellor Press, 1985.

Greenblatt, Stephen. "Psychoanalysis and Renaissance Culture." In *Literary Theory / Renaissance Texts,* ed. Patricia Parker and D. Quint, pp. 210–25. Baltimore: Johns Hopkins University Press, 1986.

———. *Renaissance Self-Fashioning: From More to Shakespeare.* Chicago: University of Chicago Press, 1980.

———. "Shakespeare and the Exorcists." In *Shakespeare and the Question of Theory,* ed. Patricia Parker and Geoffrey Hartman, pp. 163–87. London: Methuen, 1985.

———. *Shakespearean Negotiations: The Circulation of Social Energy in Renaissance England.* Oxford: Clarendon Press, 1988.

Grierson, Herbert. *The First Half of the Seventeenth Century.* Periods of European Literature Series. Edinburgh: William Blackwood & Sons, 1906.

Grundy, Joan. *The Spenserian Poets.* London: Edward Arnold, 1969.

Guidi, Guido. *Ars medicinalis.* 3 vols. Venice: Juntas, 1611.

Hagthorpe, John. *Divine Meditations, and Elegies.* London: Bernard Alsop, 1622.

———. *Visiones Rervm: The Visions of Things; or, Foure Poems, Wherein the Authour Expresseth His Invention By Way of Dreame.* London: Bernard Alsop, 1623.

Hale, John. *The Civilization of Europe in the Renaissance.* London: HarperCollins, 1993.

Hammond, Gerald, ed. *Elizabethan Poetry: Lyrical and Narrative.* A Casebook. London: Macmillan, 1984.

Hanfling, Oswald, ed. *Philosophical Aesthetics: An Introduction.* Oxford and Milton Keynes: Blackwell in association with The Open University, 1992.

Hankinson, R. J. *The Sceptics.* London and New York: Routledge, 1995.

Harcourt, Glenn. "Andreas Vesalius and the Anatomy of Antique Sculpture." *Representations* 17 (1987), 28–61.

Hardy, Barbara. "An Approach through Narrative." 1968. In *Towards a Poetics of Fiction,* ed. by Mark Spilka, pp. 31–40. Essays from *Novel: A Forum on Fiction,* 1967–1976. Bloomington: Indiana University Press, 1977.

Harré, Rom. "Some Narrative Conventions of Scientific Disourse." In *Narrative in Culture: The Use of Storytelling in the Sciences, Philosophy, and Literature,* ed. Christopher Nash, pp. 81–101. Warwick Studies in Philosophy and Literature Series. London: Routledge, 1994.

Harries, Karsten. "Metaphor and Transcendence." *Critical Inquiry* 5 (1978), 73–90.

Harris, Jonathan Gil. *Foreign Bodies and the Body Politic: Discourses of Social Pathology in Early Modern England.* Cambridge: Cambridge University Press, 1998.

Harvey, William. *The Circulation of the Blood and Other Writings.* Trans. Kenneth J. Franklin with an introduction by Andrew Wear. London: J. M. Dent & Sons, 1990.

———. *Lectures on the Whole of Anatomy.* An annotated translation of *Prelectiones Anatomiae Universalis* by C. D. O'Malley, F. N. L. Poynter, and K. F. Russell. Berkeley: University of California Press, 1961.

Hauser, Arnold. *The Social History of Art: Renaissance, Mannerism and Baroque.* 4 vols. London: Routledge & Kegan Paul, 1968.

Hausman, Carl R. *Metaphor and Art: Interactionism and Reference in the Verbal and Nonverbal Arts.* Cambridge: Cambridge University Press, 1989.

Hawkes, Terence. *Metaphor.* Critical Idiom Series, 25. London: Routledge, 1972; repr. with revised bibliography, 1989.

Hayle, W. G., and J. P. Margham. *Dictionary of Biology.* London and Glasgow: HarperCollins, 1988.

Headley, Henry, ed. *Selected Beauties of Ancient English Poetry.* 2 vols. London: T. Cadell, 1787.

Healy, Margaret. *Fictions of Disease in Early Modern England.* Basingstoke: Palgrave, 2001.

Healy, Thomas. *New Latitudes: Theory and English Renaissance Literature.* London: Edward Arnold, 1992.

———. *Richard Crashaw.* Leiden: E. J. Brill, 1986.

———. "Sound Physic: Phineas Fletcher's *The Purple Island* and the Poetry of Purgation." *Renaissance Studies* 5 (1991): 341–52.

Heckscher, William. *Rembrandt's Anatomy of Dr. Nicolaas Tulp: An Iconological Study.* New York: New York University Press, 1958.

Helman, Cecil, *Body Myths.* London: Chatto & Windus, 1991.

Henderson, John. "The Hospitals of Late-Medieval and Renaissance Florence: A Preliminary Survey." In *The Hospital in History,* ed. Lindsay Granshaw and Roy Porter, pp. 63–92. London: Routledge, 1989.

Henderson's Dictionary of Biological Terms. 9th ed. London: Longman, 1979.

Henry, John. "The Matter of Souls: Medical Theory and Theology in Seventeenth-Century England." In *The Medical Revolution of the Seventeenth Century,* ed. French and Wear, pp. 87–113.

Herbermann, Charles G., and others, eds. *The Catholic Encylopedia: An International Work of Reference on the Constitution, Doctrine, and History of the Catholic Church.* 15 vols. and index. New York: Encyclopedia Press, 1913.

Heydon, John. *A New Method of Rosie Crucian Physick; Wherein is Shewed the Cause, and Therewith Their Experienced Medicines for the Cure of All Diseases.* London: Thomas Lock, 1658.

Heywood, James, and Thomas Wright, comp. *Cambridge University Transactions during the Puritan Controversies of the 16th and 17th Centuries.* 2 vols. London: Henry G. Bohn, 1854.

Hick, John. *The Metaphor of God Incarnate.* London: SCM, 1993.

Hill, Christopher. *Change and Continuity in Seventeenth Century England.* London: Weidenfeld & Nicholson, 1974.

Hillman, David. "Visceral Knowledge: Shakespeare, Skepticism, and the Interior of the Early Modern Body." In *The Body in Parts,* ed. Hillman and Mazzio, pp. 81–105.

Hillman, David, and Carla Mazzio, eds. *The Body in Parts: Fantasies of Corporeality in Early Modern Europe.* London: Routledge, 1997.

Hobbes, Thomas. *Leviathan; or, The Matter, Forme, & Power of a Common-Wealth Ecclesiasticall and Civill.* London: Andrew Crooke, 1651.

Hodges, Devon L. *Renaissance Fictions of Anatomy.* Amherst: University of Massachusetts Press, 1985.

Honig, Edwin. *Dark Conceit: The Making of Allegory.* Oxford and New York: Oxford University Press, 1966.

Hooker, Richard. *Of the Lawes of Ecclesiasticall Politie: Eyght Bookes.* London: John Windet, [1593]–97.

Howell, James. *Epistolæ Ho-Elianæ: Familiar Letters, Domestic and Forren; Divided into Four Books,*

Partly Historical, Political, Philosophical, upon Emergent Occasions. 5th edn. 4 vols. (bound together). London: Thomas Guy, 1678.

Huffman, C. C. *Coriolanus in Context.* Cranbury: Associated University Press, 1971.

Huffman, William H. *Robert Fludd and the End of the Renaissance.* London and New York: Routledge, 1988.

Huit, Ephraim. *The Anatomy of conscience; or, The Svmme of Paul's Regeneracy; Wherein are Handled the Places of Conscience, Worship, and Scandall; With Divers Rules of Christian Practise; Very Profitable for the Weake Christian.* London: William Sheffard, 1626.

Hunter, Richard A., and Ida Macalpine. "William Harvey and Robert Boyle." *Notes and Records of the Royal Society* 13 (1958): 115–27.

Hunter, Jr., William B., ed. *The English Spenserians: The Poetry of Giles Fletcher, George Wither, Michael Drayton, Phineas Fletcher and Henry More.* Salt Lake City: University of Utah Press, 1977.

Illich, Ivan. "A Plea for Body History." *Michigan Quarterly Review* 26 (1987): 342–48.

Iser, Wolfgang. *The Implied Reader.* Baltimore: Johns Hopkins University Press, 1974.

Jackson, Henry. Μικροκοσμογραφια*; or, A Description of the Body of Man; Being a Practical Anatomy, Shewing the Manner of Anatomizing from Part to Part. . . .* London: Livewell Chatman, 1664.

James, D. Geraint, ed. *Circulation of the Blood.* Tunbridge Wells: Pitman Medical Publishing, 1978.

Janacek, Bruce. "Catholic Natural Philosophy: Alchemy and the Revivification of Sir Kenelm Digby." In *Rethinking the Scientific Revolution,* ed. Margaret J. Osler, pp. 89–118. Cambridge: Cambridge University Press, 2000.

Jardine, Lisa. *Ingenious Pursuits: Building the Scientific Revolution.* London: Little, Brown & Co., 1999.

Jefferson, Ann. "Bodymatters: Self and Others in Bakhtin, Sartre and Barthes." In *Bakhtin and Cultural Theory,* ed. Ken Hirschkop and David Shepherd. Manchester: Manchester University Press, 1989; repr. 1989, pp. 152–77.

Jenner, Mark S. R. "Body, Image, Text in Early Modern Europe." *Social History of Medicine* 12 (1999): 143–54.

Jessopp, A. *Diocesan Histories* [Norwich]. London: SPCK, 1884.

Johnson, Mark. *The Body in the Mind: The Bodily Basis of Meaning, Imagination, and Reason.* Chicago: University of Chicago Press, 1987.

Jollat, François. *Les figures et portraicts des parties du corps humain.* Paris: Iaques Kerver, 1575.

Jones, John. *The Arte and Science of Preserving Bodie and Soule in Healthe, Wisdome, and Catholike Religion; Physically, Philosophically, and Diuinely Deuised.* London: Henrie Bynneman, 1579.

Jones, Leah. *The Divine Science: The Aesthetic of Some Representative Seventeenth-Century English Poets.* New York: Columbia University Press, 1940.

Jones, Peter Murray. "Reading Medicine in Tudor Cambridge." In *The History of Medical Education in Briain,* ed. V. Nutton and R. Porter, pp. 153–83. Clio Medica 30 / The Wellcome Institute Series in the History of Medicine. Amsterdam and Atlanta, GA., 1995.

———. "Thomas Lorkyn's Dissections, 1564/5 and 1566/7." *Transactions of the Cambridge Bibliographical Society,* IX (1988): 209–29.

Jordan, Edward. *A Briefe Discourse of a Disease Called the Suffocation of the Mother. Written uppon Occasion Which Has Beene of Late Taken thereby, to Suspect Possession of an Evill Spirit, or Some Such Like Supernaturall Power. Wherein in Declared that Divers Strange Actions and Passions of the Body of Man, Which in the Common Opinion, are Imputed to the Diuell, Have their True Naturall Causes, and Do Accompanie this Disease.* 1603. Reprint in MacDonald, ed., *Witchcraft and Hysteria in Elizabethan London.*

Kastor, Frank S. *Giles and Phineas Fletcher.* Twayne's English Authors Series, 225. Boston: Twayne, 1978.

Keats, John. *The Poetical Works.* Ed. H. Buxton Forman. London: Oxford University Press, 1915.

Kellett, C. E. "A Note on Rosso and the Illustrations to Charles Estienne's *De dissectione.*" *Journal of the History of Medicine* 12 (1957): 325–36.

Kemp, Martin. "A Drawing for the *Fabrica;* and Some Thoughts upon the Vesalius Muscle-men." *Medical History* 14 (1970): 277–88.

———. "'The mark of truth': Looking and Learning in Some Anatomical Illustrations from the Renaissance and Eighteenth Century." In *Medicine and the Five Senses,* ed. W. F. Bynum and Roy Porter, pp. 85–121. Cambridge: Cambridge University Press, 1993.

Kemp, Martin, and Marina Wallace. *Spectacular Bodies: The Art and Science of the Human Body from Leonardo to Now.* London: Hayward Gallery Publishing, 2000. [Accompanies the exhibition organized by the Hayward Gallery, London, curated by Martin Kemp and Marina Wallace.]

Kerrigan, William. "The Fearful Accommodations of John Donne." In *John Donne and the Seventeenth-Century Metaphysical Poets,* ed. Harold Bloom, pp. 37–50. Modern Critical Views. New York: Chelsea House Publishers, 1986.

Keynes, Geoffrey. *A Bibliography of the Writings of Dr William Harvey 1578–1657.* Ed. Gweneth Whitteridge and Christine English. 3rd ed., rev. Winchester: St. Paul's Bibliographies, 1989.

———. *The Life of William Harvey.* Oxford: Clarendon Press, 1966.

Kimber, Diana Clifford, Carolyn E. Gray, and Caroline E. Stackpole. *Textbook of Anatomy and Physiology.* 11th edn. New York: Macmillan, 1942; repr. 1943.

Klein, Melanie. "Early Stages of the Oedipus Conflict" (1928). In *The Selected Melanie Klein,* ed. Mitchell, pp. 69–83.

Klibansky, Raymond, Erwin Panofsky, and Fritz Saxl. *Saturn and Melancholy: Studies in the History of Natural Philosophy, Religion and Art.* London: Nelson, 1964.

Knight, David M. *Natural Science Books in English, 1600–1900.* London: Portman Books, 1989. Originally published in 1972.

Kocher, Paul H. *Science and Religion in Elizabethan England.* New York: Octagon Books, 1969; c. 1953 by Henry E. Huntington Library & Art Gallery.

Kuhn, Thomas S. *The Structure of Scientific Revolutions.* Chicago: University of Chicago Press, 1962.

Lacan, Jacques. *Écrits: A Selection.* Trans. Alan Sheridan. London: Tavistock, 1977; London: Routledge, 1992.

Lakoff, George. "The Meanings of Literal." *Metaphor and Symbolic Activity* 1 (1986): 291–96.

Lakoff, George, and Mark Johnson. *Metaphors We Live By.* Chicago: University of Chicago Press, 1980.

Lambert, Samuel W. "Phineas Fletcher and Francis Quarles and *The Purple Island* of Phineas Fletcher." *Annals of Medical History,* repr. from n.s., 3 (1931): 568–78.

Langdale, Abram Barnett. *Phineas Fletcher: Man of Letters, Science and Divinity,* Columbia University Studies in English and Comparative Literature, 125. New York: Columbia University Press, 1937.

Langer, Ullrich. "Invention." In *The Cambridge History of Literary Criticism,* ed. Nisbet and Rawson, vol. III, ed. Norton, 136–44.

Laplanche, J., and J.-B. Pontalis. *The Language of Psycho-analysis.* Intr. Daniel Lagache, trans. Donald Nicholson-Smith. London: Karnac Books and the Institute of Psycho-analysis, 1988.

Laqueur, Thomas. *Making Sex: Body and Gender from the Greeks to Freud.* Cambridge: Harvard University Press, 1992.

Larkey, Sanford V. "The Vesalian Compendium of Geminus and Nicholas Udall's Translation: Their Relation to Vesalius, Caius, Vicary and De Mondeville." *Library,* 4th ser., 13, no. 4 (1933): 367–94.

Larkey, Sanford V., and Linda Tum Suden. "Jackson's English Translation of Berengarius of Carpi's 'Isagogae Breves', 1660 and 1664." *Isis* 21 (1934): 57–70.

Larkin, J. F., and P. L. Hughes, eds. *Stuart Royal Proclamations.* 2 vols. Oxford: Clarendon Press, 1973.

Larmore, Charles. "Scepticism." In *The Cambridge History of Seventeenth-Century Philosophy,* ed. David Garber and Michael Ayers, with the assistance of Roger Ariew and Alan Gabbey. 2 vols. Cambridge: Cambridge University Press, 1998. II, 1145–92.

Lawrence, Christopher, and Steven Shapin, eds. *Science Incarnate: Historical Embodiments of Natural Knowledge.* Chicago: University of Chicago Press, 1998.

Leach, Heather. "Crossing the Line: Stories on Mapping the Maze." *Times Higher Education Supplement,* 5 November 2004, pp. 18–19.

Leech, Clifford. *The Dramatist's Experience: With Other Essays in Literary Theory.* London: Chatto & Windus, 1970.

———. *Tragedy,* The Critical Idiom, vol. 1. London: Methuen, 1969; repr. 1970.

Leedham-Green. E. S., *Books in Cambridge Inventories: Book-Lists from Vice-Chancellor's Court Probate Inventories in the Tudor and Stuart Periods.* 2 vols. Cambridge: Cambridge University Press, 1986.

Leishman, J. B. *The Monarch of Wit: An Analytical and Comparative Study of the Poetry of John Donne.* 6th edn. London: Hutchinson, 1962; repr. 1967.

Lewalski, Barbara Kiefer. *Protestant Poetics and the Seventeenth-Century Religious Lyric.* Princeton: Princeton University Press, 1979.

Lind, L. R. *Studies in Pre-Vesalian Anatomy: Biography, Translations, Documents.* Philadelphia: The American Philosophical Society, 1975.

———, ed., *The Epitome of Andreas Vesalius.* Trans. from the Latin. Cambridge: M.I.T. Press, 1969.

Lindemann, Mary. *Medicine and Society in Early Modern Europe.* New Approaches to European History Series. Cambridge: Cambridge University Press, 1999.

Lippmann, Walter. *A Preface to Morals.* London: George Allen & Unwin, 1931.

Louth, Andrew. "The Body in Western Catholic Christianity." In *Religion and the Body,* ed. Coakley, pp. 111–30.

Lyas, Colin. "The Evaluation of Art." In *Philosophical Aesthetics,* ed. Hanfling, pp. 349–80.

Lyly, John. *Euphues: The Anatomy of Wit; Verie Pleasant for All Gentlemen to Reade, and Most Necessarie to Remember; Wherein are Contained the Delights that Wit Followeth in his Youth, by the Pleasantnesse of Loue; and the Happinesse He Reapeth in Age, by the Perfectnesse of Wisedome.* London: G. Eld, 1617.

Lyons, William, ed. *Modern Philosophy of Mind.* London: J. M. Dent, 1995.

MacDonald, George. *England's Antiphon.* [London]: Macmillan, [1868].

MacDonald, Michael, ed. *Witchcraft and Hysteria in Elizabethan London: Edward Jorden and the Mary Glover Case.* London: Tavistock/Routledge, 1991.

Maclean, Ian. *The Renaissance Notion of Woman: A Study in the Fortunes of Scholasticism and Medical Science in European Intellectual Life.* Cambridge Monographs on the History of Medicine. Cambridge: Cambridge University Press, 1980.

Manning, John. *The Emblem.* London: Reaktion Books, 2002.

Margolis, Joseph. *Art and Philosophy.* Brighton: Harvester Press, 1980.

Marriott, Edward. "Men and Porn." *Guardian Weekend,* 8 November 2003, pp. 44–47, 89.

Marshall, Pamela. *Wollaton Hall and the Willoughby Family.* Nottingham: Nottingham Civic Society, 1999.

Martensen, Robert. "The Transformation of Eve: Women's Bodies, Medicine and Culture in Early Modern England." In *Sexual Knowledge, Sexual Science: The History of Attitudes to Sexuality,* ed. Roy Porter and Mikulas Teich, pp. 107–33. Cambridge: Cambridge University Press, 1995.

Maus, Katharine Eisaman. *Inwardness and Theatre in the English Renaissance.* Chicago: University of Chicago Press, 1995.

———, ed. *Four Revenge Tragedies.* Oxford English Drama. Oxford: Oxford University Press, 1995.

May, Elizabeth. *Wollaton Hall: Family House and Natural History Museum.* Derby: J. H. Hall & Sons, 1989.

Mazzeo, Joseph Anthony. "Metaphysical Poetics and the Poetry of Correspondence." *Journal of the History of Ideas* 14 (1953): 221–34.

———. "Universal Analogy and the Culture of the Renaissance." *Journal of the History of Ideas* 15 (1954): 299–304.

Mazzio, Carla. "Sins of the Tongue." In *The Body in Parts,* ed. Hillman and Mazzio, pp. 52–79.

McGrath, Alister E. *Reformation Thought: An Introduction.* 2nd ed. Oxford: Blackwell, 1993; repr. 1994.

McKeown, Thomas. *The Role of Medicine: Dream, Mirage or Nemesis.* 2nd ed. Oxford: Basil Blackwell, 1989.

Merquior, J. G. *Foucault.* Fontana Modern Masters Series. London: Fontana Press, 1991.

Michael, Emily, and Fred. S. Michael. "Two Early Modern Concepts of Mind: Reflecting Substance vs. Thinking Substance." *Journal of the History of Philosophy* 27 (1989): 29–48.

Miller, David Lee. "Spenser's Poetics: The Poem's Two Bodies." *Publications of the Modern Language Association* 101 (1986): 170–85.

Mink, Louis. "History and Fiction as Modes of Comprehension." *New Literary History* 1 (1970): 541–58.

Mitchell, Juliet, ed. *The Selected Melanie Klein.* London: Penguin, 1991.

Montagu, M. F. Ashley, ed. *Studies and Essays in the History of Science and Learning Offered in Homage to George Sarton. . . .* New York: Henry Schuman, 1946.

Morton, Leslie T., and Robert J. Moore. *A Bibliography of Medical and Biomedical Biography.* Aldershot: Scolar Press, 1989.

Mulvey, Laura. "Visual Pleasure and Narrative Cinema." *Screen* 16, no. 3 (1975): 6–18.

Munk, William. *The Roll of the Royal College of Physicians of London; Comprising Biographical Sketches of All the Eminent Physicians, Whose Names Are Recorded in the Annals from the Foundation of the College in 1518 to its Removal in 1825, from Warwick Lane to Pall Mall East.* 2nd ed., revised and enlarged. Vol. I, *1518–1700.* London: The College of Physicians, 1878.

Murphy, Charles D. "John Davies's Versification of Sidney's Prose." *Philological Quarterly* 21 (1942): 410–14.

Murray, W. A. "Donne and Paracelsus: An Essay in Interpretation." In *Essential Articles for the Study of John Donne's Poetry,* ed. Roberts, pp. 122–28.

Museum of London Archaeology Service (MoLAS), *Barber-Surgeons' Hall Garden, Monkwell Square, London EC2 . . . An Archaeological Research Evaluation.* London: Museum of London, 1998.

Neal, Gilbert W. *Renaissance Concepts of Method.* New York: Columbia University Press, 1960.

Neill, Michael. *Issues of Death: Mortality and Identity in English Renaissance Tragedy.* Oxford: Clarendon Press, 1997.

Nicolson, Marjorie Hope. *The Breaking of the Circle: Studies in the Effect of the 'New Science' on Seventeenth-Century Poetry.* Rev. ed. London and New York: Columbia University Press, 1965.

Nisbet, H. B., and Claude Rawson, eds. *The Cambridge History of Literary Criticism.* 9 vols. Cambridge: Cambridge University Press, 1989–. III, *The Renaissance,* ed. Glyn P. Norton (1999).

Nixon, Anthony. *The Dignitie of Man, Both in the Perfections of His Sovle and Bodie; Shewing As Well the Faculties in the Disposition of the One: As the Senses and Organs in the Composition of the Other.* London: Edward Allde, 1612.

Nohrnberg, James. *The Analogy of "The Faerie Queene."* Princeton: Princeton University Press, 1976.

Norden, John. *A Store-Hovse of Varieties, Briefly Discoursing the Change and Alteration of Things in this World.* London: n. p., 1601.

Norford, Don Parry. "Microcosm and Macrocosm in Seventeenth-Century Literature." *Journal of the History of Ideas* 38 (1977): 409–28.

Normand, Lawrence. "Reading the Body in *The Merchant of Venice.*" *Textual Practice* 5 (1991): 55–73.

Nutton, Vivian. "Greek Science in the Sixteenth-Century Renaissance." In *Renaissance and Revolution: Humanists, Scholars, Craftsmen and Natural Philosophers in Early Modern Europe,* ed. J. V. Field and Frank A. J. L. James, pp. 15–28. Cambridge: Cambridge University Press, 1993.

———. "The Rise of Medical Humanism: Ferrara 1464–1555." *Renaissance Studies* 11 (1997): 2–19.

———. "Wittenberg Anatomy." In *Medicine and the Reformation,* ed. Ole Peter Grell and Andrew Cunningham, pp. 11–32. London: Routledge, 1993.

O'Malley, C. D. *Andreas Vesalius of Brussels, 1514–1564.* Berkeley: University of California Press, 1964.

Oates, J. C. T. *Cambridge University Library: A History.* 2 vols. Cambridge: Cambridge University Press, 1986.

Ousby, Ian. *The Cambridge Guide to Literature in English.* Cambridge: Cambridge University Press, 1993.

The Oxford English Dictionary . . . (*OED*). 12 vols. and supplement. Oxford: Clarendon Press, 1933; repr. 1970.

The Pocket Oxford Dictionary of Current English. Comp. F. G. Fowler and H. W. Fowler. 4th edn., rev. H. G. le Mesurier and E. McIntosh. Oxford: Clarendon Press, 1946; repr. 1947.

Pagel, Walter. "Religious Motives in the Medical Biology of the XVIIth Century." *Bulletin of the Institute of the History of Medicine* 3 (1935): 97–128, 213–31, 265–312. Reprint in Pagel, *Religion and Neoplatonism in Renaissance Medicine,* pp. 97–128, 213–31, 265–312.

———. *Religion and Neoplatonism in Renaissance Medicine.* Ed. Marianne Winder, Collected Studies Series, CS226. London: Variorum Reprints, 1985.

———. "The Vindication of 'Rubbish.'" In Pagel, *Religion and Neoplatonism in Renaissance Medicine,* pp. 1–14. Originally published in *Middlesex Hospital Journal* 45 (1945): 42–45.

———. "William Harvey and the Purpose of Circulation." *Isis* 42 (1951): 22–38.

Paré, Ambroise. *The Workes of that Famous Chirurgeon Ambrose Parey.* Trans. from Latin and compared with the French by Thomas Johnson. London: Thomas Cotes and R. Young, 1634.

———. *The Works of that Famous Chirurgeon Ambrose Parey.* Trans. out of Latin, and compared with the French, by Th[omas] Johnson: Together with *Three Tractates Concerning the Veins, Arteries, and Nerves: Exemplified with Large Anatomical Figures: Translated out of Adrianus Spigelius.* London: John Clark, 1678.

Parfitt, George. *English Poetry of the Seventeenth-Century.* Longman Literature in English Series. 2nd ed. London and New York: Longman, 1992.

Park, Katharine. “The Criminal and the Saintly Body: Autopsy and Dissection in Renaissance Italy.” *Renaissance Quarterly* 47 (1994): 1–33.

———. “Healing the Poor: Hospitals and Medical Assistance in Renaissance Florence.” In *Medicine and Charity before the Welfare State,* ed. Jonathan Barry and Colin Jones, pp. 26–45. London: Routledge, 1991.

———. “The Rediscovery of the Clitoris.” In *The Body in Parts,* ed. Hillman and Mazzio, pp. 170–93.

Paster, Gail Kern. “The Body and Its Passions.” *Shakespeare Studies* 29, ed. Leeds Barroll, 44–50. Madison: Fairleigh Dickinson University Press: 2001.

Patrides, C. A. *Premises and Motifs in Renaissance Thought and Literature.* Princeton: Princeton University Press, 1982.

Patterson, A. *Shakespeare and the Popular Voice.* Oxford: Basil Blackwell, 1989.

Pauw, Pieter. *Primitiae anatomicae: de humani corporis ossibus.* Leiden: Justi à Colster, 1615.

The Penguin Book of Renaissance Verse, 1509–1659. Selected and intr. David Norbrook, ed. H. R. Woudhuysen. London: Penguin Books, 1993.

Pennington, L. E., ed. *The Purchas Handbook: Studies of the Life, Times and Writings of Samuel Purchas 1577–1626,* 2 vols., with bibliographies of his books and of works about him. Works issued by the Hakluyt Society, 2nd ser., nos. 185–86. London: The Hakluyt Society, 1997.

Perry, John. *A Dialogue on Personal Identity and Immortality.* Indianapolis: Hackett Publishing, 1978.

Petherbridge, Deanna, and Ludmilla Jordanova. *The Quick and the Dead: Artists and Anatomy.* London: South Bank Centre, 1997. [Accompanies the National Touring Exhibition organized by the Hayward Gallery, London, for the Arts Council of England, curated by Deanna Petherbridge.]

Pico della Mirandola, Giovanni. *Oration on the Dignity of Man.* Trans. Elizabeth Livermore Forbes, intr. Paul Oskar Kristeller. In *The Renaissance Philosophy of Man,* ed. Cassirer, Kristeller, and Randall, pp. 213–54.

———. *A Platonick Discourse on Love, Written . . . in Explication of a Sonnet by Hieronimo Benivieni.* Trans. by Thomas Stanley. London, 1651.

Plater, Felix. *De corporis humani structura et usu . . . libri III, tabulis methodice explicati, iconibus accurate illustrati.* Basle: Ambrosius Froben, 1583.

Plato. *Timaeus.* Vol. III of *The Dialogues,* trans. Benjamin Jowett, 4th edn., 4 vols. Oxford: Clarendon Press, 1953.

———. *The Republic.* Trans. Francis MacDonald Cornford. Oxford: Clarendon Press, 1941; repr. 1966.

“A Poet’s Anatomy.” *British Medical Journal* (May 9, 1925), p. 890.

Pohlman, Augustus Grote. “*The Purple Island* by Phenias Fletcher: A Seventeenth Century Layman’s Poetical Conception of the Human Body.” *Johns Hopkins Hospital Bulletin* [vol. 18] no. 197 (August 1907): 317–21.

Pollard, A. W., and G. R. Redgrave. *A Short-Title Catalogue of Books Printed in England, Scotland, & Ireland and of English Books Printed Abroad 1475–1640.* 2nd ed. rev. and enlarged by W. A. Jackson and F. S. Ferguson, completed by Katharine F. Pantzer. 3 vols. London: Bibliographical Society, 1976–1991.

Pope, Alexander. *The Poems.* Ed. John Butt, a one-volume edition of the Twickenham text with selected annotations. London and New York: Routledge, 1963; repr. 1989.

Popkin, Richard H. *The History of Scepticism from Erasmus to Spinoza.* Berkeley, Los Angeles, and London: University of California Press, 1979.

Porter, Roy. *The Greatest Benefit to Mankind: A Medical History of Humanity from Antiquity to the Present.* London: HarperCollins, 1997.

———. *Mind-Forg'd Manacles: A History of Madness in England from the Restoration to the Regency.* Harmondsworth: Penguin, 1990.

Purchas, Samuel. *Purchas his Pilgrim. Microcosmus; or, The Historie of Man: Relating the Wonders of his Generation, Vanities in his Degeneration, Necessity of his Regeneration; Meditated on the Words of David. Psal. 39.5. Verily, Every Man at his Best State is Altogether Vanitie.* London: W.S., 1619.

———. *Purchas his Pilgrim. Microcosmus; or, The Historie of Man: Relating the Wonders of his Generation, Vanities in his Degeneration, Necessity of his Regeneration; Meditated on the Words of David. Psal. 39.5. Verily, Every Man at his Best State is Altogether Vanitie.* 1619. Reprint, The English Experience, 146, Amsterdam: Theatrum Orbis Terrarum; New York: Da Capo, 1969.

Puttenham, George. *The Arte of English Poesie 1589.* Reprint, Menston: Scolar Press, 1968.

Quarles, Francis. *The Complete Works in Prose and Verse.* With memorial-introduction, notes &c., by Alexander B. Grosart. 3 vols. Edinburgh: Edinburgh University Press, i.e.: T. & A. Constable, 1880.

Quilligan, Maureen. *The Language of Allegory: Defining the Genre.* Ithaca: Cornell University Press, 1979.

Ralegh, Sir Walter. *The History of the World.* London: [W. Stansby] for Walter Burre, 1614.

———. *Sir Walter Raleigh's Sceptick; or, Speculations, and Observations of the Magnificency and Opulency of Cities.* London: W. Bentley, 1651.

Raynalde, Thomas, trans. *The Birth of Mankinde; Otherwise Named the Womans Booke, Set Forth in English by Thomas Raynalde, Newly Translated out of Laten into Englysshe.* London: T. Adams, 1604.

———. *The Birth of Man-kinde; Othewise Named the Womas Booke.* London: James Boler, 1626.

Redman, Linda J. Flasch. "*The Purple Island:* A Critical Study." PhD dissertation, Oklahoma State University, 1974.

Reese, James M. "Song of Solomon." In *The Oxford Companion to the Bible,* ed. Bruce M. Metzer and Michael D. Coogan, pp. 708–10. Oxford: Oxford University Press, 1993.

Remmelin, Johann. *Catoptrum microcosmicum suis aere incisis visionibus splendens cum historia et pinace de nova prodit.* Ulm: Johann Görlin, 1639.

Ricci, James V. *The Genealogy of Gynaecology.* Philadelphia: Blakiston, 1950.

Richards, I. A. *The Philosophy of Rhetoric.* Oxford: Oxford University Press, 1936.

Ricoeur, Paul. "Metaphor and the Main Problem of Hermeneutics." *New Literary History* 6 (1974): 95–110.

Rigolot, François. "The Rhetoric of Presence: Art, Literature, and Illusion." In *The Cambridge History of Literary Criticism,* ed. Nisbet and Rawson, vol. III, ed. Norton, 161–67.

Rimmon-Kenan, Shlomith. *Narrative Fiction: Contemporary Poetics.* New Accents Series. London and New York: Routledge, 1983; repr. 1994.

Riva, J. G. [A series of thirty-six anatomical and pathological plates]. N. p., n. d. The British Library: Shelfmark L.R.301.bb.7.

Riverius, Lazarus. *The Practice of Physick.* London: Peter Cole, 1658.

Roberts, John R., ed. *Essential Articles for the Study of John Donne's Poetry.* Hassocks: Harvester Press, 1975.

Roberts, K. B., and J. D. W. Tomlinson. *The Fabric of the Body: European Traditions of Anatomical Illustration.* Oxford: Clarendon Press, 1992.

Rope, H. E. G. "John Davies of Hereford, Catholic and Rhymer." *The Anglo-Welsh Review* 11 (1961): 20–36.

Roper, Lyndal. *Oedipus and the Devil.* London and New York: Routledge, 1994.

Rorty, Richard. "'He had both a bubbling wit and the ability to make fun of himself.'" *Times Higher Education Supplement,* 12 November 2004, p. 16.

Rose, Jacqueline. *Sexuality in the Field of Vision.* London: Verso, 1991.

Ross, Alexander. *Medicus Medicatus; or, The Physicians Religion Cured, by a Lenitive or Gentle Potion: With Some Animadversions upon Sir Kenelme Digbie's Observations on "Religio Medici."* London: James Young, 1645.

Rowe, Katherine. "'God's handy worke': Divine Complicity and the Anatomist's Touch." In *The Body in Parts,* ed. Hillman and Mazzio, pp. 284–309.

Rupp, Jan C. C. "Matters of Life and Death: The Social and Cultural Conditions of the Rise of Anatomical Theatres, With Special Reference to Seventeenth Century Holland." *History of Science* 28 (1990): 263–87.

Russell, K. F. "A Checklist of Medical Books Published in English Before 1600." *Bulletin of the History of Medicine* 21 (1947): 922–58.

Ryle, Gilbert. *The Concept of Mind.* London: Hutchinson, 1949; repr. 1959.

Sadler, John. *The Sicke Woman's Private Looking-Glasse; Wherein Methodically Are Handled All Uterine Effects, or Diseases Arising from ye Wombe Enabling Women to Informe the Physitian about ye Cause of their Griefe.* 1636. Reprint, Amsterdam: Theatrum Orbis Terrarum; Norwood, N.J.: Walter J. Johnson, 1977.

Salgâdo, Gâmini. *Three Jacobean Tragedies.* Harmondsworth: Penguin, 1965; repr. 1978.

Sanford, Ezekial, ed. *The Works of the British Poets, With Lives of the Authors.* 50 vols. (XVIII–XXI, XXIII, XXV–L, ed. by R. Walsh). Philadelphia: Mitchell, Ames & White, 1819.

Saunders, Alison. *The Sixteenth-Century Blason Poétique.* Bern: Peter Lang, 1981.

Saunders, J. B. de S. M., and C. D. O'Malley, eds. *The Illustrations from the Works of Andreas Vesalius of Brussels.* New York: Dover Publications, 1950.

Sawday, Jonathan. "Bodies by Art Fashioned: Anatomy, Anatomists, and English Poetry 1570–1680." PhD thesis, University of London, 1988.

———. *The Body Emblazoned: Dissection and the Human Body in Renaissance Culture.* London: Routledge, 1995.

———. "The Fate of Marsyas: Dissecting the Renaissance Body." In *Renaissance Bodies: The Human Figure in English Culture, c. 1540–1660,* ed. by Lucy Gent and Nigel Llewellyn, pp. 111–35 and pp. 260–64. Critical Views Series. London: Reaktion Books, 1990; repr. 1995.

———. "Self and Selfhood in the Seventeenth Century." In *Rewriting the Self: Histories from the Renaissance to the Present,* ed. by Roy Porter, pp. 29–48. London: Routledge, 1997.

Schoenfeldt, Michael. *Bodies and Selves in Early Modern England: Physiology and Inwardness in Spenser, Shakespeare, Herbert, and Milton.* Cambridge Studies in Literature and Culture, 34. Cambridge: Cambridge University Press, 1999.

———. "Fables of the Belly in Early Modern England." In *The Body in Parts,* ed. by Hillman and Mazzio, pp. 242–61.

Scholz, Susanne. *Body Narratives: Writing the Nation and Fashioning the Subject in Early Modern England.* Basingstoke: Macmillan, 2000.

Schultz, Bernard. *Art and Anatomy in Renaissance Italy,* Studies in the Fine Arts: Art Theory, 12. Ann Arbor: UMI Research, 1985.

Schupbach, William. *The Paradox of Rembrandt's "Anatomy of Dr. Tulp."* London: Wellcome Institute for the History of Medicine, 1982.

Seebass, H., and A. C. Thiselton. "Flesh." In *The New International Dictionary of New Testament Theology,* ed. Brown, vol. I, 671–82.

Shapin, Steven. "The Philosopher and the Chicken: On the Dietetics of Disembodies Knowledge." In *Science Incarnate,* ed. Lawrence and Shapin, pp. 21–50.

Shapin, Steven, and Christopher Lawrence. "Introduction: The Body of Knowledge." In *Science Incarnate,* ed. Lawrence and Shapin, pp. 1–15.

Shakerley, Jeremy. *The Anatomy of Urania Practica; or, A Short Mathematicall Discourse; Laying Open the Errors and Impertinencies Delivered in a Treatise Lately Published by Mr. Vincent Wing, and Mr. William Leybourne, under the title of Urania Practica.* London: Thomas Brudenell, 1649.

Shakespeare, William. *The Arden Edition of the Works of William Shakespeare: Coriolanus.* Ed. Philip Brockbank. London: Routledge, 1988.

———. *The Complete Works.* Compact edn. Ed. Stanley Wells and Gary Taylor. Oxford: Clarendon Press, 1999.

Shepherd, Simon. *Spenser.* London: Harvester Wheatsheaf, 1989.

Sidney, Sir Philip. *Selected Writings: Astrophil and Stella, The Defence of Poesy and Miscellaneous Poems.* Ed. Richard Dutton. Manchester: Carcanet, 1987.

Sinfield, Alan. *Faultlines: Cultural Materialism and the Politics of Dissident Reading.* Oxford: Clarendon Press, 1992.

Singer, Charles. *The Evolution of Anatomy: A Short History of Anatomical and Physiological Discovery to Harvey; Being the Substance of the FitzPatrick Lectures Delivered at the Royal College of Physicians of London in the Years 1923 and 1924.* London: Kegan Paul, Trench, Trübner & Co., 1925.

———. "Historical Relations of Religion and Science." In *Science, Religion and Reality,* ed. Joseph Needham, pp. 85–148. London: Sheldon Press; New York and Toronto: Macmillan, 1925.

———. *Monumenta Medica.* 2 vols. Vol. II, *The Fasciculo Di Medicina, Venice 1493,* intr. by Charles Singer. Florence: R. Lier and Co., 1925.

———. *A Short History of Anatomy and Physiology from the Greeks to Harvey.* New York: Dover Publications, 1957. An unabridged republication of the first edition with a new Preface by the author; originally published as *The Evolution of Anatomy.*

———. *Vesalius on the Human Brain: Introduction, Translation of Text, Translation of Descriptions of Figures, Notes to the Translations, Figures.* London: Oxford University Press, 1952.

Sinibaldi, Giovanni Benedetto, *Rare Verities: The Cabinet of Venus Unlocked, and Her Secrets Laid Open; Being a Translation of Part of Sinibaldus His Geneanthropeia, and a Collection of Some Things out of Other Latin Authors, Never before in English.* London: P. Briggs, 1658.

Smith, John. *The Portrait of Old Age: Wherein is Contained a Sacred Anatomy Both of Soul, and Body, and a Perfect Account of the Infirmities of Age Incident to Them Both; Being a Paraphrase upon the Six Former Verses of the 12. Chapter of Ecclesiastes.* 2nd ed. corrected. London: J. Macock, 1666.

Snell, Richard S. *Clinical Anatomy for Medical Students.* 3rd ed. Boston: Little, Brown & Co., 1986.

Sontag, Susan. *Aids and Its Metaphors.* London: Penguin, 1990.

Sorri, Mari, and Jerry Gill. *A Post-Modern Epistemology: Language, Truth and Body,* Problems in Contemporary Philosophy Series, 19. Lewiston: Edwin Mellen, 1989.

Spaher, Michael, and Johann Remmelin. *A Survey of the Microcosme; or, The Anatomy of the Bodies of Man and Woman; wherein Skin, Veins, Muscles, Viscera, Bones, and Ligaments thereof Are Accurately Delineated, and So Disposed by Pasting, as that All the Parts of the Said Bodies, Both Internal and External, Are Exactly Represented in Their Proper Site.* Corrected by Clopton Havers. London: James Moxon, [1659].

Sparshott, F. E. "'As,' or The Limits of Metaphor." *New Literary History* 6 (1974): 75–94.

Spenser, Edmund. *The Faerie Queene.* Ed. A. C. Hamilton, assisted by R. J. Manning. Longman Annotated English Poets. London: Longman, 1977.

———. *The Works,* The Globe Edition. Ed. from the original editions and manuscripts by R. Morris, with a memoir by John W. Hales. London: Macmillan, 1869; repr. 1929.

Spielmann, M. H. *The Iconography of Andreas Vesalius (André Vésale) Anatomist and Physician 1514–1564: Paintings, Pictures, Engravings, Illustrations, Sculpture, Medals.* Wellcome Historical Medical Museum London, Research Studies in Medical History, 3. London: John Bale, Sons & Danielsson, 1925.

Spolsky, Ellen. *Gaps in Nature: Literary Interpretation and the Modular Mind.* Albany: SUNY Press, 1993.

———. *Satisfying Skepticism: Embodied Knowledge in the Early Modern World.* Aldershot: Ashgate, 2001.

Stallybrass, Peter. "Footnotes." In *The Body in Parts,* ed. Hillman and Mazzio, pp. 312–25.

———. "Patriarchal Territories: The Body Enclosed." In *Rewriting the Renaissance: The Discourses of Sexual Difference in Early Modern Europe,* ed. Margeret Ferguson, Maureen Quilligan, and Nancy Vickers, pp. 123–42. Women in Culture and Society Series. Chicago: University of Chicago Press, 1986.

Stedman's Medical Dictionary. 26th ed. London: Williams & Wilkins, 1995.

Steen, Gerard. *Understanding Metaphor in Literature: An Empirical Approach.* London: Longman, 1994.

Stevenson, W., ed. *Report on the Manuscripts of Lord Middleton Preserved at Wollaton Hall, Nottingham.* London: Historical Manuscripts Commission, 1911.

Strathmann, Ernest A. *Sir Walter Ralegh: A Study in Elizabethan Skepticism.* New York: Columbia University Press, 1951.

Sutherland, Stewart R. "The Concept of Revelation." In *Religion, Reason and the Self: Essays in Honour of Hywel D. Lewis,* ed. Stewart R. Sutherland and T. A. Roberts, pp. 35–45. Cardiff: University of Wales Press, 1989.

Sylvester, Joshua, trans. *Du Bartas His Devine Weekes and Workes.* London: H. Lownes, 1611.

———, trans., *Du Bartas His Devine Weekes and Workes With A Compleate Collecti*[on] *Of All the Other Delight-full Workes.* London: Robert Young, 1641.

Taylor, Charles. *Sources of the Self: The Making of the Modern Identity.* Cambridge: Cambridge University Press, 1989.

Taylor, Gary. *Reinventing Shakespeare: A Cultural History from the Restoration to the Present.* London: Hogarth Press, 1989.

Temkin, Owsei. *The Double Face of Janus and Other Essays in the History of Medicine.* Baltimore: Johns Hopkins University Press, 1977.

———. "The Elusiveness of Paracelsus." In Temkin, *The Double Face of Janus and Other Essays in the History of Medicine,* pp. 225–38. Originally published in *Bulletin of the History of Medicine* 26 (1952), 201–17.

———. "Metaphors of Human Biology." In Temkin, *The Double Face of Janus and Other Essays in the History of Medicine,* pp. 271–83. Originally published in *Science and Civilization,* ed. by Robert C. Stauffer (Madison: University of Wisconsin Press, 1949), pp. 169–94.

———. "Was Servetus Influenced by Ibn an-Nafis?" In Temkin, *The Double Face of Janus and Other Essays in the History of Medicine,* pp. 284–86

Thorndike, Lynn. *A History of Magic and Experimental Science during the First Thirteen Centuries of Our Era.* 8 vols. I–II (London: Macmillan, 1923–58), III–VIII (New York: Columbia University Press, 1947, 1960). I (1923), II (1947), III (1960).

Thornton, John L. *Medical Books, Libraries and Collectors: A Study of Bibliography and the Book Trade in Relation to the Medical Sciences.* 3rd rev. ed. Aldershot: Gower, 1966.

Thornton, John L., and Carole Reeves. *Medical Book Illustration: A Short History.* Cambridge and New York: Oleander Press, 1983.

Traub, Valerie. *Desire and Anxiety: Circulations of Sexuality in Shakespearean Drama.* London: Routledge, 1992.

———. "Gendering Mortality in Early Modern Anatomies." In *Feminist Readings of Early Modern Culture: Emerging Subjects,* ed. Valerie Traub, M. Lindsay Kaplan, and Dympna Callaghan, pp. 44–92. Cambridge: Cambridge University Press, 1996.

Tripp, David. "The Image of the Body in the Formative Phases of the Protestant Reformation." In *Religion and the Body,* ed. Coakley, pp. 131–52.

Tuve, Rosemond. *Elizabethan and Metaphysical Imagery: Renaissance Poetic and Twentieth-Century Critics.* Chicago: University of Chicago Press, 1947.

Un., Ro. *The Little World; or, A Lively Description of All the Partes and Properties of Man; For Invention Wittie, For Iudgement Learned, and For Practise Necessarie.* London: W. Jones, 1612.

Upham, Alfred Horatio. *The French Influence in English Literature from the Accession of Elizabeth to the Restoration.* New York: Columbia University Press, 1911.

Ussher, James. *Immanuel; or, The Mystery of the Incarnation of the Son of God.* London: John Parker, 1638.

Valadez, Frances. "Anatomical Studies at Oxford and Cambridge." In *Medicine in Seventeenth Century England, a Symposium Held at UCLA in Honor of C. D. O'Malley,* ed. Allen G. Debus, pp. 393–420. Berkeley: University of California Press, 1974.

Valverde da Hamusco, Juan. *Anatomia del corpo humano.* Rome: Antonio Salamanca and Antonio Lafrery, 1560.

———. *Historia de la composicion del cuerpo humano.* Rome: Antonio Salamanca and Antonio Lafrery, 1556.

———. *Vivae imagines partium corporis humani aereis formis expressae.* Antwerp: Christophe Plantin, 1566.

Van der Spieghel, Adriaan. *De formato fœtv liber singularis aeneis figvris exornatus.* Padua: Johann Bap de Martinis and Liuius Pasquatus, 1627.

———. *De formato fœtv liber singularis aeneis figvris exornatus.* Frankfurt: Matthaeus Merian, 1631.

———. *De humani corporis fabrica libri decem* [with] *Ivlii Casserii Placentini tabulae anatomicae LXXIIX.* Venice: n. p., 1627.

———. *De humani corporis fabrica libri decem* [with] *Ivlii Casserii Placentini tabulae anatomicae LXXIIX.* Frankfurt: Matthaeus Merian, 1632.

Van Noppen, Jean-Pierre, and Edith Hols, comp. *Metaphor II: A Classified Bibliography of Publications, 1985–1990.* Amsterdam Studies in the Theory and History of Linguistic Science, 5th ser.: Library & Information Sources in Linguistics, 20. Amsterdam and Philadelphia: John Benjamins Publishing, 1990.

Veeser, H. Aram. "Introduction: The New Historicism." In *The New Historicism Reader,* ed. H. Aram Veeser, pp. 1–32. London: Routledge, 1994.

Veith, Ilza. *Hysteria: The History of a Disease.* Northvale, N.J. and London: Jason Aronson, 1993; 1st ed. 1965.

Venn, J. and J. A., *Alumni Cantabrigiensis: A Biographical List of all Known Students, Graduates and Holders of Office at the University of Cambridge, from the Earliest Times to 1900,* 2 Parts: Part 1, vols. I–IV; Part II, vols. I–VI (Cambridge: Cambridge University Press, 1922–54).

Venn, John, comp. *Biographical History of Gonville and Caius College*, 8 vols., incomplete. Cambridge: Cambridge University Press, 1897–.

Venn, John, ed. *The Register of Baptisms, Marriages and Burials in St Michael's Parish, Cambridge (1538–1837)*. Cambridge: Printed for the Cambridge Antiquarian Society, 1891.

Vesalius, Andreas. *Andreae Vesalii icones anatomicae*. [Munich]: [Bremen], [1934].

———. *De humani corporis fabrica libri septem*. Basle: Johann Oporinus, 1543.

———. *De humani corporis fabrica librorum epitome*. Basle: Johann Oporinus, 1543.

Veslingus, Johann. *The Anatomy of the Body of Man . . . Published in Latin by Joh[ann] Veslingus, Reader of the Publick Anatomy in the Most Famous University of Padua . . . and Englished by Nich[olas] Culpeper*. London: Peter Cole, 1653.

Vicary, Thomas. *The Anatomie of the Bodie of Man: This Edition of 1548, As Reissued by the Surgeons of St. Bartholomew's in 1577 with a Life of Vicary*. Ed. Frederick J. Furnivall and Percy Furnivall. London: N. Trübner for the Early English Text Society, 1888.

Vickers, Nancy J. "'This heraldry in Lucrece' face.'" In *The Female Body in Western Culture: Contemporary Perspectives*, ed. Susan Rubin Suleiman, pp. 209–22. Cambridge: Harvard University Press, 1986.

———. "Members Only: Marot's Anatomical Blazons." In *The Body in Parts*, ed. Hillman and Mazzio, pp. 3–21

Walker, W. L. *The Spirit and the Incarnation in the Light of Scripture, Science, and Practical Need*. 3rd ed. Edinburgh: T. & T. Clark, 1907.

Walton, Izaak. *The Compleat Angler; or, Contemplative Man's Recreation, Being a Discourse on Rivers, Fish-Ponds, and Fishing, in Two Parts, the First Written by Mr. Izaak Walton, the Second by Charles Cotton, Esq; to Which Are Now Prefixed, the Lives of the Authors; Illustrated with Cuts . . . and Notes. . . .* London: Thomas Hope, 1760.

Ward, Richard. *The Anatomy of Warre; or, Warre with the Wofull, Fruits, and Effects thereof, Laid out to the Life: Wherein from Scripture, and Experience, These Things are Clearly Handled. . . .* London: John Dalham and Rich[ard] Lownds, [1642].

Warwick, Roger, and Peter L. Williams, eds. *Gray's Anatomy*. 35th ed. Edinburgh: Longman, 1973.

Waswo, Richard. "Theories of Language." In *The Cambridge History of Literary Criticism*, ed. Nisbet and Rawson, vol. III, ed. Norton, 25–35.

Wellcome Institute for the History of Medicine and Related Sciences, London. *Subject Catalogue of the History of Medicine and Related Sciences*. Munich: Krauss International Publications, 1980.

Wells, Robin Headlam. *Spenser's "Faerie Queene" and the Cult of Elizabeth*. London: Croom Helm; Totowa, N.J.: Barnes & Noble, 1983.

Whately, William. *The New Birth; or, A Treatise of Regeneration, Delivered in Certaine Sermons*. London: Felix Kyngston, 1619.

White, Hayden. "The Historical Text as Literary Artifact." In *The Writing of History*, ed. Canary and Kozicki, pp. 41–62.

Whitteridge, Gweneth. "Gaspard Bauhin." *Dictionary of Scientific Biography* 1 (1970): 522–55.

———. *William Harvey and the Circulation of the Blood*. London: Macdonald; New York: American Elsevier, 1971.

Whitteridge, Gweneth, and Veronica Stokes. *A Brief History of the Hospital of Saint Bartholomew*. London: The Governors of the Hospital of Saint Bartholomew, 1961.

Whitteridge, Gweneth, ed. *The Anatomical Lectures of William Harvey: Prelectiones Anatomie Universalis, De Musculis*. Edinburgh and London: Published for the Royal College of Physicians, London, by E. & S. Livingstone, 1964.

Wibbing, Siegfried. "*Sōma,* body." In *The New International Dictionary of New Testament Theology,* ed. Brown, I, 232–38

Willard, Thomas. "Donne's Anatomy Lesson: Vesalian or Paracelsian?" *John Donne Journal* 3 (1984), 34–61.

Williams, Gordon. *A Dictionary of Sexual Language and Imagery in Shakespearean and Stuart Literature.* 3 vols. London and Atlantic Highlands, N.J.: Athlone Press, 1994.

Williams, Linda. *Hard Core: Power, Pleasure, and the "Frenzy of the Visible."* Berkeley: University of California Press, 1989; London: Pandora, 1990.

Wilson, Luke. "William Harvey's *Prelectiones:* The Performance of the Body in the Renaissance Theater of Anatomy." *Representations* 17 (1987): 62–95.

Wilson, Richard. "Introduction: Historicising New Historicism." In *New Historicism and Renaissance Drama,* ed. Wilson and Dutton, pp. 1–16

Wilson, Richard, and Richard Dutton, ed. *New Historicism and Renaissance Drama.* London: Longman, 1992.

Wimsatt, W. K. *The Verbal Icon.* Lexington: University of Kentucky Press, 1954.

Wing, Vincent, and William Leybourn. *Ens Fictum Shakerlœi; or, The Annihilation of Mr. Jeremie Shakerley, his In-artificial Anatomy of Urania Practica.* London: Robert Leybourn, 1649.

Winstanley, William. *The Lives of the Most Famous English Poets; or, The Honour of Parnassus.* London: H. Clark, 1687.

Wood, A. C., ed. *The Continuation of the Willoughby Family,* by Cassandra, Duchess of Chandos, being Vol. II of the manuscript. Eton, Windsor: Shakespeare Head Press for the University of Nottingham, 1958.

Wood, Rufus. *Metaphor and Belief in "The Faerie Queene."* Basingstoke and London: Macmillan, 1997.

Wright, Louis B. *Middle-Class Culture in Elizabethan England.* Ithaca: Cornell University Press, 1958; repr. 1965; c. 1935 by The University of North Carolina; c. 1958 by the Trustees of Amherst College.

Wu, Duncan. *Romanticism: An Anthology.* 2nd edn. Oxford: Blackwell, 1998.

Yates, Frances A. *The French Academies of the Sixteenth Century.* Studies of the Warburg Institute, 15. London: Warburg Institute, University of London, 1947.

Young, Alan R., assisted by Beert Verstraete. *The English Tournament Imprese.* New York: AMS Press, 1988.

Zerner, Henri. "Observations on the Use of the Concept of Mannerism." In *The Meaning of Mannerism,* ed. Franklin W. Robinson and Stephen G. Nichols, Jr., pp. 105–21. Hanover, N.H.: University Press of New England, 1972.

Index

Page numbers in italics refer to illustrations